£49-50

MAGDALEN COLLEGE LIBRARY
SHELF No. 612.81
AND
v.1

D1092862

WITHDRAWN
From Magdalen College Library

300734406R

Neurocomputing

Neurocomputing
Foundations of Research

Edited by James A. Anderson and Edward Rosenfeld

MAGDALEN COLLEGE LIBRARY

The MIT Press
Cambridge, Massachusetts
London, England

MAGDALEN COLLEGE LIBRARY

Second printing, 1988

© 1988 Massachusetts Institute of Technology

All rights reserved. No part of this book may be
reproduced in any form by any electronic or mechanical
means (including photocopying, recording, or information
storage and retrieval) without permission in writing from
the publisher.

This book was prepared for production by combining
existing camera copy with newly typeset text set in Times
New Roman by Asco Trade Typesetting Ltd., Hong Kong,
and printed and bound by Halliday Lithograph in the
United States of America.

Library of Congress Cataloging-in-Publication Data

Neurocomputing: foundations of research.

   Includes index.
   1. Neural circuitry.   2. Computers—Circuits.
3. Higher nervous activity.   I. Anderson, James A.
II. Rosenfeld, Edward.
QP363.3.N46   1988   612'.81   87-3022
ISBN 0-262-01097-6

Contents

General Introduction
James A. Anderson

Brain-Like Machines

Neural networks, connectionist models, or, using a more recent name, neuromorphic systems are systems that are deliberately constructed to make use of some of the organizational principles that are felt to be used in the human brain. This volume contains a number of the original papers that describe many of the important ideas and techniques that are used in these models. Although there are a great many variations between authors and systems, there are also great similarities, both in the problems they try to solve with networks and in the techniques they use. At the present time, in the infancy of the field, similiarities predominate; as evolution continues, more variety will appear.

Many, if not most, of the ideas contained here were originally proposed to explain observations in neurobiology or psychology. Understanding human behavior and brain construction are perennially interesting and important questions, and the purely scientific desire to understand one of the most complex systems in nature is still the main motivation for many of the people who work with neural networks.

However, if you really understand something, you can usually make a machine do it. If you want to make machines think, act, or move like humans, a good initial strategy for most scientists is to study how humans think, act, or move. This strategy is not a guaranteed winner: extensive studies of human limb motion would probably not have suggested the wheel as the most appropriate way to move across smooth surfaces.

Humans have always wanted to build intelligent machines. Stories and legends of automata, robots, and mechanical men have been common for thousands of years, and are even more so today when we are actually building them. Current computer technology potentially allows us to build systems of a complexity that approaches the number of elements and interconnections of the brain. At the same time, we do not know how to organize this complexity or what functions to compute with it, even if we did manage to build it. Nor do we know if building a complex, brain-like system would yield intelligent behavior without other important features, some of which may still be unknown.

Somehow the brain is capable of taking neurons—the brain's basic computing elements—which are five or six orders of magnitude slower than silicon logic gates, and organizing them so as to perform some computations many times faster than the fastest digital computer now in existence. One way the brain seems to have managed to do this is by massive hardware parallelism: that is, the computing elements are arranged so that very many of them are working on a problem at the same time. Since there are huge numbers of neurons, somehow the weak computing powers of these many slow elements are combined together to form a powerful resultant. The speed of neurons has not increased much in evolution, once a few tricks like myelination were developed. The hardware and software cooperate, so the way to get more power seems to be to add more neurons, a strategy highly developed in our own massive cerebral cortex.

In the fifty-year history of digital computers, a quite different evolutionary path was followed. Conceptually, and often in reality, there is only one computing engine, the Central Processing Unit (CPU), and this unit has gotten faster and faster. At first it was made of vacuum tubes; now it is made of silicon VLSI (Very Large-Scale Integrated) circuitry; and in the future even faster compounds such as gallium arsenide will be used. Increase in computing speed has come about largely through faster hardware. Only in the last decade has good use been made of the first steps toward parallelism, where several CPUs are working at the same time on the same problem. It has turned out to be extraordinarily difficult to coordinate and program multiple CPUs of a traditional type.

Part of the reason that attempts to coordinate multiple fast CPUs have been difficult is that users and manufacturers want to run the same kinds of software, which run well on traditional computers, on parallel machines—only faster. It is unlikely that it will be possible to run traditional software on neural network machines. The appropriate software may be *very* different from traditional software.

It is worth pointing out that neural models have a narrow biological base. They are essentially all models of the newer parts of the mammalian nervous system, usually the cerebral cortex. Sometimes the earlier stages of sensory processing will be considered, but usually as a way of preprocessing information for a cortex to 'look at.' This is not necessarily bad. Because it evolved recently and is anatomically rather homogeneous, with only a few cell types and standard connection patterns to deal with, cortex may be more easily understandable than older, more highly optimized parts of the nervous system. The older parts of the vertebrate nervous system may be quite different in organization from the cortex, and invertebrates may be very different indeed.

Attempts to model cerebral cortex have an important consequence. If we model cortex, we have a good idea of its function. Most of our complex cognitive functions seem to be carried on there: speech, language, perception. Ideally, when we look at cognitive psychology or cognitive science, we are looking at software that runs on a cortical computer. This means it is possible to get reasonably good data on some of the organizational details of the output of the system by studying how humans do it.

This suggests also that a good way to find out what kind of software runs well on neural networks will be to look at what kind of software runs well on *us*. This means that cognitive science, besides its intrinsic scientific interest, might be viewed as a technique for reverse engineering the software for a parallel computer. Having an insight into writing, and running, good software for neuromorphic computers might save a great deal of unhappiness and failed expectations. As one example, expecting massively parallel neural network machines to balance a checkbook, perform logic, or keep fine detail straight might be unwise, since these are functions that are notoriously difficult for humans. Asking the same machines to make good guesses, disambiguate, resolve conflicting information, or form concepts might be reasonable.

We are so used to the constructional peculiarities of traditional computers that we have a tendency to think of them as familiar, but somewhat quirky, old friends. This familiarity often blinds experienced users to the extreme unnaturalness of the mindset required in order to use traditional computers effectively. Neural network systems might turn out to be truly 'user friendly' since they work like us!

Theoretical Themes

Several themes constantly recur in construction of neural network models: network structure, learning algorithms, and knowledge representation. Let us first sketch the generic connectionist model.

The Generic Connectionist Model

There are very many neurons, or nerve cells, in the human brain, at least ten billion. Each neuron receives inputs from other cells, integrates the inputs, and generates an output, which it then sends to other neurons, or, in some cases, to effector organs such as muscles or glands. Neurons receive inputs from other neurons by way of specialized structures called *synapses* and send outputs to other neurons by way of output lines called *axons*. A single neuron can receive on the order of hundreds or thousands of input lines and may send its output to a similar number of other neurons. A neuron is a complex electrochemical device that contains a continuous internal potential called a *membrane potential*, and, when the membrane potential exceeds a threshold, the neuron can propagate an all-or-none *action potential* for long distances down its axon to other neurons. Synapses come in a number of different forms, but two basic varieties are of particular note: *excitatory* synapses, which make it more likely that the neuron receiving them will fire action potentials, and *inhibitory* synapses, which make the neuron receiving them less likely to fire action potentials (see figure 1).

Neuroscientists usually measure the *activity* of a neuron by its firing frequency, i.e., the number of action potentials per second or something closely related to firing frequency. Biological neurons are *not* binary, that is, having only an on or off state as their output. Outputs are continuous valued and the neuron acts something like a voltage to frequency converter, converting membrane potential into firing rate (see paper 23 on the *Limulus*). Many network models use elements that are continuous valued to some extent. However, a number of neural network models assume that the basic computing elements are binary, that is, can only be on or off. The resulting binary valued systems are valuable, often give useful insights into the behavior of complex networks of whatever type, and are often easier or more convenient to analyze than systems of more complex neurons (see figure 2).

Given our degree of ignorance of nervous system function and the early stages of our modeling efforts, it would be most unreasonable to dismiss one or another assumption made by a modeler as "unbiological" until we see how the resulting system works. And of course, biological plausibility is significant only if you want to model the brain. If, instead, we desire to construct a useful device, there is no reason whatsoever to be bound by the way the brain happens to do it.

When the network is functioning, many cells can be active simultaneously. To

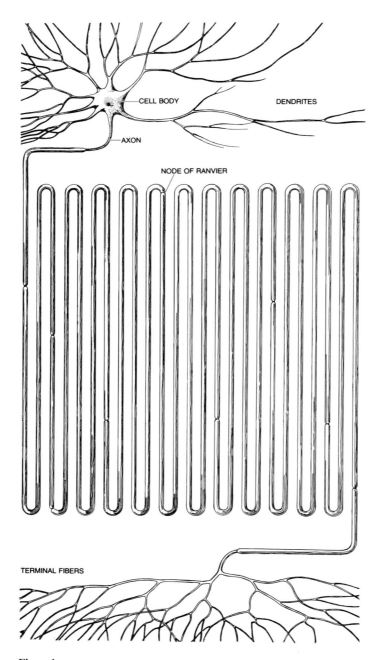

Figure 1

The typical neuron of a vertebrate animal can carry nerve impulses for a considerable distance. The neuron depicted here, with its various parts drawn to scale, is enlarged 250 times. The nerve impulses originate in the cell body, and are propagated along the axon, which may have one or more branches. This axon, which is folded for diagrammatic purposes, would be a centimeter long at actual size. Some axons are more than a meter long. The axon's terminal branches form synapses with as many as 1,000 other neurons. Most synapses join the axon terminals of one neuron with the dendrites forming a "tree" around the cell body of another neuron. Thus the dendrites surrounding the neuron in the diagram might receive incoming signals from tens, hundreds, or even thousands of other neurons. Many axons, such as this one, are insulated by a myelin sheath interrupted at intervals by the regions known as nodes of Ranvier. [Caption and figure from C. F. Stevens (1979), "The neuron," *The Brain*, Scientific American (Ed.), San Francisco: Freeman, p. 73.]

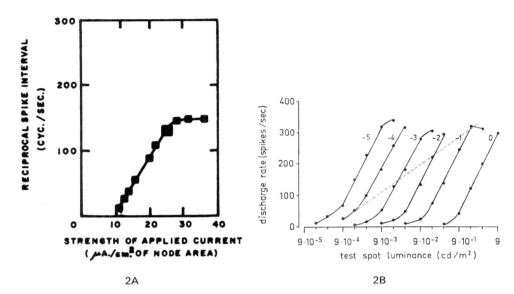

2A   2B

Figure 2
(A) Curve of the reciprocal mean spike interval in response to a maintained depolarizing current of 1-sec duration and of increasing strength for a Class I crab axon. [Adapted with permission from Reginald A. Chapman (1966), "The repetitive responses of isolated axons from the crab, *Carcinus maenas*," *Journal of Experimental Biology* 45. Copyright © 1966 by the Company of Biologists.] (B) Intensity versus response characteristics of a retinal on-center ganglion cell at different adaptation luminances: ordinate, discharge during the first 50 msec of the response; abscissa, test spot luminance (size of test spot 20-min arc). The figures on the curve are the logarithm of the adaptation luminance at which the curves were obtained. [Abridged caption and figure from O. D. Creutzfeldt, (1972), "Transfer function of the retina," *EEG Journal*, Supplement No. 31: *Recent Contributions to Neurophysiology*, J.-P. Courdeaux and P. Gloor (Eds.), Amsterdam: Elsevier.]

Overall, both these examples of real neurons display a common picture of the response of a neuron to stimulation. In one case (A) the stimulus is electrical, and in the other (B) it is light intensity. Response is taken to be firing frequency. The response displays a threshold stimulus below which there is little or no response; then, as stimulus intensity increases, there is a somewhat linear region; and finally, there is a region of saturation where there is little increase in response as stimulation increases. The response to light shown in (B) indicates that the form of the response remains almost unchanged as average illumination increases, due to a variety of adaptation phenomena. Neural network modelers sometimes approximate this response function as a sigmoid or as a linear function with clipping.

describe the system at a moment in time, we have to give the activities of all the cells in the system at that time. This set of simultaneous element activities is represented by a *state vector*, corresponding to the activities of many cells.

Neural networks have lots of computing elements connected to lots of other elements. This set of connections is often arranged in a *connection matrix*. The overall behavior of the system is determined by the structure and strengths of the connections. It is possible to change the connection strengths by various learning algorithms. It is also possible to build in various kinds of dynamics into the responses of the computing elements.

Learning

Detailed computations in neural networks are largely performed by the connection strengths—hence the name 'connectionist.' There is often a decoupling between the learning phase and the retrieval phase of operation of a network. In the *learning phase*, the connection strengths in the network are modified. Sometimes, if the constructor of the network is very clever or if the problem structure is so well defined that it allows it, it is possible to specify the connection strengths a priori. Otherwise, it is necessary to modify strengths using one of a number of useful learning algorithms, many of which are described in detail in the papers following.

There are currently a large number of learning algorithms used to set connection strengths. We shall discuss them in more detail in the introductions to the papers and in the papers themselves. Learning algorithms have been the heart of neural network research for the past three decades.

Network Operation

In the *retrieval phase*, some initial information, in the form of an initial state vector or activity pattern, is put into the system. In most connectionist networks, if no initial information is provided, nothing useful is retrieved, and the more information is provided, generally the more reliable is the output information. The initial input pattern passes through the connections to other elements, giving rise to an output pattern. If the system works properly, the output pattern contains the conclusions of the system. Given the complexity of the connections and the fact that many operations are going on simultaneously, it is often very hard to analyze exactly what is going on in the network. Although programs for traditional computers may be complicated, there is always faith that a program bug has a simple, usually localized, cause. This may not be true for a neural network, since both correct information and error may be spread out over many connections and many model neurons. This widespread distribution of computation also leads to a number of intrinsic error mechanisms such as interference between different events that are difficult, and probably impossible, to get rid of, given that many of the desirable features of networks (noise and damage tolerance, generalization) arise from the same cause.

Network Structure

There are a number of ways of organizing the computing elements in neural networks. Typically the elements are arranged in groups or layers. A single layer of neurons that connects to itself is referred to as an *autoassociative* system. Single-layer and two-layer systems, with only an input and an output layer, are easy to analyze and quite powerful, and are used extensively in many of the papers we present. More recently, learning algorithms have become available that can force the network to develop appropriate connection strengths in multilayer networks. There is currently a great deal of effort devoted to understanding multilayer systems. Such systems are potentially much more powerful than one- and two-layer systems, but they are also more complex and harder to analyze. For example, sometimes there are elements in the middle layers that are neither input nor output neurons, and are referred to as *hidden units* (see figure 3).

Representation

In the earliest days of neural networks, major research attention focused on the network structure, assumptions about the properties of the model neurons, and the learning algorithms used to change connection strengths.

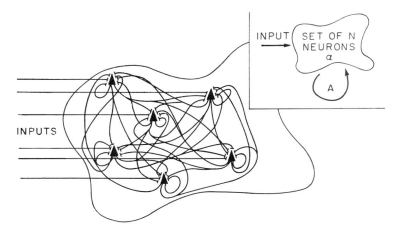

I. SET OF N NEURONS, α
2. EVERY NEURON IN α IS CONNECTED TO EVERY
   OTHER NEURON IN α THROUGH LEARNING
   MATRIX OF SYNAPTIC CONNECTIVITIES A

3A

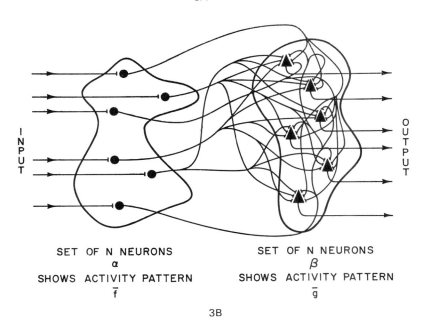

3B

Figure 3
(A) One-layer network. A group of neurons feeds back on itself. [Caption and figure from James A.
Anderson, Jack W. Silverstein, Stephen A. Ritz, and Randall Jones, "Distinctive features, categorical
perception, and probability learning: some applications of a neural model," *Psychological Review* 84,
figure 3.] (B) Two-layer network with an input set and an output set. This is a strictly feedforward
network, where the input set projects to the output set, but not vice versa. [Figure from James A.
Anderson, Jack W. Silverstein, Stephen A. Ritz, and Randall Jones, "Distinctive features, categorical
perception, and probability learning: some applications of a neural model," *Psychological Review* 84,
figure 1.] (C) Three-layer network. Schematic drawing of the network architecture. Input units are shown
on the bottom of the pyramid.... Each hidden unit in the intermediate layer receives inputs from all
of the input units on the bottom layer, and in turn sends its output to all 26 units in the output layer.
[Abridged caption and figure from Terrence J. Sejnowski and Charles R. Rosenberg, "NETtalk: a parallel
network that learns to read aloud," The Johns Hopkins University Electrical Engineering and Computer
Science Technical Report JHU/EECS-86/01.]

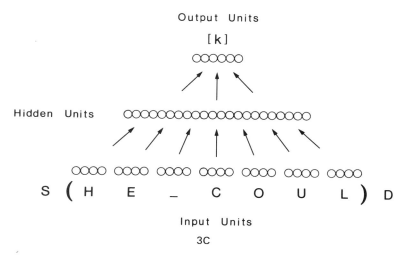

Figure 3 (continued)

Now that we have some idea of what various learning algorithms and network structures can do, there is a growing interest in the problems of representation of information in the network. The mammalian brain often follows very simple rules for representation, for example, arranging visual information into topographic maps on the cortex or mapping the surface of the body onto a cortical map that looks like a distorted map of the body. Presumably the way the brain represents information is useful for the kinds of things the brain does with it.

There is also interest in studying systems that can organize themselves, that is, develop the optimal way of arranging the information they must represent. Modelers can either make use of representations given to us by nature or improve on them.

Our feeling is that problems of representation may prove to be the major area for research in neural networks in the next few years. This is because *all* practical applications of networks are critically dependent on the way the problem is represented. With the proper description of the inputs, many learning algorithms will probably work adequately. Without the proper description, none may work. One of the major criticisms of neural networks has been the necessity up to this point to 'handcraft' the representations in order to make the systems work. There are some obvious commonsense general rules for representations that also seem to be brain-like. First, similar inputs *usually* should give rise to similar representations. Second, things to be separated should be given widely different representations. Third, if something—a sensory property or feature of the input—is important, lots of elements should be used to represent it. Fourth, do as much lower-level preprocessing as possible, so the learning and adaptive parts of the network need do as little work as possible. Build 'invariances' into the hardware and do not require the system to learn them.

The Future

We have provided an Afterword after the last paper, to summarize some of our feelings about the way the field of neural networks has developed and where it now stands.

These comments are meant to be terse. The field of neural networks has had disasters in the past, and the science as a whole has passed from great enthusiasm to rejection to, now, modest excitement. Current understanding of what networks can and cannot do is much more realistic than it was twenty years ago, and our technical resources are much greater, so it is easy to contemplate special-purpose hardware and very large systems. Perhaps in this cycle of reincarnation neural networks will take their place as a useful complement to and extension of traditional computer hardware. We hope so.

Bibliographic Note

We have collected a number of papers that we feel are particularly useful for understanding neural networks. Sometimes these are not the first published versions of the ideas, though often they are. We have usually tried to find the clearest exposition of a point of view, or the most telling example or application of it. We shall point out some of the earlier literature when we think it is appropriate or make historical comments in the introductions to the papers. We have tried to make these introductions to the individual papers reasonably self-contained. Our own experience has been that introductions to individual papers are read much more often than general introductions such as this.

The field of brain modeling and neural networks has attracted over the years a number of scientists with exceptionally strong egos. Protracted and bitter battles over priorities have been waged, remarkable for the field at a time when there were so few participants and so little respect, and still remarkable now, in better times.

Our own feeling is that all ideas have roots. It is virtually impossible to find an 'original' idea that was not presented earlier in the scientific literature, if one is willing to take the effort to search the historical record. When the need for an idea is in the air, the idea is usually discovered by several groups simultaneously. A recent example of this from the neural network literature is the simultaneous discovery of the back propagation learning algorithm in three places in 1985.

We should simply point out that science is a cooperative effort and that no scientist is an island. All are subject to the ideas of the time. To use Newton's metaphor, if one scientist happens to pick up a shinier or more valuable pebble than another, perhaps he only needed enough wisdom to be standing at the point on the beach where those pebbles were likely to be. Or, he had slightly sharper eyes. Or, he spent a lot of time at the shore. Or, perhaps, he only stubbed his toe on something and happened to pick it up through dumb luck. Now, with many scientists and others searching the neural net beaches, lots of new pebbles may be discovered. Science is unlike art and literature in that if one scientist does not find a particular idea, then another one will, and soon.

1
Introduction

(1890)
William James

Psychology (Briefer Course), New York: Holt, Chapter XVI, "Association," pp. 253–279

William James was the greatest psychologist that America has produced. *Psychology (Briefer Course)* is a Psych 1 text, designed to give undergraduates an exposure to the main ideas of scientific psychology, as it was understood in the late nineteenth century. James wrote a two-volume set, *Principles of Psychology*, which was abridged and condensed to produce the shorter version. From our point of view, this was desirable, because in the abridgment James cut out details of exposition and experiment, which have not aged well, and left the main ideas and arguments, which have.

We have reprinted only the chapter on association in this volume, because of its remarkable foreshadowing of some of the main ideas of neural networks. However, the rest of the book is equally impressive in other ways, and is well worth reading as a source of insights that are still valuable. The book is available in paperback in several editions.

It is often not appreciated that James taught physiology for a while early in his career and knew a great deal about the structure of the brain. Chapters 2–9 of *Psychology (Briefer Course)* are devoted to a description of neuroanatomy and neurophysiology as understood at the time. Chapter 9 contains the wonderful line, "The way really to understand the brain is to dissect it" (p. 81), followed by a detailed description of how to obtain a sheep's brain, where to obtain an autopsy kit, and what to look for during dissection. It is hard to overemphasize the importance of this point, both to James and to those who wish to understand or model the brain as a working organ. Unfortunately, some of the currently available paperback editions of *Psychology (Briefer Course)* cut out the chapters with the neuroanatomy and neurophysiology, and renumber the remaining chapters. Looking back from the perspective of a century it is, first, remarkable how much was known about the brain in 1890, and, second, how much we have learned since then.

As James points out emphatically in several places, the brain is not constructed to think abstractly—it is constructed to ensure survival in the world. It has many of the characteristics of a good engineering solution applied to mental operation: do as good a job as you can, cheaply, and with what you can obtain easily. If this means using ad hoc solutions, with less generality than one might like, so be it. We are living in one particular world, and we are built to work in it and with it. As James put it (pp. 3–4),

Mental facts cannot properly be studied apart from the physical environment of which they take cognizance. The great fault of the older rational psychology was to set up the soul as an absolute spiritual being with certain faculties of its own by which the several activities of remembering, imagining, reasoning, willing, etc. were explained, almost without reference to the peculiarities of the world with which these activities deal. But the richer insight of modern days perceives that our inner faculties are *adapted* in advance to the features of the world in which we dwell.... Mind and world in short have evolved together, and in consequence are something of a mutual fit.

James' comment about rational psychology could be applied, unfortunately, to some kinds of current artificial intelligence and neural network research.

This evolutionary perspective is prominent in current neurobiology as well as in James' book, and should not be lost sight of when working with neural networks. The emerging field of neuroethology demonstrates over and over how tightly coupled details of nervous system organization and species specific behavior are. Frogs effectively see only moving bugs (Lettvin, Maturana, McCulloch, and Pitts, 1959). Toads see moving worms: light worms on dark backgrounds in the summer and dark worms on light backgrounds in the winter (Ewart, 1980). In primates, many cells in areas of temporal neocortex respond preferentially to faces (Baylis, Rolls, and Leonard, 1987). The number of such examples could be extended indefinitely. Camhi (1984) gives a series of wonderful stories about the close connection between the nervous system and an animal's behavior.

One important implication of this is that we should not be surprised if brains cannot compute everything, but only a small, but useful, subset of problems. Brains are only as powerful as they have to be, and are often surprisingly special purpose. The kind of brain organization that we seem to have is very poor at doing arithmetic and formal logic; when it was evolving, there were limited opportunities for doing either. But the ability to form concepts, to see that different things were examples of the same thing, was truly important, as were the abilities to form somewhat arbitrary associations, to make good guesses, and so on. It is surprising that our brains, which evolved for purposes like these, are capable of doing arithmetic and logic at all.

The most interesting thing about the chapter on association is that it presents a detailed, mechanistic model of association that is almost identical in structure to later associative neural networks. If James had thought in terms of mathematical models and had access to a computer, it is hard to believe that he would not have developed a computational network model for association.

First, James believed that association was mechanistic and a function of the cerebral cortex.

Second, he formulated a general elementary principle of association (James' emphasis): "*When two brain processes are active together or in immediate succession, one of them, on reoccuring tends to propagate its excitement into the other*" (p. 256). If we replace "brain process" with neuron, we have a correlational learning rule, almost identical to the Hebb synapse (Hebb, paper 4).

Third, there is a summing rule for brain activity (James' emphasis): "*The amount of activity at any given point in the brain cortex is the sum of the tendencies of all other points to discharge into it, such tendencies being proportionate (1) to the number of times the excitement each other point may have accompanied that of the point in question; (2) to the intensities of such excitements; and (3) to the absence of any rival point functionally disconnected with the first point, into which the discharges might be diverted*" (p. 257).

If we replace "point in the brain cortex" with neuron, or element in the connectionist sense, then we have a model that gives neuron activity as the sum of its inputs, weighted by a connection strength given by the history of past correlations (point 1), and the current excitement of other neurons (point 2), and with an inhibitory mechanism (point 3). This structure is very close to any one of a number of network models using Hebbian synaptic modification and linear summation of synaptic inputs. This connection is made even stronger if we compare figure 57 in James with the diagram of a typical simple heteroassociative network (see, for example, Anderson et al., paper 22, figure 1).

James considers most complex events to be made up of numerous subassociations, all interconnected by "elementary nerve tracts." He also discusses, under the old psychological term *redintegration*, the ability of networks of partial associations to reconstruct the missing pieces through the cross associations. This tendency, which is shown by almost all connectionist networks and is usually held to be a virtue in them, is seen by James as a mixed blessing. He suggests what we would now call a control structure—interest—as a way to keep this tendency in check, which otherwise would lead to a mechanistic "core dump" containing excess detail.

It is worth discussing William James' connectionist model to show that modern neural networks are not as unique as sometimes claimed. The outlines of network systems were discussed in the nineteenth century. What is different today is the ability, thanks to the computer and general familiarity with mathematical analysis, to take such theoretical networks and actually construct them and use them. Our command of the detail, techniques, and application of modeling is now very much greater. However, the basic insights of James into the operation of the mind have not been fundamentally altered.

References

G. C. Baylis, E. T. Rolls, and C. M. Leonard (1987), "Functional subdivisions of the temporal lobe neocortex," *The Journal of Neuroscience* 7:330–342.

J. M. Camhi (1984), *Neuroethology*, Sunderland, MA: Sinauer.

J.-P. Ewart (1980), *Neuroethology: An Introduction to the Neurophysiological Fundamentals of Behavior*, Berlin: Springer.

J. Y. Lettvin, H. R. Maturana, W. S. McCulloch, and W. H. Pitts (1959), "What the frog's eye tells the frog's brain," *Proceedings of the I.R.E.* 47:1940–1951.

(1890)
William James

Psychology (Briefer Course), New York: Holt, Chapter XVI, "Association," pp. 253–279

Chapter XVI
Association

The Order of Our Ideas

After discrimination, association! It is obvious that all advance in knowledge must consist of both operations; for in the course of our education, objects at first appearing as wholes are analyzed into parts, and objects appearing separately are brought together and appear as new compound wholes to the mind. Analysis and synthesis are thus the incessantly alternating mental activities, a stroke of the one preparing the way for a stroke of the other, much as, in walking, a man's two legs are alternately brought into use, both being indispensable for any orderly advance.

The manner in which trains of imagery and consideration follow each other through our thinking, the restless flight of one idea before the next, the transitions our minds make between things wide as the poles asunder, transitions which at first sight startle us by their abruptness, but which, when scrutinized closely, often reveal intermediating links of perfect naturalness and propriety—all this magical, imponderable streaming has from time immemorial excited the admiration of all whose attention happened to be caught by its omnipresent mystery. And it has furthermore challenged the race of philosophers to banish something of the mystery by formulating the process in simpler terms. The problem which the philosophers have set themselves is that of ascertaining, between the thoughts which thus appear to sprout one out of the other, *principles of connection* whereby their peculiar succession or coexistence may be explained.

But immediately an ambiguity arises: which sort of connection is meant? connection *thought-of,* or connection *between thoughts?* These are two entirely different things, and only in the case of one of them is there any hope of finding 'principles.' The jungle of connections *thought of* can never be formulated simply. Every conceivable connection may be thought of—of coexistence, succession, resemblance, contrast, contradiction, cause and effect, means and end, genus and species, part and whole, substance and property, early and late, large and small, landlord and tenant, master and servant,—Heaven knows what, for the list is literally inexhaustible. The only simplification which could possibly be aimed at would be the reduction of the relations to a small number of *types,* like those which some authors call the 'categories' of the understanding. According as we followed one category or another we should sweep, from any object with our thought, in this way or in that, to others. Were *this* the sort of connection sought between one moment of our thinking and another, our chapter might end here. For the only summary description of these categories is that they are all thinkable relations, and that the mind proceeds from one object to another by some intelligible path.

Is It Determined by Any Laws?

But as a matter of fact, What determines the particular path? Why do we at a given time and place proceed to think of *b* if we have just thought of *a,* and at another time and place why do we think, not of *b,* but of *c?* Why do we spend years straining after a certain scientific or practical problem, but all in vain—our thought unable to evoke the solution we desire? And why, some day, walking in the street with our attention miles away from that quest, does the answer saunter into our minds as carelessly as if it had never been called for—suggested, possibly, by the flowers on the bonnet of the lady in front of us, or possibly by nothing that we can discover?

The truth must be admitted that thought works under strange conditions. Pure 'reason' is only one out of a thousand possibilities in the thinking of each of us. Who can count all the silly fancies, the grotesque suppositions, the utterly irrelevant reflections he makes in the course of a day? Who can swear that his prejudices and irrational opinions constitute a less bulky part of his mental furniture than his clarified beliefs? And yet, the *mode of genesis* of the worthy and the worthless in our thinking seems the same.

The Laws Are Cerebral Laws

There seem to be mechanical conditions on which thought depends, and which, to say the least, *determine the order in which the objects for her comparisons and selections are presented.* It is a suggestive fact that Locke, and many more recent Continental psychologists, have found themselves obliged to invoke a mechanical process to account for the *aberrations* of thought, the obstructive prepossessions, the frustrations of reason. This they found in the law of habit, or what we now call association by contiguity. But it never occurred to these writers that a process which could go the length of actually producing some ideas and sequences in the mind might safely be trusted to produce others too; and that those habitual associations which further thought may also come from the same mechanical source as those which hinder it. Hartley accordingly suggested habit as a sufficient explanation of the sequence of our thoughts, and in so doing planted himself squarely upon the properly *causal* aspect of the problem, and sought to treat both rational and irrational associations from a single point of view. How does a man come, after having the thought of A, to have the thought of B the next moment? or how does he come to think A and B always together? These were the phenomena which Hartley undertook to explain by cerebral physiology. I believe that he was, in essential respects, on the right track, and I propose simply to revise his conclusions by the aid of distinctions which he did not make.

Objects Are Associated, Not Ideas

We shall avoid confusion if we consistently speak as if *association,* so far as the word stands for an *effect, were between* THINGS THOUGHT OF—*as if it were* THINGS, *not ideas, which are associated in the mind.* We shall talk of the association of *objects,* not of the association of *ideas.* And so far as association stands for a *cause,* it is between *processes in the brain*—it is these which, by being associated in certain ways, determine what successive objects shall be thought.

The Elementary Principle

I shall now try to show that there is no other *elementary* causal law of association than the law of neural habit. All the *materials* of our thought are due to the way in which one elementary process of the cerebral hemispheres tends to excite whatever other elementary process it may have excited at any former time. The number of elementary processes at work, however, and the nature of those which at any time are fully effective in rousing the others, determine the character of the total brain-action, and, as a consequence of this, they determine the object thought of at the time. According as this resultant object is one thing or another, we call it a product of association by contiguity or of association by similarity, or contrast, or whatever other sorts we may have recognized as ultimate. Its *production,* however, is, in each one of these cases, to be explained by a merely quantitative variation in the elementary brain-processes momentarily at work under the law of habit.

My thesis, stated thus briefly, will soon become more clear; and at the same time certain disturbing factors, which coöperate with the law of neural habit, will come to view.

Let us then assume as the basis of all our subsequent reasoning this law: *When two elementary brain-processes have been active together or in immediate succession, one of them, on re-occurring, tends to propagate its excitement into the other.*

But, as a matter of fact, every elementary process has unavoidably found itself at different times excited in conjunction with *many* other processes. Which of these others it shall awaken now becomes a problem. Shall *b* or *c* be aroused next by the present *a*? To answer this, we must make a further postulate, based on the fact of *tension* in nerve-tissue, and on the fact of summation of excitements, each incomplete or latent in itself, into an open resultant (see p. 128). The process *b,* rather than *c,* will awake, if in addition to the vibrating tract *a* some other tract *d* is in a state of sub-excitement, and formerly was excited with *b* alone and not with *a.* In short, we may say:

The amount of activity at any given point in the brain-cortex is the sum of the tendencies of all other points to discharge into it, such tendencies being proportionate (1) *to the number of times the excitement of each other point may have accompanied that of the point in question;* (2) *to the intensity of such excitements; and* (3) *to the absence of any rival point functionally disconnected with the first point, into which the discharges might be diverted.*

Expressing the fundamental law in this most complicated way leads to the greatest ultimate simplification. Let us, for the present, only treat of spontaneous trains of thought and ideation, such as occur in revery or musing. The case of voluntary thinking toward a certain end shall come up later.

Spontaneous Trains of Thought

Take, to fix our ideas, the two verses from 'Locksley Hall':

"I, the heir of all *the ages* in the foremost files of time," and—

"For I doubt not through *the ages* one increasing purpose runs."

Why is it that when we recite from memory one of these lines, and get as far as *the ages*, that portion of the *other* line which follows and, so to speak, sprouts out of *the ages* does not also sprout out of our memory and confuse the sense of our words? Simply because the word that follows *the ages* has its brain-process awakened not simply by the brain-process of *the ages* alone, but by it *plus* the brain-processes of all the words preceding *the ages*. The word *ages* at its moment of strongest activity would, *per se*, indifferently discharge into either 'in' or 'one.' So would the previous words (whose tension is momentarily much less strong than that of *ages*) each of them indifferently discharge into either of a large number of other words with which they have been at different times combined. But when the processes of '*I, the heir of all the ages*,' simultaneously vibrate in the brain, the last one of them in a maximal, the others in a fading, phase of excitement, then the strongest line of discharge will be that which they *all alike* tend to take. '*In*' and not '*one*' or any other word will be the next to awaken, for its brain-process has previously vibrated in unison not only with that of *ages*, but with that of all those other words whose activity is dying away. It is a good case of the effectiveness over thought of what we called on p. 168 a 'fringe.'

But if some one of these preceding words—'heir,' for example—had an intensely strong association with some brain-tracts entirely disjoined in experience from the poem of 'Locksley Hall'—if the reciter, for instance, were tremulously awaiting the opening of a will which might make him a millionaire—it is probable that the path of discharge through the words of the poem would be suddenly interrupted at the word 'heir.' His *emotional interest in that word* would be such that its *own special associations would prevail* over the combined ones of the other words. He would, as we say, be abruptly reminded of his personal situation, and the poem would lapse altogether from his thoughts.

The writer of these pages has every year to learn the names of a large number of students who sit in alphabetical order in a lecture-room. He finally learns to call them by name, as they sit in their accustomed places. On meeting one in the street, however, early in the year, the face hardly ever recalls the name, but it may recall the place of its owner in the lecture-room, his neighbors' faces, and consequently his general alphabetical position: and then, usually as the common associate of all these combined data, the student's name surges up in his mind.

A father wishes to show to some guests the progress of his rather dull child in kindergarten-instruction. Holding the knife upright on the table, he says, "What do you call that, my boy?" "I calls it a *knife*, I does," is the sturdy reply, from which the child cannot be induced to swerve by any alteration in the form of question, until the father, recollecting that in the kindergarten a pencil was used and not a knife, draws a long one from his pocket, holds it in the same way, and then gets the wished-for answer, "I calls it *vertical*." All the concomitants of the kindergarten experience had to recombine their effect before the word 'vertical' could be reawakened.

Total Recall

The ideal working of the law of compound association, as Prof. Bain calls it, were it unmodified by any extraneous influence, would be such as to keep the mind in a perpetual treadmill of concrete reminiscences from which no detail could be omitted. Suppose, for example, we begin by thinking of a certain dinner-party. The only thing which all the components of the dinner-party could combine to recall would be the first concrete occurrence which ensued upon it. All the details of this occurrence could in turn only combine to awaken the next following occurrence, and so on. If a, b, c, d, e, for instance, be the elementary nerve-tracts excited by the last act of the dinner-party, call this act A, and l, m, n, o, p be those of walking home through the frosty night, which we may call B, then the thought of A must awaken that of B, because a, b, c, d, e will each and all discharge into l through the paths by which their original discharge took place. Similarly they will discharge into m, n, o, and p; and these latter tracts will also each reinforce the other's action because, in the experience B, they have already vibrated in unison. The lines in Fig. 57 symbolize the summation of discharges into each of the components of B, and the consequent strength of the combination of influences by which B in its totality is awakened.

Hamilton first used the word 'redintegration' to designate all association. Such processes as we have just described might in an emphatic sense be termed redintegrations, for they would necessarily lead, if unobstructed, to the reinstatement in thought of the *entire* content of large trains of past experience. From this complete redintegration there could be no escape save through the irruption of some new and strong present impression of the senses, or through the excessive tendency of some one of the elementary brain-tracts to discharge independently into an aberrant quarter of the brain. Such was the tendency of the word 'heir' in

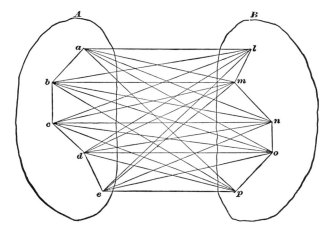

Figure 57

the verse from 'Locksley Hall,' which was our first example. How such tendencies are constituted we shall have soon to inquire with some care. Unless they are present, the panorama of the past, once opened, must unroll itself with fatal literality to the end, unless some outward sound, sight, or touch divert the current of thought.

Let us call this process *impartial redintegration*, or, still better, *total recall*. Whether it ever occurs in an absolutely complete form is doubtful. We all immediately recognize, however, that in some minds there is a much greater tendency than in others for the flow of thought to take this form. Those insufferably garrulous old women, those dry and fanciless beings who spare you no detail, however petty, of the facts they are recounting, and upon the thread of whose narrative all the irrelevant items cluster as pertinaciously as the essential ones, the slaves of literal fact, the stumblers over the smallest abrupt step in thought, are figures known to all of us. Comic literature has made her profit out of them. Juliet's nurse is a classical example. George Eliot's village characters and some of Dickens's minor personages supply excellent instances.

Perhaps as successful a rendering as any of this mental type is the character of Miss Bates in Miss Austen's 'Emma.' Hear how she redintegrates:

"'But where could *you* hear it?' cried Miss Bates. 'Where could you possibly hear it, Mr. Knightley? For it is not five minutes since I received Mrs. Cole's note—no, it cannot be more than five—or at least ten—for I had got my bonnet and spencer on, just ready to come out—I was only gone down to speak to Patty again about the pork—Jane was standing in the passage—were not you, Jane?—for my mother was so afraid that we had not any salting-pan large enough. So I said I would go down and see, and Jane said: "Shall I go down instead? for I think you have a little cold, and Patty has been washing the kitchen." "Oh, my dear," said I—well, and just then came the note. A Miss Hawkins—that's all I know—a Miss Hawkins, of Bath. But, Mr. Knightley, how could you possibly have heard it? for the very moment Mr. Cole told Mrs. Cole of it, she sat down and wrote to me. A Miss Hawkins—'"

Partial Recall

This case helps us to understand why it is that the ordinary spontaneous flow of our ideas does not follow the law of total recall. *In no revival of a past experience are all the items of our thought equally operative in determining what the next thought shall be. Always some ingredient is prepotent over the rest.* Its special suggestions or associations in this case will often be different from those which it has in common with the whole group of items; and its tendency to awaken these outlying associates will deflect the path of our revery. Just as in the original sensible experience our attention focalized itself upon a few of the impressions of the scene before us, so here in the reproduction of those impressions an equal partiality is shown, and some items are emphasized above the rest. What these items shall be is, in most cases of spontaneous revery, hard to determine beforehand. In subjective terms we say that *the prepotent items are those which appeal most to our* INTEREST.

Expressed in brain-terms, the law of interest will be: *some one brain-process is always prepotent above its concomitants in arousing action elsewhere.*

"Two processes," says Mr. Hodgson, "are constantly going on in redintegration. The one a process of corrosion, melting, decay; the other a process of renewing, arising, becoming.... No object of representation remains long before consciousness in the same state, but fades, decays, and becomes indistinct. Those parts of the object, however, which possess an interest resist this tendency to gradual decay of the whole object.... This inequality in the object—some parts, the uninteresting, submitting to decay; others, the interesting parts, resisting it—when it has continued for a certain time, ends in becoming a new object."

Only where the interest is diffused equally over all the parts is this law departed from. It will be least obeyed by those minds which have the smallest variety and intensity of interests—those who, by the general flatness and poverty of their aesthetic nature, are kept for ever rotating among the literal sequences of their local and personal history.

Most of us, however, are better organized than this,

and our musings pursue an erratic course, swerving continually into some new direction traced by the shifting play of interest as it ever falls on some partial item in each complex representation that is evoked. Thus it so often comes about that we find ourselves thinking at two nearly adjacent moments of things separated by the whole diameter of space and time. Not till we carefully recall each step of our cogitation do we see how naturally we came by Hodgson's law to pass from one to the other. Thus, for instance, after looking at my clock just now (1879), I found myself thinking of a recent resolution in the Senate about our legal-tender notes. The clock called up the image of the man who had repaired its gong. He suggested the jeweller's shop where I had last seen him; that shop, some shirt-studs which I had bought there; they, the value of gold and its recent decline; the latter, the equal value of greenbacks, and this, naturally, the question of how long they were to last, and of the Bayard proposition. Each of these images offered various points of interest. Those which formed the turning-points of my thought are easily assigned. The gong was momentarily the most interesting part of the clock, because, from having begun with a beautiful tone, it had become discordant and aroused disappointment. But for this the clock might have suggested the friend who gave it to me, or any one of a thousand circumstances connected with clocks. The jeweller's shop suggested the studs, because they alone of all its contents were tinged with the egoistic interest of possession. This interest in the studs, their value, made me single out the material as its chief source, etc., to the end. Every reader who will arrest himself at any moment and say, "How came I to be thinking of just this?" will be sure to trace a train of representations linked together by lines of contiguity and points of interest inextricably combined. This is the ordinary process of the association of ideas as it spontaneously goes on in average minds. *We may call it ordinary, or mixed, association,* or, if we like better, *partial recall.*

Which Associates Come Up, in Partial Recall?

Can we determine, now, when a certain portion of the going thought has, by dint of its interest, become so prepotent as to make its own exclusive associates the dominant features of the coming thought—can we, I say, determine *which* of its own associates shall be evoked? For they are many. As Hodgson says:

"The interesting parts of the decaying object are free to combine again with any objects or parts of objects with which at any time they have been combined before.

All the former combinations of these parts may come back into consciousness; one must, but which will?"

Mr. Hodgson replies:

"There can be but one answer: that which has been most *habitually* combined with them before. This new object begins at once to form itself in consciousness, and to group its parts round the part still remaining from the former object; part after part comes out and arranges itself in its old position; but scarcely has the process begun, when the original law of interest begins to operate on this new formation, seizes on the interesting parts and impresses them on the attention to the exclusion of the rest, and the whole process is repeated again with endless variety. I venture to propose this as a complete and true account of the whole process of redintegration."

In restricting the discharge from the interesting item into that channel which is simply most *habitual* in the sense of most frequent, Hodgon's account is assuredly imperfect. An image by no means always revives its most frequent associate, although frequency is certainly one of the most potent determinants of revival. If I abruptly utter the word *swallow*, the reader, if by habit an ornithologist, will think of a bird; if a physiologist or a medical specialist in throat-diseases, he will think of deglutition. If I say *date*, he will, if a fruit-merchant or an Arabian traveller, think of the produce of the palm; if an habitual student of history, figures with A.D. or B.C. before them will rise in his mind. If I say *bed*, *bath*, *morning*, his own daily toilet will be invincibly suggested by the combined names of three of its habitual associates. But frequent lines of transition are often set at naught. The sight of a certain book has most frequently awakened in me thoughts of the opinions therein propounded. The idea of suicide has never been connected with the volume. But a moment since, as my eye fell upon it, suicide was the thought that flashed into my mind. Why? Because but yesterday I received a letter informing me that the author's recent death was an act of self-destruction. Thoughts tend, then, to awaken their most recent as well as their most habitual associates. This is a matter of notorious experience, too notorious, in fact, to need illustration. If we have seen our friend this morning, the mention of his name now recalls the circumstances of that interview, rather than any more remote details concerning him. If Shakespeare's plays are mentioned, and we were last night reading 'Richard II.,' vestiges of that play rather than of 'Hamlet' or 'Othello' float through our mind. Excitement of peculiar tracts, or peculiar modes of general excitement in the brain, leave a sort of tenderness or exalted sensibility behind them which takes days to die away. As long as it lasts, those tracts or those modes are liable to have their activities awakened by causes which at other times might leave them in repose. Hence, *recency* in experience is a prime factor in determining revival in thought.*

Vividness in an original experience may also have the same effect as habit or recency in bringing about likelihood of revival. If we have once witnessed an execution, any subsequent conversation or reading about capital punishment will almost certainly suggest images of that particular scene. Thus it is that events lived through only once, and in youth, may come in after-years, by reason of their exciting quality or emotional intensity, to serve as types or instances used by our mind to illustrate any and every occurring topic whose interest is most remotely pertinent to theirs. If a man in his boyhood once talked with Napoleon, any mention of great men or historical events, battles or thrones, or the whirligig of fortune, or islands in the ocean, will be apt to draw to his lips the incidents of that one memorable interview. If the word *tooth* now suddenly appears on the page before the reader's eye, there are fifty chances out of a hundred that, if he gives it time to awaken any image, it will be an image of some operation of dentistry in which he has been the sufferer. Daily he has touched his teeth and masticated with them; this very morning he brushed, used, and picked them; but the rarer and remoter associations arise more promptly because they were so much more intense.

A fourth factor in tracing the course of reproduction is *congruity in emotional tone* between the reproduced idea and our mood. The same objects do not recall the same associates when we are cheerful as when we are melancholy. Nothing, in fact, is more striking than our inability to keep up trains of joyous imagery when we are depressed in spirits. Storm, darkness, war, images of disease, poverty, perishing, and dread afflict unremittingly the imaginations of melancholiacs. And those of sanguine temperament, when their spirits are high, find it impossible to give any permanence to evil forebodings or to gloomy thoughts. In an instant the train of association dances off to flowers and sunshine, and images of spring and hope. The records of Arctic or African travel perused in one mood awaken no thoughts but those of horror at the malignity of Nature; read at another time they suggest only enthusiastic reflections on the indomitable power and pluck of man. Few novels so overflow with joyous animal spirits as 'The Three Guardsmen' of Dumas. Yet it may awaken in

the mind of a reader depressed with sea-sickness (as the writer can personally testify) a most woful consciousness of the cruelty and carnage of which heroes like Athos, Porthos, and Aramis make themselves guilty.

Habit, recency, vividness, and emotional congruity are, then, all reasons why one representation rather than another should be awakened by the interesting portion of a departing thought. We may say with truth that *in the majority of cases the coming representation will have been either habitual, recent, or vivid, and will be congruous.* If all these qualities unite in any one absent associate, we may predict almost infallibly that that associate of the going object will form an important ingredient in the object which comes next. In spite of the fact, however, that the succession of representations is thus redeemed from perfect indeterminism and limited to a few classes whose characteristic quality is fixed by the nature of our past experience, it must still be confessed that an immense number of terms in the linked chain of our representations fall outside of all assignable rule. To take the instance of the clock given on page 263. Why did the jeweller's shop suggest the shirt-studs rather than a chain which I had bought there more recently, which had cost more, and whose sentimental associations were much more interesting? Any reader's experience will easily furnish similar instances. So we must admit that to a certain extent, even in those forms of ordinary mixed association which lie nearest to impartial redintegration, *which* associate of the interesting item shall emerge must be called largely a matter of accident—accident, that is, for our intelligence. No doubt it is determined by cerebral causes, but they are too subtle and shifting for our analysis.

Focalized Recall, or Association by Similarity

In partial or mixed association we have all along supposed the interesting portion of the disappearing thought to be of considerable extent, and to be sufficiently complex to constitute by itself a concrete object. Sir William Hamilton relates, for instance, that after thinking of Ben Lomond he found himself thinking of the Prussian system of education, and discovered that the links of association were a German gentleman whom he had met on Ben Lomond, Germany, etc. The interesting part of Ben Lomond as he had experienced it, the part operative in determining the train of his ideas, was the complex image of a particular man. But now let us suppose that the interested attention refines itself still further and accentuates a portion of the passing object, so small as to be no longer the image of a concrete thing, but only of an abstract quality or property. Let us moreover suppose that the part thus

---

* I refer to a recency of a few hours. Mr. Galton found that experiences from boyhood and youth were more likely to be suggested by words seen at random than experiences of later years. See his highly interesting account of experiments in his *Inquiries into Human Faculty,* pp. 191–203.

accentuated persists in consciousness (or, in cerebral terms, has its brain-process continue) after the other portions of the object have faded. *This small surviving portion will then surround itself with its own associates* after the fashion we have already seen, and the relation between the new thought's object and the object of the faded thought will be a *relation of similarity.* The pair of thoughts will form an instance of what is called '*association by similarity.*'

The similars which are here associated, or of which the first is followed by the second in the mind, are seen to be *compounds.* Experience proves that this is always the case. *There is no tendency on the part of* SIMPLE '*ideas,*' attributes, or qualities to remind us of their like. The thought of one shade of blue does not summon up that of another shade of blue, etc., unless indeed we have in mind some general purpose of nomenclature or comparison which requires a review of several blue tints.

Now two compound things are similar when some one quality or group of qualities is shared alike by both, although as regards their other qualities they may have nothing in common. The moon is similar to a gas-jet, it is also similar to a foot-ball; but a gas-jet and a foot-ball are not similar to each other. When we affirm the similarity of two compound things, we should always say *in what respect it obtains.* Moon and gas-jet are similar in respect of luminosity, and nothing else; moon and foot-ball in respect of rotundity, and nothing else. Foot-ball and gas-jet are in no respect similar—that is, they possess no common point, no identical attribute. *Similarity, in compounds, is partial identity.* When the *same* attribute appears in two phenomena, though it be their only common property, the two phenomena are similar in so far forth. To return now to our associated representations. If the thought of the moon is succeeded by the thought of a foot-ball, and that by the thought of one of Mr. X's railroads, it is because the attribute rotundity in the moon broke away from all the rest and surrounded itself with an entirely new set of companions—elasticity, leathery integument, swift mobility in obedience to human caprice, etc.; and because the last-named attribute in the foot-ball in turn broke away from its companions, and, itself persisting, surrounded itself with such new attributes as make up the notions of a 'railroad king,' of a rising and falling stock-market, and the like.

The gradual passage from total to focalized, through what we have called ordinary partial, recall may be symbolized by diagrams. Fig. 58 is total, Fig. 59 is partial, and Fig. 60 focalized, recall. *A* in each is the passing, *B* the coming, thought. In 'total recall,' all parts of *A* are equally operative in calling up *B.* In

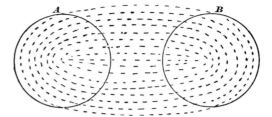

Figure 58

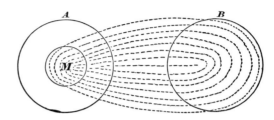

Figure 59

'partial recall,' most parts of *A* are inert. The part *M* alone breaks out and awakens *B.* In similar association or 'focalized recall,' the part *M* is much smaller than in the previous case, and after awakening its new set of associates, instead of fading out itself, it continues persistently active along with them, forming an identical part in the two ideas, and making these, *pro tanto,* resemble each other.*

Why a single portion of the passing thought should break out from its concert with the rest and act, as we say, on its own hook, why the other parts should become inert, are mysteries which we can ascertain but not explain. Possibly a minuter insight into the laws of neural action will some day clear the matter up;

* Miss M. W. Calkins (Philosophical Review, I. 389, 1892) points out that the persistent feature of the going thought, on which the association in cases of similarity hinges, is by no means always so slight as to warrant the term 'focalized.' "If the sight of the whole breakfast-room be followed by the visual image of yesterday's breakfast-table, with the same setting and in the same surroundings, the association is practically total," and yet the case is one of similarity. For Miss Calkins, accordingly, the more important distinction is that between what she calls *desistent* and *persistent* association. In 'desistent' association all parts of the going thought fade out and are replaced. In 'persistent' association some of them remain, and form a bond of similarity between the mind's successive objects; but only where this bond is extremely delicate (as in the case of an abstract relation or quality) is there need to call the persistent process 'focalized.' I must concede the justice of Miss Calkins's criticism, and think her new pair of terms a useful contribution. Wundt's division of associations into the two classes of *external* and *internal* is congruent with Miss Calkins's division. Things associated internally must have some element in common; and Miss Calkins's word 'persistent' suggests how this may cerebrally come to pass. 'Desistent,' on the other hand, suggests the process by which the successive ideas become external to each other or preserve no inner tie.

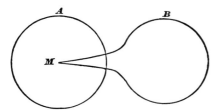

Figure 60

possibly neural laws will not suffice, and we shall need to invoke a dynamic reaction of the consciousness itself. But into this we cannot enter now.

Voluntary Trains of Thought

Hitherto we have assumed the process of suggestion of one object by another to be spontaneous. The train of imagery wanders at its own sweet will, now trudging in sober grooves of habit, now with a hop, skip, and jump, darting across the whole field of time and space. This is revery, or musing; but great segments of the flux of our ideas consist of something very different from this. They are guided by a distinct purpose or conscious interest; and the course of our ideas is then called *voluntary*.

Physiologically considered, we must suppose that a purpose means the persistent activity of certain rather definite brain-processes throughout the whole course of thought. Our most usual cogitations are not pure reveries, absolute driftings, but revolve about some central interest or topic to which most of the images are relevant, and towards which we return promptly after occasional digressions. This interest is subserved by the persistently active brain-tracts we have supposed. In the mixed associations which we have hitherto studied, the parts of each object which form the pivots on which our thoughts successively turn have their interest largely determined by their connection with some *general interest* which for the time has seized upon the mind. If we call Z the brain-tract of general interest, then, if the object abc turns up, and b has more associations with Z than have either a or c, b will become the object's interesting, pivotal portion, and will call up its own associates exclusively. For the energy of b's brain-tract will be augmented by Z's activity,—an activity which, from lack of previous connection between Z and a and Z and c, does not influence a or c. If, for instance, I think of Paris whilst I am *hungry*, I shall not improbably find that its *restaurants* have become the pivot of my thought, etc., etc.

Problems

But in the theoretic as well as in the practical life there are interests of a more acute sort, taking the form of definite images of some achievement which we desire to effect. The train of ideas arising under the influence of such an interest constitutes usually the thought of the *means* by which the end shall be attained. If the end by its simple presence does not instantaneously suggest the means, the search for the latter becomes a *problem*; and the discovery of the means forms a new sort of end, of an entirely peculiar nature—an end, namely, which we intensely desire before we have attained it, but of the nature of which, even whilst most strongly craving it, we have no distinct imagination whatever (compare pp. 241–2).

The same thing occurs whenever we seek to recall something forgotten, or to state the reason for a judgment which we have made intuitively. The desire strains and presses in a direction which it feels to be right, but towards a point which it is unable to see. In short, the *absence of an item* is a determinant of our representations quite as positive as its presence can ever be. The gap becomes no mere void, but what is called an *aching* void. If we try to explain in terms of brain-action how a thought which only potentially exists can yet be effective, we seem driven to believe that the brain-tract thereof must actually be excited, but only in a minimal and sub-conscious way. Try, for instance, to symbolize what goes on in a man who is racking his brains to remember a thought which occurred to him last week. The associates of the thought are there, many of them at least, but they refuse to awaken the thought itself. We cannot suppose that they do not irradiate *at all* into its brain-tract, because his mind quivers on the very edge of its recovery. Its actual rhythm sounds in his ears; the words seem on the imminent point of following, but fail (see p. 165). Now the only difference between the effort to recall things forgotten and the search after the means to a given end is that the latter have not, whilst the former have, already formed a part of our experience. If we first study *the mode of recalling a thing forgotten*, we can take up with better understanding the voluntary quest of the unknown.

Their Solution

The forgotten thing is felt by us as a gap in the midst of certain other things. We possess a dim idea of where we were and what we were about when it last occurred to us. We recollect the general subject to which it pertains. But all these details refuse to shoot together into a solid whole, for the lack of the missing thing, so

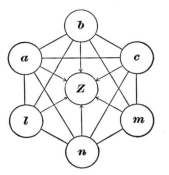

Figure 61

we keep running over them in our mind, dissatisfied, craving something more. From each detail there radiate lines of association forming so many tentative guesses. Many of these are immediately seen to be irrelevant, are therefore void of interest, and lapse immediately from consciousness. Others are associated with the other details present, and with the missing thought as well. When *these* surge up, we have a peculiar feeling that we are 'warm,' as the children say when they play hide and seek; and such associates as these we clutch at and keep before the attention. Thus we recollect successively that when we last were considering the matter in question we were at the dinner-table; then that our friend J. D. was there; then that the subject talked about was so and so; finally, that the thought came *à propos* of a certain anecdote, and then that it had something to do with a French quotation. Now all these added associates *arise independently of the will, by the spontaneous processes we know so well. All that the will does is to emphasize and linger over those which seem pertinent, and ignore the rest.* Through this hovering of the attention in the neighborhood of the desired object, the accumulation of associates becomes so great that the combined tensions of their neural processes break through the bar, and the nervous wave pours into the tract which has so long been awaiting its advent. And as the expectant, sub-conscious itching, so to speak, bursts into the fulness of vivid feeling, the mind finds an inexpressible relief.

The whole process can be rudely symbolized in a diagram. Call the forgotten thing Z, the first facts with which we felt it was related a, b, and c, and the details finally operative in calling it up l, m, and n. Each circle will then stand for the brain-process principally concerned in the thought of the fact lettered within it. The activity in Z will at first be a mere tension; but as the activities in a, b, and c little by little irradiate into l, m, and n, and as *all* these processes are somehow connected with Z, their combined irradiations upon

Z, represented by the centripetal arrows, succeed in rousing Z also to full activity.

Turn now to the case of finding the unknown means to a distinctly conceived end. The end here stands in the place of a, b, c, in the diagram. It is the starting-point of the irradiations of suggestion; and here, as in that case, what the voluntary attention does is only to dismiss some of the suggestions as irrelevant, and hold fast to others which are felt to be more pertinent—let these be symbolized by l, m, n. These latter at last accumulate sufficiently to discharge all together into Z, the excitement of which process is, in the mental sphere, equivalent to the solution of our problem. The only difference between this and the previous case is that in this one there need be no original sub-excitement in Z, coöperating from the very first. In the solving of a problem, all that we are aware of in advance seems to be its *relations*. It must be a cause, or it must be an effect, or it must contain an attribute, or it must be a means, or what not. We know, in short, a lot *about* it, whilst as yet we have no *acquaintance* with it. Our perception that one of the objects which turn up is, at last, our *quoesitum*, is due to our recognition that its relations are identical with those we had in mind, and this may be a rather slow act of judgment. Every one knows that an object may be for some time present to his mind before its relations to other matters are perceived. Just so the relations may be there before the object is.

From the guessing of newspaper enigmas to the plotting of the policy of an empire there is no other process than this. We must trust to the laws of cerebral nature to present us spontaneously with the appropriate idea, but we must know it for the right one when it comes.

It is foreign to my purpose here to enter into any detailed analysis of the different classes of mental pursuit. In a scientific research we get perhaps as rich an example as can be found. The inquirer starts with a fact of which he seeks the reason, or with an hypothesis of which he seeks the proof. In either case he keeps turning the matter incessantly in his mind until, by the arousal of associate upon associate, some habitual, some similar, one arises which he recognizes to suit his need. This, however, may take years. No rules can be given by which the investigator may proceed straight to his result; but both here and in the case of reminiscence the accumulation of helps in the way of associations may advance more rapidly by the use of certain routine methods. In striving to recall a thought, for example, we may of set purpose run through the successive classes of circumstance with which it may possibly have been connected, trusting that when the right member of the class has turned up it will help the thought's revival. Thus we may run through all the *places* in which we may have had it. We may run through the *persons* whom we remember to have conversed with, or we may call up successively all the *books* we have lately been reading. If we are trying to remember a person we may run through a list of streets or of professions. Some item out of the lists thus methodically gone over will very likely be associated with the fact we are in need of, and may suggest it or help to do so. And yet the item might never have arisen without such systematic procedure. In scientific research this accumulation of associates has been methodized by Mill under the title of 'The Four Methods of Experimental Inquiry.' By the 'method of agreement,' by that of 'difference,' by those of 'residues' and 'concomitant variations' (which cannot here be more nearly defined), we make certain lists of cases; and by ruminating these lists in our minds the cause we seek will be more likely to emerge. But the final stroke of discovery is only prepared, not effected, by them. The brain-tracts must, of their own accord, shoot the right way at last, or we shall still grope in darkness. That in some brains the tracts *do* shoot the right way much oftener than in others, and that we cannot tell why,— these are ultimate facts to which we must never close our eyes. Even in forming our lists of instances according to Mill's methods, we are at the mercy of the spontaneous workings of Similarity in our brain. How are a number of facts, resembling the one whose cause we seek, to be brought together in a list unless one will rapidly suggest another through association by similarity?

Similarity No Elementary Law

Such is the analysis I propose, first of the three main types of spontaneous, and then of voluntary, trains of thought. It will be observed that the *object called up may bear any logical relation whatever to the one which suggested it.* The law requires only that one condition should be fulfilled. The fading object must be due to a brain-process some of whose elements awaken through habit some of the elements of the brain-process of the object which comes to view. This awakening is the causal agency in the kind of association called Similarity, as in any other sort. The similarity *itself* between the objects has no causal agency in carrying us from one to the other. It is but a result—the effect of the usual causal agent when this happens to work in a certain way. Ordinary writers talk as if the similarity of the objects were itself an agent, coördinate with habit, and independent of it, and like it able to push

objects before the mind. This is quite unintelligible. The similarity of two things does not exist till both things are there—it is meaningless to talk of it as an *agent of production* of anything, whether in the physical or the psychical realms. It is a relation which the mind perceives after the fact, just as it may perceive the relations of superiority, of distance, of causality, of container and content, of substance and accident, or of contrast, between an object and some second object which the associative machinery calls up.

Conclusion

To sum up, then, we see that the *difference between the three kinds of association reduces itself to a simple difference in the amount of that portion of the nerve-tract supporting the going thought which is operative in calling up the thought which comes.* But the *modus operandi* of this active part is the same, be it large or be it small. The items constituting the coming object waken in every instance because their nerve-tracts once were excited continuously with those of the going object or its operative part. This ultimate physiological law of habit among the neural elements is what *runs* the train. The direction of its course and the form of its transitions are due to the unknown conditions by which in some brains action tends to focalize itself in small spots, while in others it fills patiently its broad bed. What these differing conditions are, it seems impossible to guess. Whatever they are, they are what separate the man of genius from the prosaic creature of habit and routine thinking. In the chapter on Reasoning we shall need to recur again to this point. I trust that the student will now feel that the way to a deeper understanding of the order of our ideas lies in the direction of cerebral physiology. The *elementary* process of revival can be nothing but the law of habit. Truly the day is distant when physiologists shall actually trace from cell-group to cell-group the irradiations which we have hypothetically invoked. Probably it will never arrive. The schematism we have used is, moreover, taken immediately from the analysis of objects into their elementary parts, and only extended by analogy to the brain. And yet it is only as incorporated in the brain that such a schematism can represent anything *causal*. This is, to my mind, the conclusive reason for saying that the order of *presentation of the mind's materials* is due to cerebral physiology alone.

The law of accidental prepotency of certain processes over others falls also within the sphere of cerebral probabilities. Granting such instability as the brain-tissue requires, certain points must always discharge more quickly and strongly than others; and this prepotency would shift its place from moment to moment by accidental causes, giving us a perfect mechanical diagram of the capricious play of similar association in the most gifted mind. A study of dreams confirms this view. The usual abundance of paths of irradiation seems, in the dormant brain, reduced. A few only are pervious, and the most fantastic sequences occur because the currents run—'like sparks in burnt-up paper'—wherever the nutrition of the moment creates an opening, but nowhere else.

The *effects of interested attention and volition* remain. These activities seem to hold fast to certain elements and, by emphasizing them and dwelling on them, to make their associates the only ones which are evoked. *This* is the point at which an anti-mechanical psychology must, if anywhere, make its stand in dealing with association. Everything else is pretty certainly due to cerebral laws. My own opinion on the question of active attention and spiritual spontaneity is expressed elsewhere (see p. 237). But even though there be a mental spontaneity, it can certainly not create ideas or summon them *ex abrupto*. Its power is limited to *selecting* amongst those which the associative machinery introduces. If it can emphasize, reinforce, or protract for half a second either one of these, it can do all that the most eager advocate of free will need demand; for it then decides the direction of the *next* associations by making them hinge upon the emphasized term; and determining in this wise the course of the man's thinking, it also determines his acts.

2
Introduction

(1943)
Warren S. McCulloch and Walter Pitts

A logical calculus of the ideas immanent in nervous activity
Bulletin of Mathematical Biophysics 5:115–133

It would be possible to write a whole book about the important aspects of this paper and the influence it has had. It is an attempt to understand what the nervous system might actually be *doing*, given primitive computing elements that are abstractions of the properties of neurons and their connections as they were known in 1943.

McCulloch and Pitts only had three references in the paper, all on mathematical logic: Carnap, *The Logical Syntax of Language*; Hilbert and Ackermann, *Foundations of Theoretical Logic*; and Whitehead and Russell, *Principia Mathematica*. In case it was thought that McCulloch did not know his physiology, a 1947 paper, "Modes of functional organization of the cerebral cortex," is about 3 pages long and has 101 references, all physiological. In 1947, McCulloch was talking about physiology; in the 1943 paper, McCulloch and Pitts are discussing what the physiology is computing—hence logic.

McCulloch and Pitts list five assumptions governing the operation of neurons. The assumptions describe what has become known as the 'McCulloch-Pitts' neuron, which is familiar to all computer scientists, often by that name. The McCulloch-Pitts neuron is a binary device; that is, it can be in only one of two possible states. Each neuron has a fixed threshold. The neuron can receive inputs from excitatory synapses, all of which have identical weights. It also can receive inputs from inhibitory synapses, whose action is absolute; that is, if the inhibitory synapse is active, the neuron cannot turn on. There is a time quantum for integration of synaptic inputs, based loosely on the physiologically observed synaptic delay.

The mode of operation of the McCulloch-Pitts neuron is simple. During the time quantum, the neuron responds to the activity of its synapses, which reflect the state of the presynaptic cells. If no inhibitory synapses are active, the neuron adds its synaptic inputs and checks to see if the sum meets or exceeds its threshold. If it does, the neuron then becomes active. If it does not, the neuron is inactive.

As an example, suppose we have a simple unit with two excitatory inputs, a and b, and with a threshold of one. Synapses connected to active cells contribute one unit. Suppose at the start of the ith time quantum, both a and b are inactive. Then, at the start of the $(i + 1)$th time quantum, the unit is inactive since the sum of two inactive synapses is zero. If a is active but b is inactive, the state of the unit at the start of the $(i + 1)$th time quantum is active since the inactive synapse plus the active synapse is one, which equals the threshold. Similarly, if b is active and a is inactive, or if both are active, the cell becomes active. This threshold-one unit with two excitatory inputs is performing the logical operation INCLUSIVE OR on its inputs, and becomes active only if a OR b OR BOTH a AND b is active.

If the unit was given a threshold of two, it would compute the logic function AND, since it would only become active if *a* AND *b* are active.

The McCulloch-Pitts neuron is therefore performing simple threshold logic. As McCulloch and Pitts put it in their Introduction, "The 'all-or-none' law of nervous activity is sufficient to insure that the activity of any neuron may be represented as a proposition. Physiological relations existing among nervous activities correspond, of course, to relations among the propositions; and the utility of the representation depends upon the identity of these relations with those of the logic of propositions" (p. 117).

As was obvious to McCulloch and Pitts, the network of connections between the simple propositions can give rise to very complex propositions. The central result of the paper is that any finite logical expression can be realized by McCulloch-Pitts neurons. This was an exciting result, since it showed that simple elements connected in a network could have immense computational power. Since the elements were based on neurophysiology, it suggested that the brain was potentially a powerful logic and computational device. The formal results about networks of logical elements also influenced others when they were thinking about the potential of digital computers, for example, John von Neumann (see paper 7).

Unfortunately, the formal proofs and the notation in this early paper are exceptionally difficult. A much better introduction to McCulloch-Pitts neurons, which also extends their results and places them in the perspective of later work in automata theory and theory of computation, is *Computation*, by Marvin Minsky (1967). This clearly written book develops a theory of computation directly from McCulloch-Pitts neurons.

One point that was clear to McCulloch and Pitts—so obvious that it was not specifically mentioned—was that a single neuron was simple, and that computational power came because simple neurons were embedded in an interacting nervous *system*. If we do not count Williams James, this paper describes perhaps the first true connectionist model, since it has simple computing elements, arranged partly in parallel, doing powerful computation with appropriately constructed connection strengths.

The question to ask next is whether McCulloch-Pitts neurons are correct approximations to real neurons, that is, whether they are a good brain model. Given the state of neurophysiology in 1943, when the ionic and electrical basis of neural activity was unclear, the approximations McCulloch and Pitts made were much more supportable than they are now. The dominant feature of observed cell activity was the 'all-or-none' action potential. It was not possible to make intracellular recordings, so it was hard to see that the postsynaptic potentials, due to presynaptic activation actually extended over a good many milliseconds, were graded, and that neurons were acting a lot more like voltage to frequency converters than like simple logic elements. (See the paper on *Limulus*, paper 23, for a more realistic model of a neural system.)

McCulloch and Pitts themselves were aware of the many continuous phenomena occurring in the nervous system, and even in the Introduction to the 1943 paper they comment on the potential importance of continuous changes in threshold brought about by adaptation and learning. It is worth mentioning this only because even now we find extended discussions of McCulloch-Pitts neurons in the scientific literature suggesting that they are adequate brain models and useful approximations to neurophysiology. This is not correct. Neurons, except in some special cases, are not simple computing devices realizing the propositions of formal logic. Binary neurons can be, however, useful approximations of underlying continuous processes in some special cases.

The immense theoretical influence of this paper was not among neuroscientists but among computer scientists. The history of this work is encouraging for theoreticians. It is not necessary to be correct in detail, or even in the original domain of application, to create an enduring work of great importance. It is possible to buy McCulloch-Pitts neurons at your local Radio Shack store, in the form of logic circuits.

A bibliographic note: McCulloch's papers are collected in a highly recommended book, *Embodiments of Mind* (1965).

References

W. McCulloch (1965), *Embodiments of Mind*, Cambridge, MA: MIT Press.

M. Minsky (1967), *Computation: Finite and Infinite Machines*, New York: Prentice-Hall.

MAGDALEN COLLEGE LIBRARY

(1943)
Warren S. McCulloch and Walter Pitts

A logical calculus of the ideas immanent in nervous activity
Bulletin of Mathematical Biophysics 5:115–133

Because of the "all-or-none" character of nervous activity, neural events and the relations among them can be treated by means of propositional logic. It is found that the behavior of every net can be described in these terms, with the addition of more complicated logical means for nets containing circles; and that for any logical expression satisfying certain conditions, one can find a net behaving in the fashion it describes. It is shown that many particular choices among possible neurophysiological assumptions are equivalent, in the sense that for every net behaving under one assumption, there exists another net which behaves under the other and gives the same results, although perhaps not in the same time. Various applications of the calculus are discussed.

I. Introduction

Theoretical neurophysiology rests on certain cardinal assumptions. The nervous system is a net of neurons, each having a soma and an axon. Their adjunctions, or synapses, are always between the axon of one neuron and the soma of another. At any instant a neuron has some threshold, which excitation must exceed to initiate an impulse. This, except for the fact and the time of its occurrence, is determined by the neuron, not by the excitation. From the point of excitation the impulse is propagated to all parts of the neuron. The velocity along the axon varies directly with its diameter, from less than one meter per second in thin axons, which are usually short, to more than 150 meters per second in thick axons, which are usually long. The time for axonal conduction is consequently of little importance in determining the time of arrival of impulses at points unequally remote from the same source. Excitation across synapses occurs predominantly from axonal terminations to somata. It is still a moot point whether this depends upon irreciprocity of individual synapses or merely upon prevalent anatomical configurations. To suppose the latter requires no hypothesis *ad hoc* and explains known exceptions, but any assumption as to cause is compatible with the calculus to come. No case is known in which excitation through a single synapse has elicited a nervous impulse in any neuron, whereas any neuron may be excited by impulses arriving at a sufficient number of neighboring

synapses within the period of latent addition, which lasts less than one quarter of a millisecond. Observed temporal summation of impulses at greater intervals is impossible for single neurons and empirically depends upon structural properties of the net. Between the arrival of impulses upon a neuron and its own propagated impulse there is a synaptic delay of more than half a millisecond. During the first part of the nervous impulse the neuron is absolutely refractory to any stimulation. Thereafter its excitability returns rapidly, in some cases reaching a value above normal from which it sinks again to a subnormal value, whence it returns slowly to normal. Frequent activity augments this subnormality. Such specificity as is possessed by nervous impulses depends solely upon their time and place and not on any other specificity of nervous energies. Of late only inhibition has been seriously adduced to contravene this thesis. Inhibition is the termination or prevention of the activity of one group of neurons by concurrent or antecedent activity of a second group. Until recently this could be explained on the supposition that previous activity of neurons of the second group might so raise the thresholds of internuncial neurons that they could no longer be excited by neurons of the first group, whereas the impulses of the first group must sum with the impulses of these internuncials to excite the now inhibited neurons. Today, some inhibitions have been shown to consume less than one millisecond. This excludes internuncials and requires synapses through which impulses inhibit that neuron which is being stimulated by impulses through other synapses. As yet experiment has not shown whether the refractoriness is relative or absolute. We will assume the latter and demonstrate that the difference is immaterial to our argument. Either variety of refractoriness can be accounted for in either of two ways. The "inhibitory synapse" may be of such a kind as to produce a substance which raises the threshold of the neuron, or it may be so placed that the local disturbance produced by its excitation opposes the alteration induced by the otherwise excitatory synapses. Inasmuch as position is already known to have such effects in the case of electrical stimulation, the first hypothesis is to be excluded unless and until it be substantiated, for the second involves no new hypothesis. We have, then, two explanations of inhibition based on the same general premises, differing only in the assumed nervous nets and, consequently, in the time required for inhibition. Hereafter we shall refer to such nervous nets as *equivalent in the extended sense*. Since we are concerned with properties of nets which are invariant under equivalence, we may make the physical assumptions which are most convenient for the calculus.

Many years ago one of us, by considerations impertinent to this argument, was led to conceive of the response of any neuron as factually equivalent to a proposition which proposed its adequate stimulus. He therefore attempted to record the behavior of complicated nets in the notation of the symbolic logic of propositions. The "all-or-none" law of nervous activity is sufficient to insure that the activity of any neuron may be represented as a proposition. Physiological relations existing among nervous activities correspond, of course, to relations among the propositions; and the utility of the representation depends upon the identity of these relations with those of the logic of propositions. To each reaction of any neuron there is a corresponding assertion of a simple proposition. This, in turn, implies either some other simple proposition or the disjunction or the conjunction, with or without negation, of similar propositions, according to the configuration of the synapses upon and the threshold of the neuron in question. Two difficulties appeared. The first concerns facilitation and extinction, in which antecedent activity temporarily alters responsiveness to subsequent stimulation of one and the same part of the net. The second concerns learning, in which activities concurrent at some previous time have altered the net permanently, so that a stimulus which would previously have been inadequate is now adequate. But for nets undergoing both alterations, we can substitute equivalent fictitious nets composed of neurons whose connections and thresholds are unaltered. But one point must be made clear: neither of us conceives the formal equivalence to be a factual explanation. *Per contra!*—we regard facilitation and extinction as dependent upon continuous changes in threshold related to electrical and chemical variables, such as afterpotentials and ionic concentrations; and learning as an enduring change which can survive sleep, anaesthesia, conclusions and coma. The importance of the formal equivalence lies in this: that the alterations actually underlying facilitation, extinction and learning in no way affect the conclusions which follow from the formal treatment of the activity of nervous nets, and the relations of the corresponding propositions remain those of the logic of propositions.

The nervous system contains many circular paths, whose activity so regenerates the excitation of any participant neuron that reference to time past becomes indefinite, although it still implies that afferent activity has realized one of a certain class of configurations over time. Precise specification of these implications by means of recursive functions, and determination of those that can be embodied in the activity of nervous nets, completes the theory.

II. The Theory: Nets without Circles

We shall make the following physical assumptions for our calculus.

1. The activity of the neuron is an "all-or-none" process.

2. A certain fixed number of synapses must be excited within the period of latent addition in order to excite a neuron at any time, and this number is independent of previous activity and position on the neuron.

3. The only significant delay within the nervous system is synaptic delay.

4. The activity of any inhibitory synapse absolutely prevents excitation of the neuron at that time.

5. The structure of the net does not change with time.

To present the theory, the most appropriate symbolism is that of Language II of R. Carnap (1938), augmented with various notations drawn from B. Russell and A. N. Whitehead (1927), including the *Principia* conventions for dots. Typographical necessity, however, will compel us to use the upright 'E' for the existential operator instead of the inverted, and an arrow ('$\rightarrow$') for implication instead of the horseshoe. We shall also use the Carnap syntactical notations, but print them in boldface rather than German type; and we shall introduce a functor S, whose value for a property P is the property which holds of a number when P holds of its predecessor; it is defined by '$S(P)(t) . \equiv . P(Kx) . t = x'$'; the brackets around its argument will often be omitted, in which case this is understood to be the nearest predicate-expression [Pr] on the right. Moreover, we shall write $S^2 Pr$ for $S(S(Pr))$, etc.

The neurons of a given net $\mathcal{N}$ may be assigned designations 'c_1', 'c_2', $\cdots$, 'c_n'. This done, we shall denote the property of a number, that a neuron c_i fires at a time which is that number of synaptic delays from the origin of time, by 'N' with the numeral i as subscript, so that $N_i(t)$ asserts that c_i fires at the time t. N_i is called the *action* of c_i. We shall sometimes regard the subscripted numeral of 'N' as if it belonged to the object-language, and were in a place for a functoral argument, so that it might be replaced by a number-variable $[z]$ and quantified; this enables us to abbreviate long but finite disjunctions and conjunctions by the use of an operator. We shall employ this locution quite generally for sequences of Pr; it may be secured formally by an obvious disjunctive definition. The predicates 'N_1', 'N_2', $\cdots$, comprise the syntactical class 'N'.

Let us define the *peripheral afferents* of $\mathcal{N}$ as the neurons of $\mathcal{N}$ with no axons synapsing upon them. Let $N_1, \cdots, N_p$ denote the actions of such neurons and $N_{p+1}, N_{p+2}, \cdots, N_n$ those of the rest. Then a *solution of* $\mathcal{N}$ will be a class of sentences of the form $S_i: N_{p+1}(z_1). \equiv . Pr_i(N_1, N_2, \cdots, N_p, z_1)$, where Pr_i contains no free variable save z_1 and no descriptive symbols save the N in the argument $[Arg]$, and possibly some constant sentences $[sa]$; and such that each S_i is true of $\mathcal{N}$. Conversely, given a $Pr_1(^1p^1{}_1, ^1p^1{}_2, \cdots, ^1p^1{}_p, z_1, s)$, containing no free variable save those in its Arg, we shall say that it is *realizable in the narrow sense* if there exists a net $\mathcal{N}$ and a series of N_i in it such that $N_1(z_1). \equiv . Pr_1(N_1, N_2, \cdots, z_1, sa_1)$ is true of it, where sa_1 has the form $N(0)$. We shall call it *realizable in the extended sense*, or simply *realizable*, if for some n $S^n(Pr_1)(p_1, \cdots, p_p, z_1, s)$ is realizable in the above sense. c_{pi} is here the realizing neuron. We shall say of two laws of nervous excitation which are such that every S which is realizable in either sense upon one supposition is also realizable, perhaps by a different net, upon the other, that they are equivalent assumptions, in that sense.

The following theorems about realizability all refer to the extended sense. In some cases, sharper theorems about narrow realizability can be obtained; but in addition to greater complication in statement this were of little practical value, since our present neurophysiological knowledge determines the law of excitation only to extended equivalence, and the more precise theorems differ according to which possible assumption we make. Our less precise theorems, however, are invariant under equivalence, and are still sufficient for all purposes in which the exact time for impulses to pass through the whole net is not crucial.

Our central problems may now be stated exactly: first, to find an effective method of obtaining a set of computable S constituting a solution of any given net; and second, to characterize the class of realizable S in an effective fashion. Materially stated, the problems are to calculate the behavior of any net, and to find a net which will behave in a specified way, when such a net exists.

A net will be called *cyclic* if it contains a circle: i.e., if there exists a chain $c_i, c_{i+1}, \cdots$ of neurons on it, each member of the chain synapsing upon the next, with the same beginning and end. If a set of its neurons $c_1, c_2, \cdots, c_p$ is such that its removal from $\mathcal{N}$ leaves it without circles, and no smaller class of neurons has this property, the set is called a *cyclic* set, and its cardinality is the *order* of $\mathcal{N}$. In an important sense, as we shall see, the order of a net is an index of the complexity of its behavior. In particular, nets of zero order have especially simple properties; we shall discuss them first.

Let us define a *temporal propositional expression* (a TPE), designating a *temporal propositional function* (TPF), by the following recursion:

1. A $^1p^1[z_1]$ is a TPE, where p_1 is a predicate-variable.
2. If S_1 and S_2 are TPE containing the same free individual variable, so are SS_1, $S_1 \vee S_2$, $S_1.S_2$ and $S_i. \varpropto S_2$.
3. Nothing else is a TPE.

THEOREM I *Every net of order* 0 *can be solved in terms of temporal propositional expressions.*

Let c_i be any neuron of $\mathcal{N}$ with a threshold $\theta_i > 0$, and let $c_{i1}, c_{i2}, \cdots, c_{ip}$ have respectively $n_{i1}, n_{i2}, \cdots, n_{ip}$ excitatory synapses upon it. Let $c_{j1}, c_{j2}, \cdots, c_{jq}$ have inhibitory synapses upon it. Let κ_i be the set of the subclasses of $\{n_{i1}, n_{i2}, \cdots, n_{ip}\}$ such that the sum of their members exceeds θ_i. We shall then be able to write, in accordance with the assumptions mentioned above,

$$N_i(z_1). \equiv . S\left\{ \prod_{m=1}^{q} \varpropto N_{jm}(z_1) \cdot \sum_{a \varepsilon \kappa_i} \prod_{s \varepsilon a} N_{is}(z_1) \right\} \tag{1}$$

where the 'Σ' and 'Π' are syntactical symbols for disjunctions and conjunctions which are finite in each case. Since an expression of this form can be written for each c_i which is not a peripheral afferent, we can, by substituting the corresponding expression in (1) for each N_{jm} or N_{is} whose neuron is not a peripheral afferent, and repeating the process on the result, ultimately come to an expression for N_i in terms solely of peripherally afferent N, since $\mathcal{N}$ is without circles. Moreover, this expression will be a TPE, since obviously (1) is; and it follows immediately from the definition that the result of substituting a TPE for a constituent $p(z)$ in a TPE is also one.

THEOREM II *Every* TPE *is realizable by a net of order zero.*

The functor S obviously commutes with disjunction, comjunction, and negation. It is obvious that the result of substituting any S_i, realizable in the narrow sense (i.n.s.), for the $p(z)$ in a realizable expression S_1 is itself realizable i.n.s.; one constructs the realizing net by replacing the peripheral afferents in the net for S_1 by the realizing neurons in the nets for the S_i. The one neuron net realizes $p_1(z_1)$ in i.n.s., and Figure 1-a shows a net that realizes $Sp_1(z_1)$ and hence SS_2, i.n.s., if S_2 can be realized i.n.s. Now if S_2 and S_3 are realizable then $S^m S_2$ and $S^n S_3$ are realizable i.n.s., for suitable m and n. Hence so are $S^{m+n} S_2$ and $S^{m+n} S_3$. Now the nets of Figures 1b, c and d respectively realize $S(p_1(z_1) \vee p_2(z_1))$, $S(p_1(z_1) . p_2(z_1))$, and $S(p_1(z_1) . \infty p_2(z_1))$ i.n.s. Hence $S^{m+n+1}(S_1 \vee S_2)$, $S^{m+n+1}(S_1 . S_2)$, and $S^{m+n+1}(S_1 . \infty S_2)$ are realizable i.n.s. Therefore $S_1 \vee S_2 S_1 . S_2 S_1 . \infty S_2$ are realizable if S_1 and S_2 are. By complete induction, all TPE are realizable. In this way all nets may be regarded as built out of the fundamental elements of Figures 1a, b, c, d, precisely as the temporal propositional expressions are generated out of the operations of precession, disjunction, conjunction, and conjoined negation. In particular, corresponding to any description of state, or distribution of the values *true* and *false* for the actions of all the neurons of a net save that which makes them all false, a single neuron is constructible whose firing is a necessary and sufficient condition for the validity of that description. Moreover, there is always an indefinite number of topologically different nets realizing any TPE.

THEOREM III *Let there be given a complex sentence* S_1 *built up in any manner out of elementary sentences of the form* $p(z_1 - zz)$ *where* zz *is any numeral, by any of the propositional connections: negation, disjunction, conjunction, implication, and equivalence. Then* S_1 *is a TPE and only if it is false when its constituent* $p(z_1 - zz)$ *are all assumed false—i.e., replaced by false sentences— or that the last line in its truth-table contains an* '**F**',—*or there is no term in its Hilbert disjunctive normal form composed exclusively of negated terms.*

These latter three conditions are of course equivalent (Hilbert and Ackermann, 1938). We see by induction that the first of them is necessary, since $p(z_1 - zz)$ becomes false when it is replaced by a false sentence, and $S_1 \vee S_2$, $S_1 . S_2$ and $S_1 . \infty S_2$ are all false if both their constituents are. We see that the last condition is sufficient by remarking that a disjunction is a TPE when its constituents are, and that any term

$$S_1 . S_2 . \ldots . S_m . \infty S_{m+1} . \infty . \ldots . \infty S_n$$

can be written as

$$(S_1 . S_2 . \ldots . S_m) . \infty (S_{m+1} \vee S_{m+2} \vee \ldots . \vee S_n),$$

which is clearly a TPE.

The method of the last theorems does in fact provide a very convenient and workable procedure for constructing nervous nets to order, for those cases where there is no reference to events indefinitely far in the past in the specification of the conditions. By way of

example, we may consider the case of heat produced by a transient cooling.

If a cold object is held to the skin for a moment and removed, a sensation of heat will be felt; if it is applied for a longer time, the sensation will be only of cold, with no preliminary warmth, however transient. It is known that one cutaneous receptor is affected by heat, and another by cold. If we let N_1 and N_2 be the actions of the respective receptors and N_3 and N_4 of neurons whose activity implies a sensation of heat and cold, our requirements may be written as

$$N_3(t) : \equiv : N_1(t - 1) . \vee . N_2(t - 3) . \infty N_2(t - 2)$$

$$N_4(t) . \equiv . N_2(t - 2) . N_2(t - 1)$$

where we suppose for simplicity that the required persistence in the sensation of cold is say two synaptic delays, compared with one for that of heat. These conditions clearly fall under Theorem III. A net may consequently be constructed to realize them, by the method of Theorem II. We begin by writing them in a fashion which exhibits them as built out of their constituents by the operations realized in Figures 1a, b, c, d: i.e., in the form

$$N_3(t) . \equiv . S\{N_1(t) \vee S[(SN_2(t)) . \infty N_2(t)]\}$$

$$N_4(t) . \equiv . S\{[SN_2(t)] . N_2(t)\}.$$

First we construct a net for the function enclosed in the greatest number of brackets and proceed outward; in this case we run a net of the form shown in Figure 1a from c_2 to some neuron c_a, say, so that

$$N_a(t) . \equiv . SN_2(t).$$

Next introduce two nets of the forms 1c and 1d, both running from c_a and c_2, and ending respectively at c_4 and say c_b. Then

$$N_4(t) . \equiv . S[N_a(t) . N_2(t)] . \equiv . S[(SN_2(t)) . N_2(t)].$$

$$N_b(t) . \equiv . S[N_a(t) . \infty N_2(t)] . \equiv . S[(SN_2(t)) . \infty N_2(t)].$$

Finally, run a net of the form 1b from c_1 and c_b to c_3, and derive

$$N_3(t) . \equiv . S[N_1(t) \vee N_b(t)]$$

$$. \equiv . S\{N_1(t) \vee S[(SN_2(t)) . \infty N_2(t)]\}.$$

These expressions for $N_3(t)$ and $N_4(t)$ are the ones desired; and the realizing net *in toto* is shown in Figure 1e.

This illusion makes very clear the dependence of the correspondence between perception and the "external world" upon the specific structural properties of the intervening nervous net. The same illusion, of course, could also have been produced under various other assumptions about the behavior of the cutaneous receptors, with correspondingly different nets.

We shall now consider some theorems of equivalence: i.e., theorems which demonstrate the essential identity, save for time, of various alternative laws of nervous excitation. Let us first discuss the case of *relative inhibition*. By this we mean the supposition that the firing of an inhibitory synapse does not absolutely prevent the firing of the neuron, but merely raises its threshold, so that a greater number of excitatory synapses must fire concurrently to fire it than would otherwise be needed. We may suppose, losing no generality, that the increase in threshold is unity for the firing of each such synapse; we then have the theorem:

THEOREM IV *Relative and absolute inhibition are equivalent in the extended sense.*

We may write out a law of nervous excitation after the fashion of (1), but employing the assumption of relative inhibition instead; inspection then shows that this expression is a *TPE*. An example of the replacement of relative inhibition by absolute is given by Figure 1f. The reverse replacement is even easier; we give the inhibitory axons afferent to c_i any sufficiently large number of inhibitory synapses apiece.

Second, we consider the case of extinction. We may write this in the form of a variation in the threshold θ_i; after the neuron c_i has fired; to the nearest integer—and only to this approximation is the variation in threshold significant in natural forms of excitation—this may be written as a sequence $\theta_i + b_j$ for j synaptic delays after firing, where $b_j = 0$ for j large enough, say $j = M$ or greater. We may then state

THEOREM V *Extinction is equivalent to absolute inhibition.*

For, assuming relative inhibition to hold for the moment, we need merely run M circuits $\mathcal{T}_1, \mathcal{T}_2, \cdots, \mathcal{T}_M$ containing respectively $1, 2, \cdots, M$ neurons, such that the firing of each link in any is sufficient to fire the next, from the neuron c_i back to it, where the end of the circuit $\mathcal{T}_j$ has just b_j inhibitory synapses upon c_i. It is evident that this will produce the desired results. The reverse substitution may be accomplished by the diagram of Figure 1g. From the transitivity of replacement, we infer the theorem. To this group of theorems also belongs the well-known

THEOREM VI *Facilitation and temporal summation may be replaced by spatial summation.*

This is obvious: one need merely introduce a suitable sequence of delaying chains, of increasing numbers of synapses, between the exciting cell and the neuron whereon temporal summation is desired to hold. The assumption of spatial summation will then give the required results. See e.g. Figure 1h. This procedure had application in showing that the observed temporal summation in gross nets does not imply such a mechanism in the interaction of individual neurons.

The phenomena of learning, which are of a character persisting over most physiological changes in nervous activity, seem to require the possibility of permanent alterations in the structure of nets. The simplest such alteration is the formation of new synapses or equivalent local depressions of threshold. We suppose that some axonal terminations cannot at first excite the succeeding neuron; but if at any time the neuron fires, and the axonal terminations are simultaneously excited, they become synapses of the ordinary kind, henceforth capable of exciting the neuron. The loss of an inhibitory synapse gives an entirely equivalent result. We shall then have

THEOREM VII *Alterable synapses can be replaced by circles.*

This is accomplished by the method of Figure 1i. It is also to be remarked that a neuron which becomes and remains spontaneously active can likewise be replaced by a circle, which is set into activity by a peripheral afferent when the activity commences, and inhibited by one when it ceases.

III. The Theory: Nets with Circles

The treatment of nets which do not satisfy our previous assumption of freedom from circles is very much more difficult than that case. This is largely a consequence of the possibility that activity may be set up in a circuit and continue reverberating around it for an indefinite period of time, so that the realizable Pr may involve reference to past events of an indefinite degree of remoteness. Consider such a net $\mathcal{N}$, say of order p, and let $c_1, c_2, \cdots, c_p$ be a cyclic set of neurons of $\mathcal{N}$. It is first of all clear from the definition that every N_s of $\mathcal{N}$ can be expressed as a *TPE*, of $N_1, N_2, \cdots, N_p$ and the absolute afferents; the solution of $\mathcal{N}$ involves then only the determination of expressions for the cyclic set. This done, we shall derive a set of expressions $[A]$:

$$N_i(z_1). \equiv . Pr_i[S^{n_{i1}}N_1(z_1), S^{n_{i2}}N_2(z_1), \cdots, S^{n_{ip}}N_p(z_1)], \qquad (2)$$

where Pr_i also involves peripheral afferents. Now if n is the least common multiple of the n_{ij}, we shall, by substituting their equivalents according to (2) in (3) for the N_j, and repeating this process often enough on the result, obtain S of the form

$$N_i(z_1) \cdot \equiv \cdot Pr_1[S^n N_1(z_1), S^n N_2(z_1), \cdots, S^n N_p(z_1)]. \tag{3}$$

These expressions may be written in the Hilbert disjunctive normal form as

$$N_i(z_1) \cdot \equiv \cdot \underset{\substack{\alpha \varepsilon \kappa \\ \beta_\alpha \varepsilon \kappa}}{\Sigma} \, S_\alpha \underset{j \varepsilon \kappa}{\Pi} S^n N_j(z_1)$$

$$\underset{j \varepsilon \beta_\alpha}{\Pi} \infty S^n N_j(z_1), \text{ for suitable } \kappa, \tag{4}$$

where S_α is a TPE of the absolute afferents of $\mathcal{N}$. There exist some 2^p different sentences formed out of the p N_i by conjoining to the conjunction of some set of them the conjunction of the negations of the rest. Denumerating these by $X_1(z_1), X_2(z_1), \cdots, X_{2^p}(z_1)$, we may, by use of the expressions (4), arrive at an equipollent set of equations of the form

$$X_i(z_1) \cdot \equiv \cdot \overset{2^p}{\underset{j=1}{\Sigma}} Pr_{ij}(z_1) \cdot S^n X_j(z_1). \tag{5}$$

Now we import the subscripted numerals i, j into the object-language: i.e., define Pr_1 and Pr_2 such that $Pr_1(zz_1, z_1) \cdot \equiv \cdot X_i(z_1)$ and $Pr_2(zz_1, zz_2, z_1) \cdot \equiv \cdot Pr_{ij}(z_1)$ are provable whenever zz_1 and zz_2 denote i and j respectively.

Then we may rewirte (5) as

$$(z_1)zz_p : Pr_1(z_1, z_3)$$

$$\cdot \equiv \cdot (Ez_2)zz_p \cdot Pr_2(z_1, z_2, z_3 - zz_n)$$

$$\cdot Pr_1(z_2, z_3 - zz_n) \tag{6}$$

where zz_n denotes n and zz_p denotes 2^p. By repeated substitution we arrive at an expression

$$(z_1)zz_p : Pr_1(z_1, zz_n zz_2)$$

$$\cdot \equiv \cdot (Ez_2)zz_p(Ez_3)zz_p \ldots (Ez_n)zz_p$$

$$\cdot Pr_2(z_1, z_2, zz_n(zz_2 - 1))$$

$$\cdot Pr_2(z_2, z_3, zz_n(zz_2 - 1)) \ldots \ldots \tag{7}$$

$Pr_2(z_{n-1}, z_n, 0) \cdot Pr_1(z_n, 0)$, for any numeral zz_2 which denotes s. This is easily shown by induction to be equipollent to

$$(z_1)zz_p : \cdot Pr_1(z_1, zz_n zz_2)$$

$$: \equiv : (Ef)(z_2)zz_2 - 1 f(z_2 zz_n)$$

$$\leq zz_p \cdot f(zz_n zz_2)$$

$$= z_1 \cdot Pr_2(f(zz_n(z_2 + 1)),$$

$$f(zz_n z_2)) \cdot Pr_1(f(0), 0) \tag{8}$$

and since this is the case for all zz_2, it is also true that

$$(z_4)(z_1)zz_p : Pr_1(z_1, z_4)$$

$$\cdot \equiv \cdot (Ef)(z_2)(z_4 - 1) \cdot f(z_2)$$

$$\leq zz_p \cdot f(z_4) = z_1 f(z_4) = z_1$$

$$\cdot Pr_2[f(z_2 + 1), f(z_2), z_2]$$

$$\cdot Pr_1[f(\text{res}(z_4, zz_n)), \text{res}(z_4, zz_n)], \tag{9}$$

where zz_n denotes n, $\text{res}(r, s)$ is the residue of $r \bmod s$ and zz_p denotes 2^p. This may be written in a less exact way as

$$N_i(t) \cdot \equiv \cdot (E\phi)(x)t - 1 \cdot \phi(x) \leq 2^p \cdot \phi(t) = i$$

$$\cdot P[\phi(x + 1), \phi(x) \cdot N_{\phi(0)}(0)],$$

where x and t are also assumed divisible by n, and Pr_2 denotes P. From the preceding remarks we shall have

THEOREM VIII *The expression* (9) *for neurons of the cyclic set of a net $\mathcal{N}$ together with certain TPE expressing the actions of other neurons in terms of them, constitute a solution of $\mathcal{N}$.*

Consider now the question of the realizability of a set of S_i. A first necessary condition, demonstrable by an easy induction, is that

$$(z_2)z_1 \cdot p_1(z_2) \equiv p_2(z_2) \cdot \rightarrow \cdot S_i \equiv S_i \begin{Bmatrix} p_1 \\ p_2 \end{Bmatrix} \tag{10}$$

should be true, with similar statements for the other free p in S_i: i.e., no nervous net can take account of future peripheral afferents. Any S_i satisfying this requirement can be replaced by an equipollent S of the form

$$(Ef)(z_2)z_1(z_3)zz_p : f \varepsilon Pr_{mi}$$

$$: f(z_1, z_2, z_3) = 1 \cdot \equiv \cdot p_{z_3}(z_2) \tag{11}$$

where zz_p denotes p, by defining

$$Pr_{mi} = \hat{f}[(z_1)(z_2)z_1(z_3)zz_p : \cdot f(z_1, z_2, z_3) = 0 \cdot \mathbf{v} \cdot f(z_1, z_2, z_3)$$

$$= 1 : f(z_1, z_2, z_3) = 1 \cdot \equiv \cdot p_{z_3}(z_2) : \rightarrow : S_i].$$

Consider now these series of classes α_i, for which

$$N_i(t) : \equiv : (E\phi)(x)t(m)q : \phi \varepsilon \alpha_i : N_m(x)$$

$$\cdot \equiv \cdot \phi(t, x, m) = 1$$

$$\cdot [i = q + 1, \cdots, M] \tag{12}$$

holds for some net. These will be called *prehensible* classes. Let us define the *Boolean ring* generated by a class of classes κ as the aggregate of the classes which can be formed from members of κ by repeated application of the logical operations; i.e., we put

$$\mathscr{R}(\kappa) = p\hat{\ }\hat{\lambda}[(\alpha, \beta) : \alpha \varepsilon \kappa$$

$$\rightarrow \alpha \varepsilon \lambda : \alpha, \beta \varepsilon \lambda \cdot \rightarrow \cdot - \alpha, \alpha \cdot \beta, \alpha \mathbf{v} \beta \varepsilon \lambda].$$

We shall also define

$$\bar{\mathscr{R}}(\kappa). = .\mathscr{R}(\kappa) - \iota'p' - "\kappa,$$

$$\mathscr{R}_e(\kappa) = p'\hat{\lambda}[(\alpha, \beta): \alpha \varepsilon \kappa \rightarrow \alpha \varepsilon \lambda . \rightarrow . - \alpha, \alpha . \beta, \alpha \mathbf{v} \beta, S"\alpha \varepsilon \lambda]$$

$$\bar{\mathscr{R}}_e(\kappa) = \mathscr{R}_e(\kappa) - \iota'p' - "\kappa,$$

and

$$\sigma(\psi, t) = \hat{\phi}[(m). \phi(t + 1, t, m) = \psi(m)].$$

The class $\mathscr{R}_e(\kappa)$ is formed from κ in analogy with $\mathscr{R}(\kappa)$, but by repeated application not only of the logical operations but also of that which replaces a class of properties $P \varepsilon \alpha$ by $S(P) \varepsilon S"\alpha$. We shall then have the

LEMMA $\quad Pr_1(\boldsymbol{p}_1, \boldsymbol{p}_2, \cdots, \boldsymbol{p}_m, \boldsymbol{z}_1)$ is a TPE if and only if

$$(\boldsymbol{z}_1)(\boldsymbol{p}_1, \cdots, \boldsymbol{p}_m)(E\boldsymbol{p}_{m+1})$$

$$: \boldsymbol{p}_{m+1} \varepsilon \bar{\mathscr{R}}_e(\{\boldsymbol{p}_1, \boldsymbol{p}_2, \cdots, \boldsymbol{p}_m\}) \tag{13}$$

$$\boldsymbol{p}_{m+1}(\boldsymbol{z}_1) \equiv Pr_1(\boldsymbol{p}_1, \boldsymbol{p}_2, \cdots, \boldsymbol{p}_m, \boldsymbol{z}_1)$$

is true; and it is a TPE not involving 'S' if and only if this holds when '$\bar{\mathscr{R}}_e$' is replaced by '$\bar{\mathscr{R}}$', and we then obtain

THEOREM IX $\quad$ *A series of classes $\alpha_1, \alpha_2, \cdots \alpha_s$ is a series of prehensible classes if and only if*

$$(Em)(En)(p)n(i)(\psi): (x)m\psi(x) = 0 \mathbf{v} \psi(x) = 1 : \rightarrow : (E\beta)$$

$$(Ey)m. \psi(y) = 0. \beta \varepsilon \mathscr{R}[\hat{\gamma}((Ei). \gamma = \alpha_i)]. \mathbf{v} . (x)m.$$

$$\psi(x) = 0. \beta \varepsilon \mathscr{R}[\hat{\gamma}((Ei). \gamma = \alpha_i)] : (t)(\phi): \phi \varepsilon \alpha_i. \tag{14}$$

$$\sigma(\phi, nt + p). \rightarrow . (Ef). f \varepsilon \beta . (w)m(x)t - 1.$$

$$\phi(n(t + 1) + p, nx + p, w) = f(nt + p, nx + p, w).$$

The proof here follows directly from the lemma. The condition is necessary, since every net for which an expression of the form (4) can be written obviously verifies it, the ψ's being the characteristic functions of the S_α and the β for each ψ being the class whose designation has the form $\Pi_{i \varepsilon \alpha} Pr_i \Pi_{j \varepsilon \beta_x} Pr_j$, where Pr_k denotes α_k for all k. Conversely, we may write an expression of the form (4) for a net $\mathscr{N}$ fulfilling prehensible classes satisfying (14) by putting for the Pr_α Pr denoting the ψ's, and a Pr, written in the analogue for classes of the disjunctive normal form, and denoting the α corresponding to that ψ, conjoined to it. Since every S of the form (4) is clearly realizable, we have the theorem.

It is of some interest to consider the extent to which we can by knowledge of the present determine the whole past of various special nets: i.e., when we may construct a net the firing of the cyclic set of whose neurons requires the peripheral afferents to have had a set of past values specified by given functions ϕ_i. In this case the classes α_i of the last theorem reduced to

unit classes; and the condition may be transformed into

$$(Em, n)(p)n(i, \psi)(Ej): . (x)m : \psi(x)$$

$$= 0 . \mathbf{v} . \psi(x) = 1 : \phi_i \varepsilon \sigma (\psi, nt + p)$$

$$: \rightarrow : (w)m(x)t - 1 . \phi_i(n(t + 1) + p, nx + p, w)$$

$$= \phi_j(nt + p, nx + p, w) : . (u, v)(w)m$$

$$. \phi_i(n(u + 1) + p, nu + p, w)$$

$$= \phi_i(n(v + 1) + p, nv + p, w).$$

On account of limitations of space, we have presented the above argument very sketchily; we propose to expand it and certain of its implications in a further publication.

The condition of the last theorem is fairly simple in principle, though not in detail; its application to practical cases would, however, require the exploration of some 2^{2^n} classes of functions, namely the members of $\mathscr{R}(\{\alpha_1, \cdots, \alpha_s\})$. Since each of these is a possible β of Theorem IX, this result cannot be sharpened. But we may obtain a sufficient condition for the realizability of an S which is very easily applicable and probably covers most practical purposes. This is given by

THEOREM X $\quad$ Let us define a set K of S by the following recursion:

1. Any TPE and any TPE whose arguments have been replaced by members of K belong to K;

2. If $Pr_1(\boldsymbol{z}_1)$ is a member of K, then $(\boldsymbol{z}_2)\boldsymbol{z}_1 . Pr_1(\boldsymbol{z}_2)$, $(E\boldsymbol{z}_2)\boldsymbol{z}_1 . Pr_1(\boldsymbol{z}_2)$, and $C_{mn}(\boldsymbol{z}_1) . s$ belong to it, where C_{mn} denotes the property of being congruent to m modulo n, $m < n$.

3. *The set* **K** *has no further members*

Then every member of K is realizable.

For, if $Pr_1(\boldsymbol{z}_1)$ is realizable, nervous nets for which

$$N_i(\boldsymbol{z}_1). \equiv . Pr_1(\boldsymbol{z}_1). SN_i(\boldsymbol{z}_1)$$

$$N_i(\boldsymbol{z}_1). \equiv . Pr_1(\boldsymbol{z}_1) \mathbf{v} SN_i(\boldsymbol{z}_1)$$

are the expressions of equation (4), realize $(\boldsymbol{z}_2)\boldsymbol{z}_1 . Pr_1(\boldsymbol{z}_2)$ and $(E\boldsymbol{z}_2)\boldsymbol{z}_1 . Pr_1(\boldsymbol{z}_2)$ respectively; and a simple circuit, $c_1, c_2, \cdots, c_n$, of n links, each sufficient to excite the next, gives an expression

$$N_m(\boldsymbol{z}_1). \equiv . N_1(0). C_{mn}$$

for the last form. By induction we derive the theorem.

One more thing is to be remarked in conclusion. It is easily shown: first, that every net, if furnished with a tape, scanners connected to afferents, and suitable efferents to perform the necessary motor-operations, can compute only such numbers as can a Turing machine; second, that each of the latter numbers can be computed by such a net; and that nets with circles can be computed by such a net; and that nets with circles can compute, without scanners and a tape, some of the numbers the machine can, but no others, and not all of them. This is of interest as affording a psychological justification of the Turing definition of computability and its equivalents, Church's λ—definability and Kleene's primitive recursiveness: If any number can be computed by an organism, it is computable by these definitions, and conversely.

IV. Consequences

Causality, which requires description of states and a law of necessary connection relating them, has appeared in several forms in several sciences, but never, except in statistics, has it been as irreciprocal as in this theory. Specification for any one time of afferent stimulation and of the activity of all constituent neurons, each an "all-or-none" affair, determines the state. Specification of the nervous net provides the law of necessary connection whereby one can compute from the description of any state that of the succeeding state, but the inclusion of disjunctive relations prevents complete determination of the one before. Moreover, the regenerative activity of constituent circles renders reference indefinite as to time past. Thus our knowledge of the world, including ourselves, is incomplete as to space and indefinite as to time. This ignorance, implicit in all our brains, is the counterpart of the abstraction which renders our knowledge useful. The role of brains in determining the epistemic relations of our theories to our observations and of these to the facts is all too clear, for it is apparent that every idea and every sensation is realized by activity within that net, and by no such activity are the actual afferents fully determined.

There is no theory we may hold and no observation we can make that will retain so much as its old defective reference to the facts if the net be altered. Tinitus, paraesthesias, hallucinations, delusions, confusions and disorientations intervene. Thus empiry confirms that if our nets are undefined, our facts are undefined, and to the "real" we can attribute not so much as one quality or "form." With determination of the net, the unknowable object of knowledge, the "thing in itself," ceases to be unknowable.

To psychology, however defined, specification of the net would contribute all that could be achieved in that field—even if the analysis were pushed to ultimate psychic units or "psychons," for a psychon can be no less than the activity of a single neuron. Since that activity is inherently propositional, all psychic events

have an intentional, or "semiotic," character. The "all-or-none" law of these activities, and the conformity of their relations to those of the logic of propositions, insure that the relations of psychons are those of the two-valued logic of propositions. Thus in psychology, introspective, behavioristic or physiological, the fundamental relations are those of two-valued logic.

Expression for the Figures

In the figure the neuron c_i is always marked with the numeral i upon the body of the cell, and the corresponding action is denoted by 'N' with i as subscript, as in the text.

Figure 1a $\quad N_2(t) . \equiv . N_1(t-1)$

Figure 1b $\quad N_3(t) . \equiv . N_1(t-1) \mathbf{v} N_2(t-1)$

Figure 1c $\quad N_3(t) . \equiv . N_1(t-1) . N_2(t-1)$

Figure 1d $\quad N_3(t) . \equiv . N_1(t-1) . \infty N_2(t-1)$

Figure 1e $\quad N_3(t) : \equiv : N_1(t-1)$
$$. \mathbf{v} . N_2(t-3) . \infty N_2(t-2)$$
$$N_4(t) . \equiv . N_2(t-2) . N_2(t-1)$$

Figure 1f $\quad N_4(t) : \equiv : \infty N_1(t-1) . N_2(t-1)$
$$\mathbf{v} N_3(t-1) . \mathbf{v} . N_1(t-1) .$$
$$N_2(t-1) . N_3(t-1)$$
$$N_4(t) : \equiv : \infty N_1(t-2) . N_2(t-2)$$
$$\mathbf{v} N_3(t-2) . \mathbf{v} . N_1(t-2) .$$
$$N_2(t-2) . N_3(t-2)$$

Figure 1g $\quad N_3(t) . \equiv . N_2(t-2) . \infty N_1(t-3)$

Figure 1h $\quad N_2(t) . \equiv . N_1(t-1) . N_1(t-2)$

Figure 1i $\quad N_3(t) : \equiv : N_2(t-1) . \mathbf{v} . N_1(t-1)$
$$. (Ex)t - 1 . N_1(x) . N_2(x)$$

Hence arise constructional solutions of holistic problems involving the differentiated continuum of sense awareness and the normative, perfective and resolvent properties of perception and execution. From the irreciprocity of causality it follows that even if the net be known, though we may predict future from present activities, we can deduce neither afferent from central, nor central from efferent, nor past from present activities—conclusions which are reinforced by the contradictory testimony of eye-witnesses, by the difficulty of diagnosing differentially the organically diseased, the hysteric and the malingerer, and by comparing one's own memories or recollections with his contemporaneous records. Moreover, systems which so respond to the difference between afferents to a regenerative net and certain activity within that net, as to reduce the difference, exhibit purposive behavior; and organisms are known to possess many such systems, subserving homeostasis, appetition and attention. Thus both the formal and the final aspects of that activity which we are wont to call *mental* are rigorously

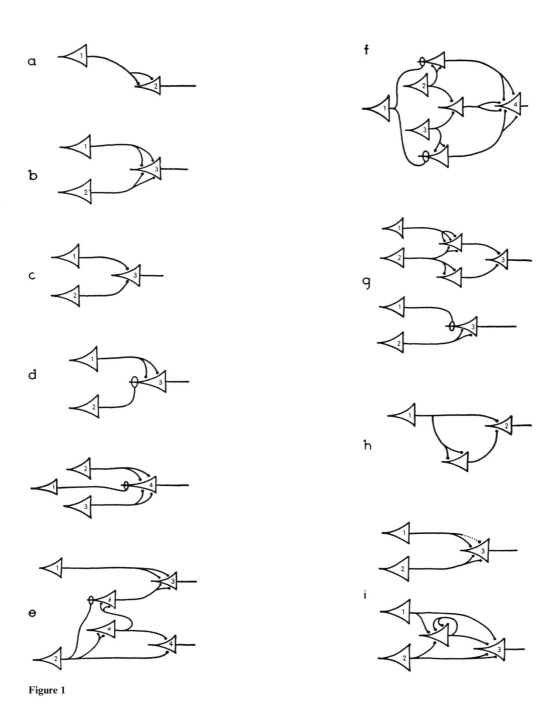

Figure 1

deduceable from present neurophysiology. The psychiatrist may take comfort from the obvious conclusion concerning causality—that, for prognosis, history is never necessary. He can take little from the equally valid conclusion that his observables are explicable only in terms of nervous activities which, until recently, have been beyond his ken. The crux of this ignorance is that inference from any sample of overt behavior to nervous nets is not unique, whereas, of imaginable nets, only one in fact exists, and may, at any moment, exhibit some unpredictable activity. Certainly for the psychiatrist it is more to the point that in such systems "Mind" no longer "goes more ghostly than a ghost." Instead, diseased mentality can be understood without loss of scope or rigor, in the scientific terms of neurophysiology. For neurology, the theory sharpens the distinction between nets necessary or merely sufficient for given activities, and so clarifies the relations of disturbed structure to disturbed function. In its own domain the difference between equivalent nets and nets equivalent in the narrow sense indicates the appropriate use and importance of temporal studies of nervous activity: and to mathematical biophysics the theory contributes a tool for rigorous symbolic treatment of known nets and an easy method of constructing hypothetical nets of required properties.

Literature

Carnap, R. 1938. *The Logical Syntax of Language.* New York: Harcourt, Brace and Company.

Hilbert, D., und Ackermann, W. 1927. *Grundüge der Theoretischen Logik.* Berlin: J. Springer.

Russell, B., and Whitehead, A. N. 1925. *Principa Mathematica.* Cambridge: Cambridge University Press.

3
Introduction

(1947)
Walter Pitts and Warren S. McCulloch

How we know universals: the perception of auditory and visual forms
Bulletin of Mathematical Biophysics 9:127–147

This paper by Pitts and McCulloch came only a short time after their famous 1943 paper. It is less well known than the first, but has many interesting ideas, and is much more in the direction in which neuroscience and network research has progressed since the 1940s. It contains some rather detailed neurophysiology, and suggests a model for the organization and operation of the superior colliculus that is very modern in character.

The basic problem addressed in the paper is one that recurs constantly in psychology, neurophysiology, artificial intelligence, and neural network research and has not yet been solved, in general. As Pitts and McCulloch state it, "We seek general methods for designing nervous nets which recognize figures in such a way as to produce the same output for every input belonging to the figure" (p. 128). The most direct realization of this problem is the construction of invariant geometrical codings of images. An example Pitts and McCulloch use is recognizing that a square is a square wherever it appears in the visual field. In its generality, however, this problem is one expression of a concept formation problem that goes back to the Greeks. How, by seeing different examples of something, can I learn that all the examples are instances of the same thing?

Pitts and McCulloch refer to a raw sense image as an *apparition*, a wonderfully suggestive name that unfortunately was not used in the later literature. In the first part of the paper, they discuss several mathematical techniques of differing complexity for transforming an apparition into the constant representation. They propose that at an early level the nervous system is specifically constructing a transformation to carry the input to a uniform representation. In the cases they discuss in detail, they assume some linearity in the transformation. They assume that we have access somehow to transformations carrying a number of different examples into the invariant representation. Then, the overall transformation is the average of the individual representations. They then discuss some extensions of this basic idea, with excursions into detailed neuroanatomy.

One of their suggestions involves having anatomically separate layers or substructures do the individual transformations, which are then brought together at another layer to form the final representation. At first glance this idea seems wasteful of neural machinery, but they suggest that it actually may be a powerful way of accomplishing the task and discuss then current cortical neuroanatomy in considerable detail in light of their idea. If we rephrase these ideas as that of using multiple converging preliminary representations to form a final representation, we can have cooperating independent computations giving rise to an overall final result.

As a way of using less machinery, they suggest the possibility of a time-space trade-off, reusing the neural hardware in a cycle of temporal processing that can replace a spatial dimension. This idea was more attractive then than now, because in 1947 there was some interest in the idea that cortical EEG (electroencephalogram) rhythms were reflections of a scanning or sweep process, like generating a television image with a beam of electrons that is swept rapidly across the screen. This clever idea has not held up.

The notion that the nervous system constructs average transformations as a way of generalizing from multiple examples has also been used in later work on concept formation. (See Knapp and Anderson, paper 36, for an example in which this technique is used explicitly.)

One of the most modern aspects of this paper is the discussion of spatial computation in the superior colliculus. As Pitts and McCulloch discuss in the paper, Julia Apter had shown a precise, though interestingly distorted, map on the collicular surface by recording the point of maximum evoked electrical potential when a spot of light was presented in the visual field. She also showed that a motor map was in register with the spatial map, as could be demonstrated by showing that an animal would move its eyes to a particular location if that point on the sensory map were excited. Later neurophysiology and neuroanatomy have demonstrated the existence of multiple maps of this kind, throughout the cortex; for example, it appears that one of the most common information representation strategies in the brain is to form a spatial map representing a quantity of interest (see Knudsen, du Lac, and Esterly, 1987). Often these maps require a fair amount of preprocessing to be formed. For example, there is a two-dimensional map of auditory space on the colliculus of the owl that requires rather complex convergence of information from interaural time delays and intensities at different frequencies to be formed. Later studies have amply confirmed the presence of the collicular maps, and the registration between sensory and motor representations, and have suggested more realistic models for the collicular control of the extraocular muscles for direction of gaze. However, the basic models are often strikingly similar in philosophy to that proposed by Pitts and McCulloch (see McIlwain, 1976).

The difference in computational strategy used between the 1943 paper and the 1947 paper is remarkable. The strong implication in the 1943 paper is that the brain is computing logic functions. The later collicular model uses very different techniques. Here, the collicular computer is taking a spatial map, computing the "weighted center of gravity" of the activity on the map, and moving the eyes to that location. This is not a logic function as customarily defined, but something that looks a great deal like spatially distributed analog computation, one that could easily be realized with realistic analog neurons. The power of logic has been replaced with the power of spatial representation and analog computation. By the standards of modern neuroscience, this was the right way to go.

A nice touch in this paper is the use of the original captions for some of the figures, in the original language. With the dropping of the language requirements in most American graduate schools, there will not be too many young American engineers or scientists able to read them.

References

E. I. Knudsen, S. du Lac, and S. D. Esterly (1987), "Computational maps in the brain," *Annual Review of Neuroscience*, W. M. Cowan, E. M. Shooter, C. F. Stevens, and R. F. Thompson (Eds.) 10:41–65.

J. T. McIlwain (1976), "Large receptive fields and spatial transformations in the visual system," *International Review of Physiology, Neurophysiology II*, R. Porter (Ed.) 10:223–248.

(1947)
Walter Pitts and Warren S. McCulloch

How we know universals: the perception of auditory and visual forms
Bulletin of Mathematical Biophysics 9 : 127–147

Two neural mechanisms are described which exhibit recognition of forms. Both are independent of small perturbations at synapses of excitation, threshold, and synchrony, and are referred to particular appropriate regions of the nervous system, thus suggesting experimental verification. The first mechanism averages an apparition over a group, and in the treatment of this mechanism it is suggested that scansion plays a significant part. The second mechanism reduces an apparition to a standard selected from among its many legitimate presentations. The former mechanism is exemplified by the recognition of chords regardless of pitch and shapes regardless of size. The latter is exemplified here only in the reflexive mechanism translating apparitions to the fovea. Both are extensions to contemporaneous functions of the knowing of universals heretofore treated by the authors only with respect to sequence in time.

To demonstrate existential consequences of known characters of neurons, any theoretically conceivable net embodying the possibility will serve. It is equally legitimate to have every net accompanied by anatomical directions as to where to record the action of its supposed components, for experiment will serve to eliminate those which do not fit the facts. But it is wise to construct even these nets so that their principal function is little perturbed by small perturbations in excitation, threshold, or detail of connection within the same neighborhood. Genes can only predetermine statistical order, and original chaos must reign over nets that learn, for learning builds new order according to a law of use.

Numerous nets, embodied in special nervous structures, serve to classify information according to useful common characters. In vision they detect the equivalence of apparitions related by similarity and congruence, like those of a single physical thing seen from various places. In audition, they recognize timbre and chord, regardless of pitch. The equivalent apparitions in all cases share a common figure and define a group of transformations that take the equivalents into one another but preserve the figure invariant. So, for example, the group of translations removes a square appearing at one place to other places; but the figure

of a square it leaves invariant. These figures are the *geometric objects* of Cartan and Weyl, the *Gestalten* of Wertheimer and Köhler. We seek general methods for designing nervous nets which recognize figures in such a way as to produce the same output for every input belonging to the figure. We endeavor particularly to find those which fit the histology and physiology of the actual structure.

The epicritical modalities map the continuous variables of sense into the neurons of a fine cortical mosaic that strikingly imitates a continuous manifold. The visual half-field is projected continuously to the *area striata*, and tones are projected by pitch along Heschl's gyrus. We can describe such a manifold, say $\mathcal{M}$, by a set of coordinates $(x_1, x_2, \cdots, x_n)$ constituting the point-vector x, and denote the distributions of excitation received in $\mathcal{M}$ by the functions $\phi(x, t)$ having the value unity if there is a neuron at the point x which has fired within one synaptic delay prior to the time t, and otherwise, the value zero. For simplicity, we shall measure time in mean synaptic delays, supposed equal, constant, and about a millisecond long. Indications of time will often not be given.

Let G be the group of transformations which carry the functions $\phi(x, t)$ describing apparitions into their equivalents of the same figure. The group G may always be taken finite, as is seen from the atomicity of the manifold; let it have N members. We shall distinguish four problems of ascending complexity:

1) The transformation T of G can be generated by transformations t of the underlying manifold $\mathcal{M}$, so that $T\phi(x) = \phi[t(x)]$; e.g., if G is the group of translations, then $T\phi(x) = \phi(x + a_T)$, where a_T is a constant vector depending only upon T. If G is the group of dilatations, $T\phi(x) = \phi(\alpha_T x)$, where α_T is a positive real number depending only upon T. All such transformations are linear:

$$T[\alpha\phi(x) + \beta\psi(x)] = \alpha\phi[t(x)] + \beta\psi[t(x)]$$

$$= \alpha T\phi(x) + \beta T\psi(x).$$

2) The transformations T of G cannot be so generated, but are still linear and independent of the time t. An

example is to take the gradient of $\phi(x)$, or to replace $\phi(x)$ by its average over a certain circle surrounding x.

3) The transformations T of G are linear, but depend also upon the time. For example, they take a moving average over the preceding five synaptic delays or take some difference as an approximation to the time-derivative of $\phi(x, t)$.

4) Not all T of G are linear.

Our special nets are essays in problem 1. The simplest way to construct invariants of a given distribution $\phi(x, t)$ of excitation is to average over the group G. Let f be an arbitrary functional which assigns a unique numerical value, in any way, to every distribution $\phi(x, t)$ of excitation in $\mathscr{M}$ over time. We form every transform $T\phi$ of $\phi(x, t)$, evaluate $t[T\phi]$, and average the result over G to derive

$$a = \frac{1}{N} \sum_{\substack{\text{all} \\ T \varepsilon G}} f[T\phi]. \tag{1}$$

If we had started with $S\phi$, S of G, instead of ϕ, we should have

$$\frac{1}{N} \sum_{T \varepsilon G} f[TS\phi] = \frac{1}{N} \sum_{\substack{\text{All } T \\ \text{such that} \\ TS^{-1} \varepsilon G}} f[T\phi] = a, \tag{2}$$

for TS^{-1} is in the group when, and only when, T is in the group; that is, the terms of the sum (1) are merely permuted.

To characterize completely the figure of $\phi(x, t)$ under G by invariants of this kind, we need a whole manifold Ξ of such numbers a for different functionals f, with as many dimensions in general as the original $\mathscr{M}$; if we describe Ξ by coordinates $(\xi_1, \xi_2, \cdots, \xi_m) = \xi$, we may fulfill this requirement formally with a single f which depends upon ξ as a parameter as well as upon the distribution ϕ which is its argument, and write

$$\phi_{f, G}(\xi) = \frac{1}{N} \sum_{T \varepsilon G} f[T\phi\xi]. \tag{3}$$

If the nervous system needs less then complete information in order to recognize shapes, the manifold Ξ may be much smaller than $\mathscr{M}$, have fewer dimensions, and indeed reduce to isolated points. The time t may be one dimension of Ξ, as may some of the x_j representing position in $\mathscr{M}$.

Suppose that G belongs to problems 1 or 2 and that the dimensions of Ξ are all spatial; then the simplest nervous net to realize this formal process is obtained in the following way: Let the original manifold $\mathscr{M}$ be duplicated on $N-1$ sheets, a manifold $\mathscr{M}_T$ for each T of G, and connected to $\mathscr{M}$ or its sensory afferents in

such a way that whatever produces the distribution $\phi(x)$ on $\mathscr{M}$ produces the transformed distribution $T\phi(x)$ on $\mathscr{M}_T$. Thereupon, separately for each value of ξ for each $\mathscr{M}_T$, the value of $f[T\phi\xi]$ is computed by a suitable net, and the results from all the $\mathscr{M}_T$'s are added by convergence on the neuron at the point ξ of the mosaic Ξ. But to proceed entirely in this way usually requires too many associative neurons to be plausible. The manifolds $\mathscr{M}_T$ together possess the sum of the dimensions of $\mathscr{M}$ and the degrees of freedom of the group G. More important is the number of neurons and fibers necessary to compute the values of $f[T\phi, \xi]$, which depends, in principle, upon the entire distribution $T\phi$, and therefore requires a separate computer for every ξ for every T of G. This difficulty is most acute if f be computed in a structure separated from the $\mathscr{M}_T$, since in that case all operations must be performed by relatively few long fibers. We can improve matters considerably by the following device: Let the manifolds $\mathscr{M}_T$ be connected as before, but raise their thresholds so that their specific afferents alone are no longer able to excite them; cause adjuvant fibers to ramify throughout each $\mathscr{M}_T$ so that when active they remedy the deficiency in summation and permit $\mathscr{M}_T$ to display $T\phi(x)$ as before. Let all the neurons with the same coordinate x on the N different $\mathscr{M}_T$'s send axons to the neuron at x on another recipient sheet exactly like them, say Q—this Q may perfectly well be one of the $\mathscr{M}_T$'s—and suppose any one of them can excite this neuron. If the adjuvant neurons are excited in a regular cycle so that every one of the sheets $\mathscr{M}_T$ in turn, and only one at a time, receives the increment of summation it requires for activity, then all of the transforms $T\phi$ of $\phi(x)$ will be displayed successively on Q. A single f computer for each ξ, taking its input from Q instead of from the $\mathscr{M}_T$'s, will now suffice to produce all the values of $f[T\phi, \xi]$ in turn as the "time-scanning" presents all the $T\phi$'s on Q in the course of a cycle. These values of $f(T\phi, \xi)$ may be accumulated through a cycle at the final Ξ-neuron in any way.

This device illustrates a useful general principle which we may call the *exchangeability of time and space*. This states that any dimension or degree of freedom of a manifold or group can be exchanged freely with as much delay in the operation as corresponds to the number of distinct places along that dimension.

Let us consider the auditory mechanism which recognizes chord and timbre independent of pitch. This mechanism, or part of it, we shall suppose situated in Heschl's gyrus, a strip of cortex two to three centimeters long on the superior surface of the temporal lobe. This strip receives afferents from lower auditory mechanisms so that the position on the cortex corresponds to the

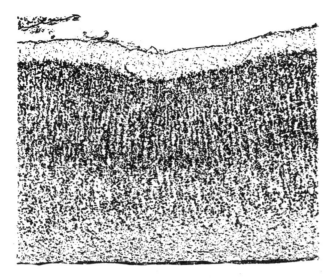

Figure 1 Vertical section of the primary auditory cortex in the long axis of Heschl's gyrus, stained by Nissl's method which stains only cell bodies. Note that the columnar cortex, typical of primary receptive areas, shows two tiers of columns, the upper belonging to the receptive layer IV and the lower, lighter stained, to layer V.

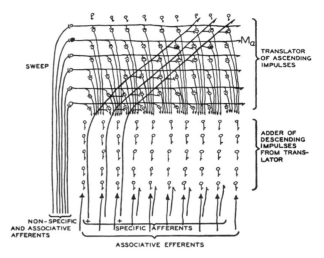

Figure 2 Impulses of some chord enter slantwise along the specific afferents, marked by plusses, and ascend until they reach the level $\mathcal{M}_a$ in the columns of the receptive layer activated at the moment by the nonspecific afferents. These provide summation adequate to permit the impulses to enter that level but no other. From there the impulses descend along columns to the depth. The level in the column, facilitated by the nonspecific afferents, moves repetitively up and down, so that the excitement delivered to the depths moves uniformly back and forth as if the sounds moved up and down together in pitch, preserving intervals. In the deep columns various combinations are made of the excitation and are averaged during a cycle of scansion to produce results depending only on the chord.

pitch of tones, low tones exciting the outer and forward end, high tones the inner and posterior. Octaves span equal cortical distances, as on the keyboard of a piano. The afferents conveying this information from the medial geniculate slant upward through the cortex, branching into telodendria in the principal recipient layer IV, which consists of vertical columns of fifty or more neurons concerning the course of whose ramifying axons there is no certain knowledge except that their activity eventually excites columns of cells situated beneath the recipient layers. Their axons converge to a layer of small pyramids whose axons terminate principally in the secondary auditory cortex or adjacent parts of the temporal lobe. To the layers above and below the receptive layers also come "associative" fibers from elsewhere in the cortex, particularly from nearby. There is no good Golgi picture of the primary auditory cortex in monkeys, but unless it is unlike all the rest of the cortex, it also receives nonspecific afferents from the thalamus, which ascend to branch indiscriminately at every level. A picture of the primary auditory cortex stained by Nissl's method is given in Figure 1, and a schematic version in Figure 2.

The secondary auditory cortex has separate specific afferents and the same structure as the primary except for possessing some large pyramids known to send axons to distant places in the cortex such as the motor face and speech areas.

In this case, the fundamental manifold $\mathcal{M}$ is a one-dimensional strip, and x is a single coordinate measuring position along it. The group G is the group of uniform translations which transform a distribution $\phi(x, t)$ of excitation along the strip into $T_a\phi = \phi(x + a, t)$. The group G is thus determined by adding the various constants to the coordinate x, and therefore belongs to problem 1. The set of manifolds $\mathcal{M}_T$ is a set of strips $\mathcal{M}_a$ that could be obtained by sliding the whole of $\mathcal{M}_a$ back and forth various distances along its length. The same effect is obtained by slanting the afferent fibers upward, as in Figure 2, and in the auditory cortex itself where the levels in the columnar receptive layer constitute the $\mathcal{M}_T$. These send axons to the deeper layer, a mass capable of reverberation and summation over time, that may well constitute the set of $f(T\phi, \xi)$ computers for the various ξ, or part of them.

To complete the parallel with our general model, we require adjuvant fibers to activate the various levels $\mathcal{M}_a$ successively. It is to the nonspecific afferents that modern physiology attributes the well-known rhythmic sweep of a sheet of negativity up and down through the cortex—the alpha-rhythm. If our model fits the facts, this alpha-rhythm performs a temporal "scanning" of the cortex which thereby gains, at the cost of time, the equivalent of another spatial dimension in its neural manifold.

According to Ramón y Cajal (1911), Lorente de Nó (1922), and J. L. O'Leary (1941), the specific visual afferents originate in the lateral geniculate body and travel upward through the calcarine cortex, to ramify horizontally for long distances in the stripe of Gennari. This is called the *granular layer* by Brodmann from Nissl stains, and is also called the *external stria of Baillarger*, from its myeloarchitecture. (Zunino, 1909). It is the fourth, or receptive, layer of Lorente de Nó. It may be divided into a superior part IVa, consisting of the larger star-cells and star-pyramids, and an inferior part IVb, consisting of somewhat smaller star-cells, arranged in columns, although the distinction of parts is not always evident (O'Leary, 1941, p. 141). The stripe of Gennari is the sole terminus of specific afferent fibers in the cat and higher mammals, although not in the rabbit. Its neurons send numerous axons horizontally and obliquely upward and downward within the layer; others ascend to the plexiform layer at the surface or descend to the subjacent fifth layer of efferent cells; and axons from the large star-pyramids even enter the subjacent white matter.

The electrical records of J. L. O'Leary and G. H. Bishop (1941) indicate that the normal response of the striate cortex to an afferent volley is triphasic, commencing in layer IV, shown by a surface-positive potential. Next it rises to the surface, making it negative; then as the surface becomes positive, it descends first to the third layer to project to other cortical areas, and then reaches the fifth layer, whence it goes to the pulvinar, the superior colliculus (Barris, Ingram, and Ranson, 1935), and tegmental oculo-motor nuclei, especially to the para-abducens nucleus, which subserves conjugate deviation of the eyes. (Personal communication from Elizabeth Crosby.) This triphasic response, having the period of the alpha-rhythm, is too long to be easily envisaged as a single cycle of purely internal reverberation in the striate cortex. This opinion is confirmed by the superimposed faster response to more intense afferent volleys. It is more reasonable to regard efferents to undifferentiated thalamic nuclei and non-specific afferents from them (Dempsey and Morrison, 1943) as responsible for the sustention of this triphasic rhythm. As in the auditory mechanism, we assign them the function of "scanning" by exciting sheets seriatim in the upper layers of the cortex.

A version of the visual cortex which agrees with these facts and which constitutes a mechanism of the present type for securing invariance to dilatation and constriction of visual forms is diagrammed in Figure 3. For comparison with this scheme some drawings by Cajal

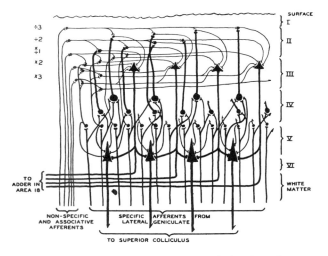

Figure 3 Impulses relayed by the lateral geniculate from the eyes ascend in specific afferents to layer IV where they branch laterally, exciting small cells singly and larger cells only by summation. Large cells thus represent larger visual areas. From layer IV impulses impinge on higher layers where summation is required from nonspecific thalamic afferents or associative fibers. From there they converge on large cells of the third layer which relay impulses to the parastriate area 18 for addition. On their way down they contribute to summation on the large pyramids of layer V which relays them to the superior colliculus.

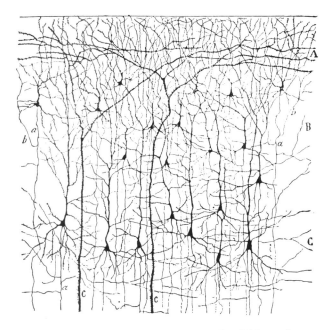

Figure 4a The following is the original caption. Kleine und mittelgrosse Pyramidenzellen der Sehrinde eines 20 tägigen Neugeborenen (Fissura calcarina). *A*, plexiforme Schicht; *B*, Schicht der kleinen Pyramiden; *C*, Schicht der mittelgrossen Pyramiden; *a*, absteigender Axencylinder; *b*, rückläufige Collateralen; *c*, Stiele von Riesenpyramiden.

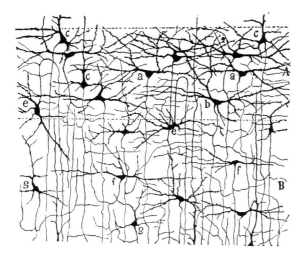

Figure 4b The following is the original caption. Schichten der Sternzellen der Sehrinde des 20 tägigen Neugeborenen (Fissura calcarina). *A*, Schicht der grossen Sternzellen; *a*, halbmondförmige Zellen; *b*, horizontale Spindelzelle; *c*, Zellen mit einem zarten radiären Fortsatz; *e*, Zelle mit gebogenem Axencylinder; *B*, Schicht der kleinen Sternzellen; *f*, horizontale Spindelzellen; *g*, dreieckige Zellen mit starken gebogenen Collateralen; *h*, Pyramiden mit gebogenem Axencylinder, an der Grenze der fünften Schicht; *C*, Schicht der kleinen Pyramiden mit gebogenem Axencylinder.

(1900) from Golgi preparations are shown in Figure 4 with the original captions.

Figure 3 is a diagram of part of the neurons in a vertical section of cortex taken radially outward from that cortical point to which the center of the fovea projects. The lowest tier of small cells in IVb is the primary receptive manifold $\mathcal{M}$; the upper tiers of internuncials in I, II, and III, to which the upper tiers of layer IVa separately project, constitute the manifolds $\mathcal{M}_a$ for uniform constriction of all the coordinates of an apparition by factors $0 < \alpha < 1$. This reduplication of the layers of IVa in additional upper internuncial tiers is of course unnecessary since the nonspecific afferents might equally well scan the layers of star-pyramids themselves. The magnifications of the apparition are represented on the internuncial tiers drawn beneath the efferents in the third layer. It is quite likely that these are in reality the small star-cells of IVb, or even the long horizontal extensions of the specific afferents within the outer stria of Baillarger. Histological sections of the visual cortex are now being cut radial to the projection of the center of the fovea and perpendicular to it. It is evident that many details of this and the other hypothetical nets of this paper might be chosen in several ways with equal reason; we have only taken the most likely in the light of present knowledge. The sheet of excitement from nonspecific

afferents sweeping up and down the upper three layers, therefore, produces all magnifications and constrictions seriatim on the efferent cells of layer III, traveling from there to the parastriate cortex where the functionals f are made of them and the results added.

It is worth observing again, when special example can fix it, that the group-invariant spatio-temporal distribution of excitations which represents a figure need not resemble it in any simple way. Thus, purely for illustration, we might suppose that the efferent pyramids in the layer III of our diagram project topographically upon another cortical mosaic, which only responds to corners, and accumulates over a cycle of scansion. A square in the visual field, as it moved in and out in successive constrictions and dilatations in Area 17, would trace out four spokes radiating from a common center upon the recipient mosaic. This four-spoked form, not at all like a square, would then be the size-invariant figure of square. In fact, Area 18 does not act like this, for during stimulation of a single spot in the parastriate cortex, human patients report perceiving complete and well-defined objects, but without definite size or position, much as in ordinary visual mental imagery. This is why we have situated the mechanism of Figure 3 in Area 17, instead of later in the visual association system. This also makes it likely that one of the dimensions of the apperceptive manifold $\varXi$, upon whose points the group-averages of various properties of the apparition are summed, is time.

This point is especially to be taken against the Gestalt psychologists, who will not conceive a figure being known save by depicting it topographically on neuronal mosaics, and against the neurologists of the school of Hughlings Jackson, who must have it fed to some specialized neuron whose business is, say, the reading of squares. That language in which information is communicated to the homunculus who sits always beyond any incomplete analysis of sensory mechanisms and before any analysis of motor ones neither needs to be nor is apt to be built on the plan of those languages men use toward one another.

Besides the mechanisms which compute invariants as averages, there is another variety of nervous net that can perceive universals. These nets we call *reflex-mechanisms*. Consider the reflex-arc from the eyes through the tectum to the oculomotor nuclei and so to the muscles which direct the gaze. We propose that the superior colliculus computes by double integration the lateral and vertical coordinates of the "center of gravity of the distribution of brightness" referred to the point of fixation as origin, and supplies impulses at a rate proportional to these coordinates to the lateral and vertical eye-muscles in such a way that these then turn the visual axis toward the center of gravity. As the center of gravity approaches the origin, its ordinate and abscissa diminish, slowing the eyes and finally stopping them when the visual axes point at the "center of brightness." This provides invariants of translation. If a square should appear anywhere in the field, the eyes turn until it is centered, and what they see is the same, whatever the initial position of the square. This is a reflex-mechanism, for it operates on the principle of the servo-mechanism, or "negative feedback."

We find considerable support for this conjecture in the profuse anatomical and physiological literature on the corpora quatrigemina anteriora. Histologically, in mammals they are arranged in nine laminae, composed alternately of grey and white matter. Aside from the central grey of the aqueduct, we may enumerate these as follows, from the most superficial inward, naming them with C. V. Ariëns-Kappers, G. C. Huber, and E. C. Crosby (1936):

1) A superficial layer of fine white myelinated fibers running antero-posteriorly. These arise in the posterior end of the middle temporal gyrus, about Area 37 of Brodmann, in the part of the temporal lobe which associates visual and auditory material. (E. Crosby, unpublished.) This is the *stratum zonale*, so called by Cajal (1911).

2) A *stratum griseum superficiale*, composed of radially directed cells of sundry types, each with dendrites ramifying near one or both of the adjacent layers, and an axon plunging down into the fourth layer.

3) The *stratum opticum*. This dense layer of myelinated fibers courses antero-posteriorly and constitutes the major afferent supply to the colliculus. The upper portion comes directly from the optic chiasm, as fibers from the nasal side of the contralateral retina and the temporal side of the ipsilateral, and pierces the rostral surface. These direct fibers diminish in number and importance in the higher mammals, giving place to fibers from the occipital cortex beneath them in the layer. These come up from the depths with the radiation from Area 17 somewhat caudal to that from Area 18 or 19 or both. (Barris, Ingram, and Ransom, 1935). There are some other cortical fibers of unknown origin in this stratum also, but none from the frontal eye-fields of Area 8 (*ibid.*), which projects directly to oculomotor nuclei (Ward and Reed, 1946). The fibers of the stratum opticum end in bushy terminal arborizations in the grey matter above and below it.

4) A *stratum griseum mediale*, which, together with the three laminae beneath—the *stratum album mediale*, and the two *strata alba et grisea profunda*—makes up Cajal's (1911) "Zone ganglionaire ou des fibres hori-

zontales." Here lie the principal bigeminate efferents. The dendrites of these cells pervade the superficial grey, the stratum opticum, and their own layer. Fibers reach their somata from all the upper strata and the commissure of the superior colliculus. Their axons course horizontally, laterally, and then somewhat caudally, descend to the stratum album profundum, and leave the tectum laterally or else pierce the medial surface as commissural fibers to the other colliculus. The former comprises the "uncrossed" bundle of tecto-pontine fibers (not tecto-spinal: *ibid.*) besides the main "voie optique reflexe" of Cajal. The latter leaves the tectum to spiral ventrad and caudad around the aqueduct and the third and fourth nerve nuclei, decussates, and passes caudad under the medial longitudinal fasciculus to the para-abducens and VI-th nerve nuclei, and to the cervical cord (Cajal, 1911). As it passes, it gives collaterals to all the oculomotor nuclei, mostly crossed at the rostral end and mostly uncrossed posteriorly (E. Crosby, unpublished). As we proceed caudally, the oculomotor nuclei innervate the ocular muscles in this order: superior rectus, medial rectus, inferior oblique, inferior rectus, and thence the superior oblique and the lateral rectus, substantiating the scheme of B. Brouwer (1917). These nuclei are interconnected by the medial longitudinal fasciculus, whereby axonal collaterals presumably inhibit antagonists and facilitate synergists. They are aided in this by modest interstitial nuclei such as the para-abducens, subserving conjugate deviation, and perhaps one between the medial recti, for convergence. Such nuclei also serve to transmit the cortical, striatal, acoustic, and vestibular impulses to the oculomotor nerves (*ibid.*, and Lorente de Nó, 1933). Some drawings by Ramón y Cajal from Golgi preparations of the superior colliculus are reproduced, with his captions, in Figure 5.

Julia Apter (1945, 1946), by illuminating small spots on the retina of the cat and finding the tectal point of maximum evoked potential, has demonstrated that each half of the visual field, seen through the nasal half of one eye and the temporal half of the other, maps point-by-point upon the contralateral colliculus. The contours of projection, in angular degrees lateral and vertical from the visual axis, are drawn on the right colliculus as dotted lines in Figure 6. Presumably the calcarine cortex would map similarly, although this has not been tried. In addition, by strychninizing a single point on the collicular surface and flashing a diffuse light on the retina, she obtained change in gaze so as to fix a certain constant point in the visual field. The points for various strychninized places are sketched in solid lines on the right colliculus of Figure 6. It is clear that they nearly coincide with the retinal points

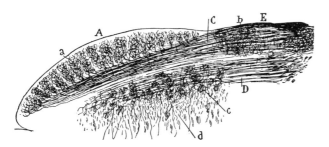

Figure 5a The following is the original caption. Coupe sagittale montrant l'ensemble des fibres optiques du tubercule quadrijumeau antérieur; souris âgée de 24 heures. Méthode de Golgi. *A*, écorce grise du tubercule antérieur; *C*, courant superficiel des fibres optiques; *D*, courant profond; *E*, région postérieure du corps genouillé externe; *b*, foyer où se terminent des collatérales des fibres optiques; *c*, nids péricellulaires formés par les fibres optiques; *d*, fibres transversales de la couche ganglionnaire.

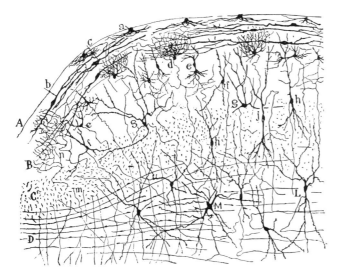

Figure 5b The following is the original caption. Coupe transversale du tubercule quadrijumeau antérieur; lapin âgé de 8 jours. Méthode de Golgi. *A*, surface du tubercule tout près de la ligne médiane; *B*, couche grise superficielle ou couche cendrée de Tartuferi comprenant les zones des cellules horizontales et des cellules fusiformes verticales; *C*, couche des fibres optiques; *D*, couche des fibres transversales ou zone blanc cendré profonde de Tartuferi; *L*, *M*, cellules de la couche ganglionnaire ou des fibres transversales; *a*, cellules marginales; *b*, cellules fusiformes transversales ou horizontales; *c*, autre cellule de même espèce, montrant bien son cylindre-axe; *d*, petites cellules à bouquet dendritique compliqué; *e*, cellules fusiformes verticales; *f*, *g*, differents types cellulaires de la couche grise superficielle; *h*, *j*, cellules fusiformes de la zone des fibres optiques; *m*, collatérale descendante allant à la substance grise centrale; *n*, arborisation terminale des fibres optiques.

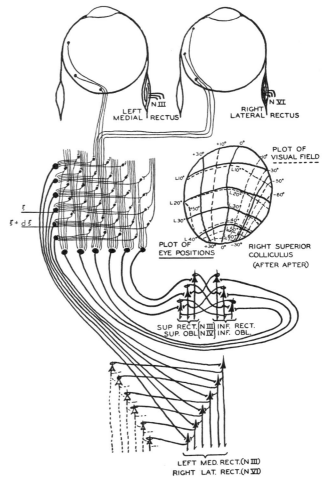

Figure 6 A simplified diagram showing occular afferents to left superior colliculus, where they are integrated anteroposteriorly and laterally and relayed to the motor nuclei of the eyes. A figure of the right superior colliculus mapped for visual and motor response by Apter is inserted. An inhibiting synapse is indicated as a loop about the apical dendrite. The threshold of all cells is taken to be one.

which project to the strychninized spot. She showed that if the diffuse light on the retina were replaced by a localized one, the response would occur if, and only if, the points projecting to the strychninized spot were illuminated—apart from certain other smaller effects from the fovea.

All these results agree well with our initial hypothesis. If x and y are respectively lateral and vertical coordinates in the visual field, and $\phi(x, y)$ is the brightness inhabiting the point (x, y)—that is, the response of the spot in the optic nerve which images (x, y)—the coordinates $\bar{x}$ and $\bar{y}$ of the center of brightness are

$$\bar{x} = \int_V dy \int x\phi(x,y)\,dx,$$

$$\bar{y} = \int_V dy \int y\phi(x,y)\,dx, \tag{4}$$

where integration is over the whole visual field V. If ξ_R, η_R are respectively sagittal and lateral coordinates measuring position on the right colliculus C_R, and ξ_L and η_L their mirror images on the left colliculus C_L, there will be a mapping

$$x = x_R = p(\xi_L, \eta_L),$$

$$y = y_R = q(\xi_L, \eta_L), \quad \text{if } x > 0, \tag{5}$$

and

$$x = x_L = p(\xi_R, \eta_R),$$

$$y = y_L = q(\xi_R, \eta_R), \quad \text{if } x \leq 0,$$

To transform equations (4) into the coordinates of the colliculus will then yield

$$\bar{x} = \bar{x}_R - \bar{x}_L,$$

$$\bar{y} = \bar{y}_R + \bar{y}_L,$$

$$\bar{x}_R = \int_{C_L} \int \Phi_L(\xi,\eta) p(\xi,\eta) J(\xi,\eta)\,d\xi d\eta,$$

$$\bar{y}_R = \int_{C_L} \int \Phi_L(\xi,\eta) q(\xi,\eta) J(\xi,\eta)\,d\xi d\eta,$$

$$\bar{x}_L = \int_{C_R} \int \Phi_R(\xi,\eta) p(\xi,\eta) J\,d\xi d\eta,$$

$$\bar{y}_L = \int_{C_R} \int \Phi_R(\xi,\eta) q(\xi,\eta) J\,d\xi d\eta;$$

where

$$J(\xi,\eta) = \begin{vmatrix} \dfrac{\partial p}{\partial \xi} & \dfrac{\partial p}{\partial \eta} \\[2mm] \dfrac{\partial q}{\partial \xi} & \dfrac{\partial q}{\partial \eta} \end{vmatrix}$$

and

$$\Phi_L(\xi,\eta) = \phi[-p(\xi,\eta), q(\xi,\eta)],$$

$$\Phi_R(\xi,\eta) = \phi[p(\xi,\eta), q(\xi,\eta)]$$

are the distributions of brightness on the surface of the colliculus.

Now it clearly makes no difference to the final result whether the true center of gravity $(\bar{x}, \bar{y})$ determines the net frequency of impulses sent into the eye-muscles, or whether it is some other pair of numbers u and v that increase monotonically with x and y respectively and

vanish with them. For in any case, the eyes must be moved in such a direction as to diminish (u, v), and *pari passu* $(\bar{x}, \bar{y})$; and finally they must remove (u, v), and therefore $(\bar{x}, \bar{y})$, to the origin at the visual axes. Thus, if the two quantities computed from $\phi(x, y)$ to determine lateral and vertical motion respectively have the form

$$u = u_R - u_L, \quad v = v_R + v_L,$$

$$u_R = \int_{C_L} \int U(\xi,\eta) \Phi_L(\xi,\eta)\,d\xi d\eta, \tag{6}$$

$$v_R = \int_{C_L} \int V(\xi,\eta) \Phi_L(\xi,\eta)\,d\xi d\eta,$$

with a similar integral with Φ_R for u_L and v_L, and any U and V fulfilling the condition that for every η, $U(\xi,\eta)$ is properly monotonic in η, and for every ξ, $V(\xi,\eta)$ is properly monotonic in η, then u and v will have the required properties that they shall vanish and vary monotonically with $\bar{x}$ and $\bar{y}$ respectively. J. Apter (1945, 1946) shows that one can write, approximately,

$$x = p(\xi),$$

$$y = q(\eta), \tag{7}$$

with $p(\xi)$ and $q(\eta)$ both properly monotonically increasing, neglecting the other variable. This would yield

$$U(\xi,\eta) = p(\xi) p'(\xi) q'(\eta),$$

$$V(\xi,\eta) = q(\eta) q'(\eta) p'(\xi),$$

so that

$$u_R = \int_{C_L} p(\xi)\,dp(\xi) \int \Phi_L(\xi,\eta) q'(\eta)\,d\eta,$$

$$v_R = \int_{C_L} q(\eta)\,dq(\eta) \int \Phi_L(\xi,\eta) p'(\xi)\,d\xi,$$

together with the corresponding expressions for u_L and v_L involving Φ_R, furnish an approximation to $(\bar{x}, \bar{y})$. Most general of this type is

$$u = u_R - u_L$$

$$= \int_C R_1(\xi)\,d\xi \int [\Phi_L(\xi,\eta) - \Phi_R(\xi,\eta)] S_1(\eta)\,d\eta, \tag{8}$$

$$v = v_R + v_L$$

$$= \int_C R_2(\eta)\,d\eta \int [\Phi_L(\xi,\eta) + \Phi_R(\xi,\eta)] S_2(\eta)\,d\eta, \tag{9}$$

where S_1 and S_2 are non-negative and R_1 and R_2 are properly monotonic and vanish at the origin. The integration is taken over the range of the collicular coordinates. If $S_1 = S_2 = 1$, $R_1(\xi) = \xi = R_2(\xi)$, this is the center of gravity of the afferent excitement upon the colliculi. The simplest schematic way of computing expressions (8) and (9) is actually to carry out the double "integration" on the colliculus, as in Figure 6, so that to compute expression (8) we first add all the afferent impulses within a thin transverse strip, $(\xi, \xi + d\xi)$, to compute

$$d\xi \int \Phi_L(\xi, \eta) \, d\eta.$$

This quantity, for most caudal or greatest ξ, is fed highest into a chain of successively exciting oculomotor neurons; for most anterior, smallest ξ, it comes in lowest. This process provides a net frequency of impulses to the right lateral and medial recti which is certainly weighted by some monotonic factor $R_1(\xi)$. Reciprocal inhibition by axonal collaterals from the nuclei of the antagonist eye-muscles, which are excited similarly by the other colliculus, serves to perform the algebraic subtraction to obtain $u = u_R - u_L$. The computation of the vertical position v of the quasi-center of gravity is done similarly. It is also possible, in whole or in part, that the difference $\Phi_L(\xi, \eta) - \Phi_R(\xi, \eta)$ in equation (8), or the sum $\Phi_L(\xi, \eta) + \Phi_R(\xi, \eta)$ in (9), is computed by commissural fibers running between contralateral tectal points with the same coordinates, instead of in the oculomotor nuclei.

We have omitted to divide the final results u and v by the total luminous flux $A = \int_V \int \Phi(x, y) \, dx \, dy$ before calling (u, v) the "quasi-center of gravity." For the reflex this makes no difference, since (u, v) finally lies at the origin, which does not change on multiplication by A. Similarly, Apter's single-point strychninizations are not relevant to the question. But if several distinct points are strychninized on the colliculus at once, then equation (8) requires gaze to deviate by a lateral distance which is the *sum* of the deviations evoked from the points separately. This may happen; but it seems more likely that the total excitation from the colliculus is in fact kept constant by compensatory variations in the background of facilitation or inhibition, produced perhaps by reverberation with the periaqueductal grey, if not internally in the tectum. H. Klüver's observation (1942) should be recalled here, that even decorticate monkeys whose corpora quadrigemina are not otherwise deafferented detect and discriminate total luminous flux.

But if the colliculus takes a "weighted center of gravity" of an impingent distribution of light, in our most general sense, for suitably chosen partially monotonic positive functions $U(\xi, \eta)$, $V(\xi, \eta)$, so dividing it by the total luminous flux, then, and only then, by a theorem of Reisz, whenever a finite (or infinite) number of points

of the colliculus are simultaneously strychninized, the consequent gaze will lie within the smallest convex polygon (or simplex) containing all the points whose projections are strychninized.

This example may be straightforwardly generalized to provide a uniform principle of design for reflex-mechanisms which secure invariance under an arbitrary group G. In some way, out of the whole series of transforms $T\phi$ of an apparition, one of them ϕ_0 is elected to be standard—e.g., one of a standard overall size—and when presented with ϕ, the mechanism computes one or more suitable parameters $a(\phi)$, $b(\phi)$, $\cdots$, which define its position within the series of $T\phi$'s in a univocal way so that their simultaneous equality $a(\phi) = a(S\phi)$, $b(\phi) = b(S\phi)$, etc., is sufficient to entail $S = I$, the identity. The errors

$$E_1(\phi) = a(\phi) - a(\phi_0),$$
$$E_2(\phi) = b(\phi) - b(\phi_0),$$

if they do not already all vanish, then impel the mechanism to perform a suitable operation $T\phi$ so determined as to diminish the parameters $E(T\phi)$ as compared to $E(\phi)$. This process may be repeated many times, reducing the $E(\phi)$ at every stage, until the $E(\phi)$'s all vanish and $\phi = \phi_0$, its standard. The mechanism is circular: it follows the scheme

$$\text{(Outside)} \rightarrow \text{(Group Operator)} \rightarrow \text{(Output)}$$

$$\begin{array}{cccc} \uparrow & \uparrow & \uparrow & \uparrow \\ a(\phi) & b(\phi) & \cdot & \cdot \\ | & | & | & | \\ \end{array}$$

$$\text{(Error Computor)} \leftarrow$$

In the case of the colliculus, the group is the two-dimensional translation-group, and the two quantities $a(\phi)$ and $b(\phi)$ are the coordinates of the "weighted center of gravity" of equation (6). For any general group of the type we are considering, quantities $a(\phi)$ of this type may always be found, as is shown in the theory of the irreducible representations of the group G.

We have focussed our attention on particular hypothetical mechanisms in order to reach explicit notions about them which guide both histological studies and experiment. If mistaken, they still present the possible kinds of hypothetical mechanisms and the general character of circuits which recognize universals, and give practical methods for their design. These procedures are a systematic development of the conception of reverberating neuronal chains, which themselves, in preserving the sequence of events while forgetting their time of happening, are abstracted universals of a kind. Our circuits extend the abstraction to a wide realm of properties. By systematic use of the principle of the exchangeability of time and space, we have enlarged the realm enormously. The adaptability of our methods to unusual forms of input is matched by the equally unusual form of their invariant output, which will rarely resemble the thing it means any closer than a man's name does his face.

The authors wish to express their great indebtedness to Professor Elizabeth Crosby for her generous assistance and more especially for permission to quote her as yet unpublished observations.

This work was aided by grants from the Josiah Macy, Jr. Foundation and the Rockefeller Foundation.

Literature

Apter, J. 1945. "The Projection of the Retina on the Superior Colliculus of Cats." *J. Neurophysiol.*, **8**, 123–134.

Apter, J. 1946. "Eye Movements Following Strychninization of the Superior Colliculus of Cats." *J. Neurophysiol.*, **9**, 73–85.

Ariëns-Kappers, C. V., G. C. Huber, and E. C. Crosby. 1936. *The Comparative Anatomy of the Nervous System of Vertebrates.* New York: The Macmillan Co.

Barris, R. W., W. R. Ingram, and S. W. Ranson. 1935. "Optic Connections of the Diencephalon and Midbrain in Cat." *J. Comp. Neur.*, **62**, 117–144.

Brouwer, B. 1917. "Klinisch-Anatomische Onderzoekingen Over de Oculomotoriuskern." Voordracht, Gehouden in de Vergadering der Amsterdamsche Neurologenvereeniging op 7 December 1916, in het Binnen-Gasthuis, Collegekamer voor Neurologie. *Neder. Tijdschrift voor Geneeskunde*, Eerste Helft, **14**, 1–11.

Dempsey, E. W., and R. S. Morison. 1943. "The Electrical Action of a Thalamo-Cortical Relay System." *Amer. Jour. of Physiology*, **138**, 2, 283–296.

Huber, G. C., E. C. Crosby, R. T. Woodburne, L. A. Gillilan, J. O. Brown, and B. Tamthai. 1943. "The Mammalian Midbrain and Isthmus Regions. I. The Nuclear Pattern." *J. Comp. Neur.*, **78**, 129–534.

Klüver, H. 1942. "Functional Significance of the Geniculo-Striate System." *Visual Mechanisms*, Pennsylvania: J. Cattell Press, 253–299.

Lorente de Nó, R. 1922. "La Corteza Cerebral del Raton." *Trab. Lab. Invest. Biol. Univ. Madr.*, **20**, 41–78.

Lorente de Nó, R. 1933. "The Vestibulo-Ocular Reflex Arc." *Arch. Neurol. and Psychiat.*, **30**, 245–291.

Morison, R. S., and E. W. Dempsey. 1942. "The Production of Rhythmically Recurrent Cortical Potentials After Localized Thalamic Stimulation." *Amer. Jour. of Physiol.*, **135**, 293–300.

Morison, R. S., and E. W. Dempsey. 1943. "Mechanism of Thalamo-Cortical Augmentation and Repetition." *Amer. Jour. of Physiol.*, **138**, 297–308.

O'Leary, J. L., and G. H. Bishop. 1941. "The Optically Excitable Cortex of the Rabbit." *J. Comp. Neur.*, **68**, 423–478.

Ramón y Cajal, S. 1911. *Histologie du Système Nerveux.* Paris: Maloine.

Ramón y Cajal, S. 1900. *Die Sehrinde.* Leipzig: Barth.

Ward, A. A. and H. L. Reed. 1946. "Mechanism of Pupillary Dilatation Elicited Cortical Stimulation." *J. Neurophysiol.*, **9**, 329–336.

Woodburne, K. T., E. C. Crosby and R. E. McCotter. 1946. "The Mammalian Midbrain and Isthmus Region. II. The Fibre Connections." *J. Comp. Neur.*, **85**, 67–92.

Zunino, G. 1909. "Die Myeloarchitektonische Differenzierung der Grosshirnrinde beim Kaninchen." *J. Psych. u. Neur.*, **14**, 38–70.

4
Introduction

(1949)
Donald O. Hebb

The Organization of Behavior, New York: Wiley, Introduction and Chapter 4, "The first stage of perception: growth of the assembly," pp. xi–xix, 60–78

Donald O. Hebb's book, *The Organization of Behavior*, is famous among neural modelers because it was the first explicit statement of the physiological learning rule for synaptic modification that has since become known as the Hebb synapse. However, the book covers a great deal more material than that, and is a thoughtful and thorough review of neuropsychology, as of 1949.

We have included Hebb's Introduction in this excerpt. It is a brief and lucid discussion of the connection between psychology and physiology, and has not dated one bit between 1949 and now.

The introduction is also notable because in it is one of the first uses of the word "connectionism" in the context of a complex brain model. The final paragraph of the Introduction contains these lines: "The theory is evidently a form of connectionism, one of the switchboard variety, though it does not deal in direct connections between afferent and efferent pathways: not an 'S-R' psychology, if R means a *muscular* response. The connections serve rather to establish autonomous central activities, which then are the basis of further learning" (p. xix). Most modern day connectionists could find little to argue with in that summation.

A more detailed description of Hebb's physiological ideas is found in chapter 4. In this chapter, Hebb has a detailed discussion of neurophysiology and neuroanatomy as it relates to his ideas. It is worth emphasizing, if it is not obvious at this point, that the early modelers really knew their neuroscience. James, McCulloch, and Hebb were highly knowledgable about the nervous system, and used their knowledge extensively in their models. Much modern work in neural networks has moved far away from its roots in the study of the brain and psychology. This is a cause for concern, both because the field is losing contact with its foundations and because it has lost a source of valuable ideas.

Hebb suggests several important ideas in chapter 4. First, and most famous, was the clear statement of what has become known as the "Hebb" synapse. To restate Hebb's description, for the nth time, "*When an axon of cell A is near enough to excite a cell B and repeatedly or persistently takes part in firing it, some growth process or metabolic change takes place in one or both cells such that A's efficiency, as one of the cells firing B, is increased*" (p. 50). This, like the other ideas in Hebb's book, is not a mathematical statement, though it is close to one. For example, Hebb does not discuss the various possible ways inhibition might enter the picture, or the quantitative learning rule that is being followed. This has meant that a number of sometimes quite different learning rules can legitimately be called "Hebb synapses." (Paper 6, an early computer simulation of Hebb's ideas, discusses the modifications one must make to this bare outline to make the system work.)

Second, Hebb is keenly aware of the "distributed" nature of the representation he is assuming the nervous system uses. The idea is that to represent something, many cells must participate in the representation. Hebb was aware of the work of Lashley (paper 5), suggesting widely distributed representations, and made some use of his ideas, though not in the strongest form of complete "equipotentiality."

Third, Hebb postulated the formation of what he called "cell assemblies," which were really the heart of the entire book. The basic idea was that there were interconnected, self-reinforcing subsets of neurons that formed the representations of information in the nervous system. Single cells might belong to more than one assembly, depending on the context. Multiple cell assemblies could be active at once, corresponding to complex perceptions or thoughts. There was a distributed representation at the functional level as well as at the anatomical level. Hebb devotes a good deal of attention to the details of the neuroanatomy and physiology that might underlie cell assemblies. The later chapters in the book contain many discussions of how cell assemblies can be used to help explain a number of psychological phenomena.

In retrospect, the idea that there exist temporarily stable, relatively long lasting neural activity patterns that are important in mental activity has reappeared in the various "attractor" models for brain activity (see, for example, Hopfield, paper 27; Grossberg, paper 24; or Anderson et al., paper 22). Details of the observed and predicted stability depend critically on learning assumptions that are nearly always based to some degree on Hebb synapses. Some of the ideas described in this book have become part of the accepted lore of the field.

(1949)
Donald O. Hebb

The Organization of Behavior, New York: Wiley, Introduction and Chapter 4,
"The first stage of perception: growth of the assembly," pp. xi–xix, 60–78

Introduction

It might be argued that the task of the psychologist, the task of understanding behavior and reducing the vagaries of human thought to a mechanical process of cause and effect, is a more difficult one than that of any other scientist. Certainly the problem is enormously complex; and though it could also be argued that the progress made by psychology in the century following the death of James Mill, with his crude theory of association, is an achievement scarcely less than that of the physical sciences in the same period, it is nevertheless true that psychological theory is still in its infancy. There is a long way to go before we can speak of understanding the principles of behavior to the degree that we understand the principles of chemical reaction.

In an undertaking of such difficulty, the psychologist presumably must seek help wherever he can find it. There have been an increasing number of attempts to develop new mathematical methods of analysis. With these, in general, I do not attempt to deal. The method of factor analysis developed by Spearman (1927) and greatly elaborated by Thurstone (1935) is well established as a powerful tool for handling certain kinds of data, though the range of its use has been limited by dependence on tests that can be conveniently given to large groups of subjects. Another method is the application of mathematics more directly to the interaction of populations of neurons, by Rashevsky, Pitts, Householder, Landahl, McCulloch, and others.* Bishop (1946) has discussed the work from the point of view of neurophysiology, and his remarks are fully concurred with here. The preliminary studies made with this method so far have been obliged to simplify the psychological problem almost out of existence. This is not a criticism, since the attempt is to develop methods that can later be extended to deal with more complex data; but as matters stand at present one must wait for further results before being sure that the attempt will succeed. Undoubtedly there is great potential value in such work, and if the right set of initial assumptions can be found it will presumably become, like factor analysis, a powerful ally of other methods of study.

However, psychology has an intimate relation with the other biological sciences, and may also look for help there. There is a considerable overlap between the problems of psychology and those of neurophysiology, hence the possibility (or necessity) of reciprocal assistance. The first object of this book is to present a theory of behavior for the consideration of psychologists; but another is to seek a common ground with the anatomist, physiologist, and neurologist, to show them how psychological theory relates to their problems and at the same time to make it more possible for them to contribute to that theory.

Psychology is no more static than any other science. Physiologists and clinicians who wish to get a theoretical orientation cannot depend only on the writings of Pavlov or Freud. These were great men, and they have contributed greatly to psychological thought. But their contribution was rather in formulating and developing problems than in providing final answers. Pavlov himself seems to have thought of his theory of conditioned reflexes as something in continual need of revision, and experimental results have continued to make revisions necessary: the theory, that is, is still developing. Again, if one were to regard Freud's theory as needing change only in its details, the main value of his work would be stultified. Theorizing at this stage is like skating on thin ice—keep moving, or drown. Ego, Id, and Superego are conceptions that help one to see and state important facts of behavior, but they are also dangerously easy to treat as ghostly realities: as anthropomorphic agents that *want* this or *disapprove* of that, *overcoming* one another by force or guile, and *punishing* or *being punished*. Freud has left us the task of developing these provisional formulations of his to the point where such a danger no longer exists. When theory becomes static it is apt to become dogma; and psychological theory has the further danger, as long as so many of its problems are unresolved, of inviting a relapse into the vitalism and indeterminism of traditional thought.

---

*Two papers by Culbertson (*Bull. Math. Biophys.*, 1948, *10*, 31–40 and 97–102), and Bishop's review article, list some of the more important of the actual titles in this field.

It is only too easy, no matter what formal theory of behavior one espouses, to entertain a concealed mysticism in one's thinking about that large segment of behavior which theory does not handle adequately. To deal with behavior at present, one must oversimplify. The risk, on the one hand, is of forgetting that one has oversimplified the problem; one may forget or even deny those inconvenient facts that one's theory does not subsume. On the other hand is the risk of accepting the weak-kneed discouragement of the vitalist, of being content to show that existing theories are imperfect without seeking to improve them. We can take for granted that any theory of behavior at present must be inadequate and incomplete. But it is never enough to say, because *we* have not yet found out how to reduce behavior to the control of the brain, that no one in the future will be able to do so.

Modern psychology takes completely for granted that behavior and neural function are perfectly correlated, that one is completely caused by the other. There is no separate soul or life-force to stick a finger into he brain now and then and make neural cells do what they would not otherwise. Actually, of course, this is a working assumption only—as long as there are unexplained aspects of behavior. It is quite conceivable that some day the assumption will have to be rejected. But it is important also to see that we have not reached that day yet: the working assumption is a necessary one, and there is no real evidence opposed to it. Our failure to solve a problem so far does not make it insoluble. One cannot logically be a determinist in physics and chemistry and biology, and a mystic in psychology.

All one can know about another's feelings and awarenesses is an inference from what he *does*—from his muscular contractions and glandular secretions. These observable events are determined by electrical and chemical events in nerve cells. If one is to be consistent, there is no room here for a mysterious agent that is defined as not physical and yet has physical effects (especially since many of the entities of physics are known only through their effects). "Mind" can only be regarded, for scientific purposes, as the activity of the brain, and this should be mystery enough for anyone: besides the appalling number of cells (some nine billion, according to Herrick) and even more appalling number of possible connections between them, the matter out of which cells are made is being itself reduced by the physicist to something quite unlike the inert stick or stone with which mind is traditionally contrasted. After all, it is that contrast that is at the bottom of the vitalist's objection to a mechanistic biology, and the contrast has lost its force (Herrick, 1929).

The mystic might well concentrate on the electron and let behavior alone. A philosophical parallelism or idealism, whatever one may think of such conceptions on other grounds, is quite consistent with the scientific method, but interactionism seems not to be.

Psychologist and neurophysiologist thus chart the same bay—working perhaps from opposite shores, sometimes overlapping and duplicating one another, but using some of the same fixed points and continually with the opportunity of contributing to each other's results. The problem of understanding behavior is the problem of understanding the total action of the nervous system, and *vice versa*. This has not always been a welcome proposition, either to psychologist or to physiologist.

A vigorous movement has appeared both in psychology and psychiatry to be rid of "physiologizing," that is, to stop using physiological hypotheses. This point of view has been clearly and effectively put by Skinner (1938), and it does not by any means represent a relapse into vitalism. The argument is related to modern positivism, emphasizes a method of correlating observable stimuli with observable response, and, recognizing that "explanation" is ultimately a statement of relationships between observed phenomena, proposes to go to the heart of the matter and have psychology confine itself to such statements *now*. This point of view has been criticized by Pratt (1939) and Köhler (1940). The present book is written in profound disagreement with such a program for psychology. Disagreement is on the grounds that this arises from a misconception of the scientific method as it operates in the earlier stages. Those apparently naïve features of older scientific thought may have had more to do with hitting on fertile assumptions and hypotheses than seems necessary in retrospect. The anti-physiological position, thus, in urging that psychology proceed now as it may be able to proceed when it is more highly developed, seems to be in short a counsel of perfection, disregarding the limitations of the human intellect. However, it is logically defensible and may yet show by its fertility of results that it is indeed the proper approach to achieving prediction and control of behavior.

If some psychologists jib at the physiologist for a bedfellow, many physiologists agree with them heartily. One must sympathize with those who want nothing of the psychologist's hair-splitting or the indefiniteness of psychological theory. There is much more certainty in the study of the electrical activity of a well-defined tract in the brain. The only question is whether a physiology of the human brain as a whole can be achieved by such studies alone. One can discover the properties of its various parts more or less in isolation; but it is a truism by now that the part may have properties that are not evident in isolation, and these are to be discovered only by study of the whole intact brain. The method then calls for learning as much as one can about what the parts of the brain do (primarily the physiologist's field), and relating behavior as far as possible to this knowledge (primarily for the psychologist); then seeing what further information is to be had about how the total brain works, from the discrepancy between (1) actual behavior and (2) the behavior that would be predicted from adding up what is known about the action of the various parts.

This does not make the psychologist a physiologist, for precisely the same reason that the physiologist need not become a cytologist or biochemist, though he is intimately concerned with the information that cytology and biochemistry provide. The difficulties of finding order in behavior are great enough to require all one's attention, and the psychologist is interested in physiology to the extent that it contributes to his own task.

The great argument of the positivists who object to "physiologizing" is that physiology has not helped psychological theory. But, even if this is true (there is some basis for denying it), one has to add the words *so far*. These has been a great access of knowledge in neurophysiology since the twenties. The work of Berger, Dusser de Barenne, and Lorente de Nó (as examples) has a profound effect on the physiological conceptions utilized by psychology, and psychology has not yet assimilated these results fully.

The central problem with which we must find a way to deal can be put in two different ways. Psychologically, it is the problem of thought: some sort of process that is not fully controlled by environmental stimulation and yet cooperates closely with that stimulation. From another point of view, physiologically, the problem is that of the transmission of excitation from sensory to motor cortex. This statement may not be as much oversimplified as it seems, especially when one recognizes that the "transmission" may be a very complex process indeed, with a considerable time lag between sensory stimulation and the final motor response. The failure of psychology to handle thought adequately (or the failure of neurophysiology to tell us how to conceive of cortical transmission) has been the essential weakness of modern psychological theory and the reason for persistent difficulties in dealing with a wide range of experimental and clinical data, as the following chapters will try to show, from the data of perception and learning to those of hunger, sleep, and neurosis.

In mammals even as low as the rat it has turned out to be impossible to describe behavior as an interaction

directly between sensory and motor processes. Something like *thinking*, that is, intervenes. "Thought" undoubtedly has the connotation of a human degree of complexity in cerebral function and may mean too much to be applied to lower animals. But even in the rat there is evidence that behavior is not completely controlled by immediate sensory events: there are central processes operating also.

What is the nature of such relatively autonomous activities in the cerebrum? Not even a tentative answer is available. We know a good deal about the afferent pathways to the cortex, about the efferent pathways from it, and about many structures linking the two. But the links are complex, and we know practically nothing about what goes on between the arrival of an excitation at a sensory projection area and its later departure from the motor area of the cortex. Psychology has had to find, in hypothesis, a way of bridging this gap in its physiological foundation. In general the bridge can be described as some comparatively simple formula of cortical transmission.* The particular formula chosen mainly determines the nature of the psychological theory that results, and the need of choosing is the major source of theoretical schism.

Two kinds of formula have been used, leading at two extremes to (1) switchboard theory, and sensori-motor connections; and (2) field theory. (Either of these terms may be regarded as opprobrium; they are not so used here.) (1) In the first type of theory, at one extreme, cells in the sensory system acquire connections with cells in the motor system; the function of the cortex is that of a telephone exchange. Connections rigidly determine what animal or human being does, and their acquisition constitutes learning. Current forms of the theory tend to be vaguer than formerly, because of effective criticism of the theory in its earlier and simpler forms, but the fundamental idea is still maintained. (2) Theory at the opposite extreme denies that learning depends on connections at all, and attempts to utilize instead the field conception that physics has found so useful. The cortex is regarded as made up of so many cells that it can be treated as a statistically homogeneous medium. The sensory control of motor centers depends, accordingly, on the distribution of the sensory excitation and on ratios of excitation, not on locus or the action of any specific cells.

Despite their differences, however, both theoretical approaches seem to imply a prompt transmission of sensory excitation to the motor side, if only by failing to specify that this is not so. No one, at any rate,

---

*The simplicity possibly accounts for the opinion expressed by an anatomist who claimed that psychologists think of the brain as having all the finer structure of a bowlful of porridge.

has made any serious attempt to elaborate ideas of a central neural mechanism to account for the delay, between stimulation and response, that seems so characteristic of thought. There have indeed been neural theories of "motor" thought, but they amount essentially to a continual interplay of proprioception and minimal muscular action, and do not provide for any prolonged sequence of intracerebral events as such.

But the recalcitrant data of animal behavior have been drawing attention more and more insistently to the need of some better account of central processes. This is what Morgan (1943) has recognized in saying that "mental" variables, repeatedly thrown out because there was no place for them in a stimulus-response psychology, repeatedly find their way back in again in one form or another. The image has been a forbidden notion for twenty years, particularly in animal psychology; but the fiend was hardly exorcised before "expectancy" had appeared instead. What is the neural basis of expectancy, or of attention, or interest? Older theory could use these words freely, for it made no serious attempt to avoid an interactionist philosophy. In modern psychology such terms are an embarrassment; they cannot be escaped if one is to give a full account of behavior, but they still have the smell of animism: and must have, until a theory of thought is developed to show how "expectancy" or the like can be a physiologically intelligible process.

In the chapters that follow this introduction I have tried to lay a foundation for such a theory. It is, on the one hand and from the physiologist's point of view, quite speculative. On the other hand, it achieves some synthesis of psychological knowledge, and it attempts to hold as strictly as possible to the psychological evidence in those long stretches where the guidance of anatomy and physiology is lacking. The desideratum is a conceptual tool for dealing with expectancy, attention, and so on, and with a temporally organized intracerebral process. But this would have little value if it did not also comprise the main facts of perception, and of learning. To achieve something of the kind, the limitations of a schema are accepted with the purpose of developing certain conceptions of neural action. This is attempted in Chapters 4 and 5; Chapters 1 to 3 try to clear the ground for this undertaking. From Chapter 6 onward the conceptions derived from schematizing are applied to the problems of learning, volition, emotion, hunger, and so on. (In general, the reader may regard Chapters 1 to 5 as mainly preparatory, unless he is particularly interested in the neurological details, or in the treatment of perception; to get the gist of the theory that is presented here one should read the two following paragraphs, and turn directly to Chapter 6.) In outline, the conceptual structure is as follows:

Any frequently repeated, particular stimulation will lead to the slow development of a "cell-assembly," a diffuse structure comprising cells in the cortex and diencephalon (and also, perhaps, in the basal ganglia of the cerebrum), capable of acting briefly as a closed system, delivering facilitation to other such systems and usually having a specific motor facilitation. A series of such events constitutes a "phase sequence"—the thought process. Each assembly action may be aroused by a preceding assembly, by a sensory event, or—normally—by both. The central facilitation from one of these activities on the next is the prototype of "attention." The theory proposes that in this central facilitation, and its varied relationship to sensory processes, lies the answer to an issue that is made inescapable by Humphrey's (1940) penetrating review of the problem of the direction of thought.

The kind of cortical organization discussed in the preceding paragraph is what is regarded as essential to adult waking behavior. It is proposed also that there is an alternate, "intrinsic" organization, occurring in sleep and in infancy, which consists of hypersynchrony in the firing of cortical cells. But besides these two forms of cortical organization there may be disorganization. It is assumed that the assembly depends completely on a very delicate timing which might be disturbed by metabolic changes as well as by sensory events that do not accord with the pre-existent central process. When this is transient, it is called emotional disturbance; when chronic, neurosis or psychosis.

The theory is evidently a form of connectionism, one of the switchboard variety, though it does not deal in direct connections between afferent and efferent pathways: not an "S-R" psychology, if R means a *muscular* response. The connections serve rather to establish autonomous central activities, which then are the basis of further learning. In accordance with modern physiological ideas, the theory also utilizes local field processes and gradients, following the lead particularly of Marshall and Talbot (1942). It does not, further, make any single nerve cell or pathway essential to any habit or perception. Modern physiology has presented psychology with new opportunities for the synthesis of divergent theories and previously unrelated data, and it is my intent to take such advantage of these opportunities as I can.

4. The First Stage of Perception: Growth of the Assembly

This chapter and the next develop a schema of neural action to show how a rapprochement can be made between (1) perceptual generalization, (2) the permanence of learning, and (3) attention, determining tendency or the like. It is proposed first that a repeated stimulation of specific receptors will lead slowly to the formation of an "assembly" of association-area cells which can act briefly as a closed system after stimulation has ceased; this prolongs the time during which the structural changes of learning can occur and constitutes the simplest instance of a representative process (image or idea). The way in which this cell-assembly might be established, and its characteristics, are the subject matter of the present chapter. In the following chapter the interrelationships between cell-assemblies are dealt with; these are the basis of temporal organization in central processes (attention, attitude, thought, and so on). The two chapters (4 and 5) construct the conceptual tools with which, in the following chapters, the problems of behavior are to be attacked.

The first step in this neural schematizing is a bald assumption about the structural changes that make lasting memory possible. The assumption has repeatedly been made before, in one way or another, and repeatedly found unsatisfactory by the critics of learning theory. I believe it is still necessary. As a result, I must show that in another context, of added anatomical and physiological knowledge, it becomes more defensible and more fertile than in the past.

The assumption, in brief, is that a growth process accompanying synaptic activity makes the synapse more readily traversed. This hypothesis of synaptic resistances, however, is different from earlier ones in the following respects: (1) structural connections are postulated between single cells, but single cells are not effective units of transmission and such connections would be only one factor determining the direction of transmission; (2) no direct sensori-motor connections are supposed to be established in this way, in the adult animal; and (3) an intimate relationship is postulated between reverberatory action and structural changes at the synapse, implying a dual trace mechanism.

The Possibility of a Dual Trace Mechanism

Hilgard and Marquis (1940) have shown how a reverberatory, transient trace mechanism might be proposed on the basis of Lorente de Nó's conclusions, that a cell is fired only by the simultaneous activity of two or more afferent fibers, and that internuncial fibers are arranged in closed (potentially self-exciting) circuits. Their diagram is arranged to show how a reverberatory circuit might establish a sensori-motor connection between receptor cells and the effectors which carry out a conditioned response. There is of course a good deal of psychological evidence which is opposed to such an oversimplified hypothesis, and Hilgard and Marquis do not put weight on it. At the same time, it is important to see that something of the kind is not merely a possible but a necessary inference from certain neurological ideas. To the extent that anatomical and physiological observations establish the possibility of reverberatory after-effects of a sensory event, it is established that such a process would be the physiological basis of a transient "memory" of the stimulus. There may, then, be a memory trace that is wholly a function of a pattern of neural activity, independent of any structural change.

Hilgard and Marquis go on to point out that such a trace would be quite unstable. A reverberatory activity would be subject to the development of refractory states in the cells of the circuit in which it occurs, and external events could readily interrupt it. We have already seen (in Chapter 1) that an "activity" trace can hardly account for the permanence of early learning, but at the same time one may regard reverberatory activity as the explanation of other phenomena.

There are memories which are instantaneously established, and as evanescent as they are immediate. In the repetition of digits, for example, an interval of a few seconds is enough to prevent any interference from one series on the next. Also, some memories are both instantaneously established and permanent. To account for the permanence, some structural change seems necessary, but a structural growth presumably would require an appreciable time. If some way can be found of supposing that a reverberatory trace might cooperate with the structural change, and *carry the memory until the growth change is made*, we should be able to recognize the theoretical value of the trace which is an activity only, without having to ascribe all memory to it. The conception of a transient, unstable reverberatory trace is therefore useful, if it is possible to suppose also that some more permanent structural change reinforces it. There is no reason to think that a choice must be made between the two conceptions; there may be traces of both kinds, and memories which are dependent on both.

A Neurophysiological Postulate

Let us assume then that the persistence or repetition of a reverberatory activity (or "trace") tends to induce lasting cellular changes that add to its stability. The assumption* can be precisely stated as follows: *When an axion of cell* A *is near enough to excite a cell* B *and repeatedly or persistently takes part in firing it, some growth process or metabolic change takes place in one or both cells such that* A's *efficiency, as one of the cells firing* B, *is increased.*

The most obvious and I believe much the most probable suggestion concerning the way in which one cell could become more capable of firing another is that synaptic knobs develop and increase the area of contact between the afferent axon and efferent soma. ("Soma" refers to dendrites and body, or all of the cell except its axon.) There is certainly no direct evidence that this is so, and the postulated change if it exists may be metabolic, affecting cellular rhythmicity and limen; or there might be both metabolic and structural changes, including a limited neurobiotaxis. There are several considerations, however, that make the growth of synaptic knobs a plausible conception. The assumption stated above can be put more definitely, as follows:

When one cell repeatedly assists in firing another, the axon of the first cell develops synaptic knobs (or enlarges them if they already exist) in contact with the soma of the second cell. This seems to me the most likely mechanism of a lasting effect of reverberatory action, but I wish to make it clear that the subsequent discussion depends only on the more generally stated proposition italicized above.

It is wise to be explicit on another point also. The proposition does not require action at any great distance, and certainly is not the same as Kappers' (Kappers, Huber, and Crosby, 1936) conception of the way in which neurobiotaxis controls axonal and dendritic outgrowth. But my assumption is evidently related to Kappers' ideas, and not inconsistent with them. The theory of neurobiotaxis has been severely criticized, and clearly it does not do all it was once thought to do. On the other hand, neurobiotaxis may still be one factor determining the connections made by neural cells. If so, it would cooperate very neatly with the knob formation postulated above. Criticism has been directed at the idea that neurobiotaxis directs axonal growth throughout its whole course, and that the process sufficiently accounts for all neural connections. The idea is not tenable, particularly in view of such work as that of Weiss (1941*b*) and Sperry (1943).

But none of this has shown that neurobiotaxis has *no* influence in neural growth; its operation, within ranges of a centimeter or so, is still plausible. Thus, in figure 6 (Lorente de Nó, 1938*a*), the multiple synaptic

* See p. 229 for a further discussion of this point and an elaboration of the assumption made concerning the nature of memory.

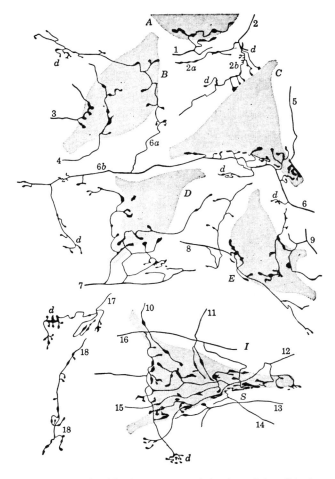

Figure 6 Relationships between synaptic knobs and the cell body. From Lorente de Nó, 1938*a*. Courtesy of Charles C. Thomas and of the author.

knobs of fiber 2 on cell *C* might be outgrowths from a fiber passing the cell at a distance, and determined by the fact of repeated simultaneous excitations in the two. Again, the course followed by fiber 7 in the neighborhood of cell *D* may include deflections from the original course of the fiber, determined in the same way.

The details of these histological speculations are not important except to show what some of the possibilities of change at the synapse might be and to show that the mechanism of learning discussed in this chapter is not wholly out of touch with what is known about the neural cell. The changed facilitation that constitutes learning might occur in other ways without affecting the rest of the theory. To make it more specific, I have chosen to assume that the growth of synaptic knobs, with or without neurobiotaxis, is the basis of the change of facilitation from one cell on another, and this is not altogether implausible. It has been demonstrated by Arvanitaki (1942) that a contiguity alone will permit the excitation aroused in one cell to be transmitted to another. There are also earlier experiments, reviewed by Arvanitaki, with the same implication. Even more important, perhaps, is Erlanger's (1939) demonstration of impulse transmission across an artificial "synapse," a blocked segment of nerve more than a millimeter in extent. Consequently, in the intact nervous system, an axon that passes close to the dendrities or body of a second cell would be capable of *helping* to fire it, when the second cell is also exposed to other stimulation at the same point. The probability that such closely timed coincidental excitations would occur is not considered for the moment but will be returned to. When the coincidence does occur, and the active fiber, which is merely close to the soma of another cell, adds to a local excitation in it, I assume that the joint action tends to produce a thickening of the fiber—forming a synaptic knob—or adds to a thickening already present.

Lorente de Nó (1938a) has shown that the synaptic knob is usually not a terminal structure (thus the term "end foot" or "end button" is misleading), nor always separated by a stalk from the axon or axon collateral. If it were, of course, some action at a distance would be inevitably suggested, if such connections are formed in learning. The knob instead is often a rather irregular thickening in the unmyelinated part of an axon near its ending, where it is threading its way through a thicket of dendrites and cell bodies. The point in the axon where the thickening occurs does not appear to be determined by the structure of the cell of which it is a part but by something external to the cell and related to the presence of a second cell. The number and size of the knobs formed by one cell in contact with a second cell vary also. In the light of these facts it is not implausible to suppose that the extent of the contact established is a function of joint cellular activity, given propinquity of the two cells.

Also, if a synapse is crossed only by the action of two or more afferent cells, the implication is that the greater the area of contact the greater the likelihood that action in one cell will be *decisive* in firing another.*

* One point should perhaps be made explicit. Following Lorente de Nó, two afferent cells are considered to be effective at the synapse, when one is not, only because their contacts with the efferent cell are close together so their action summates. When both are active, they create a larger region of *local* disturbance in the efferent soma. The larger the knobs in a given cluster, therefore, the smaller the number that might activate the cell on which they are located. On occasion, a single afferent cell must be effective in transmission. It is worth pointing this out, also, because it might appear to the reader otherwise that there is something mysterious about emphasis on the necessity of activity in two or more cells to activate the synapse. All that has really been shown is that in some circumstances two or more afferent cells are necessary. However, this inevitably implies that an increase in the number of afferent cells simultaneously active must increase the reliability with which the synapse is traversed.

Thus three afferent fibers with extensive knob contact could fire a cell that otherwise might be fired only by four or more fibers; or fired sooner with knobs than without.

In short, it is feasible to assume that synaptic knobs develop with neural activity and represent a lowered synaptic resistance. It is implied that the knobs appear in the course of learning, but this does not give us a means of testing the assumption. There is apparently no good evidence concerning the relative frequency of knobs in infant and adult brains, and the assumption does *not* imply that there should be none in the newborn infant. The learning referred to is learning in a very general sense, which must certainly have begun long before birth (see *e.g.*, the footnote on pp. 121–2).

Conduction from Area 17

In order to apply this idea (of a structural reinforcement of synaptic transmission) to visual perception, it is necessary first to examine the known properties of conduction from the visual cortex, area 17, to areas 18, 19, and 20. (In view of the criticisms of architectonic theory by Lashley and Clark [1946], it may be said that Brodmann's areas are referred to here as a convenient designation of relative cortical position, without supposing that the areas are necessarily functional entities or always histologically distinctive.)

It has already been seen that there is a topological reproduction of retinal activities in area 17, but that conduction from 17 to 18 is diffuse. Von Bonin, Garol, and McCulloch (1942) have found that a localized excitation in 17 is conducted to a large part of 18, a band lying along the margins of 17. There is no point-to-point correspondence of 17 and 18. Excitation from 18 is conducted back to the nearest border region of 17; to all parts of area 18 itself; and to all parts of the contralateral 18, of area 19 (lying anterior to 18), and of area 20 (in the lower part of the temporal lobe).

The diffusity of conduction from area 17 is illustrated by the diagram of figure 7. Cells lying in the same part of 17 may conduct to different points in 18. The cells in 18, thus stimulated, also lead to points in 18 itself which are widely separated; to any part of the ipsilateral areas 19 and 20; and, though one synapse, to any part of the contralateral 19 and 20. Conversely, *cells lying in different parts of* 17 *or* 18 *may have connections with the same point in* 18 *or* 20.

Thus there is convergence as well as spread of excitation. The second point illustrated by figure 7 is a selective action in 18, depending on the convergence of fibers from 17. In the figure, *F* and *G* are two cells in area 18 connecting the same macroscopic areas. *F*,

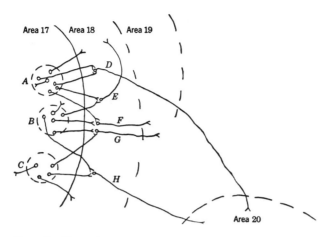

Figure 7 Illustrating convergence of cells in Brodmann's area 17 upon cells in area 18, these cells in turn leading to other areas. *A, B, C*, three grossly distinct regions in area 17; *D, E, F, G, H*, cells in area 18. See text.

however, is one that happens to be exposed to excitations from both *A* and *B* (two different regions in area 17). When an area-17 excitation includes both *A* and *B*, *F* is much more likely to be fired than *G*. The figure does not show the short, closed, multiple chains which are found in all parts of the cortex and whose facilitating activity would often make it possible for a single fiber from *B* to fire *G*. But the same sort of local bombardment would also aid in firing *F*; and the cell which receives excitations from two area-17 fibers simultaneously would be more likely to fire than that which receives excitation from only one.

On the other hand, when *B* and *C* (instead of *A* and *B*) are excited simultaneously, *G* would be more likely to fire than *F*. Any specific region of activity in area 17 would tend to excite specific cells in area 18 which would tend not to be fired by the excitation of another region in 17. These specific cells in 18 would be diffusely arranged, as far as we know at random. They would be usually at some distance from one another and would always be intermingled with others which are not fired by the same afferent stimulation, but because of their lasting structural connections would tend always to be selectively excited, in the same combination, whenever the same excitation recurs in area 17. This of course would apply also in areas 19 and 20. Since a single point in 18 fires to many points throughout 19 and 20, excitation of any large number of area-18 cells means that convergence in 19 and 20 must be expected. How often it would happen is a statistical question, which will be deferred to a later section.

The tissues made active beyond area 17, by two different visual stimuli, would thus be (1) grossly the same,

(2) histologically distinct. A difference of stimulating pattern would not mean any gross difference in the part of the brain which mediates perception (except in the afferent structures up to and including area 17, the visual cortex). Even a completely unilateral activity, it should be noted, would have diffuse effects throughout areas 18, 19, and 20 not only on one side of the brain but on both. At the same time, a difference of locus or pattern of stimulation would mean a difference in the particular cells in these areas that are consistently or maximally fired.

Mode of Perceptual Integration: The Cell-Assembly

In the last chapter it was shown that there are important properties of perception which cannot be ascribed to events in area 17, and that these are properties which seem particularly dependent on learning. That "identity" is not due to what happens in 17 is strongly implied by the distortions that occur in the projection of a retinal excitation to the cortex. When the facts of hemianopic completion are also considered, the conclusion appears inescapable. Perception must depend on other structures besides area 17.

But we now find, at the level of area 18 and beyond, that all topographical organization in the visual process seems to have disappeared. All that is left is activity in an irregular arrangement of cells, which are intertangled with others that have nothing to do with the perception of the moment. We know of course that perception of simple objects is unified and determinate, a well-organized process. What basis can be found for an integration of action, in cells that are anatomically so disorganized?

An answer to this question is provided by the structural change at the synapse which has been assumed to take place in learning. The answer is not simple; perceptual integration would not be accomplished directly, but only as a slow development, and, for the purposes of exposition, at least, would involve several distinct stages, with the first of which we shall now be concerned.

The general idea is an old one, that any two cells or systems of cells that are repeatedly active at the same time will tend to become "associated," so that activity in one facilitates activity in the other. The details of speculation that follow are intended to show how this old idea might be put to work again, with the equally old idea of a lowered synaptic "resistance," under the eye of a different neurophysiology from that which engendered them. (It is perhaps worth while to note that the two ideas have most often been combined only in the special case in which one cell is associated with

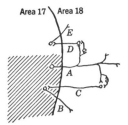

Figure 8 Cells A and B lie in a region of area 17 (shown by hatching) which is massively excited by an afferent stimulation. C is a cell in area 18 which leads back into 17. E is in area 17 but lies outside the region of activity. See text.

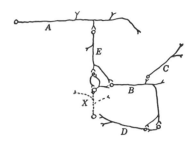

Figure 9 A, B, and C are cells in area 18 which are excited by converging fibers (not shown) leading from a specific pattern of activity in area 17. D, E, and X are, *among the many cells with which* A, B, *and* C *have connections*, ones which would contribute to an integration of their activity. See text.

another, or a higher level or order in transmission, which it fires; what I am proposing is a possible basis of association of two afferent fibers of the same order—in principle, a sensori-sensory association,* in addition to the linear association of conditioning theory.)

The proposal is most simply illustrated by cells A, B, and C in figure 8. A and B, visual-area cells, are simultaneously active. The cell A synapses, of course, with a large number of cells in 18, and C is supposed to be one that happens to lead back into 17. Cells such as C would be those that produce the local wedge-shaped area of firing in 17 when a point in 18 is strychninized (von Bonin, Garol, and McCulloch, 1942). The cells in the region of 17 to which C leads are being fired by the same massive sensory excitation that fires A, and C would almost necessarily make contact with some cell B that also fires into 18, or communicate with B at one step removed, through a short-axon circuit. With repetition of the same massive excitation in 17 the same firing relations would recur and, according to the assumption made, growth changes would take place at synapses AC and CB. This means that A and B, both afferent neurons of the same order, would no longer act independently of each other.

At the same time, in the conditions of stimulation that are diagrammed in figure 8, A would also be likely to synapse (directly, or *via* a short closed link) with a cell D which leads back into an unexcited part of 17, and there synapses with still another cell E of the same order as A and B. The synapse DE, however, would be unlikely to be traversed, since it is not like CB exposed to concentrated afferent bombardment. Upon frequent repetition of the particular excitation in area 17, a functional relationship of activity in A and B would increase much more than a relationship of A to E.

* It should be observed, however, that some theorists have continued to maintain that "S-S" (sensori-sensory) associations are formed in the learning process, and have provided experimental evidence that seems to establish the fact. See, *e.g.*, Brogden, *J. Exp. Psychol.*, 1947, *37*, 527–539, and earlier papers cited therein.

The same considerations can be applied to the activity of the enormous number of individual cells in 18, 19, and 20 that are simultaneously aroused by an extensive activity in 17. Here, it should be observed, the evidence of neuronography implies that there are anatomical connections of every point with every other point, within a few millimeters, and that there is no orderly arrangement of the cells concerned.

Figure 9 diagrams three cells, A, B, and C, that are effectively fired in 18 by a particular visual stimulation, frequently repeated (by fixation, for example, on some point in a constant distant environment). D, E, and X represent possible connections which might be found between such cells, directly or with intervening links. Supposing that time relations in the firing of these cells make it possible, activity in A would contribute to the firing of E, and that in B to firing C and D. Growth changes at the synapses AE, BC, BD, and so on, would be a beginning of integration and would increase the probability of coordinated activity in each pair of neurons.

The fundamental meaning of the assumption of growth at the synapse is in the effect this would have on the timing of action by the efferent cell. The increased area of contact means that firing by the efferent cell is more likely to follow the lead of the afferent cell. A fiber of order n thus gains increased control over a fiber $n + 1$, making the firing of $n + 1$ more predictable or determinate. The control cannot be absolute, but "optional" (Lorente de Nó, 1939), and depends also on other events in the system. In the present case, however, the massive excitation in 17 would tend to establish constant conditions throughout the system during the brief period of a single visual fixation; and the postulated synaptic changes would also increase the degree of this constancy. A would acquire an increasing control of E, and E, with each repetition of the visual stimulus, would fire more consistently at the same time that *B* is firing (*B*, it will be recalled, is directly controlled by the area-17 action). Synaptic changes *EB* would therefore result. Similarly, *B* acquires an increasing control of *D*; and whenever a cell such as *D* happens to be one that connects again with *B*, through *X*, a closed cycle (*BDXB*) is set up.

It is, however, misleading to put emphasis on the coincidences necessary for the occurrence of such a simple closed circuit. Instead of a ring or hoop, the best analogy to the sort of structure which would be set up or "assembled" is a closed solid cage-work, or three-dimensional lattice, with no regular structure, and with connections possible from any one intersection to any other. Let me say explicitly, again, that the *specificity of such an assembly of cells in 18 or 20, to a particular excitation in 17, depends on covergences.* Whenever two cells, directly or indirectly controlled by that excitation, converge on another cell (as *E* and *X* converge on *B* in figure 9) the essential condition of the present schematizing is fulfilled; the two converging cells need not have any simple anatomical or physiological relation to one another, and physiological integration would not be supposed to consist of independent closed chains.

This has an important consequence. Lorente de Nó (1938*b*) has put stress on the fact that activity in a short closed circuit must be rapidly extinguished, and could hardly persist as long as a hundredth of a second. It is hard, on the other hand, to see how a long, many-linked chain, capable of longer reverberation, would get established as a functional unit. But look now at figure 10, which diagrams a different sort of possibility. Arrows represent not neurons, but multiple pathways, of whatever complexity is necessary so that each arrow stands for a functional unit. These units fire in the order 1, 2, 3, ⋯ 15. The pathway labeled (1, 4) is the first

to fire, and also the fourth; (2, 14) fires second and fourteenth; and so on. The activity 1–2–3–4 is in a relatively simple closed circuit. At this point the next unit (2, 14) may be refractory, which would effectively extinguish reverberation in that simple circuit. But at this point, also, another pathway (5, 9) may be excitable and permit activity in the larger system to continue in some way as that suggested by the numbers in the figure. The sort of irregular three-dimensional net which might be the anatomical basis of perceptual integration in the association areas would be infinitely more complex than anything one could show with a diagram and would provide a large number of the multiple parallel (or alternate) units which are suggested by figure 10. If so, an indefinite reverberation in the structure might be possible, so long as the background activity in other cells in the same gross region remained the same. It would not of course remain the same for long, especially with changes of visual fixaton; but such considerations make it possible to conceive of "alternating" reverberation which might frequently last for periods of time as great as half a second or a second.

(What I have in mind, in emphasizing half a second or so as the duration of a reverberatory activity, is the observed duration of a single content in perception (Pillsbury, 1913; Boring, 1933]. Attention wanders, and the best estimate one can make of the duration of a single "conscious content" is of this time-order.)

This then is the cell-assembly. Some of its characteristics have been defined only by implication, and these are to be developed elsewhere, particularly in the remainder of this chapter, in the following chapter, and in Chapter 8 (see pp. 195–7). The assembly is thought of as a system inherently involving some equipotentiality, in the presence of alternate pathways each having the same function, so that brain damage might remove some pathways without preventing the system from functioning, particularly if the system has been long established, with well-developed synaptic knobs which decrease the number of fibers that must be active at once to traverse a synapse.

Statistical Considerations

It must have appeared to the reader who examined figures 8 and 9 carefully that there was something unlikely about its being arranged at the Creation to have such neat connections exactly where they were most needed for my hypothesis of perceptual integration. The answer of course is statistical: the neurons diagrammed were those which happen to have such connections, and, given a large enough population of connecting fibers distributed at random, the improbable

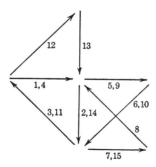

Figure 10 Arrows represent a simple "assembly" of neural pathways or open multiple chains firing according to the numbers on each (the pathway "1, 4" fires first and fourth, and so on), illustrating the possibility of an "alternating" reverberation which would not extinguish as readily as that in a simple closed circuit.

connection must become quite frequent, in absolute numbers. The next task is to assess the statistical element in these calculations, and show that probability is not stretched too far.

The diagrams and discussion of the preceding section require the frequent existence of two kinds of coincidence: (1) synchronization of firing in two or more converging axons, and (2) the anatomical fact of convergence in fibers which are, so far as we know, arranged at random. The necessity of these coincidences sets a limit to postulating functional connections *ad lib.* as the basis of integration. But this is not really a difficulty, since the psychological evidence (as we shall see) also implies that there are limits to perceptual integration.

Consider first the enormous frequency and complexity of the actual neural connections that have been demonstrated histologically and physiologically. One is apt to think of the neural cell as having perhaps two or three or half a dozen connections with other cells, and as leading from one minute point in the central nervous system to one other minute point. This impression is far from the truth and no doubt is due to the difficulty of representing the true state of affairs in a printed drawing.

Forbes (1939) mentions for example an estimate of 1300 synaptic knobs on a single anterior horn cell. Lorente de Nó's drawings (1943, figures 71–73, 75) show a complexity, in the ramification of axon and dendrite, that simply has no relation whatever to diagrams (such as mine) showing a cell with one or two connections. The gross extent of the volume of cortex infiltrated by the collaterals of the axon of a single neuron is measured in millimeters, not in microns; it certainly is not a single point, microscopic in size. In area 18, the strychnine method demonstrates that each tiny area of cortex has connections with the whole region. (These areas are about as small as 1 sq. mm., according to McCulloch, 1944*b*.) It puts no great strain on probabilities to suppose that there would be, in area 18, some anatomical connection of any one cell, excited by a particular visual stimulation, with a number of others excited in the same way.

There is, therefore, the *anatomical* basis of a great number of convergences among the multitude of cortical cells directly or indirectly excited by any massive retinal activity. This is to be kept in mind as one approaches the physiological question of synchronization in the converging fibers. In the tridimensional, lattice-like assembly of cells that I have supposed to be the basis of perceptual integration, those interconnecting neurons which synapse with the same cell would be functionally in parallel. Figure 10 illustrates this. The

pathways labeled (1, 4), (8), and (13), converging on one synapse, must have the same function in the system; or the two-link pathway (5, 9)–(6, 10) the same function as the single link (2, 14). When impulses in one such path are not effective, those in another, arriving at a different time, could be.

Once more, the oversimplification of such diagrams is highly misleading. At each synapse there must be a considerable dispersion in the time of arrival of impulses, and in each individual fiber a constant variation of responsiveness; and one could never predicate a determinate pattern of action in any small segment of the system. In the larger system, however, a statistical constancy might be quite predictable.

It is not necessary, and not possible, to define the cell-assembly underlying a perception as being made up of neurons all of which are active when the proper visual stimulation occurs. One can suppose that there would always be activity in some of the group of elements which are in functional parallel (they are not of course geometrically parallel). When for example excitation can be conducted to a particular point in the system from five different directions, the activity characteristic of the system as a whole might be maintained by excitation in any three of the five pathways, and no one fiber would have to be synchronized with any other one fiber.

There would still be some necessity of synchronization, and this has another aspect. In the integration which has been hypothesized, depending on the development of synaptic knobs and an increasing probability of control by afferent over efferent fibers, there would necessarily be a gradual change of the frequency characteristics of the system. The consequence would be a sort of fractionation and recruitment, and some change in the neurons making up the system. That is, some units, capable at first of synchronizing with others in the system, would no longer be able to do so and would drop out: "fractionation." Others, at first incompatible, would be recruited. *With perceptual development there would thus be a slow growth in the assembly,* understanding by "growth" not necessarily an increase in the number of constituent cells, but a change. How great the change would be there is no way of telling, but it is a change that may have importance for psychological problems when some of the phenomena of association are considered.

This then is the statistical approach to the problem. It is directly implied that an "association" of two cells in the same region, or of two systems of cells, would vary, in the probability of its occurrence, over a wide range. If one chose such pairs at random one would find some between which no association was possible, some in which association was promptly and easily established when the two were simultaneously active, and a large proportion making up a gradiation from one of these extremes to the other. The larger the system with a determinate general pattern of action, the more readily an association could be formed with another system. On a statistical basis, the more points at which a chance anatomical convergence could occur, the greater the frequency of effective interfacilitation between the two assemblies.

Psychologically, these ideas mean (1) that there is a prolonged period of integration of the individual perception, apart from associating the perception with anything else; (2) that an association between two perceptions is likely to be possible only after each one has independently been organized, or integrated; (3) that, even between two integrated perceptions, there may be a considerable variation in the ease with which association can occur. Finally, (4) the apparent necessity of supposing that there would be a "growth," or fractionation and recruitment, in the cell-assembly underlying perception means that there might be significant differences in the properties of perception at different stages of integration. One cannot guess how great the changes of growth would be; but it is conceivable, even probable, that if one knew where to look for the evidence one would find marked differences of identity in the perceptions of child and adult.

The psychological implications of my schematizing, as far as it has gone, have been made explicit in order to show briefly that they are not contrary to fact. We are not used to thinking of a simple perception as slowly and painfully learned, as the present chapter would suggest; but it has already been seen, in the discussion of the vision of the congenitally blind after operation, that it actually is. The slowness of learning, and the frequent instances of total failure to learn at all in periods as great as a year following operation (Senden, 1932), are extraordinary and incredible (if it were not for the full confirmation by Riesen, 1947). The principles of learning to be found in psychological textbooks are derived from the behavior of the half-grown or adult animal. Our ideas as to the readiness with which association is set up apply to the behavior of the developed organism, as Boring (1946) has noted; there is no evidence whatever to show that a similarly prompt association of separate perceptions can occur at birth—that it is independent of a slow process in which the perceptions to be associated must first be integrated.

As to the wide range in difficulty of associating two ideas or perceptions, even for the adult, this is psychologically a matter of common experience. Who has not had trouble remembering, in spite of repeated efforts, the spelling or pronunciation of some word, or the name of some acquaintance? The fact of the unequal difficulty of associations is not stressed in the literature, probably because it does not fit into conditioned-reflex theory; but it is a fact. My speculations concerning the nature of the trace and the aboriginal development of perception thus are not obviously opposed to the psychological evidence. Further evaluation can be postponed until the speculations have been fully developed.

Introduction

(1950)
K. S. Lashley

In search of the engram

Society of Experimental Biology Symposium, No. 4: *Psychological Mechanisms in Animal Behavior*
Cambridge: Cambridge University Press, pp. 454–455, 468–473, 477–480

Karl Lashley was a Harvard neuropsychologist who studied the formation and storage of memory. This famous paper, which summarized years of work on the biology of memory, addresses directly the important question of distributed versus localized representation of information, which is central to modeling of the nervous system. We have included Lashley's summary, and a couple of small excerpts that bear on this issue.

As has often been pointed out, metaphors for the brain are usually based on the most complex device currently available: in the seventeenth century the brain was compared to a hydraulic system, and in the early twentieth century to a telephone switchboard. Now, of course, we compare the brain to a digital computer. One of the editors (JAA) remembers a number of talks, reports, and papers in psychology in the 1960s and 1970s that presented block diagrams of mental functions that looked for all the world like machine code for a PDP-8. Perhaps in the late 1980s we shall invert this process and make metaphors for computer design based on brain architecture!

The telephone switchboard analogy in its crudest form had a sensory input joined to a motor output by way of specific connections in the central exchange, i.e., the cerebral cortex. Memory was largely concerned with setting up the internal switching. That is, if x is seen, then do y, because the signal is properly routed. Real models of this type were, of course, more subtle, but they tended to predict that if you destroyed a chunk of brain containing the discrete paths between input and output, the memory would vanish along with the connection.

When Lashley tried to do this by lesioning various parts of rat brains, it did not work. It turned out that quite large areas of rat cortex could be removed and the animal was still capable of demonstrating learned behavior. The robustness of these results led Lashley to suggest at first that there was relatively little localization of function in cerebral cortex. More precise modern techniques have shown that this is untrue, and that there is a great deal of localization in cerebral cortex: for example, there are numerous topographic maps of visual space onto the surface of our cortex, maps of the body surface in areas of cortex associated with the somatosensory system, maps of frequency in the auditory system, and so on. In fact, fairly precise topographic localization of function is ubiquitous in cortex, and in other regions of the brain as well.

However, it would be equally in error to suppose that stored information is located only in a very small set of cells and their connections. For example, given the nature of the visual world, any complex image is liable to excite a large number of neurons in the visual system. Exactly how many is a matter for debate. Those believing in localization might estimate as few as a few hundred discharging neurons are adequate to represent complex perceptions. Those believing that information is more spread out would estimate a few percent of the relevant parts of cortex are necessary for adequate representation. A fascinating pair of papers by Barlow (1971, 1985) discusses this issue.

This debate is alive and well in the connectionist literature. Models have appeared assuming extreme element selectivity, where one active element corresponds to a complex, high level perception, and others have assumed widely distributed representations, where activity of a single element tells little about the item being represented. These different assumptions are important, because they tend to lead to different kinds of computations and have quite different statistical properties.

It should be pointed out that one natural way to realize semantic networks with connectionist hardware often requires highly selective elements, if one makes the natural assumption that single nodes in the network correspond to single active elements in a connectionist system. If one wants to use connectionist networks as computing devices—say, in artificial intelligence—there is absolutely no reason not to use highly selective, highly meaningful primitive elements (see Feldman and Ballard, paper 29).

However, if one is trying to model brain or psychological function, the evidence seems to us to be compelling that information is distributed, in that many elements must be simultaneously active to represent information.

Lashley's summary states strongly his conclusions opposing localization and in favor of distribution. The modern picture is somewhat less clear. Some low level reflex pathways seem to be quite localized. There is also evidence that several kinds of memory, apparently associated with different brain structures, exist (Thompson, 1986).

However for the cerebral cortex, even though Lashley overstates some of his conclusions and modern evidence indicates a greater degree of localization than he thought existed, we think that there is compelling evidence that in the brain "... every instance of recall requires the activity of literally millions of neurons" and that "... the reservation of individual synapses for special associative reactions is impossible." Lashley conceives of brain operation as large scale patterns of activation involving a great many active neurons leading to other large patterns of activity. Many, though not all, connectionist models incorporate this assumption explicitly or implicitly.

References

H. Barlow (1971), "Single units and sensation: a neuron doctrine for perceptual psychology," *Perception* 1:371–394.

H. Barlow (1985), "The role of single neurons in the psychology of perception," *Quarterly Journal of Experimental Psychology* (The Twelfth Bartlett Memorial Lecture) A37:121–145.

R. F. Thompson (1986), "The neurobiology of learning and memory," *Science* 233:941–947.

(1950)
K. S. Lashley

In search of the engram

Society of Experimental Biology Symposium, No. 4: *Psychological Mechanisms in Animal Behavior*
Cambridge: Cambridge University Press, pp. 454–455, 468–473, 477–480

I. Introduction

'When the mind wills to recall something, this volition causes the little [pineal] gland, by inclining successively to different sides, to impel the animal spirits toward different parts of the brain, until they come upon that part where the traces are left of the thing which it wishes to remember; for these traces are nothing else than the circumstance that the pores of the brain through which the spirits have already taken their course on presentation of the object, have thereby acquired a greater facility than the rest to be opened again the same way by the spirits which come to them; so that these spirits coming upon the pores enter therein more readily than into the others.'

So wrote Descartes just three hundred years ago in perhaps the earliest attempt to explain memory in terms of the action of the brain. In the intervening centuries much has been learned concerning the nature of the impulses transmitted by nerves. Innumerable studies have defined conditions under which learning is facilitated or retarded, but, in spite of such progress, we seem little nearer to an understanding of the nature of the memory trace than was Descartes. His theory has in fact a remarkably modern sound. Substitute nerve impulse for animal spirits, synapse for pore and the result is the doctrine of learning as change in resistance of synapses. There is even a theory of scanning which is at least more definite as to the scanning agent and the source of the scanning beam than is its modern counterpart.

As interest developed in the functions of the brain the doctrine of the separate localization of mental functions gradually took form, even while the ventricles of the brain were still regarded as the active part. From Prochaska and Gall through the nineteenth century, students of clinical neurology sought the localization of specific memories. Flechsig defined the association areas as distinct from the sensory and motor. Aphasia, agnosia and apraxia were interpreted as the result of the loss of memory images either of objects or of kinaesthetic sensations of movements to be made. The theory that memory traces are stored in association areas adjacent to the corresponding primary sensory areas seemed reasonable and was supported by some clinical evidence. The extreme position was that of Henschen, who speculated concerning the location of single ideas or memories in single cells. In spite of the fact that more critical analytic studies of clinical symptoms, such as those of Henry Head and of Kurt Goldstein, have shown that aphasia and agnosia are primarily defects in the organization of ideas rather than the result of amnesia, the conception of the localized storing of memories is still widely prevalent (Nielsen, 1936).

While clinical students were developing theories of localization, physiologists were analysing the reflex arc and extending the concept of the reflex to include all activity. Bechterew, Pavlov and the behaviourist school in America attempted to reduce all psychological activity to simple associations or chains of conditioned reflexes. The path of these conditioned reflex circuits was described as from sense organ to cerebral sensory area, thence through associative areas to the motor cortex and by way of the pyramidal paths to the final motor cells of the medulla and cord. The discussions of this path were entirely theoretical, and no evidence on the actual course of the conditioned reflex are was presented.

In experiments extending over the past 30 years I have been trying to trace conditioned reflex paths through the brain or to find the locus of specific memory traces. The results for different types of learning have been inconsistent and often mutually contradictory, in spite of confirmation by repeated tests. I shall summarize to-day a number of experimental findings. Perhaps they obscure rather than illuminate the nature of the engram, but they may serve at least to illustrate the complexity of the problem and to reveal the superficial nature of many of the physiological theories of memory that have been proposed.

VI. The Engram within Sensory Areas (Equipotential Regions)

The experiments reported indicate that performance of habits of the conditioned reflex type is dependent

Copyright © 1950 The Society for Experimental Biology. Reprinted with the permission of Cambridge University Press.

upon the sensory areas and upon no other part of the cerebral corte. What of localization within the sensory areas? Direct data upon this question are limited, but point to the conclusion that so long as some part of the sensory field remains intact and there is not a total loss of primary sensitivity, the habit mechanism can still function. Thus, in a series of experiments attempting to locate accurately the visual cortex of the rat, parts of the occipital lobes were destroyed in a variety of combinations. In these experiments it appeared that, so long as some part of the anterolateral surface of the striate cortex (the projection field of the temporal retina corresponding to the macula of primates) remained intact, there was no loss of habit. Any small part of the region was capable of maintaining the habits based on discrimination of intensities of light (Lashley, 1935b).

In a later experiment an attempt was made to determine the smallest amount of visual cortex which is capable of mediating habits based upon detail vision. The extent of visual cortex remaining after operation was determined by counting undegenerated cells in the lateral geniculate nucleus. Discrimination of visual figures could be learned when only one-sixtieth of the visual cortex remained (Lashley, 1939). No comparable deta are available on postoperative retention, but from incidental observations in other experiments I am confident that retention would be possible with the same amount of tissue.

In an early study by Franz (1911) the lateral surfaces of the occipital lobes of the monkey were destroyed after the animals had been trained in pattern and colour discrimination. These operations involved the greater part of what is now known to be the projection field of the macula. There was no loss of the habits. I have destroyed the cortex of the retrocalcarine fissure (the perimacular field) without destroying visual memories. The results with monkeys thus support the more ample data for the rat; the visual memory traces survive any cortical lesion, provided some portion of the field of acute vision remains intact.

This lack of definite habit localization might really have been predicted from psychological data alone. Analysis of the effective stimuli in discriminative learning reveals that the association is independent of particular sensory nerve fibres. It is a response to a pattern of excitation which may vary widely in position on the sensory surface and consequently in cortical projection. The reactions involved in motor habits show the same sort of functional equivalence; a motor habit is not a predetermined set of muscular contractions but is a series of movements in relation to bodily posture and to the complex pattern of the environment. The

writing of one's name, for example, is not a stereotyped series of contractions of particular muscles but is a series of movements in relation to the body planes which can be performed with any motor organ and with any degree of amplitude.

I have not time here to report in detail the experiments which justify the conclusion that neither the afferent path nor the efferent is fixed by habit. The mass of evidence accumulated by gestalt psychologists shows conclusively that it is the pattern and not the localization of energy on the sense organ that determines its functional effect. Similar motor equivalence is demonstrated by a variety of less systematic evidence. The psychological studies, like the more limited direct experiments on the brain, point to the conclusion that the memory trace is located in all parts of the functional area; that various parts are equipotential for its maintenance and activation.

VII. Facilitative Functions in Learning and Retention (Mass Action)

The experiments thus far reported have been concerned almost entirely with discriminative habits requiring only an association between a single sensory stimulus and a motor response. A very different picture develops in experiments with other types of learning. If rats are trained in the maze and then have portions of the cortex removed, they show more or less loss of the habit. If a small amount of cortex is destroyed, 5–10%, the loss may be scarcely detectable. If large amounts, say 50% or more, are destroyed, the habit is completely lost, and relearning may require many times as much practice as did initial learning. The amount of loss, measured in terms of the practice required for relearning, is, on the average, closely proportional to the amount of cortex destroyed. Text-fig. 8 shows the relation for one group of rats on a relatively difficult maze with eight culs de sac. There is some evidence that the more difficult the task, the greater the relative effect of the larger lesions (Lashley, 1929; Lashley & Wiley, 1933). Similar results have been obtained with latchbox learning and retention (Lashley, 1935a). So far as it is possible to analyse the data from more than 200 diverse operations, the amount of loss from a given extent of cortical destruction is about the same, no matter what part of the cerebral hemispheres is destroyed, provided that the destruction is roughly similar in both hemispheres.

The explanation of this quantitative relationship is difficult. In learning the maze the rat certainly employs a variety of sensory cues, visual, tactile, kinaesthetic, olfactory, possibly auditory. Brain injuries destroy vari-

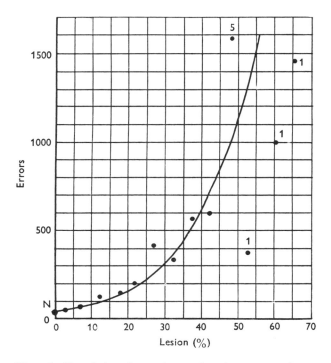

Figure 8 The relation of errors in maze learning to extent of cerebral damage in the rat. The extent of brain injury is expressed as the percentage of the surface area of the isocortex destroyed. Data from 60 normal and 127 brain-operated animals are averaged by class intervals of 5% destruction. The curve is the best fitting one of logarithmic form. For lesions above 45% the number of cases (indicated by numerals on the graph) is too small for reliability. (After Lashley & Wiley, 1933.)

ous sensory fields and the larger the lesion the greater the reduction in available sense data. The production of different amounts of sensory deficit would thus appear to be the most reasonable explanation of the quantitative relation between habit loss and extent of lesion (Hunter, 1930; Finley, 1941). Sensory deficit certainly plays a role in it. In the experiment on effects of incisions through the cortex, which was described earlier, the severity of loss of the maze habit correlated highly with the interruption of sensory pathways, as determined from degeneration of the thalamus.

However, sensory loss will not account for all of the habit deterioration. There is evidence which shows that another more mysterious effect is involved. In the first place, destruction of a single sensory area of the cortex produces a far greater deficit in maze or latch-box performance than does loss of the corresponding sense modality. A comparison was made of the effects on retention of the latch-box habits of combined loss of vision, vibrissae touch, and the anaesthesia to touch and movement produced by sectioning the dorsal half of the spinal cord at the third cervical level. This latter operation severs the columns of Gall and Burdoch, which convey tactile and kinaesthetic impulses, and also severs the pyramidal tracts which have a dorsal position in the rat. The combined peripheral sense privation and section of the pyramids produced less loss of the latch-box habits than did destruction of a single sensory area of the cortex (Lashley, 1935a). Secondly, when blind animals are trained in the maze, the removal of the primary visual cortex produces a severe loss of the habit with serious difficulty in re-learning, although the animals could have used no visual cues during the initial learning (Lashley, 1943).

A possible explanation of this curious effect was that the rat forms concepts of spatial relations in visual terms, as man seems to do, and that the space concepts are integrated in the visual cortex. The visual cortex might then function in the formation of spatial habits, even when the animal loses its sight. To test this Tsang (1934) reared rats blind from birth, trained them as adults in the maze, then destroyed the visual cortex. The resultant loss of the maze habit by these animals was as severe as in animals which had been reared with vision. The hypothesis concerning the formation of visual space concepts was not confirmed.

Our recent studies of the associative areas of the monkey are giving similar results to those gained with rats. Visual and tactile habits are not disturbed by the destruction singly, either of the occipital, parietal, or lateral temporal regions, so long as the primary sensory fields remain. However, combined destruction of these regions, as shown in Text-fig. 9, does produce a loss of the habits with retarded relearning. Higher level functions, such as the conditional reaction, delayed reaction, or solution of the multiple stick problem, show deterioration after extensive damage in any part of the cortex. The capacity for delayed reaction in monkeys, for example (to remember in which of two boxes food was placed), may be seriously reduced or abolished by removal either of the prefrontal lobes or of the occipital associative cortex or of the temporal lobes. That is, small lesions, embracing no more than a single associative area, do not produce loss of any

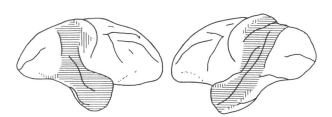

Figure 9 Minimal lesion which produces disturbances in tactile or visual memory in the monkey.

habit; large lesions produce a deterioration which effects a variety of habits, irrespective of the sensori-motor elements involved.

Results such as these have led me to formulate a theory of mass action or mass facilitation. It is, essentially, that performance of any function depends upon two variables in nervous activity. The reaction mechanism, whether of instinctive or of learned activity, is a definite pattern of integrated neurons with a variable threshold of excitability. The availability of such patterns, the ease with which they can be activated, is dependent upon less specific facilitative effects. This facilitation can come from a variety of sources. Some instinctive behaviour seems to require hormonal activation, probably a direct chemical effect upon specific nervous elements. Emotional facilitation may produce a temporary activation. Continued activity of related mechanisms may facilitate the whole group of associated reactions; a sort of warming-up effect.

There are indications (Krechevsky, 1936), although little systematic evidence, that the severity of postoperative amnesia varies with the intensity of motivation. Rats trained in a discrimination without punishment with electric shock for errors may show loss of the habit after lesions which do not produce loss in animals which were trained with punishment. The greater effects of cortical lesions in monkeys than in rats may be in part a result of the greater difficulty in getting consistent motivation in the higher animals. In man an amnesia often seems to be a difficulty rather than impossibility of recall; recall may be possible but only with extreme effort and fatigue. I believe that the evidence strongly favours the view that amnesia from brain injury rarely, if ever, is due to the destruction of specific memory traces. Rather, the amnesias represent a lowered level of vigilance, a greater difficulty in activating the organized patterns of traces, or a disturbance of some broader system of organized functions.

In interpreting apparent loss of memory after cerebral damage, extreme caution is necessary. The poor performance in tasks may be due to the destruction of specific associative connexions, but is instead generally, I believe always, the result rather of interference with a higher level functional patterning. Some experiments of Dr Klüver's (personal communication) illustrate this point. Monkeys were trained in a variety of discriminative reactions calling for use of different sense modalities by a method that required them to pull in the stimulus objects by attached strings. Extensive lesions in different cortical areas all caused loss of these habits. The monkeys simply pulled the strings at random. They were retrained in the discrimination of

weights. When this was learned, the habits based on other sense modalities (reactions to intensities of light, for example) returned spontaneously. What had been disturbed by all the operations was the set or attitude to compare stimuli, not the specific memory of which one was correct.

This example perhaps illustrates at a primitive level the characteristic of amnesias as seen clinically. Apparent loss of memory is secondary to a disorder in the structuring of concepts. Some physiological mode of organizing or integrating activity is affected rather than specific associative bonds.

X. Summary

This series of experiments has yielded a good bit of information about what and where the memory trace is not. It has discovered nothing directly of the real nature of the engram. I sometimes feel, in reviewing the evidence on the localization of the memory trace, that the necessary conclusion is that learning just is not possible. It is difficult to conceive of a mechanism which can satisfy the conditions set for it. Nevertheless, in spite of such evidence against it, learning does sometimes occur. Although the negative data do not provide a clear picture of the nature of the engram, they do establish limits within which concepts of its nature must be confined, and thus indirectly define somewhat more clearly the nature of the nervous mechanisms which must be responsible for learning and retention. Some general conclusions are, I believe, justified by the evidence.

(1) It seems certain that the theory of well-defined conditioned reflex paths from sense organ via association areas to the motor cortex is false. The motor areas are not necessary for the retention of sensori-motor habits or even of skilled manipulative patterns.

(2) It is not possible to demonstrate the isolated localization of a memory trace anywhere within the nervous system. Limited regions may be essential for learning or retention of a particular activity, but within such regions the parts are functionally equivalent. The engram is represented throughout the region.

(3) The so-called associative areas are not storehouses for specific memories. They seem to be concerned with modes of organization and with general facilitation or maintenance of the level of vigilance. The defects which occur after their destruction are not amnesias but difficulties in the performance of tasks which involve abstraction and generalization, or conflict of purposes. It is not possible as yet to describe these defects in the present psychological terminology. Goldstein (1940) has expressed them in part as a shift from the abstract

to the concrete attitude, but this characterization is too vague and general to give a picture of the functional disturbance. For our present purpose the important point is that the defects are not fundamentally those of memory.

(4) The trace of any activity is not an isolated connexion between sensory and motor elements. It is tied in with the whole complex of spatial and temporal axes of nervous activity which forms a constant substratum of behaviour. Each association is oriented with respect to space and time. Only by long practice under varying conditions does it become generalized or dissociated from these specific co-ordinates. The space and time co-ordinates in orientation can, I believe, only be maintained by some sort of polarization of activity and by rhythmic discharges which pervade the entire brain, influencing the organization of activity everywhere. The position and direction of motion in the visual field, for example, continuously modifies the spinal postural adjustments, but, a fact which is more frequently overlooked, the postural adjustments also determine the orientation of the visual field, so that upright objects continue to appear upright, in spite of changes in the inclination of the head. This substratum of postural and tonic activity is constantly present and is integrated with the memory trace (Lashley, 1949).

I have mentioned briefly evidence that new associations are tied in spontaneously with a great mass of related associations. This conception is fundamental to the problems of attention and interest. There are no neurological data bearing directly upon these problems, but a good guess is that the phenomena which we designate as attention and interest are the result of partial, subthreshold activation of systems of related associations which have a mutual facilitative action. It seems impossible to account for many of the characters of organic amnesias except in such general terms as reduced vigilance or reduced facilitation.

(5) The equivalence of different regions of the cortex for retention of memories points to multiple representation. Somehow, equivalent traces are established throughout the functional area. Analysis of the sensory and motor aspects of habits shows that they are reducible only to relations among components which have no constant position with respect to structural elements. This means, I believe, that within a functional area the cells throughout the area acquire the capacity to react in certain definite patterns, which may have any distribution within the area. I have elsewhere proposed a possible mechanism to account for this multiple representation. Briefly, the characteristics of the nervous network are such that, when it is subject to any pattern of excitation, it may develop a pat-

tern of activity, reduplicated throughout an entire functional area by spread of excitations, much as the surface of a liquid develops an interference pattern of spreading waves when it is disturbed at several points (Lashley, 1942a). This means that, within a functional area, the neurons must be sensitized to react in certain combinations, perhaps in complex patterns of reverberatory circuits, reduplicated throughout the area.

(6) Consideration of the numerical relations of sensory and other cells in the brain makes it certain, I believe, that all of the cells of the brain must be in almost constant activity, either firing or actively inhibited. There is no great excess of cells which can be reserved as the seat of special memories. The complexity of the functions involved in reproductive memory implies that every instance of recall requires the activity of literally millions of neurons. The same neurons which retain the memory traces of one experience must also participate in countless other activities.

Recall involves the synergic action or some sort of resonance among a very large number of neurons. The learning process must consist of the attunement of the elements of a complex system in such a way that a particular combination or pattern of cells responds more readily than before the experience. The particular mechanism by which this is brought about remains unknown. From the numerical relations involved, I believe that even the reservation of individual synapses for special associative reactions is impossible. The alternative is, perhaps, that the dendrites and cell body may be locally modified in such a manner that the cell responds differentially, at least in the timing of its firing, according to the pattern of combination of axon feet through which excitation is received.

6
Introduction

(1956)
N. Rochester, J. H. Holland, L. H. Haibt, and W. L. Duda

Tests on a cell assembly theory of the action of the brain, using a large digital computer
IRE Transactions on Information Theory IT-2: 80–93

When McCulloch, or Hebb, or James had an idea about the nervous system, they could think about it and describe it in words or, in a few fortunate cases, general mathematical formulas, but that was it. Nowadays, of course, the first response to an idea about almost anything complex is, "Let's simulate it on the computer." (Unfortunately, this sometimes takes the form of "Why think when you can simulate?" but, mostly, the effects have been beneficial.)

During the 1950s the world entered the age of computer simulation. We shall not emerge from it soon. When someone proposes a learning rule, or a theory for brain organization, it is now possible to test it to see if it works. With very complex systems it is often difficult or impossible to predict how they will behave beforehand, no matter how intelligent the analyst. Behavior is sometimes explainable in retrospect, but it is necessary to do the numerical experiments to see if ideas are actually workable, or if unforeseen problems appear. They often do. As only one example, there are a number of learning rules that can be proved to work mathematically. Unfortunately, when simulations are done, learning times are found to be enormous, totally outside the bounds of practicality. Or the results are immensely sensitive to noise, or error, or to values of particular parameters.

It is sometimes not appreciated by the public at large, including most scientists, how difficult and tricky it can be to get a computer simulation to work. The response to a correctly running simulation is often something along the lines of "It took you six months just to do that?" With 'real' experiments in neuroscience or psychology, there are usually a large number of dead animals or a depleted subject account to show that the experimenter has, indeed, been working. With failed simulations, there is merely a bill for CPU charges, frustration, and the problem of explaining to others why you have been wasting your time.

In spite of this, simulations of brain theories have been done from the earliest days of the digital computer. One early and well known computer simulation of the nervous system was by Farley and Clark (1954), who studied in some detail the properties of interconnected, neuron-like elements.

A less well known, but more interesting, computer simulation from our point of view in this paper. Rochester et al. did their work at the IBM research laboratories in Poughkeepsie. They studied the learning system proposed by Hebb in 1949, and kept in close touch with Hebb's group at McGill (specifically Hebb and Peter Milner) during the process. This paper represents perhaps the first attempt to test a well formulated, detailed neural theory with a computer simulation.

The authors may also have been the first to discover the nearly universal finding for computer simulations designed to check brain models: the first attempt did not work.

They made the appropriate adaptive response. In conjunction with Hebb and Milner, they tried again, with a modification of the learning rule, and this time it worked much better.

There are a number of interesting technical aspects to this work. Because it was necessary actually to realize a system, they had to make a number of important assumptions that were not present in Hebb's original theory. Simulations require great precision, and details cannot be ignored or discussed qualitatively.

Neurons in the first simulation were on-off binary devices, with a threshold. The authors made the discovery that, as first described, as time increased the strength of the Hebb synapses would grow without bound. In order to avoid this, they proposed a normalization rule, where the sum of all the synaptic strengths equaled a constant. Strong synapses grow strong at the expense of others. Some variant of this rule is common in later work on neural networks. They also found it necessary to incorporate a fatigue factor, where firing makes the cell less likely to fire in the immediate future. This could cause one cell assembly state to decay after a while and others to grow.

When the system was first simulated, it did not show signs of developing cell assemblies. Though it did show complex activity in response to inputs, it did not seem to display the self-organization that had been predicted.

The second try made assumptions that sound more modern, given the perspective of over thirty years. Based on a suggestion by Peter Milner, they made two major changes in the 'physiology.' First, they incorporated inhibition. The argument was that a cell assembly is self-exciting, and one active cell assembly inhibits others from becoming active. Synapses could range in strength from $+1$ to -1, instead of only from 0 to $+1$. Second, they modified the model neurons—they were no longer binary, but output was related to frequency of firing; that is, the average frequency of immediate past discharge was used as the activity measure of the neuron, and neuron activity was graded (from 0 to 15). The Hebb learning rule was modified so that if the frequencies of pre- and postsynaptic activities were correlated, the synapse increased, and if one cell fired and the other was inactive, the synaptic strength decreased. This is a natural generalization of the Hebb synapse that is commonly used now, in many variants. The activity of the postsynaptic cell was obtained by summing the presynaptic activity weighted by the synaptic strength, again a simple integrating rule commonly used now, though there were some complexities introduced by the fatigue factor and normalization. There was also significant local anatomy, in that neurons only connected to nearby neurons. They used quite a large system—512 neurons.

Now the simulation showed signs of working. The authors commented that cell assemblies did seem to form, with excitatory connections among members of an assembly and inhibitory connections between assemblies. There were still problems, but the paper concluded on an encouraging note.

This early paper makes the important point that neural network research is technology driven. When a computer is used, the form and precision of the assumptions and the definitions of success change radically. Before the computer, theories could be proposed and argued about, and even analyzed, but not really tested. Now we know whether or not they *work* and can quickly get an idea of how well. Many small details that could be safely ignored can no longer be ignored. Success or failure depends on getting the details right as much as getting the overall structure right.

It is hard to improve on the summary in last paragraph of the paper: "This kind of investigation cannot prove how the brain works. It can, however, show that some models are unworkable and provide clues as to how to revise the models to make them work. Brain theory has progressed to the point where it is not an elementary problem to determine whether or not a model is workable. Then, when a workable model is achieved, it may be that a definitive experiment can be devised to test whether or not the workable model corresponds to a detail of the brain" (p. 88).

Reference

B. Farley and W. A. Clark (1954), "Simulation of self-organizing systems by digital computer," *IRE Transactions on Information Theory* 4:76–84.

(1956)
N. Rochester, J. H. Holland, L. H. Haibt, and W. L. Duda

Tests on a cell assembly theory of the action of the brain, using a large digital computer
IRE Transactions on Information Theory IT-2: 80–93

Theories by D. O. Hebb and P. M. Milner on how the brain works were tested by simulating neuron nets on the IBM Type 704 Electronic Calculator. The formation of cell assemblies from an unorganized net of neurons was demonstrated, as well as a plausible mechanism for short-term memory and the phenomena of growth and fractionation of cell assemblies. The cell assemblies do not yet act just as the theory requires, but changes in the theory and the simulation offer promise for further experimentation.

Introduction

The problem of how the brain works can be approached by investigating the elementary components, the neurons, and then seeing how larger and larger assemblies of these operate. Or it can be approached by observing the behavior of the entire organism and working back to determine what the components must be. The former activity is called neurophysiology and the latter is called psychology. Before we can say that the problem is well in hand, these two approaches must meet in the middle so that we have a single consistent picture that firmly connects psychology and neurophysiology.

As the neurophysiologist considers more and more complicated structures of neurons he gets into problems that are less and less related to his normal way of thinking. Curiously, however, some of these problems do not begin to resemble parts of psychology. What is happening is that the neurophysiologist is beginning to think about information handling machines that are too complex to be understood without the specialized knowledge of other disciplines. These other disciplines are information theory, computer theory, and mathematics. People in these other fields need to augment the work of the neurophysiologists and psychologists before the brain can be properly understood.

In the experimental study of the brain it is not yet possible to observe well the electrical interconnections among neurons. No one has yet been able to simultaneously record input and output signals of a single neuron in the brain. For this reason it has not yet been possible to test certain theories about how the brain works by experimentation on animals.

It is possible to measure the electrical characteristics of an isolated neuron in some circumstances.[1,2] One can imagine an elaborate network of such neurons and conjecture on the behavior of the network. The analytical treatment of these networks has proved that one can construct any desired kind of logical machine from elements that are probably much less powerful than neurons.[3,4]

The analytical approach has not been very effective in actually describing the behavior of complicated networks of neurons. However, it has proved effective to simulate such networks and to draw conclusions from the behavior of the simulated network of neurons.

Two sets of simulation experiments were made and another is in progress. In the first of these it was possible to simulate a network of up to 99 neurons and a test was made of part of the theory advanced by D. O. Hebb in his monograph, *The Organization of Behavior*.[5] The second set tested an unpublished revision of P. M. Milner of part of Hebb's theory with a network of 512 neurons. The third set is to test a further revision. In each case the original neurophysiological theory had to be interpreted in order to get something definite enough to simulate, and these interpretations were done by the present authors.

The 69-Neuron Discrete Pulse Simulation

In this paper the term "neuron" will generally be used as an abbreviation for the term "simulated neuron". Likewise the term "synapse" will be used to stand for the term "simulated synapse", in other words for the simulation of the coupling mechanism that enables one neuron to send signals to another. Where ambiguity could arise, qualifying adjectives will be used.

The basic idea of the simulation can be seen by reference to Fig. 1. The large rectangle in Fig. 1 stands for all of the 2048-word high speed electrostatic memory of the Type 701 calculator. The memory was divided into 70 parts, one for each neuron and one for the program. In the area reserved for each simulated

Copyright 1956 IEEE. Reprinted with permission.

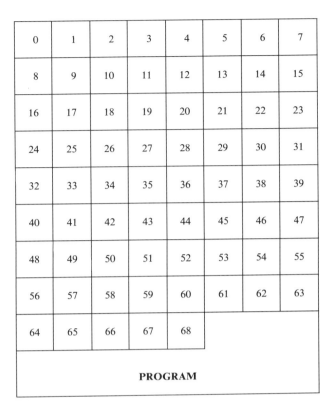

0	1	2	3	4	5	6	7
8	9	10	11	12	13	14	15
16	17	18	19	20	21	22	23
24	25	26	27	28	29	30	31
32	33	34	35	36	37	38	39
40	41	42	43	44	45	46	47
48	49	50	51	52	53	54	55
56	57	58	59	60	61	62	63
64	65	66	67	68			

PROGRAM

Figure 1   Allocation of memory

neuron were some numbers that might theoretically be measured on a corresponding living neuron. These numbers gave all of the information that was needed about each neuron. Specifically, the things that were known about each neuron, either from its location in memory or from the numbers stored there, were:

1. It's number (name)

2. How long since it had fired

3. How tired it was from having been fired excessively

4. For each of 10 output (efferent) synapses:

4.1. The number of the (efferent) neuron that it simulated

4.2. The magnitude of the signal that it sent to that (efferent) neuron when this (afferent) neuron fired.

Under control of the program, the calculator repeatedly scanned the 69 neurons and, by making calculations, caused these numbers to change as they would have changed if the network had actually been constructed. Therefore, after each pass over the data in memory, the data represented, in great detail, the state of each neuron and synapse in the network at the next instant of time.

In this model, time was quantized into time steps. A

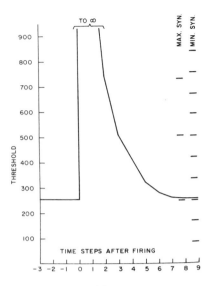

Figure 2   Threshold curve.

neuron could fire at any time step, but not between. A time step corresponded approximately to the interval between the firing of one neuron in a chain to the firing of the next. In the simulation, the average length of time required for a single time step was about 5.3 seconds and this corresponded to perhaps 0.7 milliseconds in the brain. Therefore, the simulation was slower by a factor of 7600.

At any given time step a neuron was either fired or in some state of recovery from being fired. Various recovery curves were used and the one shown in Fig. 2 was typical.

During any given run on the calculator, the neurons were interconnected in a particular net. Each neuron was connected so as to stimulate 10 other neurons. Usually the net was designed by the calculator. It would make a random choice of the neuron to be stimulated by each of the 10 output synapses of each neuron. It would record these choices on punched cards and retain them for the rest of the run.

If a neuron fired at time step (n − 1) it would stimulate 10 neurons so as to tend to cause them to fire at time step n. The size of the signals sent to the 10 neurons would depend only upon the fact that the original neuron fired and upon the magnitudes of the interconnecting synapses. To say this another way, the input signals to a neuron, together with its threshold, would determine whether or not the neuron would fire, but if it did fire, the strength of firing would not depend on the input signals.

The input situation of a typical neuron is shown in Fig. 3 with some possible values of synapse magnitudes. The behavior of neuron x is shown in the following table.

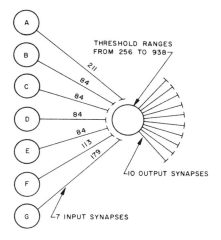

THRESHOLD RANGES
FROM 256 TO 938

10 OUTPUT SYNAPSES

7 INPUT SYNAPSES

Figure 3 Example of a simulated neuron and its connections.

Neurons that fired on step (n − 1)	Mag	Threshold of x	Would x fire on step n?
AB	295	256	Yes
BCE	252	256	No
BCDE	336	256	Yes
ABCDEFG	839	839	Yes
ABCDEFG	839	938	No
ACF	408	376	Yes
EFG	376	376	Yes

It can be seen that the input circuits to such a neuron can provide quite sophisticated switching.

Not all of the properties of the simulated neurons have been described. However, to make the exposition easier to follow, it is convenient to skip ahead and show some observations on the behavior of networks of neurons. Except for some minor difficulties, this behavior would be obtained with neurons like those already described. The discussion of these minor difficulties will be clearer after showing these results.

Fig. 4 shows an example of what will be called diffuse reverberation. Each row in this figure indicates with a 1 those neurons that fired and with an 0 those neurons that did not fire in a particular time step. Each column, of the 64 columns at the right, shows the history of a single neuron. The right hand 64 columns of Fig. 4 show, therefore, the complete firing history of 64 neurons for 50 time steps.

Fig. 5 shows, as a function of time, the number of neurons that were simultaneously fired. The time covered here is a little larger than in Fig. 4 and shows the complete history beginning with a quiescent net and continuing until the activity died out.

We propose this diffuse reverberation as a plausible mechanism for short term memory, the kind of memory that is involved in remembering the inter-

mediate results in mental arithmetic. We will discuss later some conjectures as to how the brain can make use of such a memory mechanism.

Now another property of the neurons will be described. When neuron A participated in firing neuron B, the synapse that enabled A to stimulate B was increased in magnitude unless it already had reached the limit of 938, in which case it remained constant. This characteristic was our version of Hebb's basic *neurophysiological* postulate. Hebb postulated that, *"When an axon of cell A is near enough to excite cell B and repeatedly or persistently takes part in firing it, some growth process or metabolic change takes place in one or both cells such that A's efficiency, as one of the cells firing B, is increased."* [6]

This property of simulated neurons is somewhat curious. No process of just this sort has been observed in living tissue. However, it has not been possible to demonstrate, by measurement, that the Hebb postulate is false. Nothing else has been observed that could account for learning and memory in a plausible way. The Hebb postulate suggests a plausible machine that does not contradict experiment.

The purpose of the assumption about the growth of synapses is to get a mechanism for the retention of long term memory. When an animal experiences some event there will be activity in its brain. This activity will consist of a spacial and temporal pattern of firing of neurons. During the experience, the synapses involved will be strengthened, according to Hebb's postulate. Therefore, the same, or a similar, sequence of neural events is more likely to take place later than it would have been if the animal had not had that experience. A repetition of some part of the neural events that were associated with an experience is assumed to be the act of recalling the experience. It is evident that the mechanism that Hebb postulated would tend to cause recollections. The question of whether or not the postulate is sufficient is, in a sense, the main topic of this paper.

If no additional rule were made, the Hebb postulate would cause synapse values to rise without bound. Therefore, an additional rule was established: The sum of the synapse values should remain constant. This meant that, if a synapse was used by one neuron to help cause another to fire, the synapse would grow. On the other hand, if a synapse was not used effectively, it would degenerate and become even less effective, because active synapses would grow and then, to obey the rule about a constant sum of magnitudes, all synapses would be reduced slightly, so the inactive synapses would decrease.

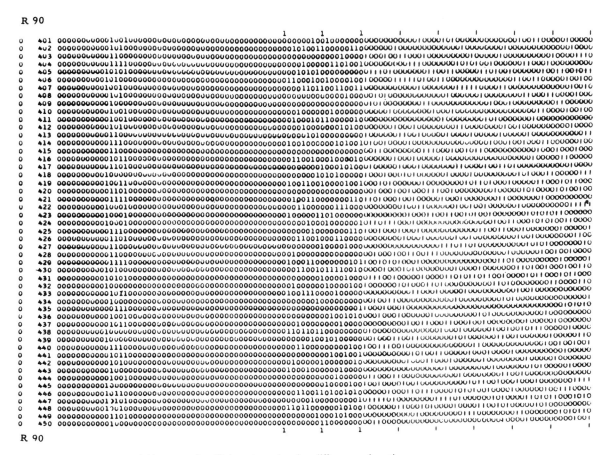

Figure 4   Firing pattern of 64 neurons for 50 time steps showing diffuse reverberation.

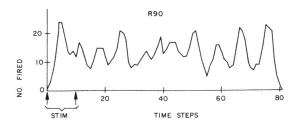

Figure 5   Number of neurons firing at each time step showing diffuse reverberation.

Before discussing network action further, another property of the neurons will be mentioned. A neuron fired at too high a frequency becomes less sensitive, so that more stimulation is required to fire it. The effect of this is shown in Fig. 6, which shows the threshold as a function of time when the neuron is fired repeatedly with a constant level of stimulation. As with a living neuron, this simulated neuron fires rapidly at first and then settles down to a lower rate of firing.

This process is called fatigue because of the obvious analogy to living neurons. A significant aspect of fatigue is that it is a form of memory and, as such, may plan an important part in the operation of the brain.

The concept of cell assembly occupies a key position in Hebb's theory. A cell assembly is a group of neurons that are interconnected in a very complex fashion and within which diffuse reverberation can take place. Fig. 4 shows just such a situation.

Parts of the cortex are imagined to consist of a large number of cell assemblies, each of which contains a large number of neurons. Only a small fraction of the cell assemblies are aroused at any one time. In other

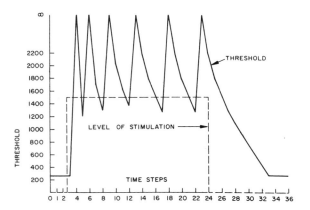

Figure 6 Threshold as a function of time.

words signals are reverberating in only a few cell assemblies at once. Just which cell assemblies would be aroused at any one time would depend in large part upon what cell assemblies had been aroused at a previous instant of time, and in small part upon signals from elsewhere.

In the language of information theory, this part of the brain can be considered to be a finite state transducer, in which the internal state is determined by noting which cell assemblies are aroused and which are quiescent. In other words, the brain should exhibit a kaleidoscopic sequence of patterns of cell assembly arousal. It is outside the scope of this paper to expound Hebb's theory, so it will be assumed henceforth that the reader either understands the significance of a finite state transducer or has read Hebb's book.

In passing, it is worthwhile to point out how appropriate the finite state transducer description is for Craik's "hypothesis on the nature of thought."[8]

Hebb's theory required that it be possible for a neuron to belong to several different cell assemblies and that not all of these assemblies be aroused at once. Hebb's theory also required that it be possible for a neuron to change its affiliation from one cell assembly to another. It may be possible to devise a theory that has only the second requirement, but no further consideration of this possibility will take place in this paper.

The problem of how cell assemblies can arise and how they become modified, is vital to this theory. It will be shown that Hebb's scheme is unlikely to work with neurons of the type described so far. It will also be shown that, by suitably improving the neurons and by making the network more complex, cell assemblies can be made to form spontaneously. It will further be shown that these cell assemblies are not entirely

satisfactory but that there is a plausible course for further investigation.

Suppose that there is initially some activity in a network of neurons and that input signals are impinging on the network. Suppose also that from time to time a particular input signal, S, arrives. When S first arrives, it will impinge on some internal state, I_j. In other words it will impinge upon some particular configuration of states of individual neurons. The particular sequence of internal states, $I_{j+1}, I_{j+2}, I_{j+3}, \ldots,$ that follows will strengthen certain synapses in such a way that the sequence $I_j, I_{j+1}, I_{j+2}, I_{j+3}, \ldots$ is more likely to occur again. It was conjectured that the next time S occurred, some part of I_j would be in existence and that some part of the sequence $I_j, I_{j+1}, I_{j+2}, I_{j+3}, \ldots$ would be reinforced. As S appeared repeatedly some characteristic response to S gradually would become sufficiently reinforced as to be identifiable. As the characteristic sequence was arising there would appear, in it, points where diffuse reverberation could occur. In other words there would be some internal state I_{j+k} which would repeat some part of an earlier state in the sequence. As soon as this happened the rate of reinforcement of the connections would increase, because each time the stimulus S occurred the sequence of states would be such as to give several reinforcements to some of the connections instead of just one reinforcement. It was conjectured that cell assemblies corresponding to some common stimuli would arise in the brain this way.

In order to test this conjecture about the manner in which cell assemblies form, a program was written to generate an appropriate environment for the neuron network, and an arrangement was set up for the network to receive signals from the environment. To receive the signals, six neurons were chosen to act as receptors. It was arranged that no neurons would stimulate these receptors. Instead, they could be fired only by an external program to enable the calculator to reach in and modify the one bit on each of the six neurons that indicated whether or not it had just received enough stimulation to fire. The synapses from the receptors spread out diffusely through the network.

The neuron net was stimulated once every ten time steps with a 6-bit signal that could define the state of each of the six receptors. The signals were chosen by a program whose action is illustrated in Fig. 7. It is a Markov process in which there is some probability that the input will be random but mixed in with the random signals are frequent occurrences of certain sequences. The network then had the opportunity to develop a characteristic response to each of the three sequences.

The network did not develop any characteristic

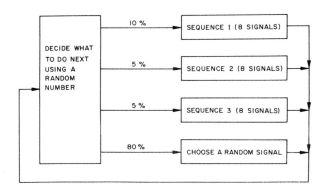

Figure 7 Environment.

responses and there was no sign of development of cell assemblies. A number of variations on this experiment were tried, all with the same result. Then the reason for the difficulty was realized and a simulation experiment was run to verify the explanation.

In such a neuron network, the idea that a detailed temporal-spacial pattern of firing can be effectively reinforced by a partial repetition is false. The reason can be seen from the following experiment. A simulation experiment was run to a convenient point where diffuse reverberation was taking place. Then all the data was punched on tabulating cards. These cards contained all relevant information so that, if they were read by the calculator, the simulation would go on from where it left off. Before the cards were read by the calculator, however, they were reproduced to give four identical decks of cards. Then three of these decks were slightly modified, each in a different way. In each case the modification was to choose some neuron that was about to be fired and manually change the number that specified its state of recovery so that it wouldn't fire quite so soon. Then four simulation experiments were run, one with each deck.

The four sets of results were compared and it was found that the detailed patterns of firing diverged rapidly. In just ten time steps, in each case, over 30 per cent of the neurons firing were different. This result is shown in Fig. 8. This shows that even slight differences rapidly grow to be large differences so there is little chance that a detailed pattern of firing can be effectively reinforced.

It was concluded from this work that some additional structure was needed within a network to allow all assemblies to form. A plausible model of a short term memory had been demonstrated but rather convincing evidence had been found to show that Hebb's postulate was not enough to make cell assemblies form.

Some other experiments were run which coincided

Time Step	Total Neurons Firing				Different Neurons Firing		
	C	A	B	D			
	Cntl. Run	N40 Sup.	N61 Sup.	N70 Sup.	A·C	B·C	D·C
151	31	30	30	30	1	1	1
152	30	30	28	28	0	2	2
153	28	29	27	27	1	3	5
154	30	31	29	31	1	9	5
155	31	31	28	30	2	19	9
156	33	34	26	30	3	27	17
157	30	29	31	31	5	37	19
158	31	26	32	29	9	37	16
159	32	30	32	34	14	32	22
160	34	32	33	38	20	35	22

Figure 8 Divergence after suppressing one firing of one neuron. Three separate runs are represented in this chart in addition to the control run, C. In run A, neuron 40 was suppressed; in B, N61 was suppressed; and in D, N70 was suppressed.

in time with the work of Farley and Clark[9] and which reached essentially identical results. However, these did not seem to throw any light on the central problem of how the brain works, so this line of investigation was dropped.

512-Neuron F.M. Simulation

At this point we conferred with D. O. Hebb and one of his people, P. M. Milner. Milner had been working on a revision of part of Hebb's theory to introduce more recent neurophysiological data. The essence of Milner's idea was that inhibitory synapses, as well as excitatory synapses, are needed and that within a cell assembly most synapses are excitatory, while between cell assemblies most synapses are inhibitory. This idea sounded to us like a plausible cure for the troubles in the first model. It made engineering sense.

The significance of the idea can be seen by considering two cell assemblies. These will act like an Eccles-Jordan Flip Flop circuit. Suppose one is aroused. It keeps itself going by its internal excitatory connections and keeps the other quiescent by the inhibitory interconnections. Finally it begins to fatigue. As it begins to falter, it inhibits the other less strongly, so sporadic residual activity in the other begins to increase. This in turn inhibits the aroused cell assembly, causing it to falter more. This feedback condition causes an abrupt switching so that the aroused one becomes quiescent and the quiescent one becomes aroused. A more detailed discussion of this can be found in Appendix 1.

It seemed certain that the switching action would take place, but it was not clear whether the possibility of having inhibitory synapses would be enough to allow cell assemblies to arise or whether some cell assembly structure would have to be built in at the start.

Experiments with the discrete pulse model indicated that diffuse reverberation was a fairly reliable sort of thing in a net of 63 neurons, but quite erratic in a net with 21 neurons. Therefore it was felt that in a new experiment there should be a larger number of neurons in a net. A major obstacle to this was that the calculator was not fast enough to manage a very much larger net, even though this was to be done on the Type 704 which is faster than the 701. Something had to be sacrificed.

It was decided to sacrifice the knowledge of exactly when an individual neuron fired. All that the machine or the experimenter could know was the frequency at which a neuron was firing, and not the exact instants of time at which it did fire. The frequency would vary from time to time, so this was called the FM model.

One particular version of the FM model will be described here. There were 512 neurons, each with 6 input (afferent) synapses and a number of output (efferent) synapses that varied from one neuron to another. The synapse magnitude lay between -1 and $+1$ and changed as long term learning took place. The frequency of a neuron varied from 0 to 15. Equations are given in Appendix 2 to specify precisely how these quantities varied from time to time, and a qualitative description is given below in the text.

The magnitude of a synapse was much like a correlation coefficient between the two neurons that it connected. If the frequencies of the two neurons usually went up and down together, the synapse magnitude would grow toward $+1$. If, on the other hand, one neuron was usually inactive while the other was active, the synapse magnitude would approach -1. This is the FM version of Hebb's basic neurophysiological postulate.[6]

The frequency of a neuron was obtained essentially by calculating, for each synapse, the product of the synapse magnitude and the frequency of the stimulating (afferent) neuron, adding these products, and normalizing. It was further bounded by not being allowed to go negative. Therefore a neuron could have a high frequency only if it was stimulated through positive synapses by neurons with large frequencies and not simultaneously stimulated through negative synapses by neurons with large frequencies.

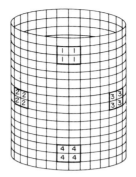

Figure 9 Arrangement of neuron in F.M. model.

The fatigue increased if the frequency was high; stayed constant if the frequency was intermediate; and decreased if the frequency was low. Furthermore, it was not allowed beyond the bounds of 0 and 7. A fatigue of 7 could nearly stop a neuron while a fatigue of 3 did little to it.

An important change was made in the nature of the connections in the net. A distance bias was introduced so that two nearby neurons were more likely than two remote neurons to be connected together through a synapse. In the experiment described in this paper, the neurons were visualized as being arranged in a cylinder, as shown in Fig. 9. The cylinder was 16 neurons high and 32 neurons around. If two neurons were within eight of each other, they were as likely to be connected by a synapse as any other two neurons that were within eight of each other. However, no neurons that were farther apart were connected by synapses.

Four blocks of four neurons each were selected to act as receptors. These four blocks are shown in Fig. 9. The procedure that was used most of the time was that receptor areas 1 and 4 were controlled to have maximum activity for three successive time steps and then the net was allowed to operate with no external stimulation for three time steps. Then areas 2 and 3 were controlled to have maximum activity for three time steps and then the net was again left alone for three time steps. This cycle was repeated many times. The cycle was considered to be the equivalent of about 0.2 seconds in an animal and took about 160 seconds on the calculator.

Cell assemblies did actually build up around each of the receptor areas. Within a cell assembly the interconnections were largely excitatory and between cell assemblies they were largely inhibitory.

The activity of each neuron at each time step for one complete cycle is given in Table 1.

Table 1 Activity during one complete cycle of simulation

1.

```
0040000000006000001403141003 00
00400003100100000077010400000000
000000207000000000770100000120 00
000000000000050010000050003000000
001000530110040000061040000200000
00000101000400000005001100000000
00000000001000040001000011000000
0000001000000000001100010000000
0000040000 30000 3000000000000001
0000000100000004105000000004000110
0000000000000006000000041000000 00
0000000000000600000000000000000
000003300000000100000000010100004
00000000050010100377000000010000
0000000000000000077000 000001000
0 00000 0005010000010000010000010
```

4.

```
00200010000010500047007000000100 0
000010021000001500250000000001000
000000000 3002030003640 5001000000
000000000000013075563011004410000
00100000000000000000000000000000
0003010000100000100005 0000000000
00000000010000100020000000000000
0000000000000000000000000 000000
00000000000000000040400000000 00000
00000000000010000010001000600040 00
000300000000010000500000000000020
00020000010000010000000577000000
02001000000000000100000007001000
000003100055010000120000000060002
00000060007000000070700030030000 4
000000000001000001310 30000000000
```

2.

```
004002000000005000160000010000300
00 3020042002000200770 60 30001000 0
00000000500000200077507000000000
00000000300000200050000001000 200
0000004100000000000000000000100000
00000103100000000002020100000000
0004000000100010000000001000000
00000010100000000000000000000000
00000000001000010007000000000000
000000011000001002000004 00000000
000000000100004100410000000000000
000000000000310013000002000000 00
0000023000000000000000000000000
0000001000031000047700000000000000
00010200000000000477000000200000
0 000010000000000000000000 0000000
```

5.

```
000000002100305200050050000000000
000010010000003500241000001031 00
000000000510002005 35000003000000
00000000000000104305402000340010
000000000000000000000010000000000
000000000010000010006000010000000
00000000000000000060000000000000
000000000000000001000000000000000
0000200000000000012000000000000000
000000000001000000000010060 05000
00000100000000000000000000004130
00000000 30000000000300657000000
0010100100000000010200000010000000
0000030000700000000200300000000010
1000006000700000000030004000100004
000010000000000001000000000000000
```

3.

```
003001110000012000670 07000000100 0
002020033100 000020077010020000000
000000000000010300077507000000000
000000012000002053730001013 0000
000000000000000000000000000000000
000300030010000010000 301000000000
000300000000010000000 00000000000
00000010000000000000000000000000 00
000000000000000130700000000000000
0000000100001001001000110600000000
0003000000000200004 0000010000000
0000000000000000011000067 0000000
00000020000000000000000107000000
000001100005000011770000000000000
0000013000500100077700000030000 3
00000010 0000000001300 30000000000
```

6.

```
0000000054010025000030000000000000
000100000020010010020000000000 3000
001000000310112105 2500000150 00000
0000000000000000010030100004 0020
3000000000000000000010000000000
000000000000030 00006001010 0000
00000000000010007010000000000000
00000000000000000000000010000000
000 300000000000030000003030 00
0000000000000000000 0C1000005001
0 00000100000000010000000015230
0100110103000000000040 45000010 00
00000001000000000050000010030 00
010000000700100000000400000000 02
00000001100000 000000005400110 20 0
13002000000000000000000400000000
```

Table 1 (continued)

7.

```
0 000000002 30 3001400 00000000000000
0001000000100010011000000000000
0010000002001210200000035000000
0 0000000000000000000000000010010
5000000000000200000000000000010000
0100000000100003000001000000000000
000000000000000000007010000100010
0000000000000770000000000077000000
0000000000007770000200300776030 00
0001101000000000000000000000000002
0010000000000000300000000003400 0
0010000100000000000405001003000
100000000000000000005400000000000
0100000000010000000022000000000
100000001000000000000005015000000
0 30 00010000000000000005000 00000
```

8.

```
000000000020000110 0000000000000000
0000 00000000010002000000000100000
0000000000000000060110000750060000
0000000000100000000000000010010000
600000000000200500101000000700000
0000000000000301300000000000000000
000000000000000006000000000101010
1000000000057705070000007700000 0
00000000000077100000433077500010
0000000001000000000000000000000000
0040000000000000300000000004 0000
00000000000701000000014100000000 0
000000000000007000004000006010 00
010000000000000200000400000600000
0000000000000000000000000340100 00
00000000000000200000000010000000 0
```

9.

```
000000000000000000000000000000000
00000000000000006000000030070000
0000000000000007000000700360040
000000000000000000000000030005000
0000000000101005001010000073 0000
00000000001006420010000000000000
00000000000000011000070000000000
00000000057705070000007701 0
0000010000777000004144770000 30
0000000000001000010000000000000
106000000000000100000000140007
0000000000704070000024000000 0000
000000000001100700010000000700050
000000000000000050000000005600000
000000000004001400000010500100 00
00000000000003050000000000001 000
```

10.

```
0 00000000000000001000000000001006
0 00000000000000600000060000 70000
000000000000000007000000710450 040
000000000000005000000000000005000
000 300000 3000360200000040750000
0000000010006530020000000000000000
4000000000000000 31000070020000000
0000000000067706070000102000000 0
000 0006000000370000004131700500 30
00000000000004000060000000000100
100000000000000001000700000000007
00000000105 3507000000000000000000
0 00000000013305000600000040600061
0000000000000006000000020005501300
0000100000040047000020000 30020000
0000000000 3300405000000000000013000
```

11.

```
000000000000000041000204001030 05
001000000000000070000060020600 00
0000002000000000020 0002020051005 0
00100000000200600000000000000500 0
000000003005004420402 3000 30 060 00
00000000130371400013003000000000
5000000000000005 30020700300200 00
000000000001160102001000400010 00
00000070000003400000102200050000
000000000 000004000060000004400 00
200000000000100201201705001100 005
00000000100 320600000000000010100
0000001000336000005000107000 0140
000000000700000400000200040143 00
000000000000400560000000020013000
000000007400206000000010000300 00
```

12.

```
00000000000010 0640004060 3005006
0010000000000000100000540 30000 00
00000004000000100004000011 0010
0010000000 30060000 40000000 00
0000004500400540040404403050160000
00000000000510100104005000010000
4000000010100044000000 30050100
00000000200022010 00020013000000 0
00000040003000010000002000304 02
0000000000002000031002007400 00
00000000000004014026063001000 00
000000000003040000000000130 032
0000031001 3120000030003040140004
0000000070000000000301000133 00
00000000000003110001000001 0000
0000000052103000000010000 30001
```

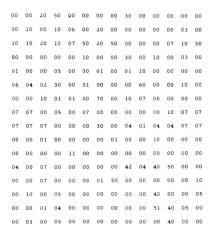

00	00	20	50	00	00	00	00	30	00	00	00	00	00
00	10	00	10	06	00	20	00	00	00	00	00	03	00
10	10	20	10	07	50	20	50	00	00	00	07	10	50
00	00	00	00	00	10	00	00	30	00	10	00	03	00
01	00	00	05	00	00	01	00	01	10	00	00	00	00
06	04	02	30	00	01	00	00	00	60	00	00	10	00
00	00	10	01	01	00	70	00	10	07	00	00	00	00
07	07	00	05	00	07	00	00	00	00	00	10	07	07
07	07	07	00	00	30	00	04	01	04	04	07	07	
00	00	01	00	00	00	00	01	00	00	10	00	00	00
00	00	00	00	11	00	00	00	00	00	00	00	00	00
04	00	07	00	00	00	00	00	42	04	40	50	00	00
00	00	07	00	00	00	01	50	00	00	00	00	00	10
00	10	00	05	00	00	00	00	00	00	40	00	00	05
00	00	01	04	00	00	00	00	00	51	40	05	00	
00	03	00	05	00	00	00	00	00	00	40	00	00	

Figure 10 Illustration of cell assemblies.

In table 1 the numbers are just half of the frequencies of the neurons. This was done to reduce the cost of printing. Since it is difficult to sec what is going on without practice and effort, a small section of the sixth and ninth steps have been reproduced side by side in Fig. 10. These times were chosen to contrast the arousal of 1 and 4 which suppress 2 and 3, with the arousal of 2 and 3 which suppress 1 and 4. Nearly every neuron has chosen allegiance to one cell assembly or another. Only 3 of the 224 neurons shown are active at both times.

Examination of the synapses also showed that cell assemblies had formed. A dividing line was found between area 1 and area 2. Synapses that crossed this line were predominately negative while synapses that failed to cross it were predominately positive.

There is no doubt that cell assemblies did form. A very detailed statistical study of the allegiance of neurons of cell assemblies was not made because, as will soon be evident, the model still needs improvements, and the statistics of the improved model would be different.

A further significant characteristic needed by the Hebb-Milner theory was evident. Over sequences of 100 or so time steps (perhaps the equivalent of 17 minutes) neurons were observed to change allegiance from one cell assembly to another. In other words, "fractionation" and "recruiting" were observed.

Some evidence was found to indicate that one cell assembly tended to arouse another. However, the tendency was weak. It can be seen that the only possible excitatory synapses between cell assemblies would be those involved with neurons of dubious allegiance. Apparently this was insufficient to allow spontaneous activity in the net. The only activity was caused by the

input signals and the arousal of cell assemblies was completely determined by the input. The theory requires that the preceding central activity (set) be much more influential than the input stimuli, so clearly some changes are needed.

Plans for the Future

After studying the detailed results of this experiment, we arrived at a conjecture as to what should be done next. This conjecture was based on our intuition gained from experience in designing computing machines. We felt that the inhibitory synapses should be separated from the excitatory synapses and should follow different rules. Appendix 1 describes some detailed considerations of the transmission of activity from cell assembly to cell assembly.

We then consulted again with P. W. Milner and learned that he had just produced a further revision of the theory that had just this property of synapses with differing characteristics. His new model appears also to have the characteristic that the cell assemblies would be much more diffuse than in the FM Model described here. This would correspond better to what is expected in the brain and would make a better machine because one cell assembly could directly affect a larger number of others. It is not within the scope of this paper to discuss this new scheme because we have not yet reduced it to our terminology and tested it. However, the work is proceeding.

Summary

The first set of experiments, designed to test parts of the theory advanced in *The Organization of Behavior*, by D. O. Hebb, simulated a network of 69 neurons with a "Discrete Pulse Model." This set of experiments clearly illustrated the diffuse reverberation that is advanced as an explanation of short term memory. There was, however, no tendency for neurons to group into cell assemblies.

The second set of experiments were designed to test P. M. Milner's revision of Hebb's theory with an "F. M. Model" which kept track of the frequency of firing of 512 neurons but ignored the precise timing of individual firings. Cell assemblies formed and exhibited the "fractionation" and "recruiting" required by the theory. The cell assemblies, however, were not able to arouse one another, so this model was too heavily dominated by environment.

A third set of experiments is in progress. It is hoped that this set will get around the next major obstacle in producing a model that will do what the neurophysiological theory requires.

This kind of investigation cannot prove how the brain works. It can, however, show that some models are unworkable and provide clues as to how to revise the models to make them work. Brain theory has progressed to the point where it is not an elementary problem to determine whether a model is workable. Then, when a workable model has been achieved, it may be that a definitive experiment can be devised to test whether or not the workable model corresponds to a detail of the brain.

Appendix 1. The Interaction of Cell Assemblies

Suppose that all synapses within a cell assembly are excitatory and that both excitatory and inhibitory synapses go between all assemblies. Suppose also that the effect of stimulation at a synapse rises suddenly when the preceding (afferent) neuron fires, and then dies out more slowly. For example, a model of this could be a chemical transmitter that was discharged on the stimulated (efferent) neuron and that was destroyed at an exponential rate. Suppose also that the effect of an excitatory synapse fades more slowly than the effect of an inhibitory synapse. In terms of the chemical transmitter, this could mean that two different chemicals were used for inhibition and excitation, and that these were destroyed at different rates. Finally, suppose that the total inhibitory stimulation of an aroused cell assembly on a quiescent cell assembly dominates the total excitatory stimulation.

While a cell assembly is firing actively it will suppress its neighbors. However, when its neurons tire and it begins to falter, the inhibition will drop more rapidly than the excitation. When the level of inhibition drops below the level of excitation, switching will take place.

This sort of interaction between neurons is being built into the third set of experiments.

Appendix 2. Equations Describing the F.M. Model

The structure of the net is given by

$$j = g(h, i)$$

where i is the number of the efferent neuron, j is the number of the afferent neuron, and h is the number of the afferent synapse for the ith neuron. g(h, i) is determined at the beginning of an experiment, and remains constant.

The following quantities for all i, j determine the state of the model at any time t.

Symbol	Number of bits	Description
x(i, t)	4	Frequency of neuron i at time t
$\bar{x}$(i, t)	4	Average frequency of neuron i at time t
d(i, t)	3	Fatigue of neuron i at time t
r(i, j, t)	8	Magnitude of the synapse at time t coupling stimulation from neuron i to neuron j
R(i, t)	8	A function of x(i, t), x(i, t − 1), ...

Initial conditions for the net are given by the values x(i, 0), $\bar{x}$(i, 0), d(i, 0), r(i, j, 0), and R(i, 0).

The quantities S(i, j, t) and x′(i, t) are intermediate results in the calculation. A single time step consists of the successive evaluation of the following formulas:

1) $S(i, j, t) = r(i, j, t)\sqrt{R(i, t)R(j, t)}$

2) $R(i, t + 1) = \left(1 - \dfrac{1}{m}\right)R(i, t) + (x(i, t) - \bar{x}(i, t))^2$

$R(j, t + 1) = \left(1 - \dfrac{1}{m}\right)R(j, t) + (x(j, t) - \bar{x}(j, t))^2$

$m = 32$

3) $S(i, j, t + 1) = \left(1 - \dfrac{1}{m}\right)S(i, j, t) + (x(i, t) - \bar{x}(i, t))$

$\cdot (x(j, t) - \bar{x}(j, t))$

$m = 32$

4) $r(i, j, t + 1) = \dfrac{S(i, j, t + 1)}{\sqrt{R(i, t + 1)R(j, t + 1)}}$

5) We define p(i, t) = j such that r(i, j, t) ⩾ 0, and q(i, t) = j such that r(i, j, t) < 0, including only values of j such that the synapse (i, j) exists.

Then

$x'(i, t + 1) =$

$$k_0 \left[\frac{\sum\limits_{p(i,t+1)} [r(i,j,t+1) + k_1]x(j,t)}{\sum\limits_{p(i,t+1)} r(i,j,t+1) + k_1} - \frac{\sum\limits_{q(i,t+1)} [r(i,j,t+1) - k_1]x(j,t)}{\sum\limits_{q(i,t+1)} r(i,j,t+1) - k_1)} \right]$$

$k_0 = 1.25, \quad k_1 = 1/512$

6) $d(i, t + 1) = f(d(i, t), x'(i, t + 1))$

$x'(i, t+1)$ / $d(i,t)$	0–4	5–9	10–15
0	0	0	2
1	0	1	3
2	1	2	4
3	2	3	5
4	3	4	6
5	4	5	7
6	5	6	7
7	6	7	7

7)
$$x(i, t+1) = \begin{cases} \text{an externally controlled value, when the neuron i is a stimulated receptor} \\ X(x'(i, t+1), d(i, t+1)) \quad \text{when neuron i is not a stimulated receptor} \end{cases}$$

Table of $x(i, t) = X(x'(i, t), d(i, t))$

x' \ a	0	1	2	3	4	5	6	7
0	0	0	0	0	0	0	0	0
1	1	1	1	1	1	1	1	0
2	2	2	2	2	1	1	1	0
3	3	3	3	2	1	1	1	0
4	4	4	4	3	3	2	1	1
5	5	5	4	4	4	3	2	1
6	7	6	6	4	4	3	2	1
7	7	7	7	6	5	4	2	1
8	8	8	8	7	5	5	2	1
9	9	9	9	7	5	5	2	1
10	10	10	9	8	6	5	3	1
11	11	11	11	8	6	5	3	1
12	12	12	12	9	6	6	3	2
13	13	13	12	9	7	6	3	2
14	14	13	13	10	7	6	4	2
15	15	15	13	10	7	6	4	2

8) $\bar{x}(i, t+1) = \left(1 - \dfrac{1}{m}\right)\bar{x}(i, t) + \dfrac{x}{m}(i, t+1)$

References

1. Brink, F. Jr. "Excitation and Conduction in the Neuron" and "Synaptic Mechanisms". pp. 50–120 in *Handbook of Experimental Psychology*, Ed. by S. S. Stevens, John Wiley and Sons, Inc., New York; 1951.

2. Eccles, J. C., *The Neurophysiological Basis of the Mind*, Oxford: The Clarendon Press, 1953.

3. McCulloch, W. S., and Pitts, W., "A Logical Calculus of the Ideas Immanent in Nervous Activities", Bull. Math. Bio-Physics, vol. 5, pp. 115–133, 1943.

4. Kleene, S. C., "Representation of Events in Nerve Nets in Finite Automata", in *Automata Studies, Annuals of Mathematics Studies*, No. 34. Ed. by C. E. Shannon and J. McCarthy. Princeton: Princeton University Press, 1956.

5. Hebb, D. O., *The Organization of Behavior*, New York: John Wiley and Sons, Inc., 1949.

6. Ref. (5), p. 62.

7. Ref. 2, and notice that Hebb's postulate (Ref. 6) is not necessarily related closely to Eccles "post-tetanic potentiation". On p. 196 Eccles shows the effect of a million volleys (Fig. 61, 36 minute curve) and this is much more severe than is relevant for the present discussion.

8. Craik, K. J. W., *The Nature of Explanation*, Cambridge: The University Press, 1952.

9. Farley, D. G., and Clark, W. A., Proceedings of the Western Joint Computer Conference, 1955.

7
Introduction

(1958)
John von Neumann

The Computer and the Brain, New Haven: Yale University Press, pp. 66–82

John von Neumann was one of the great figures in science in the first part of the twentieth century. He did major work in quantum mechanics and mathematical physics. He was active in the Manhattan Project, and he played an important role in the development of the digital computer, first as a mathematical consultant to the ENIAC project, the first electronic digital computer, at the Moore School of Engineering at the University of Pennsylvania, and second as one of the designers of an early computer at the Institute for Advanced Studies at Princeton.

A draft report that von Neumann wrote in 1945 (von Neumann, 1982) has become famous as the first place where the idea of a stored program, which resided in the computer's memory along with the data it was to operate on, was clearly stated. Interestingly, von Neumann discusses McCulloch and Pitts in several places (von Neumann, 1982):

4.2. It is worth mentioning that the neurons of the higher animals are (relay-like) elements in the above sense. They have all-or-none character, that is two states: Quiescent and excited.... An excited neuron emits the standard stimulus along many lines (axons). Such a line can, however, be connected in two different ways to the next neuron: First, in an *excitatory synapsis*, so that the stimulus causes excitation of that neuron. Second, in an *inhibitory synapsis*, so that the stimulus absolutely prevents the excitation of that neuron by any stimulus on any other (excitatory) synapsis. The neuron also has a definite reaction time, between the reception of a stimulus and the emission of the stimuli caused by it, the *synaptic delay*.

Following W. Pitts and W. S. McCulloch ... we ignore the more complicated aspects of neuron functioning: Thresholds, temporal summation, relative inhibition, ... etc.... It can easily be seen that these simplified neuron functions can be imitated by telegraph relays or by vacuum tubes.

Clearly the potential analogy between digital computers and the brain was present in von Neumann's mind, from the earliest days of the device.

Just before von Neumann's death in 1957 he prepared some material for the Silliman Lectures at Yale that was later published as a book, *The Computer and the Brain*. Much of the book is review material that is more familiar to us now than it would have been to von Neumann's audience in 1957. However, there are a number of gems scattered throughout the book.

A few of the points von Neumann makes in our short excerpt introduce themes that keep recurring throughout the later papers presented in this volume.

He mentions the importance of memory in biological nervous systems (as in electronic ones) and comments that the exact means or mode of storage is not known, which is still true, though we are closing in. He concludes with an order of magnitude calculation that human memory capacity must be huge. He suggests a capacity on the order of 10^{20} bits. The only recent paper we know of to consider this problem, which is of importance to the study of the mind in all its aspects, is by Tom Landauer (1986), who concludes, using primarily psychological arguments from three separate directions, that long term memory capacity is around a billion (10^9) bits. In any case, it is enormous, and von Neumann suggests some possible ways that biological memory could work. He suggests that 'active' implementations, where storage elements are constantly electrically active (such as reverberating loops of activity), are too wasteful of the computing elements. He gives as an alternative a sketch for a simple 'strengthening by use' synaptic learning rule to form a permanent physical change in the brain. The idea of synaptic modification as the basis for memory has been prevalent in neuroscience since the nineteenth century, and is generally accepted by neuroscientists, though details differ greatly from theory to theory.

Von Neumann makes an important argument in the short section entitled "The logical structure of the nervous system." He points out that neurons have very low precision, not more than a few bits. Certainly, they are far less precise than the decimal precision of 10 or more places easily available on an electronic computer. He then points out that this lack of accuracy has major effects on the *kinds* of computation that the nervous system can do. Roundoff errors will make it impossible for the 'arithmetic' performed by neurons to be very deep; that is, operations cannot be strung together in long chains—only a few operations can be carried out in succession before errors overwhelm the calculation. He suggests that a more reliable mode of operation would use 'logical' (for example, relational) calculations, which should be more resistant to roundoff errors than arithmetic.

The observation that the *kinds* of computations performed by the nervous system must be closely bound to its physical realization must be true. The brain has to do what it does, with what it has got to do it with. This important point is sometimes forgotten when analogies are drawn between computers and the nervous system, even today.

References

T. Landauer (1986), "How much do people remember? Some estimates of the quantity of learned information in long-term memory," *Cognitive Science* 10:477–493.

J. von Neumann (1945/1982), "First draft of a report on the EDVAC," *The Origins of Digital Computers: Selected Papers*, B. Randall (Ed.), Berlin: Springer (3rd edition).

(1958)
John von Neumann

The Computer and the Brain, New Haven: Yale University Press, pp. 66–82

Analogies with Artificial Computing Machines

Lastly, I would like to mention that systems of nerve cells, which stimulate each other in various possible cyclical ways, also constitute memories. These would be memories made up of active elements (nerve cells). In our computing machine technology such memories are in frequent and significant use; in fact, these were actually the first ones to be introduced. In vacuum-tube machines the "flip-flops," i.e. pairs of vacuum tubes that are mutually gating and controlling each other, represent this type. Transistor technology, as well as practically every other form of high-speed electronic technology, permit and indeed call for the use of flip-floplike subassemblies, and these can be used as memory elements in the same way that the flip-flops were in the early vacuum-tube computing machines.

The Underlying Componentry of the Memory Need Not Be the Same as That of the Basic Active Organs

It must be noted, however, that it is a priori unlikely that the nervous system should use such devices as the main vehicles for its memory requirements; such memories, most characteristically designated as "memories made up from basic active organs," are, in every sense that matters, extremely expensive. Modern computing machine technology started out with such arrangements—thus the first large-scale vacuum tube computing machine, the ENIAC, relied for its primary (i.e. fastest and most directly available) memory on flip-flops exclusively. However, the ENIAC had a very large size (22,000 vacuum tubes) and by present-day standards a very small, primary memory (consisting of a few dozens of ten-digit decimal numbers only). Note that the latter amounts to something like a few hundred bits—certainly less than 10^3. In present-day computing machines the proper balance between machine size and memory capacity (cf. above) is generally held to lie around something like 10^4 basic active elements, and a memory capacity of 10^5 to 10^6 bits. This is achieved by using forms of memory which are technologically entirely different from the basic active organs of the machine. Thus a vacuum tube or

transistor machine might have a memory residing in an electrostatic system (a cathode ray tube), or in suitably arranged large aggregates of ferromagnetic cores, etc. I will not attempt a complete classification here, since other important forms of memory, like the acoustic delay type, the ferro-electric type, and the magnetostrictive delay type (this list could, indeed, be increased), do not fit quite so easily into such classifications. I just want to point out that the componentry used in the memory may be entirely different from the one that underlies the basic active organs.

These aspects of the matter seem to be very important for the understanding of the structure of the nervous system, and they would seem to be as yet predominantly unanswered. We know the basic active organs of the nervous system (the nerve cells). There is every reason to believe that a very large-capacity memory is associated with this system. We do most emphatically *not* know what type of physical entities are the basic components for the memory in question.

Digital and Analog Parts in the Nervous System

Having pointed out in the above the deep, fundamental, and wide-open problems connected with the memory aspect of the nervous system, it would seem best to go on to other questions. However, there is one more, minor aspect of the unknown memory subassembly in the nervous system, about which a few words ought to be said. These refer to the relationship between the analog and the digital (or the "mixed") parts of the nervous system. I will devote to these, in what follows, a brief and incomplete additional discussion, after which I will go on to the questions *not* related to the memory.

The observation I wish to make is this: processes which go through the nervous system may, as I pointed out before, change their character from digital to analog, and back to digital, etc., repeatedly. Nerve pulses, i.e. the digital part of the mechanism, may control a particular stage of such a process, e.g. the contraction of a specific muscle or the secretion of a specific chemical. This phenomenon is one belonging to the analog class, but it may be the origin of a train of nerve pulses

Copyright © 1958 Yale University Press.

which are due to its being sensed by suitable inner receptors. When such nerve pulses are being generated, we are back in the digital line of progression again. As mentioned above, such changes from a digital process to an analog one, and back again to a digital one, may alternate several times. Thus the nerve-pulse part of the system, which is digital, and the one involving chemical changes or mechanical dislocations due to muscular contractions, which is of the analog type, may, by alternating with each other, give any particular process a mixed character.

Role of the Genetic Mechanism in the Above Context

Now, in this context, the genetic phenomena play an especially typical role. The genes themselves are clearly parts of a digital system of components. Their effects, however, consist of stimulating the formation of specific chemicals, namely of definite enzymes that are characteristics of the gene involved, and, therefore, belong in the analog area. Thus, in this domain, a particular specific instance of the alternation between analog and digital is exhibited, i.e. this is a member of a broader class, to which I referred as such above in a more general way.

Codes and Their Role in the Control of the Functioning of a Machine

Let me now pass on to the questions involving other aspects than those of memory. By this I mean certain principles of organizing logical orders which are of considerable importance in the functioning of any complicated automaton.

First of all, let me introduce a term which is needed in the present context. A system of logical instructions that an automaton can carry out and which causes the automaton to perform some organized task is called a *code*. By logical orders, I mean things like nerve pulses appearing on the appropriate axons, in fact anything that induces a digital logical system, like the nervous system, to function in a reproducible, purposive manner.

The Concept of a Complete Code

Now, in talking about codes, the following distinction becomes immediately prominent. A code may be *complete*—i.e., to use the terminology of nerve pulses, one may have specified the sequence in which these impulses appear and the axons on which they appear. This will then, of course, define completely a specific behavior of the nervous system, or, in the above comparison, of the corresponding artificial automaton involved. In computing machines, such complete codes

are sets of orders, given with all necessary specifications. If the machine is to solve a specific problem by calculation, it will have to be controlled by a complete code in this sense. The use of a modern computing machine is based on the user's ability to develop and formulate the necessary complete codes for any given problem that the machine is supposed to solve.

The Concept of a Short Code

In contrast to the complete codes, there exists another category of codes best designated as *short codes*. These are based on the following idea.

The English logician A. M. Turing showed in 1937 (and various computing machine experts have put this into practice since then in various particular ways) that it is possible to develop code instruction systems for a computing machine which cause it to behave as if it were another, specified, computing machine. Such systems of instructions which make one machine *imitate* the behavior of another are known as *short codes*. Let me go into a little more detail in the typical questions of the use and development of such short codes.

A computing machine is controlled, as I pointed out above, by codes, sequences of symbols—usually binary symbols—i.e. by strings of bits. In any set of instructions that govern the use of a particular computing machine it must be made clear which strings of bits are orders and what they are supposed to cause the machine to do.

For two different machines, these *meaningful* strings of bits need not be the same ones and, in any case, their respective effects in causing their corresponding machines to operate may well be entirely different. Thus, if a machine is provided with a set of orders that are peculiar to another machine, these will presumably be, in terms of the first machine, at least in part, *nonsense*, i.e. strings of bits which do not necessarily all belong to the family of the *meaningful* ones (in terms of the first-mentioned machine), or which, when "obeyed" by the first-mentioned machine, would cause it to take actions which are not part of the underlying organized plan toward the solution of a problem, the solution of which is intended, and, generally speaking, would not cause the first-mentioned machine to behave in a purposive way toward the solution of a visualized, organized task, i.e. the solution of a specific and desired problem.

The Function of a Short Code

A code, which according to Turing's schema is supposed to make one machine behave as if it were another specific machine (which is supposed to make the former *imitate* the latter) must do the following things. It must contain, in terms that the machine will understand and (purposively obey), instructions (further detailed parts of the code) that will cause the machine to examine every order it gets and determine whether this order has the structure appropriate to an order of the second machine. It must then contain, in terms of the order system of the first machine, sufficient orders to make the machine cause the actions to be taken that the second machine would have taken under the influence of the order in question.

The important result of Turing's is that in this way the first machine can be caused to imitate the behavior of *any* other machine. The order structure which it is thus caused to follow may be entirely different from that one characteristic of the first machine which is truly involved. Thus the order structure referred to may actually deal with orders of a much more complex character than those which are characteristic of the first machine: every one of these orders of the secondary machine may involve the performing of several operations by the first-mentioned machine. It may involve complicated, iterative processes, multiple actions of any kind whatsoever; generally speaking, anything that the first machine can do in any length of time and under the control of all possible order systems of any degree of complexity may now be done as if only "elementary" actions—basic, uncompounded, primitive orders—were involved.

The reason for calling such a secondary code a *short code* is, by the way, historical: these short codes were developed as an aid to coding, i.e. they resulted from the desire to be able to code more briefly for a machine than its own natural order system would allow, treating it as if it were a different machine with a more convenient, fuller order system which would allow simpler, less circumstantial and more straightforward coding.

The Logical Structure of the Nervous System

At this point, the discussion is best redirected toward another complex of questions. These are, as I pointed out previously, not connected with the problems of the memory or with the questions of complete and short codes just considered. They relate to the respective roles of logics and arithmetics in the functioning of any complicated automaton, and, specifically, of the nervous system.

Importance of the Numerical Procedures

The point involved here, one of considerable importance, is this. Any artificial automaton that has been constructed for human use, and specifically for the control of complicated processes, normally possesses a purely logical part and an arithmetical part, i.e. a part in which arithmetical processes play no role, and one in which they are of importance. This is due to the fact that it is, with our habits of thought and of expressing thought, very difficult to express any truly complicated situation without having recourse to formulae and numbers.

Thus an automaton which is to control problems of these types—constancy of temperature, or of certain pressures, or of chemical isostasy in the human body—will, if a human designer has to formulate its task, have that task defined in terms of numerical equalities or inequalities.

Interaction of Numerical Procedures with Logic

On the other hand, there may be portions of this task which can be formulated without reference to numerical relationships, i.e. in purely logical terms. Thus certain qualitative principles involving physiological response or nonresponse can be stated without recourse to numbers by merely stating qualitatively under what combinations of circumstances certain events are to take place and under what combinations they are not desired.

Reasons for Expecting High Precision Requirements

These remarks show that the nervous system, when viewed as an automaton, must definitely have an arithmetical as well as a logical part, and that the needs of arithmetics in it are just as important as those of logics. This means that we are again dealing with a computing machine in the proper sense and that a discussion in terms of the concepts familiar in computing machine theory is in order.

In view of this, the following question immediately presents itself: when looking at the nervous system as at a computing machine, with what precision is the arithmetical part to be expected to function?

This question is particularly crucial for the following reason: all experience with computing machines shows that if a computing machine has to handle as complicated arithmetical tasks as the nervous system obviously must, facilities for rather high levels of precision must be provided. The reason is that calculations are likely to be long, and in the course of long calculations not only do errors add up but also those committed early in the calculation are amplified by the latter parts of it; therefore, considerably higher precision is needed than the physical nature of the problem would by itself appear to require.

Thus one would expect that the arithmetical part

of the nervous system exists and, when viewed as a computing machine, must operate with considerable precision. In the familiar artificial computing machines and under the conditions of complexity here involved, ten- or twelve-decimal precision would not be an exaggeration.

This conclusion was well worth working out just because of, rather than in spite of, its absolute implausibility.

Nature of the System of Notations Employed: Not Digital but Statistical

As pointed out before, we know a certain amount about how the nervous system transmits numerical data. They are usually transmitted by periodic or nearly periodic trains of pulses. An intensive stimulus on a receptor will cause the latter to respond each time soon after the limit of absolute refractoriness has been underpassed. A weaker stimulus will cause the receptor to respond also in periodic or nearly periodic way, but with a somewhat lower frequency, since now not only the limit of absolute refractoriness but even a limit of a certain relative refractoriness will have to be underpassed before each next response becomes possible. Consequently, intensities of quantitative stimuli are rendered by periodic or nearly periodic pulse trains, the frequency always being a monotone function of the intensity of the stimulus. This is a sort of frequency-modulated system of signaling; intensities are translated into frequencies. This has been directly observed in the case of certain fibers of the optic nerve and also in nerves that transmit information relative to (important) pressures.

It is noteworthy that the frequency in question is not directly equal to any intensity of stimulus, but rather that it is a montone function of the latter. This permits the introduction of all kinds of scale effects and expressions of precision in terms that are conveniently and favorably dependent on the scales that arise.

It should be noted that the frequencies in question usually lie between 50 and 200 pulses per second.

Clearly, under these conditions, precisions like the ones mentioned above (10 to 12 decimals!) are altogether out of question. The nervous system is a computing machine which manages to do its exceedingly complicated work on a rather low level of precision: according to the above, only precision levels of 2 to 3 decimals are possible. This fact must be emphasized again and again because no known computing machine can operate reliably and significantly on such a low precision level.

Another thing should also be noted. The system described above leads not only to a low level of precision, but also to a rather high level of reliability. Indeed, clearly, if in a digital system of notations a single pulse is missing, absolute perversion of meaning, i.e. nonsense, may result. Clearly, on the other hand, if in a scheme of the above-described type a single pulse is lost, or even several pulses at lost—or unnecessarily, mistakenly, inserted—the relevant frequency, i.e. the meaning of the message, is only inessentially distorted.

Now, a question arises that has to be answered significantly: what essential inferences about the arithmetical and logical structure of the computing machine that the nervous system represents can be drawn from these apparently somewhat conflicting observations?

Arithmetical Deterioration. Roles of Arithmetical and Logical Depths

To anyone who has studied the deterioration of precision in the course of a long calculation, the answer is clear. This deterioration is due, as pointed out before, to the *accumulation* of errors by superposition, and even more by the *amplification* of errors committed early in the calculation, by the manipulations in the subsequent parts of the calculation; i.e. it is due to the considerable number of arithmetical operations that have to be performed in series, or in other words to the great *arithmetical depth* of the scheme.

The fact that there are many operations to be performed in series is, of course, just as well a characteristic of the *logical* structure of the scheme as of its arithmetical structure. It is, therefore, proper to say that all of these deterioration-of-precision phenomena are due to the great *logical depth* of the schemes one is dealing with here.

Arithmetical Precision or Logical Reliability, Alternatives

It should also be noted that the message-system used in the nervous system, as described in the above, is of an essentially *statistical* character. In other words, what matters are not the precise positions of definite markers, digits, but the statistical characteristics of their occurrence, i.e. frequencies of periodic or nearly periodic pulse-trains, etc.

Thus the nervous system appears to be using radically different system of notation from the ones we are familiar with in ordinary arithmetics and mathematics: instead of the precise systems of markers where the position—and presence or absence—of every marker counts decisively in determining the meaning of the message, we have here a system of notations in which the meaning is conveyed by the *statistical* properties of the message. We have seen how this leads to a lower level of arithmetical precision but to a higher level of logical reliability: a deterioration in arithmetics has been traded for an improvement in logics.

Other Statistical Traits of the Message System That Could Be Used

This context now calls clearly for the asking of one more question. In the above, the frequencies of certain periodic or nearly periodic pulse-trains carried the *message*, i.e., the *information*. These were distinctly *statistical* traits of the message. Are there any other statistical properties which could similarly contribute as vehicles in the transmission of information?

So far, the only property of the message that was used to transmit information was its frequency in terms of pulses per second, it being understood that the message was a periodic or nearly periodic train of pulses.

Clearly, other traits of the (statistical) message could also be used: indeed, the frequency referred to is a property of a single train of pulses whereas every one of the relevant nerves consists of a large number of fibers, each of which transmits numerous trains of pulses. It is, therefore, perfectly plausible that certain (statistical) relationships between such trains of pulses should also transmit information. In this connection it is natural to think of various correction-coefficients, and the like.

The Language of the Brain Not the Language of Mathematics

Pursuing this subject further gets us necessarily into questions of *language*. As pointed out, the nervous system is based on two types of communications: those which do not involve arithmetical formalisms, and those which do, i.e. communications of orders (logical ones) and communications of numbers (arithmetical ones). The former may be described as language proper, the latter as mathematics.

It is only proper to realize that language is largely a historical accident. The basic human languages are traditionally transmitted to us in various forms, but their very multiplicity proves that there is nothing absolute and necessary about them. Just as languages like Greek or Sanskrit are historical facts and not absolute logical necessities, it is only reasonable to assume that logics and mathematics are similarly historical, accidental forms of expression. They may have essential variants, i.e. they may exist in other forms than the ones to which we are accustomed. Indeed, the nature of the central nervous system and of the message systems that it transmits indicate positively that this is

so. We have now accumulated sufficient evidence to see that whatever language the central nervous system is using, it is characterized by less logical and arithmetical depth than what we are normally used to. The following is an obvious example of this: the retina of the human eye performs a considerable reorganization of the visual image as perceived by the eye. Now this reorganization is effected on the retina, or to be more precise, at the point of entry of the optic nerve by means of three successive synapses only, i.e. in terms of three consecutive logical steps. The statistical character of the message system used in the arithmetics of the central nervous system and its low precision also indicate that the degeneration of precision, described earlier, cannot proceed very far in the message systems involved. Consequently, there exist here different logical structures from the ones we are ordinarily used to in logics and mathematics. They are, as pointed out before, characterized by less logical and arithmetical depth than we are used to under otherwise similar circumstances. Thus logics and mathematics in the central nervous system, when viewed as languages, must structurally be essentially different from those languages to which our common experience refers.

It also ought to be noted that the language here involved may well correspond to a short code in the sense described earlier, rather than to a complete code: when we talk mathematics, we may be discussing a *secondary* language, built on the *primary* language truly used by the central nervous system. Thus the outward forms of *our* mathematics are not absolutely relevant from the point of view of evaluating what the mathematical or logical language *truly* used by the central nervous system is. However, the above remarks about reliability and logical and arithmetical depth prove that whatever the system is, it cannot fail to differ considerably from what we consciously and explicitly consider as mathematics.

8
Introduction

(1958)
F. Rosenblatt

**The perceptron: a probabilistic model
for information storage and organization in the brain**
Psychological Review 65:386–408

The perceptron created a sensation when it was first described. It was the first precisely specified, computationally oriented neural network, and it made a major impact on a number of areas simultaneously. Rosenblatt was originally a psychologist, and the things that the perceptron was computing were things that a psychologist would think were important. The fact that it was a learning machine that was potentially capable of complex adaptive behavior made it irresistibly attractive to engineers. There were immense practical benefits to be gained from a real learning machine. The device was mathematically complex as well. It and variants of it were challenging to analyze, and it led to some very important general insights into the powers and limitations of learning machines. Much of the later work on perceptrons and successors was done by engineers and physicists, a situation still true today in research on neural networks.

It must be acknowledged that this important paper is hard to read. Perceptrons are described in a number of variations, each labeled with its own arbitrary name. The analysis does not proceed easily. There are so many options, variables, and learning rules introduced that it becomes confusing. The figures are often difficult to follow.

However, if we compare the perceptron with its successors, we see that the machinery Rosenblatt proposed is still used today, in one form or another. Rosenblatt used as his basic architecture one layer of model neurons projecting to another layer of model neurons by way of parallel bundles of connections. He started with a sensory surface— a "retina"—projecting to higher areas. The retina, the first layer, connected to a second layer, an "association area" with random, but localized, connectivity in the simplest perceptron. That is, a number of cells in a region of the retina projected to a single A-unit (Association Unit) in the higher layer. The association unit layer was reciprocally connected to a third layer of R-units (Response Units), that is, the A-unit connected to the R-unit and vice versa. The activation of the appropriate R-unit for a given input pattern or class of input patterns was the goal of the operation of the perceptron. It was deemed undesirable that more than one R-unit be on at a time. To prevent this, a set of reciprocal inhibitory connections was used, so that an R-unit inhibited all the association units that did *not* connect to it. Therefore, when an R-unit was activated, it indirectly suppressed competitors. A system with an extreme form of this behavior is today sometimes called a "Winner-Take-All" system and appears in many current network models. (See Feldman and Ballard, paper 29). The simple perceptron, as described here, was a three-layer device (see figure 2A), but Rosenblatt also considered systems with two layers of association units, a four-layer system (see figure 1).

Probably even more important than the anatomy was what Rosenblatt chose to compute. It was clear from the introductory section that he felt that what earlier theorists of the nervous system had done was fundamentally flawed. He was particularly harsh on theorists who made the brain compute logic functions, which he said amounted to "logical contrivances" (p. 387). He reserved praise for those who were aware that the noise and randomness present in the nervous system were not merely inconveniences because of poor design and construction, but were essential to the kinds of computations brains performed. The language of "symbolic logic and Boolean analysis" was not suitable for the brain.

Rosenblatt felt the appropriate way to view brain operation was as a learning associator. The goal was to couple classification responses to stimuli. Since the environment was noisy and variable, and internal connections in the nervous system were not completely prespecified, it was necessary for the system to learn to make the associations by rewiring itself. The serious technical problem then became not logical realizability, but separability; i.e., whether or not it is possible to separate the input stimuli that need to be discriminated and to make the same response to stimuli that need to be classified together. This meant that the structure of the stimuli had to be investigated. If there are clusters of events in the world that belong together, and if the stimuli coding these events are similar in the well defined sense that they tend to activate the same units and not be "too far" apart, then it is possible to classify them together. The system is very dependent for operation on the internal representation of the world. Perceptrons were found to be rather poor at discriminating arbitrary random patterns, but demonstrably quite good at separating items in a "differentiated environment" where "each response is associated to a distinct class of mutually correlated ... stimuli" (p. 405).

Learning in Rosenblatt's first paper is not analyzed with the depth that was found in later work (see Block, paper 11). The learning rules were largely of a simple reinforcement kind. After a brief preliminary period one of the R-units, randomly determined, would become active. The active R-unit would supress A-units not connected to it. If an A-unit was active when the stimulus was presented, its activity was increased. Then, after learning, if the stimulus appeared again, the A-units it activated would show stronger activity, would drive the R-units harder, and, consequently, the appropriate R-unit would be more likely to respond correctly.

This simplest learning rule was "self-organizing," because a stimulus and a random R-unit became coupled. Rosenblatt also mentioned systems where the responses were "forced." This meant activating the appropriate R-unit and then modifying the activities of the A-units. Rosenblatt showed that the perceptron was capable of learning random associations, but that there was a limit on the capacity, in that if too many random stimuli were associated with a response, accuracy decreased. If, on the other hand, the system was learning structured sets of stimuli, so that members of a class activated similar sets of A-units, then the capacity of the system could increase, and accuracy could potentially become asymptotically high.

It was in this kind of "concept forming" mode that perceptrons displayed their greatest theoretical weaknesses. As much subsequent work showed, it was *very* hard to arrange connections so that stimuli that belong together according to some rule would activate the same units (see Minsky and Papert, paper 13). Spatial arrangements of cells on a retina would not work for many interesting equivalence classes. And simple overlap clustering, as was soon discovered, was not adequate either. But even with severe limitations, the perceptron displayed the invaluable ability to generalize; that is, it could respond appropriately to patterns it had never seen because they were similar to patterns it had seen.

Rosenblatt spent some time discussing the ability of the perceptron to work in the presence of noise and with damaged or missing connections or units. He points out in this context that the memory is distributed, that is, resident in many connection strengths, and therefore resistant to damage.

By the standards of papers only a few years later, the analysis in this first paper is sketchy. There was no proof, or even awareness, of the famous Perceptron Convergence Theorem, which proved that perceptrons could learn many classifications. There were only a few statistical calculations indicating plausibility of learning, but overlooking the detailed limitations on learnability that proved so important later.

Some of the excitement of the discovery of the perceptron still comes through this early work. Here was a brain model that could *do* something. The potential for progress seemed very great. As Rosenblatt put it, "The question may well be raised at this point of where the perceptron's capabilities actually stop.... the system described is sufficient for pattern recognition, associative learning, and such cognitive sets as are necessary for selective attention and selective recall. The system appears to be potentially capable of temporal pattern recognition.... with proper reinforcement it will be capable of trial and error learning, and can learn to emit ordered sequences of responses ..." (p. 404).

Although these claims were ambitious, they were correct in that variants of the basic perceptron architecture were capable of performing them, though perhaps not as easily as it first appeared. It is important to realize that Rosenblatt was quite aware of some more serious computational limitations on the perceptron that he felt might prove very difficult to solve. Specifically, he mentioned "relative judgments" and "symbolic behavior." He mentioned that the perceptron acted in many respects like a brain damaged patient, being literal, inflexible, and unable to handle abstractions.

In fact, these are exactly areas where current neural networks still are inadequate. At present, our networks, even after thirty years of progress, still act "brain damaged." It is an open question as to whether severe modifications of network theory will have to be made to handle these highest cognitive functions.

In spite of the preliminary level of analysis in this early paper, it was clear that Rosenblatt had a good idea of both the strengths and the weaknesses of his approach. If the model had been allowed a less controversial infancy, we might have made more progress than we have done up to now.

As one final point, Rosenblatt uses the word "connectionist" as a descriptor of the perceptron, just as Hebb did for cell assemblies.

(1958)
F. Rosenblatt

The perceptron: a probabilistic model
for information storage and organization in the brain
Psychological Review 65:386–408

If we are eventually to understand the capability of higher organisms for perceptual recognition, generalization, recall, and thinking, we must first have answers to three fundamental questions:

1. How is information about the physical world sensed, or detected, by the biological system?
2. In what form is information stored, or remembered?
3. How does information contained in storage, or in memory, influence recognition and behavior?

The first of these questions is in the province of sensory physiology, and is the only one for which appreciable understanding has been achieved. This article will be concerned primarily with the second and third questions, which are still subject to a vast amount of speculation, and where the few relevant facts currently supplied by neurophysiology have not yet been integrated into an acceptable theory.

With regard to the second question, two alternative positions have been maintained. The first suggests that storage of sensory information is in the form of coded representations or images, with some sort of one-to-one mapping between the sensory stimulus

The development of this theory has been carried out at the Cornell Aeronautical Laboratory, Inc., under the sponsorship of the Office of Naval Research, Contract Nonr-2381(00). This article is primarily an adaptation of material reported in Ref. 15, which constitutes the first full report on the program.

and the stored pattern. According to this hypothesis, if one understood the code or "wiring diagram" of the nervous system, one should, in principle, be able to discover exactly what an organism remembers by reconstructing the original sensory patterns from the "memory traces" which they have left, much as we might develop a photographic negative, or translate the pattern of electrical charges in the "memory" of a digital computer. This hypothesis is appealing in its simplicity and ready intelligibility, and a large family of theoretical brain models has been developed around the idea of a coded, representational memory (**2, 3, 9, 14**). The alternative approach, which stems from the tradition of British empiricism, hazards the guess that the images of stimuli may never really be recorded at all, and that the central nervous system simply acts as an intricate switching network, where retention takes the form of new connections, or pathways, between centers of activity. In many of the more recent developments of this position (Hebb's "cell assembly," and Hull's "cortical anticipatory goal response," for example) the "responses" which are associated to stimuli may be entirely contained within the CNS itself. In this case the response represents an "idea" rather than an action. The important feature of this approach is that there is never any simple mapping of the stimulus into memory, according to some code which would permit its later reconstruction. Whatever in-

Copyright 1958 by the American Psychological Association, Inc. Reprinted by permission of the publisher and author.

formation is retained must somehow be stored as a *preference for a particular response;* i.e., the information is contained in *connections* or *associations* rather than topographic representations. (The term *response,* for the remainder of this presentation, should be understood to mean any distinguishable state of the organism, which may or may not involve externally detectable muscular activity. The activation of some nucleus of cells in the central nervous system, for example, can constitute a response, according to this definition.)

Corresponding to these two positions on the method of information retention, there exist two hypotheses with regard to the third question, the manner in which stored information exerts its influence on current activity. The "coded memory theorists" are forced to conclude that recognition of any stimulus involves the matching or systematic comparison of the contents of storage with incoming sensory patterns, in order to determine whether the current stimulus has been seen before, and to determine the appropriate response from the organism. The theorists in the empiricist tradition, on the other hand, have essentially combined the answer to the third question with their answer to the second: since the stored information takes the form of new connections, or transmission channels in the nervous system (or the creation of conditions which are functionally equivalent to new connections), it follows that the new stimuli will make use of these new pathways which have been created, automatically activating the appropriate response without requiring any separate process for their recognition or identification.

The theory to be presented here takes the empiricist, or "connectionist" position with regard to these ques-

tions. The theory has been developed for a hypothetical nervous system, or machine, called a *perceptron.* The perceptron is designed to illustrate some of the fundamental properties of intelligent systems in general, without becoming too deeply enmeshed in the special, and frequently unknown, conditions which hold for particular biological organisms. The analogy between the perceptron and biological systems should be readily apparent to the reader.

During the last few decades, the development of symbolic logic, digital computers, and switching theory has impressed many theorists with the functional similarity between a neuron and the simple on-off units of which computers are constructed, and has provided the analytical methods necessary for representing highly complex logical functions in terms of such elements. The result has been a profusion of brain models which amount simply to logical contrivances for performing particular algorithms (representing "recall," stimulus comparison, transformation, and various kinds of analysis) in response to sequences of stimuli—e.g., Rashevsky (14), McCulloch (10), McCulloch & Pitts (11), Culbertson (2), Kleene (8), and Minsky (13). A relatively small number of theorists, like Ashby (1) and von Neumann (17, 18), have been concerned with the problems of how an imperfect neural network, containing many random connections, can be made to perform reliably those functions which might be represented by idealized wiring diagrams. Unfortunately, the language of symbolic logic and Boolean algebra is less well suited for such investigations. The need for a suitable language for the mathematical analysis of events in systems where only the gross organization can be characterized, and the precise structure is unknown, has led the author to formulate the current model in terms of probability theory rather than symbolic logic.

The theorists referred to above were chiefly concerned with the question of how such functions as perception and recall might be achieved by a deterministic physical system of any sort, rather than how this is actually done by the brain. The models which have been produced all fail in some important respects (absence of equipotentiality, lack of neuroeconomy, excessive specificity of connections and synchronization requirements, unrealistic specificity of stimuli sufficient for cell firing, postulation of variables or functional features with no known neurological correlates, etc.) to correspond to a biological system. The proponents of this line of approach have maintained that, once it has been shown how a physical system of any variety might be made to perceive and recognize stimuli, or perform other brainlike functions, it would require only a refinement or modification of existing principles to understand the working of a more realistic nervous system, and to eliminate the shortcomings mentioned above. The writer takes the position, on the other hand, that these shortcomings are such that a mere refinement or improvement of the principles already suggested can never account for biological intelligence; a *difference in principle* is clearly indicated. The theory of statistical separability (Cf. **15**), which is to be summarized here, appears to offer a solution in principle to all of these difficulties.

Those theorists—Hebb (**7**), Milner (**12**), Eccles (**4**), Hayek (**6**)—who have been more directly concerned with the biological nervous system and its activity in a natural environment, rather than with formally anal-

ogous machines, have generally been less exact in their formulations and far from rigorous in their analysis, so that it is frequently hard to assess whether or not the systems that they describe could actually work in a realistic nervous system, and what the necessary and sufficient conditions might be. Here again, the lack of an analytic language comparable in proficiency to the Boolean algebra of the network analysts has been one of the main obstacles. The contributions of this group should perhaps be considered as suggestions of what to look for and investigate, rather than as finished theoretical systems in their own right. Seen from this viewpoint, the most suggestive work, from the standpoint of the following theory, is that of Hebb and Hayek.

The position, elaborated by Hebb (**7**), Hayek (**6**), Uttley (**16**), and Ashby (**1**), in particular, upon which the theory of the perceptron is based, can be summarized by the following assumptions:

1. The physical connections of the nervous system which are involved in learning and recognition are not identical from one organism to another. At birth, the construction of the most important networks is largely random, subject to a minimum number of genetic constraints.

2. The original system of connected cells is capable of a certain amount of plasticity; after a period of neural activity, the probability that a stimulus applied to one set of cells will cause a response in some other set is likely to change, due to some relatively long-lasting changes in the neurons themselves.

3. Through exposure to a large sample of stimuli, those which are most "similar" (in some sense which must be defined in terms of the particular physical system) will tend

to form pathways to the same sets of responding cells. Those which are markedly "dissimilar" will tend to develop connections to different sets of responding cells.

4. The application of positive and/ or negative reinforcement (or stimuli which serve this function) may facilitate or hinder whatever formation of connections is currently in progress.

5. *Similarity*, in such a system, is represented at some level of the nervous system by a tendency of similar stimuli to activate the same sets of cells. Similarity is not a necessary attribute of particular formal or geometrical classes of stimuli, but depends on the physical organization of the perceiving system, an organization which evolves through interaction with a given environment. The structure of the system, as well as the ecology of the stimulus-environment, will affect, and will largely determine, the classes of "things" into which the perceptual world is divided.

THE ORGANIZATION OF A PERCEPTRON

The organization of a typical photo-perceptron (a perceptron responding to optical patterns as stimuli) is shown in Fig. 1. The rules of its organization are as follows:

1. Stimuli impinge on a retina of sensory units (S-points), which are assumed to respond on an all-or-nothing basis, in some models, or with a pulse amplitude or frequency proportional to the stimulus intensity, in other models. In the models considered here, an all-or-nothing response will be assumed.

2. Impulses are transmitted to a set of association cells (A-units) in a "projection area" (A_I). This projection area may be omitted in some models, where the retina is connected directly to the association area (A_{II}).

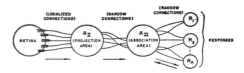

FIG. 1. Organization of a perceptron.

The cells in the projection area each receive a number of connections from the sensory points. The set of S-points transmitting impulses to a particular A-unit will be called the *origin points* of that A-unit. These origin points may be either *excitatory* or *inhibitory* in their effect on the A-unit. If the algebraic sum of excitatory and inhibitory impulse intensities is equal to or greater than the threshold (θ) of the A-unit, then the A-unit fires, again on an all-or-nothing basis (or, in some models, which will not be considered here, with a frequency which depends on the net value of the impulses received). The origin points of the A-units in the projection area tend to be clustered or focalized, about some central point, corresponding to each A-unit. The number of origin points falls off exponentially as the retinal distance from the central point for the A-unit in question increases. (Such a distribution seems to be supported by physiological evidence, and serves an important functional purpose in contour detection.)

3. Between the projection area and the association area (A_{II}), connections are assumed to be random. That is, each A-unit in the A_{II} set receives some number of fibers from origin points in the A_I set, but these origin points are scattered at random throughout the projection area. Apart from their connection distribution, the A_{II} units are identical with the A_I units, and respond under similar conditions.

4. The "responses," R_1, R_2, . . . , R_n are cells (or sets of cells) which

respond in much the same fashion as the A-units. Each response has a typically large number of origin points located at random in the A_{II} set. The set of A-units transmitting impulses to a particular response will be called the source-set for that response. (The source-set of a response is identical to its set of origin points in the A-system.) The arrows in Fig. 1 indicate the direction of transmission through the network. Note that up to A_{II} all connections are forward, and there is no feedback. When we come to the last set of connections, between A_{II} and the R-units, connections are established in both directions. The rule governing feedback connections, in most models of the perceptron, can be either of the following alternatives:

(*a*) Each response has excitatory feedback connections to the cells in its own source-set, or

(*b*) Each response has inhibitory feedback connections to the complement of its own source-set (i.e., it tends to prohibit activity in any association cells which do not transmit to it).

The first of these rules seems more plausible anatomically, since the R-units might be located in the same cortical area as their respective source-

sets, making mutual excitation between the R-units and the A-units of the appropriate source-set highly probable. The alternative rule (*b*) leads to a more readily analyzed system, however, and will therefore be assumed for most of the systems to be evaluated here.

Figure 2 shows the organization of a simplified perceptron, which affords a convenient entry into the theory of statistical separability. After the theory has been developed for this simplified model, we will be in a better position to discuss the advantages of the system in Fig. 1. The feedback connections shown in Fig. 2 are inhibitory, and go to the complement of the source-set for the response from which they originate; consequently, this system is organized according to Rule *b*, above. The system shown here has only three stages, the first association stage having been eliminated. Each A-unit has a set of randomly located origin points in the retina. Such a system will form similarity concepts on the basis of *coincident areas* of stimuli, rather than by the similarity of contours or outlines. While such a system is at a disadvantage in many discrimination experiments, its capability is still quite impressive, as will be demonstrated presently. The system shown in Fig. 2 has only two responses, but there is clearly no limit on the number that might be included.

The responses in a system organized in this fashion are mutually exclusive. If R_1 occurs, it will tend to inhibit R_2, and will also inhibit the source-set for R_2. Likewise, if R_2 should occur, it will tend to inhibit R_1. If the total impulse received from all the A-units in one source-set is stronger or more frequent than the impulse received by the alternative (antagonistic) response, then the first response will

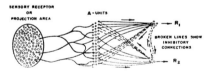

FIG. 2A. Schematic representation of connections in a simple perceptron.

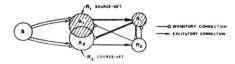

FIG. 2B. Venn diagram of the same perceptron (shading shows active sets for R_1 response).

tend to gain an advantage over the other, and will be the one which occurs. If such a system is to be capable of learning, then it must be possible to modify the A-units or their connections in such a way that stimuli of one class will tend to evoke a stronger impulse in the R_1 source-set than in the R_2 source-set, while stimuli of another (dissimilar) class will tend to evoke a stronger impulse in the R_2 source-set than in the R_1 source-set.

It will be assumed that the impulses delivered by each A-unit can be characterized by a value, V, which may be an amplitude, frequency, latency, or probability of completing transmission. If an A-unit has a high value, then all of its output impulses are considered to be more effective, more potent, or more likely to arrive at their endbulbs than impulses from an A-unit with a lower value. The value of an A-unit is considered to be a fairly stable characteristic, probably depending on the metabolic condition of the cell and the cell membrane, but it is not absolutely constant. It is assumed that, in general, periods of activity tend to increase a cell's value, while the value may decay (in some models) with inactivity. The most interesting models are those in which cells are assumed to compete for metabolic materials, the more active cells gaining at the expense of the less active cells. In such a system, if there is no activity, all cells will tend to remain in a relatively constant condition, and (regardless of activity) the net value of the system, taken in

TABLE 1

COMPARISON OF LOGICAL CHARACTERISTICS OF α, β, AND γ SYSTEMS

	α-System (Uncompensated Gain System)	β-System (Constant Feed System)	γ-System (Parasitic Gain System)
Total value-gain of source set per reinforcement	N_{a_r}	K	0
ΔV for A-units active for 1 unit of time	$+1$	K/N_{a_r}	$+1$
ΔV for inactive A-units outside of dominant set	0	K/N_{A_r}	0
ΔV for inactive A-units of dominant set	0	0	$\dfrac{-N_{a_r}}{N_{A_r}-N_{a_r}}$
Mean value of A-system	Increases with number of reinforcements	Increases with time	Constant
Difference between mean values of source-sets	Proportional to differences of reinforcement frequency $(n_{s_{r_1}}-n_{s_{r_2}})$	0	0

Note: In the β and γ systems, the total value-change for any A-unit will be the sum of the ΔV's for all source-sets of which it is a member.

N_{a_r} = Number of active units in source-set
N_{A_r} = Total number of units in source-set
$n_{s_{r_j}}$ = Number of stimuli associated to response r_j
K = Arbitrary constant

its entirety, will remain constant at all times. Three types of systems, which differ in their value dynamics, have been investigated quantitatively. Their principal logical features are compared in Table 1. In the alpha system, an active cell simply gains an increment of value for every impulse, and holds this gain indefinitely. In the beta system, each source-set is allowed a certain constant rate of gain, the increments being apportioned among the cells of the source-set in proportion to their activity. In the gamma system, active cells gain in value at the expense of the inactive cells of their source-set, so that the total value of a source-set is always constant.

For purposes of analysis, it is convenient to distinguish two phases in the response of the system to a stimulus (Fig. 3). In the *predominant phase*, some proportion of A-units (represented by solid dots in the figure) responds to the stimulus, but the R-units are still inactive. This phase is transient, and quickly gives way to the *postdominant phase*, in which one of the responses becomes active, inhibiting activity in the complement of its own source-set, and thus preventing the occurrence of any alternative response. The response which happens to become dominant is initially random, but if the A-units are reinforced (i.e., if the active units are allowed to gain in value), then when the same stimulus is presented again at a later time, the same response will have a stronger tendency to recur, and learning can be said to have taken place.

ANALYSIS OF THE PREDOMINANT PHASE

The perceptrons considered here will always assume a fixed threshold, θ, for the activation of the A-units. Such a system will be called a *fixed-threshold model*, in contrast to a *continuous transducer model*, where the response of the A-unit is some continuous function of the impinging stimulus energy.

In order to predict the learning curves of a fixed-threshold perceptron, two variables have been found to be of primary importance. They are defined as follows:

P_a = the expected proportion of A-units activated by a stimulus of a given size,

P_c = the conditional probability that an A-unit which responds to a given stimulus, S_1, will also respond to another given stimulus, S_2.

It can be shown (Rosenblatt, **15**) that as the size of the retina is increased, the number of S-points (N_s) quickly ceases to be a significant parameter, and the values of P_a and P_c approach the value that they would have for a retina with infinitely many points. For a large retina, therefore, the equations are as follows:

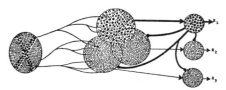

FIG. 3A. Predominant phase. Inhibitory connections are not shown. Solid black units are active.

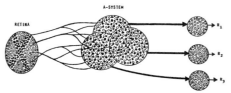

FIG. 3B. Postdominant phase. Dominant subset suppresses rival sets. Inhibitory connections shown only for R₁.

FIG. 3. Phases of response to a stimulus.

$$P_a = \sum_{e=\theta}^{x} \sum_{i=\theta}^{\min (y,\, e-\theta)} P(e,i) \qquad (1)$$

where

$$P(e,i) = \binom{x}{e} R^e (1-R)^{x-e}$$
$$\times \binom{y}{i} R^i (1-R)^{y-i}$$

and

R = proportion of S-points activated by the stimulus

x = number of excitatory connections to each A-unit

y = number of inhibitory connections to each A-unit

θ = threshold of A-units.

(The quantities e and i are the excitatory and inhibitory components of the excitation received by the A-unit from the stimulus. If the algebraic sum $\alpha = e + i$ is equal to or greater than θ, the A-unit is assumed to respond.)

$$P_c = \frac{1}{P_a} \sum_{e=\theta}^{x} \sum_{i=e-\theta}^{y} \sum_{l_e=0}^{e} \sum_{l_i=0}^{i}$$
$$\sum_{g_e=0}^{x-e} \sum_{g_i=0}^{y-i} P(e,i,l_e,l_i,g_e,g_i) \quad (2)$$

$$(e - i - l_e + l_i + g_e - g_i \geq \theta)$$

where

$P(e,i,l_e,l_i,g_e,g_i)$

$$= \binom{x}{e} R^e (1-R)^{x-e}$$
$$\times \binom{y}{i} R^i (1-R)^{y-i}$$
$$\times \binom{e}{l_e} L^{l_e} (1-L)^{e-l_e}$$
$$\times \binom{i}{l_i} L^{l_i} (1-L)^{i-l_i}$$
$$\times \binom{x-e}{g_e} G^{g_e} (1-G)^{x-e-g_e}$$
$$\times \binom{y-i}{g_i} G^{g_i} (1-G)^{y-i-g_i}$$

and

L = proportion of the S-points illuminated by the first stimulus, S_1, which are not illuminated by S_2

G = proportion of the residual S-set (left over from the first stimulus) which is included in the second stimulus (S_2).

The quantities R, L, and G specify the two stimuli and their retinal overlap. l_e and l_i are, respectively, the numbers of excitatory and inhibitory origin points "lost" by the A-unit when stimulus S_1 is replaced by S_2; g_e and g_i are the numbers of excitatory and inhibitory origin points "gained" when stimulus S_1 is replaced by S_2. The summations in Equation 2 are between the limits indicated, subject to the side condition $e - i - l_e + l_i + g_e - g_i \geq \theta$.

Some of the most important characteristics of P_a are illustrated in Fig. 4, which shows P_a as a function of the retinal area illuminated (R). Note that P_a can be reduced in magnitude by either increasing the threshold, θ, or by increasing the proportion of inhibitory connections (y). A comparison of Fig. 4b and 4c shows that if the excitation is about equal to the inhibition, the curves for P_a as a function of R are flattened out, so that there is little variation in P_a for stimuli of different sizes. This fact is of great importance for systems which require P_a to be close to an optimum value in order to perform properly.

The behavior of P_c is illustrated in Fig. 5 and 6. The curves in Fig. 5 can be compared with those for P_a in Fig. 4. Note that as the threshold is increased, there is an even sharper reduction in the value of P_c than was the case with P_a. P_c also decreases as the proportion of inhibitory connections increases, as does P_a. Fig. 5, which is

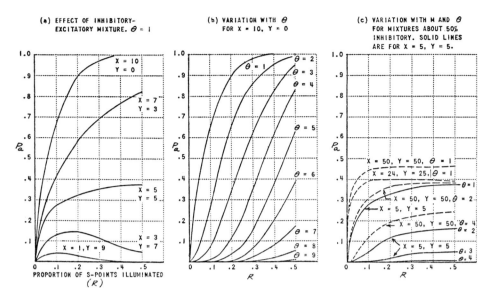

FIG. 4. P_a as function of retinal area illuminated.

calculated for nonoverlapping stimuli, illustrates the fact that P_c remains greater than zero even when the stimuli are completely disjunct, and illuminate no retinal points in common. In Fig. 6, the effect of varying amounts of overlap between the stimuli is shown. In all cases, the value of P_c goes to unity as the stimuli approach perfect identity. For smaller stimuli (broken line curves), the value of P_c is lower than for large stimuli. Similarly, the value is less for high thresholds than for low thresholds. The minimum value of P_c will be equal to

$$P_{cmin} = (1 - L)^x (1 - G)^y. \quad (3)$$

In Fig. 6, P_{cmin} corresponds to the curve for $\theta = 10$. Note that under these conditions the probability that the A-unit responds to both stimuli (P_c) is practically zero, except for stimuli which are quite close to identity. This condition can be of considerable help in discrimination learning.

MATHEMATICAL ANALYSIS OF LEARNING IN THE PERCEPTRON

The response of the perceptron in the predominant phase, where some fraction of the A-units (scattered throughout the system) responds to the stimulus, quickly gives way to the postdominant response, in which activity is limited to a single source-set, the other sets being suppressed. Two possible systems have been studied for the determination of the "dominant" response, in the postdominant phase. In one (the mean-discriminating system, or μ-system), the response whose inputs have the greatest mean value responds first, gaining a slight advantage over the others, so that it quickly becomes dominant. In the second case (the sum-discriminating system, or Σ-system), the response whose inputs have the greatest net value gains an advantage. In most cases, systems which respond to mean values have an advantage over systems which respond to sums, since the means are

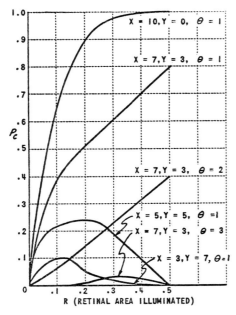

FIG. 5. P_c as a function of R,
for nonoverlapping stimuli.

less influenced by random variations in P_a from one source-set to another. In the case of the γ-system (see Table 1), however, the performance of the μ-system and Σ-system become identical.

We have indicated that the perceptron is expected to learn, or to form associations, as a result of the changes in value that occur as a result of the activity of the association cells. In evaluating this learning, one of two types of hypothetical experiments can be considered. In the first case, the perceptron is exposed to some series of stimulus patterns (which might be presented in random positions on the retina) and is "forced" to give the desired response in each case. (This forcing of responses is assumed to be a prerogative of the experimenter. In experiments intended to evaluate trial-and-error learning, with more sophisticated perceptrons, the experimenter does not force the system to

respond in the desired fashion, but merely applies positive reinforcement when the response happens to be correct, and negative reinforcement when the response is wrong.) In evaluating the learning which has taken place during this "learning series," the perceptron is assumed to be "frozen" in its current condition, no further value changes being allowed, and the same series of stimuli is presented again in precisely the same fashion, so that the stimuli fall on identical positions on the retina. The probability that the perceptron will show a bias towards the "correct" response (the one which has been previously reinforced during the learning series) in preference to any given alternative response is called P_r, the probability of correct choice of response between two alternatives.

In the second type of experiment, a learning series is presented exactly as before, but instead of evaluating the perceptron's performance using the same series of stimuli which were shown before, a new series is presented, in which stimuli may be drawn from the same *classes* that were previ-

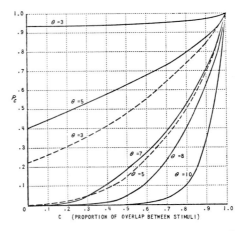

FIG. 6. P_c as a function of C. $X = 10$, $Y = 0$. Solid lines: R = .5; broken lines: R = .2.

ously experienced, but are not necessarily identical. This new test series is assumed to be composed of stimuli projected onto random retinal positions, which are chosen independently of the positions selected for the learning series. The stimuli of the test series may also differ in size or rotational position from the stimuli which were previously experienced. In this case, we are interested in the probability that the perceptron will give the correct response for the *class* of stimuli which is represented, regardless of whether the particular stimulus has been seen before or not. This probability is called P_g, the probability of correct generalization. As with P_r, P_g is actually the probability that a bias will be found in favor of the proper response rather than any one alternative; only one pair of responses at a time is considered, and the fact that the response bias is correct in one pair does not mean that there may not be other pairs in which the bias favors the wrong response. The probability that the correct response will be preferred over *all* alternatives is designated P_R or P_G.

In all cases investigated, a single general equation gives a close approximation to P_r and P_g, if the appropriate constants are substituted. This equation is of the form:

$$P = P(N_{a_r} > 0) \cdot \phi(Z) \qquad (4)$$

where

$$P(N_{a_r} > 0) = 1 - (1 - P_a)^{N_e}$$

$\phi(Z)$ = normal curve integral
from $-\infty$ to Z

and

$$Z = \frac{c_1 n_{s_r} + c_2}{\sqrt{c_3 n_{s_r}^2 + c_4 n_{s_r}}}.$$

If R_1 is the "correct" response, and R_2 is the alternative response under consideration, Equation 4 is the probability that R_1 will be preferred over

R_2 after n_{s_r} stimuli have been shown for each of the two responses, during the learning period. N_e is the number of "effective" A-units in each source-set; that is, the number of A-units in either source-set which are not connected in common to both responses. Those units which are connected in common contribute equally to both sides of the value balance, and consequently do not affect the net bias towards one response or the other. N_{a_r} is the number of active units in a source-set, which respond to the test stimulus, $S_t \cdot P(N_{a_r} > 0)$ is the probability that at least one of the N_e effective units in the source-set of the correct response (designated, by convention, as the R_1 response) will be activated by the test stimulus, S_t.

In the case of P_g, the constant c_2 is always equal to zero, the other three constants being the same as for P_r. The values of the four constants depend on the parameters of the physical nerve net (the perceptron) and also on the organization of the stimulus environment.

The simplest cases to analyze are those in which the perceptron is shown stimuli drawn from an "ideal environment," consisting of randomly placed points of illumination, where there is no attempt to classify stimuli according to intrinsic similarity. Thus, in a typical learning experiment, we might show the perceptron 1,000 stimuli made up of random collections of illuminated retinal points, and we might arbitrarily reinforce R_1 as the "correct" response for the first 500 of these, and R_2 for the remaining 500. This environment is "ideal" only in the sense that we speak of an ideal gas in physics; it is a convenient artifact for purposes of analysis, and does not lead to the best performance from the perceptron. In the ideal environment situation, the constant c_1 is always equal to zero, so that, in the case of P_g (where c_2 is also zero), the value of Z will be zero, and P_g can never be any better than the random expectation of 0.5. The evaluation of P_r for these conditions, however, throws some interesting light on the differences between the alpha, beta, and gamma systems (Table 1).

First consider the alpha system, which has the simplest dynamics of the three. In this system, whenever an A-unit is active for one unit of time, it gains one unit of value. We will assume an experiment, initially, in which N_{s_r} (the number of stimuli associated to each response) is constant for all responses. In this case, for the sum system,

$$\left.\begin{array}{l} c_1 = 0 \\ c_2 = (1 - P_a)N_e \\ c_3 = 2P_a\omega \\ c_4 \approx 0 \end{array}\right\} \quad (5)$$

where ω = the fraction of responses connected to each A-unit. If the source-sets are disjunct, $\omega = 1/N_R$, where N_R is the number of responses in the system. For the μ-system,

$$\left.\begin{array}{l} c_1 = 0 \\ c_2 = (1 - P_a)N_e \\ c_3 = 0 \\ c_4 = 2\omega \end{array}\right\} \quad (6)$$

The reduction of c_3 to zero gives the μ-system a definite advantage over the Σ-system. Typical learning curves for these systems are compared in Fig. 7 and 8. Figure 9 shows the effect of variations in P_a upon the performance of the system.

If n_{s_r}, instead of being fixed, is treated as a random variable, so that the number of stimuli associated to each response is drawn separately from some distribution, then the per-

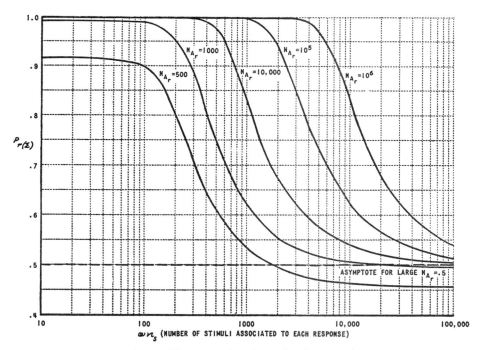

FIG. 7. $P_r(\Sigma)$ as function of ωn_s, for discrete subsets. ($\omega_c = 0$, $P_a = .005$. Ideal environment assumed.)

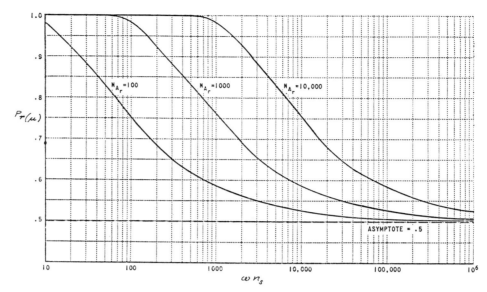

FIG. 8. $P_{r(\mu)}$ as function of ωn_s. (For $P_a = .07$, $\omega_c = 0$. Ideal environment assumed.)

formance of the α-system is considerably poorer than the above equations indicate. Under these conditions, the constants for the μ-system are

$$\left.\begin{aligned} c_1 &= 0 \\ c_2 &= 1 - P_a \\ c_3 &= 2P_a^2 q^2 \left[\frac{(\omega N_R - 1)^2}{N_R - 2} + 1 \right] \\ c_4 &= \frac{2(1 - P_a)N_R}{(1 - \omega_c)N_A} \end{aligned}\right\} \quad (7)$$

where

q = ratio of $\sigma_{n_{s_r}}$ to $\bar{n}_{s_r}$
N_R = number of responses in the system
N_A = number of A-units in the system
ω_c = proportion of A-units common to R_1 and R_2.

For this equation (and any others in which n_{s_r} is treated as a random variable), it is necessary to define n_{s_r} in Equation 4 as the expected value of this variable, over the set of all responses.

For the β-system, there is an even greater deficit in performance, due to the fact that the net value continues to grow regardless of what happens to the system. The large net values of the subsets activated by a stimulus tend to amplify small statistical differences, causing an unreliable performance. The constants in this case (again for the μ-system) are

$$\left.\begin{aligned} c_1 &= 0 \\ c_2 &= (1 - P_a)N_e \\ c_3 &= 2(P_a N_e q \omega N_R^2)^2 \\ c_4 &= 2(1 - P_a)\omega N_R N_e \end{aligned}\right\} \quad (8)$$

In both the alpha and beta systems, performance will be poorer for the sum-discriminating model than for the mean-discriminating case. In the gamma-system, however, it can be shown that $P_{r(\Sigma)} = P_{r(\mu)}$; i.e., it makes no difference in performance whether the Σ-system or μ-system is used. Moreover, the constants for the γ-system, with variable n_{s_r}, are identical to the constants for the alpha μ-sys-

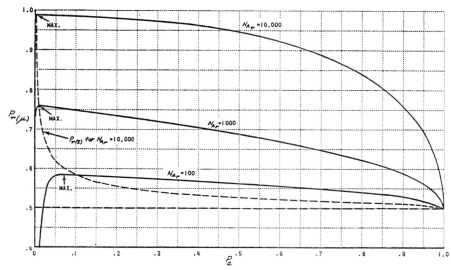

FIG. 9. $P_{r(\mu)}$ as function of P_a. (For $n_{s_r} = 1,000$, $\omega_c = 0$. Ideal environment assumed.)

tem, with n_{s_r} fixed (Equation 6). The performance of the three systems is compared in Fig. 10, which clearly demonstrates the advantage of the γ-system.

Let us now replace the "ideal en-

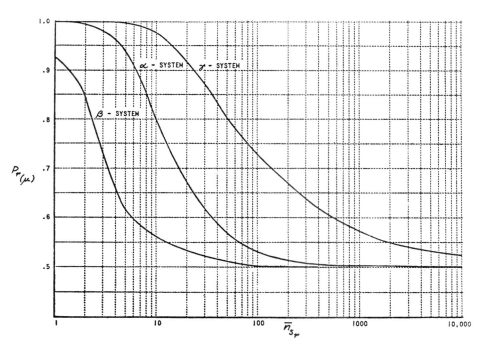

FIG. 10. Comparison of α, β, and γ systems, for variable n_{s_r}
($N_R = 100$, $\sigma_{n_{rs}} = .5\bar{n}_{s_i}$, $N_A = 10,000$, $P_a = .07$, $\omega = .2$).

vironment" assumptions with a model for a "differentiated environment," in which several distinguishable classes of stimuli are present (such as squares, circles, and triangles, or the letters of the alphabet). If we then design an experiment in which the stimuli associated to each response are drawn from a different class, then the learning curves of the perceptron are drastically altered. The most important difference is that the constant c_1 (the coefficient of n_{s_r} in the numerator of Z) is no longer equal to zero, so that Equation 4 now has a nonrandom asymptote. Moreover, in the form for P_g (the probability of correct generalization), where $c_2 = 0$, the quantity Z remains greater than zero, and P_g actually approaches the same asymptote as P_r. Thus the equation for the perceptron's performance after infinite experience with each class of stimuli is identical for P_r and P_g:

$$P_{r\infty} = P_{g\infty} = [1 - (1 - P_a)^{N_e}]$$
$$\times \phi \left(\frac{c_1}{\sqrt{c_3}} \right) \quad (9)$$

This means that *in the limit it makes no difference whether the perceptron has seen a particular test stimulus before or not; if the stimuli are drawn from a differentiated environment, the performance will be equally good in either case.*

In order to evaluate the performance of the system in a differentiated environment, it is necessary to define the quantity $P_{c\alpha\beta}$. This quantity is interpreted as the expected value of P_c between pairs of stimuli drawn at random from classes α and β. In particular, P_{c11} is the expected value of P_c between members of the same class, and P_{c12} is the expected value of P_c between an S_1 stimulus drawn from Class 1 and an S_2 stimulus drawn from Class 2. P_{c1x} is the expected value of P_c between members of Class 1 and

stimuli drawn at random from all other classes in the environment.

If $P_{c11} > P_a > P_{c12}$, the limiting performance of the perceptron ($P_{g\infty}$) will be better than chance, and learning of some response, R_1, as the proper "generalization response" for members of Class 1 should eventually occur. If the above inequality is not met, then improvement over chance performance may not occur, and the Class 2 response is likely to occur instead. It can be shown (15) that for most simple geometrical forms, which we ordinarily regard as "similar," the required inequality can be met, if the parameters of the system are properly chosen.

The equation for P_r, for the sum-discriminating version of an α-perceptron, in a differentiated environment where n_{s_r} is fixed for all responses, will have the following expressions for the four coefficients:

$$\left. \begin{aligned}
c_1 &= P_a N_e (P_{c11} - P_{c12}) \\
c_2 &= P_a N_e (1 - P_{c11}) \\
c_3 &= \sum_{r=1,2} P_a (1 - P_a) N_e \\
&\quad \times [P_{c1r}^2 + \sigma_s^2(P_{c1r}) \\
&\quad + \sigma_j^2(P_{c1r}) + (\omega N_R - 1)^2 \\
&\quad \times (P_{c1x} + \sigma_s^2(P_{c1x}) \\
&\quad + \sigma_j^2(P_{crx})) + 2(\omega N_R - 1) \\
&\quad (P_{c1r} P_{c1x})] + P_a^2 N_e^2 \\
&\quad \times [\sigma_s^2(P_{c1r}) + (\omega N_R - 1)^2 \\
&\quad \times \sigma_s^2(P_{c1x}) + 2(\omega N_R - 1)\epsilon] \\
c_4 &= \sum_{r=1,2} P_a N_e [P_{c1r} - P_{c1r}^2 \\
&\quad - \sigma_s^2(P_{c1r}) - \sigma_j^2(P_{c1r}) \\
&\quad + (\omega N_R - 1)(P_{c1x} - P_{c1x}^2 \\
&\quad - \sigma_j^2(P_{c1x}))]
\end{aligned} \right\} \quad (10)$$

where

$\sigma_s^2(P_{c1r})$ and $\sigma_s^2(P_{c1x})$ represent the variance of P_{c1r} and P_{c1x} measured over the set of possible test stimuli, S_t, and

$\sigma_j^2(P_{c1r})$ and $\sigma_j^2(P_{c1x})$ represent the variance of P_{c1r} and P_{c1x} measured over the set of all A-units, a_j.

ϵ = covariance of $P_{c1r}P_{c1x}$, which is assumed to be negligible.

The variances which appear in these expressions have not yielded, thus far, to a precise analysis, and can be treated as empirical variables to be determined for the classes of stimuli in question. If the sigma is set equal to half the expected value of the variable, in each case, a conservative estimate can be obtained. When the stimuli of a given class are all of the same shape, and uniformly distributed over the retina, the subscript s variances are equal to zero. $P_{g(\Sigma)}$ will be represented by the same set of coefficients, except for c_2, which is equal to zero, as usual.

For the mean-discriminating system, the coefficients are:

$$
\left.
\begin{aligned}
c_1 &= (P_{c11} - P_{c12}) \\
c_2 &= (1 - P_{c11}) \\
c_3 &= \sum_{r=1,2} \left[\frac{1}{P_a(N_e - 1)} - \frac{1}{N_e - 1} \right] \\
&\quad \times [\sigma_j^2(P_{c1r}) + (\omega N_R - 1)^2 \\
&\quad \times \sigma_j^2(P_{c1x})] + [\sigma_s^2(P_{c1r}) \\
&\quad + (\omega N_R - 1)^2 \sigma_s^2(P_{c1x})] \\
c_4 &= \sum_{r=1,2} \frac{1}{P_a N_e} [P_{c1r} - P_{c1r}^2 \\
&\quad - \sigma_s^2(P_{c1r}) - \sigma_j^2(P_{c1r}) \\
&\quad + (\omega N_R - 1)(P_{c1x} - P_{c1x}^2 \\
&\quad - \sigma_s^2(P_{c1x}) - \sigma_j^2(P_{c1x}))]
\end{aligned}
\right\} \quad (11)
$$

Some covariance terms, which are considered negligible, have been omitted here.

A set of typical learning curves for the differentiated environment model is shown in Fig. 11, for the mean-discriminating system. The parameters are based on measurements for a

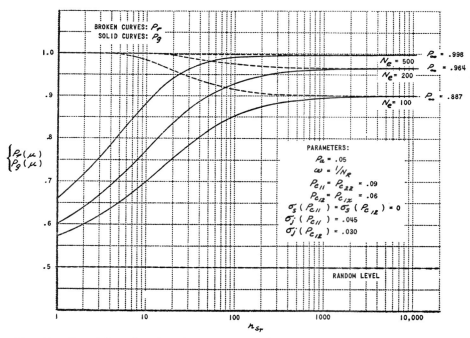

FIG. 11. P_r and P_g as function of n_{s_r}. Parameters based on square-circle discrimination.

square-circle discrimination problem. Note that the curves for P_r and P_g both approach the same asymptotes, as predicted. The values of these asymptotes can be obtained by substituting the proper coefficients in Equation 9. As the number of association cells in the system increases, the asymptotic learning limit rapidly approaches unity, so that for a system of several thousand cells, the errors in performance should be negligible on a problem as simple as the one illustrated here.

As the number of responses in the system increases, the performance becomes progressively poorer, if every response is made mutually exclusive of all alternatives. One method of avoiding this deterioration (described in detail in Rosenblatt, **15**) is through the binary coding of responses. In this case, instead of representing 100 different stimulus patterns by 100 distinct, mutually exclusive responses, a limited number of discriminating features is found, each of which can be independently recognized as being present or absent, and consequently can be represented by a single pair of mutually exclusive responses. Given an ideal set of binary characteristics (such as dark, light; tall, short; straight, curved; etc.), 100 stimulus classes could be distinguished by the proper configuration of only seven response pairs. In a further modification of the system, a single response is capable of denoting by its activity or inactivity the presence or absence of each binary characteristic. The efficiency of such coding depends on the number of independently recognizable "earmarks" that can be found to differentiate stimuli. If the stimulus can be identified only in its entirety and is not amenable to such analysis, then ultimately a separate binary response pair, or bit, is required to

denote the presence or absence of each stimulus class (e.g., "dog" or "not dog"), and nothing has been gained over a system where all responses are mutually exclusive.

BIVALENT SYSTEMS

In all of the systems analyzed up to this point, the increments of value gained by an active A-unit, as a result of reinforcement or experience, have always been *positive*, in the sense that an active unit has always *gained* in its power to activate the responses to which it is connected. In the gamma-system, it is true that some units lose value, but these are always the *inactive units*, the active ones gaining in proportion to their rate of activity. In a *bivalent system*, two types of reinforcement are possible (positive and negative), and an active unit may either gain or lose in value, depending on the momentary state of affairs in the system. If the positive and negative reinforcement can be controlled by the application of external stimuli, they become essentially equivalent to "reward" and "punishment," and can be used in this sense by the experimenter. Under these conditions, a perceptron appears to be capable of trial-and-error learning. A bivalent system need not necessarily involve the application of reward and punishment, however. If a binary-coded response system is so organized that there is a single response or response-pair to represent each "bit," or stimulus characteristic that is learned, with positive feedback to its own source-set if the response is "on," and negative feedback (in the sense that active A-units will lose rather than gain in value) if the response is "off," then the system is still bivalent in its characteristics. Such a bivalent system is particularly efficient in reducing some of the bias effects (preference for the wrong response due to greater size or frequency of its associated stimuli) which plague the alternative systems.

Several forms of bivalent systems have been considered (15, Chap. VII). The most efficient of these has the following logical characteristics.

If the system is under a state of positive reinforcement, then a positive ΔV is added to the values of all active A-units in the source-sets of "on" responses, while a negative ΔV is added to the active units in the source-sets of "off" responses. If the system is currently under negative reinforcement, then a negative ΔV is added to all active units in the source-set of an "on" response, and a positive ΔV is added to active units in an "off" source-set. If the source-sets are disjunct (which is essential for this system to work properly), the equation for a bivalent γ-system has the same coefficients as the monovalent α-system, for the μ-case (Equation 11).

The performance curves for this system are shown in Fig. 12, where the asymptotic generalization probability attainable by the system is plotted for the same stimulus parameters that were used in Fig. 11. This is the probability that all bits in an n-bit response pattern will be correct. Clearly, if a majority of correct responses is sufficient to identify a stimulus correctly, the performance will be better than these curves indicate.

In a form of bivalent system which utilizes more plausible biological assumptions, A-units may be either excitatory or inhibitory in their effect on connected responses. A positive ΔV in this system corresponds to the incrementing of an excitatory unit, while a negative ΔV corresponds to the incrementing of an inhibitory unit.

Such a system performs similarly to the one considered above, but can be shown to be less efficient.

Bivalent systems similar to those illustrated in Fig. 12 have been simulated in detail in a series of experiments with the IBM 704 computer at the Cornell Aeronautical Laboratory. The results have borne out the theory in all of its main predictions, and will be reported separately at a later time.

IMPROVED PERCEPTRONS AND SPONTANEOUS ORGANIZATION

The quantitative analysis of perceptron performance in the preceding sections has omitted any consideration of *time* as a stimulus dimension. A perceptron which has no capability for temporal pattern recognition is referred to as a "momentary stimulus perceptron." It can be shown (15) that the same principles of statistical separability will permit the perceptron to distinguish velocities, sound sequences, etc., provided the stimuli leave some temporarily persistent trace, such as an altered threshold,

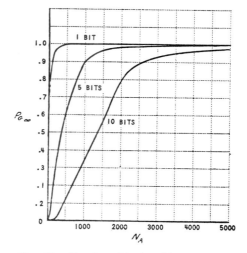

FIG. 12. P_{G_∞} for a bivalent binary system (same parameters as Fig. 11).

which causes the activity in the A-system at time t to depend to some degree on the activity at time $t - 1$.

It has also been assumed that the origin points of A-units are completely random. It can be shown that by a suitable organization of origin points, in which the spatial distribution is constrained (as in the projection area origins shown in Fig. 1), the A-units will become particularly sensitive to the location of contours, and performance will be improved.

In a recent development, which we hope to report in detail in the near future, it has been proven that if the values of the A-units are allowed to decay at a rate proportional to their magnitude, a striking new property emerges: the perceptron becomes capable of "spontaneous" concept formation. That is to say, if the system is exposed to a random series of stimuli from two "dissimilar" classes, and all of its responses are automatically reinforced without any regard to whether they are "right" or "wrong," the system will tend towards a stable terminal condition in which (for each binary response) the response will be "1" for members of one stimulus class, and "0" for members of the other class; i.e., the perceptron will spontaneously recognize the difference between the two classes. This phenomenon has been successfully demonstrated in simulation experiments, with the 704 computer.

A perceptron, even with a single logical level of A-units and response units, can be shown to have a number of interesting properties in the field of selective recall and selective attention. These properties generally depend on the intersection of the source sets for different responses, and are elsewhere discussed in detail (15). By combining audio and photo inputs, it is possible to associate sounds, or audi-tory "names" to visual objects, and to get the perceptron to perform such selective responses as are designated by the command "Name the object on the left," or "Name the color of this stimulus."

The question may well be raised at this point of where the perceptron's capabilities actually stop. We have seen that the system described is sufficient for pattern recognition, associative learning, and such cognitive sets as are necessary for selective attention and selective recall. The system appears to be potentially capable of temporal pattern recognition, as well as spatial recognition, involving any sensory modality or combination of modalities. It can be shown that with proper reinforcement it will be capable of trial-and-error learning, and can learn to emit ordered sequences of responses, provided its own responses are fed back through sensory channels.

Does this mean that the perceptron is capable, without further modification in principle, of such higher order functions as are involved in human speech, communication, and thinking? Actually, the limit of the perceptron's capabilities seems to lie in the area of relative judgment, and the abstraction of relationships. In its "symbolic behavior," the perceptron shows some striking similarities to Goldstein's brain-damaged patients (5). Responses to definite, concrete stimuli can be learned, even when the proper response calls for the recognition of a number of simultaneous qualifying conditions (such as naming the color if the stimulus is on the left, the shape if it is on the right). As soon as the response calls for the recognition of a relationship between stimuli (such as "Name the object left of the square." or "Indicate the pattern that appeared before the circle."), however, the problem generally becomes excessively difficult for the perceptron. Statistical separability alone does not provide a sufficient basis for higher order abstraction. Some system, more advanced in principle than the perceptron, seems to be required at this point.

CONCLUSIONS AND EVALUATION

The main conclusions of the theoretical study of the perceptron can be summarized as follows:

1. In an environment of random stimuli, a system consisting of randomly connected units, subject to the parametric constraints discussed above, can learn to associate specific responses to specific stimuli. Even if many stimuli are associated to each response, they can still be recognized with a better-than-chance probability, although they may resemble one another closely and may activate many of the same sensory inputs to the system.

2. In such an "ideal environment," the probability of a correct response diminishes towards its original random level as the number of stimuli learned increases.

3. In such an environment, no basis for generalization exists.

4. In a "differentiated environment," where each response is associated to a distinct class of mutually correlated, or "similar" stimuli, the probability that a learned association of some specific stimulus will be correctly retained typically approaches a better-than-chance asymptote as the number of stimuli learned by the system increases. This asymptote can be made arbitrarily close to unity by increasing the number of association cells in the system.

5. In the differentiated environment, the probability that a stimulus which has not been seen before will be correctly recognized and associated to its appropriate class (the probability of correct generalization) approaches the same asymptote as the probability of a correct response to a previously reinforced stimulus. This asymptote will be better than chance if the inequality $P_{c12} < P_a < P_{c11}$ is met, for the stimulus classes in question.

6. The performance of the system can be improved by the use of a contour-sensitive projection area, and by the use of a binary response system, in which each response, or "bit," corresponds to some independent feature or attribute of the stimulus.

7. Trial-and-error learning is possible in bivalent reinforcement systems.

8. Temporal organizations of both stimulus patterns and responses can be learned by a system which uses only an extension of the original principles of statistical separability, without introducing any major complications in the organization of the system.

9. The memory of the perceptron is *distributed*, in the sense that any association may make use of a large proportion of the cells in the system, and the removal of a portion of the association system would not have an appreciable effect on the performance of any one discrimination or association, but would begin to show up as a general deficit in *all* learned associations.

10. Simple cognitive sets, selective recall, and spontaneous recognition of the classes present in a given environment are possible. The recognition of relationships in space and time, however, seems to represent a limit to the perceptron's ability to form cognitive abstractions.

Psychologists, and learning theorists in particular, may now ask: "What has the present theory accomplished, beyond what has already been done in the quantitative theories of Hull, Bush and Mosteller, etc., or physiological theories such as Hebb's?" The present theory is still too primitive, of course, to be considered as a full-fledged rival of existing theories of human learning. Nonetheless, as a first approximation, its chief accomplishment might be stated as follows:

For a given mode of organization (α, β, or γ; Σ or μ; monovalent or bivalent) the fundamental phenomena of *learning, perceptual discrimination, and generalization can be predicted entirely from six basic physical parameters*, namely:

x: the number of excitatory connections per A-unit,

y: the number of inhibitory connections per A-unit,

θ: the expected threshold of an A-unit,

ω: the proportion of R-units to which an A-unit is connected,

N_A: the number of A-units in the system, and

N_R: the number of R-units in the system.

N_s (the number of sensory units) becomes important if it is very small. It is assumed that the system begins with all units in a uniform state of value; otherwise the initial value distribution would also be required. *Each of the above parameters is a clearly defined physical variable, which is measurable in its own right, independently of the behavioral and perceptual phenomena which we are trying to predict.*

As a direct consequence of its foundation on physical variables, the present system goes far beyond existing learning and behavior theories in three main points: parsimony, verifiability, and explanatory power and generality. Let us consider each of these points in turn.

1. *Parsimony.* Essentially all of the basic variables and laws used in this system are already present in the structure of physical and biological science, so that we have found it necessary to postulate only one hypothetical variable (or construct) which we have called V, the "value" of an association cell; this is a variable which must conform to certain functional characteristics which can clearly be stated, and which is assumed to have a potentially measurable physical correlate.

2. *Verifiability.* Previous quantitative learning theories, apparently without exception, have had one important characteristic in common: they have all been based on measurements of *behavior*, in specified situations, using these measurements (after theoretical manipulation) to predict *behavior* in other situations. Such a procedure, in the last analysis, amounts to a process of curve fitting and extrapolation, in the hope that the constants which describe one set of curves will hold good for other curves in other situations. While such extrapolation is not necessarily circular, in the strict sense, it shares many of the logical difficulties of circularity, particularly when used as an "explanation" of behavior. Such extrapolation is difficult to justify in a new situation, and it has been shown that if the basic constants and parameters are to be derived anew for any situation in which they break down empirically (such as change from white rats to humans), then the basic "theory" is essentially irrefutable, just as any successful curve-fitting equation is irrefutable. It has, in fact, been widely conceded by psychologists that there is little point in trying to "disprove" any of the major learning theories in use today, since by extension, or a change in parameters, they

have all proved capable of adapting to any specific empirical data. This is epitomized in the increasingly common attitude that a choice of theoretical model is mostly a matter of personal aesthetic preference or prejudice, each scientist being entitled to a favorite model of his own. In considering this approach, one is reminded of a remark attributed to Kistiakowsky, that "given seven parameters, I could fit an elephant." This is clearly *not* the case with a system in which the independent variables, or parameters, can be measured *independently* of the predicted behavior. In such a system, it is not possible to "force" a fit to empirical data, if the parameters in current use should lead to improper results. In the current theory, a failure to fit a curve in a new situation would be a clear indication that either the theory or the empirical measurements are wrong. Consequently, if such a theory *does* hold up for repeated tests, we can be considerably more confident of its validity and of its generality than in the case of a theory which must be hand-tailored to meet each situation.

3. *Explanatory power and generality.* The present theory, being derived from basic physical variables, is not specific to any one organism or learning situation. It can be generalized in principle to cover any form of behavior in any system for which the physical parameters are known. A theory of learning, constructed on these foundations, should be considerably more powerful than any which has previously been proposed. It would not only tell us what behavior might occur in any known organism, but would permit the *synthesis* of behaving systems, to meet special requirements. Other learning theories tend to become increasingly qualitative as they are generalized.

Thus a set of equations describing the effects of reward on T-maze learning in a white rat reduces simply to a statement that rewarded behavior tends to occur with increasing probability, when we attempt to generalize it from any species and any situation. The theory which has been presented here loses none of its precision through generality.

The theory proposed by Donald Hebb (7) attempts to avoid these difficulties of behavior-based models by showing how psychological functioning might be derived from neurophysiological theory. In his attempt to achieve this, Hebb's philosophy of approach seems close to our own, and his work has been a source of inspiration for much of what has been proposed here. Hebb, however, has never actually achieved a model by which behavior (or any psychological data) can be *predicted* from the physiological system. His physiology is more a suggestion as to the *sort* of organic substrate which might underlie behavior, and an attempt to show the plausibility of a bridge between biophysics and psychology.

The present theory represents the first actual completion of such a bridge. Through the use of the equations in the preceding sections, it is possible to predict learning curves from neurological variables, and likewise, to predict neurological variables from learning curves. How well this bridge stands up to repeated crossings remains to be seen. In the meantime, the theory reported here clearly demonstrates the feasibility and fruitfulness of a quantitative statistical approach to the organization of cognitive systems. By the study of systems such as the perceptron, it is hoped that those fundamental laws of organization which are common to all information handling systems, machines and men included, may eventually be understood.

REFERENCES

1. ASHBY, W. R. *Design for a brain.* New York: Wiley, 1952.
2. CULBERTSON, J. T. *Consciousness and behavior.* Dubuque, Iowa: Wm. C. Brown, 1950.
3. CULBERTSON, J. T. Some uneconomical robots. In C. E. Shannon & J. McCarthy (Eds.), *Automata studies.* Princeton: Princeton Univer. Press, 1956. Pp. 99–116.
4. ECCLES, J. C. *The neurophysiological basis of mind.* Oxford: Clarendon, 1953.
5. GOLDSTEIN, K. *Human nature in the light of psychopathology.* Cambridge: Harvard Univer. Press, 1940.
6. HAYEK, F. A. *The sensory order.* Chicago: Univer. Chicago Press, 1952.
7. HEBB, D. O. *The organization of behavior.* New York: Wiley, 1949.
8. KLEENE, S. C. Representation of events in nerve nets and finite automata. In C. E. Shannon & J. McCarthy (Eds.), *Automata studies.* Princeton: Princeton Univer. Press, 1956. Pp. 3–41.
9. KÖHLER, W. Relational determination in perception. In L. A. Jeffress (Ed.), *Cerebral mechanisms in behavior.* New York: Wiley, 1951. Pp. 200–243.
10. McCULLOCH, W. S. Why the mind is in the head. In L. A. Jeffress (Ed.), *Cerebral mechanisms in behavior.* New York: Wiley, 1951. Pp. 42–111.

11. McCULLOCH, W. S., & PITTS, W. A logical calculus of the ideas immanent in nervous activity. *Bull. math. Biophysics*, 1943, **5**, 115–133.
12. MILNER, P. M. The cell assembly: Mark II. *Psychol. Rev.*, 1957, **64**, 242–252.
13. MINSKY, M. L. Some universal elements for finite automata. In C. E. Shannon & J. McCarthy (Eds.), *Automata studies.* Princeton: Princeton Univer. Press, 1956. Pp. 117–128.
14. RASHEVSKY, N. *Mathematical biophysics.* Chicago: Univer. Chicago Press, 1938.
15. ROSENBLATT, F. *The perceptron: A theory of statistical separability in cognitive systems.* Buffalo: Cornell Aeronautical Laboratory, Inc. Rep. No. VG-1196-G-1, 1958.
16. UTTLEY, A. M. Conditional probability machines and conditioned reflexes. In C. E. Shannon & J. McCarthy (Eds.), *Automata studies.* Princeton: Princeton Univer. Press, 1956. Pp. 253–275.
17. VON NEUMANN, J. The general and logical theory of automata. In L. A. Jeffress (Ed.), *Cerebral mechanisms in behavior.* New York: Wiley, 1951. Pp. 1–41.
18. VON NEUMANN, J. Probabilistic logics and the synthesis of reliable organisms from unreliable components. In C. E. Shannon & J. McCarthy (Eds.), *Automata studies.* Princeton: Princeton Univer. Press, 1956. Pp. 43–98.

9
Introduction

(1958)
O. G. Selfridge

Pandemonium: a paradigm for learning

Mechanisation of Thought Processes: Proceedings of a Symposium Held at the National Physical Laboratory, November 1958, London: HMSO, pp. 513–526

This well known paper by Oliver Selfridge has been influential in several domains. It foreshadows the mixing of motivations that characterizes both cognitive science and neural modeling at present. It makes some suggestions for solving a practical problem (Morse code translation) with techniques suggested by both human information processing (the Pandemonium model) and engineering (hill climbing). Neuroscience is present in the emphasis on parallel arrays of computing elements with modifiable connections. This eclectic mix is typical of much present day work.

The Pandemonium model is notable for its simplicity and its parallelism. The idea of having multiple independent systems simultaneously looking at the input and responding according to their own bias is simple and powerful. The processing elements can be simple, but they are not simplistic, since each is looking for something specific in the outside world. It is also wasteful of processing power if a traditional computer is used to model it. But, with appropriate memory and computing elements present in great profusion, as they would be in a biological realization, this problem evaporates.

Selfridge also proposes a simple "hill climbing" learning scheme. The response of any particular unit is given by its input weights, i.e., the strengths connecting that unit to the input or to units in earlier layers. If we knew what inputs we wanted the units to classify, then we could modify the input weights until we maximized the effectiveness of the system. Selfridge suggested that one way to do that would be by modifying the weights a little bit in all directions and choosing the direction that gives the largest increase in effectiveness. Then do this again from the new point, and so on. This ensures that the response of the system will be improving or, at worst, not decreasing, at all times.

The problem with this technique is the same one that haunts a number of later network schemes. The weight changing technique ensures that the system will climb a hill of effectiveness, and find a maximum. But this might be a response foothill and the true maximum might be on a peak in a different range entirely. Local search may not find you the true global maximum. Of course, in many simple cases it does. Studying the topography of function landscapes keeps a number of neural network modelers usefully occupied.

The second type of adaptive change he suggests is one that has not been used as often as weight changing, but has great potential. He suggests that it might be possible to generate new units, or completely rewire old ones, that currently serve no useful function. Specifically, he suggests that new units be formed from logical combinations of the most successful old units in order to generate the most useful new units. In current terminology, this would consist of adaptively restructuring the "representation" of the input. There seems to us to be no question that this technique is used extensively by the nervous system, in that much of the computational power of the system is gained not from the rather weak computing power of the neural network, but from the representation of the input by the nervous system. Details of this representation are under the control of experience in the individual to some extent and are the product of great evolutionary pressure in the longer term.

We must point out the impact of this model on psychology. Ulrich Neisser in an important book, *Cognitive Psychology* (1967), discusses Pandemonium extensively as a suggestive model for human pattern recognition and information processing. About the best introductory psychology text of the '70s, Lindsay and Norman's *Human Information Processing* (1972, 2nd edition 1977) based much of its discussion of perception on Pandemonium. This book was notable for its wonderful figures that captured the essence of Pandemonium. Unfortunately, this text required too much thought for most Psych 1 students, and made the commercial error of discussing science rather than the more entertaining parts of psychology.

References

Peter H. Lindsay and Donald A. Norman (1972), *Human Information Processing: An Introduction to Psychology*, New York: Academic Press.

Ulrich Neisser (1957), *Cognitive Psychology*, New York: Appleton-Century-Crofts.

(1958)
O. G. Selfridge

Pandemonium: a paradigm for learning

Mechanisation of Thought Processes: Proceedings of a Symposium Held at the National Physical Laboratory, November 1958, London: HMSO, pp. 513–526

Introduction

We are proposing here a model of a process which we claim can adaptively improve itself to handle certain pattern recognition problems which cannot be adequately specified in advance. Such problems are usual when trying to build a machine to imitate any one of a very large class of human data processing techniques. A speech typewriter is a good example of something that very many people have been trying unsuccessfully to build for some time.

We do not suggest that we have proposed a model which can learn to typewrite from merely hearing speech. Pandemonium does not, however, seem on paper to have the same kinds of inherent restrictions or inflexibility that many previous proposals have had. The basic motif behind our model is the notion of parallel processing. This is suggested on two grounds: first, it is often easier to handle data in a parallel manner, and, indeed, it is usually the more "natural" manner to handle it in; and, secondly, it is easier to modify an assembly of quasi-independent modules than a machine all of whose parts interact immediately and in a complex way.

We are not going to apologize for a frequent use of anthropomorphic or biomorphic terminology. They seem to be useful words to describe our notions.

What we are describing is a process, or, rather, a model of a process. We shall not describe all the reasons that led to its particular formulation, but we shall give some reasons for hoping that it does in fact possess the flexibility and adaptability that we ascribe to it.

The Problem Environment for Learning

Pandemonium is a model which we hope can learn to recognize patterns which have not been specified. We mean that in the following sense: we present to the model examples of patterns taken from some set of them, each time informing the model which pattern we had just presented. Then, after some time the model guesses correctly which pattern has just been presented before we inform it. For a large class of pattern recognition ensembles there has never existed any adequate written or statable description of the distinctions between the patterns. The only requirement we can place on our model is that we want it to behave in the same way that men observably behave in. In an absolute sense this is a very unsatisfactory definition of any task, but it may be apparent that it is the way in which most tasks are defined for most men. Lucky is he whose job can be exactly specified in words without any ambiguity or necessary inferences. The example we shall illustrate in some detail is translating from manually keyed Morse code into, say, typewritten messages. Now it is true that when one learns Morse code one learns that a dash should be exactly three times are length of a dot and so on, but it turns out that this is really mostly irrelevant. What matters is only what the vast army of people who use Morse code and with whom one is going to have to communicate understand and practise when they use it. It turns out that this is nearly always very different from school book Morse.

In the same way the only adequate definition of the pattern of a spoken word, or one hand-written, must be in terms of the consensus of the people who are using it.

We use the term pattern recognition in a broad sense to include not only that data processing by which images are assigned to one or another pattern in some set of patterns, but also the processes by which the patterns and data processing are developed by the organism or machine; we generally call this latter "learning".

Pandemonium, Idealized and Practical

We first construct an idealized pandemonium (*fig. 1*). Each possible pattern of the set, represented by a demon in a box, computes his similarity with the image simultaneously on view to all of them, and gives an output depending monotonically on that similarity. The decision demon on top makes a choice of that

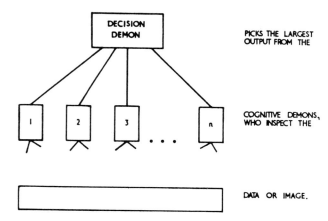

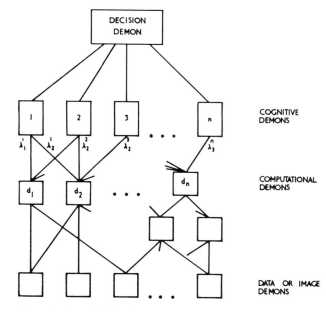

Figure 1   Idealized Pandemonium.

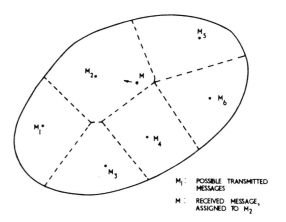

Figure 2   Signal space.

Figure 3   Amended Pandemonium.

pattern belonging to the demon whose output was the largest.*

Each demon may, for example, be assigned one letter of the alphabet, so that the task of the A-demon is to shout as loud of the amount of 'A-ness' that he sees in the image.† Now it will usually happen that with a reasonable collection of categories—like the letters of the alphabet—the computations performed by each of these ideal cognitive demons will be largely the same.

---

*This is an exact correlate of a communications system wherein given a received message $M(T)$ and a number of possible transmitted messages $M_i(T)$, that M_i is chosen, that is, deemed to have been transmitted, which minimizes $\int |M(T) - M_i(T)|^2 dT$. (Such a procedure is optimum under certain conditions). This integral is, as it were, the square of a distance in a signal phase space—fig. 2— and thus that transmitted message is selected that is most similar to the received one.

†It is possible also to phrase it so that the A-demon is computing the distance in some phase of the image from some ideal A; it seems to me unnecessarily platonic to postulate the existence of 'ideal' representatives of patterns, and, indeed, there are often good reasons for not doing so.

In many instances a pattern is nearly equivalent to some logical function of a set of features, each of which is individually common to perhaps several patterns and whose absence is also common to several other patterns.*

We therefore amend our idealized Pandemonium. The amended version—fig. 3—has some profound advantages, chief among which is its susceptibility to that kind of adaptive self-improvement that I call learning.

The difference between fig. 1 and fig. 3 is that the common parts of the computations that each cognitive demon carries out in fig. 1 have in fig. 3 been assigned instead to a host of subdemons. At this state the organization has four levels. At the bottom the data demons serve merely to store and pass on the data. At the next level the computational demons or subdemons perform certain more or less complicated computations on the data and pass the results of these up to the next level, the cognitive demons who weigh the evidence, as it were. Each cognitive demon computes a shriek, and from all the shrieks the highest level demon of all, the decision demon, merely selects the loudest.

The Conception of Pandemonium

We cannot ab initio know the ideal construction of our Pandemonium. We try to assure that it contains the

---

* See, for example, Jerome Bruner, "A study of Thinking",

seeds of self-improvement. Of the four levels in *fig. 3*, all but the third, the subdemons, which compute, are specified by the task. For the third level, therefore, we collect a large number of possible useful functions, eliminating a priori only those which could not conceivably be relevant, and make a reasonable selection of the others, being bound by economy and space. We then guess reasonable weights for them. The behavior at this point may even be acceptably good, but usually it must be improved by means of the adaptive changes we are about to discuss.

The Evolution of Pandemonium
There are several kinds of adaptive changes which we will discuss for our γ Pandemonium. They are all essentially very similar, but they may be programmed and discussed separately. The first may be called "Feature Weighting".

Although we have not yet specified what the cognitive demons compute, the sole task at present is to add a weighted sum of the outputs of all the computational demons or subdemons; the weightings will of course differ for the different cognitive demons, but the weightings will be the only difference between them. Feature weighting consists of altering the weights assigned to the subdemons by the cognitive demons so as to maximize the score of the entire γ Pandemonium. How then can we do this?

The Score
What we mean by the score here is how well the machine is doing the task we want it to do. This presumes that we are monitoring the machine and telling it when it makes an error and so on, and for the rest of the discussion we shall be assuming that we have available some such running score. Now at some point we shall be very interested in having the machine run without that kind of direct supervision, and the question naturally arises whether the machine can meaningfully monitor its own performance. We answer that question tentatively yes, but delay discussing it till a later section.

Feature Weighting and Hill-Climbing
The output of any cognitive demon is

$$d_i = \sum \lambda_j^i d_j$$

so that the complete set of weights for all the cognitive demons over all the subdemons is a vector:

$$\Lambda = \{\Lambda^i\} = \{\lambda_1^1, \lambda_2^1 .. \lambda_1^2, ... \lambda_m^n\}$$

Now for some (unknown) set of weights Λ, the behaviour of this whole Pandemonium is optimum, and

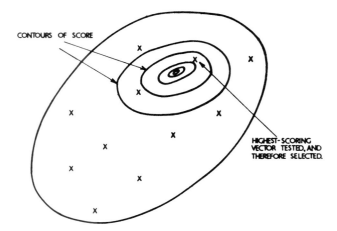

Figure 4 First hill-climbing technique: pick vectors at random (points in the space), score them, and select the one that scores highest.

the problem of feature weighting is to find that set. This may be described as a hill-climbing problem. We have a space (of Λ) and a function on the space (the score), which we may consider an altitude, and which we wish to maximize by a proper search through Λ. One possible technique is to select weighting vectors at random, score them, and finally to select the vector that scored highest (see *fig. 4*). It will usually, however, turn out to be profitable to take advantage of the continuity properties of the space, which usually exist in some sense, in the following way: select vectors at random until you find one that scores perceptibly more than the others. From this point take small random steps in all directions (that is, add small random vectors) until you find a direction that improves your score. When you find such a step, take it and repeat the process. This is illustrated in *fig. 5*, and is the case of a blind man trying to climb a hill. There may be, of course, many false peaks on which one may find oneself trapped in such a procedure (*fig. 6*).

The problem of false peaks in searching techniques is an old and familiar one. In general, one may hope that in spaces of very high dimensionality the interdependence of the components and the score is so great as to make very unlikely the existence of false peaks completely isolated from the main or true peak. If must be realized, however, that this is a purely experimental question that has to be answered separately for every hill-climbing situation. It does turn out in hill-climbing situations that the choice of starting point is often very important. The main peak may be very prominent, but unless it has wide-spread foot-hills it may take a very long time before we ever begin to gain altitude.

This may be described as one of the problems of

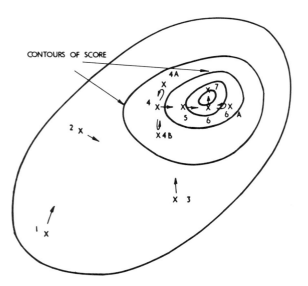

Figure 5 Second hill-climbing technique: pick vectors until one of them (Number 4) outscores the previous ones. Then take short random steps, retracing any that decrease the score.

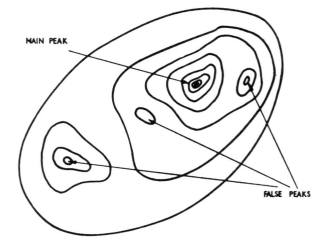

Figure 6 General space showing false peaks. One of the false peaks is quite isolated from the main or true peak.

training, namely, to encourage the machine or organism to get enough on the foot-hills so that small changes in his parameters will produce noticeable improvement in his altitude or score. One can describe learning situations where most of the difficulty of the task lies in finding any way of improving one's score, such as learning to ride a unicycle, where it takes longer to stay on for a second than it does to improve that one second to a minute; and others where it is easy to do a little well and very hard to do very well, such as learning to play chess. It is also true that often the main peak is a plateau rather than an isolated spike. That is to say, optimal behaviour of the mechanism, once reached, may be rather insensitive to the change of some of the parameters.

Subdemon Selection

The second kind of adaptive change that we wish to incorporate into our Pandemonium is subdemon selection. At the conception of our demoniac assembly we collected somewhat arbitrarily a large number of subdemons which we guessed would be useful and assigned them weights also arbitrarily. The first adaptive change, feature weighting, optimized these weights, but we have no assurance at all that the particular subdemons we selected are good ones. Subdemon selection generates new subdemons for trial and eliminates inefficient ones, that, is, ones that do not much help improve the score.

We propose to do this initially by two different techniques, which may be called "mutated fission"

and "conjugation". The first point to note is that it is possible to assign a worth to each of the subdemons. It may be done in several ways, and we may, for example, write the worth W_i of the ith demon

$$W_i = \sum_j |\lambda_i^j|,$$

so that the worthy demons are those whose outputs are likely to affect most strongly the decisions made.

We assume that feature weighting has already run so long that the behaviour of the machine has been approximately optimized, and that scores and worths of machine and its demons have been obtained. First we eliminate those subdemons with low worths. Next we generate new subdemons by mutating the survivors and reweighting the assembly. At present we plan to pick one subdemon and alter some of his parameters more or less at random. This will usually require that we reduce the subdemon himself to some canonical form so that the random changes we insert do not have the effect of rewriting the problem so that it will not run or so that it will enter a closed loop without any hope of getting out of it.*

Besides mutated fission, we are proposing another method of subdemon improvement called "conjugation". Our purpose here is two-fold: first to provide a logical variety in the functions computed by the subdemons, and, secondly, to provide length and complexity in them.

What we do is this: given two 'useful' subdemons, we generate a new subdemon which is the continuous analogue of one of the ten nontrivial binary two-

---

*We are at present running our Pandemonium on an IBM 704. The analogues for kinds of simulation are obvious.

Table 1 Non-trivial binary functions on two variables

A.B	A ∨ B
A.~B	A ∨ ~B
~A.~B	~A ∨ ~B
~A.B	~A ∨ ~B
A.B ∨ ~A.~B	A.~B ∨ ~A.B

variable functions on them. For example, the product of two subdemon outputs, corresponding to the logical product, would suggest the simultaneous presence of two features. The ten non-trivial such functions are listed in *Table 1*.

Control Adaptation

The first two levels of adaptation are directly concerned with immediate improvement of behaviour and the score. We should also like to improve the entire organization, and in the same way. We shall deal with this point somewhat cursorily, being reluctant to specify things too far in advance of experiment. In principle, we propose that the control operations should themselves be handled by demons subject to changes like feature weighting and subdemon selection. It is obviously a little more difficult and perhaps impossible here to define the usefulness or worth of a particular demon. It is also clear that it will sometimes take much longer to check the usefulness of some change in some control demon—for example, in one of those which control the mutations in subdemon selection. Furthermore, at this level, some of the demons, presumably, will be in a position to change themselves, for otherwise we should need another level of possible change, and so on. This raises the possibility of irreversible changes, and it is not obvious that *all* parts of the machine should be subject to adaptive change. But these are largely heuristic questions.

The Evolutionary Process

The adaptive changes mentioned above will tend, we hope, to promote a kind of evolution in our Pandemonium. The scheme sketched is really a natural selection on the processing demons. If they serve a useful function they survive, and perhaps are even the source for other subdemons who are themselves judged on their merits.

It is perfectly reasonable to conceive of this taking place on a broader scale—and in fact it takes place almost inevitably. Therefore, instead of having but one Pandemonium we might have some crowd of them, all fairly similarly constructed, and employ natural selection on the crowd of them. Eliminate the relatively

poor and encourage the rest to generate new machines in their own images.

Unsupervised Operation

So far all of the operation of the machine has been on the basis of constant monitoring by a person who is telling the machine when it makes an error. A very valid question is whether the machine can form any independent opinion of its own on how well it is doing. I suggest that it can in the following way: one criterion of correct decisions will be that they are fairly unequivocal, that is, that there is one and only one cognitive demon whose output far outshines the rest. Some running average of the degree to which this is so would presumably somewhat reflect the score of the machine. Note that it would be vital that the machine be trained first to do well enough before it is left to its own resources and supervision.

A Real-Life Example: Morse Translation

As we mentioned before, the entire notion of Pandemonium was conceived as a practical way of automatically improving data-processing for pattern recognition. Our initial model task is to distinguish dots and dashes in manually keyed Morse code, so that our Pandemonium can be illustrated in *fig. 7*. Note that the functions and behaviour of all demons have been specified except for the computing subdemons. We shall reiterate those specifications.

(1) The decision demon's output is 'dot' or 'dash' according as the dot demon's output is greater or less than the dash demon's.

(2) The cognitive demons, dot and dash, each compute a weighted sum of the outputs of some 150 comput-

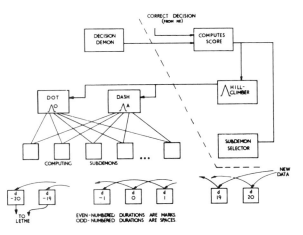

Figure 7 First Morse Pandemonium.

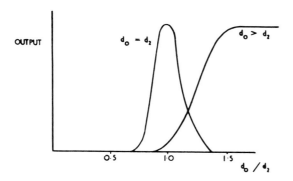

Figure 8 Operation of the subdemons $d_0 = d_2$, $d_0 > d_2$.

ing subdemons. Initial weights we have assigned arbitrarily, but, we hope, reasonably.

(3) The data-handling demons receive data in the form of durations, alternatively of marks and spaces, and they pass them down the line.

The computing subdemons are constructed from only a very few operational functions, which are carefully non-binary. For example the subdemons $d_0 = d_2$ and $d_0 > d_2$ have their outputs shown in *fig. 8*. The operational functions follow:

(1) '$=$'. This function computes the degree of equality of some set of variables (see *fig. 8*).

(2) '$<$', '$>$', compute the degree to which some variable is less than or is greater than some other variable (see *fig. 8*).

(3) '*max*', '*rmax*', compute the degree to which some variable is the largest of an arbitrary set of variables or an arbitrary set of consecutive variables.

(4) 'O_i', 'A_i', store the degree to which the ith duration *has been* identified as a dot or dash.

(5) '*Av*' computes an average of some set of variables.

(6) '*M*' is a family of tracking means. For example, it might compute

$$M^\alpha(c) = \alpha M(i - 1) + (1 - \alpha)$$

(7) '*Ox*', '*Ax*'. $Ox1$ is the last duration identified as a dot. $Ax3$ is the third last duration identified as a dash, etc.

The above is the present functional vocabulary of the computing operations for our subdemons. The subdemons themselves are built with a simple syntax. For the initial set, at conception, we merely select a set of operational functions and follow them with the numbers of some particular data demons.

Conclusion

What I shall present at the meeting in November will be the details of the progress of Pandemonium on the Morse code translation problem. The initial problem we have given the machine is to distinguish dots and dashes. When the behaviour of the machine has improved itself to the point where little further improvement seems to be occurring, we shall add three more cognitive demons, the symbol space, the letter space, and the word space. Presumably after some further time this new Pandemonium will settle down to some unimprovable state. Then we shall replace the senior or decision-making demon with a row of some forty or so character demons with a new decision-making demon above them, letting the new cognitive demons for the character demons use all the inferior demons, cognitive and otherwise, for their inputs. It is probably also desirable that previous decisions be available for present decisions, so that a couple of new functional operations might be added. There need be little concern about logical circularity, because we have no requirement for logical consistency, but are merely seeking agreeable Morse translation.

How much of the whole program will have been run and tested by November I cannot be sure of. At the present (July) we have had some fair testing of hill-climbing procedures.

Acknowledgment

I should like to acknowledge the valuable contributions of discussions with many of my friends, including especially M. Minsky, U. Neisser, F. Frick and J. Lettvin.

10
Introduction

(1960)
Bernard Widrow and Marcian E. Hoff

Adaptive switching circuits
1960 IRE WESCON Convention Record, New York: IRE, pp. 96–104

Based the correctness or incorrectness of its immediate past classification, the perceptron changed its coupling coefficients ("synaptic weights"). The Perceptron Convergence Theorem and other work showed that, given some known constraints (for example, linear separability), some modification rules would eventually lead to correct classification of items in the input set.

Many learning rules, however, took unacceptably long to converge to a set of weights that classified correctly. Even if a set of successful weights were found, it was not clear whether they were optimal in any sense.

This paper by Widrow and Hoff proposed an adaptive system that could potentially learn quickly and accurately. Although not specifically a perceptron, it was related to perceptrons. It was called an "adaptive neuron," which was a Threshold Logic Unit with variable connection strengths. The adaptive neuron computed a weighted sum of activities of the inputs times the synaptic weight, plus a bias element. If the sum was greater than zero, the output became $+1$. If it was equal to or less than zero, output was -1.

The learning rule suggested by Widrow and Hoff was elegant in its simplicity, and it, and variants of it, have been extensively used in neural networks in recent years. Widrow and Hoff assumed that there is an input pattern, and an output classification of the input pattern by the adaptive neuron, which can take values of either a $+1$ or a -1. They assumed that there was a "teacher" (which they call a "boss" in the paper) that knew what the answer was supposed to be for that input. The output of the summer was also available. They assumed the system was able to form an *error signal* between what the output is supposed to be and what the summer computed; that is, it formed the difference between them. The synaptic weights were adjusted, and the sum recomputed, so the error signal became zero; that is, the system response was exactly correct for that input, having a value of either a $+1$ or a -1. Notice that the response was not simply a correct classification, in which case the sum merely had to be greater than or less than 0, but exactly correct, equal to $+1$ or -1. This means that learning, that is, change of synaptic weights, can continue even if the classification is correct because the output of the summer may not be exactly $+1$ or -1. There still may be a nonzero error signal allowing continued learning, even though the system is correctly classifying every input. This continued learning is important, because one reason perceptrons took so long to learn was that synapses were changed only when the gross classification was incorrect. One can show this kind of continued learning in humans. In many experiments a subject's reaction time to a stimulus continues to drop even though the subject is making every response correctly.

Widrow and Hoff assume we are interested in reducing the square of the error signal to its smallest possible value. All possible values of the input weighting coefficients give rise to an error value, that is, an error surface in a *weight space* is formed. We do not know where the minimum of the error surface is, exactly, since we cannot see the entire surface, but we can measure the local topography, so we can see what directions of adjustment will decrease the error the most. A good way to adjust the weights, therefore, is to use a *gradient descent* method; that is, always adjust the weights so that change in weights moves the system down the error surface in the direction of the locally steepest descent. Problems with gradient descent in general involve getting trapped in local minima, i.e., the bottom of valleys, which are not the lowest global error. However, minimizing square error means, as Widrow and Hoff show, that there is a simple quadratic error surface, with only one, global, minimum. Computing the gradient at a point involves computing partial derivatives of the square of the error with respect to the weights. Widrow and Hoff show that this derivative is proportional to the error signal. By measuring the error signal, we can tell which direction to move to correct the error.

In applying this technique to the ADALINE (which first stood for ADAptive LInear NEuron, and then, ADAptive LINear Element, as neural models became less popular—Widrow, 1987), Widrow and Hoff were helped because only values of $+1$ or -1 were allowed at the inputs of the system. Therefore, the proper weight correction rule was that every one of the n input weighted input lines was to be adjusted to eliminate $1/n$ of the error signal. This was the optimal gradient descent method of error correction.

It is easy to generalize from a single adaptive neuron to a set of adaptive neurons, where a pattern on the input can give rise to a pattern of output element activities. This is because a particularly important aspect of the learning rule proposed by Widrow and Hoff is that it is *local*. To change the synapse, all that is required is information that is potentially present at the single adaptive element: the input, the output, and what the output should have been, which allows computation of the error signal. The adaptive neuron need not care what other neurons are doing in a single-layer multineuron system.

One important assumption of the error correction idea is the existence of the "teacher," which has perfect knowledge of the appropriate response. Algorithms with a teacher are referred to as *supervised*. In the more recent neural network literature, "back propagation" is the best known supervised algorithm (Rumelhart, Hinton, and Williams, papers 41 and 42). Back propagation is a generalization of the simple Widrow-Hoff rule.

It seems to be an obvious general rule that the more information is known about the nature of the error, the more quickly the system will learn. If only global correctness or incorrectness of the categorizations of a set of weights is known, the system will learn, but slowly, as in simple perceptron rules. If an error signal can be formed that is an exact representation of the error, rapid learning rules can be followed, such as those described by Widrow and Hoff. Barto, Sutton, and Anderson (paper 32) discuss some alternative learning rules that have different amounts of information available about the error.

It is quite fascinating to note how much concern there is in this technical report with practical implementations and applications for these learning rules. Very concrete suggestions for devices to realize the rules are made, and some pilot hardware was made for demonstrations. At a recent Neural Networks for Computation Conference, Widrow gave a demonstration of one of the original ADALINEs, which has worked perfectly for twenty-five years, to the delight of the audience. It is only recently that there has occurred a flowering of hardware designs and implementations similar to as that which flourished in the first golden age of neural networks, around 1960.

This error correction algorithm is also known as the LMS (Least Mean Squares) algorithm in signal processing, and a number of detailed formal results on it and variants of it have been derived. More details about the LMS algorithm, its variants and extensions, and their use in signal processing can be found in Widrow and Stearns (1985).

References

B. Widrow (1987), historical talk at Neural Networks for Computation Conference, Snowbird, UT, April 1987.

B. Widrow and S. D. Stearns (1985), *Adaptive Signal Processing*, New York: Prentice-Hall.

(1960)
Bernard Widrow and Marcian E. Hoff

Adaptive switching circuits
1960 IRE WESCON Convention Record, New York: IRE, pp. 96–104

A. Introduction

The modern science of switching theory began with work by Shannon[1] in 1938. The field has developed rapidly since then, and at present a wealth of literature exists[2] concerning the analysis and synthesis of logical networks which might range from simple interlock systems to telephone switching systems to large-scale digital computing systems.

An example illustrating the use of switching theory is that of the design of an interlock system for the control of traffic in a railroad switch yard. The first step is the preparation of a "truth table", an exhaustive listing of all input possibilities (the positions of all incoming and outgoing trains), and what the desired system output should be (what the desired control signals should be) for each input situation. The next step is the construction of a Boolean function, and the following steps are algebraic reduction and design of the logical control system.

The design of the traffic control system is an example wherein the truth table must be followed precisely and reliably. Errors would be destructive. The design of the arithmetic element of a digital computer is another example wherein the truth table must be followed precisely.

There are other situations in which some errors are inevitable, however, and here errors are usually costly but not catastrophic. These situations call for statistically optimum switching circuits. A common performance objective is the minimization of the average number of errors. An example is that of prediction of the next bit in a correlated stochastic binary number sequence. The predictor output is to be a logical combination of a finite number of previous input sequence bits. An optimum system is a sequential switching circuit that predicts with a minimum number of errors.

Suppose that a record of the binary sequence is printed on tape and cut up into pieces (with indication of the positive direction of time preserved), say 25 bits long. Place all pieces where the most recent event is ONE in one pile, and the remainder in another pile. Delete the most recent bit on each piece of tape. If the statistical scheme could be discovered by which the pieces of tape are classified, this would lead to a prediction scheme. It is apparent that prediction is a certain kind of classification.

Assuming statistical regularity, a reasonable way to proceed might be to form a truth table, and let the data from each piece of tape be an entry in the table. It might be expected that with the data of 100 pieces of tape, a fairly good predictor could be developed. The truth table would have only 100 entries however, out of a total of 2^{24}. The "best" way to fill in the remainder of the truth table depends upon the nature of the sequence statistics and the error cost criteria. Filling in the table is a difficult and a crucial part of the problem. Even if the truth table were filled in, however, the designer would have the difficult task of realizing a logical network to satisfy a truth table with 2^{24} entries.

An approach to such problems is taken in this paper which does not require an explicit use of the truth table. The design objective is the minimization of the average number of errors, rather than a minimization of the number of logical components used. The nature of the logical elements is quite unconventional. The system design procedure is adaptive, and is based upon an iterative search process. *Performance feedback* is used to achieve automatic system synthesis, i.e., the selection of the "best" system from a restricted but useful class of possibilities. The designer "trains" the system to give the correct responses by "showing" it examples of inputs and respective desired outputs. The more examples "seen", the better is the system performance. System competence will be directly and quantitatively related to amount of experience.

B. A Neuron Element

In Fig. 1, a combinatorial logical circuit is shown which is a typical element in the adaptive switching circuits to be considered. This element bears some resemblance to a "neuron" model introduced by von Neurman[3], whence the name.

The binary input signals on the individual lines have values of $+1$ or -1, rather than the usual values of 1

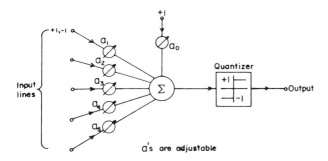

Figure 1 An adjustable neuron.

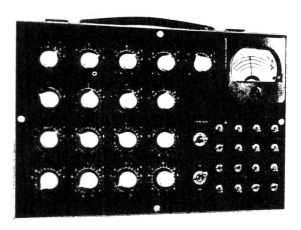

Figure 2 Adaline.

or 0. Within the neuron, a linear combination of the input signals is formed. The weights are the gains a_1, $a_2, \ldots$, which could have both positive and negative values. The output signal is $+1$ if this weighted sum is greater than a certain threshold, and -1 otherwise. The threshold level is determined by the setting of a_0, whose input is permanently connected to a $+1$ source. Varying a_0 varies a constant added to the linear combination of input signals.

For fixed gain settings, each of the 2^5 possible input combinations would cause either a $+1$ or -1 output. Thus, all possible inputs are classified into two categories. The input-output relationship is determined by choice of the gains $a_0, \ldots, a_5$. In the adaptive neuron, these gains are set during the "training" procedure.

In general, there are 2^{2^5} different input-output relationships or truth functions by which the five input variables can be mapped into the single output variable. Only a subset of these, *the linearly separated truth functions*[4], can be realized by all possible choices of the gains of the neuron of Fig. 1. Although this subset is not all-inclusive*, it is a useful subset, and it is "searchable", i.e., the "best" function in many practical cases can be found iteratively without trying all functions within the subset.

Application of this neuron in adaptive pattern classifiers was first made by Mattson.[5,6] He has shown that complete generality in choice of switching function could be had by combining these neurons. He devised an iterative digital computer routine for finding the best set of a's for the classification of noisy geometric patterns. An iterative procedure having similar objectives has been devised by these authors and is described in the next section. The latter procedure is quite simple to implement, and can be analyzed by statistical methods that have already been developed for the analysis of adaptive sampled-data systems.

*It becomes a vanishingly small fraction of all possible switching functions as the number of inputs gets large.

C. An Adaptive Pattern Classifier

An adaptive pattern classification machine (called "Adaline", for *ada*ptive *lin*ear) has been constructed for the purpose of illustrating adaptive behavior and artificial learning. A photograph of this machine, which is about the size of a lunch pail, is shown in Fig. 2.

During a training phase, crude geometric patterns are fed to the machine by setting the toggle switches in the 4×4 input switch array. Setting another toggle switch (the reference switch) tells the machine whether the desired output for the particular input pattern is $+1$ or -1. The system learns something from each pattern and accordingly experiences a design change. The machine's total experience is stored in the values of the weights $a_0 \ldots a_{16}$. The machine can be trained on undistorted noise-free patterns by repeating them over and over until the iterative search process converges, or it can be trained on a sequence of noisy patterns on a one-pass basis such that the iterative process converges statistically. Combinations of these methods can be accommodated simultaneously. After training, the machine can be used to classify the original patterns and noisy or distorted versions of these patterns.

A block schematic of Adaline is shown in Fig. 3. In the actual machine, the quantizer is not built in as a device but is accomplished by the operator in viewing the output meter. Different quantizers (2-level, 3-level, 4-level) are realized by using the appropriate meter scales (see Fig. 2). Adaline can be used to classify patterns into several categories by using multi-level quantizers and by following exactly the same adaptive procedure.

The following is a description of the iterative searching routine. A pattern is fed to the machine, and the reference switch is set to correspond to the desired output. The error (see Fig. 3) is then read (by switching

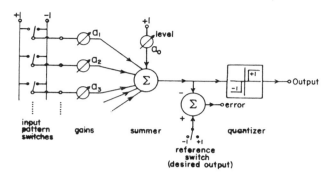

Figure 3 Schematic of Adaline.

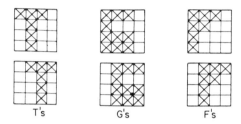

Figure 4 Patterns for classification experiment.

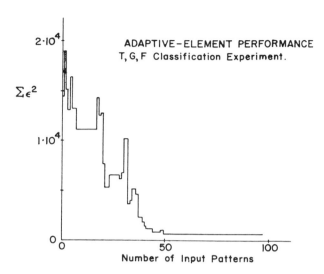

Figure 5 Adaptive-element performance curve.

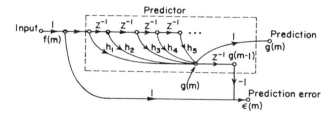

Figure 6 An adjustable sampled-data predictor.

the reference switch; the error voltage appears on the meter, rather than the neuron output voltage). All gains including the level are to be changed by the same absolute magnitude, such that the error is brought to zero. This is accomplished by changing each gain (which could be positive or negative) in the direction which will diminish the error by an amount which reduces the error magnitude by $1/17$. The 17 gains may be changed in any sequence, and after all changes are made, the error for the present input pattern is zero. Switching the reference back, the meter reads exactly the desired output. The next pattern, and its desired output, is presented and the error is read. The same adjustment routine is followed and the error is brought to zero. If the first pattern were reapplied at this point, the error would be small but not necessarily zero. More patterns are inserted in like manner. Convergence is indicated by small errors (before adaption), with small fluctuations about a stable root mean-square value. The iterative routine is purely mechanical, and requires no thought on the part of the operator. Electronic automation of this procedure will be discussed below.

The results of a typical adaption on six noiseless patterns is given in Figs. 4 and 5. The patterns were selected in a random sequence, and were classified into 3 categories. Each T was to be mapped to $+60$ on the meter dial, each G to 0, and each F to -60. As a measure of performance, after each adaptation, all six patterns were read in (without adaptation) and six

errors were read. The sum of their squares denoted by $\sum \varepsilon^2$ was computed and plotted. Fig. 5 shows the learning curve for the case in which all gains were initially zero.

D. Statistical Theory of Adaption for Sampled-Data Systems

This section is a summary of the portions of Widrow's statistical theory of adaption for sampled-data systems[7,8] that is useful in the analysis of adaptive switching circuits.

Consider the general linear sampled-data system formed of a tapped delay line, shown in Fig. 6. This system is intended to be a statistical predictor. The individual inpulses of the impulse response may be adjusted in the following manner. Apply a mean square reading meter to $\varepsilon(m)$, the difference between the present input and the delayed prediction. This meter will measure mean square error in prediction. Adjust h_1, h_2, h_3, ..., until the meter reading is minimized.

The problem of adjusting the h's is not trivial, because their effects upon performance interact. Suppose

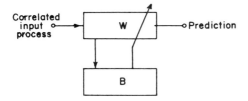

Figure 7 An adaptive predictor.

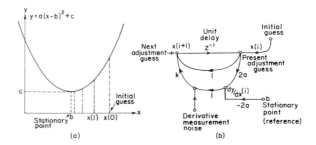

Figure 8 One-dimensional surface searching.

that the predictor has only two impulses in its impulse response, h_1 and h_2. The mean square error for any setting of h_1 and h_2 can be readily derived:

$$\varepsilon(m) = f(m) - h_1 f(m-1) - h_2 f(m-2)$$

$$\overline{\varepsilon^2}(m) = \phi_{ff}(0)h_1^2 + \phi_{ff}(0)h_1^2 - 2\phi_{ff}(1)h_1 - 2\phi_{ff}(2)h_2$$

$$+ 2\phi_{ff}(1)h_1 h_2 + \phi_{ff}(0) \qquad (1)$$

The discrete autocorrelation function of the input is $\phi_{ff}(j)$.

The mean square error given by equations (1) is what the mean square meter would read if it were to average over very large sampled size. The mean square error is a parabolic function of the predictor adjustments h_1 and h_2, and, in general, can easily be shown to be a quadratic function of such adjustments, regardless of how many there are.

The optimum n-impulse predictor can be derived analytically be setting the partial derivatives of $\overline{\varepsilon^2}$ of equation (1) equal to zero. This is the discrete analogue of Wiener's optimization[7] of continuous filters. Finding the optimum system experimentally is the same as finding a minimum of a paraboloid in n dimensions. This could be done manually by having a human operator read the meter and set the adjustment, or it could be done automatically by making use of any one of several iterative gradient methods for surface-searching, as devised by numerical analysts. When either of these schemes is employed, an adaptive system results that consists essentially of a "worker" and a "boss". The worker in this case predicts, whereas the boss has the job of adjusting the worker.

Figure 7 is a block-diagram representation of such a basic adaptive unit. The boss continually seeks a better worker by trial and error experimentation with the structure of the worker. Adaption is a multi-dimensional *performance feedback* process. The "error" signal in the feedback control sense is the gradient of mean square error with respect to adjustment.

Many of the commonly used gradient methods search surfaces for stationary points by making changes in the independent variables (starting with an initial guess) in proportion to measured partial derivatives to obtain

the next guess, and so forth. These methods give rise to geometric (exponential) decays in the independent variables as they approach a stationary point for second-degree or quadratic surfaces. One-dimensional surface-searching is illustrated in Fig. 8.

The surface being explored in Fig. 8 is given by Eq. (2). The first and second derivatives are given by Eqs. (3) and (4).

$$y = a(x - b)^2 + c \qquad (2)$$

$$\frac{dy}{dx} = 2a(x - b) \qquad (3)$$

$$\frac{d^2 y}{dx^2} = 2a \qquad (4)$$

A sampled-data feedback model of the iterative process is shown in Fig. 8(b). This flow-graph can be reduced, and the resulting characteristic equation is

$$-(2ak + 1)z^{-1} + 1 = 0 \qquad (5)$$

The iterative process is stable when $-1/a < k < 0$, and transients decay completely in one step when $k = -1/2a$.

"Noise" in the measurements of the mean square error surface due to small sample size causes noisy derivative measurements. These noises enter the adaption process, as indicated in Fig. 8(b), and cause noisy system adjustments.

Variance in x about the optimum value causes the average of y to be greater than the minimum value c. The increase in $\overline{y}$ equals the variance in x multiplied by a, as can be seen from Eq. (2). It is useful to define a dimensionless parameter M the "misadjustment", as the ratio of the mean increase in mean square error to the minimum mean square error. It is a measure of how the system performs on the average, after adapting transients have died out, compared with the fixed optimum system. With regard to the curve of Fig. 8,

$$M = \frac{\overline{y} - c}{c} \qquad (6)$$

More detailed derivations of misadjustment formulas covering several different methods of surface searching and derivative measurement are presented in Refs. 7 and 8. The particular formulas which can be applied to the analysis of adaptive switching circuits are the following.

When derivatives are measured by data repeating, i.e., when the same system input data is applied for both N "forward" and N "backward" measurements of mean square error, the misadjustment is given by

$$M = \frac{1}{2(N\tau)} \qquad (7)$$

τ is the time constant of the iterative process of Fig. 8, and is equal to $-1/2ak$. A unit time constant means that the adjustment error decreases by a factor $1/e$ per iteration cycle. Equation (7) is conservative, and appreciably so only for small values of τ, less than 1. In the limiting case of one-step adaption, $\tau = 0$ and the appropriate misadjustment formula is

$$M = \frac{1}{N} \qquad (8)$$

In deriving Formulas (7) and (8), it has been assumed that the error samples are Gaussian distributed, with zero mean, and are uncorrelated. It can be shown that these results are highly insensitive to this distribution density shape, and are appreciably affected by correlation only when it exceeds 0.8. It is interesting to note that the quality of adaption depends only on the number of samples "seen" by the system in adapting.

The expressions (7) and (8) are based on the supposition that fresh data is brought in for each cycle of iteration. If the system adapts on a fixed body of N error samples the misadjustment is given by Formula (8). When there are m interacting adjustments instead of just one, Expressions (7) and (8) may be generalized by multiplication by m.

E. Statistical Theory of Adaption for the Adaptive Neuron Element

The error signal measured and used in adaption of the neuron of Fig. 1 is the difference between the desired output and the sum before quantization. This error is indicated by ε in Fig. 9. The actual neuron error, indicated by ε_n in Fig. 9, is the difference between the neuron output and the desired output.

The objective of adaption is the following. Given a collection of input patterns and the associated desired outputs, find the best set of weights $a_0, a_1, \ldots a_m$ to minimize the mean square of the neuron error, $\overline{\varepsilon_n^2}$.

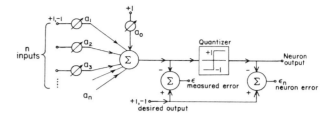

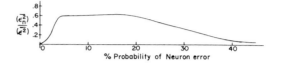

Figure 9 Relations between neuron errors and measured errors.

Individual neuron errors could only have the values of $+2$, 0, and -2 with a two-level quantizer. Minimization of $\overline{\varepsilon_n^2}$ is therefore equivalent to minimizing the average number of neuron errors.

The simple adaption procedure described in this paper minimizes $\overline{\varepsilon^2}$ rather than $\overline{\varepsilon_n^2}$. The measured error ε has zero mean (a consequence of the minimization of $\overline{\varepsilon^2}$) and will be assumed to be Gaussian-distributed. By making use of certain geometric arguments or by using a statistical theory of amplitude quantization,[10] it can be shown that $\overline{\varepsilon_n^2}$ is a monotonic function of $\overline{\varepsilon^2}$, and that minimization of $\overline{\varepsilon^2}$ is equivalent to minimization of $\overline{\varepsilon_n^2}$ and to minimization of the probability of neuron error. The ratio of these mean squares has been calculated and is plotted in Fig. 9 as a function of the neuron error probability.

Given any collection of input patterns and the associated desired outputs, the measured mean square error $\overline{\varepsilon^2}$ must be a precisely parabolic function of the gain settings, $a_0, \ldots a_n$. Let the kth pattern be indicated as the vector $S(k) = s_1(k), s_2(k), \ldots s_n(k)$. The s's have values of $+1$ or -1, and represent the n input components numbered in a fixed manner. The kth error is

$$\varepsilon(k) = d(k) - a_0 - a_1 s_1(k) - a_2 s_2(k) - \cdots - a_n s_n(k) \qquad (9)$$

For simplicity, let the neuron have only two input lines and a level control. The square of the error is accordingly

$$\varepsilon^2(k) = d^2(k) + a_0^2 + s_1^2(k)a_1^2 + s_2^2(k)a_2^2$$
$$- 2d(k)a_0 - 2d(k)s_1(k)a_1$$
$$- 2d(k)s_2(k)a_2$$
$$+ 2s_1(k)a_0 a_1 + 2s_2(k)a_0 a_2$$
$$+ 2s_1(k)s_2(k)a_1 a_2 \qquad (10)$$

The mean square error averaged over k is

$$\overline{\varepsilon^2} = a_0^2 + \phi(s_1, s_1)a_1^2 + \phi(s_2, s_2)a_2^2 - \overline{d}a_0$$
$$- 2\phi(d, s_1)a_1 - 2\phi(d, s_2)a_2$$
$$+ 2\overline{s}_1 a_0 a_1 + 2\overline{s}_2 a_0 a_2$$
$$+ 2\phi(s_1, s_2)a_1 a_2 + \phi(d, d) \tag{11}$$

The ϕ's are *spatial* correlations. $\phi(s_1, s_2) = \overline{s_1 s_2}$, etc. Note that $\phi(s_j, s_j) = \overline{s_j s_j} = 1$.

Adjusting the a's to minimize $\overline{\varepsilon^2}$ is equivalent to searching a parabolic stochastic surface (having as many dimensions as there are a's) for a minimum. How well this surface can be searched will be limited by sample size, i.e., by the number of patterns seen in the searching process.

The method of searching that has proven most useful is the method of steepest descent. Vector adjustment changes are made in the direction of the gradient. Transients consist of sums of geometric sequence components (there are as many natural "frequencies" as the number of adjustments, as can be seen from generalization of the flow graph of Fig. 9—see Ref. 9). It can be shown that the method of steepest descent will be stable when the proportionality constant k between partial derivative and size of change is less than the reciprocal of the second partial derivative. It can also be shown that when k is small, transients can be approximately represented as being of the single time constant 1/2k.

The method of adaption that has been used requires an extremely small sample size per iteration cycle, namely one pattern. One-pattern-at-a-time adaption has the advantages that derivatives are very easy to measure and that no storage is required within the adaptive machinery except for the gain values (which contain the past experience of the neuron).

The square of the error for a single pattern (the mean square error for a sample size of one) is given by Eq. (10). The partial derivatives are

$$\frac{\partial \varepsilon^2(k)}{\partial a_0} = [-2d(k) + 2a_0 + 2s_1(k)a_1 + 2s_2(k)a_2]$$

$$\frac{\partial \varepsilon^2(k)}{\partial a_1} = s_1(k)[-2d(k) + 2a_0 + 2s_1(k)a_1 + 2s_2(k)a_2]$$

$$\frac{\partial \varepsilon^2(k)}{\partial a_2} = s_2(k)[-2d(k) + 2a_0 + 2s_1(k)a_1 + 2s_2(k)a_2] \tag{12}$$

Comparison of the Eqs. (12) with Eq. (9) shows that the derivatives are simply related to the measured error, and suggest that the *derivatives could be measured without squaring and averaging and without actual differentiation*. The jth partial derivative is given by

$$\frac{\partial \varepsilon^2(k)}{\partial a_j} = -2s_j(k)\varepsilon(k) \tag{13}$$

It follows that all partial derivatives have the same magnitude, and have signs determined by the error sign and the respective input signal signs. The procedure described in Sec. C for bringing $\varepsilon(k)$ to zero with each successive input pattern gives the constant k a value of $1/2(n + 1)$. From the previous discussion we see that the time constant of the iterative process is therefore $\tau = (n + 1)$ patterns. On the 4×4 adaline, there are $n = 16$ input line gains plus a level control. Therefore, the time constant should be roughly 17 patterns (for verification, see the learning curve of Fig. 5). The search procedure could be readily modified to speed up or slow down the adaption process by adjusting k.

The misadjustment Formulas (7) and (8) when applied to the adaptive neuron give the per unit increase in measured mean square error as a result of adapting on a finite number of patterns. Since the ratio of probability of neuron error to the mean square error $\overline{\varepsilon^2}$ is essentially constant over a wide range of error probabilities (Fig. 9), the misadjustment as expressed by Formulas (7) and (8) may be interpreted in terms of the ratio of the increase in error probability to the minimum error probability.

If adaption is accomplished by injection of a fresh pattern each iteration cycle, the misadjustment, as derived from Eq. (7), is

$$M = \frac{(n + 1)}{2\tau} \tag{14}$$

Following the procedure of bringing $\varepsilon(k)$ to zero each iteration cycle, the misadjustment is

$$M = \frac{(n + 1)}{2\tau} = \frac{(n + 1)}{2(n + 1)} = \frac{1}{2} \tag{15}$$

If adaption is accomplished by taking a fixed collection of N patterns and repeating them over and over for several time constants (where the time constant is long, several times N), the misadjustment, as derived from Eq. (8), is

$$M = \frac{(n + 1)}{N} \tag{16}$$

Simulation tests have shown that the misadjustment formulas are highly accurate over a very wide range of pattern and noise characteristics. A description of a typical experiment and its results is given in Fig. 10.

Noisy 3×3 patterns were generated by randomly injecting errors in ten percent of the the positions of the four "pure" patterns, X, T, C, J.

The best system, arrived at by slow precise adaption

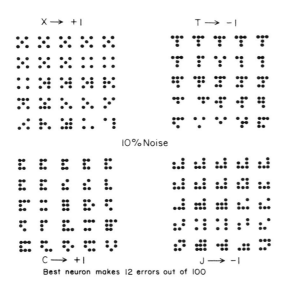

Figure 10 Experimental adaption on 10 noisy 3 × 3 patterns.

EXPERIMENT #	PATTERNS ADAPTED ON	NUMBER OF ERRORS	MISADJUSTMENT
1	95, 79, 07, 60 73, 61, 08, 02, 72, 26	25	$M = \frac{25-12}{12} = 108\%$
2	70, 69, 52, 55, 32 97, 30, 38, 87, 01	19	$M = \frac{19-12}{12} = 58\%$
3	65, 12, 84, 83, 34 38, 71, 66, 13, 80	20	$M = \frac{20-12}{12} = 67\%$
4	07, 42, 85, 88, 63 35, 37, 92, 79, 22	28	$M = \frac{28-12}{12} = 133\%$

on the full body of 100 noisy patterns, we able to classify these patterns as desired except for twelve errors. The gains were then set to zero and ten patterns were chosen at random. The best system for these patterns was arrived at and tested on the full body of 100 patterns. Twenty-five classification errors out of 100 were made. The misadjustment was 108 percent. The experiment was repeated three more times, and the misadjustments that resulted, in order, were 58 percent, 67 percent and 133 percent. Since N = 10 patterns and n = 9 input lines, the theoretical misadjustment was

$$M = \frac{n+1}{N} = 100 \text{ percent}$$

An average taken over the four experiments gives a measured misadjustment of 91.5 percent.

The adaptive classifier can adapt after seeing remarkably few patterns. A misadjustment of 20 percent should be acceptable in most applications. To achieve this, all one has to do is supply the adaptive classifier with a number of patterns equal to five times the number of input lines, regardless of how noisy the patterns are and how difficult the "pure" patterns are to separate. Although the misadjustment formulas have been derived for the specific classifier consisting

of a single adaptive neuron, it is suspected that the following "rule of thumb" will apply fairly well to all adaptive classifiers: *the number of patterns required to train an adaptive classifier is equal to several times the number of bits per pattern.*

F. Networks of Adaptive Neurons

Linearly separable* pure patterns and noisy versions of them are readily classified by the single neuron. Non-linearly separable pure patterns and their noisy equivalents can also be separated by a single neuron, but absolute performance can be improved and the generality of the classification scheme can be greatly increased by using more than one neuron.

Two Adalines were combined by using the following adaption procedure: if the desired output for a given input pattern applied to both machines was −1, then both machines were adapted in the usual manner to ensure this; if the desired output was +1, the machine with the smallest measured error ε was assigned to adapt to give a +1 output while the other machine remained unchanged. If either or both machines gave outputs of +1, the pattern was classified as +1. If both machines gave −1 outputs, the pattern was classified as −1.

This procedure assigns specific "responsibility" to the neuron that can most easily assume it. If, at the beginning of adaption, a given neuron takes responsibility for producing a +1 with a certain input pattern, it will invariably take this responsibility each time the pattern is applied during training. Notice that it is not necessary for a teacher to assign responsibility. The combination does this automatically and requires only input patterns and the associated desired outputs, like the single neuron.

Various classification problems could be solved simultaneously by multiplexing neurons or combinations of neurons. One neuron might be trained to decide whether the man in a given picture does or does not have a green tie, while another neuron or combination could be trained to decide whether or not the man has a checkered shirt. Each neuron or combination has its own output line, and each is fed the appropriate desired output signal during training. The input signals are common to all neurons. In this manner, it is possible to form adaptive classifiers that can separate with great accuracy large quantities of complicated patterns into many output categories. All that is needed is large quantities of adaptive neurons.

*A more complete discussion of linear separability is given in references 4 and 5.

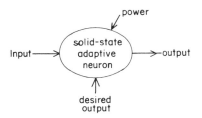

Figure 11 Electronic automatically adapted neuron.

G. Adaptive Microelectronic Systems

The structure of the neuron described in this paper and its adaption procedure is sufficiently simple that an effort is under way to develop a physical device which is an all-electronic fully automatic Adaline. The objective is a self-contained device, like the one sketched in Fig. 11, that has a signal input line, a "desired output" input line (actuated during training only), an output line, and a power supply. The device itself should be suitable for mass production, should contain few parts, should be reliable, and probably should consist of solid-state components.

To have such an adaptive neuron, it is necessary to be able to store the gain values, which could be positive or negative, in such manner that these values could be changed electronically.

Present efforts have been based on the use of multi-aperture magnetic cores (MAD elements[11]). The special characteristics of these cores permit multilevel storage with continuous, non-destructive read out. In addition, the stored levels are easily changed by small controlled amounts, with the direction of the change being determined by logic performed by the MAD element. The results of this work have shown that macroscopic adaptive neurons made of MAD elements will soon exist, and that with the use of thin ferromagnetic films, adaptive microelectronic neurons will ultimately exist.

H. Applications for Adaptive Logical Circuit Elements

If a computer were built of adaptive neurons, details of structure could be imparted by the designer by training (showing its examples of what he would like to do) rather than by direct designing. This design concept becomes more significant as size and complexity of digital systems increase. The demands of modern technology are such that larger and more complex digital systems are continually being contemplated, and in step with this, progress in microelectronics makes such systems physically and economically possible.

The problem of reliability is greatly aggravated by increase in size and complexity.

Shannon and Moore[12] and von Neumann[3] have proposed schemes for increasing the reliability of fixed digital systems by using redundancy. The reliability of systems may be increased further by combining adaption and redundancy. Consider a multiplex consisting of three machines solving the same problem with the same input data. Let the output of each machine be a single binary number, expressed as $+1$ or -1. If these machines were perfectly reliable, their outputs would always agree. If not, then von Neumann proposed that the majority should rule. The neuron shown in Fig. 1 with a_0 set to zero, and the other gains set to $+1$ would give a majority output. Each machine has equal vote. Unequal vote (higher vote going to the more reliable machine) is possible by making the a's adjustable, and causing these adjustments to be made automatically to optimize performance. The adaptive vote taker, identical to the adaptive neuron, can be trained by periodically injecting a certain input when the desired output is known. The adaptive vote taker could ideally give the correct outcome with only a single correct machine by giving it a heavy vote and attenuating the votes of the unreliable machines. This is in effect an adaptive routing procedure for information flow, and allows systems in a small measure to be self-healing.

One of the most promising areas of research in computer system theory is that of problem-solving machines, theorem-proving machines, and artificially "intelligent" machines. The earliest proponents of this research were Turing and Shannon.[13] Their suggestions were put to practice with some success by Newell, Simon, and Shaw,[14] by Samuel,[15] and by others.

An automatic problem-solving computer should have a memory system from which information could be extracted according to classification rather than according to address number. The use of stored games, or "rote learning",[15] would be considerably more powerful if it were possible to extract from the memory previous situations that are *similar* and not necessarily *identical* to the current situation. Far less experience and storage would be needed to adapt to a given level of competence. The extent of classification before storing should be slight (e.g., is the pattern of checkers or of chess?), and a consistent scheme for the arrangement of the pattern bits should be established before storing. Final classification should be done within the memory itself. Each storage register might contain an Adaline or a network of Adalines.

A request from a "central control" for a certain type of information would be sent to every register in the memory simultaneously. This has the effect of setting the adjustments of all the Adalines. Only the registers whose classifiers respond properly (e.g., give $+1$ outputs) answer the request and transmit their information back to the "central control."

Very sophisticated learning procedures would become possible if one had such recall-by-association parallel-access memory systems. The simplicity of Adaline and the progress being made in microelectronics gives a strong indication that such memory systems will come into existence in the not too distant future.

Acknowledgment

This research was supported by the Office of Naval Research under contract with Stanford University.

References

1. C. E. Shannon, "A symbolic analysis of relay and switching circuits," Trans. of AIEE, Vol. 37, 1938, pp. 713–723.

2. S. H. Caldwell, Switching Circuits and Logical Design, Wiley, 1958.

3. J. von Neuman, "Probabilistic Logic and Synthesis of Reliable Organisms from Unreliable Components", in Automata Studies, Princeton University Press, 1956.

4. Robert McNaughton, "Unate truth functions", Tech. Report No. 4, Appl. Math. and Stat. Lab., Stanford University, October 21, 1957.

5. R. L. Mattson, "The design and analysis of an adaptive system for statistical classification", S. M. Thesis, MIT, May 22, 1959.

6. R. L. Mattson, "A self-organizing logical system", 1959 Eastern Joint Computer Conference Convention Record, Inst. of Radio Eng., N.Y.

7. B. Widrow, "Adaptive sampled-data systems—a statistical theory of adaptation", 1959 WESCON Convention Record, Part 4.

8. B. Widrow, "Adaptive sampled-data systems", IFAC Moscow Congress Record, Butterworth Publications, London, 1960.

9. Wiener, N., Extrapolation, Interpolation, and Smoothing of Stationary Time Series with Engineering Applications, New York, Wiley, 1949.

10. B. Widrow, "A study of rough amplitude quantization by means of Nyquist sampling theory", IRE Transactions, PGCT, Vol. CT-3, Number 4, Dec. 1956.

11. H. D. Crane, "A high-speed logic system using magnetic elements and connecting wire only", Proc. IRE, pp. 63–73, Jan. 1959.

12. E. F. Moore and C. E. Shannon, "Reliable circuits using less reliable relays", Bell Telephone System Technical Publications, Monograph 2696, January 1957.

13. C. E. Shannon, "Programming a computer for playing chess", Phil. Magn., 41, p. 256, March 1950.

14. A. Newell, J. C. Shaw, and H. A. Simon, "Chess-playing programs and the problem of complexity", IBM Journal of Res. and Dev., 2, p. 320, Oct. 1958.

15. A. L. Samuel, "Some studies in machine learning using the game of checkers", IBM Journal of Res. and Dev., Vol. 3, No. 3, pp. 211–229, July 1959.

11
Introduction

(1962)
H. D. Block

The Perceptron: a model for brain functioning. I
Reviews of Modern Physics 34:123–135

This paper, when compared with the earlier description of the simple perceptron given in paper 8 (Rosenblatt, 1958), demonstrates the great increase in the level of analysis that had been achieved in a few years.

Work had progressed using the three techniques used for research on neural networks today: mathematical analysis, computer simulation, and special purpose hardware. Rosenblatt and coworkers at Cornell had built a "Mark I" perceptron with a retina containing 400 photosensitive receptors, on a 20 by 20 grid. The "Mark I" had 512 associator units and 8 response units (R-units) for the final classifications. These numbers are comparable with those that would be used routinely today for network simulations on a scientific workstation of roughly VAX power. A number of results from the Mark I are presented in the paper.

It was now possible to ask, and answer, the appropriate questions. First, given a set of patterns that the perceptron was supposed to classify, was such a classification possible at all? Rosenblatt had found experimentally in the 1958 paper that the perceptron did not classify well with random vectors, but did better with classes with correlative structure. The basic classification element in the perceptron is the R-unit, which forms a weighted sum of the active elements times the connection strengths. The R-unit had a threshold. If the sum was greater than threshold, the R-unit took value 1; if less than threshold, the R-unit took value -1. Suppose there are n inputs to the R-unit. Then all possible inputs to the R-unit are points in an n-dimensional space. Consider the points in this space where the output of the summer equals the threshold. This occurs when the sum of the product of the synaptic weights times the coordinates of the inputs equals the threshold. This equation is the equation of a hyperplane in the n-dimensional space. A classification can be learned precisely only if this hyperplane can separate all the points that are to have value 1 in the R-unit from those with value -1. If a hyperplane cannot separate the points, then the classification cannot be learned exactly. This property is called linear separability. The concept is very intuitive. Figure 1 shows two sets of points in the plane, one set linearly separable, the other not.

Unfortunately, many important classifications are not linearly separable. It was limitations of this kind that led to a loss of interest in perceptrons and related devices because, as Minksy and Papert showed (paper 13), *simple* perceptrons could not compute a number of important quantities that a brain was clearly able to compute.

Second, if a solution existed, could the perceptron find it? For some learning rules, the answer could be proved to be yes. This paper contains a concise proof of the famous "Perceptron Convergence Theorem" for a simple perceptron. The proof analyzes a simple perceptron using an error correcting learning rule: the R-unit was given an input with a known correct classification. If the perceptron correctly classified the

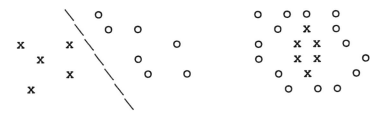

Figure 1
(Left) Crosses and circles are linearly separable. (Right) Crosses and circles are not linearly separable.

input, none of the weights were changed. If there was an error, then the active units, since they were wrong, had their weights decremented (if the R-unit was supposed to be $+1$) or incremented (if the R-unit was supposed to be -1). Inactive connections were untouched. The detailed proof is a little involved. The important result, however, is that the connection weights of the perceptron could indeed learn to classify correctly, if classification was possible. This was a striking and powerful result.

The result was sensitive to the learning rule. Block shows that if instead of the rule given above, we use "forced" learning, where weights are changed regardless of the response of the R-unit, the theorem does not hold. For best results, it was necessary to use the "If it works, don't change it" rule. This requirement has a deleterious effect on the learning rate. As the classification gets more and more accurate, weight changes occur less and less frequently. Error free classification can, therefore, take a very long time. Some error correction rules do not suffer this limitation (Widrow and Hoff, paper 10) since the error signal can be reduced by weight changes even if the binary classification was correct.

The paper concludes with a brief discussion of the problems of generalization. Given the structure of the inputs to the R-units, it can be concluded that stimuli that overlap a great deal with correctly classified inputs will tend to be responded to similarly. Generalization of responses to inputs other than what the system saw during learning is one of the most important properties of a neural network and can be accomplished by a number of different and important mechanisms such as simple correlation (Cooper, paper 16), prototype formation (Knapp and Anderson, paper 36), or formation of attractor basins (see, in particular, Amari, paper 21; Anderson et al., paper 22; Hopfield, paper 27).

The best and clearest review of learning systems from this period is found in Nilson's classic book, *Learning Machines* (1965). To see the complexity of analysis of function possible using variants of the perceptron, it is worth looking at the paper by Block, Knight, and Rosenblatt (1962) that immediately succeeded this one in *Reviews of Modern Physics*. It analyzes a four-layer series coupled perceptron using differential equations. This perceptron variant was capable of remarkable feats of adaption and could learn some temporal sequences.

References

H. D. Block, B. W. Knight, Jr., and F. Rosenblatt (1962), "Analysis of a four-layer series coupled perceptron, II," *Reviews of Modern Physics* 34:135–142.

N. J. Nilson (1965), *Learning Machines—Foundations of Trainable Pattern Classifying Systems*, New York: McGraw-Hill.

(1962)
H. D. Block

The Perceptron: a model for brain functioning. I
Reviews of Modern Physics 34:123–135

THE Perceptron is a self-organizing or adaptive system proposed by Rosenblatt.[1] Its primary purpose is to shed some light on the problem of explaining brain function in terms of brain structure. It also has technological applications as a pattern-recognizing device, but here our emphasis is on the brain function-structure problem. The technological aspects are not completely irrelevant however, since a model, no matter how appealing it may appear from the point of view of structural similarity, must also be judged on the basis of its performance.

In brief a perceptron consists of a *retina* of *sensory units* (for example photocells); these are connected (for example by wires) to *associator units*. The connections are many to many and random. The associator units may be connected to each other or to *response units*. When a *stimulus* is presented to the retina (for example as a pattern of illumination) impulses are conducted from the activated sensory units to the associator units. If the total signal arriving at an associator unit exceeds a certain *threshold* then the associator becomes *active* and sends an impulse to units to which it is connected. The magnitude of the impulse carried by certain connections depends on the past activity of their termini according to certain preassigned *reinforcement rules*. Thus the device changes its internal functional properties. The resultant behavior exhibits, as is shown in the text, interesting aspects of learning, discrimination, generalization, and memory.

We present here a survey of the work to date. We also give in detail the proofs of certain theorems which illustrate some of the methods of analysis and which, in view of their central position, illuminate a wide area. Further elaboration and details are given in the references cited, particularly in Rosenblatt's summary report[1] which presents a detailed and comprehensive exposition of the entire subject.

BACKGROUND AND MOTIVATION

1. Structure and Function

For most of the organs of the body (e.g., the heart, lungs, kidneys, stomach, intestines, liver, spleen, blood stream, bones, skin, peripheral nervous system, etc.) we have some idea of the functions each performs and some explanation of how the structure operates to achieve the function.

Conversely, for most of the functions necessary to sustain life (e.g., locomotion, sensitivity, respiration, reproduction, digestion, nutrition, excretion, etc.) we have some idea of which structures are involved and the manner in which they implement the function.

It is now generally believed that the brain is the principal organ involved in thought, but there is no reasonably precise explanation of how the action of the brain structure produces the "higher functions."

Admittedly there are still open problems for the other organs. We do not know enough to build a real lung. But we do know in a general way that the function of respiration is to replace carbon dioxide in the blood by oxygen. The lung offers a large thin surface area where air can get on one side and blood on the other, and the exchange can take place. Admittedly, for a specialist in this field the really interesting problems start at this point. Nevertheless we take it as evident that at the present time the degree of our ignorance on the brain "function-structure problem" is of a higher order than for the corresponding "respiration-lung problem." To pose the problem precisely, we should now define brain function and brain structure.

2. Brain Function

Since many psychological phenomena have not yet been investigated, any description of "brain function" is necessarily incomplete. Moreover, even those phenomena which have been intensively studied often admit a multiplicity of interpretations.[2,3] It therefore seems pointless to attempt to make a precise definition of "brain function." Clearly there is such a phenomenon. Clearly it is related to perception, memory, discrimination, recognition, association, comparison, learning, communication, reasoning, and attention. An operational definition of these terms might be given in terms of the relation between (1) a sequence of inputs, which might be suitable physical stimuli; e.g., a light pattern on the eye, sound on the ear, pressure or heat on the skin, etc., and (2) a sequence of outputs, which might be the observable response of the subject.

Such a definition might be criticized as being too narrow in that it neglects *thought* which is not triggered by an observable stimulus or displayed in an observable response. This objection might be answered by the argument that, in principle, thought must be accompanied by physical, hence observable, changes some-

---

Research sponsored by the Office of Naval Research.
[1] F. Rosenblatt, *Principles of Neurodynamics: Perceptrons and the Theory of Brain Mechanisms* (Spartan Books, Washington, D. C., 1961).

[2] E. G. Boring, H. S. Langfeld, and H. P. Weld, *Foundations of Psychology* (John Wiley & Sons, Inc., New York, 1948).
[3] E. R. Hilgard, *Introduction to Psychology* (Harcourt Brace & Company, Inc., New York, 1957), Chaps. 10–31.

where in the brain and these might be interpreted as the inputs and the outputs.

We are not concerned here with the difficult problems of determining which behavior patterns are *innate* and which ones are *learned*. While questions of innate behavior, such as the navigation systems of certain birds and the mechanisms of their genetic transmission, are clearly formidable, nevertheless most scientists expect that the explanations, when they come, will be based on conventional physics, chemistry, and mathematical analysis and could, in principle, be duplicated by engineering techniques a few orders of ingenuity beyond our present state. For learned behavior on the other hand, we believe, as will be elaborated subsequently, that the physics and chemistry involved may be straightforward, but the organization of information and the mode of operation are based on radically novel principles requiring entirely new concepts of analysis. Thus from the viewpoint of the present paper we would not be particularly concerned with neuronal circuitry yielding an "unconditioned reflex" behavior pattern, such as the pupillary contraction of the eye under bright light. We would, on the other hand, be very interested in learning what physical changes occur in a child's brain when he learns to recognize the letter "A" and how, in precise terms, these changes account for the learning.

Similarly it is possible to arrange circuits so that a machine will perform tasks which appear to have much in common with thinking. For example patter-recognizers,[4,5] chess players,[6] and other "thinking machines" have been built.[7] Although such special purpose machines clarify the nature of the logical problem to be solved, they contribute little to the understanding of how the brain is organized to perform these functions. Even the "heuristic reasoning machines,"[8] while they reveal a great deal about the nature of human reasoning, are several orders of abstraction beyond the basic mechanisms with which we are concerned here.

Let us leave the definition of "brain function" in this rather vague state. These concepts are very slippery and will probably never be formulated in a way that will satisfy everybody.

3. Gross Brain Structure

The gross anatomy of the brain has been well studied.[9–12] Certain regions appear to have well defined functions; in particular, stimulation of points in the post central gyrus causes the subject to feel sensations, while stimulation of the precentral gyrus causes motor action. The body is precisely mapped (with distortions in scale) on these regions, so that the response is quite specific. These facts do not shed any light on our problem, however, since they serve only to move the input and output terminals from the receptor and motor organs onto the brain surface. The main question remains, "what happens between the input and the output?" While other localized areas appear to be concerned with specific functions,[13] the localization usually implies a predominance of function rather than an absolute localization. Furthermore it is also true that there is a certain equi-potentiality involved in brain functions, in which the functions of extirpated parts can be taken over by other parts and the loss of function varies as the mass of brain removed.[14] Indeed large sections of the brain can be removed with no apparent permanent loss in function. The search for specific structures performing specific functions has been generally without success. It seems clear that memory and the other higher functions are distributed in the fine structure of the brain. It is not known however to what extent different functions have structural units in common.[15]

Elsasser[16] says, "When the histologist looks at the brain he sees something which is very reminiscent of large electronic computers. He sees a small number of basic components repeated over and over again. All the complexity lies in the innumerable interconnections, not in the variety of basic components. So far as we know, the brain consists exclusively of neurons. Again, so far as we know, a neuron does nothing but conduct electrochemical pulses from its head end to its tail end. Some of the neurons leave the brain (efferent nerves), others enter it (afferent nerves), but apart from this the head and tail ends of neurons make synaptic connections with other neurons. Thus if one is to study the physiological background of memory one might start with such a model of interconnected

[4] W. K. Taylor, Proc. Inst. Elec. Engrs., (London), **106**, 198 (1959).

[5] R. L. Grimsdale, F. H. Sumner, C. J. Tunis, and T. Kilburn, Proc. Inst. Elec. Engrs., **106**, Part B (1959).

[6] R. M. Friedberg, IBM J. Research Develop. **2**, 2 (1958).

[7] *Symposium on the Design of Machines to Simulate the Behavior of the Human Brain*, I.R.E. Trans. on Electronic Computers, EC-5 (1956).

[8] A. Newell, J. C. Shaw, and H. A. Simon, Psych. Rev. **65**, 151 (1958).

[9] George W. Gray, Sci. Am. **179**, 4 (1948).

[10] J. F. Fulton, *Physiology of the Nervous System* (Oxford University Press, New York, 1943).

[11] See, M. Singer, in *Histology*, edited by R. O. Greep (Blakiston Company, New York, 1954).

[12] W. G. Walter, *The Living Brain* (W. W. Norton and Company, Inc., New York, 1953).

[13] W. Penfield and T. Rasmussen, *The Cerebral Cortex of Man* (The MacMillan Company, New York, 1950).

[14] K. S. Lashley, *Brain Mechanisms and Intelligence* (University of Chicago Press, Chicago, Illinois, 1929).

[15] K. S. Lashley, Research Publs., of the Assoc. Research Nervous Mental Diseases, **36**, 1 (1958).

[16] W. M. Elsasser, *The Physical Foundation of Biology* (Pergamon Press, New York, 1958), p. 138.

neurons. We do not claim that this model is altogether true, but it is simple, and presents itself on the basis of anatomical data. *There is no anatomical evidence for a storage organ* used to file away the immense amount of information which every person retains in his memory. Also, *brain physiology has not brought to light any evidence for* the existence of the highly complicated *special scanning devices.*"

4. Neurons

The doctrine expressed in this quotation, that the neurons are the functional units of the brain, is largely due to Ramon y Cajal and is now widely held among neurophysiologists.[17] A considerable amount is known about the action of individual neurons.[18–20] A grossly oversimplified description is as follows. When the cell body of a neuron is sufficiently stimulated, an electrochemical pulse travels from the cell body down along the axon, out along the branches to the end feet which impinge (synapse) on the cell bodies of other neurons, thus tending to stimulate or inhibit those neurons. This general description applies to the "internuncial" neurons. The action of a sensory (afferent) neuron differs from this in that its cell body is excited by an external stimulus (for example light impinging on a retinal receptor cell, or pressure on a special-purpose capsule in the skin). The action of a motor (efferent) neuron differs from the general description in that its end feet terminate in a muscle fiber (or a gland); electrochemical impulses arriving at these end feet tend to activate or inhibit the muscle contraction (or the gland output). There are perhaps 10^8 of these input and output neurons, constituting about 1% of the 10^{10} neurons in the brain.

The speed of conduction of the pulses in the neurons varies from about 5 m/sec in the fine neurons up to about 125 m/sec in the large ones. The time for a pulse to be conducted along the length of the neuron is of the order of 3×10^{-4} sec. The time to cross a synapse is of the order of 10^{-3} sec. After a neuron fires there is an absolute refractory period of the order 10^{-2} sec during which the neuron cannot fire again. There is also a relative refractory period, of increased threshold.

We reiterate that the above description is grossly oversimplified. It does, however, furnish a general idea of the manner in which neurons operate.

5. Organization of Neurons in the Brain

To establish detailed anatomical information regarding the connections of the 10^{10} neurons in the brain presents a formidable laboratory task. By ingenious and painstaking techniques, such as microelectrode stimulation, or degeneration and staining, some information has been obtained.[21–23] Some neurons are long, some are short. Some make contacts with nearby neurons, others wander the length of the brain before contacting another neuron. Some neurons connect with only a few others, others contact thousands. A simplified scheme is shown in Fig. 1. In the words of M. Singer, "Almost any type of connection scheme that can be imagined can be found in the brain." It seems impossible to map the entire topology of the neural network. Moreover, even if we accomplished this, we would then face the disheartening task of analyzing the performance of such a network. Now in a digital computer every connection must be exact or the answer can be entirely wrong. If it were also true for the brain that the misplacement or malfunction of a single connection could completely destroy the function, then we could not hope to understand how the brain operates until we have accomplished the impossible tasks of determining the exact wiring diagram of the neural net and analyzing it. However, it is clearly not true that the connections must be exactly right for the brain to function at all. This is proved by the fact that, although neurons do not regenerate, functions which are temporarily lost after extirpation of sections of the brain, are later recovered. Furthermore it seems unlikely that the genes would carry the information to specify every one of 10^{13} connections. It seems more plausible that only certain parameters of growth are specified and the fine connections are grown in a more or less random manner, subject to these constraints. Thus the detailed connection scheme would be unique to each individual. If it is true that individuals, with connection schemes specified only by certain parameters of growth, function

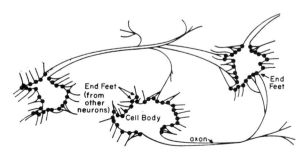

FIG. 1. Neurons (schematic).

[17] T. H. Bullock, Science **129**, 997 (1959).

[18] See, Frank Brink, in *Handbook of Experimental Psychology* edited by S. S. Stevens (John Wiley & Sons, Inc., New York, 1951).

[19] J. C. Eccles, *The Physiology of Nerve Cells* (Johns Hopkins Press, Baltimore, Maryland, 1957).

[20] Revs. Modern Phys. (*Biophysical Science*) **31**, 1–598 (1959). See also *Biophysical Science*, edited by J. L. Oncley (John Wiley & Sons, Inc., New York, 1959).

[21] D. A. Sholl, *Organization of the Cerebral Cortex* (Methuen and Company, Ltd., London, 1956).

[22] A. D. Adrian, *The Physical Background of Perception* (Oxford University Press, New York, 1947).

[23] J. C. Eccles, *The Neurophysiological Basis of Mind* (Oxford University Press, New York, 1953).

in similar ways, then there is hope that the performance of such a system might be analyzed in terms of such parameters. This also implies that the operation of the brain is radically different in principle from the logical circuitry of digital computers.[24] The discovery of these principles poses some challenging mathematical questions.

6. The Prospects for a Model

We have seen that our description of brain function is vague and our knowledge of brain structure is very sketchy indeed. It might well be argued that we have not formulated precisely a brain "function-structure" problem at all. Precisely formulated or not, it is clear that the problem exists. Again it is entirely possible that the areas of our ignorance of brain structure and brain function cover items which are essential for an explanation, and, until these are revealed, no understanding is possible. On the other hand, based on the above description of the brain structure, we can, as suggested in the quotation from Elsasser,[16] consider a model of interconnected neurons and study its behavior. Such networks can be arranged to perform any logical function,[25–28] but these arrangements are contrived and do not appear to resemble the biological organization at all. Some early experiments[29–31] were performed on straightforward neural networks, but these systems were very small and simple and the results were difficult to interpret. More plausible, but descriptive models have also been proposed in recent years.[32,33] While the verbal description of the conjectured functioning of such systems is quite attractive, the vagueness in the specifications of these models precluded the possibility of reasonably rigorous analysis or verification. With the Perceptron,[34,35] Rosenblatt offered for the first time a model which was: (a) specified in terms precise enough to permit testing of asserted performance, (b) sufficiently complex to offer the hope that its behavior would be interesting, (c) sufficiently simple to suggest that its performance might be analyzed and predicted, and (d) consistent with the known biological facts. Admittedly the model represents an enormous simplification of even the known brain structure; but if it does not violate the biological constraints (such as the number of units, the organization of connections, the reliability of components, the mechanism of signal transmission, the speed of response, the stability of the performance with respect to component malfunction or extirpation, the capacity for information storage, etc.) and if it exhibits even rudimentary brain functions, then, even if it does not in fact operate in the same manner as the brain does, it still provides at least a possible explanation of how the brain structure, as we know it at this time, *might* be organized to perform these functions.

PERCEPTRONS

7. General Description

The term *Perceptron* refers to a class of theoretical brain models, such as illustrated in Fig. 2.

A stimulus S (for example a pattern of light) is presented to the sensory retina. The illuminated sensory elements send pulses with varying time delays to the associators. Some of the pulses are positive (excitatory) and some are negative (inhibitory). If the algebraic sum of the pulses arriving at an associator in a suitable time interval exceeds a certain threshold (which need not be the same for all associators), that associator sends out pulses as indicated by the arrows to other associators and/or to the response units. Each unit may have its own refractory period. Each connection may have its own transmission time, pulse magnitude and sign, or frequency and phase. The response units also have an activation threshold and may have excitatory and inhibitory connections with some associators and/or each other.

So far we have made no provision for change (learning or memory) in the system. On this the anatomical, histological, or physiological findings offer no clue. The general belief is that pathways through the network are somehow, as a consequence of being used, facilitated for future conduction. Thus, Hebb[32] says, "When an axon of cell A is near enough to excite a cell B and repeatedly or persistently takes part in firing it, some growth process or metabolic change takes place in one or both of the cells such that A's efficiency, as one of the cells firing B, is increased." This might be brought about in the biological system by the growth of additional end feet, or by chemical changes in the neurons such as the production of enzymes in the cell body which alter the threshold in a small region of the cell body, or by several other plausible means.[36] "Long term memory" in humans, which can survive for a century in spite of severe shocks, must be stored in some fairly permanent

[24] J. von Neumann, *The Computer and the Brain* (Yale University Press, New Haven, Connecticut, 1958).

[25] J. T. Culbertson, *Consciousness and Behavior* (William C. Brown Company, Dubuque, Iowa, 1950).

[26] W. Pitts and W. S. McCulloch, Bull. Math. Biophys. **9**, 127 (1947).

[27] W. S. McCulloch and W. Pitts, Bull. Math. Biophys. **5**, 115 (1943).

[28] D. A. Sholl and A. M. Uttley, Nature **171**, 387 (1953).

[29] W. A. Clark and B. G. Farley, Proceedings of the Western Joint Computer Conference, p. 86 (1955).

[30] B. G. Farley and W. A. Clark, I.R.E. Trans. Professional Group on Inform. Theory 4, 76 (1954).

[31] N. Rochester, J. H. Holland, L. H. Haibt, and W. L. Duda, I.R.E. Trans. on Inform. Theory, **IT-2**, 80–93 (1956).

[32] D. O. Hebb, *The Organization of Behavior* (John Wiley & Sons, Inc., New York, 1949).

[33] R. L. Beurle, Trans. Roy. Soc. (London), **B240**, 55 (1956).

[34] F. Rosenblatt, Cornell Aeronautical Laboratory Report No. VG-1196-G-1 (January, 1958).

[35] F. Rosenblatt, Psych. Rev. **65**, 386 (1958).

[36] F. Rosenblatt, Cornell Aeronautical Laboratory, Project PARA Technical Memorandum No. 10 (December, 1959).

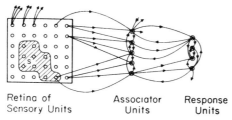

Fᴵɢ. 2. Organization of Perceptron.

structure; "short-term memory" might be stored by means of a transient state of activity. In any case many of the conjectures are functionally equivalent to the rule that when the two ends of a connection are sequentially active the connection is strengthened, i.e., the pulse it carries is increased. This description of the *reinforcement rule* is intentionally vague; it can be realized in various ways, some of which are given in precise terms below.

Parameters which must be specified to define the perceptron of Fig. 2 are: The number of sensory elements, the number (or probability distribution) of excitatory and inhibitory connections at each level and the geometrical constraints on them, the number of associators and the number of responses; the thresholds, refractory periods, summation intervals, and transmission times. For studying the behavior of such a perceptron we would also have to specify the set of stimulus patterns, the order and times of their presentation, and the observations to be made on the responses. The reinforcement rule must, of course, also be defined.

We shall not pursue further here the arguments showing that the above model is consistent with the biological constraints.[1,34]

8. Techniques of Investigation

For studying the behavior of perceptrons, three general techniques are available.

(a) *Mathematical analysis.* When it is successful, this approach offers many advantages, such as the predictability of the performance of classes of perceptrons, the effects of variations in the parameters, and so forth. For a model of the complexity of the general perceptron of Fig. 2 the analysis is quite complicated (see Sec. 6 of the paper which follows[37]). For certain simplified cases as in the simple perceptron of Fig. 4 which is discussed later, the analysis is fairly complete. In Sec. 9 we prove some theorems and illustrate the analytical techniques for such systems. In the paper which follows, a more complicated system is analyzed.

(b) *Simulation on a digital computer.* The principal advantage of this method is that it can always be done, subject, of course, to time, storage, and cost limitations. A considerable amount of data has been obtained in

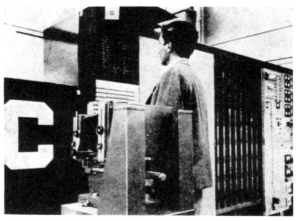

Fɪɢ. 3. Mark I Perceptron at Cornell Aeronautical laboratory. (a) Overall view with sensory input at left, association units in center, and control panel and response units at far right. The sensory to associator plugboard, shown in (b) is located behind the closed panel to the right of the operator. The image of the letter "C" on the front panel is a repeater display, for monitoring sensory inputs.

this way.[1,38] Some of these will be described in Sec. 9 below.

(c) *Construction of an actual machine.* This has an enormous advantage in speed over the digital computer, since essentially all the action goes on in parallel simultaneously and the response appears almost immediately, while in the digital simulation all computations are done in sequence. While an actual machine enjoys certain types of flexibility, such as the ease with which the experimenter can vary the stimulus patterns, it is a serious task to change the wiring diagram (in the digital computer this can be generated quickly by a suitable program) and it is impossible to alter certain basic features of the network. There is also the complicating factor of the inexact performance of hardware. A machine of the complexity of Fig. 2 has not yet been built, but one having the organization of Fig. 4 (but with eight binary-response units) has been built, and is known as the Mark I, (Fig. 3).[39-41] The retina is a 20×20 grid of photocells mounted in the picture plane of a camera to which the stimulus pictures are shown. There are 512 associator units and eight binary-response units. Each sensory unit can have up to forty connections to the associator units.

[37] H. D. Block, B. W. Knight, Jr., and F. Rosenblatt. Revs. Modern Phys. **34**, 135 (1962).

[38] F. Rosenblatt, Proc. I.R.E. **48**, 301 (1960).
[39] J. C. Hay, F. C. Martin, and C. W. Wightman, Record of I.R.E. 1960 National Convention, Part 2, New York, (1960).
[40] C. W. Wightman, Cornell Aeronautical Laboratory, Project PARA Technical Memorandum No. 4 (February, 1959).
[41] J. C. Hay and A. E. Murray, Cornell Aeronautical Laboratory Report VG-1196-G-5 (February, 1960).

This wiring is normally made according to a table of random numbers. The associator to response connections are varied by motor-driven potentiometers. Many interesting experimental results have been obtained.[39] Some of these are mentioned in Sec. 9(f) below.

For purposes of building a machine having more units and organized more in the direction of Fig. 2, the principal requirement is for inexpensive, compact, low power associator units and connections having the desired variability of weights. Neither precision nor reliability of the components is important. [This is contrasted with the usual engineering situation in which as the number of components increases the reliability of each component must also increase, for the failure of a single element results in the failure of the entire system; the probability of system failure thus increases rapidly as the number of components increases. For the perceptron it is the other way around; cf. the extirpation experiments described in 9(f) below.] Another need is for an inexpensive method of making connections. Again these need not be precise, but there must be a great many of them. Recent developments[42–46] are encouraging and lead to reasonable hope for success.

9. Analysis

a. A Simple Perceptron.

Consider the simple[47] perceptron shown in Fig. 4.

Let there be N_s sensory units, N_a associator units, and n stimulus patterns (each stimulus pattern is a specified set of activated retinal points). We denote typical sensory units by s_σ, typical associators by a_μ, and typical stimuli by S_i. Let us represent the connection between s_σ and a_μ by the real number $C_{\sigma\mu}$; in particular the $C_{\sigma\mu}$ might be random numbers having the possible values $+1$, -1, 0. When the stimulus S_i is applied to the retina, the signal

$$\alpha_\mu{}^i = \sum_{\substack{\sigma \\ s_\sigma \epsilon S_i}} C_{\sigma\mu}$$

is transmitted instantly to the associator a_μ. If $\alpha_\mu{}^i \geqq \theta$, where θ is an arbitrary, but fixed real number, the

[42] K. R. Shoulders, *Simulation of Neural Networks by Optical-Photographic Methods* (Stanford Research Institute, Menlo Park, California, December, 1959).

[43] K. R. Shoulders, *Research in Microelectronics Using Electron-beam-activated Machining Techniques* (Stanford Research Institute, Menlo Park, California, September, 1960).

[44] B. Widrow, Stanford Electronics Laboratory Technical Report 1553-2, Stanford, California, (1960).

[45] J. K. Hawkins and C. J. Munsey, *A Magnetic Integrator for the Perceptron Program* (Aeroneutronics, Newport Beach, California, 1960).

[46] A. E. Brain, *The Simulation of Neural Elements by Electrical Networks Based on Multi-Aperture Magnetic Cores* (Stanford Research Institute, Menlo Park, California, 1960).

[47] We use the term "simple" here in the colloquial sense. The term is also used with a technical meaning, a precise definition of which is given in Rosenblatt[1]. The Perceptron of Fig. 4 is also "simple" in the technical sense.

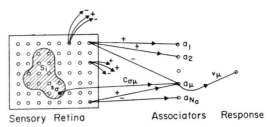

Fig. 4. Simple perceptron.

associator a_μ is said to be *active* and instantly transmits a signal v_μ to the response unit. An inactive associator transmits no signal. The total signal arriving at the response unit is $u = \sum'_\mu v_\mu$, where $\sum'$ is taken over the active associator units. If $u > \Theta$, where Θ is an arbitrary but fixed non-negative number, the response output is $+1$. If $u < -\Theta$ the response is -1. If $|u| \leqq \Theta$ the response is 0.

In this model the connections $C_{\sigma\mu}$ do not change. Therefore $A(S_i)$, the set of associators activated by stimulus S_i, does not change. Thus once the numbers $C_{\sigma\mu}$ have been determined (more will be said about this later) we may disregard the sensory retina altogether and start with the Venn diagram of Fig. 5. Let

$$e_{\mu i} = \begin{cases} 1 & \text{if } a_\mu \in A(S_i) \\ 0 & \text{if } a_\mu \notin A(S_i). \end{cases}$$

The input to the response unit, when stimulus S_i is presented to the retina is then

$$u_i = \sum_\mu v_\mu e_{\mu i}. \tag{1}$$

This model does not use the known biological facts of delay, refractory period, and variability of neurons. We shall see later that these features can be very helpful indeed, but we now show that even without them we get very interesting performance.

b. Discrimination: Learning by Error Correction

Let us assign each stimulus to one of two classes, which we denote by $+1$ and -1. Say stimulus S_i is assigned to class ρ_i, where ρ_i is $+1$ or -1. This dichotomization is then represented by $\rho = (\rho_1, \rho_2, \cdots, \rho_i, \cdots, \rho_n)$. We would like the perceptron, in its terminal state, to give the correct response to each stimulus. From Eq. (1) we see that the response to S_i is correct if and only if

$$\rho_i u_i = \sum_\mu v_\mu e_{\mu i} \rho_i > \Theta. \tag{2}$$

Let B denote the matrix with elements $b_{\mu i} = e_{\mu i} \rho_i$. It may happen that no choice of numbers for y_μ

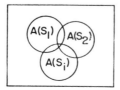

Fig. 5. Associator units.

$(\mu=1,2,\cdots,N_a)$ will yield the inequalities

$$\sum_\mu y_\mu b_{\mu i}>\Theta, \quad i=1, 2, \cdots, n. \tag{3}$$

If this is the case, then the discrimination problem (ρ) cannot possibly be solved (perfectly) by this perceptron, no matter what reinforcing arrangement is used. However, as will be discussed later, for reasonable dichotomies in most systems of interest there will be numbers y_μ satisfying (3) and hence positive numbers ζ_i such that

$$\sum_\mu y_\mu b_{\mu i}\equiv\zeta_i>\Theta, \quad i=1, 2, \cdots, n. \tag{4}$$

In particular if the Venn diagram (Fig. 5) has, in each set $A(S_i)$, an associator $a_{\mu(i)}$ which is in no other $A(S_j)$, then we can simply set $y_{\mu(i)}=(1+\theta)\rho_i$ for each $i=1, 2, \cdots, n$ and the other y_μ's $=0$ and satisfy (4). More generally if the matrix B is of full rank and $N_a\geqq n$ then, by the standard theorem in algebra there will be a solution $y=(y_1,\cdots,y_\mu,\cdots y_{N_a})$ to (4). For the remainder of this section we assume that there is a solution y to (4).

The *error correction procedure* is as follows. A stimulus S_i is shown and the perceptron gives a response. If this response is correct then no reinforcement is made. If the response is incorrect then the v_μ for active associators a_μ is incremented[48,49] by $\eta\rho_i$.[50] The inactive associators are left alone. The initial values are arbitrary, say $(v_1^0,\cdots,v_\mu^0,\cdots,v_{N_a}^0)$.

Suppose the stimuli are shown in an arbitrary sequence, such that each stimulus recurs infinitely often. We shall show that after a certain finite number of steps the machine will thereafter give the correct response to all the stimuli, so that no further changes take place. The proof given here is a distillation of a succession of proofs by Rosenblatt, Joseph, Kesten, and the author.[49,51,52]

(1) Let $\xi=\begin{bmatrix}\xi_1\\\xi_2\\\vdots\\\xi_n\end{bmatrix}$, where the ξ_i are real numbers;

$$F(\xi)=(B\xi,B\xi)\equiv\sum_\mu\left(\sum_i b_{\mu i}\xi_i\right)^2$$

[48] R. D. Joseph, Cornell Aeronautical Laboratory Report No. VG-1196-G-7. See also Ph.D. Thesis, Cornell University, Ithaca, New York, 1961.

[49] F. Rosenblatt, Cornell Aeronautical Laboratory Report No. VG-1196-G-4 (February, 1960).

[50] Other schemes for the amount of reinforcement have been investigated.[1,48] In particular, the system in which the inactive connections suffer a decrement in such a way that the sum of the input connections at each response unit remains constant is called a "γ system" in contrast to the "α system" described in the text. The "Γ system" on the other hand, conserves the sum of the output connections at each associator. Another system which has been studied is the "λ system" in which the v_μ cannot exceed a certain bound. Another modification, in which each connection v_μ suffers a decay proportional to v_μ will be used in the paper which follows. In the interest of simplicity of presentation we do not go into any of these here.

[51] R. D. Joseph, Cornell Aeronautical Laboratory, Project PARA Technical Memorandum No. 12 (May, 1960).

[52] R. D. Joseph, Cornell Aeronautical Laboratory, Project PARA Technical Memorandum No. 13 (July, 1960).

and, for $\xi\neq0$, $f(\xi)=F(\xi)/\sum_i\xi_i^2$. Note that $f(\xi)$ is constant along each ray, i.e., $f(\lambda\xi)=f(\xi)$ for any $\lambda\neq0$. Thus all the values that $f(\xi)$ takes on are assumed on the unit sphere $\sum_i\xi_i^2=1$. Since this is a compact set, we have for all $\xi\neq0$, $0\leqq f(\xi)\leqq M$.

Now choose any ξ^* such that $\sum_i\xi_i^{*2}=1$ and each component $\xi_i^*\geqq0$. If $f(\xi^*)=0$ then $B\xi^*=0$ and

$$0=(B\xi^*,y)=(\xi^*,By)=\sum_i\xi_i^*\sum_\mu b_{\mu i}y_\mu=\sum_i\xi_i^*\zeta_i>0,$$

by (4), which is a contradiction. Therefore $f(\xi)$ does not vanish on that portion of the unit sphere which lies in the closed first orthant. Hence $f(\xi)$ assumes a positive minimum value on that set, say $m>0$. Since $f(\xi)$ is constant along each ray, it follows that, for any vector $\xi\neq0$ having each component $\xi_i\geqq0$, $f(\xi)\geqq m$.

Therefore we have proved, on the basis of the assumption that there exists a solution to Eq. (4), that there is a constant $m>0$ with the following property. For any nonzero vector ξ having all its components non-negative:

$$0<m\leqq f(\xi)\leqq M.$$

(2) At any time t, let $x_i(t)(i=1,2,\cdots,n)$ be the number of times the machine has incorrectly identified stimulus S_i (and hence has been reinforced, so far, x_i times by amount $\rho_i\eta$). Then $v_\mu(t)=v_\mu^0+\sum_i e_{\mu i}\rho_i\eta x_i(t)$. If the stimulus S_j is now shown to the machine the input to the response unit is

$$u_j(t)=\sum_\mu v_\mu^0 e_{\mu j}+\sum_\mu\left[\sum_i e_{\mu i}\rho_i\eta x_i(t)e_{\mu j}\right]. \tag{5}$$

Hence

$$\sum_\mu\left[\rho_j e_{\mu j}\sum_i e_{\mu i}\rho_i x_i(t)\right]\eta=\rho_j u_j(t)-\rho_j u_j^0, \tag{6}$$

where $u_j^0=\sum_\mu v_\mu^0 e_{\mu j}$.

Reinforcement occurs at this stage if and only if the response determined by (5) is incorrect, i.e., if $\rho_j u_j(t)\leqq\Theta$. Thus if reinforcement occurs we have, from (6),

$$\sum_\mu\left[b_{\mu j}\sum_i b_{\mu i}x_i(t)\right]\leqq(\Theta-\rho_j u_j^0)/\eta\equiv D_j. \tag{7}$$

Suppose that reinforcement takes place. Let us consider the change in $F(\xi)$ as ξ goes from

$$P_1=(x_1,x_2,\cdots,x_j,\cdots,x_n)$$

to

$$P_2=(x_1,x_2,\cdots,x_j+h,\cdots,x_n),$$

for $0\leqq h\leqq1$. We have

$$\left.\frac{\partial F}{\partial\xi_j}\right]_{P_1}=2\sum_\mu\left(b_{\mu j}\sum_i b_{\mu i}x_i\right),$$

$$\left.\frac{\partial F}{\partial\xi_j}\right]_{P_2}=2\sum_\mu b_{\mu j}\sum_i b_{\mu i}(x_i+h\delta_{ji})$$
$$=2\sum_\mu b_{\mu j}\sum_i b_{\mu i}x_i+2h\sum_\mu b_{\mu j}^2,$$

and in the interval from P_1 to P_2 we have, using (7)

$$\frac{\partial F}{\partial \xi_j} \leq 2 \sum_\mu b_{\mu j} \sum_i b_{\mu i} x_i + 2hN_j \leq 2D_j + 2hN_j,$$

where N_j is the number of associators activated by S_j. Therefore the change in $F(\xi)$ as ξ_j is varied from x_j to x_j+1 is

$$\Delta F = \int_{x_j}^{x_j+1} \frac{\partial F}{\partial \xi_j} \partial \xi_j \leq 2 \int_0^1 (D_j + N_j h) dh$$

$$\leq 2D_j + N_j \leq \max_j(2D_j + N_j) \equiv D. \quad (8)$$

The total change in F from the beginning of the training, $x^0 = (0,0,\cdots,0)$ to the state where

$$x = (x_1, x_2, \cdots, x_i, \cdots, x_n)$$

is

$$F(x) - F(0) = F(x) = \sum \Delta F \leq D \sum_i x_i.$$

Hence

$$0 < m \leq f(x) = \frac{F(x)}{\sum_i x_i^2} \leq \frac{D \sum_i x_i}{\sum_i x_i^2} \leq \frac{Dn}{\sum_i x_i},$$

where the last inequality follows from Schwartz's inequality:

$$n \sum_i x_i^2 = \sum_i 1^2 \sum_i x_i^2 \geq (\sum_i x_i)^2.$$

Therefore

$$\sum_i x_i \leq (n/m\eta) \max_i[2(\Theta - \rho_i u_i^0) + \eta N_i]. \quad (9)$$

After at most this many corrections there will be no more; i.e., the machine will thereafter give only correct responses. It has learned the dichotomy.

The above can be generalized so that instead of the corrections all having the same magnitude η each time, they have magnitudes $h_1, h_2, \cdots, h_\nu, \cdots$, where the h_ν are bounded and

$$\sum_{\nu=1}^\infty h_\nu$$

diverges. The analysis is analogous, with $\eta x_i(t)$ in Eq. (5) replaced by $X_i(t)$, the absolute magnitude of reinforcement applied up to time t as a result of incorrect responses to stimulus S_i. Analogous to (8) we get

$$\Delta F \leq 2(\Theta - \rho_j u_j^0) h_\nu + h_\nu^2 N_j \leq (C + Nh) h_\nu,$$

where $C = \max_j 2(\Theta - \rho_j u_j^0)$, $N = \max_j N_j$, $h =$ the maximum h_ν used to date. Then, summing as before, we get after K corrections

$$m \leq \frac{F(X)}{\sum_i X_i^2} \leq \frac{(C+Nh)\sum h_\nu}{\sum_i X_i^2} \leq \frac{(C+Nh)n \sum h_\nu}{(\sum_i X_i)^2}$$

$$= \frac{(C+Nh)n}{\sum_{\nu=1}^K h_\nu}.$$

Hence

$$\sum_{\nu=1}^K h_\nu \leq \frac{(C+Nh)n}{m}. \quad (9')$$

Thus the process terminates if

$$\sum_{\nu=1}^\infty h_\nu$$

diverges and the h_ν are bounded. This last condition can clearly be weakened to the condition that for arbitrarily large values of K:

$$\frac{\max(h_\nu)}{(\nu=1,\cdots,K)} \leq \frac{rm}{Nn} \quad \text{where} \quad r < 1;$$

for then we get from $(9')$

$$\sum_{\nu=1}^K h_\nu \leq \frac{C}{m(1-r)}.$$

Various modifications and generalizations of the fundamental theorem expressed in Eq. (9) have been obtained.[49,51,52] We confine ourselves here to the following remark.

The condition for the existence of a solution is, from (3):

$$\text{sgn} \sum_\mu e_{\mu i} y_\mu = \rho_i.$$

Hence, with a fixed Venn diagram, the number of dichotomies which the machine will be able to learn is equal to the number of orthants (in n space) which can be entered by linear combinations of the N_a row vectors of the matrix $e_{\mu i}$. An upper bound for this number[53] shows that it may be considerably less than the 2^n possible dichotomizations. However most of these dichotomizations are "unreasonable" and could not be learned by humans either. As an example of the power of the machine, the Mark I was shown the twenty horizontal bars (4×20 retinal units) and the twenty vertical bars (20×4 retinal units) with the dichotomy being that alternate bars were in opposite classes. This should be a difficult dichotomy, since bars with the greatest overlap are in opposite classes (compare the discussion below). The Mark I learned this dichotomy perfectly after seeing 214 stimuli, requiring 600 sec of reinforcement in all. (In this experiment the reinforcement rule was to hold on the reinforcement until the sign of the response changed).[49] It would be interesting to compare this performance with that of a human subject on the same problem.

[53] R. D. Joseph and L. Hay, Cornell Aeronautical Laboratory, Project PARA Technical Memorandum No. 8 (1960).

c. Forced Learning.

In the *forced-learning* reinforcement rule the v_μ of each active associator is incremented by $\eta\rho_i$ each time S_i is shown, regardless of the machine's response. The input to the response unit when stimulus S_j is presented to the retina is again given by

$$u_j(t)=\mu_j{}^0+\eta\sum_\mu e_{\mu j}\sum_i e_{\mu i}\rho_i x_i(t),\qquad(10)$$

where $x_i(t)$ now denotes the number of times the stimulus S_i has been shown, up to time t. Suppose that at some time T, $x_i(T)=p_iT$, for $i=1, 2, \cdots, n$. The vector

$$p=\begin{bmatrix}p_1\\p_2\\\vdots\\p_i\\\vdots\\p_n\end{bmatrix},\quad p_i\geqq0,\quad \sum_i p_i=1,$$

corresponds to the relative frequencies of occurrence of the various stimuli

$$\begin{bmatrix}S_1\\S_2\\\vdots\\S_i\\\vdots\\S_n\end{bmatrix}.$$

The condition for the response to S_j to be correct is $\rho_j u_j>\Theta$, i.e.,

$$\rho_j u_j{}^0+\eta T\sum_\mu \rho_j e_{\mu j}\sum_i e_{\mu i}\rho_i p_i>\Theta.\qquad(11)$$

Inequality (11) will hold in general for large T, if and only if

$$\sum_\mu \rho_j e_{\mu j}\sum_i e_{\mu i}\rho_i p_i>0.\qquad(12)$$

Thus, if the perceptron can learn this dichotomy under forced learning, for some choice of p, we can take $y_\mu=\sum_i e_{\mu i}\rho_i p_i$ to satisfy (3). Therefore it will learn under error correction. Conversely if the machine can learn the dichotomy under error correction then from (5) starting with zero initial values, the x_i obtained by the error correction procedure will yield a frequency vector

$$p_i=x_i/\sum_j x_j$$

satisfying (12). To summarize: If the perceptron can learn a dichotomy under forced learning, it will learn it under error correction. If it can learn it under error correction it can learn it under forced learning for some choices of the frequency vector, but not in general (see below) for others. Thus the error correction method is a more generally effective method than forced learning.

Let $\mathfrak{N}$ denote the matrix whose elements are $n_{ij}=n_{ji}=\sum_\mu e_{\mu j}e_{\mu i}$. Note that n_{ij} is the number of associators activated by both S_i and S_j; i.e., the number of elements in $A(S_i)\cap A(S_j)$. The matrix B^TB

has for its i–jth element $\rho_i n_{ij}\rho_j$. The desired dichotomy having been specified, we can relabel the stimuli without loss of generality, so that all those in the $+1$ category come before all those in the -1 category. Then the matrix B^TB has the appearance

$$U=\begin{bmatrix}n_{11} & n_{12} & n_{13} & -n_{14} & -n_{15}\\n_{21} & n_{22} & n_{23} & -n_{24} & -n_{25}\\n_{31} & n_{32} & n_{33} & -n_{34} & -n_{35}\\\hline -n_{41} & -n_{42} & -n_{43} & n_{44} & n_{45}\\-n_{51} & -n_{52} & -n_{53} & n_{54} & n_{55}\end{bmatrix},$$

where $\quad U=B^TB,$

and where we have assumed, for purposes of illustration that there are three stimuli in the first class and two in the second. Clearly if there is any negative entry in U then there is some frequency vector p for which (12), which now reads

$$\sum_i U_{ji}p_i>0,\qquad(13)$$

will fail to hold for some j. If the stimuli are "equally likely" to occur, $(p_i=1/n)$ then (13) is the requirement that each row sum of U is positive; or roughly, that each stimulus has a greater intersection (in the associator set) with its own class than with the opposite class.

If the stimuli are presented in random order, with p_i the probability of occurrence of S_i, then, in (10), $x_i(t)$ and $u_i(t)$ are random variables, with the expected value of $u_j(T)$ again given by the left side of (11). Using Tchebycheff's inequality, rigorous bounds on the probability of error as a function of time can be obtained.[48] By using a normal approximation, an estimate of the learning curves (Probability of correct response vs t) have been found.[48] The success of the perceptron at this type of learning has led to its being applied as an engineering pattern-recognizing device.

The entries of the $\mathfrak{N}$ matrix, for a given set of stimuli, stem from the connections $C_{\sigma\mu}$ and the threshold θ. By fixing these suitably we could "gimmick" the $\mathfrak{N}$ matrix to fit our needs. [For example by taking $C_{\sigma1}=1$ for those sensory units s_σ which are activated by S_1; $C_{\sigma1}=-1$ if s_σ is not activated by S_1; and setting $\theta_1=\frac{1}{2}\sum_\sigma (C_{\sigma1}+|C_{\sigma1}|)$ we can be sure that associator a_1 responds to stimulus S_1 and only to S_1. If we now want an associator a_μ to respond to S_1 and to S_2 and to no other stimulus we could use a preliminary layer of associators and the connection scheme (Fig. 6), where a_2 responds only to S_2 and $\theta_\mu=1$. Similarly we can deal with the other logical functions. Thus in fact, using the preliminary layer of n associators, we can fix the number of elements in each subset of the Venn

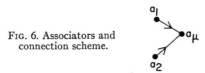

Fig. 6. Associators and connection scheme.

diagram (Fig. 5), and thus control the $\mathfrak{N}$ matrix.] However we take the position that we want the perceptron to react equally well when the set of stimuli $S_1, \cdots, S_n$ are not specified in advance. Thus an $\mathfrak{N}$ matrix that leads to interesting results for a preassigned set of stimuli may be poor for a different set. For this reason the $C_{\sigma\mu}$ are taken at random. The resulting entries in the $\mathfrak{N}$ matrix become random variables which have been studied[1] and tabulated in some detail.[54] From these data it is possible to select design parameters and to predict performance of perceptrons constructed in this way.[55] These results have been verified and elaborated by simulation testing of Perceptrons[1,56] as well as by experiments on Mark I.[39] (Actually, by the use of a reinforcement scheme applied to the $C_{\sigma\mu}$ analogous to that used on the v_μ, the perceptron can improve its sensory connection scheme for a given problem.[1] We shall not go into the analysis of such systems here).

d. Generalization.

The simple perceptron we are here analyzing exhibits a good deal of generalization, i.e., correct response to a stimulus it has not seen before. Examination of inequality (13) reveals the reason for this. If the stimulus has considerable overlap with some of the members of the same class and very little overlap with members of the opposite class it will give the correct response on the basis of having seen the like stimuli. To illustrate with an extreme example suppose the $\mathfrak{N}$ matrix has the following appearance

$$
\begin{bmatrix}
1 & 1 & 0 & 0 & 1 & 0 & 0 & 0 & 0 & 0 \\
1 & 1 & 1 & 0 & 0 & 0 & 0 & 0 & 0 & 0 \\
0 & 1 & 1 & 1 & 0 & 0 & 0 & 0 & 0 & 0 \\
1 & 0 & 0 & 1 & 1 & 0 & 0 & 0 & 0 & 0 \\
0 & 0 & 0 & 0 & 0 & 1 & 1 & 0 & 0 & 1 \\
0 & 0 & 0 & 0 & 0 & 1 & 1 & 1 & 0 & 0 \\
0 & 0 & 0 & 0 & 0 & 0 & 1 & 1 & 1 & 0 \\
0 & 0 & 0 & 0 & 0 & 0 & 0 & 1 & 1 & 1 \\
0 & 0 & 0 & 0 & 0 & 1 & 0 & 0 & 1 & 1
\end{bmatrix} \quad (14)
$$

In the matrix each stimulus overlaps (in the associator set) with its two nearest neighbors but with no others. (The retina in this case is conceived as being

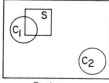

Retina

FIG. 7. Stimulus pattern of circles and squares.

---

[54] F. Rosenblatt, Cornell Aeronautical Laboratory Report No. VG-1196-G-6 (May, 1960).

[55] R. D. Joseph, I.R.E. 1960 Convention Record, 2, New York, (1960).

[56] F. Rosenblatt, Proc. I.R.E. 48, 301 (1960).

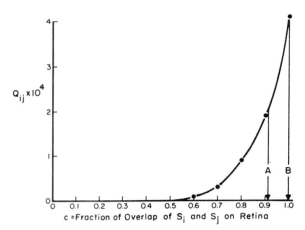

FIG. 8. Q_{ij} as a function of c.

cylindrical; patterns which go off the right edge reappear at the corresponding point at the left edge. In this way each stimulus has exactly two nearest neighbors.) Then, starting with zero initial values, once the machine sees one stimulus and is reinforced it will give the correct response to its two neighboring stimuli even if it has not been shown these previously. While this type of $\mathfrak{N}$ matrix is a gross oversimplification of those obtained in practice, it illustrates a basic mechanism which is operating here as we show next.

Suppose that we are interested in discriminating squares from circles, all of unit area. In Fig. 7 it is clear that the circle C_1 in the upper left of the retina has many more sensory points in common with the square S than it has with the other circle C_2. Why should the machine tend to classify the first circle C_1 with the second circle C_2 rather than with the square S, with which it has the greater retinal overlap?

To answer this, let us use some typical data. Suppose the connections $C_{\sigma\mu}$ are made by taking at each associator 5 inhibitory and 5 excitatory inputs connected randomly to the retina. Further suppose that $\theta = 5$ and the total retina is of area 10/3. Then the probability Q_{ij} of the associator being activated by both of two stimuli S_i and S_j is a function of the retinal overlap c of the stimuli, S_i and S_j as shown in Fig. 8 (the data for Fig. 8 are taken from the tables[54] mentioned earlier).

With an extremely fine lattice spacing on the retina the overlap between any one of the circles and any one of the squares is less than 91%, as can be verified by elementary geometry. Thus for a circle and a square we are always to the left of point A in Fig. 8, and the probability of an associator being activated by both is less than 2×10^{-4}. If the circles can be displaced by very small amounts we will have close to 100% retinal overlap between a circle and a nearby circle; consequently we are near point B in Fig. 8 with the probability of the associator being activated by both circles greater than 4×10^{-4}. If there are, say, 3000 associators, then the expected number of associators activated by any specified circle and square is less than 0.6. On the other hand, each circle has neighbors for which the expected number of associators activated by it and a specified neighbor is greater than 1.2. Therefore it is plausible that the matrix $\mathfrak{N}$ should have an appearance not unlike that indicated in Eq. (14) and the possibility for generalization is present. We reiterate that this is a gross oversimplification of what happens in the actual examples that have been dealt with in practice, but it illustrates a mechanism which makes some dichotomies "natural" and others "unnatural."

The effect operating here depends on the maximal overlap between stimuli of opposite classes being less than the maximal overlap between members of the same class. This effect can be sharpened, for example, by replacing the solid stimulus figures by their boundaries (with some small width). For example the ratio of the amount of overlap between any circular ring and a square ring to the overlap between neighboring circular rings clearly tends to zero with the ring thickness, instead of the 91% of the solid figures in the above example. The process of "contour extraction" can be realized by simple neuronal circuitry. For example if the output of each sensory unit is nullified by the stimulation of its four nearest neighbors then the figure will be replaced by its boundary. Another arrangement is to take account of a relatively long absolute refractory period and small rapid random motions of the retina, such as the human eye makes. After a short instant, the figure will again be replaced by its contour. Other similar pattern property filters[25,57] can be used to organize information at the sensory level, resulting in enhancing the effect under discussion.

The performance of the perceptron can also be improved by modifying the two-valued nature of the output of the associators by taking into account the magnitude of the input in excess of the threshold.[48,58] We shall not go into this here.

The above discussion is concerned with "perfect" or 100% performance. The perceptron will emit a response in any case and, under much more general circumstances, the performance will be "better than chance." By putting several "better than chance" machines in parallel and using, say, a "majority decision" rule for the final response, considerable improvements in performance are obtained. Often, however, better performance results from combining all the associators of the parallel systems into a larger single set of associators. "Learning curves," giving the probability of a correct response as a function of training time, have been obtained by analysis as

indicated above and also by simulation and by Mark I experiments.[1,48] The results justified the approximations of the analysis as well as indicating the ability of the machine.

The simple perceptron discussed here generalizes on the basis of retinal overlap. The four layer system which will be described and analyzed in the paper which follows[37] generalizes also on the basis of temporal contiguity.

e. Spontaneous Organization.

The *spontaneous-organization* program consists of showing the machine stimuli, letting it compute its own response, and reinforce in accordance with that response. The only contact between the experimenter and the machine is the presentation of the stimuli. Although it is true that for certain special cases and with certain modifications[38] the simple perceptrons described here do make interesting dichotomies, they do not do so in general.[56,59,60] The four-layer systems described in the next paper (and more generally the cross coupled systems[1]) do make interesting spontaneous classifications, as will be shown there. Since it is the spontaneous classification that corresponds to the machine having an "original concept," we see that we shall have to look to the paper which follows for this.

f. Psychological Testing.

While it is possible on the basis of the analysis indicated to estimate the performance of simple perceptrons, experimentation with Mark I has suggested various "psychological" experiments on that machine. We cite some of these results.[39] Figure 9 compares the "forced learning" with the "error-correction" procedure. The same perceptron is "trained" by each method and "tested" at various times. During the testing, of course, no reinforcement is applied.

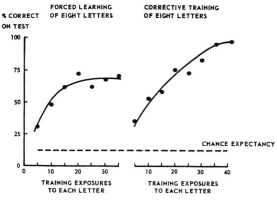

FIG. 9. Learning curves for eight letter identification task (each letter upright, but in a variety of locations).

[57] M. Babcock, A. Inselberg, L. Löfgren, H. von Foerster, P. Weston, and G. Zopf, Tech. Rept. No. 2, University of Illinois, 1960.

[58] R. D. Joseph, Cornell Aeronautical Laboratory, Project PARA Technical Memorandum No. 11 (March, 1960).

[59] See, F. Rosenblatt, in *The Mechanization of Thought Processes* (Her Majesty's Stationery Office, London, England, 1959).

[60] F. Rosenblatt, Cornell Aeronautical Laboratory, Project PARA Technical Memorandum No. 2 (October, 1958).

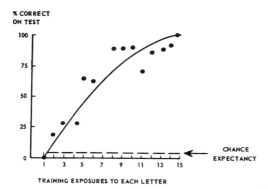

FIG. 10. Learning curve for 26 letters, each in standard position; corrective training.

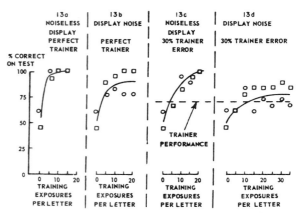

FIG. 12. Effects of noisy display and imperfect trainer on learning of "E"–"X" discrimination.

Figure 10 shows a learning curve for recognition of all 26 letters of the alphabet in one standard position. For this experiment the coding of the outputs of five binary-response units was selected in a quasi-optimal manner.

To study the effects of random noise in the stimulus pattern, the target stimuli were taken as the letters E and X, with a small amount of retinal shift allowed. Noise in the target display is illustrated in Fig. 11. The effect of the noise on performance is illustrated in Figs. 12(a) and 12(b). An additional disturbance, in which the trainer makes errors (when deciding what reinforcement to apply, he misidentifies the stimuli) at random 30% of the time, is introduced in Figs. 12(c) and 12(d). In Fig. 12(c) the perceptron does better than the trainer and in fact rises to the same level of performance as in Fig. 12(a).

In Fig. 13 damage to the machine is simulated by removing association units at random from a perceptron already trained on E–X discrimination. The decline in performance is gradual rather than sudden and varies with the amount removed. Since the memory does not operate by comparison with a stored file of patterns, but rather is distributed throughout the structure we get behavior here analogous to Lashleys' law of "Equipotentiality and Mass Action."

More advanced perceptual problems, such as figure ground determination, relations among objects in complex fields ('the square is inside the circle,' 'the

tree is behind the dog') and so on, are beyond the capacity of the simple perceptron considered here, but it seems possible that the richer models will be able to perform these functions.[1]

g. Stimulus Modalities.

We have described the stimuli as visual patterns on a retina. Other interpretations of the input patterns are equally possible.

If, for example, each sensory unit has as its source of activation, the output of some "property filter,"[57] then a "stimulus pattern" on the retina represents a listing of the presence or absence of the various properties. Similarly as far as the logic and functioning of the simple perceptron are concerned the stimulus patterns could represent, for example, the magnitudes of the Fourier components of a sound wave or the combinations of taste sensors activated by particular food preparations. The analysis would be similar to that given above. Indeed a simple perceptron had considerable success in selecting medical diagnoses where the sensory input was the patient's coded clinical signs and symptoms.[61]

By putting several perceptrons in parallel one may, with cross connections, obtain *conditioned reflexes*, *association* between stimulus patterns of different modalities and so on.[1]

The perceptron might also be used as the perceptual input to the first stage of a "heuristic logic" machine.[8]

10. Current Research

In all of the problems discussed so far, we have been concerned with an *instantaneous* pattern. If on the other hand our interest is in the properties of a temporal sequence such as is involved, e.g., in speech recognition or the sequence of nerve impulses being fed back to the brain as a muscle movement is performed, then the

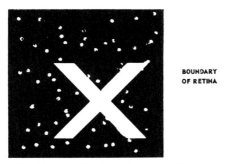

BOUNDARY OF RETINA

FIG. 11. Example of a noisy target display.

[61] A. E. Murray, Cornell Aeronautical Laboratory Report VE-1446-G-1 (November, 1960).

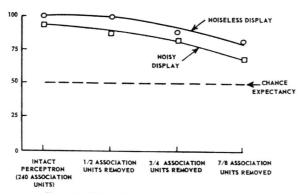

FIG. 13. Effect of association unit removal on trained "E"–"X" discrimination.

simple perceptron of Fig. 4 is no longer adequate. We shall find certain temporal effects in the paper which follows, but for others it is necessary to introduce time delays into the system.[1] A speech recognizing perceptron which utilizes such delays is currently being built at Cornell University.

Other activities now in progress[1] include quantitative studies of cross-coupled and multi-layer systems (by means of analysis and digital simulation), studies of selective attention mechanisms, the effects of geometric constraints on network organization, new types of reinforcement rules, and attempts at relating this research to biological data. Work is also in progress on development of electrolytic and other low-cost integrating devices and additional electronic components necessary for the construction of large-scale physical models.

It is clear that we are still far from the point of understanding how the brain functions. It is equally clear, we believe, that a promising road is open for further investigation.

12
Introduction

(1969)
D. J. Willshaw, O. P. Buneman, and H. C. Longuet-Higgins

Non-holographic associative memory
Nature 222:960–962

This short paper is an elegant and concise presentation of two ingenious network models. It contains a simple optical system realizing a correlation memory, called a correlograph, and a closely related neural network suggested by the optical memory.

The correlograph is a pattern associator, and not a classifier, as the perceptron was. This is an important point. The perceptron had a set of discrete output units, only one of which, by design, was active when a pattern was presented. This is sometimes called a "grandmother cell" system, that is, a cell (or R-unit) that fires when, and only when, grandmother appears. A correlograph is used to associate two patterns with many nonzero elements, not to recognize grandmother. The patterns consist of holes in an opaque screen. (The diagram shown in figure 3 of the paper can be realized with pencil and paper; the lenses are only used to focus the rays of light on the correct places.) The first pattern is lit from behind, giving a number of spots of light. Light from the spots of light in the first pattern pass through the holes in the second pattern. Since there are many holes in the second pattern, there are many possible rays from A through the hole, given by the number of spots in A times the number in B. The rays illuminate a screen at C. The screen is marked where the dots appear.

To retrieve the first pattern, the apparatus is run backward. Rays are run from all the spots on C, through the holes in B. The problem is that the spots corresponding to those in the original pattern, A, are illuminated, but there are also a number of spurious spots, corresponding to a beam from a spot through an inappropriate hole. The solution to the problem is that the original spots of A receive more rays in total than the spurious points because there is a ray to a spot in A from all the holes in B, but spurious points will only receive a ray from one or a few of the holes in B. If we set a decision threshold so we only look at the brightest spots on A, we can reconstruct A.

The correlograph therefore can be made to act as a pattern associator; i.e., given pattern B it can construct pattern A. It is so easy to make one of these devices using a piece of paper and a ruler, following figure 3, and it is so instructive in demonstrating the nature of distributed associative memories of all types that doing so is strongly recommended.

It is possible to store several sets of patterns and successfully reconstruct them, using the same correlograph. If too many associations are stored, there will be too many spurious spots, and errors will be made; that is, the reconstruction of A will be noisy. Estimating the number of patterns that can be stored is a statistical question. The authors get estimates for the useful capacity and the information capacity of the correlograph in bits. The capacity of the system is shown to be quite large, nearly as great as if the system was using a simple random access memory. Of course there is always the statistical potential for error, which is not the case for a traditional memory.

The authors then propose a neural network model, based on the correlograph, which struck them as somewhat more physiologically plausible than the correlograph. Input-lines (axons?) run horizontally and output lines run vertically (dendrites?), forming a matrix of connections. It is assumed that connections are not graded, but only on or off. Initially all the connections are zero. When two patterns are to be associated, junctions between input and output are formed. During learning, the input pattern and output pattern are simultaneously presented. A synaptic connection is made if both input and output units are active. This is a Hebbian learning rule, because the junction is strengthened (i.e., turned on) only when both input and output units are active at the same time. For reconstruction, the input pattern is presented and the activity on an output line is the sum of the active synapses in that column. The output is then thresholded, as before. The storage capacity and information density can also be computed for this network, with results similar to those for the correlograph. Both have information storage capacities proportional to the number of potential possible connections. (For the correlograph, it must be assumed that the output storage plane is composed of discrete points to do the calculation.)

This associative network is close to a purely digital device. Input lines can only be on or off, and connections are only made or not made. Only the output unit needs a graded response, giving the sum of the active synapses, but this graded output is then thresholded. For this reason, it is exceptionally easy to implement this model using integrated circuits. It and variants of it have been analyzed and proposed numerous times in the neural network literature. Potential connection densities with modern VLSI techniques are extremely high.

It might be a fair observation that many VLSI implementations of neural nets, particularly the first attempts at them, end up looking very much like this associative network. Simple networks of discrete elements may have important practical uses. More extended discussions of these models and related work can be found in the references, which repay reading.

References

D. J. Willshaw (1972), "A simple model capable of inductive generalization," *Proceedings of the Royal Society, Series B* 182:233–247.

D. J. Willshaw (1981), "Holography, associative memory, and inductive generalization, *Parallel Models of Associative Memory*, G. Hinton and J. A. Anderson (Eds.), Hillsdale, NJ: Erlbaum.

(1969)
D. J. Willshaw, O. P. Buneman, and H. C. Longuet-Higgins

Non-holographic associative memory
Nature 222:960–962

The features of a hologram that commend it as a model of associative memory can be improved on by other devices.

The remarkable properties of the hologram as an information store have led some people[1,2] to wonder whether the memory may not work on holographic principles. There are, however, certain difficulties with this hypothesis if the holographic analogy is pressed too far; how could the brain Fourier-analyse the incoming signals with sufficient accuracy, and how could it improve on the rather feeble signal-to-noise ratio[3] of the reconstructed signals? Our purpose here is to show that the most desirable features of holography are manifested by another type of associative memory, which might well have been evolved by the brain. A mathematical investigation of this non-holographic memory shows that in optimal conditions it has a capacity which is not far from the maximum permitted by information theory.

Our point of departure is Gabor's observation[4,5] that any physical system which can correlate (or for that matter convolve) pairs of patterns can mimic the performance of a Fourier holograph. Such a system, which could be set up in any school physics laboratory, is shown in Fig. 1. The apparatus is designed for making "correlograms" between pairs of pinhole patterns, and then using the correlogram and one of the patterns for reconstructing its partner. One of the pinhole patterns is mounted at A, and the other at B. The distance between them equals f, the focal length of the lens L. A viewing screen is placed at C, at a distance f from the lens, and a diffuse light source is mounted behind A. The pattern of bright dots appearing at C is the correlogram between the pattern at A and the pattern at B. Formally, $C = \bar{A} * B$, where the asterisk stands for convolution and $\bar{A}$ is the result of rotating the pattern A through half a turn round the optical axis. If A and B were interchanged, the pattern at C would be $\bar{B} * A = A * \bar{B} = \bar{C}$, so that the correlogram would be inverted. This is clear enough if B is a pinhole, and shows that the order of the patterns is important.

Copyright © 1969 Macmillan Magazines Limited. Reprinted by permission.

To recover pattern A from pattern B we convert the correlogram into a pattern of pinholes in a black card and place the light source behind it, so that the light shines through C and B on to a viewing screen at A (Fig. 2) A pattern of spots now appears on the viewing screen. All the spots of the original pattern A are present, but a number of spurious spots as well. If the pinholes were infinitesimal and there were no diffraction effects the reconstructed pattern would be $\bar{C} * B = A * \bar{B} * B$, just as in Fourier holography. If B were a random pattern, one could argue, $\bar{B} * B$ would approximate to a delta function at the origin, so that the reconstructed pattern would look like a slightly bespattered version of the original pattern A. How can we pick out the genuine spots from the others?

To solve this problem let us simplify the set-up by removing the lens (Fig. 3). Suppose, for example, that A has two holes and B has three. Then the pattern C will consist of six bright spots (barring coincidences). When these spots are converted into pinholes and illuminated from the right, a total of 18 ($=6 \times 3$) rays will emerge from B and impinge on the screen at A. But we shall not see eighteen spots on this screen, because six of the rays will converge, in sets of three, on to the two points of the original pattern. The other twelve rays will give rise to spurious spots, but (again barring coincidences) these spots will be fainter than the genuine ones. We can therefore expect to be able to pick out the wheat from the chaff with a detector with a threshold slightly less than three units of brightness.

This reasoning applies equally to the "correlograph", with lens, illustrated in Figs. 1 and 2. So, having found how to get rid of the unwanted background in reconstructing A from B and C, we can now envisage the possibility of constructing multiple correlograms, comprising all the spots present in $C_1 = \bar{A}_1 * B_1$ or in $C_2 = \bar{A}_2 * B_2$, and so on. The presentation of B_1 should evoke A_1, presentation of B_2 should evoke A_2, and so on, up to the limit set by the information capacity of the system. But what is this limit?

To answer this question let us evade the complicated (and basically irrelevant) issues raised by the finite wavelength of light, edge effects and so on, and pose the question in terms of a discrete, and slightly more

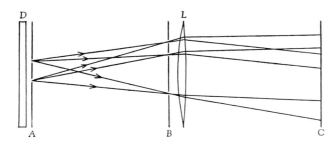

Figure 1 Constructing a correlogram. *D* is a diffuse light source, *L* a lens, and *C* the plane of the correlogram of *A* with *B*.

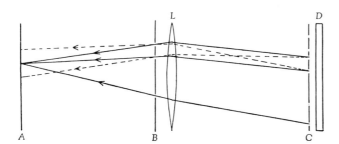

Figure 2 Reconstructing a pattern: ———, paths traversed in Fig. 1; ---, paths not traversed in Fig. 1.

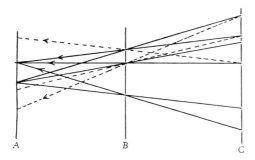

Figure 3 Showing that original spots are generally brighter.

abstract, model. We suppose *A*, *B* and *C* to be discrete spaces, each containing *N* points, a_1 to a_N, b_1 to b_N, and c_1 to c_N. The point-pair (a_i, b_j) is mapped on to the point c_k if $i - j = k$ or $k - N$. Conversely, the point-pair (c_k, b_j) is mapped on to a_i if the same condition is met. Imagine now that we have *R* pairs of patterns which we wish to associate together, each pair consisting of *M* points selected from *A* and another *M* selected from *B*. The total number of point-pairs determined by all the pairs of patterns will be RM^2, and we may think of this number of "rays" striking *C*. If they impinge at random, the probability of any point c_k not being struck will be

$$\exp(-RM^2/N) = 1 - p, \text{ say}$$

The correlogram for the whole set of *R* pairs will then consist of the remaining *pN* points of *C*.

Now consider the reconstruction process. One of the *B*-patterns, comprising *M* of the points b_1 to b_N, is selected, and combined with the correlogram to produce *pNM* "rays" impinging on *A*. Each point of the original *A*-pattern will receive exactly *M* rays, so that we should set the threshold of our detector at *M* if we want to pick up all the original points. Now consider any one of the $N - M$ other points in *A*. It may receive a ray through any one of the *M* "holes" in *B*; the probability that it receives a ray through a given hole is just *p*, for this is the chance that the point on *C* "behind" the hold belongs to the correlogram. The chance of an unwanted point reaching the threshold is thus p^M, and the probable number of spurious points of brightness *M* is consequently $(N - M)p^M$. If *M* is a fairly large number, this will be a sensitive function of *p*, and for given *N* and *M* the critical value of *p* above which spurious points begin to appear may be found from the relation

$$(N - M)p^M = 1$$

Alternatively, this may be viewed as a relation which sets a lower limit to the value of *M* for given values of *N* and *p*. A slightly safer estimate is given by

$$Np^M = 1, \text{ or } M = -\log N/\log p$$

If *M* falls below this value, the reconstruction will be marred by spurious points.

Next we enquire about the amount of information stored in the memory when *R* pairs have been memorized and *M* satisfies the aforementioned condition for accurate retrieval. We can evoke any one of *R* *A*-patterns by presenting the appropriate *B*-pattern. There are $\binom{N}{M}$ possible *A*-patterns altogether, so the amount of information needed to store any one of them is $\log \binom{N}{M}$, which is roughly $M \log N$ natural units of information. The total amount of information stored is, therefore, approximately

$$I = RM \log N \text{ natural units}$$

But according to our original calculation of *p*

$$RM^2 = -N \log(1 - p)$$

and if we are working at the limit of accurate retrieval

$$M = -\log N/\log p \simeq \log_2 N \text{ (see below)}$$

It follows immediately that

$$I = N \log p \log(1 - p)$$

As one might have anticipated, this expression has its

maximum value when p is $0\cdot5$—when the correlogram occupies about half of C.

What is remarkable is the size of I_{max}.

$I_{max} = N(\log 2)^2$ natural units $= N\log 2$ bits. The maximum amount of information that could possibly be stored in C is N bits. So the correlograph, in this discrete realization, stores its information nearly ($\log_e 2 = 69$ per cent) as densely as a random access store with no associative capability.

As described, the discrete correlograph, like the holograph, will "recognize" displaced patterns. If an A-pattern $\{a_i\}$ and a B-pattern $\{b_j\}$ have been associated, then presentation of the displaced B-pattern $\{b_{j+d}\}$ will evoke the displaced A-pattern $\{a_{i+d}\}$.

But the resemblance does not cease there. Just as in holography, the information to be stored is laid down (i) in parallel, (ii) non-locally and (iii) in such a way that it can survive local damage. In parallel, because each mapping $(a_i, b_j) \to c_k$ can be effected without reference to any other; the same applies to the reconstructive mappings $(c_k, b_j) \to a_i$. Non-locally, because the presence of a_i in an A-pattern is registered at M separate points on the correlogram, one for each point of the B-pattern. And robustly, because if the system is not stretched to its theoretical limit it can (as we shall show elsewhere) be used for the accurate reconstruction of A-patterns even when some of the correlogram is "ablated" and/or the B-patterns are inaccurately presented. But it can only be made secure against such contingencies by sacrificing storage capacity—as one would expect.

In our discussion of the process of reconstruction we had occasion to note that a point c_k might owe its presence on the correlogram to the joint occurrence of (a_i, b_j); but that if a pattern were presented containing the point b_{j+d}, the 'ray' (c_k, b_{j+d}) would light up the point a_{i+d}, which might never have occurred in any A-pattern. It was this feature which underlay the ability of the system to recognize displaced patterns; but the same feature is a slight embarrassment when one comes to consider how a discrete correlograph, with the reconstructive facility, could be realized in neural tissue. We will not dwell on this point, except to acknowledge that it was drawn to our attention by Dr F. H. C. Crick, to whom H. C. L.-H. is indebted for provocative comments. But it led us on to a further refinement of our model, in which a given point c_k is admitted to the correlogram only if the particular pair (a_i, b_j) occurs in one of the pairs of patterns, and not otherwise. On this assumption there might be as many as N^2 separate point-pairs to take into account, and a correspondingly large number of points in the space C.

In this form our associative memory model ceases to be a correlograph, having lost the ability to recognize displaced patterns, but its information capacity is now potentially far greater than before. To show this, we will adopt a rather different type of representation, in which the points of A become N_A parallel lines, and those of B become N_B parallel lines. The points of C are the $N_A N_B$ intersections between the lines a_i and the lines b_j.

In this network model, as before, a particular point of C is included in the active set if the pair of lines (a_i, b_j) which pass through it have been called into play in at least one association of an A-pattern with a B-pattern. Let us suppose that R pairs of patterns have been associated in this way, each pair comprising a selection of M_A lines from A and M_B lines from B. Then the chance that a given point of C has not been activated by the recording is

$$\exp(-RM_A M_B/N_C) = 1 - p, \text{ say}$$

where we have written N_C for $N_A N_B$. If B-patterns are being used to recall A-patterns, then there will be a minimum value of M_B such that if the threshold on the A-lines is set at M_B (so as to detect all the genuine lines) spurious lines will begin to be detected as well. (The argument is just the same as that applied to the correlograph earlier on.) This minimum value of M_B is given by

$$N_A p^{M_B} = 1$$

or

$$M_B = -\log N_A/\log p \simeq \log_2 N_A$$

Now the amount of information stored in the memory when R pairs of A-patterns have been memorized is roughly

$$I_A = RM_A \log N_A$$

But from our equation for $1 - p$

$$RM_A M_B = -N_C \log(1 - p)$$

therefore

$$I_A = N_C \log p \log(1 - p)$$

showing that, as in the correlograph, the density with which the associative net stores information is 69 per cent of the theoretical maximum value. We may note, in passing, that I_B, defined as $RM_B \log N_B$, is also equal to $N_C \log p \log(1 - p)$.

An associative network of this kind also operates (i) in parallel (ii) non-locally and (iii) in such a way that local damage or inaccuracy is not necessarily disastrous.

Output lines

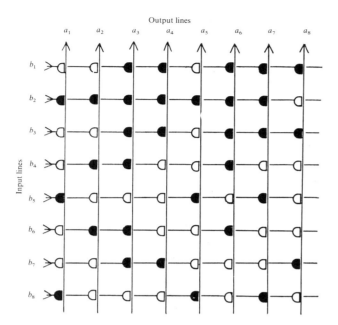

Figure 4 An associative net.

We intend to go into the details of (iii) elsewhere. We now succumb to the temptation of indicating how such an associative memory might be realized in neural tissue though, as Brindley has pointed out[6], function need not determine structure uniquely.

The system we have in mind is represented diagrammatically in Fig. 4. The horizontal lines are axons of the N_B input neurones $b_1, b_2, ..$, while the vertical lines are dendrites of the N_A output neurones $a_1, a_2, ..,$. At the intersection of b_j with a_i is a modifiable synapse c_{ij}. This synapse is initially inactive, but becomes active after a coincidence in which a_i and b_j are made to fire at the same time by some external stimulus. Such a coincidence is supposed to occur if an A-pattern containing a_i is presented in association with a B-pattern containing b_j. After the activation of c_{ij} (which we regard as a permanent effect) the firing of b_j will locally depolarize the membrane of a_i. The output neurone a_i is then supposed to fire if M_B or more input cells depolarize it simultaneously.

In Fig. 4 we indicate what the state of the network would be after it had learned to associate the following pairs of patterns:

B-pattern	A-pattern
1,2,3	4,6,7
2,5,8	1,5,7
2,4,6	2,3,6
1,3,7	3,4,8

The synapses indicated by solid semicircles would be active, those indicated by open semicircles being still inactive. In this particular example, N_A and N_B are both 8, and M_A ($\simeq \log_2 N_B$) and M_B ($\simeq \log_2 N_A$) are both 3. R, the number of pairs of patterns associated, has been chosen so as to make p, the proportion of synapses active, close to 0·5; in fact p equals 0·5 exactly. These various numbers illustrate the system working near its maximum capacity. The reader may verify that every B-pattern except the first evokes the correct A-pattern at a threshold of 3; the only mistake the system makes is that when supplied with the B-pattern 1,2,3 it responds with an A-pattern 3,4,6,7 containing four elements.

To summarize, we have attempted to distil from holography the features which commend it as a model of associative memory, and have found that the performance of a holograph can be mimicked and actually improved on by discrete non-linear models, namely the correlograph and the associative net just described. Quite possibly there is no system in the brain which corresponds exactly to our hypothetical neural network; but we do attach importance to the principle on which it works and the quantitative relations which we have shown must hold if such a system is to perform, as it can, with high efficiency.

Notes and References

[1] Van Heeden, P. J., *App. Optics*, 2, 393 (1963).

[2] Pribram, K. H., *Sci. Amer.*, 220, 73 (1969).

[3] Willshaw, D., and Longuet-Higgins, H. C., *Machine Intelligence* 4 (edit. by Michie, D.) (Edinburgh University Press, 1969).

[4] Gabor, D., *Nature*, **217**, 1288 (1968).

[5] Gabor, D., *Nature*, **217**, 584 (1968).

[6] Brindley, G. S., *Proc. Roy. Soc.*, B, **168**, 361 (1967).

13

Introduction

(1969)
Marvin Minsky and Seymour Papert

Perceptrons, Cambridge, MA: MIT Press, Introduction, pp. 1–20, and p. 73 (figure 5.1)

In the popular history of neural networks, first came the classical period of the perceptron, when it seemed as if neural networks could do anything. A hundred algorithms bloomed, a hundred schools of learning machines contended. Then came the onset of the dark ages, where, suddenly, research on neural networks was unloved, unwanted, and, most important, unfunded.

A precipitating factor in this sharp decline was the publication of the book *Perceptrons* by Minsky and Papert. Yet *Perceptrons* is a brilliant book. It was simply and clearly written in an elegant and informal style. Why did it have such apparently devastating effects?

There were many causes and it would be unfair to blame *Perceptrons* alone. According to Bernard Widrow, a pioneer in the field, there had always been great resistance to the whole idea of actually building an artificial "chunk of brain" (Widrow, 1987). The area was suspect scientifically from the beginning. Other problems included excessive hype, especially when commercial exploitation seemed likely. Few things get other scientists more irritated than the appearance of extravagant claims in the newspapers. Brain models lend themselves well to dramatic news stories and reporters for daily newspapers in the 1960s were not noted, with a few honorable exceptions, for their scientific understanding or their understatement. Rosenblatt quotes a headline from an Oklahoma newspaper in his introduction to *Principles of Neurodynamics* (1962): "Frankenstein Monster Designed by Navy Robot That Thinks." Nowadays reporting of science and technology is generally much better. So far, the news stories that have been written during the current renaissance of neural networks are much more responsible than they were twenty years ago.

Perceptrons is a book on mathematics and the theory of computation. We have reprinted here the entire Introduction, which conveniently summarizes many of the results and techniques used in the rest of the book. We have reprinted, in addition, a single page that contains an interesting figure.

Perceptrons discusses limits. Rosenblatt and others were aware of problems with perceptrons, in that they did not seem to compute certain things very well. Some theoretical limitations were known from the earliest days, for example, the requirement for linear separability of the data points for perfect classification by an output unit. There were problems if too much generalization was required. Some of these difficulties are mentioned in the concluding chapter of Rosenblatt's book, but it would be fair to say that there was little appreciation for the magnitude of the theoretical limitations of perceptrons.

Minsky and Papert placed the study of computational limitations of perceptrons on a solid foundation. The simple perceptron (Rosenblatt, paper 8) contains three layers: a retina, a set of "Association" or "A-units," and a set of "Response" or "R-units." The units of the perceptron are threshold logic units. The output of a threshold logic unit can be only one of two states. Therefore, threshold logic units can be assumed to be computing a logical predicate of their inputs: true for some things and false for others. The simple perceptron has many A-units receiving inputs from a retina, and projecting to a single R-unit. Rosenblatt considered the R-unit to be computing a "classification," but Minsky and Papert said it was actually computing a logical predicate, based on information fed to it by the A-units connected to it. Intuitively, the A-units form a family of local predicates, called ψ, computing a set of local properties or features. The family of local predicates is called Ψ. Minsky and Papert assumed that the local predicates, the A-units, and the family of local predicates were all looking at a 'retina,' R, which consisted of points on the plane. This allowed them to make extensive use of geometrical arguments for their proofs. In the simple "linear" perceptron, analyzed in the most depth in the book, the R-unit looks at the local predicates from the A-units, takes their sum, weighted by a multiplicative constant, compares it with a threshold, and then responds with a value for the overall predicate, $\Psi(X)$, which is either true or false.

With this structure, the critical question becomes, "What logical functions can $\Psi(X)$ successfully compute?"

This representation of the output of the perceptron as a logical predicate has a different significance than the more modest kinds of computation that Rosenblatt and others stressed in their work. A logical predicate seems to be a much more forbidding function than simply trying to get classifications correct most of the time. Logical functions can be of great generality and power.

The potential utility of perceptrons as computing devices rested on their structural simplicity. To be simple, they had to have limitations on the number of connections. If it was possible for the local predicates to look at every point in the entire retina, anything whatsoever could be computed by having a predicate computed for every possible pattern in the plane. This would be impractical, to say the least, since the number of possible patterns on the retina grows exponentially with the size of the retina.

Minsky and Papert used two important classes of limitations on the local predicates in their proofs: *order limited*, where only a certain maximum number of retinal points could be connected to the local decision unit computing the local predicate, and *diameter limited*, where only a geometrically restricted region of retina was connected to the local predicate.

The most famous mathematical results in the book come from Minsky and Papert's discussion of the geometrical predicate *connectedness*. A connected figure is in one piece; i.e., it can be drawn without lifting the pencil from the paper. Immediately, given the initial definition of a diameter limited perceptron, it is possible to show that a diameter limited perceptron *cannot* compute connectedness. The proof is simple, intuitive, and given in the Introduction in Section 0.8. Minsky and Papert later show that the same limitation holds true for order limited perceptions, though the proof is more complicated. There was also extended discussion of some other predicates, for example, parity, which required counting the active points on the plane and deciding whether the number of points was odd or even. Parity could also not be computed with limited perceptrons.

Were these important but somewhat technical results enough to terminate an entire active line of research? It seems unlikely. Perceptrons and related early network models were in decline for several years before this book because they had failed to achieve much beyond their initial successes. Practical results failed to materialize. Experience with simulations had indicated problems scaling up small systems to large ones. Small systems worked very well; larger ones tended not to work at all. (That scaling was a true theoretical difficulty was confirmed by some of the results in *Perceptrons*.) And the new field of Artificial Intelligence, in which Minsky and Papert were prominent, was generating good results itself using very different approaches to the problem of machine intelligence.

So the appearance of *Perceptrons* was the final step in a process that had gone on for several years. Minsky and Papert probably express a general irritation and disappointment with perceptrons when they make such comments as, "... most of this writing [on perceptrons] is without scientific value" (p. 4) and "The results of these hundreds of projects and experiments [on perceptrons] were generally disappointing, and the explanations inconclusive" (p. 19).

Perceptrons may have summed up a general feeling in the scientific community in the late '60s. However, it delivered the final, near-fatal blow to the whole field of neural networks. Part of the reason was its obvious brilliance and the focus and lucidity of its attack. Allan Newell started his long review of *Perceptrons* in *Science* with this sentence: "This is a great book" (Newell, 1969). And the book is, in fact, a model for how to do analysis of a complex system. Perhaps most serious in its impact on future work were the conjectures in the last chapter. There, the authors expressed the strong belief that limitations of the kind they discovered for simple perceptrons would be held to be true for perceptron variants, more specifically, multilayer systems. As they put it, "... we consider it to be an important research problem to elucidate ... our intuitive judgement that the extension [to multilayer systems] is sterile" (p. 232).

This conjecture, following over two hundred pages of brilliant analysis, became telling indeed. It thoroughly dampened the enthusiasm of granting agencies to support further research. Why bother, since more complex versions would have the same problems? Unfortunately, this conjecture now seems to be wrong, and more advanced neural networks are capable of computing some logical predicates in efficient ways that perceptrons could not. One example is the predicate "parity," which is briefly discussed in the paper by Rumelhart, Hinton, and Williams on the back propagation algorithm for multilayered networks (paper 41, p. 334).

There was a point about perceptrons and learning machines that Minsky and Papert did not discuss but that became quite important in the resurrection of neural networks. That was their ability to aet as psychological models. It is true, and proven, that simple perceptrons cannot compute the predicate connectedness. However, a brief glance at figure 5.1 (p. 73) will indicate something interesting: *we cannot compute it either*. The only way to tell which spiral is connected is to trace out the contours. This simple experiment demonstrates that in our immediate perceptions, we act as if we had some of the same computational limitations that a perceptron does. Of course, we have the ability, which the perceptron did not, to convert ourselves into an inefficient serial algorithm for computing connectedness, by tracing the contours of the figure. But subjectively, this feels like a very different mode of operation of our mental apparatus. Perceptron-like models can be quite successful at modeling a number of aspects of human perception and cognition. A body of research on human psychology supports this conclusion. Many of these psychological results are summarized in the two-volume work *Parallel Distributed Processing* (Rumelhart and McClelland, 1986, Vol. 1; McClelland and Rumelhart, 1986, Vol. 2).

References

J. L. McClelland and D. E. Rumelhart (Eds.) (1986), *Parallel Distributed Processing: Explorations in the Microstructures of Cognition*, Vol. 2, Cambridge, MA: MIT Press.

A. Newell (1969), "A step toward the understanding of information processing," *Science* 165:780–781.

F. E. Rosenblatt (1962), *Principles of Neurodynamics*, New York: Spartan.

D. E. Rumelhart and J. L. McClelland (Eds.) (1986), *Parallel Distributed Processing: Explorations in the Microstructures of Cognition*, Vol. 1, Cambridge, MA: MIT Press.

B. Widrow (1987), historical talk at Neural Networks for Computation Conference, Snowbird, UT, April 1987.

(1969)
Marvin Minsky and Seymour Papert

Perceptrons, Cambridge, MA: MIT Press, Introduction, pp. 1–20, and p. 73 (figure 5.1)

0
Introduction

0.0 Readers

In writing this we had in mind three kinds of readers. First, there are many new results that will interest specialists concerned with "pattern recognition," "learning machines," and "threshold logic." Second, some people will enjoy reading it as an essay in abstract mathematics; it may appeal especially to those who would like to see geometry return to topology and algebra. We ourselves share both these interests. But we would not have carried the work as far as we have, nor presented it in the way we shall, if it were not for a different, less clearly defined, set of interests.

The goal of this study is to reach a deeper understanding of some concepts we believe are crucial to the general theory of computation. We will study in great detail a class of computations that make decisions by weighing evidence. Certainly, this problem is of great interest in itself, but our real hope is that understanding of its mathematical structure will prepare us eventually to go further into the almost unexplored theory of parallel computers.

The people we want most to speak to are interested in that general theory of computation. We hope this includes psychologists and biologists who would like to know how the brain computes thoughts and how the genetic program computes organisms. We do not pretend to give answers to such questions—nor even to propose that the simple structures we shall use should be taken as "models" for such processes. Our aim—we are not sure whether it is more modest or more ambitious—is to illustrate how such a theory might begin, and what strategies of research could lead to it.

It is for this third class of readers that we have written this introduction. It may help those who do not have an immediate involvement with it to see that the theory of pattern recognition might be worth studying for other reasons. At the same time we will set out a simplified version of the theory to help readers who

have not had the mathematical training that would make the later chapters easy to read. The rest of the book is self-contained and anyone who hates introductions may go directly to Chapter 1.

0.1 Real, Abstract, and Mythological Computers

We know shamefully little about our computers and their computations. This seems paradoxical because, physically and logically, computers are so lucidly transparent in their principles of operation. Yet even a school boy can ask questions about them that today's "computer science" cannot answer. We know very little, for instance, about how much computation a job should require.

As an example, consider one of the most frequently performed computations: *solving a set of linear equations*. This is important in virtually every kind of scientific work. There are a variety of standard programs for it, which are composed of additions, multiplications, and divisions. One would suppose that such a simple and important subject, long studied by mathematicians, would by now be thoroughly understood. But we ask, How many arithmetic steps are absolutely required? How does this depend on the amount of computer memory? How much time can we save if we have *two* (or n) identical computers? Every computer scientist "knows" that this computation requires something of the order of n^3 multiplications for n equations, but even if this be true no one knows—at this writing—how to begin to prove it.

Neither the outsider nor the computation specialist seems to recognize how primitive and how empirical is our present state of understanding of such matters. We do not know how much the speed of computations can be increased, in general, by using "parallel" as opposed to "serial"—or "analog" as opposed to "digital"—machines. We have no theory of the situations in which "associative" memories will justify their higher cost as compared to "addressed" memories. There is a great deal of folklore about this sort of contrast, but much of this folklore is mere supersition; in the case we have studied carefully, the common beliefs turn out to be not merely "unproved"; they are often drastically wrong.

The immaturity shown by our inability to answer questions of this kind is exhibited even in the language used to formulate the questions. Word pairs such as "parallel" vs. "serial;" "local" vs. "global," and "digital" vs. "analog" are used as if they referred to well-defined technical concepts. Even when this is true, the technical meaning varies from user to user and context to context. But usually they are treated so loosely that the species of computing machine defined by them belongs to mythology rather than science.

Now we do not mean to suggest that these are mere pseudo-problems that arise from sloppy use of language. This is not a book of "therapeutic semantics"! For there *is* much content in these intuitive ideas and distinctions. The problem is how to capture it in a clear, sharp theory.

0.2 Mathematical Strategy

We are not convinced that the time is ripe to attempt a very general theory broad enough to encompass the concepts we have mentioned and others like them. Good theories rarely develop outside the context of a background of well-understood real problems and special cases. Without such a foundation, one gets either the vacuous generality of a theory with more definitions than theorems—or a mathematically elegant theory with no application to reality.

Accordingly, our best course would seem to be to strive for a *very thorough* understanding of well-chosen particular situations in which these concepts are involved.

We have chosen in fact to explore the properties of the simplest machines we could find that have a clear claim to be "parallel"—for they have no loops or feedback paths—yet can perform computations that are nontrivial, both in practical and in mathematical respects.

Before we proceed into details, we would like to reassure non-mathematicians who might be frightened by what they have glimpsed in the pages ahead. The mathematical methods used are rather diverse, but they seldom require advanced knowledge. We explain most of that which goes beyond elementary algebra and geometry. Where this was not practical, we have marked as *optional* those sections we feel might demand from most readers more mathematical effort than is warranted by the topic's role in the whole structure. Our theory is more like a tree with many branches than like a narrow high tower of blocks; in many cases one can skip, if trouble is encountered, to the beginning of the following chapter.

The reader of most modern mathematical texts is made to work unduly hard by the author's tendency to cover over the intellectual tracks that lead to the discovery of the theorems. We have tried to leave visible the lines of progress. We should have liked to go further and leave traces of all the false tracks we followed; unfortunately there were too many! Nevertheless we have occasionally left an earlier proof even when we later found a "better" one. Our aim is not so much to prove theorems as to give insight into methods and to encourage research. We hope this will be read not as a chain of logical deductions but as a mathematical novel where characters appear, reappear, and develop.

0.3 Cybernetics and Romanticism

The machines we will study are abstract versions of a class of devices known under various names; we have agreed to use the name "perceptron" in recognition of the pioneer work of Frank Rosenblatt. Perceptrons make decisions—determine whether or not an event fits a certain "pattern"—by adding up evidence obtained from many small experiments. This clear and simple concept is important because most, and perhaps all, more complicated machines for making decisions share a little of this character. Until we understand it very thoroughly, we can expect to have trouble with more advanced ideas. In fact, we feel that the critical advances in many branches of science and mathematics began with good formulations of the "linear" systems, and these machines are our candidate for beginning the study of "parallel machines" in general.

Our discussion will include some rather sharp criticisms of earlier work in this area. Perceptrons have been widely publicized as "pattern recognition" or "learning" machines and as such have been discussed in a large number of books, journal articles, and voluminous "reports." Most of this writing (some exceptions are mentioned in our bibliography) is without scientific value and we will not usually refer by name to the works we criticize. The sciences of computation and cybernetics began, and it seems quite rightly so, with a certain flourish of romanticism. They were laden with attractive and exciting new ideas which have already borne rich fruit. Heavy demands of rigor and caution could have held this development to a much slower pace; only the future could tell which directions were to be the best. We feel, in fact, that the solemn experts who most complained about the "exaggerated claims" of the cybernetic enthusiasts were, in the balance, much more in the wrong. But now the time has come for maturity, and this requires us to match our speculative enterprise with equally imaginative standards of criticism.

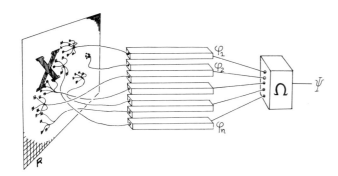

Figure 0.1

0.4 Parallel Computation

The simplest concept of parallel computation is represented by the diagram in Figure 0.1. The figure shows how one might compute a function $\psi(X)$ in two stages. First we compute *independently* of one another a set of functions $\varphi_1(X), \varphi_2(X), \ldots, \varphi_n(X)$ and then combine the results by means of a function Ω of n arguments to obtain the value of ψ.

To make the definition meaningful—or, rather, productive—one needs to place some restrictions on the function Ω and the set Φ of functions $\varphi_1, \varphi_2, \ldots$. If we do not make restrictions, we do not get a theory: any computation ψ could be represented as a parallel computation in various trivial ways, for example, by making one of the φ's be ψ and letting Ω do nothing but transmit its results. We will consider a variety of restrictions, but first we will give a few concrete examples of the kinds of functions we might want ψ to be.

0.5 Some Geometric Patterns; Predicates

Let R be the ordinary two-dimensional Euclidean plane and let X be a geometric figure drawn on R. X could be a circle, or a pair of circles, or a black-and-white sketch of a face. In general we will think of a figure X as simply a subset of the points of R (that is, the black points).

Let $\psi(X)$ be a function (of figures X on R) that can have but two values. We usually think of the two values of ψ as 0 and 1. But by taking them to be FALSE and TRUE we can think of $\psi(X)$ as a predicate, that is, a variable statement whose truth or falsity depends on the choice of X. We now give a few examples of predicates that will be of particular interest in the sequel.

$$\psi_{\text{CIRCLE}}(X) = \begin{cases} 1 & \text{if the figure } X \text{ is a circle,} \\ 0 & \text{if the figure is not a circle;} \end{cases}$$

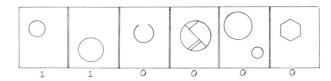

$$\psi_{\text{CONVEX}}(X) = \begin{cases} 1 & \text{if } X \text{ is a convex figure,} \\ 0 & \text{if } X \text{ is not a convex figure;} \end{cases}$$

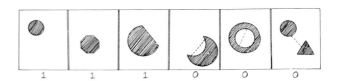

$$\psi_{\text{CONNECTED}}(X) = \begin{cases} 1 & \text{if } X \text{ is a connected figure,} \\ 0 & \text{otherwise.} \end{cases}$$

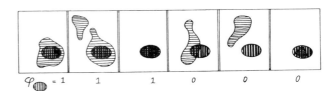

We will also use some very much simpler predicates.* The very simplest predicate "recognizes" when a particular single point is in X: let p be a point in the plane and define

$$\varphi_p(X) = \begin{cases} 1 & \text{if } p \text{ is in } X, \\ 0 & \text{otherwise.} \end{cases}$$

Finally we will need the kind of predicate that tells when a particular set A is a subset of X:

$$\varphi_A(X) = \begin{cases} 1 & \text{if } A \subset X, \\ 0 & \text{otherwise.} \end{cases}$$

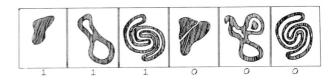

0.6 One Simple Concept of "Local"

We start by observing an important difference between $\psi_{\text{CONNECTED}}$ and ψ_{CONVEX}. To bring it out we state a fact about convexity:

* We will use "φ" instead of "ψ" for those very simple predicates that will be combined later to make more complicated predicates. No absolute logical distinction is implied.

DEFINITION A set X fails to be convex if and only if there exist three points such that q is in the line segment joining p and r, and

$$\left\{ \begin{array}{ll} p & \text{is in } X, \\ q & \text{is not in } X, \\ r & \text{is in } X. \end{array} \right.$$

Thus we can test for convexity by examining triplets of points. If all the triplets pass the test then X is convex; if any triplet fails (that is, meets all conditions above) then X is not convex. Because all the tests can be done independently, and the final decision made by such a logically simple procedure—unanimity of all the tests—we propose this as a first draft of our definition of "local."

DEFINITION A predicate ψ is <u>conjunctively local</u> of order k if it can be computed, as in §0.4, by a set Φ of predicates φ such that

$$\left\{ \begin{array}{l} \text{Each } \varphi \text{ depends upon no more than } k \text{ points of } R; \\ \psi(X) = \left\{ \begin{array}{ll} 1 & \text{if } \varphi(X) = 1 \text{ for every } \varphi \text{ in } \Phi, \\ 0 & \text{otherwise.} \end{array} \right. \end{array} \right.$$

Example ψ_{CONVEX} is conjunctively local of order 3.

The property of a figure being *connected* might not seem at first to be very different in kind from the property of being convex. Yet we can show that:

THEOREM 0.6.1 $\psi_{\text{CONNECTED}}$ is not conjunctively local of any order.

Proof Suppose that $\psi_{\text{CONNECTED}}$ has order k. Then to distinguish between the two figures

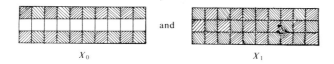

there must be some φ which has value 0 on X_0 which is not connected. All φ's have value 1 on X_1, which is connected. Now, φ can depend on at most k points, so there must be at least one middle square, say S_j, that does not contain one of these points. But then, on the figure X_2,

which is connected, φ must have the same value, 0, that it has on X_0. But this cannot be, for all φ's must have value 1 on X_2.

Of course, if some φ is allowed to look at *all* the points of R then $\psi_{\text{CONNECTED}}$ can be computed, but this would go against any concept of the φ's as "local" functions.

0.7 Some Other Concepts of Local

We have accumulated some evidence in favor of "conjunctively local" as a geometrical and computationally meaningful property of predicates. But a closer look raises doubts about whether it is broad enough to lead to a rich enough theory.

Readers acquainted with the mathematical methods of topology will have observed that "conjunctively local" is similar to the notion of "local property" in topology. However, if we were to pursue the analogy, we would restrict the φ's to depend upon all the points inside small circles rather than upon fixed numbers of points. Accordingly, we will follow two parallel paths. One is based on *restrictions on numbers of points* and in this case we shall talk of predicates of *limited order*. The other is based on restrictions of distances between the points, and here we shall talk of *diameter-limited predicates*. Despite the analogy with other important situations, the concept of local based on diameter limitations seems to be less interesting in our theory—although one might have expected quite the opposite.

More serious doubts arise from the narrowness of the "conjunctive" or "unanimity" requirement. As a next step toward extending our concept of *local*, let us now try to separate essential from arbitrary features of the definition of *conjunctive localness*. The intention of the definition was to divide the computation of a predicate ψ into two stages:

STAGE I *The computation of many properties or features φ_α which are each easy to compute, either because each depends only on a small part of the whole input space R, or because they are very simple in some other interesting way.*

STAGE II *A decision algorithm Ω that defines ψ by combining the results of the Stage I computations. For the division into two stages to be meaningful, this decision function must also be distinctively homogeneous, or easy to program, or easy to compute.*

The particular way this intention was realized in our example ψ_{CONVEX} was rather arbitrary. In Stage I we made sure that the φ_α's were easy to compute by requiring each to depend only upon a few points of R. In Stage II we used just about the simplest imaginable decision rule; if the φ's are *unanimous* we accept the figure; we reject it if even a single φ disagrees.

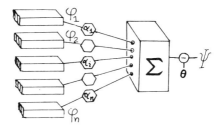

Figure 0.2

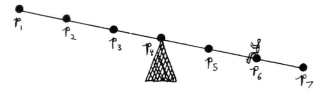

Figure 0.3

We would prefer to be able to present a perfectly precise definition of our intuitive local-vs.-global concept. One trouble is that phrases like "easy-to-compute" keep recurring in our attempt to formulate it. To make this precise would require some scheme for comparing the complexity of different computation procedures. Until we find an intuitively satisfactory scheme for this, and it doesn't seem to be around the corner, the requirements of both Stage I and Stage II will retain the heuristic character that makes formal definition difficult.

From this point on, we will concentrate our attention on a particular scheme for Stage II—"weighted voting," or "linear combination" of the predicates of Stage I. This is the so-called perceptron scheme, and we proceed next to give our final definition.

0.8 Perceptrons

Let $\Phi = \{\varphi_1, \varphi_2, \ldots, \varphi_n\}$ be a family of predicates. We will say that

ψ is linear with respect to Φ

if there exists a number θ and a set of numbers $\{\alpha_{\varphi_1}, \alpha_{\varphi_2}, \ldots, \alpha_{\varphi_n}\}$ such that $\psi(X) = 1$ if and only if $\alpha_{\varphi_1}\varphi_1(X) + \cdots + \alpha_{\varphi_n}\varphi_n(X) > \theta$. The number θ is called the threshold and the α's are called the coefficients or weights. (See Figure 0.2.) We usually write more compactly

$$\psi(X) = 1 \text{ if and only if } \sum_{\varphi \in \Phi} \alpha_\varphi \varphi(X) > \theta.$$

The intuitive idea is that each predicate of Φ is supposed to provide some evidence about whether ψ is true for any figure X. If, on the whole, $\psi(X)$ is strongly correlated with $\varphi(X)$ one expects α_φ to be positive, while if the correlation is negative so would be α_φ. The idea of correlation should not be taken literally here, but only as a suggestive analogy.

Example Any conjunctively local predicate can be expressed in this form by choosing $\theta = -1$ and $\alpha_\varphi = -1$ for every φ. For then

$$\sum (-1)\varphi(X) > -1$$

exactly when $\varphi(X) = 0$ for every φ in Φ. (The senses of TRUE and FALSE thus have to be reversed for the φ's, but this isn't important.)

Example Consider the seesaw of Figure 0.3 and let X be an arrangement of pebbles placed at *some* of the equally spaced points $\{p_1, \ldots, p_7\}$. Then R has seven points. Define $\varphi_i(X) = 1$ if and only if X contains a pebble at the ith point. Then we can express the predicate

"The seesaw will tip to the right"

by the formula

$$\sum (i - 4)\varphi_i(X) > 0,$$

where $\theta = 0$ and $\alpha_i = (i - 4)$.

There are a number of problems concerning the possibility of infinite sums and such matters when we apply this concept to recognizing patterns in the Euclidean plane. These issues are discussed extensively in the text, and we want here only to reassure the mathematician that the problem will be faced. Except when there is a good technical reason to use infinite sums (and this is sometimes the case) we will make the problem finite by two general methods. One is to treat the retina R as made up of discrete little squares (instead of points) and treat as equivalent figures that intersect the same squares. The other is to consider only bounded X's and choose Φ so that for any bounded X only a finite number of φ's will be nonzero.

DEFINITION   A perceptron is a device capable of computing all predicates which are linear in some given set Φ of partial predicates.

That is, we are given a set of φ's, but can select freely their "weights," the α_φ's, and also the threshold θ. For reasons that will become clear as we proceed, there is little to say about all perceptrons in general. But, by imposing certain conditions and restrictions we will find much to say about certain particularly interesting *families* of perceptrons. Among these families are

1. Diameter-limited perceptrons: for each φ in Φ, the set of points upon which φ depends is restricted not to exceed a certain *fixed diameter* in the plane.

2. Order-restricted perceptrons: we say that a perceptron has order $\leq n$ if no member of Φ depends on more than n points.

3. Gamba perceptrons: each member of Φ may depend on all the points but must be a "linear threshold function" (that is, each member of Φ is itself computed by a perceptron of order 1, as defined in 2 above).

4. Random perceptrons: These are the form most extensively studied by Rosenblatt's group: the φ's are random Boolean functions. That is to say, they are order-restricted and Φ is generated by a stochastic process according to an assigned distribution function.

5. Bounded perceptrons: Φ contains an infinite number of φ's, but all the α_φ lie in a finite set of numbers.

To give a preview of the kind of results we will obtain, we present here a simple example of a theorem about diameter-restricted perceptrons.

THEOREM 0.8   No diameter-limited perceptron can determine whether or not all the parts of any geometric figure are connected to one another! That is, no such perceptron computes $\psi_{\text{CONNECTED}}$.

The proof requires us to consider just four figures

and a diameter-limited perceptron ψ whose support sets have diameters like those indicated by the circles below:

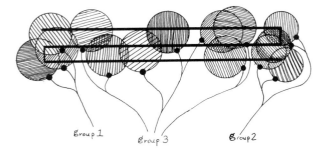

It is understood that the diameter in question is given at the start, and we *then* choose the X_{ij}'s to be several diameters in length. Suppose that such a perceptron could distinguish disconnected figures (like X_{00} and X_{11}) from connected figures (like X_{10} and X_{01}), according to whether or not

$$\sum \alpha_\varphi \varphi > \theta$$

that is, according to whether or not

$$\left[\sum_{\text{group 1}} \alpha_\varphi \varphi(X) + \sum_{\text{group 2}} \alpha_\varphi \varphi(X) + \sum_{\text{group 3}} \alpha_\varphi \varphi(X) - \theta\right] > 0$$

where we have grouped the φ's according to whether their support sets lie near the left, right, or neither end of the figures. Then for X_{00} the total sum must be negative. In changing X_{00} to X_{10} only $\sum_{\text{group 1}}$ is affected, and its value must *increase* enough to make the total sum become positive. If we were instead to change X_{00} to X_{01} then $\sum_{\text{group 2}}$ would have to increase. But if we were to change X_{00} to X_{11}, both $\sum_{\text{group 1}}$ and $\sum_{\text{group 2}}$ will have to increase by these same amounts since (locally!) the same changes are seen by the group 1 and group 2 predicates, while $\sum_{\text{group 3}}$ is unchanged in every case. Hence, net change in the $X_{00} \rightarrow X_{11}$ case must be even more positive, so that if the perceptron is to make the correct decision for X_{00}, X_{01}, and X_{10}, it is forced to accept X_{11} as connected, and this is an error! So no such perceptron can exist.

Readers already familiar with perceptrons will note that this proof—which shows that diameter-limited perceptrons cannot recognize connectedness—is concerned neither with "learning" nor with probability theory (or even with the geometry of hyperplanes in n-dimensional hyperspace). It is entirely a matter of relating the geometry of the patterns to the algebra of weighted predicates. Readers concerned with physiology will note that—insofar as the presently identified functions of receptor cells are all diameter-limited—this suggests that an animal will require more than neurosynaptic "summation" effects to make these cells compute connectedness. Indeed, only the most advanced animals can apprehend this complicated visual concept. In Chapter 5 this theorem is shown to extend also to order-limited perceptrons.

0.9 Seductive Aspects of Perceptrons

The purest vision of the perceptron as a pattern-recognizing device is the following:

The machine is built with a fixed set of computing elements for the partial functions φ, usually obtained by a random process. To make it recognize a particular pattern (set of input figures) one merely has to set the coefficient α_φ to suitable values. Thus "programming" takes on a pleasingly homogeneous form. Moreover since "programs" are representable as points $(\alpha_1, \alpha_2, \ldots, \alpha_n)$ in an n-dimensional space, they inherit a metric which makes it easy to imagine a kind of automatic programming which people have been tempted to call *learning*: by attaching feedback devices to the parameter controls they propose to "program" the machine by providing it with a sequence of input patterns and an "error signal" which will cause the coefficients to change in the right direction when the machine makes an inappropriate decision. The *perceptron convergence theorems* (see Chapter 11) define conditions under which this procedure is guaranteed to find, eventually, a correct set of values.

0.9.1 Homogeneous Programming and Learning

To separate reality from wishful thinking, we begin by making a number of observations. Let Φ be the set of partial predicates of a perceptron and $L(\Phi)$ the set of all predicates linear in Φ. Thus $L(\Phi)$ is the repertoire of the perceptron—the set of predicates it can compute when its coefficients α_φ and threshold θ range over all possible values. Of course $L(\Phi)$ could in principle be the set of *all* predicates but this is impossible in practice, since Φ would have to be astronomically large. So any physically real perceptron has a limited repertoire. The ease and uniformity of programming have been bought at a cost! We contend that the traditional investigations of perceptrons did not realistically measure this cost. In particular they neglect the following crucial points:

1. The idea of thinking of classes of geometrical objects (or programs that define or recognize them) as classes of n-dimensional vectors $(\alpha_1, \ldots, \alpha_n)$ loses the geometric individuality of the patterns and leads only to a theory that can do little more than *count* the number of predicates in $L(\Phi)$! This kind of imagery has become traditional among those who think about pattern recognition along lines suggested by classical statistical theories. As a result not many people seem to have observed or suspected that there might be *particular* meaningful and intuitively simple predicates that belong to *no* practically realizable set $L(\Phi)$. We will extend our analysis of $\psi_{\text{CONNECTED}}$ to show how deep this problem can be. At the same time we will show that certain predicates which might intuitively seem to be difficult for these devices *can*, in fact, be recognized by low-order perceptrons: ψ_{CONVEX} already illustrates this possibility.

2. Little attention has been paid to the size, or more precisely, the information content, of the parameters $\alpha_1, \ldots, \alpha_n$. We will give examples (which we believe are typical rather than exceptional) where the ratio of the largest to the smallest of the coefficients is meaninglessly big. Under such conditions it is of no (practical) avail that a predicate be in $L(\Phi)$. In some cases the information capacity needed to store $\alpha_1, \ldots, \alpha_n$ is even greater than that needed to store the whole class of figures defined by the pattern!

3. Closely related to the previous point is the problem of *time of convergence* in a "learning" process. Practical perceptrons are essentially finite-state devices (as shown in Chapter 11). It is therefore vacuous to cite a "perceptron convergence theorem" as assurance that a

learning process will eventually find a correct setting of its parameters (if one exists). For it could do so trivially by cycling through all its states, that is, by trying all coefficient assignments. The significant question is how fast the perceptron learns relative to the time taken by a completely random procedure, or a completely exhaustive procedure. It will be seen that there are situations of some geometric interest for which the convergence time can be shown to increase even faster than exponentially with the size of the set R.

Perceptron theorists are not alone in neglecting these precautions. A perusal of any typical collection of papers on "self-organizing" systems will provide a generous sample of discussions of "learning" or "adaptive" machines that lack even the degree of rigor and formal definition to be found in the literature on perceptrons. The proponents of these schemes seldom provide any analysis of the range of behavior which can be learned nor do they show much awareness of the price usually paid to make some kinds of learning easy: they unwittingly restrict the device's total range of behavior with hidden assumptions about the environment in which it is to operate.

These critical remarks must not be read as suggestions that we are opposed to making machines that can "learn." Exactly the contrary! But we do believe that significant learning at a significant rate presupposes some significant prior structure. Simple learning schemes based on adjusting coefficients can indeed be practical and valuable when the partial functions are reasonably matched to the task, as they are in Samuel's checker player. A perceptron whose φ's are properly designed for a discrimination known to be of suitably low order will have a good chance to improve its performance adaptively. Our purpose is to explain why there is little chance of much good coming from giving a high-order problem to a quasi-universal perceptron whose partial functions have not been chosen with any particular task in mind.

It may be argued that *people* are universal learning machines and so a counterexample to this thesis. But our brains are sufficiently structured to be programmable in a much more general sense than the perceptron and our *culture* is sufficiently structured to provide, if not actual program, at least a rather complex set of interactions that govern the course of whatever the process of self-programming may be. Moreover, it takes time for us to become universal learners: the sequence of transitions from infancy to intellectual maturity seems rather a confirmation of the thesis that the rate of acquisition of new cognitive structure (that is, learning) is a sensitive function of the level of existing cognitive structure.

0.9.2 Parallel Computation

The perceptron was conceived as a parallel-operation device in the physical sense that the partial predicates are computed simultaneously. (From a formal point of view the important aspect is that they are computed independently of one another.) The price paid for this is that *all* the φ_i must be computed, although only a minute fraction of them may in fact be relevant to any particular final decision. The *total amount* of computation may become vastly greater than that which would have to be carried out in a well planned sequential process (using the same φ's) whose decisions about what next to compute are conditional on the outcome of earlier computation. Thus the choice between parallel and serial methods in any particular situation must be based on balancing the increased value of reducing the (total elapsed) time against the cost of the additional computation involved.

Even low-order predicates may require large amounts of wasteful computation of information which would be irrelevant to a serial process. This cost may sometimes remain within physically realizable bounds, especially if a large tolerance (or "blur") is acceptable. High-order predicates usually create a completely different situation. An instructive example is provided by $\psi_{\text{CONNECTED}}$. As shown in Chapter 5, *any* perceptron for this predicate on a 100×100 toroidal retina *needs* partial functions that *each* look at many hundreds of points! In this case the concept of "local" function is almost irrelevant: the partial functions are themselves global. Moreover, the fantastic number of possible partial functions with such large supports sheds gloom on any hope that a modestly sized, randomly generated set of them would be sufficiently dense to span the appropriate space of functions. To make this point sharper we shall show that for certain predicates and classes of partial functions, the *number* of partial functions that have to be used (to say nothing of the size of their coefficients) would exceed physically realizable limits.

The conclusion to be drawn is that the appraisal of any particular scheme of parallel computation cannot be undertaken rationally without tools to determine the extent to which the problems to be solved can be analyzed into local and global components. The lack of a *general* theory of what is global and what is local is no excuse for avoiding the problem in particular cases. This study will show that it is not impossibly difficult to develop such a theory for a limited but important class of problems.

0.9.3 The Use of Simple Analogue Devices

Part of the attraction of the perceptron lies in the possibility of using very simple physical devices—"analogue computers"—to evaluate the linear threshold functions. It is perhaps generally appreciated that the utility of this scheme is limited by the sparseness of *linear* threshold functions in the set of *all* logical functions. However, almost no attention has been paid to the possibility that the set of linear functions which are *practically* realizable may be rarer still. To illustrate this problem we shall compute (in Chapter 10) the range and sizes of the coefficients in the linear representations of certain predicates. It will be seen that certain ratios can increase faster than exponentially with the number of distinguishable points in R. It follows that for "big" input sets—say, R's with more than 20 points—no simple analogue storage device can be made with enough information capacity to store the whole range of coefficients!

To avoid misunderstanding perhaps we should repeat the qualifications we made in connection with our critique of the perceptron as a model for "learning devices." We have no doubt that analogue devices of this sort have a role to play in pattern recognition. *But we do not see that any good can come of experiments which pay no attention to limiting factors that will assert themselves as soon as the small model is scaled up to a usable size.*

0.9.4 Models for Brain Function and Gestalt Psychology

The popularity of the perceptron as a model for an intelligent, general-purpose learning machine has roots, we think, in an image of the brain itself as a rather loosely organized, randomly interconnected network of relatively simple devices. This impression in turn derives in part from our first impressions of the bewildering structures seen in the microscopic anatomy of the brain (and probably also derives from our still-chaotic ideas about psychological mechanisms).

In any case the image is that of a network of relatively simple elements, randomly connected to one another, with provision for making adjustments of the ease with which signals can go across the connections. When the machine does something bad, we will "teach" it not to do it again by weakening the connections that were involved; perhaps we will do the opposite to reward it when it does something we like.

The "perceptron" type of machine is one particularly simple version of this broader concept; several others have also been studied in experiments.

The mystique surrounding such machines is based in part on the idea that when such a machine learns the information stored is not localized in any particular spot but is, instead, "distributed throughout" the structure of the machine's network. It was a great disappointment, in the first half of the twentieth century, that experiments did not support nineteenth century concepts of the localization of memories (or most other "faculties") in highly local brain areas. Whatever the precise interpretation of those not particularly conclusive experiments should be, there is no question but that they did lead to a search for nonlocal machine-function concepts. This search was not notably successful. Several schemes were proposed, based upon large-scale fields, or upon "interference patterns" in global oscillatory waves, but these never led to plausible theories. (Toward the end of that era a more intricate and substantially less global concept of "cell-assembly"—proposed by D. O. Hebb [1949]—lent itself to more productive theorizing; though it has not yet led to any conclusive model, its popularity is today very much on the increase.) However, it is not our goal here to evaluate these theories, but only to sketch a picture of the intellectual stage that was set for the perceptron concept. In this setting, Rosenblatt's [1958] schemes quickly took root, and soon there were perhaps as many as a hundred groups, large and small, experimenting with the model either as a "learning machine" or in the guise of "adaptive" or "self-organizing" networks or "automatic control" systems.

The results of these hundreds of projects and experiments were generally disappointing, and the explanations inconclusive. The machines usually work quite well on very simple problems but deteriorate very rapidly as the tasks assigned to them get harder. The situation isn't usually improved much by increasing the size and running time of the system. It was our suspicion that even in those instances where some success was apparent, it was usually due more to some relatively small part of the network, and not really to a global, distributed activity. Both of the present authors (first independently and later together) became involved with a somewhat therapeutic compulsion: to dispel what we feared to be the first shadows of a "holistic" or "Gestalt" misconception that would threaten to haunt the fields of engineering and artificial intelligence as it had earlier haunted biology and psychology. For this, and for a variety of more practical and theoretical goals, we set out to find something about the range and limitations of perceptrons.

It was only later, as the theory developed, that we realized that understanding this kind of machine was important whether or not the system has practical applications in particular situations! For the same kinds of problems were becoming serious obstacles to the progress of computer science itself. As we have already remarked, we do not know enough about what makes some algorithmic procedures "essentially" serial, and to what extent—or rather, at what cost—can compu-

tations be speeded up by using multiple, overlapping computations on larger more active memories.

0.10 General Plan of the Book

The theory divides naturally into three parts. In Part I we explore some very general properties of liner predicate families. The theorems in Part I apply usually to all perceptrons, independently of the kinds of patterns considered; therefore the theory has the quality of algebra rather than geometry. In Part II we look more narrowly at interesting geometric patterns, and get sharper but, of course, less general, theorems about the geometric abilities of our machines. In Part III we examine a variety of questions centered around the potentialities of perceptrons as *practical* devices for pattern recognition and learning. The final chapter traces some of the history of these ideas and proposes some plausible directions for further exploration.

Figure 5.1

14, 15
Introduction

(1972)
Teuvo Kohonen
Correlation matrix memories
IEEE Transactions on Computers C-21:353–359

(1972)
James A. Anderson
A simple neural network generating an interactive memory
Mathematical Biosciences 14:197–220

These two papers propose the same model for associative memory and were published simultaneously. The authors were working independently and were not aware of each other until considerably after these papers appeared. The authors had quite different backgrounds: Anderson was a neurophysiologist and Kohonen was an electrical engineer. A similar situation occurred more recently in the development of the "back-propagation" algorithm (Rumelhart, Hinton, and Williams, paper 41), which was developed simultaneously and independently in three places. Independent simultaneous discovery is common in science. Here are two examples among the modest number of papers in this collection.

Described here is what has become known as the "linear associator" model for associative memory. This model is a natural extension and generalization of the kind of anatomy seen in the perceptron. It is interesting to see how differently the two authors develop the same basic idea: Kohonen is concerned with the mathematical structure, while Anderson is concerned with physiological plausibility.

There are several important assumptions in this model. First, it is a linear model of the neuron. (Kohonen does not make the neural analogies as explicitly as does Anderson, but the mathematics is identical.) This is where the linear associator deviates most markedly from the perceptron and most previous models. Many neurons, particularly sensory neurons, are not binary, with only two output possibilities, but are 'analog'; that is, they respond to changes in inputs with changes in firing rate. In some cases, particularly the *Limulus* eye (see paper 23), the cell is approximated very well by a linear system. This meant that "Threshold Logic Units," of the kind used by Rosenblatt in the perceptron, and, indeed, the kind of analysis of neurons in terms of logical predicates used by Minsky and Papert in *Perceptrons*, was inappropriate for the nervous system. There was enough evidence for small signal linearity to investigate models using it in some detail.

The basic neural element then became a slightly generalized *Limulus* neuron: it took its presynaptic firing rates, multiplied them by the synaptic weights, and added them up, with the output of the neuron proportional to the sum. The model neuron became an analog integrator with a continuous valued output. All the terms—input activities, synaptic weights, and output activities—were continuous valued. The model assumed that addition of inputs and the transduction were sufficiently linear so that simple linear algebra would approximate the output of the system; thus, given an input state vector, the output state vector was given by the connection matrix times the input state vector. The input-output relations of the system were specified as a simple vector-matrix multiplication.

Whatever one says about the correctness of models of this class, they are certainly easy to analyze!

These models were memory models, which meant that a learning rule had to be specified. The learning rule proposed was a generalization of the Hebb synapse: the synaptic change was proportional to the product of pre- and postsynaptic activities (Anderson) or to the correlation between elements in the input and output vectors (Kohonen); these, of course, are the same thing, plus or minus a neuron model. The connection matrix then became the outer product between the input and output vectors. It was easy to show that such a system acted as a good associator. Suppose that an input and output vector were stored and the connection (outer product) matrix was formed. Then later presentation of the input vector and the resultant multiplication of the input by the connection matrix would generate the output vector. If more than one association were stored in the connection matrix, the association would not be perfect, and a number of calculations of retrieval accuracy are presented in both papers.

Note the emphasis on association between large activity patterns, that is, input and output vectors. This emphasis was also found in Willshaw, Buneman, and Longuet-Higgins (paper 12) and is different from grandmother cell systems such as the simple perceptron, where only a single highly meaningful classification cell is active. Having many cells active at both input and output at the same time allows multiple correlations between different elements at the synaptic level. This strengthens the network's abilities to generalize, average out noise, and form prototypes. It also removed an asymmetry between input and output representations. For example, if both input and output are vectors, it is easy to think of feeding back the output back into the input, for another pass through the system. Kohonen calls this an "autoassociative" system, and it serves as the basis for several feedback network models, which also add some strong signal nonlinearities (see, for example, Hopfield, paper 27, or Anderson et al., paper 22). If the input and output vectors are different, Kohonen calls the system "heteroassociative."

The only case where association is perfect is when the input vectors is orthogonal. This puts an upper limit of the dimensionality on the number of vectors that can be stored. Later experience has indicated that the system works very well for random vectors if the maximum number of vectors stored is 10–20% of the dimensionality, that is, the number of model neurons. This model, in common with all neural networks, is wasteful of memory, both in the form of synaptic weights and in model neurons. However, its manifold virtues may make the trade-off acceptable in practical systems and in the brain.

The most complete analyses of this class of models are found in two books by Kohonen (1977, 1984), where a great many extensions and examples are presented.

One of the predictions of the linear associator is the existence of "cross-talk" if nonorthogonal vectors are learned. This strong interaction between separate memories gives rise to a whole class of behavioral predictions based on interactions, both constructive and destructive, between stored memories. Knapp and Anderson (paper 36) give a detailed analysis of such a system. Error correcting methods such as the Widrow-Hoff gradient descent algorithm (paper 10) are effective with the linear associator, and much of the discussion in Kohonen's books is concerned with error correcting systems.

The major criticism of this class of models is their linearity. Certainly, neurons are not linear for large signals, and most later work building on the linear associator as a starting point has included large signal nonlinearities. But the useful associative properties, generated by the Hebb synapse, carry over to more complicated systems, and the possibility for a degree of linear interaction between inputs and memory sometimes is quite useful.

References

T. Kohonen (1977), *Associative Memory—A System Theoretic Approach*, Berlin: Springer-Verlag.

T. Kohonen (1984), *Self-Organization and Associative Memory*, Berlin: Springer-Verlag.

(1972)
Teuvo Kohonen

Correlation matrix memories
IEEE Transactions on Computers C-21:353–359

A new model for associative memory, based on a correlation matrix, is suggested. In this model information is accumulated on memory elements as products of component data. Denoting a *key vector* by $q^{(p)}$, and the *data* associated with it by another vector $x^{(p)}$, the pairs $(q^{(p)}, x^{(p)})$ are memorized in the form of a matrix

$$c \sum_p x^{(p)} q^{(p)T} = M_{xq}$$

where c is a constant. A randomly selected subset of the elements of M_{xq} can also be used for memorizing. The recalling of a particular datum $x^{(r)}$ is made by a transformation $x^{(r)} = M_{xq} q^{(r)}$. This model is failure tolerant and facilitates associative search of information; these are properties that are usually assigned to holographic memories. Two classes of memories are discussed: a *complete correlation matrix memory* (CCMM), and randomly organized *incomplete correlation matrix memories* (ICMM). The data recalled from the latter are stochastic variables but the fidelity of recall is shown to have a deterministic limit if the number of memory elements grows without limits. A special case of correlation matrix memories is the *auto-associative* memory in which any part of the memorized information can be used as a key. The memories are selective with respect to accumulated data. The ICMM exhibits adaptive improvement under certain circumstances. It is also suggested that correlation matrix memories could be applied for the classification of data.

Index Terms—Associative memory, associative net, associative recall, correlation matrix memory, nonholographic associative memory, pattern recognition.

1. Introduction

For the associative search of memorized information, optical holography has been suggested [1]–[4]. Some specific mathematical models for simulated holographic memories have recently been presented [5]–[8]. It has also been assumed that the recording of information in biological memories might be based on holography [9]–[11]. Steinbuch [12], [13], and Willshaw, Buneman, and Longuet-Higgins [14], [15] suggested a nonholographic associative memory for the same purpose, based on a switching matrix.

In this paper we replace the "Lernmatrix" of Stein-buch by a correlation matrix of component signals. It is assumed that products of signals can be formed and memorized by network elements. The possibility of the formation of products of neural signals has been analyzed by Rapoport [16], Jenik [17], [18], and Küpfmüller and Jenik [19]. In this paper we discuss some analytical properties of matrix transformations used for associative recall but not the possible role of such networks in neural systems.

The correlation matrix might become immensely large with a large number of input signals. We can show that it will then suffice to take a set of random samples of its elements for the representation of the information stored in the matrix, and the rest of the elements are put equal to zero. In addition to such a randomly sampled matrix, we discuss a randomly generated associative network in which a set of memory elements is interconnected at random to two input elements. If the number of memory elements is sufficiently large, this model still reconstructs information stored in it. Such a matrix is both failure tolerant, and completely randomly organized.

In this paper, after setting up the structure of the model, we will make a formal mathematical approach to the problem in which the statistical nature of this model is analyzed.

The Model

The correlation matrix model has the structure depicted in Fig. 1. Here we have an input field which consists of two parts: the set of input elements denoted by an index set I comprises a *key field* used for the encoding of data, and the set denoted by an index set J is a *data field*. All input elements, called *receptors* in what follows, receive a set of simultaneous input signals. For simplicity, we are working in discrete-time representation in which the signals are assumed to be present at sampling instants. All signals of the key field taken together form a *key vector* denoted by $q^{(p)}$; here the superscript p is a discrete-time index, or the label of a particular *pattern*. All signals of the data field taken

Copyright © 1972 The Institute of Electrical and Electronics Engineers, Inc.

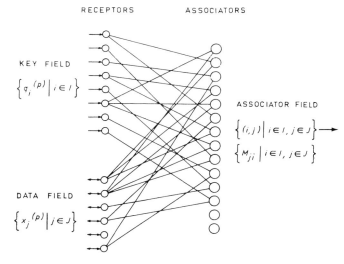

RECEPTORS   ASSOCIATORS

KEY FIELD

$\left\{ q_i^{(p)} \mid i \in I \right\}$

ASSOCIATOR FIELD

$\left\{ (i,j) \mid i \in I, j \in J \right\} \longrightarrow$

$\left\{ M_{ji} \mid i \in I, j \in J \right\}$

DATA FIELD

$\left\{ x_j^{(p)} \mid j \in J \right\}$

Figure 1 Associative network.

together form a *datum vector* denoted by $x^{(p)}$ where p labels the pattern. In *component form*, elements of the set $\{ q_i^{(p)} | i \in I \}$ constitute the components of the key vector whereas $\{ x_j^{(p)} | j \in J \}$ is the set of data signals.

Yet another field of *memory elements* or *associators* consists of elements labeled by a pair (i,j) corresponding to the ith element of the key field and the jth element of the data field to which the associator is connected. If there are connections for all possible pairs (i,j) and only one of a type, we speak of a *complete correlation matrix memory* (CCMM). If connections exist only for a randomly selected subset of all possible pairs (i,j), or if they are created at random without prior examination for the existence of a pair of connections, we speak of *incomplete correlation matrix memories* (ICMM's). The *memorization* of data is made by increasing the value of every associator element M_{ji} by an amount directly proportional to $q_i^{(p)} x_j^{(p)}$: for a set

$$P = \{ 1, 2, \ldots, p, \ldots \}$$

of patterns,

$$M_{ji} = c \sum_{p \in P} q_i^{(p)} x_j^{(p)} \qquad (1)$$

where c is a normalizing constant. The *recall* of a particular $x_j^{(r)}$ associated with a key vector $q^{(r)}$ is made by transformation

$$\hat{x}_j^{(r)} = \sum_i M_{ji} q_i^{(r)}. \qquad (2)$$

The memorization-recall-transformation defined by (1) and (2) has a bearing on the well-known Gauss-Markov estimator.

First we show that the recalled data indeed have

a certain structural conformity to the memorized information.

II. Complete Correlation Matrix Memory (CCMM)

The complete correlation matrix is an array that has one and only one memory element for every pair (i,j) of indices, for $i \in I, j \in J$. The contents of the associator field are described by the matrix

$$M_{xq} = c \sum_p x^{(p)} q^{(p)T} \qquad (3)$$

where c is a constant and T denotes the transpose. A particular datum $x^{(r)}$ will be recalled by a transformation

$$\hat{x}^{(r)} = M_{xq} q^{(r)}. \qquad (4)$$

Let us substitute M_{xq} from (3) to (4):

$$\hat{x}^{(r)} = c \sum_p x^{(p)} q^{(p)T} q^{(r)}$$

$$= cx^{(r)} [\|q^{(r)}\|^2] + c \sum_{p \neq r} x^{(p)} q^{(p)T} q^{(r)} \qquad (5)$$

where $\| \cdot \|$ denotes the Euclidean norm of a vector. If the inner products of any two key vectors $q^{(p)}$ and $q^{(r)}$ are zero, i.e., if all key vectors are *orthogonal*, we see that $\hat{x}^{(r)}$ is directly proportional to $x^{(r)}$. If the Euclidean norms of different key vectors are equal, c can further be selected as

$$c = \|q^{(r)}\|^{-2} \qquad \text{(constant with } r\text{)} \qquad (6)$$

in which case the recall is perfect for every $x^{(r)}$,

$$\hat{x}^{(r)} = x^{(r)}.$$

The nonorthogonality of key vectors gives rise to crosstalk that may have the same polarity as the recorded data or the opposite. The relative crosstalk level for a particular datum is defined as

$$L^{(p,r)} = \frac{q^{(p)T} q^{(r)}}{\|q^{(r)}\|^2}. \qquad (7)$$

Equation (7) gives a measure to *selectivity*.

III. Incomplete Correlation Matrix Memories (ICMM)

A. Stochastically Sampled Correlation Matrix

The number of matrix elements, mn in total, may grow impractically large if the dimensions of q and x are increased. Let us first discuss in this section a hypothetical case in which a randomly selected subset of the elements of M_{xq} is used to represent the complete correlation matrix. Let us define *sampling coefficients* s_{ij} that take the value 1 at all sampled elements of the correlation matrix and 0 elsewhere. The *sampled correlation matrix* is then defined as

$$(M_{xq})_{ji} = c \sum_p s_{ij} q_i^{(p)} x_j^{(p)} \tag{8}$$

and the recalled pattern reads, in analogy with (5), for all j,

$$\begin{aligned}
\hat{x}_j^{(r)} &= c \sum_p \sum_{i=1}^m s_{ij} q_i^{(p)} q_i^{(r)} x_j^{(p)} \\
&= c \left[\sum_{i=1}^m s_{ij} (q_i^{(r)})^2 \right] x_j^{(r)} + c \sum_{p \neq r} \sum_{i=1}^m s_{ij} q_i^{(p)} q_i^{(r)} x_j^{(p)} \\
&= K_j^{(r)} x_j^{(r)} + \sum_{p \neq r} K_j^{(p,r)} x_j^{(p)}.
\end{aligned} \tag{9}$$

The gain factors $K_j^{(r)}$ and $K_j^{(p,r)}$ are *stochastic variables*. Notice that if we were dealing with estimation problems or the classification of statistically distributed input signals, we should discuss the input signals $q_i^{(p)}$ and $x_j^{(p)}$ as stochastic variables. This, however, is not the case with the present study. In our model every signal has a unique value at every sampling instant, i.e., a value that must be memorized as such. To put it in another way, because the signals are not regarded as stochastic variables, there is no need to discuss their statistical distributions or correlations between these signals. In the following we thus assume that the sampling of matrix elements is independent of all $q_i^{(p)}$ in which case the $q_i^{(p)}$ can be regarded as *constant parameters* with values attained in a particular realization, and only the $s_{ij} \in \{0, 1\}$ are stochastic variables; the probability for s_{ij} being 1 is denoted by w ($0 \leqslant w \leqslant 1$) where

$$w = \frac{s}{mn} \tag{10}$$

and

s Number of sampled matrix elements.
m Number of components in q.
n Number of components in x.

Because it is our main objective to analyze what sort of noise (or statistical error) is introduced by the use of a randomly sampled incomplete matrix instead of a complete one, we shall derive expressions for the expectation values and variances of signals recalled from the memory, whereby these statistical operations refer only to the sampling process.

It is known from elementary statistics that if $Y_1, Y_2, \ldots, Y_m$ are independent stochastic variables with means $M_1, M_2, \ldots, M_m$ and variances $\sigma_1^2, \sigma_2^2, \ldots, \sigma_m^2$, respectively, we have for a linear combination X of the Y's,

$$X = \sum_{i=1}^m a_i Y_i$$

$$E(X) = \sum_{i=1}^m a_i M_i$$

$$\operatorname{var}(X) = \sum_{i=1}^m a_i^2 \sigma_i^2.$$

Because all s_{ij} can be assumed independent since n is a large number, the probability distribution $\Pr(s_{ij} = 1)$, as is well known, is *binomial*. The *mean* of s_{ij} is then

$$E(s_{ij}) = w \tag{11}$$

and the *variance* of s_{ij} is

$$\operatorname{var}(s_{ij}) = w(1 - w). \tag{12}$$

$K_j^{(r)}$ and $K_j^{(p,r)}$ we have now, regarding $q_i^{(p)}$ as parameters, the means and variances

$$E(K_j^{(r)}) = cw \sum_{i=1}^m (q_i^{(r)})^2 \tag{13}$$

$$\operatorname{var}(K_j^{(r)}) = c^2 w(1 - w) \sum_{i=1}^m (q_i^{(r)})^4. \tag{14}$$

The fidelity of recall is conveniently described in terms of a *relative standard deviation* which for the noise due only to the searched pattern is

$$\frac{\sqrt{\operatorname{var}(K_j^{(r)})}}{E(K_j^{(r)})} = \sqrt{\frac{1 - w}{w}} \frac{\sqrt{\sum_{i=1}^m (q_i^{(r)})^4}}{\sum_{i=1}^m (q_i^{(r)})^2}. \tag{15}$$

The crosstalk from another pattern (p) is expressed by the average crosstalk level,

$$E(K_j^{(p,r)}) = cw \sum_{i=1}^m q_i^{(p)} q_i^{(r)} \tag{16}$$

and the variance of this level is

$$\operatorname{var}(K_j^{(p,r)}) = c^2 w(1 - w) \sum_{i=1}^m (q_i^{(p)} q_i^{(r)})^2. \tag{17}$$

The relative standard deviation of the recalled data due to all patterns is obtained from (9) and the previous considerations, for all j,

$$\begin{aligned}
&\frac{\sqrt{\operatorname{var}(\hat{x}_j^{(r)})}}{E(\hat{x}_j^{(r)})} \\
&= \sqrt{\frac{1 - w}{w}} \frac{\sqrt{\sum_{i=1}^m \left(\sum_p q_i^{(p)} q_i^{(r)} x_j^{(p)} \right)^2}}{\sum_p \sum_{i=1}^m q_i^{(p)} q_i^{(r)} x_j^{(p)}}.
\end{aligned} \tag{18}$$

If other parameters are finite, the relative standard deviation approaches zero for $w \to 1$.

Example 1 Let us take a representative case with one memorized pattern only and $q_i^{(p)} \in \{+1, -1\}$ for which we obtain

$$\frac{\sqrt{\operatorname{var}(\hat{x}_j^{(r)})}}{E(\hat{x}_j^{(r)})} = \sqrt{\frac{1-w}{mw}} = \sqrt{\frac{mn-s}{ms}}. \qquad (19)$$

Since (19) gives the relative standard deviation of a recalled pattern, we obtain an order of magnitude rule. If w is much smaller than 1, as is usually the case, if the maximum allowed relative standard deviation is N, and if the number of components in the key vector is $m \gg 1$, then the minimum number of sampled matrix elements must be

$$s = mnw \geqslant \frac{n}{N^2} \qquad (20)$$

which thus depends on the number n of components in the data vector x only. For example, if $N = 0.1$, we must have $s \geqslant 100n$.

It is not difficult to generalize the above results and show that the number of sampled matrix elements for a given percentage noise is obviously directly proportional to the number of components in the data vector, for any amount of them. This, of course, is advantageous with a large number of elements in q.

Example 2 If we again consider the recall of a memorized single pattern, and $q_i^{(p)} \in \{+1, 0, -1\}$, we have a distribution of $q_i^{(p)}$ which is less uniform than in Example 1. Denoting the fraction of nonzero components in the key vectors by $\alpha (0 \leqslant \alpha \leqslant 1)$ we have

$$\frac{\sqrt{\operatorname{var}(\hat{x}_j^{(r)})}}{E(\hat{x}_j^{(r)})} = \sqrt{\frac{1-w}{\alpha mw}} = \sqrt{\frac{mn-s}{\alpha ms}}. \qquad (21)$$

In this case the necessary number of sampled matrix elements for relative standard deviation N is

$$s \geqslant \frac{n}{\alpha N^2}. \qquad (22)$$

For example, if there are 80 percent zeros in the key and $N = 0.1$, we must have $s \geqslant 500n$.

B. Randomly Generated Correlation Matrix Memory

We will now make a more realistic approach and assume that connections are generated stochastically *without prior examination for the existence of connections*. A matrix M_{xq} formed of the M_{ji} may thus include *multiple* elements if two associators have identical connections. The only difference with respect to the cases of Section III-A is that we must now use *occupation numbers* z_{ij} of matrix elements (instead of s_{ij}), $z_{ij} \in \{0, 1, 2, \ldots\}$.

$$(M_{xq})_{ji} = c \sum_p z_{ij} q_i^{(p)} x_j^{(p)} \qquad (23)$$

where the distribution of z_{ij} is *Poissonian*,

$$\Pr(z_{ij} = \zeta) = \frac{\mu^\zeta}{\zeta!} e^{-\mu} \qquad (24)$$

and the parameter of this distribution is

$$\mu = \frac{s}{mn} \qquad (25)$$

where s is the number of associators.

The distribution defined by (24) is obviously the same as the distribution of hits in s stochastic throws on squares of a board with mn squares. The recalled pattern is for all j,

$$\hat{x}_j^{(r)} = \left[c \sum_{i=1}^m z_{ij} (q_i^{(r)})^2 \right] x_j^{(r)} + c \sum_{p \neq r} \sum_{i=1}^m z_{ij} q_i^{(p)} q_i^{(r)} x_j^{(p)}$$

$$= K_j^{(r)} x_j^{(r)} + \sum_{p \neq r} K_j^{(p,r)} x_j^{(p)}. \qquad (26)$$

Now we have

$$E(z_{ij}) = \mu = \frac{s}{mn} \qquad (27)$$

$$\operatorname{var}(z_{ij}) = \mu \qquad (28)$$

and then

$$E(K_j^{(r)}) = c\mu \sum_{i=1}^m (q_i^{(r)})^2 \qquad (29)$$

$$\operatorname{var}(K_j^{(r)}) = c^2 \mu \sum_{i=1}^m (q_i^{(r)})^4 \qquad (30)$$

and

$$E(K_j^{(p,r)}) = c\mu \sum_{i=1}^m q_i^{(p)} q_i^{(r)} \qquad (31)$$

$$\operatorname{var}(K_j^{(p,r)}) = c^2 \mu \sum_{i=1}^m (q_i^{(p)} q_i^{(r)})^2. \qquad (32)$$

The relative standard deviation of gain factor $K_j^{(r)}$ due to the searched pattern only is

$$\frac{\sqrt{\operatorname{var}(K_j^{(r)})}}{E(K_j^{(r)})} = \mu^{-1/2} \frac{\sqrt{\sum_{i=1}^m (q_i^{(r)})^4}}{\sum_{i=1}^m (q_i^{(r)})^2} \qquad (33)$$

and the relative standard deviation of a recalled signal, due to all patterns, is

$$\frac{\sqrt{\operatorname{var}(\hat{x}_j^{(r)})}}{E(\hat{x}_j^{(r)})} = \mu^{-1/2} \frac{\sqrt{\sum_{i=1}^m \left(\sum_p q_i^{(p)} q_i^{(r)} x_j^{(p)} \right)^2}}{\sum_p \sum_{i=1}^m q_i^{(p)} q_i^{(r)} x_j^{(p)}}. \qquad (34)$$

If other parameters are finite, the relative standard deviation approaches zero for $s \to \infty$, i.e., $\mu \to \infty$.

Example 3 Let us repeat the problem of Example 1 for randomly generated connections. The only difference is in the expression for variance and we obtain

$$\frac{\sqrt{\text{var}(K_j^{(r)})}}{E(K_j^{(r)})} = \sqrt{\frac{1}{\mu m}}. \tag{35}$$

For any value of μ, we must now have

$$s \geqslant \frac{n}{N^2}. \tag{36}$$

Example 4 Repeating the problem of Example 2 for randomly generated connections, we have

$$\frac{\sqrt{\text{var}(K_j^{(r)})}}{E(K_j^{(r)})} = \sqrt{\frac{1}{\alpha \mu m}}. \tag{37}$$

Thus, without any restriction on μ, we must have

$$s \geqslant \frac{n}{\alpha N^2}. \tag{38}$$

IV. Recall by an Incomplete Key

On account of the fact that information in a correlation matrix memory is stored in redundant form, the recalling of a stored item can also be made by an incomplete key which has a high correlation with the key used during memorization. To show this fact by a specific example, we define deterministic *projection parameters* $P_i \in \{0, 1\}$ which for known elements of $q^{(r)}$ are 1 and otherwise 0. The memorization is still defined by (1) but during recall, the key vector has the components

$$q_i^{(r)} = P_i q_i^{(p_0)}, \qquad p_0 \in P. \tag{39}$$

Then we have for an ICMM

$$\begin{aligned}
\hat{x}_j^{(r)} &= c \sum_p \sum_{i=1}^m z_{ij} P_i q_i^{(p)} x_j^{(p)} q_i^{(p_0)} \\
&= \left[c \sum_{i=1}^m z_{ij} P_i q_i^{(p_0)} q_i^{(p_0)} \right] x_j^{(p_0)} \\
&\quad + c \sum_{p \neq p_0} \sum_{i=1}^m z_{ij} P_i q_i^{(p)} q_i^{(p_0)} x_j^{(p)}.
\end{aligned} \tag{40}$$

The expectation value of recalled data is

$$\begin{aligned}
E(\hat{x}_j^{(r)}) &= c\mu \sum_{i=1}^m P_i (q_i^{(p_0)})^2 x_j^{(p_0)} \\
&\quad + c\mu \sum_{p \neq p_0} \sum_{i=1}^m P_i q_i^{(p)} q_i^{(p_0)} x_j^{(p)}
\end{aligned} \tag{41}$$

and the variance is

$$\text{var}(\hat{x}_j^{(r)}) = c^2 \mu \sum_{i=1}^m P_i \left(\sum_p q_i^{(p)} q_i^{(p_0)} x_j^{(p)} \right)^2. \tag{42}$$

(Note $P_i^2 = P_i$.)

V. Auto-Associative Correlation Matrix Memory

There is nothing in the foregoing which would restrict us from implementing an auto-associative memory in which the pattern itself or any part of it could be used as the key. In this case during memorization $q^{(p)} = x^{(p)}$ and thus

$$M_{xq} = c \sum_p x^{(p)} x^{(p)T} \tag{43}$$

is the correlation matrix. Every memory element is now connected to two different elements of an input field. (The key and data fields are thus mixed up.) During recall, $q^{(r)}$ is replaced by a part of $x^{(p_0)}$. The projection parameters P_i are now 1 for the known elements of the key pattern and otherwise 0. Then for the CCMM

$$\hat{x}_j^{(r)} = c \sum_p \sum_{i=1}^m P_i x_i^{(p)} x_j^{(p)} x_i^{(p_0)}. \tag{44}$$

For the randomly generated ICMM

$$\hat{x}_j^{(r)} = c \sum_p \sum_{i=1}^m P_i z_{ij} x_i^{(p)} x_j^{(p)} x_i^{(p_0)}. \tag{45}$$

Now we obtain for the recalled data in ICMM the expectation value

$$\begin{aligned}
E(\hat{x}_j^{(r)}) &= c\mu \sum_{i=1}^m P_i (x_i^{(p_0)})^2 x_j^{(p_0)} \\
&\quad + c\mu \sum_{p \neq p} \sum_{i=1}^m P_i x_i^{(p)} x_j^{(p)} x_i^{(p_0)}
\end{aligned} \tag{46}$$

and the variance

$$\text{var}(x_j^{(r)}) = c^2 \mu \sum_{i=1}^m P_i \left(\sum_p x_i^{(p)} x_j^{(p)} x_i^{(p_0)} \right)^2 \tag{47}$$

where

$\mu = s/n(n-1)$.
s Number of associators.
n Number of input elements.

If all memorized patterns, or at least the key parts of them are orthogonal, we can recall a pattern by its part with true conformity, and the noise is given by (47). Unless the key parts of the patterns are orthogonal, there will be biased crosstalk from other patterns given by the last terms in (46). Notice that if there are many zeros in the pattern, the requirement of approximate orthogonality is usually fulfilled.

VI. Unsupervised Learning in the ICMM

When the key vectors contain many zero components, there will be an appreciable probability that a large number of the matrix elements M_{ji} are zero or very small. In the recall, the contribution of these elements is small and, therefore, we may guess that without causing appreciable errors, all elements smaller than a certain limit could be deleted from the set of associators. Now we can stochastically generate a corresponding amount of new connections (associators) and obviously we will have found new large matrix elements of the complete correlation matrix. Using these for memorization and recall, we can deduce, e.g., from Example 5 that the relative noise will have been reduced. By intermittently breaking old connections and generating new ones during the memorization, unsupervised learning in the stochastically generated memory takes place. Achievable results depend strongly on the statistics of key vectors.

Example 5 Let us take an illustrative simplified example, the memorization and recall of a repeating single pattern. The unnormalized correlation matrix of a single pattern is (with $c = 1$)

$$M_{xq} = xq^T. \tag{48}$$

Let us assume that $q_i, x_j \in \{0,1\}$ and denote the relative fraction of 1's in the key by α and in the data by β. We shall inspect the recall of a component x_j:

$$\hat{x}_j = \sum_{i=1}^{m} z_{ij} q_i^2 x_j. \tag{49}$$

In the beginning,

$$E(\hat{x}_j) = m\alpha\mu x_j \tag{50}$$

$$\text{var}(\hat{x}_j) = m\alpha\mu x_j. \tag{51}$$

(Note that $x_j^2 = x_j \in \{0,1\}$.)

Now we start a process of rejecting all $M_{ji} = 0$. For further simplicity, we clear the contents of the surviving M_{ji}, generate new connections, and repeat the memorization of (q, x). At the νth stage of this process, denote the number of associators with the contents $M_{ji} = 1$ by $\gamma^\nu (\gamma^0 = \alpha\beta s)$. At the $(\nu + 1)$th stage, we reject $s - \gamma^\nu$ associators and generate an equal number of new ones. Of these, $\alpha\beta(s - \gamma^\nu)$ will hit places at which the complete correlation matrix would have 1's. Therefore,

$$\gamma^{\nu+1} - \gamma^\nu = \alpha\beta(s - \gamma^\nu). \tag{52}$$

The solution of (52) reads for M processes,

$$\gamma^M = [1 - (1 - \alpha\beta)^{M+1}]s. \tag{53}$$

Here

$$\lim_{M \to \infty} \gamma^M = s. \tag{54}$$

Finally, after infinitely many renewing processes, all associators will have connections for which the complete correlation matrix would have 1's. There are now s elements distributed over $\alpha\beta mn$ possible places corresponding to components of q that are 1. Therefore, the mean and variance of a recalled pattern at j are, respectively,

$$E(\hat{x}_j) = \alpha m \cdot \frac{s}{\alpha\beta mn} x_j = \frac{s}{\beta n} x_j \tag{55}$$

$$\text{var}(\hat{x}_j) = \frac{s}{\beta n} x_j. \tag{56}$$

Thus comparing (50), (51), (55), and (56), we see that the relative standard deviation has been reduced by a factor $\sqrt{\alpha\beta}$. This example shows us that the incomplete correlation matrix memory is capable of unsupervised learning, and the result is the better, the more zeros there are in M_{xq}. On the other land, the speed of learning is slower in this case.

VII. Use of Correlation Matrix Memories for Pattern Classification

By the classification of patterns we mean that a set of patterns consists of subsets each of which is mapped on a single element. Classification with the aid of correlation matrix memories means here that we assign one element j in the x field to each occurring class and the patterns to be classified are used as the q vectors. The memory is "taught" by assigning a value $x_h = 1$ for the element h if the exposed figre belongs to the class h, but 0 otherwise (supervised learning). During recalling, a pattern q is used as the key and $\hat{x}$ is formed as before. The class h is now found by the decision

$$\hat{x}_h^{(r)} = \max_j x_j^{(r)}. \tag{57}$$

VIII. Computer Simulations

The noise level of recalled patterns, due to the random structure of the associative network, can easily be computed from (18), (34), (42), and (47) for any type of memorized patterns. It might be desirable to have a direct demonstration of the reliability and efficiency of these memories. Therefore, some computer simulations on the auto-associative model have been performed and are shown in Fig. 2. The number of input elements was 140 in this experiment and these elements formed a two-dimensional retina; the number of memory ele-

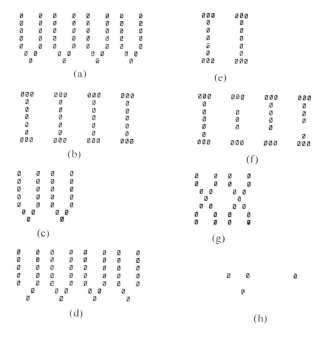

Figure 2 (a) First memorized pattern, shown in a retina of 140 elements. Signals denoted by the symbol φ have the value 1 whereas signals denoted by blanks have the value 0. (b) Second memorized pattern, the memory traces of which have been superimposed on those of case (a). (c) First key pattern used for recall when the same symbols as in case (a) have been used. (d) Result of recall when all $\hat{x}_j^{(r)} \geqslant 5.5$ have been denoted by the symbol φ, and recalled signals smaller than 5.5 by a blank. (e) Second key pattern used for recall when the same symbols as in case (a) have been used. (f) Result of recall when the same symbols as in case (d) have been used. (g) Third key pattern which is uncorrelated with the memorized items. (h) Result of recall when the same symbols as in case (d) have been used.

ments was 4000 and every element was connected to two elements of the input retina using a random number generator for the designation of connections. The source data consisted to binary patterns. The stored items were recalled by key patterns using the transformation defined by (45). To standardize the output signals, all $\hat{x}_j^{(r)} \geqslant 5.5$ were displayed as binary 1's whereas smaller output signals were displayed as binary 0's. In Fig. 2(c) and (e), the key pattern had a high correlation with a stored item so that the stored information was reconstructed as shown in Fig. 2(d) and (f), respectively. The key pattern used in Fig. 2(g) had a low correlation with stored patterns so that only minor crosstalk from memorized items is present. The missing portions of the recalled patterns in Fig. 2(d) and (f) are due to the incompleteness of the correlation matrix. This effect manifests itself in all finite networks but its contribution decreases with increasing size of the memory.

Acknowledgment

The author wishes to thank M. Kilpi for programming the computer simulations.

References

1 P. J. van Heerden, "A new optical method of storing and retrieving information," *Appl. Opt.*, vol. 2, pp. 387–392, 1963.

2 ———, "Theory of optical information storage in solids." *Appl. Opt.*, vol. 2, pp. 393–400, 1963.

3 ———, *The Foundation of Empirical Knowledge, with a Theory of Artificial Intelligence.* Wassenaar, Netherlands: Wistik, 1968.

4 R. J. Collier. "Some current views on holography," *IEEE Spectrum*, vol. 3, pp. 67–74, July 1966.

5 H. C. Longuet-Higgins, "Holographic model of temporal recall," *Nature*, vol. 217, p. 104, 1968.

6 P. T. Chopping, "Holographic model of temporal recall," *Nature*, vol. 217, pp. 781–782, 1968.

7 D. J. Willshaw and H. C. Longuet-Higgens, in *Machine Intelligence*, vol. 4, B. Meltzer and D. Michie, Eds. Edinburgh. Great Britain: Edinburgh University Press, 1969, pp. 349–357.

8 D. Gabor, "Associative holographic memories," *IBM J. Res. Develop.*, vol. 13, pp. 156–159, 1969.

9 I. T. Diamond and K. L. Chow, in *Psychology, A Study of a Science*, vol. 4, S. Koch, Ed. New York: McGraw-Hill, 1962, pp. 158–241.

10 E. B. Carne, *Artifical Intelligence Techniques.* Washington, D. C.: Spartan 1965, p. 18.

11 P. R. Westlake, "The possibilities of neural holographic processes within the brain," *Kybernetik*, vol. 7, pp. 129–153, 1970.

12 K. Steinbuch, "Die Lernmatrix, *Kybernetik*, vol. 1, pp. 36–45, 1961.

13 ———, *Automat und Mensch.* Berlin: Springer-Verlag, 1963, pp. 213–241.

14 D. J. Willshaw, O. P. Buneman, and H. C. Longuet-Higgins, "Non-holographic associative memory," *Nature*, vol. 222, pp. 960–962, 1969.

15 D. J. Willshaw and H. C. Longuet-Higgins, in *Machine Intelligence*, vol. 5, B. Meltzer and D. Michie, Eds. Edinburgh, Great Britain: Edinburgh University Press, 1970, pp. 351–359.

16 A. Rapoport, "Addition and multiplication theorems for the input of two neurons converging on a third," *Bull. Math. Biophys.*, vol. 13, pp. 179–188, 1951.

17 F. Jenik, in *Ergebnisse der Biologie*, H. Autrum, Ed. Berlin: Springer-Verlag, 1961, pp. 206–245.

18 ———, in *Neural Theory and Modelling*, R. F. Reiss, Ed. Stanford, Calif.: Stanford University Press, 1964, pp. 190–212.

19 K. Küpfmüller and F. Jenik, "Ueber die Nachrichtenverarbeitung in der Nervenzelle," *Kybernetik*, vol. 1, pp. 1–6, 1961.

20 T. Kohonen, "A class of randomly organized associative memories," *Acta Polytech. Scandinavica*, no. El 25, 1971.

(1972)
James A. Anderson

A simple neural network generating an interactive memory
Mathematical Biosciences 14:197–220

A model of a neural system where a group of neurons projects to another group of neurons is discussed. We assume that a trace is the simultaneous pattern of individual activities shown by a group of neurons. We assume synaptic interactions add linearly and that synaptic weights (quantitative measure of degree of coupling between two cells) can be coded in a simple but optimal way where changes in synaptic weight are proportional to the product of pre- and post-synaptic activity at a given time. Then it is shown that this simple system is capable of "memory" in the sense that it can (1) recognize a previously presented trace and (2) if two traces have been associated in the past (that is, if trace $\bar{f}$ was impressed on the first group of neurons and trace $\bar{g}$ was impressed on the second group of neurons and synaptic weights coupling the two groups changed according to the above rule) presentation of $\bar{f}$ to the first group of neurons gives rise to $\bar{g}$ plus a calculable amount of noise at the second set of neurons. This kind of memory is called an "interactive memory" since distinct stored traces interact in storage. It is shown that this model can effectively perform many functions. Quantitative expressions are derived for the average signal to noise ratio for recognition and one type of association. The selectivity of the system is discussed. References to physiological data are made where appropriate. A sketch of a model of mammalian cerebral cortex which generates an interactive memory is presented and briefly discussed. We identify a trace with the activity of groups of cortical pyramidal cells. Then it is argued that certain plausible assumptions about the properties of the synapses coupling groups of pyramidal cells lead to the generation of an interactive memory.

All that we are is the result of what we have thought: it is founded on our thoughts, it is made up of our thoughts.—Dhammapada, I.I.

Introduction

Memory is a mental function which seems in its generality to be a central problem in neurophysiology and neuropsychology. Experiments on memory are difficult to perform and good data is scanty, although considerable information about memory function in higher mammals has accumulated in recent years.

In two previous papers I discussed a very simple model for the organization of long term memory storage [3]. Basically, the model assumes, first, that memory "traces" (the items which are to be stored—the basic elementary unit of memory) are composed of a complex pattern of individual activities shown by a large spatial array of elements, and, second, that memory storage is formed by constructing a storage array which is the sum of many of the basic "traces."

More precisely, the trace was represented by an N element vectors, $\bar{f}_k$, where N is very large. Elements of $\bar{f}_k$ can take any values. (They are not binary elements.) Assume we have K traces ($\bar{f}_1, \bar{f}_2, \ldots, \bar{f}_k$) to be stored. Then, we represented the memory array in this model by a storage vector, $\bar{s}$, defined as

$$\bar{s} = \sum_{k=1}^{k=K} \bar{f}_k.$$

$\bar{s}$ represents our sole information about the system.

Storage of traces is simple in such a system and could be performed by a simple physiological mechanism; for example, strengthening of synaptic contacts by an amount proportional to activity in the presynaptic cell would be sufficient in some neuron models. It is not clear, however, that it is possible to extract much information from such a system. The retrieval problem was discussed in the previous papers and it was shown that it was indeed possible for such a simple system to perform many of the functions which would be expected of a biological memory. Some of the properties shown by the system, particularly the kinds of mistakes and distortions it makes, are reminiscent of those made by a human memory [3].

The following discussion is an extension of this simple idea to another model. It is an attempt to model a very common kind of anatomical configuration in the central nervous system where one large group of neurons projects to another large group of neurons, or projects to itself via recurrent collaterals having a significantly

Copyright © 1972 American Elsevier Publishing Company.

long conduction time. Examples of this type of highly parallel projection are legion: there are many projections of the thalamic nuclei to cortex, for example, and extensive intracortical projection systems.

If a group of neurons projects to another we shall show that strengthening or weakening the synaptic connection between the two groups according to a simple multiplicative function of activity in pre- and postsynaptic cells automatically generates an interactive memory akin to those discussed in the previous paper. The properties of this model will be discussed and a model of mammalian cortex embodying some of these ideas will be sketched in the last section of the paper.

The Model

Let us consider two groups of neurons, α and β, where α sends projections to β (Fig. 1). We shall assume that (1) Exactly M neurons of α project to each neuron of β, (2) Exactly M neurons of β are projected to by each neuron of α, (3) Synaptic weights add linearly and the activity of a neuron in β is proportional to the sum of the synaptic weights times activity of neurons in α, and (4) α and β have the same number of neurons, N. Assumptions (1), (2) and (4) are approximations of the real situation for mathematical convenience.

Assumption (3) requires justification. There is an extensive literature on neural models which assumes that the nervous system is basically digital, that is, the presence or absence of a spike at a given moment is the matter of most importance to the nervous system. This approach to neural modeling is interesting mathematically, and has led to some elegant and important results

in automata theory (Ref. 23, chap. 3). However, a great deal of experimental evidence now suggests that an analog model of the mammalian central nervous system is correct in many cases.

Evidence indicates that for most systems in mammals, particularly the "higher" systems, what is significant to the system is the behavior of the cell over a short period of time—the average firing frequency, for example—and not the presence or absence of a single spike. Perkel and Bullock [26] have made an extensive list of biological systems and the apparent kinds of neural codes used, many of which depend on temporal spike patterns over a period long in relation to the duration of a single spike.

One carefully studied sensory system in primates shows this kind of coding. Mountcastle has studied the tactile sensitivity of the monkey hand [24]. A stimulator probe indented the skin of the thenar eminence of the palm of a monkey. The number of impulses in 600 msec. was recorded from myelinated axons of the palmar branch of the median nerve and was found to be linearly related to the indentation of skin produced by the stimulator probe. The same linear relation to skin indentation was found in units in the ventrobasal thalamus to stimuli applied to the glabrous skin of the monkey hand if the spontaneous activity level of the cell was subtracted. The same linearity of response was found in a cortical neuron in the post-central gyrus of an unanesthetized monkey. Human observers gave a similar linear relation between subjective intensity and depth of mechanical stimulation by a probe tip applied to the pad of the middle finger.

A strictly linear relation between a stimulus (depth of indentation) and neuronal response is rare, however, what is often found and which encouraged Mountcastle to formulate it as a general rule for sensory systems, is that there is a linear relation between the output of the first order afferent fiber and sensory response of the nervous system. Most receptors are markedly nonlinear transducers; for example, response of single tactile receptors (average number of impulses to a stimulus) in the hairy skin areas of the hand gives a power function response to indentation with an exponent of about 0.5. Power functions as transducer outputs are common, and generally the exponent is preserved up to primary sensory cortex.

Maffei et al. have shown [20, 21] that the firing rate of lateral geniculate body (LGB) cells follows the sinusoidal intensity modulation of a light stimulus in quite linear fashion. Maffei has also shown that the LGB may use spatial averaging to improve signal to noise ratio. Their data indicates that spatial averaging pre-

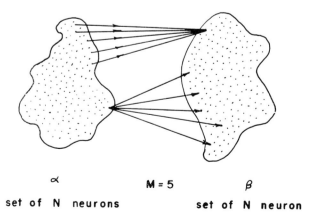

α M = 5 β

set of N neurons   set of N neuron

Figure 1 A group of neurons, α, projects to another group of neurons, β. M neurons of α project to each neuron of β, and M neurons of β are projected to by each neuron of α. Both groups have N neurons.

serves the linearlity of the response of the cell, in consonance with our assumption (3).

It would be unrealistic to claim that neurons do not introduce substantial nonlinearities. (See Ref. 7 for examples of some of the significant nonlinearities encountered in LGB cells under conditions similar to those of Maffei). However, the overall picture is of much less nonlinearity than might at first be expected from a system incorporating such nonlinear threshold devices as neurons.

At this point it might be wise to make the distinction often made in circuit analysis between large and small signal characteristics. We would expect a linear addition assumption to be quite accurate for cortical neurons if the stimulus increment was small, whereas significant nonlinearities would occur if the increment was large. The particular system under consideration is of paramount importance. Burns (Ref. 6, chap. 6) has shown that interactions between eyes can be quite nonlinear. Thus Burns finds that in primary visual cortex, where cells respond to stimuli applied to each eye, presentation of the same stimulus to both eyes in spatial register gives enhancement of firing rate far beyond what would be expected from the sum of the response to stimulation of either eye alone.

Before we go further, let us make some comments about the linear assumption. First, the model assumes, in general, small signal linearity in storage. Second, retrieval, in the model presented here, assumes as well large-signal linearity, but the model is not very sensitive to the details of the large-signal transfer function as long as monotonicity is preserved, that is, as long as increasing excitatory synaptic activity increases postsynaptic cell firing or increasing inhibitory activity decreases cell firing.

We will assume that what is of interest in our model is the simultaneous activity of the entire group of neurons, thus we define a trace to be this pattern of individual activities. We further assume that traces are "large" in that a single trace contains a good deal of information. We will make calculations with "traces" as our elementary units.

Since traces are assumed to be the simultaneous activities of large groups of neurons, we can formally represent a trace as a vector of N elements, where N is the number of neurons present. Thus, if $\bar{f}$ is a trace, we define the "power" of $\bar{f}$, P, as the vector dot product,

$$P = \bar{f} \cdot \bar{f}.$$

We assume that traces add together in storage, they are not separated. Thus, our initial approach to calculations will be to place some statistical constraints on the set of allowable traces. We assume they have equal power, P. Since power, in some intuitive sense, stands for the "importance" of a trace, assumption of equal trace power seems unnatural. The calculations to follow could equally will have been carried out using P as the "average" power of trace in the set of allowable traces. However, the results to be discussed are not basically changed and the additional complications in exposition seemed to make the equal power assumption a simplifying convenience.

We assume that different traces in a sum are uncorrelated. We assume that, on the average over sets of sums of allowable traces, that the statistics of every element will be the same.

By the Central Limit Theorem, we predict that the value of the sum of many uncorrelated traces approximates a normally distributed random variable.

Since we know nothing of the details of the traces involved in any particular sum, values we calculate are "averages," calculated over many sets of sums of allowable traces.

As discussed in Ref. 3, we assume that the mean value of elements in a trace is zero. This assumption can be shown to give nearly optimal properties (optimal in the sense that the signal power to noise power ratio is maximized) to the retrieval system to be discussed and greatly simplifies calculations. The zero mean assumption implies both positive and negative values for elements in a trace. However, neurons can only have positive average firing frequencies. We can meet the zero mean requirement by assuming neurons have a spontaneous activity level and then defining the activity comprising a trace as deflections, plus or minus, from this resting level. Other realizations of this requirement are possible, but this definition seems natural. Much cortical data shows transduction in both a positive and negative direction which often seems to be referred to the spontaneous activity level (see Ref. 24, p. 398), the well known data of Fox and O'Brien [13], as well as the data in other experimental papers. This implies that the neuron would be expected to act something like a limiter with less dynamic range in the negative than positive direction. Freeman [14, 15] in a well-developed and experimentally supported electrophysiological model of cat prepyriform cortex has developed a basically linear model with clipping to explain the results of his extensive experiments.

Calculations will be made as follows. We will be interested in the behavior of one trace in the sum. We will approximate the sum of the other traces by the values taken by a normally distributed random variable and with this additive "noise" can then make simple calculations.

Let us now return to the model shown in Fig. 1. Let

us consider an input trace, $\bar{f}$, to α. We have assumed that the activity of a neuron in β will be given by the sum of the activities coupled to the neuron by the M neurons from α projecting to it [Assumption (3)]. If we denote by a_{ij} the value of the synapse connecting the ith element in α with the jth element in β, and if $g(j)$ is the value taken by the jth element in β, then we have as the fundamental relation for our system,

$$g(j) = \sum_{i=1}^{i=N} a_{ij} f(i).$$

We see that if $\bar{f}$ is represented by a column vector and if a_{ij} are elements, representing synaptic weights in a "connectivity matrix", A, then a vector $\bar{g}$, representing activity in β is generated according to

$$\bar{g} = A\bar{f}.$$

We are interested in the following problem. We have trace $\bar{f}_1$, representing activity in α, which we wish to have associated with another, trace $\bar{g}_1$, representing activity in β, thus we wish to construct a connectivity matrix A so that

$$\bar{g}_1 = A\bar{f}_1,$$

as shown in Fig. 2. In general, we wish to couple many pairs of traces by way of our projection system. Let us assume we have a set of K pairs of traces

$$[(\bar{f}_1, \bar{g}_1); (\bar{f}_2, \bar{g}_2); \dots ; (\bar{f}_K, \bar{g}_K)],$$

that we wish to couple, that is, if we present pattern of activity $\bar{f}_k$ to the set of neurons α, we wish the set of neurons, β, to show a pattern of activity "close to" $\bar{g}_k$. Let us first try to construct a matrix A which is somehow optimal and, second, calculate just how accurately the projection system reconstructs $\bar{g}_k$ at β when $\bar{f}_k$ is presented at α.

The first problem is considered here in a slightly different manner than in Ref. 3. The second problem is considered later.

We will assume that we code the incoming trace,

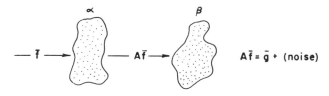

Figure 2 The general scheme for associating two distinct traces, $\bar{f}$, and $\bar{g}$. Trace $\bar{f}$ is impressed on group of neurons α. The connectivity matrix, A, coupling α and β gives rise to trace $\bar{g}$ (which was associated with $\bar{f}$ in the past) plus noise due to the presence of other associations in A.

$\bar{f}$, according to a well defined set of rules, generating a vector $h(\bar{f})$. We assume $h(\bar{f})$ is a reasonable function—nonzero, continuous. Since $[\bar{f}_j]$ are uncorrelated, $[h(\bar{f}_j)]$ will be uncorrelated. Since we have assumed that synaptic increments or decrements due to different traces add together, we can write for the activity of a single element in β when $\bar{f}$ is impressed on α,

$$g(i) = \bar{f} \cdot \Sigma_j h(\bar{f}_j).$$

Let us assume that a trace corresponding to $\bar{f}$ is present, that is, $\bar{f}$ has appeared to the system and the coding of $\bar{f}$, $g(\bar{f})$ is represented in the synapses. Then

$$g(i) = \bar{f} \cdot h(\bar{f}) + \bar{f} \cdot \Sigma_j h(\bar{f}_j).$$

Remembering that $h(\bar{f}_j)$ are vectors, we see the second term on the average over allowable sets of traces is a function of the average length of $h(\bar{f}_j)$ since the $\bar{f}_j$ are uncorrelated. We will assume an optimal coding, $h(\bar{f})$ is one which on the average maximizes the first term while keeping the second term as small as possible. The geometry of the dot product indicates that this maximum is obtained when $\bar{f}$ and $h(\bar{f})$ point in the same direction, that is

$$h(\bar{f}) = c\bar{f},$$

where c is a constant.

Let us point out the implications of this simple result. Many models of long term memory assume that permanent changes in synaptic weight are made when the trace is laid down [12]. *We will show that if, when an association is learned by a set of neurons projecting to another set of neurons, synaptic weight is changed by an amount proportional to the product of activity in the presynaptic neuron and of activity in the postsynaptic neuron, a memory formed of interacting traces is generated which (1) is optimal in the sense just considered and (2) can be shown to effectively recognize and associate traces.* It is thus possible to generate a psychologically global memory system which is formed by physiologically local changes produced according to simple local rules, requiring the synapse to be affected only by the product of pre- and postsynaptic activities. Note that this does not correspond to a simple strengthening-by-use learning change, although synaptic change dependent only on the activity of the presynaptic cells is capable of generating a simple recognition memory model of an interactive type [3].

Recognition

We will first consider the problem of recognition in this model where recognition is defined as the ability

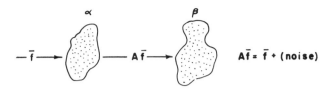

Figure 3 The general scheme for recognition by self-association. Presentation of trace $\bar{f}$ to set of neurons α gives rise to trace $\bar{f}$ at set of neurons β, plus noise due to other stored associations in A.

to state, with some calculatable degree of certainty, whether or not a trace presented to α has been presented to the system before. For this calculation we will assume that previous storage implies that the trace is associated with itself, that is, presentation of the trace to α gives rise to the trace in β, thus

$$\bar{f} = A\bar{f},$$

(see Fig. 3). This may seem like an artificial definition if α and β are assumed to be two distinct sets of neurons, but if we were to consider a system, common in cortex, where a set of neurons projects to itself, we see that this scheme for recognition arises naturally.

We assume we have K pairs of associated traces stored in our system according to the rule

$$A_k \bar{f}_k = \bar{g}_k,$$

where A_k is the connectivity matrix generated when only $(\bar{f}_k, \bar{g}_k)$ are associated by the system. Then we form the sum

$$A = \sum_{k=1}^{k=K} A_k.$$

We assume a trace $\bar{f}_0$ is presented whose connectivity matrix may be present in A.

We wish to calculate a statistic which will let us decide whether $(\bar{f}_0, \bar{f}_0)$ is present or absent from the associations stored in A. A reasonable kind of statistic to use, since it is exceptionally easy to calculate with strictly local interactions, and which is also the optimal linear filter is the "matched filter" given in this case by

$$V = A\bar{f}_0 \cdot \bar{f}_0.$$

We will now proceed to calculate the part of V due to the presence of $(\bar{f}_0, \bar{f}_0)$ and the amount due to the noise generated by the other stored associations. We will use as our parameter of interest the output signal to noise ratio $[(S/N)_0]$ defined as is usual by

$(S/N)_0 = $ (output due to signal)^2/
          (mean square output due to noise).

We assume M neurons in α project to each neuron in β. We further assume diagonal elements of the matrix

are zero. To assume otherwise for the recognition problem would lead to trouble since self-associations will always give rise to positive diagonal elements. For now we will assume that the correlation between elements in a trace and between traces is zero.

By our optimal coding scheme we know that nonzero elements of the elementary matrix coupling two traces must be proportional to a constant times the value of the input to α, $\bar{f}_k$, thus for the association $(\bar{f}_k, \bar{g}_k)$ we obtain the following representation:

$$A_k = c \begin{bmatrix} g_k(1)\bar{f}_k^{(1)} \\ g_k(2)\bar{f}_k^{(2)} \\ \vdots \\ g_k(N)\bar{f}_k^{(N)} \end{bmatrix}.$$

Here we have assumed that M is large enough so that the normalizing constant for each element can be assumed to be the same. (This is equivalent to assuming that the trace power is approximately the same for every group of neurons in α that projects to a single neuron in β.) The $\bar{f}_k^{(i)}$ are row vectors defined so that $\bar{f}_k^{(i)}(j) = 0$ if neuron j in α does not connect with neuron i in β and

$$\bar{f}_k^{(i)}(j) = \bar{f}_k(j),$$

if neuron (j) in α is connected to neuron (i) in β. We can easily calculate the normalizing constant so that for K pairs of associations.

$$A = \sum_{k=1}^{k=K} A_k,$$

$$A = \frac{N}{MP} \sum_{k=1}^{k=K} \begin{bmatrix} g_k(1)\bar{f}_k^{(1)} \\ g_k(2)\bar{f}_k^{(2)} \\ \vdots \\ g_k(N)\bar{f}_k^{(N)} \end{bmatrix}.$$

We will assume that K is sufficiently large so that we can make our random variable approximation. We can see that individual terms of this matrix can be approximated by the sum of K of the products of two uncorrelated random variables of mean zero and variance P/N. Thus, variance of an element of A_k averaged over sets of allowable traces is

$$E[f(i)^2 f(j)^2] = E[f(i)^2]E[f(j)^2] = P^2/N^2.$$

K of these elementary connectivity matrices go to form A with

$$\text{var}(a_{ij}) = KP^2/N^2.$$

We can now form the $(S/N)_0$ for recognition. We will assume we have $K + 1$ pairs of associated traces stored in A. We present a trace $\bar{f}_k$ which is assumed to be identical to a stored self-association. Then

$$(S/N)_0 = \frac{(A_k \bar{f_k} \cdot \bar{f_k})^2}{E\{[(A - A_k)\bar{f_k} \cdot \bar{f_k}]^2\}},$$

where the average is to be taken allowable sets of traces. Then,

$$(S/N)_0 = \frac{P^2}{E\{[(A - A_k)\bar{f_k} \cdot \bar{f_k}]^2\}}.$$

We now evaluate

$$E\{[(A - A_k)\bar{f_k} \cdot \bar{f_k}]^2\}.$$

We know that $(A - A_k)$ is a matrix composed of elements either zero or random variables approximated above. We have the useful formula, if X is a random variable

$$\text{var}(cX) = c^2 \text{ var } X. \tag{1}$$

Then the vector approximating $(A - A_k)\bar{f_k}$ is given by a set of random variables of mean zero with

$$\text{var}[(A - A_k)\bar{f_k}(j)] = (MP/N)(KP^2/N^2) = (KMP^3/N^3).$$

This vector is multiplied by a constant N/MP. Then

$$\text{var}[(A - A_k)\bar{f_k} \cdot \bar{f_k}] = \sum_{i=1}^{i=N} f(i)X = \text{var } X \sum_{i=1}^{i=N} f(i)X$$

$$= (KMP^4)/N^3.$$

And for the $(S/N)_0$ we find, squaring the normalizing constant according to Eq. 1

$$(S/N)_0 = \frac{P^2}{(N^2/M^2P^2)(KMP^4/N^3)} = MN/K.$$

This result should now be discussed. Let us first note that this result is consistent with those obtained in previous papers, in particular, it shares with them the important property that an interactive memory works better as it gets larger, and, in this case, more highly interconnected. This finding is suggestive since it is known that in mammals the amount of cortex associated with a function appears to be determined by the relative importance of the function in the animal's behavior. Two examples are given by Thompson (Ref. 32, p. 317). First, he notes that anatomical studies indicate that in the dog, a relatively small amount of cortical tissue is devoted to representation of forepaw. In the racoon, which makes extensive use of its forepaws, there is a far larger amount of cortex devoted to forepaw representation. Second, and perhaps the best known example of this, is the grotesque little man representing the relative size of parts of motor cortex (determined by noting response to electrical stimulation) in humans (Ref. 32, p. 318). He has huge

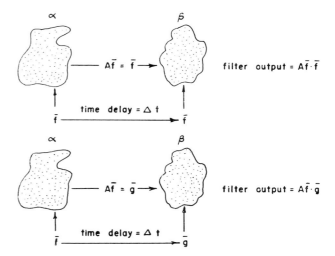

Figure 4 A system recognizing traces by self-association also allows recognition of temporal sequences. When trace $\bar{f}$ is impressed on set of neurons α it may give rise to output $\bar{f}$ (plus noise) if the self-association $(\bar{f}, \bar{f})$ has been stored in A. The filter output $A\bar{f} \cdot \bar{f}$ indicates whether or not the self-association $(\bar{f}, \bar{f})$ was stored. If we assume that the connection between α and β requires a time delay, Δt, then if the input to the system, $\bar{f}$, changes to $\bar{g}$ during the same period that the association $A\bar{f}$ is being generated at β, the filter output is identical to that for recognition.

fingers and face and relatively tiny body and feet. Also, the simple increase in size of the human brain during evolution suggests the presence of a significant mass effect where the cortex appears to work better (generating a selective advantage presumably due to more complex or appropriate behavior) as it gets larger (Ref. 22, p. 634).

Second, let us point out that the recognition model postulated here, where a self-association is assumed to be stored, is also capable of recognizing temporal sequences (Fig. 4). If we assume that there is a significant delay between the arrival of a pattern of activity $\bar{f}$ at α and the generation of pattern of activity $\bar{g}$ at β, we can see that if the input pattern changes, say, from $\bar{f}$ to $\bar{g}$ during this time delay, the formal scheme for recognition of the sequence and for recognizing a self-association are identical, giving rise to identical $(S/N)_0$. (Systems with long conduction times (20–200 msec) are characteristic of cortex.)

Third, let us briefly try to justify the choice of a matched filter recognition scheme. Although it has the maximum $(S/N)_0$ of any simple filter and requires only local operations to form, this, of course, is no guarantee that it is used by the nervous system. The presence of a matched filter would have some implications.

A simple matched filter (a variant of a template matching scheme) runs into the problem of the so-called perceptual invariants. The problem can be avoided by assuming that storage takes place after invariant transformations have occurred. However, even a simple template scheme is more realistic than it might seem at first. In the visual system, it is clear that matched filter detection of a trace that still retains a good deal of topographic organization is not rotation invariant. Although some literature on visual pattern recognition makes the assumption that human pattern recognition is largely rotation invariant, this is not the case, as sensory psychologists have known since the last century [18]. (A simple experiment will demonstrate the effect: it is very difficult to recognize even familiar faces if the observer's head is tilted 45°.) Relatively little quantitative work has been done on this effect. Dearborn (1899, described in Ref. 18) required his subjects to detect repeated presentations of forms displayed with various degrees of rotation. Dearborn reported that cards repeated in their original orientation were recognized 70% of the time while forms rotated by 9° were recognized only 43% of the time.

The matched filter is translation invariant if even elementary centering systems are permitted, as head or eye movements in vision. It is pertinent to point out that animals and humans have an extensive repertoire of orienting and gaze directing behaviors. Thus a snake, with an eye with a slit pupil, can be seen to keep its pupil opening vertical, preserving a constant orientation of the retinal image, no matter what the angle of the snake's body. Size invariance is a more difficult problem. Simple schemes have been proposed (Ref. 6, p. 110) which preserve size invariance over a limited range in ways which are compatible with a matched filter detection system. It is known [31] that brain damage which leaves a large central scotoma ("blind spot") does not interfere with recognition of figures, such as a large triangle, whose contours ring the scotoma, a result which might be expected from a matched filter. In any case, there often appears to be an optimal size of retinal image for many complex items. Thus the size of type will unconsciously determine the distance at which printed matter will be held, both very large and very small type being "hard to read". In vision, a complex system is present to give an output which is directly related to size invariance.

A More Complex System

More detailed calculations are possible on the model. The most interesting involve associations between unrelated traces. In this section an example of such a system will be considered.

Assuming that the trace, $\bar{g}$, associated with $\bar{f}$, is a

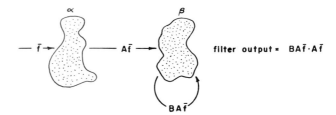

Figure 5 A more complex system allows the reconstruction of output traces uncorrelated with the input traces. It combines a projection system A, coupled with a projection system which couples β with itself.

trace that can be recognized by itself allows us to use a recognition and an association system together to improve the $(S/N)_0$. Figure 5 shows such a system. We assume that K_A traces are stored in the synapses connecting α and β and that K_B traces are stored in the recognition system attached to β. We assume that α and β each have N neurons and that all stored traces have power P. A is the connectivity matrix associated with connections between α and β; B is the connectivity matrix representing the recognition system associated with β.

We will first calculate the equivalent of the recognition $(S/N)_0$ for this system and then show how this result can be generalized to partially reconstruct the associated trace.

We wish to know whether a trace $\bar{f}$ presented to α generates a trace, $\bar{g}$, which is one of the traces recognized by the recognition system.

For these calculations we will introduce the notation $\bar{n}_A$ to indicate the noise added to the associated trace $\bar{g}$ due to the presence of the other stored associations, and $\bar{n}_B$ to represent the added noise produced by the recognition system, thus,

$$A\bar{f} = \bar{g} + \bar{n}_A, \qquad B\bar{g} = \bar{g} + \bar{n}_B,$$

Since we only have noisy information about $\bar{g}$, we form our statistic (equivalent to the recognition statistic discussed previously when $A = I$, the matrix with all diagonal elements equal to one) as

$$V = A\bar{f} \cdot BA\bar{f}.$$

Then

$$A\bar{f} \cdot BA\bar{f} = (\bar{g} + \bar{n}_A) \cdot (B\bar{g} + B\bar{n}_A)$$
$$= (\bar{g} + \bar{n}_A) \cdot (\bar{g} + \bar{n}_B + B\bar{n}_A)$$
$$= \bar{g} \cdot \bar{g} + \bar{g} \cdot \bar{n}_B + \bar{g} \cdot B\bar{n}_A + \bar{g} \cdot \bar{n}_A$$
$$+ \bar{n}_A \cdot \bar{n}_B + \bar{n}_A - B\bar{n}_A.$$

We assume as before that traces are uncorrelated, trace

elements are uncorrelated and the mean of elements in a trace is zero. We will sketch the calculation of these quantities, averaged over sets of allowable traces.

We wish to calculate the variance of $\bar{g} \cdot \bar{n}_B$. In the calculation in the previous section we calculated $(S/N)_0$ for recognition, thus

$$\bar{g} \cdot B\bar{g} = \bar{g} \cdot \bar{g} + \bar{g} \cdot \bar{n}_B = P + \bar{g} \cdot \bar{n}_B.$$

We found

$$(S/N)_0 = (\bar{g} \cdot \bar{g})^2 / \mathrm{var}(\bar{g} \cdot \bar{n}_B) = M_B N / K_B.$$

Solving

$$\mathrm{var}(\bar{g} \cdot \bar{n}_B) = (K_B P^2)/(M_B N).$$

Similarly,

$$\mathrm{var}(\bar{g} \cdot \bar{n}_A) = (K_A P^2)/(M_A N).$$

The term $\mathrm{var}(\bar{n}_A \cdot \bar{n}_B)$ is also easy to calculate. We can show that the power

$$P_{n_A} = \bar{n}_A \cdot \bar{n}_A = (K_A P)/M_A,$$

$$P_{n_B} = \bar{n}_B \cdot \bar{n}_B = (K_B P)/M_B.$$

Then we see that since the variance of an element of $\bar{n}_A = K_A P/M_A N$ and of $\bar{n}_B = K_B P/M_B N$ and there are N elements,

$$\mathrm{var}(\bar{n}_A \cdot \bar{n}_B) = (K_A K_B P^2)/(M_A M_B N).$$

We must go to the details of matrix B for calculation of the variance of the last two terms. We can approximate elements of B by random variables with variance $K_B P^2/N^2$; $\bar{n}_A$ is approximated by elements with variance $K_A P/M_A N$. Variance of the product vector is given by a vector with elements with variance

$$(N/M_B P)^2 M_B (K_B P^2/N^2)(K_A P/M_A N)$$

$$= (K_A K_B P/M_A M_B N).$$

The first term in the above expression arises from the normalizing constant which multiplies B and which appears as the square in the variance.

We see that

$$\mathrm{var}(\bar{g} \cdot B\bar{n}_A) = (K_A K_B P^2)/(M_A M_B N)$$

$$\mathrm{var}(\bar{n}_A \cdot B\bar{n}_A) = (K_A^2 K_B P^2)/(M_A^2 M_B N).$$

Now we can find $(S/N)_0$:

$$(S/N)_0$$

$$= \cfrac{P^2}{\cfrac{K_B P^2}{M_B N} + \cfrac{K_A K_B P^2}{M_A M_B N} + \cfrac{K_A P^2}{M_A N} + \cfrac{K_A K_B P^2}{M_A M_B N} + \left(\cfrac{K_A}{M_A}\right)^2 \cfrac{K_B P^2}{M_B N}}$$

$$= \cfrac{N}{(K_B/M_B)(K_A/M_A + 1)^2 + K_A/M_A}.$$

If we assume $K_A = K_B = K$; $M_A = M_B = M$ as a simplifying approximation, then

$$(S/N)_0 = \frac{MN}{K[K/M + 1)^2 + 1]}.$$

If many traces are stored (K/M is large)

$$(S/N)_0 = M^3 N/K^3.$$

This result would signify that $(S/N)_0$ would drop off rapidly as K became very large, but indicates that there is still a linear dependence of $(S/N)_0$ on N. However, the $(S/N)_0$ is very dependent on M, indicating that a highly interconnected system should work far better than a weakly interconnected system.

This $(S/N)_0$ is related to the question as to whether a trace presented to the system has an associated trace present in the system but says nothing about the details of the associated trace, a result of more interest. In general we would wish to extract as much information as possible from the association, in particular, we wish to have an idea of the structure of the associated trace— enough detail to be able to use it in further processing in the nervous system.

If we have N elements, a good strategy to recover information about the associated trace is to process the output of groups of elements, increasing or decreasing the number of elements in a group as needed to effect a compromise between amount of recovered detail and noise added by the memory.

As an example of this approach, we can use the formula just derived in a grouping scheme. We will assume that we are most interested in recovering the magnitude of the sum of squares of the values taken by elements. If we take V,

$$V = A\bar{f} \cdot BA\bar{f},$$

and note that the dot product can be taken over groups of R elements in β, we see that for a group of R elements the $(S/N)_0$ is given by

$$(S/N)_0 = \frac{R}{(K_B/M_B)(K_A/M_A + 1)^2 + K_A/M_A}.$$

(This result holds if $\bar{g} \cdot \bar{g}$ over the R elements has the average value,

$$E(\bar{g} \cdot \bar{g}) = RP/N.$$

The $(S/N)_0$ for a given region depends on the power of $\bar{g} \cdot \bar{g}$ in the region.)

We see from the above that V will be an estimate of the sum of squares of the trace summed over the R chosen elements. By increasing R we obtain the intuitive conclusion that we can increase the $(S/N)_0$

while simultaneously losing detail on the level of single elements.

It should be noted that a system like this apparently exists in the retina where the size of the spatio-temporal light integrating area can be increased or decreased depending on the average light intensity, sacrificing acuity for detectability in low light and attaining very high acuity in high light intensities (Ref. 8, chaps. IV, V, VI; Ref. 4).

By grouping elements, the $(S/N)_0$ of an average output can be varied over a range

$$\frac{1}{(K_B/M_B)(K_A/M_A + 1) + K_A/M_A}$$

$$\leqslant (S/N)_0$$

$$\leqslant \frac{N}{(K_B/M_B)(K_A/M_A + 1) + K_A/M_A}.$$

If we assume that groups of neurons will be chosen spatially close to each other, we see that by assuming even weak topographic organization of the neurons involved, this will decrease the high spatial frequency response of the recovered associated trace while the low frequency outline will be more stable. Detail will be lost but major structure will be preserved. This is in accord with common sense.

Selectivity

An important problem for a model where traces can easily be confused is the question of selectivity, that is, how "close" must a trace be to the stored trace to be recognized.

We can do some simple geometrical calculations to get an insight into this question. We have assumed all traces have power P, that is, they are vectors with tips lying on an N-dimensional hypersphere of radius $P^{1/2}$ (Fig. 6).

Let us consider $\bar{f}$ perturbed by a vector $\Delta \bar{f}$ so that

$$\bar{f} + \Delta \bar{f} = \bar{f}'; \qquad \bar{f} \cdot \bar{f} = \bar{f}' \cdot \bar{f}' = P.$$

Considering a simple recognition system with only one stored trace, $\bar{f}$, if we present our perturbed vector $\bar{f}'$, then

$$A\bar{f}' \cdot \bar{f}' = (A\bar{f} + A\Delta\bar{f}) \cdot \bar{f}'$$
$$= (A\bar{f} + A\Delta\bar{f}) \cdot (\bar{f} + \Delta\bar{f})$$
$$= P + \bar{f} \cdot \Delta\bar{f} + \bar{f} \cdot A\Delta\bar{f} + A\Delta\bar{f} \cdot \Delta\bar{f}.$$

We see $A\Delta\bar{f}$ can be calculated if we assume M is large enough so that for all (j, k)

$$\bar{f}^{(j)} \cdot \Delta\bar{f} = \bar{f}^{(k)} \cdot \Delta\bar{f} = (M/N)(\bar{f} \cdot \Delta\bar{f}),$$

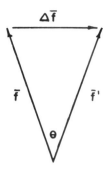

$$\bar{f} \cdot \bar{f} = \bar{f}' \cdot \bar{f}' = P$$

Figure 6 A trace $\bar{f}$ is perturbed by a vector $\Delta\bar{f}$, generating a new trace $\bar{f}'$. Both $\bar{f}$ and $\bar{f}'$ have the same power.

is an adequate approximation. Then

$$A\Delta\bar{f} = (N/MP)(M/N)(\bar{f} \cdot \Delta\bar{f})\bar{f},$$

and

$$A\Delta\bar{f} \cdot \bar{f} = \bar{f} \cdot \Delta\bar{f}.$$

Similarly,

$$A\Delta\bar{f} \cdot \Delta\bar{f} = (\bar{f} \cdot \Delta\bar{f})^2/P.$$

By geometry (Fig. 6) we see

$$\bar{f} \cdot \Delta\bar{f} = P(\cos\theta - 1).$$

If we form a ratio of filter output as a function of θ over filter output when $\theta = 0$, where θ is the angle between $\bar{f}$ and $\bar{f}'$, then,

$$(A\bar{f}' \cdot \bar{f}')/(A\bar{f} \cdot \bar{f}) = \cos^2\theta.$$

We wish to find the angle θ at which the output of the filter is reduced by half; we see $\theta = 45°$. This corresponds to the half-power point of engineering filter theory.

We now ask what are the chances of a random trace falling into the region where the filter output is greater than half of its maximum value. Restated, this is the ratio of the surface content of the N-dimensional hypersphere within $45°$ of a given trace to the surface content of the N-dimensional hypersphere [30]. We can calculate this quantity by observing we can establish a recursion formula, when S_N is the surface-content of an N-dimensional hypersphere and V_N is the volume-content of an N-dimensional hypersphere and when R is the radius,

$$S_N = 4 \int_0^{\pi/2} V_{N-2} R \, d\theta.$$

By a similar argument, if SC_N is the surface-content of

the hypersphere contained in a cone of base angle θ and VC_N is the volume content of such a cone (assuming the cone extends in positive and negative direction, like an hourglass)

$$SC_N = 4 \int_0^\theta VC_{N-2} R \, d\theta.$$

Thus the desired ratio is given, for N elements in the vector by

$$SC_N/S_N = \frac{4RVC_{N-2} \int_0^\theta d\theta}{4RV_{N-2} \int_0^{\pi/2} d\theta} = (2\theta/\pi)(VC_{N-2}/V_{N-2}).$$

If N is even

$$(SC_N/S_N) = (2\theta/\pi)^{N/2},$$

and if N is odd,

$$(SC_N/S_N) = (2\theta/\pi)^{(N+1)/2}.$$

When $\theta = 45°$, $(2\theta/\pi) = \frac{1}{2}$. This quantity becomes extremely small when N is large. Thus we see that selectivity as well as $(S/N)_0$ increases as N increases.

Cortical Model

The preceding discussion suggests a simple cortical model which would be capable of generating an interactive memory. The model will be sketched here.

Cortex is similar in histological structure over much of its extent. The most frequently occurring kind of cell—up to 80% of neurons in rabbit cortex [16]—are variants of what are called "pyramidal cells" with characteristic pyramid-shaped cell bodies and dendrites perpendicular to the cortical surface, extending almost to it, and then branching extensively parallel and close to the surface. These cells show rich synaptic contacts on all parts of the dendrites and have an array of "spines" which apparently correspond to synaptic contacts on the dendrites, although not all synapses are associated with spines. These cells also show extensive collateral branches of their axons, often projecting both intra- and extracortically. It is hard to estimate the number of different axons synapsing on pyramidal cells. Measurements quoted by Brodal (Ref. 5, p. 658) indicate that in monkey as many as 60,000 synapses are found on a motor area neuron, in visual cortex, about 7,000. In any case, several thousand cells may synapse on a single pyramidal cell. Origins of these synapses differ: some are sensory afferents, others intracortical fibers of various types—callosal, short projections, long projections. Much is known about

the details of interconnection, for this brief discussion we will merely assert that the cortex is very highly interconnected.

Most electrophysiological recordings from cortex have been made from pyramidal cells, simply because they are the largest and most numerous cells present. This data suggests that cells behave in many respects as unique individuals, each cell taking its particular sample of the surrounding afferent inputs. Since afferents are organized topographically in some cases (vision, for example) this gives common features to neighboring cells, but when cells are investigated in detail, each cell appears different from its neighbors in its particular blend of properties. Creutzfeldt and Ito [10] and Creutzfeldt [9] found that primary visual cortical cells appeared to receive large inputs from only a few lateral geniculate fibers (2 to 4). They suggested that the variety of forms of receptive field displayed by primary visual cortex cells could be explained by the possible permutations of the types of lateral geniculate cells. Goldstein et al. [17] found that auditory cortex is weakly organized tonotopically with nearby cells showing quite different tuning curves. Single units in primary auditory cortex responded in an "individualistic and variegated" manner. Hubel and Wiesel [19] suggested that in monkey visual cortex, many overlapping mosaics of sensory parameters are present, the sample received by a single cell depending on the afferents in its particular area.

The viewpoint that cortical cells respond as individuals finds confirmation in a paper by Noda and Adey [25]. They recorded single units chronically from cat association cortex (parietal cortex). In over 70 cases they recorded two units with the same electrode. They separated and analyzed these units. They found that when the animal was alert and awake, the cells, spatially close together, were uncorrelated in their discharge. This was true as well when the animal was in REM sleep, presumably a time of intense subjective experience. When the animal was awake but drowsy, there was a weak correlation between cells and when the animals were in deep sleep, cell discharges were highly correlated.

Let us now assume that we can identify the "traces" discussed earlier with patterns of increased or decreased pyramidal cell discharge in a given cortical area.

We assumed in the mathematical model that elements of a single trace were uncorrelated. The Noda and Adey findings [25] lend direct support to this assumption. Observations on EEG suggest a similar conclusion more indirectly. The generators apparently giving rise to the EEG tend to desynchronize (i.e. the resulting EEG amplitude distribution approaches a normal distribution more closely) during REM sleep and the awake state than in deep sleep [2]. This evidence indicates that during times when memory may be presumed to be functioning (alert, awake state and dream state) cortical elements tend to show uncorrelated activity.

We must consider the interconnections of the cortex. I would like to suggest that there is some evidence indicating the presence of two major classes of synapses significant for memory in the pyramidal cells. More probably, there may be a spectrum of types of synapses with the classes to be described forming two ends of a spectrum, but the analysis is not affected by assuming two discrete classes instead of a spectrum.

First, there is the familiar type of synapse, with relatively large PSPs with relatively short rise times (1.5–10 msec). This type of synapse would be characteristic of incoming sensory afferents or strong intracortical projection systems. The work of Rall [27] suggests that PSPs can be classified on the basis of shape as to their electrical distance from the cell body, since distant synapses will give slowly rising long PSPs due to the cable properties of the dendrites. Closer synapses will give more rapidly rising and falling PSPs, other properties of the PSPs being identical. Thus the first system would consist primarily of synapses close to the cell body.

However, we have noted that many thousand synapses occur on pyramidal cells. Since there are generally only a few clearly recognizable PSPs [11], one might wonder what the other synapses are doing.

Intracellular recordings from cortical neurons generally show spontaneous fluctuations of the membrane potential [1]. Although spontaneous PSPs are often present, this activity usually appears superimposed on some other fluctuating activity. Elul has determined the frequency spectrum of neuronal intracellular activity and finds it to have a frequency spectrum similar to that of the EEG with most energy in the frequencies below 10 cps. Although the distribution of amplitudes of the intracellular slow wave activity is not Gaussian (it is somewhat asymmetrical skewed in the positive direction) it is sufficiently close to the histogram expected of a Gaussian process to suggest some interesting modeling possibilities to see if the known asymmetries of the cortical neurons coupled with various input probability distributions might lead to the observed distribution.

In any case, I would like to suggest, as a second system, that this slow activity is generated by the activity of other synapses on the pyramidal cells. These synapses would be electrically remote from the cell body (on spines?) and would be severely low-pass filtered. Their influence would appear as a biasing of cell activity where the activity of any given input would be unnoticeable but the action of many thousand weak, long time course inputs would be highly significant. Since we know that most cells in cortex are spontaneously active and are interconnected, assuming weak, extensive interconnections seems the most reasonable way of explaining these ubiquitous low frequency membrane fluctuations.

Burns (Ref. 6, p. 64) has presented evidence which suggests that isolated cat forebrain acts very much like a highly interconnected random network.

Another line of evidence indicates the presence of a system electrically remote from the cell body. Smith and Smith [29] in an extensive study of the statistics of spontaneous cell activity in cortex detected the presence of two distinct systems giving rise to the spontaneous activity of cortical cells in their preparation, the unanesthetized, isolated cat forebrain. They found that the statistics of spontaneous activity could be explained as the result of two Poisson processes, one a "shower" of spikes following Poisson statistics and the other a process which "gated" the shower on and off at random intervals. They found when they passed a weak polarizing current through the tip of their electrode that the "gating" process was greatly affected but the average frequency of the Poisson shower was not changed drastically. Their interpretation of this is that the gating process is located near the cell body and the system giving rise to the shower is located farther from the soma. One might like to identify the "shower" process with the extensive weak interactions and the "gating" process with the first system but this is premature. Creutzfeldt, et al. [11] suggest in interpreting their stimulation data that "nonspecific" thalamic afferents are electrically more remote than "specific" thalamic afferents and that most inhibitory synapses appear to be located on the soma. Scheibel and Scheibel [28] suggest on anatomical grounds that nonspecific afferents exert a temporally diffuse biasing control on the pyramidal cell because of the pattern and location of the terminal arborization of these fibers.

Let us consider how an idealized model which assumes a relatively strong, fast system, and a weak, slow, highly interconnected system might work to generate memory. We have assumed that in learning, as we assumed in the mathematical model, synaptic strength coupling two neurons in the weak system is increased or decreased proportional to the product of pre- and postsynaptic activity during a given period—when the two groups of cells are "associating" the desired traces—then we can see that the weak system will give rise to an interactive memory system of the type discussed in previous sections.

The system would work roughly like this. A sensory input is impressed on the first group of neurons by the fast, strong system. A pattern of activity, corresponding to the input trace, is established in the first group of neurons. This pattern then "filters through" the weak system generating associations in a second group of neurons by the mechanisms discussed in the earlier sections of this paper. These associations can interact with the input trace to produce the recognition statistics. The input trace might be made available to the second group of neurons by a connection between first and second groups by the fast strong system or by another means, such as a direct sensory projection. Or the trace produced by the weak system could then be processed further by the second group of neurons and its connections.

A slow, weak system seems like a good candidate for memory since it would have the properties of extensive interconnection (high "M") and long time course (allowing complex temporal interactions at times consistent with the time courses of psychological events).

A common pattern of response of cortical neurons is to show a brief burst of activity when a new stimulus is presented and then to show a long inhibition. It might be possible that this inhibition serves to quiet the cells for the memory readout so that the diffuse and noisy memory trace trickling through the weak system would not be submerged by a strong sensory input.

Acknowledgments

I would like to thank Professor W. Ross Adey and the Space Biology Laboratory, Brain Research Institute, UCLA, for providing financial support for this work. I would also like to thank Professor Alan Grinnell, Department of Zoology, UCLA for providing greatly appreciated assistance.

References

1 W. R. Adey, *Neurosciences Research Program Bulletin* **7(2)**, 75 (1969).

2 W. R. Adey, *The Neurosciences*, Vol. 2 (F. O. Schmitt, Ed.). Rockefeller University, New York (1970), p. 224.

3 J. A. Anderson, *Kybernetik* **5**, 113 (1969); *Matheatical Biosciences* **8**, 137 (1970).

4 M. A. Bouman, in *Sensory Communication* (W. A. Rosenblith, Ed.) MIT Press, Cambridge, Mass. (1961), p. 377.

5 A. Brodal, *Neurological Anatomy*, 2nd ed., Oxford University Press, New York (1969).

6 B. D. Burns, *The Uncertain Nervous System*, Arnold, London (1968).

7 B. Cleland and C. Enroth-Cugell, *Acta Physiol. Scand.* **68**, 365 (1966).

8 T. N. Cornsweet, *Visual Perception*, Academic, New York (1970).

9 O. D. Creutzfeldt, in *The Neurosciences*, Vol. 2 (F. O. Schmitt, Ed.) Rockefeller University, New York (1970).

10 O. D. Creutzfeldt and M. Ito, *Experimental Brain Research* **6**, 324 (1968).

11 O. D. Creutzfeldt, K. Maekawa and L. Hösli, in *Progress in Brain Research*, Vol. 31 (K. Akert and P. G. Waser, Eds.) Elsevier, Amsterdam (1969).

12 J. C. Eccles, in *Brain and Conscious Experience* (J. C. Eccles, Ed.) Springer, Berlin (1966), p. 314.

13 S. S. Fox and J. H. O'Brien, *Science* **147**, 888 (1965).

14 W. J. Freeman, *J. Neurophysiol.* **31**, 337 (1968).

15 W. J. Freeman, *Mathematical Biosciences* **2**, 181 (1968).

16 A. Globus and A. B. Scheibel, *J. Comp. Neurol.* **131**, 155 (1966).

17 M. H. Goldstein, Jr., J. L. Hall, II, and B. O. Butterfield, *J. Acoust. Soc. Am.* **42**, 444 (1968).

18 H. W. Hake, quoted in *Pattern Recognition* (L. Uhr, Ed.) Wiley, New York (1966), p. 151.

19 D. H. Hubel and T. N. Wiesel, *J. Physiol.* **195**, 215 (1968).

20 L. Maffei, *J. Neurophysiol.* **31**, 283 (1968).

21 L. Maffei and G. Rizzolatti, *J. Neurophysiol.* **30**, 333 (1967).

22 E. Mayr, *Animal Species and Evolution*, Harvard, Cambridge, Mass. (1963).

23 M. L. Minsky, *Computation: Finite and Infinite Machines*, Prentice-Hall, Englewood Cliffs, New Jersey (1967).

24 V. B. Mountcastle, in *The Neurosciences* (G. C. Quarton, T. Melnechuk and F. O. Schmitt, Eds.) Rockefeller University, New York (1967).

25 H. Noda and W. R. Adey, *J. Neurophysiol.* **33**, 572 (1970).

26 D. H. Perkel and T. H. Bullock, *Neurosciences Research Program Bulletin* **6**, 221 (1968).

27 W. Rall, in *Neural Theory and Modeling* (R. F. Reiss, Ed.) Stanford University, Stanford, Calif. (1964), p. 73.

28 M. E. Scheibel and A. B. Scheibel, in *The Neurosciences*, Vol. 2 (F. O. Schmitt, Ed.), Rockefeller University, New York (1970), p. 443.

29 D. R. Smith and G. K. Smith, *Biophys. J.* **5**, 47 (1965).

30. D. M. Y. Sommerville, *An Introduction to the Geometry of N-Dimensions*, Dutton, New York (1929), chap. 8.

31 H. L. Teuber, in *Brain and Conscious Experience* (J. C. Eccles, Ed.) Springer, Berlin (1966), p. 182.

32 R. F. Thompson, *Foundations of Physiological Psychology*, Harper and Row, New York (1967).

16
Introduction

(1973)
L. N. Cooper

A possible organization of animal memory and learning
Proceedings of the Nobel Symposium on Collective Properties of Physical Systems,
B. Lundquist and S. Lundquist (Eds.), New York: Academic Press, pp. 252–264

Leon Cooper is another in the long line of physicists who have become interested in brain function and who have done significant work in the field. Cooper received a Nobel Prize for his work in superconductivity; he is the 'C' in the BCS theory of superconductivity. He has done distinguished work in quantum mechanics. He first became interested in memory because of the importance of the observer in many formulations of quantum mechanics. Observers can be true observers only if they can retain observations long enough to report them. This means they must have a memory, hence the initial interest in memory.

One of the editors of this volume (JAA) was at Rockefeller University in New York in 1971. One of Cooper's graduate students, Menasche Nass, burst into my office one day and said that he wanted to learn everything there was to know about memory, because he understood I was supposed to know something about it. At that time, the work described in paper 15 had been completed, and I was able to provide a couple of reprints and manuscripts along with a lot of half-baked speculation and a few psychological references. Later, Cooper and I had extensive and fascinating discussions about the way the brain might work. (Some of these results were presented at a AAAS symposium in Washington, DC, in December 1972; see the references at the end of this introduction.) At the 1972 Gordon Conference, some of our more exciting discussions were conducted while driving very rapidly over the twisty roads of New Hampshire in Cooper's gray Camaro.

In conjunction with the 1973 Nobel Symposium, Cooper presented this well known paper. It takes the linear associator model (papers 14 and 15) and presents a number of intriguing speculations based on it. Cooper suggests ways of using the cross-talk terms, that is, terms representing interactions between inputs if the network learns nonorthogonal inputs, as a way of generating weaker associations than the main association, and for building chains of associations. He also discusses the formation of *communities* of similar vectors, a notion that gets picked up later in some of the models for simple concept formation discussed in, for example, Knapp and Anderson (paper 36). A community is defined by the magnitude of the inner product between the members, which must exceed a certain value. The natural measure of similarity in the simple linear associator is the inner product between stored and new vectors, because inner products appear in the coefficients multiplying the output vector. Using the inner product, then, this model gives a natural way of forming equivalence classes of input vectors.

Cooper uses this clustering to make an important point, which echoes the comment made by von Neumann that "the logic of the brain is not the language of mathematics" (paper 7). Cooper points out the obvious fact that animals, at least according to associative models, do not behave according to rules of formal logic but instead follow "animal logic," which jumps to conclusions, based on an idiosyncratic blend of past experiences. This may seem to be an obvious observation, and it is true of virtually all neural network models. Yet it stands in explicit contradiction to many rule based theories of cognition found in Artificial Intelligence.

Cooper points out that when error correcting rules or multilayer networks are used, it is possible for the internals of the networks to vary greatly from one realization of a network to another, but their responses to a set of test inputs may be identical. It is often surprising to realize the large degree of gross anatomical variability between human brains. One requirement of a properly functioning educational system must be to mask the natural variability in its pupils, by teaching the appropriate, agreed upon, cognitive structures!

Cooper finishes the paper with some speculations on the organization of visual cortex. This particular question was a subtheme in theoretical neurobiology throughout the 1970s. When it became clear that visual cortex was both exquisitely organized and capable of adaptive modification in unusual environments, it seemed an ideal target for learning neural networks. In this volume, papers by von der Malsburg (paper 17), Grossberg (paper 19), and Bienenstock, Cooper, and Munro (paper 26) address this problem in detail. The paper by Nass and Cooper (1975) contains some of the first results on this problem from Cooper's laboratory.

References

J. A. Anderson, L. N. Cooper, W. Freiberger, U. Grenander, and M. M. Nass (1972), AAAS Symposium on Theoretical Biology and Biomathematics, Washington, DC.

M. M. Nass and L. N. Cooper (1975), "A theory for the development of feature detecting cells in visual cortex," *Biological Cybernetics* 19:1–18.

(1973)

L. N. Cooper

A possible organization of animal memory and learning

Proceedings of the Nobel Symposium on Collective Properties of Physical Systems,
B. Lundquist and S. Lundquist (Eds.), New York: Academic Press, pp. 252–264

Summary

A brief account is given of some of the properties of a neural model which displays on a primitive level features which suggest some of the mental behavior associated with animal memory and learning. The model as well as the basic passive procedure by which the neural network modifies itself with experience is consistent with known neurophysiology as well as with what information might be available in the neuron network. One must, however, assume that there exists a means of communication (electrical or chemical) between the cell body and dendrite ends—communication in a direction opposite to the flow of electrical signals. This same modification procedure could also lead to the formation of cells of the type observed by Hubel & Wiesel in the cat's visual cortex. It is suggestive that a network modification procedure that could produce such early processing cells might also be responsible, in cortex, for 'higher' mental processes. The explicit mathematics employed here is that of linear transformations on a vector space. However, as only certain topological properties are used, it is possible to construct a more general non-linear theory.

We have been analyzing a class of neural models that display, on a primitive level, features such as recognition, association and generalization, which suggest some of the mental behavior associated with animal memory and learning [1]. The mechanisms employed seem to be biologically plausible and are not inconsistent with known neurophysiology. In addition the networks that result seem to be a reasonable outcome of evolutionary development under the pressure of survival. Some of the ideas discussed are related to or are generalizations of earlier concepts such as perceptrons or similar models [2–4]. In addition non-local memories have been explored previously among others by Longuet-Higgins [5, 6].

This work was supported in part by the US National Science Foundation.

Quasi-Random Network

Because of the enormous complexity of the neural network, it seems reasonable to assume that an animal's central nervous system is not completely predetermined genetically. This view is reinforced by the observed fact that in mammals a functioning system is relatively resistant to the death or malfunction of individual units. A system is required, therefore, which is either highly redundant, in which the exact placement of individual units is not critical, or in some other way can continue functioning in spite of the failure of individual units.

A quasi-random network is in some ways analogous to a many-body system as treated in statistical mechanics. There the number of degrees of freedom is large compared to the number of parameters such as temperature or pressure, usually specified to characterize any given ensemble. The specification of the few parameters ensures that any actual configuration is a member of a particular ensemble.

We consider the possibility that genetic information determines a small number of overall parameters governing the growth of neurons. These might be, for example, density or types of neurons in various regions, approximate regions of projection, general direction of growth, extent of dendritic arborization, number and type of synaptic connections, etc. There is some evidence that this may be the case, particularly in cerebral cortex. In visual cortex there has long been known to be strong topographic organization, but there also appears to be local randomness of connection [7]. Auditory cortex seems to show even less obvious organization [8], and the olfactory and taste systems show no easily understood organization at all [9]. The model we consider accepts such local randomness as a fundamental principle of organization. Such a system could arise naturally in evolution by allowing progressive modification of previously completely specified neural systems.

Linear Mappings

Neuron behavior is complex and highly non-linear. However, in spite of the short time non-linear behavior of neurons, there is presently justification for considering linear or quasi-linear models in which the neuron potentials are averaged over short periods of time as surprisingly good approximations of some aspects of neural response [10]. Anderson explored the properties of a simple neural model based on a linear mapping between one set of neurons and another [11, 12]. This mapping contains a non-local memory with some of the properties of the holographic-type memory discussed by Longuet-Higgins [5, 6], but in contrast to the model of Longuet-Higgins, this mapping seems easily realizable with known physiology. Anderson showed that his model would act as a matched filter and be capable of recognition of events to which it is previously exposed. We have developed further the consequences of such systems and have shown that they are capable of recognition, association and a form of generalization which seems possible to relate in a primitive way to some animal mental behavior.

Network Modification

In addition we have been able to introduce a method of network modification in which learning takes place in an effortless or what we call a passive manner. The system is placed in an environment and, without any search procedure, forms an internal representation of its external world—an internal representation which enables it to recognize and associate.

The means of modifying the neural network employed here has a long history. It involves changing the strength of the synaptic junction (the ratio of output to input spiking frequencies) according to products of pre- and postsynaptic activity. This is physiologically possible (as will be discussed in more detail) if there is a means of communicating information between the cell bodies and dendrite ends—communication in a direction opposite to the flow of electrical signals. If such communication exists, alterations in the synaptic junctions could be made according to what information is locally available. (One can easily add to this the possibility of overall (global) controls affecting all of the network.)

Though there is no difficulty imagining a variety of possible mechanisms by which such changes might take place, experimental evidence bearing on which of these might be operative (or whether in fact synaptic modifications occur) is very sparse.

The modification procedure could also account for the formation of cells which respond to particular patterns as observed, for example, by Hubel & Wiesel in visual cortex of the cat [13]. The suggestive possibility emerges that the action of a single adaptive mechanism designed originally, perhaps, for the formation of cells to facilitate processing in the outer portions of the brain, can account also for 'higher' mental processes such as memory.

Space of Events and Representations

The duration and extent of an 'event' should be defined self-consistently by the interaction between the environment and the system itself. We proceed initially, however, as though an event is a well-defined objective happening and envision a space of events, E, labelled $e^1, e^2, e^3 \ldots$ Imagine that these are mapped by the sensory and early processing devices of the system through the mapping P (processing) into signal distributions in the neuron space $f^1, f^2, f^3 \ldots$ The mapping, P, is denoted by the double arrow, fig. 1. For the moment we maintain the fiction that this mapping is not modified by experience. (What seems actually to be the case is that such early processing systems are at least partially constructed in the youth of the animal and become 'hardened' at some relatively early stage in its development [14]).

Although we do not discuss the mapping P in any detail, it can be very complex; it must be rich and detailed enough so that a sufficient amount of information is preserved to be of interest. *We assume that the mapping P from E to F has the fundamental property of preserving in a sense (not yet completely defined) the closeness or separateness of events.*

Two events e^ν and e^μ map into f^ν and f^μ whose separation is related to the separation of the original events. In a vector representation we imagine that two events as similar as a white cat and a grey cat map into vectors which are close to parallel while two events as different as the sound of a bell and the sight of food map into vectors which are close to orthogonal to each other.

Given the signal distribution in F which is the result of an event in E, we imagine that the signal distribution f is mapped onto another set of neurons (or onto the same set) by a mapping, A, denoted by the single arrow, fig. 1. This latter type of mapping is modifiable, and we propose that it is in such mappings that animal memory is contained.

The cortex of higher animals is of course very complex, and if any such systems of neurons exist they would be expected not only to be complicated but to be interspersed with neuron assemblies which do specific processing. Further one would imagine that large numbers of such systems arranged serially or in parallel would be required to reproduce even a portion of the complexity and variety of mental processes.

As will be seen later, the behavior of such neuron sub-networks is dependent on parameters which could easily be imagined to vary from sub-network to sub-network resulting in quite different characteristics. It is possible that sequences of such sub-networks with different values of the biological parameters could be involved in actual mental behavior.

In what follows we construct an idealized model of a network which incorporates a modi-

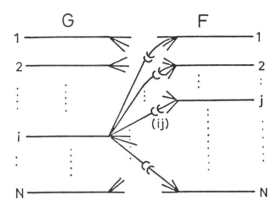

Fig. 2. The ideal associator unit. Each of the N incoming neurons in F is connected to each of the N outgoing neurons in G by a single ideal junction. (Only the connections to i are drawn.) We assume that the firing rate of neuron i in G, g_i, is mapped from the firing rates of all of the neurons in F by: $g_i = \Sigma_j A_{ij} f_j$.

fiable mapping and explore some of its properties.

Consider N neurons $1, 2 \dots N$, each of which has some spontaneous firing rate r_{j0}. (This need not be the same for all of the neurons nor need it be constant in time.) We can then define an N-tuple whose components are the difference between the actual firing rate r_j of the jth neuron and the spontaneous firing rate r_{j0}.

$$f_j \equiv r_j - r_{j0}$$

By constructing two such banks of neurons connected to one another (or even by the use of a single bank which feeds signals back to itself), we arrive at a simplified model as illustrated in fig. 1.

The actual connections between one neuron and another are generally complex and redundant; we idealize the network by replacing this multiplicity of connections between axon and dendrites by a single ideal junction which summarizes logically the effect of all of the synaptic contacts between the incoming axon branch from neuron j in the F bank and the dendrite tree of the outgoing neuron i in the G bank (fig. 2). Each of the N incoming neurons, in F, is connected to each of the N outgoing neurons, in G, by a single ideal junction. We then assume that: *the firing rate of neuron i in G, g_i, is mapped*

Fig. 1. The N neurons in the F bank are connected via synaptic junctions to the N neurons of the G bank.

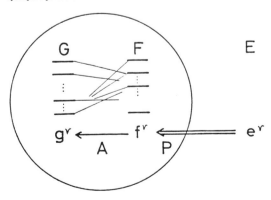

from the firing rates of all of the neurons, f_j, in F by:

$$g_i = \sum_{j=1}^{N} A_{ij} f_j$$

This is the fundamental linear assumption which gives the influence of firing rates in *F* on those in *G*. Although most of the results we obtain below do not require so strong an assumption, the simplicity of this hypothesis makes worthwhile an exploration of its consequences. In making this hypothesis, we are focusing our attention on firing rates, on time averages of the instantaneous signals in a neuron (or perhaps a small population of neurons). We are further using the known integrative properties of dendrite branches. Although we have available both excitatory or inhibitory synaptic junctions, such a module could be built if necessary of excitatory neurons alone since a decrease of incoming signal (if $r_j - r_{j0} < 0$) would decrease the output (less excitation).

The Associative Mapping, Memory and Mental Processes

It is in modifiable mappings of the type *A* that the experience and memory of the system are stored. In contrast with machine memory which is at present local (an event stored in a specific place) and addressable by locality (requiring some equivalent of indices and files) animal memory is likely to be distributed and addressable by content or by association. In addition for animals there need be no clear separation between memory and 'logic'. We show below that the mapping *A* can have the properties of a memory that is non-local, content addressable and in which 'logic' is a result of association and an outcome of the nature of the memory itself.

The suggestion that animal memory is non-local goes back at least to Lashley [15]; it is implied in Perceptron-like devices [2–4]. The holographic memory of Longuet-Higgins [5, 6] is non-local and content addressable but difficult to realize physiologically. The form we employ here was introduced by Anderson [11, 12].

Anderson's associative mapping is most easily written in the basis of the mapped vectors the

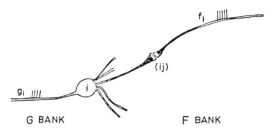

Fig. 3. The ideal junction.

system has experienced:

$$A = \sum_{\mu\nu} c_{\mu\nu} g^\mu \times f^\nu$$

Although this is a transparent mathematical form, its meaning as a mapping among neurons deserves some discussion. The ij'th element of *A* gives the strength of the ideal junction between the incoming neuron j in the *F* bank and the outgoing neuron i in the *G* bank (fig. 3).

Thus, if only f_j is non-zero

$$g_i = A_{ij} f_j.$$

Since

$$A_{ij} = \sum_{\mu\nu} c_{\mu\nu} g_i^\mu f_j^\nu$$

the ij'th junction strength is composed of a sum of the entire experience of the system as reflected in firing rates of the neurons connected to this junction. Each experience or association $(\mu\nu)$, however, is stored over the entire array of $N \times N$ junctions. This is the essential meaning of a distributed memory: Each event is stored over a large portion of the system, while at any particular local point many events are superimposed.

Recognition and Recollection

The fundamental problem posed by a distributed memory is the address and accuracy of recall of the stored events. Consider first the 'diagonal' portion of *A*

$$(A)_{\text{diagonal}} \equiv \mathcal{R} \equiv \sum_{\nu} c_{\nu\nu} g^\nu \times f^\nu$$

An arbitrary event, *e*, mapped into the signal, *f*, will generate the response in *G*

$$g = Af$$

If we equate recognition with the strength of this response g, say the value of

(g, g),

then the mapping A will distinguish between those events it contains (the f^ν, $\nu = 1, 2, ..., k$) and other events separated from these.

The word 'separated' in the above context requires definition. In a type of argument given previously [11, 12], the vectors f^ν are thought to be independent of one another and to satisfy the requirements that on the average

$$\sum_{i=1}^{N} f_i^\nu = 0$$

$$\sum_{i=1}^{N} (f_i^\nu)^2 = 1.$$

Any two such vectors have components which are random with respect to one another so that a new vector, f, presented to $\mathcal{R}$ above gives a noise-like response since on the average (f^ν, f) is small. The presentation of a vector seen previously, f^λ, however, gives the response

$$\mathcal{R}f^\lambda = c_{\lambda\lambda} f^\lambda + \text{noise}$$

It is then shown that if the number of imprinted events, k, is small compared to N, the signal to noise ratios are reasonable.

If we define separated events as those which map into orthogonal vectors, then clearly a recognition matrix composed of k orthogonal vectors $f^1 f^2, ..., f^k$

$$\mathcal{R} = \sum_{\nu=1}^{k} c_{\nu\nu} g^\nu \times f^\nu$$

will distinguish between those vectors contained, $f^1 ... f^k$, and all vectors separated from (perpendicular to) these. Further the response of $\mathcal{R}$ to a vector previously recorded is unique and completely accurate

$$\mathcal{R}f^\lambda = c_{\lambda\lambda} g^\lambda$$

In this special situation the distributed memory is as precise as a localized memory.

In addition this type of memory has, as has been pointed out before [5, 6], the interesting property of recalling an entire associated vector g^λ

even if only part of f^λ is presented. Let

$$f^\lambda = f_1^\lambda + f_2^\lambda$$

If only part of f^λ, say f_1^λ, is presented, we obtain

$$\mathcal{R}f_1^\lambda = c_{\lambda\lambda}(f_1^\lambda, f_1^\lambda) g^\lambda + \text{noise}$$

The result is thus the entire response to the full f^λ with a reduced coefficient plus noise.

Association

If we now take the point of view that presentation of the event e^ν which generates the vector f^ν is recognized if

$$\mathcal{R}f^\nu = cg^\nu + \text{noise}$$

Then the off-diagonal terms

$$A \equiv \sum_{\mu \neq \nu} c_{\mu\nu} g^\mu \times f^\nu$$

may be interpreted as leading to association of events initially separated from one another

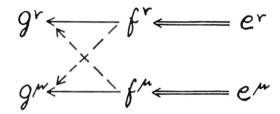

where $(f^\nu, f^\mu) = 0$.

For with such terms the presentation of the event e^ν will generate not only g^ν (which is equivalent to recognition of e^ν) but also (perhaps more weakly) g^μ which should result with the presentation of e^μ. Thus, for example, if g^μ will initiate some response (originally a response to e^μ) the presentation of e^ν when $c_{\mu\nu} \neq 0$ will also initiate this response.

We, therefore, can write the association matrix

$$A = \sum_{\mu\nu} c_{\mu\nu}' g^\mu \times f^\nu = \mathcal{R} + \mathcal{A}$$

where

$$\mathcal{R} = (A)_{\text{diagonal}} \equiv \sum_{\nu} c_{\nu\nu} g^\nu \times f^\nu \quad \text{[recognition]}$$

and

$$\mathcal{A} = (A)_{\text{off-diagonal}} \equiv \sum_{\mu \neq \nu} c_{\mu\nu} g^\mu \times f^\nu \quad \text{[association]}$$

The $c_{\mu\nu}$ are then the 'direct' recognition and association coefficients.

"Logic"

In actual experience the events to which the system is exposed are not in general highly separated nor are they independent in a statistical sense. There is no reason, therefore, to expect that all vectors, f^ν, printed into A would be orthogonal or even very far from one another. Rather it seems likely that often large numbers of these vectors would lie close to one another. Under these circumstances a distributed memory of the type contained in A will become confused and make errors. It will 'recognize' and 'associate' events never in fact seen or associated before.

To illustrate, assume that the system has been exposed to a class of non-separated events $\{e^1 \dots e^k\} : \{e^\alpha\}$ which map into the k vectors $\{f^1 \dots f^k\} : \{f^\alpha\}$.

The closeness of the mapped events can be expressed in a linear space by the concept of community. We define the community of a set of vectors such as $\{f^\alpha\}$ above as the lower bound of the inner products (f^s, f^t) of any two vectors in this set. *The community of the set of vectors $\{f^\alpha\}$ is Γ, $C(f^\alpha) = \Gamma$, if Γ is the lower bound of (f^s, f^t) for all f^s and f^t in $\{f^\alpha\}$.*

If each exposure results in an addition to A (or to $\mathcal{R}$) of an element of the form

$$c_{\nu\nu} g^\nu \times f^\nu,$$

then the response to an event f^s from this class $f^s \in \{f^\alpha\}$ is

$$\mathcal{R}f^s = g = \sum_\nu c_{\nu\nu} g^\nu (f^\nu, f^s) = c_{ss} g^s + \sum_{\nu \neq s} (f^\nu, f^s) c_{\nu\nu} g^\nu$$

where $(f^\nu, f^s) \geqslant \Gamma$. If Γ is large enough the response to f^s is, therefore, not very clearly distinguished from that of any other f contained in $\{f^\alpha\}$. (In the next section we discuss how such an A might be constructed.)

If a new event, e^{k+1}, not seen before is presented to the system and this new event is close to the others in the class α (for example, suppose that e^{k+1} maps into f^{k+1} which is a member of the community $\{f^\alpha\}$) then $\mathcal{R}f^{k+1}$ will produce a response not too different from that produced for one of the vectors $f^s \in \{f^\alpha\}$. Therefore, the event e^{k+1} will be recognized though not seen before.

This, of course, is potentially a very valuable error. For the associative memory recognizes and then attributes properties to events which fall into the same class as events already recognized. If in fact the vectors in $\{f^\alpha\}$ have the form

$$f^\nu = f^0 + n^\nu$$

where n^ν varies randomly, f^0 will eventually be recognized more strongly than any of the particular f^ν actually presented.

We have here an explicit realization of what might loosely be called 'animal logic'—which, of course, is not logic at all. Rather what occurs might be described as the result of a built-in directive to 'jump to conclusions'. The associative memory by its nature takes the step

$$f^0 + n^1, f^0 + n^2 \dots f^0 + n^k \dots \to f^0$$

which one perhaps attempts to describe in language as passing from particulars: cat[1], cat[2], cat[3] ... to the general: cat.

How fast this step is taken depends (as we will see in the next section) on the parameters of the system. By altering these parameters, it is possible to construct mappings which vary from those which retain all particulars to which they are exposed, to those which lose the particulars and retain only common elements—the central vector of any class.

In addition to 'errors' of recognition, the associative memory also makes errors of association. If, for example, all (or many) of the vectors of the class $\{f^\alpha\}$ associate some particular g^β so that the mapping A contains terms of the form

$$\sum_{\nu=1}^{k} c_{\beta\nu} g^\beta \times f^\nu$$

with $c_{\beta\nu} \neq 0$ over much of $\nu = 1, 2, \dots, k$, then the new event e^{k+1} which maps into f^{k+1} as in the previous example will not only be recognized

$$\mathcal{R}f^{k+1}, \mathcal{R}f^{k+1}) \quad \text{large}$$

but will also associate g^β

$$Af^{k+1} = cg^\beta + \dots$$

as strongly as any of the vectors in $\{f^\alpha\}$.

If errors of recognition lead to the process described in language as going from particulars to the general, errors of association might be described as going from particulars to a universal: cat[1] meows, cat[2] meows ... → all cats meow.

There is, of course, no 'justification' for this animal process. It is performed as a consequence of the nature of the system. Whatever efficacy it has will depend on the order of the world in which the animal system finds itself. If the world is properly ordered, an animal system which 'jumps to conclusions' in the sense above may be better able to adapt and react to the hazards of its environment and thus survive. The animal philosopher sophisticated enough to argue 'the tiger ate my friend but that does not allow me to conclude that he might want to eat me' might then be a recent development whose survival depends on other less sophisticated animals who jump to conclusions.

By a sequence of mappings of the form above (or by feeding the output of A back to itself) one obtains a fabric of events and connections

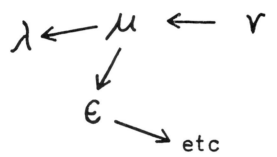

which is rich as well as suggestive. One easily sees the possibility of a flow of electrical activity influenced both by internal mappings of the form A and the external input. This flow is governed not only by direct association coefficients $c_{\mu\nu}$ (which can be explicitly learned as described next) but also by indirect associations due to the overlapping of the mapped events as indicated in fig. 4. In addition one can easily imagine situations arising in which direct access to an event, or a class of events, has been lost ($c_{\gamma\gamma} = 0$ in fig. 4) while the existence of this event or class of events in A influences the flow of electrical activity.

One serious problem in making the identifications suggested above is a direct consequence of the assumption of the linearity of the system. Any state is generally a superposition of various vectors. Thus one has to find a means by which events—or the entities into which they are mapped are distinguished from one another.

There are various possibilities; neurons are so non-linear that it is not at all difficult to imagine

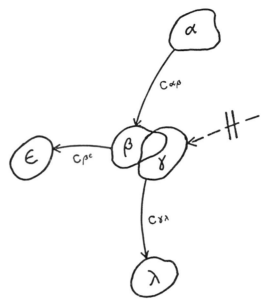

Fig. 4. There will be a flow from $\{\alpha\}$ to $\{\lambda\}$ even though no directly learned coefficient $c_{\alpha\lambda}$ exists, due to the overlap between the $\{\beta\}$ and $\{\gamma\}$ classes. In the particular situation above direct access to $\{\gamma\}$ has been lost.

non-linear or threshold devices that would separate one vector from another. But the occurrence of a vector in the class $\{f^\alpha\}$ in a distributed memory is a set of signals over a large number of neurons each of which is far from threshold. A basic problem, therefore, is how to associate the threshold of a single cell or a group of cells with such a distributed signal. How this might come about will be described in a later section.

In addition to the appearance of such 'pontifical' cells, there will be a certain separation of mapped signals due to actual localization of the areas in which these signals occur. For example, optical and auditory signals are subjected to much processing before they actually meet in cortex. It is possible to imagine that identification of optical or auditory signals (as optical or auditory) occurs first from where they appear and their immediate cluster of associations. Connections between an optical and an auditory event might occur as suggested in fig. 5.

I need hardly mention that even assuming that the physiological assumptions that underlie these constructions are possible (or even correct), there is a distance to be travelled before it is

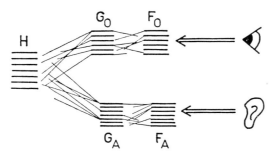

Fig. 5. A Model Optical-Auditory System.

shown that combinations of such elements as introduced above, working together could really reproduce animal mental behavior. The most important step is to make contact between theory and experiment. Some attempts in this direction will be discussed below.

Network Modification, Learning

We now ask how a mapping of the type A might be put into the network. The ij'th element of the associative mapping A

$$A_{ij} = \sum_{\mu\nu} c_{\mu\nu} g_i^\mu f_j^\nu$$

is a weighted sum over the j components of all mapped signals, f^ν, and the i components of the responses, g^μ, appropriate for recollection or association. Such a form could be obtained by additions to the element A_{ij} of the following type:

$$\delta A_{ij} \sim g_i f_j$$

This δA_{ij} is proportional to the product of the differences between the actual and the spontaneous firing rates in the pre and post synaptic neurons i and j[16]. The addition of such changes in A for all associations $g^\mu \times f^\nu$ results finally in a mapping with the properties discussed in the previous section.

For such modifications to occur, there must be a means of communication between the cell body and dendrite ends in order that the necessary information be available at the appropriate junctions; this information must move in a direction opposite to the flow of electrical signals [17]. The junction ij, for example, must have informa-

tion of the firing rate f_j (which is locally available) as well as the firing rate g_i which is somewhat removed (fig. 6). One possibility would be that the integrated electrical signals from the dendrites produce a chemical or electrical response in the cell body which controls the spiking rate of the axon and at the same time communicates to the dendrite ends the information of the integrated slow potential.

There are a variety of means by which the coefficient A_{ij} might be modified, given that the necessary information is available at the ij'th junction. Among these might be growth of additional dendritic spines, adding new synaptic junctions, activation of synaptic junctions previously inactive, changes in membrane resistivity and/or changes in the amount of transmitter in a synapse. Although some structural changes have been observed, there is little evidence yet to choose among the possibilities mentioned above or in fact little evidence that such processes take place at all in the cortex of an adult animal.

To make the modifications

$$\delta A \sim g^\mu \times f^\nu$$

by any of the mechanisms suggested above, the system must have the signal distribution f^ν in its F bank and g^μ in its G bank. It is easy to obtain f^ν since this is mapped in from the event e^ν by P. But to get g^μ in the G bank is more difficult since this in effect is what the system is trying to learn.

In what we denote as active learning (which has been much explored in the past) the system is presented with some f^λ, searches for a response, and is given some indication of when it is coming

Fig. 6. In order that the junction (ij) be modified in proportion to $g_i f_j$, a means is needed for communicating the firing rate g_i which is the result of signals incoming from all the dendrites $g_i = \Sigma_j A_{ij} f_j$ back to the junction (ij).

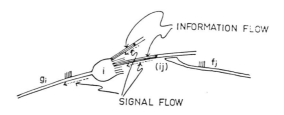

closer. When (after what could be a long time) by some procedure or another it finds the "right" response, say g^ω, it is "rewarded" and responds to the reward by printing into A the information:

$$\delta A_{ij} = \eta g_i^\omega f_j^\lambda$$

(The information is available at the time of the reward since at that time the system is mapping f^λ, responding g^ω, and thus has just the desired spiking frequencies in the F and G banks of neurons.) Active learning probably describes a type of learning in which a system response to an input is matched against an expected or desired response and judged correct or incorrect.

However, there is a type of animal learning which does not seem from visible external indications to require this type of a search procedure. It is the type of learning in which, as far as can be seen, an animal is placed in an environment and seems to learn to recognize and to recollect in a passive manner.

To arrive at an algorithm which produces passive learning, we utilize a distinction between forming an internal representation of events in the external world as opposed to producing a response to these events which is matched against what is expected or desired in the external world.

The simple but important idea is that *the internal electrical activity which in one mind signals the presence of an external event is not necessarily (or likely to be) the same electrical activity which signals the presence of that same event for another mind.* There is nothing that requires that the same external event be mapped into the same neural patterns by different animals. The event e^ν which for one animal is mapped into the signal distributions f^ν and g^ν, in another animal is mapped into f'^ν and g'^ν. What is required for eventual agreement between animals in their description of the external world is not that electrical signals mapped be identical but rather that the relation of the signals to each other and to events in the external world be the same (fig. 7).

Passive Learning

Call $A^{(t)}$ the A matrix after the presentation of t events. We write

$$A^{(t)} = \gamma A^{(t-1)} + \eta g^t \times f^t$$

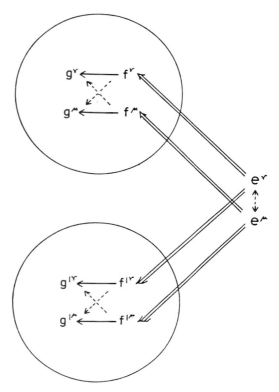

Fig. 7. Representations in two different systems of the same external fabric of events. The two representations are not identical but they each stand in a one-to-one relation to the external fabric and to each other.

In the equation γ is dimensionless and is a measure of the uniform decay of information at every site (a type of forgetting). One would expect that

$$0 \leqslant \gamma \leqslant 1$$

It turns out that values of γ close to one are of most interest. Since f or g are firing rates (spikes/second), η or $\varepsilon = \eta/\gamma$ have the dimensions of s^2. In what follows we normalize all vectors $(f,f) = (g,g) = 1$ so that η and ε become dimensionless.

If we now say that g^t is

$$g^t = A^{(t-1)} f^t + g_R^t + g_A^t$$

We see that the total post-synaptic potentials are composed of three terms: a passive response, $A^{(t-1)} f^t$, an active but random term, g_R^t, and an active response, g_A^t. For purely passive learning we consider only the first term so that

$$\delta A = \eta g^t \times f^t = \eta A^{(t-1)} f^t \times f^t.$$

Here the post synaptic potentials are just those produced by the existing mapping, $A^{(t-1)}$, when the vector f^t in F is mapped into G.

$$g^t = A^{(t-1)} f^t$$

The passive learning algorithm is then

$$A^{(t)} = A^{(t-1)}(\gamma + \eta f^t \times f^t)$$

$$= \gamma A^{(t-1)}(1 + \varepsilon f^t \times f^t)$$

where in general $\varepsilon = n/\gamma$ is presumably much smaller than one. Before any external events have been presented, A has the form $A^{(0)}$ which could be random. This will contain among other things the connectivity of the network.

With this algorithm, after k events A has the form

$$A^{(k)} = \gamma^k A^{(0)} \prod_{\nu=1}^{k} (1 + \varepsilon f^\nu \times f^\nu)$$

where $\prod_0$ is an ordered product in which the factors with lower indices stand to the left:

$$\prod_{\nu=1}^{k} f(\nu) = f(1) f(2) \dots f(k)$$

This can also be written:

$$A^{(k)} = \gamma^k A^{(0)} \Bigg[1 + \varepsilon \sum_{\nu=1}^{k} |f^\nu\rangle \langle f^\nu |$$

$$+ \varepsilon^2 \sum_{\nu<\mu} |f^\nu\rangle \langle f^\nu | f^\mu\rangle \langle f^\mu | + \dots$$

$$+ \varepsilon^k |1\rangle \langle 1|2\rangle \langle 2|3\rangle \langle 3|4\rangle \dots \langle k-1|k\rangle \langle k| \Bigg]$$

It is striking that the passive learning algorithm generates its own response $A^{(0)} f^\nu$ to the incoming vector f^ν, a response that depends on the original configuration of the network through $A^{(0)}$ and on the vector f^ν mapped from the event e^ν. For example if f^ν is the only vector presented, A eventually takes the form

$$A \sim g^\nu \times f^\nu$$

where

$$g^\nu = A^{(0)} f^\nu.$$

Special cases of A

We now display the form of A in several special cases; in all of these ε is assumed to be constant and small.

(1) If the k vectors are orthogonal, A becomes

$$A^{(k)} = \gamma^k A^{(0)} \left(1 + \varepsilon A^{(0)} \sum_{\nu=1}^{k} f^\nu \times f^\nu \right).$$

Letting $A^{(0)} f^\nu = g^\nu$, the second term takes the form of the "diagonal" part of A

$$(A)_{\text{diagonal}} \equiv \mathcal{R} = \sum_{\nu=1}^{k} g^\nu \times f^\nu$$

and will serve for the recognition of the vectors $f^1 \dots, f^k$. (It should be observed that the associated vectors g^ν are not given in advance; they are generated by the network.) If ε is small however this seems inadequate for recognition since the recognition term would be weak. Further one would expect recognition to build up only after repeated exposure to the same event.

(2) The following example demonstrates that the passive learning algorithm does build up recognition coefficients for repeated inputs of the same event. If the same vector f^0 is presented l times, A becomes eventually

$$A^{(l)} \simeq \gamma^l_A{}^{(0)} (1 + e^{l\varepsilon} f^0 \times f^0).$$

If l is large enough so that $e^{l\varepsilon} \gg 1$, the recognition term will eventually dominate. Presumably when $e^{l\varepsilon}$ becomes large enough there should be no further increase in the coefficient. This can be accomplished by making ε a function of the response to the incoming vector in some way so that beyond some maximum value there is no further increase of the coefficient.

(3) The presentation of m orthogonal vectors $l_1, l_2, \dots l_m$ times results in a simple generalization of the second result. When $\gamma = 1$ (for simplicity)

$$A^{(l_1+l_2+\dots+l_m)} = A^{(0)} \left(1 + \sum_{\nu=1}^{m} e^{l_\nu \varepsilon} f^\nu \times f^\nu \right)$$

which is just a separated associative recognition and recall matrix

$$A \simeq \sum_{\nu=1}^{m} c_{\nu\nu} g^\nu \times f^\nu$$

if

$$e^{l_\nu \varepsilon} \equiv c_{\nu\nu} \gg 1.$$

(4) Some of the effect of non-orthogonality can be displayed by calculating the result of an input consisting of l noisy vectors distributed randomly around a central f^0

$$f^\nu = f^0 + n^\nu$$

Here n^ν is a stochastic vector whose magnitude is small compared to that of f^0. We obtain

$$A^{(l)} \simeq \gamma^l A^{(0)} \exp\left(l\varepsilon \frac{n^2}{N}\right)(1 + e^{l\varepsilon} f^0 \times f^0)$$

where n is the average magnitude of n^ν. We see that the generated $A^{(l)}$, with the additional factor due to the noise, is just of the form for recognition of f^0. Thus the repeated application of a noisy vector of the form above results in an A which recognizes the central vector f^0.

Association Terms

Off-diagonal or associative terms can be generated as follows. Assume that A has attained the form

$$A = \sum_{\nu=1}^{k} A^{(0)} f^\nu \times f^\nu = \sum_{\nu=1}^{k} g^\nu \times f^\nu.$$

Now present the events e^α and e^β so that they are associated so that the vectors f^α and f^β map together. (The precise conditions which result in such a simultaneous mapping of f^α and f^β will depend on the construction of the system. The simplest situation to imagine is that in which $(f^\alpha + f^\beta)/\sqrt{2}$ is mapped if e^α and e^β are presented to the system close enough to each other in time.) We may assume that e^α and e^β are separated so that $(f^\alpha, f^\beta) = 0$. In the F bank of neurons we then have $(f^\alpha - f^\beta)/\sqrt{2}$, where the vector is normalized for convenience.

After one such presentation of e^α and e^β, A becomes:

$$A^{(1)} \simeq \sum_{\nu=1}^{k} g^\nu \times f^\nu + \frac{\eta}{2}(g^\beta \times f^\alpha + g^\alpha \times f^\beta).$$

The second term gives the association between α and β with the coefficient

$$c_{\alpha\beta} = c_{\beta\alpha} = \eta/2$$

which presumably (except in special circumstances) would be small. If f^α and f^β do not occur again

in association, $c_{\alpha\beta}$ or $c_{\beta\alpha}$, although they do grow upon the presentation of f^α or f^β separately, always remain small compared to the respective recognition coefficients $c_{\beta\beta}$ or $c_{\alpha\alpha}$. However if $(f^\alpha + f^\beta)/\sqrt{2}$ is a frequent occurrence (appearing for example l times) the coefficient of the cross term becomes

$$c_{\alpha\beta} \simeq \frac{\gamma^l}{2} e^{l\varepsilon}$$

as large as the recognition coefficient.

We are at present persuing analytical calculations and machine simulations to explore further the development of A in a variety of situations.

Hubel-Wiesel Type Cells

It is most important to make some contact between theory and experiment. We have been working in several directions in such attempts; I will describe one below.

In a classic series of experiments Hubel & Wiesel [13] have shown that cells in the cat's visual cortex respond in a very specific way to external visual patterns to which the cat is exposed. Such patterns can for example be vertical or horizontal lines of specific sizes, in specific positions or with definite motions. Recently Hirsch & Spinelli [18] have shown that depriving the cat of visual experience (such as the experience of horizontal lines) results in the absence of Hubel-Wiesel cells which respond to such lines in the adult. We outline below one means by which such cells might develop, assuming that network modification can occur in a similar manner to that employed previously for the construction of the mapping A.

Since these cells occur in the early processing regions we divide the mapping P so that

$$P = M_2 \times M_1$$

$$c^\alpha <\!\!\!\sim\!\!\!\sim d^\alpha <\!\!=\!\!= e^\alpha$$
$$\quad\quad M_2 \quad\quad M_1$$

where now M_1 is taken to be prewired but M_2 is modifiable (such modification being possible presumably only in the early development of the animal). As M_2 is to be modified by the same mechanism used previously we write:

$$\delta M_2 \sim g^\mu \times f^\nu.$$

We now assume that the spontaneous firing rate of the cells in the left hand bank of M_2 (other than the random spikes we admit below) is very low or zero and that $M_2^{(0)}$ (the initial value of the mapping M_2) is small, so that the cells in the left hand bank, at least in the development stages, are in general below threshold and do not respond strongly to external signals. These cells therefore are in a pre-threshold and highly non-linear region.

We suppose further (as is in fact the case) that the cells in the left hand bank fire occasional spikes spontaneously and at random. We can then write:

$$M_2^{(t)} = \gamma M_2^{(t-1)} + \eta P_n M_2^{t-1} d^t \times d^t + \eta c^n \times d^t$$

where $P_n M_2$ is the projection of M_2 onto the n'th outgoing cell

$$P_n M_2 = \sum_i \delta_{ni}(M_2)_{ij} \quad \text{for all} \quad j.$$

As before the first term gives the uniform loss of memory, the second the passive response, while the third gives active additions to M_2 due to the simultaneous arrival of the mapped vector d^t and the spontaneous and random firing of the n'th cell in the left hand bank

$$c_i^n = 0 \quad i \neq n$$

$$c_i^n = 1 \quad i = n$$

Since M_2 is initially thought to be small, an incoming vector d^t produces no change in M_2 unless one of the cells in the left hand bank is firing simultaneously. If, however, the entry of the pattern d^α coincides with the firing of the n'th cell we add a term to M_2 of the form

$$\delta M_2 = \eta P_n M_2 d^\alpha \times d^\alpha + \eta c^n \times d^\alpha$$

It is reasonable to assume that a single such entry is not sufficient to fire the n'th cell upon a second presentation of d^α. However, with l coincidences of d^α and the firing of c^n we obtain

$$M_2^{(l)} \simeq \gamma^l e^{l\varepsilon} c^n \times d^\alpha$$

Therefore, there is a build-up of such recognition cells and, depending upon the parameters, they will eventually fire upon the presentation of the pattern with which they have been associated.

There are several problems. One would expect several cells to respond to the same pattern. This is very likely to be the case in fact. However, the addition of a mechanism, such as exists in the visual processing system of Limulus, by which a firing cell suppresses the activity of its neighbours would reduce the number of cells which would respond to the same pattern. In addition it might occur that the same cell would pick up several patterns. How likely this is is being tested by computer simulation with various values of the parameters. Whether this happens in fact is an open question which could be answered experimentally. Such multiple patterns could also eventually be diminished by the addition of a Limulus-like suppression.

All of the prior considerations of picking central vectors out of noisy entries would apply; the details would depend on the parameters chosen. Such cells, linked to F or G type neuron banks, might also be of use for the separation of one vector from another as suggested at the end of the second section.

If these considerations in some measure correspond to facts, it becomes intriguing to speculate that a form of network modification which arose originally in evolutionary development to assist in the formation of special cells for early processing—to increase the flexibility of systems so that they required less genetic pre-programming—might (functioning in a region of cortex which retains for much or all of a lifetime its ability for modification) be responsible also for what we like to call "higher" mental processes.

The author wishes to express his appreciation for the hospitality offered to him by the Laboratoires de Physique Théorique et Hautes Energies, Universités Paris VI and Paris XI (Orsay) where part of this article was written. He would also like to thank those colleagues, in particular Professors Jack Cowen, Charles Elbaum and Bruce Knight, who have offered helpful criticism.

References

1. Anderson, J, Cooper, L, Nass, M, Freiberger, W & Grenander, U, AAAS symposium, theoretical biology and bio-mathematics (1972).
2. Block, H D, Rev mod phys 1962, 34, 123.
3. Block, H D, Knight, B W, Jr & Rosenblatt, F, Rev med phys 1962, 34, 135.

4. Minsky, M & Papert, S, Perceptrons: An introduction to computational geometry. MIT Press (1969).

5. Longuet-Higgins, H C, Nature, London 1968, 217, 104.

6. — Proc roy roc London B 1968, 171, 327.

7. Creutzfeldt, O D & Ito, M, Exptl brain res 1968, 6, 324.

8. Goldstein, M, Jr, Hall, J L, II & Butterfield, B O, J acoustical society of America 1968, 42, 444.

9. Lettvin, J Y & Gesteland, R C, Cold Spring Harbor symp quant biol 1965, 30.

10. Mountcastle, V B, The neurosciences (ed G C Quarton, T Melnechuk & F O Schmitt) p. 393. Rockefeller University Press, New York, 1967.

11. Anderson, J A, Math biol-sci 1970, 8, 137.

12. — Ibid 1972, 14, 197.

13. Hubel, D H & Wiesel, T N, J physiol 1962, 160, 106.

14. Wiesel, T N & Hubel, D H, J neurophysiol 1965, 28, 1029.

15. Lashley, K S, Arch neurol psychiat, Chicago 1924, 12, 249.

16. Alterations in junction strengths proportional to f_j or the immediate dendrite response to f_j would also seem to be physiologically possible and in some situations might be useful. However, such modifications do not result in the various properties discussed here.

17. Such 'retrograde signalling' has also been postulated by J P Changeux & A Danchin. Private communication.

18. Hirsch, H V & Spinelli, D N, Exptl brain res 1971, 12, 509. See also Blakemore, C & Cooper, G F, Nature 1970, 228, 477.

17
Introduction

(1973)
Chr. von der Malsburg

Self-organization of orientation sensitive cells in the striata cortex
Kybernetik 14:85–100

In the early 1970s experimental data on the organization of cerebral cortex, particularly primary visual cortex (also known as area 17 or V1), were accumulating. Instead of being "equipotential" as Lashley had suggested, cortex was highly organized. In visual cortex it had been known for a number of years that there was a topographic map of the visual field onto the surface of cortex, with some intriguing systematic distortions, for example, a great overrepresentation of the foveal retina. However, closer examination revealed much more detail. Many easily recorded cells in the visual cortex respond preferentially to oriented line segments. Hubel and Wiesel (1962) showed that cortical cells were arranged in what they called "columns," perpendicular to the cortical surface. All the oriented cells in a single column would respond to the same orientation. More striking was later work (Hubel and Wiesel, 1968, 1974) indicating that orientation preferences of nearby cortical columns were highly correlated, and that preference varied smoothly across the surface of the ocrtex. (See Peters and Jones, 1985, for more detail.)

Along with the study of cortical organization, there was concurrently a series of dramatic experiments suggesting that cortical organization could be modified when an animal is raised in a particular environment. Hirsh and Spinelli (1970), in perhaps the prototypical experiment of this type, raised kittens wearing goggles so that they saw horizontal line segments in one eye and vertical line segments to the other. The kittens appeared to rewire their visual cortex so as to conform to the environment; that is, the cells connected to the horizontal eye only had horizontal receptive fields, and the cells connected to the vertical eye only had vertical orientation. There were no longer any binocular cells that responded to both eyes.

A common human visual pathology appears to arise from a mechanism of this type. If the images from the two eyes cannot be made to coincide because of defects in the extraocular muscles ("squint" or "wall eye") or if one eye is covered during development, the cortex appears to modify itself so that one eye is functionally disconnected from the cortex and becomes useless, even though it is biologically perfect. The cortex seems to suppress conflicting information by suppressing one eye. This problem is called *amblyopia* and is a common visual abnormality, found in several percent of the population. (For a review of some of the human and animal data, see Aslin, Alberts, and Petersen, 1981).

These results indicated that cortex was modifiable in predictable ways. Since there were a number of learning network theories being studied at this time, it seemed natural to see whether networks combined with synaptic modification rules could produce a model cortex that displayed modifiability and organization of the observed type.

One of the first attempts to do so is this paper from von der Malsburg. Technically, von der Malsberg incorporates a more complex, and more realistic, local anatomy into the kinds of learning systems we have already seen. There are excitatory and inhibitory cells that interact strongly, and there is significant local structure. Connection strengths are a function of distance between cells. Excitatory cells have a shorter range of connections than inhibitory cells. In the simulations, excitatory cells excite their neighbors, both excitatory and inhibitory. Inhibitory cells inhibit only the next-to-immediate-neighbor excitatory cells. This is a simplification of the center-surround organization prominent in the mammalian visual system; there is a ring of inhibition surrounding a central core of excitation. Von der Malsburg requires the local connectivity to organize the columns in his simulations.

Von der Malsburg assumed there was a model retina that projected randomly to the excitatory cortical cells. Initial connection strengths were also random.

Local network structure has not played much of a role in modeling up to this point. Previous network models (linear associator, most perceptrons, correlograph models) had either complete connectivities (i.e., everything connected to everything) or a uniform distribution of random connections, though some perceptrons had random connections over a restricted area. Structured local interactions are obvious and important in the brain. Von der Malsburg's model, which bases its assumptions on connection patterns found in the visual system, is a step in the direction of greater realism.

Only afferent (input) synapses on excitatory cells learn by a variant of a Hebb synapse. Coincident activation of the excitatory cell and its input causes an increment in strength of the connection. A renormalization rule is used, so that the total sum of strengths of afferent synaptic strengths remains a constant. This avoids the problem of unlimited growth of synaptic strengths, so when one connection is strengthened, the others are weakened. We have met a rule of this kind before in an early computer model (Rochester, et al., paper 6) and in one perceptron variant (Rosenblatt, paper 8).

The response of the neuron to its inputs is more complex than the simple summing element used up to now. The response of the cell is given by a nonlinear differential equation. The rate of change of output activity of the cell (table 1 in paper 17) is influenced by the usual weighted sum of synaptic inputs, but also includes a term containing the activity from the immediate past state, multiplied by a decay constant. Left to itself, the activity of the cell would exponentially decay to zero. More important, there is a threshold nonlinearity in the response of cells. There is a distinction made between the internal excitatory state of the cell and its output. The cell does not give an output if its excitation is below a certain value. It is no longer a simple summer, but is of such dynamic complexity that it is necessary to resort to computer simulation to study its behavior.

Von der Malsburg is now using a more complex model for the neuron than we have previously seen. Before, we have had simple threshold logic units, which computed an internal variable (say, the linear summation of synaptic inputs) and responded either on (above a threshold) or off (below a threshold). In the models of Willshaw, Buneman, and Longuet-Higgins (paper 12) or in the linear associator, the output of the unit was a faithful reflection of its internal state. Now, the two are combined. There is a threshold below which the output of the cell is zero, and above which the cell gives an analog output reflecting its internal state. Almost all network models now use some limiting function that has this property. The PDP (Parallel Distributed Processing) group (for example, see McClelland and Rumelhart, paper 25, or Rumelhart, Hinton, and Williams, paper 41) refers to the limiting function as a "squashing function"; that is, the output of the cell is kept within a limited range by a nonlinear relation between internal state and output. Grossberg refers to it as a "sigmoidal" nonlinearity (paper 19).

During the simulations, the system was presented with oriented stimuli, line segments, on its model retina. Von der Malsburg shows that the excitatory cells tend to develop "receptive fields" that look like the oriented line segments that the retina has been seeing. The cortical cells tend to develop "columns," i.e., groups of nearby cells that fire as a cluster. Also, and strikingly displayed in figures 12 and 13 of paper 17, adjacent cells tend to develop similar orientation preferences, and a track across the cortex sometimes displays a smoothly changing preferred orientation, which has some similarity to the actual experimental results.

This is an interesting paper for a number of reasons. It is one of the first to incorporate detailed cortical anatomy into its simulated network and to check the computer simulations of the model against physiological data. Overall, the results, while preliminary, were encouraging and suggested that it was possible to make some contact between theoretical networks and cortical organization.

References

R. N. Aslin, J. R. Alberts, and M. R. Petersen (1981), *Development of Perception*, Vol. 2: *The Visual System*, New York: Academic Press.

H. V. B. Hirsh and D. N. Spinelli (1970), "Visual experience modifies distribution of horizontally and vertically oriented receptive fields in cats," *Science* 168:869–871.

D. H. Hubel and T. N. Wiesel (1962), "Receptive fields, binocular interaction and functional architecture in the cat's visual cortex," *Journal of Physiology* 160:106–154.

D. H. Hubel and T. N. Wiesel (1968), "Receptive fields and functional architecture of monkey striate cortex," *Journal of Physiology* 195:215–243.

D. H. Hubel and T. N. Wiesel (1974), "Sequence regularity and geometry of orientation columns in the monkey striate cortex," *Journal of Comparative Neurology* 158:267–294.

A. Peters and E. G. Jones (1985), *Cerebral Cortex*: Vol. 3: *Visual Cortex*, New York: Plenum.

(1973)
Chr. von der Malsburg

Self-organization of orientation sensitive cells in the striata cortex

Kybernetik 14:85–100

Abstract

A nerve net model for the visual cortex of higher vertebrates is presented. A simple learning procedure is shown to be sufficient for the organization of some essential functional properties of single units. The rather special assumptions usually made in the literature regarding preorganization of the visual cortex are thereby avoided. The model consists of 338 neurones forming a sheet analogous to the cortex. The neurones are connected randomly to a "retina" of 19 cells. Nine different stimuli in the form of light bars were applied. The afferent connections were modified according to a mechanism of synaptic training. After twenty presentations of all the stimuli individual cortical neurones became sensitive to only one orientation. Neurones with the same or similar orientation sensitivity tended to appear in clusters, which are analogous to cortical columns. The system was shown to be insensitive to a background of disturbing input excitations during learning. After learning it was able to repair small defects introduced into the wiring and was relatively insensitive to stimuli not used during training.

I. Introduction

The task of the cortex for the processing of visual information is different from that of the peripheral optical system. Whereas eye, retina and lateral geniculate body (LGB) transform the images in a "photographic" way, i.e. preserving essentially the spatial arrangement of the retinal image, the cortex transforms this geometry into a space of concepts.

Within the last decade electrophysiology took the first steps into discovering the way in which the visual cortex performs this transformation. This paper is mainly concerned with the following features, which have been found in the primary visual cortex (area 17) of cat and monkey (Hubel and Wiesel, 1962, 1963, 1968).

1. There are neurones, which are selectively sensitive to the presentation of light bars and edges of a certain orientation (Hubel and Wiesel, 1962).

2. The neurones seem to be organized in "functional columns", i.e. the neurones lying within one cylinder vertical to the cortical surface are sensitive to the same orientation (Hubel and Wiesel, 1963).

3. Neighbouring columns tend to respond to stimuli of similar orientation (Hubel and Wiesel, 1963, 1968).

Although these findings are interesting in themselves, they will yield their full potential profit only if two questions are answered:

I. For what reason and to what end is area 17 organized in this way?

II. By which mechanisms are these neuronal properties determined?

Ad I: The fibres of the optic radiation are most sensitive to phasic changes of light intensity. Such intensity changes are brought about by eye movements, which scan the receptive fields of individual retinal and geniculate neurones over the light and dark contours of an image. Moving light edges and bars are therefore the most important stimuli which lead to geniculate output, i.e. cortical input. This may be considered one of the conditions for the existence of edge and bar detectors at the first cortical levels.

Ad II: The only mechanism proposed so far in the literature is genetical predetermination of the required circuitry (Hubel and Wiesel, 1963). This view has several disadvantages.

First, it would cost the system an immense volume of genetic information to tell all the terminal branches of the afferent axons with which cortical neurone they have to make contact.

Second, a rigid, genetically determined circuit would not have a high degree of plasticity. Such plasticity was demonstrated by experiments, in which the trigger features of visual cortical cells of young kitten could be determined in various ways by visual experience (Hirsch and Spinelli, 1970; Blakemore and Cooper, 1970; Blakemore and Mitchell, 1973).

Finally, plasticity, i.e. a process of self-organization should be possible in later stages of information processing by the brain, when it has to deal with situations not foreseen by nature.

© by Springer-Verlag 1973

The aim of this paper is to propose a mechanism of self-organization of the visual cortex which is able to explain in a simple way the facts 1 to 3 above and which also reduces the genetical problem to a reasonable level.

II. The Model

We describe now a model structure for the visual cortex and its specific afferents, which is in principle in accord with the known anatomical data and which has the minimum degree of complication required for the purpose of this paper.

a) The Elements

The model consists of a network of cells. The information transmitting signal is thought to be the discharge rate of the cells, averaged over a small interval of time. Thus the signal we employ here is a smooth function of time. This avoids the problem of artificial pulse — synchronization caused by the quantization of time, which is necessary in computer simulations. The cells make contact via connections (synapses). The connections can differ in weight, and are characterized by a number called strength of connection. The strength of connection from a cell A to a cell B will be denoted by p_{AB}. This term would include excitatory effects such as the sum of the postsynaptic potentials caused in a cell B by all the synapses of cell A on the dendrites and cell body of B. No assumptions are being made about the morphological variables which might determine the strength of connection. It may be different effectiveness or position of single synapses or merely a variable number of synapses between cells A and B (probably both). There are excitatory (E) and inhibitory (I) cells which have positive and negative strengths of connection respectively.

It is a simplification to characterize the connection between two cells by one number (or actually two numbers, as there are two directions, $A \rightarrow B$ and $B \rightarrow A$). In reality there is at least the complication of a variable synaptic delay and of different time-courses of excitation in cell B caused by one pulse of cell A. But the simplification of one number is sufficient for our purpose. The excitation per second caused by A in B is equal to the output of A (i.e. its firing frequency) times the strength of connection $A \rightarrow B$. This may, of course, be negative or positive.

It is furthermore assumed that the excitation and inhibition of one cell caused by all its presynaptic elements are summed linearly. The resulting quantity constitutes the input to the cell. The internal state of the cell is described by a quantity $H(t)$, which is called "excitatory state" (ES). As output signal of the cell we take that part of its ES, which exceeds a threshold. It shall be denoted by $H^*(t)$ (see Table 1). The cells influence each other via their output signals. The total input into a cell k is $\sum_l p_{lk} H_l^*(t)$, the p_{lk} being the strength of connection from cell l to cell k.

In the absence of input to a cell its ES decays exponentially with time. This can be described by the differential equation $d/dt\, H(t) = -\alpha H(t)$. The decay constant α is introduced to represent two phenomena, the decay of the postsynaptic potentials and the postexcitatory polarization following each action potential. A single decay constant is a crude but adequate approximation to these two precesses. The cells used here are very simple models of real neurones. The most conspicuous simplification is the linear dependency of the output signal on the input, as long as the threshold is exceeded. This implies in particular, that there is no intrinsic upper limitation to the output signal, as is imposed in reality by an absolute refractive period. Such an unlimited output can lead to instability in a network, which contains circular excitatory pathways. Therefore, in this model, instabilities can only be avoided by a properly devised inhibitory system within the network. In real systems there seems also to exist a limiting mechanism apart from the absolute refractive period, as cells rarely fire with maximum frequency for an extended period of time (above 100 msec).

b) The Wiring of the Model

The cells of the model form a two dimensional arrangement, the "cortical plane". Their distribution is uniform, E- and I-cells have equal density, although the relative proportions do not matter. The strength of connection between any two cells is a function $f(x)$ of their distance x. This function should be monotonically decreasing, e.g. bell-shaped. It is characterized by a range R and an amplitude A. There are functions $f(x)$ with different range R and amplitude A according to the different types of pairs of pre- and postsynaptic elements: $E \rightarrow E$, $E \rightarrow I$, $I \rightarrow E$ and $I \rightarrow I$, (see Fig. 1):

$f_{EE}(x)$ with amplitude $A_{EE} > 0$ and range R_{EE},

$f_{EI}(x)$ with amplitude $A_{EI} > 0$ and range $R_{EI} = R_{EE}$,

$f_{IE}(x)$ with amplitude $A_{IE} < 0$ and range $R_{IE} > R_{EE}$.

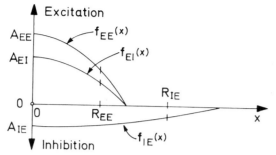

Fig. 1. Schematic representation of the dependency of the intra-cortical connection strengths on cell distance x. Explanation of symbols in the text

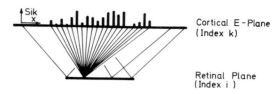

Fig. 2. Schematic drawing to show the organization of the afferents. The lower horizontal line represents the retinal plane or a cross section through the bundle of afferent fibres. The upper horizontal line represents a section of the cortical plane. The vertical bars symolize the different strengths of connection from one fibre i to many of the cortical E-cells k. The distribution of the heights of the bars is random. The connection of the other afferent fibres to one cortical cell are not shown

No connection is assumed here between the I-cells ($A_{II} = 0$). This assumption and the restriction to $R_{EI} = R_{EE}$ do not lead to a loss of features essential for this paper. The intracortical connections described will serve to organize columns. For this it is essential that R_{IE} be larger than R_{EE}, as will become evident later on.

How does this model cortex compare with the cortex found in higher vertebrates? Firstly, it has only two kinds of cells, E- and I-cells. We do not try here to identify them with two of the many classes of neurones described in actual cortex. This identification may, in fact, be impossible, as perhaps no single cell of the cortex has the field of innervation we look for. The model could in this case be saved by the existence of multicellular units: clusters of cells integrated by strong mutual excitation. The individual arborizations of these cells could then add up to give the postulated field of innervation. It was Colonnier, who introduced the concept of local fields of innervation to explain columnar organization. He also discussed histological evidence for this scheme (Colonnier, 1966).

It should be emphasized, that the model described here certainly corresponds to only a part of the real cortex; for example long-range excitatory connections within the cortex are not represented. Also there may be several systems like the one described here, which occupy the same space and which are weakly linked to each other.

c) The Afferent Organization

There is a set of afferent fibres which provide the input to the model cortex. The fibres have circular receptive fields within a small area of the retina. Where the retina is hit by the light of a stimulus, it switches its optical fibres to an active state, i.e. a state of constant firing. All the other fibres are silent. The model retina is thus again a simplification as only one type of cells (on-cells) without a center-surround organization and no off-cells are assumed. This simplification is possible, as only static light bars will be used as stimuli. Each afferent fibre projects to an area of the cortical plane and connects to all the E-cells within that area. A possible connection to I-cells is left out for the sake of simplicity.

Up to this point the wiring of the system is homogeneous: Each element is entirely equivalent to its neighbours of the same class, if one considers only relative coordinates. We now add an element of irregularity. Let s_{ik} be the strength of connection between the fibre i and the E-cell k. It is chosen to be an element of a set of random numbers (Fig. 2). This set was arbitrarily assumed to have uniform distribution in an interval $[0, s]$. No correlation is assumed between the numbers s_{ik} for different k. (We will introduce later a correlation between the s_{ik} for different values of i.) If $A_i^*(t)$ denotes the signal on the afferent fibre i, then the afferent input into the cell k is $\sum_i s_{ik} A_i^*(t)$.

A word has to be said about retinotopic organization here: It is well known, that there is a continuous mapping from the retina onto the visual cortex. Suppose, that this is also true for the model in the sense that the receptive field position on the retina and the geometrical centers of the fields of projection of the afferent fibres correspond to a continuous mapping. But the probabilistic distribution of connection strengths within the field of projection will lead to a random scatter in the "centers of gravity" of the fields of projection and the retinotopic organization will be upset on a small scale. For the present study one can forget retinotopic organization altogether since the fields of projection of neighbouring retinal cells overlap considerably. Suppose that the small

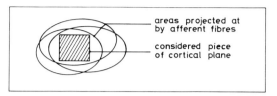

Fig. 3. View onto the cortical plane, showing the overlap of the fields of projection of different fibres. The hatched area lies entirely within the region of overlap and so the information about the exact positions of the receptive fields of the afferent fibres is lost within this area

piece of cortex considered here lies entirely within the region of overlap of the fibres coming from the small piece of relevant retina, as is illustrated by Fig. 3. Within this small cortical area all afferent fibres can then be regarded as equivalent in spite of their different position on the retina.

Several experimental findings on cat and monkey suggest, that also in reality retinotopical organization gives way to random scatter on a small scale: If one records during one cortical electrode penetration from successive neurones, one will find a large random scatter superimposed on the slow systematic displacement due to retinotopic organization (Hubel and Wiesel, 1962; Albus, 1973). In addition, if one maps the two receptive fields of binocular units, one finds a disparity in their positions which changes unsystematically from cell to cell (Joshua and Bishop, 1970). This suggests a still larger individual scatter in the course of single afferent fibres, as the position of the receptive field corresponds to an average position of the fibres constituting it. An additional argument is the rather irregular course of afferent fibres, seen on Golgi pictures (Ruiz-Marcos and Valverde, 1970; Ramon y Cajal, 1955, p. 613).

d) The Learning Principle

The system as it is described up to now is not yet able to explain the experimental facts stated in the introduction. For this, a process of self-organization is required, i.e. the system has to have the possibility to modify itself. This is done in the following way: if during a stimulation the afferent fibre i is active and if the stimulus leads to the firing of the E-cell k, then s_{ik}, the strength of connection between fibre i and cell k is increased by an increment Δs_{ik}. This corresponds to synaptic learning as it was proposed earlier in one form or another (Hebb, 1949; Uttley, 1970; Brindley, 1969; Marr, 1971). The learning principle as defined here leaves the question open by which mechanism

such type of synaptic learning may be brought about: by a chemical change within the existing synapses, by a change of their position, by an increase in the number of synapses or by a change of their dimensions. There are morphological data which support the last two alternatives (Cragg, 1968; Møllgaard et al., 1971).

The principle, as it was just stated, leaves one main problem. If s_{ik} is increased by a constant amount Δs at each time when a coincidence $i - k$ takes place, then this will lead to synaptic strengths which will grow forever and eventually will cause instability of the circuit. One way out would be to let the s_{ik} saturate: the increments get smaller and smaller as s_{ik} approaches a maximum value. For this model we choose a different solution, which is stated in the form of a learning principle:

if there is a coincidence of activity in an afferent fibre i and a cortical E-cell k, then s_{ik}, the strength of connection between the two, is increased to $s_{ik} + \Delta s$, Δs being proportional to the signal on the afferent fibre i and to the output signal of the E-cell k. Then all the s_{jk} leading to the same cortical cell k are renormalized to keep the sum $\sum_j s_{jk}$ constant.

This last step could correspond to the idea, that the total synaptic strength converging on one neurone is limited by the dendritic surface available. It means, that some s_{ik} are increased at the expense of others.

III. The Function of the Model

a) Basic Equations of Evolution

What functional states are there for the model network described in the last section and how will it be influenced by stimulation? To answer these questions we have to write down the equations, which gowern the evolution of the system. They are summarized in Table 1 (compare Grossberg, 1972). At first sight they look like linear differential equations. But H_k^* is a nonlinear function of H_k and this nonlinearity is essential, i.e. one cannot get rid of it by approximations. There are no mathematical ways to solve these equations in a closed form and that is the reason we had to do numerical calculations on a computer.

In this paper we are interested only in static stimuli, i.e. stimuli which are switched on for a moment and switched off again. Ideally the network's response will be an initial transient settling down to a

Table 1. Equations of evolution in time

$H_k(t)$	Excitatory state (ES) of cell k at time t
θ_k	Threshold of cell k
$H_k^*(t)$	Signal of cell k

$$H_k^*(t) = \begin{cases} H_k(t) - \theta_k & \text{if } H_k(t) > \theta_k \\ \text{zero} & \text{otherwise} \end{cases}$$

$A_i^*(t)$	Signal of afferent fibre i
N	Number of cortical cells
M	Number of afferent fibres
α_k	Decay constant of ES of cortical cell k
s_{ik}	Strength of connection between fibre i and cell k
p_{lk}	Strength of connection from cell l to cell k

$$\frac{d}{dt} H_k(t) = -\alpha_k H_k(t) + \sum_{l=1}^{N} p_{lk} H_l^*(t) + \sum_{i=1}^{M} s_{ik} A_i^*(t), \quad k = 1, \dots, N$$

Table 2. Stationary equations

E_k	ES of E-cell number k
I_k	ES of I-cell number k
E_k^*, I_k^*	Corresponding signals
N	Number of E-cells and number of I-cells
M	Number of afferent fibres

Strengths of Connections:

$p_{lk} > 0$	from E-cell l to E-cell k
$q_{lk} > 0$	from I-cell l to E-cell k
$r_{lk} > 0$	from E-cell l to I-cell k
$s_{ik} > 0$	from afferent fibre i to E-cell k

$$\text{a)} \quad E_k = \sum_{\substack{l=1 \\ l \neq k}}^{N} p_{lk} E_l^* - \sum_{l=1}^{N} q_{lk} I_l^* + \sum_{i=1}^{M} s_{ik} A_i^*$$

$$\text{b)} \quad I_k = \sum_{l=1}^{N} r_{lk} E_l^* \qquad \qquad k = 1, \dots, N$$

steady state, which lasts until the end of the stimulus period. As the details of the switching-on and -off periods are irrelevant for this paper, we will restrict our argument to the steady state, however short it might be in reality.

The specialization of the equation of Table 1 to the steady state, or $dH_k/dt = 0$ reads:

$$\alpha_k H_k = \sum_{l=1}^{N} p_{lk} H_l^* + \sum_{i=1}^{M} s_{ik} A_i^*, \quad k = 1, \dots, N.$$

Here one can divide by α_k and absorb the factor $1/\alpha_k$ on the right side into the definition of the coefficients p_{lk} and s_{ik} giving p_{lk}' and s_{ik}':

$$H_k = \sum_{l=1}^{N} p_{lk}' H_l^* + \sum_{i=1}^{M} s_{ik}' A_i^*, \quad k = 1, \dots, N.$$

These equations can be made more explicit, as according to the conventions of the model many of the p_{lk} and s_{ik} vanish and others are negative. By the introduction of more specialized symbols the final set of equations of Table 2 is obtained. (The primes are dropped again, as no confusion can arise.)

b) Specification of Details

Many of the definitions used in the description of the model above were unprecise, because too many details would have obscured the principle. These details have now to be specified. Some of these features will still be found to be oversimplified, but their special form was dictated by the necessity to economize computer time and space.

A hexagonal array was chosen for E- and I-cells (see Fig. 4). For every E-cell there is a corresponding I-cell occupying the same place. The hexagonal arrangement has the advantage of giving to each cell an almost circular surround of neighbouring cells. The total number of cells was chosen to be $2 \times 169 = 338$ cells, giving a network of a major diameter of 15 cells (Fig. 6). The threshold of all the cells was made equal to 1.

The wiring of the network is explained in Fig. 4: Each active E-cell directly excites the immediately neighbouring E-cells with strength p and the immediately neighbouring I-cells (including the one occupying the same place as the active cell) with strength r. Each active I-cell inhibits with strength q its next-to-immediate neighbours amongst the E-cells only.

These distributions are rather crude representations of the bell shaped curves of Fig. 1. The fact, that

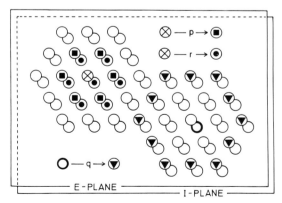

Fig. 4. A small part of the simulated cortex, showing the hexagonal array of the E-cells (upper plane) and the I-cells (lower plane). The different symbols are used to designate those cells which are connected with strengths, p, r and q. Every cell is connected with its neighbors in the same way (except at the borders)

Table 3. Numerical parameters (Definitions see text and Table 2)

$N = 169$	
$M = 19$	
$p = 0.4$	
$q = 0.3$	
$s = 0.25$	
$r = 0.286$	
$h = 0.05$	

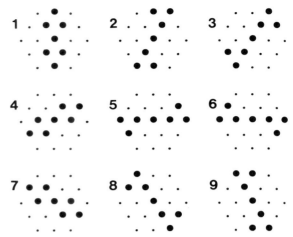

Fig. 5. The standard set of stimuli used on the model "retina". Large and small dots represent active and non-active fibres respectively

the I-cells do not inhibit their directly neighbouring E-cells has no major consequences on the function of the model, as was checked in separate calculations. It is a measure of economy.

The afferents are made up to 19 fibres. Each fibre is thought to fan out into 169 branches to contact the 169 E-cells. Correspondingly, a matrix of 19×169 numbers representing the strengths of connection had to be specified. This matrix was derived from a set s'_{ik} ($i = 1, ..., 19$; $k = 1, ..., 169$) of random numbers with a flat distribution within an interval $[0, s]$. (For the values of s and all the other numerical parameters see Table 3.)

With the numbers s'_{ik} the total afferent synaptic strength s'_k leading to the E-cell k can be written $s'_k = \sum_{i=1}^{19} s'_{ik}$. This number has to be a constant during

learning. For the sake of simplicity, all the s'_k were changed to their mean value, which is $19 \cdot s/2$. This was done by renormalizing the s'_{ik} according to:

$$s_{ik} = s'_{ik} \cdot 19 \cdot \frac{s}{2} / s'_k, \quad i = 1, ..., 19; \quad k = 1, ..., 169.$$

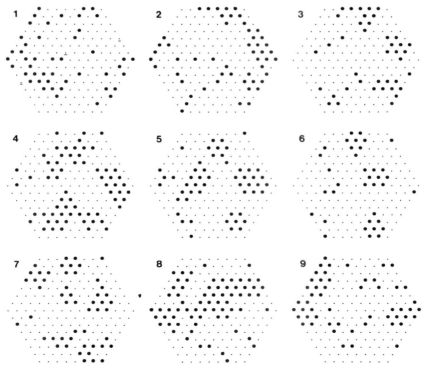

Fig. 6. The reactions of the cortical E-cells to the stimuli of Fig. 5. Large dots represent firing cells. The numbers to the upper left correspond to those of Fig. 5

The s_{ik} thus derived are the starting values of the afferent strengths of connection. All the simulations of this paper are based on the same initial set of synaptic strengths.

In each learning step the s_{ik} values are increased to

$$s'_{ik} = s_{ik} + h \cdot A_i^* \cdot E_k^*$$

after the stimulation. Then the s'_{ik} are renormalized in the way just described to give the s_{ik} of a next generation.

The stimuli always consisted of seven active afferent fibres, i.e. seven of the 19 A_i^* in equation a) of Table 2 were set to 1, the others to zero.

There was a standard set of nine stimuli, which was employed mainly. This set is shown in Fig. 5. It was chosen to represent light bars of different orientation.

It should however be emphasized that in the absence of retinotopical organization the important property of the stimuli in Fig. 5 is not their geometrical arrangement but rather their relationships established by mutual overlap.

c) The Procedure of the Numerical Calculations

The solutions to the equations of Table 2 were found by numerical calculation on a UNIVAC 1108. The method we employed was stepwise approximation by an iterative procedure. That means, the equations we really used were those of Table 1, the different steps of the approximation corresponding to their solution at consecutive time steps. Not every set of parameters p, q, r and s lead to stable solutions. Those finally employed were found partly by trial and error.

The solution is not reached in a monotone and quick way. The ES of the cells will first follow a course of damped oscillations and then approach slowly their final values. To save time, this slow approximation was stopped after 20 iterative ("time"-) steps and the result taken as approximate solution to the equation of Table 2. By then the change from step to step in the ES of most cells was smaller than 0.5%. After the solution was found, learning was done by the described manipulations on the s_{ik}. All the other parameters were held fixed.

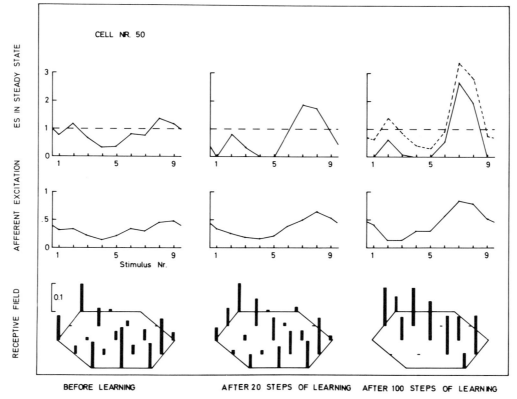

Fig. 7. Receptive field organization, afferent excitation and ES of the E-cell No. 50 in the steady state response to the nine stimuli. Left, middle and right column correspond to the system without learning, with 20 steps and with 100 steps of learning, respectively. The heights of the vertical bars in the hexagon in lower row represent the connection strengths s_{ik} of the 19 retinal fibres to the cell. Their arrangement corresponds to Fig. 5. The position of this cell is underlined in the first diagram of Fig. 6 in the fifth line from the top

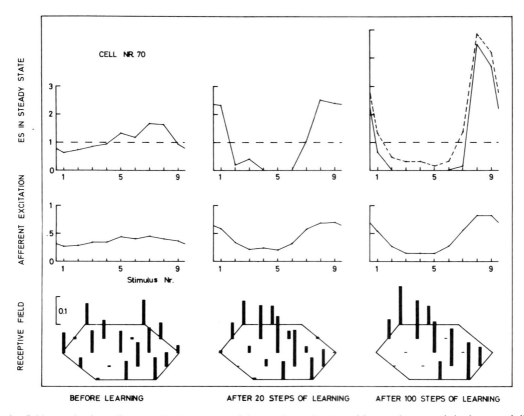

Fig. 8. Receptive field organization, afferent excitation and ES of the *E*-cell No. 70. Its position on the cortex is in the seventh line from the top (see Fig. 6, 1). For explanation see Fig. 7

d) The Results without Learning

Figure 6 shows in a qualitative way the reaction of the network to the nine stimuli (Fig. 5), before any learning took place. The small and large dots represent *E*-cells with ES below and above threshold respectively. The *I*-cells are not shown. As can be seen, the cells have already the clear tendency to fire in clusters. To get a more quantitative impression of the reaction of the cells consider the left column of diagrams in Figs. 7, 8 and 9, which summarize the behaviour of three typical cells, Nos. 50, 70 and 120 (for their positions see Fig. 6). The vertical bars in the hexagon in the bottom row show the connection strengths s_{ik} of the 19 afferent fibres to the cell. Their hexagonal arrangement corresponds to the one in Fig. 5. For each stimulus the afferent excitation (which corresponds to the sum of seven of the bars) is plotted in the middle row of Figs. 7–9 against stimulus number. Notice, that the points in these diagrams tend to form a continuous line, i.e. neighbouring points are correlated. This is a consequence of the retinal overlap between neighbouring stimuli.

The upper graphs of the left columns in Figs. 7–9 show the ES of the particular cells in response to the nine stimuli. This plot could also be called the cells orientational tuning curve. Its details are determined roughly by the afferent excitation, although it is modified by what goes on in the cortical neighbourhood.

Table 4. Classification of cells according to reaction type

a) Classification of tuning curves

	No response	Unimodal	Multimodal
Before learning	12	87	70
20 steps of learning	43	118	8
100 steps of learning	21	147	1

b) Width of unimodal tuning curves (*n* is the number of neighbouring stimuli to which the cells responded)

n	1	2	3	4	5	6	7
Before learning	20	24	18	19	5	—	1
20 steps of learning	24	19	45	25	5	—	—
100 steps of learning	8	43	64	25	7	—	—

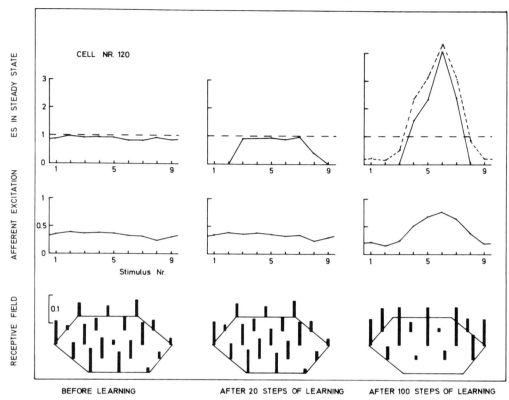

Fig. 9. Receptive field organization, afferent excitation and ES of the *E*-cell No. 120. Position in the eleventh line from the top (Fig. 6, 1). For explanation see Fig. 7

A summary of the main features of the 169 tuning curves is shown in Table 4. Twelve cells never reach threshold. Seventy of them react to stimuli in a multimodal fashion, i.e. they fire within separate regions. This is exemplified by the cell in Fig. 7 (left column). A large fraction, eighty-seven, of the cells, could already be called orientation sensitive, as their tuning curves are unimodal, although not very sharp (e.g. cell number 70, Fig. 8). In summary one can say, that although there is no systematics in the organization of the afferents, we already get a tendency to firing in clusters ("columnar organization") and a considerable fraction of the cells (51%) with unimodal orientation tuning ("orientation specific units"), although the tuning curves may still be comparatively flat.

e) The Results after a Learning Phase

A step of learning consists of one presentation of all the nine different stimuli and the learning manipulations on the s_{ik} subsequent to each presentation.

During the learning phase the sequence of presentation of the stimuli was 1, 6, 2, 7, 3, 8, 4, 9 instead of the natural one (1, 2, 3, …). This was to avoid special effects resulting from the consecutive presentation of two maximally overlapping stimuli.

The behaviour of the system after 20 steps of learning is shown in Fig. 10. The cells now show a much stronger tendency to fire in clusters than they did before learning (compare Fig. 6). This can be interpreted in the following way: the intracortical connections give the cells a natural tendency to fire in separate clusters. At first the cells are disturbed in this tendency by the afferent excitation, which at the beginning does not favour clusters at all. Nevertheless the ES of the cortical cells will be highest, if they can exchange maximal intracortical excitation and minimal intracortical inhibition, which is the case with firing in patterns of small clusters. Now, the higher the ES of a cortical cell, the stronger will be its influence on the afferent organization via learning. Finally those patterns of afferent organization will persist, which favour cortical clusters.

If the statistics of the afferent stimulations are stationary, the learning process in the system will saturate after some time. This saturation is not yet reached after 20 learning steps. Therefore the reactions

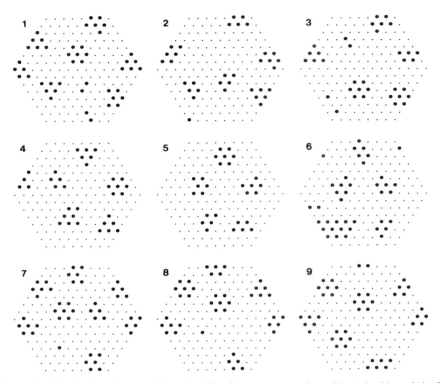

Fig. 10. Reaction of the cortical *E*-cells after 20 steps of learning. The figure corresponds to Fig. 6. In this and the following figure there is no learning between the nine stimulations shown

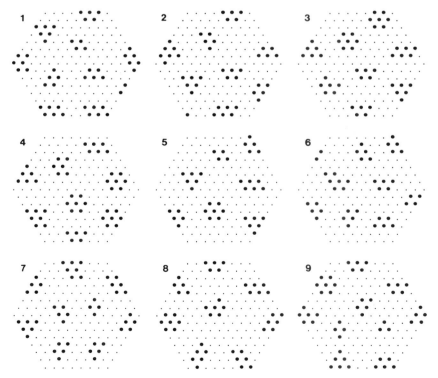

Fig. 11. Reaction of the cortical *E*-cells after 100 steps of learning

of the system continue to change. Figure 11 shows the responses of the E-cells after 100 steps of learning, of which the last 40 were accelerated by doubling the learning constant h (its value being 0.1 instead of 0.05).

To see the effect of learning on the level of the single cells, consider the middle and right columns in Figs. 7–9. In many cases the tuning curves are now much steeper, the cells being either strongly excited or little excited (e.g. Figs. 7 and 8). This behaviour is more pronounced with more learning (see right columns). The cell of Fig. 7 had at first two sensitive regions, one of which disappeared within the first 20 steps of learning. In the case of Fig. 9 there was no reaction of the cell to any of the stimuli up to learning step 20. Later on, however, it was occasionally excited to the firing level by its neighbours. This led to a sudden modulation of its tuning curve, which is very steep after 100 learning steps.

The changes in the tuning curves of the cells with learning are the result of positive feedback: Whenever a cell fires, it will strengthen those afferent connections which were exciting the cell. This leads to increased afferent excitation of the cell, when the same stimulus is applied the next time. Increased afferent excitation will in its turn make the cell fire more strongly in response to the stimulus, and this will lead to accelerated learning as long as saturation is far. The corresponding modifications of the afferent organizations and of the excitation curves are apparent in Figs. 7–9, bottom row. In the end, only those s_{ik} persist which are used by the effective stimuli (compare Fig. 5 for the stimuli). The decrease of the unused connection strengths is a consequence of the condition that their sum be constant.

The chain of positive feedback described in the last paragraph is modified by the intracortical excitation and inhibition, which are added to the afferent excitation. This is the point, where intracortical dynamics enters and leads to the clustering of firing.

The behaviour of all the E-cells with learning is summed up in Table 4. It shows, that less and less cells have multimodal tuning curves (70 before learning, 8 after 20 steps and one after 100 steps). Note also, that very sharp tuning curves, i.e. reaction of a cell to only a single stimulus, seems not to be favoured by the system after extended learning (24 after 20 steps and only 8 after 100 steps).

The broken lines in the upper right graphs of Figs. 7–9 show the excitation of the cells alone, no inhibition being subtracted. The inhibition is the difference between the two curves of each graph. As one can see, the tuning curves are made a bit

narrower by the inhibition. In 20 cases a second separate sensitive region of a cell was suppressed by the inhibition, as in the example of the cell of Fig. 7. Therefore in this model the inhibition takes part in the organization of the "cortex" in an essential way (apart from its stabilizing function), although it is homogeneously distributed and not modified by learning at all.

Two of the three aims we set for the model in the introduction of this paper are now accomplished: First, clusters of cortical activity are brought about by intracortical dynamics. Second, organization of orientation specific units is brought about by a learning strategy rather than by genetical determination.

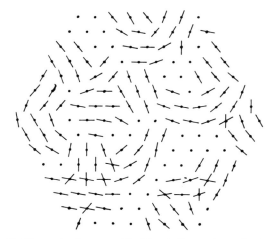

Fig. 12. View onto the cortex. Each bar indicates the optimal orientation of the E-cell (for definition see text). Dots without a bar are cells which never reacted to the standard set of stimuli. Two bars indicate two separate sensitive regions

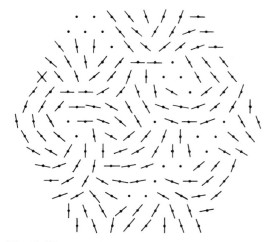

Fig. 13. View onto the cortex after 100 steps of learning

The third task was an explanation of the fact, that neighbouring cells have the tendency to react to stimuli of neighbouring orientation (Hubel and Wiesel, 1963, 1968). This now turns out to be a natural consequence of the existence of clusters and of their influence on the afferent synaptic strength, as can be seen directly by inspection of Fig. 12.

Each bar in this figure indicates the median of the orientations to which the corresponding cell responded. If, for instance, a cell fired in response to stimuli 1, 2 and 3, the orientation corresponding to stimulus 2 is plotted, regardless of the magnitude of the three answers. If the cell fired in response to stimuli 1 and 2, the plotted orientation lies halfway between those for 1 and 2. Two crossing bars indicate a "bimodal" reaction of the cell, i.e. responses to two orientations separated by one or more uneffective orientations.

It can be seen that the probability of similar orientations in adjacent cells is high. This tendency is emphasized in Fig. 13, which shows the optimal orientations after 100 steps of learning. There are even series of cells with continuously turning orientations (e.g. the seventh line in Fig. 13) as described in the literature (Hubel and Wiesel, 1968).

f) The Effect of Non-Standard Stimuli

An important question one can pose now is how the trained system will react to stimuli which it does not know yet. To answer this question, the system was tested with 45 different stimuli m_i, $i = 1, ..., 45$. As the standard stimuli n_k, $k = 1, ..., 9$ of Fig. 5, the m_i consisted of seven retinal points each. The m_i were charaterized by the maximal overlap $V_{i\max}$ they had with any of the n_k:

$$V_{i\max} = \max_{k=1,\,...,\,9} (m_i \cap n_k).$$

The m_i were chosen to form five groups of equal $V_{\max}$, containing nine stimuli each. $V_{\max}$ varied from 2 in the first group to 6 in the fifth group. As the model retina is so small, no stimuli with $V_{\max} = 1$ could be found. Within one group the stimuli were chosen to be as different as possible.

To judge the effect of a stimulus on the system, the mean output signal E of the E-cells was computed:

$$E = \frac{1}{N} \sum_k E_k^*.$$

In Fig. 14 the average of E for the nine stimuli in one group is plotted against the maximum overlap $V_{i\max}$ with the standard stimuli. The overlap 7 means the standard set itself. The flat curve, ($\circ - \circ$) is the effect of the stimuli on the "naive" network, before learning of any stimuli took place. No significant

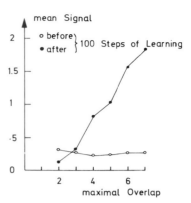

Fig. 14. The mean output signal of the E-cells in response to different sets of stimuli. The stimuli of each set have a maximal overlap with stimuli of the standard set. This overlap is calculated as number of fibres the stimuli have in common and it is shown on the abscissa. An overlap of seven corresponds to the standard set

differences between the groups are found. The steeper curve ($\bullet - \bullet$) shows the effect of the different stimulus sets after the network went through 100 steps of learning the standard stimuli, which, in Fig. 14, corresponds to the group of overlap 7. The stimuli most similar to the stimuli used for training are clearly favoured. The response becomes smaller the less similar the test stimuli are to the standard set. In the case of overlap 2 there is even a suppression in comparison with the sensitivity before training. This indicates, that the system responds less to "new" stimuli once it has been tought a given set.

It should be noted, that the mean output signal of the cells increased after learning (e.g. from 0.25 to 1.8 for the standard stimuli in Fig. 14). A more realistic model should be able to keep the mean excitation of the network constant in spite of learning. This could be done by an adapting inhibitory system. The trained network would then actively suppress the response to all stimuli, with which it was not trained. With this modification, even more than in the present model, the network could be regarded as an effective filter.

g) The Sensitivity to Nonspecific Input

Up to now the task of the model was fairly simple: a set of stimuli, which could be characterized by one parameter (their orientation) was presented. The problem for each cortical cell was to become selectively sensitive to a small range of the stimulus parameter only. In reality there are certainly different sources of perturbations:

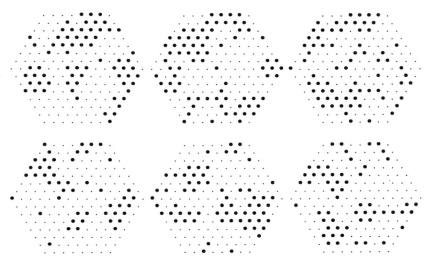

Fig. 15. Six reactions of the E-cells to the same stimulus (1 of Fig. 5) before learning. The differences are produced by extra random input

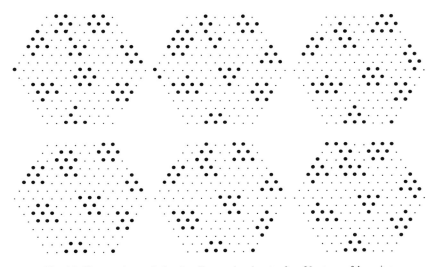

Fig. 16. Six reactions of the E-cells to stimulus 1 after 20 steps of learning

1. There is a very large number of different stimuli, which make up a large percentage of the information flow, but each individual stimulus occurs so rarely that no cortical cell would get specialized to it.

2. Those stimuli occurring frequently have small variations of composition, which can be regarded as a disturbance.

3. There is a nonspecific input to the cells, which is not related to the visual information.

To test the ability of the model to work in the presence of perturbations the following experiment was done. A random number t_k was added to the specific afferent excitation $\sum_i s_{ik} A_i^*$ (see Table 2) received by cell number k ($k = 1, \ldots, 169$). There was a new set of random numbers for every stimulation.

The first generation of the s_{ik} was chosen from the interval [0, 0.175]; the t_k from the interval [0, 0.525], three times as large. Consequently the mean of the expressions $\sum_i s_{ik} A_i^*$ is 0.613 ± 0.095 (root mean square deviation) and the mean of the t_k is 0.263 ± 0.153. Their sum, $\sum_i s_{ik} A_i^* + t_k$, has a mean of 0.875 (as before) and its r.m.s. deviation is ± 0.180, which is largely determined by the perturbation t_k.

The most important feature of the afferent excitatory input to the cortical cells is its differentiation between different cells during one stimulation and for one cell between different stimulations. These differences have to be detected and enhanced by learning. However in the present experiment they are completely buried under the differences produced by the perturbative random excitation t_k.

How serious the perturbation is can be assessed from Fig. 15, which shows the reactions of the untrained system to six times the same stimulus (number one of Fig. 5). Without the perturbation the reactions would be identical. After 20 steps of learning (involving nine stimuli each and with $h = 0.1$, compare Table 3) the picture is quite different: Fig. 16, also six presentations of stimulus one, shows much less variations.

To measure the decrease in variability from Fig. 15 to Fig. 16 by a number, the entropy of these variations was calculated. If E-cell k fired n times during m presentations of the same stimulus, then the probability of k to react can be defined as $p_k = n/m$, and h_k, the corresponding entropy can be calculated

$$h_k = -p_k \operatorname{ld} p_k - (1 - p_k) \operatorname{ld}(1 - p_k)$$

(ld is the binary logarithm). If k fires half the time, then the entropy is maximum, $h_k = 1.0$. The mean entropy of all the E-cells then is

$$H = \frac{1}{N} \sum_k h_k ,$$

where N is the number of cells.

The entropy of the variability of reactions to one and the same stimulus before learning (Fig. 15) is $H = 0.674$ and after 20 steps of learning (Fig. 16) it is $H = 0.203$, considerably reduced.

In conclusion one can say: If there is a systematic structure in the information arriving at the cortical cells, it can be detected and enhanced by the learning system even if it is buried in nonstructured, random excitation.

h) Redundancy of Information Storage

It has been demonstrated, that the connectivity between two points of nervous tissue can change after simultaneous stimulation of these points (e.g. Bliss and Gardner-Medwin, 1970). These changes may be interpreted as synaptic learning. Unfortunately they were never longlasting and after some hours or sometimes days the connectivity had decayed to its previous value. This fact, which was also found in other preparations, has always been a serious argument against the interpretation of synaptic conditioning as a basis for permanent memory.

With our model we performed an experiment, which may be relevant to this question. In the system as it was left after 100 steps of learning, twelve

arbitrarily chosen afferent fibres were strengthened by increasing the corresponding synaptic strengths s_{ik} to triple value, and the synaptic input to the cortical cells was renormalized to keep their total synaptic input constant (s. Section IIIb). This reduced the increased numbers s_{ik} to a somewhat lower value. Then the system was allowed to learn for 40 more steps (with $h = 0.1$). It was found, that most of the connection strengths were brought back to a level, which was close to the value before the added increase. Two of the increased s_{ik} were connected to cells which never reacted before, during or after this experiment. Consequently no learning took place and these s_{ik} stayed high. The sum of the remaining 10 s_{ik} was 0.963 before the experiment. After the increase and renormalization it was 2.351, and after the 40 steps of learning it was back to 1.026, although saturation had not yet been reached.

The explanation of this result is that the characteristics of a pattern, to which a cell responds, are determined by all the effective connections leading to the cell. If the strength of only one of these connections is changed, the optimal pattern of the cell will not be changed very much. This one connection will then readapt by the process of learning to its previous value.

The experiment shows, that the system has enough redundancy in its information storage to make it insensitive to such small defects as an arbitrary change in the strength of some connections. This kind of argument could be able to explain the failure of the experiments mentioned above: The conductivity changes artificially produced in the experiment may be considered as disturbances of the normal function of the neural structure and are therefore "repaired" by the nervous system.

IV. Discussion

It was the aim of this paper to show, that there is at least one way to explain a large part of the functional organization in the visual cortex of cat and monkey without depending completely on a genetically predetermined connectivity between the cortex and its afferent fibres.

Most of the principles used here have been described before in the literature: the form of intracortical connectivity (Fig. 1) (Colonnier, 1966; Wilson and Cowan, to be published), mechanisms of synaptical conditioning (Hebb, 1949; Rosenblatt, 1961; Brindley, 1969; Grossberg, 1972; Uttley, 1970; Marr, 1971), and random connections (Beurle, 1956; Rosenblatt, 1961; Marr, 1971). The equations used here are very similar to those of (Grossberg, 1972).

a) The Structure of the Model and Generalizations

This paper proposed two principal mechanisms: the development of pattern sensitive cortical cells by a self-organizing process involving synaptic learning, and the arrangement of functional columns as a consequence of intracortical connections rather than due to a predetermined distribution of afferents.

The two main directions of local intracortical fibre systems are tangential and vertical to the cortical surface. The model takes these two directions into account in a simplified way. The vertical connections are implicit in the representation of all the cells of one vertical cylinder by only two cells, an excitatory and an inhibitory. The underlying assumption is that all cells of this cylinder are so strongly connected by vertical fibres, that they fire virtually simultaneous under most conditions.

The horizontal connections, on the other hand, have been represented explicitly. In contrast to the histological picture they spread symmetrically in all directions. This can be justified by the fact, that each connection between two functional units represents an average over many fibres, which connect a multitude of individual cells. The firing of cortical cells in clusters is due to the excitatory horizontal connections. As a consequence of the homogeneity of intracortical connection the borders between such clusters are not fixed and may shift slightly from one stimulation to the next. The inhibitory and excitatory interaction between cortical neurones makes it necessary to think in terms of collective networks rather than of isolated neurones. This means, that the functional behaviour of a cortical neurone is not only a function of its receptive field in terms of its afferent input, but also of its intracortical connections.

To organize orientation specificity of the cortical cells a mechanism of adaptation, namely synaptic conditioning, is introduced. This mechanism is applied to a network with nonspecific, random interconnections and is able to transform it into a highly specialized system. Some of the special features of the "learning principle" employed here need discussion.

The total sum of synaptic strength converging onto one cell was kept constant: while some synapses grew stronger, others became weaker. This was introduced for several reasons. One is stability: a system with only growing excitatory synapses is unstable. A second reason is the requirement of high specificity of the cells: they should become insensitive to all stimuli for which they were not trained. Both of these functions can probably also be realized by replacing the principle of constant

synaptic strengths by additional training of inhibitory connections.

Some of the limitations of the proposed model are caused not by the underlying principles but rather by the restricted number of cells and connections and the highly restricted sensory inputs used in the computer simulations. More cells would give the cortex more degrees of freedom and it could adapt to a greater range of stimuli. An obvious generalization would be the inclusion of moving stimuli. The important stimulus property selected for would then no longer be retinal position but rather a temporal sequence of positions. Another generalization would be the organization of a hierarchical system of feature detectors.

b) Comparison with Experiments

The learning principle was applied only to synapses of the afferent fibres. This may be taken as a special case of learning in early development such as the "learning" of binocularity (Wiesel and Hubel, 1965) or orientation tuning (Blakemore and Cooper, 1970).

Many neurones in the visual cortex of very young kittens are orientation sensitive before any visual experience has occurred, but they are not so sharply tuned as in adult animals (Hubel and Wiesel, 1963; Pettigrew, 1972). In the light of these findings it is interesting to note that also in the proposed model a high percentage of orientation sensitive cells is found before any training took place (Table 4). Narrowly tuned neuronal orientation specificity found during experiments on unexperienced kittens may be a consequence of fast learning during the experiments, a property also shown by the model (see below).

The model leads to one important "prediction". If only a restricted set of stimuli is presented during the training period, the cortical neurones will specialize to these stimuli and will thereafter become insensitive to all other stimuli. This corresponds to experimental findings on training of young animals: if kittens are raised in an environment consistling entirely of horizontal or vertical stripes, they become virtually blind for contours perpendicular to the orientation they had experienced during the training period (Blakemore and Cooper, 1970). When testing the visual cortex cells of such animals a highly significant anisotropy in the distribution of preferred orientations was found. It was later shown that an exposure time as short as one hour during a sensitive period was sufficient to induce the anisotropy of preferred orientations (Blakemore and Mitchell, 1973). This compares well with the quick convergence of the self-organization in the model network, which became relatively insensitive to untrained stimuli after as little as a hundred presentations of the set of training stimuli. This phenomenon like the corresponding experiments on unexperienced kittens may be an analogue to imprinting rather then learning in the definition of ethologists.

Recent experiments in this laboratory make it doubtful, that many geniculate fibres, the receptive fields of which are arranged in a line parallel to the optimal orientation, converge on individual cortical "simple" cells. The excitatory input of such cells appears to have a receptive field the form and size of which correspond to those of individual retinal ganglion cells rather than to lines (Benevento, Creutzfeldt and Kuhnt, 1972). It seems that the temporal sequence of retinal excitation, i.e. the stimulus movement is a more important aspect of cortical organization. In order to test this, the dynamic aspects rather than steady state conditions may have to be investigated with our model.

The results of this theoretical study are encouraging as they show, that such simple assumptions about intracortical connections and the mechanism of synaptic conditioning as made in this model proved to be sufficient to explain some of the most striking functional properties of the visual cortex.

Acknowledgements. The author would like to thank Professor Dr. Otto Creutzfeldt for providing excellent working conditions and for helpful suggestions, Dr. H. Wässle for critical remarks on the manuscript and Dr. Jean Ennever for correcting my English. The numerical calculations have been done on the UNIVAC 1108 computer of the Gesellschaft für wissenschaftliche Datenverarbeitung, Göttingen.

References

Albus, K.: Topology of orientation sensitivity in the cortical areas 17 and 18 of the cat. Pflügers Arch. Suppl. to **339**, R 91 (1973)

Benevento, L. A., Creutzfeldt, O. D., Kuhnt, U.: Significance of intracortical inhibitions in the visual cortex. Nature (Lond.) **238**, 124—126 (1972)

Beurle, R. L.: Properties of a mass of cells capable of regenerating pulses. Phil. Trans. Roy. Soc. (Lond.) B **240**, 55—94 (1956)

Blakemore, C., Cooper, G. F.: Development of the brain depends on the visual environment. Nature (Lond.) **228**, 477—478 (1970)

Blakemore, C., Mitchell, D. E.: Environmental modification of the visual cortex and the neural basis of learning and memory. Nature (Lond.) **241**, 467—468 (1973)

Bliss, T. V. P., Gardner-Medwin, A. R.: Long-lasting increases of synaptic influence in the unanaesthetized hippocampus. J. Physiol. (Lond.) **216**, 32—33 P (1971)

Brindley, G. S.: Nerve net models of plausible size that perform many simple learning tasks. Proc. Roy. Soc. (Lond.) B **174**, 173—191 (1969)

Colonnier, H. L.: Structural design of the neocortex. In: Eccles, J. C. (Ed.): Brain and conscious experience, p. 1—21. Berlin-Heidelberg-New York: Springer 1966

Cragg, B. G.: Are there structural alterations in synapses related to functioning? Proc. Roy. Soc. B **171**, 319—323 (1968)

Grossberg, S.: Neural expectation: cerebellar and retinal analogs of cells fired by learnable or unlearned pattern classes. Kybernetik **10**, 49—57 (1972)

Hebb, D. O.: Organization of Behaviour. New York: John Wiley 1949

Hirsch, H. V. B., Spinelli, D. N.: Visual experience modifies distribution of horizontally and vertically oriented receptive fields in cats. Science **168**, 869—871 (1970)

Hubel, D. H., Wiesel, T. N.: Receptive fields, binocular interaction and functional architecture in the cat's visual cortex. J. Physiol. (Lond.) **160**, 106—154 (1962)

Hubel, D. H., Wiesel, T. N.: Receptive fields of cells in striate cortex of very young, visually inexperienced kittens. J. Neurophysiol. **26**, 994—1002 (1963)

Hubel, D. H., Wiesel, T. N.: Receptive fields and functional architecture of monkey striate cortex. J. Physiol. (Lond.) **195**, 215—243 (1968)

Joshua, D. E., Bishop, P. O.: Binocular single vision and depth discrimination. Receptive field disparities for central and peripheral vision and binocular interactions on peripheral single units in cat striate cortex. Exp. Brain. Res. **10**, 389—416 (1970)

Marr, D.: Simple memory. Phil. Trans. Roy. Soc. (Lond.) B **262**, 23—81 (1971)

Møllgaard, K., Diamond, M. C., Bennett, E. L., Rosenzweig, M. R., Lindner, B.: Quantitative synaptic changes with differential experience in rat brain. Int. J. Neurosci. **2**, 113—128 (1971)

Pettigrew, J. D.: The importance of early visual experience for neurones of the developing geniculostriate system. Invest. Ophthal. **11**, 386—392 (1972)

Ramon y Cajal, S.: Histologie du système nerveux, Vol. II. Madrid: Consejo Superior de Investigationes Cientificas, Instituto Ramon y Cajal 1955

Rosenblatt, F.: Principles of neurodynamics: Perceptrons and the theory of brain mechanisms. Washington D.C.: Spartan Books 1961

Ruiz-Marcos, A., Valverde, F.: Dynamic architecture of the visual cortex. Brain Res. **19**, 25—39 (1970)

Uttley, A. M.: The informon: a network for adaptive pattern recognition. J. theor. Biol. **27**, 31—67 (1970)

Wiesel, T. N., Hubel, D. H.: Comparison of the effects of unilateral and bilateral eye closure on cortical unit responses in kittens. J. Neurophysiol. **28**, 1029—1040 (1965)

18

Introduction

(1975)
W. A. Little and Gordon L. Shaw

A statistical theory of short and long term memory
Behavioral Biology 14:115–133

This paper by Little and Shaw describes a neural network model that makes effective use of some ideas that have become prominent in recent years. In particular, they use a probabilistic, rather than deterministic, model of the neuron and assume that memory for short time periods is bound up with the development of stable states in the nervous system.

The paper describes an integrated and subtle theory. Little and Shaw start by approximating a neuron as a two-state element, either firing or not firing an action potential. However, instead of describing the threshold behavior of the neuron by a fixed threshold, as was done in previous work, they observed that there are many sources of noise in the real neuron; for example, there is present a Poisson process governing the number of quanta (synaptic vesicles) released when a synapse is activated, as well as a spontaneous quantal discharge rate. Quantal transmitter release is not identical from event to event. Therefore, given a certain depolarization, the chance of an action potential actually occurring is a function of both the threshold of the cell and the internal variability. The cell could fire when the depolarization is below threshold, though it is less likely to fire then. Or it might be silent when it is above threshold, though it is more likely to fire then. A probabilistic neuron model is important in later work, for example, in Boltzmann models (Ackley, Hinton, and Sejnowski, paper 38).

Suppose the system is in a certain state at one time. Then it is possible to calculate the *probability* of going to another state, given the synaptic weights and estimates of the internal variability. If there are N neurons, each of which can be on or off, then there are 2^N possible states. We can then consider a very large matrix (2^N dimensions) of probabilities, P, which gives the transition probabilities between all different states. An element in the matrix P, $P_{\alpha\beta}$, gives the transition probability from state α to state β. A state will be long lasting if the chance of a transition between a state and itself, that is, a diagonal term $P_{\alpha\alpha}$, is close to one. The closer to one $P_{\alpha\alpha}$ is, the longer the duration of the state. Little and Shaw show that such long lasting states exist, and that there may be a number of them comparable to the number of neurons in the system.

Once they have shown that long lasting states exist, the next problem is to make the stable states take the desired form, that is, act as a memory for particular patterns. They suggest two ways to do this. The first is by manipulation of the threshold of firing of the cell. The second is by use of the familiar Hebb synaptic modification rule. Little and Shaw argue that Hebbian modification of the active elements in a state α gives rise to an increase in the diagonal element $P_{\alpha\alpha}$; that is, the duration of the state will lengthen. Therefore Hebbian modification tends to make the system develop meaningful long lasting states in response to appropriate input patterns. This model is similar in character to later attractor models such as Hopfield's (paper 27).

Hopfield, Little, and Shaw are physicists. Their metaphors for the behavior of the nervous system are taken from solid state physics, in particular, a system of atomic spins in a crystal. The mathematics for systems of interacting spins has been studied for many years (in, for example, the Ising model). In such systems, long range order can be generated from local interactions, as in this paper.

(1975)
W. A. Little and Gordon L. Shaw

A statistical theory of short and long term memory
Behavioral Biology 14:115–133[1]

We present a theory of short, intermediate and long term memory of a neural network incorporating the known statistical nature of chemical transmission at the synapses. Correlated pre- and post-synaptic facilitation (related to Hebb's Hypothesis) on three time scales are crucial to the model. Considerable facilitation is needed on a short time scale both for establishing short term memory (active persistent firing pattern for the order of a sec) and the recall of intermediate and long term memory (latent capability for a pattern to be re-excited). Longer lasting residual facilitation and plastic changes (of the same nature as the short term changes) provide the mechanism for imprinting of the intermediate and long term memory. We discuss several interesting features of our theory: nonlocal memory storage, large storage capacity, access of memory, single memory mechanism, robustness of the network and statistical reliability, and usefulness of statistical fluctuations.

Introduction

We have developed a theory of memory based on a model (Little, 1974; Shaw and Vasudevan, 1974) of a neural network which incorporates the known behavior of the constituent neurons and, in particular, the experimentally observed probabilistic nature of the chemical transmission at the synapses (Katz, 1966). The model shows under what conditions persistence of the firing pattern for the network as a whole for periods of the order of seconds can occur, and how these patterns are related to the synaptic parameters. The persistent states or patterns we identify as representing the *active* or short term memory. Intermediate and long term memory (McGaugh, 1966; McGaugh and Dawson, 1971), on the other hand, we associate with the connectivity of the network as a whole and the values of the synaptic parameters. In the model, intermediate and long term memory are shown to reside in the latent capability of the network to gen-

erate, when stimulated, certain time correlated firing patterns.[2]

A striking feature of the model is that, given the facilitation of certain pre- and postsynaptic parameters, a large number of different firing patterns may be induced in the network as the result of different stimuli. Thus, information contained in the values of the synaptic parameters and connectivity of the network can be transferred from this static, long term storage to the active state by such stimuli. In addition, one can show that assuming the enhancement as a result of use, of certain pre- and postsynaptic parameters, information contained in short term memory becomes imprinted in the values of the synaptic parameters. This occurs in such a way that later, upon suitable stimulus, this information may be recalled or transferred back to the active form to reproduce the identical firing pattern which previously was contained in short term memory during the imprinting state.

To avoid any confusion we will define the following important concepts as they will be used in this paper:

Strong stimulus. We take this to be a strong input signal which will override the internal firing of the network forcing it to have a specific firing pattern *a* (or set of patterns). This pattern *a* (or patterns) will persist as long as the stimulus is present.

Facilitation. We consider facilitation to be an enhancement with firing of certain neuronal parameters.

Following McGaugh (1966, 1971) we identify three time scales of memory:

(a) *Short term* or image like memory lasting from a fraction of a second to seconds. In our model, this corresponds to a specific firing pattern *a* (or more generally a cyclic group of patterns) persisting in the *active* state for this time period after excitation by a strong stimulus which had induced considerable facilitation. The pattern will persist for the order of a second before synaptic depression (Katz, 1966) sets in.

[1] This research was supported in part by the Research Corporation. We thank E. M. Harth, R. K. Josephson, R. D. Luce, J. L. McGaugh, D. H. Perkel, C. L. Stephens, and M. Verzeano for helpful discussions.

Copyright © 1975 Academic Press, Inc.

[2] This is precisely described mathematically in terms of the eigenvalues and eigenvectors of the probability matrix $\underline{P}$ described below. See, e.g., Bartlett (1966). A complete description for the neuronal system will be given by us in a future publication.

(b) *Intermediate memory* lasting up to hours during which time the imprinting into long term memory can be affected by drugs or electric shock (McGaugh, 1966, 1971). We see this as a period during which residual facilitation of synaptic parameters, after process (a), remains sizable so that the pattern *a* is easily re-excited in the network and persists as before.

(c) *Long term* or permanent memory. We see this as long term or plastic changes (i.e., permanent changes in synaptic strength and/or growth of new synapses) in neuronal parameters which enable patterns *a* to be re-excited readily as in (b).

We find that in order to have learning in our model, we require correlated pre- and post-synaptic facilitation in a manner similar to that postulated by Hebb (1949). In essence, we have a mathematical realization of Hebb's postulate of learning. Important consequences of the model are that long term memory is associated with the network as a whole, is relatively insensitive to local damage, and has large storage capacity. Moreover, the statistical nature of the synaptic transmission plays an important role in both increasing the available storage and the accessibility of this storage. Note that we use *statistical* to describe the probabilistic nature of chemical transmission at the synapses and *not* to imply any random nature of the axonal connections between neurons.

The principal requirement of our model is that in addition to the known pre-synaptic increases in the amount of chemical transmitter released (Dudel and Kuffler, 1961; Mallart and Martin, 1967), which is dependent on the recent firing history of the incoming *axon*, there should also be a changed response (either pre or post synaptically) dependent on the recent firing history of the *neuron*. Given *both* of these factors, then the features of learning and recall discussed above follow from the model.

Alternately, the requirements of the model might also be satisfied by suitable feedback loops from the neuron back to its afferents via axoaxonal synapses. This could yield an effective increase or decrease of postsynaptic potential as a result of this neuron firing. Pre-synaptic inhibition *is* known to play a role in the afferent level of the brain but is not so prevalent in the higher levels (Eccles, 1973).

The important concept of synaptic changes being involved in learning was emphasized by Hebb (1949) and Eccles (1953). There have been many theoretical papers written describing models of memory based on synaptic changes. See, e.g., Anderson (1972), Brindley (1969), Griffith (1971), Harth (1970), Marr (1970), Uttley (Burns, 1968), and Wilson and Cowan (1972). The two key elements that we stress in our model are the known

statistical fluctuations in the synaptic transmission (Katz, 1966) and the assumed three time scales of facilitation: large short term and longer lasting residual facilitation and plastic changes. Both features are crucial in order to have a large storage capacity. Furthermore, we stress that the use of the formalism of a $2^N \times 2^N$ probability matrix related *explicitly* to neuronal parameters allows one to gain intuitive understanding of the memory behavior in our model. Using the formalism, we give a simple derivation of the relation between learning for the network and synaptic facilitation.

We wish to emphasize strongly the importance of combining information from behavioral learning experiments and electrophysiological neuronal measurements. A useful theory should help to bridge the enormous gap between single neuron and behavioral phenomena by predicting the properties of a network of neurons. For example, we stress the significance of correlating the several time scales in memory with similar time scales in neuronal facilitation. The key feature, of course, is to determine experimentally if the correlated pre-post facilitation exists. Most viable theories of memory require some form of this Hebb hypothesis. Recent preliminary results of Decima (1974) (also see Black-Cleworth (1974)) indicate possible long lasting facilitation in the spinal motorneurons of cats and rats as the result of orthodromic excitation of the neurons followed 200 msec later by antidromic activation of the same motor pool. This interesting result must of course be verified. We wish to note, however, the clever feature in delaying the antidromic firing by some fraction of a second. Presumably, this was motivated by the well known delay necessary in learning experiments in which the conditioning stimulus must precede the unconditoned stimulus.

We present briefly our theory and examine its implications.

Review of Model

Consider a network of N neurons. We picture the state of the system in terms of the neurons firing or not firing at a given time and look at the evolution of such a state in discrete time steps τ of the order of a few msec, which is of the order of the refractory period and also the decay time of a below threshold post-synaptic potential. At each fixed time $\ell\tau$ then there are 2^N possible firing states of the N neurons denoted by a:

$$a = \{S_1, S_2, \ldots, S_N\}, \tag{1}$$

where $S_i = 1$ if the ith neuron fires, -1 if the ith neuron does not fire. We need to calculate the probability $P_{a'a}$ that a given firing state a at time $\ell\tau$ goes to $a' =$

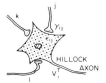

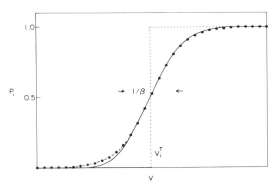

Figure 1 Schematic drawing of the ith neuron showing synaptic junctions from the j, k and l^{th} neurons. At the jth synapse to the ith neuron, the number of quanta of chemical transmitter released when the jth neuron fires is given by a Poisson process with mean γ_{ij}. The factor ε_{ij} represents the conversion factor for the chemical transmitter producing a given potential change at the hillock. The neuron fires when the sum of the inhibitory and excitatory post-synaptic potentials exceed V_i^T at the hillock.

Figure 2 Plot of p_i, the probability of the ith neuron firing vs V, the total post synaptic potential where V_i^T is the nominal threshold potential. The dashed line (with points) is the probabilistic expression Eq. (22) of (Shaw and Vasudevan, 1974) and the continuous line is the approximate expression $p_i = (\exp(-\beta(V - V_i^T)) + 1)^{-1}$ of (Little, 1974). The dotted step function corresponds to the limit that the statistical fluctuations in the synaptic transmission of chemical transmitter are ignored.

$\{S_1', S_2' \ldots \ldots S_N'\}$ at $t = (\ell + 1)\tau$. This can be obtained in a simple way. We concentrate on the probability p_i that the ith neuron receives a depolarization $V > V_i^T$, its threshold, and thus fires. As in Fig. 1 the ith neuron has a synaptic junction, either excitatory or inhibitory, at an ending of the axon from the jth neuron. We include (Shaw and Vasudevan, 1974) three known sources of fluctuation in the post-synaptic potential: (a) The number of quanta of chemical transmitter released when the jth neuron fires is given by a Poisson process[3] (Katz, 1966) with mean γ_{ij}. (b) There is also a low spontaneous rate γ_{ij}^{SP}. (c) The quanta have a Gaussian distribution of chemical with mean v and standard deviation σ (which may depend on i and j) (Katz, 1966). We define a factor ε_{ij} which represents the conversion factor for the chemical transmitter producing a given potential change at the hillock. The factor depends on the sensitivity of the post-synaptic membrane at the jth synapse and on the spatial geometry of the neuron (which determines the spatial decay of a post-synaptic potential produced in traveling to the hillock). This can be quite complicated. See, e.g., Rall (1970). The term ε_{ij} will be positive for an excitatory synapse and negative for an inhibitory synapse. Thus at a given time step $(\ell + 1)\tau$ there are two mean contributions to the potential change at the hillock of the ith neuron: the spontaneous contribution $\gamma_{ij}^{SP} v \varepsilon_{ij}$, and the potential

$$(S_j + 1)/2 V_{ij} = \gamma_{ij} v \varepsilon_{ij}(S_j + 1)/2 \qquad (2)$$

containing the factor $(S_j + 1)/2$ (Eq. (1)), which may be 1 or 0 and represents the firing (or not firing) of the jth neuron at the time $\ell\tau$. Folding in the Poisson rates γ with the Gaussian distribution, we showed (Shaw and Vasudevan, 1974) that the probability of the ith neuron firing is the smooth distribution illustrated in Fig. 2.

Thus, given the connectivity of the neural system, the synaptic parameters γ and ε, and the thresholds V^T, we may calculate directly, using products of factors given in Fig. 2, the probabilities $P_{a'a}$ of going from firing state a at $\ell\tau$ to a' at $(\ell + 1)\tau$. These elements form a $2^N \times 2^N$ (nonsymmetric) probability matrix $\underline{P}$ (with elements $P_{a'a} \geq 0$ and $\Sigma_{a'} P_{a'a} = 1$). *For the moment* we assume that the γ_{ij}, ε_{ij} and V_i^T are *independent of the recent firing history* of the system, i.e., we ignore facilitation. The P is thus independent of time and the probability of going to state a' after r time steps is

$$(P^r)_{a'a}. \qquad (3)$$

We are interested in the "long" time behavior[4] of Eq. (3) for the firing patterns: $t = r\tau$ for r about 100 or t of the order of seconds (Hebb, 1949). In particular, we are interested in the situation in which a given pattern a when excited persists for r time steps. This implies that $(P_{aa})^r$ is sizable, or

$$P_{aa} \gtrsim 1 - \frac{1}{r}. \qquad (4)$$

---

[3] The exact details of these distributions are not important to our model. It is known that, e.g., in certain cases, the number of vesicles released might behave more like a binomial distribution rather than a Poisson one.

[4] This is governed by (Bartlett, 1966) the eigenvalues λ of $\underline{P}$ (elements of $\underline{P}$ when it has been transformed to diagonal form). There are 2^N such λ's of a given $\underline{P}$. They range in magnitude from 0 to 1 and there is always one $\lambda = 1$. In order to have a dependence of firing pattern a' on pattern a at time $r\tau$ previous, we must in general have another λ with magnitude $\sim [1 - (1/r)]$. Corresponding to a given λ is an eigenvector φ which is a linear combination of the 2^N firing patterns a. An eigenvector ϕ_i has the property that it reproduces itself diminished in amplitude by its eigenvalue λ_i with each time step: $P\phi_i = \lambda_i\phi_i$. In our model, long term memory storage of a "piece" of information corresponds to having a λ large but not necessarily ≈ 1 as we discuss below.

In our model, long term memory storage of a "piece" of information corresponds to having a P_{aa} large but not necessarily ≈ 1 as we discuss below. Many P_{aa}'s can be large corresponding to many pieces of storage. The number of the 2^N P_{aa}'s which can be simultaneously large is expected to be greater than N, the number of neurons. We suggest that it is roughly the number of variable parameters which is the order of the number of synapses (which if all the neurons in the network were directly interconnected would be of order N^2) See Griffith (1971).

Justification of Discrete Time

In the model, we have used discrete time steps to study the time evolution of the firing patterns. It is essential to make this approximation if one is to use the matrix method. However, as we have discussed elsewhere through a close analogy with other systems (Little, 1974), one can show that certain features of the discrete step model are retained by the continuum system.

Montroll (1952) has discussed the mathematical machinery for expressing the long term behavior of the probability amplitude of a Markoff process in terms of a transfer matrix taken between states at discrete time intervals. Essentially the same approach may be used to describe the continuous time behavior of the neural network in terms of an effective transfer matrix at discrete time intervals. However, the details of this are rather complicated and go beyond what we wish to discuss in this paper. Instead we will argue for the plausibility of the result by considering the analogy between the neural network and the behavior of an amorphous and a crystalline spin system. The problem of determining the configuration of atomic spins in a crystalline lattice may be formulated in a manner closely similar to that used here for the neural network. Given a particular configuration of spins, $a \equiv \{S_1, S_2 \ldots\}$, where $S_i = +1 \, (-1)$ refers to spin i being "up" (down), one can calculate the probability of finding a configuration $a' \equiv \{S_1', S_2' \ldots\}$ in the next row in the crystal. This probability defines an element of a matrix $P_{a'a}$. From an analysis of the properties of the matrix one can show that such a crystal undergoes a phase transition from the paramagnetic state to the ferromagnetic state at a sharply defined temperature, the Curie temperature. At the Curie point the properties of the matrix change so that in the ferromagnetic state information contained in the first row of the crystal propagates throughout the crystal. We have shown (Little, 1974) that the analogous property to this in the neural network is the capability of the network of sustaining a persistent firing pattern.

In the spin problem it is essential that the crystalline lattice be strictly regular in order to use the matrix method, for the matrix describes the propagation of order from row to row. In a disordered or amorphous solid one cannot use the matrix method to determine the behavior of the spins, for the atoms in the lattice do not follow one another in a regular manner. Nevertheless, such systems still exhibit a clearly defined phase transition and in the ferromagnetic state order persists throughout the crystal. A much more sophisticated mathematical procedure would be needed to describe completely the behavior of such an amorphous spin system. Yet, certain key features relating to the order of this system are contained in the behavior of the simple regular crystal. In view of the almost exact one-to-one correspondence between the continuum model of the neural network and the amorphous lattice-spin problem, one should likewise expect certain key features of the discrete time model to occur also in the continuum system. It is for this reason that we believe that the features relating to the propagation of firing pattern of the neural network, which maps into the propagation of order in the spin problem, will be retained in the continuum model.

A second example (Little, 1974; Huang, 1963) of a discrete model which gives the essential features of the continuum system is the lattice-gas model of the real liquid-gas phase transition. In the lattice gas, one considers the behavior of a gas in which the particles are only allowed to move on the mesh points of a regular lattice. Again, one may formulate the problem by a matrix method in which one describes the interaction between particles at the mesh points of one row with the particles on the next row. One finds that this model gives a remarkably good description of the liquid-gas transition of real fluids. Yet the particles of a real fluid, of course, are not confined to move on the mesh points of a lattice but move in a continuous manner over all space. Nevertheless, the key features of the discrete model are again contained in the full continuum model.

In addition, to the above arguments for determining the long-term behavior of the neural system from a discrete time model, one may also argue that there is some physiological evidence for the synchronous firing of some of the neurons. This is related to the existence of an interesting phenomenon observed some time ago by Erlanger and Gasser (1937) and Hodgkin (1938). A supernormal threshold was found in certain axons under special conditions: after an action potential, the relative refractory period is followed by a supernormal period and this in turn by another relative refractory period before returning to normal. This supernormal period *if present* in cortical cells would tend to recruit the firing pattern of a group of neurons into synchrony. This would provide some physiological support for the discrete time assumption.

The actual parameters which we use in the discrete step model are related to, but are not identical to, those of a continuum model. In addition, we recognize the fact that the continuum system, with variability of propagation times, diffusion times, etc., contains a greater richness and complexity than does our model. It is possible that the key element of memory lies in a subtle coding of the firing frequency of the individual neurons, rather than in the overall firing pattern of the network. In this case, our discrete step model would not contain the essential features of memory. We are presuming that such frequency coding is not the key element.

Facilitation and the Requirements of Learning

In the above, we have treated the synaptic parameters γ_{ij} and ε_{ij} as if they were time independent, i.e., we ignored both facilitation and permanent plastic changes. Now we discuss the crucial point of how the requirements of storage and recall of memory in our model require *specific* variation of the γ and ε in Eq. (2) with the recent firing history of the neurons.

We need to consider how in our model an input stimulus can generate a temporary, *active* persistent firing pattern or short term memory trace. Then we must consider how this information can be transferred to a more permanent form such that, with the aid of suitable stimuli, the original firing pattern of short term memory may be reactivated.

For simplicity in the following analysis we consider a strong input stimulus consisting of a rapid firing sequence of certain input neurons. We assume this signal will override the internal firing of the network forcing it to have a specific firing pattern. This pattern, a will persist as long as the stimulus is present. We then ask what conditions the matrix P must satisfy so that a will persist for several seconds after the stimulus is switched off. There are two situations of interest:

(i) The connectivity of the network and neuronal parameters may be such that in the absence of any facilitation the element P_{aa} may directly satisfy Eq. (4). Under these circumstances the pattern a will persist for several seconds as discussed earlier. This situation is not of the greatest interest because relatively few patterns a can be expected to satisfy this condition. On the other hand, many more patterns may be caused to persist if the following occurs:

(ii) Under the action of the input stimulus and the forced excitation of the pattern a facilitation of some of the synapses may occur so that the new P_{aa} satisfies

Eq. (4). During such facilitation and prior to the onset of depression, the pattern a will persist. This persistence then depends on the characteristic times of the various kinds of facilitation.

We need to determine what form the facilitation must take in order for the given stimulated pattern to persist. This can be done by examining in detail the elements of the probability matrix, P. We have shown previously that in our model the element of the matrix $P_{a'a}$ which gives the probability of going from a pattern a to a' in one step is proportional[5] to

$$\exp\left(\sum_i \beta_i S_i' \left\{\sum_j V_{ij}\left[\frac{(S_j + 1)}{2}\right] - V_i^T\right\}\right) \tag{5}$$

where the set $\{S_j\}$ describes the pattern or state, a, Eq. (1) and the set $\{S_i'\}$ describes the state a'. V_{ij} is given by Eq. (2) and β_i is a measure of the "width" of the threshold as illustrated in Fig. 2. In order to simplify the following discussion we have omitted the spontaneous contributions γ^{SP} and we will treat β_i as if it were essentially a constant and independent of i.[6]

The input stimulus puts the network into a firing pattern a. In order for such a pattern to persist after termination of the input stimulus, P_{aa} must be large. Thus, for a diagonal element P_{aa} (i.e., $\{S_i'\} = \{S_i\}$) we wish to *maximize* the exponent of (5):

$$\sum_i S_i \left[\sum_j V_{ij}\left(\frac{S_j + 1}{2}\right) - V_i^T\right] \tag{6}$$

This can occur if either or both of the following types of facilitation occur.

(a) First, if the threshold V_i^T of a cell i decreases as a result of firing or increases gradually as a result of disuse then those neurons which have just fired will tend to fire more readily in the future. We assume that over a *limited range*, the threshold may then change from V_i^T to a new value

$$V_i^T - W_i S_i, \tag{7}$$

where W_i is a positive quantity whose magnitude is a function of the recent firing history of the neuron. The sign of the facilitation is given by S_i for the particular firing pattern a.

As we shall show later, this type of threshold modification (7) does not appear to be sufficient in itself to yield a system capable of associative recall of events which had been learned together. On the other hand,

---

[5] This follows from Eq. (6) of (Little, 1974). We have dropped the denominator in that equation, which is simply a normalizing factor, and brought the product over i within the exponent in Eq. (5), making it a sum over i.
[6] See Eq. (22) of (Shaw and Vasudevan, 1974) for the actual form of β_i.

the correlated pre- and post-synaptic facilitation discussed in (b) does have this feature.

(b) We assume, again over a limited range, that the postsynaptic potential, V_{ij} changes as a result of the ij-synapse firing *and* the neuron i also firing such that V_{ij} becomes

$$V_{ij} + S_i U_{ij}(S_j + 1)/2, \tag{8}$$

where U_{ij} is a positive quantity whose magnitude is a function of the recent firing history of the jth axon and of the ith neuron. Then again the diagonal element P_{aa} will be strongly enhanced. This can be seen as follows. If the set $\{S_j\}_a$ is identical to the original set $\{S_j\}$ for which S_i, S_j were used in Eq. (8), then every term in the sum over j in Eq. (6) will be increased by $U_{ij}((S_j + 1)/2)^2$, which is positive. In addition, every term in the sum over i will be multiplied by S_i^2, which is likewise positive. Thus, the effect of change in V_{ij} described by the second term of Eq. (8) will be to increase the magnitude of P_{aa}. In general, all other elements $P_{a'a}$ for $a' \neq a$ will be enhanced by a lesser amount because for these the sum over i will contain some negative elements $S_i' S_i$, rather than all positive as with P_{aa}.

The above argument is strictly true only if all the neurons connect to all other neurons so that every V_{ij} and thus U_{ij} is finite. In an actual network such complete interconnections do not occur. The absence of these connections gives rise to an extremely important statistical aspect of the model. Consider a state a' which is identical to a except for a small subset of the S_i's. If for this subset no interconnections occur, i.e., the appropriate V_{ij} and thus U_{ij} in Eq. (8) are absent, then as far as Eq. (8) contributes, this $P_{a'a}$ will be enhanced as much as is P_{aa}. Thus, instead of the stimulus exciting the persistence of the identical pattern, a, it will excite any one of an ensemble of almost identical patterns. In a similar way, if the input signals terminate on a subset of all the neurons then they will not be able to excite a single pattern, a in general but only one or other of an ensemble of similar patterns which may differ from one another by the state of those neurons not closely connected to be input neurons.

Our argument is thus generalized so that the facilitation of the types discussed above will enhance the probability of a class or ensemble of patterns persisting for some seconds. This ensemble of patterns is thus a representation or image of the input stimuli.

Thus, facilitation of the pre- and post-synaptic factors described by one or both of Eqs. (7) and (8) will result in a persistent trace being formed in the network which carries an image of the input stimuli. This trace will

persist until synaptic depression sets in or other stimuli activate other patterns.

With the onset of synaptic depression (Katz, 1966) the factors W_i and U_{ij} will decrease in magnitude and P_{aa} will subside and become more comparable to the other terms $P_{a'a}$. The facilitation of some other neurons from a new input stimulus (or through heterosynaptic facilitation (Kandel and Tauc, 1965a,b)) may cause some other element P_{aa} to become large and thus drive the network to a new pattern, γ. Suppose, however, that as a result of facilitation W_i and U_{ij} do not return to zero but, rather, to a slightly enhanced value. Then, a subsequent weaker stimulus together with these residual contributions could cause the pattern a to be reactivated. Likewise, an input stimulus not identical to a, but with some elements in common with a could, with the aid of the slightly enhanced values of W_i and U_{ij}, also reactivate the pattern a. This may be shown by the same arguments as applied above to Eqs. (6)–(8). Thus there is a set of such patterns $\{a_c\}$ which lead to a, i.e., P_{aa_c} is also large.

These plastic changes of W_i and U_{ij} would thus retain information of the initial input pattern a for long periods of time. Under appropriate stimulus at a later time, this information could result in the reactivation of the same pattern in the network. Such plastic changes would thus represent a repository of earlier traces and would constitute the storage location of intermediate and long term memory.

Our model thus provides a means by which the input trace can be stored temporarily during which time physiological changes would have a chance to occur. In addition, it shows that in order to recall a previous memory trace, it is necessary for the input stimulus to have elements in common with the pattern we wish to recall.

Thus far we have discussed only the simplest type of input stimulus—that of a steady train of pulses on certain input neurons. If the input is a more complicated form of stimulus consisting of a time sequence of varying strengths, then this would stimulate a firing pattern which would follow this time variation. These may be included in our model, by generalizing the above arguments to include such time varying or cyclic patterns in networks with appropriate connections. We might have cycles $(a \rightarrow a' \rightarrow a)$ among a subset of neurons so that this subset has firing frequency different from the others. This leads to an expression similar to (8) for the learning criterion but involves certain combinations of the S_i's and S_j's of the neurons in the cycle instead of the single S_i or S_j terms.

Related Work on Facilitation

We have seen that in our model, we were led by our learning criteria to specific facilitation assumptions. Eqs. (7) and (8), for the neural parameters V_i^T and V_{ij}. We now briefly examine the physiological bases for these facilitations. We refer the reader to the very extensive reviews of Kandel and Spencer (1968) and Kandel *et al.* (1970).

The general concept of synaptic changes being involved in learning dates back to the discovery of the synapse by Sherrington around 1900. See, e.g., Eccles (1953). Hebb (1949) stated his neurophysiological postulate: "When an axon of cell *A* is near enough to excite a cell *B* and repeatedly or persistently takes part in firing it, some growth process or metabolic change takes place in one or both cells such that *A*'s efficiency as one of the cells firing *B* is enhanced." As mentioned previously, many theoretical models have been based on this postulate. Stent (1974) has suggested an interesting physiological mechanism whereby a gradual change in the number of post-synaptic neurotransmitter receptors is related to synchronous (within 20 msec) firing behavior of the pre- and postsynaptic neurons.

Our *correlated* pre-post synaptic facilitation as described by Eq. (8) has the property that if a pre-synaptic axon fires and the post-synaptic neuron fires (does not fire) synchronously, then that synapse is enhanced if excitatory (if inhibitory). Thus Eq. (8) is exactly a mathematical representation of Hebb's postulate. The proof in the previous section demonstrates mathematically how Hebb's postulate yields a mechanism whereby learning can occur in the neural network.

There is considerable data on the time behavior of presynaptic facilitation. We note the following features found by Mallart and Martin (1967) for the transmitter release at the neuromuscular junction of the frog:

(1) The magnitude of the facilitation caused by a single previous firing was large (approximately 100%).

(2) There were two components of the time decay. One short one (with a time constant of about 35 msec) and a second much slower one.

(3) During repetitive firings (up to 100/sec) the individual firing facilitatory effects summed linearly.

There is no direct evidence for post-synaptic changes in the mammalian brain. There is some indirect evidence (Deutsch, 1973) from studies of the effects of pharmacological agents that such modifications of the postsynaptic neuron do occur in learning. Postsynaptic facilitation has been looked for in Aplysia, and not found (Kandel and Spencer, 1968; Kandel *et al.*, 1970). However, again in Aplysia, Stephens (1973) has observed a decrement in postsynaptic neuron activity

and suggested a relationship to habituation (Groves and Thompson, 1970), and Parnes *et al.* (1974) have observed long lasting ($\sim$ hr) effects of brief inputs on the firing frequencies in a pacemaker neuron. Decima (1973), in experiments carried out in the spinal cord of cats, has been the correlated effects of the post-synaptic neuron firing activity acting back on the pre-synaptic axon via the electric fields set up during firing. As mentioned in the introduction, long lasting effects were observed when he delayed (Decima, 1974) the antidromic firing by some 200 msec. See, also, Black-Cleworth (1974).

A very interesting consequence of Eq. (8) is that if the facilitation factor U_{ij} is sufficiently large in magnitude, the synapse can change from excitatory to inhibitory. For example, consider an excitatory input from the jth neuron to the ith neuron so that V_{ij} is positive. However, if the jth neuron repeatedly fires but the ith neuron does not fire so that S_i is negative, then for large U, the facilitation term will be large and negative. Hence the synapse will change from excitatory to inhibitory for the time scale associated with U. There is no experimental evidence for this correlated pre-post phenomena. However, we note that Wachtel and Kandel (1967) have found a synapse (in the L7 neuron in Aplysia) which changes from excitatory to inhibitory depending on the firing frequency of the input axon.

We fully expect that V_i^T and V_{ij} do have a *more* complicated dependence on S_i and S_j than Eqs. (7) and (8). We have not investigated these more general possibilities. In our model we have considered only the simplest forms which exhibit learning. We assume only that (7) and (8) are valid over a limited range.

We are not aware of any evidence concerning the experimental validity of the facilitation Eq. (7) for the threshold V_i^T.

Consistent with the experimental features (1)–(3) listed above for presynaptic facilitation (Mallart and Martin, 1967), we have in mind that U_{ij} (and W_i) might have, roughly, the following time dependence following stimulation of a firing pattern: U increases with a magnitude related to the duration of stimulus. After firing repeatedly for $\sim$ sec, synaptic depression (Katz, 1966) sets in and the magnitude of U rapidly decreases. There remains a residual quantity which decays slowly with time. This time scale is related to the time scale for intermediate term memory. (Finally, permanent plastic changes can take place in V_{ij} corresponding to long term memory.)

The model may readily accommodate the fact that in the central nervous system, appreciable activity occurs without action potential spike generation. Such effects may be treated within the model as yet another form of enhancement or inhibition in which the activity of certain neurons may be controlled by certain input axons or other neurons. Similarly, electrically induced post-synaptic potentials may be treated in the same manner. In such a treatment, these effects serve to set the values of the parameters of the model. However, we assume in our model that the principal activity remains in the action potential.

Numerical Calculations

We have studied a number of the properties of our model by simulating the behavior of a simple network on a digital computer. Some of the properties relating to the generation of persistent patterns have been described previously (Little, 1974). More recently, we have studied those features related to facilitation and learning as described by Eqs. (7) and (8).

One particularly interesting property of the model lies in "noise" due to the probabilistic fluctuations (Katz, 1966) (i.e., the statistical nature) of the release of the chemical transmitter. We considered a four-neuron network, so that there are $2^4 = 16$ possible firing patterns a for the network. Choosing various values for the neuronal parameters γ_{ij}, ε_{ij} and V_i^T we calculated the 16×16 probability matrix P. Referring to Eq. (2) for V_{ij}, we keep the mean amount of chemical released $\gamma_{ij}v$ fixed but vary the size v of the vesicle (and hence vary γ_{ij}). As $v \to 0$, the number γ gets large and the fluctuations (or "noise") approach zero; for v large, γ is small and the noise is large. We found that the number c of diagonal elements P_{aa} which had appreciable magnitude varied with v or the amount of noise. There is a value of v for which this number c had a maximum. This is illustrated in Fig. 3. At this value for v, the largest number of patterns can be stimulated and persist in the network for a given degree of facilitation. Thus there is some optimum value for v or the noise as far as the storage capacity of the network is concerned. Having some noise serves a useful purpose by increasing the storage capacity over that of a comparable network which is free of such statistical fluctuations.

A second property we studied relates to the effects of facilitation in the computer model. By modifying the synaptic parameters according to the prescription given in Eq. (8), we found that the appropriate diagonal element P_{aa} becomes large and that as a result the pattern a dominates the long time behavior of the network. One particularly interesting extension of this relates to what one may call associative recall: Suppose we have an input stimulus which is the sum of signals which separately would stimulate two different patterns a and γ. In our model, facilitation will occur

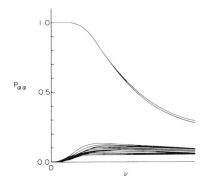

Figure 3 Plot of diagonal elements of matrix $\underline{P}$ as a function of the vesicle size v which is a measure of the statistical fluctuations or "noise" in the network. We note that more elements P_{aa} have appreciable strength in the presence of this noise than in its absence ($v = 0$). There is an "optimum" value of v for which several P_{aa} are "large." In this four neuron example, for large v all elements of $\underline{P}$ approach $1/2^4 = 0.06$.

so that the synaptic parameters change according to Eq. (8). However, the pattern $\{S_i\}$ is replaced by $\{\bar{S}_i\}$, where $\bar{S}_i$ is the average of S_i for the superposition of the two patterns a and γ. We find that stimulus of the network by either a or γ results in the generation of the combined pattern a plus γ. This is not what occurs if the two patterns are "learned" separately. Then a leads to a and γ to γ. This difference results from the cross terms between the pre- and post-synaptic factors in Eq. (8). The model then predicts that if one observes two events at the same time, a stimulus which recalls one will recall the other as well.

We note that this feature is unique to the learning criterion (8) and is not given by the contribution of an effect on the threshold alone such as Eq. (7). We believe this is a strong argument for some form of pre- and post-synaptic facilitation in learning, for, associative recall appears to be a characteristic feature of memory in the brain.

Discussion

We list and discuss several of the interesting features of our theory:

1) *Nonlocal memory storage.* In our model long-term memory lies in the latent capability to generate, when stimulated, persistent (for ~sec) firing patterns of the network. Thus, even though this capability is related to the neuronal parameters, the storage of information is associated with the network as a whole rather than to a local part of it. This is to be contrasted with the local type of information storage used in an electronic computer. This nonlocal feature of our model seems to

be in agreement with evidence for such diffuse storage of information in the brain.

2) *Large storage capacity.* The capacity in our model is associated with the number of P_{aa}'s which can be large. It seems plausible that this number is related to the number of synaptic parameters for this represents the number of variables in the problem. Thus the capacity is determined roughly by the number of synapses in the network rather than by the much smaller number of neurons (10^{10} neurons vs 10^{14} synapses in the human brain). This is an extremely important point and it would be worthwhile to investigate more precisely the maximum storage capacity of the model.

3) *Access or recall of memory.* A piece of information can be recalled to the active firing state directly by exciting one of any of a set of patterns $\{a_c\}$ which have "pieces" in common with a. This direct accessibility to pattern a by a whole class of stimuli is in contrast with the usual method of recall of information in an electronic computer where one must "interrogate" each set of storage elements to find the relevant pieces. Our model would thus account for the remarkable speed with which an enormous range of information may be located in the human brain in spite of the brain's relatively low cycle speed compared to an electronic computer. In our model it is only necessary to have the appropriate stimulus for facilitation to cause essentially immediate recall of any one of the latent patterns of the network.

Our model also has the property that the retrieval cue is an important if not essential element in recall of memory. This is in agreement with much work in this area such as that described by Tulving (1974) in the human brain.

4) *Single memory mechanism.* Even though we distinguish three kinds of memory storage—short ($\sim$ sec), intermediate ($\sim$ hr) and long ("permanent")—there is just one basic mechanism for all three in our model. These three associated time scales for memory reflect three different time scales of facilitation of the *same* neuronal parameters.

On the other hand, in our model memory has two representations. The first is the actual, active firing pattern of the network following a stimulus. This occurs when the naive network receives a strong input stimulus or later when another stimulus reactivates this firing pattern. The second form is in the latent capability of the network to generate a given pattern upon suitable stimulus. This latent capability lies in the values of the various synaptic parameters. These parameters thus indirectly store the same information as the pattern the network is capable of generating.

5) *Robustness of the network and statistical reliability.* The enormous storage capacity (related to the number of synapses) of our model makes it possible to build in a great deal of redundancy. We might imagine that there are a number of similar "small" networks storing the same information and thus ensuring statistical reliability of the firing patterns. Perhaps some 10^4 heavily interconnected neurons could play the role of such a small network having a storage capacity of 10^8. Furthermore the connectivity of each network might be such that axons from a number of neurons converge in an output stage. This would also ensure that limited local damage (e.g., supposedly many neurons die at random and are not replaced) will not destory the stored information.

6) *Usefulness of statistical fluctuations.* We have seen that the statistical nature of synaptic transmission plays an important role in our model in increasing the storage capacity over that of a similar network which is free of such "noise." (We stress, however, there was an optimum value of the noise in this regard.) A second useful purpose of the statistical feature is to enhance the ability to switch back and forth between different firing patterns (1), or in other words, increase the accessibility of the stored information.

In the model "noise" or statistical fluctuations plays an important role in increasing the number of accessible patterns which the system can store. From Fig. 3 one sees that the effect of fluctuations is to increase the value of the diagonal elements of the probability matrix P for a large number of terms. Any one of these patterns, upon stimulation, will thus persist for slightly longer than it would in the absence of such fluctuations. During this persistence, facilitation may enhance the diagonal element for this pattern, increasing its persistent lifetime so that it can persist until synaptic depression sets in. Thus the fluctuations allow a large number of elements to be accessible for enhancement through facilitation and thus accessible for recall.

A given stimulus could in a naive neural network lead to seemingly uncorrelated firing patterns. However, the same stimulus after being matched repeatedly with an unconditioned stimulus could then give a precise, sharp pattern as described above. This is in agreement with the experiments of John (1974).

John (1974), also, gives evidence for the idea of having a number of repeating small networks as described in (5) above. We emphasize that the seemingly contradictory concepts of considerable statistical fluctuations and large reliability are made compatible in our model via the mechanisms of short duration facilitation. This short, but large amplitude facilitation during recall selects the one associated pattern from the many others.

Summary

We present a theory of short term and long term memory which incorporates, in an essential manner, the statistical nature of synaptic transmission. We require that the neuronal firings lead to specific pre- and postsynaptic facilitation. Then information from an input stimulus is transferred to short term memory and ultimately becomes imprinted in long term memory by changes in the synaptic parameters. The model exhibits: a large storage capacity of a non-local character; a single mechanism for short, intermediate and long term memory based on facilitation of the neuronal parameters; associative recall of memory; a robustness of the network against local damage, and a statistical reliability.

The model presented does not take into account the details of the organization of the brain. For example, we have not included any features of the sensory inputs or the role of the thalamus or hippocampus in controlling the overall behavior. We hope to be able to include such important aspects into our model of memory to bring it into closer contact with behavioral experiments.

References

Anderson, J. A. (1972). A simple neural network generating an interactive memory. *Math. Biosci.* **14**, 197–220.

Bartlett, M. A. (1966). "An Introduction to Stochastic Processes," Chapter 2. Cambridge: Cambridge University Press.

Black-Cleworth, P. A. (1974). "Conditioned blink acquired by pairing click and electrical stimulation of the facial nerve." *In* C. D. Woody, K. Brown, T. Crow, and J. Knispel (Eds.), "Cellular Mechanisms Subserving Changes in Neuronal Activity." pp. 111–118. Los Angeles: U.C.L.A.-Brain Information Service.

Brindley, G. S. (1969). Nerve net models of plausible size that performs many simple learning tasks. *Proc. Roy. Soc. London Ser.* **174**, 173–191.

Burns, B. D. (1968). "The Uncertain Nervous System," Chapter 7. London: Edward Arnold Publishers.

Decima, E. E., and Goldberg, L. J. (1973). Antidromic electrical interaction between alpha motorneurons and presynaptic terminals. *Brain Res.* **57**, 1–14.

Decima, E. E. (1974). Plastic synaptic changes induced by orthodromic -antidromic pairing. *In* "Mechanism of Synaptic Action," Abstract, p. 14. Jerusalem Symposium, XXVI International Congress of Physiological Sciences.

Deutsch, J. A. (1973). The cholinergic synapse and the site of memory. *In* J. A. Deutsch (Ed.), "The Physiological Basis of Memory," Chapter 3. New York: Academic Press.

Dudel, J., and Kuffler, S. W. (1961). Mechanism of facilitation at the crayfish neuromuscular junction. *J. Physiol.* **155**, 530–542.

Eccles, J. C. (1953). "The Neurophysiological Basis of Mind," Oxford: Clarendon Press.

Eccles, J. C. (1973). "The Understanding of the Brain," p. 93. New York: McGraw-Hill.

Griffith, J. S. (1971). "Mathematical Neurobiology," Chapter 6. New York: Academic Press.

Groves, P. M. and Thompson, R. F. (1970). Habituation: A dual-process theory. *Psychol. Rev.* **77**, 419–450.

Harth, E. M., Csermely, T. J., Beek, B., and Lindsay, R. D. (1970). Brain functions and neural dynamics. *J. Theort. Biol.* **26**, 93–120.

Hebb, D. O. (1949). "The Organization of Behavior." New York: Wiley.

Hodgkin, A. L. (1938). The subthreshold potentials in a crustacean nerve fibre. *Proc. Roy. Soc. London Ser.* **842**, 87–121.

Huang, K. (1963). "Statistical Mechanics." New York: Wiley.

John, E. R. (1972). Switchboard versus statistical theories of learning and memory, *Science* **177**, 850–864.

Kandel, E. R., and Tauc, L. (1965a). Heterosynaptic facilitation in neurons of the abdominal ganglion of Aplysia depilans. *J. Physiol. London* **181**, 1–27.

Kandel, E. R., and Tauc, L. (1965b). Mechanism of heterosynaptic facilitation in the giant cell of the abdominal ganglion of Aplysia depilans. *J. Physiol. London* **181**, 28–47.

Kandel, E. R., and Spencer, W. A. (1968). Cellular neurophysiological approaches in the study of learning. *Physiol. Rev.* **48**, 65–134.

Kandel, E., Castellucci, V., Pinsker, H., and Kupfermann, I. (1970). The role of synaptic plasticity in the short-term modification of behavior. *In* G. Horn and R. A. Hinde (Eds.), "Short-Term Changes in Neural Activity and Behavior," pp. 281–322. Cambridge: Cambridge University Press.

Katz, B. (1966). "Nerve, Muscle and Synapse," Chapter 9. New York: McGraw-Hill.

Little, W. A. (1974). The existence of persistent states in the brain. *Math. Biosci.* **19**, 101–120.

Mallart, A. and Martin, A. R. (1967). An analysis of facilitation at the neuromuscular junction of the frog. *J. Physiol. London* **193**, 679–694.

Marr, D. (1970). A theory for cerebral neocortex. *Proc. Roy. Soc. London Ser.* **176**, 161–234.

McGaugh, J. L. (1966). Time dependent processes in memory storage. *Science* **153**, 1351–1358.

McGaugh, J. L., and Dawson, R. G. (1971). Modification of memory storage processes. *Behav. Sci.* **16**, 45–63.

Montroll, E. W. (1952). Markoff chains, Wiener integrals and quantum theory. *Comm. Pure Appl. Math.* **V**, 415–453.

Parnas, I., Armstrong, D., and Strumwasser, F. (1974). Prolonged excitatory and inhibitory synaptic modulation of a bursting pacemaker neuron. *J. Neurophysiol.* **37**, 594–608.

Rall, W. (1970). Dendritic neuron theory and dendrodendritic synapses in a simple cortical system. *In* F. O. Schmitt (Ed.), "The Neurosciences, Second Study Program," pp. 552–565. New York: Rockefeller University Press.

Shaw, G. L., and Vasudevan, R. (1974). Persistent states of neural networks and the random nature of synaptic transmission. *Math. Biosci.* **21**, 207–218.

Stent, G. S. (1973). A physiological mechanism for Hebb's postulate of learning. *Proc. Nat. Acad. Sci. USA* **70**, 997–1001.

Stephens, C. L. (1973). Progressive decrements in the activity of Aplysia neurones following repeated intracellular stimulation: Implications for habituation. *J. Exp. Biol.* **58**, 411–421.

Tulving, E. (1974). Cue-dependent forgetting. *Amer. Sci.* **62**, 74–82.

Wachtel, H. and Kandel, E. R. (1967). A direct synaptic connection mediating both excitation and inhibition. *Science* **158**, 1206–1208.

Wilson, H. R. and Cowan, J. D. (1972). Excitatory and inhibitory interactions in localized populations of model neurons. *Biophys. J.* **12**, 1–23.

Introduction

(1976)
S. Grossberg

Adaptive pattern classification and universal recoding: I. Parallel development and coding of neural feature detectors
Biological Cybernetics 23:121–134

Stephen Grossberg has been one of the most visible scientists working in neural networks for nearly twenty years. His productivity has been immense, and he has written dozens of papers and several books. At the same time, he has generated considerable controversy because of his frequent vigorous arguments in favor of his approach to the mathematical analysis of brain function.

Grossberg's work has a reputation for difficulty. There are several reasons for this. First, he has been so productive and written so many papers, that his papers are rarely self-contained, but make constant reference to previous results appearing elsewhere. This makes it hard to judge a part of his work in isolation from the whole. Second, his papers are written at an unusually high level of mathematical abstraction. He proves theorems and writes nonlinear differential equations to describe the behavior of his theoretical constructions. He and his group did not start to use computer simulations routinely until recently. This means that many of his theoretical results are presented as brief qualitative statements about the behavior of differential equations and complex networks. Many members of the natural target audience for his papers do not have the degree of mathematical sophistication to understand or judge results presented in this form. Third, he tends to use a self-consistent but unusual vocabulary in his papers.

This 1976 paper by Grossberg is a theoretical analysis inspired by the developmental physiology of cortical organization. Some of these data were also discussed and simulated by von der Malsburg (paper 17). As mentioned in that paper, there is strong evidence for the development and modification of feature detectors in visual cortex in response to a particular environment. Raising kittens or monkeys in a warped environment can produce demonstrable changes in the kinds and distributions of cell types found in cortex. The experimental results sometimes suggest that cortex tunes itself to pick up the most useful features of the environment and ignores or suppresses other information.

Much of Grossberg's work starts with a nonlinear extension of lateral inhibition (see paper 23), which functions as an automatic gain control. This network was originally proposed as an answer to the scaling problem of how to maintain relative intensities in an input pattern over a wide range of average intensities. By use of nonlinear feedback, the network "normalizes" the input patterns, so they fit the dynamic range of the neurons, but maintains relative intensities intact. "Saturation," where cells are only at an upper or lower firing limit, is seen by Grossberg as an important problem because then relative intensities are lost. A "center-surround" gain control system is used frequently in Grossberg's work. He first pointed out its value, and its consistency with center-surround receptive field organization, in the late 1960s.

This concern for relative intensities is in contrast to network models composed of neurons using threshold logic, or feedback models that deliberately use saturation, where relative intensities in the spatial input pattern are lost.

Grossberg sees short term memory (STM) as a way of keeping patterns active after the original input has vanished. A short term memory is a persistent activity pattern in a set of neurons, maintained by nonlinear feedback system. Grossberg has developed a theory of short term memory, briefly described in section 2 of the paper, where a spatial input pattern is processed in several possible ways: by formation of a quenching threshold, below which activity is suppressed, and by being contrast enhanced in various ways if activity is above threshold. By appropriate setting of the network parameters the network can respond only to the maximum input amplitude or can contrast enhance part or all of the spatial input pattern. One of the ways this is accomplished is by incorporating what Grossberg calls a "sigmoid" (S-shaped) output function into the way the neuron responds to its inputs. This simple and useful idea is now used by many modelers (McClelland and Rumelhart, paper 25, for example). The activation level of a cell is a function of its inputs. However, the output amplitude of the cell need not reflect exactly the activation level, as it would in a linear model; rather, it is a nonlinear sigmoidal function of it. At low and high levels of activation, changes in activation lead to very small changes in output, while at intermediate levels of activation, changes in activation lead to larger changes in output. This kind of activation function respects the limited dynamic range of a cell.

The remainder of the paper is spent discussing how input patterns can be picked up by synapses at particular cells. As a simple example, if only one cell in the short term memory is active in response to a particular pattern (by appropriate normalizing, thresholding, and contrast enhancement), then the synapses to that cell will pick up the pattern (theorem 1, section 8). Grossberg discusses what happens when multiple patterns are present, and how it is possible to keep the "meaning" of a feature detector stable as the learning network is bombarded with a constantly changing set of inputs. Once a code—a particular set of discharging elements—has been established, it would be desirable to make it permanent.

One way of doing this would be to keep learning under tight biochemical control, so that learning occurs for a brief period and then ceases, once responses to the features of a particular environment have been learned. This solution lacks elegance, but it is almost certainly a major mechanism used by the early stages of the visual system, as Grossberg points out. The earlier levels of the cortex (area 17) are modifiable for a while in early development, and then no longer seem to be plastic. Fortunately, other parts of cortex seem to be more modifiable, since many adult humans are still capable of learning. It is hard to avoid the conclusion that a learning network is a dangerous thing. Biology has conspired to keep modifiability in the real nervous system under tight control.

Some of Grossberg's papers have been reprinted in a book (1982), which conveniently brings together his major ideas. References to his journal articles are given in the references in this paper and those in the other article by Grossberg in this collection (paper 24).

Reference

S. Grossberg (1982), *Studies in Mind and Brain*, New York: Reidel.

(1976)

S. Grossberg

Adaptive pattern classification and universal recoding: I.
Parallel development and coding of neural feature detectors
Biological Cybernetics 23:121–134

Abstract. This paper analyses a model for the parallel development and adult coding of neural feature detectors. The model was introduced in Grossberg (1976). We show how experience can retune feature detectors to respond to a prescribed convex set of spatial patterns. In particular, the detectors automatically respond to average features chosen from the set even if the average features have never been experienced. Using this procedure, any set of arbitrary spatial patterns can be recoded, or transformed, into any other spatial patterns (universal recoding), if there are sufficiently many cells in the network's cortex. The network is built from short term memory (STM) and long term memory (LTM) mechanisms, including mechanisms of adaptation, filtering, contrast enhancement, tuning, and nonspecific arousal. These mechanisms capture some experimental properties of plasticity in the kitten visual cortex. The model also suggests a classification of adult feature detector properties in terms of a small number of functional principles. In particular, experiments on retinal dynamics, including amacrine cell function, are suggested.

1. Introduction

This paper analyses a model for the development of neural feature detectors during an animal's early experience with its environment. The model also suggests mechanisms of adult pattern discrimination that remain after development has been completed. The model evolved from earlier experimental and theoretical work. Various data showed that there is a critical period during which experimental manipulations can alter the patterns to which feature detectors in the visual cortex are tuned (e.g., Barlow and

Pettigrew, 1971; Blakemore and Cooper, 1970; Blakemore and Mitchell, 1973; Hirsch and Spinelli, 1970, 1971; Hubel and Wiesel, 1970; Wiesel and Hubel, 1963, 1965). This work led Von der Malsburg (1973) and Pérez et al. (1974) to construct models of the cortical tuning process, which they analysed using computer methods. Their models are strikingly similar. Both use a mechanism of long term memory (LTM) to encode changes in tuning. This mechanism learns by classical, or Pavlovian, conditioning (Kimble, 1967) within a neural network. Such a concept was qualitatively described by Hebb (1949) and was rigorously analysed in its present form by Grossberg (e.g., 1967, 1970a, 1971, 1974). The LTM mechanism in a given interneuronal pathway is a plastic synaptic strength which has two crucial properties: (a) it is computed from a time average of the product of presynaptic signals and postsynaptic potentials; (b) it multiplicatively gates, or shunts, a presynaptic signal before it can perturb the postsynaptic cell.

Given this LTM mechanism, both models invoke various devices to regulate the retinocortical signals that drive the tuning process. On-center off-surround networks undergoing additive interactions, attenuation of small retinocortical signals at the cortex, and conservation of the total synaptic strength impinging on each cortical cell are used in both models. Grossberg (1976) realized that all of these mechanisms for distributing signals could be replaced by a minimal model for parallel processing of patterns in noise, which is realized by an on-center off-surround recurrent network whose interactions are of shunting type (Grossberg, 1973). Three crucial properties of this model are: (a) normalization, or adaptation, of total network activity; (b) contrast enhancement of input patterns; and (c) short term memory (STM) storage of the contrast-enhanced pattern. Using these properties, Grossberg (1976) eliminates the conservation of total synaptic strength—which is incompatible with

Supported in part by the Advanced Research Projects Agency under ONR Contract No. N00014-76-C-0185

© by Springer-Verlag 1976

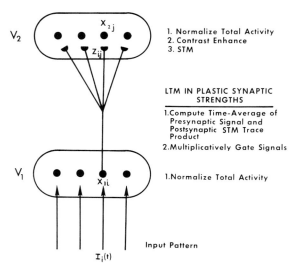

Fig. 1. Minimal model of developmental tuning using STM and LTM mechanisms

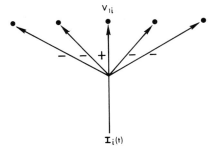

Fig. 2. Nonrecurrent, or feedforward, on-center off-surround network

classical conditioning—and shows that the tuning process can be derived from *adult* STM and LTM principles. The model is schematized in Figure 1. It describes the interaction via plastic synaptic pathways of two network regions, V_1 and V_2, that are separately capable of normalizing patterns, but V_2 can also contrast enhance patterns and store them in STM. In the original models of Von der Malsburg and Pérez et al., V_1 was interpreted as a "retina" or "thalamus" and V_2 as "visual cortex". In Part II, an analogous anatomy for V_1 as "olfactory bulb" and V_2 as "prepyriform cortex" will be noted. In Section 5, a more microscopic analysis of the model leads to a discussion of V_1 as a composite of retinal receptors, horizontal cells, and bipolar cells, and of V_2 as a composite of amacrine cells and ganglion cells. Such varied interpretations are possible because the same functional principles seem to operate in various anatomies.

Using this abstract structure, it was suggested in Grossberg (1976) how hierarchies of cells capable of discriminating arbitrary spatial patterns can be synthesized. Also a striking analogy was described between the structure and properties of certain reaction-diffusion systems that have been used to model development (Gierer and Meinhardt, 1972; Meinhardt and Gierer, 1974) and of reverberating shunting networks. This paper continues this program by rigorously analysing mathematical properties of the model, which thereupon suggest other developmental and adult STM and LTM mechanisms that are related to it. The following sections will describe these connections with a minimum of mathematical detail. Mathematical proofs are contained in the Appendix.

2. The Tuning Process

This section reviews properties of the model that will be needed below. Suppose that V_1 consists of n states (or cells, or cell populations) v_{1i}, $i = 1, 2, ..., n$, which receive inputs $I_i(t)$ whose intensity depends on the presence of a prescribed feature, or features, in an external pattern. Let the population response (or activity, or average potential) of v_{i1} be $x_{1i}(t)$. The relative input intensity $\Theta_i = I_i I^{-1}$, where $I = \sum_{k=1}^{n} I_k$, measures the relative importance of the feature coded by v_i in any given input pattern. If the Θ_i's are constant during a given time interval, the inputs are said to form a *spatial pattern*. How can the laws governing the $x_{1i}(t)$ be determined so that $x_{1i}(t)$ is capable of accurately registering Θ_i? Grossberg (1973) showed that a bounded, linear law for x_{1i}, in which x_{1i} returns to equilibrium after inputs cease, and in which neither input pathways nor populations v_{1i} interact, does not suffice; cf., Grossberg and Levine (1975) for a review. The problem is that as the total input I increases, given *fixed* Θ_i values, each x_{1i} saturates at its maximal value. This does not happen if off-surround interactions also occur. For example, let the inputs I_i be distributed via a nonrecurrent, or feedforward, on-center off-surround anatomy undergoing shunting (or mass action, or passive membrane) interactions, as in Figure 2. Then

$$\dot{x}_{1i} = -A x_{1i} + (B - x_{1i})I_i - x_{1i} \sum_{k \neq i} I_k \qquad (1)$$

with $0 \leq x_{1i}(0) \leq B$. At equilibrium (namely, $\dot{x}_{1i} = 0$),

$$x_{1i} = \Theta_i \frac{BI}{A + I}, \qquad (2)$$

which is proportional to Θ_i no matter how large I becomes. Since also $BI(A + I)^{-1} \leq B$, the total activity

$$x_1 \equiv \sum_{k=1}^{n} x_{1k}$$ never exceeds B; it is normalized, or adapts, due to automatic gain control by the in-

hibitory inputs. The normalization property in (2) shows that x_{1i} codes Θ_i rather than instantaneous fluctuations in I.

To store patterns in STM, recurrent or feedback pathways are needed to keep signals active after the inputs cease. Again the problem of saturation must be dealt with, so that some type of recurrent on-center off-surround anatomy is suggested. The minimal solution is to let V_2 be governed by a system of the form

$$\dot{x}_{2j} = -Ax_{2j} + (B - x_{2j})[f(x_{2j}) + I_{2j}] - x_{2j} \sum_{k \neq j} f(x_{2k}),$$
$$(3)$$

where $f(w)$ is the average feedback signal produced by an average activity level w, and I_{2j} is the total excitatory input to v_{2j} (Fig. 3a). In particular, v_{2j} excites itself via the term $(B - x_{2j})f(x_{2j})$, and v_{2k} inhibits v_{2j} via the term $-x_{2j}f(x_{2k})$, for every $k \neq j$. The choice of $f(w)$ dramatically influences how recurrent interactions within V_2 transform the input pattern $I^{(2)} = (I_{21}, I_{22}, \ldots, I_{2N})$ through time. Grossberg (1973) shows that a sigmoid, or S-shaped, $f(w)$ can reverberate important inputs in STM after contrast-enhancing them, yet can also suppress noise.

Various generalizations of recurrent networks have been studied, such as

$$\dot{x}_{2j} = -Ax_{2j} + (B - x_{2j})\left[\sum_{k=1}^{N} f(x_{2k})C_{kj} + I_{2j}\right]$$
$$- (x_{2j} + D)\sum_{k=1}^{N} f(x_{2k})E_{kj},$$
$$(4)$$

$D \geq 0$, where the excitatory coefficients C_{kj} ("on-center") decrease with the distance between populations v_{2k} and v_{2j} more rapidly than do the inhibitory coefficients E_{kj} ("off-surround"). Levine and Grossberg (1976) show that, in such cases, the inhibitory off-surround signals $\sum_{k=1}^{N} f(x_{2k})E_{kj}$ to v_{2j} can be chosen strong enough to offset the saturating effects of inputs I_{2j} plus excitatory on-center signals $\sum_{k=1}^{N} f(x_{2k})C_{kj}$. Ellias and Grossberg (1975) study generalizations of (4) in which inhibitory interneurons interact with their excitatory counterparts.

Below we will consider networks in which the excitatory signals I_{2j} to V_2 are sums of signals from many populations in V_1. Moreover, the synaptic strengths of these signals can be trained. This fact suggests another reason for making V_2 recurrent. A recurrent anatomy is needed within V_2 to prevent saturation in response to trainable signals. To see this, note in the nonrecurrent network (1) that each excitatory input to v_{1i} is replicated as an inhibitory input to all v_{1k}, $k \neq i$. The size of a trainable signal to v_{2j}

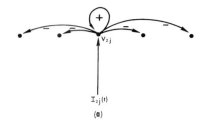

(a)

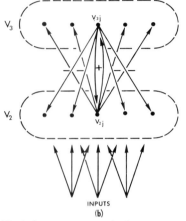

INPUTS
(b)

Fig. 3. Some recurrent, or feedback, on-center off-surround networks

depends on the activity at v_{2j}. This signal therefore cannot be replicated at populations v_{2k}, $k \neq j$, unless recurrent interactions within V_2 exist. Moreover, whether or not signals are trainable, whenever I_{2j} is a sum of signals from many populations, recurrent signals within V_2 prevent saturation at a large saving of extra signal pathways to the populations v_{2k}, $k \neq j$.

A related scheme for marrying sums of (trainable) signals with pattern normalization is illustrated in Figure 3b. Here a sum of signals I_{2j} from V_1 perturbs each v_{2j}. Population v_{2j} thereupon excites an on-center of cells near v_{3j}, and inhibits a broad off-surround of populations centered at v_{3j}. Thus, when a pattern $I^{(2)}$ arrives at V_2, it is normalized at V_3 before saturation can take place across V_2. Then feedback signals from V_3 to V_2 prevent saturation at V_2 from setting in as follows. Each population v_{3j} that receives a large net excitatory signal from V_2 excites its on-center of cells near v_{2j}, and inhibits a broad off-surround of populations centered at v_{2j}. This feedback inhibition prevents the pattern $I^{(2)}$ from saturating V_2, much as recurrent inhibition in Equation (4) works. Figure 3b can also be expanded to explicitly include inhibitory interneurons, as in Ellias and Grossberg (1975).

Normalization in V_1 by (1) occurs gradually in time, as each x_{1i} adjusts to its new equilibrium value, but it will be assumed below to occur instantaneously with x_{1i} approaching Θ_i rather than $\Theta_i BI(A+I)^{-1}$. These simplifications yield theorems about the tuning process that avoid unimportant details. The assumption that normalization occurs instantaneously is tenable because the normalized pattern at V_1 drives slow changes in the strength of connections from V_1 to V_2. Instantaneous normalization means that the pattern at V_1 normalizes itself before the connection strengths have a chance to substantially change.

Let the synaptic strength of the pathway from v_{1i} to the jth population v_{2j} in V_2 be denoted by $z_{ij}(t)$ (see Fig. 1). Let the total signal to v_{2j} due to the normalized pattern $\Theta = (\Theta_1, \Theta_2, ..., \Theta_n)$ at V_1 and the vector $z^{(j)}(t) = (z_{1j}(t), z_{2j}(t), ..., z_{nj}(t))$ of synaptic strengths be

$$S_j(t) \equiv \Theta \cdot z^{(j)}(t) \equiv \sum_{k=1}^{n} \Theta_k z_{kj}(t); \qquad (5)$$

that is, each $z_{kj}(t)$ *gates* the signal Θ_k from v_{1k} on its way to v_{2j}, and these gated signals combine additively at v_{2j} (cf., Grossberg, 1967, 1970a, 1971, 1974). Since $z^{(j)}(t)$ determines the size of the input to v_{2j}, given any pattern Θ, it is called the *classifying vector* of v_{2j} at time t. Every v_{2j}, $j = 1, 2, ..., N$, in V_2 receives such a signal when Θ is active at V_1. In this way, Θ creates a pattern of activity across V_2.

Given any activity pattern across V_2, it can be transformed in several ways as time goes on. Two main questions about this process are: (a) will the *total* activity of V_2 be suppressed, or will some of its activities be stored in STM? and (b) which of the *relative* activities across V_2 will be preserved, suppressed, or enhanced? Several papers (Ellias and Grossberg, 1975; Grossberg, 1973; Grossberg and Levine, 1975) analyse how the parameters of a reverberating shunting on-center off-surround network determine the answers to these questions. Below some of these facts are cited as they are needed. In particular, if all the activities are sufficiently small, then they will not be stored in STM. If they are sufficiently large, then they will be contrast enhanced, normalized, and stored in STM. Figure 4 schematizes two storage possibilities. Figure 4a depicts a pattern of activity across V_2 before it is transformed by V_2. Given suitable parameters, if some of the initial activities exceed a quenching threshold (QT), then V_2 will *choose* the population having maximal initial activity for storage in STM, as in Figure 4b. Under other circumstances, all initial activities below the QT are suppressed, whereas *all* initial activities above QT are contrast enhanced, normalized, and stored in STM (Fig. 4c); that is, *partial* contrast in STM is possible. Grossberg (1973) shows that partial contrast can occur if the signals between populations in a recurrent shunting on-center off-surround network are sigmoid (S-shaped) functions of

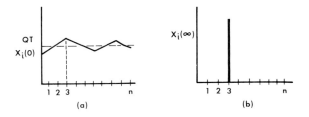

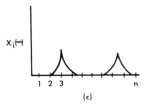

Fig. 4. Contrast enhancement and STM by recurrent network: (a) initial pattern; (b) choice; (c) partial contrast

their activity levels. Ellias and Grossberg (1975) show that partial contrast can occur if the self-excitatory signals of populations in V_2 are stronger than their self-inhibitory signals, and moreover if the excitatory signals between populations in V_2 decrease with inter-population distance faster than the inhibitory signals.

The enhancement and STM storage processes also occur much faster than the slow changes in connection strengths z_{ij}; hence, it is assumed below that these processes occur instantaneously in order to focus on the slow changes in z_{ij}.

The slow changes in z_{ij} are assumed to be determined by a time averaged product of the signal from v_{1i} to v_{2j} with the cortical response at v_{2j}; thus

$$\dot{z}_{ij} = -C_{ij} z_{ij} + D_{ij} x_{2j},$$

where C_{ij} is the decay rate (possibly variable) of z_{ij}, and D_{ij} is the signal from v_{1i} to v_{2j}. For example, if $C_{ij} = 1$, the V_1 and V_2 patterns are normalized, and V_2 chooses only the population v_{2j} whose initial activity is maximal for storage in STM (Fig. 4b), then while v_{2j} is active,

$$\dot{z}_{ij} = -z_{ij} + \Theta_i, \quad \text{for all } i = 1, 2, ..., n.$$

It remains to determine how these z_{ij} and all other z_{ik}, $k \neq j$, change under other circumstances. To eliminate conceptual and mathematical difficulties that arise if z_{ij} can decay even when V_1 and V_2 are inactive, we let *all* changes in each z_{ij} be determined by which populations in V_2 have their activities chosen for storage in STM. In other words, all changes in z_{ij} are driven by the *feedback* within the excitatory recurrent loops of V_2 that establish STM storage. Then

$$\dot{z}_{ij} = (-z_{ij} + \Theta_i)x_{2j} \tag{6}$$

where $\sum_{k=1}^{N} x_{2k}(t) = 1$ if STM in V_2 is active at time t,

whereas $\sum_{k=1}^{N} x_{2k}(t) = 0$ if STM in V_2 is inactive at time t.

If V_2 *chooses* a population for storage in STM, as in Figure 4b, then

$$x_{2j} = \begin{cases} 1 & \text{if } S_j > \max\{\varepsilon, S_k : k \neq j\} \\ 0 & \text{if } S_j < \max\{\varepsilon, S_k : k \neq j\}, \end{cases} \tag{7}$$

where as in (5), $S_j = \Theta \cdot z^{(j)}$ with $\Theta_i = I_i \left(\sum_{k=1}^{n} I_k \right)^{-1}$. Equation (7) omits the cases where two or more signals S_j are equal, and are larger than all other signals and ε. In these cases, the x_{2j}'s of such S_j's are equal and add up to 1. Such a normalization rule for equal maximal signals will be tacitly assumed in all the cases below, but will otherwise be ignored to avoid tedious details. Equation (6) shows that z_{ij} can change only if $x_{2j} > 0$. Equation (7) shows that V_2 chooses the maximal activity for storage in STM. This activity is normalized ($x_{2j} = 0$ or 1), and it corresponds to the population with largest initial signal ($S_j > \max\{S_k : k \neq j\}$). No changes in z_{ij} occur if all signals S_j are too small to be stored in STM (all $S_j \leq \varepsilon$).

If partial contrast in STM holds, as in Figure 4c, then the dynamics of a reverberating shunting network can be approximated by a rule of the form

$$x_{2j} = \begin{cases} f(S_j) \left[\sum_{S_k > \varepsilon} f(S_k) \right]^{-1} & \text{if } S_j > \varepsilon \\ 0 & \text{if } S_j < \varepsilon \end{cases} \tag{8}$$

where $f(w)$ is an increasing nonnegative function of w such that $w = 0$; e.g., $f(w) = w^2$. In (8), the positive constant ε represents the QT; the function $f(w)$ controls how suprathreshold signals S_j will be contrast enhanced; and the ratio of $f(S_j)$ to $\sum\{f(S_k) : S_k > \varepsilon\}$ expresses the normalization of STM.

3. Ritualistic Pattern Classification

After developmental tuning has taken place, the above mechanisms describe a model of pattern classification in the "adult" network. These mechanisms will be described first as interesting in themselves, and as a helpful prelude to understanding the tuning process. They are capable of classifying arbitrarily complicated spatial patterns into mutually nonoverlapping, or partially overlapping, sets depending on whether (7) or (8) holds. These mechanisms realize basic principles of pattern discrimination using shunting interactions.

An alternative scheme of pattern discrimination using a mixture of shunting and additive mechanisms has already been given (Grossberg, 1970b, 1972). Together these schemes suggest numerous anatomical and physiological variations that embody the same small class of functional principles. Since particular anatomies imply that particular physiological rules should be operative, intriguing questions about the dynamics of various neural structures, such as retina, neocortex, hippocampus, and cerebellum, are suggested.

First consider what happens if V_2 chooses a population for storage in STM. After learning ceases (that is, $\dot{z}_{ij} \equiv 0$), all classifying vectors $z^{(j)}$ are constant in time, and Equations (6) and (7) reduce to the statement that population v_{2j} is stored in STM if

$$S_j > \max\{\varepsilon, S_k : k \neq j\}. \tag{9}$$

In other words, v_{2j} codes all patterns Θ such that (9) holds; alternatively stated, v_{2j} is a *feature detector* in the sense that all patterns

$$P_j = \{\Theta : \Theta \cdot z^{(j)} > \max(\varepsilon, \Theta \cdot z^{(k)} : k \neq j)\} \tag{10}$$

are classified by v_{2j}. The set P_j defines a *convex cone* C_j in the space of nonnegative input vectors $J = (I_1, I_2, ..., I_n)$, since if two such vectors $J^{(1)}$ and $J^{(2)}$ are in C_j, then so are all the vectors $\alpha J^{(1)}$, $\beta J^{(2)}$, and $\gamma J^{(1)} + (1 - \gamma)J^{(2)}$, where $\alpha > 0$, $\beta > 0$, and $0 < \gamma < 1$. The convex cone C_j defines the *feature* coded by v_{2j}.

The classification rule in (10) has an informative geometrical interpretation in n-dimensional Euclidean space. The signal $S_j = \Theta \cdot z^{(j)}$ is the inner product of Θ and $z^{(j)}$ (Greenspan and Benney, 1973). Letting $\|\xi\| = \sqrt{\sum_{k=1}^{n} \xi_k^2}$ denote the Euclidean length of any real vector $\xi = (\xi_1, \xi_2, ..., \xi_n)$, and $\cos(\eta, \omega)$ denote the cosine between two vectors η and ω, it is elementary that

$$S_j = \|\Theta\| \, \|z^{(j)}\| \cos(\Theta, z^{(j)}).$$

In other words, the signal S_j is the length of the projection of the normalized pattern Θ on the classifying vector $z^{(j)}$ times the length of $z^{(j)}$. Thus if all $z^{(j)}$, $j = 1, 2, ..., N$, have equal length, then among all patterns with the same length, (10) classifies all patterns Θ in P_j whose angle with $z^{(j)}$ is smaller than the angles between Θ and any $z^{(k)}$, $k \neq j$, and is small enough to satisfy the ε-condition. In particular, patterns Θ that are *parallel* to $z^{(j)}$ are classified in P_j. The choice of classifying vectors $z^{(j)}$ hereby determines how the patterns Θ will be divided up. Section 8 will show that the tuning mechanism (6)–(7) makes the $z^{(j)}$ vectors more parallel to prescribed patterns Θ, and thereupon changes the classifying sets P_j. In summary

(i) the number of populations in V_2 determines the maximum number N of pattern classes P_j;

(ii) the choice of classifying vectors $z^{(j)}$ determines

how different these classes can be; for example, choosing all vectors $z^{(j)}$ equal will generate one class that is redundantly represented by all v_{2j}; and

(iii) the size of ε determines how similar patterns must be to be classified by the same v_{2j}.

If the choice rule (7) is replaced by the partial contrast rule (8), then an important new possibility occurs, which can be described either by studying STM responses to all Θ at fixed v_{2j}, or to a fixed Θ at all v_{2j}. In the former case, each v_{2j} has a *tuning curve*, or *generalization gradient*; namely, a maximal response to certain patterns, and submaximal responses to other patterns. In the latter case, each pattern Θ is *filtered* by V_2 in a way that shows how close Θ lies to *each* of the classifying vectors $z^{(j)}$. The pattern will only be classified by v_{2j}—that is, stored in STM—if it lies sufficiently close to $z^{(j)}$ for its signal S_j to exceed the quenching threshold of V_2.

For example, suppose that some of the classifying vectors $z^{(j)}$ are chosen to create large signals at V_2 when vertical lines perturb V_1, and that other $z^{(j)}$ create large signals at V_2 when horizontal lines perturb V_1. If a pattern containing both horizontal and vertical lines perturbs V_1, then the population activities in V_2 corresponding to both types of lines can be stored in STM, unless competition between their populations drives all activity below the QT. Now let V_3 be another "cortex" that receives signals from V_2, in the same fashion that V_2 receives signals from V_1. Given an appropriate choice of classifying vectors for V_3, there can exist cells in V_3 that fire in STM only if horizontal *and* vertical lines perturb a prescribed region of V_1; e.g., hypercomplex cells. The existence of tuning curves in a given cortex V_i hereby increases the discriminative capabilities of the next cortex V_{i+1} in a hierarchy; cf., Grossberg (1976).

The above mechanisms will now be discussed as cases of a general scheme of pattern classification. This is done with two goals in mind: firstly, to emphasize that these mechanisms might well exist in other than "retinocortical" analogs; and secondly, to generate explicit experimental directives in a variety of neural structures. One such directive will be described in Section 5.

4. Shunts vs. Additive Interactions as Mechanisms of Pattern Classification

The processing stages utilized in Section 3 are the following:

A) Normalization

Input patterns are normalized in V_1 by an on-center off-surround anatomy undergoing shunting interactions.

B) Partial Filtering by Signals

The signals S_j generated at V_2 by a normalized pattern on V_1 create the data base on which later computations are determined. The signal generating rule (5), for example, has the following important property. Suppose that an input $I_i(t) = \Theta_i I(t)$ is normalized to x_{1i}, as in (2), rather than to the approximate value Θ_i. The signal from V_1 to v_{2j} becomes

$$\tilde{S}_j = BI(A+I)^{-1}S_j$$

and (9) is replaced by the analogous rule

$$\tilde{S}_j > \max\{\varepsilon, \tilde{S}_k : k \neq j\} \ .$$

Then V_2 will classify a given pattern into the same class P_j no matter how large I is chosen. In other words, the signal generating rule is invariant under suprathreshold variations of the total activity at V_1. If I_i is the transduced receptor response to an external input J_i—that is, $I_i = g(J_i)$—then the signal-generating rule is invariant, given *any* $z^{(j)}$'s, if $g(w) = w^p$ for some $p > 0$.

C) Contrast Enhancement of Signals

The signals S_j are contrast enhanced by the recurrent on-center off-surround anatomy within V_2, and either a choice (Fig. 4b) or a tuning curve (Fig. 4c) results.

Two successive stages of lateral inhibition are needed in this model. The first stage normalizes input patterns. The second stage sharpens the filtering of signals.

Additive mechanisms can also achieve classification of arbitrarily complicated spatial patterns. These mechanisms also employ three successive stages A)–C) of pattern processing, with stage A) normalizing input patterns, stages A) and C) using inhibitory interactions, and stage C) completing the pattern classification that is begun by the signal generating rules of stage B). The additive model can differ in several respects from the shunting model:

(i) its anatomy can be feedforward; that is, there need not be a recurrent network in stage C);

(ii) threshold rules replace the inner product signal-generating rule (5) to determine partial filtering of signals; and

(iii) the responses in time of stages A)–C) to a sustained pattern at V_1 are not the same in the additive model. For example, sustained responses in the shunting model can be replaced by responses to the onset and offset of the pattern in the additive model (Grossberg, 1970b).

Mixtures of additive and shunting mechanisms are also possible. The additive mechanisms will now be summarized to illustrate the basic stages A)–C).

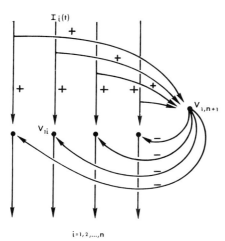

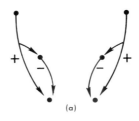

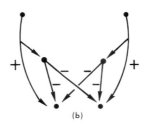

Fig. 5. Normalization and low-band filtering by subtractive non-specific interneuron and signal threshold rules

Fig. 6. (a) Specific subtractive inhibitory interneurons; (b) Non-specific inhibitory interneurons

An additive nonspecific inhibitory interneuron normalizes patterns at V_1 (Fig. 5). Many variations on this theme exist (Grossberg, 1970b) in which such parameters as the lateral spread of inhibition, the number of cell layers, and the rates of excitatory and inhibitory decay can be varied. The idea in its simplest form is this. The excitatory input I_i excites a bifurcating pathway. One branch of the pathway is specific, and the other branch is nonspecific. The lateral inhibitory interneuron $v_{1,n+1}$ lies in the nonspecific branch. It sums the excitatory inputs I_i, and generates a nonspecific signal back to all the specific pathways if a signal threshold Γ is exceeded. Each input I_i also generates a specific signal from v_{1i} that is a linear function of I_i above a signal threshold. Each pathway from v_{1i} in V_1 to v_{2j} in V_2 has its own signal threshold Γ_{ij}. The net signal from v_{1i} to v_{2j} is

$$K_{ij} = [I_i - \Gamma_{ij}]^+ - \left[\sum_{k=1}^{n} I_k - \Gamma\right]^+,$$

where the notation $[u]^+ = \max(u, 0)$ defines the threshold rule. Define $\Theta_{ij} = \Gamma_{ij}\Gamma^{-1}$ and let the spatial pattern $I_i = \Theta_i I$ perturb V_1. Then

$$K_{ij} = [\Theta_i I - \Theta_{ij}\Gamma]^+ - [I - \Gamma]^+. \tag{11}$$

The net signal K_{ij} has the following properties:
  (i) $K_{ij} \leq 0$ for all values of $I > 0$ if $\Theta_i \leq \Theta_{ij}$;
  (ii) $K_i > 0$ for $I > \Theta_{ij}\Theta_i^{-1}$ if $\Theta_i > \Theta_{ij}$; and
  (iii) $K_{ij} \leq (\Theta_i - \Theta_{ij})\Gamma$ for all $I > 0$.
In other words, by (i), no signal is emitted from v_{1i} to v_{2j} if $\Theta_i < \Theta_{ij}$; by (ii), if $\Theta_i > \Theta_{ij}$, a signal is emitted from v_{1i} if I exceeds a threshold depending on Θ_i and Θ_{ij}; and by (iii), the total activity in the cells v_{1i} is normalized. Partial filtering of signals is thus achieved by

the choice of threshold pattern $\Theta^{(j)} = (\Theta_{1j}, \Theta_{2j}, \ldots, \Theta_{nj})$ rather than by the choice of classifying vector $z^{(j)} = (z_{1j}, z_{2j}, \ldots, z_{nj})$.

Stage C) is needed because the total signal to v_{2j} can be maximized by patterns Θ which are very different from the threshold pattern $\Theta^{(j)}$. This problem arises because the signals K_{ij} continue to grow linearly as a function of I after the threshold value $\Theta_{ij}\Theta_i^{-1}$ is exceeded. Grossberg (1970b) shows that the problem can be avoided by inhibiting each signal K_{ij} if it gets too large. For example, let the net signal from v_{1i} to v_{2j} be

$$S_{ij}^* = K_{ij} - \alpha[K_{ij} - \beta]^+, \tag{12}$$

where $\alpha > 1$ and $0 < \beta \ll 1$. This mechanism inhibits the signal from v_{1i} to v_{2j} if it represents a Θ_i which is too much larger than Θ_{ij}. Equation (12) can be realized by any of the several inhibitory mechanisms: a specific subtractive inhibitory interneuron (Fig. 6a), a switchover from net excitation to net inhibition when the spiking frequency in the pathway from v_{1i} to v_{2j} becomes too large (Bennett, 1971; Blackenship et al., 1971; Wachtel and Kandel, 1971), or postsynaptic blockade of the v_{2j} cell membrane at sufficiently high spiking frequencies. Signal S_{ij}^* is positive only if Θ_i is sufficiently close to Θ_{ij} in value. Stage C) is completed by choosing the signal threshold of v_{2j} so high that v_{2j} only fires if *all* signals S_{ij}^*, $i = 1, 2, \ldots, n$, are positive; that is, only if the input pattern Θ is close to the threshold pattern $\Theta^{(j)}$. The second stage of inhibition hereby completes the partial filtering process by choosing a population v_{2j} in V_2 to code $\Theta^{(j)}$, as in Figure 4b. If the specific inhibitory interneurons if Figure 6a are replaced by a lateral spread of inhibition, as in Figure 6b, then a tuning curve is generated, as in Figure 4c.

5. What Do Retinal Amacrine Cells Do?

This section illustrates how the principles A)–C) can generate interesting questions about particular neural processes. Grossberg (1970b, 1972) introduces a retinal model in which shunting and additive interactions both occur. In this model, retinal amacrine cells are examples of the inhibitory interaction in stage C). We will note that amacrine cells have *opposite* effects on signals if they realize a shunting rather than an additive model. In the retinal model of Grossberg (1972), normalization is accomplished by an on-center off-surround anatomy undergoing shunting interactions. Analogously, in vivo receptors excite bipolar cells (on-center) as well as horizontal cells, and the horizontal cells inhibit bipolar cells via their lateral interactions (off-surround). Partial filtering of the normalized inputs is accomplished by signal thresholds; for example, using the normalized x_{1i} activities in (2), the simplest signal function from v_{1i} to v_{2j} is $K_{ij} = [x_{1i} - \Gamma_{ij}]^+$. Stage C) is then accomplished by a mechanism such as (12), by which large signals are inhibited. Whether a choice (Fig. 4b) or a tuning curve (Fig. 4c) is generated depends, in part, on how broadly these lateral inhibitory signals that complete stage C) are distributed. This second stage of inhibition is identified with the inhibition that amacrine cells, fed by bipolar cell activity, generate at ganglion cells. Grossberg (1972) notes data that support the idea that stage C) is realized by an additive mechanism such as (12). In particular, amacrine cells often respond when an input pattern is turned on, or off, or both. Two questions about amacrine cells now suggest themselves.

(i) If this interpretation of amacrine cells is true, then they will shut off signals from the bipolar cells to the ganglion cells when these signals become too *large*; that is, they act as high-band filters. By contrast, inhibition in stage C) of the shunting model shuts off signals if they become too *small*. Opposite effects due to the second inhibitory stage can hereby create a similar functional transformation of the input pattern! If a shunting role for amacrine cells is sought, then the following types of anatomy would be anticipated: inhibitory bipolar-to-amacrine-to-bipolar cell feedback that contrast enhances the receptor-to-bipolar signals, or inhibitory ganglion-to-amacrine-to-ganglion cell feedback that contrast enhances the bipolar-to- ganglion cell signals, or some functionally similar

feedback loop. To decide between these two possible roles for amacrine cells, one must test whether amacrine cells suppress large signals or small ones; in either case, if the model is applicable, contrast enhancement of the normalized and filtered retinal pattern is the result, so that this property cannot be used as a criterion.

(ii) Does the spatial extent of lateral amacrine interaction determine the amount of contrast, or the breadth of the tuning curves, in ganglion cell responses, as in Figures 4b and 4c? For example, there exist narrow field diffuse amacrine cells, wide field diffuse amacrine cells, stratified diffuse amacrine cells, and unstratified amacrine cells (Boycott and Dowling, 1969). Do these specializations guarantee particular tuning characteristics in the corresponding ganglion cells?

Grossberg (1972) also suggests a cerebellar analog based on the same principles. Thus at least formal aspects of various neural structures seem to be emerging as manifestations of common principles. These results suggest a program of classifying seemingly different anatomical and physiological data according to whether they realize similar functional transformations of patterned neural activity, such as total activity normalization, partial filtering by signals, and contrast enhancement of the signal pattern. Below are described certain properties of the shunting mechanism that will be needed when development is discussed.

6. Arousal as a Tuning Mechanism

The recurrent networks in V_2 all have a quenching threshold (QT); namely, a criterion activity level that must be exceeded before a population's activity can reverberate in STM. Changing the QT or, equivalently, changing the size of signals to V_2, can retune the responsiveness of populations in V_2 to prescribed patterns at V_1. For example, suppose that an unexpected, or novel, event triggers a nonspecific arousal input to V_2, which magnifies all the signals from V_1 to V_2 (see Part II). Then certain signals, which could not otherwise be stored in STM, will exceed the QT and be stored. For example, if V_2 is capable of partial contrast in STM and also receives a nonspecific arousal input, then (8) can be replaced by

$$x_{2j} = \begin{cases} f(\phi S_j) \left[\sum_{\phi S_k > \varepsilon} f(\phi S_k) \right]^{-1} & \text{if} \quad \phi S_j > \varepsilon \\ 0 & \text{if} \quad \phi S_j < \varepsilon \end{cases} \tag{13}$$

where ϕ is an increasing function of the arousal level. Note that an increase in ϕ allows more V_2 populations to reverberate in STM; cf., Grossberg (1973) for mathematical proofs. In a similar fashion, if an unexpected event triggers nonspecific shunting inhibition of the inhibitory interneurons in the off-surrounds of V_2, then the QT will decrease (Grossberg, 1973; Ellias and Grossberg, 1975), yielding an equivalent effect. Equation (8) can then be changed to

$$
x_{2j} = \begin{cases} f(S_j) \left[\sum_{S_k > \phi^* \varepsilon} f(S_k) \right]^{-1} & \text{if} \quad S_j > \phi^* \varepsilon \\ 0 & \text{if} \quad S_j < \phi^* \varepsilon \end{cases} \tag{14}
$$

where ϕ^* is a decreasing function of the arousal level.

Reductions in arousal level have the opposite effect. For example, if (13) holds, and arousal is lowered until only one population in V_2 exceeds the QT, then a choice will be made in STM, as in Figure 4b. Thus a choice in STM can be due either to *structural* properties of the network, such as the rules for generating signals between populations in V_2 [cf. the faster-than-linear signal function in Grossberg (1973)], or to an arousal level that is not high enough to create a tuning curve. Similarly, if arousal is too small, then all functions x_{2j} in (13) will always equal zero, and no STM storage will occur.

Changes in arousal can have a profound influence on the time course of LTM, as in (6), because they change the STM patterns that drive the learning process. For example, if during development arousal level is chosen to produce a choice in STM, then the tuning of classifying vectors $z^{(j)}$ will be sharper than if the arousal level were chosen to generate partial contrast in STM.

The influence of arousal on tuning of STM patterns can also be expressed in another way, which suggests a mechanism that will be needed in Part II when universal recoding is discussed.

7. Arousal as a Search Mechanism

Suppose that arousal level is fixed during learning trials, and that a given pattern Θ at V_1 does not create any STM storage at V_2 because all the inner products $\Theta \cdot z^{(j)}$ are too small. If arousal level is then increased in (13) until some $x_{2j} > 0$, STM storage will occur. In other words, changing the arousal level can facilitate *search* for a suitable classifying population in V_2.

Why does arousal level increase if no STM storage occurs at V_2? This is a property of the expectation mechanism that is developed in Part II. Also in Part II a pattern Θ at V_1 that is not classified by V_2 will use this mechanism to release a subliminal search routine that terminates when an admissible classification occurs.

8. Development of an STM Code

System (6)–(7) will be analysed mathematically because it illustrates properties of the model in a particularly

simple and lucid way. The first result describes how this system responds to a single pattern that is iteratively presented through time.

Theorem 1 (One Pattern)

Given a pattern Θ, suppose that there exists a unique j such that

$$
S_j(0) > \max \{\varepsilon, S_k(0) : k \neq j\} . \tag{15}
$$

Let Θ be practiced during a sequence of nonoverlapping intervals $[U_k, V_k]$, $k = 1, 2, \dots$. Then the angle between $z^{(j)}(t)$ and Θ monotonically decreases, the signal $S_j(t)$ is monotonically attracted towards $\|\Theta\|^2$ and $\|z^{(j)}\|^2$ oscillates at most once as it pursues $S_j(t)$. In particular, if $\|z^{(j)}(0)\| \leq \|\Theta\|$, then $S_j(t)$ is monotone increasing. Except in the trivial case that $S_j(0) = \|\Theta\|^2$, the limiting relations

$$
\lim_{t \to \infty} \|z^{(j)}(t)\|^2 = \lim_{t \to \infty} S_j(t) = \|\Theta\|^2 \tag{16}
$$

hold if and only if

$$
\sum_{k=1}^{\infty} (V_k - U_k) = \infty . \tag{17}
$$

Remark. If $z^{(j)}(0)$ is small, in the sense that $\|z^{(j)}(0)\| \leq \|\Theta\|$, then by Theorem 1, as time goes on, the learning process maximizes the inner product signal $S_j(t) = \Theta \cdot z^{(j)}(t)$ over all possible choices of $z^{(j)}$ such that $\|z^{(j)}\| \leq \|\Theta\|$. This follows from the obvious fact that

$$
\sup \{\Theta \cdot \psi : \|\psi\| \leq \|\Theta\|\} = \|\Theta\|^2 .
$$

Otherwise expressed, learning makes $z^{(j)}$ parallel to Θ, and normalizes the length of $z^{(j)}$.

What happens if several different spatial patterns $\Theta^{(k)} = (\Theta_1^{(k)}, \Theta_2^{(k)}, \dots, \Theta_n^{(k)})$, $k = 1, 2, \dots, M$, all perturb V_1 at different times? How are changes in the z_{ij}'s due to one pattern prevented from contradicting changes in the z_{ij}'s due to a different pattern? The choice-making property of V_2 does this for us; it acts as a sampling device that prevents contradictions from occurring. A heuristic argument will now be given to suggest how sampling works. This argument will then be refined and made rigorous. For definiteness, suppose that M spatial patterns $\Theta^{(k)}$ are chosen, $M \leq N$, such that their signals at time $t = 0$ satisfy

$$
\Theta^{(k)} \cdot z^{(k)}(0) > \max \{\varepsilon, \Theta^{(k)} \cdot z^{(j)}(0) : j \neq k\} \tag{18}
$$

for all $k = 1, 2, \dots, M$. In other words, at time $t = 0$, $\Theta^{(k)}$ is coded by v_{2k}. Let $\Theta^{(1)}$ be the first pattern to perturb V_1. By (18), population v_{21} receives the largest signal from V_1. All other populations v_{2j}, $j \neq 1$, are thereupon inhibited by the off-surround of v_{21}, whereas v_{21} reverberates in STM. By (6), none of the synaptic strengths $z^{(j)}(t)$, $j \neq 1$, can learn while $\Theta^{(1)}$ is presented. As in Theorem 1, presenting $\Theta^{(1)}$ makes $z^{(1)}(t)$ more parallel

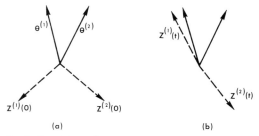

Fig. 7. Practicing $\Theta^{(1)}$ brings $z^{(1)}(t)$ closer to $\Theta^{(1)}$ and $\Theta^{(2)}$ than $z^{(2)}(0)$

to $\Theta^{(1)}$ as t increases. Consequently, if a different pattern, say $\Theta^{(2)}$, perturbs V_1 on the next learning trial, then it will excite v_{22} more than any other v_{2j}, $j \neq 2$: it cannot excite v_{21} because the coefficients $z^{(1)}(t)$ are more parallel to $\Theta^{(1)}$ than before; and it cannot excite any v_{2j}, $j \neq 1, 2$, because the v_{2j} coefficients $z^{(j)}(t)$ still equal $z^{(j)}(0)$. In response to $\Theta^{(2)}$, v_{22} inhibits all other v_{2j}, $j \neq 2$. Consequently none of the v_{2j} coefficients $z^{(j)}(t)$ can learn, $j \neq 2$; learning makes the coefficients $z^{(2)}(t)$ become more parallel to $\Theta^{(2)}$ as t increases. The same occurs on all learning trials. By inhibiting the post-synaptic part of the learning mechanism in all but the chosen V_2 population, the on-center off-surround network in V_2 samples one vector $z^{(j)}(t)$ of trainable coefficients at any time. In this way, V_2 can learn to distinguish as many as N patterns if it contains N populations.

This argument is almost correct. It fails, in general, because by making (say) $z^{(1)}(t)$ more parallel to $\Theta^{(1)}$, it is also possible to make $z^{(1)}(t)$ more parallel to $\Theta^{(2)}$ than $z^{(2)}(0)$ is. Thus when $\Theta^{(2)}$ is presented, it will be coded by v_{21} rather than v_{22}. In other words, practicing one pattern can recode other patterns. A typical example of this property is illustrated in Figure 7. Figure 7a depicts the two dimensional patterns $\Theta^{(1)}$ and $\Theta^{(2)}$ as solid vectors, and the two classifying vectors $z^{(1)}(0)$ and $z^{(2)}(0)$ as dotted vectors. Clearly (18) holds for $j = 1, 2$. As a result of practicing $\Theta^{(1)}$ during a fixed interval, Figure 7b is produced. Note that $\Theta^{(2)} \cdot z^{(1)}(t) > \Theta^{(2)} \cdot z^{(2)}(t)$ after the practice interval terminates. Consequently, v_{21}, rather than v_{22}, codes $\Theta^{(2)}$ when $\Theta^{(2)}$ is practiced. This property can be iterated to show how systematic trends in the sequence of practiced patterns can produce systematic drifts in recoding. Consider Figure 8. Again two dimensional patterns are denoted by solid vectors and classifying vectors are denoted by dotted vectors. Let the patterns be practiced in the order $\Theta^{(1)}, \Theta^{(2)}, ..., \Theta^{(M)}$, where $M \gg N$. By successively practicing $\Theta^{(1)}, \Theta^{(2)}, ..., \Theta^{(r-1)}$, the vector $z^{(1)}(t)$ is dragged along clockwise until it almost reaches $\Theta^{(r-1)}$. Then $\Theta^{(r)}$ is practiced, and since $\Theta^{(r)}$ is coded by v_{22}, $z^{(1)}(t)$ stops moving and $z^{(2)}(t)$ begins

to move clockwise; $z^{(2)}(t)$ continues to move clockwise while $\Theta^{(r+1)}, \Theta^{(r+2)}, ..., \Theta^{(2r-1)}$ are practiced. Then $z^{(3)}(t)$ begins to move clockwise, and so on. The clockwise drift in the practice schedule hereby shifts each $z^{(j)}(t)$, $j = 1, 2, ..., M-1$, to a position that is close to the one $z^{(j+1)}(0)$ occupied. In other words, essentially all vectors in V_2 are reclassified. If the same practice schedule $\Theta^{(1)}, \Theta^{(2)}, ..., \Theta^{(M)}$ is repeated on a second learning trial, then essentially all v_{2i} are recoded by $v_{2,i+2}$, and so on. Each learning trial recodes V_2 until all the N populations in V_2 code one of the N most clockwise vectors $\Theta^{(k)}$. This asymptotic coding of V_2 is stable, except for a wild oscillation in the coding of v_{21} on each learning trial, if the same practice schedule is always repeated. If, however, a counter-clockwise drift in practiced patterns is then imposed, all of V_2 will be recoded until the N most counter-clockwise vectors $\Theta^{(k)}$ are coded. In general, if there are many patterns relative to the number of populations in V_2, and if the statistical structure of the practice sequences continually changes, then there need not exist a stable coding rule in V_2. This is quite unsatisfactory.

By contrast, if there are few, or sparse, patterns relative to the number of populations in V_2, then a stable coding rule does exist, and the STM choice rule in V_2 does provide an effective sampling technique. Such a situation is approximated, for example, when the network is exposed to a "visually deprived" environment, in imitation of experiments on young animals. A theorem concerning this case will now be stated, if only to suggest what auxiliary mechanisms will be needed to establish a stable coding rule in the general case. This theorem shows how populations learn to code convex regions of features. In particular, if v_{2j} learns to code a certain set of features, then it automatically codes *average* features derived from this set.

The following nomenclature will be needed to state the theorem. A *partition* $\bigoplus_{k=1}^{K} \mathscr{P}_k$ of a finite set $\mathscr{P}$ is a subdivision of $\mathscr{P}$ into nonoverlapping and exhaustive subsets $\mathscr{P}_j$. The *convex hull* $\mathscr{H}(\mathscr{P})$ of a finite set $\mathscr{P}$ is the set of all convex combinations of elements in $\mathscr{P}$; for example, if $\mathscr{P} = \{\Theta^{(1)}, \Theta^{(2)}, ..., \Theta^{(M)}\}$, then

$$\mathscr{H}(\mathscr{P}) = \left\{ \sum_{k=1}^{M} \lambda_k \Theta^{(k)} : \text{each } \lambda_k \geq 0 \text{ and } \sum_{k=1}^{M} \lambda_k = 1 \right\}.$$

Given a set $\mathscr{P}$ with subset $\mathscr{Q}$, let $\mathscr{R} = \mathscr{P} \backslash \mathscr{Q}$ denote the set of elements in $\mathscr{P}$ that are not in $\mathscr{Q}$. If the classifying vector $z^{(j)}(t)$ codes the set of patterns $\mathscr{P}_j(t)$, let $\mathscr{P}_j^*(t) = \mathscr{P}_j(t) \cup \{z^{(j)}(t)\}$. The *distance* between a vector P and a set of vectors $\mathscr{Q}$, denoted by $\|P - \mathscr{Q}\|$, is defined by

$$\|P - \mathscr{Q}\| = \inf\{\|P - Q\| : Q \in \mathscr{Q}\}.$$

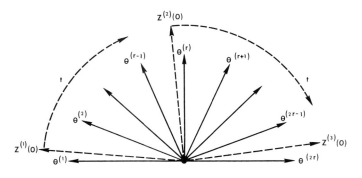

Fig. 8. Practicing a sequence of spatial patterns can recode all the populations

Theorem 2 (Sparse Patterns)

Let the network practice any set $\mathscr{P} = \{\Theta^{(i)}: i = 1, 2, ..., M\}$ of patterns for which there exists a partition

$$\mathscr{P} = \bigoplus_{k=1}^{N} \mathscr{P}_k(0)$$ such that

$$\min \{u \cdot v : u \in \mathscr{P}_j(0), v \in \mathscr{P}_j^*(0)\} > \max \{u \cdot v : u \in \mathscr{P}_j(0),$$

$$v \in \mathscr{P}^*(0) \backslash \mathscr{P}_j^*(0)\} \tag{19}$$

for all $j = 1, 2, ..., N$. Then $\mathscr{P}_j(t) = \mathscr{P}_j(0)$ and the functions

$$D_j(t) = \|z^{(j)}(t) - \mathscr{H}(\mathscr{P}^{(j)}(t))\| \tag{20}$$

are monotone decreasing for $t \geqq 0$ and $j = 1, 2, ..., N$. If moreover the patterns in $\mathscr{P}^{(j)}(0)$ are practiced in intervals $[U_{jm}, V_{jm}]$, $m = 1, 2, ...,$ such that

$$\sum_{m=1}^{\infty} (V_{jm} - U_{jm}) = \infty \tag{21}$$

then

$$\lim_{t \to \infty} D_j(t) = 0 . \tag{22}$$

Remarks. In other words, if the classifying vectors initially code the patterns into sparse classes, in the sense of (19), then this code persists through time, and the classifying vectors approach a convex combination of their coded patterns. As (20) and (22) show, learning permits each v_{2j} to respond as vigorously as possible to its class of coded patterns.

The above results indicate that, given a fixed number of patterns, it becomes easier to establish a stable code for them as the number of populations in V_2 increases. Once V_2 is constructed, however, it is not possible to increase its number of populations at will. Moreover, *in vivo*, an enormous variety of patterns typically barrages the visual system. How can a stable code be guaranteed no matter how many patterns perturb V_1?

One way is to assume that a biochemically determined *critical period* exists during which the z_{ij}'s are capable of learning; once the critical period terminates,

some chemical factor is removed and the z_{ij}'s remain fixed in the last code to be established. The existence of a critical period has been reported (Hubel and Wiesel, 1970), but whether it is due to a chemical factor, or *merely* to a chemical factor, is as yet unknown. From a formal point of view, such a mechanism suffers from several significant related disadvantages. The most obvious one is that all the coded information that is learned throughout the critical period can be obliterated if its last phase exhibits an unlikely statistical trend. In addition, a repetitive statistical trend can prevent many patterns from being coded at all. For example, in Figure 8, once the classifying vectors code the N most clockwise patterns, many of the other $M - N$ patterns might be too far away from $z^{(1)}$ to satisfy the ε-condition in (7); they will then not be coded by any population. Yet each of these $M - N$ patterns has been presented as frequently as the N patterns that are coded. More generally, because populations which are already coded can be recoded so easily, it is hard to search for as yet uncommitted populations to code as yet uncoded patterns. This problem prevents a universal recoding from being achieved (see Part II).

These negative remarks can be supplemented by intriguing positive observations. Stabilizing the code seems to require the same formal machinery that is needed in models of adult attention and discrimination learning (Grossberg, 1975). This machinery, in turn, is highly evokative of data concerning attentional modulation of olfactory patterns by the prepyriform cortex of cats (Freeman, 1974). Auxiliary mechanisms for stabilizing the code will therefore be motivated below. It is understood that a biochemically triggered critical period can coexist with these mechanisms, or indeed can preempt them in sufficiently primitive organisms.

Various mechanisms can be contemplated which partially stabilize the code, but which are not sufficient. A satiation mechanism will be sketched below to

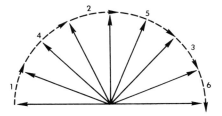

Fig. 9. Practicing in the order 1, 2, 3, 4, 5, 6 can recode all the populations even if satiation exists

clarify what is needed. Consider (6) with

$$x_{2j}(t) = \begin{cases} G_j(t) & \text{if } S_j(t)G_j(t) > \max\{\varepsilon, S_k(t)G_k(t) : k \neq j\} \\ 0 & \text{if } S_j(t)G_j(t) < \max\{\varepsilon, S_k(t)G_k(t) : k \neq j\} \end{cases} \tag{23}$$

where

$$G_j(t) = g\left(1 - \int_0^t x_{2j}(v)K(t-v)dv\right). \tag{24}$$

In (24), $g(w)$ is a monotone increasing function such that $g(0) = 0$ and $g(1) = 1$. $K(w)$ is a monotone decreasing function such that $K(0) = 1$ and $K(\infty) = 0$; for example, $K(w) = e^{-w}$. Equation (23) says that persistent activation of v_{2j} causes its STM response to satiate, or adapt; if v_{2j} is active during a sufficiently long interval, its activity approaches zero. Correspondingly, $z^{(j)}$'s fluctuations are damped within a time interval of fixed length. Such a mechanism is inadequate if the training schedule allows v_{2j} to recover its maximal strength. Figure 9 shows, for example, an ordering of patterns that permits recoding of essentially all populations in V_2.

This problem is only made worse by replacing the choice rule in (23) by a partial contrast rule such as

$$x_{2j} = \begin{cases} \dfrac{f(S_jG_j)}{\displaystyle\sum_{S_kG_k > e} f(S_kG_k)} & \text{if } S_jG_j > \varepsilon \\ 0 & \text{if } S_jG_j < \varepsilon. \end{cases}$$

Here, if a prescribed pattern Θ causes a maximal STM response at v_{2j}, then the activity x_{2j} is suppressed by G_j more rapidly than the activities of other Θ-activated populations. There can consequently be a shift in the locus of maximal responsiveness even to a single pattern—that is, recoding—in addition to the difficulty cited in Figure 9.

Such examples clarify what is essential:

(A) *Before* $z^{(j)}(t)$ learns a pattern, or class of related patterns, it must be able to fluctuate freely in response to pattern inputs in search of a classification.

(B) *After* $z^{(j)}(t)$ learns a pattern, it must be prevented from coding very different patterns, no matter what the training schedule is. In particular, satiating $z^{(j)}$'s

ability to change through time does not suffice, since a very different pattern can still be coded by $z^{(j)}$ if this pattern elicits a larger signal at v_{2j}, say due to the size of $\|z^{(j)}\|$ rather than the direction of vector $z^{(j)}$, than at any of the uncommited populations.

Requirements (A) and (B) constrain the interaction of STM and LTM mechanisms, given that (6) holds. For example, by (6), if a pattern Θ creates signals while v_{2j} is active in STM, then $z^{(j)}(t)$ will change. Suppose that a sequence $\Theta^{(1)}$, $\Theta^{(2)}$ of two very different patterns is successively presented to V_1, and that $z^{(1)}(t)$ codes $\Theta^{(1)}$. In response to $\Theta^{(1)}$, v_{21} is activated, but $z^{(1)}(t)$ does not substantially change because it already codes $\Theta^{(1)}$. Now let $\Theta^{(2)}$ perturb V_1. By requirement (B), $z^{(1)}(t)$ must not be allowed to change. By (6), $z^{(1)}(t)$ will change unless either no signal is emitted from V_1 when v_{21} is active, or a signal is emitted from V_1 only after v_{21} is inactivated. These two cases will be separately considered in the next two paragraphs.

In the former case, some type of feedback to V_1 must suppress the V_1-to-V_2 signals that would otherwise be generated by $\Theta^{(2)}$. This feedback somehow tells V_1 that $\Theta^{(2)}$ is very different from the pattern $\Theta^{(1)}$ that is presently coded in STM. By (A), however, $\Theta^{(2)}$ can generate V_1-to-V_2 signals at *some* time, either to search for a classifying vector, or to activate its already learned STM representation. Thus after V_1-to-V_2 signals are suppressed long enough for STM activity in v_{21} to also be suppressed, then V_1-to-V_2 signals are reactivated.

In the latter case, changing $\Theta^{(1)}$ to $\Theta^{(2)}$ somehow suppresses the STM activity that codes $\Theta^{(1)}$; in particular, somehow the network can tell when the spatial patterns that perturb V_1 are changed. In both cases, the same general issue is raised: how does the network process a temporal succession $\Theta^{(1)}$, $\Theta^{(2)}$, ..., $\Theta^{(k)}$, ... of spatial patterns $\Theta^{(k)} = (\Theta_1^{(k)}, \Theta_2^{(k)}, ..., \Theta_n^{(k)})$; that is, a *space-time pattern*. Space-time patterns are the typical inputs to a receptive field *in vivo*. The problem of stabilizing the STM code forces us to consider their processing in some detail. Part II of this paper considers this problem.

Appendix

Proof of Theorem 1. Consider the case in which

$$|\Theta|^2 > S_j(0) > \max\{\varepsilon, S_k(0) : k \neq j\}. \tag{A1}$$

The case in which $S_j(0) \geq |\Theta|^2$ can be treated similarly. First it will be shown that if the inequalities

$$|\Theta|^2 > S_j(t) > \max\{\varepsilon, S_k(t) : k \neq j\} \tag{A2}$$

hold at any time $t = T \in \bigcup_{m=1}^{x} [U_m, V_m]$, then they hold at all times $t \in [T, \infty) \cap \bigcup_{m=1}^{x} [U_m, V_m]$. By (A2), $x_{2j}(T) = 1$ and $x_{2k}(T) = 0$, $k \neq j$.

Consequently, by (6),

$$\dot{z}_{ij}(T) = -z_{ij}(T) + \Theta_i \tag{A3}$$

and

$$\dot{z}_{ik}(T) = 0 \tag{A4}$$

for $k \neq j$ and $i = 1, 2, \ldots, n$. By (A2)–(A4),

$$\dot{S}_j(T) = -S_j(T) + |\Theta|^2$$
$$> 0 = \dot{S}_k(T), \tag{A5}$$

$k \neq j$. Thus (A2) holds for all $t \in [T, \infty) \cap \bigcup_{m=1}^{\infty} [U_m, V_m]$. By (A2) and (A5), for all $t \in \bigcup_{m=1}^{\infty} [U_m, V_m]$, $S_j(t)$ increases monotonically towards $|\Theta|^2$ and (16) holds if and only if (17) holds. For $t \notin \bigcup_{m=1}^{\infty} [U_m, V_m]$, all $\dot{S}_k(t) = 0$, $k = 1, 2, \ldots, n$.

Letting $N_j = |z^{(j)}|^2$ and $C_j = \cos(z^{(j)}, \Theta) \equiv S_j N_j^{-1/2} |\Theta|^{-1}$, it readily follows from (A5) that for all $t \in \bigcup_{m=1}^{\infty} [U_m, V_m]$,

$$\dot{N}_j = 2(-N_j + S_j) \tag{A6}$$

and

$$\dot{C}_j = |\Theta| N_j^{-1/2}(1 - C_j^2). \tag{A7}$$

Equation (A7) shows that the angle between $z^{(j)}(t)$ and Θ closes monotonically as Θ is practiced. Since $S_j(t)$ is a monotonic function, (A6) shows that $N_j(t)$ oscillates at most once.

In particular, suppose $\|z^{(j)}(0)\| \leq \|\Theta\|$. Then $S_j(0) \leq \|\Theta\|^2$, since otherwise

$$\Theta \cdot z^{(j)}(0) > \Theta \cdot \Theta \geq z^{(j)}(0) \cdot z^{(j)}(0)$$

which implies

$$1 \geq C_j(0) > \|\Theta\| \, \|z^{(j)}(0)\|^{-1} \geq \|z^{(j)}(0)\| \, \|\Theta\|^{-1},$$

and thus

$$\|z^{(j)}(0)\| > \|\Theta\| > \|z^{(j)}(0)\|,$$

which is a contradiction. By (A5), therefore $\|z^{(j)}(0)\| \leq \|\Theta\|$ implies that $S_j(t)$ is monotone increasing ∎

Proof of Theorem 2. Inequality (19) is based on the fact that, if a fixed set of patterns $\Theta^{(j_1)}, \Theta^{(j_2)}, \ldots, \Theta^{(j_k)}$ is classified by $z^{(j)}(t)$ for all $t \geq 0$, then

$$z^{(j)}(t) \in \mathcal{H}(\Theta^{(j_1)}, \Theta^{(j_2)}, \ldots, \Theta^{(j_k)}, z^{(j)}(0)), \tag{A8}$$

for all $t \geq 0$. For example, suppose that the patterns are practiced in the order $\Theta^{(j_1)}, \Theta^{(j_2)}, \ldots, \Theta^{(j_k)}$ during the nonoverlapping intervals $[U_1, V_1], [U_2, V_2], \ldots, [U_k, V_k]$. Except during these intervals, $\dot{z}^{(j)} = 0$. Thus for $t \in [U_1, V_1]$,

$$\dot{z}^{(j)} = -z^{(j)} + \Theta^{(j_1)},$$

or

$$z^{(j)}(t) = z^{(j)}(0)e^{-(t-U_1)} + \Theta^{(j_1)}(1 - e^{-(t-U_1)}),$$

so that

$$z^{(j)}(t) \in \mathcal{H}(\Theta^{(j_1)}, z^{(j)}(0)) \subset \mathcal{H}(\Theta^{(j_1)}, \ldots, \Theta^{(j_k)}, z^{(j)}(0)).$$

For $t \in [U_2, V_2]$,

$$z^{(j)}(t) = [z^{(j)}(0)e^{-(V_1-U_1)} + \Theta^{(j_1)}(1 - e^{-(V_1-U_1)})]e^{-(t-U_2)}$$
$$+ \Theta^{(j_2)}(1 - e^{-(t-U_2)}). \tag{A9}$$

Hence

$$z^{(j)}(t) \in \mathcal{H}(\Theta^{(j_1)}, \Theta^{(j_2)}, z^{(j)}(0)) \subset \mathcal{H}(\Theta^{(j_1)}, \ldots, \Theta^{(j_k)}, z^{(j)}(0)),$$

and so on.

Condition (19) is then applied using the fact that, for any $U \in P_j(0)$, $V \in \mathcal{H}(P_j^*(0))$, and $W \in \mathcal{H}(P^*(0) \backslash P_j^*(0))$,

$$U \cdot V > \max\{\varepsilon, U \cdot W\} \tag{A10}$$

because

$$U \cdot V \geq \min\{u \cdot v : u \in P_j(0), v \in P_j^*(0)\}$$

and

$$\max\{u \cdot v : u \in P_j(0), v \in P^*(0) \backslash P_j^*(0)\} \geq U \cdot W.$$

Until a pattern is reclassified, however, (A8) shows that $z^{(j)}(t) \in \mathcal{H}(P_j^*(0))$ and that $z^{(k)}(t) \in \mathcal{H}(P^*(0) \backslash P_j^*(0))$ for any $k \neq j$. But then, by (A10), reclassification is impossible.

That $D_j(t)$ in (20) is monotone decreasing follows from iterations of (A9). That (21) implies (22) follows just as in the proof of Theorem 1. ∎

References

Barlow, H. B., Pettigrew, J. D.: Lack of specificity of neurones in the visual cortex of young kittens. J. Physiol. (Lond.) **218**, 98–100 (1971)

Bennett, M. V. L.: Analysis of parallel excitatory and inhibitory synaptic channels. J. Neurophysiol. **34**, 69–75 (1971)

Blackenship, J. E., Wachtel, H., Kandel, E. R.: Ionic mechanisms of excitatory, inhibitory, and dual synaptic actions mediated by an identified interneuron in abdominal ganglion of Aplysia. J. Neurophysiol. **34**, 76–92 (1971)

Blakemore, C., Cooper, G. F.: Development of the brain depends on the visual environment. Nature (Lond.) **228**, 477–478 (1970)

Blakemore, C., Mitchell, D. E.: Environmental modification of the visual cortex and the neural basis of learning and memory. Nature (Lond.) New Biol. **241**, 467–468 (1973)

Boycott, B. B., Dowling, J. E.: Organization of the primate retina: light microscopy. Phil. Trans. roy. Soc. B. **255**, 109–184 (1969)

Ellias, S. A., Grossberg, S.: Pattern formation, contrast control, and oscillations in the short term memory of shunting on-center off-surround networks. Biol. Cybernetics **20**, 69–98 (1975)

Freeman, W. J.: Neural coding through mass action in the olfactory system. Proceeding IEEE Conference on biologically motivated automata theory 1974

Gierer, A., Meinhardt, H.: A theory of biological pattern formation. Kybernetik **12**, 30–39 (1972)

Greenspan, H. P., Benney, D. J.: Calculus. New York: McGraw-Hill 1973

Grossberg, S.: Nonlinear difference-differential equations in prediction and learning theory. Proc. nat. Acad. Sci. (Wash.) **58**, 1329–1334 (1967)

Grossberg, S.: Some networks that can learn, remember, and reproduce any number of complicated space-time patterns, II. Stud. appl. Math. **49**, 135–166 (1970a)

Grossberg, S.: Neural pattern discrimination. J. theor. Biol. **27**, 291–337 (1970b)

Grossberg, S.: Pavlovian pattern learning by nonlinear neural networks. Proc. nat. Acad. Sci. (Wash.) **68**, 828–831 (1971)

Grossberg, S.: Neural expectation: cerebellar and retinal analogs of cells fired by learnable or unlearned pattern classes. Kybernetik **10**, 49–57 (1972)

Grossberg, S.: Contour enhancement, short term memory, and constancies in reverberating neural networks. Stud. appl. Math. **52**, 213–257 (1973)

Grossberg, S.: Classical and instrumental learning by neural networks. In: Rosen, R. and Snell, F. (Eds.): Progress in Theoretical Biology, pp. 51–141. New York: Academic Press 1974

Grossberg, S.: A neural model of attention, reinforcement, and discrimination learning. Int. Rev. Neurobiol. **18**, 263–327 (1975)

Grossberg, S.: On the development of feature detectors in the visual cortex with applications to learning and reaction-diffusion systems. Biol. Cybernetics **21**, 145–159 (1976)

Grossberg, S., Levine, D. S.: Some developmental and attentional biases in the contrast enhancement and short term memory of recurrent neural networks. J. theor. Biol. **53**, 341–380 (1975)

Hebb, D. O.: The organization of behavior. New York: Wiley 1949

Hirsch, H. V. B., Spinelli, D. N.: Visual experience modifies distribution of horizontally and vertically oriented receptive fields in cats. Science **168**, 869–871 (1970)

Hirsch, H. V. B., Spinelli, D. N.: Modification of the distribution of receptive field orientation in cats by selective visual exposure during development. Exp. Brain Res. **12**, 509–527 (1971)

Hubel, D. H., Wiesel, T. N.: The period of susceptibility to the physiological effects of unilateral eye closure in kittens. J. Physiol. (Lond.) **206**, 419–436 (1970)

Kimble, G. A.: Foundations of conditioning and learning. New York: Appleton-Century-Crofts 1967

Levine, D. S., Grossberg, S.: Visual illusions in neural networks: line neutralization, tilt aftereffect, and angle expansion. J. theor. Biol., in press (1976)

Meinhardt, H., Gierer, A.: Applications of a theory of biological pattern formation based on lateral inhibition. J. Cell. Sci. **15**, 321–346

Pérez, R., Glass, L., Shlaer, R.: Development of specificity in the cat visual cortex. J. math. Biol. (1974)

Von der Malsburg, C.: Self-organization of orientation sensitive cells in the striate cortex. Kybernetik **14**, 85–100 (1973)

Wachtel, H., Kandel, E. R.: Conversion of synaptic excitation to inhibition at a dual chemical synapse. J. Neurophysiol. **34**, 56–00 (1971)

Wiesel, T. N., Hubel, D. H.: Single-cell responses in striate cortex of kittens deprived of vision in one eye. J. Neurophysiol. **26**, 1003–1017 (1963)

Wiesel, T. N., Hubel, D. H.: Comparison of the effects of unilateral and bilateral eye closure on cortical unit responses in kittens. J. Neurophysiol. **28**, 1029–1040 (1965)

20
Introduction

(1976)
D. Marr and T. Poggio

Cooperative computation of stereo disparity
Science 194:283–287

This paper by Marr and Poggio applies a neural network to a realistic problem in computational vision: putting the images from the two eyes together to form a depth map of the visual world. Although Marr later felt that this model was inadequate (Marr, 1982, sections 3.2 and 3.3), it still stands as one of the best known examples of a 'cooperative' algorithm, one where a number of elements, each seeing only a small portion of the problem, interact with each other and agree on a common global solution.

Stereopsis is a favorite problem in computational vision because the task is understandable and constrained. We have two eyes and they see two slightly different scenes. The nearer an object is to us, the greater the angular difference, or the *disparity* between the points on the two retinas representing the same location on the object. If the world contained only single points to see, then we could measure depth by triangulation, using disparity, and there would be no problem. Unfortunately, real images are highly complex, and it is not obvious what the corresponding points are in the different images seen by the two eyes. If we cannot reliably detect corresponding points, then we cannot measure disparity accurately. Even if the world is simplified so it is made up of small numbers of discrete points, there are still many false targets; that is, triangulation becomes ambiguous because of spurious coincidences between different dots (see figure 1 in the paper). Even this seemingly simple problem quickly becomes very difficult. It is fair to say that the correct general solution to it is still not known.

The strategy that Marr and Poggio used to disambiguate dot positions is one that is used frequently in computational vision. It starts from the observation that the world is highly constrained and that scenes are not all equally likely. The implicit claim is that the visual system evolved to work best in one particular environment.

Marr and Poggio assume that there are two environmental constraints operating: First, a point on a surface can have only one disparity value at a time; that is, a point cannot be at two depths at once. Second, the world has lots of smooth surfaces, and relatively few boundaries between the surfaces. The second constraint implies that nearby points are likely to be at similar depths.

There is evidence for cells in visual cortex that respond to disparities in the visual field; i.e., they respond best if the same point in the image falls on different retinal locations in the two eyes. Marr and Poggio set up several layers of disparity neurons, each corresponding to one value of disparity. In their largest simulation they used seven layers of neurons, representing shifts of corresponding relative locations between the two eyes from $+3$ to -3. If there are many discrete points in the image, then there are large numbers of ambiguous possible matches where a point in one eye is paired with an inappropriate point in the other eye. So a cell might be excited inappropriately.

Marr and Poggio assume cells are connected together to form a network to compute the correct interpretation of the two images. The assumed constraints allow a connection matrix to be formed. The continuity of surfaces implies that there are excitatory connections among cells within a disparity layer; i.e., nearby locations on the image tend to lie at similar disparities. Depending on its distance from the observer, a point along a line of view corresponds to different disparities between the two eyes. The first constraint says that only one cell that corresponds to a particular line of view can be on, and the rest must be off. Therefore the connection matrix contains inhibitory connections between cells along a particular line of view. Since there are two lines of view from the two eyes, there are two sets of inhibitory connections. Each cell therefore receives a number of inhibitory and excitatory inputs from nearby cells.

Given the connection matrix and an image, an interative procedure can be started. Cells are familiar threshold logic units; only outputs of 0 and 1 are allowed. An initial guess (say, points randomly distributed in disparity) is provided. A cell sums excitation, sums inhibition, subtracts excitation from inhibition, and, based on whether the sum is greater than or less than threshold, computes its output. The system than takes the new state and iterates it again. After a few passes, the system reaches a stable state, where repeated iterations do not change the pattern. (Note the similarity of these dynamics to those suggested by Hopfield in paper 27.)

In the Marr and Poggio model we see an initial state vector repeatedly iterated according to simple rules, with the outcome of the repeated iterations being a global stable state. Systems like this have become common in the neural network literature. In some of these systems, as here, the connection strengths can be inferred from the computational structure of the problem. A well known example of this, which has some interesting similarities in structure to Marr and Poggio's model, is in the paper by Hopfield and Tank (1985), which uses a neural network to estimate the solution to the "Traveling Salesman" problem. In other systems of this type, the connection strengths are formed by the system, using a learning rule.

References

J. Hopfield and D. Tank (1985), "'Neural' computation," *Biological Cybernetics* 52:141–152.

D. Marr (1982), *Vision*, San Francisco: W. H. Freeman.

(1976)
D. Marr and T. Poggio

Cooperative computation of stereo disparity
Science 194:283–287

A cooperative algorithm is derived for extracting disparity information from stereo image pairs.

Perhaps one of the most striking differences between a brain and today's computers is the amount of "wiring." In a digital computer the ratio of connections to components is about 3, whereas for the mammalian cortex it lies between 10 and 10,000 (*1*).

Although this fact points to a clear structural difference between the two, this distinction is not fundamental to the nature of the information processing that each accomplishes, merely to the particulars of how each does it. In Chomsky's terms (*2*), this difference affects theories of performance but not theories of competence, because the nature of a computation that is carried out by a machine or a nervous system depends only on a problem to be solved, not on the available hardware (*3*). Nevertheless, one can expect a nervous system and a digital computer to use different types of algorithm, even when performing the same underlying computation. Algorithms with a parallel structure, requiring many simultaneous local operations on large data arrays, are expensive for today's computers but probably well-suited to the highly interactive organization of nervous systems.

The class of parallel algorithms includes an interesting and not precisely definable subclass which we may call cooperative algorithms (*3*). Such algorithms operate on many "input" elements and reach a global organization by way of local, interactive constraints. The term "cooperative" refers to the way in which local operations appear to cooperate in forming global order in a well-regulated manner. Cooperative phenomena are well known in physics (*4, 5*), and it has been proposed that they may play an important role in biological systems as well (*6–10*). One of the earliest suggestions along these lines was made by Julesz (*11*), who maintains that stereoscopic fusion is a cooperative process. His model, which consists of an array of dipole magnets with springs coupling the tips of adjacent dipoles, represents a suggestive metaphor for this idea. Besides its biological relevance, the extraction of stereoscopic information is an important and yet unsolved problem in visual information processing (*12*). For this reason—and also as a case in point—we describe a cooperative algorithm for this computation.

In this article, we (i) analyze the computational structure of the stereo-disparity problem, stating the goal of the computation and characterizing the associated local constraints; (ii) describe a cooperative algorithm that implements this computation; and (iii) exhibit its performance on random-dot stereograms. Although the problem addressed here is not directly related to the question of how the brain extracts disparity information, we shall briefly mention some questions and implications for psychophysics and neurophysiology.

Computational Structure of the Stereo-Disparity Problem

Because of the way our eyes are positioned and controlled, our brains usually receive similar images of a scene taken from two nearby points at the same horizontal level. If two objects are separated in depth from the viewer, the relative positions of their images will differ in the two eyes. Our brains are capable of measuring this disparity and of using it to estimate depth.

Three steps (S) are involved in measuring stereo disparity: (S1) a particular location on a surface in the scene must be selected from one image; (S2) that same location must be identified in the other image; and (S3) the disparity in the two corresponding image points must be measured.

If one could identify a location beyond doubt in the two images, for example by illuminating it with a spot of light, steps S1 and S2 could be avoided and the problem would be easy. In practice one cannot do this (Fig. 1), and the difficult part of the computation is solving the correspondence problem. Julesz found that we are able to interpret random-dot stereograms, which are stereo pairs that consist of random dots when viewed monocularly but fuse when viewed stereoscopically to yield patterns separated in depth. This might be thought surprising, because when one tries to

Copyright © 1976 AAAS.

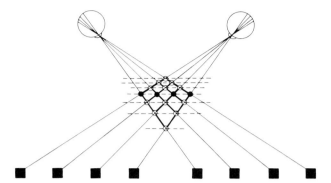

Figure 1 Ambiguity in the correspondence between the two retinal projections. In this figure, each of the four points in one eye's view could match any of the four projections in the other eye's view. Of the 16 possible matchings only four are correct (closed circles), while the remaining 12 are "false targets" (open circles). It is assumed here that the targets (closed squares) correspond to "matchable" descriptive elements obtained from the left and right images. Without further constraints based on global considerations, such ambiguities cannot be resolved. Redrawn from Julesz (*11*, figure 4.5-1).

set up a correspondence between two arrays of random dots, false targets arise in profusion (Fig. 1). Even so, we are able to determine the correct correspondence. We need no other cues.

In order to formulate the correspondence computation precisely, we have to examine its basis in the physical world. Two constraints (C) of importance may be identified (*13*): (C1) a given point on a physical surface has a unique position in space at any one time; and (C2) matter is cohesive, it is separated into objects, and the surfaces of objects are generally smooth compared with their distance from the viewer.

These constraints apply to locations on a physical surface. Therefore, when we translate them into conditions on a computation we must ensure that the items to which they apply there are in one-to-one correspondence with well-defined locations on a physical surface. To do this, one must use surface markings, normal surface discontinuities, shadows, and so forth, which in turn means using predicates that correspond to changes in intensity. One solution is to obtain a primitive description [like the primal sketch (*14*)] of the intensity changes present in each image, and then to match these descriptions. Line and edge segments, blobs, termination points, and tokens, obtained from these by grouping, usually correspond to items that have a physical existence on a surface.

The stereo problem may thus be reduced to that of matching two primitive descriptions, one from each eye. One can think of the elements of these descriptions as carrying only position information, like the white

squares in a random-dot stereogram, although in practice there will exist rules about which matches between descriptive elements are possible and which are not. The two physical constraints C1 and C2 can now be translated into two rules (R) for how the left and right descriptions are combined:

R1) *Uniqueness*. Each item from each image may be assigned at most one disparity value. This condition relies on the assumption that an item corresponds to something that has a unique physical position.

R2) *Continuity*. Disparity varies smoothly almost everywhere. This condition is a consequence of the cohesiveness of matter, and it states that only a small fraction of the area of an image is composed of boundaries that are discontinuous in depth.

In real life, R1 cannot be applied simply to gray-level points in an image. The simplest counterexample is that of a goldfish swimming in a bowl: many points in the image receive contributions from the bowl and from the goldfish. Here, and in general, a gray-level point is in only implicit correspondence with a physical location, and it is therefore impossible to ensure that gray-level points in the two images correspond to exactly the same physical position. Sharp changes in intensity are usually due either to the goldfish, to the bowl, or to a reflection, and therefore define a single physical position precisely.

A Cooperative Algorithm

By constructing an explicit representation of the two rules, we can derive a cooperative algorithm for the computation. Figure 2a exhibits the geometry of the rules in the simple case of a one-dimensional image. L_x and R_x represent the positions of descriptive elements on the left and right images. The thick vertical and horizontal lines represent lines of sight from the left and right eyes, and their intersection points correspond to possible disparity values. The dotted diagonal lines connect points of constant disparity.

The uniqueness rule R1 states that only one disparity value may be assigned to each descriptive element. If we now think of the lines in Fig. 2a as a network, with a node at each intersection, this means that only one node may be switched on along each horizontal or vertical line.

The continuity rule R2 states that disparity values vary smoothly almost everywhere. That is, solutions tend to spread along the dotted diagonals.

If we now place a "cell" at each node (Fig. 2b) and connect it so that it inhibits cells along the thick lines in the figure and excites cells along the dotted lines,

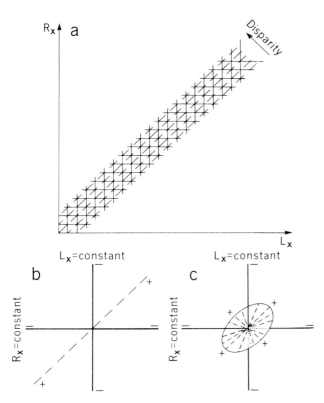

Figure 2 The explicit structure of the two rules R1 and R2 for the case of a one-dimensional image is represented in (a), which also shows the structure of a network for implementing the algorithm described by Eq. 2. Solid lines represent "inhibitory" interactions, and dotted lines represent "excitatory" ones. The local structure at each node of the network in (a) is given in (b). This algorithm may be extended to two-dimensional images, in which case each node in the corresponding network has the local structure shown in (c). Such a network was used to solve the stereograms exhibited in Figs. 3 to 6.

then, provided the parameters are appropriate, the stable states of such a network will be precisely those in which the two rules are obeyed. It remains only to show that such a network will converge to a stable state. We were able to carry out a combinatorial analysis [as in (9, 15)] which established its convergence for random-dot stereograms (16).

This idea may be extended to two-dimensional images simply by making the local excitatory neighborhood two dimensional. The structure of each node in the network for two-dimensional images is shown in Fig. 2c.

A simple form of the resulting algorithm (3) is given by the following set of difference equations:

$$\mathbf{C}^{(n+1)} = \sigma\{\Xi(\mathbf{C}^{(n)}) + \mathbf{C}^{(0)}\} \tag{1}$$

that is,

$$
C_{xyd}^{(n+1)} = \sigma \left\{ \sum_{x'y'd' \in S(xyd)} C_{x'y'd'}^{(n)} - \varepsilon \sum_{x'y'd' \in O(xyd)} C_{x'y'd'}^{(n)} + C_{xyd}^{(0)} \right\} \tag{2}
$$

where $C_{xyd}^{(n)}$ represents the state of the node or cell at position (x, y) with disparity d at iteration n, Ξ is the linear operator that embeds the local constraints (S and O are the circular and thick line neighborhoods of the cell xyd in Fig. 2c), ε is the "inhibition" constant, and σ is a sigmoid function with range $[0, 1]$. The state $C_{xyd}^{(n+1)}$ of the corresponding node at time $(n + 1)$ is thus determined by a nonlinear operator on the output of a linear transformation of the states of neighboring cells at time n.

The desired final state of the computation is clearly a fixed point of this algorithm; moreover, any state that is inconsistent with the two rules is not a stable fixed point. Our combinatorial analysis of this algorithm shows that, when σ is a simple threshold function, the process converges for a rather wide range of parameter values (16). The specific form of the operator is apparently not very critical.

Noniterative local operations cannot solve the stereo problem in a satisfactory way (11). Recurrence and nonlinearity are necessary to create a truly cooperative algorithm that cannot be decomposed into the superposition of local operations (17). General results concerning such algorithms seem to be rather difficult to obtain, although we believe that one can usually establish convergence in probability for specific forms of them.

Examples of Applying the Algorithm

Random-dot stereograms offer an ideal input for testing the performance of the algorithm, since they enable one to bypass the costly and delicate process of transforming the intensity array received by each eye into a primitive description (14). When we view a random-dot stereogram, we probably compute a description couched in terms of edges rather than squares, whereas the inputs to our algorithm are the positions of the white squares. Figures 3 to 6 show some examples in which the iterative algorithm successfully solves the correspondence problem, thus allowing disparity values to be assigned to items in each image. Presently, its technical applications are limited only by the preprocessing problem.

This algorithm can be realized by various mechanisms, but parallel, recurrent, nonlinear interactions, both excitatory and inhibitory, seem the most natural. The difference equations set out above would then rep-

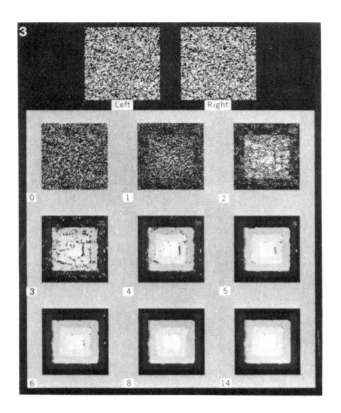

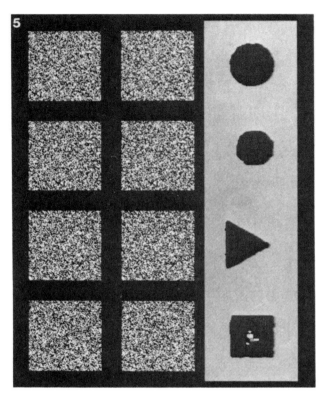

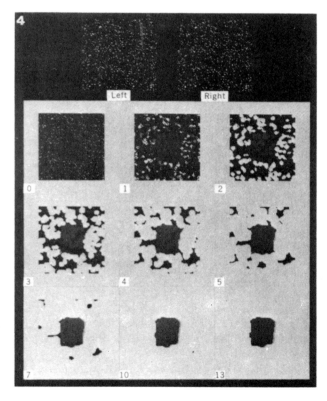

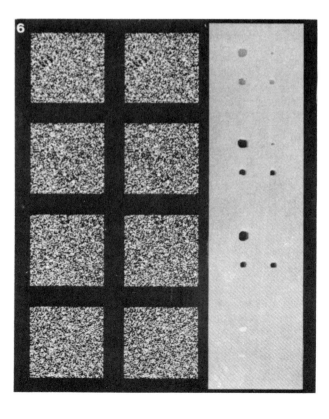

resent an approximation to the differential equations that describe the dynamics of the network.

Implications for Biology

We have hitherto refrained from discussing the biological problem of how stereopsis is achieved in the mammalian brain. Our analyses of the computation, and of the cooperative algorithm that implements it, raise several precise questions for psychophysics and physiology. An important preliminary point concerns the relative importance of neural fusion and of eye movements for stereopsis. The underlying question is whether there are many disparity "layers" (as our algorithm requires), or whether there are just three "pools" (18)—crossed, uncrossed, and zero disparity. Most physiologists and psychologists seem to accept the existence of numerous, sharply tuned binocular "disparity detectors," whose peak sensitivities cover a wide range of disparity values (19, 20). We do not believe that the available evidence is decisive (21), but an answer is critical to the biological relevance of our analysis. If, for example, there were only three pools or layers with a narrow range of disparity sensitivities, the problem of false targets is virtually removed, but at the expense of having to pass the convergence plane of the eyes across a surface in order to achieve fusion. Psychophysical experiments may provide some insight into this problem, but we believe that only physiology is capable of providing a clear-cut answer.

If this preliminary question is settled in favor of a "multilayer" cooperative algorithm, there are several obvious implications of the network (Fig. 2) at the physiological level: (i) the existence of many sharply tuned disparity units that are rather insensitive to the nature of the descriptive element to which they may refer; (ii) organization of these units into disparity layers (or stripes or columns); (iii) the presence of reciprocal excitation within each layer; and (iv) the presence of reciprocal inhibition between layers along the two lines of sight. Ideally, the inhibition should exhibit the characteristic "orthogonal" geometry of the thick lines in Fig. 2, but slight deviations may be permissible (16).

At the psychophysical level, several experiments (under stabilized image conditions) could provide critical evidence for or against the network: (i) results about the size of Panum's area and the number of disparity "layers"; (ii) results about "pulling" effects in stereopsis (20); and (iii) results about the relationship between disparity and the minimum fusable pattern size (Fig. 6).

Discussion

Our algorithm performs a computation that finds a correspondence function between two descriptions, sub-

---

◁ **Figures 3–6** The results of applying the algorithm defined by Eq. 2 to two random-dot stereograms. Fig. 3. The initial state of the network $C^{(0)}$ is defined by the input such that a node takes the value 1 if it occurs at the intersection of a 1 in the left and right eyes (Fig. 2), and it has the value 0 otherwise. The network iterates on this initial state, and the parameters used here, as suggested by the combinatorial analysis, were $\theta = 3.0$, $\varepsilon = 2.0$, and $M = 5$, where θ is the threshold and M is the diameter of the "excitatory" neighborhood illustrated in Fig. 2c. The stereograms themselves are labeled *Left* and *Right*, the initial state of the network as 0, and the state after n iterations is marked as such. To understand how the figures represent states of the network, imagine looking at it from above. The different disparity layers in the network lie in parallel planes spread out horizontally, so that the viewer is looking down through them. In each plane, some nodes are on and some are off. Each of the seven layers in the network has been assigned a different gray level, so that a node that is switched on in the top layer (corresponding to a disparity of +3 pixels) contributes a dark point to the image, and one that is switched on in the lowest layer (disparity of −3) contributes a lighter point. Initially (iteration 0) the network is disorganized, but in the final state stable order has been achieved (iteration 14), and the inverted wedding-cake structure has been found. The density of this stereogram is 50 percent. Fig. 4. The algorithm of Eq. 2, with parameter values given in the legend to Fig. 3, is capable of solving random-dot stereograms with densities from 50 percent to less than 10 percent. For this and smaller densities, the algorithm converges increasingly slowly. If a simple homeostatic mechanism

is allowed to control the threshold θ as a function of the average activity (number of "on" cells) at each iteration [compare (15)], the algorithm can solve stereograms whose density is very low. In this example, the density is 5 percent and the central square has a disparity of +2 relative to the background. The algorithm "fills in" those areas where no dots are present, but it takes several more iterations to arrive near the solution than in cases where the density is 50 percent. When we look at a sparse stereogram, we perceive the shapes in it as cleaner than those found by the algorithm. This seems to be due to subjective contours that arise between dots that lie on shape boundaries. Fig. 5. The disparity boundaries found by the algorithm do not depend on their shapes. Examples are given of a circle, an octagon (notice how well the difference between them is preserved), and a triangle. The fourth example shows a square in which the correlation is 100 percent at the boundary but diminishes to 0 percent in the center. When one views this stereogram, the center appears to shimmer in a peculiar way. In the network, the center is unstable. Fig. 6. The width of the minimal resolvable area increases with disparity. In all four stereograms the pattern is the same and consists of five circles with diameters of 3, 5, 7, 9, and 13 dots. The disparity values exhibited here are +1, +2, +3, and +6, and for each pattern we show the state of the network after ten iterations. As far as the network is concerned, the last pair (disparity of +6) is uncorrelated, since only disparities from −3 to +3 are present in our implementation. After ten iterations, information about the lack of correlation is preserved in the two largest areas.

ject to the two constraints of uniqueness and continuity. More generally, if one has a situation where allowable solutions are those that satisfy certain local constraints, a cooperative algorithm can often be constructed so as to find the nearest allowable state to an initial one. Provided that the constraints are local, use of a cooperative algorithm allows the representation of global order, to which the algorithm converges, to remain implicit in the network's structure.

The interesting difference between this stereo algorithm and standard correlation techniques is that one is not required to specify minimum or maximum correlation areas to which the analysis is subsequently restricted. Previous attempts at implementing automatic stereocomparison through local correlation measurement have failed in part because no single neighborhood size is always correct (12). The absence of a "characteristic scale" is one of the most interesting properties of this algorithm, and it is a central feature of several cooperative phenomena (22). We conjecture that the matching operation implemented by the algorithm represents in some sense a generalized form of correlation, subject to the a priori requirements imposed by the constraints. The idea can easily be generalized to different constraints and to other forms of equations 1 or 2, and it is technically quite appealing.

Cooperative algorithms may have many useful applications [for example, to make best matches for associative retrieval problems (15)], but their relevance to early processing of information by the brain remains an open question (23). Although a range of early visual processing problems might yield to a cooperative approach ["filling-in" phenomena, subjective contours (24), grouping, figural reinforcement, texture "fields," and the correspondence problem for motion], the first important and difficult task in problems of biological information processing is to formulate the underlying computation precisely (3). After that, one can study good algorithms for it. In any case, we believe that an experimental answer to the question of whether depth perception is actually a cooperative process is a critical prerequisite to further attempts at analyzing other perceptual processes in terms of similar algorithms.

Summary

The extraction of stereo-disparity information from two images depends upon establishing a correspondence between them. In this article we analyze the nature of the correspondence computation and derive a cooperative algorithm that implements it. We show that this algorithm successfully extracts information from random-dot stereograms, and its implications for the psychophysics and neurophysiology of the visual system are briefly discussed.

References and Notes

1. D. A. Sholl, *The Organisation of the Cerebral Cortex* (Methuen, London, 1956). The comparison depends on what is meant by a component. We refer here to the level of a gate and of a neuron, respectively.

2. A. N. Chomsky, *Aspects of the Theory of Syntax* (MIT Press, Cambridge, Mass., 1965).

3. D. Marr and T. Poggio, *Neurosci. Res. Prog. Bull.*, in press (also available as *Mass. Inst. Technol. Artif. Intell. Lab. Memo 357*).

4. H. Haken, Ed. *Synergetics-Cooperative Phenomena in Multi-component Systems* (Teuber, Stuttgart, 1973).

5. H. Haken, *Rev. Mod. Phys.* **47**, 67 (1975).

6. J. D. Cowan, *Prog. Brain Res.* **17**, 9 (1965).

7. H. R. Wilson and J. D. Cowan, *Kybernetik* **13**, 55 (1973).

8. M. Eigen, *Naturwissenschaften* **58**, 465 (1971).

9. P. H. Richter, paper contributed to a competition of the Bavarian Academy of Science, Max-Planck-Institut für Biophysikalische Chemie, 1974.

10. A. Gierer and H. Meinhardt, *Kybernetik* **12**, 30 (1972).

11. B. Julesz, *Foundations of Cyclopean Perception* (Univ. of Chicago Press, Chicago, 1971).

12. K. Mori, M. Kidode, H. Asada, *Comp. Graphics Image Process.* **2**, 393 (1973).

13. D. Marr, *Mass. Inst. Technol. Artif. Intell. Lab. Memo 327* (1974).

14. ——, *Philos. Trans. R. Soc. London Ser. B* **275**, 483 (1976).

15. ——, *ibid.* **252**, 23 (1971). See especially section 3.1.2.

16. —— and T. Poggio, in preparation.

17. T. Poggio and W. Reichardt, *Q. Rev. Biophys.*, in press.

18. W. Richards, *J. Opt. Soc. Am.* **62**, 410 (1971).

19. H. B. Barlow, C. Blakemore, J. D. Pettigrew, *J. Physiol.* (*London*) **193**, 327 (1967); J. D. Pettigrew, T. Nikara, P. O. Bishop, *Exp. Brain Res.* **6**, 391 (1968); C. Blakemore, *J. Physiol.* (*London*) **209**, 155 (1970).

20. B. Julesz and J.-J. Chang, *Biocybernetics* **22**, 107 (1976).

21. D. H. Hubel and T. N. Wiesel, *Nature* (*London*) **225**, 41 (1970).

22. K. G. Wilson, *Rev. Mod. Phys.* **47**, 773 (1975).

23. Julesz (*11*), Cowan (*6*), and Wilson and Cowan (*7*) were the first to discuss explicitly the cooperative aspect of visual information processing. Much has been published recently on possible cooperative processes in nervous systems, ranging from the "catastrophe" literature [E. C. Zeeman, *Sci. Am.* **234**, 65 (April (1976)] to various attempts of more doubtful credibility. There has hitherto been no careful study of a cooperative algorithm in the context of a carefully defined computational problem [but see (*15*)], although algorithms that may be interpreted as cooperative were discussed, for instance, by P. Dev [*Int. J. Man-Mach. Stud.* **7**, 511 (1975)] and by A. Rosenfeld, R. A. Hummel, and S. W. Zucker [*Syst. Man. Cybern.* **6**, 420 (1976)]. In particular neither Dev nor J. I. Nelson [*J. Theor. Biol.* **49**, 1 (1975)] formulated the computational structure of the stereo-disparity problem. As a consequence, the resulting geometry of the inhibition be-

tween their disparity detectors does not correspond to ours (Fig. 2c) and apparently fails to provide a satisfactory algorithm.

24. S. Ullmann, *Mass. Inst. Technol. Artif. Intell. Lab. Memo 367* (1967); also in *Biol. Cybern.*, in press.

25. We thank W. Richards for valuable discussions, H. Lieberman for making it easy to create stereograms, and K. Prendergast for preparing the figures. The research described was done at the Artificial Intelligence Laboratory of the Massachusetts Institute of Technology. Support for the laboratory's artificial intelligence research is provided in part by the Advanced Research Projects Agency of the Department of Defense under Office of Naval Research contract N00014-75-C-0643; T. P. acknowledges the support of the Max-Planck-Gesellschaft during his visit to the Massachusetts Institute of Technology.

21
Introduction

(1977)
S.-I. Amari

Neural theory of association and concept-formation
Biological Cybernetics 26:175–185

This paper by Amari is terse, and it is easy to overlook the many interesting ideas it contains. It discusses both pattern associators, where an input pattern gives rise to an appropriate, different, output pattern, and what Amari calls "concept-forming" nets, which are recurrent, autoassociative nets, where the input and the output are the same pattern, and the output can be fed back into the input.

One unusual aspect of this paper is that it does not take an abstraction of a single neuron as its elementary unit. Amari considers small groups of mutually connected neurons, which he calls "neuron pools," to be the basic units of his model. In earlier work, he analyzed such neuron pools, and showed that they would act as two-state devices, having an "excited" state, when many neurons in the pool were active, and a "resting" state, when most of the neurons in the pool were off. The pool could be switched from excited to resting and back again by changing the average input activity to the pool. This bistability lets Amari approximate the pool by a McCulloch-Pitts neuron (paper 2) that switchs on when its inputs are above threshold, and is off otherwise.

It is usually assumed that neurons, or an approximation of them, are the basic elements in neural networks. Except for a certain attractive simplicity, there is no experimental support for this idea. A number of network modelers use neutral terms such as "element" or "unit" to denote the basic computing device in neural networks, rather than use "model neuron," which makes a strong biological statement. Amari makes this point explicitly, by using a higher level construction as the basic computing element. The resulting mathematics turns out to be identical to that obtained with McCulloch-Pitts neurons, but the point being made is worth remembering.

For his discussion of associative nets, Amari considers a number of variations on simple associative systems. Association is defined as pattern association, as in the linear associator or that given by Willshaw, Buneman, and Longuet-Higgins (paper 12), where an input pattern x was transformed into an output pattern y by using a transformation T; that is, $y = Tx$. Amari assumes association nets have another possible input besides the input pattern, an input from a "teacher," which is only provided during a learning phase. Modification of connection strength constructs the appropriate transformation T so that the correct input-output relations will be formed.

Amari also discusses "concept forming" nets, which are recurrent; that is, the output is fed back into the input. He defines a "concept" as a stable state of the network; that is, given a network input-output transformation T, and an input pattern x, he is interested in stable states of the form $x = Tx$. He shows that there can be a number of them in a given system, and the connection strengths can be formed in the same way as in the associative systems.

The way to change the transformation **T** is to adjust its weights properly. We have already seen several learning rules proposed in this collection. Amari discusses the problem at a high level of generality. He defines two special functions; R, a function of weights, input, and teacher; and L, a function only of the weights, which is the expected value of R averaged over all input patterns. He then shows, given his assumptions, that the weights will change so as to minimize function L, and the learning rule will have change in weights adjusted so as to move down the gradient of L; that is, the system is doing gradient descent in L. An example we have already seen that fits into this framework is the Widrow-Hoff learning rule, where L would be the average mean square error and the weights adjust themselves so as to minimize L.

Amari constructs several different R functions for different learning rules, and shows that they give rise to different final sets of weights. Of particular interest, in light of later developments, is Amari's section 3.2C, "Learning based on potential." The form of L is the "energy" function that plays so prominent a role in neural network research in the 1980s. Amari mentions that the "minimum energy" solution is found when the weights are pointing in the direction of the eigenvector with maximum eigenvalue of the covariance matrix formed from the inputs.

In the linear associator and related pattern associators, orthogonality relations are particularly important. It is easy to show that if the input patterns are mutually orthogonal, then the associator will work perfectly, and there will be no cross-talk between different inputs. Cross-talk, one input evoking the association of another input, can be viewed as either a blessing, because it allows some possibilities for novel chains of associations, or a curse, because it is something unlearned, and therefore is noise. Both Kohonen (1977) and Amari have discussed this problem from the viewpoint of noise reduction, that is, constructing the matrix so as to make the best approximation to perfect association. (The other point of view, that cross-talk can be used constructively, is presented in Knapp and Anderson, paper 36.) The last part of the Amari paper discusses orthogonalizing as a way of eliminating interference between patterns.

Amari discusses "noise immunity," that is, the ability of a network, when given a learned input corrupted by noise, to generate the appropriate output with less added noise. In light of later developments it is interesting to note that Amari uses "concept forming" autoassociative networks to correct errors in patterns. Given an input pattern that is close to a stable concept pattern, the state of the system after repeated iterations will be the stable concept and the noise will have been eliminated. He also discusses some of the conditions allowing formation of stable attractor states. Many later network models use a similar formulation of network activity.

Reference

T. Kohonen (1977), *Associative Memory: A System-Theoretical Approach*, Berlin: Springer.

(1977)
S.-I. Amari

Neural theory of association and concept-formation
Biological Cybernetics 26:175–185

Abstract. The present paper looks for possible neural mechanism underlying such high-level brain functioning as association and concept-formation. Primitive neural models of association and concept-formation are presented, which will elucidate the distributed and multiply superposed manner of retaining knowledge in the brain. The models are subject to two rules of self-organization of synaptic weights, orthogonal and covariance learning. The convergence of self-organization is proved, and the characteristics of these learning rules are shown. The performances, especially the noise immunity, of the association net and concept-formation net are analyzed.

I. Introduction

A typical "neural theory" builds a model of a specific portion of the brain in the beginning, and predicts the functioning of that portion by computer-simulated experiments. We, however, do not intend to build a model of a specific portion at the present stage. We rather intend to figure out possible neural mechanisms of learning, storing and using knowledge which might be found widely in the brain in various versions. We search for such mechanisms of distributed and multiply-superposed information processing as might underlie neural association and concept-formation. These mechanisms will help us in building more realistic models at the next stage.

The present work is a development of the model, consisting of mutually connected bistable neuron pools, proposed by Amari (1971) and partly analyzed in Amari (1972a). In order to make mathematical analysis tractable, the model is kept as simple as possible as long as the essential features are not missed. This is a common attitude of the author's neural researches (Amari, 1971; 1972a, b; 1974a, b; 1975; 1977a, b).

We consider the following association net. The net, learning from k pairs of stimulus patterns $(x_1, z_1), ..., (x_k, z_k)$, self-organizes in such a manner that, when the net receives a key pattern $x_\alpha (\alpha = 1, ..., k)$, it correctly outputs the associated partner z_α. The self-organization is carried into effect through modification of the synaptic weights of the net. This type of neural association has been investigated by many authors (e.g., Nakano, 1972; Kohonen, 1972; Anderson, 1972; Amari, 1972a; Uesaka and Ozeki, 1972; Wigström, 1973), where the so-called "correlation" of pattern components is memorized in the synaptic weights. The correlational association works well when the patterns $x_1, ..., x_k$ are mutually orthogonal, but otherwise it does not work well by virtue of the mutual interference of superposed patterns. The present paper considers two methods of neural self-organization, orthogonal and covariance learning, which eliminate the above interference. Orthogonal learning is closely related to Kohonen's generalized inverse approach (Kohonen, 1974; Kohonen and Oja, 1976).

The concept-formation net, on the other hand, is a sequential net having recurrent connections. The net, receiving many stimulus patterns which are distributed in k clusters, self-organizes in such a manner that the net forms an equilibrium state corresponding to each cluster. The equilibrium states correspond to the concept patterns which the net retains. After the learning is completed, the net, receiving a pattern belonging to a cluster, falls into the equilibrium state corresponding to the cluster. The net keeps and reproduces any concept pattern in this manner (short-term memory), while forming many equilibrium states by changing synaptic weights yields the long-term memory. Orthogonal learning and covariance learning again play an important role.

This research was supported in part by a grant by the Sloan Foundation to the Center for Systems Neuroscience, University of Massachusetts

© by Springer-Verlag 1977

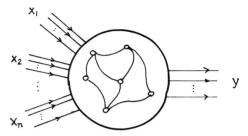

Fig. 1. Neuron pool

The main results of the present paper are: i) a mathematical analysis of the performance, especially noise immunity, of association and concept-formation nets, ii) the formulation of orthogonal and covariance learning, and iii) proof of convergence of neural self-organization including covariance and orthogonal learning.

II. Neural Models of Association and Concept-formation

2.1. Neuron Pool

We first consider the behavior of a neuron pool, which is composed of mutually connected neurons. Neuron pools play the role of building blocks of the models in the present paper. A neuron pool has a bundle of output axons. The output signal y of the neuron pool is represented by the average pulse rate over this bundle of axons. Let $x_1, x_2, ..., x_n$ be n signals entering into the neuron pool. A signal x_i is also carried by a bundle of axons (Fig. 1), representing the average pulse rate of the corresponding bundle.

The behavior of a neuron pool in which many neurons are connected in a random manner has been studied in detail (e.g., Amari, 1971, 1972b, 1974; Amari et al., 1977). Let us consider a bistable neuron pool. It has two states, an excited state which emits a high pulse rate output and a resting state which emits a low pulse rate output. It can be in either state. It enters the excited state when a weighted sum

$$u = \sum_{\tau=1}^{n} w_i x_i$$

of the input signals exceeds a threshold value h_1, and stays in that state. It enters the resting state, when the weighted sum u becomes lower than another threshold value h_2 ($h_2 < h_1$).

We represent the high pulse rate output by $y = 1$, and the low pulse rate by $y = 0$. The behavior of a bistable neuron pool, then, can approximately be described by the following equation

$$y = \varphi[\sum w_i x_i + (h_1 - h_2)y - h_1], \qquad (2.1)$$

where φ is the step-function defined by

$$\varphi(u) = \begin{cases} 1, & u > 0 \\ 0, & u \leqq 0 \end{cases}.$$

When $h_1 = h_2$ holds, the equation is simplified to yield the input-output relation

$$y = \varphi(\sum w_i x_i - h). \qquad (2.2)$$

This is the behavior of the so-called McCulloch-Pitts formal neuron. If we attach a feedback connection to a McCulloch-Pitts neuron with weight $h_1 - h_2$, (2.2) coincides with (2.1).

We hereafter use McCulloch-Pitts formal neurons as the constituent elements of our neuron nets, where an element may be a bistable neuron pool. A bundle of input or output axons of a neuron pool will also be treated as a single input or output line. We call w_i the connection weight of input x_i, and the weighted sum u the potential.

2.2. Model Net for Association

Let us consider a net of non-recurrent connections consisting of m elements (m neuron pools). The net receivers n input signals $x_1, ..., x_n$ and transforms them into m output signals $y_1, ..., y_m$, where y_i is the output from the i-th element (Fig. 2). Every input x_j is connected with all the elements. Let w_{ij} be the connection weight of x_j entering into the i-th element. Let h_i be the threshold of the i-th element.

The output y_i is determined by

$$y_i = \varphi\left(\sum_{j=1}^{n} w_{ij} x_j - h_i\right), \qquad i = 1, ..., m. \qquad (2.3)$$

This determines the transformation from $x_1, ..., x_n$ to $y_1, ..., y_m$. It is convenient to use vector-matrix notation. We represent the input and output signals by column vectors, e.g.,

$$x = (x_1, x_2, ..., x_n)^T$$

is the input vector, where the superscript T denotes transposition. The connection weights are also represented by a matrix

$$W = (w_{ij}),$$

which we call the connection matrix. The transformation (2.3) can then be rewritten in the form

$$y = \varphi(Wx - h), \qquad (2.4)$$

where function φ operates component-wise. We symbolically write

$$y = Tx \qquad (2.5)$$

to show that x is transformed into y.

An input vector x is called an input pattern. It represents a set of n features $x_1, x_2, ..., x_n$ of a signal pattern which the net processes. An output vector y is also called an output pattern.

The net has another set of inputs $z_1, ..., z_m$ (Fig. 2). Signal z_i arrives directly at the i-th element, and plays the role of a "teacher" of this element. Since the teacher signal

$$z = (z_1, z_2, ..., z_m)^T$$

is strong, the output signal y is always set equal to z, when z arrives. The teacher signal z arrives at the net in the learning phase only. The net usually works without the teacher input.

Let us consider k pattern signals $x_1, ..., x_k$. When pattern z_α is associated to pattern x_α, we have k pairs of patterns (x_α, z_α)'s. When the net, receiving input x_α, outputs the associated pattern z_α, i.e., when $z_\alpha = Tx_\alpha$ holds, we say that the net associates z_α with x_α. The net self-organizes such that z_α is associated with x_α by receiving k given pattern pairs (x_α, z_α)'s repeatedly, x_α from the ordinary input and z_α from the second input. This requires that $z_\alpha = Tx_\alpha$ eventually holds when the second input is finally allowed to revert to zero.

This is a primitive model of a neural association mechanism. We can construct more complicated and plausible models of association based on the mechanism of this simple model. The mechanism can be incorporated with lateral-inhibition networks such as those of Spinelli (1970) and Malsburg (1973).

2.3. Model Net for Concept Formation

A recurrent net can be obtained by connecting the outputs of a net to its inputs. We consider the net shown in Figure 3, where the net consists of n elements. Let $x_1(t), ..., x_n(t)$ be the outputs of the net at time t. We denote the outputs by

$$x(t) = [x_1(t), ..., x_n(t)]^T$$

and call it the state of the net at time t. The output $x(t+1)$ at time $t+1$ is determined by

$$x(t+1) = \varphi[Wx(t) - h] \qquad (2.6)$$

which we write symbolically as

$$x(t+1) = Tx(t), \qquad (2.7)$$

and call T the state-transition operator. The second input z plays the role of setting the net in the initial state specified from the outside. When signal z arrives at the second input, the net is set in the state z. The net, then, begins its state transition, if no further signals come from the second input.

A state x which satisfies $Tx = x$ is called an equilibrium state of the net. The net can have many equi-

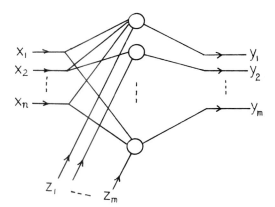

Fig. 2. Net for association

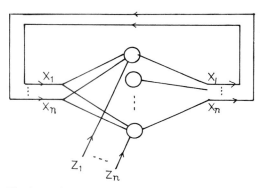

Fig. 3. Net for concept formation

librium states. Equilibrium states represent information patterns which the net can retain persistently (the short term memory). We regard an equilibrium state as a concept pattern which the net has formed. The net can thus retain many concept patterns.

Many patterns will arrive at the net in the learning phase from the outside. When the patterns consists of a number of clusters, we consider that there is a concept corresponding to each cluster of patterns. We show in Figure 4 an example of a pattern distribution, where patterns are distributed around x_1, x_2, and x_3, forming three clusters. The abscisa really is an n-dimensional pattern space and the ordinate represents the relative frequency of patterns.

Receiving a distribution of patterns one by one repeatedly, the net must self-organize by modifying the connection weights such that it automatically finds the clusters and forms equilibrium states. Moreover, when a pattern x' belonging to the cluster of x_α is input later, the net, first entering state x', must tend to the equilibrium state x_α thus formed. The net may be regarded as a pattern generator in which patterns $x_1, ..., x_k$ are coded in the form of equilibria.

frequency

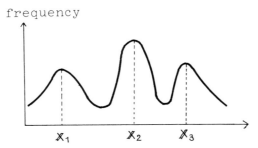

Fig. 4. Distribution of input patterns

III. Self-organization of Nerve Elements

3.1. Convergence of Learning

It is widely believed that a nerve net self-organizes by modifying its synaptic weights. Hebb (1949) proposed the rule that simultaneous firing of presynaptic and postsynaptic neurons brings an increase in that synaptic weight. This hypothesis has widely been accepted in various versions (e.g., Caianiello, 1961; Rosenblatt, 1961; Grossberg, 1969). We also adopt similar but somewhat extended rules of modification of connection weights.

A connection weight w_i of an element increases in proportion to a reinforcement signal r_i in the learning phase. It decays slowly at the same time. The rule of modification of a weight is, hence, formulated as

$$w_i(t+1) = (1-c)w_i(t) + dr_i(t), \tag{3.1}$$

where c and d are constants, and $w_i(t)$ and $r_i(t)$ are respectively the connection weight and reinforcement signal at time t.

The reinforcement signal $r_i(t)$ is determined generally depending on the input signal $x(t)$, the connection weight $w(t)$, the output signal $y(t)$ and the teacher signal $z(t)$ (when it exists). Nerve nets can realize various possibilities of information processing if provided with appropriate reinforcement signals.

When $r_i(t)$ is given by $r_i(t) = x_i(t)y(t)$, it realizes the Hebbian rule. When a teacher signal $z(t)$ arrives at the element, we can consider the rule $r_i(t) = x_i(t)z(t)$. This may be called forced learning, because the teacher signal comes from outside. When the reinforcement signal is given by $r_i(t) = x_i(t)u(t)$, where $u(t)$ is the potential evoked by the input $x(t)$

$$u(t) = \sum w_j(t)x_j(t),$$

we obtain a learning rule based on the potential of the neuron. This plays an important role in orthogonal learning defined later.

We again use the vector notation, $w = (w_1, ..., w_n)$, $r(r_1, ..., r_n)$, where w and r are row vectors, so that wx gives the inner product of w and x. The reinforce-

ment signal r is a function of input x, w, and z. We represent it by

$$r = r(w, x, z). \tag{3.2}$$

Let $\Delta w(t)$ be the increment of the connection weight at time t

$$\Delta w(t) = w(t+1) - w(t).$$

Then, the rule of weight modification can be written as

$$\Delta w(t) = -cw(t) + dr[w(t), x(t), z(t)] \tag{3.3}$$

for suitable positive constants c and d. This shows the direction of modification of w.

We consider the case when there exists a function $R(w, x, z)$ for which

$$\frac{\partial R}{\partial w} = w - \frac{d}{c} r(w, x, z) \tag{3.4}$$

holds[1], where $\partial R/\partial w = (\partial R/\partial w_1, ..., \partial R/\partial w_n)$ is the gradient vector. Then

$$\Delta w = -c \frac{\partial R}{\partial w}. \tag{3.5}$$

This means that w is modified in the negative direction of the gradient of R with respect to w.

The connection weight is modified depending on the $x(t)$ and $z(t)$, $t = 1, 2, 3, ...$. (We do not have any teacher signals in the case of concept formation.) We consider the situation where $[x(t), z(t)]$'s are independent samples from a fixed distribution. A typical case is that there are k input signals $x_1, ..., x_k$, x_α coming with probability or relative frequency p_α. The teacher signal z_α is paired with input signal x_α, in the case of association. We show the convergence of the connection weights.

Let $L(w)$ be the expected value of R for possible inputs (x, z),

$$L(w) = \langle R(w, x, z) \rangle, \tag{3.6}$$

where $\langle \rangle$ denotes the expectation with respect to x and z. In the above case,

$$L(w) = \sum_{\alpha=1}^{k} p_\alpha R(w, x_\alpha, z_\alpha).$$

By taking the expectation of (3.5), we have

$$\langle \Delta w \rangle = -c \partial L(w)/\partial w.$$

This shows that w is modified on the average in the direction of the steepest descent of $L(w)$. Hence, it is expected that w eventually converges to the value which minimizes $L(w)$ to within small fluctuations.

---

[1] The necessary and sufficient condition for the existence of R is given by $\partial r_i/\partial w_j = \partial r_j/\partial w_i$. As will be shown later, the condition is satisfied for most interesting cases

Assume that $L(w)$ is a function which has a unique minimum and satisfies some regularity conditions. Then, we can apply the theory of stochastic approximation (e.g., Wasan, 1969). When c is small, we can show that the expected value $\langle w(t)\rangle$ of the connection weights converges to the minimum of $L(w)$. Moreover, the deviation of $w(t)$ from the expected value can be made as small as desired, by letting c be sufficiently small.

We assume in the following that the decay constant c is sufficiently small so that the connection weights converge to the minimum of $L(w)$. The minimum is attained at w satisfying

$$cw = d\langle r(w, x, z)\rangle . \tag{3.7}$$

3.2. Examples of Simple Learning

A) Hebbian Learning. In the case of Hebbian learning, r_i is given by $r_i = yx_i$, where $y = \varphi(wx - h)$. Let $\Phi(u)$ be the integral of φ

$$\Phi(u) = \int_0^u \varphi(x)dx .$$

The function

$$R(w, x) = \tfrac{1}{2}ww^T - \frac{d}{c}\Phi(wx - h)$$

satisfies (3.4). Therefore, by Hebbian learning, the connection weight converges to w satisfying

$$w = \frac{d}{c}\langle \varphi(wx - h)x^T\rangle . \tag{3.8}$$

B) Forced Learning. The reinforcement signal is given by $r_i = zx_i$ with teacher signal z. It is easy to see that

$$R(w, x, z) = \tfrac{1}{2}ww^T - \frac{d}{c}zwx$$

gives the forced rule (3.4). From

$$2L(w) = \left(\left\|w - \frac{d}{c}\langle zx^T\rangle\right\|\right)^2 - \left(\frac{d}{c}|\langle zx^T\rangle|\right)^2 ,$$

the minimum is attained at

$$w = \frac{d}{c}\langle zx^T\rangle .$$

This is sometimes called the "correlation" of z and x.

C) Learning Based on Potential. The learning rule based on potential is given by

$$R(w, x) = \tfrac{1}{2}\left[ww^T - \frac{d}{c}(wx)^2\right],$$

where the reinforcement signal is $r = (wx)x$. We have

$$L(w) = \tfrac{1}{2}\left(ww^T - \frac{d}{c}w\langle xx^T\rangle w^T\right),$$

where $\langle xx^T\rangle$ is the matrix whose (i, j) entry is given by $\langle x_i x_j\rangle$. (It should be remembered that x is a column vector while w is a row vector.)

The function $L(w)$, however, has no minimum in this case. Therefore, the connection weight $w(t)$ does not converge under the above rule of learning. If the connection weight is subject to the subsidiary condition $ww^T = \text{const}$, so that $w(t)$ is normalized after each step of learning, we can prove that $w(t)$ converges to the minimum of $L(w)$ under the subsidiary condition. It is the direction of the eigenvector of the matrix $\langle xx^T\rangle$ corresponding to the maximum eigenvalue.

3.3. Orthogonal Learning

An element may have both excitatory and inhibitory channels from the inputs. An input signal, therefore, not only excites it but also inhibits it at the same time through inhibitory interneurons. The connection weight w_i can be decomposed into $w_i = w_i^+ - w_i^-$, where w_i^+ and w_i^- represent, respectively, the excitatory and inhibitory efficiency.

Excitatory and inhibitory neurons may have different self-organization rules. We consider the case where the excitatory w_i^+ is modified subject to forced learning with teacher signal, while the inhibitory weight w_i^- is modified subject to learning based on potential;

$$\Delta w^+ = -cw^+ + dzx^T ,$$

$$\Delta w^- = -c'w^- + d'(wx)x^T .$$

The excitatory weight converges to

$$w^+ = \frac{d}{c}\langle zx^T\rangle . \tag{3.9}$$

Define

$$R(w^-, w^+, x) = \tfrac{1}{2}\left[|w^-|^2 + \frac{d'}{c'}|(w^+ - w^-)x|^2\right]. \tag{3.10}$$

We then have $\Delta w^- = -c'\partial R/\partial w^-$. The function

$$L(w^-, w^+) = \langle R\rangle$$

$$= \tfrac{1}{2}\left[|w^-|^2 + \frac{d'}{c'}(w^+ - w^-)\langle xx^T\rangle(w^+ - w^-)\right]$$

is quadratic in w^- and its coefficient matrix is positive definite. Hence, it has a unique minimum, and w^- converges to the minimum. Since w^+ converges to

(3.9) independently of w^-, the minimum w^- is given by the equation

$$w^- = \frac{d'}{c'}[(w^+ - w^-)\langle xx^T\rangle].$$

The total synaptic weight $w = w^+ - w^-$, therefore, converges to one satisfying

$$w = \frac{d}{c}\langle zx^T\rangle - \frac{d'}{c'}w\langle xx^T\rangle. \tag{3.11}$$

or

$$w = e\langle zx^T\rangle(\varepsilon E + \langle xx^T\rangle)^{-1}, \tag{3.12}$$

where E is the unit matrix and $e = c'd/cd'$, $\varepsilon = c'/d'$.

Let us consider the case where k signals $x_1, ..., x_k$ come to the input $(k < n)$ with paired teacher signals z_α. We then have

$$\langle zx\rangle = \sum_\alpha p_\alpha z_\alpha x_\alpha, \quad \langle xx^T\rangle = \sum_\alpha p_\alpha x_\alpha x_\alpha^T.$$

We assume that k vectors $x_1, ..., x_k$ are linearly independent.

Let us introduce the set of dual orthogonal vectors $x_1^*, x_2^*, ..., x_k^*$ associated with $x_1, ..., x_k$. The x_α^* is a row vector satisfying the two conditions:

1) x_α^* is a linear combination of $x_1^T, ..., x_k^T$,

2) $x_\alpha^* x_\beta = \begin{cases} 1, & \alpha = \beta \\ 0, & \alpha \neq \beta. \end{cases}$

In other words, x_α^* belongs to the subspace spanned by $x_1^T, ..., x_k^T$ and is orthogonal to all the x_β's except x_α. The dual orthogonal vector is uniquely determined by

$$x_\alpha^* = \frac{1}{n}\sum_{\beta=1}^k G_{\alpha\beta}^{-1}x_\beta^T, \tag{3.13}$$

where $(G_{\alpha\beta}^{-1})$ is the inverse matrix of $(G_{\alpha\beta}) = (x_\alpha^T x_\beta)/n$.

Since $\langle xx^T\rangle$ is a singular matrix, it is impossible to expand (3.12) in a power series in ε, even when ε is small. By the method given in Appendix I,

$$w = e\sum_\alpha z_\alpha x_\alpha^* - \frac{e\varepsilon}{n}\sum_{\alpha,\beta}\frac{1}{p_\beta}z_\alpha G_{\alpha\beta}^{-1}x_\beta^* + O(\varepsilon^2), \tag{3.14}$$

where $O(\varepsilon^2)$ represents the second-order term in ε.

3.4. Covariance Learning

We propose another possible neural rule of self-organization. The excitatory weight w^+ is modified subject to forced learning as before, while the reinforcement signal r_i for the inhibitory synapse is given, in this case, by

$$r_i = e(bx_i + az - ab),$$

where a and b are constants. We then have

$$w = e'\langle(x-a)(z-b)\rangle, \tag{3.15}$$

where $e' = d/c = (d'/c')e$ and $a = (a, a, ..., a)^T$.

When a and b are, respectively, the expected values of x_i and z, $\langle(x_i - a)(z - b)\rangle$ is called the covariance of x_i and z. We call this rule of self-organization covariance learning.

When k input signals x_α's come with z_α's we have

$$w = e'\sum_\alpha p_\alpha(z_\alpha - b)(x_\alpha - a)^T. \tag{3.16}$$

When a and b are put equal to zero, this learning reduces to simple forced learning or correlation learning.

IV. Self-organization under Noisy Situations

4.1. Self-organization of Neural Nets

A non-recurrent net model for association receives k pairs of patterns (x_α, z_α) repeatedly, and the connection weight w_i of the i-th element converges to

$$w_i = e\langle z_i x^T\rangle(\varepsilon E + \langle xx^T\rangle)^{-1}$$

under orthogonal learning, where z_i is the teacher signal entering the i-th element. Hence, the connection matrix $W = (w_{ij})$ of the net converges to

$$W = e\langle zx^T\rangle(\varepsilon E + \langle xx^T\rangle)^{-1}. \tag{4.1}$$

This can be evaluated as

$$W = e\sum_\alpha z_\alpha x_\alpha^* - \frac{e\varepsilon}{n}\sum_{\alpha,\beta}\frac{1}{p_\beta}z_\alpha G_{\alpha\beta}^{-1}x_\beta^* + O(\varepsilon^2). \tag{4.2}$$

When ε is sufficiently small, we have

$$Wx_\beta \doteq ez_\beta, \quad \beta = 1, ..., k.$$

Therefore, ε represents the inaccuracy of the orthogonalization effect which suppresses the interference among many patterns.

Under covariance learning, the connection matrix converges to

$$W = e'\sum_\alpha p_\alpha(z_\alpha - b)(x_\alpha - a)^T, \tag{4.3}$$

where a and b are constant vectors.

A recurrent net, on the other hand, is forced to be in state x, when it receives an input x from the second or the initial-state setting terminals. The net immediately outputs the x, which is fed back to the ordinary input terminals of the net. Learning takes place under this situation, so that x plays the roles of the input signal and the teacher signal at the same time.

The connection matrix of the net converges to

$$W = e\langle xx^T\rangle(\varepsilon E + \langle xx^T\rangle)^{-1} \tag{4.5}$$

by orthogonal learning. It converges to

$$W = e'\langle(x-a)(x-a)^T\rangle, \tag{4.5}$$

by covariance learning.

4.2. Learning under Noisy Situation

We have so far considered noiseless learning, where patterns x_α and z_α arrive without noise disturbances. In reality, the net receives noisy versions of x_α's and z_α's. Let $\tilde{x}_\alpha$ be a noisy version of x_α

$$\tilde{x}_\alpha = x_\alpha + n,$$

where n is a noise vector. When the noise disturbs the i-th component of x_α, the i-th component n_i of n is equal to 1 or -1, according as $x_i = 0$ or 1. The n_i is otherwise equal to 0. We consider a random noise, which independently disturbs every component of a pattern with probability δ. We call δ the noise rate of noisy pattern $\tilde{x}_\alpha$. The noisy $\tilde{x}_\alpha$ differs from x_α in $n\delta$ components on the average.

We consider the case where the net receives noisy pattern pairs $(\tilde{x}_\alpha, \tilde{z}_\alpha)$. The noise independently disturbs the pattern pair (x_α, z_α) each time. Let $\bar{x}_\alpha$ be the average of noisy versions of x_α, i.e., the expectation of $\tilde{x}_\alpha$ over all the possible noise. We easily have

$$\bar{x}_\alpha = (1 - 2\delta)x_\alpha + 1, \tag{4.6}$$

where 1 is the column vector defined by $1 = (1, 1, ..., 1)^T$.

Similarly, by taking the expectation over all possible patterns and noises,

$$\langle zx^T \rangle = \sum_\alpha p_\alpha \bar{z}_\alpha \bar{x}_\alpha^T,$$

$$\langle xx^T \rangle = \sum p_\alpha \bar{x}_\alpha \bar{x}_\alpha^T + \sigma E,$$

where $\sigma = \delta(1 - \delta)$. Therefore, we see from (4.1) that the connection weight converges, under noisy orthogonal learning, to

$$W = e \sum \bar{z}_\alpha \bar{x}_\alpha^* + \frac{e\varepsilon'}{n} \sum \bar{z}_\alpha G_{\alpha\beta}^{-1} \frac{1}{p_\beta} \bar{x}_\beta^* + O(\varepsilon'^2), \tag{4.7}$$

where $\varepsilon' = \varepsilon + \sigma$ and the $\bar{x}_\alpha^*$'s are the dual orthogonal vectors of the $\bar{x}_\alpha$'s.

Noisy learning results in changing x_α and z_α to $\bar{x}_\alpha$ and $\bar{z}_\alpha$, respectively. This is not so important, because they are similar. The important effect of the noise is that it apparently increases ε to $\varepsilon' = \varepsilon + \sigma$, thus increasing the inaccuracy of orthogonalization.

A cluster of patterns can be considered as being composed of noisy versions $\tilde{x}_\alpha$'s of the representative patterns x_α of that cluster. Let δ_α be the noise rate of $\tilde{x}_\alpha$. A typical pattern $\tilde{x}_\alpha$ in the cluster is different from x_α in $n\delta_\alpha$ components on the average, so that δ_α represents the dispersion of the cluster. Given a distribution of patterns consisting of k clusters, we have

$$\langle x \rangle = \sum p_\alpha \bar{x}_\alpha,$$

$$\langle xx^T \rangle = \sum p_\alpha \bar{x}_\alpha \bar{x}_\alpha^T + \sigma E,$$

where $\sigma = \sum p_\alpha \delta_\alpha (1 - \delta_\alpha)$.

The connection weight matrix of a recurrent net converges, under orthogonal learning of the above distribution, to

$$W = e(\sum p_\alpha \bar{x}_\alpha \bar{x}_\alpha^T + \sigma E)[(\sigma + \varepsilon)E + \sum p_\alpha \bar{x}_\alpha \bar{x}_\alpha^T]^{-1}. \tag{4.8}$$

When both ε and σ are small, we can expand it into

$$W = e\left(\frac{\sigma}{\sigma + \varepsilon}E + \frac{\varepsilon}{\sigma + \varepsilon}\sum \bar{x}_\alpha \bar{x}_\alpha^*\right) + O(\varepsilon) \tag{4.9}$$

(see Appendix II). When $\sigma \ll \varepsilon$, i.e., the dispersion of clusters is small compared to ε, we have

$$W \doteq e \sum \bar{x}_\alpha \bar{x}_\alpha^*.$$

However, when $\sigma \gg \varepsilon$, we have $W \doteq eE$ which carries no information about the distribution of patterns. Therefore, orthogonal learning works well only when the dispersion σ of patterns is smaller than ε. Hence, ε represents the degree of noise immunity for orthogonalization.

Covariance learning is not so sensitive to noise disturbances. The connection weight converges to

$$W = e' \sum p_\alpha (\bar{z}_\alpha - b)(\bar{x}_\alpha - a)^T \tag{4.10}$$

for a non-recurrent net, and to

$$W = e'[\sum p_\alpha (\bar{x}_\alpha - a)(\bar{x}_\alpha - a)^T + \sigma E] \tag{4.11}$$

for a recurrent net.

V. Behavior of Association and Concept-formation Net

5.1. Simple Illustration of Association Mechanism

The matrix

$$W = e' \sum p_\alpha \bar{z}_\alpha \bar{x}_\alpha^T \tag{5.1}$$

is sometimes called the correlation matrix of k pattern pairs, and many researchers have studied associative memory models by the use of this matrix (e.g., Nakano, 1972; Amari, 1972a; Kohonen, 1972; Anderson, 1972). Consider the very special case where k patterns $x_1, ..., x_k$ are mutually orthogonal,

$$x_\alpha^T x_\beta = 0 \quad (\alpha \neq \beta). \tag{5.2}$$

In this case, for given input x_β, the net produces the potential vector

$$u = Wx_\beta = (e'p_\beta|x_\beta|^2)z_\beta$$

which is proportional to the associated pattern z_β. Hence, if the threshold h is adequately chosen, the net outputs the associated pattern z_β from input x_β $(\beta = 1, ..., k)$. The orthogonality condition (5.2) makes it possible to extract the necessary pattern from W in which many pattern pairs are superposed. Most studies of the so-called correlation memory have used mutually orthogonal or nearly orthogonal patterns to take advantage of this fact.

When the orthogonality condition does not exactly hold, and when a noisy version $x_\beta + n$ is input as a key pattern, the induced potential vector is written as

$$W(x_\beta + n) = (e'p_\beta|x_\beta|^2)z_\beta + N,$$

where

$$N = e' \sum_{\alpha \neq \beta} p_\alpha(x_\alpha^T x_\alpha)z_\alpha + Wn$$

represents the term due to the interference from the other superposed patterns and the noise. When N is small enough, the interference is completely eliminated by operating the non-linear threshold function φ. The neural association is realized by the two mechanisms: 1) elimination of interference of patterns by orthogonalization; and 2) noise elimination by operating the non-linear function φ. Amari (1972a) has studied the condition of noise elemination by the use of the stability concept of threshold-element nets. Uesaka and Ozeki (1972) have shown the probability of noise elimination by using randomly generated patterns x_α's, where each component takes on 1 and -1 with equal probability. In this case, two vectors are orthogonal on the average. Kohonen (1972) has studied the linear aspects of the correlation matrix memory. He also proposed the generalized-inverse approach, which uses the matrix W minimizing

$$L(W) = \sum_{\alpha=1}^{k} |Wx_\alpha - z_\alpha|^2 .$$

An iterative method of obtaining W was also shown (Kohonen, 1974). Orthogonal learning of the present paper has a close relation with the generalized inverse approach.

It should be noted that holography-like memory such as the holophone (Willshaw and Longuet-Higgins, 1969; Pfaffelhuber, 1975) also uses correlation memory. However, the dimensionality of the memory is as small as that of the patterns, so that it is difficult to superpose many patterns in the holography-like case.

5.2. Noise Immunity of Association Net

Let n be a noise disturbing a pattern x_α into $x'_\alpha = x_\alpha + n$. We call

$$\delta = \frac{1}{n} \sum_{i=1}^{n} |x'_{\alpha,i} - x_{\alpha,i}| = \frac{1}{n} \sum |n_i| \tag{5.3}$$

the noise rate of x'_α. The noise rate of an output pattern z'_α is defined similarly. An association net is said to have a high noise immunity, if the net, receiving a noisy input x'_α, outputs the correct or less noisy pattern z'_α. The behavior of an association net is evaluated by the

noise immunity shown by the relation between the noise rates of the input and output.

This relation, of course, depends on the whole set of pattern pairs (x_α, z_α) in a very complicated manner, so that it is almost impossible to obtain the relation. We, therefore, use the stochastic technique to evaluate the noise immunity. To this end, we assume that k pattern pairs (x_α, z_α) are randomly generated. Let a be the probability that every component $x_{\alpha,i}$ of every x_α is set equal to 1. Let b be the probability of $z_{\beta,j}$ being set equal to 1. All these components are independently determined.

We treat the case where n, m, and k are sufficiently large. Then, the law of large numbers holds and we can get an asymptotic evaluation of the noise immunity. The net must memorize k patterns z_α which altogether include mk components. The net has nm modifiable connection weights w_{ij}. Therefore,

$$s = mk/nm = k/n \tag{5.4}$$

represents the ratio of the number of memorized components to that of the modifiable synapses.

We first consider a net subject to covariance learning of k pattern pairs, with $p_\alpha = 1/k$. By evaluating the probability distribution of the interference term N (where we use the central limit theorem, the law of large numbers and assume that h_i's are adequately chosen), we have the following relation between the input noise ratio δ and the output noise ratio δ':

$$\delta' = F\left[\frac{\sqrt{a(1-a)(1-2\delta)}}{2\sqrt{sb(1-b)(a+\delta-2a\delta)}}\right] \tag{5.5}$$

where

$$F(s) = \frac{1}{\sqrt{2\pi}} \int_s^\infty \exp(-u^2/2)du .$$

The relation is shown in Figure 5, where $a = b = 0.2$. When $s = 0.1$, i.e., when the number of superposed pattern pairs is 10% of the dimensionality of input patterns, the net has an ability of eliminating an input noise of $\delta = 0.1$ almost completely, outputting the precise associated pattern. When $s = 0.3$, noise reduction also takes place, but the output pattern includes a noise of rate 2%, even when the input pattern is noiseless. When $s = 0.4$, there is no noise reduction. The output noise rate is similar to that of the input. Therefore, the association net with covariance learning works well for $s < 0.3$ or $k < 0.3\,n$. It should be noted that the net works better for smaller a and b. Mere correlation learning (5.1) does not work well, even if the thresholds are adequately tuned.

Association by covariance learning, however, has the following shortcomings: 1) Even if $\delta = 0$, i.e., the correct cue pattern is input, the net outputs a noisy pattern, $\delta' \neq 0$. 2) When the set of pattern pairs cannot be regarded as a typical sample from randomly generated patterns, it does not work well.

Association by orthogonal learning does not share these shortcomings. It works well for an arbitrary (non-random) set of pattern pairs, as long as the x_α's are linearly independent. When $\delta = 0$, the output patterns are noiseless $\delta' = 0$. However, the above performances are guaranteed only when ε is neglisibly small. When the netself-organizes under the influence of noise, ε increases to $\varepsilon' = \varepsilon + \sigma$. Therefore, it is not realistic to ignore the term of order ε.

Since x_α^*'s depend on all the x_α's in a very complicated manner, it is difficult to evaluate the noise immunity in the general situation. We again consider the case where the (x_α, z_α)'s are randomly generated, though the net works well for non-randomly generated pattern sets. We treat the case where ε and δ are so small that their higher order terms can be neglected. We then have the following noise relation

$$\delta' \leq F\left[\frac{a(1-a) - \varepsilon's + \delta}{2\sqrt{abs\delta(1-a+s+sa)}}\right], \qquad (5.6)^2$$

which is shown in Figure 6, where $a = b = 0.2$, $\varepsilon' = 0.1$. For small δ, the performance is very good, even when $s = 0.5$. In the case of $s = 0.1$, even the fifteen percent noise is completely reduced. However, this evaluation holds only when $\varepsilon' = \varepsilon + \sigma$ is not so large.

5.3. Dynamics of a Concept Formation Net

When each cluster of patterns is so sharply distributed that $\sigma \doteq 0$, $\bar{x}_\alpha \doteq x_\alpha$ hold, the connection matrix (4.9) reduces to

$$W = e \sum x_\alpha x_\alpha^* + O(\varepsilon).$$

Therefore, for small ε, $Tx_\alpha = x_\alpha$ ($\alpha = 1, \ldots, k$) holds. This shows that each pattern x_α becomes an equilibrium state of the net. By receiving an input pattern x, the net will tend to the state x_α which is closest to it. The net can store k patterns x_α in this manner. It can reproduce any of them correctly from a noisy input. The net may be regarded as a pattern generator, in which k patterns are coded.

In the case where the dispersion σ of the clusters is not neglisibly small, the net works only when $\sigma < \varepsilon$. Otherwise, the term $[\sigma/(\sigma + \varepsilon)]E$ dominates in W, which then gives no information about the necessary patterns. We may say in this sense that ε gives the degree of resolution of patterns: Only a pattern cluster whose dispersion σ is smaller than ε is treated as comprising versions of only one pattern. It is difficult for

---

2 The derivation is complicated and involves some stochastic assumptions

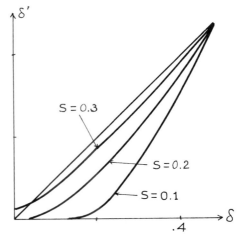

Fig. 5. Noise reduction rate of an association net with covariance learning

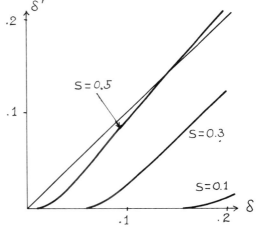

Fig. 6. Noise reduction rate of an association net with orthogonal learning

the net to recognize a cluster of patterns whose dispersion is larger than ε. The parameter ε, on the other hand, must be small, because otherwise we cannot neglect the harmful interference term $O(\varepsilon)$. Therefore, orthogonal learning works well for a trainable pattern generator in which the patterns to be generated are precisely taught. The cerebellum might have this mechanism in which the patterns are taught by the cerebrum.

When the net treats clusters of largely dispersed patterns, covariance learning works better. We hence analyze the behavior of the net with covariance learning. We consider again the case where k patterns x_α are randomly generated. We moreover assume for the sake of simplicity in calculation that the connection weight w_{ii} has a fixed self-inhibition term of amount $-e'/4$.

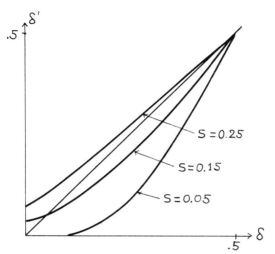

Fig. 7. Noise reduction rate of a concept-formation net with covariance learning

Assume that, in the neighborhood of a pattern x_α, the relation

$$\delta' = g(\delta),\tag{5.7}$$

holds, where δ and δ' are, respectively, the noise rates of states before and after state transition. Then, the noise rate δ_t at time t obeys the dynamical equation

$$\delta_{t+1} = g(\delta_t),\tag{5.8}$$

because the state $x(t+1)$ of the net is determined from $x(t)$. We can analyze the dynamical behavior of the net through the change in the noise rate. When δ_t of (5.8) converges to an equilibrium δ_0, the net tends to and remains in states close to x_α whose noise rate is δ_0. We can say in this case that the net can retain pattern x_α persistently within an error of noise rate δ_0. Unfortunately, there does not exist an exact relation like (5.7). We obtain instead the relation

$$\delta' \leq g(\delta),$$

where

$$g(\delta) = F\left[\frac{kp_\alpha(0.5-\delta)(0.5-\delta_\alpha)^2}{\sqrt{s\gamma}(0.5+\delta)}\right]\tag{5.9}$$

for patterns around x_α, where we put $a = 0.5$,

$$\gamma = k\sum_\beta p_\beta^2(0.5-\delta_\beta)^4$$

and δ_α represents the dispersion of patterns around x_α.

We show the function $g(\delta)$ in Figure 7, where $p_\alpha = 1/k$, $\delta_\alpha = 0.2$ for all α. Since we have put $a = b = 0.5$, the result is not so good compared to covariance association learning. For $s = 0.1$, dynamical Equation (5.8) has a stable equilibrium at $\delta = 0.01$. Therefore,

the net forms the corresponding concept within a 1% noise rate. For $s = 0.15$, the equation has a stable equilibrium at $\delta = 0.06$, so that the concept pattern is formed in the net within a 6% noise rate. However, for $s = 0.2$, there are no equilibria except for $\delta = 0.5$. This means that the net fails to form concept patterns in the case when $s > 0.2$, or $k > 0.2\, n$. The situation seems better for for small a and b.

Appendix I. Derivation of (3.14)

Let X be a $k \times n$ matrix consisting of k column vectors x

$$X = [x_1, x_2, ..., x_k].$$

Let P be a $k \times k$ diagonal matrix, whose diagonal entries are p_α. We denote by $\sqrt{P}$ the diagonal matrix whose diagonal entries are $\sqrt{p_\alpha}$. Let z be the row vector

$$z = (z_1, z_2, ..., z_k).$$

We easily have

$$\langle xx^T \rangle = XPX^T = (X\sqrt{P})(X\sqrt{P})^T,$$
$$\langle zx^T \rangle = zPX^T = z\sqrt{P}(X\sqrt{P})^T,$$
$$G = X^TX/n.$$

Equation (3.12) is rewritten as

$$w = ezPX^T(\varepsilon E + XPX^T)^{-1}.$$

We can use the identity

$$(X\sqrt{P})^T(\varepsilon E + XPX^T)^{-1} = (\varepsilon E_k + \sqrt{P}X^TX\sqrt{P})^{-1}\sqrt{P}X^T,$$

where E_k is the $k \times k$ identity matrix, and the expansion formula

$$(\varepsilon E_k + A)^{-1} = A^{-1} - \varepsilon A^{-2} + O(\varepsilon^2),$$

where A is a non-singular $k \times k$ matrix. (Since $k < n$, the $n \times n$ matrix XPX^T is singular, while the $k \times k$ matrix $\sqrt{P}X^TX\sqrt{P}$ is non-singular.) We obtain

$$w = ez(X^TX)^{-1}X^T - e\varepsilon z(X^TX)^{-1}P^{-1}(X^TX)^{-1} + O(\varepsilon^2).$$

Taking account of $(X^TX)^{-1}X^T = G^{-1}X^T$, and rewriting the above w in the vector notation, we have (3.14). It should be noted that the $k \times n$ matrix $X^+ = G^{-1}X^T$, which is composed of k row vectors x_α^*'s, is the generalized inverse of X [Albert (1972), see also Kohonen (1972)].

Appendix II. Derivation of (4.9)

Let $\bar{X}$ be the matrix defined by $\bar{X} = [\bar{x}_1, ..., \bar{x}_k]$. We have

$$W = e(\sigma E + \bar{X}P\bar{X}^T)[(\sigma + \varepsilon)E + \bar{X}P\bar{X}^T]^{-1}.$$

Let $\lambda_1, \lambda_2, ..., \lambda_k$ be the non-zero eigenvalues of the matrix XPX^T, and let $e_1, ..., e_k$ be the corresponding eigenvectors. We add $n - k$ vectors $e_{k+1}, ..., e_n$ such that $\{e_i\}\ i = 1, ..., n$ forms a set of orthonormal basis vectors. We then have

$$\bar{X}P\bar{X}^T = \sum_{i=1}^{k} \lambda_i e_i e_i^T,$$

$$E = \sum_{i=1}^{n} e_i e_i^T.$$

By the use of the relations

$$\left(\sum_{i=1}^{n} \mu_i e_i e_i^T \right)^{-1} = \sum_{i=1}^{n} \frac{1}{\mu_i} e_i e_i^T ,$$

$$\left(\sum \mu_i e_i e_i^T \right)\left(\sum v_i e_i e_i^T \right) = \sum \mu_i v_i e_i e_i^T ,$$

we obtain

$$W = e\left(\sum_{i=1}^{k} \frac{\sigma + \lambda_i}{\sigma + \varepsilon + \lambda_i} e_i e_i^T + \sum_{i=k+1}^{n} \frac{\sigma}{\sigma + \varepsilon} e_i e_i^T \right).$$

Expanding the above, where $\sigma + \varepsilon$ is assumed to be small, we obtain

$$W = e \sum_{i=1}^{n} \frac{\sigma}{\sigma + \varepsilon} e_i e_i^T + e \sum_{i=1}^{k} \frac{\varepsilon}{\sigma + \varepsilon} e_i e_i^T - \varepsilon e \sum_{i=1}^{k} \frac{1}{\lambda_i} e_i e_i^T + O[\varepsilon(\varepsilon + \sigma)] .$$

Here, we put $\sigma = 0$ and compare the result carefully with (4.2) or (4.7). We then have (4.9).

Conclusion

We have used two types of nets with recurrent connections and non-recurrent connections as models of association and concept-formation, respectively. We proposed two types of self-organization, one called orthogonal learning and the other covariance learning. We have analyzed the behavior of these nets and elucidated possible neural mechanism of information processing including association and concept-formation.

The present models are still too simple to be compared with the actual brain. However, the actual brain might have the mechanisms analyzed here commonly in its various portions. The present analysis will help in building more realistic models of the brain because it provides a possible neural logic.

Acknowledgement. This research was initiated at the University of Tokyo and completed during the author's visit to the Center for Systems Neuroscience of the University of Massachusetts on sabbatical leave from the University of Tokyo. The author would like to thank Professor M. A. Arbib for helpful comments on the manuscript.

References

Albert, A.: Regression and the Moore-Penrose pseudoinverse. New York: Academic Press 1972

Amari, S.: Characteristics of randomly connected threshold element networks and network systems. Proc. IEEE **59**, 35—47 (1971)

Amari, S.: Learning patterns and pattern sequences by self-organizing nets of threshold elements. IEEE Trans. Comp. C-21, 1197—1206 (1972a)

Amari, S.: Characteristics of random nets of analog neuron-like elements. IEEE Trans. on Systems, Man, and Cybernetics SMC-**2**, 643—657 (1972b)

Amari, S.: A method of statistical neurodynamics. Kybernetik **14**, 201—215 (1974a)

Amari, S.: A mathematical theory of nerve nets. In: Kotani, M. (Ed.): Advances in biophysics, Vol. 6, pp. 75—120. Tokyo: University Press 1974b

Amari, S.: Homogeneous nets of neuron-like elements. Biol. Cybernetics **17**, 211—220 (1975)

Amari, S.: Dynamics of pattern formation in lateral-inhibition type neural fields. To be published (1977a)

Amari, S.: A mathematical approach to neural systems. In: Arbib, M. A., Metzler, J. (Eds.): Systems neuroscience. I. New York: Academic Press 1977b

Amari, S., Yoshida, K., Kanatani, K.: Mathematical foundation of statistical neurodynamics. SIAM J. Appl. Math. **33** (1977)

Anderson, J. A.: A simple neural networks generating interactive memory. Math. Biosc. **14**, 197—220 (1972)

Caianiello, E. R.: Outline of a theory of thought processes and thinking machines. J. theor. Biol. **1**, 204—235 (1961)

Grossberg, S.: Onlearning, information lateral inhibition, and transmitters. Math. Biosc. **4**, 255—310 (1969)

Hebb, D. O.: The organization of behavior. New York: Wiley 1949

Kohonen, T.: Correlation matrix memories. IEEE Trans. Comp. C-21, 353—359 (1972)

Kohonen, T.: An adaptive associative memory principle. IEEE Trans. Comp. C-23, 444—445 (1974)

Kohonen, T., Oja, E.: Fast adaptive formation of orthogonalizing filters and associative memory in recurrent networks of neuron-like elements. Biol. Cybernetics **21**, 85—95 (1976)

Malsburg von der, C.: Self-organization of orientation sensitive cells in the striate cortex. Kybernetik **14**, 85—100 (1973)

Nakano, K.: Associatron—a model of associative memory. IEEE Trans. on Systems, Man, and Cybernetics SMC-**2**, 380—388 (1972)

Pfaffelhuber, E.: Correlation memory models—a first approximation in a general learning scheme. Biol. Cybernetics **18**, 217—223 (1975)

Rosenblatt, F.: Principles of Neurodynamics. Washington: Spartan 1961

Spinelli, N.: OCCAM: a computer model for a content-addressable memory in the central nervous systems. Biology of memory 293—306. New York: Academic Press 1970

Uesaka, Y., Ozeki, K.: Some properties of association-type memories. JIEECE of Japan (in Japanese), **55**-D, 323—330 (1972)

Wasan, M. T.: Stochastic approximation. Cambridge: University Press 1969

Wigström, H.: A neuron model with learning capability and its relation to mechanism of association. Kybernetik **12**, 204—215 (1973)

Willshaw, D. J., Longuet-Higgins, H. C.: The holophone—recent developments. In: Metzler, B., Michie, D. (Eds.): Machine intelligence, Vol. 4, pp. 349—357. Edinburgh: University Press 1969

Introduction

(1977)
James A. Anderson, Jack W. Silverstein, Stephen A. Ritz, and Randall S. Jones

Distinctive features, categorical perception, and probability learning: some applications of a neural model
Psychological Review 84:413–451

[Introductory comments by JAA] This paper arose from my firsthand experience with the cultural differences between psychologists and neuroscientists. One of the most entertaining things about neural network research for the participants is the chance it gives to learn lots of things in different scientific environments. But every area of science develops its own culture, sets its own problems, and uses its own jargon, and sometimes these differences can be unsettling.

My Ph.D. was in neurophysiology from MIT, with a thesis on the neurobiology of *Aplysia*. When I arrived at UCLA for a postdoc, I had the chance to work alone for a while, not entirely voluntarily, because the person I was to work with left almost as soon as I arrived. The time was at the end of the '60s, when the *Zeitgeist* was "What does it all mean, anyway?" Although work on *Aplysia* was interesting, and they do have colorful insides, there seemed no immediate possibility of using a very stupid animal like *Aplysia* to figure out what I thought were the important questions: How does the human brain work and what does it do? Two years of solitude on the seventh floor of the Brain Research Institute, spiced with far more outside political involvement and excitement than I expected when I went to California, led to the ideas presented in paper 15 and its precursors.

In 1971 I had a chance to go to Rockefeller University and work on pigeon cerebellum. That year it was hard to get a job doing anything in science, and pigeons sounded wonderful in comparison to unemployment. While I was at Rockefeller dissecting pigeon semicircular canals, I had a chance to talk to some of the mathematical psychologists in the laboratory of Professor William K. Estes, at that time a couple of floors above my office in Theobald Smith Hall. Networks like the linear associator had strong behavioral predictions because everything mixed together at the synapses in the act of storage. For example, they immediately suggested that psychological effects like interference and enhancement between separate memories could be produced easily (see Knapp and Anderson, paper 36). They also suggested an entire range of effects related to reaction time. If items are superimposed in memory, and if a test item has to be compared to stored items in memory, there cannot be an item by item comparison of test and memory items. The entire memory has to be checked for a match simultaneously. This analysis eventually led to some neural network models for psychological reaction time experiments (see Anderson, 1973).

I convinced Bill Estes that I would make a suitable postdoc in his laboratory in order to learn some mathematical psychology. I found it a fascinating and instructive year. Psychologists were constantly talking about "features," which seemed to be some kind of perceptual atom, from which we constructed complicated perceptions. I thought I knew what features were because I had read Hubel and Wiesel: they were selective neurons. Unfortunately, it became clear that psychological and linguistic features were *much* more complicated that the selective neurons that most physiologists called "feature detecting" cells. To my knowledge, no true psychological "feature" detecting cells have been found, then or now; that is, there seem to be no grandmother cells for features. To me, that meant that features had to be distributed patterns of discharge of many cells. In networks, patterns could be just as selective as individual cells and would have all the virtues of distribution, for example, damage and noise resistance.

I had become fascinated at UCLA with the network of intracortical recurrent connections that were prominent in cortical neuroanatomy. It seemed the most natural thing in the world to combine the linear associator with the recurrent connections in cortex and form an autoassociative system. As a learning feedback system, it should display all kinds of interesting temporal properties. And, best of all, the resulting connection matrix would have orthogonal eigenvectors that should have just the properties that I wanted for features: they would be patterns (since they were orthogonal), they would act highly selectively in the network with no cross-talk, and, due to feedback, they would tend to analyze the inputs in terms of the eigenvectors with large positive eigenvalues. There were some problems with instability, but they could be solved with a simple nonlinearity.

This paper applies the resulting model to speech perception and a phenomenon in experimental psychology called probability learning. The theoretical section starts with a review of the simple heteroassociative linear associator. This segues into the autoassociator, where a set of neurons feeds back onto itself, so input and output patterns are the same. It impossible to avoid the notion of feedback when an auto-associative system is proposed.

The generalized Hebb synapse is used as the synaptic learning rule. In the hetero-associative system, this rule gives rise to a pattern associator, as we have seen. In the autoassociative system, it gives rise to a system with pronounced dynamics, governed by positive feedback, and a symmetric connection matrix.

A linear version of the autoassociative system is discussed first. Because of the linearity, it is easy to carry through an analysis in terms of the eigenvectors and eigenvalues of the connection matrix that is formed by the Hebb rule. Jack Silverstein, one of my collaborators, pointed out that Hebbian learning in an autoassociative system gives a final connection matrix that is related to the sample covariance matrix of factor analysis. This implies that the eigenvectors with largest positive eigenvalues account for most of the variance in the learned input set.

The amount of feedback is governed by the eigenvalues, which act as a feedback gain control. In the dynamics of feedback, eigenvectors with large positive eigenvalues grow faster than ones with smaller eigenvalues. The power method in numerical analysis, which is used for the calculation of eigenvectors, uses this technique. If an input pattern is a mixture of eigenvectors, then after a while the components of eigenvectors with larger positive eigenvalues are enhanced relative to those smaller eigenvalues. Because of their connection with the sample covariance matrix, the "most important" eigenvectors for making discriminations will grow most rapidly. The claim is that these "important" eigenvectors act like the "distinctive features" mentioned above since inputs tend to be represented in terms of combinations of them.

Unfortunately, a linear system with positive feedback is potentially unstable. However, it is easy to stabilize the network by introducing a simple nonlinearity: limits are placed on the firing rate of individual neurons. This has the effect of containing the state vector within a "box" of limits (a hypercube), so the model is sometimes referred to as the "brain-state-in-a-box" (BSB) model. The dynamics of the system are linear within the box and nonlinear once neural discharges start to limit. This is a discontinuous version of the "sigmoid" used by Grossberg (paper 19) or the "squashing functions" described Rumelhart and McClelland (1986), and it contains a linear region. Since the BSB model is small signal linear, it is easy to analyze.

Formally, the BSB model is a dynamical system with attractors. Starting points anywhere in large regions of state space can end up in the same stable attractor points, usually corners of the hypercube, with all elements either fully on or fully off. It is possible to get attractors internal to the hypercube as well (see Hopfield, papers 27 and 35).

Like many "one-shot" neural networks, once the network enters a stable state, it will stay there forever. Using Hebbian "antilearning" (outer product learning with the opposite sign), it is possible to destabilize a stable state and make the system move to a new state. This effect has been used to make a model for multistable perception, for example, the way ambiguous figures like the Necker cube oscillate into and out of the paper on which they are drawn (see Kawamoto and Anderson, 1985).

The psychological applications make use of this attractor geometry. The first is categorical perception of consonants. Experiments indicate that when consonants, particularly stop consonants (/p/, /t/, /b/, /d/, etc.), are perceived, it is difficult or impossible to hear small differences between particular examples of a phoneme. If, for example, a continuum of examples of two stops is constructed (say, from /pa/ to /ba/) by varying the onset of voicing, subjects seem to be unable to discriminate between different examples that are perceived as the same phoneme, but can discriminate between equally close examples that are perceived as different phonemes. There is a large peak in discrimination ability at the indentification boundary between the phonemes.

In an attractor model, this can be interpreted as losing information about the starting point and retaining only the final, stable state, which for all examples perceived is the same phoneme. This analysis also predicts some other effects. For example, reaction times should be longest to stimului near the decision boundaries, where discrimination is best—which is what is found.

The other application of the BSB model is less familiar to nonpsychologists. The model is applied to an effect in experimental psychology called *probability learning*. Suppose either one of two lights can be turned on at a trial. Which of the two lights is chosen for a particular trial is random, and given by a certain probability. If subjects are asked to guess which of the lights will appear, they match the fraction of their guesses with the probability of occurrence of the light; i.e., if the left light is randomly turned on for 70% of the trials, the subjects will guess that light about 70% of the time.

Probability matching, which is shown by animals as well as humans in many tasks, does not maximize the probability of correct response.

Probability matching is a strong prediction of several learning models in psychology, for example, statistical learning theory. Therefore there is a large amount of good data available on the effect. The BSB model was applied to it by teaching the network the observed stimuli. This generates a matrix with a structure reflecting the probabilities of appearance of the stimuli. Then, when a prediction is needed, random points are chosen in state space, the system iterates until an attractor associated with one or the other stimulus is reached, and an appropriate response is made. The model does a reasonable job of accounting for the observed experimental effects.

Recent work on the BSB model has focused on its concept forming abilities and its abilities to perform "cognitive computation," i.e., to do the kinds of computation that humans seem to do well (see Anderson, 1986; Anderson and Murphy, 1986). The BSB model can be shown to be an energy minimizing system, in the sense used by Hopfield (paper 27; see also Golden, 1986).

References

J. A. Anderson (1973), "A theory for the recognition of items from short memorized lists," *Psychological Review* 80:417–438.

J. A. Anderson (1986), "Cognitive capabilities of a parallel system," *Disordered Systems and Biological Organization*, E. Bienenstock, F. Fogelman Soulie, and G. Weisbuch (Eds.), Berlin: Springer.

J. A. Anderson and G. M. Murphy (1986), "Psychological concepts in a parallel system," *Evolution, Games, and Learning*, D. Farmer, A. Lapedes, N. Packard, and B. Wendrofff (Eds.), Amsterdam: North-Holland.

R. M. Golden (1986), "The 'brain-state-in-a-box' neural model is a gradient descent algorithm," *Journal of Mathematical Psychology* 30:73–80.

A. H. Kawamoto and J. A. Anderson (1985), "A neural network model of multistable perception," *Acta Psychologica* 59:35–65.

D. E. Rumelhart and J. L. McClelland (Eds.) (1986), *Parallel Distributed Processing: Explorations in the Microstructures of Cognition*, Vol. 1, Cambridge, MA: MIT Press.

(1977)
James A. Anderson, Jack W. Silverstein, Stephen A. Ritz, and Randall S. Jones

Distinctive features, categorical perception, and probability learning: some applications of a neural model
Psychological Review 84:413–451

A previously proposed model for memory based on neurophysiological considerations is reviewed. We assume that (a) nervous system activity is usefully represented as the set of simultaneous individual neuron activities in a group of neurons; (b) different memory traces make use of the same synapses; and (c) synapses associate two patterns of neural activity by incrementing synaptic connectivity proportionally to the product of pre- and postsynaptic activity, forming a matrix of synaptic connectivities. We extend this model by (a) introducing positive feedback of a set of neurons onto itself and (b) allowing the individual neurons to saturate. A hybrid model, partly analog and partly binary, arises. The system has certain characteristics reminiscent of analysis by distinctive features. Next, we apply the model to "categorical perception." Finally, we discuss probability learning. The model can predict overshooting, recency data, and probabilities occurring in systems with more than two events with reasonably good accuracy.

In the beginner's mind there are many possibilities, but in the expert's there are few.

—Shunryu Suzuki
1970

I. Introduction

If we knew some of the organizational principles of brain tissue, it might be possible to make a few general statements about how the brain works in a psychological sense. There is a close relation between available building blocks and the performance that can be realized easily with those component parts in most systems, and we should expect the same to be true for the nervous system.

Let us consider an example of some importance to psychologists. In current theories of cognition and perception one often finds explanations of phenomena in terms of what are essentially little computer programs, complete with flow charts and block diagrams. Certainly, the desire to decompose complex mental events into simpler basic units follows the strategy that has been triumphantly successful in the physical sciences.

Even a poor understanding of brain organization might be of value in placing some kinds of limits on these elementary operations.

All of us are somewhat familiar with digital computers and, particularly, with computer programs. In some psychological models, the elementary instructions that we use to tell a computer what to do seem to serve as the model for brain function. Assumed are "comparisons," "scans," "lists," "decisions," and other computer-like operations.

However, computers are made with fast, reliable, binary electronic components that are designed to operate by executing very quickly a long series of simple operations. When an elementary operation takes a fraction of a microsecond and when internal noise is not a

Copyright 1977 by the American Psychological Association, Inc. Reprinted by permission of the publisher and author.

major problem, this is a very successful technique.

But most neuroscientists agree at present that the brain is a slow, intrinsically parallel, analog device that contains a great deal of internal noise from a variety of sources. It is very poorly suited to the accurate execution of a long series of simple operations. What is it good at?

The brain is best adapted to interacting, highly complex, spatially distributed parallel operations. Adjectives such as "distributed," "parallel," and "holographic" are sometimes used to describe operations of this type.

Since neurons are very slow compared to electronic devices—on the order of a few milliseconds at the very fastest—we should expect there to be time for only a few of the elementary operations that compose the instructions for a "brain computation." There would be neither the speed, equipment, nor accuracy for enormous strings of simple operations.

We may conclude that the elementary operations used by the brain are of a much more powerful and different kind than the simpler instructions familiar to us from our experience with computers.

In this article we will present the outlines of a theory that is suggested by the anatomy and physiology of the brain and that is realized by arrays of simple, parallel, analog elements that are meant to be an oversimplification of the neurons in a mammalian neocortex. The model

is associative and distributed, and we shall see that its elementary operations are of a very different type from simple logical operations. Systems that are intrinsically parallel have a number of pronounced and unfamiliar properties, as well as some impressive capabilities.

When we propose a model that tries to join together information from several fields, we have difficulties when we try to verify it. At present, we simply do not know enough about the detailed connectivity and synaptic properties of the brain to do more than make our models in qualitative agreement with what is known of the neurobiology. Similarly, the psychological data are useful when it comes to testing general approaches, but they often do not allow unequivocal tests of the details of a theory. Thus, it seems to us best to try to make our neurally based models refined enough to fit, in detail, a few experiments—just to show it can be done. But we would also like to point out, in a more impressionistic way, areas of agreement between theory and observation elsewhere. We have deliberately made our theories extremely simple, perhaps unrealistically so, because if we can make adequate models with very simple theories, surely it will be possible to do better when more complex and/or more realistic assumptions are made. We feel that many different versions of distributed memories can be made to give results similar to those we describe here.

We will first discuss some necessary theoretical background. We will present a simple version of a distributed, associative memory. We will then modify the simple, linear model to incorporate positive feedback of a set of neurons on itself. We will show that feedback gives rise to behavior that is reminiscent of the analysis of an input in terms of what are called "distinctive features," a type of analysis that is commonly held to be of great importance in perception. We will then apply the model to two widely differing psychological phenomena. We shall discuss in detail the classic set of experiments generally described as "probability learning," and we shall discuss more generally the perceptual phenomenon called "categorical perception."

These diverse effects, which at first sight might seem to be very complicated and involve much information processing, may be

This work was supported in part by grants from the Ittleson Family Foundation to the Center for Neural Studies, Brown University, and by a Public Health Service General Research Support Grant from the Division of Biological and Medical Sciences, Brown University. Jack W. Silverstein was supported by the National Science Foundation Grant MCS75-15153-A01.

Some material derived from this paper has been presented at the "Workshop on Formal Theories in Information Processing," Stanford University, Stanford, California, June 1974. Some material was also presented at the 7th Mathematical Psychology meeting, Ann Arbor, Michigan, August 1974; at the annual meeting of the Society for Neuroscience, St. Louis, Missouri, October 1974, and at the 9th Mathematical Psychology Meeting, New York, August 1976.

Requests for reprints should be sent to James A. Anderson, Center for Neural Studies, Department of Psychology, Box 1853, Brown University, Providence, Rhode Island 02912.

explainable by a single, rather simple, set of assumptions.

II. Theoretical Development

In the past few years a number of related realizations of distributed memories applied to brain models have been put forward by several groups of investigators (Anderson, 1970, 1972; Cooper, 1974; Grossberg, 1971; Kohonen, 1972, 1977; Little & Shaw, 1975; Willshaw, Buneman, & Longuet–Higgins, 1969). One form of the central learning assumptions of these models was first proposed by Hebb (1949) but, as is often the case in an active field of science, many of the fundamental assumptions of the models have been arrived at independently by different groups.

In our development here, we will follow the notation and basic assumptions of Anderson (1968, 1970, 1972, 1977; Cooper, 1974). This version of a distributed, associative memory is formally exceptionally simple. It is easy to work with and may give a first approximation to some of the common properties of many distributed models.

We start by making two central assumptions. First, nervous system activity can be most usefully represented as the set of simultaneous individual neuron activities in a group of neurons. Neuron "activity" is considered to be related to a continuous variable, the average firing frequency. Patterns of *individual* activities are stressed, because properties of particular neuron activities need not be related to each other for the system to function. Indeed, some evidence (Noda & Adey, 1970) indicates that interneuronal spike activity correlations of nearby cells recorded with the same electrode in parietal ("association") cortex may be quite low when the brain is doing "interesting" things. They found that in an animal in REM (rapid eye movement) sleep or in the awake, alert state, correlations were very low, whereas the same pairs of cells had highly correlated discharges in deep sleep. The same appears to be true in hippocampus (Noda, Manohar, & Adey, 1969).

Recent work by Creutzfeldt, Innocenti, and Brooks (1974) seems to suggest that most cells in primary visual cortex, even those close to one another in the same cortical column, are not strongly coupled together, again implying a good deal of individuality of cell response. The individuality of cell responses in auditory cortex has been remarked upon (Goldstein, Hall, & Butterfield, 1968). Morrell, Hoeppner, and de Toledo (1976) studied single units in cat and rabbit parastriate cortex. They found that nearby units responding to the same stimulus (recorded with the same electrode) often differed in the type and direction of alteration of their discharges when a stimulus configuration and a cutaneous shock were paired in a Pavlovian conditioning paradigm. They comment that their data "provide little support for the notion of coherent changes in large neuronal populations" (p. 448).

Thus we are dealing, to a first approximation, with a system of individualistic cells, each with its own properties. Although cells near to one another may show similar response properties (for example, orientation or binocularity in visual cortex), each cell behaves differently from its neighbors when studied in detail. We will be concerned with the behavior of large groups of cells, but this does not mean that each cell is doing the same thing. Indeed, most distributed systems do not work well if cells all respond to the same inputs in the same ways, since diversity of cell properties allows for better operation. Simple redundancy is the most uninteresting way to provide reliability.

This assumption allows us to represent these large-scale activity patterns as vectors of high dimensionality with independent components. All the models we work with will use these vectors as the elementary units. We will show that it is possible to develop theories where these complex activity patterns, representing discharges of very many neurons, can act as basic units that combine and interact in relatively simple ways.

As our second major assumption, we hold that different memory traces (sometimes called "engrams"), corresponding to these large patterns of individual neuron activity, interact strongly at the synaptic level so that different traces are not separate in storage. Considerable physiological evidence supports this idea. As Sir John Eccles comments, "each neurone and even each synaptic junction are built into many engrams. The systematic study

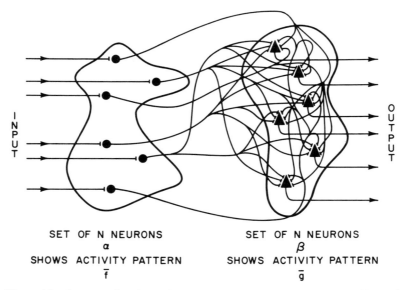

SET OF N NEURONS   SET OF N NEURONS
α   β

SHOWS ACTIVITY PATTERN   SHOWS ACTIVITY PATTERN
$\bar{f}$ $\bar{g}$

Figure 1. We consider the properties of sets of N neurons, α and β. Close inspection of this spaghetti-like drawing will reveal that every neuron in α projects to (i.e., has a synapse with) every neuron in β. Since this drawing, where $N = 6$, understates the size and connectivity of the nervous system by several orders of magnitude, it can be seen that single neurons and single synapses may have little effect on the discharge patterns of the group as a whole. Properties of such large, interconnected systems can sometimes be modeled simply with the techniques of linear algebra, the approach taken in the text. ($\bar{f}$ and $\bar{g}$ indicate vectors.)

of the responses of individual neurones in the cerebrum, cerebellum, and in the deeper nuclei of the brain is providing many examples of this multiple operation" (Eccles, 1972, p. 59).

Association Model

Let us assume we have two groups of N neurons, α and β (see Figure 1). We will assume that every neuron in α projects to every neuron in β. This clearly unrealistic assumption is not made in Anderson (1972). A detailed discussion of some of the mathematical aspects of this model are found in Cooper (1974) and Nass and Cooper (1975).

To proceed further, we must describe how the activity of a neuron reflects its synaptic inputs. For many cells, this can be extremely complex. However for some systems rather simple relations are found. We shall assume at first that neurons are simple linear analog integrators of their inputs, and we shall see what kinds of models evolve from this assumption. There is evidence supporting this assumption in a few well-studied systems. In the

lateral inhibitory system of *Limulus*, rather good linear integration is found (Knight, Toyoda, & Dodge, 1970; Ratliff, Knight, Dodge, & Hartline, 1974). Linear transmission, according to Mountcastle (1967), holds for many mammalian sensory systems, once beyond an initial nonlinear transduction.

We assume there is a synaptic strength, a_{ij}, which couples the jth neuron in α with the ith neuron in β. Thus, subject to our assumption of linear integration, we can write the following: If $f(j)$ is the activity shown by neuron j in α and if $g(i)$ is the activity shown by neuron i in β at a given time, then

$$g(i) = \sum_j a_{ij} f(j). \qquad (1)$$

Let us now consider the following situation: The set of neurons α shows an activity pattern, a vector, $\mathbf{f}_1$, the set of all the $f(j)$. The set of neurons β shows an activity pattern, a vector, $\mathbf{g}_1$. We wish to associate the pattern $\mathbf{f}_1$ with the pattern $\mathbf{g}_1$ so that later presentation of $\mathbf{f}_1$ alone will give rise to $\mathbf{g}_1$ in the set β. Let us assume that initially our set of synaptic connectivities a_{ij} is zero.

We can ask what detailed local information could influence a synaptic junction to allow storage of memory. Locally available information includes the presynaptic activity and, we shall assume, the postsynaptic activity. Let us make about the simplest assumption for synaptic learning that allows for pre- and postsynaptic interaction at the synaptic level. Let us postulate, as our essential learning assumption, that *to associate pattern $\mathbf{f}_1$ in α with pattern $\mathbf{g}_1$ in β we need to change the set of synaptic weights according to the product of presynaptic activity at a junction with the activity of the postsynaptic cell.*

For convenience, let

$$\|\mathbf{f}_1\| = 1, \quad \text{where} \quad \|\mathbf{f}\| = \left[\sum_{i=1}^{N} f(i)^2\right]^{1/2}.$$

This quantity is usually called the length of the vector. The change in the ijth synapse is given by $f_1(j)g_1(i)$. The set of connections form a matrix $\mathbf{A}_1$ given by

$$\mathbf{A}_1 = \mathbf{g}_1\mathbf{f}_1^T$$
$$= [\mathbf{g}_1 f_1(1), \mathbf{g}_1 f_1(2) \cdots \mathbf{g}_1 f_1(N)], \quad (2)$$

where T is the transpose operation (throughout this paper all vectors will be assumed to be N-dimensional column vectors).

Assume that after we have "printed" the set of connectivities $\mathbf{A}_1$, pattern of activity $\mathbf{f}_1$ arises in α. Then we see that activity in β is given by

$$\mathbf{A}_1\mathbf{f}_1 = \mathbf{g}_1(\mathbf{f}_1^T\mathbf{f}_1) = \|\mathbf{f}_1\|^2\mathbf{g}_1 = \mathbf{g}_1; \quad (3)$$

so we have $\mathbf{g}_1$ appearing as the pattern of activity in β, which corresponds to our definition of association.

It is very unlikely that these sets of neurons exist only to associate a single set of activity patterns. Let us assume that we have K sets of associations, $(\mathbf{f}_1,\mathbf{g}_1), (\mathbf{f}_2,\mathbf{g}_2), \ldots, (\mathbf{f}_K,\mathbf{g}_K)$, each generating a matrix of synaptic increments, $\mathbf{A}_k$. Then, since we have assumed that a single synapse participates in storing many traces, we form an overall connectivity matrix $\mathbf{A}$ given by

$$\mathbf{A} = \sum_k \mathbf{A}_k.$$

Let us assume that the $\mathbf{f}_i$ are mutually orthogonal, that is, that the inner product of $\mathbf{f}_i$ with $\mathbf{f}_j$, defined as

$$\mathbf{f}_i^T\mathbf{f}_j = \sum_{s=1}^{N} f_i(s)f_j(s),$$

is zero for $i \neq j$. Then, if pattern $\mathbf{f}_j$ is impressed on the set of neurons α, we have

$$\mathbf{A}\mathbf{f}_j = \sum_k \mathbf{A}_k\mathbf{f}_j$$
$$= \mathbf{A}_j\mathbf{f}_j + \sum_{k \neq j} \mathbf{A}_k\mathbf{f}_j$$
$$= \mathbf{g}_j + \sum_{k \neq j} \mathbf{g}_k(\mathbf{f}_k^T\mathbf{f}_j)$$
$$= \mathbf{g}_j.$$

Thus, the system associates perfectly. If the $\mathbf{f}$s are not orthogonal the system will produce noise as well as the correct association, but the system is often quite usable. Actually, the "mistakes" made by this system are often as interesting as the "correct" responses (see Anderson, 1977).

Orthogonality is an effective way of dealing with the notion of independence of inputs. If two inputs are orthogonal, then there is no interaction between them; that is, the response of the system to one input is in no way influenced by the other input. It is as if the inputs are going through completely different mechanisms.

A Numerical Example

To show how this system works, let us construct a simple, eight-dimensional system. Assume we have the four orthogonal input vectors, $\mathbf{f}_1$, $\mathbf{f}_2$, $\mathbf{f}_3$, and $\mathbf{f}_4$ shown in Table 1. These vectors are normalized Walsh functions, which are digital versions of sine and cosine functions. We choose arbitrary output vectors, $\mathbf{g}_1$, $\mathbf{g}_2$, $\mathbf{g}_3$, and $\mathbf{g}_4$ and wish to make the associations between pairs of vectors $(\mathbf{f}_i,\mathbf{g}_i)$. Note that we need place no restrictions on the $\mathbf{g}$s. Thus $\mathbf{g}_1$ and $\mathbf{g}_2$ are orthogonal, and $\mathbf{g}_3$ and $\mathbf{g}_4$ are very close together. They also vary considerably in length: $\mathbf{g}_1$ is 2.24 units long, $\mathbf{g}_2$ is 3.61 units long, and $\mathbf{g}_3$ and $\mathbf{g}_4$ are 4.47 units long. The $\mathbf{A}_1$ matrix, the association matrix between $\mathbf{f}_1$ and $\mathbf{g}_1$, is shown in Table 2. We have not shown the other three matrices, $\mathbf{A}_2$, $\mathbf{A}_3$, and $\mathbf{A}_4$, but their construction should be clear from $\mathbf{A}_1$. The resulting sum of all four matrices is shown in the second part of Table 2.

Table 1
Input and Output Vectors Used in the Numerical Example Given in the Text

Type of vectors

Input

$$\mathbf{f}_1 = \frac{1}{\sqrt{8}}\begin{bmatrix} 1 \\ 1 \\ 1 \\ 1 \\ -1 \\ -1 \\ -1 \\ -1 \end{bmatrix} \quad \mathbf{f}_2 = \frac{1}{\sqrt{8}}\begin{bmatrix} -1 \\ -1 \\ 1 \\ 1 \\ -1 \\ -1 \\ 1 \\ 1 \end{bmatrix} \quad \mathbf{f}_3 = \frac{1}{\sqrt{8}}\begin{bmatrix} 1 \\ -1 \\ 1 \\ -1 \\ 1 \\ -1 \\ 1 \\ -1 \end{bmatrix} \quad \mathbf{f}_4 = \frac{1}{\sqrt{8}}\begin{bmatrix} -1 \\ -1 \\ 1 \\ 1 \\ 1 \\ 1 \\ -1 \\ -1 \end{bmatrix}$$

Output

$$\mathbf{g}_1 = \begin{bmatrix} 1 \\ 0 \\ -1 \\ 0 \\ 1 \\ -1 \\ -1 \\ 0 \end{bmatrix} \quad \mathbf{g}_2 = \begin{bmatrix} -1 \\ 2 \\ 0 \\ -1 \\ -1 \\ -1 \\ -1 \\ 2 \end{bmatrix} \quad \mathbf{g}_3 = \begin{bmatrix} 3 \\ 0 \\ -1 \\ -1 \\ -2 \\ 0 \\ -1 \\ 2 \end{bmatrix} \quad \mathbf{g}_4 = \begin{bmatrix} 4 \\ 0 \\ -1 \\ -1 \\ -1 \\ 0 \\ 0 \\ 1 \end{bmatrix}$$

Inspection of this matrix shows little sign of the component parts that went to make it up, and asking about the value of any particular element is pointless in relation to the information stored in the matrix. However, calculation will show that

$$\mathbf{A}\mathbf{f}_i = \mathbf{g}_i$$

for all four input vectors, which shows that indeed such a simple matrix can "learn" four essentially arbitrary associations.

One of the most important aspects of these models is their similarity to a filter in the strict sense of a system which responds weakly to an input that has not been learned (i.e., to

Table 2
Matrices Associating Pairs of Vectors in the Numerical Example

Matrix

$$\mathbf{A}_1 = \mathbf{g}_1 \mathbf{f}_1^T$$

$$\mathbf{A}_1 = \frac{1}{\sqrt{8}}\begin{bmatrix} 1 & 1 & 1 & 1 & -1 & -1 & -1 & -1 \\ 0 & 0 & 0 & 0 & 0 & 0 & 0 & 0 \\ -1 & -1 & -1 & -1 & 1 & 1 & 1 & 1 \\ 0 & 0 & 0 & 0 & 0 & 0 & 0 & 0 \\ 1 & 1 & 1 & 1 & -1 & -1 & -1 & -1 \\ -1 & -1 & -1 & -1 & 1 & 1 & 1 & 1 \\ -1 & -1 & -1 & -1 & 1 & 1 & 1 & 1 \\ 0 & 0 & 0 & 0 & 0 & 0 & 0 & 0 \end{bmatrix}$$

$$\mathbf{A} = \mathbf{A}_1 + \mathbf{A}_2 + \mathbf{A}_3 + \mathbf{A}_4$$

$$\mathbf{A} = \frac{1}{\sqrt{8}}\begin{bmatrix} 1 & -5 & 7 & 1 & 7 & 1 & -3 & -9 \\ -2 & -2 & 2 & 2 & -2 & -2 & 2 & 2 \\ -1 & 1 & -3 & -1 & -1 & 1 & 1 & 3 \\ 1 & 3 & -3 & -1 & -1 & 1 & -1 & 1 \\ 1 & 5 & -3 & 1 & -3 & 1 & -3 & 1 \\ 0 & 0 & -2 & -2 & 2 & 2 & 0 & 0 \\ -1 & 1 & -3 & -1 & 1 & 3 & -1 & 1 \\ -1 & -5 & 5 & 1 & 1 & -3 & 3 & -1 \end{bmatrix}$$

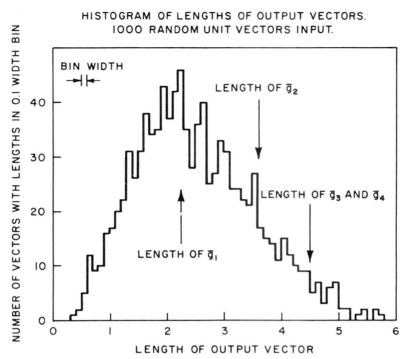

Figure 2. One thousand random unit vectors were input to the matrix constructed as a numerical example in the bottom half of Table 2. The figure shows a histogram of the lengths of the 1,000 resulting output vectors. The arrows point to the lengths of the actual output vectors associated with the four inputs used in the matrix. ($\bar{g}_1$, $\bar{g}_2$, $\bar{g}_3$, and $\bar{g}_4$ indicate vectors.)

which the filter is not "tuned") but which responds strongly to an input that has been seen before. We can demonstrate this property with our matrix. Suppose we have a random unit vector as input, that is, a vector generated by the uniform distribution on the unit sphere in eight-dimensional space. We should expect, if the system is filter-like, that there will be a short vector appearing at the output, on the average. We used a computer to generate 1,000 random unit vectors and looked at the lengths of the resulting output vectors. The distribution of output lengths is shown in Figure 2. The lengths of the actual stored associations—the **g**s—are shown by arrows in the figure. Strictly on the basis of length, almost no outputs due to random vectors were as long as **g**$_3$ and **g**$_4$ (96% were shorter), and 85% of the outputs were shorter than **g**$_2$. Even **g**$_1$, which was half the length of **g**$_3$ and **g**$_4$, was longer than 45% of the random outputs. Thus, even with as crude a measure of filter characteristic as output length, in a very low dimensionality

system, a fairly good job of discrimination can be made between old and new inputs. Introduction of even a very few elementary mechanisms to "sharpen up" the response from the system, such as cascades of filters or positive feedback (discussed in Section IV), can make these matrices good filters with an interesting "cognitive" structure to them.

We should observe, however, that the system can make mistakes. Noise is inherent in the system. The histogram shows that there are a few inputs that can give a larger output than given by any of the inputs the system has learned. Thus, we can see that a useful type of analysis for such systems is statistical, and many aspects of their behavior can be studied well with the techniques of communication theory and decision theory. If such a distributed system is indeed present in our cortex, possibly we can see why such statistical methods are so successful in accounting for many of the phenomena found for the psychology of even our highest mental functions.

III. Biological and Psychological "Features"

Introduction

The association model just presented is a good associator, and it has been applied elsewhere to several sets of experimental data (see Anderson, 1977; Nass & Cooper, 1975). A simple variant of the model, relying on the filter characteristics of a similar system, has been used to propose an explanation for some of the data arising in the Sternberg list-scanning experiment (Anderson, 1973).

We hope now to show that an extension of the model has interesting qualitative similarities to some important characteristics of human perception. We hope to show that we can represent noisy inputs to the system in terms of their "distinctive features" and that the features that the model generates are both cognitively significant, in that they are the most useful for discriminating among members of a stimulus set, and, at the same time, are most strongly represented in the output from the system. Thus, what appears to be a highly structured and analytical approach to perception—distinctive feature analysis—can be explained as the result of the operation of a highly parallel, analog system with feedback.

We shall interweave the theoretical discussion with a very brief discussion of the psychological theory and a little data from the neurosciences. This occasionally awkward means of presentation is intended to convey the close interdependence between theory and data from several fields.

Let us make clear all we hope to do. At this time, we do not have the data from either psychology or neuroscience to decisively test the theory we are to present. However, we hope to show that there are striking qualitative similarities between the structure of the theory and the structure that many feel typifies some kinds of perception.

What is a Distinctive Feature?

That there are entities called distinctive features and that these entities are somehow of importance in perception is a commonly accepted belief is psychology at present. A recent elementary textbook (Lindsay & Norman, 1972) builds a large part of the first few chapters around the idea of features, and this approach is central to Neisser's very influential book, *Cognitive Psychology* (1967). Since the early 1950s, there has been strong evidence from linguistics that phonemes could be characterized as being represented by the presence or absence of a small set (12 in the original analysis) of distinctive features (Jakobsen, Fant, & Halle, 1961). Each phoneme was uniquely represented by its own particular set of features. The use of a good set of features is an excellent type of pre-processing and is a commonly used practical pattern recognition technique. The great reduction in dimensionality of the stimulus allows the system to discard irrelevant information and eliminate noise, while making later stages simpler and more reliable, since they have to cope with less complex inputs.

How Might the Brain Do Feature Analysis?

Feature analysis seems to be a strategy used by the brain. How is this analysis performed?

The simplest way might be to have, somewhere, neurons that respond when, and only when, a particular distinctive feature appears. These would then be true "feature detecting" neurons. We might point out that in this case, discharges corresponding to different features would be orthogonal in the sense discussed in the previous section, since different features would give rise to activity patterns with non-zero values in different sets of elements.

Barlow (1972) has argued strongly that this is truly the way the brain works. He stated a number of what he calls "dogmas" about the relation between brain and perception, proposing that sensory systems are organized so as to achieve as complete a representation of the stimulus as possible with the smallest number of discharging neurons. He estimated that as few as 1,000 cells in visual cortex may fire in response to a complex visual stimulus. He proposed that what we call "perception" may correspond to the activity of a small number of high level, very selective neurons, each of which "says something of the order of complexity of a word" (p. 385). Informal discussions indicate to us that many neurophysiologists are sympathetic to this point of view.

Biology and Psychology: Audition

Suppose we look at published feature lists for spoken language (see Lindgren, 1965) and for letter perception (Gibson, 1969; Laughery, 1969). The features proposed by these and other writers are psychological features. They are quite complex. Lindgren describes the acoustic characteristics of the distinctive features proposed for spoken language. Almost every feature is characterized by complex changes in wide, often ill-defined bands of frequencies. For example, the "vocalic versus non-vocalic" feature is characterized by "presence versus absence of a sharply defined formant structure" (Lindgren, 1965, p. 55). The "nasal versus oral" feature is described as "spreading the available energy over wider (versus narrower) frequency regions by a reduction in the intensity of certain (primarily the first) formants and introduction of additional (nasal) formants" (Lindgren, 1965, p. 55). The only feature that seems to correspond to a well-defined set of frequencies is the "voiced versus voiceless" feature, which describes the presence or absence of vocal cord vibration. Even here, though, men, women, and children have characteristic vocal cord frequencies differing over a two-octave range, all presumably capable of exciting the voicing feature. The complicated structure of the vocal tract would usually be expected to give rise to equally complicated variations in frequencies of resonances with changes in geometry.

The neurophysiology of the auditory system is very complex and not well understood at present. Although the lower levels of the auditory pathway seem to be primarily frequency analyzers, neurons in auditory cortex are highly individualistic and variable in their responses. Many cells in primary auditory cortex have very sharply tuned responses, but others have quite wide bandwidths.

If species-specific vocalizations are used to stimulate cells, the picture is no simpler. Wollberg and Newman (1972) recorded from the auditory cortex of squirrel monkeys, using recordings of species-specific calls as stimuli. They reported, "some cells responded with temporally complex patterns to many vocalizations. Other cells responded with simpler patterns to only one call. Most cells lay between these two extremes" (p. 212).

Funkenstein and Winter (1973; Winter & Funkenstein, 1973) did a similar experiment with a wider range of squirrel monkey vocalizations. They also found a variety of cell responses, from a small percentage of cells that responded only to particular calls, to cells that responded to particular frequencies in any context—noise, pure tones or calls.

Evans (1974) reviews a number of experiments and describes the bewildering variety of response types observed. Even in research on an animal as unintelligent as a bullfrog, which has a behaviorally important mating call with two spectral peaks, Frishkopf, Capranica, and Goldstein (1968) did not uncover cells that responded only to both peaks presented simultaneously, a property which would be required of a "mating call detector." Although the frequency responses of cells in the frog auditory system were commonly tuned to one or the other spectral peak, the investigators found no cells in the midbrain and medulla that put the two peaks together.

Biology and Psychology: Vision

The feature lists that are proposed for recognition of capital letters are deceptively simple. In the lists, one typically finds proposed features such as vertical line segments at left, center, or right; horizontal line segments at top, middle, or bottom; or curve slants or parallel lines (Laughery, 1969).

With a list of this type, it is only a small conceptual leap to identify these psychological features with groups of particular cells of the type known to exist in primary visual cortex which show orientation and edge sensitivity (Hubel & Wiesel, 1962, 1968). It should be emphasized that single cells in primary visual cortex do *not* show the requisite selectivity in the sense that they respond to features and only features. Cells in primary visual cortex are quite selective, but they respond to many aspects of the stimulus. Schiller, Finlay, and Volman (1976), in perhaps the most careful quantitative study of single-cell response properties in monkey striate cortex, make the following comments in the abstract of the summary paper:

1. Several statistical analyses were performed on 205 S-type and CX-type cells which had been completely analyzed on 12 response variables: orientation tuning, end stopping, spontaneous activity, response variability, direction selectivity, contrast selectivity for flashed or moving stimuli, selectivity for interaction of contrast and direction of stimulus movement, spatial-frequency selectivity, spatial separation of subfields responding to light increment or light decrement, sustained/transient response to flash, receptive-field size, and ocular dominance.
2. Correlation of these variables showed that within any cell group, these response variables vary independently. (p. 1362)

(S and CX cells correspond roughly to the familiar simple versus complex distinction, but with more precise definition.)

Besides finding support of our earlier assumption of the great individuality of cortical neurons, we can see that single cells can have their discharges modified by a wide range of aspects of the stimulus. Here also we do not seem to find cells that respond only to psychological features, although there are cells with very pronounced selectivities.

A Regrettable Misapprehension

It is apparent from reading the literature in this area that the word "feature" as used by psychologists and by neuroscientists has come to mean different things. When a psychologist discusses features, what seems to be meant is a complex kind of perceptual atom which is independent of other atoms and constitutes an elementary unit out of which perception is built. The feature lists that have been proposed for both letter perception and speech perception involve many different aspects of the input stimulus. Their simplicity is deceiving when considered in light of the properties of the single cells of the nervous system.

It is also apparent, regrettably, that when a neurobiologist refers to a "feature detector," he is typically referring to a single neuron which displays a certain amount of selectivity in its discharge, often for the biologically important and relevant aspects of the stimulus. This does not mean that this cell has the specificity of response to be a detector of the psychological feature. Something more is involved.

IV. A Theoretical Approach to Psychological Features

There is an implicit neural model—that formulated clearly by Barlow—in the identification of single-cell properties with features. There is an alternative point of view that uses the observed single-cell selectivities to provide selectivity to the psychological features, but now as part of well-defined activity patterns that *use* cell properties.

McIlwain (1976) makes the important point that the large receptive fields often seen in the higher levels of the visual system and the response of the cells to many aspects of a stimulus do not necessarily mean that the overall system lacks precision. Our distributed system using the output of the entire group of cells can be very precise, in that the discharge pattern due to one input can be reliably differentiated from that arising from a different stimulus, even though there are many cells in the group that may respond to both inputs. As a relevant example, the associative network presented in Section II of this article is completely interconnected. A cell in the second set of neurons can respond to any cell in α, the first set. A single cell has a large receptive field and is very unselective. However, we showed that the output patterns of all the cells in β can be made to respond strongly only to particular inputs, and thus the system displays considerable selectivity.

Can we make a neural model analyze its inputs in terms of features? It is clear that distinctive feature analysis has an important learned component. In the perception of written letters, this is obvious. In spoken language, which is much more biologically determined than are written letters, Eimas, Siqueland, Jusczyk, and Vigorito (1971) have shown that some aspects of linguistic distinctive features are both built in and modifiable. For example, a category boundary for voice onset time, which appears to correspond to the voicing–voiceless feature, is present in the human infant. Yet this feature boundary can be modified in the adult, depending on the phonetic structure of the language spoken (Eimas & Corbit, 1973).

Our theoretical aim is to reduce the dimensionality of the stimulus so that a very com-

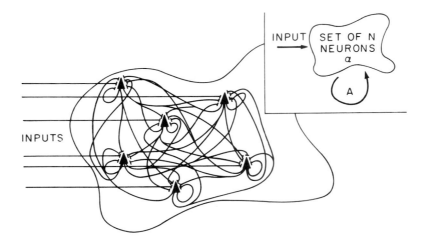

I. SET OF N NEURONS, α

2. EVERY NEURON IN α IS CONNECTED TO EVERY
OTHER NEURON IN α THROUGH LEARNING
MATRIX OF SYNAPTIC CONNECTIVITIES A

Figure 3. A group of neurons feeds back on itself. Again, note that $N = 6$, as in Figure 1. Note that each cell feeds back to itself as well as to its neighbors.

plicated stimulus, exciting perhaps millions of selective cells, could act as if only a small number of independent elements were involved.

A Model

Assume we have an associative system (Section II) which couples a set of neurons, α, to itself, instead of to a different set of neurons. We again make the approximation that every neuron projects to every other neuron. Let us assume that this feedback connection is through a matrix, **A**, of synaptic connectivities. Figure 3 shows this situation.

Let us consider the case where a pattern of activity on the set α is coupled to itself. The increment in synaptic strength, Δa_{ij} is proportional to the product of the activity shown by the ith neuron, $f(i)$, and the jth neuron, $f(j)$. Note that

$$\Delta a_{ij} = \Delta a_{ji}. \tag{4}$$

This means that **A** is what is called a symmetric matrix.

This implies, in turn, the existence of N mutually orthogonal vectors $\mathbf{e}_1, \ldots, \mathbf{e}_N$ such that

$$\mathbf{A}\mathbf{e}_i = \lambda_i \mathbf{e}_i, \quad i = 1, \ldots, N,$$

where each λ_i is a real number. The $\mathbf{e}_i$s are called eigenvectors of **A**, while each λ_i is called the eigenvalue associated with $\mathbf{e}_i$. Since there are N mutually orthogonal eigenvectors, every vector is a linear combination of the eigenvectors. An important consequence of this is that **A** is completely determined by its sets of eigenvectors and corresponding eigenvalues.

In many systems, the eigenvalues and eigenvectors are of great importance. Our system is no exception. Let us consider how a matrix, starting from zero, would develop.

Assume we present K orthonormal inputs, that is, mutually orthogonal unit vectors, each input $\mathbf{f}_i$ appearing k_i times. Then by our associative model, each $\mathbf{f}_i$ is an eigenvector of **A** with corresponding eigenvalue k_i, since

$$\mathbf{A} = \sum_{i=1}^{K} k_i \mathbf{f}_i \mathbf{f}_i^T$$

and

$$\mathbf{A}\mathbf{f}_j = \left[\sum_{i=1}^{K} k_i \mathbf{f}_i \mathbf{f}_i^T \right] \mathbf{f}_j = k_j \mathbf{f}_j, \quad j = 1, \ldots, K.$$

If $K < N$, the remaining $N - K$ eigenvectors of **A** have zero eigenvalues. We see then, in this case, that the eigenvectors of **A** with large eigenvalues will tend to correspond to commonly presented patterns, and the eigenvalue (in a more complex system) will be at least a rough estimate of the frequency of presentation.

Suppose now the inputs are arbitrary except that the average input

$$\sum_{i=1}^{K} k_i \mathbf{f}_i / \sum_{i=1}^{K} k_i$$

is zero. This important assumption is acknowledgment of the fact that inhibition and excitation are equally important and equally prominent in the operation of the nervous system. We make the above assumption in the form it takes because it is very convenient for our mathematical interpretation, but the exact conditions that occur in the nervous system are beyond our present knowledge.

With this assumption, $\mathbf{A}$ is then a scalar multiple of the sample covariance matrix of the inputs, since the latter is given by

$$\sum_{i=1}^{K} p_i \mathbf{f}_i \mathbf{f}_i^T - [\sum_{i=1}^{K} p_i \mathbf{f}_i]^T [\sum_{i=1}^{K} p_i \mathbf{f}_i], \quad (5)$$

where $p_i = k_i / \sum k_i$, $i = 1, \ldots, K$. We should point out that this matrix is positive semidefinite, that is, that all the eigenvalues are greater than or equal to zero.

From principal components analysis, we find that the eigenvectors of $\mathbf{A}$ are related very strongly to the inputs in the following way. We will use terms from probability theory. Let $\mathbf{f}$ be the random vector which takes the values $\mathbf{f}_i$ with probability p_i, $i = 1, \ldots, K$. Let cov($\mathbf{f}$) denote the covariance matrix of $\mathbf{f}$, given by Equation 5. The main result from principal components analysis states that any unit vector $\mathbf{u}$ that maximizes the variance of the random variable $c = \mathbf{u}^T \mathbf{f}$ over all unit vectors must be an eigenvector $\mathbf{u}_1$ of cov($\mathbf{f}$) with the largest eigenvalue λ_1. The variance of $c_1 = \mathbf{u}_1^T \mathbf{f}$ turns out to be λ_1. The maximum variance of $\mathbf{u}^T \mathbf{f}$ over all unit vectors orthogonal to $\mathbf{u}_1$ is the second largest eigenvalue λ_2 of cov($\mathbf{f}$), and it must occur at the corresponding eigenvector. This maximal principle follows through for all the eigenvalues where at the jth step we maximize all of the unit vectors orthogonal to the $j - 1$ eigenvectors already established. The random vector $\mathbf{f}$ can then be expressed as

$$\mathbf{f} = \sum_{i=1}^{N} c_i \mathbf{u}_i, \quad c_i = \mathbf{u}_i^T \mathbf{f}.$$

The c_i values turn out to be mutually uncorrelated, and we see that

$$\text{var}\|\mathbf{f}\|^2 = \text{var} \sum_{i=1}^{N} c_i \mathbf{u}_i^T [\sum_{i=1}^{N} c_i \mathbf{u}_i]$$

$$= \sum_{i=1}^{N} \text{var } c_i = \sum_{i=1}^{N} \lambda_i.$$

Each input then is a linear combination of the eigenvectors of $\mathbf{A}$, that is,

$$\mathbf{f}_i = \sum_{j=1}^{N} c_j{}^i \mathbf{u}_j,$$

where, qualitatively, the eigenvectors with large eigenvalues account for most of the differences between inputs. Furthermore, no correlation exists between $c_{j_1}{}^i$ and $c_{j_2}{}^i$, where $j_1 \neq j_2$. The eigenvectors of $\mathbf{A}$ can therefore be considered as the basic components of the inputs.

As an additional comment, we know from linear algebra that the eigenvectors give the maximum responses from the system. Over all vectors of unit length (that is, all vectors $\mathbf{x}$, such that $\|\mathbf{x}\| = 1$), the largest value of $\|\mathbf{Ax}\|$ occurs when $\mathbf{x}$ is the eigenvector $\mathbf{e}_i$ corresponding to the largest eigenvalue. The next largest value of $\|\mathbf{Ax}\|$ over unit vectors orthogonal to $\mathbf{e}_i$ occurs when $\mathbf{x}$ is the eigenvector $\mathbf{e}_j$ corresponding to the second largest eigenvalue, and so on.

Significance of Feedback

Let us consider what this system might do to an input to the set of neurons from the sensory receptors or from an earlier stage of processing.

Suppose the input is composed of one of the eigenvectors of the feedback matrix that has a large positive eigenvalue. This means that the activity pattern will pass through the feedback matrix unchanged in direction. It will add algebraically to what is already going on in the set of neurons. Since the eigenvalue is positive, it will add to ongoing activity, that is, to the eigenvector. The larger amplitude pattern will be fed back again and the output from the feedback matrix will again add to the activity in the set of neurons. This is positive feedback, and the amplitude of the eigenvector will grow. Depending on the details of the system, it may grow without bound or merely show a longer and stronger response, but it will be increased in strength relative to other patterns. Consider an input which contains a contribution from an eigenvector with small or zero eigenvalue. Positive feedback will not significantly enhance this pattern, and the amplitude may increase very slowly or not at all.

Thus the input pattern, after a while, will tend to be composed of only the components of the original input that have large positive eigenvalues, and only these components will participate in further processing. We have just seen that the eigenvalue is in some sense a measure of the importance of the particular eigenvector in discriminating among different members of the items the system has learned, so the patterns with large eigenvalues are the most important patterns for the system. Since this is exactly the behavior we want from distinctive features, *let us specifically identify the eigenvectors of the feedback matrix with large positive eigenvalues as the distinctive features of the system.* We see that in all important aspects of their behavior they act as we would like distinctive features to act.

Let us note as well that a similar technique is used in pattern recognition and statistics. We have previously mentioned the similarity of this analysis to principal components analysis, and many pattern recognition tasks use very similar techniques because of their theoretical optimality (Young & Calvert, 1974, Chap. 6).

The operation of this system gives us good insight into the profound practical differences between the brain and a digital computer. The reason such pattern recognition techniques are not used more widely is economic: Excessive computation time is required because of the large matrix operations involved. However, we see that an adaptive parallel feedback system with highly interconnected analog elements can process an input so it most strongly weights its features, according to importance, in only one step. We shall discuss this process further in the next few sections.

Some Comments on Neurobiology

The idea of a distributed feedback network is consistent with much we know about the physiology and anatomy of cerebral cortex. We should emphasize as well that the actual pathways may sometimes be more complex than our model. For example, inhibition seems often to be accomplished in mammals by inhibitory interneurons. This extra neuron need not affect the mathematics of the model.

There are a number of feedback systems in cortex and thalamus. Perhaps the most attractive candidate to perform operations like those we discuss in this article is the very rich network of recurrent collaterals of cortical pyramidal cells. Recurrent collaterals are axonal fibers which branch off the axon of a pyramidal cell, loop back into the nearby gray matter, and synapse extensively with the dendrites of nearby pyramids over a range of several millimeters. Globus and Scheibel (1967) comment that the recurrent collaterals of pyramids are the most common class of fibers in neocortex.

Freeman and his collaborators (see Freeman, 1975, for a detailed review) have worked extensively, both experimentally and theoretically, on the electrical activity and connections of prepyriform cortex (olfactory cortex) and olfactory bulb in cats. This primitive cortex may show in simple form the connections that are more highly developed in neocortex. Freeman has had considerable success in applying linear systems analysis to these networks. He has incorporated in some of his models the type of excitatory feedback that we have suggested as a basis for feature analysis. An excitatory collateral system from prepyriform pyramids onto nearby pyramids has been described anatomically, as has a similar collateral system in hippocampus, another primitive cortex. The anatomy of these connections, and other recurrent systems in neocortex, as well as some of the physiology, is reviewed in Shepherd (1974).

Higher order loops, from one region of cortex to another and back, or from thalamus to cortex and back, are also common in the brain and may also be candidates for the psychologically significant feedback loops we would like to find. However, the physiological data on these systems, other than those suggesting their existence and importance, are presently sparse. Further discussion of evidence for physiological mechanisms that may participate in feedback interactions is given elsewhere (Anderson, 1977, Note 1).

More Detailed Study of Feedback

In the study described in detail in Anderson (1977), we used linear systems analysis to obtain exact solutions of an interesting case. We showed that the response of the feedback system had the properties we claimed for it. Suppose we represent an input as a weighted sum of eigenvectors of the feedback matrix, **A**. Since the eigenvectors are orthogonal, they can serve as a basis set. Suppose the input is then presented to the system. We showed that after a period of time, the relative weights of the eigenvectors changed, and that the eigenvectors with large positive eigenvalues were much more heavily represented in the activity of the set of neurons than were the eigenvectors with smaller eigenvalues. We also showed that the response of the system to eigenvectors with large positive eigenvalues lasts longer. Some variants of the model have regions of stability as well; that is, the response of the system dies back to zero or remains bounded as long as the largest eigenvalue does not exceed a certain critical value. Above this value, the system "blows up"; that is, the amplitude of the activity pattern increases without bound. In this calculation, the time constants of the feedback system of the brain were quite long relative to the duration of the neural activity representing the sensory input.

In this article, we consider a slightly different model. We assume, essentially, that the time constant of the feedback is fast compared to the sensory input. Thus, feedback and current activity directly add in the same time period.

As a speculative comment, we observe that the first model bears a certain impressionistic relation to the way visual processing in reading has been conjectured to be performed. A visual input is received and coded with a great burst of activity from the sensory neurons, which then become relatively quiet; however, processing continues for 200 or 300 msec before the eyes jump to a new location in a saccade. Then the process is repeated. We might conjecture

that the auditory system—particularly in speech perception—is following a somewhat different strategy. Phonemes follow one another in quick succession, tens of milliseconds apart, and the nervous system must respond more quickly to a constantly changing input.

There are many ways we could set up the feedback model we shall discuss for the remainder of this article, but we initially chose one that was exceptionally convenient for computer simulations. Our general philosophy of modeling is always to try simple things first. We suspect that similar but more complex variants will not show very different qualitative behavior.

The dynamics of the system are assumed to occur in discrete time. Let $\mathbf{x}(t)$ denote the activity vector (the "state vector") at time t. Integer values are taken by t. The activity at time $t + 1$ is assumed to be the sum of the activity at time t and the action of the feedback matrix on the activity at time t. The summing of the output of the feedback system and the activity at this time is assumed to occur in the same time quantum, which implies that their time courses are comparable. Thus, we have

$$\mathbf{x}(t + 1) = \mathbf{x}(t) + \mathbf{A}\mathbf{x}(t) = (\mathbf{I} + \mathbf{A})\mathbf{x}(t), \quad (6)$$

where **I** is the identity matrix. Throughout this discussion, we let **A** be fixed. The system is, indeed, a positive feedback system, due to the fact that all eigenvalues of **A** are nonnegative. To see this, let $\mathbf{e}_1, \ldots, \mathbf{e}_N$ denote the orthonormal eigenvectors of **A** with corresponding nonnegative eigenvalues $\lambda_i, i = 1, \ldots, N$. Then **A** can be written as

$$\mathbf{A} = \sum_{i=1}^{N} \lambda_i \mathbf{e}_i \mathbf{e}_i^T.$$

Let

$$\mathbf{x}(t) = \sum_{i=1}^{N} x_i \mathbf{e}_i.$$

Then

$$\mathbf{x}(t + 1) = (\mathbf{I} + \mathbf{A})\mathbf{x}(t)$$
$$= \left[\sum_{i=1}^{N} (1 + \lambda_i) \mathbf{e}_i \mathbf{e}_i^T \right] \sum_{j=1}^{N} x_j \mathbf{e}_j$$
$$= \sum_{i=1}^{N} x_i (1 + \lambda_i) \mathbf{e}_i, \quad (7)$$

so that

$$\|\mathbf{x}(t+1)\|^2$$

$$= \left[\sum_{i=1}^{N} x_i (1 + \lambda_i) \mathbf{e}_i \right]^T \sum_{j=1}^{N} x_j (1 + \lambda_j) \mathbf{e}_j$$

$$= \sum_{i=1}^{N} x_i^2 (1 + \lambda_i)^2 \geq \sum x_i^2 = \|\mathbf{x}(t)\|^2. \quad (8)$$

Thus, the length of the activity vector is non-decreasing at every step. If $\mathbf{x}(t)$ is made up only of eigenvectors of $\mathbf{A}$ with zero eigenvalues, then from Equation 6 we see that $\mathbf{x}(t+1) = \mathbf{x}(t)$, so if the system starts at one of these points it stays there for all time. All other vectors will respond to the system. In fact, for these vectors strict inequality holds in Equation 8.

At this point we must introduce an important feature into the model, one that will break with the linearity that we have assumed up to now. We pointed out that in one version of the feedback model certain positive eigenvalues of $\mathbf{A}$ were stable, in that the system activity did not grow indefinitely large as time progressed. In the present version of the model, the same is not true; activity will grow without bound. Unfortunately, the desirable features of positive feedback are exactly the ones that cause catastrophes. This is inappropriate behavior for a system that requires unquestioned stability at all times. The cases of unstable neuronal discharge that we know of give rise to highly pathological seizure states. The normally functioning brain seems to be extremely stable and resistant to "runaway," which is a very important observation in view of the powerful excitatory mechanisms that the brain contains.

The simplest way of containing the activity of the system is to use the fact that neurons have limits on their activities: They cannot fire faster than some frequency (usually around several hundred spikes per sec, and in some auditory units, as fast as 1,000 spikes per sec), and they cannot fire slower than 0 spikes per sec. Thus, there are positive and negative limits on firing rate.

Suppose we incorporate this property into our model. A particular activity pattern is a point in a very high dimensionality space. The coordinate axes correspond to activities of individual neurons. Thus, putting limits on firing frequency corresponds to putting the allowable activity patterns into a box. Since large, high-dimensionality vectors describing the system are often called "state vectors," we name this model the "brain-state-in-a-box" model, with the associated image of a state vector, a point in space, buzzing around like a bee under the influence of input and feedback.

We formalize this situation by assuming (possibly unrealistically) that the limits are symmetrical around the origin. That is, saturation in the system is achieved by keeping the system on or inside the cube in N-dimensional space defined by $x_i = \pm C$; $i = 1, \ldots, N$, where x_i is the activity of the ith neuron.

Applying this assumption to our model, at each time step, the activity vector first undergoes the change given by Equation 6, and then each coordinate which is either greater than C or less than $-C$ is replaced by C or $-C$, respectively. Using the maximum (max) and minimum (min) functions, we can write the dynamics of the saturating system for the ith element of $\mathbf{x}$, at time $t + 1$, as

$$x(t+1)_i$$
$$= \max\left(-C, \min\left(C, [(\mathbf{I} + \mathbf{A})x(t)]_i\right)\right),$$
$$i = 1, \ldots, N.$$

Our primary interest in the cube is the corners, that is, points of the form

$$C(\pm 1, \pm 1, \ldots, \pm 1)^T.$$

There are 2^N different corners. Suppose $\mathbf{x}_0$ is a corner with the property that each coordinate of $\mathbf{A}\mathbf{x}_0$ is nonzero and carries the same sign as the corresponding coordinate of $\mathbf{x}_0$. Then it follows that there exists a neighborhood N of $\mathbf{x}_0$ (for our purposes, a neighborhood of $\mathbf{x}_0$ can be thought of as a ball centered about $\mathbf{x}_0$) such that if the activity vector ever lands in the intersection of N and the cube, it eventually reaches $\mathbf{x}_0$ and stays there for all future time. Points of this type are called stable. If some of the coordinates of $\mathbf{A}\mathbf{x}_0$ are zero, then $\mathbf{x}_0$ can still be stable if there is a neighborhood N of $\mathbf{x}_0$ such that each element $\mathbf{x}$ of N, where $\mathbf{x} \neq \mathbf{x}_0$, satisfies the above condition. It is easy to see that if a corner is stable its antipodal corner is stable.

If an eigenvector of $\mathbf{A}$ lies along a diagonal

of the cube, then the two corresponding corners are stable. Moreover, all points (except 0) sufficiently close to the eigenvector will wind up in either corner. The collection of points in the cube that are attracted to a stable corner $\mathbf{x}_0$ is called the region of stability of $\mathbf{x}_0$. We will use these regions (where all points are considered equivalent, since they all end in the same corner) in our applications in the next two sections.

Our experience with computer simulations and our intuitions suggest that the qualitative behavior of the system is straightforward. Suppose we start off with an activity vector which is receiving powerful positive feedback; that is, without limits the vector would grow indefinitely. The vector lengthens until it reaches one of the walls of the box; that is, one of its component neurons reaches the firing limit. The vector will try to get longer, but it cannot escape from the box. Thus, it will head for a corner, where it will stay if the corner is stable. It can be shown that in many cases only some corners are stable.

An Important Special Case

In some of the calculations we shall perform in the next two sections, we must specify how the eigenvalues change with time. In one important special case, this is extremely simple.

We observe that in a saturating model, the final state of the system is always a corner. Thus, if we increment matrix $\mathbf{A}$ by the final state given by that corner, the incremental set of synaptic changes will always be the same.

If the corners are eigenvectors, which could be the case after a long time learning only corners, then the eigenvalues will be increased by a constant amount every time that the corner corresponding to that eigenvector appears. This can be seen easily. If $\mathbf{f}$ is both an eigenvector and a corner, then $\mathbf{Af} = \lambda\mathbf{f}$. The incremental change in $\mathbf{A}$ due to learning is given by $\Delta\mathbf{A} = \eta\mathbf{f}\,\mathbf{f}^T$, where η is a learning parameter. If $\mathbf{f}$ is normalized so $\|\mathbf{f}\| = 1$, then

$$\Delta\mathbf{Af} = \eta\mathbf{f}(\mathbf{f}^T\mathbf{f}) = \eta\mathbf{f}$$

and

$$(\mathbf{A} + \Delta\mathbf{A})\mathbf{f} = (\lambda + \eta)\mathbf{f}. \qquad (9)$$

A Further Comment

Mathematically, the covariance matrix defined in Equation 5 is positive semidefinite; that is, all the eigenvalues are either zero or positive. One reason for this is the presence of large positive values on the main diagonal, the a_{jj}. This is so, since if we have K stored inputs with each input f_i appearing k_i times, then

$$a_{jj} = \sum_{i=1}^{K} k_i f_i^2(j).$$

This term corresponds to the feedback of a neuron on itself. Although so-called "autapses" are found occasionally in cortical pyramidal cells (van der Loos & Glaser, 1972), most of the physiologically studied cases where a cell feeds back on itself involve special neural circuits, often inhibitory. A well-known example is the Renshaw cell system, a special class of cells providing inhibitory feedback to spinal motor neurons. We need not be restricted to the values given above for the main diagonal of the matrix. We can let these values be zero, if we wish, as we show in Appendix C.

Zero Activity Level

The model presented previously predicts that almost all neurons will be firing at either maximum or minimum rate. Clearly, in a real nervous system, many, if not most, neurons probably will not respond to a given stimulus, although a sizable fraction may participate in the activity pattern. There are several ways we can have a number of zero elements in our activity vectors. Even in the model given here, where feedback is very powerful and connectivity is complete, input patterns will occasionally give rise to a zero activity level of a neuron in the set if all the eigenvectors with nonzero eigenvalues have zero in the coordinate corresponding to that neuron. If the space is even moderately filled with eigenvectors (i.e., if K is somewhat comparable to N and many eigenvalues are nonzero) or if noise is introduced into the system, then this is very uncommon. However, if we assume any one of several mechanisms—thresholds of feedback, adaptation, restricted connectivity—we can easily produce a system where many elements in the N-dimensional space remain at zero, even in the final state.

In the previously mentioned paper by Barlow (1972), it was pointed out that cortical neurons seem to respond in a characteristic way. A linear model would make no restrictions on the activity pattern. That is, we could have activity patterns containing a great many small changes in activity or a few large ones. But the nervous system seems to have chosen the latter kind of response pattern. We see, if we make appropriate assumptions, that we may restrict saturation to a set of the most "important" neurons for the discrimination of the input stimulus.

Cells in cortex are rather selective in that they do not respond to most inputs but have a strong response to some. Our brain-state-in-a-box model has something of this aspect to it, in that in the final state cells may be fully on, fully off, or not participating in the activity pattern at all. It would be of interest to look at cortical neurons in the awake, behaving animal in light of our proposed model.

V. Categorical Perception

The best evidence for distinctive feature analysis comes from linguistics. Therefore, it seemed natural to us to try to apply this model to find if it agreed with the kind of perceptual analysis that occurs in speech perception.

Preprocessing

Distinctive features are usually viewed as a system for efficient preprocessing, whereby a noisy stimulus is reduced to its essential characteristics and decisions are made on these. We showed that the feature vectors arising in the model are akin to those found in principal components analysis and are indeed the most useful patterns for these kinds of discriminations. More, however, is suggested by the model. The brain-state-in-a-box model suggests that the final output of such a system is a stable state corresponding to a corner of the high-dimensional hypercube formed by saturating neural activity patterns. A good preprocessor should put a noisy input into a more or less noise-free standard form for use in later stages of perception. Analysis and noise-free resynthesis of the signal is not necessary if the whole system gives directly as output a noise-free standard form. Features may never occur by themselves, for example, but only serve the function of "steering" the noisy input into the appropriate corner.

Categorical Perception

One of the characteristics that seems to be found in speech perception is "categorical perception." When stimuli of an artificially constructed set vary continuously across a feature dimension, the listener is experimentally found to have difficulties making discriminations within categories. Thus, voice onset time—the time when the vocal cords start to vibrate— is the physical feature used in making the /p/ versus /b/ discrimination. It is found that the listener cannot discriminate well among stimuli classified as /p/ even though voice onset times may vary considerably. Conversely, discriminations across a category boundary are very good. Two stimuli differing only slightly acoustically, with small differences in voice onset time, are well discriminated if they happen to fall on different sides of the category boundary. It has been suggested (see review by Studdert-Kennedy, 1975, for references to the literature) that incoming speech stimuli are analyzed by both a "categorizer" and a "precategorical acoustic store" (PAS) analogous to iconic memory in vision. It is found experimentally that consonants, particularly stop consonants, display much stronger categorical perception than do vowels. The reason for this, it is suggested, is that consonants are so short in duration that the sensory information contained in the PAS decays too quickly to be used to make discriminations, while the categorical information is much more stable. This hypothesis predicts that vowels, for example, will show more aspects of categorical perception if they are degraded in noise or shortened in duration, both of which would primarily distort the PAS. The predicted result is found. Thus categorical perception may be a very important—indeed, characteristic—property of language perception.

The Brain-State-in-a-Box Model

The brain-state-in-a-box model presented here is a model of categorical perception in a rather pure sense. All points in a region are classified together (differences vanish) and all points in different regions, even if initially very near each other, are classified apart. Thus, no discriminations are possible within the regions, and perfect discrimination occurs between regions.

In previous work (Anderson, 1977; Anderson, Silverstein, & Ritz, in press) we showed that it was possible to use the saturating model to categorically perceive "vowels,"—that is, vectors corresponding to the outputs from a bank of frequency filters when spoken vowels were the input. For the simulation, we used an eight-dimensional space (i.e., an eight-dimensional vector representing each vowel) and showed that if we let our feedback matrix **A** learn according to the rules presented earlier, the system would eventually come to act as a good preprocessor. By this, we meant that nine initial vowel codings, derived from experimental data and often starting close to one another, would, after the operation of feedback, be associated with separate stable corners. The system "learned" to do this with about 20,000 total presentations of the nine vowels in the set of stimuli. Initially the system classified different vowels in the same corner and took many computer iterations (i.e., passages through the feedback system according to Equation 6), but after 20,000 learning trials, all vowels had their own final corner and the system performed correct classification after only seven iterations.

This simultaneous increase in both accuracy and speed of classification struck us as a good demonstration of what we feel are the practical virtues of such a system for perception.

A Computer Simulation

Testing categorical perception experimentally involves two parts: identification and discrimination. A set of artificial stimuli are first constructed which vary smoothly from one speech sound to the other. Sometimes it is easy to construct a continuously graded set of stimuli—for example, voice onset time, where

the physical feature is quite clear-cut—but often it is more difficult, as for some of the consonant-vowel formant transitions. The essence of the effect is that subjects do not *hear* the continuous variation, but instead perceive an abrupt shift from one sound to the other.

For the *identification* experiment, subjects are presented with different stimuli and asked to say which phoneme they hear. Most of the published *discrimination* data have relied on the psychophysical technique called "ABX discrimination." Two stimuli—A and B—which differ by a few milliseconds in voice onset time are presented. The subject is then presented with a third stimulus—either A or B—and is required to say whether the third stimulus matches the first or the second. Thus, if the subject cannot make a discrimination he must guess, which means he will be right 50% of the time, on the average.

We decided that the most straightforward computer simulation we could do would be to simply duplicate these experiments, using the outputs from the saturating neural model, and to show that it behaved in a way which looked like the human data. Since we know very little, to say the least, about the neurophysiology of speech perception, it seems to us premature to do more than point out general similarities of the model with the phenomenon of categorical perception.

We used an eight-dimensional system again, which seemed to us large enough to be indicative of the behavior of a real system, yet small enough to be manageable and of reasonable cost. We assumed we had two eigenvectors of the feedback matrix with nonzero eigenvalues pointing toward two corners. Corner A was $(1,1,1,1,-1,-1,-1,-1)$, and Corner B was $(1,1,-1,-1,1,1,-1,-1)$. This situation is shown in Figure 4. To perform our simulations, we had to have a set of stimuli which varied smoothly from one feature to the other. We picked 16 equally spaced points along the unit sphere through the plane containing the two eigenvectors and the origin.

In a real system, of course, there would be noise as well as other eigenvectors. Our simulations would show perfect categorization if there were no noise. We added zero-mean Gaussian noise to each of the eight components of the initial position. In the figures, *SD* represents the value of the standard deviation of this noise. Figure 4 shows the average length of the noise vector added to the starting point. Note that an *SD* of .4 corresponds to an average added noise about 1 unit long, or a distance as long as that from the origin to the starting point. This is a great deal of added noise. An *SD* of .1 corresponds roughly to the distance between initial positions two steps apart. This seemed to us to be a reasonable range of lengths to cover, and one that might be found in a perception system under normal operating conditions. The noise vector is not constrained to lie on the plane containing the eigenvectors and the origin. The geometry of this model is very complicated.

Results of the Simulations

We did a Monte Carlo simulation of 100 presentations of each starting point with added random noise. When the final state of the system was Corner A, we called it Response A,

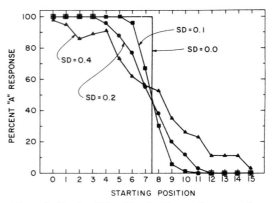

Figure 5. In the "identification" experiment, a "response" corresponds to the corner in which an input ends. The curves show the results of adding various amounts of random noise to a given starting position. One hundred trials were used for each point.

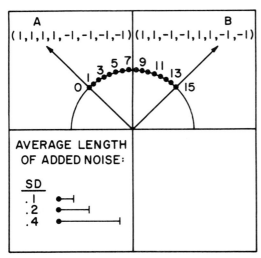

Figure 4. In the simulation of categorical perception presented in Section V, an eight-dimensional system is constructed with two eigenvectors which point toward two corners. Limits of saturation are two units from the origin. Starting positions for test inputs are equally spaced points along the unit circle, numbered 0 to 15. Zero-mean Gaussian noise of standard deviation (*SD*) is added to each of the eight components. The average resulting lengths of the noise vectors corresponding to different *SD*s are shown in the lower-left quadrant. This figure portrays a two-dimensional slice of an eight-dimensional space.

and similarly for B. Figure 5 shows the results of this simulation for several values of *SD*. If there is no added noise (i.e., *SD* = 0), the system categorizes perfectly. If there is noise of *SD* equal to .4, there is a nearly linear decrease in the probability of Response A as the starting point moves from 0 to 15. Intermediate values of noise produced curves which look very much like the published data presented in Studdert–Kennedy (1975), Pisoni (1971), or Eimas and Corbit (1973).

The real test of a categorical perceiver is the difficulty it has performing discriminations within categories. Since experiments often use an ABX paradigm, we simply did so in our simulation, although it was expensive in terms of computer time. We took as an initial input a starting point at, say, point *n*. We then added random noise and noted which corner appeared as the final state. We repeated the process for an input at point *n* + 4 or *n* + 6. We again noted the final state. Then we randomly chose the first or second starting point, added different noise, repeated the process, and attained a third final state. Finally we determined whether the final state agreed with what it was supposed to ("correct") or whether it was in error. For several combinations of corners, it was necessary to require the program to guess, corresponding to a forced choice. In this case, it was correct or incorrect with a probability of .50. We did Monte Carlo simulations using only 40 trials per point to save

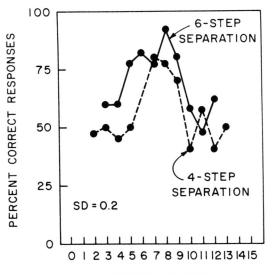

Figure 6. The "discrimination" experiment had the computer perform an ABX experiment. Two starting points four or six steps apart corresponded to A and B. The program noted which final state appeared for each input. A third input—X—was chosen from either A or B, and the computer decided whether it was classified correctly, incorrectly, or whether guessing was necessary. Added Gaussian noise had standard deviation of .2 units. For each point, 40 responses were averaged.

computer money. The results are given in Figure 6.

We have shown only data for an added noise of *SD* equal to .2 units, since this value had identification functions that looked appropriate to us in light of the experimental data. The discrimination functions look similar to what is seen—discrimination shows a pronounced peak when the category boundary separates the starting points and shows a drop if both points start on the same side of the boundary.

An interesting by-product of the identification simulation was an estimate of the "reaction time" required for the system to classify an input as one or the other corner. We simply counted the number of iterations required to saturate the system and attain the final state. This number was averaged over the 100 trials for each point and is plotted in Figure 7 for *SD* equal to .2. Decreasing noise seemed to slightly increase average categorization time.

Maximum change in the number of steps required between conditions was around 20%, and the shape of the curve was quite similar in all cases.

The required number of steps was about twice as great when the stimulus started from a point near the category boundary. An effect like this has been observed. Data from Pisoni and Tash (1974) for an experiment using voice onset time showed a change in reaction time from about 475 msec in the centers of the categories to about 575 msec at the category boundary, as determined by the identification function. The distribution of reaction times was symmetrical around the category boundary, as is ours.

Adaptation

Since we potentially have a learning system, and since we have already shown in our previous discussion how the eigenvalues change when the system is learning eigenvectors pointed toward corners (see Equation 9), it seemed an obvious extension of our simulations to look briefly at the effects of "adaptation" on categorical perception.

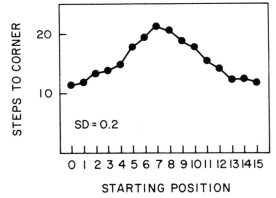

Figure 7. The number of computer iterations required for all components of the system to saturate is plotted here. Added noise had a standard deviation of .2. As starting position varied, the number of steps required to saturate increased near the boundary. Different noise conditions showed very similar patterns. If each iteration or its equivalent takes about the same time to perform in a real nervous system, then this graph could be interpreted as a rough indicator of the pattern of reaction times that would be observed in categorical perception as an input stimulus is moved from one category to another across a boundary. Each point is the average of 100 responses.

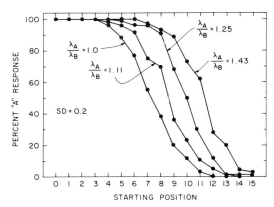

Figure 8. When the eigenvalues associated with the eigenvectors are not equal, the identification function shifts. If we assume repeated presentation of an adapting stimulus causes synaptic "antilearning," then this simulation corresponds to how we would expect our system to behave if one feature vector is adapted. Each point is the average of 100 responses.

Cooper (1975) has reviewed some of the adaptation literature and points out some of the general features experimentally observed. Adaptation is typically produced experimentally by having the subject listen to a minute of the adapting stimulus at a presentation rate of 2 per sec. After adaptation, the identification function for the unadapted stimulus (A) indicates a shift toward the adapted stimulus (B). The curve for the identification function is displaced parallel to its initial position; as Cooper comments, "the slopes of the identification functions obtained after adaptation were as steep as the slopes of functions obtained in the unadapted state" (p. 26). There was evidence of crossed series adaptation; that is, subjects who adapted to voicing in one pair of phonemes also adapted (to a slightly lesser extent) to another pair of phonemes differing in the same feature.

Our simple simulation could not test crossed series adaptation because our system only had two eigenvectors. However, we can easily check the shift in identification function.

We assumed that "adaptation" was equivalent to synaptic "antilearning," or learning according to Equation 2 with negative sign. This is *not* simple fatigue but a more complex and subtle process involving precise synaptic change at many synapses. By Equation 9, we see that adaptation causes the eigenvalue

of the adapted stimulus to decrease slightly. The eigenvectors did not shift direction. In our initial simulation (see Figure 5) we assumed the two eigenvalues were equal. For the simulation of adaptation we used exactly the same program but decreased the eigenvalue λ_B. The results for several different ratios of eigenvalue are shown in Figure 8. The noise had *SD* equal to .2 units. The ratio λ_A/λ_B = 1.0 was included to correspond to the initial simulation. It can be seen that the data look very much like parallel displacements of the unadapted curve toward Response B, exactly as seen in the experimental data.

It might be mentioned that we did virtually no searching for parameters in most of these simulations, and the simulations used many of the same programs as those used for probability learning in the next section.

Conclusions

We suggest that the simulations presented here are quite good replicas of the major experimental findings of categorical perception. Our previous work with the simulation of vowel preprocessing, coupled with the work discussed here, suggests that models for speech perception might consider using some such ideas as positive feedback, saturation, and synaptic learning, which seem to be responsible for the interesting effects in our simulations.

This simulation shows clearly some of the features we feel may be typical of natural systems constructed with distributed, parallel arrays of interconnected analog elements. The system acts like an adaptive filter, where the filter characteristics are determined by the past history of the filter. The system does not "analyze" its inputs in the sense that a computer or logician might analyze them, by dissecting them into component parts, but it simply responds to them. However, the response of the system is determined by its past, so its analysis becomes meaningful in terms of this past.

VI. Probability Learning

We shall consider here the set of experiments usually called "probability learning." We shall make a direct application of the model previously described and show that it can provide a model for this seemingly remote application.

The ability to estimate the probability of occurrence of an event with a random component—whether or not it will rain, who will win an election, what the stock market will do—is obviously important in daily life. It has also served as the basis of a large body of work in experimental psychology. Probability learning has been studied in a number of ways. A classic experimental technique involves a prediction by the subject as to which of two lights will be turned on. Typically, a subject will sit facing the lights. He will be asked, usually immediately after a signal, which light will turn on after a brief interval. The two lights are usually turned on randomly.

The general results are quite consistent from experiment to experiment. The subjects will tend to "match" the probabilities of the events; that is, if the left light occurs with a probability of .8, the subject, after a number of trials, will predict that light about 80% of the time. This result is not what one would expect if one were to assume that the subject was trying to maximize his chances of successful prediction. If he were, then he would choose the most probable light all of the time. Of course, by providing appropriately large payoffs, it is possible to encourage the subjects to change strategies, but, if left to themselves, there is a strong tendency to match probabilities in most simple experimental situations.

Statistical Learning Theory

The prediction of probability matching is a simple consequence of statistical learning theory, which is one of the reasons for the great interest in the effect. Derivation of this result is given in a number of places (see Estes & Straughan, 1954, Estes, 1957, and the other papers collected in Neimark & Estes, 1967).

Let us briefly sketch some of the important aspects of the derivation. We assume there are a number of alternative responses, A_1, A_2, $\cdots A_r$, which can be made by the subject predicting, for example, which of several lights will turn on. The response is followed by an event—E_1, E_2, $\cdots E_r$—which is the actual turning on of one of the lights. Learning theory assumes that the actual event will change the future probability of predicting that event. Suppose the probability of a response, A_j, on the nth-trial is given by $p_{n,j}$. If an event, E_j, occurs, then the probability of Response A_j is increased. In early formulations of the learning rule, the correctness or incorrectness of a prediction determined the subsequent learning. However, it seems that only the occurrence of a particular event is required to increase the associated response probability. As Estes (1972) says in a review article, "the mathematical operator applied on each trial is determined solely by the information received by the subject" (p. 82) as long as special rewards for success and failure are not present. This was shown directly by Reber and Millward (1968). They showed that mere observation of the event lights, in the absence of prediction, produced probability matching in the subjects.

The quantitative rule that the increase in probability of A_j will follow if Event E_j occurs can be derived from statistical sampling theory, or can arise from other assumptions, and follows the form,

$$p_{j,n+1} = (1 - \theta)p_{j,n} + \theta. \qquad (10)$$

The parameter, θ, is an important learning parameter in statistical learning theory, and $0 \leq \theta \leq 1$. If Event E_j does not occur, then the probability of making Response A_j in the future will be given by

$$p_{j,n+1} = (1 - \theta)p_{j,n}. \qquad (11)$$

Qualitatively, these expressions make the probability of a response tend to *increase* after the associated event, and tend to *decrease* after another event. If we consider the behavior of the probabilities, the expected value for $p_{i,n+1}$ is given by

$$E(p_{i,n+1}) = (1 - \theta)E(p_{i,n}) + \theta\pi_{i,n}, \qquad (12)$$

where $\pi_{i,n}$ is the current value of the actual probability of occurrence of the Event E_i. Thus, after the first trial,

$$E(p_{i,2}) = (1 - \theta)E(p_{i,1}) + \theta\pi_{i,1},$$

after the second trial,

$$\begin{aligned}
E(p_{i,3}) &= (1 - \theta)E(p_{i,2}) + \theta\pi_{i,2} \\
&= (1 - \theta)^2 E(p_{i,1}) \\
&\qquad + \theta(1 - \theta)\pi_{i,1} + \theta\pi_{i,2},
\end{aligned}$$

and so on.

For the important special case, often used in experiments, where the probability of an Event E_i is fixed at some value, π_i, the expected value of the probability of making Response A_i after n trials is given by

$$E(p_{i,n}) = \pi_i - (\pi_i - p_{i,1})(1 - \theta)^{n-1}. \quad (13)$$

The second term of this expression will go to zero as n increases if θ is greater than zero, that is, if the system learns at all. Thus, the asymptotic value of $E(p_{i,n})$ is given by the probability of the event, π_i, which is essentially what is observed. This result is completely independent of the initial probability of the response and the learning parameter, θ.

Problems

This elegant result agrees well with many aspects of the data. However, the actual data present some difficulties for the simple theory, which has led to a number of attempts to modify or extend the simple model.

In some of these extensions, subjects are assumed to memorize past sequences and to predict the event that occurred in the past. There are other approaches, involving "hypothesis testing" or other, generally "intelligent" behavior on the part of the subject.

We will assume here that the tendency of the subjects to look for regularities and sequencies in the data interferes with, and is separate from, an underlying straightforward learning phenomenon with no cognitive component.

There are two problems with the simple model, which can be seen in most of the experimental data.

First, when there are only two alternative events, the subjects match probabilities closely but they do not do so exactly. Myers (1976) says in a recent review that the response probability "consistently overshoots" exact matching, and the overshooting has been seen "by almost anyone who has run subjects for more than 300 trials" (p. 173). Overshooting is not large, usually at most a few percent in the two-choice experiment, but it is significant and thus a problem for theorists. The overshooting becomes much larger when there are more than two alternatives (Neimark & Estes, 1967, p. 261). Estes (1964) suggests that over-

shooting may be a function, to a certain extent, of the instructions given the subject. Myers comments that "investigators who are determined to do so can produce overshooting in a variety of ways; undershooting is somewhat more difficult to achieve, but possible" (p. 175).

Exact probability matching is a good first approximation. However, it is clear that a model should have sufficient flexibility to explain the small but consistent deviations above and below exact matching that are a feature of the experimental data.

Second, a matter that could either be considered aesthetic or substantive, depending on perspective, is the value of the learning parameter, θ. Although θ does not appear in the expression for the asymptote, it determines the rate and trajectory with which the response probability will approach its asymptotic value, because the factor $(1 - \theta)^{n-1}$ appears in the second term of Equation 13. Given the time course of response probabilities, it is possible to estimate θ. When different probability conditions are run with the same subjects, θ does not appear to be constant (Neimark & Estes, 1967, pp. 257–258). The change in θ may be considerable, even in very similar experiments.

A Neural Model for Probability Learning

We have at hand, in our brain-state-in-a-box model, the means for automatically generating discrete responses. We also have, if we introduce noise into the system, a means of introducing probabilities naturally into the model. Although we can handle any number of events, we will restrict ourselves to two at first.

Let us observe here that there are two parts to this model. The first is the *learning feedback matrix*. We have discussed the properties of this matrix in some detail previously. The second part involves the *dynamics* of the system. The initial state moves, under the influence of feedback, into a corner. How fast it moves and into which corner it moves are given by the properties of the feedback matrix at that time. We assume here that the feedback matrix changes only *after* a stable corner has been reached. This means that the matrix does not learn while the state vector is changing; that

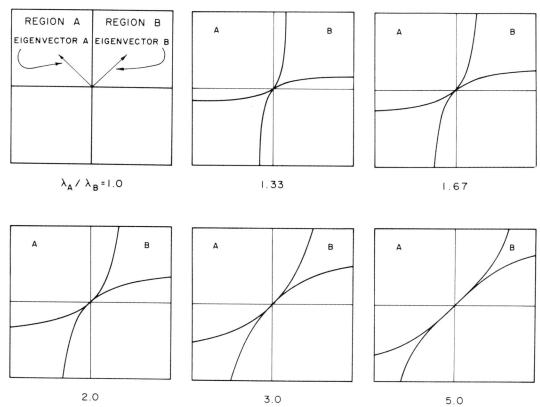

Figure 9. A two-dimensional, saturating neural system. Each coordinate axis corresponds to a single neuron, or the amplitude of a pattern of neuron activity. This activity saturates; thus activity is confined to the interior and edges of the square. There are two eigenvectors in this system, one pointing to each pair of diagonally opposite corners. When feedback is positive, all inputs to points in Region A cause the activity vector to end up in the corner to which eigenvector **A** points, or in the opposite corner. As the ratio of eigenvalue *A* to eigenvalue *B* increases, the size of Region A increases.

is, the time constant of learning is long compared to the dynamics of the system.

Suppose we have a two-dimensional system with the dimensions coupled by a feedback matrix. Thus the two-dimensional limits of saturation form a square. Suppose that the two eigenvectors of the feedback matrix point toward the corners of the square. If we start off at some initial point of the square, positive feedback will drive the system toward one of the corners (see Figure 9).

Let us identify one pair of diagonal corners, associated with one eigenvector, with one response. Let us identify the response associated with corners $(-1, +1)$ and $(+1, -1)$ as Response A and the other pair of corners $(1, 1)$ and $(-1, -1)$ as Response B. Identifying two diagonally opposite corners with one

response is made primarily for convenience, and could be avoided if necessary.

This two-dimensional system is not so restrictive as it seems. Assume that one response is associated with one large pattern of neural activity, as it might be in a real system, and the other response is associated with another large pattern, orthogonal to the first. These patterns give rise to two orthogonal vectors in a high-dimensional space. Then by considering the plane through the vectors and the origin, we have a two-dimensional system, with corners associated with eigenvectors. The axes of our two-dimensional system might correspond to amplitudes of complex activity patterns interacting with each other. The tractable case we will discuss might be a reasonable approximation to a much more complicated, high-dimensionality system.

As we have mentioned, the plane will be divided into two regions. The origin is unstable and need not be considered further. One region will be associated with one response, and the other region with the other. This is shown in Figure 9.

Suppose the subject wishes to make a prediction (i.e., to make one response or the other). There is no lack of random noise in the nervous system. We will assume that the subject initiates the prediction process by starting at some random point on the plane. If the chances of starting at all points on the plane are equal, then by simply calculating the areas associated with each region and dividing by the total area, we know the probabilities of each response.

The calculation of these probabilities is straightforward, though involved, and is presented in Appendix A. The boundaries between regions are simple curves, and the areas of the regions can be calculated exactly, with a simple resulting expression. If λ_A and λ_B are the eigenvalues of the feedback matrix associated with Response A and Response B, respectively, and if

$$\lambda_0 = \frac{\lambda_A}{\lambda_B},$$

then the probability of Response A is given by

$$p_A = \frac{3\lambda_0^2 + \lambda_0^3}{(\lambda_0 + 1)^3}. \tag{14}$$

As the ratio increases, Response A becomes more and more probable. The shapes of the regions for different values of the ratio λ_0 are shown in Figure 9. The probability of Response A versus the ratio of the eigenvalues is shown in Figure 10. Note that p_A is monotonically increasing for $\lambda_A/\lambda_B \geq 1$.

Learning Rules

To make predictions of response probabilities, we must specify only how the eigenvalues change with time. We then have a rule for turning the ratio of eigenvalues into response probabilities.

We observe that in the saturating model, the final state of the system is always a corner.

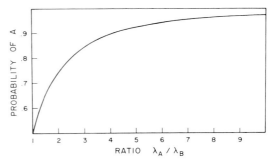

Figure 10. In the model for probability learning presented in the text, it is possible to calculate the probability of each response if the eigenvalue associated with each response is known. Their ratio then gives the probability according to the graph.

Thus, as we showed at the end of Section IV, the eigenvalues follow a very simple learning rule: They increment by a constant amount every time a corner is reached, as shown in Equation 9.

Let us allow that memory decays with time. We also assume that there is a stable part of the eigenvalue which is not affected by either decay or learning. This might correspond to the knowledge, for example, that there are *two* possible responses. Thus, we get a formula for the eigenvalues that is similar to, but not identical with, that used for the probabilities in statistical learning theory.

Let g be a decay factor, $0 \leq g \leq 1$, and let η be the amount of increment produced when the system learns. Then, if Event A has just occurred, and if the set of synaptic changes associated with that corner has been added to the feedback matrix, that is, A has been learned, the two eigenvalues for the $n + 1$st trial are given by

$$\lambda_A(n + 1) = 1 + g[\lambda_A(n) - 1] + \eta \tag{15a}$$

and

$$\lambda_B(n + 1) = 1 + g[\lambda_B(n) - 1]. \tag{15b}$$

The first term, "1," is the constant part; the second term is the decay term; and the third part is the increment. If Event B occurs, λ_B receives the increment. Note that the success or failure of the prediction made by the subject does not appear in this scheme.

Since we are concerned only with response probabilities, we are concerned only with the ratio of λ_A and λ_B. In other applications—

prediction of reaction time, for example, in related tasks—we cannot make this assumption and must know the true values of the eigenvalues.

The general similarity of this model to many aspects of stimulus sampling theory should be emphasized. The learning scheme for eigenvalues is a variant of the "linear" learning model. Thus, many of the predictions will be somewhat similar to those of statistical learning theory. The *essential* difference is that here the eigenvalues are what are changing— the size of an eigenvalue becomes a sophisticated measure of trace strength—and that changes in probability of response occur as a result of the operation of a complex process instead of directly.

Asymptotic Behavior

It can be seen that the model is completely defined once we specify the constants and have the sequence of events. We can derive asymptotic behavior of the eigenvalues quite easily in some cases. If there is a constant value of probability of Events A and B, with probabilities π_A and π_B, and if the eigenvalues start at λ_{A_0} and λ_{B_0}, then average values of λ_A and λ_B after n trials are given by

$$E[\lambda_A(n)] = 1 + \frac{\pi_A \eta (1 - g^n)}{1 - g} + g^n(\lambda_{A_0} - 1) \quad (16a)$$

and

$$E[\lambda_B(n)] = 1 + \frac{\pi_B \eta (1 - g^n)}{1 - g} + g^n(\lambda_{B_0} - 1); \quad (16b)$$

and after a very large number of trials, if $g < 1$, the asymptotic values of the averages are given by

$$E(\lambda_A) = 1 + \frac{\pi_A \eta}{1 - g}$$

$$E(\lambda_B) = 1 + \frac{\pi_B \eta}{1 - g}.$$

Since our formula for response probability (Equation 14) requires knowledge of the ratio of the two eigenvalues, it is easy to get an estimate of this value. We can approximate the expected value of the ratio well enough for

our purposes by letting

$$E\left(\frac{\lambda_A}{\lambda_B}\right) \cong \frac{E(\lambda_A)}{E(\lambda_B)}. \quad (17)$$

At asymptote, we see that

$$E\left(\frac{\lambda_A}{\lambda_B}\right) \cong \frac{1 - g + \pi_A \eta}{1 - g + \pi_B \eta}.$$

For the values of parameters used here, the approximation is good to better than 1%.

This relation does not predict simple probability matching but a complex relation involving η and g. Using simple calculus, we can see the largest expected value of overshoot will occur when there is no forgetting, that is, when $g = 1$. Then

$$E(\lambda_A/\lambda_B) \approx \frac{\pi_A}{\pi_B}.$$

We can then calculate the estimate of the maximum expected value of probability from Equation 14. The maximum expected overshoot in this case is about 10% when π_A is around .75.

The smallest average asymptotic value of λ_A/λ_B will occur when the second term is zero, that is, when there is no learning. Then

$$\lambda_A/\lambda_B = 1,$$

and the probability of both Response A and B will be .5. Thus, we can predict values of asymptotic probability both above and below probability matching.

It does not seem to be possible to derive exact expressions for the important characteristics of the random variable λ_A/λ_B, such as the expected value and the variance. However, characteristics of the response probability (Equation 14) can, somewhat surprisingly, be calculated, a result shown in Appendix B.

Application to Data

Let us see if we can fit some experimental data with the model as it stands. Our purpose in this section, it must be emphasized, is not primarily to provide a different or quantitatively more satisfactory model for probability learning, but to establish that our neurally based model is capable of handling a wide range of interesting phenomena. We are not concerned with explaining every aspect of the available data.

A good set of data to use is the series of experiments reported by Friedman et al. (1964). They used an exceptionally large group of subjects—80 Indiana University undergraduates—in a 3-day series of probability learning experiments. In the first 2 days, subjects received sequences of 48 trial blocks. The probabilities of an event changed from block to block ("variable $- \pi$ series"). Odd numbered blocks had both π_A and π_B equal to .5. Probabilities in the even numbered blocks varied from .1 to .9 in steps of .1, excluding .5. Each subject received all probability conditions during the 2 days. Two different sequences of block probabilities were used, but the first, last, and alternate intermediate blocks were assigned .5 probabilities. On the 3rd day, eight 48-trial blocks were given. The first and last blocks had π_A equal to .5; the middle six blocks had π_A equal to .8.

Friedman et al. (1964) were able to fit the data reasonably well by assuming the basic equations of statistical learning theory, Equations 10 and 11, with a changing θ. The parameter, θ, varied over a considerable range. To fit the transition between a block with π_A equal to .1, to the succeeding block with π_A equal to .5, the best θ was found to be .62. To fit the transition between π_A equal to .4 and π_A equal to .5, θ was found to be .07. Other transitions fell in between, with θ increasing as the difference in probability between blocks increased.

We felt it would be of interest to see if our model would fit the data with a single set of parameters. Our model requires, as does statistical learning theory, detailed knowledge of the stimulus sequences and ordering of blocks to make the best predictions. This data was not available in most cases in the Friedman et al. (1964) paper, although a wealth of carefully gathered and computed averaged data was presented.

Strong learning effects were demonstrated to be present between the beginning and the end of the series, and it was stated that the ordering of blocks had a "highly significant effect" which "severely limits the analyses that can usefully be accomplished with the

variable $- \pi$ sequence" (Friedman et al., 1964, p. 260).

Friedman et al. (1964) were primarily concerned, in the analysis of the variable $- \pi$ series, with the transitions between a variable probability block and the following .5 probability block, since this allowed them to estimate θ and to get some idea of the general behavior of the statistical learning theory model with respect to the data.

Our model was sufficiently complex and probabilistic to make it difficult to generate simple expressions for some of the averages. Since we had access to a large computer, the most direct way of initially checking the fit of our model to Friedman et al.'s (1964) data was to do a simulation, using 80 computer-generated pseudosubjects and then to compare the results with the real data.

We at first assumed that our pseudosubjects would receive the sequence of stimuli in Friedman et al.'s (1964) Summary Table 7, which has the following sequence of block probabilities: .5, .1, .5, .2, .5, .3, .5, .4, .5, .6, .5, .7, .5, .8, .5, .9, .5. Every pseudosubject received an individual set of events generated with probabilities given by the probability sequences. Eigenvalues associated with responses of the pseudosubjects were calculated according to the learning scheme, and the probabilities of response of each subject were calculated and averaged across subjects. This meant the pseudosubjects did not actually make pseudoresponses, which would be then processed as the responses of the real subjects were, so the data have lost a significant source of variance. The statistics can also be calculated directly by use of the formulas in Appendix B. The resulting average probabilities were then compared with the real data in Friedman et al.'s (1964) Table 7. Figure 11 shows the results.

A crude search was made in order to see which set of parameters produced a simulation most resembling the actual data. The best fitting parameters were found to be $\eta = .3$ and $g = .90$. Parameters were not especially critical. The value of the decay factor, .9, was such as to make the contribution to the current eigenvalue from trials over 48 trials (i.e., a block) in the past negligible, so this simulation could be viewed as a sequence of fits of transi-

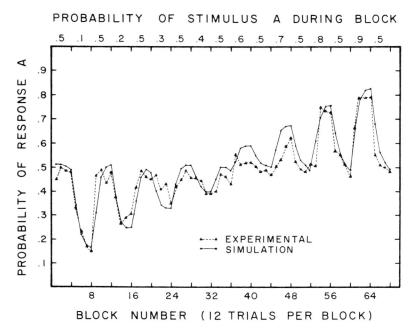

PROBABILITY OF STIMULUS A DURING BLOCK

Figure 11. Computer simulation and real data of the first two days of the experiment performed by Friedman et al. (1964). The solid line gives the results of a computer simulation with 80 pseudosubjects receiving random sequences according to the probability schedule. The learning parameter, η, was .3; and the decay parameter, g, was .90. The pseudosubjects' calculated average response probabilities are plotted. The dashed line is the actual experimental data, with 80 subjects making real responses. Subjects did not receive blocks in the sequence given.

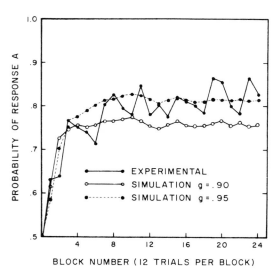

Figure 12. Two computer simulations and real data of the 3rd day of the experiment in Friedman et al. (1964). The learning parameter, η, of the simulation was .3. The decay parameter, g, takes two values, .90 and .95. The experimental data are given by the solid line.

tions to the .5 blocks. Friedman et al. (1964) plotted these transitions (and fitted curves to them) individually in their Figure 4.

Data from the 3rd Day

The fit of the simulated to the actual data was encouraging, although the actual data from the first 2 days of the experiment concealed a demonstrated multitude of extraneous effects, such as event sequence, and subject learning.

The probability sequence on the 3rd day was the same for all the subjects, now well practiced, and provided a good test of the model. Figure 12 shows the fits for data from the 3rd day. The same set of parameters provided a reasonable fit, although the fit was made better by increasing the decay parameter, g, from .90 to .95. Note that both the data and the simulation show a small, but definite, overshoot above .8. The response probability over the last two 48-trial blocks

(12 trial blocks, 25–32) was .82, as opposed to probability matching. The same average for the simulation with g equal to .95 was .81.

In a previous paper (Anderson, 1973), a model similar in many respects to this was proposed for the learning of short lists and for a choice reaction time experiment. A quantity, γ, was defined, and is equal to $\eta/(1 - g)$ in the present section. This quantity was found, in fitting the data, to vary from about 2 to about 7. In the probability learning experiments, with η equal to .3 and g equal to .95, $\gamma = 6$; and with η equal to .3 and g equal to .90, $\gamma = 3$, which falls in the same range. Since list learning and probability learning seem at first glance to be very different tasks, this coincidence is interesting.

Recency Curves

Friedman et al.'s (1964) article, and other articles as well, pay special attention to the behavior of response probability when a continuous sequence of a particular event ("a run") occurs. Both statistical learning theory and our theory predict a so-called "positive recency" effect, where the response probability for an event *increases* after a run of that event. When subjects are well practiced, they show this effect very strongly, although they do not always do so during the first few blocks of trials.

Friedman et al. (1964) provide extensive recency data for their subjects during the 3rd day, when $\pi_A = .8$ and $\pi_B = .2$, and at the end of the 2nd day, when π_A and $\pi_B = .5$.

When they fit the data with the simple statistical learning theory model, they find fair fits, but θ again is not constant. In the data from the 3rd day, θ was estimated to be .058 during the transitions to and from the $\pi_A = .8$ condition, while the fit during a run suggested $\theta = .17$.

If we assume the eigenvalues start off, in our model, at asymptote, calculated from the formula and parameters given previously, we can calculate (exactly) the eigenvalues after a particular sequence by use of Equations 15a and 15b and can then calculate the associated probabilities using Equation 14. When this was done with the data given in Table 11 of Friedman et al. (1964) for runs

in the $\pi_A = .8$, 3rd-day condition, and with the data read from Figure 11 for the last block (.5 probability) during Day 2, the fit was not very good for runs of 2, 3, or 4 events, although the match was quite good after a run of 15 or 20 A events.

Our model predicts that response probability can never equal 1.00, even after an infinite run of a particular event. As Reber and Millward (1971) point out, published recency data usually indicate asymptotes well below 1.00. In Reber and Millward's (1971) experiment, which involved a continuously variable event probability which was "tracked" by the subjects, the probability of Response A after a run of up to 10 A events was extraordinarily low and seemed stable at about .78.

In the data from runs of Event A in the $\pi_A = .8$ condition of the Friedman et al. (1964) study, values of response probability also seemed well away from 1.0. Since there were, as might be expected, relatively few very long runs, even in a .8 condition, the individual data points are of dubious significance beyond runs of about 10. However, the average probability of Response A for runs of length 17 through 20 in the 48 trial blocks 3, 4, 5, and 6 is .91; that is, there were 211 A responses out of 231 total responses. We can predict what the model with parameters determined previously would give us after 20 trials of the same event simply by following the formula. We find that, with η equal to .3 and g equal to .95, the probability of Response A after a run of 20 is .91, indicating good fit for long runs. However, the failure of the fit for short runs is intriguing and seems also to be implied by the Friedman et al. finding of wide differences in θ.

Short-term Memory and Probability Learning

Perhaps the simplest explanation of this short-run divergence from a theory satisfactory for the long run is simply that there is a highly weighted contribution from the past few events. Possibly, we are seeing an effect that is due to the limited capacity, short-term store, which is nearly universally accepted to exist in human memory.

Millward and Reber (1968) asked subjects about past trials in a probability learning

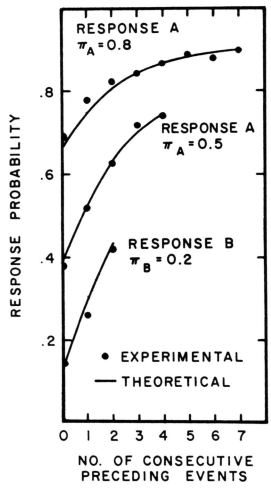

RESPONSE A
$\pi_A = 0.8$

RESPONSE A
$\pi_A = 0.5$

RESPONSE B
$\pi_B = 0.2$

● EXPERIMENTAL

— THEORETICAL

NO. OF CONSECUTIVE
PRECEDING EVENTS

Figure 13. Experimental recency data and theoretical calculations for the experiment in Friedman et al. (1964). Response probabilities are calculated for "runs" of events. A run of zero length means that the preceding event was the other event. These data were simulated by postulating a rapidly decaying short-term trace, in addition to the more slowly decaying trace assumed previously (see text for details).

experiment and found, in most conditions, that "memory for past events goes back about five events and/or four event runs" (p. 988).

Since limited capacity, in our model, can be modeled by assuming it is synonymous with "rapid decay," suppose we postulate that there is a rapidly decaying short-term representation of the preceding three or four events.

There seemed no point in making extensive parameter fits, since we would now have two more parameters to add to our basic model

As is well known, with a large computer, enough parameters, and enough persistence, any given set of data can be fitted, so we simply picked some reasonable parameters. We let the short-term decay factor, g_s, be .65, which corresponds to a memory span of three or four. Since we know that a g of .95 fitted data quite well when averaged over 12 trial blocks (which would wash out contributions from a process with a decay factor of .65), we assumed the long-term decay factor, $g_1 = .95$.

We assume the two processes are separate, decay separately, and add together to give the magnitude of the eigenvalue, each following an equation like Equation 15a or 15b. Thus, at asymptote, for the eigenvalue associated with Response A, we have

$$\lambda_A = 1 + \frac{\pi_A \eta_s}{1 - g_s} + \frac{\pi_A \eta_1}{1 - g_1}.$$

As a starting point, let us assume that at asymptote there are equal contributions from the short- and long-term processes; that is, the second term equals the third term. We then find that the learning parameter of the short-term process has η_s equal to 1.067 and that of the long-term process, η_1, equal to .1525. This is all the information we need to calculate the average probabilities of the subjects during runs of a particular event. We show the theoretical and experimental curves in Figure 13. The fit is quite good. All these points are fitted with the above parameters, and only the starting points vary. There is a value plotted for a run of zero length, which gives the probability of one response when a run of the other event is about to start. (A run of one event has to start with the other event.) We assume that the eigenvalues at the "−1" trial started at asymptote.

Comments About Variance

Analysis of the variance in these models is rather complicated, since several mechanisms can contribute to the observed experimental variance. First, the raw data is generated by a probabilistic process. Even if we knew the subject's response probabilities exactly, the resulting data from a relatively small number of responses would contribute a significant amount of variance. Second, in our model

Table 3

Comparison Between Published Three-Choice Experimental Data and a Computer Simulation

Source of data	Event p			Simulation			Experiment		
	π_A	π_B	π_C	A	B	C	A	B	C
Gardner[a]	.60	.30	.10	.75	.21	.04	.68	.24	.08
Cotton & Rechtshaffen[b]							.66		
Gardner	.60	.20	.20	.79	.10	.11	.68	.16	.16
Cotton & Rechtshaffen							.66		
Gardner	.70	.20	.10	.84	.11	.05	.80	.13	.07
Cotton & Rechtshaffen							.80		
Gardner	.70	.15	.15	.86	.06	.08	.80	.10	.10
Cotton & Rechtshaffen							.81		
Cole[c]	.44	.33	.22	.52	.32	.16	.51	.31	.17
Cole	.67	.22	.11	.82	.13	.05	.84	.11	.05

[a] Data taken from Gardner (1957). Average of response probabilities for Trials 286–450.

[b] Data taken from Cotton and Rechtshaffen (1958). Average of response probabilities for Trials 286–450.

[c] Date taken from Cole (1965). Average of response probabilities on Trials 501–1,000. A correction procedure was used in this experiment.

there is a contribution due to the intrinsic variance in the probability of a subject's response. This rather involved calculation is given in Appendix B. Third, in our model, intersubject differences in parameters have a profound effect on asymptotic probability, causing a large increase in the variance of the pooled data. This effect is largest when deviations from equal probability are large.

Our guess is that intersubject variation in parameters will turn out to be a major cause of experimental variance, and since we have no idea what kind of distribution of parameters might exist for subjects, we felt it best to do no detailed calculations of variance at this time.

Extension to More Events

The derivation, sketched earlier, of the simplest form of statistical learning theory says that probability matching is independent of the number of stimuli. However, experimentally, overshooting of the most probable stimulus increases greatly as the number of alternatives increases (Estes, 1972). Some data are available for three-choice experiments (Cole, 1965; Cotton & Rechtshaffen, 1958; Gardner, 1957). All these experiments show pronounced overshooting above matching, generally many percent.

Since our model becomes difficult to work with analytically in higher dimensional boxes,

we resorted to a Monte Carlo simulation to estimate the response probabilities in a three-choice system. We considered three orthogonal vectors pointing toward corners in a four-dimensional space. (There are orthogonal vectors pointing toward corners only when the dimensionality of the space is divisible by a power of 2.) The orthogonal vectors were $e_1 = (1, -1, -1, 1)$, $e_2 = (1, 1, -1, -1)$, and $e_3 = (1, -1, 1, -1)$. The eigenvalues could be calculated from our asymptotic formula. With eigenvectors and eigenvalues known, the feedback matrix was constructed. Then a random point in the box was chosen (with a uniform distribution) as a starting point. The feedback then forced the system into a corner (see Table 3).

Occasionally stable corners not associated with an eigenvector appeared when the eigenvalues were nearly equal. These corners were ignored in the calculation of probabilities. It is not clear what these extraneous corners might correspond to psychologically—possibly paralyzed uncertainty.

In each Monte Carlo simulation, 1,000 initial random vectors were used, and the number of times a particular corner appeared was counted. This gave an estimate of the probabilities of a particular response. We used the same parameter values we used previously, $\eta = .3$ and $g = .95$. These two values produced only very slight overshooting in the two-choice system but very pronounced overshooting in the three-choice system.

Table 3 shows the results of the simulations and compares them with the limited experimental data. We find agreement with the actual data is surprisingly good in several cases. In a couple of the cases where the fit is not so good (the 60–20–20 and 60–30–10 conditions of Gardner, 1957, and of Cotton & Rechtshaffen, 1958), the data in their published figures clearly indicate that asymptote had not been reached and the probabilities of the most probable response were still increasing. Cole used the last 500 trials of a 1,000-trial experiment, while the other two workers used average probabilities of Trials 286–450 in a shorter experiment. Cole's experiment incorporated a correction procedure, where subjects continued to predict until they were correct. This should not greatly affect our prediction for the first response probabilities in this experiment.

One might wonder why it would not be possible in the three-choice case to consider Responses B and C, say, to be a single "response" and then to apply the two-choice analysis to them. We showed in our simulations that the two-choice model does not give rise to as much overshooting (for the same parameters) as the multiple-choice model, so this must be an incorrect approach. One reason for this is that we are not using a linear system but one with a high degree of nonlinearity, both in geometry and dynamics. Inputs and outputs cannot be freely combined in a nonlinear system because the superposition principle does not hold. Many "obvious" approaches are incorrect when applied to nonlinear systems, a point to be careful of when analyzing the operation of a system with such spectacular nonlinearities as the brain.

VII. Conclusions

Our aim in this article has been to present a relatively detailed and precise model, which was suggested by the anatomy and physiology of the brain, and to consider some interesting psychological applications. Although it was necessary to oversimplify reality to get a model that we could work with easily, the resulting model was sufficiently rich to have

a pronounced structure, and, with a very modest amount of manipulation, gave rise to some testable predictions. We feel that the best way to work with such a very wide-ranging model is to try to fit with reasonable success many different phenomena, rather than to try to explain every detail of a limited body of experimental data.

We have argued that our model provided a theoretical framework for entities that acted and behaved very much like the distinctive features of psychology. A set of neurons with positive feedback tends to analyze its inputs by most heavily weighing the eigenvectors of the feedback matrix with large positive eigenvalues and by suppressing the rest. We also pointed out that these particular eigenvectors are often the most meaningful in terms of the discrimination to be performed, since they contain most of the information allowing discriminations to be made among the stimulus set.

When we introduced a saturating nonlinearity—the "brain-state-in-a-box" model—we were able to suggest directly a model for probability learning, which fitted some actual experimental data in reasonable detail. As an application of the model in a different area, we showed that the same model, with slight modifications, acted as a categorical perceiver with properties similar to those seen in recent data.

The model has some obvious shortcomings. Among these are some grievous oversimplications of the physiology, the assumption of a high degree of linearity in some parts of the system, and, later, the assumption of "hard" saturation, with no transition from linearity to saturation. Other necessary details that are needed for an actual operating system are simply ignored. For example, a mechanism must be provided to get the brain state *out* of a corner, once it has gotten *in*. There are a multitude of ways this could be accomplished, all of which could be used together. There could be selective adaptation or habituation of rapidly firing cells and of their synapses; there could be large amounts of noise which could force the system from corner to corner; and, quite probably, there could be a special neural circuit, where rapidly firing cells generate recurrent inhibition to turn themselves off. We preferred to ignore this problem here, since we did not need to consider it for the problems we discussed. However, eventually we must make concrete assumptions about how brain states decay as well as how they grow.

Some of the quantitative predictions for probability learning may be fairly sensitive to the assumptions about the geometry of the system. It is quite unlikely that our hypothetical enclosing "box" is a hypercube, with all sides equal. We are not sure how a more general box would affect the behavior of the system. Many of the qualitative properties would remain the same, but the system is sufficiently complex that we cannot say for certain.

There are many immediate extensions of the model. As one example, the probability learning model serves also as a model for choice reaction time. If we assume that instead of starting randomly, we start at a point in the box that is determined to some extent by the stimulus— by initially passing the stimulus through a memory filter of the kind we have been discussing here, for example—we can make a model for choice reaction time. In fact, the model as stated has similarities, which can be made precise and explicit, with random walk models for reaction time, which are very successful in explaining many of the quantitative aspects of reaction time experiments (Link, 1975). If we consider the corner as an absorbing state and the noisy memory filter output evolving in time (with additive noise) as the directed random walk, then the resemblance becomes striking.

As another point, the model strongly suggests that the brain would rather be wrong than undecided; that is, a misperception (the wrong corner) is better than no perception at all. Some misperceptions are clearly more likely than others, reflecting the past learning of the system. Also, the model suggests that brain states in perception move from stable state to stable state, with abrupt transitions between stable states.

All these suggestions can be made precise relatively easily and can be compared with the rich body of data and concepts in experimental psychology. We have tried to show, with theoretical discussion and with several detailed examples, that such an effort may be worthwhile.

Reference Note

1. Anderson, J. A. *What is a distinctive feature?* (Tech. Rep. 74-1). Providence, R.I.: Brown University, Center for Neural Studies, 1974.

References

Anderson, J. A. A memory storage model utilizing spatial correlation functions. *Kybernetik*, 1968, *5*, 113–119.

Anderson, J. A. Two models for memory organization using interacting traces. *Mathematical Biosciences*, 1970, *8*, 137–160.

Anderson, J. A. A simple neural network generating an interactive memory. *Mathematical Biosciences*, 1972, *14*, 197–220.

Anderson, J. A. A theory for the recognition of items from short memorized lists. *Psychological Review*, 1973, *80*, 417–438.

Anderson, J. A. Neural models with cognitive implications. In D. LaBerge & S. J. Samuels (Eds.), *Basic processes in reading: Perception and comprehension.* Hillsdale, N.J.: Erlbaum, 1977.

Anderson, J. A., Silverstein, J. W., & Ritz, S. A. Vowel pre-processing with a neurally based model. In *Conference Record: 1977 I.E.E.E. International Conference on Acoustics, Speech, and Signal Processing.* Hartford, Conn.: in press.

Barlow, H. B. Single units and sensation: A neuron doctrine for perceptual psychology. *Perception*, 1972, *1*, 371–394.

Cole, M. Search behavior: A correction procedure for three choice probability learning. *Journal of Mathematical Psychology*, 1965, *2*, 145–170.

Cooper, L. N. A possible organization of animal memory and learning. In B. Lundquist & S. Lundquist (Eds.), *Proceedings of the Nobel Symposium on Collective Properties of Physical Systems.* New York: Academic Press, 1974.

Cooper, W. E. Selective adaptation to speech. In F. Restle, R. M. Shiffrin, N. J. Castellan, H. R. Lindman, & D. B. Pisoni (Eds.), *Cognitive theory* (Vol. 1). Hillsdale, N.J.: Erlbaum, 1975.

Cotton, J. W., & Rechtshaffen, A. Replication report: Two- and three-choice verbal conditioning phenomena. *Journal of Experimental Psychology*, 1958, *56*, 96–97.

Creutzfeldt, O., Innocenti, G. M., & Brooks, D. Vertical organization in the visual cortex (area 17) in the cat. *Experimental Brain Research*, 1974, *21*, 315–336.

Eccles, J. C. Possible synaptic mechanisms subserving learning. In A. G. Karczmar & J. C. Eccles (Eds.), *Brain and human behavior.* New York: Springer, 1972.

Eimas, P. D., & Corbit, J. D. Selective adaptation of linguistic feature detectors. *Cognitive Psychology*, 1973, *4*, 99–109.

Eimas, P. D., Siqueland, E. R., Jusczyk, P., & Vigorito,

J. Speech perception in infants. *Science*, 1971, *171*, 303–306.

Estes, W. K. Theory of learning with constant, variable, or contingent probabilities of reinforcement. *Psychometrika*, 1957, *22*, 113–132.

Estes, W. K. Probability learning. In A. W. Melton (Ed.), *Categories of human learning*. New York: Academic Press, 1964.

Estes, W. K. Research and theory on the learning of probabilities. *Journal of the American Statistical Association*, 1972, *67*, 81–102.

Estes, W. K., & Straughan, J. H. Analysis of a verbal conditioning situation in terms of statistical learning theory. *Journal of Experimental Psychology*, 1954, *47*, 225–234.

Evans, E. F. Neural processes for the detection of acoustic patterns and for sound localization. In F. O. Schmitt & F. G. Worden (Eds.), *The neurosciences: Third study program*. Cambridge, Mass.: MIT Press, 1974.

Freeman, W. J. *Mass action in the nervous system*. New York: Academic Press, 1975.

Friedman, M. P., et al. Two-choice behavior under extended training with shifting probabilities of reinforcement. In R. C. Atkinson (Ed.), *Studies in mathematical psychology*. Stanford, Calif.: Stanford University Press, 1964.

Frishkopf, L. S., Capranica, R. R., & Goldstein, M. H., Jr. Neural coding in the bullfrog's auditory system: A teleological approach. *Proceedings of the I.E.E.E.*, 1968, *56*, 969–980.

Funkenstein, H. H., & Winter, P. Responses to acoustic stimuli of units in the auditory cortex of awake squirrel monkeys. *Experimental Brain Research*, 1973, *18*, 464–488.

Gardner, R. A. Probability learning with two and three choices. *American Journal of Psychology*, 1957, *70*, 174–185.

Gibson, E. J. *Principles of perceptual learning and development*. New York: Meredith, 1969.

Globus, A., & Scheibel, A. B. Pattern and field in cortical structure: The rabbit. *Journal of Comparative Neurology*, 1967, *131*, 155–172.

Goldstein, M. H., Jr., Hall, J. L. II, & Butterfield, B. O. Single-unit activity in the primary auditory cortex of unanesthetized cats. *Journal of the Acoustical Society of America*, 1968, *43*, 444–455.

Grossberg, S. Pavlovian pattern learning by nonlinear neural networks. *Proceedings of the National Academy of Sciences*, 1971, *68*, 828–831.

Hebb, D. O. *The organization of behavior*. New York: Wiley, 1949.

Hubel, D. H., & Wiesel, T. N. Receptive fields, binocular interaction, and functional architecture in the cat's visual cortex. *Journal of Physiology*, 1962, *160*, 106–154.

Hubel, D. H., & Wiesel, T. N. Receptive fields and functional architecture of monkey striate cortex. *Journal of Physiology*, 1968, *195*, 215–243.

Jakobson, R., Fant, G. G. M., & Halle, M. *Preliminaries to speech analysis: The distinctive features and their correlates*. Cambridge, Mass.: MIT Press, 1961.

Knight, B. W., Toyoda, J. I., & Dodge, F. A., Jr. A quantitative description of the dynamics of excitation and inhibition in the eye of *Limulus*. *Journal of General Physiology*, 1970, *56*, 421–437.

Kohonen, T. Correlation matrix memories. *I.E.E.E. Transactions on Computers*, 1972, *C-21*, 353–359.

Kohonen, T. *Associative memory: A system theoretic approach*. Berlin: Springer-Verlag, 1977.

Laughery, K. R. Computer simulation of short-term memory: A component decay model. In G. H. Bower & J. R. Spence (Eds.), *The psychology of learning and motivation* (Vol. 3). New York: Academic Press, 1969.

Lindgren, N. Machine recognition of human language: Part 2. Theoretical models of speech perception and language. *I.E.E.E. Spectrum*, 1965, *2*, 45–59.

Lindsay, P. H., & Norman, D. A. *Human information processing: An introduction to psychology*. New York: Academic Press, 1972.

Link, S. W. The relative judgment theory of two choice reaction time. *Journal of Mathematical Psychology*, 1975, *12*, 114–135.

Little, W. A., & Shaw, G. L. A statistical theory of short and long term memory. *Behavioral Biology*, 1975, *14*, 115–135.

McIlwain, J. T. Large receptive fields and spatial transformations in the visual system. In R. Porter (Ed.), *International review of physiology: Neurophysiology II* (Vol. 10). Baltimore, Md.: University Park Press, 1976.

Millward, R. B., & Reber, A. S. Event-recall in probability learning. *Journal of Verbal Learning and Verbal Behavior*, 1968, *7*, 980–989.

Morrell, F., Hoeppner, T. J., & de Toledo, L. Mass action re-examined: Selective modification of single elements within a small population. *Neuroscience Abstracts*, 1976, *2*, 448.

Mountcastle, V. B. The problem of sensing and the neural coding of sensory events. In G. C. Quarton, T. Melnechuk, & F. L. Schmitt (Eds.), *The neurosciences*. New York: Rockefeller University Press, 1967.

Myers, J. L. Probability learning and sequence learning. In W. K. Estes (Ed.), *Handbook of learning and cognitive processes* (Vol. 3). Hillsdale, N.J.: Erlbaum, 1976.

Nass, M. M., & Cooper, L. N. A theory for the development of feature detecting cells in visual cortex. *Biological Cybernetics*, 1975, *19*, 1–18.

Neimark, E. D., & Estes, W. K. *Stimulus sampling theory*. San Francisco: Holden-Day, 1967.

Neisser, U. *Cognitive psychology*. New York: Appleton-Century-Crofts, 1967.

Noda, H., & Adey, W. R. Firing of neuron pairs in cat association cortex during sleep and wakefulness. *Journal of Neurophysiology*, 1970, *33*, 672–684.

Noda, H., Manohar, S., & Adey, W. R. Correlated firing of hippocampal neuron pairs in sleep and wakefulness. *Experimental Neurology*, 1969, *24*, 232–247.

Pisoni, D. B. *On the nature of categorical perception of speech sounds*. Unpublished doctoral dissertation, University of Michigan, 1971.

Pisoni, D. B., & Tash, J. Reaction time to comparisons within and across phonetic categories. *Perception & Psychophysics*, 1974, *15*, 285–290.

Ratliff, F., Knight, B. W., Dodge, F. A., Jr., & Hartline, H. K. Fourier analysis of dynamics of excitation and inhibition in the eye of *Limulus*: Amplitude, phase, and distance. *Vision Research*, 1974, *14*, 1155–1168.

Reber, A. S., & Millward, R. B. Event observation in probability learning. *Journal of Experimental Psychology*, 1968, *77*, 317–327.

Reber, A. S., & Millward, R. B. Event tracking in probability learning. *American Journal of Psychology*, 1971, *84*, 85–99.

Schiller, P. H., Finlay, B. L., & Volman, S. F. Quantitative studies of single-cell properties in monkey striate cortex: V. Multivariate statistical analyses and models. *Journal of Neurophysiology*, 1976, *39*, 1362–1374.

Shepherd, G. M. *The synaptic organization of the brain.* New York: Oxford University Press, 1974.

Studdert-Kennedy, M. The nature and function of phonetic categories. In F. Restle, R. M. Shiffrin, N. J. Castellan, H. R. Lindman, & D. B. Pisoni (Eds.), *Cognitive theory* (Vol. 1). Hillsdale, N.J.: Erlbaum, 1975.

Suzuki, S. *Zen mind, beginner's mind.* New York: Weatherhill, 1970.

van der Loos, H., & Glaser, E. M. Autapses in neocortex cerebri: Synapses between a pyramidal cell's axon and its own dendrites. *Brain Research*, 1972, *48*, 355–360.

Willshaw, D. J., Buneman, O. P., & Longuet-Higgins, H. C. Non-holographic associative memory. *Nature*, 1969, *222*, 960–962.

Winter, P., & Funkenstein, H. H. The effect of species specific vocalization on the discharge of auditory cortical cells in the awake squirrel monkey. *Experimental Brain Research*, 1973, *18*, 489–504.

Wollberg, Z., & Newman, J. D. Auditory cortex of squirrel monkey: Response patterns of single cells to species-specific vocalizations. *Science*, 1972, *175*, 212–214.

Young, T. Z., & Calvert, T. W. *Classification, estimation, and pattern recognition.* New York: American Elsevier, 1974.

Appendix A

The calculation of size of the regions corresponding to each corner is reasonably straightforward. In our system we have two orthogonal eigenvectors which point in the $(1, 1)$ and $(-1, 1)$ directions. Rotate the system by $-\pi/4$ radians to obtain the normalized eigenvectors

$$\mathbf{e}_x = \begin{pmatrix} 1 \\ 0 \end{pmatrix}, \quad \mathbf{e}_y = \begin{pmatrix} 0 \\ 1 \end{pmatrix}.$$

The eigenvalues are unchanged by a rotation; that is,

$$\lambda_A = \lambda_x, \quad \lambda_B = \lambda_y.$$

A point with coordinates, $\mathbf{v} = \begin{pmatrix} x_0 \\ y_0 \end{pmatrix}$ in this rotated frame, then may be expressed as

$$\mathbf{v} = x_0 \mathbf{e}_x + y_0 \mathbf{e}_y. \tag{A1}$$

If we consider that a feedback cycle occurs once every unit time, t_i, $i = 0, 1, 2, \ldots$, then

$$\mathbf{v}(t_0) = x_0 \mathbf{e}_x + y_0 \mathbf{e}_y$$

$$\begin{aligned} \mathbf{v}(t_1) &= x_0(\mathbf{e}_x \lambda_x) + y_0(\mathbf{e}_y \lambda_y) + \mathbf{v}(t_0) \\ &= x_0 \mathbf{e}_x(1 + \lambda_x) + y_0 \mathbf{e}_y(1 + \lambda_y) \end{aligned}$$

$$\begin{aligned} \mathbf{v}(t_2) &= x_0 \mathbf{e}_x(1 + \lambda_x)\lambda_x + y_0 \mathbf{e}_y(1 + \lambda_y)\lambda_y \\ &\qquad\qquad\qquad\qquad\qquad + \mathbf{v}(t_1) \\ &= x_0 \mathbf{e}_x(1 + \lambda_x)^2 + y_0 \mathbf{e}_y(1 + \lambda_y)^2. \end{aligned}$$

By induction we get

$$\mathbf{v}(t_n) = x_0 \mathbf{e}_x(1 + \lambda_x)^n + y_0 \mathbf{e}_y(1 + \lambda_y)^n.$$

Since $\mathbf{e}_x = \begin{pmatrix} 1 \\ 0 \end{pmatrix}$, $\mathbf{e}_y = \begin{pmatrix} 0 \\ 1 \end{pmatrix}$, we have

$$\mathbf{v}(t_n) = \begin{pmatrix} x_0(1 + \lambda_x)^n \\ y_0(1 + \lambda_y)^n \end{pmatrix}. \tag{A2}$$

In order to get a reasonably simple, general solution for the sizes of the regions, we must make the approximation, which should be quite good if step size is not too large, that time is continuous.

For continuous time we have

$$dx/dt = \lambda_x x$$

$$dy/dt = \lambda_y y.$$

These two equations may be solved to determine the "motion" of a point (x, y) as it is fed back through the system. The solutions are

$$\lambda_x^{-1} \ln x = t + k_1$$
$$\lambda_y^{-1} \ln y = t + k_2.$$

Eliminating t, we find

$$\lambda_x \ln y - \lambda_y \ln x = \text{constant}$$

$$\ln \frac{y^{\lambda_x}}{x^{\lambda_y}} = \text{constant} \quad (A3)$$

$$y^{\lambda_x} = k x^{\lambda_y} \quad \text{or} \quad y = k' x^{\lambda_y/\lambda_x}, \quad (A4)$$

where k' is a constant. This family of curves provides the trajectories that will be followed by an initial point under the influence of feedback. A point with initial coordinates (x_0, y_0) will follow the curve expressed by Equation A4 with

$$k' = \frac{y_0}{x_0^{\lambda_y/\lambda_x}}. \quad (A5)$$

We must now consider the behavior of these curves at the boundary of the square. A point will travel along the curve given by Equation A4 until it reaches the boundary. At this point all motion in the direction normal to the boundary will be prevented. Only the component of the predicted motion that is tangent to the boundary will contribute to the actual motion. If $\mathbf{t}$ is a normalized tangent vector along the boundary, then

$$\left. \frac{d\mathbf{v}}{dt} \right|_{\substack{\text{along} \\ \text{boundary}}} = \left\{ \left(\frac{dx}{dt} \right) \mathbf{e}_x + \left(\frac{dy}{dt} \right) \mathbf{e}_y \right\} \cdot \mathbf{t}. \quad (A6)$$

We are interested in finding the curve represented by Equation A4 for which a point will reach the boundary and stop. This will occur when the scalar product (dot product) of Equation A6 of the point reached on the boundary is zero. Thus we wish to find the point at which

$$\left(\frac{dx}{dt} \mathbf{e}_x + \frac{dy}{dt} \mathbf{e}_y \right) \cdot \mathbf{t} = 0.$$

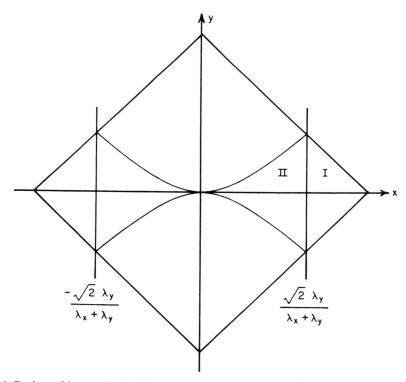

Figure A1. Regions of integration for the calculation of response probabilities for the two-dimensional neural system. The eigenvectors point along the coordinate axes.

Consider only the positive quadrant, with a normalized tangent vector,

$$\mathbf{t} = \frac{-\mathbf{e}_x + \mathbf{e}_y}{\sqrt{2}}.$$

We have then

$$0 = \{x\lambda_x \mathbf{e}_x + y\lambda_y \mathbf{e}_y\} \cdot \left\{\frac{-\mathbf{e}_x + \mathbf{e}_y}{\sqrt{2}}\right\},$$

so that

$$0 = -x\lambda_x + y\lambda_y.$$

Since $x + y = \sqrt{2}$ on the boundary, we have

$$0 = x\lambda_x + x\lambda_y - \sqrt{2}\lambda_y,$$

and therefore

$$x = \frac{\sqrt{2}\,\lambda_y}{\lambda_x + \lambda_y}$$

$$y = \frac{\sqrt{2}\,\lambda_x}{\lambda_x + \lambda_y}.$$

The lines, $x = \sqrt{2}\,\lambda_y/(\lambda_x + \lambda_y)$ and $x = -\sqrt{2}\,\lambda_y/(\lambda_x + \lambda_y)$ determine by their intersection with the boundary four (unstable) equilibrium points. The curves representing all interior points of the square which will move so as to intersect the boundary at one of these equilibrium points are given by Equation A4 with

$$k' = \pm \frac{y_0}{x_0^{\lambda_y/\lambda_x}}$$

$$= \pm \left(\frac{\sqrt{2}\,\lambda_x}{\lambda_x + \lambda_y}\right)\left(\frac{\lambda_x + \lambda_y}{\sqrt{2}\,\lambda_y}\right)^{\lambda_y/\lambda_x}.$$

The regions of the two-dimensional system are thus determined (see Figure A1).

The area of the sector labelled I and II in Figure A1 can be found by integrating $y = k'x^{\lambda_y/\lambda_x}$ from 0 to $x = \sqrt{2}\,\lambda_y/(\lambda_x + \lambda_y)$ and adding the area of the triangular segment, I. This area times 4 will give the amount of the total area for which a vector will end up in corners $(1, 1)$ or $(-1, -1)$ in the nonrotated system. Since the total area of the square is 4, the area I-II corresponds to the probability of Response A.

If P is the probability of ending up in corners $(1, 1)$ or $(-1, -1)$ then

$$P = \int_0^{\sqrt{2}\lambda_y/(\lambda_x + \lambda_y)} k'x^{\lambda_y/\lambda_x}dx$$

$$+ \frac{1}{2}\left(\sqrt{2} - \frac{\sqrt{2}\,\lambda_y}{\lambda_x + \lambda_y}\right)\left(\frac{\sqrt{2}\,\lambda_x}{\lambda_x + \lambda_y}\right)$$

$$= \frac{k'x^{\lambda_y/\lambda_x+1}}{1 + \lambda_y/\lambda_x}\Bigg|_0^{\sqrt{2}\lambda_y/(\lambda_x+\lambda_y)} + \left(\frac{\lambda_x}{\lambda_x + \lambda_y}\right)^2$$

$$= \left(\frac{\sqrt{2}\,\lambda_x}{\lambda_x + \lambda_y}\right)\left(\frac{\lambda_x + \lambda_y}{\sqrt{2}\,\lambda_y}\right)^{\lambda_y/\lambda_x}\left(\frac{\lambda_x}{\lambda_x + \lambda_y}\right)$$

$$\times \left(\frac{\sqrt{2}\,\lambda_y}{\lambda_x + \lambda_y}\right)^{\lambda_y/\lambda_x+1} + \left(\frac{\lambda_x}{\lambda_x + \lambda_y}\right)^2$$

$$= \sqrt{2}\left(\frac{\lambda_x}{\lambda_x + \lambda_y}\right)^2\left(\frac{\sqrt{2}\,\lambda_y}{\lambda_x + \lambda_y}\right) + \left(\frac{\lambda_x}{\lambda_x + \lambda_y}\right)^2$$

$$= \frac{3\lambda_x^2\lambda_y + \lambda_x^3}{(\lambda_x + \lambda_y)^3}$$

$$= \frac{3\lambda^2 + \lambda^3}{(\lambda + 1)^3}, \qquad (A7)$$

where $\lambda = \lambda_x/\lambda_y$. This is the formula used to calculate response probabilities in the section on probability learning.

Appendix B

We can write the equation governing λ_A as

$$\lambda_A(n + 1) = 1 + P_{n+1}, \qquad (B1)$$

where

$$P_{n+1} = gP_n + \eta I_{n+1}, \quad P_0 = 0, \qquad (B2)$$

where the I_ns are independent random variables taking the value 1 with probability π and 0 with probability $1 - \pi$. The solution for Equation B2 is given by

$$P_n = \eta \sum_{j=1}^{n} g^{n-j}I_j \qquad (B3)$$

and can easily be verified. However it is virtually impossible to express the distribution of P_n in a workable form. We can calculate moments of P_n using Equation B3. For example the expected value of P_n is

$$E(P_n) = \eta \sum_{j=1}^{n} g^{n-j}E(I_j) = \eta\pi \sum_{j=0}^{n-1} g^j$$

$$= \eta\pi \frac{1 - g^n}{1 - g}, \qquad (B4)$$

while the second moment is

$$E(P_n{}^2) = \eta^2 E\Big[\Big(\sum_{j=1}^{n} g^{n-j} I_j\Big)\Big(\sum_{k=1}^{n} g^{n-k} I_k\Big)\Big] = \eta^2 \sum_{j,k=1}^{n} g^{n-j+n-k} E(I_j I_k)$$

$$= \eta^2\Big[\pi \sum_{j=1}^{n} g^{2(n-j)} + \sum_{j\neq k} g^{2n-j-k} E(I_j) E(I_k)\Big]$$

$$= \eta^2\Big[\pi \frac{1-g^{2n}}{1-g^2} + \sum_{j=1}^{n} g^{n-j} E(I_j) \sum_{k=1}^{n} g^{n-k} E(I_k) - g^{n-j} E(I_j)\Big]$$

$$= \eta^2\Big[\pi \frac{1-g^{2n}}{1-g^2} + \pi^2\Big(\frac{1-g^n}{1-g}\Big)^2 - \pi^2 \frac{1-g^{2n}}{1-g^2}\Big]$$

$$= \eta^2\pi\Big[(1-\pi)\frac{1-g^{2n}}{1-g^2} + \pi\Big(\frac{1-g^n}{1-g}\Big)^2\Big] \qquad \text{(B5)}$$

The variance, then, of P_n is

$$\text{var}(P_n) = E(P_n{}^2) - E(P)^2$$
$$= \eta^2\pi(1-\pi)\frac{1-g^{2n}}{1-g^2}. \quad \text{(B6)}$$

The higher moments can be found in the same way, although the calculations become more tedious.

As n approaches infinity, P_n cannot converge to a real number or to another random variable, since at each step we are adding a non-attenuating random element. However the distribution functions $F_n(x) \equiv P(P_n \leq x)$ converge to the distribution function of the random variable

$$P = \eta \sum_{j=0}^{\infty} g^j I_j, \qquad \text{(B7)}$$

where the I_js are defined as before. Since the distributions of P_n are defined on the interval between 0 and $\eta/(1-g)$, the moments of P_n converge to the moments of P. The moments of P are easy to get by using an iterative scheme. We can write P as

$$P = gP' + \eta I, \qquad \text{(B8)}$$

where $P' = \sum_{j=0}^{\infty} g^j I_{j+1}$, and $I = I_1$. Since P and P' have the same distribution, and since P' and I are independent we have for all $n \geq 1$

$$E(P^n) = E(gP' + \eta I)^n = g^n E(P^n)$$
$$+ \pi \sum_{j=1}^{n} \binom{n}{j} g^{n-j} \eta^j E(P^{n-j}), \quad \text{(B9)}$$

where (n, r) are the binomial coefficients [that is, $(n, r) = n!/r!(n-r)!$]. Therefore we have

$$E(P^n) = \frac{1}{1-g^n}\Big[\pi \sum_{j=1}^{n} \binom{n}{j} g^{n-j} \eta^j E(P^{n-j})\Big]. \qquad \text{(B10)}$$

It is easy to see that

$$\lambda_B(n) = 1 + \eta \sum_{j=1}^{n} g^{n-j}(1-I_j)$$
$$= 1 + \frac{\eta(1-g^n)}{1-g} - P_n, \quad \text{(B11)}$$

so that

$$\lambda(n) = \frac{\lambda_A(n)}{\lambda_B(n)} = \frac{1+P_n}{1 + \frac{\eta(1-g^n)}{1-g} - P_n}$$
$$= \frac{1+a_n}{a_n - P_n} - 1, \quad \text{(B12)}$$

where $a_n = 1 + [(1-g^n)/(1-g)]$. However, the random variable $1/(a_n - P_n)$ is difficult to deal with even in the limiting case. All moments of $\lambda(n)$ must be derived from infinite series of moments of P_n. It is computationally feasible to get approximations, but we made no attempt to do so.

Fortunately, we have a remarkably different situation for the probability of selection. We write

$$\text{Prob}[\lambda(n)] = \frac{\lambda(n)[3\lambda(n) + \lambda(n)^2]}{[1+\lambda(n)]^3}$$
$$= 1 - \frac{1+3\lambda(n)}{[1+\lambda(n)]^3}, \quad \text{(B13)}$$

and it can easily be verified that

$$\frac{1+3\lambda(n)}{[1+\lambda(n)]^3} = \frac{1}{(1+a_n)^3}$$
$$\times [a_n{}^3 + 3a_n{}^2 - 6a_n P(n)$$
$$+ 3(1-a_n)P(n)^2 + 2P^3(n)], \quad \text{(B14)}$$

so that

$$\text{Prob}[\lambda(n)] = \frac{1+3a_n}{(1+a_n)^3} - \frac{1}{(1+a_n)^3}$$

$$\times [2P^3(n) + 3(1 - a_n)P^2(n) - 6a_nP(n)]. \quad (B15)$$

Therefore in order to determine the expected value of $\text{Prob}[\lambda(n)]$ we need only to compute the first three moments of $P(n)$, while the variance of $\text{Prob}[\lambda(n)]$ requires $P(n)$'s first six moments.

As n approaches infinity the distributions of $\text{Prob}[\lambda(n)]$ approach the distribution of the random variable $\text{Prob}(\lambda)$, where λ is given by

$$\lambda = \frac{1 + a}{a - P} - 1, \quad a = 1 + \frac{\eta}{1 - g}. \quad (B16)$$

Equation B15 holds true for λ with $P(n)$ replaced by P, so that that expected value and variance of $\text{Prob}(\lambda)$ can be derived as in the finite case.

Appendix C

We would like to see if we must be restricted to the values given in Equation 5 for the feedback system to have its desirable properties. Suppose we let the main diagonal be zero. This means a cell influences only its neighbors, and not itself. We see that the value of the diagonal element given by Equation 5 is related to the mean square of the activities $f(i)$ over all the traces. If we make the assumption—in harmony with our general approach, but presently untestable—that, on the average, cells are equally active across the total stimulus set, that is, they are all equally "important" in some sense, then the diagonal elements will be almost the same.

Let us consider the matrix we shall call $\mathbf{D}$, which contains only diagonal elements. Then,

$$\mathbf{D} \cong c\mathbf{I},$$

where c is a positive constant. If the covariance matrix given by Equation 5 is denoted $\mathbf{V}$ and we require the feedback matrix, $\mathbf{A}$, to have zeroes along the diagonal, then

$$\mathbf{A} \cong \mathbf{V} - \mathbf{D} \cong \mathbf{V} - c\mathbf{I}. \quad (C1)$$

Suppose $\mathbf{e}_i$ is an eigenvector of the covariance matrix with eigenvalue λ_i and with all the important discriminative properties we have discussed. Let us assume that the equality in Equation C1 holds. Then if

$$\mathbf{A} = \mathbf{V} - c\mathbf{I}$$

$$\mathbf{Ae}_i = \mathbf{Ve}_i - c\mathbf{e}_i$$

$$= \lambda_i\mathbf{e}_i - c\mathbf{e}_i$$

$$= (\lambda_i - c)\mathbf{e}_i. \quad (C2)$$

Thus, $\mathbf{e}_i$ is an eigenvector of $\mathbf{A}$ and its eigenvalue is $(\lambda_i - c)$. We see that all the eigenvalues are reduced by an amount c. Thus some eigenvalues can be negative. If the equality does not hold, then for small deviations of $\mathbf{D}$ from a constant times the identity matrix, the eigenvectors of $\mathbf{A}$ will be close to the eigenvectors of the covariance matrix and will have about the same properties.

As a special case of interest, assume the eigenvectors pointed toward the corners of an N-dimensional hypercube. Suppose the system only learns these corners. Since the cube has equal sides, then all the $f_k^2(i)$ will be the same (i.e., the square of the saturation limit) and we see that the exact equality will hold). Thus, we can avoid the awkward necessity to have large positive elements along the main diagonal and lose little, if any, of the information processing power of the system.

Note that negative eigenvalues, as long as they are greater than -1, may cause the final state of the system to converge to zero, as can be seen from Equation 8. This will occur if the initial input is a linear combination of eigenvectors with eigenvalues in the interval $(-1, 0)$.

23

Introduction

(1978)

Scott E. Brodie, Bruce W. Knight, and Floyd Ratliff

The response of the *Limulus* retina to moving stimuli: a prediction by Fourier synthesis
Journal of General Physiology 72:129–154, 162–166

A collection of papers on neural networks without an article on *Limulus* would be unthinkable. The study, analysis, and understanding of the lateral eye of *Limulus* stands as one of the true triumphs of neurobiology. It is also full of lessons for theoreticians trying to understand brain function in more complex organisms.

Limulus polyphemus is also known as the horseshoe crab and is familiar to anyone who has spent time at the beach in New England. It is a large, but harmless, crab-like animal found in great profusion in many locations on the East Coast of the United States. The name is misleading; it is an arthropod and not a crab. It is a very old species, and seems not to have changed much in the fossil record for two hundred million years. (Perhaps it is doing something right.) H. K. Hartline first studied the lateral eye of *Limulus* in the 1930s, work which eventually lead to a Nobel Prize. The laboratory of Hartline and Floyd Ratliff at Rockefeller University produced a series of papers in the '60s and '70s that almost completely 'explained' the *Limulus* eye, in terms of its input-output relationships.

This paper is one of the final products of this work. Many of the earlier papers are collected in a volume of reprints (Ratliff, 1974) or are described in Ratliff's well known classic, *Mach Bands* (1965).

The great advantage of *Limulus* for physiology is the ease with which it is possible to do experiments on the lateral eye. The compound *Limulus* eye is made up of a large number of independent 'little eyes' called *ommatidia*, an optical system very unlike the familiar vertebrate eye, where a lens projects an image on a retina. There are 800 to 1,000 ommatidia, each one of which is electrically and mechanically isolated from its neighbors. Each ommatidium has a restricted field of view. In their totality, fields of the ommatidia cover almost the entire hemisphere, though with poor spatial resolution.

The phenomenon of "lateral inhibition," ubiquitous in sensory systems, and elsewhere in the brain as well, was first described in *Limulus*. When light is shone on a single ommatidium, the optic nerve fiber connecting that ommatidium to the *Limulus* brain shows an initial burst of activity and then a steady regular discharge frequency, a monotonic function of light intensity. If a neighboring ommatidium is illuminated, it reduces the firing rate of the first cell, i.e. inhibits it. Since the inhibition occurs at the same level, it is referred to as *lateral inhibition*. Lateral inhibition of a slightly more complex kind is found in our own retina in the well known "center-surround" receptive field organization. A number of visual effects such as Mach Bands and the illusions seen with Hermann grids can be explained simply by this mechanism.

Lateral inhibition is a powerful signal processing technique. Most prominently, it serves as a simple automatic gain control mechanism and serves to enhance edge contrast by increasing the firing rate differential between cells on the two sides of the edge.

Although these effects are important, it should not be forgotten, especially by connectionist modelers, that the *Limulus* eye is a true neural network. It consists of a group of neurons connected by synapses of varying weights. To understand its dynamics we need detailed models for the individual neurons and their interconnections and interactions.

The Rockefeller group has provided us with that information. The most profound lesson that their work has taught us is that the *Limulus* eye is *simple*. When one thinks of the horrendous potential complexities of a set of interconnected threshold elements with potentially highly nonlinear synaptic interactions, it is remarkable that our best understood nervous system can be understood by analyzing it as a linear system. As Brodie, Knight, and Ratliff comment, "Our characterization task is greatly simplified by the fact that the *Limulus* eye is, to an excellent approximation, a 'linear system'; that is, its response to the sum of two stimuli is the sum of its responses to each stimulus presented separately."

This is a truly remarkable finding, and is the kind of result that gives hope to those interested in the understanding of the nervous system. It is possible that the brain is not so complicated as it might have been, and that some systems may actually be quite simple if analyzed in the right way.

If the *Limulus* eye is truly a linear system, then it should be possible to obtain its transfer function, in both the time domain and the space domain. This paper discusses in considerable detail the mathematical and experimental techniques and approximations necessary in order to obtain the transfer functions.

Once the transfer function is obtained, it should be possible to predict the output of the retina (in terms of the pattern of activities of fibers in the optic nerve) for *any* moving or stationary image. When this was tried, the fit between prediction and experiment was quite good for several arbitrary input patterns. Of special note is the arbitrary moving pattern shown in figure 14B, which is, in origin, a silhouette of the skyline of Wesleyan University.

It is often hard to convince those who are not neuroscientists that there are actually areas where we know a great deal about the detailed interactions of real nervous networks. In many cases these are much simpler than they have any right to be. To paraphrase a famous quote from physics, let us make our models as complex as they have to be, *but no more so.*

Bibiographic note: To save space, we removed some lengthy discussions of the optics used in the experiments, which were described in appendix A. There are more details of the experiments and analysis techniques provided in the immediately succeeding paper in this issue of the *Journal of General Physiology* (Brodie, Knight, and Ratliff, 1978).

References

S. E. Brodie, B. W. Knight, and F. Ratliff (1978), "The spatiotemporal transfer function of the *Limulus* lateral eye," *Journal of General Physiology* 78:166–190.

F. Ratliff (1965), *Mach Bands: Quantitative Studies on Neural Networks in the Retina*, San Francisco: Holden-Day.

F. Ratliff (Ed.) (1974), *Studies on Excitation and Inhibition in the Retina*, New York: Rockefeller University Press.

(1978)
Scott E. Brodie, Bruce W. Knight, and Floyd Ratliff

The response of the *Limulus* retina to moving stimuli: a prediction by Fourier synthesis
Journal of General Physiology 72:129–154, 162–166

The *Limulus* retina responds as a linear system to light stimuli which vary moderately about a mean level. The dynamics of such a system may conveniently be summarized by means of a spatiotemporal transfer function, which describes the response of the system to moving sinusoidal gratings. The response of the system to an arbitrary stimulus may then be calculated by adding together the system's responses to suitably weighted sinusoidal stimuli. We have measured such a spatiotemporal transfer function for the *Limulus* eye. We have then accurately predicted, in a parameter-free calculation, the eye's response to various stimulus patterns which move across it at several different velocities.

Introduction

The retina of the lateral eye of the horseshoe crab *Limulus polyphemus* has been the subject of extensive study (for reviews, see Hartline and Ratliff, 1972; Ratliff, 1974). In recent years, many of these studies have been addressed to the resolution of the eye's neural response into the actions of individual physiological subsystems at the cellular level. To this end, these studies have exploited techniques such as the optical isolation of single visual units and the electrical manipulation of single cells by means of intracellular microelectrodes. These methods are intended to suppress the broad integrative actions of the retina so that the actions of small components may be studied. The present study is complementary to this earlier work. Our goal is to develop a methodology which describes the neural action of the *Limulus* retina as an integrated whole. In principle, such a description should permit the quantitative prediction of the response of the retina to an arbitrary stimulus, as well as permit the further elucidation of the component cellular processes which underly the response.

In the present paper, we describe the application of such a methodology to the *Limulus* retina. We have developed a set of standard visual stimuli which generate responses which should completely characterize the retina. We also show how this characterization may be used to predict the response of the retina to light stimuli which vary both in space and in time. In the following paper (Brodie et al., 1978) we relate this description of the retina's integrative action to the underlying electrophysiological processes.

Our characterization task is greatly simplified by the fact that the *Limulus* retina is, to an excellent approximation, a "linear system;" that is, its response to the sum of two stimuli is the sum of its responses to each stimulus presented separately (Knight et al., 1970). Although various nonlinear effects have been observed in the *Limulus* eye (Hartline and Ratliff, 1957; Barlow and Lange, 1974), they appear to be more prominent in the action of isolated visual units than in the response of the retina as a whole. Ultimately, the validity of our linear description should be judged by the extent of the agreement between the response of the eye and the predictions derived from the linear characterization.

Theory

In this section we develop the theoretical basis for our analysis of the response of the *Limulus* retina. We begin with a very general discussion of linear systems analysis, and then specialize this treatment for application to the present measurements. We first make three fundamental assumptions about the system under study, which we may refer to as stationarity, linearity, and continuity.

By stationarity, we refer to the assumption that the properties of the system are stable with respect to time. In other words, we require that the system give the same response each time it is presented with the same stimulus. If we denote the stimulus as a function of time by $\mathscr{S}(t)$, and the corresponding response by $\mathscr{R}(t)$, and use an arrow ($\rightarrow$) to denote the action of the system under study, we may express the stationarity conditions as:

$$\mathscr{S}(t) \rightarrow \mathscr{R}(t) \text{ implies } \mathscr{S}(t - \tau) \rightarrow \mathscr{R}(t - \tau), \qquad (1)$$

where τ is any constant shift in time. Of course, as nearly every biological system ages, and most neuro-

Copyright © 1978 The Rockefeller University Press.

physiological preparations deteriorate, this assumption is necessarily an approximation.

Linearity refers to the assumption that the system obeys the superposition rule, that the response to a sum of inputs is the sum of the responses to the inputs taken separately. In symbols,

$$\mathscr{S}_1(t) \to \mathscr{R}_1(t),\ \mathscr{S}_2(t)$$
$$\to \mathscr{R}_2(t) \text{ implies } \mathscr{S}_1(t) + \mathscr{S}_2(t) \to \mathscr{R}_1(t) + \mathscr{R}_2(t). \quad (2)$$

This is a very strong assumption, whose consequences will be vigorously exploited below. In general, many biological systems saturate when presented with very strong stimuli, but operate nearly linearly when presented with stimuli consisting of small fluctuations about a mean value.

The assumption of continuity states that small changes in the stimulus presented to the system produce only small changes in the response. This is a mild assumption for most systems in the middle of their operating range, but it often does not hold for systems at the extremes of their range. The effect of this assumption is to justify various mathematical manipulations below. For example, if x is some parameter and $\mathscr{S}_x(t)$ varies smoothly with x, then the continuity assumption allows us to assert the following continuous analog of Eq. 2:

$$\mathscr{S}_x(t) \to \mathscr{R}_x(t) \text{ implies } \int \mathscr{S}_x(t)\,dx \to \int \mathscr{R}_x(t)\,dx. \quad (3)$$

The assumptions of linearity and continuity together imply "weighted" versions of Eqs. 2 and 3:

$$\mathscr{S}_1(t) \to \mathscr{R}_1(t),\ \mathscr{S}_2(t)$$
$$\to \mathscr{R}_2(t) \text{ implies } a\mathscr{S}_1(t) + b\mathscr{S}_2(t) \quad (2')$$
$$\to a\mathscr{R}_1(t) + b\mathscr{R}_2(t);$$
$$\mathscr{S}_x(t) \to \mathscr{R}_x(t) \text{ implies } \int a(x)\mathscr{S}_x(t)\,dx \to \int a(x)\mathscr{R}_x(t)\,dx, \quad (3')$$

where a and b are numbers, and $a(x)$ is a function of x.

In general, a stationary, linear, continuous system (a "linear system," for short) can be completely characterized in several equivalent ways.

One such characterization consists of measuring the system's response to an "impulse," a stimulus of finite strength delivered within an arbitrarily short time. We will denote such a stimulus as a Dirac delta-function, $\delta(t)$. The response to such an impulse delivered at time $t = 0$ will be called the "impulse response," denoted $\mathscr{I}(t)$. The impulse response provides a complete characterization of a linear system by virtue of the following identity, which constitutes a fundamental property of the delta-function:

$$\mathscr{S}(t) = \int \mathscr{S}(u)\delta(t - u)\,du, \quad (4)$$

which expresses an arbitrary stimulus $\mathscr{S}(t)$ as a weighted combination of impulses occurring at different times; the weighting function is simply the stimulus function itself. The stationarity assumption implies that the response to the stimulus $\delta(t - u)$ is $\mathscr{I}(t - u)$; we may now apply Eq. 3' to conclude

$$\mathscr{S}(t) \to \mathscr{R}(t) = \int \mathscr{S}(u) \cdot \mathscr{I}(t - u)\,du. \quad (5)$$

This is a formula for the response of the system to an arbitrary stimulus $\mathscr{S}$, given in terms of the impulse response $\mathscr{I}$.

An alternative characterization of a linear system can be obtained from Eq. 5 by considering the response of the system to a sinusoidal input. We may greatly simplify the calculations by adopting the complex-exponential notation for sinusoidal functions; we thus identify $\cos \omega t$ with the real part of the complex exponential $e^{i\omega t} = \cos \omega t + i \sin \omega t$. In general, any complex quantity is to be interpreted as representing its real part. This is valid, because "taking the real part" of a complex quantity obeys the superposition principle. With this convention, we choose for our stimulus $\mathscr{S}(t) = e^{i\omega t}$, a sinusoid of angular frequency ω. According to Eq. 5, for this stimulus, we obtain the response:

$$\mathscr{R}(t) = \int e^{i\omega u}\mathscr{I}(t - u)\,du$$
$$= \int e^{i\omega(t-u)}\mathscr{I}(u)\,du = \left(\int e^{-i\omega u}\mathscr{I}(u)\,du\right) \cdot e^{i\omega t}. \quad (6)$$

Thus, the response of a linear system to a sinusoidal input $\mathscr{S}(t) = e^{i\omega t}$ is a sinusoid of the same frequency, multiplied by some (complex) number, which depends on the frequency, ω, and upon the impulse response, $\mathscr{I}$, of the system. We refer to this coefficient (considered as a function of ω) as the "transfer function" of the system, and denote it by $\mathscr{F}(\omega)$. We then have

$$\mathscr{R}(t) = \mathscr{F}(\omega) \cdot e^{i\omega t}, \quad (6')$$

where

$$\mathscr{F}(\omega) = \int e^{-i\omega t}\mathscr{I}(t)\,dt. \quad (7)$$

$\mathscr{F}(\omega)$ is the Fourier transform of the impulse response, $\mathscr{I}(t)$.

The transfer function $\mathscr{F}(\omega)$ provides a complete characterization of a linear system as a consequence of the following Fourier inversion formula:

$$\mathscr{S}(t) = \frac{1}{2\pi} \int e^{i\omega t} \tilde{\mathscr{S}}(\omega) \, d\omega, \tag{8}$$

where

$$\tilde{\mathscr{S}}(\omega) \equiv \int e^{-i\omega t} \mathscr{S}(t) \, dt. \tag{9}$$

This expresses an arbitrary stimulus $\mathscr{S}(t)$ as a weighted sum of sinusoids $e^{i\omega t}$; the weighting function $\tilde{\mathscr{S}}(\omega)$ is the Fourier transform of the stimulus. Applying Eqs. 3′ and 6′ to Eq. 8 yields

$$\mathscr{R}(t) = \frac{1}{2\pi} \int e^{i\omega t} \cdot \mathscr{F}(\omega) \cdot \tilde{\mathscr{S}}(\omega) \, d\omega. \tag{10}$$

This is an expression for the response of the system to an arbitrary stimulus $\mathscr{S}(t)$ in terms of the transfer function $\mathscr{F}(\omega)$. The response is the inverse Fourier transform of the product of the transfer function of the system and the Fourier transform $\tilde{\mathscr{S}}(\omega)$ of the arbitrary stimulus $\mathscr{S}(t)$.

It is appropriate to note here that although these two mathematical characterizations of a linear system are informationally equivalent, and each can be readily obtained from the other (by Eq. 7 and its Fourier inversion formula), the transfer function is often the more suitable characterization for direct laboratory measurement. This is because an impulse stimulus, though of finite total strength, has extremely large intensity (even in a laboratory realization of the theoretical infinite intensity). Such large signals may easily drive the system out of its linear range, or even damage it irreversibly (consider, for example, the study of a skin pressure receptor; in this case, an impulse takes the form of a sharp blow). Even when an impulse stimulus might not saturate the system, it may be difficult or impossible to provide a satisfactory impulse stimulus, especially if one studies a transduction whose only accessible input is the output of another transduction. In such a case, sinusoidal inputs are readily obtained, but impulses are unavailable.

We now proceed to formulate the analysis given above in terms appropriate to our particular experimental situation. We introduce coordinates on the *Limulus* lateral eye, with the x-axis horizontal, the y-axis vertical, and the origin centered on the test ommatidium (the ommatidium whose eccentric cell action potentials are being recorded). A completely general stimulus takes the form $\mathscr{S} = \mathscr{S}(x, y, t)$; for computational convenience, we restrict all further discussion to stimuli which depend only on x and t, that is, to stimuli which, at any time t, are constant along each vertical line. This reduces our problem to one dimension of time, and one of space. To further facilitate

the analysis, we ignore the discrete structure of the *Limulus* eye, and assume instead that it is made of a continuum of photosensitive elements, each possessing the same dynamical properties as the test ommatidium (Kirschfield and Reichardt, 1964). The ommatidia are sufficiently numerous and homogeneous so that this approximation is reasonably innocuous. Our final specialization is to restrict our calculations to predictions of the response to arbitrary intensity patterns which move uniformly across the eye. Although the Fourier methods outlined above are perfectly adequate to predict the response to an arbitrary time-varying stimulus, the restriction to stimuli of the form $\mathscr{S}(x, t) = \mathscr{S}(x - vt)$, where $\mathscr{S}(x)$ is some spatial pattern of illumination, and v is the drift velocity, greatly facilitates the calculation of the necessary Fourier transforms. Such stimuli, with $\mathscr{S}(x)$ arbitrary, and v at our disposal, are sufficiently general to provide a rigorous test of the adequacy of our characterization of the response of the eye.

This characterization is given in terms of a spatio-temporal transfer function, which generalizes the temporal transfer function of Eq. 7, above. We consider the response as a function of space and time, $\mathscr{R} = \mathscr{R}(x, t)$, and ask what is the response to a traveling spatial sinusoid $\mathscr{S}(x, t) = e^{i(\xi x + \omega t)}$. We may put this expression in the form $\mathscr{S}(x - vt)$ by writing

$$e^{i(\xi x + \omega t)} = e^{i\xi(x + (\omega/\xi)t)} = e^{i\xi(x - vt)},$$

where $v = -\omega/\xi$; this is the equation of a sinusoid of spatial frequency ξ moving with velocity $v = \omega/\xi$. In addition to our previous assumptions, we now assume that the response of the eye is invariant under translation (change of origin). This is the analogue for space of the stationarity assumption in time. With this assumption, an argument strictly analogous to that given above implies that the response $\mathscr{R}(x, t)$ to a sinusoidal input $\mathscr{S}(x, t) = e^{i(\xi x + \omega t)}$ is again a sinusoidal function of space and time, with the same spatial frequency ξ and the same temporal frequency ω. In other words, we have the input-output relation:

$$\mathscr{S}(x, t) = e^{i(\xi x + \omega t)} \quad \text{implies} \quad \mathscr{R}(x, t) = \mathscr{F}(\xi, \omega) \cdot e^{i(\xi x + \omega t)}, \tag{11}$$

where

$$\mathscr{F}(\xi, \omega) = \int \int e^{-i(\xi x + \omega t)} \mathscr{I}(x, t) \, dx \, dt \tag{12}$$

is a (complex) number depending on ξ and ω, given by the spatiotemporal Fourier transform of $\mathscr{I}(x, t)$, the response of the system to a spatiotemporal impulse (a vertical line at $x = 0$ flashed at the instant $t = 0$).

We now fit the pieces together to calculate the response of the system to an arbitrary moving pattern. We first obtain the Fourier transform of the spatial pattern of the stimulus $\mathcal{S}(x,t) = \mathcal{S}(x - vt)$:

$$\mathcal{S}(\xi) = \int e^{-i\xi x}\mathcal{S}(x)\,dx. \tag{13}$$

We also note the corresponding inversion formula:

$$\mathcal{S}(u) = \frac{1}{2\pi}\int e^{+i\xi u}\mathcal{S}(\xi)\,d\xi. \tag{14}$$

Taking $u = x - vt$, we obtain a Fourier representation for the stimulus:

$$\mathcal{S}(x - vt) = \frac{1}{2\pi}\int e^{i(\xi x - \xi vt)}\mathcal{S}(\xi)\,d\xi$$
$$= \frac{1}{2\pi}\int e^{i(\xi x + \omega t)}\mathcal{S}(\xi)\,d\xi, \tag{15}$$

where $\omega = \omega(\xi) = -\xi v$.

Applying Eq. 3' to Eqs. 11 and 15 gives the final input-output relation:

$$\mathcal{S}(x - vt) \to \mathcal{R}(x,t)$$
$$= \frac{1}{2\pi}\int e^{i(\xi x + \omega t)} \cdot \mathcal{F}(\xi,\omega) \cdot \mathcal{S}(\xi)\,d\xi$$
$$= \frac{1}{2\pi}\int e^{i(\xi x - \xi vt)}\mathcal{F}(\xi,-\xi v)\mathcal{S}(\xi)\,d\xi. \tag{16}$$

Eqs. 11, 13, and 16 give a complete scheme for the characterization of our system: the transfer function $\mathcal{F}(\xi,\omega)$ may be obtained by measurement of the responses of the system to sinusoidal stimuli. Given $\mathcal{F}(\xi,\omega)$, the response to an arbitrary moving stimulus $\mathcal{S}(x - vt)$ may be obtained by taking the Fourier transform of the stimulus spatial pattern, multiplying by the transfer function, and taking the inverse Fourier transform. In this way, knowledge of the transfer function $\mathcal{F}(\xi,\omega)$ serves to completely characterize the system.

It is useful to assume that, in addition to being homogeneous (spatially invariant), the eye under study is isotropic, in the sense that the eye is indifferent to reflections about the test ommatidium ($x = 0$). Equivalently, the impulse response function shows the symmetry $\mathcal{I}(-x,t) = \mathcal{I}(x,t)$. This induces certain useful symmetries in $\mathcal{F}(\xi,\omega)$. For example, we have, from Eq. 12 that

$$\mathcal{F}(-\xi,\omega) = \iint e^{-i(-\xi x + \omega t)}\mathcal{I}(x,t)\,dx\,dt$$
$$= \iint e^{-i(\xi(-x) + \omega t)}\mathcal{I}(x,t)\,dx\,dt$$

$$= \iint e^{-i(\xi u + \omega t)}\mathcal{I}(-u,t)\,du\,dt \tag{17}$$
$$= \iint e^{-i(\xi u + \omega t)}\mathcal{I}(u,t)\,du\,dt = \mathcal{F}(\xi,\omega).$$

Now consider

$$\overline{\mathcal{F}(-\xi,-\omega)} = \overline{\iint e^{-i(-\xi x - \omega t)}\mathcal{I}(x,t)\,dx\,dt}$$
$$= \overline{\iint e^{i(\xi x + \omega t)}\mathcal{I}(x,t)\,dx\,dt}$$
$$= \iint e^{-i(\xi x + \omega t)}\mathcal{I}(x,t)\,dx\,dt \tag{18}$$
$$= \mathcal{F}(\xi,\omega),$$

where the horizontal bars indicate complex conjugates, and where we have used the fact that $\mathcal{I}(x,t)$ is real.

Together Eqs. 17 and 18 imply

$$\mathcal{F}(\xi,-\omega) = \overline{\mathcal{F}(\xi,\omega)}. \tag{19}$$

The symmetry (Eq. 17) allows an important experimental simplification. We have:

$$e^{i(\xi x + \omega t)} \to \mathcal{F}(\xi,\omega)e^{i(\xi x + \omega t)}$$
$$e^{i(-\xi x + \omega t)} \to \mathcal{F}(-\xi,\omega)e^{i(-\xi x + \omega t)}$$
$$= \mathcal{F}(\xi,\omega)e^{i(-\xi x + \omega t)}.$$

Adding, and dividing by 2, we obtain (using Eq. 2')

$$\mathcal{S}(x,t) = e^{i\omega t} \cdot \cos\xi x$$
$$\to \mathcal{F}(\xi,\omega)e^{i\omega t}\cos\xi x = \mathcal{R}(x,t). \tag{20}$$

Examining the output at $x = 0$ (the test ommatidium), we have

$$\mathcal{S}(x,t) = e^{i\omega t}\cos\xi x$$
$$\to \mathcal{R}(t) \equiv \mathcal{R}(0,t) = \mathcal{F}(\xi,\omega)e^{i\omega t}. \tag{11'}$$

The implication of this equation is that we may determine the transfer function $\mathcal{F}(\xi,\omega)$ by examining the response of the eye to the stationary counterphase grating stimulus $e^{i\omega t} \cdot \cos\xi x$, instead of the drifting sinusoidal grating stimulus $e^{i(\xi x + \omega t)}$. The counterphase stimulus, which consists of a spatial sinusoidal grating, placed with a peak centered over the test ommatidium, modulated by multiplication by a time-varying sinusoidal signal, is especially well suited for experimental use (see below).

Again, by restricting our attention to the output at the test ommatidium, we obtain an analaogous version of Eq. 16:

$$\mathcal{S}(x, vt) \rightarrow \mathcal{R}(t)$$

$$\equiv \mathcal{R}(0, t) = \frac{1}{2\pi} \int e^{i\omega t} \mathcal{F}(\xi, \omega) \mathcal{S}(\xi) \, d\xi. \qquad (16')$$

$$= \frac{1}{2\pi} \int e^{-i\xi vt} \mathcal{F}(\xi, -\xi v) \mathcal{S}(\xi) \, d\xi.)$$

The assumption of linearity is also useful for reasons of experimental convenience. For example, it is unnecessary to present the stimuli of Eq. 11' one at a time. Instead, we may form the linear combination stimulus

$$\mathcal{S}(x, t) = \sum_n c_n e^{i\omega_n t} \cos \xi x,$$

where the c_n are constants at our disposal and the ω_n are the temporal frequencies at which $\mathcal{F}(\xi, \omega)$ is sought. By Eq. 2', we have, for this stimulus,

$$\mathcal{S}(x, t) = \left(\sum_n c_n e^{i\omega_n t} \right) \cdot \cos \xi x$$

$$\rightarrow \mathcal{R}(t) = \sum_n c_n \mathcal{F}(\xi, \omega_n) e^{i\omega_n t}. \qquad (21)$$

For the fixed spatial frequency ξ, the transfer function can easily be recovered from this composite response with the aid of Fourier methods (see below). Indeed, the frequencies ω_n can be so chosen that all the second harmonics are distinct from the input frequencies. With such a choice of the ω_n, the (presumably negligible) response detected at second harmonic frequencies can be used as a simple monitor of the linearity of the system (Victor et al., 1977). Such a choice of test stimulus has two advantages in the present situation. First, it substantially reduces the data acquisition time necessary to characterize the system response over the grid of spatiotemporal frequency points. Second, because such a stimulus contains modulation over a substantial range of temporal frequencies, it prevents the occurrence of "phase locking," a distinctly nonlinear phenomenon which affects many neural encoders driven predominantly at a single frequency (Knight, 1972 a). By choosing the weighting coefficients c_n roughly reciprocal to the anticipated magnitude $|\mathcal{F}(\xi, \omega_n)|$ of the transfer function at each frequency ω_n, one produces a response with roughly equal output power at each frequency ω_n. Under such circumstances, phase locking becomes exceedingly unlikely. Furthermore, this procedure optimizes the signal-to-noise ratio at each frequency.

In the preceding analysis, we have described the prediction of a response, $\mathcal{R}(t)$ from the knowledge of the stimulus $\mathcal{S}(x, t)$ and a spatiotemporal transfer function $\mathcal{F}(\xi, \omega)$, which may be determined by measuring the response to particular stimuli. For such analysis, the output function $\mathcal{R}(t)$ is in general a continuous function of time, and the input is a continuous function of space and time. In our experimental situation, the input variable is straightforward. $\mathcal{S}(x, t)$ specifies the illumination incident on the eye at the position x, and time t. The output variable is more problematical. As described above, we limit our attention to the response of a single "test ommatidium," whose electrical activity is monitored as a train of discrete action potentials. In order to interpret this sequence of discrete events as a continuous function of time, we invoke a notion of "mean impulse density." To frame the definition of this output variable, we consider an ensemble of N statistically independent replicas of the test ommatidium, each presented with the same stimulus. Over any brief time interval $(t, t + dt)$, we can determine the number, $m_N(t, t + dt)$, of impulses occuring in the entire ensemble of N elements. As N increases, we expect to find at least a few impulses over even very short time intervals, at least in the absence of phase-locking. We define the mean impulse density $r(t)$ as a normalized limit of such impulse counts:

$$r(t) \, dt \equiv \lim_{N \to \infty} \frac{1}{N} m_N(t, t + dt). \qquad (22)$$

Under our assumptions that the eye is dynamically homogeneous, the mean impulse density can be identified with the "population firing rate" (Knight, 1972 a).

In practice, only one specimen of the test ommatidium is available, precluding direct application of the definition (Eq. 22). We may assume however, that the responses of this single unit to successive presentations of the same stimulus are statistically independent, and replace the average in Eq. 22 with an average over repetitions of the same stimulus. Even in this case, however, direct calculation of $r(t)$ from the definition requires a great deal of data, and many procedures have been advocated for its optimization (see Appendix B). Fortunately, for our purposes, an explicit calculation of $r(t)$ is unnecessary; we need only its Fourier components at various frequencies. A least-squares method for estimating these parameters directly from the list of impulse occurrence times is discussed in Appendix B. With this method, we obtain our transfer function $\mathcal{F}(\xi, \omega)$ in terms of the transduction from light intensity to mean impulse density.

For the purpose of displaying the measured response to moving patterns, however, an explicit function of time is needed. Here, we have found an alternative output variable, the "mean instantaneous rate" to be convenient. We begin by defining, for any one sample of impulse train data, an "instantaneous rate" function $s(t)$. (This function, which we define for all times t, generalizes the usual definition, which applies only those times at which impulses occur.) At any time t, $s(t)$ is defined as the reciprocal of the duration of the interval between impulses in which the time t falls. We have

$$s(t) = \sum_n \chi_{(t_n, t_{n+1}]}(t) \cdot \frac{1}{t_{n+1} - t_n}, \qquad (23)$$

where t_n is the time of occurrence of the n'th impulse, and

$$\chi_{(a, b]}(t) = \begin{cases} 1, & a < t \leq b \\ 0, & t \leq a \quad \text{or} \quad t > b. \end{cases}$$

Note that, for any time t, only one term in the summation in Eq. 23 is non-zero. Considering now M presentations of the identical stimulus, and denoting by $s_m(t)$ the instantaneous rate function observed in response to the m'th stimulus presentation, we define $\sigma(t)$, the mean instantaneous rate:

$$\sigma(t) \equiv \langle s_m(t) \rangle = \frac{1}{M} \sum_{m=1}^{M} s_m(t). \qquad (24)$$

Thus, the mean instantaneous rate at any time t is the average, over all repetitions of the stimulus, of the reciprocals of the lengths of the intervals between impulses in which the timepoint t happens to fall. If the overall impulse rate, v, is very fast compared to the time scale of the response under study, the alternative output functions $r(t)$ and $\sigma(t)$ will be very similar; however, at low impulse rates or high modulation frequencies, the functions differ. One may nevertheless compare data given as $\sigma(t)$ with predictions derived from a transfer function obtained in terms of $r(t)$ by incorporating a transfer function which describes the transduction from one output variable to the other.

We now calculate such a transfer function, under the assumption that the modulation is sufficiently small that we may employ a perturbation analysis. It is convenient to break the overall transduction $r(t) \rightarrow \sigma(t)$ into two pieces: first, the transduction from mean impulse density to instantaneous rate, and then from instantaneous rate to mean instantaneous rate: $r(t) \rightarrow s(t) \rightarrow \sigma(t)$. The first transduction, from a population rate to a single unit rate, has been treated in great generality elsewhere (Knight, 1972 a); we simply cite the result: if each of an ensemble of identical neurons encodes a continuous signal into a sequence of impulses according to a "deterministic" law such that, for any given stimulus, t_{n+1} is a monotonic (steadily increasing) function of t_n, then the transduction from population rate to the instantaneous rate of a single unit corresponds to the transfer function

$$B(\omega, v) \equiv \frac{1 - e^{-i\omega/v}}{i\omega/v}. \qquad (25)$$

To analyze the second transduction, from instantaneous rate to a mean instantaneous rate, we consider an encoder producing a sequence of impulses with sinusoidally modulated instantaneous rate. If such a device produces an impulse at time t, we have

$$s(t) = v + \varepsilon e^{i\omega t}, \qquad (26)$$

where v is the mean rate, and $\varepsilon \ll v$. If an impulse does not occur at the time t, $s(t)$ takes its value from the following impulse:

$$s(t) = v + \varepsilon e^{i\omega t_{n+1}}, \qquad t_n < t \le t_{n+1}. \qquad (26')$$

To obtain $\sigma(t)$, we average Eq. 26' over all those times t_{n+1}, at which an impulse could immediately follow the time t. To first order, we have

$$\sigma(t) = \langle s(t) \rangle = \frac{1}{T} \int_t^{t+T} (v + \varepsilon e^{i\omega t_{n+1}}) \, \mathrm{d}t_{n+1}$$

$$= v + \frac{\varepsilon}{T} \int_0^T e^{i\omega(t+\tau)} \, \mathrm{d}\tau \qquad (27)$$

$$= v + \varepsilon e^{i\omega t} \cdot \frac{1}{T} \int_0^T e^{i\omega\tau} \, \mathrm{d}\tau,$$

where $T \equiv 1/v$ is the mean interval between impulses. Subtracting the constant term v and dividing by the input $e^{i\omega t}$ yields the transfer function from $s(t)$ or $\sigma(t)$:

$$\frac{1}{T} \int_0^T e^{i\omega\tau} \, \mathrm{d}\tau = \frac{e^{i\omega T} - 1}{i\omega T} = \frac{e^{i\omega/v} - 1}{i\omega/v} = \overline{B(\omega, v)},$$

namely the complex conjugate of $B(\omega, v)$, as defined in Eq. 25.

Now, if we start with a signal of the form $r(t) = A e^{i\omega t}$, where A is any (complex) number, then the corresponding $s(t)$ signal will be given by $s(t) = A \cdot B(\omega, v) e^{i\omega t}$, and $\sigma(t) = A \cdot B(\omega, v) \cdot \overline{B(\omega, v)} e^{i\omega t} = A \cdot |B(\omega, v)|^2 e^{i\omega t}$. Taking A to be the spatiotemporal transfer function $\mathscr{F}(\xi, \omega)$ which relates stimulus $\mathscr{S}$ to the output variable r, we conclude

$$\mathscr{S}(x, t) = e^{i(\xi x + \omega t)} \rightarrow r(t) = \mathscr{F}(\xi, \omega) e^{i\omega t} \quad \text{implies}$$
$$\mathscr{S}(x, t) = e^{i(\xi x + \omega t)} \rightarrow \sigma(t) = |B(\omega, v)|^2 \mathscr{F}(\xi, \omega) e^{i\omega t}. \qquad (28)$$

Thus, if the transfer function in terms of $r(t)$ is $\mathscr{F}(\xi, \omega)$, then the transfer function in terms of $\sigma(t)$ is simply $|B(\omega, v)|^2 \cdot \mathscr{F}(\xi, \omega)$.[1]

For our *Limulus* data, the factor $|B(\omega, v)|^2$ has only a very slight effect on the results of the Fourier synthesis calculations. This is a consequence of the fact that our moving stimuli contained very little spectral power in the high frequency regime, where $|B(\omega, v)|^2$ differs significantly from unity. In other circumstances, this correction would have a more significant effect.

We note that this linear analysis applies, strictly speaking, only to the variations of the various functions about their mean level (light intensity, on input; and impulse rate, on output). This restriction is necessary, in part, because our experimental variables do not take on negative values. Furthermore, when considered as a function of absolute stimulus intensity, the response of most sensory transducers, including the *Limulus* eye, is decidedly nonlinear, and exhibits a nearly logarithmic response to time-independent stimuli which vary over many orders of magnitude (Stevens, 1975).

Finally, over periods comparable to our experimental trials, many transducers, including the *Limulus* eye, show considerable fatigue or adaptation, in violation of our assumption of stationarity. Such systems may be treated by linear methods only in terms of their fluctuations about a mean operating level which has been adjusted to include the effects of short-term

---

[1] Our notation here recognizes the fact that in this paper we deal only with real ω; more generally, for complex ω, the transfer function is given by the analytic expression $B(\omega, v) B(-\omega, v) \mathscr{F}(\xi, \omega)$.

adaptation. Thus, in the present study, to compare responses with linear predictions, the mean level and adaptation rate were measured, subtracted from the output signal, and then added back to the results of the linear calculation. We made no attempt to predict these mean output levels.[2]

Materials and Methods

Stimulus

Patterns of light, varying in space and time, were formed on the screen of a large oscilloscope (Hewlett-Packard model 1321A High Speed Graphic Display, Hewlett-Packard Co., Palo Alto, Calif.) using analog voltages produced by a system of circuits designed for this purpose (Shapley and Rossetto, 1976), and digital-to-analog converters (DAC) controlled by a PDP 11/45 computer (Digital Equipment Corps., Maynard, Mass.). This time-varying pattern was then imaged by a high-quality camera lens (Nikon Nikkor f/1.2, focal length 55 mm, Nikon, Garden City, N. Y.) onto the flat surface of a fiber-optic taper, which was glued to the cornea of a *Limulus* eye. This fiber-optic device, supplied according to our specifications by Walter P. Siegmund of the American Optical Corp., Southbridge, Mass., conveyed the visual stimulus to the curved array of *Limulus* photoreceptors. The details of the optical system are discussed in Appendix A. The overall effect of this optical system was to convert a pattern 15 cm wide and 2 cm high on the oscilloscope face to a stimulus 1 cm wide and 0.13 cm high on the corneal surface. The height of the image, roughly one-fourth of the height of the eye, was chosen as a compromise between the theoretically desirable goal of illuminating the entire eye, and the need to illuminate a sufficiently narrow band to get an acceptably high impulse rate. (Larger stimuli produce low impulse rates by producing more lateral inhibition, and by spreading the same photon flux over a greater number of ommatidia).

For all stimuli produced, a high-frequency triangle-wave was applied to the Y-input of the display oscilloscope; the X-input was driven with a sawtooth waveform generated by the computer, and the Z-input (intensity) was driven by computer-generated signals synchronized to the X-input sawtooth. This arrangement produced a rectangle of light on the screen, whose intensity varied with horizontal position and time, but whose intensity was independent of vertical position. Three types of stimuli were used: a "setup" stimulus to align the stimulus coordinates with the test ommatidium; an "analysis" stimulus for measuring the spatiotemporal transfer function; and a "synthesis" stimulus which consisted of a uniformly drifting pattern of illumination.

The "setup" stimulus consisted of a single bright vertical line at $x = 0$ surrounded by a uniform dim background; this stimulus did not vary with time. This pattern was manually

moved across the face of the oscilloscope until the bright line was centered on the test ommatidium, as indicated by monitoring its impulse-train discharge. This adjustment was generally reproducible to within ± 0.004 eyewidths,[3] and proved more than adequate for the purposes of these experiments.

The "analysis" stimulus was produced by using for the Z-input of the display oscilloscope a signal which consisted of a constant offset plus the analog product of three computer-generated signals: a constant voltage (to control the total contrast); a "temporal modulation" signal whose value was changed before each sweep of the X-input sawtooth (each sweep lasted 0.01536 s; this yielded a temporal sampling rate of 65.1 Hz); and a "spatial modulation" signal, a rapid sinusoidal modulation synchronized to the X-input sawtooth so as to produce a $\cos \xi x$ spatial pattern. (This is shown schematically as a block diagram in Fig. 1.) The X-sawtooth was obtained by rapidly producing 256 successive equally spaced voltage values with a DAC. This provided a spatial sampling mesh of 256 points/eye width (which corresponds to eight points per cycle at the highest spatial frequency used). The temporal modulation signal was a sum of eight sinusoids with frequencies 0.1, 0.233, 0.5, 1.033, 2.1, 4.233, 8.5, and 17.033 Hz.[4] The relative amplitudes of these components were 60, 50, 45, 30, 15, 10, 20, and 40, respectively.

The "systhesis" stimulus was generated in a similar manner, by holding the "temporal modulation" signal constant, and by producing the spatial modulation by sending to the DAC successive numbers from a list of intensities which described the arbitrary spatial pattern, synchronized to the X-sawtooth. By progressively shifting the phase of the intensity list with respect to the sawtooth, the pattern was made to drift across the screen at any desired rate.

By reversal of the order in which the X-sawtooth voltages were read out by the DAC, it was possible to reflect the synthesis stimulus about the test ommatidium, in order to verify our assumptions concerning the symmetry of the eye, and also to check the accuracy of the alignment of the stimulus origin with the test ommatidium. For the synthesis stimuli, spatial resolution was upgraded to 512 points/eye width, at the expense of lowering the sweep rate to 39.1 Hz. (For comparison, the comparable temporal resolution for commercial television is 30 Hz in the U.S., 25 Hz in Europe; movies are typically shown at a frame rate of 24 Hz.) Spatial patterns were fixed at 2.0 eye widths in length, and were presented in a periodic fashion through a "window" one eye width wide, so that at high drift velocities, the pattern

---

[2] In the *Limulus* visual system, the use of signals consisting of perturbations about a mean level also allows us to ignore in the analysis the phenomenon of inhibitory thresholds (Hartline and Ratliff, 1957), a decidedly nonlinear effect. See also below.

[3] In our experimental apparatus, the width of the *Limulus* eye under study is the most natural unit for the horizontal coordinate. For a typical eye, this may be converted as follows: 1.0 eye widths = 1.0 cm = 40 ommatidial diameters (in the horizontal direction). The conversion to visual angle (in the horizontal direction) is somewhat complicated, because even though the ommatidia are rather evenly spaced, their optical axes diverge nonuniformly. Thus, in the center of the eye, an ommatidium subtends $\sim 6°$ of visual angle, but the whole eye (40 ommatidia in width) covers a total visual angle of only $\sim 200°$.

[4] $(2^{n+1} - 1)/30$, in Hz.

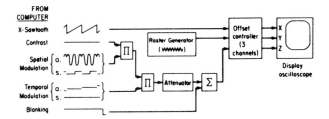

Figure 1 Block diagram of spatiotemporal stimulus generation. Π indicates analog multipliers; Σ indicates analog summer. The raster generator produces a free-running 100 KHz triangle wave for the Y-axis. The attenuator is a voltage divider set by hand to adjust overall contrast level. The offset controller adds analog signals to manually adjusted constant voltages. Three channels are provided: X and Y offset moves the stimulus origin on the oscilloscope face; Z offset adjusts mean illumination level. Two cycles of computer-generated input are shown at left (a. denotes analysis episodes; s. denotes synthesis episodes). Note spatial and temporal modulation are synchronized to X-sawtooth. Constant "contrast" voltage allows computer control of contrast level. "Blanking" signal darkens oscilloscope between experimental episodes.

was seen several times during the course of an experimental episode.

The response of the display oscilloscope to these analog control signals was calibrated with a silicon photocell. At typical mean operating levels, the response was linear up to ~40% contrast;[5] experiments were performed at total contrasts of <35%. No attempt was made to calibrate the absolute intensity of the stimulus, because of the high variability of the optical density of the *Limulus* cornea between specimens.

The Biological Preparation

Adult horseshoe crabs, *Limulus polyphemus*, measuring 15–20 cm across the carapace, were obtained from Gulf Specimens Inc., Panacea, Fla. The animals were kept in filtered artificial seawater at 10° C. They were generally used within 6 wk of delivery, during which time they were not fed. Animals selected for use had "clear" eyes, with no perceptible abrasion of the cornea, and in an informal "neurological exam" they demonstrated brisk, vigorous flexion of the hinge muscle after noxious stimulation of the gill. In general, the speed and strength of the hinge muscle reflex appeared to correlate well with the health of the retina.

Recordings of neural activity were made using an *in situ* preparation (Corning et al. 1965; Biederman-Thorson and Thorson, 1971; Adolph, 1971; Kaplan and Barlow, 1975). In brief, the animal was secured to a wooden board on top of a manipulator which allowed the animal to be rotated and tilted. The gills were placed on a paper towel moistened with the seawater in which the animal had been living. The animals were always capable of vigorous motion when removed

[5] Contrast = {(peak intensity) − (trough intensity)}/{(peak intensity) + (trough intensity)}.

from the apparatus, often after as long as 18 h. A surgical trephine was used to cut a hole 2 cm in diameter in the carapace, about 3 cm anterior to the animal's right eye, above the optic nerve. The nerve was transected, dissected free, and pulled into a small recording chamber, which was then screwed into the carapace. The chamber was filled with seawater, and the nerve dissected with glass needles until a strand which contained a single functioning axon was obtained (Hartline and Graham, 1932). This strand was laid on a cotton wick-silver/silver chloride electrode. The signal from the electrode was amplified and filtered by a differential amplifier, and monitored via oscilloscope and loudspeaker.

The temperature of the crab was measured by means of a thermistor probe (Yellow Springs Instrument Co., Yellow Springs, Ohio) inserted in a hole placed medial to the animal's left eye. The animal's temperature was controlled by coupling it to a constant-temperature circulator (Lauda Div., Brinkmann Instruments, Inc. Westbury, N.Y.) with a modified ice-bag. Eye temperature was held at 22° C, and typically varied <0.5° C over the course of a 10-h experiment. This elevated temperature was chosen because it raises the mean impulse rate (Adolph, 1973) and enhances the response to flickering light (Brodie, 1978).

Data Acquisition

The amplified signal from the wick electrode was converted to a train of uniform pulses by a discriminator. These pulses served as input to the PDP 11/45 computer, which recorded the successive intervals between impulses in a file on magnetic disk for later analysis. Resolution was 10^{-4} s, and the same clock was used to time impulse arrivals as was used to generate the time-varying stimuli. The uniform pulses were also used to drive a "hyperbolic sweep" monitor, which gave a visual indication of instantaneous rate. (This device employed a new digital design by M. Rossetto, and replaced the analog circuit of MacNichol and Jacobs, 1955).

Protocol

The experimental schedule consisted of 60-s periods of illumination in alternation with 90-s periods of darkness. This episode pattern was designed to maintain the eye in a uniform state of light adaptation, over the duration of the experiment, as estimated by the total number of neural impulses produced in each episode. Successive episodes alternated between the analysis stimulus (a sinusoidal grating in space modulated temporally by a sum-of-sinusoids signal) and the synthesis stimulus (a pattern of light drifting across the eye at a constant speed). A stimulus cycle consisted of 16 episodes: analysis episodes at each of eight spatial frequencies (1/10, 1, 2, 4, 8, 16, 20, and 32 cycles/eye width) interleaved with two presentations of the synthesis stimulus at each of four drift velocities, one presentation in each direction.

These stimulus cycles (which lasted 40 min) were repeated indefinitely until the nerve fiber ceased conducting impulses. Experiments typically lasted at least 6 h, and occasionally as long as 10 h. Nerve conduction failures were as a rule the result of the drying out of the exposed portion of the optic nerve; activity was readily obtained from more proximal portions of the nerve. We are thus confident that the retina did not significantly deteriorate over the course of an experiment.

Computations

All computations were based on data from the last 50 s of each 60-s episode, well after the initial on-transient had decayed. The spatiotemporal transfer function $\mathscr{F}(\xi, \omega)$ was obtained from the analysis episodes by means of a least-squares fitting algorithm, as described in Appendix B. This procedure is equivalent to ordinary discrete Fourier analysis of binned (histogram) data, with arbitrarily narrow bins; equivalently, it yields the spectrum of the impulse train interpreted as a series of δ-functions. The algorithm is particularly suited to the handling of pooled data from episodes with identical stimuli.

For each spatial frequency, the algorithm determines real numbers r_n so that the function $f(t) = \Sigma r_n f_n(t)$ best approximates the response $r(t)$ (in a certain least-squares sense; see Appendix B), where the f_n are the functions 1, $t - t_m$ (t_m is the midpoint of the data collection period), $\sin \omega_m t$, $\cos \omega_m t$, $\sin 2\omega_m t$, and $\cos 2\omega_m t$ (where the ω_m are the input frequencies). The coefficient of the function 1 gives the mean impulse rate over the episode; the coefficient of the function $t - t_m$ (the "ramp slope") describes the slow decay of the impulse rate over the course of an episode. As described above, these parameters are ignored in the remaining analysis, but are added back to the Fourier synthesis at the end of the calculation. The coefficients of the $\sin \omega_m t$ and $\cos \omega_m t$ terms determined the value of $\mathscr{F}(\xi, \omega_m)$ (where ξ is the spatial frequency of the stimulus which produced the particular data being analyzed): the amplitude of $\mathscr{F}(\xi, \omega_m)$ is the square root of the sum of the squares of these two coefficients, while the tangent of the phase of $\mathscr{F}(\xi, \omega_m)$ is given by their quotient. The coefficients of $\sin 2\omega_m t$ and $\cos 2\omega_m t$ measure nonlinearities in the response of the system. The entire set of coefficients was determined for the pooled data from all the analysis episodes at each spatial frequency.

The phase information for the transfer function at each spatiotemporal frequency pair (ξ_m, ω_m) was obtained, as described above, by taking the arctangent of a quotient; the computer expressed these results as real numbers between $-\pi$ and π. These phase data were individually adjusted by a multiple of 2π so as to obtain continuous phase curves. Thus, values of $\mathscr{F}(\xi, \omega)$, as amplitude and phase, were obtained at the 64 spatiotemporal frequency points corresponding to all the possible combinations of the eight spatial and eight temporal frequencies present in the analysis stimuli. In order to estimate $\mathscr{F}(\xi, \omega)$ for (ξ, ω) between the points of the input lattice, a two-dimensional cubic spline was used. For this purpose, the transfer function data were expressed in terms of two separate real functions (log amplitude, and phase) of the variables $\log \xi$ and $\log \omega$. The complex transfer function $\mathscr{F}(\xi, \omega)$ was then reconstructed from the amplitude and phase.

To avoid artifacts due to abrupt frequency cutoffs, the transfer function was extrapolated beyond the spatiotemporal frequency lattice at which it had been measured. The extrapolation to high spatial frequency extended the observed attenuation of amplitude seen at spatial frequencies above 20 cycles/eye width; for each spatial frequency, the amplitude was fixed as a small constant multiple of the amplitude observed at the highest spatial frequency where measurements were made. Phases were extrapolated by setting them equal to the phases measured at the highest spatial frequency used. The Fourier syntheses were insensitive to the details of these high-frequency extrapolations. It was unnecessary to extrapolate to low spatial frequency, as the data extended down to 0.1 cycles/eye width. The high temporal frequency extrapolations were provided as approximate continuations of the typical observed high-frequency roll-off in amplitude and phase.

For the extrapolation to low temporal frequencies, the low-frequency transfer-function measurements of Biederman-Thorson and Thorson (1971) were used as a guide. Under experimental conditions quite similar to ours, they measured temporal transfer functions from 0.4 Hz down to 0.004 Hz. In this regime they found that the transfer function could be expressed as $\mathscr{F}(\omega) = K \cdot (i\omega)^p$ where p is a real exponent between 0.18 and 0.27 (mean 0.23), and K is a real constant of proportionality. For simplicity we adopted the exponent $p = 0.25$, and extrapolated the transfer functions accordingly, fixing the proportionality constant by the amplitude observed at the lowest frequency where measurements were available (0.1 Hz), and extrapolating the phase to the low-frequency phase lead of $\pi/8$ radians implied by the exponent $p = 1/4$. Although the very low frequency features of the Fourier syntheses were not insensitive to the details of the low temporal frequency extrapolation, this parameter-free procedure produced no systematic discrepancies between experiment and prediction in the very low frequency range.

The synthesis episodes were treated differently. The data from episodes with identical stimuli were averaged together by computing the instantaneous rate function $s(t)$ for each episode, and then averaging these functions on a mesh of 1,024 equally spaced points covering the 50-s episode length. The resultant averaged response function, $\sigma(t)$, was plotted on a digital plotter (CalComp 565, California Computer Products, Inc., Anaheim, Calif.) for later comparison with the Fourier syntheses.

The synthesis stimuli were presented in two directions at each velocity. As these pairs of stimuli consisted of reflections of each other about the test ommatidium, they served to verify our assumptions about the symmetry of the inhibitory fields, and to verify the accurate placement of the stimulus origin so as to coincide with the test ommatidium. In all cases but one, the responses to the two mirror-image stimuli appeared nearly identical (see Results, below). This observation permitted us to further increase the signal-to-noise ratio by averaging together all the synthesis episodes at each velocity, thus combining the response to each synthesis stimulus with the response to its mirror image.

The Fourier synthesis computations were done in Fortran complex arithmetic with the Fast Fourier Transform algorithm (FFT) on a mesh of 1,024 equally spaced points covering the response to one full period of the periodic synthesis stimulus. (Thus, for rapidly drifting patterns, each experimental record provided several repetitions of the response to the moving stimulus. Except for the slow drift in mean impulse rate, these repetitions should, in principle, be identical.) The translation of the synthesis formula (Eq. 16′) (derived above in terms of Fourier integrals) into a form suitable for use with the (discrete) FFT is essentially straightforward. The computation consisted of filling an array with the transform of the stimulus pattern (obtained either analytically or by FFT), multiplying by the transfer function, and inverting by FFT.[6] The time-stationary (periodic) portion of the response was then available as the real part of this result. The mean impulse rate and "ramp" (describing the slow drift of the impulse rate over the course of an episode) were added to the periodic response, and the sum was plotted in a form compatible with the plots of the averaged synthesis episodes for direct visual comparison. This calculation was repeated for each stimulus drift velocity.

We wish to emphasize that this entire calculation, from the measurements of the transfer function to the calculation of the Fourier synthesis prediction, allowed no adjustment of free parameters. The prediction is explicitly and unambiguously determined by the measured transfer function, mean impulse rate, and ramp slope.

Results

The outcome of the analysis portion of the protocol is depicted in Figs. 2 and 3. Fig. 2 shows the average instantaneous rate function $\sigma(t)$ for the response to a typical analysis episode. It is important to note that, contrary to its noisy appearance, such a record is in fact a definite response to a fixed temporal signal, albeit a harmonically rich signal. In Fig. 3, the marked dependence of the response on the spatial frequency of the analysis stimulus is illustrated. At low spatial frequencies, a moderate response to the temporal modulation of the stimulus is observed. As the spatial frequency is increased, an increase in the response to flicker is apparent. The peak sensitivity is around four cycles/eye width. At high spatial frequencies, the response decreases, until, at 32 cycles/eye width it is essentially undetectable. It may be noted that, as all the analysis episodes share a common temporal modulation signal, the records at different spatial frequencies

[6] Because of the phenomenon of "aliasing," the FFT may be interpreted equivalently as operating either on the array of frequencies $0, \xi, 2\xi, \ldots, (2^N - 1)\xi$, or on the array $0, \xi, 2\xi, \ldots, 2^{N-1}\xi, -(2^{N-1} - 1)\xi, -(2^{N-1} - 2)\xi, \ldots, -2\xi, -\xi$. For our purposes, the second interpretation is the correct one, because the power in the negative-frequency components greatly exceeds that in the positive high-frequency components, due to the high-frequency cutoff of both the stimulus $\mathscr{S}(\xi)$ and the transfer function $\mathscr{F}(\xi, -\xi v)$. The transfer function for negative frequencies is obtained from the symmetry relations (Eqs. 17–19).

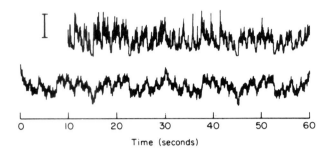

Figure 2 Response of test ommatidium to temporally modulated sine-wave grating. The bottom record shows the sum-of-sinusoids temporal signal used to modulate a sinusoidal grating stimulus over each 60-s episode. The top record shows the average instantaneous rate response $\sigma(t)$ from 14 repetitions of such an analysis stimulus, with a spatial frequency of 4 cycles/eye width. The data from the first 10 s of each episode were discarded to avoid the effect of the initial on-transient. Scale marker: 10 impulses/s.

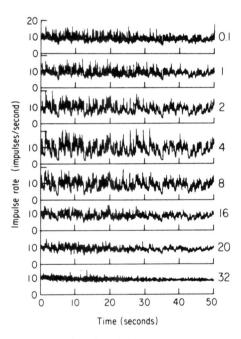

Figure 3 Effect of spatial frequency on response to temporally modulated sine-wave gratings. Each record is the average of 14 episodes. The stimulus considered of a sinusoidal grating (spatial frequencies shown at right) placed with a peak centered over the test ommatidium, modulated according to the temporal signal shown in Fig. 2.

show corresponding features at corresponding time points.

These effects are specified quantitatively in Fig. 4, which shows the full spatiotemporal transfer function derived from the same preparation as in Fig. 3. Though the general trend of the curves agrees with the description above, the following features may be noted. The relative sensitivity of the eye to sine-wave gratings of differing spatial frequency depends strongly on temporal frequency. Thus, at low temporal frequency, the response is greatest at intermediate spatial frequency falling off gently at low spatial frequency, and sharply at high spatial frequency. At intermediate temporal frequency, the eye is most responsive at low spatial frequency, with response decreasing monotonically with increasing spatial frequency. At high temporal frequency, there is little dependence on spatial frequency, except for the ultimate high-frequency cutoff. These findings may be considered the analog, with sine-wave gratings, of the more familiar "small spot/large spot" experiments (Ratliff et al., 1967; Ratliff et al., 1969; Knight et al., 1970).

The spatial dependence of the phase of the transfer function is more subtle, with detail concentrated at the lower temporal frequencies. The low-frequency phase lead is slightly greater at low spatial frequencies and persists to higher temporal frequencies. Once it starts, however, the rate of increase of the phase lag with temporal frequency is greater at low spatial frequency, so that there is little difference in the phase lags seen at all spatial frequencies at high temporal frequency. The full implications of the transfer function measurements are discussed at greater length in the following article (Brodie et al., 1978).

As was discussed above, the Fourier syntheses were computed in terms of the mean instantaneous rate output variable $\sigma(t)$. For this purpose, the transfer function amplitudes, measured in terms of the mean impulse density, $r(t)$, were multiplied by the correction factor $|B(\omega, v)|^2$. The corrected amplitudes are shown in Fig. 4C. The effect of the correction is mainly to attenuate the response at frequencies above the mean impulse rate. The phases are, of course, unchanged.

Typical averaged responses from the synthesis portion of the protocol are shown at the top of Fig. 5. For the experiment shown, the synthesis stimulus consisted of a square wave of spatial frequency 0.5 cycles/eye width, which was moved slowly across the eye. Because the stimulus was viewed by the animal through an effective "window" one eye width across, the stimulus had the appearance of a "step" of light intensity advancing across the screen. As the records from such mirror-image presentation of the stimulus are, in gen-

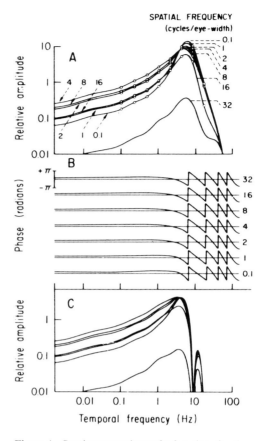

Figure 4 Spatiotemporal transfer functions for the preparation of Fig. 3. (A) Bode plots (log amplitude vs. log frequency) of the fractional modulation of the mean impulse density $r(t)$ for each spatial frequency. The points indicated (○) were obtained from experimental measurements; the remaining portions of the curve were extrapolated as described in the text. (This preparation produced no detectable response at 32 cycles/eye width (see Fig. 2). As a curve for this spatial frequency was needed for computational purposes, it was extrapolated by setting the amplitude at 32 cycles/eye width equal to 10% of the amplitude measured at 20 cycles/eye width.) (B) Phase vs. log frequency is indicated (modulo 2π) on a separate axis for each spatial frequency. The curves were extrapolated in the same regions as the amplitudes, above. (C) The transfer function amplitudes for the same preparation, in terms of the mean instantaneous rate function $\sigma(t)$, obtained by multiplying the transfer function in (A) by the transfer function $|B(\omega, v)|^2$ (see text). The small undulations of the amplitude curves (A) and (C) at low frequency are artifacts of the extrapolation procedure.

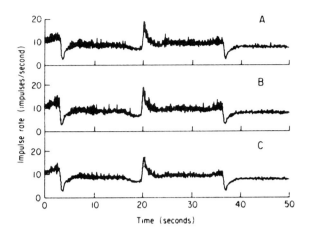

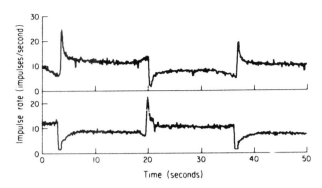

Figure 5 Comparison of the response to mirror-image stimuli. The top two records show the average instantaneous rate response $\sigma(t)$ obtained from 14 presentations of a drifting edge stimulus moving with drift velocity (A) $+0.06$ eye widths/s or (B) -0.06 eye widths/s. The record (C) is the averaged response of all 28 episodes. The preparation is the same as used in Figs. 3 and 4.

Figure 6 Step-transient responses of an ommatidium with an asymmetric inhibitory field. The test ommatidium in this preparation was located within a few ommatidia of the posterior edge of the eye. The anticipatory Mach bands were much more pronounced when the step stimulus moved toward that edge of the eye (top record) than when the stimulus moved away from the edge (bottom record).

eral, nearly identical, we have deemed it appropriate to average together all such responses. Such an averaged response is shown in the bottom of Fig. 5; the improvement in signal-to-noise ratio is evident. The features seen in this record, while to some extent dependent on the drift velocity of the stimulus (here, 0.06 eye widths/s), are common to such step-responses (see below). Of particular interest are the anticipatory "Mach bands"[7] of excitation or inhibition that precede the crossing of the test ommatidium by the moving edge. The crossing itself is seen as a clear on- or off-transient, which then decays, sometimes with a small overshoot, as here. In the intervals between the step transients, the impulse rate settles to a steady-state value. This value is nearly the same, regardless of whether the steady state is a response to the bright or dim region of the step-pattern stimulus.

An example of the effect of an asymmetrical inhibitory field on such records is seen in Fig. 6. In this preparation, the test ommatidium was located within 2 mm of an edge of the eye, and thus received manifestly asymmetrical input from the rest of the retina. Thus, a Mach band typical of the drift velocity (0.06 eye widths/s) is seen when the steps drift toward the edge of the eye, while steps drifting away from the edge can affect very few ommatidia before encountering the test ommatidium, and thus they are scarcely anticipated.

[7] Strictly speaking, the term "Mach bands" refers to the maxima and minima seen in a static stimulus pattern consisting of a gradient of intensity between two uniform areas of different intensity (see Ratliff, 1965). We use the term loosely here to include the maxima and minima in neural responses to stepwise changes in intensity.

Such ommatidia were scrupulously avoided in the rest of the study. Hence, all further figures depicting responses to moving stimuli display the average of responses to mirror-image stimuli without further comment.

The records from synthesis episodes at four different drift velocities are shown in Fig. 7. The responses show a marked dependence on the drift velocity of the stimulus. At very low speeds the step in light intensity takes a significant time to cross the test ommatidium, and there is only a modest transient response to the step stimulus. This transient decays monotonically to a steady response. As the velocity is increased, the transient responses increase dramatically, momentarily driving the unit at over three times its average impulse rate. The inhibitory precursors are somewhat strengthened, but, of course, occupy a shorter interval of time. Immediately after the on-transient responses, the impulse rate falls rapidly, overshooting the subsequent steady response to the bright portion of the step. At high velocity, the off-transients, which would have to extend to "negative" impulse rates to mirror the observed on-transients, are severely truncated.

The predictions of the Fourier synthesis procedure (using the transfer-function data from Fig. 4 B and C) are shown above the experimental records. The agreement between the Fourier predictions and the experimental records is, on the whole, excellent. The linear theory successfully predicts the form and height of the step transients (a sensitive function of drift velocity), and the width and strength of the Mach bands. The limited dependence of the "steady state" response on the intensity of the illumination between step transients

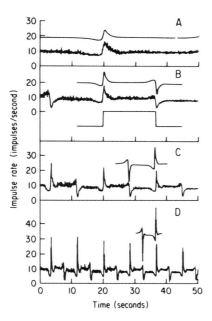

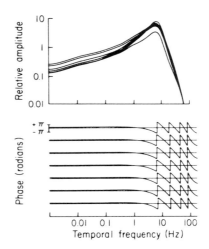

Figure 8 Transfer function for a "sick" *Limulus*. Bode plots of measured spatiotemporal transfer function, plotted as in Fig. 4. At the peak, amplitudes decrease monotonically with increasing spatial frequency.

Figure 7 Fourier synthesis of step-transient responses. The figure shows the observed averaged response $\sigma(t)$ to drifting steps of light intensity (same preparation as Figs. 3, 4, and 5). Drift velocities were (A) 0.03, (B) 0.06, (C) 0.12, and (D) 0.24 eye widths/s. The episodes at the slowest speed provided only one on-transient; the others provided at least one full cycle of the stimulus. The curves offset immediately above the observed records are the predictions, for one cycle of the stimulus, of the Fourier synthesis procedure described in the text, applied to the spatiotemporal transfer function shown in Fig. 4. The intensity of the stimulus at the test ommatidium is shown for one stimulus cycle in (B).

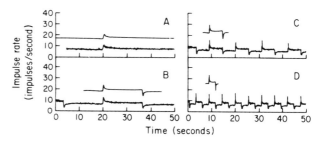

Figure 9 Fourier synthesis for "sick" *Limulus* (same preparation as Fig. 8). Predicted and measured responses (averages of eight episodes) to drifting step stimuli are plotted as in Fig. 7. Drift velocities were (A) 0.03, (B) 0.06, (C) 0.18, and (D) 0.36 eye widths/s.

is also correctly predicted by the Fourier synthesis calculation.

The synthesis shows only a few systematic discrepancies from the actual responses. The slight overshoot of the response at intermediate velocities is somewhat underestimated, and the sculpturing of the on-transient at the lowest velocity is slightly distorted. The biggest discrepancy is the truncation of the off-transients at high drift rates. This highly nonlinear phenomenon is beyond the scope of our linear theory. The truncation also produces secondary effects, such as the absence of overshoot after high-speed off-transients, which likewise are not predicted by the Fourier synthesis.

We have obtained results comparable to those shown above on several other preparations. Further evidence of the extent to which our transfer function measurements characterize the response of a *Limulus* eye to moving stimuli was also obtained.

Figs. 8 and 9 show the results of an analysis-synthesis experiment performed on a *Limulus* with weak and sluggish reflexes; this specimen would not have been used had a healthier one been available. The trans-

fer function shows better optical resolution than that of Fig. 4, with a readily measurable response at 32 cycles/eye width, but very little dependence on spatial frequency. This apparent lack of lateral inhibition is confirmed by the synthesis records, which show virtually no Mach band effects at all. Nonetheless, the agreement between the Fourier synthesis and the experimental records is striking, at velocities ranging over an order of magnitude. It thus appears that our transfer function measurements accurately describe the dynamics of even a somewhat pathological eye.

We have also performed syntheses of the response to drifting patterns other than steps. Figs. 10 and 11 show the results of a synthesis of the response to the "step complement" stimulus of Ratliff and Sirovich (1978). This stimulus, which is composed of a sinusoid of the lowest possible frequency (0.5 cycles/eye width) plus a fast exponential decay superposed on a step,

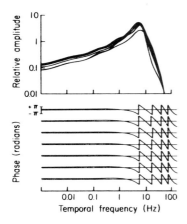

Figure 10 Transfer function for experiment of Fig. 11. At peak, amplitudes decrease monotonically with increasing spatial frequency.

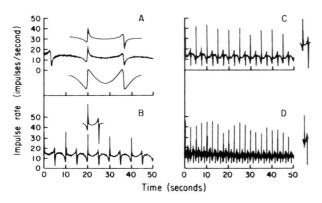

Figure 11 Predicted and measured responses to moving "step complement" stimulus. Once cycle of the stimulus is reproduced as the bottom record in (A). Drift velocities were (A) 0.06, (B) 0.20, (C) 0.40, and (D) 0.80 eye widths/s. Measured responses are the average of 12 episodes.

is designed to resemble the step stimulus as little as possible, yet produce similar visual responses. Even though the stimulus possesses no sharp dicontinuities, the response clearly resembles the typical response to true drifting steps, especially at low velocities. At higher speeds, the eye readily perceives the 0.5 cycles/eye-width sinusoid. All of these responses, up to a velocity of 0.8 eye widths/s, are well predicted by the Fourier synthesis. With the measured spatiotemporal transfer function, we can perform a Fourier synthesis to predict the response of this preparation to a true step stimulus, such as the one used for the experiments of Figs. 7 and 9. These predictions are shown in Fig. 12, for comparison with the response to the step-complement stimulus.

As a final test of the ability of our procedure to

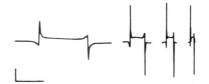

Figure 12 Predicted step responses for preparation of Figs. 10 and 11. Drift velocities, from left to right: 0.06, 0.20, 0.40, 0.80 eye widths/s. Scale marker horizontal, 10 s; vertical, 10 impulses/s.

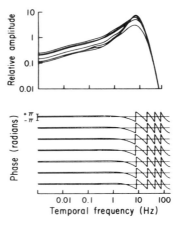

Figure 13 Transfer function for experiment of Fig. 14.

handle "arbitrary" stimuli, we produced a visual stimulus whose light-intensity profile resembles a row of buildings. The results of an experiment using this stimulus are shown in Figs. 13 and 14. The agreement between the predicted and measured responses is again excellent.

Discussion

The extensive agreement between the measured responses to moving stimuli and the predictions of our Fourier-synthetic calculations demonstrates the essential validity of our program of linear systems analysis. Though the major assumption of linearity is known to hold for many aspects of the *Limulus* visual transduction, especially in the vicinity of a fixed operating point, there are important known exceptions, such as the dependence of inhibitory coupling on excitation levels, and the phenomenon of inhibitory thresholds. Our results confirm that, in spite of these potential complications, the *Limulus* system responds with linear behavior well beyond the range of small perturbations.

The only striking nonlinear effect demonstrated in our study is the truncation of off-transients corresponding to the limitation of the pulse-coding scheme.

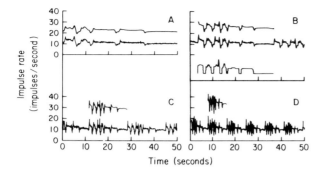

Figure 14 Predicted and measured responses to moving "arbitrary" stimulus. One cycle of the stimulus is reproduced as bottom record in (B). Drift velocities were (A) 0.03, (B) 0.06, (C) 0.12, and (D) 0.24 eye widths/s. Measured responses are the average of 22 episodes.

Our data appear to be consistent with the hypothesis that, under the conditions of widespread illumination and moderate impulse rate, the effective threshold for lateral inhibition (at least for inhibitory transients) is the absolute threshold: the absence of impulses in the inhibiting units. This may be a small limitation in practice: those stimuli in which features which greatly exceed the intensity of the mean illumination are brief and well separated from each other (such as the stimulus of Fig. 14) produce little or none of this truncation effect, especially at low stimulus velocities.

This study provides verification of the behavior of large numbers of interacting neurons in the dynamic situation. We confirm the presence of Mach bands which precede waves of excitation moving across the eye. The relative insensitivity of the *Limulus* retina to slowly changing, or slowly moving, stimuli has been definitively demonstrated. As has often been stated elsewhere, such response characteristics have the effect of accentuating contours and movements in the information passed by the eye to the brain. On the other hand, it should be pointed out that for moderately rich stimuli, the *Limulus* eccentric cell provides a fairly accurate depiction of the stimulus at even rather leisurely drift velocities.

We thus conclude that the spatiotemporal transfer functions, as measured above, in fact serve as concise complete characterizations of the *Limulus* visual system in the regime studied, and therefore that the properties of a *Limulus* retina are well specified by the retina's response to sinusoidal gratings modulated sinusoidally in time. In principle, then, the task of the *Limulus* visual physiologist may be reduced, to a large extent, to predicting and explaining the various features of the measured transfer functions.

Appendix B: On the Choice of Output Variable for the Fourier Analysis of Impulse Train Data

The measurement of the harmonic content of various output variables has proved to be one of the most useful techniques for studying the dynamics of biological systems. When the output variable is a continuous function of time, such as muscle tension or an intracellular "slow potential," the application of these Fourier methods is straightforward. On the other hand, when the output is a train of neural impulses, the information carried by the signal is presumably contained in the impulse occurrence times. In this case no function of continuous time is directly available for Fourier analysis. Several procedures for obtaining such a function have been proposed in the past (for example, French and Holden, 1971; Knight, 1972 *b*; Fohlmeister et al., 1977). We have recently adopted a procedure which determines the Fourier coefficients directly from the impulse occurrence times, without the intermediate calculation of a continuous time function. The method is equivalent to the use of post-stimulus-time histograms ("binning") with arbitrarily narrow bins. This procedure, its advantages, and its weaknesses, are discussed below.

In general, all calculations of Fourier coefficients may be interpreted as the result of a "least-squares" best-fitting procedure. Such a procedure determines those coefficients c_n that minimize a quadratic error estimator of the form:

$$\Delta = \int \{f(t) - \sum_n c_n f_n(t)\}^2 \, dt, \qquad (B1)$$

where the f_n are the functions with which we are attempting to approximate the data $f(t)$. (Here, for simplicity, we suppress the limits of integration and the corresponding division by the length of the interval of integration.) If the f_n are sines and cosines, the c_n are the usual Fourier coefficients, but other choices for f_n are equally suitable. In this context, the various procedures for Fourier analyzing impulse-train data amount to different explicit choices of the algorithm for obtaining $f(t)$, a function of a continuous variable, from the sequence of impulse occurrence times.

Once such a choice has been made, one may readily differentiate Eq. B1 with respect to the c_n; setting the partial derivatives $\partial \Delta / \partial c_m$ equal to zero yields a system of simultaneous equations for the c_n:

$$\int f(t) \cdot f_m(t) \, dt = \sum_n c_n \cdot \int f_n(t) \cdot f_m(t) \, dt, \qquad m = 1, 2 \cdots. \quad (B2)$$

If the f_n are orthogonal (as are the sines and cosines, if integrated over an integral number of periods), the correlation matrix $\int f_n(t) \cdot f_m(t)\mathrm{d}t$ reduces to $\delta_{n,m}$, and we retrieve the usual formula for the coefficients c_n.

The system of Eqs. B2 makes perfectly good sense if we allow $f(t)$ to be any suitably integrable function, even a sequence of Dirac delta-functions,

$$f(t) = \sum_k \delta(t - t_k),$$

where t_k is the time of occurrence of the kth impulse. For this choice of f, we obtain the form:

$$\sum_k f_m(t_k) = \sum_n c_n \cdot \int f_n(t) \cdot f_m(t)\mathrm{d}t, \qquad m = 1, 2, \cdots. \tag{B3}$$

Unfortunately, such an f is not admissible in Eq. B1, because the δ-function is not square-integrable. Nonetheless, a careful limiting argument (in which one approximates the δ-function by a sequence of increasingly taller and narrower rectangular pulses) demonstrates that Eq. B3 correctly calculates the spectral components associated with the impulse shape-independent structure of the impulse train. Under the assumption that only the impulse occurrence times convey information of interest, this is no limitation. We will refer to the set of Fourier components c_n obtained from Eq. B3 as the "delta-function spectrum."

This calculation of the delta-function spectrum has several computational advantages. First the correlation matrix $\int f_n(t)f_m(t)\mathrm{d}t$ can be calculated in advance; it depends only on the functions f_n and the interval over which data are collected. Second, the calculation is linear in the data; thus, if data from several episodes with identical stimuli are to be pooled, one may simply add together the function values $f_m(t_k)$ from all episodes. Third, the algorithm is well suited to on-line data acquisition: if tables of the functions f_n are stored in memory, a pointer incremented in real time can provide rapid access to the function value $f_n(t_k)$, so that whenever an impulse occurs, the current function value is immediately available for addition to a running total. The same function table may even serve to provide a list of successive stimulus values. Such a scheme has recently been implemented with a microprocessor-driven device in our laboratory (Milkman et al., 1978).

Because the delta function spectrum calculation is linear, as described above, it follows immediately that, as the number of pooled impulse trains grows large, the Eqs. B3 approach the continuous system (Eqs. B2), if we make the choice $f(t) = r(t)$, where $r(t)$ is the mean impulse density function as defined in Eq. 22, above.

Another important feature of the formulation above is the provision for nonorthogonal functions f_n. In general, a "ramp" function $f_n(t) = t - t_0$ will not be orthogonal to both a sine and cosine function over any time interval. Similarly, sinusoids of incommensurate periods fail to be orthogonal, as do commensurate sinusoids except over carefully selected time intervals. The flexibility of the system (Eqs. B3) in dealing with such functions greatly facilitates the selection of episode lengths.

The various alternative choices for the function $f(t)$ generally fall into two classes. "Binning" methods are particularly simple to apply. They divide the episode into short successive equal time periods ("bins"), and assign to each such period the number of impulses which occur within it. "Instantaneous rate" methods, which are often useful for impulse trains with few impulses, assign function values equal to the reciprocals of the time intervals between impulses. It is also possible to combine these approaches. For impulse trains varying about a mean carrier rate, all of these functions convey the same information and are simply related by transfer functions, such as that calculated in Eqs. 25–28. Nonetheless, the delta-function procedure outline above has several relative advantages in experimental situations.

First, the delta-function spectrum is highly insensitive to discrimination errors, which may result in erroneously short intervals between impulses. These experimental artifacts greatly distort calculations based on reciprocal intervals, but scarcely perturb the running sums from which the delta-function spectrum is calculated. Indeed, any number of spurious impulses uncorrelated with the periodic stimulus have no systematic effect on the computed spectrum. Second, unlike the results of binning procedures, the delta-function spectrum contains no nulls due to the interaction of a modulation frequency component with the bin width (a parameter which is entirely external to the system under study). Finally, both binning and reciprocal interval methods suffer from phase errors, because the procedures which produce the function $f(t)$ somewhat distort the time at which an impulse is reflected in the spectral estimation; in contrast, the delta-function procedure accurately reflects each impulse at the time when it actually occurs.

In summary, the delta-function spectrum, as determined by the system of Eqs. B3, provides an excellent and easily computed characterization of the harmonic content of impulse train data. The procedure is free from many of the artifacts which affect other methods when applied to laboratory data, and it imposes no arbitrary structure of its own on the data.

Acknowledgment

We thank James Gordon for his assistance with the photographs which appear in Appendix A. This work was supported, in part, by National Institutes of Health grants EY 188, EY 1428, EY 1472, and GM 1789.

References

ADOLPH, A. R. 1971. Recording of optic nerve spikes underwater from freely-moving horseshoe crab. *Vision Res.* **11**:979–983.

ADOLPH, A. R. 1973. Thermal sensitivity of lateral inhibition in *Limulus* eye. *J. Gen. Phsyiol.* **62**:392–406.

BARLOW, R. B. 1967. Inhibitory fields in the *Limulus* lateral eye. Thesis, The Rockefeller University, New York.

BARLOW, R. B., and G. D. LANGE. 1974. A nonlinearity in the inhibitory interactions in the lateral eye of *Limulus*. *J. Gen. Physiol.* **63**:579–589.

BIEDERMAN-THORSON, M., and J. THORSON. 1971. Dynamics of excitation and inhibition in the light-adapted *Limulus* eye *in situ*. *J. Gen. Physiol.* **58**:1–19.

BRODIE, S. E. 1978. Temperature dependence of the dynamic response of the *Limulus* retina. *Vision Res.* In press.

BRODIE, S. E., B. W. KNIGHT, and F. RATLIFF. 1978. The spatiotemporal transfer function of the *Limulus* lateral eye. *J. Gen. Physiol.* **72**.

CORNING, W. C., D. A. FEINSTEIN, and J. R. HAIGHT. 1965. Arthropod preparation for behavioral, electrophysiological, and biochemical studies. *Science (Wash. D.C.).* **148**:394–395.

FOHLMEISTER, J. F., R. E. POPPELE, and R. L. PURPLE. 1977. Repetitive firing: a quantitative study of feedback in model encoders. *J. Gen. Physiol.* **69**:815–848.

FRENCH, A. S., and A. V. HOLDEN. 1971. Alias-free sampling of neuronal spike trains. *Kybernetik.* **8**:165–171.

GAUSS, C. F. 1828. Disquitiones generales circa superficies curvas. *Commentat. Soc. R. Sci. Gott. Recent.* **6**:99–146.

HARTLINE, H. K., and C. H. GRAHAM. 1932. Nerve impulses from single receptors in the eye. *J. Cell. Comp. Physiol.* **1**:227–295.

HARTLINE, H. K., and F. RATLIFF. 1957. Inhibitory interaction of receptor units in the eye of *Limulus*. *J. Gen. Physiol.* **40**:357–376.

HARTLINE, H. K., and F. RATLIFF. 1972. Inhibitory interaction in the retina of Limulus. *In* Handbook of Sensory Physiology, Vol. VII/2, Physiology of Photoreceptor Organs, M. G. F. Fuortes, editor. Springer-Verlag, Berlin. 381–447.

KAPANY, N. S. 1967. Fiber Optics: Principles and Applications. Academic Press, Inc. New York. 429 pp.

KAPLAN, E., and R. B. BARLOW, JR. 1975. Properties of visual cells in the lateral eye of *Limulus in situ*. *J. Gen. Physiol.* **66**:303–326.

KIRSCHFELD, K., and W. REICHARDT. 1964. Die Verarbeitung stationärer optischer Nachrichten im Komplexauge von Limulus. Ommatidien-Sehfeld und räumliche Verteilung der Inhibition. *Kybernetik.* **2**:43–61.

KNIGHT, B. W., J. TOYODA, and F. A. DODGE. 1970. A quantitative description of the dynamics of excitation and inhibition in the eye of Limulus. *J. Gen. Physiol.* **56**:421–437.

KNIGHT, B. W. 1972 *a*. Dynamics of encoding in a population of neurons. *J. Gen. Physiol.* **59**:734–766.

KNIGHT, B. W. 1972 *b*. The relationship between the firing rate of a single neuron and the level of activity in a population of neurons. *J. Gen. Physiol.* **59**:767–778.

MACNICHOL, E. F., and J. A. H. JACOBS. 1955. Electronic device for measuring reciprocal time intervals. *Rev. Sci. Instrum.* **26**:1176–1180.

MILKMAN, N., R. SHAPLEY, and G. SCHICK. 1978. A microcomputer based visual stimulator. *Behav. Res. Methods Instrum.* In press.

RATLIFF, F. 1965. Mach Bands: Quantitative Studies on Neural Networks in the Retina. Holden-Day, Inc., San Francisco. 365 pp.

RATLIFF, F., editor. 1974. Studies on Excitation and Inhibition in the Retina. The Rockefeller University Press, New York. 668 pp.

RATLIFF, F., B. W. KNIGHT, and N. GRAHAM. 1969. On tuning and amplification by lateral inhibition. *Proc. Natl. Acad. Sci., U. S. A.* **62**:733–740.

RATLIFF, F., B. W. KNIGHT, J. TOYODA, and H. K. HARTLINE. 1967. Enhancement of flicker by lateral inhibition. *Science. (Wash. D. C.).* **158**:392–393.

RATLIFF, F., and L. SIROVICH. 1978. Equivalence classes of visual stimuli. *Vision Res.* In press.

SHAPLEY, R., and M. ROSSETTO. 1976. An electronic visual stimulator. *Behav. Res. Methods Instrum.* **8**:15–20.

STEVENS, S. S. 1975. Psychophysics. John Wiley and Sons, New York. 329 pp.

VICTOR, J. D., R. M. SHAPLEY, and B. W. KNIGHT. 1977. Nonlinear analysis of cat retinal ganglion cells in the frequency domain. *Proc. Natl. Acad. Sci., U. S. A.* **74**:3068–3072.

24
Introduction

(1980)
Stephen Grossberg

How does a brain build a cognitive code?
Psychological Review 87:1–51

This long paper gives access to much of Stephen Grossberg's work. It is effectively two papers. In the body of the paper, Grossberg gives qualitative descriptions of many of his ideas. Then, a series of mathematical appendices provides more precise formal descriptions.

Grossberg develops a series of detailed mechanisms in this paper, but most of them are first used in, and arise from, a consideration of error correction. Widrow and Hoff (paper 10) discuss one error correction technique in some detail, and we have seen others discussed more briefly, for example, some of the perceptron variants. Most of these techniques require the use of teachers who know the right answers. The teacher is not discussed in detail except insofar as it assumed both to exist and to have access to certain kinds of information.

One of Grossberg's key points is that the neural network must generally do error correction by itself, without outside help. This places strong constraints on the way the system *can be* wired up. The initial discussion assumes two systems in series, one communicating with the other. (Lateral geniculate projecting to visual cortex is suggested as an example.) A spatial activity pattern on the first set of cells gives rise to another spatial activity pattern on the second set of cells. Suppose that a different activity pattern on the first set gives rise to the *same* pattern on the second set of neurons as the first pattern did. Let us assume this corresponds to one kind of error. Grossberg asks how the system could know an error was made, if we cannot assume an external teacher with detailed knowledge of the correct responses.

Grossberg makes the reasonable suggestion that there are reciprocal connections between the two sets of cells, so a pattern of activity going upward provokes "learned feedback" going downward. Presumably in the past the descending system learned the output-input association. Then, since both are simultaneously present, the expected input pattern (from the association) and the actual input pattern interact. The details of their interaction are discussed in considerable detail, along with general principles for nervous system organization.

Many supervised error correcting techniques described in this collection (Widrow and Hoff, paper 10; Rumelhart, Hinton, and Williams, paper 41) compute difference signals between the desired and actual outputs. Grossberg computes the sum. Effective use of the sum in an error correcting system requires some assumptions about the dynamics of the network. One mechanism suggested involves network dynamics that predict the existence of what Grossberg calls a "Quenching Threshold" (QT). These nonlinear network mechanisms were suggested in some of Grossberg's earlier papers as good ways for biological information processing systems to work (see appendix C). They have the effect of suppressing small signals and enhancing large signals.

We can then use the nonlinear dynamics to test for match or mismatch of the input and the learned feedback from higher levels. If the learned feedback signal and the input pattern match, then the sum is larger than the input pattern, and has the same pattern; that is, there has been enhancement because expectation matches input pattern. If there is a mismatch between the learned feedback and the input pattern, the mismatch will lead to a more uniform summed signal with lower peak values. The dynamics of the first set of neurons that is forming the sum can be tuned so it will suppress the activity if the Quenching Threshold is set correctly. Grossberg makes the point that a noise suppression system suggested for other reasons, effective information processing in noisy environments, can be used for another purpose, such as here, checking for matches or mismatches in a system that is trying to code information about the input patterns.

The rest of the theoretical discussion considers extensions of these ideas. Grossberg suggests that recurrent feedback can maintain activity in a set of cells if necessary, even if lower level inputs are shut off due to mismatches or the disappearance of the input stimuli. Recurrent feedback allows the maintenance of patterns in a "Short Term Memory" (STM). Recurrent feedback is provided by "shunting on-center, off-surround" networks (nonlinear lateral inhibition). This particular center-surround anatomy has been a consistent theme in Grossberg's work for many years, and he has shown it to have a number of interesting and useful properties, some of which are discussed in this paper in appendix D.

Given a recurrent network capable of forming a persistent short term memory pattern, Grossberg discusses in section 15 what might happen if the feedback and input pattern match. Then, strong signals reinforce each other, giving rise to a dynamical state that Grossberg calls an *adaptive resonance*. These long lasting states become the essential elements of the cognitive apparatus. These resonant states give rise to what Grossberg calls "cognitive codes," which represent states of the system and which can be provoked by appropriate inputs.

It is worth mentioning that long lasting or stable states, generated by recurrent interactions either involving learning or a favorable anatomy, are used in many neural networks in this collection. Systems differ considerably in the details of the feedback mechanisms, but the feedback dynamics often give rise to similar behavior. Identifying stable or long lasting states as key parts of mental information processing is natural in network theory and may be a useful insight into the workings of the brain.

Grossberg spends most of the rest of the paper applying the mechanisms he described to a number of different systems in the brain. These sections contain some intriguing speculations about brain organization.

It is typical of Grossberg's work to bring together in one theory several large scale organizing principles, their effects, and their interactions. His belief is that the same set of techniques for network computation are reused over and over in different contexts. This leads to applications of his models to areas that initially seem far afield from the initial domain of application. However, if the nervous system shows any overall organizing principles and is not just a bundle of ad hoc techniques, then one should be able to see similar principles at work in many places.

(1980)
Stephen Grossberg

How does a brain build a cognitive code?
Psychological Review 87:1–51

This article indicates how competition between afferent data and learned feedback expectancies can stabilize a developing code by buffering committed populations of detectors against continual erosion by new environmental demands. The gating phenomena that result lead to dynamically maintained critical periods, and to attentional phenomena such as overshadowing in the adult. The functional unit of cognitive coding is suggested to be an adaptive resonance, or amplification and prolongation of neural activity, that occurs when afferent data and efferent expectancies reach consensus through a matching process. The resonant state embodies the perceptual event, or attentional focus, and its amplified and sustained activities are capable of driving slow changes of long-term memory. Mismatch between afferent data and efferent expectancies yields a global suppression of activity and triggers a reset of short-term memory, as well as rapid parallel search and hypothesis testing for uncommitted cells. These mechanisms help to explain and predict, as manifestations of the unified theme of stable code development, positive and negative aftereffects, the McCollough effect, spatial frequency adaptation, monocular rivalry, binocular rivalry and hysteresis, pattern completion, and Gestalt switching; analgesia, partial reinforcement acquisition effect, conditioned reinforcers, underaroused versus overaroused depression; the contingent negative variation, P300, and ponto-geniculo-occipital waves; olfactory coding, corticogeniculate feedback, matching of proprioceptive and terminal motor maps, and cerebral dominance. The psychophysiological mechanisms that unify these effects are inherently nonlinear and parallel and are inequivalent to the computer, probabilistic, and linear models currently in use.

How do internal representations of the environment develop through experience? How do these representations achieve an impressive measure of global self-consistency and stability despite the inability of individual nerve cells to discern the behavioral meaning of the representations? How are coding errors corrected, or adaptations to a changing environment effected, if individual nerve cells do not know that these errors or changes have occurred? This article describes how limitations in the types of information available to individual cells can be overcome when the cells act together in suitably designed feedback schemes. The designs that emerge have a natural neural interpretation, and enable us to explain and predict a large variety of psychological and physiological data as manifestations of mechanisms that have evolved

This work was supported in part by the National Science Foundation (NSF MCS 77-02958).

Copyright 1980 by the American Psychological Association, Inc. Reprinted by permission of the publisher and author.

to build stable internal representations of a changing environment. In particular, various phenomena that might appear idiosyncratic or counterintuitive when studied in isolation seem plausible and even inevitable when studied as a part of a design for stable coding.

Some of the themes that will arise in our discussion have a long history in psychology. To achieve an exposition of reasonable length, the article is built around a thought experiment that shows us in simple stages how cells can act together to achieve the stable self-organization of evironmentally sensitive codes. If nothing else, the thought experiment is an efficient expository device for sketching how organizational principles, mechanisms, and data are related from the viewpoint of code development, using a minimum of technical preliminaries. On a deeper level, the thought experiment provides hints for a future theory about the types of developmental events that can generate the neural structures in which the codes are formed. It does this by correlating the types of environmental pressures to which the developmental mechanisms are sensitive with the types of neural structures that have evolved to cope with these pressures. References to previous theories and data have been chosen to clarify the thought experiment, to contrast its results with alternative viewpoints, to highlight areas in which more experimentation can sharpen or disconfirm the theory, or to refer to more complete expositions that should be consulted for a thorough understanding of particular results. The thought experiment and its consequences do not, however, depend on these references, and the reader will surely know many other references that can be used to confront and interpret the thought experiment.

1. A Historical Watershed

Some of the themes that will arise were already adumbrated in the work of Helmholtz during the last half of the 19th century (Boring, 1950; Koenigsberger, 1906). Unfortunately, the conceptual and mathematical tools needed to cast these themes as rigorous science were not available until recently. This fact helped to precipitously terminate the productive interdisciplinary activity between physics and psychology that had existed until Helmholtz's time, as illustrated by the perceptual contributions of Mach and Maxwell (Boring, 1950; L. Campbell & Garnett, 1882; Ratliff, 1965) in addition to those of Helmholtz (1866, 1962); to create a schism between psychology and physics that has persisted to the present day; and to unleash a century of controversy and antitheoretical dogma within psychology that led Hilgard and Bower (1975) to write the following first sentence in their excellent review of *Theories of Learning:* "Psychology seems to be constantly in a state of ferment and change, if not of turmoil and revolution" (p. 2).

One illustrative type of psychological data that Helmholtz studied concerned color perception. Newton had noted that white light at a point in space is composed of light of all visible wavelengths in approximately equal measure. Helmholtz realized, however, that the light we perceive to be white tends to be the average color of a whole scene (Beck, 1972). Thus perception at each point is nonlocal; it is due to a psychological process that averages data from many points to define the perceived color at each point. Moreover, this averaging process must be nonlinear, since it is more concerned with relative than absolute light intensities. Unfortunately, most of the mathematical tools that were available to Helmholtz were local and linear.

There is a good evolutionary reason why the light that is perceived to be white tends to be the average color of a scene. We rarely see objects in perfectly white light. Thus our eyes need the ability to average away spurious coloration due to colored light sources, so that we can see the "real" colors of the objects themselves. In other words, we tend to see the "reflectances" of objects, or the relative amounts of light of each wavelength that they reflect, not the total amount of light reaching us from each point. This observation is still a topic of theoretical interest and is the starting point of the modern theory of lightness (Cornsweet, 1970; Grossberg, 1972a; Land 1977).

A more fundamental difficulty faced Helmholtz when he considered the objects of perception. Helmholtz was aware that cognitive factors can dramatically influence our perceptions and that these factors can evolve or be learned through experience. He referred to all such factors as *unconscious inferences*, and developed his belief that a raw sensory datum, or *perzeption*, is modified by previous experience via a learned imaginal increment, or *vorstellung*, before it becomes a true perception, or *anschauung* (Boring, 1950). In more modern terms, sensory data activate a feedback process whereby a learned template, or expectancy, deforms the sensory data until a consensus is reached between what the data "are" and what we "expect" them to be. Only then do we "perceive" anything.

The struggle between raw data and learned expectations also has an evolutionary rationale. If perceptual and cognitive codes are defined by representations that are spread across many cells, with no single cell knowing the behavioral meaning of the code, then some buffering mechanism is needed to prevent previously established codes from being eroded by the flux of experience. It will be shown below how feedback expectancies establish such a buffer.

Unfortunately, Helmholtz was unable to theoretically represent the nonstationary, or evolutionary, process whereby the expectancy is learned, the feedback process whereby it is read out, or the competitive scheme whereby the afferent data and efferent expectancy struggle to achieve consensus. Helmholtz's conceptual and mathematical tools were linear, local, and stationary.

Section 4 begins to illustrate how nonlinear, nonlocal, and nonstationary concepts can be derived as principles of organization for adapting to a fluctuating environment. The presentation is nontechnical, but it will become apparent as we proceed that without a rigorous mathematical theory as a basis, the heuristic summary would have been impossible, since some of the properties that we will need are not intuitively obvious consequences of their underlying principles, and were derived by mathematical analysis. Furthermore, it will emerge that several design principles for adapting to different aspects of the environment operate together in the same structure. One of the facts that we must face about evolutionary systems is that their simple organizational principles can imply extraordinarily subtle properties. Indeed, part of

the dilemma that many students of mind now face is not that they do not know enough facts on which to base a theory, but rather they do not know which facts are principles and which are epiphenomena, and how to derive the multitudinous consequences that occur when a few principles act together. A rigorous theory is indispensable for drawing such conclusions.

The next two sections summarize some familiar experiments whose properties will reappear from a deeper perspective in the thought experiment. These experiments are included to further review one of the themes that Helmholtz confronted, and to prepare the reader for the results of the thought experiment. The sections can be skipped on a first reading.

2. Overshadowing: A Multicomponent Adult Phenomenon With Developmental Implications

Psychological data are often hard to analyze because many processes are going on simultaneously in a given experiment. This point is illustrated below in a classical conditioning paradigm that will be clarified by the theoretical development. Classical conditioning is considered by many to be the most passive type of learning and to be hopelessly inadequate as a basis for cognitive studies. The overshadowing phenomenon illustrates the fact that even classical conditioning is often only one component of a multicomponent process in which attention, expectation, and other "higher order" feedback processes play an important role (Kamin, 1969; Trabasso & Bower, 1968; Wagner, 1969).

Consider the four experiments depicted in Figure 1. Experiment 1 summarizes the simplest form of classical conditioning. An unconditioned stimulus (UCS), such as shock, elicits an unconditioned response (UCR), such as fear, and autonomic signs of fear. The conditioned stimulus (CS), such as a briefly ringing bell, does not initially elicit fear, but after preceding the UCS by a suitable interval on sufficiently many conditioning trials, the CS does elicit a conditioned response (CR) that closely resembles the UCR. In this way, persistently pairing an indifferent cue with a

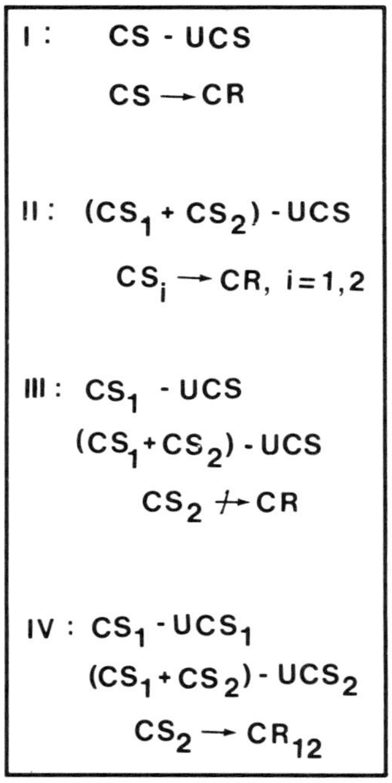

I : CS - UCS

CS → CR

II : $(CS_1 + CS_2)$ - UCS

CS_i → CR, i = 1,2

III : CS_1 - UCS

$(CS_1 + CS_2)$ - UCS

CS_2 ↛ CR

IV : CS_1 - UCS_1

$(CS_1 + CS_2)$ - UCS_2

CS_2 → CR_{12}

Figure 1. Four experiments illustrate overshadowing. (Experiment I summarizes the standard classical conditioning paradigm: conditioned stimulus–unconditioned stimulus [CS–UCS] pairing enables the CS to elicit a conditioned response (CR). Experiment II shows that joint pairing of two CSs with the UCS can enable each CS separately to elicit a CR. Experiment III shows that prior CS_1–UCS pairing can block later conditioning of CS_2 to the CR. Experiment IV shows that CS_2 can be conditioned if its UCS differs from the one used to condition CS_1. The CR that CS_2 elicits depends on the relationship between both UCSs, hence the notation CR_{12}.)

significant cue can impart some of the effects of the significant cue to the indifferent cue.

In Experiment 2, two CSs, CS_1 and CS_2, occur simultaneously before the UCS on a succession of conditioning trials; for example, a ringing bell and a flashing light both precede shock. It is typical in vivo for many cues to occur simultaneously, or in parallel, and the experimental question is, Is each cue separately conditioned to the fear reaction or is just the entire cue combination conditioned? If the cues are equally salient to the organism and are in other ways matched, then the answer is yes. If either cue CS_1 or CS_2 is presented separately after the conditioning trials, then it can elicit the CR.

Experiment 3 modifies Experiment 2 by performing the conditioning part of Experiment 1 on CS_1 before performing Experiment 2 on CS_1 and CS_2. In other words, first condition CS_1 until it can elicit the CR. Then present CS_1 and CS_2 simultaneously on many trials using the same UCS as was used to condition CS_1. Despite the results of Experiment 2, the CS_2 does not elicit the CR if it is presented after conditioning trials. Somehow prior pairing of CS_1 to the CR "blocks" conditioning of CS_2 to the CR.

The meaning of Experiment 3 is clarified by Experiment 4, which is the same as Experiment 3, with one exception. The UCS that follows CS_1 is not the same UCS that follows the stimulus pair CS_1 and CS_2 taken together. Denote the first UCS by UCS_1 and the second UCS by UCS_2. Suppose, for example, that UCS_1 and UCS_2 are different shock levels. Does CS_2 elicit a CR in this situation? The answer is yes if the two shock levels are sufficiently different. If the shock UCS_2 exceeds UCS_1 by a sufficient amount, then CS_2 elicits fear, or a negative reaction. If, however, the shock level UCS_1 exceeds UCS_2 by a sufficient amount, then CS_2 elicits relief, or a positive reaction.

How can the difference between Experiments 3 and 4 be summarized? In Experiment 3, CS_2 is an irrelevant or uninformative cue, since adding it to CS_1 does not change the expected consequence UCS. In Experiment 4, by contrast, CS_2 is informative because it predicts a change in the UCS. If the change is for the worse, then CS_2 eventually elicits a negative reaction (Bloomfield, 1969). If the change is for the better, then CS_2 eventually elicits a positive reaction (Denny, 1970).

Thus many learners are minimal adaptive predictors. If a given set of cues is followed by expected consequences, then all other cues are treated as irrelevant, as is CS_2 in Experiment 3. Each of us can define a given object using different sets of cues without ever realizing that our private sets are different, so long as the situations in which each of us uses the object always yield expected consequences. By contrast, if unexpected consequences occur, as in Experiment 4, then we somehow enlarge the set of relevant cues to include cues that were erroneously disregarded.

Several important qualitative conclusions can be drawn from these remarks. First, what is conditioned depends on our expectations, and these in turn help to regulate the cues to which we pay attention. Second, cues are conditioned, and indeed codes that interrelate these cues are built up, only if we pay attention to these cues because of their potential informativeness. Third, the mismatch between expected consequences and real events occurs only after attention has been focused on certain cues that thereupon generate the expectancy. Somehow this mismatch "feeds backwards in time" to amplify cues that have previously been overshadowed but that must have contained relevant information that we have erroneously ignored. Fourth, whenever we are faced with unexpected consequences, we do not know which cues have erroneously been ignored. The feedback process must be capable of amplifying all of the cues that are still being stored, albeit in a suppressed state. In other words, the feedback process is nonspecific. Finally, the nonspecific feedback process that is elicited by unexpected events competes with the specific consummatory channels that have focused our attention on the wrong set of cues. This competition between specific and nonspecific mechanisms helps us to reorganize our attentional focus until expected consequences are once again achieved.

This brief discussion reveals several basic processes working together in the overshadowing paradigm:

(a) classical conditioning, (b) attention, (c) learned expectancies, (d) matching between expectancies and sensory data, and (e) a nonspecific system that is activated by unexpected or novel events and competes with the specific consummatory system that focuses attention on prescribed cues.

Thus even classical conditioning is not a passive process when it occurs in realistic behavioral situations. Furthermore, its understanding requires the analysis of such teleological concepts as expectancy and attention. Helmholtz's doctrine of unconscious inference is readily called to mind.

Attention is to many individuals a holistic, if not unscientific, concept that does not mesh well with recent technological advances, say in microelectrode recording from individual nerve cells. Perhaps for this reason the fact that attentional variables can significantly influence what codes will be learned seems to have been ignored by some neurophysiologists who study the development of the visual cortex. For example, Stryker and Sherk (1975) were unable to replicate the Blakemore and Cooper (1970) study of visual code development in kittens. In the Blakemore and Cooper study, kittens were raised in a cylindrical chamber whose walls were painted with vertical black and white bars. The visual cortices of the kittens were reported to possess abnormally small numbers of horizontally tuned feature detectors. Hirsch and Spinelli (1970) performed experiments that did replicate in later experiments. In their experiments, the cats wore goggles, one lens with vertical stripes and the other with horizontal stripes. The corresponding visual cortices were reported to possess abnormally small numbers of feature detectors that were tuned to the orthogonal orientation. The entire controversy focused on such technical details as possible sampling errors due to Blakemore and Cooper's method of placing their electrodes. It is obvious, however, that the two experimental paradigms are attentionally inequivalent. Even perfect experimental technique would not necessarily imply similar experimental results.

3. Parallel Processing and the Persistence of Learned Meanings

The fact that classical conditioning, and for that matter any form of code development or learning, cannot be divorced from feedback processes that are related to attention is also made clear by the example illustrated by Figure 2. In Figure 2a, two classical conditioning experiments are depicted, one in which stimulus S_2 is the UCS for response R_2 and S_1 is its CS, and one in which S_1 is the UCS for R_1 and S_2 is its CS. What would happen if each cue S_1 and S_2 is conditioned to its own response R_1 or R_2, respectively, before a classical conditioning experiment occurs in which S_1 and S_2 are alternately scanned? This is the typical situation in real life, when we scan many cues in parallel, or intermittently, and many of these cues already have their own associations. If classical conditioning were a

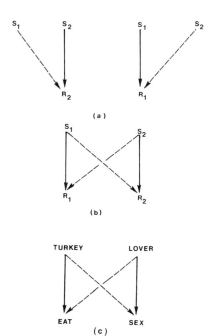

Figure 2. Classical conditioning cannot be a passive feed-forward process during real behavior. (In (a), S_1 acts as a conditioned stimulus (CS) for S_2, whereas S_2 acts as a CS for S_1. In (b), parallel processing of S_1 and S_2, each previously conditioned to responses R_1 and R_2, would yield cross-conditioning. In (c), some of the disastrous consequences of cross-conditioning are illustrated.)

passive feed-forward process, then cross-conditioning from S_1 to R_2 and from S_2 to R_1 would rapidly occur, as in Figure 2b.

However, this is absurd, as the particular example in Figure 2c vividly illustrates. Figure 2c schematizes the situation that would occur due to having a turkey dinner with one's lover. One alternately looks at lover and turkey, with lover associated with sexual responses (among others!) and turkey associated with eating responses. Why do we not come away from dinner wanting to eat our lover and to have sex with turkeys? Somehow the persistence of learned meanings can endure despite the fact that cues that are processed in parallel often generate incompatible responses. This is not always true, however, since if we, for example, consistently use a turkey as a discriminative cue for shock, or even sex, then turkeys might well become associated with fear or sexual arousal. Figure 2 depicts a situation in which the free reorganization of

attention, rather than a forced pairing of a CS with a UCS, maintains the learned persistence of meanings. Grossberg (1975) developed a thought experiment in which overcoming the environmentally imposed dilemma of Figure 2 leads to attentional mechanisms that imply the overshadowing phenomena in Figure 1.

Before leaving the subject of overshadowing, we might ask why this adult attentional phenomenon is related to the development of sensory and cognitive codes, even in infants. This article argues that feedback is necessary to stabilize the development of behaviorally meaningful codes in a rich input environment. The feedback processes include attentional mechanisms, and the stabilization of developing codes leads to gating phenomena, or the emergence of critical periods, that are dynamically maintained by the feedback processes.

From this perspective, the structure of an environmentally adaptive tissue is a dynamic scheme whose parameters change very slowly only because of the nature of its maintaining feedback. Death itself is a dramatic example of how seemingly persistent structures can rapidly disintegrate when maintaining feedback is disturbed. When the development of a structure is driven by a particular type of experience, one of the structure's maintaining factors is that variety of experience. A subtle feature of such a developing structure is its ability to selectively amplify those experiences that tend to maintain its structure. Next I

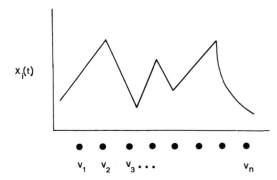

Figure 3. Each cell (or cell population) v_i possess an activity or potential $x_i(t)$, at every time t, $i = 1$, 2, ..., n. (The vector $(x_1(t), x_3(t), ..., x_n(t))$ of all these activities is a spatial pattern of activity.)

will discuss how feedback expectancies help to accomplish this end.

4. A Thought Experiment: The Need for Learned Feedback Expectancies

We now start to build a framework in which to discuss environmentally driven and behaviorally meaningful code development. Wherever possible, mathematical details will be suppressed, and the *minimal* structure capable of achieving our ends will be defined. This procedure will clarify what mathematical problems have to be solved, what their relationship is to each other, and what types of thematic variations on the minimal structures can be anticipated in different species and different neural locations in the same individual.

The central theoretical theme will be, How can a coding error be corrected if no individual cell knows that one has occurred? The importance of this issue becomes clear when we realize that erroneous cues can accidentally be incorporated into a code when our interactions with the environment are simple and will only become evident when our environmental expectations become more demanding. Even if our code perfectly matched a given environment, we would certainly make errors as the environment itself fluctuates. Furthermore, we never have an absolute criterion of whether our understanding of a fixed environment is faulty, or the environment that we thought we understood is no longer the same. The problem of error correction is fundamental whenever either the environment fluctuates or the individual keeps testing ever-deepening interpretations of the environment using ever sharper criteria of behavioral success.

We begin by introducing the functional elements on which our argument will build. Figure 3 depicts a collection of cells or cell populations, $v_1, v_2, \ldots, v_n$, each of which has an activity, or potential, $x_1(t), x_2(t), \ldots, x_n(t)$ at every time t. The activity $x_i(t)$ or v_i is imagined to be due to inputs $I_i(t)$ to v_i from a prior stage of neural processing, or the external environment, or endogenous sources within v_i itself. At every time t, these activities form a *pattern* $x(t) = (x_1(t), x_2(t), \ldots, x_n(t))$ across the cells $v_1, v_2, \ldots, v_n$, to which we will refer collectively

as a *field* of cells F. Henceforth, the time variable t will often be suppressed, since we will always take for granted that we are studying the system at a prescribed time.

Now consider two successive fields $F^{(1)}$ and $F^{(2)}$ of cells. Suppose that a pattern $x^{(1)}$ is active across $F^{(1)}$ (Figure 4). At this point the reader might wish to give $F^{(1)}$ and $F^{(2)}$ a concrete interpretation to help fix ideas. For example, one might think of $F^{(1)}$ as an idealization of the lateral geniculate nucleus (LGN) and $F^{(2)}$ as an idealization of visual cortex. The LGN processes visual data on its way to visual cortex, and it is the way station closest to the visual receptors at which our argument might hold in some species. I emphasize, however, that the results will be generally applicable to all neural stages at which behaviorally meaningful environmental inputs can drive code development. The fact that a significant fraction of visual development seems to be genetically prewired in the geniculo-cortical pathways of higher mammals like the monkey (Hubel & Wiesel, 1977) will not weaken the general conclusions that we will reach, and in fact various predictions and recent data about LGN, among other structures, will emerge from the analysis.

Suppose that the signal-carrying pathways from $F^{(1)}$ to $F^{(2)}$ act to filter the pattern $x^{(1)}$, and that due to prior developmental experience, this filter "codes" pattern $x^{(1)}$ by eliciting pattern $x^{(2)}$ across $F^{(2)}$. Knowing the detailed structure of this code is unnecessary to make our argument. However, we must be able to show how signal pathways can act as a filter that can be tuned by experience. This is done in Appendix A.

Suppose after the system learns to code $x^{(1)}$ by $x^{(2)}$ that another pattern is presented to $F^{(1)}$ and is erroneously coded at $F^{(2)}$ by $x^{(2)}$. To describe this situation conveniently, I introduce some subscripts. Denote $x^{(1)}$ and $x^{(2)}$ by $x_1^{(1)}$ and $x_1^{(2)}$, respectively, and denote the erroneously coded pattern at $F^{(1)}$ by $x_2^{(1)}$. In Figure 5 we draw the pattern $x_2^{(2)}$ that codes $x_2^{(1)}$ to equal $x_1^{(2)}$. Equality is meant to imply functional equivalence rather than actual identity. We now ask the central question, How can this coding error be corrected if no individual cell knows that an error has occurred?

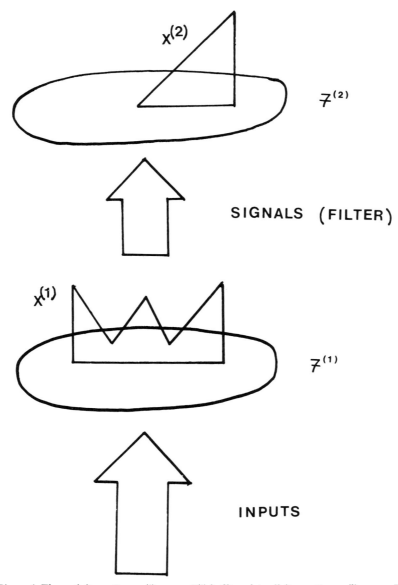

Figure 4. The activity pattern $x^{(1)}$ across $F^{(1)}$ is filtered to elicit a pattern $x^{(2)}$ across $F^{(2)}$.

Our first robust conclusion is now apparent: Whatever the mechanism is that corrects this error, it cannot exist within $F^{(2)}$, since by definition $x_1^{(2)}$ and $x_2^{(2)}$ are functionally equivalent. In principle, $F^{(2)}$ does not have the ability to distinguish the fact that $x_1^{(1)}$, and not $x_2^{(1)}$, should elicit $x_1^{(2)}$, since so far as $F^{(2)}$ knows, $x_1^{(1)}$ *is* active at $F^{(1)}$ rather than $x_2^{(1)}$.

It is important to realize that this argument is independent of coding details. It is based only on the type of information that $F^{(2)}$ cannot, in principle, possess. Much of our argument will be based on similar limitations in the types of information that particular processing stages can, in principle, possess. The robustness of this argument suggests why the design that overcomes these limitations seems to occur ubiquitously, in one form or another, in so many neural structures.

Where in the network can this error be detected in principle? At the time when $x_2^{(1)}$ elicits $x_1^{(2)}$, there exists no trace within the network that during prior learning trials it was $x_1^{(1)}$ that elicited $x_1^{(2)}$, not $x_2^{(1)}$. Somehow this fact must be represented within the network dynamics. Otherwise, $x_1^{(2)}$ could become associated with $x_2^{(1)}$, just as $x_1^{(1)}$ was on previous developmental trials. The only times that $x_1^{(1)}$ was active in the network were the developmental trials during which the filter from $F^{(1)}$ to $F^{(2)}$ was learning to code $x_1^{(1)}$ by $x_1^{(2)}$. To be, in principle, capable of testing whether the correct pattern $x_1^{(1)}$ elicits $x_1^{(2)}$ on later trials when $x_1^{(1)}$ is not presented, it must be true that during the developmental trials, $x_1^{(2)}$ activates a feedback pathway from $F^{(2)}$ to $F^{(1)}$ that is capable of learning the active pattern $x_1^{(1)}$ at $F^{(1)}$. Then when $x_2^{(1)}$ erroneously activates $x_1^{(2)}$ on future trials, $x_1^{(2)}$ can read out the correct pattern $x_1^{(1)}$ across $F^{(1)}$. When this happens, the two patterns $x_1^{(1)}$ and $x_2^{(1)}$ will be simultaneously active across $F^{(1)}$, and they can be compared, or matched, to test whether or not the correct pattern has activated $x_1^{(2)}$ (Figure 6).

In summary, if in principle it is possible to correct a coding error at $F^{(2)}$, then there

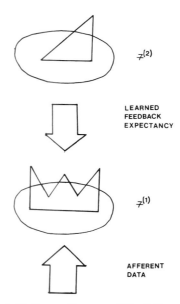

Figure 6. Pattern $x_1^{(2)}$ across $F^{(2)}$ elicits a feedback pattern $x_1^{(1)}$ to $F^{(1)}$, which is the pattern that it sampled across $F^{(1)}$ during previous developmental trials. (Field $F^{(1)}$ becomes an interface where afferent data and learned feedback expectancies are compared.)

must exist learned feedback from $F^{(2)}$ to $F^{(1)}$. This learned feedback represents the pattern that $x_1^{(2)}$ *expects* to be at $F^{(1)}$ due to prior developmental trials. The feed-forward data to $F^{(1)}$ and the learned feedback expectancy, or template, from $F^{(2)}$ to $F^{(1)}$ are thereupon compared at $F^{(1)}$. Figure 7 illustrates this sequence of events as a series of snapshots that can occur at a very fast rate, for example, on the order of hundreds of milliseconds. Helmholtz's doctrine of unconscious inference is readily called to mind.

The general nature of the preceding argument strongly suggests that feedback pathways will ubiquitously occur from "higher" neural centers to the relay stations that excite them. In fact, reciprocal thalamocortical connections seem to exist in all thalamo–neocortical systems (Macchi & Rinvik, 1976; Tsumoto, Creutzfeldt, & Legéndy, 1978).

At this point, we also recognize two more design problems for mathematics. The first problem is, How do feedback pathways from $F^{(2)}$ learn a pattern of activity across $F^{(1)}$? (See Appendix B for a summary of this mechanism.)

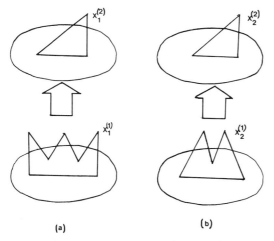

Figure 5. In (a), pattern $x_1^{(1)}$ at $F^{(1)}$ elicits the correct pattern $x_1^{(2)}$ across $F^{(2)}$. In (b), pattern $x_2^{(1)}$ elicits the incorrect pattern $x_2^{(2)}$, which is functionally equivalent to $x_1^{(2)}$ across $F^{(2)}$.

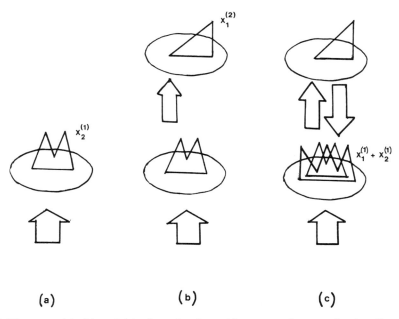

Figure 7. The stages (a), (b), and (c) schematize the rapid sequence of events whereby afferent data is filtered and activates a feedback expectancy that is matched against itself.

5. Noise Suppression, Pattern Matching, and Spatial Frequency Detection

The second design problem that we must face is this: Somehow the mismatch between the patterns $x_1^{(1)}$ and $x_2^{(1)}$ must rapidly shut off activity across $F^{(1)}$. Otherwise, $x_1^{(2)}$ would learn to code $x_2^{(1)}$ much as $x_1^{(2)}$ learned to code $x_1^{(1)}$ on the preceding developmental trials. Pattern $x_1^{(2)}$ must also be rapidly shut off if only to prevent behavioral consequences of $x_1^{(2)}$ from being triggered by further network processing. Moreover, $x_1^{(2)}$ must be shut off in such a fashion that $x_2^{(1)}$ can thereupon be coded by a more suitable pattern across $F^{(2)}$.

The only basis on which these changes can occur is the mismatch of $x_1^{(1)}$ and $x_2^{(1)}$ across $F^{(1)}$. We must therefore ask, How does the mismatch of patterns across a field $F^{(1)}$ of cells inhibit activity across $F^{(1)}$? The mathematical details are summarized in Appendix C. Here, however, it is useful to make the important distinction between mechanisms that develop due to evolutionary pressures and properties that are merely consequences of these mechanisms. One might well worry that the design of a mismatch mechanism is a rather sophisticated evolutionary task. We

now indicate that such a mechanism is a consequence of a more basic property, namely noise suppression, and that noise suppression is itself a variation of a basic evolutionary principle. Moreover, other useful properties follow from noise suppression, such as spatial frequency detection and edge enhancement.

The environmental problem out of which the noise suppression property emerges is the *noise-saturation dilemma*. This dilemma has been discussed in detail elsewhere (e.g., Grossberg, 1977, 1978d). The dilemma confronts all noisy cellular systems that process input patterns, as in Figure 3. If the inputs are too small, they can get lost in the noise. If the inputs are amplified to avoid the noisy range, they can saturate all the cells by activating all of their excitable sites, and thereby reduce to zero the cells' sensitivity to differences in the input intensities. Appendix C reviews how competitive interactions among the cells automatically retune their sensitivity to overcome the saturation problem. In a neural context, the competitive interactions are said to be shunting interactions, and they are carried by an on-center off-surround anatomy. The retuning of sensitivity is due to automatic gain control by the inhibitory

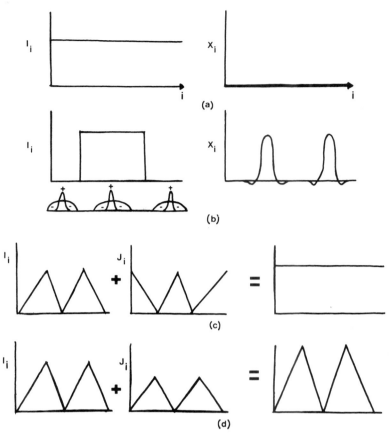

Figure 8. In (a), noise suppression converts a uniform input pattern into a zero activity pattern. In (b), a rectangular input pattern elicits differential activity at its edges because the cells within its interior and beyond its boundary perceive uniform fields. (This is a special case of spatial frequency detection.) In (c), two mismatched patterns add to generate an approximately uniform total input pattern, which will be suppressed by the mechanism of (a). In (d), two matched patterns add to yield a total input pattern that can elicit more vigorous activation than either input pattern taken separately.

off-surround signals. This fundamental property does not exist in additive models of lateral inhibition, such as the Hartline–Ratliff model (Ratliff, 1965). Appendix C shows how the automatic gain control mechanism can inhibit a uniform pattern of inputs, no matter how intense the inputs are. This is the property of noise suppression that we seek.

Figure 8a depicts this noise suppression property. A uniform pattern does not distinguish any cell from any other cell. For example, when the cells are feature detectors of one kind or another, a uniform input pattern contains no information that can distinguish one feature from any other feature. Noise suppression eliminates this irrelevant

activity and allows the network to focus on informative discriminations.

Once noise suppression is guaranteed, several consequences automatically follow. For example, Figure 8b shows that such a network responds to the edges of a rectangular input, or to spatial gradients in more general input patterns. This is because cells whose inhibitory surrounds fall outside the rectangle perceive a uniform field, and cells with inhibitory surrounds that are near the center of the rectangle also perceive a uniform field. Both types of cells suppress their inputs. Only cells near the edges of the rectangle do not perceive a uniform pattern. Consequently only the edges of the rectangle elicit large activation.

This argument tacitly supposes that the lateral inhibitory interactions affecting each cell have a prescribed spatial extent, and that the width of the rectangle exceeds this spatial scale. More generally, spatial gradients in an input pattern are matched against the spatial scale of each cell's excitatory and inhibitory interactions. Only those spatial gradients in the input pattern that are nonuniform with respect to the cell's interaction scales generate large activities. By varying the inhibitory scales across cells, one can tune different cells to respond to different spatial frequencies. Thus spatial frequency detectors are a natural consequence of noise suppression properties within cells having a prescribed inhibitory scale. Since all networks in which shunting inhibition occurs have such scales, the existence of spatial frequency detectors should come as no surprise and does not imply that neural networks are Fourier analyzers in the spatial domain (Robson, 1976). Indeed, Fourier analyzers are linear mechanisms. By contrast, shunting networks that are capable of short-term memory contain feedback pathways, and all such networks must be nonlinear to be stable (Grossberg, 1973, 1978d).

Finally, Figure 8c and 8d indicate how a noise suppression mechanism can accomplish pattern matching. Figure 8c supposes that two mismatched patterns feed into $F^{(1)}$, where they add before coupling into the shunting dynamics. Because of the mismatch, the peaks of I_i fill in the troughs of J_i. The total input pattern is approximately uniform and is consequently quenched as noise. By contrast, in Figure 8d, the two patterns match. Their peaks and troughs mutually reinforce each other, so the resultant activities can be amplified beyond the effect of just one pattern. In summary, mismatched input patterns quench activity, whereas matched patterns amplify activity across a field $F^{(1)}$ that is capable of noise suppression.

A subsidiary mathematical question is now evident: How uniform must a pattern be for it to be suppressed? Part of the answer is determined by the choice of structural parameters, such as the strength and spatial distribution of lateral inhibitory coefficients (Appendix C). However, the field $F^{(1)}$ can also be dynamically tuned, or sensitized, by fluctuations in the level of nonspecific arousal that perturbs it through time. An arousal increment can, for example, act by inhibiting the inhibitory interneurons of the network (Ellias & Grossberg, 1975; Grossberg, 1973, 1978e; Grossberg & Levine, 1975). Such a tuning mechanism can simultaneously alter the spatial frequency properties of the network by multiplicatively strengthening or weakening the inhibitory interactions of the cells (Barlow & Levick, 1969a, 1969b). Such mechanisms will arise in a natural fashion as our argument continues.

6. Triggering of Nonspecific Arousal by Unexpected Events

Having suppressed $x_2^{(1)}$ at $F^{(1)}$ due to mismatch with the feedback expectancy $x_1^{(1)}$, we must now use this suppression to inhibit $x_1^{(2)}$ at $F^{(2)}$, since the mismatch at $F^{(1)}$ is the only mechanism in the network that can, in principle, distinguish that an error has occurred at $F^{(2)}$. Moreover, until $x_1^{(2)}$ is quenched, it will continue to read out the template $x_1^{(1)}$ to $F^{(1)}$, which will prevent $x_1^{(2)}$ from eliciting a new signal to $F^{(2)}$.

We were led to the mismatch mechanism at $F^{(1)}$ by noting that $F^{(2)}$ could not discriminate whether an error had occurred. Now we note that $F^{(1)}$s information is also limited At $F^{(1)}$ it cannot be discerned *which* pattern across $F^{(2)}$ caused the mismatch at $F^{(1)}$. It could have been any pattern whatsoever. All $F^{(1)}$ knows is that a mismatch has occurred. Whatever pattern across $F^{(2)}$ caused the mismatch must be inhibited. Consequently, a mismatch at $F^{(1)}$ must have a *nonspecific* effect on all of $F^{(2)}$, since any of the cells in $F^{(2)}$ might be one of the cells that must be inhibited.

We are therefore led to the following questions: How does mismatch and subsequent quenching of activity across $F^{(1)}$ elicit a nonspecific signal (arousal!) to $F^{(2)}$? Where does the activity that drives this nonspecific arousal pulse come from?

Before answering these questions, we should realize that we have been led to a familiar conclusion: Unexpected or novel events are arousing. (To forcefully remind yourself of this basic fact, test a friend's reaction by unexpectedly slamming your hand on a table).

Now we will consider how such arousal is initiated and how it contributes to attentional processing.

Where does the activity that drives the arousal come from, and why is it released when quenching of activity at $F^{(1)}$ occurs? There are two possible answers to the first part of the question, but only one of them survives closer inspection. The activity is either endogenous (internally and persistently generated) or the activity is elicited by the sensory input. If the activity were endogenous, then arousal would occur whenever $F^{(1)}$ was inactive, whether this inactivity was due to active quenching by mismatched feedback from $F^{(2)}$ or to the absence of sensory inputs. This leads to the unpleasant conclusion that $F^{(2)}$ would be tonically flooded with arousal whenever nothing interesting was happening at $F^{(1)}$ or $F^{(2)}$. Therefore, sensory inputs to $F^{(1)}$ bifurcate before they reach $F^{(1)}$. One pathway is *specific:* It delivers information about the sensory event $F^{(1)}$. The other pathway is *nonspecific:* It activates the arousal mechanism that is capable of nonspecifically influencing $F^{(2)}$. The idea that cues have both informative (specific) and arousal (nonspecific) functions has been empirically known at least since the work of Moruzzi and Magoun on the reticular formation (Hebb, 1955; Moruzzi & Magoun, 1949).

Given that the sensory inputs to $F^{(1)}$ also activate an arousal pathway, what prevents this pathway from being activated except when activity at $F^{(1)}$ is quenched? The answer is now clear: Activity at $F^{(1)}$ inhibits the arousal pathway, and quenching of this activity disinhibits the arousal pathway. Figure 9 schematizes the (very rapid) sequence of events to which we have been led. First, a sensory event elicits a pattern $x_2^{(1)}$ across $F^{(1)}$ as it begins to activate the arousal pathway α. This activation at α is inhibited by activity from $F^{(1)}$. Simultaneously, pattern $x_2^{(1)}$ activates pathways to $F^{(2)}$ that act as a filter that erroneously activates $x_1^{(2)}$. Pattern $x_1^{(2)}$ reads out the learned feedback expectancy $x_1^{(1)}$ to $F^{(1)}$. Mismatch of $x_1^{(1)}$ and $x_2^{(1)}$ at $F^{(1)}$ quenches activity across $F^{(1)}$. The inhibitory signal from $F^{(1)}$ to α is also quenched, and the arousal pathway is disinhibited. A nonspecific arousal pulse is hereby unleashed on $F^{(2)}$.

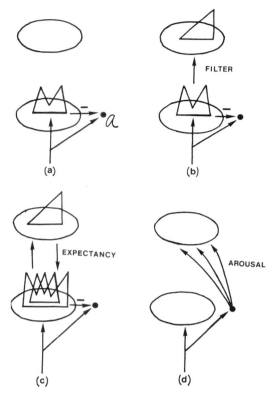

Figure 9. In (a), afferent data elicit activity across $F^{(1)}$ and an input to the arousal source α that is inhibited by $F^{(1)}$. In (b), the pattern at $F^{(1)}$ maintains inhibition of α as it is filtered and activates $F^{(2)}$. In (c), the feedback expectancy from $F^{(2)}$ is matched against the pattern at $F^{(1)}$. In d, mismatch attenuates activity across $F^{(1)}$ and thereby disinhibits α, which releases a nonspecific arousal signal to $F^{(2)}$.

7. Parallel Hypothesis Testing in Real Time: The Probabilistic Logic of Complementary Categories

The next design problem is now clearly before us: How does the increment in nonspecific arousal differentially shut off the active cells in $F^{(2)}$? The active cells are the cells that elicited the feedback expectancy to $F^{(1)}$, and since mismatch occurred at $F^{(1)}$, these cells must have been erroneously activated. Consequently, they should be shut off. Furthermore, inactive cells at $F^{(2)}$ should not be inhibited, because these cells must be available for possible coding of $x_2^{(1)}$ during the next time interval. Thus a differential suppression of cells is required: The cells that are most active when arousal occurs should be most

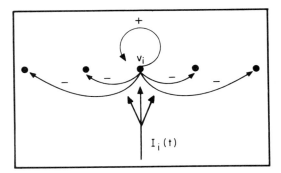

Figure 10. A recurrent shunting on-center off-surround network is capable of contrast-enhancing its input pattern, normalizing its total activity, and storing the contrast-enhanced pattern in short-term memory (STM). (If its feedback signals are properly chosen— e.g., sigmoid, or **S**-shaped signals—then a quenching threshold exists that defines the activity level below which activity is treated as noise and quenched, and above which activity is contrast enhanced and stored in STM.)

inhibited. This property realizes a kind of probabilistic logic in real time. If activating cell v_i in $F^{(2)}$ to a given degree leads to a certain degree of error or mismatch at $F^{(1)}$, then cell v_i should be inhibited to a degree that is commensurate both with its prior activation and with the size of the arousal increment, or the amount of error. If saying "yes" at v_i leads to error, then change the "yes" to "no," and do it in a graded fashion across the field $F^{(2)}$. Since cells that were only minimally active could have contributed only a small effect to the feedback expectancy, their inhibition will consequently be less, and they can contribute more to the correct coding of $x_2^{(1)}$ during the next time interval.

The arousal-initiated inhibition of cells across $F^{(2)}$ must be enduring as well as selective. Otherwise, as soon as $x_1^{(2)}$ is inhibited, the feedback expectancy, $x_1^{(1)}$, would be shut off, and $x_2^{(1)}$ would be free to reinstate $x_1^{(2)}$ across $F^{(2)}$ once again. The error would perseverate, and the network would be locked into an uncorrectable error. The inhibited cells must therefore stay inhibited long enough for $x_2^{(1)}$ to activate a different pattern across $F^{(2)}$ during the next time interval. The inhibition is therefore slowly varying compared to the time scale of filtering, feedback expectancy, and mismatch.

Once this selective and enduring inhibition is accomplished, the network has a capability for rapid hypothesis testing. By enduringly and selectively inhibiting $x_1^{(2)}$, the network "renormalizes" or "conditionalizes" the field $F^{(2)}$ to respond differently to pattern $x_2^{(1)}$ during the next time interval. If the next pattern elicited by $x_2^{(1)}$ across $F^{(2)}$ also creates a mismatch at $F^{(1)}$, then it will be suppressed, and $F^{(2)}$ will be renormalized again. In this fashion, a sequence of rapid pattern reverberations between $F^{(1)}$ and $F^{(2)}$ can successively conditionalize $F^{(2)}$ until either a match occurs or a set of uncommitted cells is found with which $x_2^{(1)}$ can build a learned filter from $F^{(1)}$ to $F^{(2)}$, and a learned expectancy from $F^{(2)}$ to $F^{(1)}$.

8. The Parallel Dynamics of Recurrent Competitive Networks: Contrast Enhancement, Normalization, Quenching Threshold, Tuning

At this point one can justifiably wonder how $x_2^{(1)}$ elicits a supraliminal pattern across $F^{(2)}$ after $x_1^{(2)}$ is inhibited? If $x_1^{(2)}$ is the pattern that $x_2^{(1)}$ originally excites, and $x_1^{(2)}$ is inhibited, then won't the next pattern elicited by $x_2^{(1)}$ across $F^{(2)}$ have very small activity? In other words, why was the second pattern not also active when $x_1^{(2)}$ was active?

It would have been if the anatomy within $F^{(2)}$ contained only feedforward, or non-recurrent pathways. Thus we are forced to conclude that the anatomy within $F^{(2)}$ contains feedback, or recurrent pathways. Since all cellular systems face the noise-saturation dilemma, these pathways are distributed in a competitive geometry, or an on-center off-surround anatomy (Figure 10). Mathematical analysis demonstrates that the normalization property holds when recurrent pathways are distributed in a competitive geometry. When these competitive networks are designed to overcome noise amplification and saturation, they enjoy several properties that we need (Appendix D). First, they are capable of contrast-enhancing small differences in initial pattern activities into large and easily discriminable differences that are thereupon stored in short-term memory (STM; see

Figure 11). This property is necessary to build up the codes for $F^{(1)}$ patterns at $F^{(2)}$. Before an $F^{(1)}$ pattern is coded by $F^{(2)}$, it might elicit an almost uniform activity pattern across $F^{(2)}$. The recurrent dynamics within $F^{(2)}$ quickly contrast enhances and stores the contrast-enhanced pattern in STM, where it can be sampled and stored in long-term memory (LTM) by the pathways from $F^{(1)}$ to $F^{(2)}$. When the next occurrence of the same pattern at $F^{(1)}$ occurs, these pathways therefore elicit a more differentiated pattern across $F^{(2)}$, which is again contrast enhanced and stored in STM. The feedback enhancement between STM and LTM continues until the two processes equilibrate other things being equal.

Another property of such a network is its tendency to conserve, or adapt, the total activity that it stores in STM. This is the normalization property that we seek. If certain cells in the network are prevented from sharing the STM activity, say due to arousal-initiated inhibition, then the total activity is renormalized by being distributed to the other cells. Thus after $x_1^{(2)}$ is inhibited across $F^{(2)}$, the network will respond to the signals due to $x_2^{(1)}$ by differentially amplifying them in a way that tends to preserve the total STM activity across $F^{(2)}$. This new STM pattern will inherit much of the STM activity that $x_1^{(2)}$ had before it was suppressed, but the new STM pattern across $F^{(2)}$ will be a quite different pattern than $x_1^{(2)}$, since it is built from $F^{(1)}$ signals that previously fared poorly in the competition for STM activity. The normalization property manifests itself in a large class of psychological data, notably data about behavioral contrast and ratio scales in choice behavior (Grossberg, 1975, 1978a).

These recurrent networks also possess a *quenching threshold* (QT), which is a parameter whose size determines what activities will be suppressed, or quenched, and what activities will be stored in STM (Grossberg, 1973). Activities in populations that start below the QT will be suppressed; activities that exceed the QT will be contrast enhanced and stored in STM. Thus the QT is the cutoff point that defines noise in a recurrent network. All networks that possess a QT can be tuned; that is, by varying the QT, the criterion of

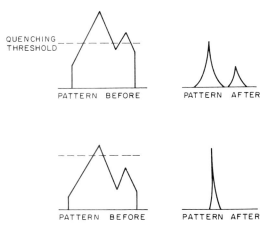

Figure 11. If the quenching threshold is variable, for example, due to shunting signals that nonspecifically control the size of the network's inhibitory feedback signals, then the network's sensitivity can be tuned to alter the ease with which inputs are stored in short-term memory.

which data shall be stored in STM and which data shall be quenched can be altered through time. Several parameters work together to determine QT size, notably the strength of recurrent lateral inhibitory pathways within the network. For example, if a nonspecific arousal pulse multiplicatively inhibits or shunts the inhibitory interneurons of a recurrent network, then its QT will momentarily decrease—the network's inhibitory "gates" will open—to facilitate STM storage.

The normalization property also helps us to understand the relevance of probabilistic models of hypothesis testing to cognitive processing. Normalization plays the role of summing all the probabilities to equal 1. Shunting, or multiplicative network dynamics, plays the role of multiplying the probabilities of independent events. However, probabilistic concepts only approximately describe some aspects of shunting competitive dynamics. A most serious difference is that although the network's hypothesis testing mechanism might produce a serial sequence of renormalizations in time, these operations are performed by parallel, rather than serial mechanisms. Serial mechanisms of hypothesis testing are not equivalent to the parallel theory.

More generally, serial behavioral properties do not imply that the control processes that subserve them are also serial. In particular, various serial, notably computer, models of memory and cognitive processing have been shown to be fundamentally inequivalent to parallel neural interactions. This inequivalence is noted for the Atkinson and Shiffrin (1968, 1971) theory of free recall in Grossberg (1978a) and for the Schneider and Shiffrin (1976) theory of automatic versus controlled visual information processing in Grossberg (1978e, Section 61). Different predictions of the two types of theory are also described in these articles, and some data that are inexplicable by the serial theories are explained using parallel properties, such as normalization, in a basic way.

9. Antagonistic Rebound Within On-Cell Off-Cell Dipoles

We are now faced with a subtle design problem: How can a nonspecific event, such as arousal, have specific consequences of any kind, let alone generate an exquisitely graded, enduring, and selective suppression of active cells? Here again mathematical analysis was absolutely essential, since the theory could not progress beyond this step had not the answer already been derived (to my own surprise) during work on reinforcement mechanisms (Grossberg, 1972b, 1972c, 1975). In this work, the mechanism helped explain such nontrivial effects as learned helplessness, vicious circle behavior, superconditioning, overshadowing, asymptotically nonchalant avoidance, and peak shift with behavioral contrast when it was joined with suitable conditioning and cognitive mechanisms that were all derived as evolutionary solutions to prescribed environmental pressures. These results extend such popular learning theories as those of Irwin (1971), Kamin (1969), Rescorla and Wagner (1972), and Seligman, Maier, and Solomon (1971) by explicating mechanisms, conceptual distinctions, and predictions in a psychophysiological framework that are invisible to descriptive theories. One might wish to know what reinforcement mechanisms have to do with the development of cognitive codes. The answer is that the property in question occurs whenever optimally designed chemical transducers, or transmitters, occur in competing

network channels, or dipoles, whether these channels arise in reinforcement mechanisms, attentional mechanisms (Grossberg, 1975), developmental mechanisms (Grossberg, 1976b), or mechanisms of motor control (Grossberg, 1978e). The property is a robust consequence of a ubiquitous neural design principle, and it guarantees a type of rapid hypothesis testing and error correction wherever this principle is used.

First let us consider some familiar behavioral facts that help to motivate the mechanism. Suppose that I wish to press a lever in response to the offset of a light. If light offset simply turned off the cells that code for light being on, then there would exist no cells whose activity could selectively elicit the lever-press response after the light was turned off. Clearly, offset of the light not only turns off the cells that are turned on by the light, but it also selectively turns on cells that will transiently be active after the light is shut off. The activity of these "off"-cells—namely the cells that are turned on by light offset—can then activate the motor commands leading to the lever press. Let us call the transient activation of the off-cell by cue offset *antagonistic rebound*.

Antagonistic rebound also occurs in a variety of other behavioral situations. For example, shock can unconditionally elicit the emotion of fear and various autonomic consequences of fear (Dunham, 1971; Estes, 1969; Estes & Skinner, 1941). Offset of shock is (other things equal) capable of eliciting relief or a complementary emotional reaction (Denny, 1970; Masterson, 1970; McAllister & McAllister, 1970). In a similar fashion I suggest that when motor command cells are organized in agonist–antagonist pairs, offset of the agonist input can elicit a rebound in the antagonist command cell that acts to rapidly brake the motion in the muscles controlled by the agonist command cell.

When such on-cell off-cell interactions are modeled, one finds examples akin to Figure 12. In Figure 12a, a nonspecific, or adaptation level, input I is delivered equally to both channels, whereas a test input J is delivered to the on-cell channel. These inputs create signals S_1 and S_2 in both channels, and the signals are multiplicatively gated by slowly varying chemical transmitters z_1 and z_2, respectively. The gated signals S_1z_1 and S_2z_2 thereupon compete and yield the on-cell off-cell responses that are depicted in Figure 12a. Appendix E describes the details that are needed for a better understanding, but the main idea behind antagonistic rebound is easy to describe. Consider Figure 12a. Here the transmitters z_1 and z_2 are depleted by being released at rates proportional to S_1z_1 and S_2z_2, respectively. More depletion of z_1 than z_2 occurs if the signal S_1 exceeds S_2. While the test input J is on, the on-channel receives a larger input than the off-channel, since its total input is J plus the nonspecific input I, whereas the off-cell channel only receives the input I. Consequently, $S_1 > S_2$, so that depletion of transmitter leads to the inequality $z_1 < z_2$. Despite this fact, one can prove that the gated signals satisfy the inequality $S_1z_1 > S_2z_2$. Consequently, the on-channel receives a larger gated signal than the off-channel, so that after competition takes place, there is a net on-reaction.

What happens when the test input is shut off? Both channels receive only the equal nonspecific input I. The signals S_1 and S_2 rapidly equalize until $S_1 = S_2$. However, the transmitters are more slowly varying in time so that the inequality $z_1 < z_2$ continues to hold. The gated signals therefore satisfy $S_1z_1 < S_2z_2$. Now the off-channel receives a larger signal. After competition takes place there is an antagonistic rebound in response to offset of the test input.

Why is the rebound transient in time? The equal signals S_1 and S_2 continue to drive the depletion of the transmitters z_1 and z_2. Gradually the amounts of z_1 and z_2 also

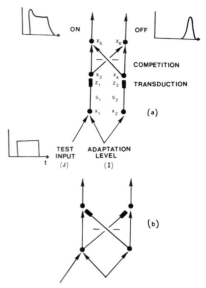

Figure 12. Two examples of on-cell off-cell dipoles. (In (a), the test input J and adaptation level input I add in the on-channel. The adaptation level input perturbs the off-channel. Each input is gated by a slowly varying excitatory transmitter [square synapses]. Then the channels compete before eliciting a net on-response or off-response. In (b), the slowly varying transmitters are inhibitory, and participate in the competition process.)

equalize so that S_1z_1 and S_2z_2 gradually equalize. As the gated signals equalize, the competition shuts off both the on-channel and the off-channel. These facts are summarized in Table 1.

10. Analgesia, Escape, Partial Reward, and Underaroused Versus Overaroused Depression

In Figure 12a, the two transmitters are excitatory and generate gated signals before competition occurs. Similar effects occur in Figure 12b in which the transmitters are inhibitory and act both as gates and as competing channels. There exist many variations on this theme in vivo. For example, by analyzing more complex learning situations, in particular, experiments on secondary conditioning phenomena, or on transfer between instrumental and classical conditioning, one can show that feedback pathways must exist within the channels that subserve incentive

Table 1
Antagonistic Rebound at Offset of Phasic Input

Test input J is on	Right after offset of J	After dipole equilibrates to offset of J
$I + J > I$	$I = I$	$I = I$
$x_1 > x_2$	$x_1 = x_2$	$x_1 = x_2$
$S_1 > S_2$	$S_1 = S_2$	$S_1 = S_2$
$z_1 < z_2$	$z_1 < z_2$	$z_1 = z_2$
$S_1z_1 > S_2z_2$	$S_1z_1 < S_2z_2$	$S_1z_1 = S_2z_2$
$x_3 > x_4$	$x_3 < x_4$	$x_3 = x_4$
$x_5 > 0 = x_6$	$x_5 = 0 < x_6$	$x_5 = 0 = x_6$

motivation. These feedback channels lead to meaningful comparisons with psychophysiological data when they are interpreted as a formal analogue of the medial forebrain bundle (Grossberg, 1972c, 1975).

Even the feed-forward networks already have surprising and important properties, however. For example, consider a network in which the on-channel supplies negative incentive motivation ("fear") and the off-channel supplies positive incentive motivation ("relief") in a conditioning paradigm. Choose shock reduction as the experimental manipulation. Let shock excite the on-channel, and suppose that the size of the positive rebound after shock terminates is monotonically related to the rewarding effect of the manipulation. Then one can derive a quantitative formula for rebound size (Grossberg, 1972c) that orders infinitely many possible experiments in terms of how rewarding they will be. In particular, reducing J units of shock to $J/2$ units is less rewarding than reducing $J/2$ units of shock to 0 units, despite the fact that shock reduction equals $J/2$ units in both cases. This analgesic effect is due to intracellular adaptation of the chemical transmitters. Analogous data have been reported by Campbell (1968); B. Campbell and Kraeling (1953); Gardner, Licklider, and Weisz (1961); and Myers (1969). Moreover, it is predicted that three indices should all covary as a function of the reticular formation arousal level, which is interpreted to be a source of nonspecific input to the incentive motivational dipoles. These indices are (a) the rewarding effect due to switching J units to $J/2$ units of shock, (b) the ability of an animal to learn to escape from presentation of a discrete fearful cue, and (c) the relative advantage of partial reward over continuous reward (Grossberg, 1972c).

One also finds that two types of depressed emotional affect exist in the dipole: an under-aroused syndrome and an overaroused syndrome. These syndromes are manifestations of the dramatic changes in the net incentive motivation that occur when the arousal level is parametrically changed (Grossberg, 1972c). The two syndromes are the endpoints in an inverted U of net incentive as a function of arousal level. At underaroused levels, the behavioral threshold is abnormally high, but

the system is hyperactive after this threshold is exceeded. At overaroused levels, the behavioral threshold is abnormally low, but the system is so hypoactive that little net incentive is ever generated. Parkinson's patients and certain hyperactive children seem to exhibit the underaroused syndrome (Fuxe & Ungerstedt, 1970; Ladisich, Volbehr, & Matussek, 1970; Ricklan, 1973), which is paradoxical because behavioral threshold is inversely related to suprathreshold reactivity. Such underaroused individuals can be brought "down" behaviorally by a drug that acts as an "up"; that is, it raises the adaptation level to the normal range. In Parkinson's patients, this up is L-dopa, and in certain hyperactive children, it is amphetamine.

A general question now presents itself: Do *all* neural dipoles share these properties whether they occur in motivational, sensory, or motor representations? This question is considered for the case of cortical red–green dipole responses to white light in Section 12.

11. Arousal Elicits Antagonistic Rebound: Surprise and Counterconditioning

A surprising feature of the on-cell off-cell dipole is its reaction to rapid temporal fluctuations in arousal, or adaptation level. This reaction allows us to answer the following question posed in Section 9: How can a nonspecific event, such as arousal, selectively suppress active on-cells? Appendix E shows that arousal fluctuations can reset the dipole, despite the fact that they generate equal inputs to the on-cell and off-cell channels. In particular, a sudden increment in arousal can, by itself, cause an antagonistic rebound in the relative activities of the dipole. Moreover, the size of the arousal increment that is needed to cause rebound can be independent of the size of the test input that is driving the on-channel. When this occurs, an arousal increment that is sufficiently large to rebound any dipole will be large enough to rebound all dipoles in a field. In other words, if the mismatch is "wrong" enough to trigger a large arousal increment, then all the errors will be simultaneously corrected. This cannot, in principle, happen in a serial processor. Moreover, the size of the rebound is an increasing

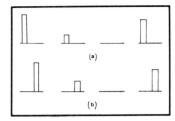

Figure 13. A rebound from on-cell activation to off-cell activation can be elicited by a rapid increment in the arousal or adaptation level of the dipole. (The size of the rebound is determined by the size of the on-cell activation. In (a) are depicted the on-responses of four cells. In (b) are depicted possible rebounds by their off-cells in response to a nonspecific increment.)

function of the size of the on-cell test input (Figure 13). Thus the amount of antagonistic rebound is precisely matched to the amount of on-cell activation that is to be inhibited. Finally, in previously inactive dipoles no rebound occurs, but the arousal increment can sensitize the dipole to future signals by changing by equal amounts the gain, or temporal averaging rate, of the on-cell and off-cell. In summary, the on-cell off-cell dipole is superbly designed to selectively reset $F^{(2)}$, and to do so in an enduring fashion because of the slow fluctuation rate of the transmitter gates.

In a reinforcement context, the rebound due to arousal shows how surprising or unexpected events can reverse net incentive motivation and thereby drive counterconditioning of a behavior's motivational support (Grossberg, 1972b, 1972c). Once the rebound capabilities of surprising events are recognized, one must evaluate with caution such general claims as "the surprising omission of . . . shock . . . can hardly act as a reinforcing event to produce excitatory conditioning" (Dickinson, Hall, & Mackintosh, 1976, p. 321).

The above mechanisms indicate how dynamical critical periods might be laid down by learned feedback expectancies. These expectancies modulate an arousal mechanism that buffers already coded populations by shutting them off so rapidly in response to erroneous STM coding that LTM recording is impossible. In other words, the mechanism helps to stabilize the LTM code against continual erosion by environmental fluctuations.

The thought experiments from which these conclusions follow are purely abstract. One experiment describes how limitations in the types of information available to individual cells can be overcome when the cells act together in suitably designed feedback schemes. Another experiment describes a solution to the noise-saturation dilemma, and yet another experiment describes how to design a chemical transducer and how dipoles formed when such transducers compete in parallel channels can achieve antagonistic rebound. As the thought experiments proceed, however, the resultant network designs take on increasingly neural interpretations. To test the theory by psychophysiological experiments, these empirical connections must be made more explicit. The next three sections discuss three of the major design features in more detail to suggest that some psychophysiological designs are examples of our abstract designs, and to explain and predict some psychophysiological phenomena using formal properties of the abstract designs as a guide. These examples are hardly exhaustive, but they will perhaps be sufficient to enable the reader to continue making new connections. Further details are in the articles of Grossberg (1972b, 1972c, 1975, 1976b, 1978e). The next three sections can be skipped on a first reading if the reader wishes to immediately study Section 15 to find out what happens when the patterns at $F^{(1)}$ and $F^{(2)}$ mutually reinforce each other.

12. Dipole Fields: Positive and Negative Aftereffects, Spatial Frequency Adaptation, Rivalry, and the McCollough Effect

Section 8 noted that $F^{(2)}$ possesses a recurrent on-center off-surround anatomy that is capable of normalizing its total STM activity within its functional channels. Section 9 showed that the cells in this recurrent anatomy are the on-cells of on-cell off-cell dipoles. I therefore conclude that $F^{(2)}$ consists of a field of on-cell off-cell dipoles such that the on-cells interact within a recurrent on-center off-surround anatomy and the off-cells also interact within a recurrent on-center off-surround anatomy. Denote by $F_+^{(2)}$ the recurrent subfield of on-cells, and by $F_-^{(2)}$ the recurrent subfield of off-cells (Figure 14). The

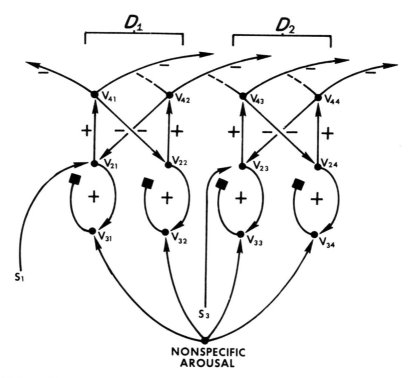

Figure 14. A possible anatomy of two dipoles (D_1 and D_2) is depicted, embedded in recurrent subfields of on-cells and off-cells. (The nonspecific arousal signal is gated by transmitters in the pathways $v_{3i} \rightarrow v_{2i}$, $i = 1, 2, \ldots$. The transmitter gates are depicted by square synapses. The arousal level hereby determines an overall level of transmitter adaptation across the dipole field. The signal S_1 turns on the cell v_{21}, which inhibits its off-cell v_{22} via the inhibitory interneuron v_{41}. Simultaneously, the on-cell v_{21} begins to differentially deplete its transmitter gate via the feedback pathway $v_{21} \rightarrow v_{31} \rightarrow v_{21}$. The interneurons v_{4i}, $i = 1, 2, \ldots$, also activate the recurrent interactions among on-cells and among off-cells that normalize their respective subfields.)

existence of neural, in particular, cortical on-cells and off-cells, and the joining together of nerve cells in on-center off-surround anatomies are familiar neural facts. Moreover, these facts have often been used to explain psychophysiological data (Carterette & Friedman, 1975; Cornsweet, 1970). The present treatment is novel in several respects, however. That a dipole field is a major tool to reset an error and to search for a correct code is, to the best of my knowledge, a new insight. Moreover, the way in which arousal fluctuations interact with slowly varying, competing transmitter gates to cause rebound or a shift in adaptation level, and the way in which shunting interactions define a quenching threshold, normalize field activity, and regulate contrast enhancement also seem to be new insights.

There exists a basic difference between the recurrent inhibition within a subfield and the dipole inhibition between on-cells and their off-cells. Dipole inhibition creates a balance between mutually exclusive categories or features. Intrafield inhibition normalizes and tunes its subfield. For example, suppose that the on-cells in a given field respond to white bars of prescribed orientation on a black field, and their corresponding off-cells respond to black bars of similar orientation on a white field. A continuous shift in the position of a white bar can induce a continuous shift of activity within the on-field, but at each position there can exist either a white bar on a black field or a black bar on a white field, but not both. Next are summarized some of the phenomena that are due to continuous changes within subfields and complementary changes when dipole rebounds cause a flip between subfields. The goal of this summary is to clarify some of the properties through which dipole fields manifest themselves in perceptual data, and to suggest that these properties are manifestations of code stabilizing mechanisms. The summary will not attempt to describe the global schemata in which these properties are embedded during a live perceptual event, although the article makes clear that interfield signaling processes, such as filtering and expectancy matching, will be important ingredients in the classification of such schemata.

An important property of a dipole field is this: If a test input excites a particular on-cell, then the on-cell inhibits its off-cell. The inhibited off-cell can, in turn, disinhibit a nearby off-cell due to the tonic arousal input and the recurrent anatomy within the off-cell field. The disinhibited off-cell thereupon inhibits its on-cell via dipole interactions. Suppose that the test input is shut off after it has been on long enough to deplete its transmitter gate. (To make this argument quantitative, we must carefully control the duration of experimental inputs relative to the transmitter depletion rate.) Then antagonistic rebound within its dipole can turn on its off-cell, which inhibits the nearby off-cell, whose on-cell is hereby disinhibited and responds by rebounding onward. Negative aftereffects are hereby generated. For example, suppose that the on-cells are orientationally selective such that nearby orientations recurrently excite each other, whereas more distinct orientations inhibit each other (Figure 15a). Then persistent inspection of a field with radial symmetry (Figure 15a) can elicit an aftereffect with circular symmetry (Figure 15c), as MacKay (1957) has reported.

In Section 5 I noted that the noise suppression properties of shunting lateral inhibition also imply spatial frequency properties. Consequently, dipole fields whose subfield inhibition is of shunting type are capable of spatial frequency adaptation. A grating with a sinusoidal luminance profile of prescribed spatial frequency will excite a band of cell types whose inhibitory fields permit maximal excitation by the input. If the input stays on for awhile, the activated transmitter gates will be differentially depleted. Test inputs

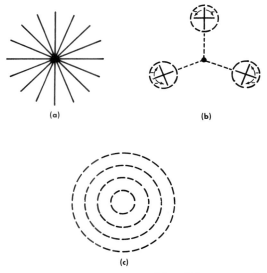

Figure 15. In (a), a pattern with radial symmetry is inspected for a long time. In (b), the net inhibitory interactions among mutually perpendicular orientations at each position are depicted. In (c), offset of the radial pattern elicits antagonistic rebounds across the field that differentially activate the perpendicular orientations.

with similar spatial frequencies share some of these gated pathways, so the overall sensitivity of response to these inputs will be less (Grossberg, 1976b). This view of spatial frequency adaptation contrasts with the view developed by Wilson (1975) that spatial frequency adaptation is due to classical conditioning of an inhibitory transmitter. It is often assumed that a slowly varying effect implies a conditioned change. The alternative notions that "fatigue" and antagonistic activity can yield perceived changes are also very old (see Brown, 1965, for a review).

The present theory refines the latter view by noting how slowly varying changes can follow from dipole adaptation without any conditioning taking place. In particular, even if the adaptational differences decay until they are very similar, contrast enhancement due to fast recurrent competitive interactions can bootstrap these differences into the perceivable range. An interaction between slow transmitters and fast recurrent interactions can hereby create behavioral effects that are much more enduring than the transmitter decay rate would suggest. This suggestion is made again later for the McCollough effect. The Wilson model differs from the present theory in its STM properties as well as in its description of slow adaptation effects. Wilson used the Wilson–Cowan equations to describe fast intercellular interactions. Among other differences, these equations do not incorporate automatic gain control by lateral inhibitory signals (Grossberg 1973). Consequently, the Wilson–Cowan equations cannot retune their sensitivity in response to shifts in background input intensity, a difficulty that also occurs in all additive models of lateral inhibition.

Pattern-contingent colored aftereffects can also be generated in a dipole field. Suppose that a prescribed field of feature detectors is color coded. Let the on-cells be maximally turned on by red light and the off-cells be maximally turned on by green light for definiteness. Then white light will excite both on-cells and off-cells; that is, white light acts like an adaptation level in this situation. Suppose that a red input whose features are extracted by the field is turned on long enough to substantially deplete its transmitter. What happens if a white input replaces the red input on test trials? The depletion caused by the prior red input now causes the white adaptation level to generate a larger gated signal to the green channel, so a green pattern-contingent aftereffect will be generated.

How enduring will this aftereffect be? Here we must recall that the anatomies of $F_+^{(2)}$ and $F_-^{(2)}$ are recurrent, and that one property of such recurrent anatomies is their ability to contrast enhance small differences in net input into large differences that can then be stored in STM (Section 8). Thus, even if the large initial differences in transmitter depletion within the on-cell off-cell dipoles decay steadily to small differences, the recurrent anatomy can contrast enhance these small differences into a perceptually visible aftereffect when the white test pattern is presented. For this to happen, however, the feature field must be protected from new inputs that can disrupt the pattern of small differences until the test trial occurs. Sleep can hereby prolong the apparent duration of the aftereffect. These properties are familiar ones in the McCollough effect (MacKay & MacKay, 1975; McCollough, 1965).

Various authors have suggested that the long duration of the McCollough effect implicates classical conditioning mechanisms. Montalvo (1976) presented a particularly ingenious application of this idea. This approach seems to trade-off one paradox for another, since the classical conditioning must produce a negative aftereffect during test trials, rather than the positive effect that was experienced during learning trials. Unless one can isolate a large class of phenomena in which classical conditioning reverses the effect on test trials, this explanation is hard to understand from the viewpoint of basic neural design. The present theory points out that slowly varying transmitter gates supplemented by rapid contrast enhancement and STM storage in a recurrent anatomy can also generate long-term effects whose duration is much longer than the transmitter decay rate would suggest. Such long-term effects must unambiguously be ruled out before classical conditioning is invoked as a unitary explanation.

Dipole field structure also helps to explain monocular rivalry (Rauschecker, Campbell, & Atkinson, 1973), whereby two superimposed gratings with the same sinusoidal luminance profile, one vertically oriented and one horizontally oriented, and each illuminated by white light or by different (say complementary) colors, are seen to alternate through time. The tendency toward rivalry can be explained by the recurrent inhibition across orientationally tuned on-cells and across orientationally tuned off-cells; the vertical on-cells tend to inhibit the horizontal on-cells, and conversely. The tendency to alternate can be explained by the fact that persistent STM reverberation of the active vertical on-cells tends to deplete their transmitter gates, thereby weakening their reverberation and providing a relative advantage to the inhibited, and therefore relatively undepleted, horizontal on-cells. When the vertical on-cell depletion reaches a critical value, the horizontal on-cells are sufficiently disinhibited to allow the recurrent dynamics to contrast enhance the horizontally coded inputs into STM. The horizontally coded on-cells thereupon reverberate in STM until the cycle repeats itself. Thus the main effect can be ascribed to combined effects of slow transmitter depletion, recurrent inhibition across orientations, and the contrast-enhancing capabilities of the recurrent network, even if there are no changes in gaze.

Of particular interest is the fact that the alternation rate depends on the color of the gratings. Two white and black gratings, or two monochromatic gratings, alternate up to three times slower than gratings that are illuminated by complementary colors. This can be discussed in terms of the rebound behavior that occurs between subfields that are orientationally coded and whose dipoles code for complementary colors. When two white and black gratings of sufficient contrast are used, the white inputs can excite both on-cells and off-cells of the color-coded dipoles, thereby inhibiting them. It is therefore assumed that apart from altering their gain, intense black and white gratings cause net excitation primarily in feature fields whose on-cells respond unselectively to light-on and whose off-cells respond to light-off. In such a feature field, the horizontal and vertical white bars excite the same subfield, and the horizontal and vertical black bars excite the complementary subfield. Each subfield tends to adapt or conserve its total STM activity (within its functional channels!) so that there exists a tendency for the horizontal and vertical inputs to compete for STM activity, and to thereby decrease the transmitter depletion rate in active cells.

By contrast, consider what happens in a color-coded dipole field in response to two gratings that use the field's complementary colors, say red-vertical and green-horizontal. Here the red-vertical bars deplete only the red field, and the green-horizontal bars deplete only the green field. There is no direct inhibition within a given subfield between horizontal and vertical orientations. Thus, other things equal, greater STM activation of red-verticals or green-horizontals is possible than in the black–white case because less intrafield competition for STM activity occurs. Greater STM activation implies faster transmitter depletion and faster alternation rates. If this explanation is correct, then it is a special case of a more general phenomenon; namely, that the frequency of perceptual oscillations can be pattern contingent due to the intrafield normalization property.

Other aftereffects provide more direct evidence for the existence of slowing varying transmitter gates. In particular, the effects of changing background illumination, or the *secondary field*, on aftereffects are remarkably similar to the effects of changing arousal level on the rebound. If a secondary field is turned on during the observation of a positive afterimage in darkness, then a rapid transition to a negative afterimage can be generated (Brown, 1965, p. 483; Helmholtz, 1866, 1924). If the secondary field is then turned off, the afterimage can revert in appearance to that of the stage when the secondary field was first turned on. In a dipole, an increase of adaptation level tends to rebound the relative dipole activities. If the arousal level is then decreased, the slowly varying transmitter levels can still be close to their original values, so that the original relative dipole activities are rapidly restored. The higher the luminance of the secondary field, the shorter the afterimage latency, and the more rapidly the afterimage is extinguished (Juhasz, 1920). In a dipole, a higher adaptation level more rapidly equalizes the amounts of transmitter in the two dipole channels by depleting them both at a faster, more uniform rate. When approximately equal levels of transmitter are achieved, the inhibitory interneurons between the dipole's populations kill any relative advantage of one population over the other. The duration of an afterimage increases with an increase in primary stimulus luminance (Brown, 1965, p. 493). In a dipole, increasing the intensity of an input to one population increases the rebound at the other population when the input terminates, much as termination of a more intense shock causes greater relief, other things being equal (Grossberg, 1972c, 1976b).

The preceding considerations lead to some experimental predictions. Some of these concern red and green cortical dipoles. For example, suppose that a red stimulus has activated a red-cell long enough to substantially deplete the transmitter. Does an increment in white light cause a green-cell rebound? Does a decrement in red light from J units to $J/2$ units cause a smaller rebound when white light is on than a decrement from $J/2$ units to 0 units? Is there an inverted $\cup$ in dipole responsiveness as a function of the arousal level or the intensity of white light? Does the relative rebound size increase as a function of arousal level size for intermediate levels of arousal? In other words, are visual dipoles designed the same way as motivational dipoles?

Another set of predictions concerns the McCollough effect. For example, how does the McCollough effect depend on the intensity of white light during test trials? A more intense white light should yield an initially larger aftereffect unless the white is so intense that overarousal occurs. Moreover, more intense white should equalize the relative transmitter stores more rapidly than less intense white. This suggests an experiment in which a double test is made. The first test uses prolonged inspection of white bars whose intensity differs across subjects. Before the second test is made, some visual experience should occur to blot out whatever small differences in transmitter storage might still exist after the bright white bars are examined. Then a second test with white bars is given. Subjects who saw less intense bars on the first test should perceive a larger aftereffect.

An experiment concerning spatial frequency adaptation is also suggested. This experiment is analogous to the experiment on aftereffects due to changes in the secondary field. Speaking generally, if spatial frequency adaptation and certain other aftereffects are all due to dipole depletion, albeit in different fields of feature detectors, then they should undergo similar transformations in response to analogous experimental manipulations, other things being equal, notably the persistence with which each feature field is disrupted by uncontrolled inputs. Suppose that when a series of vertical sinusoids drifts horizontally across the visual field, those on-cells and off-cells whose recurrent inhibitory signals collide with visually induced inputs will have their activities suppressed. Consider the on-cells and off-cells that can be activated by the prescribed spatial frequency. What happens as the contrast of the visual pattern is parametrically increased across subjects?

This is a delicate question because more than one dipole field in the coding hierarchy can be activated by such an input. Let us consider what would happen if only one dipole

field is activated. In the limit of absolute black and very bright white verticals, both the on-cells and the off-cells would be almost equally excited on the average, albeit at different times, as the light and dark verticals drift over their receptive fields. Neither on-cell nor off-cell would gain a large *relative* advantage, but both would have their transmitter stores significantly depleted by the persistence of the horizontally drifting input. Hence, significant spatial frequency adaptation would occur, but not due to large relative imbalances in the dipoles. What happens as the contrast between the white and black verticals is decreased? Then other things being equal, the off-cells will be depleted more than the on-cells. Hence, a greater relative depletion within the dipoles can be induced at smaller contrast levels than at larger contrast levels. How can this conclusion be tested? Consider two groups of subjects. Let Group 1 be adapted and tested using high contrast gratings. Let Group 2 be adapted on a lower contrast grating and tested using the same higher contrast grating used to test Group 1. The net on-responses at a black–white interface as the test grating slowly drifts across the visual field should be greater in Group 2 than in Group 1. Can such differentially enhanced boundaries between the trailing edge of black and the leading edge of white be perceived? If the answer is yes, then one can properly claim that the effect is a functional analogue within the visual system of the partial reinforcement acquisition effect in the motivational system (Grossberg, 1975).

13. Reset Wave: Reaction Time, P300, and Contingent Negative Variation

The nonspecific arousal that is triggered by unexpected events (or mismatch) selectively and enduringly inhibits active population across $F^{(2)}$. In vivo, do there exist broadly distributed inhibitory waves that are triggered by unexpected events? In average evoked potential experiments, one often finds such a wave, namely the P300 (Rohrbaugh, Donchin, & Eriksen, 1974; Squires, Wickens, Squires, & Donchin, 1976). The theory's relationship to P300 is discussed in Grossberg (1978e), in which the following properties of P300 are shown to be analogous to properties of the resetting wave: Reaction time is an increasing function of P300 size (Squires et al., 1976); P300 is not the same average evoked potential as the contingent negative variation (CNV) (Donchin, Tueting, Ritter, Kutas, & Heffley, 1975; cf. Section 16); P300 can be elicited in the absence of motor activity (Donchin, Gerbrandt, Leifer, & Tucker, 1972); resetting the STM codes of longer sequences of events can take longer than resetting the STM codes of shorter sequences of events, and due to the relationship between reaction time and P300 size, longer sequences will elicit larger P300s (Remington, 1969; Squires et al., 1976). Moreover, Chapman, McCrary, and Chapman (1978) showed that in a number- and letter-comparison task, there existed an evoked potential component with a poststimulus peak at about 250 msec that is related to the storage of cue-related information in STM. This latency fits well with the idea that STM storage occurs if the feedback expectancy does not create a mismatch. The extra 50 or so msec needed to generate a P300 would also be necessary in the network to trigger the reset wave if a mismatch does occur.

If the P300 is indeed a reset wave of the type that the thought experiment describes, then several types of experiments can be undertaken to test this hypothesis. On the anatomical side, Where does the expectancy matching take place? What pathways subserve the arousal? On the physiological side, Do dipole rebounds cause the inhibition? On the psychophysiological side (e.g., average evoked potential experiments), Is there a more direct experimental paradigm for testing whether P300 directly inhibits STM? In particular, Can a succession of P300s be reliably triggered when information is disconfirmed in successive stages? On a deeper functional level, Does the P300 act to buffer committed cells against continual recording by the flux of experience? If P300 is inhibited, can previously committed cells be recoded? In other words, when we consider cognitive coding, does a chemical switch contribute to code stability, or is code stability entirely dependent on buffering by dynamic reset mechanisms?

As was noted in Section 4, feedback expectancies that trigger STM reset mechanisms should

occur in many thalamocortical systems, so that there should exist different reset waves corresponding to each functionally distinct system. In Grossberg (1978e), the preceding scheme is generalized to a variety of examples in which competition occurs between attentional, or consummatory, pathways and novelty, or orienting, pathways. A matching process goes on within the attentional system and computes such information as follows: Are the sensory cues the ones that are expected? Do the proprioceptive motor cues match the terminal motor map that is guiding the limb? If the answer is yes, then goal-oriented arousal systems are activated to support the matching process and its consequences, such as posture. If the answer is no, then complementary arousal systems are activated that support rapid reset and orienting reactions aimed at acquiring new information with which to correct the error. Given that the P300 helps to reset sensory STM in response to unexpected events, does there exist a complementary wave that occurs along with expected events? The CNV would appear to be such a wave (Cohen, 1969), since it is associated with an animal's expectancy, decision (Walter, 1964), motivation (Cant & Bickford, 1967; Irwin, Rebert, McAdam, & Knott, 1966), volition (McAdam, Irwin, Rebert, & Knott, 1966), preparatory set (Low, Borda, Frost, & Kellaway, 1966), and arousal (McAdam, 1969).

If the P300 and the CNV are indeed complementary waves, then experiments should be undertaken to determine the neural loci at which the generators of these waves compete. For example, Section 16 suggests that the hippocampus provides output that contributes to the CNV. Does expectancy mismatch occur within the hippocampus, or in a cell nucleus that activates hippocampus, and thereby release a P300 by disinhibiting its generator?

Having noted the existence of reset and attentional waves that are triggered by sensory events, it is natural to ask whether there exist analogous waves that are triggered by motor events? To answer this question, the next section considers how eye movements can modulate the LGN's sensitivity to afferent visual signals and the related questions of whether the LGN has a dipole field organization and whether feedback from visual cortex to LGN can selectively attenuate or amplify afferent visual signals. This discussion leads to a reinterpretation of LGN data and to some predictions. These predictions concern the possible existence of a reset motor wave and the timing of certain developmental events relative to the end of the critical period for plasticity in the primary visual cortices.

14. Template Matching and Reset: PGO Wave, Geniculate Dipoles, and Corticogeniculate Feedback

An example of an "attentional" motor wave seems to be the ponto-geniculo-occipital (PGO) wave whose effects on the LGN are admirably reviewed by Singer (1977). Singer (1977) distinguished at least two types of inhibitory interneurons in his discussion of LGN dynamics:

There apparently are two inhibitory mechanisms with two different functions. One is based on intrinsic interneurons and presumably conveys the retinotopically organized and highly selective inhibitory interactions between adjacent retinocortical channels. . . . This inhibition seems to be mainly of the feed-forward type. . . . The second inhibitory pathway is exclusively of the recurrent type and is relayed via cells in nucleus reticularis thalami. . . . This extrinsic inhibitory loop is probably involved in more global modifications of LGN excitability as they occur during changes in the animal's state of alertness and during orienting responses associated with eye movements. (p. 394)

Singer noted that mesencephalic reticular formation (MRF) stimulation leads to field potentials in the LGN and the visual cortex that closely resemble PGO waves. LGN transmission is facilitated during PGO waves and during the analogous negative field potential that occurs after MRF stimulation.

One mechanism of MRF facilitation is inhibition of the cells in the nucleus reticularis thalami, which are recurrent inhibitory interneutrons between LGN relay cells. From a theoretical viewpoint, this type of disinhibition would be expected to have nonspecific effects like decreasing the quenching threshold of an entire recurrent subfield of cells, and thereby facilitating transmission of signals through these cells (Grossberg, 1973; Grossberg & Levine, 1975). Such an effect seems to occur in LGN. Since MRF stimulation can completely suppress inhibitory postsynaptic po-

tentials elicited from optic nerve or optic radiation, Singer (1977) concluded

that the intrinsic inhibitory pathways also get inactivated. However, it cannot yet be decided whether the inhibitory interneurons in the main laminae are also subject to direct reticular inhibition as is the case for cells in nucleus reticularis thalami. (p. 409)

Singer went on to suggest that corticogeniculate feedback could partially accomplish the intrinsic cell inhibition.

For present purposes, the main point is Singer's (1977) functional interpretation of the MRF-induced LGN disinhibition. He claimed

that the brief phase of disinhibition serves to reset the thalamic relay each time the point of fixation is changed. . . . To assure a bias-free initial processing of the pattern viewed after a saccade . . . inhibitory gradients ought to be erased before the eyes come to rest on the new fixation point . . . the concomitant disinhibition occurs only towards the end of the saccade right before the eyes come to rest. (p. 411)

Singer's remarks can be mechanistically interpreted as follows: As the proprioceptive coordinates of the eye muscles approach the terminal motor coordinates that control the saccade, the two sets of coordinates match, a PGO wave is initiated, it disinhibits LGN relay cells, and prepares the LGN to transmit retinal signals to the visual cortex. If the PGO wave is indeed elicited by a matching process between the terminal motor map and proprioceptive coordinates of the eye muscles, then this matching process should be capable of exciting cells that inhibit the LGN interneurons within the nucleus reticularis thalami. In what neural structure does this matching process take place? One component of this structure might already have been discovered by Tsumoto and Suzuki (1976), who report a pathway from the frontal eye fields to the perigeniculate nucleus in which are found the LGN inhibitory interneurons. Electrical stimulation of the frontal eye fields inhibits the perigeniculate cells and facilitates LGN transmission.

Singer (1977) claimed that the PGO wave resets the LGN so that it can respond to retinal signals without bias. However, nonspecifically reducing the quenching threshold is not the type of selective reset that I have discussed earlier. Indeed, Singer's discussion of LGN dynamics emphasizes the wiping away of all inhibitory gradients as a reset mechanism. But what if excitatory activities already exist in the LGN when this happens? Why do these activities not get amplified and thereupon maximally bias LGN activity in response to the next retinal input volley? I suggest that the LGN reset that is due to the nucleus thalami reticularis occurs while the eye is moving and the extrinsic inhibitory interneurons are active. This extrinsic inhibitory feedback resets the LGN by generating a high quenching threshold and thereby wiping out the LGN's excitatory patterns. As the eye comes to rest at its intended position, I suggest that matching occurs between the terminal and proprioceptive motor maps of the eye muscles, thereby activating the attentional system, in particular the PGO wave, which sensitizes the LGN to retinal and cortical signals.

Even if the preceding interpretation of Singer's argument is correct, it discusses a nonspecific effect on the QT and the sensitivity of visual pattern processing, but not the selective reset that aims at reorganizing attention in response to an error, or other unexpected event.

Is there a wave that is functionally complementary to the PGO wave, that can precede it, and that drives a selective reset of LGN dynamics in response to unexpected events? If such a wave does exist, it would be functionally analogous to the P300. In this regard, Singer (1977) parenthetically mentions the work of Foote, Manciewicz, and Mordes (1974) to explain the inhibition of LGN transmission that sometimes occurs shortly after MRF stimulation but before the facilitatory phase. Foote et al. suggest that this inhibitory pathway is due to serotonergic fibers originating in the dorsal raphe nucleus. Are these fibers the pathway over which selective reset can occur?

For a selective reset wave to exist, it must operate on on-cell off-cell dipoles. Do such dipoles exist in the LGN? Much of the data discussed by Singer was collected in the cat LGN. Singer (1977) reports here

that reciprocal inhibitory connections exist between adjacent neurons driven by the same eye that have the same receptive field center characteristics; i.e., between on-center cells and between off-center cells, respectively. (p. 390)

These interneurons are analogous to the intrafield lateral inhibition that was postulated within $F_+^{(2)}$ and $F_-^{(2)}$, but which we now recognize as a prerequisite for total activity adaptation and quenching threshold tuning in any recurrent network. In addition, there exist "reciprocal inhibitory interactions between neurons with antagonistic field center characteristics—that is, between on- and off-center units with spatially overlapping receptive fields" (Singer, 1977, p. 390). These cells would appear to form dipoles. If they are dipoles of the type discussed, then the arousal system that triggers their rebounds will feed into them—from the dorsal raphe nucleus— and activating this arousal system will rebound their relative activities.

These hypotheses should be easier to test in the monkey than the cat, because Schiller and Malpeli (1978) have reported that of the four parvocellular layers in the monkey, the two layers committed to the left eye are subdivided into an on-cell layer and an off-cell layer, and the two layers committed to the right eye are also subdivided into an on-cell layer and an off-cell layer. Do dipole interactions occur between the on-cell and off-cell layers of each eye representation? Does a suitable arousal increment rebound the relative activities of these dipoles? If so, we will have found an elegant functional reason for the existence of this structure in the monkey: Each eye has its own dipole field to carry out its selective reset modes. We will also have found an elegant reason for the existence of intrinsic and extrinsic inhibitory systems: Attentional reduction of the quenching threshold is functionally distinct from, and even complementary to, selective reset.

Another important point of Singer's (1977) article concerns the role of corticogeniculate feedback.

In a highly selective way the cortex permits transmission of binocular information that can be fused and evaluated in terms of disparity depth cues while it leaves it to the intrinsic LGN circuits to cancel transmission of signals that give rise to disturbing double images. (p. 398)

In other words, the corticogeniculate feedback acts as a template that selectively enhances the type of data that the cortex is capable of coding in a globally self-consistent way.

In summary, the LGN seems to enjoy a dipole field structure whose sensitivity to afferent sensory signals is modulated both by corticogeniculate feedback, which acts like a sensory expectancy-matching mechanism, and by MRF arousal, which lowers the LGN QT in response to proprioceptive-terminal map matching within the eye movement system.

If we interpret the geniculocortical relay as an example of our thought experiment, then several experimental predictions arise. These predictions are made with caution, since a significant part of visual development seems to be genetically prewired in the geniculocortical pathways of higher mammals (Hubel & Wiesel, 1977). It is still not clear, however, to what extent corticogeniculate feedback does help to terminate the visual critical period in these animals. Nor is it clear whether the same neural design that is used in some species, or in individual neural relays, to terminate a critical period using feedback is also used in others wherein a chemical switch or other prewired mechanisms are appended. The predictions flow from the observation that if the geniculocortical system is an example of the thought experiment, *albeit* vestigially, then its reset and search mechanisms must develop before the end of the visual critical period. In particular, if lateral inhibition within the LGN is used to help match cortical and retinal data, then these inhibitory connections must develop before the end of the critical period. The dipole field structure of the cortex must also develop before the end of the critical period. Moreover, mismatch within the LGN system should disinhibit an arousal system capable of rebounding the cortical dipoles. There exists a catecholamine arousal system to neocortex, among other structures (Fuxe, Hökfelt, & Ungerstedt, 1970; Ungerstedt, 1971; Jacobowitz, 1973; Lindvall & Björklund, 1974; Stein, 1974). Is this the arousal system being sought? Does it develop before the end of the critical period? Is this arousal system capable of driving antagonistic rebound in cortical dipoles? Is a catecholamine transmitter always used in arousal systems that drive antagonistic rebound, for example, the catecholamine system originating in the dorsal raphe nucleus that was described by Foote et al. (1974)? Finally, is there a structural similarity between all pairs of attentional and selective reset waves, as between CNV and P300, or PGO and its hypothesized complementary reset wave?

15. Adaptive Resonance, Code Stability, and Attention

The preceding sections discuss some of the network events that occur when feedforward data mismatch feedback expectancies. What happens if an approximate match occurs? Then the activity patterns at $F^{(1)}$ and $F^{(2)}$ elicit interfield signals that mutually reinforce each other, and activities at both levels are amplified and locked into STM. Because the STM activities can now persist much longer in time than the passive decay rates of individual cells, the slowly varying feed-forward filters and feedback expectancies have sufficient time to sample the STM patterns and store them in LTM. I call this dynamical state an *adaptive resonance*. The resonant state provides a global interpretation of the afferent data, or a context-dependent code, that explicates in neural terms the idea that the network is paying attention to the data. The resonance idea suggests that many individual neural events, such as cell potentials and axonal signals, are behaviorally irrelevant until they are bound together by resonant feedback. Of special importance is the observation that unless resonance occurs, no coding in LTM can take place. This observation clarifies from a mechanistic viewpoint the psychological fact that a relationship exists between paying attention to an event and coding it in LTM (Craik & Lockhart, 1972; Craik & Tulving, 1975).

The resonant state provides a context-dependent code due to several factors acting together. For example, the pattern of expectancy feedback can alter, through a matching process, the activity that a given feature detector would have experienced if only afferent signals were operative. Similarly, competitive interactions within a subfield can rapidly alter the net input pattern before storing it in STM. Thus when an activity pattern at a field $F^{(1)}$ is projected by interfield signaling to a field $F^{(2)}$, feedback from $F^{(3)}$ to $F^{(2)}$ can deform this pattern before it is further reorganized by competition within $F^{(2)}$. Because the resonant state provides a context-dependent code whose resultant patterns in STM and LTM depend on all active components of the system, it is impossible to determine the code from the measurements taken by any single microelectrode, no matter how precise its calibration. I claim that adaptive resonances are the functional units of cognitive coding, and that classification of the resonances that occur in prescribed situations is a central problem for cognitive psychology. The structural substrates of these cognitive units are nonlinear feedback modules involving whole fields of cells rather than individual nerve cells.

The technical details needed to rigorously build up the resonance idea are derived in Grossberg (1976a, 1976b), in which a summary of related coding models is also given, and further developed in Grossberg (1978e). In particular, Grossberg (1976a) points out that a coding theory that depends on a feedforward anatomy with any fixed number of cells is faced with a crippling dilemma: Either a chemical switch turns off code development at a prescribed time, but then the code will be behaviorally meaningless with a high likelihood, or the code is unstable through time whenever the number of patterns in the environment significantly exceeds the number of coding cells. It is also proved that a developing code can be stable in a sparse input environment, but this does not address the typical situation in vivo, where a continuous visual flow, and therefore a nondenumerable series of visual patterns, must be dealt with. Computer models of code development missed these basic points because they typically used small numbers of inputs and small numbers of coding cells.

Once the main point was vividly made, one could see that feedback was essential to stabilize a developing code in a rich input environment, and that the types of feedback that were needed resembled attentional mechanisms that had previously been derived from different considerations, namely classical and instrumental conditioning postulates (Grossberg, 1975). The examples in Sections 2 and 3 illustrate these attentional phenomena. Two pleasing conclusions were thereby drawn:

Adult mechanisms, in this case attentional mechanisms, are often continuations along a developmental continuum of infantile mechanisms, in this case code development mechanisms; and the rather mysterious rubrics of "paying attention" and "expectancy" could be attached to the more substantial theme of "code stability and consistency," and the establishment of dynamically maintained critical periods.

Anderson, Silverstein, Ritz, & Jones (1977) also recognized the importance of feedback in defining the functional units of neural network. Their model differs, however, from the present theory in several notable respects. The recurrent STM interactions in their model are defined by linear feedback signals. Grossberg (1973, 1978d) shows how linear feedback signals among cells that are capable of saturating create unphysical instabilities such as noise amplification and compression of an input pattern. Furthermore, known neural nonlinearities, such as sigmoid signals between cells, overcome these instabilities and contrast enhance the input pattern (Appendix D). The LTM interactions in the Anderson et al. model are described by summing up a large number of mutually orthogonal LTM vectors

$$z = \sum_{k=1}^{n} z_k$$

to form the total LTM trace across a field $F^{(1)}$ of cells. When a signal pattern S from $F^{(1)}$ to $F^{(2)}$ is gated by the total LTM trace, as in

$$S \cdot z = \sum_{k=1}^{n} S \cdot z_k$$

(see Appendix A), it might be perpendicular to all but one of the increments, say z_1. Consequently, the net signal is $S \cdot z_1$. This concept gets into difficulty because in vivo the total LTM trace z must be composed of small quantities (e.g., transmitter concentrations). Each of the summands z_k must therefore be a very small quantity unless n is also very small, but then the theory is powerless. If each z_k is very small, then the net signal depends on gating by a sum of very small quantities. This creates an unstable situation. Furthermore, the LTM trace z_{ij} from cell v_i to cell v_j in the

Anderson et al. model is assumed to equal the LTM trace z_{ji} from v_j to v_i. This symmetry assumption is too restrictive for our purposes, since we do not want the filter from $F^{(1)}$ to $F^{(2)}$ to necessarily equal the expectancy from $F^{(2)}$ to $F^{(1)}$; this would limit the tendency to achieve greater abstractness of feature extraction in a hierarchy of fields. A more serious problem for the LTM symmetry assumption is its implication that the signal from every cell be proportional to its STM trace. This follows because the growth rate of LTM trace z_{ij} is proportional to the product of signal S_{ij} from v_i and v_j times STM trace x_j of v_j. To achieve $z_{ij} \equiv z_{ji}$, it is necessary for $S_{ij}x_j \equiv S_{ji}x_i$, which is possible if $S_{ij} = \alpha x_i$ and $S_{ji} = \alpha x_j$. In particular, the recurrent signals must be linear functions of the STM traces, and the usual instabilities that recurrent linear signals generate among cells will be generated.

The next three sections summarize a few resonant schemes and predictions pertaining thereto that suggest the scope of the resonance phenomenon. Other resonances, notably the olfactory resonance that is described by the distinguished work of Freeman (1975), are discussed more completely in Grossberg (1976b, 1978e). Freeman discovered a resonant phenomenon by performing parallel electrode experiments on the cat prepyriform cortex. When the cat smells an expected scent, its cortical potentials are amplified until a synchronized oscillation of activity is elicited across the cortical tissue. The oscillation organizes the cortical activity into a temporal sequence of spatial patterns. The spatial patterns of activity across cortical cells carry the olfactory code. By contrast, when the cat smells an unexpected scent, then the cortical activity is markedly suppressed. Freeman traces the differences in cortical activity after expected versus unexpected scents to gain changes within the cortical tissue. Appendix C shows how a matching mechanism in a shunting network simultaneously changes gains as it amplifies or attenuates network activity. Freeman also notes a tendency for the most active populations to phase-lead less active populations. This also occurs automatically in a shunting network due to the correlation between gain and asymptote. Because the cortex oscillates, Freeman models his data

using second-order differential equations whose coefficients are changed by expectations in a manner that is descriptively stated, but not dynamically explained, by his model. I suggest that the oscillations are caused by feedback between cells that obey first-order differential equations whose gains are changed by signals coupled to the shunting mechanism.

Grossberg (1978b) claimed that adaptive resonances also occur in nonneural tissues, where they are suggested to be a basic design principle in a universal developmental code. Syncytium formation during sea urchin gastrulation is identified as a possible adaptive resonance phenomenon—in particular, the law whereby pseudopods from the mesenchymal cells adhere to ectodermal cells to form a syncytium has the same form as the law for an LTM trace—and some predictions are made to test this hypothesis.

16. Are Conditioned Reinforcer Pathways and Conditioned Incentive Motivational Pathways Reciprocal Pathways in an Adaptive Resonance?

Grossberg (1975) has described a psychophysiological theory of attention in which an adaptive resonance occurs. This resonance helps to explain why the dilemma of crossconditioning that is depicted in Figure 2 does not routinely occur. Figure 16 idealizes this resonance. Speaking intuitively, the internal representations of external cues elicit signals that are, before conditioning takes place, distributed nonspecifically across the various internal drive representations. During conditioning trials, the pattern of reinforcement and drive levels strengthens the LTM traces within certain of these signal pathways and weakens the LTM traces within other pathways. These conditioned changes in the signal pathways endow the external cues with conditioned reinforcer properties. On recall trials, these conditioned signals combine with internal drive inputs to determine whether or not feedback signals will be elicited. The feedback signals play the role of incentive motivation in the network. Incentive motivation is released in a given feedback pathway only if the momentary balance of conditioned reinforcer signals plus drive inputs compete

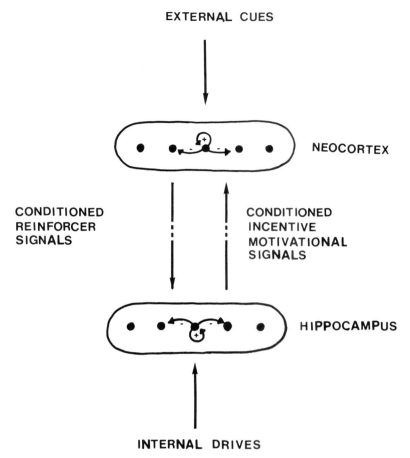

EXTERNAL CUES

NEOCORTEX

CONDITIONED
REINFORCER
SIGNALS

CONDITIONED
INCENTIVE
MOTIVATIONAL
SIGNALS

HIPPOCAMPUS

INTERNAL DRIVES

Figure 16. An adaptive resonance between neocortex and hippocampus is suggested to occur when external cues are compatible with internal needs. (The conditionable feedback pathways in this module subserve conditioned reinforcer properties and conditioned incentive motivational properties.)

favorably against these factors within the other feedback pathways.

Before conditioning occurs, each of the incentive motivational channels nonspecifically projects to the external cue representations. As in the case of the conditioned reinforcers, the incentive motivational channels are conditionable, and their LTM traces can be strengthened when their signals are large and contiguous to active external cue representations. Thus after conditioning occurs, an internal drive representation can deliver incentive motivational signals preferentially to those external cue representations with which it was previously associated. In this way, activating a given external cue representation can sensitize an ensemble of motivation-

ally related external cue representations via incentive motivational feedback. A type of subliminal psychological set is hereby formed. Since the external cue representations compete among themselves for storage in STM, the conditioned incentive motivational feedback abets the storage of compatible cues and tends to overshadow the storage of incompatible cues. Thus, during alternate scanning of incompatible cues, attentional switching between resonances compatible with one cue class and then the other class buffers each class against indiscriminate cross-conditioning with incompatible cues. Various data and related theories about reinforcement and attention are analyzed in the light of these concepts in the articles of Grossberg (1972b,

1972c, 1975). Herein I suggest a neural substrate for this resonance and a psychophysiological experiment to test its existence.

In Figure 16, $F^{(1)}$ and $F^{(2)}$ both possess recurrent on-center off-surround interactions, and both the $F^{(1)} \to F^{(2)}$ and $F^{(2)} \to F^{(1)}$ pathways are conditionable. Region $F^{(1)}$ contains external cue representations, and region $F^{(2)}$ contains internal drive representations. When this network is embedded into a more complete system of interactions, an interpretation of $F^{(1)}$ as neocortex and of $F^{(2)}$ as hippocampus is suggested. Given this interpretation, the conditioned reinforcer pathways $F^{(1)} \to F^{(2)}$ should have a final common pathway at hippocampal pyramidal cells, and their LTM traces should be sensitive to the balance of drives and reinforcements through time. Relevant data have been collected by Berger and Thompson (1977), who describe neural plasticity at the hippocampal pyramids during classical conditioning of the rabbit nictitating membrane response.

The conditioned incentive motivational pathways $F^{(2)} \to F^{(1)}$ should have a final common pathway at neocortical pyramidal cells, and their LTM traces should be sensitive to the balance of motivation and cue saliency through time. The CNV is a conditionable neocortical potential shift that has been associated with an animal's motivational state (Cant & Bickford, 1967; Irwin et al., 1966), and Walter (1964) has hypothesized that the CNV shifts the average baseline of the cortex by depolarizing the apical dendrites of its pyramidal cells. If the conditioned incentive motivational feedback is indeed realized by the CNV and if adaptive resonances between conditioned reinforcers and conditioned incentives do exist, then there should exist neural feedback loops between neocortex and hippocampus such that while conditioned reinforcer properties are being established with the hippocampal pyramid cells as a final common pathway, simultaneously conditioned incentive properties are being conditioned with the apical dendrites of neocortical pyramid cells as a final common pathway. Experiments to test this prediction would require either simultaneous measurement from electrodes in the neocortical and hippocampal loci of the resonant circuit, or correlation of electrode measurements in the hippocampus simultaneous with CNV measurements.

17. Pattern Completion, Hysteresis, and Gestalt Switching

Consider what happens to an adaptive resonance as its afferent data are slowly and continuously deformed through time, say from the letter O to the letter D. By "slowly" I mean slowly relative to the rate with which resonant feedback can be exchanged. Recall that feedback from $F^{(2)}$ to $F^{(1)}$ can deform what "is" perceived into what "is expected to be" perceived. Otherwise expressed, the feedback is a prototype, or higher order Gestalt, that can deform and even complete activity patterns across lower order feature detectors. For example, suppose that a sensory event is coded by an activity pattern across the feature detectors of a field $F^{(1)}$. The $F^{(1)}$ pattern is then coded by certain populations in $F^{(2)}$. If the sensory event has never before been experienced, then the $F^{(2)}$ populations that are chosen are those whose codes most nearly match the sensory event because the pattern at $F^{(1)}$ is projected onto $F^{(2)}$ by the positional gradients in the $F^{(1)} \to F^{(2)}$ pathways (Appendix A). If no approximate match is possible, then mismatch at $F^{(1)}$ will trigger a reset wave that selectively inhibits $F^{(2)}$ and elicits a search routine. If an approximate match is possible, however, then the feedback signals from $F^{(2)}$ to $F^{(1)}$ will elicit the template of the sensory events that are optimally coded by the $F^{(2)}$ pattern. These feedback signals rapidly deform the $F^{(1)}$ pattern until this STM pattern is a mixture of feedforward codes and feedback templates. Otherwise expressed, $F^{(2)}$ tries to complete the $F^{(1)}$ pattern using the prototype, or template, that its active populations release.

In Grossberg (1978e, Section 40), another completion mechanism is also suggested, namely a normative drift. This mechanism generalizes the line neutralization phenomenon that was described by Gibson (1937). In suitably designed feature fields, STM activity at a particular coding cell can spontaneously drift toward the "highest order" coding cell in its vicinity, due either to the existence of more cell sites, or to larger and spatially more broadly distributed feedback signals, at the highest order cells. After STM activity drifts to its local norm, the highest order cell can thereupon release its feedback template. It was shown in Levine and Grossberg (1976) that such drifts are a type of lateral masking due to the recurrent interactions within the feature field. I suggest that many Gestalt-like pattern completions are manifestations of intrafield competitive transformations, such as normative drifts, and the deformation by feedback expectancies of lower order STM patterns. Such global dynamical transformations transcend the capabilities of classical pattern discrimination models (e.g., Duda & Hart, 1973).

Two important manifestations of the completion property are hysteresis and Gestalt switching. For example, once an STM resonance is established in response to the letter O, the resonance resists changing its codes when small changes in the sensory event occur—this is hysteresis. Hysteresis occurs because the active $F^{(2)} \rightarrow F^{(1)}$ template keeps trying to deform the shifting $F^{(1)}$ STM pattern back to one that will continue to code the $F^{(2)}$ populations that originally elicited this template.

If, however, the sensory event changes so much that the mismatch of test and template patterns becomes too great, then the arousal-and-reset mechanism is triggered. This event inhibits the old code at $F^{(2)}$ and forces a search for a distinct code. A dramatic switch between global percepts can hereby be effected. The global nature of the switch is due not only to the rapid suppression of the previously active $F^{(2)}$ code but also to the fact that $F^{(2)}$ contains populations that can synthesize data from many feature detectors in $F^{(1)}$, and the feedback templates of these populations can reorganize large segments of the $F^{(1)}$ field. I suggest that an analogous two-stage process of hysteresis and reset is operative in various visual illusions, such as Necker's cube (Graham, 1965). When ambiguous figures are presented, these mechanisms can elicit spontaneous switches of perceptual interpretation due either to shifts of gaze or to the input-induced cyclic rates of transmitter depletion that can occur even if the gaze remains relatively fixed (Section 12).

If Gestalt switching is a two-stage process, then at the moment of switching, a reset wave should occur. Does a P300 occur at the moment of perceived switching? If not, can this paradigm be used to discover what average evoked potential, if any, parallels activation of the reset mechanism?

18. Binocular Resonance and Rivalry

My final example indicates that adaptive resonances need not be hierarchically organized, and points to a class of resonances of particular importance. Hysteresis can occur between two reciprocally connected fields even if they are not hierarchically organized, since the individual cells do not know whether they are in a hierarchy. For example, suppose that each eye activates a field of monocularly coded feature detectors. Suppose that each monocular field is endowed with a recurrent on-center off-surround anatomy, indeed with a recurrent dipole field of on-cell and off-cells. Let the on-cells in each monocular field be capable of exciting corresponding on-cells in the other monocular field. In other words, signals from a given monocular field act as a template for the other monocular field. It does not matter what features are coded by these detectors to draw the following conclusion. Once a resonance is established between the two monocular fields, hysteresis will prevent small and slow changes in the input patterns from changing the coded activity. Julesz (1971) introduced a field of physical dipoles to model the binocular hysteresis that he and Fender described (Fender & Julesz, 1967). Resonance between two recurrent on-center off-surround anatomies undergoing shunting dynamics provides a neural model of the phenomenon. Such a binocular resonance will generate properties of binocular rivalry, since competition within each subfield of the recurrent networks will inhibit feature detectors that do not participate in the resonance.

The construction of monocular representations whose binocular resonances code globally self-consistent invariants of stereopsis is presently being undertaken. Although this construction is not yet complete, some observations can be made in broad strokes to guide the reader who is interested in pursuing the elucidation of perceptual and motor resonances.

Before the two eyes can fixate on the scenes that will drive binocular development, there must already exist enough prewired visual feature detectors to direct the eye movement system to lock in the fixation process. Thus the existence of prewired visual feature fields does not argue against the need for visual tuning by experience. Such tuning seems necessary to achieve accurate stereopsis in the face of significant variations in bodily parameters due to individual differences and growth (Daniels & Pettigrew, 1976). An effort should be made to correlate individual and species differences in the motor mechanisms that are used to accumulate visual data and to act on the visual environment with corresponding differences in prewired sensory feature detectors and the ultimate feature fields that can be synthesized (Arbib, 1972; Creutzfeldt & Nothdurft, 1978).

Even if feature development can continue in the absence of visual experience, this does not imply that visual experience does not alter visual development. Just as imprinting can be driven by endogenous drive sources that are later supplanted by environmentally reactive drive sources (Sluckin, 1964), an effort should be made to test whether visual development is driven by endogenous arousal sources before these sources are supplanted by visual experience, in particular by visually reactive arousal sources.

Binocular visual resonances seem to be a special case of bilateral resonances that are due to the bilateral organization of the body, for example, binaural auditory resonances. As in the case of binocular corticogeniculate feedback (Singer, 1977), bilateral interactions at each of several anatomical stages help to select the activity patterns that elicit and are modulated by hierarchical signals. The hierarchical signals are supplemented by environmental feedback signals to complete the sensorimotor loops that control the circular reactions of a developing individual (Piaget, 1963). An effort should be made to correlate the structures that emit the environmental signals with those that receive them, for example, the algebraic properties of motor speech commands with the corresponding

properties of auditory feedback patterns (Grossberg, 1978e).

19. Symmetry and Symmetry Breaking in Sensory and Motor Systems

An important theme in the design of adaptive resonances will be the analysis of their symmetry and symmetry-breaking properties. This theme is unavoidable when sensory resonances are studied side by side with their motor counterparts, as Section 18 suggests. For example, the system schematized in Figure 9 shows a manifest asymmetry in the construction of its arousal and pattern analysis components. However, this system forms only one part of a larger system that enjoys a much more symmetric structure in which two subsystems compete, namely an attentional and an orienting subsystem (Lynn, 1966). The component in Figure 9 is part of the attentional system, which also includes incentive motivational and CNV components (Grossberg, 1975). This subsystem focuses attention on cues that are expected to generate prescribed consequences of behavior. It can overshadow irrelevant cues, as in Section 2, by selectively amplifying certain patterns at the expense of others. The complementary orienting system is also capable of selectively amplifying patterns, but these are not the patterns that code for sensory or cognitive events. They are, rather, the motor maps that are capable of directing the subject toward sources of unexpected environmental events.

The dichotomous but interdependent nature of these subsystems is illustrated by the existence of X-cells and Y-cells in mammalian retinas and by the neural pathways that these cells excite. The X-cells project primarily to the LGN, where their signals are processed as visual data, whereas the Y-cells have axons that bifurcate to send branches both to the LGN and the superior colliculus (Fukuda & Stone, 1974; Robson, 1976), which has been identified as an area in which a visuomotor map for eye movements is elaborated (Wurtz & Goldberg, 1972). The competitive nature of these two subsystems is illustrated by considering how different our motor reactions can be when a loud sound to the left is unexpected versus when it is a learned discriminative cue for rapid button pushing that will be highly rewarded if it is sufficiently rapid. In the former case, our eyes and head rotate rapidly to the left. In the latter case, rotation can be inhibited and supplanted by a rapid button push.

Competition between attentional and orienting subsystems may clarify certain paradoxes about mental illness. As just summarized, the attentional system focuses attention on cues that are expected to generate prescribed consequences of behavior and can thereby overshadow irrelevant cues. The competing system is triggered by unexpected events (novelty) and allows the network to redefine the set of relevant cues to avoid unexpected consequences. Overarousal of either subsystem can yield attentional deficits (Grossberg, 1972c; Grossberg & Pepe, 1970, 1971), but the exact nature of the deficit and its proper treatment depends on the particular subsystem that is overaroused. For example, a schizophreniclike syndrome of reduced attentional span and contextual collapse can be elicited by overarousal of the incentive-motivational system, but would not necessarily be cured by a depressant that acted differentially on the novelty (reticular formation) system. In fact, depressing the wrong arousal system can cause a paradoxical deterioration of a syndrome by disinhibiting the hyperactive competing arousal system that caused the syndrome. Complicating the situation further is the inverted-U in responsiveness that can be caused by parametrically exciting either of the arousal systems separately (Section 10).

Alternation between attentional and orienting reactions seems also to occur, and in a cyclic fashion, within the motor system during the performance of a familiar sequence of skilled movements. Grossberg (1978e, Sections 48–54) used a thought experiment concerning the information available to a behaving infant to derive a minimal network for the learning of circular reactions. A central mechanism in this network is the matching or mismatching of a terminal motor map, or where the end organ expects to go, and a proprioceptive motor map, or where the end organ now is. Proprioceptive-terminal map matching is the analogue within motor systems of expectancy matching in sensory systems (Tanji & Evarts,

1976). Proprioceptive-terminal map matching means that the end organ has reached the location where it expects to be. I suggest that such matching is capable of eliciting signals that not only support the motor postures and perceptual sensitivity needed to pay attention—reflected in the PGO wave—but also release from STM the next motor command in a goal-directed motor sequence. The new motor command instates a new terminal motor map that mismatches the current proprioceptive map, thereby inhibiting the attentional arousal and releasing the new orienting reaction. Thus the matching process seems to cyclically sow the seeds of its own destruction, at least until the entire motor plan is executed. An effort should be made to test whether proprioceptive-terminal matching does indeed elicit signals that reset motor commands in a goal-directed motor plan.

The minimal dimension of the symmetry that is needed to design bilateral hierarchical resonances between competing subsystems is a 16-fold symmetry, since each subsystem contains at least two levels capable of matching their patterns, and each level contains a pair of dipole fields to compute a bilateral resonance.

Despite the greater symmetry that manifests itself by studying competing subsystems side by side, it is inevitable that neural system design will exhibit substantial symmetry breaking. In addition to the asymmetry between excitatory and inhibitory configurations that supports neural development and evolution (e.g., on-center off-surround anatomy), such environmental asymmetries as between light versus dark and between up versus down must be reflected in the neural machinery that has adapted to them. Some insights concerning this neural machinery are suggested in terms of the preceding discussion. For example, if certain off-cells are tonically on in darkness, and if offset of a light triggers a transient output signal from the corresponding off-cell, then why does the tonic activity of this off-cell in the dark not drive a tonic output signal? If the off-field is normalized, then when all the off-cells are on in the dark, none of them is sufficiently active to exceed the output threshold, which is chosen higher than the quenching threshold. After a light is turned off, a particular off-cell's activity is differentially rebounded for a short time during which its activity exceeds the output threshold. Tonic activity and transient outputs are hereby reconciled. This example illustrates the importance of carefully tuning the relative levels of overall network activity and output threshold.

By contrast, suppose that the output threshold is lowered by disinhibiting the output cells' axon hillocks, or that the overall network activity is enhanced by lowering its quenching threshold—perhaps as in the nucleus reticularis thalami. Then the off-field can deliver tonic output signals to its target cells. If, for example, the target cells control the contraction of muscles, then the tonic muscle signals can maintain a posture that resists the effects of gravity, for example, standing. In this situation, periodic phasic inputs to the on-cells, whether due to external sources or to feedback signals from the off-cells, can cause an oscillatory motor reaction during every cycle of which agonist contraction is followed by an antagonist rebound, for example, walking. Thus, differential tuning of output threshold and normalized activity can convert transient off-cell output signals, as in phasic sensory responses, into tonic off-cell output signals that either balance a persistent asymmetry in environmental influences, as in standing, or energize rhythmic output bursts, as in walking.

20. Cerebral Dominance: The Anatomy of Temporal Versus Spatial Encoding

A more profound type of symmetry breaking occurs between the attentional and orienting subsystems, due to the different nature of cognitive and motor data, and within the attentional subsystem itself, due to the different processing of data about space and time. A pattern of activity across a field of populations at a given time is inherently ambiguous. Does the pattern code a single event in time, such as the features in a visual scene, or does it code the order information in a series of events? Because of this fundamental ambiguity, it is suggested in Grossberg (1978e) that different STM reset mechanisms are needed to reset spatial versus temporal data. The spatial reset mechanism is a match-

ing mechanism such as I have just discussed. The temporal reset mechanism is derived from a study of free recall and serial learning. The output signal from a population in a temporal processor is suggested to activate a self-destructive inhibitory feedback signal. This feedback inhibition prevents perseverative performance of the same item, and conditionalizes the order information among the populations that remain active, with the most active population performed first, since its reaction time for generating an output signal is smallest. The readout of order information from a field of active populations is suggested to be accomplished by either a nonspecific decrease in all the output thresholds or a nonspecific amplification of the total STM activity in the field. Again the relative size of these two levels is a crucial parameter in determining network performance. Thus, the readout of sensory order information is suggested to be mechanistically analogous to the activation of a sequential motor program. By this scheme, a list of items can be performed in a perfect serial ordering despite the fact that all the mechanisms in the network are parallel mechanisms. Serial properties do not imply serial processes.

I suggest that the cortical microanatomy that subserves spatial versus temporal processing will be found to exhibit these different STM reset mechanisms. Consequently, to unambiguously decode temporal versus spatial data, somehow the populations that code the different types of data must be spatially segregated so that they can be endowed with their disparate STM reset mechanisms. The ambiguous meaning of spatial patterns hereby suggests the need to spatially segregate the processing of sequential, including language-like, codes from codes concerning themselves with spatial integration. This dichotomy might be one reason for the emergence of cerebral dominance (Gazzaniga, 1970, chap. 8), despite the fact that a typical speech act can include both spatial and temporal coding elements, and thus requires cyclic resetting of both types of codes. Visual and auditory processing are sensory prototypes of higher codes that emphasize spatial and temporal processing, respectively. Since visual and auditory representations are bilateral, the trend toward segregation of spatial versus temporal processing in separate hemispheres can be viewed as a symmetry-breaking operation with a drift of visuallike processing into the non-dominant hemisphere and auditorylike process-ing into the dominant hemisphere. The symmetry between bilateral resonances in these regions should be correspondingly broken, leading to a generalized avalanche or command structure between the two hemispheres to coordinate the temporal unfolding of spatial representations. An effort should be made to test whether the cortical microanatomy in spatial versus temporal processors exhibits traces of different reset mechanisms in the anatomy of inhibitory feedback interneurons.

21. Conclusion: How to Understand Evolutionary Data?

The thought experiment in this article illustrates a general method for discovering the mechanisms behind psychological data. Many psychological phenomena are facets of the evolutionary process—variously called chunking, unitization, or automation—whereby behavioral fragments are grouped into new control units that become the fragments of still higher behavioral units in a continuing process of hierarchical organization and command synthesis. By its very nature, this evolutionary process hides the mechanistic substrate on which it is built, so that we can behave in a world of percepts, feelings, and plans rather than of cells, signals, and trans-mitters. Because our brains are these evolu-tionary devices, we have immediate introspec-tive evidence about basic psychological processes, and can consensually define concepts like reward, punishment, frustration, expec-tation, memory, and plan even without a scientific understanding of their mechanistic substrates. To represent these consensual concepts in our scientific work by processes that mirror their introspective properties is, however, a fundamental mistake. Then the consensual impression of events blinds us to their functional representation.

For example, language processes whose properties seem discrete and serial are often realized by continuous and parallel control processes (Grossberg, 1978a, 1978e). The two types of representation are not fundamentally equivalent and generate different predictions. Similarly, behavioral properties that seem linear are often controlled by nonlinear processes (Grossberg, 1978d). Again the two types of description are fundamentally not equivalent. When a theory is erroneously built on consensual properties, it soon meets data that it finds paradoxical. Then the theory either collapses or is decorated with a succes-sion of ad hoc hypotheses. Theoretical epi-cycles soon crowd the scientific landscape, and theory gets a bad name even though we cannot live without it.

An alternative procedure is to respect the wisdom of evolution by trying to imitate it. To do this, at each stage of theory construc-tion, prescribed environmental pressures are identified that force adaptive designs on the behaving brain. Most of us know these pressures; they are familiar precisely because they are among the constraints to which we have successfully adapted. Thus the theory is grounded on a firm basis. By contrast with the consensual method, these pressures are proper-ties of the environment rather than of our behavior. The thought experiments show how these environmental constraints generate ex-plicit minimal mechanisms for coping with them. Such experiments include information that eludes experimental techniques for several reasons. For example, they show how many system components work together, and they compress into a unified description environ-mental pressures that act over long, or at least nonsimultaneous, times. Most importantly, the thought experiments explicate design constraints that are needed to adapt in a real-time setting. These real-time constraints are often the most crucial ones, and they are invisible to descriptive or purely formal theories.

Once the minimal mechanisms that realize several environmental pressures are con-structed, mathematical analysis shows how they work together to generate data and predictions whose complexity and subtlety transcend the apparent simplicity of the environmental pressures, as well as unaided intuition. This procedure defines new con-ceptual categories into which to divide the data, and also points to important environ-mental pressures that have been overlooked, by clearly delineating what the mechanisms can and cannot do. In this way, a small number of principles and mechanisms is organized in an evolutionary progression, and large bodies of data are hierarchically grouped as manifestations of these principles.

In the present article a thought experiment shows how limitations in the types of information available to individual cells can be overcome when the cells act together in suitably designed feedback schemes. The explication of these schemes in a rigorous setting (see the appendices) forces us to study a series of general design problems whose complete solution includes many examples that go beyond the thought experiment; for example, competitive systems (their decision schemes, self-tuning, adaptation, fast pattern transformations, and STM), nonstationary prediction systems (their filtering, pattern learning, and LTM), dipole systems (their transduction and rebound properties), and resonant systems (their hysteresis, deformation, and reset properties). This thought experiment is just one in a series that has helped to unravel psychological mechanisms and to generate as yet untested predictions.

An early thought experiment used the simplest classical conditioning postulates, interpreted in real time (see Grossberg, 1974, for a review), to derive explicit neural networks. When, for example, these networks are exposed serially to long lists, a variety of serial learning properties automatically occur, such as bowing, skewing, anchoring, primacy dominating recency, anticipatory and perseverative generalization gradients, and response oscillation (Grossberg, 1969b; Grossberg & Pepe, 1970, 1971). In addition, mathematical analysis unexpectedly showed how overarousal can cause an attentional deficit with reduced attentional span and collapsed contextual constraints. This overaroused syndrome includes a change toward less skewing of the bowed error curve and toward recency dominating primacy. These formal properties have not yet been empirically tested.

Using these results on classical conditioning, another thought experiment about classical conditioning became necessary. The time intervals between CS and UCS presentations on successive learning trials are not always

the same. In a real-time theory, this trivial fact creates a severe synchronization problem whose solution unexpectedly led to explicit mechanisms of instrumental conditioning (Grossberg, 1971a, 1972b, 1972c). Many insights about instrumental mechanisms and their relationship to Pavlovian mechanisms were hereby derived. One of them is especially pertinent to this article. A dipole mechanism was forced on the theory to control net incentive motivation through time. Mathematical analysis of the dipole revealed several unexpected properties (Sections 10 and 11) including the ability of arousal, and hence of unexpected events, to adapt or rebound the dipole. The detailed understanding of dipole dynamics helped to clarify many novelty-related phenomena, such as learned helplessness, superconditioning, and vicious circle behavior. It also forced on the theory the realization that cognitive events, via expectancy matching, can directly influence reinforcement, via the dipole. In summary, a simple environmental pressure concerning a real-time synchronization problem in classical conditioning was solved by mechanisms of instrumental conditioning and led to a role for cognitive processing in the direct evaluation of reinforcement.

With these results in hand, a thought experiment about feature fields came into view. The parallel activation of many cells by external cues can easily destroy decision rules that regulate the balance of net incentive through time. The minimal solution of this difficulty is to impose a normalization property at the processing stages where cues are stored in STM (Grossberg, 1972c). This normalization property had already been noticed as a property by which competitive shunting networks solve the saturation problem (Grossberg, 1970). These results from reinforcement theory made it clear that further progress concerning feature extraction and related perceptual phenomena required a frontal attack on the mathematics of competitive systems. The early results in this direction (Grossberg, 1973) eventually led to many surprising properties, the most general being that every competitive system induces a decision scheme that can be used to predict its behavior through time (Grossberg, 1978c). For present purposes, the normalization and quenching threshold properties are particularly important, since they show how arousal can tune STM, and thereby help to control what cues are overshadowed vs. what cues are processed. Another role for cognitive events, again acting on arousal via expectancies, was hereby discerned.

Once the normalization and quenching threshold properties were discovered, a thought experiment was suggested that joins together facets of perceptual and motivational processing: How can cues with incompatible motivational consequences be processed in parallel without causing chaotic cross-conditioning (Figure 2)? This thought experiment showed how incentive motivational feedback can influence STM storage to yield stable self-consistent coding and, as side benefits, explanations of attentional data such as overshadowing and discrimination-learning data such as peak shift and behavioral contrast (Grossberg, 1975). Several other theoretical stages then followed as the attentional phenomena were recognized to be special cases of the resonance idea. It became possible to build a theory of stable code development (Grossberg, 1976a, 1976b), which, in turn, suggested a psychophysiological foundation for cognitive theory (Grossberg, 1978e), one of whose facets is heuristically summarized by the present thought experiment.

The evolutionary procedure thus embodies a program of real-time theory construction in psychological studies that underscores the need to understand the collective properties of hierarchically organized nonlinear neural networks. Because the rigorous analysis of such networks is well under way, we can anticipate an emergent resonance between experimental psychology and psychophysiological theory during our generation.

References

Anderson, J. A., Silverstein, J. W., Ritz, S. A., & Jones, R. S. Distinctive features, categorical perception, and probability learning: Some applications of a neural model. *Psychological Review*, 1977, *84*, 413–451.

Arbib, M. A. *The metaphorical brain*. New York: Wiley, 1972.

Atkinson, R. C., & Shiffrin, R. M. Human memory: A proposed system and its control processes. In K. W. Spence & J. T. Spence (Eds.), *Advances in the psychology of learning and motivation research and theory* (Vol. 2). New York: Academic Press, 1968.

Atkinson, R. C., & Shiffrin, R. M. The control of short-term memory. *Scientific American*, 1971, *225*, 82–90.

Barlow, H. B., & Levick, W. R. Changes in the maintained discharge with adaptation level in the cat retina. *Journal of Physiology*, 1969, *202*, 699–718. (a)

Barlow, H. B., & Levick, W. R. Three factors limiting the reliable detection of light by retinal ganglion cells of the cat. *Journal of Physiology*, 1969, *200*, 1–24. (b)

Beck, J. *Surface color perception*. Ithaca, N.Y.: Cornell University Press, 1972.

Berger, T. W., & Thompson, R. F. Limbic system interrelations: Functional division among hippocampal–septal connections. *Science*, 1977, *197*, 587–589.

Blakemore, C., & Cooper, G. F. Development of the brain depends on the visual environment. *Nature*, 1970, *228*, 477–478.

Bloomfield, T. M. Behavioral contrast and the peak shift. In R. M. Gilbert & N. S. Sutherland (Eds.), *Animal discrimination learning*. New York: Academic Press, 1969.

Boring, E. G. *A history of experimental psychology* (2nd ed.). New York: Appleton-Century-Crofts, 1950.

Brown, J. L. Afterimages. In C. H. Graham (Ed.), *Vision and visual perception*. New York: Wiley, 1965.

Campbell, B. A. Interaction of aversive stimuli: Summation or inhibition? *Journal of Experimental Psychology*, 1968, *78*, 181–190.

Campbell, B. A., & Kraeling, D. Response strength as a function of drive level and amount of drive reduction. *Journal of Experimental Psychology*, 1953, *45*, 97–101.

Campbell, F. W., & Howell, E. R. Monocular alternation: A method for the investigation of pattern vision. *Journal of Physiology*, 1972, *225*, 19–21.

Campbell, L., & Garnett, W. *The life of James Clerk Maxwell*. London: Macmillan, 1882.

Cant, B. R., & Bickford, R. G. The effect of motivation on the contingent negative variation (CNV). *Electroencephalography and Clinical Neurophysiology*, 1967, *23*, 594.

Carterette, E. C., & Friedman, M. P. (Eds.). *Handbook of perception: Seeing* (Vol. 5). New York: Academic Press, 1975.

Chapman, R. M., McCrary, J. W., & Chapman, J. A. Short-term memory: The "storage" component of human brain response predicts recall. *Science*, 1978, *202*, 1211–1214.

Cohen, J. Very slow brain potentials relating to expectancy: The CNV. In E. Donchin & D. B. Lindsley (Eds.), *Average evoked potentials*. Washington, D.C.: National Aeronautics and Space Administration, 1969.

Cornsweet, T. N. *Visual perception*. New York: Academic Press, 1970.

Craik, F. I. M. & Lockhart, R. S. Levels of processing: A framework for memory research. *Journal of Verbal Learning and Verbal Behavior*, 1972, *11*, 671–684.

Craik, F. I. M., & Tulving, E. Depth of processing and the retention of words in episodic memory. *Journal of Experimental Psychology: General*, 1975, *104*, 268–294.

Creutzfeldt, O. D., & Northdurft, H. C. Representation of complex visual stimuli in the brain. *Naturwissenschaften*, 1978, *65*, 307–318.

Daniels, J. D., & Pettigrew, J. D. Development of neuronal responses in the visual system of cats. In G. Gottlieb (Ed.), *Neural and behavioral specificity* (Vol. 3). New York: Academic Press, 1976.

Denny, M. R. Relaxation theory and experiments. In F. R. Brush (Ed.), *Aversive conditioning and learning*. Academic Press: New York, 1970.

Dickinson, A., Hall, G., & Mackintosh, N. J. Surprise and the attenuation of blocking. *Journal of Experimental Psychology: Animal Behavior Processes*, 1976, *4*, 313–322.

Donchin, E., Gerbrandt, L. A., Leifer, L., & Tucker, L. Is the contingent negative variation contingent on a motor response? *Psychophysiology*, 1972, *9*, 178–188.

Donchin, E., Tueting, P., Ritter, W., Kutas, M., & Heffley, E. *Electroencephalography and Clinical Neurophysiology*, 1975, *38*, 1–13.

Duda, R. O., & Hart, P. E. *Pattern classification and scene analysis*. New York: Wiley, 1973.

Dunham, P. J. Punishment: Method and theory. *Psychological Review*, 1971, *78*, 58–70.

Ellias, S. A., & Grossberg, S. Pattern formation, contrast control, and oscillations in the short term memory of shunting on-center off-surround networks. *Biological Cybernetics*, 1975, *20*, 69–98.

Estes, W. K. Outline of a theory of punishment. In B. A. Campbell & R. M. Church (Eds.), *Punishment and aversive behavior*. New York: Appleton-Century-Crofts, 1969.

Estes, W. K., & Skinner, B. F. Some quantitative properties of anxiety. *Journal of Experimental Psychology*, 1941, *29*, 390–400.

Fender, D., & Julesz, B. Extension of Panum's fusional area in binocularly stabilized vision. *Journal of the Optical Society of America*, 1967, *57*, 819–830.

Foote, W. E., Manciewicz, R. J., & Mordes, J. P. Effect of midbrain raphe and lateral mesencephalic stimulation on spontaneous and evoked activity in the lateral geniculate of the cat. *Experimental Brain Research*, 1974, *19*, 124–130.

Freeman, W. J. *Mass action in the nervous system*. New York: Academic Press, 1975.

Fukuda, Y., & Stone, J. Retinal distribution and central projections of X-, Y-, and W-cells of the cat's retina. *Journal of Neurophysiology*, 1974, *37*, 749–772.

Fuxe, K., Hökfelt, T., & Ungerstedt, U. Morphological and functional aspects of central monoamine neurons. *International Review of Neurobiology*, 1970, *13*, 93–126.

Fuxe, K., & Ungerstedt, U. Histochemical, biochemical, and functional studies on central monoamine neurons after acute and chronic amphetamine administration. In E. Costa & S. Garattini (Eds.), *Amphetamines and related compounds*. New York: Raven Press, 1970.

Gardner, W. J., Licklider, J. C. R., & Weisz, A. Z. Suppression of pain by sound. *Science*, 1961, *132*, 32–33.

Gazzaniga, M. S. *The bisected brain*. New York: Appleton-Century-Crofts, 1970.

Gibson, J. J. Adaptation with negative aftereffect. *Psychological Review*, 1937, *44*, 222–244.

Graham, C. H. Visual form perception. In C. H. Graham (Ed.), *Vision and visual perception*. New York: Wiley, 1965.

Grossberg, S. *The theory of embedding fields with applications to psychology and neurophysiology*. New York: Rockefeller Institute for Medical Research, 1964.

Grossberg, S. On learning and energy–entropy dependence in recurrent and nonrecurrent signed networks. *Journal of Statistical Physics*, 1969, *1*, 319–350. (a)

Grossberg, S. On the serial learning of lists. *Mathematical Biosciences*, 1969, *4*, 201–253. (b)

Grossberg, S. On the production and release of chemical transmitters and related topics in cellular control. *Journal of Theoretical Biology*, 1969, *22*, 325–364. (c)

Grossberg, S. Neural pattern discrimination. *Journal of Theoretical Biology*, 1970, *27*, 291–337.

Grossberg, S. On the dynamics of operant conditioning. *Journal of Theoretical Biology*, 1971, *33*, 225–255. (a)

Grossberg, S. Pavlovian pattern learning by nonlinear neural networks. *Proceedings of the National Academy of Sciences*, 1971, *68*, 828–831. (b)

Grossberg, S. Neural expectation: Cerebellar and retinal analogs of cells fired by learnable or unlearned pattern classes. *Kybernetik*, 1972, *10*, 49–57. (a)

Grossberg, S. A neural theory of punishment and avoidance. I. Qualitative theory. *Mathematical Biosciences*, 1972, *15*, 39–67. (b)

Grossberg, S. A neural theory of punishment and avoidance, II. Quantitative theory. *Mathematical Biosciences*, 1972, *15*, 253–285. (c)

Grossberg, S. Pattern learning by functional-differential neural networks with arbitrary path weights. In K. Schmitt (Ed.), *Delay and functional-differential equations and their applications*. New York: Academic Press, 1972. (d)

Grossberg, S. Contour enhancement, short-term memory, and constancies in reverberating neural networks. *Studies in Applied Mathematics*, 1973, *52*, 217–257.

Grossberg, S. Classical and instrumental learning by neural networks. In R. Rosen & F. Snell (Eds.), *Progress in theoretical biology* (Vol. 3). New York: Academic Press, 1974.

Grossberg, S. A neural model of attention, reinforcement, and discrimination learning. *International Review of Neurobiology*, 1975, *18*, 263–327.

Grossberg, S. Adaptive pattern classification and universal recoding, I: Parallel development and coding of neural feature detectors. *Biological Cybernetics*, 1976, *23*, 121–134. (a)

Grossberg, S. Adaptive pattern classification and universal recording, II: Feedback, expectation, olfaction, and illusions. *Biological Cybernetics*, 1976, *23*, 187–202. (b)

Grossberg, S. Pattern formation by the global limits of a nonlinear competitive interaction in n dimensions. *Journal of Mathematical Biology*, 1977, *4*, 237–256.

Grossberg, S. Behavioral contrast in short-term memory: Serial binary memory models or parallel continuous memory models? *Journal of Mathematical Psychology*, 1978, *17*, 199–219. (a)

Grossberg, S. Communication, memory, and development. In R. Rosen & F. Snell (Eds.), *Progress in theoretical biology* (Vol. 5). New York: Academic Press, 1978. (b)

Grossberg, S. Decisions, patterns, and oscillations in the dynamics of competitive systems with applications to Volterra-Lotka systems. *Journal of Theoretical Biology*, 1978, *73*, 101–130. (c)

Grossberg, S. Do all neural models really look alike? A comment on Anderson, Silverstein, Ritz, and Jones. *Psychological Review*, 1978, *85*, 592–596. (d)

Grossberg, S. A theory of human memory: Self-organization and performance of sensory-motor codes, maps, and plans. In R. Rosen & F. Snell (Eds.), *Progress in theoretical biology* (Vol. 5). New York: Academic Press, 1978. (e)

Grossberg, S., & Levine, D. S. Some developmental and attentional biases in the contrast enhancement and short term memory of recurrent neural networks. *Journal of Theoretical Biology*, 1975, *53*, 341–380.

Grossberg, S., and Pepe, J. Schizophrenia: Possible dependence of associational span, bowing, and primacy vs. recency on spiking threshold. *Behavioral Science*, 1970, *15*, 359–362.

Grossberg, S., & Pepe, J. Spiking threshold and overarousal effects on serial learning. *Journal of Statistical Physics*, 1971, *3*, 95–125.

Hebb, D. O. Drives and the CNS (conceptual nervous system). *Psychological Review*, 1955, *62*, 243–254.

Helmholtz, H. von. *Handbuch der physiologischen optik* (1st. ed.). Hamburg, Leipzig: Voss, 1866.

Helmholtz, H. von. *Physiological optics* (Vol. 2) (J. P. C. Southall, Ed.). New York: Dover, 1962.

Hilgard, E. R., & Bower, G. H. *Theories of learning* (4th ed.). Englewood Cliffs, N.J.: Prentice-Hall, 1975.

Hirsch, H. V. B., & Spinelli, D. N. Visual experience modifies distribution of horizontally and vertically oriented receptive fields in cats. *Science*, 1970, *168*, 869–871.

Hubel, D. H., & Wiesel, T. N. Functional architecture of macaque monkey visual cortex. *Proceedings of the Royal Society of London* (B), 1977, *198*, 1–59.

Irwin, F. W. *Intentional behavior and motivation: A cognitive theory*. Philadelphia, Pa.: Lippincott, 1971.

Irwin, D. A., Rebert, C. S., McAdam, D. W., & Knott, J. R. Slow potential change (CNV) in the human EEG as a function of motivational variables. *Electroencephalography and Clinical Neurophysiology*, 1966, *21*, 412–413.

Jacobowitz, D. M. Effects of 6-hydroxydopa. In E. Usdin & H. S. Snyder (Eds.), *Frontiers in catecholamine research*. New York: Pergamon Press, 1973.

Juhasz, A. Über die komplementärge-färbten nachbilder. *Zeitschrift fur Psychologie*, 1920, *51*, 233–263.

Julesz, B. *Foundations of cyclopean perception*. Chicago: University of Chicago Press, 1971.

Kamin, L. J. Predictability, surprise, attention, and conditioning. In B. A. Campbell & R. M. Church (Ed.), *Punishment and aversive behavior*. New York: Appleton-Century-Crofts, 1969.

Koenigsberger, L. *Hermann von Helmholtz*. (F. A. Welby, trans.). Oxford, England: Clarendon, 1906.

Ladisich, W., Volbehr, H., & Matussek, N. Paradoxical effect of amphetamine on hyperactive states in correlation with catecholamine metabolism in brain. In E. Costa & S. Garattini (Eds.), *Amphetamines and related compounds*. New York: Raven Press, 1970.

Land, E. H. The retinex theory of color vision. *Scientific American*, 1977, *237*, 108–128.

Levine, D. S., & Grossberg, S. Visual illusions in neural networks: Line neutralization, tilt aftereffect, and angle expansion. *Journal of Theoretical Biology*, 1976, *61*, 477–504.

Lindvall, O., & Björklund, A. The organization of the ascending catecholamine neuron systems in the rat brain as revealed by the glyoxylic acid fluorescence method. *Acta Physiologia Scandinavia Supplement*, 1974, *412*, 1–48.

Low, M. D., Borda, R. P., Frost, J. D., & Kellaway, P. Surface negative slow potential shift associated with conditioning in man. *Neurology*, 1966, *16*, 711–782.

Lynn, R. *Attention, arousal, and the orientation reaction*. New York: Pergamon Press, 1966.

Macchi, G., & Rinvik, E. Thalamo-telencephalic circuits: A neuroanatomical survey. In A. Rémond (Ed.), *Handbook of electroencephalography and clinical neurophysiology* (Vol. 2, Pt. A). Amsterdam: Elsevier, 1976.

MacKay, D. M. Moving visual images produced by regular stationary patterns. *Nature*, 1957, *180*, 849–850.

MacKay, D. M., & MacKay, V. What causes decay of pattern-contingent chromatic aftereffects? *Vision Research*, 1975, *15*, 462–464.

Masterson, F. A. Is termination of a warning signal an effective reward for the rat? *Journal of Comparative and Physiological Psychology*, 1970, *72*, 471–475.

McAdam, D. W. Increases in CNS excitability during negative cortical slow potentials in man. *Electroencephalography and Clinical Neurophysiology*, 1969, *26*, 216–219.

McAdam, D. W., Irwin, D. A., Rebert, C. S., & Knott, J. R. Conative control of the contingent negative variation. *Electroencephalography and Clinical Neurophysiology*, 1966, *21*, 194–195.

McAllister, W. R., & McAllister, D. E. Behavioral measurement of conditioned fear. In F. R. Brush (Ed.), *Aversive conditioning and learning*. New York: Academic Press, 1970.

McCollough, C. Color adaptation of edge-detectors in the human visual system. *Science*, 1965, *149*, 1115–1116.

Montalvo, F. S. A neural network model of the McCollough effect. *Biological Cybernetics*, 1976, *25*, 49–56.

Moruzzi, G., & Magoun, H. W. Brain stem reticular formation and activation of the EEG. *Electroencephalography and Clinical Neurophysiology*, 1949, *1*, 455–473.

Myers, A. K. Effects of continuous loud noise during instrumental shock-escape conditioning. *Journal of*

Comparative and Physiological Psychology, 1969, *68*, 617–622.

Piaget, J. *The origins of intelligence in children*. New York: Norton, 1963.

Ratliff, F. *Mach bands: Quantitative studies of neural networks in the retina*. San Francisco: Holden-Day, 1965.

Rauschecker, J. P. J., Campbell, F. W., & Atkinson, J. Colour opponent neurones in the human visual system. *Nature*, 1973, *245*, 42–45.

Remington, R. J. Analysis of sequential effects in choice reaction times. *Journal of Experimental Psychology*, 1969, *2*, 250–257.

Rescorla, R. A., & Wagner, A. R. A theory of Pavlovian conditioning: Variations in the effectiveness of reinforcement and nonreinforcement. In A. Black & W. F. Prokasy (Eds.), *Classical conditioning II*. New York: Appleton-Century-Crofts, 1972.

Ricklan, M. *L-dopa and parkinsonism: A psychological assessment*. Springfield, Ill.: Charles C Thomas, 1973.

Robson, J. G. Receptive fields: Neural representation of the spatial and intensive attributes of the visual image. In E. C. Carterette & M. P. Friedman (Eds.), *Handbook of perception* (Vol. 5). New York: Academic Press, 1976.

Rohrbaugh, J., Donchin, E., & Eriksen, C. Decision making and the P300 component of the cortical evoked response. *Perception & Psychophysics*, 1974, *15*, 368–374.

Schiller, P. H., & Malpeli, J. G. Functional specificity of lateral geniculate nucleus laminae of the rhesus monkey. *Journal of Neurophysiology*, 1978, *41*, 788–797.

Schneider, W., & Shiffrin, R. M. Automatic and controlled information processing in vision. In D. LaBarge & S. J. Samuels (Eds.), *Basic processes in reading: Perception and comprehension*. Hillsdale, N.J.: Erlbaum, 1976.

Seligman, M. E. P., Maier, S. F., & Solomon, R. L. Unpredictable and uncontrollable aversive events. In F. R. Brush (Ed.), *Aversive conditioning and learning*. New York: Academic Press, 1971.

Singer, W. Control of thalamic transmission by corticofugal and ascending reticular pathways in the visual system. *Physiological Review*, 1977, *57*, 386–420.

Sluckin, W. *Imprinting and early learning*. London: Methuen, 1964.

Squires, K., Wickens, C., Squires, N., & Donchin, E. The effect of stimulus sequence on the waveform of the cortical event-related potential. *Science*, 1976, *193*, 1142–1146.

Stein, L. Norepinephrine reward pathways: Role in self-stimulation, memory consolidation, and schizophrenia. In J. K. Cole & T. B. Sonderegger (Eds.), *Nebraska Symposium on Motivation* (Vol. 22). Lincoln: University of Nebraska Press, 1974.

Stryker, M., & Sherk, H. Modification of cortical orientation selectivity in the cat by restricted visual experience: A reexamination. *Science*, 1975, *190*, 904–905.

Tanji, J., & Evarts, E. V. Anticipatory activity of motor cortex neurons in relation to direction of an intended movement. *Journal of Neurophysiology*, 1976, *39*, 1062–1068.

Thomas, G. B., Jr. *Calculus and analytic geometry*. Reading, Mass.: Addison-Wesley, 1968.

Trabasso, T., & Bower, G. H. *Attention in learning: Theory and research*. New York: Wiley, 1968.

Tsumoto, T., Creutzfeldt, O. D., & Legéndy, C. R. Functional organization of the corticofugal system from visual cortex to lateral geniculate nucleus in the cat. *Experimental Brain Research*, 1978, *32*, 345–364.

Tsumoto, T., & Suzuki, D. A. Effects of frontal eye field stimulation upon activities of the lateral geniculate body of the cat. *Experimental Brain Research*, 1976, *25*, 291–306.

Ungerstedt, U. Stereotoxic mapping of the monoamine pathways in the rat brain. *Acta Physiologica Scandinavia*, 1971, *82* (Supplement 367), 1–48.

Wagner, A. R. Frustrative nonreward: A variety of punishment. In B. A. Campbell & R. M. Church (Eds.), *Punishment and aversive behavior*. New York: Appleton-Century-Crofts, 1969.

Walter, W. G. Slow potential waves in the human brain associated with expectancy, attention, and decision. *Arch. Psychiat. Nervenkr.*, 1964, *206*, 309–322.

Wilson, H. A synaptic model for spatial frequency adaptation. *Journal of Theoretical Biology*, 1975, *50*, 327–352.

Wurtz, R. H., & Goldberg, M. E. The role of the superior colliculus in visually evoked eye movement. In J. Dichgans & E. Bizzi (Eds.), *Cerebral control of eye movement and motion perception*. Basel, Switzerland: Karger, 1972.

Appendix A

This section summarizes some of the mechanisms whereby an activity pattern across $F^{(1)}$ elicits signals to $F^{(2)}$ that filter the pattern. The filtered signal pattern is then rapidly transformed by recurrent competitive interactions within $F^{(2)}$ before the resultant pattern is stored in STM. The STM pattern endures long enough to alter the interfield path strengths that define the filter. This is an LTM change. Then the process repeats itself until STM and LTM equilibrate.

Filter

Denote the cells of $F^{(1)}$ by $v_i^{(1)}$, $i = 1, 2, \ldots$, n, and the cells of $F^{(2)}$ by $v_j^{(2)}$, $j = 1, 2, \ldots, N$. Let the activity of $v_i^{(1)}$ at time t be $x_i^{(1)}(t)$. Suppose that the activity $x_i^{(1)}(t)$ elicits a signal $S_i = S_i(x_i^{(1)}(t))$ in the pathways from $v_i^{(1)}$ to $F^{(2)}$. Let the net signal from $v_i^{(1)}$ to $v_j^{(2)}$ be $S_i z_{ij}$, where z_{ij} provides a measure of the efficiency of the pathway e_{ij} from $v_i^{(1)}$ to $v_j^{(2)}$. In other words, z_{ij} *gates* signal S_i on its way to $v_j^{(2)}$. Then the *total* signal from $F^{(1)}$ to $v_j^{(2)}$ is

$$T_j = \sum_{i=1}^{n} S_i z_{ij}.$$

This equation for T_j has an informative geometrical interpretation in terms of the vectors $S = (S_1, S_2, \ldots, S_n)$ of signals and $z_j = (z_{1j}, z_{2j}, \ldots, z_{nj})$ of path strengths from $F^{(1)}$ to $v_j^{(2)}$. Namely, T_j is the dot product, or inner product, of S and z_j, which is written $T_j = S \cdot z_j$ (Thomas, 1968). The dot product can be evaluated in terms of the vector lengths

$$\|S\| = \sqrt{\sum_{i=1}^{n} S_i^2}$$

and

$$\|z_j\| = \sqrt{\sum_{i=1}^{n} z_{ij}^2},$$

and the cosine of the angle between S and z_j by the formula $T_j = \|S\| \, \|z_j\| \cos(S, z_j)$. In particular, if all $\|z_j\|$ are equal, then the cell $v_j^{(2)}$ in $F^{(2)}$ that receives the largest signal is the cell whose $\cos(S, z_j)$ is maximal. The cosine can be increased by choosing the coefficients of z_j more proportional, or parallel, to S, and can be decreased by choosing the coefficients more perpendicular, or orthogonal, to S. Thus each z_j *filters* S by producing a net signal T_j whose size depends on how parallel z_j is to S. Otherwise expressed, z_j *projects* S onto $v_j^{(2)}$. The pattern $T = (T_1, T_2, \ldots, T_N)$ of inputs to $F^{(2)}$ represents pattern S by projecting S onto all the cells $F^{(2)}$ with relative input sizes that depend on the choice of all the vectors $z_1, z_2, \ldots, z_N$.

Contrast Enhancement

When S is first presented to $F^{(1)}$, the pattern T of inputs that it elicits across $F^{(2)}$ might be approximately uniform. Recurrent on-center off-surround interactions within $F^{(2)}$ rapidly contrast enhance this input pattern in order to produce a sharper pattern of STM activities across $F^{(2)}$ (Grossberg, 1976a, 1976b). I illustrate this concept with the simplest case: Suppose that $F^{(2)}$ can choose the cell whose initial input is maximal for storage in STM. Denoting the activity of $v_j^{(2)}$ by $x_j^{(2)}$, this law says that

$$\begin{aligned} x_j^{(2)} &= 1 \quad \text{if} \quad T_j > \max\{\epsilon, T_k, k \neq j\} \\ x_j^{(2)} &= 0 \quad \text{if} \quad T_j < \max\{\epsilon, T_k, k \neq j\}. \end{aligned} \quad \text{(A1)}$$

The coefficient ϵ designates a quenching threshold that must be exceeded before any STM storage is possible. Suppose for definiteness that $T_1 > \max\{\epsilon, T_k, k \neq 1\}$. Then the activity of $v_1^{(2)}$ is rapidly contrast enhanced and stored in STM, whereas all other activities across $F^{(2)}$ are suppressed.

Coding

The path strengths z_{ij} are LTM traces that can slowly adapt to the signal pattern S from $F^{(1)}$ and the STM pattern across $F^{(2)}$. In the simplest case, z_{ij} changes only if $x_j^{(2)} > 0$. Then

$$\frac{d}{dt} z_{ij} = (-z_{ij} + S_i) x_j^{(2)}. \quad \text{(A2)}$$

For example, if $x_1^{(2)} = 1$ and all $x_j^{(2)} = 0$, $j \neq 1$, then this LTM law causes the signal $T_1 = S \cdot z_1$ to be maximized as S is practiced by making z_1 become parallel to S. In this way, presentation of S at $F^{(1)}$ can induce a differentiated STM pattern across $F^{(2)}$ by changing the LTM vectors $z_1, z_2, \ldots, z_N$.

Grossberg's (1976a, 1976b, 1978b, 1978e) articles describe these mechanisms in greater detail. They also show how to generalize the mechanisms to include more complex STM and LTM interactions. Despite these generalizations, the mechanisms are there shown to be unstable in a complex input environment. A precise understanding of this difficulty forced the use of learned feedback expectancies.

Appendix B

This section indicates how a single cell population in $F^{(2)}$ can learn a spatial pattern of activity across $F^{(1)}$. Analogous arguments then show how many simultaneously active cell populations across $F^{(2)}$ can learn a spatial pattern across $F^{(1)}$, albeit not necessarily the same spatial pattern that would have excited a single cell in $F^{(2)}$ by interfield signaling from $F^{(1)}$ to $F^{(2)}$.

Associative Learning

Our laws for associative learning appeared in a monograph by Grossberg (1964), and were mathematically analyzed in a series of articles, leading to a universal theorem of associative learning in Grossberg (1969a, 1971b, 1972d). The universal theorem proves that if these associative learning laws were invented at a prescribed time during the evolutionary process, then they could be used to guarantee unbiased associative learning in essentially any later evolutionary specialization. That is, the laws are capable of learning arbitrary spatial patterns in arbitrarily many, simultaneously active sampling channels that are activated by arbitrary continuous data preprocessing in an essentially arbitrary anatomy. Learning of arbitrary space-time patterns is also guaranteed given modest requirements on the temporal regularity of stimulus sampling. (See Grossberg, 1974, for a review.) Herein I summarize the fact that the unit of LTM is a *spatial pattern*. This is done by considering the *minimal* anatomy that is capable of classical conditioning.

STM and LTM Laws That Factor Pattern from Activity

Let presentation of a CS create an input $I_0(t)$ that activates the cell population v_0. Let the UCS create an input pattern $(I_1(t), I_2(t), \ldots, I_n(t))$ that activates the cell populations $v_1, v_2, \ldots, v_n$, whose outputs elicit the UCR. Let the STM trace of v_j be $x_j(t)$, $j = 0, 1, \ldots, n$, and let the LTM trace of the axon pathway e_{0i} from v_0 to v_i be $z_{0i}(t)$, $i = 1, 2, \ldots, n$ (Figure A1). Suppose that the STM and LTM traces obey the laws

$$\frac{d}{dt}x_0 = -A_0x_0 + I_0(t) \tag{A3}$$

$$\frac{d}{dt}x_i = -Ax_i + Bz_{0i} + I_i(t), \tag{A4}$$

and

$$\frac{d}{dt}z_{0i} = -Cz_{0i} + Dx_i, \tag{A5}$$

$i = 1, 2, \ldots, n$. The terms A_0 and A are STM decay rates. The term C is the LTM decay rate. The terms B and D are signals from v_0 along all the pathways e_{0i}, $i = 1, 2, \ldots, n$; for example, $B(t) = f(x_0(t - \tau))$, where $f(w)$ is a sigmoid function of w. The LTM trace z_{0i} is computed at the interface of the synaptic knob S_{0i} (at the end of e_{0i}) and the postsynaptic cell v_i—that is, at the synaptic knob and/or postsynaptic membrane —where it can gate the signals B on their way to v_i, as in term Bz_{0i} of Equation A4, and simultaneously time-average (term $-Cz_{0i}$) the product of signals D and postsynaptic STM trace x_i (term Dx_i), in A5. In particular, A2 is a special case of A5.

A *spatial pattern* is a UCS whose *relative* activities remain fixed, even though their absolute activities can fluctuate through time,

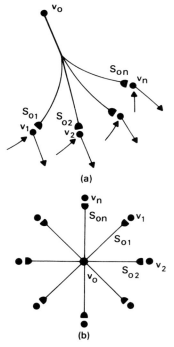

(a)

(b)

Figure A1. In (a) the conditioned stimulus (CS)-activated population v_0 samples the unconditioned stimulus (UCS)-activated populations $v_1, v_2, \ldots, v_n$; in (b) the *outstar* is the minimal network capable of classical conditioning.

namely, $I_i(t) = \theta_i I(t)$, $i = 1, 2, \ldots, n$, where θ_i is the fixed relative activity and $I(t)$ is the total UCS activity. The convention

$$\sum_{i=1}^{n} \theta_i = 1$$

guarantees the normalization

$$I(t) = \sum_{i=1}^{n} I_i(t).$$

The relative values $\theta = (\theta_1, \theta_2, \ldots, \theta_n)$ are like generalized "reflectances" that carry the information in the UCS pattern, whereas $I(t)$ provides the UCS activity that drives system changes in response to θ. It is shown below how this system, which I call an *outstar* (Figure A1), can factorize pattern information θ from information about total activity $I(t)$. This property has many important implications. For example, θ is a probability distribution, since each $\theta_i \geq 0$ and

$$\sum_{k=1}^{n} \theta_k = 1.$$

The system learns probabilities despite the fact that it can generate deterministic behavior. There exists a type of "wave-particle" dualism in these systems that helps to explain the partial successes of statistical learning models, and provides an interesting vantage point from which to think about the wave-particle dualism of quantum theory. Also, since there is no evolutionary advantage in perceptually discriminating data that cannot, in principle, be learned, we can expect the neural perceptual apparatus also to process spatial patterns. The brightness and hue constancies of vision illustrate this fact. These observations clarified how perceptual and learning mechanisms are matched to each other, and suggested study of the minimal neural networks that are capable of discriminating a spatial pattern θ; that is, reflectances. Some of these networks were constructed in Grossberg (1970, 1972a) and, not surprisingly, have an anatomy that is remarkably retinal.

System A3–A5 *factorizes* θ and $I(t)$ in the following sense. Equation A3 can be explicitly solved for $x_0(t)$ by integration, and the result used to solve for $B(t)$ and $D(t)$ as functions of time t. Then A4 and A5 can be rewritten in terms of the relative STM traces

$$X_i = x_i \left(\sum_{k=1}^{n} x_k \right)^{-1}$$

and relative LTM traces

$$Z_i = z_{0i} \left(\sum_{k=1}^{n} z_{0k} \right)^{-1}$$

as follows:

$$\frac{d}{dt} X_i = E(Z_i - X_i) + F(\theta_i - X_i) \quad (A6)$$

and

$$\frac{d}{dt} Z_i = G(X_i - Z_i). \quad (A7)$$

The coefficients E, F, and G depend only on $I(t)$, on the total STM activity

$$x = \sum_{k=1}^{n} x_k,$$

and on the total LTM activity

$$z = \sum_{k=1}^{n} z_{0k}.$$

By A4 and A5,

$$\frac{d}{dt} x = -Ax + Bz + I \quad (A8)$$

and

$$\frac{d}{dt} z = -Cz + Dx. \quad (A9)$$

Equations A8 and A9 are independent of θ; they depend only on the total activity $I(t)$. These equations *decouple* total activity data (I, x, z) from pattern data (θ, X, Z), where $X = (X_1, X_2, \ldots, X_n)$ and $Z = (Z_1, Z_2, \ldots, Z_n)$ are also probability distributions. The total activity data influence the pattern data only via the coefficients E, F, and G, which are always nonnegative. No matter how wildly the CS input $I_0(t)$ and the UCS input $I(t)$ oscillate through time, these coefficients influence only the *rates* with which X and Z are influenced by θ, but not the *directions* in which X and Z can change in response to θ. It is this property that generalizes to yield the universal theorem cited above.

In particular, term $F(\theta_i - X_i)$ in A6 says that X_i approaches θ_i as learning proceeds (UCS read into STM). Term $E(Z_i - X_i)$ in A6 says X_i approaches Z_i (readout of LTM into STM). The net effect of these two terms shows how present demands of the UCS, expressed via θ, and past memories, expressed via Z, compete to change STM via X. Equation A7 shows that Z_i approaches X_i (transfer from STM to LTM). As X approaches θ, and Z approaches X, Z learns the spatial pattern θ. On later performance trials, a CS input to v_0 activates x_0, which in turn activates the signal B. Signal B reads the pattern Z into STM via the terms Bz_{0i} in A4. Since $Z \cong \theta$, A4 shows that the x_is that are activated in this fashion are proportional to the θ_is, as desired.

Many aspects of associative learning can be understood using these STM and LTM laws in more complex anatomies. In particular, the Z_is are stimulus sampling probabilities whose properties explain in a neural setting the partial successes of statistical learning models. The distributions of STM and LTM traces also mimic and predict various data about serial learning, paired associate learning, and free recall experiments. See Grossberg (1974, 1978a, 1978e) for additional discussion.

Appendix C

This section summarizes how feedforward competitive interactions solve the saturation problem using automatic gain control by inhibitory signals, and how properties such as noise suppression, pattern matching, edge enhancement, and spatial frequency sensitivity follow as special cases.

Noise-Saturation Dilemma

All cellular systems face the following dilemma. If their inputs are too small, they can get lost in noise. If the inputs are too large, they can turn on all excitable sites, thereby saturating the system and rendering it insensitive to input differences across the cells. For example, suppose that the ith cell v_i receives an input I_i that can turn on some of its B excitable sites by mass action. Let $x_i(t)$ be the number of excited sites and $B - x_i(t)$ be the number of unexcited sites at time t. The simplest mass action law for turning on unexcited sites and letting excited sites spontaneously turn off is

$$\frac{d}{dt}x_i = -Ax_i + (B - x_i)I_i, \quad (A10)$$

$i = 1, 2, \ldots, n$. Term $(B - x_i)I_i$ says that the input I_i turns on unexcited sites $B - x_i$ by mass action. Term $-Ax_i$ says that excited sites spontaneously becomes unexcited by mass action at rate A. Hence, when $I_i \equiv 0$, x_i can decay to the equilibrium point 0.

System A10 is inadequate for the following reason: Let the inputs form a spatial pattern $I_i = \theta_i I$. Given a fixed pattern $\theta = (\theta_1, \theta_2, \ldots, \theta_n)$, choose a background intensity I and let the system reach equilibrium. This equilibrium is found by setting $(d/dt)x_i = 0$ and solving for x_i:

$$x_i = \frac{B\theta_i I}{A + \theta_i I}. \quad (A11)$$

Now keep θ fixed and increase I. That is, process the same pattern with different background activity. Then all x_i in A11 approach B even if the relative input intensity θ_i is small. This is saturation. How can the system preserve its sensitivity to θ even as I increases? In other words, how does the ith cell v_i compute its "reflectance" θ_i in response to a spatial pattern $I_i = \theta_i I$, $i = 1, 2, \ldots, n$, of inputs? Since

$$\theta_i = I_i I^{-1} = I_i \left(\sum_{k=1}^{n} I_k \right)^{-1},$$

cell v_i needs to know what all the inputs $I_1, I_2, \ldots, I_n$ are in order to compute θ_i. Since

$$\theta_i = I_i \left(I_i + \sum_{k \neq i} I_k \right)^{-1},$$

increasing the ith input I_i "excites" v_i (increases θ_i), whereas increasing any input I_k, $k \neq i$, "inhibits" v_i (decreases θ_i). When this intuition is most simply modeled by a cellular mass action network, we find the system

$$\frac{d}{dt}x_i = -Ax_i + (B - x_i)I_i - x_i \sum_{k \neq i} I_k, \quad (A12)$$

$i = 1, 2, \ldots, n$. In Equation A12, I_i excites v_i via term $(B - x_i)I_i$, just as in A10. The new term

$$-x_i \sum_{k \neq i} I_k$$

describes how the inputs I_k, $k \neq i$, inhibit (note the minus sign) the excited sites of v_i (which number x_i) by mass action. The *gain* of x_i is its decay rate. This is found by grouping together all the terms that multiply x_i. The sum of these terms is $A + I$, where

$$I = \sum_{k=1}^{n} I_k.$$

Thus the inputs automatically change the gain of x_i. In A10 the gain of x_i is $A + I_i$. The two gains differ by the sum

$$\sum_{k \neq i} I_k$$

of inhibitory signals. We now note how automatic gain control by the inhibitory signals overcomes the saturation problem.

Present a spatial pattern $I_i = \theta_i I$ to A12 and let each x_i reach equilibrium. Setting $(d/dt)x_i = 0$, we find

$$x_i = \theta_i \frac{BI}{A + I}. \qquad (A13)$$

In A13, x_i remains proportional to θ_i no matter how intense I is, and $BI(A + I)^{-1}$ has the form of a Weber-Fechner law. The saturation problem is hereby overcome using automatic gain control by inhibitory signals.

Noise Suppression

In A12, the passive equilibrium point, due to term $-Ax_i$, and the inhibitory saturation point, due to term

$$-x_i \sum_{k \neq i} I_k,$$

are both zero. This is not always true in vivo, where a cell potential can sometimes be actively inhibited below the passive equilibrium point. How does this fact alter pattern processing? Consider the system

$$\frac{d}{dt}x_i = -Ax_i + (B - x_i)I_i$$
$$- (x_i + C) \sum_{k \neq i} I_k, \qquad (A14)$$

which differs from A12 only in that x_i can fluctuate between B and $-C$, rather than B and 0, where $-C < 0$. Often in vivo B represents the saturation point of a Na^+ channel, $-C$ represents the saturation point of a K^+ channel, and B is much larger than C.

To see how the inhibitory saturation point C influences pattern processing, let A14 equilibrate to the spatial pattern $I_i = \theta_i I$. Setting $(d/dt)x_i = 0$, we find the equilibrium activities

$$x_i = \frac{(B + C)I}{A + I}\left(\theta_i - \frac{C}{B + C}\right). \qquad (A15)$$

By A15, $x_i > 0$ only if $\theta_i > C(B + C)^{-1}$. The constant $C(B + C)^{-1}$ is an *adaptation level* that θ_i must exceed in order to excite x_i. For simplicity, suppose that the ratio CB^{-1}

matches the ratio of the number of cells excited by each I_i, namely 1, to the number of cells inhibited by I_i, namely $(n - 1)$. If $CB^{-1} = (n - 1)^{-1}$, then $C(B + C)^{-1} = 1/n$. Since, in response to a uniform spatial pattern of inputs, all $\theta_i = 1/n$, no matter how intense I is, it then follows by A15 that all $x_i = 0$. This is noise suppression in its simplest form. It is due to a matched symmetry-breaking between the intracelluar excitatory versus inhibitory parameters (B, C) and the intercellular spread of off-surround versus on-center pathways.

Edge Enhancement, Spatial Frequency Detection, and Pattern Matching

The noise suppression property generalizes to systems whose excitatory and inhibitory interactions can depend on intercellular distances, as in

$$\frac{d}{dt}x_i = -Ax_i + (B - x_i) \sum_{k=1}^{n} I_k C_{ki}$$
$$- (x_i + D) \sum_{k=1}^{n} I_k E_{ki}, \qquad (A16)$$

where C_{ki} (E_{ki}) is the excitatory (inhibitory) coefficient from v_k to v_i. Noise suppression at v_i (i.e., $x_i \leq 0$) occurs in response to a uniform pattern (all $\theta_i = 1/n$) in A16 if

$$B \sum_{k=1}^{n} C_{ki} \leq D \sum_{k=1}^{n} E_{ki}, \qquad (A17)$$

which generalizes $CB^{-1} = (n - 1)^{-1}$ in A15. If a rectangular pattern perturbs such a network, then a cell's activity x_i will be suppressed either if its interactions fall so far outside the rectangle or so far inside it that the pattern looks uniform to its interaction coefficients C_{ki} and E_{ki}. Consequently, only activities near the edge of the rectangle will be enhanced. More generally, the spatial gradients of activity in any input pattern are matched against the spatial gradients in each cell's interaction coefficients to enhance the activity of only those cells to whom the input pattern looks nonuniform. In recurrent networks, this property is supplemented by active contrast-enhancing, disinhibitory, and STM processes that can join together cells with similar interaction gradients into a dynamically coherent subfield that is sensitive to a band of spatial frequencies in the input patterns.

Pattern matching is illustrated as follows. Suppose in A14 that each input I_i is a sum of two inputs J_i and K_i whose patterns $J = (J_1, J_2, \ldots, J_n)$ and $K = (K_1, K_2, \ldots, K_n)$ are to be matched. If J and K mismatch each other's peaks and troughs to form an almost uniform total pattern $I = (I_1, I_2, \ldots, I_n)$, then by A15 all x_i will be inhibited if $CB^{-1} \geq (n-1)^{-1}$. By contrast, if the two patterns reinforce each other, say $J_i = \alpha K_i$, then by

(A15),

$$x_i = \frac{(B+C)(1+\alpha)\bar{K}}{A+(1+\alpha)\bar{K}}\left[\theta_i - \frac{C}{B+C}\right], \quad \text{(A18)}$$

where

$$\bar{K} = \sum_{i=1}^{n} K_i$$

and $\theta_i = K_i(\bar{K})^{-1}$. In other words, matching J and K amplifies each x_i without changing the pattern θ_i.

Appendix D

This section summarizes some properties of recurrent on-center off-surround networks, including normalization, contrast enhancement, quenching threshold, and STM properties.

To see how recurrent networks normalize their STM activity, we first note by Appendix C that these networks need competitive interactions to solve the noise-saturation dilemma. The simplest recurrent on-center off-surround network is defined by

$$\frac{d}{dt}x_i = -Ax_i + (B - x_i)[f(x_i) + I_i] \\ - x_i\left[\sum_{k \neq i} f(x_k) + J_i\right], \quad \text{(A18)}$$

$i = 1, 2, \ldots, n$. As usual, x_i is the STM activity of v_i, term $(B - x_i)f(x_i)$ describes the self-excitation of v_i via a positive feedback signal $f(x_i)$—the recurrent on-center—and term

$$-x_i \sum_{k \neq i} f(x_k)$$

describes the inhibition of v_i via negative feedback signals $f(x_k)$, $k \neq i$—the recurrent off-surround. Term I_i is the ith excitatory input, and term J_i is the ith inhibitory input, for example,

$$J_i = \sum_{k \neq i} I_k$$

in A12.

Contrast Enhancement, Normalization, and Quenching Threshold

An important problem in system A18 is to choose the feedback signal function $f(w)$ as a function of activity level w in such a way as to suppress noise but contrast enhance and store in STM behaviorally important patterns. This problem was solved in Grossberg (1973).

The solution is reviewed in Grossberg (1978e, Sections 14 and 15).

To understand the simplest STM properties, A18 is transformed into pattern variables $X_i = x_i x^{-1}$ and total activity variables

$$x = \sum_{k=1}^{n} x_k$$

using the notation $g(w) = w^{-1}f(w)$ and supposing that all $I_i = J_i = 0$. Then

$$\frac{d}{dt}X_i = BX_i \sum_{k=1}^{n} X_k[g(X_i x) - g(X_k x)] \quad \text{(A19)}$$

and

$$\frac{d}{dt}x = -Ax + (B - x) \sum_{k=1}^{n} f(X_k x). \quad \text{(A20)}$$

For example, if $f(w)$ is linear, namely, $f(w) = Cw$, then $g(w) = C$ and all $(d/dt)X_i = 0$ in A19. In other words, A19 can perfectly remember *any* initial pattern of reflectances. However, by A20 if $A \geq B$, then $x(t)$ approaches zero as $t \to \infty$, whereas if $B > A$, then $x(t)$ approaches $B - A$ as $t \to \infty$, whether or not a prior input pattern occurs. Thus if STM storage is ever possible, then $B > A$, and consequently noise will be amplified as vigorously as inputs. A linear signal amplifies noise, and is therefore inadequate despite its perfect memory of reflectances.

A slower-than-linear signal $f(w)$, for example, $f(w) = Cw(D + w)^{-1}$ or more generally, any $f(w)$ such that $g(w)$ is monotone decreasing, is even worse. By A19, if $X_i > X_k$, $k \neq i$, then $(d/dt)X_i < 0$ and if $X_i < X_k$, $k \neq i$, then $(d/dt)X_i > 0$. All differences in reflectances are hereby erased by the reverberation, and noise amplification also occurs. The whole network experiences a type of seizure.

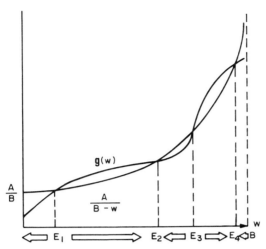

Figure A2. The even solutions E_0, E_2, ... of $g(w)$ $= A(B - w)^{-1}$ are stable equilibrium points of $x(\infty)$ $= \lim_{t \to \infty} x(t)$. (Since $g(w) = w^{-1}f(w)$, these points are solutions of $f(w) = Aw(B - w)^{-1}$. If $x(0) < E_1$, then $x(\infty) = 0$; thus E_1 defines the level below which $x(t)$ is treated as noise and quenched. All equilibrium points satisfy $E_i \leq B$; hence, short-term memory is normalized.)

If $f(w)$ is faster than linear, then the situation is better; for example, $f(w) = Cw^n$, $n > 1$, or more generally any $f(w)$ such that $g(w)$ is monotone increasing. In this case, if $X_i > X_k$, $k \neq i$, then $(d/dt)X_i > 0$, and if $X_i < X_k$, $k \neq i$, then $(d/dt)X_i < 0$. Consequently, this network chooses the population with the initial maximum in activity and totally suppresses activity in all other populations. This network behaves like a finite state, or binary choice machine. The same is true for total activity, since as $t \to \infty$, A20 becomes approximately

$$(d/dt)x \cong x[- A + (B - x)g(x)]. \quad (A21)$$

Thus the equilibrium points of $x(t)$ as $t \to \infty$ are $E_0 = 0$ and all the solutions of the equation

$$g(x) = A(B - x)^{-1}. \quad (A22)$$

If $g(0) < A/B$, then the smallest solution E_1 of A22 is unstable (Figure A2) so that small activities $x(t)$ are suppressed as $t \to \infty$. This is noise suppression due to recurrent competition. Every other solution E_2, E_4, ... of A22 is a stable equilibrium point of $x(t)$ as $t \to \infty$ (total activity quantization) and all equilibria are smaller than B (normalization).

The faster-than-linear signal contrast enhances the pattern so violently that the good property of noise suppression is joined to the extreme property of binary choice. This latter property is weakened by contructing a hybrid signal function that is chosen faster than linear at small activities to achieve noise suppression, but which levels off at high activities if only because all signal functions must be bounded. In the simplest case, $f(w)$ is a sigmoid, or **S**-shaped signal function. Then there exists a quenching threshold (QT). If v_is initial activity $x_i(0)$ falls below the QT, then its STM activity is quenched, or laterally masked: $x_i(\infty) = 0$. All the $x_i(0)$s that exceed the QT are contrast enhanced and stored in STM. Simultaneously, the total STM activity is normalized. Speaking intuitively, the QT exists because the faster-than-linear range starts to contrast enhance the pattern. Simultaneously, normalization shifts the activities into the intermediate linear range that stores any pattern, in particular the partially contrast-enchanced pattern. Because a QT exists, the network is a tunable filter. For example, a nonspecific arousal signal that multiplicatively inhibits all the recurrent inhibitory interneurons will lower the QT and facilitate storage of inputs in STM. Grossberg and Levine (1975) mathematically studied how such attentional shunts alter the resultant STM pattern by differentially sensitizing prescribed subfields of feature detectors that are joined together by competitive feedback interactions. The privileged subfields mask the activities in less sensitive subfields.

Such examples, either taken separately or linked together by feedback, provide insight into how interactions between continuously fluctuating quantities can sometimes generate discrete collective properties of the system as a whole. More generally, Grossberg (1978c) proves that every competitive system induces a decision scheme that can be used to globally characterize its pattern transformations as time goes on.

Appendix E

This section summarizes how the simplest transduction law realizable by a depletable chemical generates properties of antagonistic rebound due to specific cue offset and to nonspecific arousal onset when two parallel transduction pathways compete.

Transmitters as Gates

The transmitter law that we need can be derived in two ways. Originally, it was derived as the minimal law that was compatible with psychological postulates of classical conditioning (Grossberg, 1969c, Section 20; Grossberg, 1972c, Section 2). I now show that the law is the simplest transduction rule that can be computed using a depletable chemical transducer.

The simplest transduction rule converts an input I into a proportional signal S, namely,

$$S = BI, \qquad (A23)$$

where $B > 0$ is some proportionality constant. Equation A23 says that I is *gated* by B to yield S. If we interpret B as the amount of transducer and BI as the rate with which transducer is released to create signal S, then A23 says that the input I activates the transducer B in a statistically independent, or mass action, way.

When the transducer is released to activate another cell, there must exist a mechanism whereby it can be replenished, so that A23 can be maintained, at least approximately, through time.

Let $z(t)$ be the amount of transducer at time t. How can we keep $z(t) \cong B$ for all $t \geq 0$ so that the transduction rule

$$S = Iz(t) \qquad (A24)$$

approximately agrees with A23? This question leads to the following law for the temporal evolution of the amount $z(t)$ of available transducer

$$\frac{dz}{dt} = A(B - z) - Iz. \qquad (A25)$$

The term $A(B - z)$ in A25 says that $z(t)$ accumulates until it attains level B. The term does this by accumulating transducer at rate AB, that is proportional to B, and by feedback inhibition of the production rate at a rate $-Az(t)$ that is proportional to $z(t)$. The term $-Iz(t)$ in A25 indicates that transducer is depleted at a rate proportional to its rate of elimination, which is due to gating of I by $z(t)$. When $z(t) \cong B$, term $-Iz$ is proportional to $-BI$, as required by A23. Thus A25 is the law that "corresponds" to the law $S = BI$ when depletion of transducer can occur. It describes four effects working together: production, feedback inhibition, gating, and depletion.

Rebound Due to Cue Offset

Suppose that the adaptation level is I and that the cue input is J. Consider the simplest case in which the total signal in the on-channel is $S_1 = I + J$ and in the off-channel is $S_2 = I$. Let the transmitter z_1 in the on-channel satisfy the equation

$$\frac{d}{dt}z_1 = A(B - z_1) - S_1 z_1, \qquad (A26)$$

and the transmitter z_2 in the off-channel satisfy the equation

$$\frac{d}{dt}z_2 = A(B - z_2) - S_2 z_2. \qquad (A27)$$

After z_1 and z_2 equilibrate to S_1 and S_2, $(d/dt)z_1 = (d/dt)z_2 = 0$. Thus by A26 and A27,

$$z_1 = \frac{AB}{A + S_1} \qquad (A28)$$

and

$$z_2 = \frac{AB}{A + S_2}. \qquad (A29)$$

Since $S_1 > S_2$, it follows that $z_1 < z_2$; that is, z_1 is depleted more than z_2. However, the gated signal in the on-channel is $S_1 z_1$, and the gated signal in the off-channel is $S_2 z_2$. Since

$$S_1 z_1 = \frac{ABS_1}{A + S_1} \qquad (A30)$$

and

$$S_2 z_2 = \frac{ABS_2}{A + S_2}, \qquad (A31)$$

it follows from $S_1 > S_2$ that $S_1 z_1 > S_2 z_2$ despite the fact that $z_1 < z_2$. Thus the on-channel gets a bigger signal than the off-channel. After the two channels compete, the cue input J produces a sustained on-response whose size is proportional to

$$S_1 z_1 - S_2 z_2 = \frac{A^2 BJ}{(A + I + J)(A + I)}. \qquad (A32)$$

Now shut J off. Then the cell potentials rapidly adjust until new signal values $S_1{}^* = I$ and $S_2{}^* = I$ obtain. However, the transmitters z_1 and z_2 change much more slowly, so that A28 and A29 are approximately valid in a time interval that follows J offset. Thus the net signals are approximately

$$S_1{}^* z_1 = \frac{ABI}{A + S_1} \qquad (A33)$$

and

$$S_2{}^* z_2 \cong \frac{ABI}{A + S_2}. \qquad (A34)$$

Since $S_1 > S_2$, $S_1{}^*z_1 < S_2{}^*z_2$. The off-channel now gets the bigger signal, so an antagonistic rebound occurs whose size is approximately

$$S_2{}^*z_2 - S_1{}^*z_1 = \frac{ABIJ}{(A+I+J)(A+I)}. \quad (A35)$$

The rebound is transient because the equal signals $S_1{}^* = S_2{}^* = I$ gradually equalize the z_1 and z_2 levels until they both approach $AB(A+S_1{}^*)^{-1}$. Then $S_1{}^*z_1 - S_2{}^*z_2$ approaches zero, so the competition between channels shuts off both of their outputs.

Rebound due to Arousal Onset

Suppose that the on-channel and off-channel have equilibrated to the input levels I and J. Now increase I to I^*, thereby changing the signals to $S_1{}^* = I^* + J$ and $S_2{}^* = I^*$. The transmitters z_1 and z_2 continue to obey A28 and A29 for awhile, with $S_1 = I + J$ and $S_2 = J$. A rebound occurs if $S_2{}^*z_2 > S_1{}^*z_1$. This inequality is true if

$$I^* > I + A, \quad (A36)$$

since

$$S_2{}^*z_2 - S_1{}^*z_1 = \frac{ABJ(I^* - I - A)}{(A+I)(A+I+J)}. \quad (A37)$$

In particular, a rebound will occur if I^* exceeds $I + A$ no matter how J is chosen. In other words, if the mismatch is great enough to increment the adaptation level by more than amount A, then all dipoles will simultaneously rebound, and by an amount that increases as a function of J, as in Equation A37. This is not true in all versions of the dipole model, since the signals S_i, $i = 1, 2$, are not always linear functions of their inputs. There exist examples in which the most active dipoles can be rebounded even though less intensely activated dipoles are amplified without being rebounded. Moreover, if the signals are sigmoid functions of input size, then inverted-**U** effects occur in both the on- and off-responses to cue and arousal increments (Grossberg, 1972b, 1972c, 1975).

25
Introduction

(1981)
James L. McClelland and David E. Rumelhart

**An interactive activation model of context effects in letter perception: part 1.
An account of basic findings**
Psychological Review 88:375–407

This is a paper by the two most prominent members of what has become known as the PDP group. PDP stands for Parallel Distributed Processing, and was the informal title of a research group at the University of California, San Diego, led by Rumelhart and McClelland, who were both there at that time. *Parallel Distributed Processing* also became the title of the two-volume set of papers that is currently the most detailed single source of information on neural networks (Rumelhart and McClelland, 1986, Vol. 1; McClelland and Rumelhart, 1986, Vol. 2).

One essential property of most neural networks is that computation is done by the whole network on the entire input. This means that different parts of a state vector can contribute to the efficiency and accuracy of the overall computation. For the better part of a century, effects due to context of this type have been seen in the perception of words and letters. It is suprisingly hard to do the experiments unambiguously, but there now is no question that it is easier to recognize individual letters embedded in words than in nonwords. Presumably this advantage is due to our familiarity with words, which somehow enhances perceptibility of their components.

In addition to enhancement of perceptibility with whole words, there are also effects seen with pronounceable nonwords. There are strong constraints on which letters can appear next to one another if a word is to be pronounceable. Almost as much improvement in recognition is seen in letters embedded in pronounceable nonwords as in words.

McClelland and Rumelhart propose a multilayer network model for word and letter perception, which they call the "Interactive Activation" model. It is not a learning model; the connection strengths are set by the modelers. For their simulations, McClelland and Rumelhart use a three-layer network and allow four-letter strings of characters. The elementary computational units—the model neurons—in the Interactive Activation model are also called *nodes*, which have interpretations depending on where they are in the network.

The levels follow the kinds of analysis that have been suggested for word perception: First is the *feature level*, which extracts features from the visual image—say, oriented line segments or corners. Each of the four allowable locations has its own set of feature analyzers. Second is the *letter level*. Each of the four letter positions has individual nodes representing a specific letter appearing at that position. Third is the *word level*. There is one node appearing at this level for every word in the system's vocabulary.

It is important to note that the representation of information is not widely distributed at the word and letter levels. When a word is finally recognized, at the word level, one, and only one, node is active. Similarly, only one letter node can be ultimately active at a single position, though, of course, four active letters are required to spell a word. This is in contrast to models that allow a distributed pattern of activity across many nodes to represent a letter or a word. Questions about representation are currently of great importance. Feldman and Ballard (paper 29) discuss some of the things that can be done with localized representations like the one used here.

The network allows for complex interactions between levels and within levels. Within a level, interactions are inhibitory. That is, the output of the computation when a word is recognized is a single active node. The active node suppresses its neighbors at the word level. Activation of a letter at a position inhibits other letters at that position. Connections between layers can be excitatory or inhibitory. The letter nodes spelling a word obviously have excitatory connections to that word node. Similarly, if a word node is active, it makes excitatory contact with the letters that spell it. Most of the connection patterns are fairly clear, given the computational function of the network, but the exact connection strengths require a bit of trial and error. The real usefulness of a grandmother cell system arises here; it would be much more difficult to set connection strengths a priori in a network using a distributed coding, because of the arbitrary and unpredicable nature of the interactions between units. The use of powerful learning algorithms is almost essential for a network using distributed representations.

The units at the nodes act very much like familiar model neurons. Activation of a unit is a real number, and can be positive or negative. When the activation level of a node is positive, the node is said to be *active*. In the absence of inputs, units have a resting level that is related to word frequency. When nodes connected to it become active, a unit takes on a value given by the weighted sum of its inputs. There is a decay term, so at each time step the previous state of the unit decays a bit and the inputs from active units are added in. A sigmoidal output function is built into the dynamics (see Grossberg, paper 19) so that the output from a node will not exceed a certain maximum value, set to 1.0 in the model, or a minimum value, set to -0.20.

The Interactive Activation model is simple in conception and dynamics. Because it is realized as a computer program, it is possible to simulate directly many psychological experiments. McClelland and Rumelhart spend most of the paper applying the model to various tasks, to see if it fits the patterns of responses, reaction times, and errors seen in the human data. It is fair to say that the model is remarkably successful in doing so.

They made a reasonably large network containing 1,179 four-letter words for their simulations. The feature level extracted features probabilistically, so effects such as poor contrast, brief displays, or degraded images could be built in easily by changing the chances of extracting a feature at the feature level from the characters presented.

In common with many neural models, the behavior of the system when it operates is quite intuitive, though hard to analyze exactly. They give one example, in which if a degraded image is presented, multiple possible candidate words are activated. The likeliest words then excite their component letters through return connections, which inhibit other competing letters by suppressing them at the letter levels. Eventually a single word wins, and remains active, while the other words are inactive. The winning word is usually the best fit, i.e., the word that agrees best with the degraded input data.

Words then have an advantage in perceptibility over nonwords because they are activated faster and more strongly. Everything cooperates to enhance their activation, including feedback from the word level. Nonwords, even if the correct letters are eventually activated, are slow to achieve high activation levels. This translates into high error rate and slow response time. Pronounceable nonwords have an advantage over random letter strings for an interesting reason. Although there is no single word node associated with them, because they are pronouncable they activate partially a number of word nodes. As one of the examples used in the paper, MAVE excites sixteen word nodes for MOVE, HAVE, GAVE, etc., because they have pairs or triples of letters in common with MAVE. This means that pronounceable nonwords get a significant boost in speed of activation and accuracy with a little help from their *friends*, which McClelland and Rumelhart define as the group of words that share three letters in common with a pronounceable nonword.

The detailed analysis of a number of word recognition experiments in this paper and its successor (Rumelhart and McClelland, 1982) is fascinating and, overall, quite convincing. This paper has become a landmark in mathematical psychology.

References

J. L. McClelland and D. E. Rumelhart (Eds.) (1986), *Parallel Distributed Processing: Explorations in the Microstructures of Cognition*, Vol. 2, Cambridge, MA: MIT Press.

D. E. Rumelhart and J. L. McClelland (1982), "An interactive activation model of context effects in letter perception: part 2. The contextual enhancement effect and some tests and extensions of the model," *Psychological Review* 89:60–94.

D. E. Rumelhart and J. L. McClelland (Eds.) (1986), *Parallel Distributed Processing: Explorations in the Microstructures of Cognition*, Vol. 1, Cambridge, MA: MIT Press.

(1981)

James L. McClelland and David E. Rumelhart

An interactive activation model of context effects in letter perception: part 1.
An account of basic findings
Psychological Review 88:375–407

A model of context effects in perception is applied to the perception of letters in various contexts. In the model, perception results from excitatory and inhibitory interactions of detectors for visual features, letters, and words. A visual input excites detectors for visual features in the display. These excite detectors for letters consistent with the active features. The letter detectors in turn excite detectors for consistent words. Active word detectors mutually inhibit each other and send feedback to the letter level, strengthening activation and hence perceptibility of their constituent letters. Computer simulation of the model exhibits the perceptual advantage for letters in words over unrelated contexts and is consistent with the basic facts about the word advantage. Most importantly, the model produces facilitation for letters in pronounceable pseudowords as well as words. Pseudowords activate detectors for words that are consistent with most of the active letters, and feedback from the activated words strengthens the activations of the letters in the pseudoword. The model thus accounts for apparently rule-governed performance without any actual rules.

As we perceive, we are continually extracting sensory information to guide our attempts to determine what is before us. In addition, we bring to perception a wealth of knowledge about the objects we might see or hear and the larger units in which these objects co-occur. As one of us has argued for the case of reading (Rumelhart, 1977), our knowledge of the objects we might be perceiving works together with the sensory information in the perceptual process. Exactly how does the knowledge that we have interact with the input? And how does this interaction facilitate perception?

In this two-part article we have attempted to take a few steps toward answering these questions. We consider one specific example of the interaction of knowledge and perception—the perception of letters in words and other contexts. In Part 1 we examine the main findings in the literature on perception of letters in context and develop a model called the interactive activation model to account for these effects. In Part 2 (Rumelhart & McClelland, in press) we extend the model in several ways. We present a set of studies introducing a new technique for studying the perception of letters in context, independently varying the duration and timing of the context and target letters. We show how the model fares in accounting for the results of these experiments and discuss how the model may be extended to account for a variety of phenomena. We also present an experiment that tests—and supports—a

Preparation of this article was supported by National Science Foundation Grants BNS-76-14830 and BNS-79-24062 to J. L. McClelland and Grant BNS-76-15024 to D. E. Rumelhart, and by the Office of Naval Research under contract N00014-79-C-0323. We would like to thank Don Norman, James Johnston, and members of the LNR research group for helpful discussions of much of the material covered in this article.

Copyright 1981 by the American Psychological Association, Inc. Reprinted by permission of the publisher and author.

counterintuitive prediction of the model. Finally, we consider how the mechanisms developed in the course of exploring our model of word perception might be extended to perception of other sorts of stimuli.

Basic Findings on the Role of Context in Perception of Letters

The notion that knowledge and familiarity play a role in perception has often been supported by experiments on the perception of letters in words (Bruner, 1957; Neisser, 1967). It has been known for nearly 100 years that it is possible to identify letters in words more accurately than letters in random letter sequences under tachistoscopic presentation conditions (Cattell, 1886; see Huey, 1908, and Neisser, 1967, for reviews). However, until recently such effects were obtained using whole reports of all of the letters presented. These reports are subject to guessing biases, so that it was possible to imagine that familiarity did not determine how much was seen but only how much could be inferred from a fragmentary percept. In addition, for longer stimuli, full reports are subject to forgetting. We may see more letters than we can actually report in the case of nonwords, but when the letters form a word, we may be able to retain as a single unit the item whose spelling may simply be read out from long-term memory. Thus, despite strong arguments to the contrary by proponents of the view that familiar context really does influence perception, it has been possible until recently to imagine that the context in which a letter was presented influences only the accuracy of postperceptual processes and not the process of perception itself.

The perceptual advantage of letters in words. The seminal experiment of Reicher (1969) suggests that context does actually influence perceptual processing. Reicher presented target letters in words, unpronounceable nonwords, and alone, following the presentation of the target display with a presentation of a patterned mask. The subject was then tested on a single letter in the display, using a forced choice between two alternative letters. Both alternatives fit the context to form an item of the type pre-

sented, so that, for example in the case of a word presentation, the alternative would also form a word in the context.

Forced-choice performance was more accurate for letters in words than for letters in nonwords or even for single letters. Since both alternatives made a word with the context, it is not possible to argue that the effect is due to postperceptual guessing based on equivalent information extracted about the target letter in the different conditions. It appears that subjects actually come away with more information relevant to a choice between the alternatives when the target letter is a part of a word. And, since one of the control conditions was a single letter, it is not reasonable to argue that the effect is due to forgetting letters that have been perceived. It is hard to see how a single letter, once perceived, could be subject to a greater forgetting than a letter in a word.

Reicher's (1969) finding seems to suggest that perception of a letter can be facilitated by presenting it in the context of a word. It appears, then, that our knowledge about words can influence the process of perception. Our model presents a way of bringing such knowledge to bear. The basic idea is that the presentation of a string of letters begins the process of activating detectors for letters that are consistent with the visual input. As these activations grow stronger, they begin to activate detectors for words that are consistent with the letters, if there are any. The active word detectors then produce feedback, which reinforces the activations of the detectors for the letters in the word. Letters in words are more perceptible, because they receive more activation than representations of either single letters or letters in an unrelated context.

Reicher's basic finding has been investigated and extended in a large number of studies, and there now appears to be a set of important related findings that must also be explained.

Irrelevance of word shape. The effect seems to be independent of the familiarity of the word as a visual configuration. The word advantage over nonwords is obtained for words in lowercase type, words in uppercase type, or words in a mixture of upper-

and lowercase (Adams, 1979; McClelland, 1976).

Role of patterned masking. The word advantage over single letters and nonwords appears to depend upon the visual masking conditions used (Johnston & McClelland, 1973; Massaro & Klitzke, 1979; see also Juola, Leavitt, & Choe, 1974; Taylor & Chabot, 1978). The word advantage is quite large when the target appears in a distinct, high-contrast display followed by a patterned mask of similar characteristics. However, the word advantage over single letters is actually reversed, and the word advantage over nonwords becomes quite small when the target is indistinct, low in contrast, and/or followed by a blank, nonpatterned field.

Extension to pronounceable pseudowords. The word advantage also applies to pronounceable nonwords, such as *REET* or *MAVE*. A large number of studies (e.g., Aderman & Smith, 1971; Baron & Thurston, 1973; Spoehr & Smith, 1975) have shown that letters in pronounceable nonwords (also called pseudowords) have a large advantage over letters in unpronounceable nonwords (also called unrelated letter strings), and three studies (Carr, Davidson, & Hawkins, 1978; Massaro & Klitzke, 1979; McClelland & Johnston, 1977) have obtained an advantage for letters in pseudowords over single letters.

Absence of effects of contextual constraint under patterned-mask conditions. One important finding, which rules out several of the models that have been proposed previously, is the finding that letters in highly constraining word contexts have little or no advantage over letters in weakly constraining word contexts under the distinct-target/patterned-mask conditions that produce a large word advantage (Johnston, 1978; see also Estes, 1975). For example, if the set of possible stimuli contains only words, the context _*HIP* constrains the first letter to be either an *S*, a *C*, or a *W*; whereas the context _*INK* is compatible with 12 to 14 letters (the exact number depends on what counts as a word). We might expect that the former, more strongly constraining context would produce superior detection of a target letter. But in a very carefully controlled and executed study, Johnston (1978) found no such effect. Although constraints do influence performance under other conditions (e.g., Broadbent & Gregory, 1968), they do not appear to make a difference under the distinct-target/patterned-mask conditions of the Johnston study.

To be successful, any model of word perception must provide an account not only for Reicher's (1969) basic effect but for these related findings as well. Our model accounts for all of these effects. We begin by presenting the model in abstract form. We then focus on the specific version of the model implemented in our simulation program and consider some of the details. Subsequently, we turn to detailed considerations of the findings we have discussed in this section.

The Interactive Activation Model

We approach the phenomena of word perception with a number of basic assumptions that we want to incorporate into the model. First, we assume that perceptual processing takes place within a system in which there are several levels of processing, each concerned with forming a representation of the input at a different level of abstraction. For visual word perception, we assume that there is a visual feature level, a letter level, and a word level, as well as higher levels of processing that provide "top–down" input to the word level.

Second, we assume that visual perception involves parallel processing. There are two different senses in which we view perception as parallel. We assume that visual perception is spatially parallel. That is, we assume that information covering a region in space at least large enough to contain a four-letter word is processed simultaneously. In addition, we assume that visual processing occurs at several levels at the same time. Thus, our model of word perception is spatially parallel (i.e., capable of processing several letters of a word at one time) and involves processes that operate simultaneously at several different levels. Thus, for example, processing at the letter level presumably occurs simultaneously with processing at the word level and with processing at the feature level.

Third, we assume that perception is fundamentally an *interactive* process. That is,

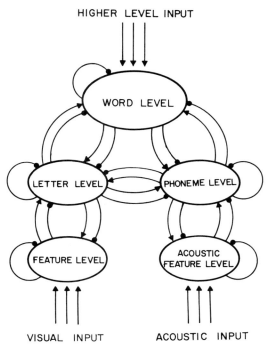

Figure 1. A sketch of some of the processing levels involved in visual and auditory word perception, with interconnections.

we assume that "top–down" or "conceptually driven" processing works simultaneously and in conjunction with "bottom-up" or "data driven" processing to provide a sort of multiplicity of constraints that jointly determine what we perceive. Thus, for example, we assume that knowledge about the words of the language interacts with the incoming featural information in codetermining the nature and time course of the perception of the letters in the word.

Finally, we wish to implement these assumptions by using a relatively simple method of interaction between sources of knowledge whose only "currency" is simple excitatory and inhibitory activations of a neural type.

Figure 1 shows the general conception of the model. Perceptual processing is assumed to occur in a set of interacting levels, each communicating with several others. Communication proceeds through a spreading activation mechanism in which activation at one level spreads to neighboring levels. The communication can consist of both excit-

atory and inhibitory messages. Excitatory messages increase the activation level of their recipients. Inhibitory messages decrease the activation level of their recipients. The arrows in the diagram represent excitatory connections, and the circular ends of the connections represent inhibitory connections. The intralevel inhibitory loop represents a kind of lateral inhibition in which incompatible units at the same level compete. For example, since a string of four letters can be interpreted as at most one four-letter word, the various possible words mutually inhibit one another and in that way compete as possible interpretations of the string.

It is clear that many levels are important in reading and perception in general, and the interactions among these levels are important for many phenomena. However, a theoretical analysis of all of these interactions introduces an order of complexity that obscures comprehension. For this reason, we have restricted the present analysis to an examination of the interaction between a single pair of levels, the word and letter levels. We have found that we can account for the phenomena reviewed above by considering only the interactions between letter level and word level elements. Therefore, for the present we have elaborated the model only on these two levels, as illustrated in Figure 2. We have delayed consideration of the effects of higher level processes and phonological processes, and we have ignored the reciprocity of activation that may occur between word and letter levels and any other levels of the system. We consider aspects of the fuller model including these influences in Part 2 (Rumelhart & McClelland, in press).

Specific Assumptions

Representation assumptions. For every relevant unit in the system we assume there is an entity called a *node.* We assume that there is a node for each word we know, and that there is a node for each letter in each letter position within a four-letter string.

The nodes are organized into levels. There are *word level* nodes and *letter level* nodes. Each node has connections to a number of

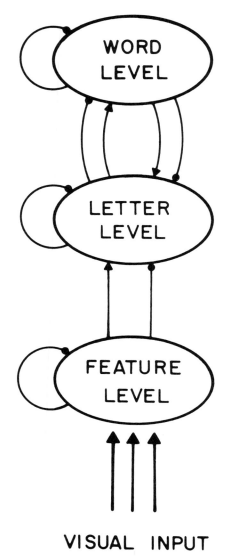

Figure 2. The simplified processing system.

tify nodes according to the units they detect, printing them in italics; stimuli presented to the system are in uppercase letters.

Connections may occur within levels or between adjacent levels. There are no connections between nonadjacent levels. Connections within the word level are mutually inhibitory, since only one word can occur at any one place at any one time. Connections between the word level and letter level may be either inhibitory or excitatory (depending on whether the letter is a part of the word in the appropriate letter position). We call the set of nodes with excitatory connections to a given node its *excitatory neighbors* and the set of nodes with inhibitory connections to a given node its *inhibitory neighbors*.

A subset of the neighbors of the letter *t* is illustrated in Figure 3. Again, excitatory connections are represented by the arrows ending with points, and inhibitory connections are represented by the arrows ending with dots. We emphasize that this is a small subset of the neighborhood of the initial *t*. The picture of the whole neighborhood, including all the connections among neighbors and their connections to their neighbors, is much too complicated to present in a two-dimensional figure.

Activation assumptions. There is associated with each node a momentary activation value. This value is a real number, and for node i we will represent it by $a_i(t)$. Any node with a positive activation value is said to be *active*. In the absence of inputs from its neighbors, all nodes are assumed to decay back to an inactive state, that is, to an activation value at or below zero. This resting level may differ from node to node and corresponds to a kind of a priori bias (Broadbent, 1967) determined by frequency of activation of the node over the long term. Thus, for example, the nodes for high-frequency words have resting levels higher than those for low-frequency words. In any case, the resting level for node i is represented by r_i. For units not at rest, decay back to the resting level occurs at some rate Θ_i.

When the neighbors of a node are active, they influence the activation of the node by either excitation or inhibition, depending on their relation to the node. These excitatory and inhibitory influences combine by a sim-

other nodes. The nodes to which a node connects are called its *neighbors*. Each connection is two-way. There are two kinds of connections: *excitatory* and *inhibitory*. If two nodes suggest each other's existence (in the way that the node for the word *the* suggests the node for an initial *t* and vice versa), then the connections are excitatory. If two nodes are inconsistent with one another (in the way that the node for the word *the* and the node for the word *boy* are inconsistent), then the relationship is inhibitory. Note that we iden-

ple weighted average to yield a net input to the unit, which may be either excitatory (greater than zero) or inhibitory. In mathematical notation, if we let $n_i(t)$ represent the net input to the unit, we can write the equation for its value as

$$n_i(t) = \sum_j \alpha_{ij} e_j(t) - \sum_k \gamma_{ik} i_k(t), \quad (1)$$

where $e_j(t)$ is the activation of an active excitatory neighbor of the node, each $i_k(t)$ is the activation of an active inhibitory neighbor of the node, and α_{ij} and γ_{ik} are associated weight constants. Inactive nodes have no influence on their neighbors. Only nodes in an active state have any effects, either excitatory or inhibitory.

The net input to a node drives the activation of the node up or down, depending on whether it is positive or negative. The degree of the effect of the input on the node is modulated by the node's current activity level to keep the input to the node from driving it beyond some maximum and minimum values (Grossberg, 1978). When the net input is excitatory, $n_i(t) > 0$, the effect on the node, $\epsilon_i(t)$, is given by

$$\epsilon_i(t) = n_i(t)(M - a_i(t)), \quad (2)$$

where M is the maximum activation level of the unit. The modulation has the desired effect, because as the activation of the unit approaches the maximum, the effect of the input is reduced to zero. M can be thought

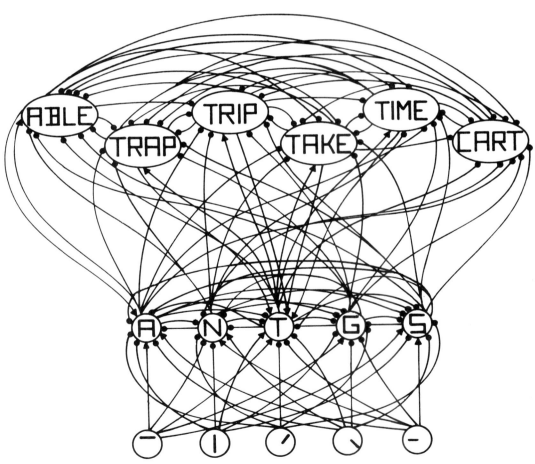

Figure 3. A few of the neighbors of the node for the letter T in the first position in a word, and their interconnections.

of as a basic scale factor of the model, and we have set its value to 1.0.

In the case where the input is inhibitory, $n_i(t) < 0$, the effect of the input on the node is given by

$$\epsilon_i(t) = n_i(t)(a_i(t) - m), \qquad (3)$$

where m is the minimum activation of the unit.

The new value of the activation of a node at time $t + \Delta t$ is equal to the value at time t, minus the decay, plus the influence of its neighbors at time t:

$$a_i(t + \Delta t)$$
$$= a_i(t) - \Theta_i(a_i(t) - r_i) + \epsilon_i(t). \qquad (4)$$

Input assumptions. Upon presentation of a stimulus, a set of featural inputs is made available to the system. Each feature in the display will be detected with some probability p. For simplicity it is assumed that feature detection occurs, if it is to occur at all, immediately after onset of the stimulus. The probability that any given feature will be detected is assumed to vary with the visual quality of the display. Features that are detected begin sending activation to all letter nodes that contain that feature. All letter level nodes that do not contain the extracted feature are inhibited.

It is assumed that features are binary and that we can extract either the presence or absence of a particular feature. So, for example, when viewing the letter R we can extract, among other features, the presence of a diagonal line segment in the lower right corner and the absence of a horizontal line across the bottom. In this way the model honors the conceptual distinction between knowing that a feature is absent and not knowing whether a feature is present.

Presentation of a new display following an old one results in the probabilistic extraction of the set of features present in the new display. These features, when extracted, replace the old ones in corresponding positions. Thus, the presentation of an E following the R described above would result in the replacement of the two features described above with their opposites.

On making responses. One of the more problematic aspects of a model such as this one is a specification of how these relatively complex patterns of activity might be related to the content of percepts and the sorts of response probabilities we observe in experiments. We assume that responses and perhaps the contents of perceptual experience depend on the temporal integration of the pattern of activation over all of the nodes. The integration process is assumed to occur slowly enough that brief activations may come and go without necessarily becoming accessible for purposes of responding or entering perceptual experience. However, as the activation lasts longer and longer, the probability that it will be reportable increases. Specifically, we think of the integration process as taking a running average of the activation of the node over previous time:

$$\bar{a}_i(t) = \int_{-\infty}^{t} a_i(x)e^{-(t - x)r}dx. \qquad (5)$$

In this equation, the variable x represents preceding time, varying between $-\infty$ and time t. The exponential portion of the expression weights the contribution of the activation of the node in previous time intervals: Essentially, its effect is to reduce the contribution of prior activations as they recede further back in time. The parameter r represents the relative weighting given to old and new information and determines how quickly the output values change in response to changes in the activations of the underlying nodes. The larger the value of r, the more quickly the output values change. *Response strength*, in the sense of Luce's choice model (Luce, 1959), is an exponential function of the running average activation:

$$s_i(t) = e^{\mu \bar{a}_i(t)}. \qquad (6)$$

The parameter μ determines how rapidly response strength grows with increases in activation. Following Luce's formulation, we assume that the probability of making a response based on node i is given by

$$p(R_i, t) = \frac{s_i(t)}{\sum_{j \epsilon L} s_j(t)}, \qquad (7)$$

where L represents the set of nodes competing at the same level with node i.

Most of the experiments we will be considering test subjects' performance on one of the letters in a word or other type of display. In accounting for these results, we have adopted the assumption that responding is always based on the output of the letter level, rather than the output of the word level or some combination of the two. The forced choice is assumed to be based only on this letter-level information. The subject compares the letter selected for the appropriate position against the forced-choice alternatives. If the letter selected is one of the alternatives, then that alternative is chosen in the forced choice. If it is not one of the alternatives, then the model assumes that one of the alternatives would simply be chosen at random.

One somewhat problematical issue involves deciding when to read out the results of processing and select response letters for each letter position. When a target display is simply turned on and left on until the subject responds, and when there is no pressure to respond quickly, we assume that the subject simply waits until the output strengths have reached their asymptotic values. However, when a target display is presented briefly followed by a patterned mask, the activations produced by the target are transient, as we shall see. Under these conditions, we assume that the subject learns through experience in the practice phase of the experiment to read out the results of processing at a time that allows the subject to optimize performance. For simplicity, we have assumed that readout occurs in parallel for all four letter positions.

The Operation of the Model

Now, consider what happens when an input reaches the system. Assume that at time t_0 all prior inputs have had an opportunity to decay, so that the entire system is in its quiescent state, and each node is at its resting level. The presentation of a stimulus initiates a process in which certain features are extracted and excitatory and inhibitory pressures begin to act upon the letter-level nodes. The activation levels of certain letter nodes are pushed above their resting levels. Others receive predominantly inhibitory inputs and are pushed below their resting levels. These letter nodes, in turn, begin to send activation to those word-level nodes they are consistent with and inhibit those word nodes they are not consistent with. In addition, within a given letter position channel, the various letter nodes attempt to suppress each other, with the strongest ones getting the upper hand. As word-level nodes become active, they in turn compete with one another and send feedback down to the letter-level nodes. If the input features were close to those for one particular set of letters and those letters were consistent with those forming a particular word, the positive feedback in the system will work to rapidly converge on the appropriate set of letters and the appropriate word. If not, they will compete with each other, and perhaps no single set of letters or single word will get enough activation to dominate the others. In this case the various active units might strangle each other through mutual inhibition.

At any point during processing, the results of perceptual processing may be read out from the pattern of activations at the letter level into a buffer, where they may be kept through rehearsal or used as the basis for overt reports. The accuracy of this process depends on a running average of the activations of the correct node and of other competing nodes.

Simulations

Although the model is in essence quite simple, the interactions among the various nodes can become complex, so that the model is not susceptible to a simple intuitive or even mathematical analysis. Instead, we have relied on computer simulations to study the behavior of the model and to see if it is consistent with the empirical data. A description of the actual computer program is given in the Appendix.

For purposes of these simulations, we have made a number of simplifying assumptions. These additional assumptions fall into three classes: (a) discrete rather than continuous time, (b) simplified feature analysis of the input font, and (c) a limited lexicon.

The simulation operates in discrete time slices, or ticks, updating the activations of all of the nodes in the system once each cycle on the basis of the values on the previous cycle. Obviously, this is simply a matter of computational convenience and not a fundamental assumption. We have endeavored to keep the time slices "thin" enough so that the model's behavior is continuous for all intents and purposes.

Any simulation of the model involves making explicit assumptions about the appropriate featural analysis of the input font. We have, for simplicity, chosen the font and featural analysis employed by Rumelhart (1970) and by Rumelhart and Siple (1974), illustrated in Figure 4. Although the experiments we have simulated employed different type fonts, we assume that the basic results do not depend on the particular font used. The simplicity of the present analysis recommends it for the simulations, though it obviously skirts several fundamental issues about the lower levels of processing.

Finally, our simulations have been restricted to four-letter words. We have equipped our program with knowledge of 1,179 four-letter words occurring at least two times per million in the Kucera and Francis (1967) word count. Plurals, inflected forms, first names, proper names, acronyms, abbreviations, and occasional unfamiliar entries arising from apparent sampling flukes

Figure 4. The features used to construct the letters in the font assumed by the simulation program, and the letters themselves. (From "Process of Recognizing Tachistoscopically Presented Words" by David E. Rumelhart and Patricia Siple, *Psychological Review*, 1974, *81*, 99–118. Copyright 1974 by the American Psychological Association. Reprinted by permission.)

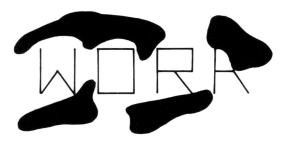

Figure 5. A hypothetical set of features that might be extracted on a trial in an experiment on word perception.

have been excluded. This sample appears to be sufficient to reflect the essential characteristics of the language and to show how the statistical properties of the language can affect the process of perceiving letters in words.

An Example

Let us now consider a sample run of our simulation model. The parameter values employed in the example are those used to simulate all the experiments discussed in the remainder of Part 1. These values are described in detail in the following section. For the purposes of this example, imagine that the word *WORK* has been presented to the subject and that the subject has extracted those features shown in Figure 5. In the first three letter positions, the features of the letters *W, O,* and *R* have been completely extracted. In the final position a set of features consistent with the letters *K* and *R* have been extracted, with the features that would disambiguate the letter unavailable. We wish now to chart the activity of the system resulting from this presentation. Figure 6 shows the time course of the activations for selected nodes at the word and letter levels, respectively.

At the word level, we have charted the activity levels of the nodes for the words *work, word, wear,* and *weak.* Note first that *work* is the only word in the lexicon consistent with all the presented information. As a result, its activation level is the highest and reaches a value of .8 through the first 40 time cycles. The word *word* is consistent with the bulk of the information presented and therefore first rises and later is pushed back down below its resting level, as a result of competition with *work*. The words *wear* and *weak* are consistent with the only letter active in the first letter position, and one of the two active in the fourth letter position. They are also inconsistent with the letters active in Positions 2 and 3. Thus, the activation they receive from the letter level is quite weak, and they are easily driven down well below zero, as a result of competition from the other word units. The activations of these units do not drop quite as low, of course, as the activation level of words such as *gill*, which contain nothing in common with the presented information. Although not shown in Figure 6, these words attain near-mini-

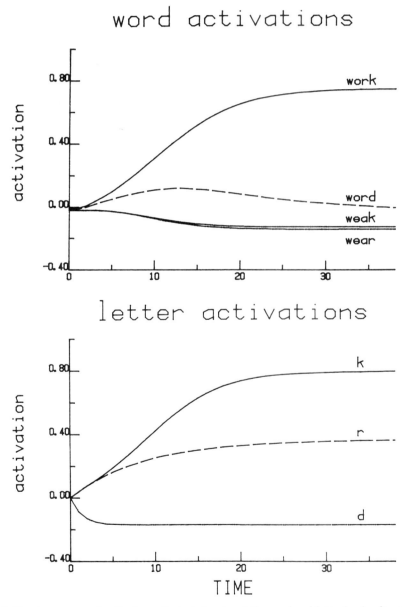

Figure 6. The time course of activations of selected nodes at the word and letter levels after extraction of the features shown in Figure 5.

mum activation levels of about −.20 and stay there as the stimulus stays on. Returning to *wear* and *weak*, we note that these words are equally consistent with the presented information and thus drop together for about the first 9 time units. At this point, however, the word *work* has clearly taken the upper hand at the word level, and produces feedback that reinforces the activation of the final *k* and not the final *r*. As a result, the word *weak* receives more activation from the letter level than the word *wear* and begins to gain a slight advantage over *wear*. The strengthened *k* continues to feed activation into the word level and strengthen consistent words. The words that contain an *R* continue to receive activation from the *r* node also, but they receive stronger inhibition from the words consistent with a *K* and are therefore ultimately weakened, as illustrated in the lower panel of Figure 6.

The strong feature–letter inhibition ensures that when a feature inconsistent with a particular letter is detected, that letter will receive relatively strong net bottom–up inhibition. Thus in our example, the information extracted clearly disconfirms the possibility that the letter *D* has been presented in the fourth position, and thus the activation level of the *d* node decreases quickly to near its minimum value. However, the bottom–up information from the feature level supports either a *K* or an *R* in the fourth position. Thus, the activation of each of these nodes rises slowly. These activations, along with those for *W, O,* and *R*, push the activation of *work* above zero, and it begins to feed back; by about Time Cycle 4, it is beginning to push the *k* above the *r* (because *WORR* is not a word). Note that this separation occurs just before the words *weak* and *wear* separate. It is the strengthening of *k* due to feedback from *work* that causes them to separate.

Ultimately, the *r* reaches a level well below that of *k* where it remains, and the *k* pushes toward a .8 activation level. As discussed below, the word-to-letter inhibition and the letter-to-letter inhibition have both been set to 0. Thus, *k* and *r* both co-exist at moderately high levels, the *r* fed only from the bottom up, and the *k* fed from both bottom up and top down.

Finally, consider the output values for the

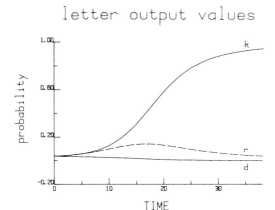

Figure 7. Output values for the letters *r, k,* and *d* after presentation of the display shown in Figure 5.

letter nodes *r, k,* and *d*. Figure 7 shows the output values for the simulation. The output value is the probability that if a response was selected at time *t*, the letter in question would be selected as the output or response from the system. As intended, these output values grow somewhat more slowly than the values of the letter activations themselves but eventually, as they reach and hold their asymptotic values, come to reflect the activations of the letter nodes. Since in the absence of masking subjects can afford to wait to read out a response until the output values have had a chance to stabilize, they would be highly likely to choose the letter *K* as the response.

Although this example is not very general in that we assumed that only partial information was available in the input for the fourth letter position, whereas full information was available at the other letter positions, it does illustrate many of the important characteristics of the model. It shows how ambiguous sensory information can be disambiguated by top–down processes. Here we have a very simple mechanism capable of applying knowledge of words in the perception of their component letters.

Parameter Selection

Once the basic simulation model was constructed, we began a lengthy process of attempting to simulate the results of several representative experiments in the literature.

Only two parameters of the model were allowed to vary from experiment to experiment: (a) the probability of feature extraction and (b) the timing of the presentation of the masking stimulus if one was used.

The probability of feature extraction is assumed to depend on the visual characteristics of the display. In most of the experiments we will consider, a bright, high-contrast target was used. Such a target would produce perfect performance if not followed by a patterned mask. In these cases probability of feature extraction was fixed at 1.0 and the timing of the target offset and coincident mask onset typically was adjusted to achieve 75% correct performance over the different experimental conditions of interest. In simulating the results of these experiments, we likewise varied the timing of the target offset/mask onset to achieve the right average correct performance from the model.

In some experiments no patterned mask was used, and performance was kept below perfect levels by using a dim or otherwise degraded target display. In these cases the probability of feature extraction was set to a value less than 1.0, which produces about the right overall performance level.

The process of exploring the behavior of the model amounted to an extended search for a set of values for all the other parameters that would permit the model to simulate, as closely as possible, the results of all of the experiments to be discussed later in Part 1, as well as those to be considered in Part 2 (Rumelhart & McClelland, in press). To constrain the search, we adopted various restrictive simplifications. First, we assumed that all nodes have the same maximum activation value. In fact, the maximum was set to 1.0, and served to scale all activations within the model. The minimum activation value for all nodes was set at $-.20$, a value that permits rapid reactivation of strongly inhibited nodes. The decay rate of all nodes was set to the value of .07. This parameter effectively serves as a scale factor that determines how quickly things are allowed to change in a single time slice. The .07 value was picked after some exploration, since it seemed to permit us to run our simulations with the minimum number of time slices per trial, at the same time as it minimized a kind

of reverberatory oscillation that sets in when things are allowed to change too much on any given time cycle. We also assigned the resting value of zero to all of the letter nodes. The resting value of nodes at the word level was set to a value between $-.05$ and 0, depending on word frequency.

We have assumed that the weight parameters, α_{ij} and γ_{ij} depend only on the processing levels of nodes i and j and on no other characteristics of their identity. This means, among other things, that the excitatory connections between all letter nodes and all of the relevant word nodes are equally strong, independent of the identity of the words. Thus, for example, the degree to which the node for an initial t excites the node for the word *tock* is exactly the same as the degree to which it excites the node for the word *this*, in spite of a substantial difference in frequency of usage. To further simplify matters, the word-to-letter inhibition was also set to zero. This means that feedback from the word level can strengthen activations at the letter level but cannot weaken them.

The output from the detector network has essentially two parameters. The value .05 was used for the parameter r, which determines how quickly the output values change in response to changes in the activations of the underlying nodes. This value is small enough that the output values change relatively slowly, so that transient activations can come and go without much effect on the output. The value 10 was given to the parameter μ in Equation 6 above. The parameter is essentially a scale factor relating activations in the model to response strengths in the Luce formulation.

The values of the remaining parameters were fixed at the values given in Table 1. It is worth noting the differences between the feature–letter influences and the letter–word influences. The feature–letter inhibition is 30 times as strong as the feature–letter excitation. This means that all of the features detected must be compatible with a particular letter before that letter will receive net excitation (since there are only 14 possible features, there can only be a maximum of 13 excitatory inputs whenever there is a single inhibitory input). The main reason for choosing this value was to permit the pre-

Table 1
Parameter Values Used in the Simulations

Parameter	Value
Feature–letter excitation	.005
Feature–letter inhibition	.15
Letter–word excitation	.07
Letter–word inhibition	.04
Word–word inhibition	.21
Letter–letter inhibition	0
Word–letter excitation	.30

sentation of a mask to clear the previous pattern of activation. On the other hand, the letter–word inhibition is actually somewhat less than the letter–word excitation. When only one letter is active in each letter position, this means that the letter level will produce net excitation of all words that share two or more letters with the target word. Because of these multiple activations, strong word–word inhibition is necessary to "sharpen" the response of the word level, as we will see. In contrast, no such inhibition is necessary at the letter level. For these reasons, the letter–letter inhibition has been set to 0, whereas the word–word inhibition has been set to .21.

Comments on Related Formulations

Before turning to the application of the model to the experimental literature, some comments on the relationship of this model to other models extant in the literature is in order. We have tried to be synthetic. We have taken ideas from our own previous work and from the work of others in the literature. In what follows, we attempt to identify the sources of most of the assumptions of the model and to show in what ways our model differs from the models we have drawn on.

First of all, we have adopted the approach of formulating the model in terms similar to the way in which such a process might actually be carried out in a neural or neural-like system. We do not mean to imply that the nodes in our system are necessarily related to the behavior of individual neurons. We will, however, argue that we have kept the kinds of processing involved well within the bounds of capability for simple neural

circuits. The approach of modeling information processing in a neural-like system has recently been advocated by Szentagothai and Arbib (1975) and is represented in many of the articles presented in the volume by Hinton and Anderson (1981) as well as many of the specific models mentioned below.

One case in point is the work of Levin (1976). He proposed a parallel computational system capable of interactive processing that employs only excitation and inhibition as its currency. Although our model could not be implemented exactly in the format of their system (called Proteus), it is clearly in the spirit of their model and could readily be implemented within a variant of the Proteus system.

In a recent article McClelland (1979) has proposed a cascade model of perceptual processing in which activations on each level of the system drive those at the next higher level. This model has the properties that partial outputs are continuously available for processing and that every level of the system processes the input simultaneously. The present model certainly adopts these assumptions. It also generalizes them, permitting information to flow in both directions simultaneously.

Hinton (Note 1) has developed a *relaxation* model for visual perception in which multiple constraints interact by means of incrementing and decrementing real numbered strengths associated with various interpretations of a portion of the visual scene in an attempt to attain a maximally consistent interpretation of the scene. Our model can be considered a relaxation system in which activation levels are manipulated to get an optimal interpretation of an input word.

James Anderson and his colleagues (Anderson, 1977; Anderson, Silverstein, Ritz, & Jones, 1977) and Kohonen and his colleagues (Kohonen, 1977) have developed a pattern recognition system which they call an *associative memory* system. Their system shares a number of commonalities with ours. One feature the models share is the scheme of adding and subtracting weighted excitation values to generate output patterns that represent cleaned-up versions of the input patterns. In particular, our α_{ij} and γ_{ij} correspond to the matrix elements of the associative memory models. Our model differs in that it has multiple levels and employs a nonlinear cumulation function similar to one suggested by Grossberg (1978), as mentioned above.

Our model also draws on earlier work in the area of word perception. There is, of course, a strong similarity between this model and the logogen model of Morton (1969). What we have implemented might be called a hierarchical, nonlinear, logogen model with feedback between levels and inhibitory interactions among logogens at the same level. We have also added dynamic assumptions that are lacking from the logogen model.

The notion that word perception takes place in a hierarchical information-processing system has, of course, been advocated by several researchers interested in word perception (Adams, 1979; Estes, 1975; Johnston & McClelland, 1980; LaBerge & Samuels, 1974; McClelland, 1976). Our model differs from those proposed in many of these papers in that processing at different levels is explicitly assumed to take place in parallel. Many of the models are not terribly explicit on this topic, although the notion that partial information could be passed along from one level to the next so that processing could go on at the higher level while it was continuing at the lower level had been suggested by McClelland (1976). Our model also differs from all of these others, except that of Adams (1979), in assuming that there is feedback from the word level to the letter level. The general formulation suggested by Adams (1979) is quite similar to our own, although she postulates a different sort of mechanism for handling pseudowords (excitatory connections among letter nodes) and does not present a detailed account.

Our mechanism for accounting for the perceptual facilitation of pseudowords involves, as we will see below, the integration of feedback from partial activation of a number of different words. The idea that pseudoword perception could be accounted for in this way was inspired by Glushko (1979), who suggested that partial activation and synthesis of word pronunciations could ac-

count for the process of constructing a pronunciation for a novel pseudoword.

The feature-extraction assumptions and the bottom–up portion of the word recognition model are nearly the same as those employed by Rumelhart (1970, Note 2) and Rumelhart and Siple (1974). The interactive feedback portion of the model is clearly one of the class of models discussed by Rumelhart (1977) and could be considered a simplified control structure for expressing the model proposed in that paper.

Application of the Simulation Model to Several Basic Findings

We are finally ready to see how well our model fares in accounting for the findings of several representative experiments in the literature. In discussing each account, we will try to explain not only how well the simulation works but why it behaves as it does. As we proceed through the discussion, we will have occasion to describe several interesting synergistic properties of the model that we did not anticipate but discovered as we explored the behavior of the system. As mentioned previously, the actual parameters used in both the examples that we will discuss and in the simulation results we will report are those summarized in Table 1. We will consider the robustness of the model, and the effects of changes in these parameters, in the discussion section at the end of Part 1.

The Word Advantage and the Effects of Visual Conditions

As we noted previously, word perception has been studied under a variety of different visual conditions, and it is apparent that different conditions produce different results. The advantage of words over nonwords appears to be greatest under conditions in which a bright, high-contrast target is followed by a patterned mask with similar characteristics. The word advantage appears to be considerably less when the target presentation is dimmer or otherwise degraded and is followed by a blank white field.

Typical data demonstrating these points (from Johnston & McClelland, 1973) are

Table 2

Effect of Display Conditions on Proportion of Correct Forced Choices in Word and Letter Perception (From Johnston & McCielland, 1973)

| | Display type | |
	Word	Letter with number signs
Visual condition		
Bright target/patterned mask	.80	.65
Dim target/blank mask	.78	.73

presented in Table 2. Forced-choice performance on letters in words is compared to performance on letters embedded in a row of number signs (e.g., *READ* vs. *#E##*). The number signs serve as a control for lateral facilitation or inhibition. This factor appears to be important under dim-target/blank-mask conditions.

Target durations were adjusted separately for each condition, so that it is only the pattern of differences within display conditions that is meaningful. The data show that a 15% word advantage was obtained in the bright-target/patterned-mask condition and only a 5% word advantage in the dim-target/blank-mask condition. Massaro and Klitzke (1979) obtained about the same size effects. Various aspects of these results have also been corroborated in two other studies (Juola et al., 1974; Taylor & Chabot, 1978).

To understand the difference between these two conditions it is important to note that in order to get about 75% correct performance in the no-mask condition, the stimulus must be highly degraded. Since there is no patterned mask, the iconic trace presumably persists considerably beyond the offset of the target. It is our assumption that the effect of the blank mask is simply to reduce the contrast of the icon by summating with it. Thus, the limit on performance is not so much the amount of time available in which to process the information as it is the quality of the information made available to the system. In contrast, when a patterned mask is employed, the mask produces spurious inputs, which can interfere with the processing of the target. Thus, in the bright-target/patterned-mask conditions, the pri-

mary limitation on performance is the amount of time that the information is available to the system in relatively legible form rather than the quality of the information presented. This distinction between the way in which blank masks and patterned masks interfere with performance has previously been made by a number of investigators, including Rumelhart (1970) and Turvey (1973). We now consider each of these sorts of conditions in turn.

Word perception under patterned-mask conditions. When a high-quality display is followed by a patterned mask, we assume that the bottleneck in performance does not come in the extraction of feature information from the target display. Thus, in our simulation of these conditions, we assume that all of the features presented can be extracted on every trial. The limitation on performance comes from the fact that the activations produced by the target are subject to disruption and replacement by the mask before they can be translated into a permanent form suitable for overt report. This general idea was suggested by Johnston and McClelland (1973) and considered by a number of other investigators, including Carr et al. (1978), Massaro and Klitzke (1979), and others. On the basis of this idea, a number of possible reasons for the advantage for letters in words have been suggested. One is that letters in words are for some reason translated more quickly into a non-maskable form (Johnston & McClelland, 1973; Massaro & Klitzke, 1979). Another is that words activate representations removed from the direct effects of visual patterned masking (Carr et al., 1978; Johnston & McClelland, 1973, 1980; McClelland, 1976). In the interactive activation model, the reason letters in words fare better than letters in nonwords is that they benefit from feedback that can drive them to higher activation levels. As a result, the probability that the activated letter representation will be correctly encoded is increased.

To understand in detail how this account works, consider the following example. Figure 8 shows the operation of our model for the letter *E* both in an unrelated (#) context and in the context of the word *READ* for a visual display of moderately high quality.

We assume that display conditions are sufficient for complete feature extraction, so that only the letters actually contained in the target receive net excitatory input on the basis of feature information. After some number of cycles have gone by, the mask is presented with the same parameters as the target. The mask simply replaces the target display at the feature level, resulting in a completely new input to the letter level. This input, because it contains features incompatible with the letter shown in all four positions, immediately begins to drive down the activations at the letter level. After only a

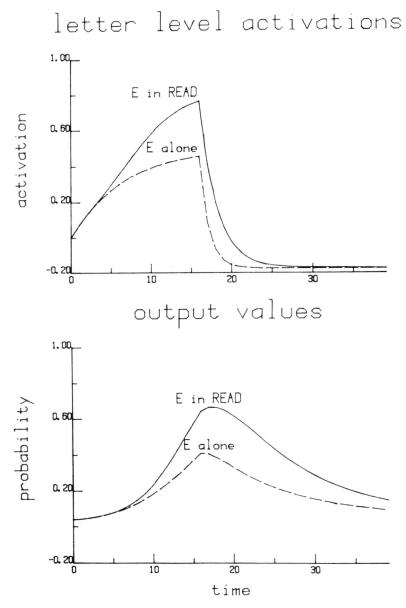

Figure 8. Activation functions (top) and output values (bottom) for the letter *E*, in unrelated context and in the context of the word *READ*.

few more cycles, these activations drop below resting level in both cases. Note that the correct letter was activated briefly, and no competing letter was activated. However, because of the sluggishness of the output process, these activations do not necessarily result in a high probability of correct report. As shown in the top half of Figure 8, the probability of correct report reaches a maximum after 16 cycles at a performance level far below the ceiling.

When the letter is part of the word (in this case, *READ*), the activation of the letters results in rapid activation of one or more words. These words, in turn, feed back to the letter level. This results in a higher net activation level for the letter embedded in the word.

Our simulation of the word advantage under patterned-mask conditions used the stimulus list that was used for simulating the blank-mask results. Since the internal workings of the model are completely deterministic as long as probability of feature extraction is 1.0, it was only necessary to run each item through the model once to obtain the expected probability that the critical letter would be encoded correctly for each item under each variation of parameters tried.

As described previously, we have assumed that readout of the results of processing occurs in parallel for all four letter positions and that the subject learns through practice to choose a time to read out in order to optimize performance. We have assumed that readout time may be set at a different point in different conditions, as long as they are blocked so that the subject knows in advance what type of material will be presented on each trial in the experiment. Thus, in simulating the Johnston and McClelland (1973) results, we allowed for different readout times for letters in words and letters in unrelated contexts, with the different times selected on the basis of practice to optimize performance on each type of material.

A final feature of the simulation is the duration of the target display. This was varied to produce an average performance on both letters embedded in number signs and letters in words that was as close as possible to the average performance on these two conditions in the 1973 experiment of John-

ston and McClelland. The value used for the run reported below was 15 cycles. As in the Johnston and McClelland study, the mask followed the target immediately.

The simulation replicated the experimental data shown in Table 2 quite closely. Accuracy on the forced choice was 81% correct for the letters embedded in words and 66% correct for letters in an unrelated (#) context.

It turns out that it is not necessary to allow for different readout times for different material types. A repetition of the simulation produced a 15% word advantage when the same readout time was chosen for both single letters and letters in words, based on optimal performance averaged over the two material types. Thus, the model is consistent with the fact that the word advantage does not depend on separating the different stimulus types into separate blocks (Massaro & Klitzke, 1979).

Perception of letters in words under conditions of degraded input. In conditions of degraded (but not abbreviated) input, the role of the word level is to selectively reinforce possible letters that are consistent with the visual information extracted and that are also consistent with the words in the subject's vocabulary. Recall that the task requires the subject to choose between two letters, both of which (on word trials) make a word with the rest of the context. There are two distinct cases to consider. Either the featural information extracted from the to-be-probed letter is sufficient to distinguish between the alternatives, or it is not. Whenever the featural information is consistent with both of the forced-choice alternatives, any feedback will selectively enhance both alternatives and will not permit the subject to distinguish between them. When the information extracted is inconsistent with one of the alternatives, the model produces a word advantage. The reason is that we assume forced-choice responses are based not on the feature information itself but on the subject's best guess about what letter was actually shown. Feedback from the word level increases the probability of correct choice in those cases where the subject extracts information that is inconsistent with the incorrect alternative but consistent with the correct alternative and a number of others. Thus, feedback would have the effect of helping the subject select the actual letter shown from several possibilities consistent with the set of extracted features. Consider again, for example, the case of the presentation of *WORD* discussed above. In this case, the subject extracted incomplete information about the final letter consistent with both *R* and *K*. Assume that the forced choice the subject was to face on this trial was between a *D* and a *K*. The account supposes that the subject encodes a single letter for each letter position before facing the forced choice. Thus, if the features of the final letter had been extracted in the absence of any context, the subject would encode *R* or *K* equally often, since both are equally compatible with the features extracted. This would leave the subject with the correct response some of the time. But if *R* were chosen instead, the subject would enter the forced choice between *D* and *K* without knowing the correct answer directly. When the whole word display is shown, the feedback generated by the processing of all of the letters greatly strengthens the *K*, increasing the probability that it will be chosen over the *R* and thus increasing the probability that the subject will proceed to the forced choice with the correct response in mind.

Our interpretation of the small word advantage in blank-mask conditions is a specific version of the early accounts of the word advantage offered by Wheeler (1970) and Thompson and Massaro (1973) before it was known that the effect depends on masking. Johnston (1978) has argued that this type of account does not apply under patterned-mask conditions. We are suggesting that it does apply to the small word advantage obtained under blank-mask conditions like those of the Johnston and McClelland (1973) experiment. We will see below that the model offers a different account of performance under patterned-mask conditions.

We simulated our interpretation of the small word advantage obtained in blank-mask conditions in the following way. A set of 40 pairs of four-letter words that differed by a single letter was prepared. The differing letters occurred in each position equally often. From these words corresponding con-

trol pairs were generated in which the critical letters from the word pairs were presented in nonletter contexts (#s). Because they were presented in nonletter contexts, we assumed that these letters did not engage the word processing system at all.

Each member of each pair of items was presented to the model four times, yielding a total of 320 stimulus presentations of word stimuli and 320 presentations of single letters. On each presentation, the simulation sampled a random subset of the possible features to be detected by the system. The probability of detection of each feature was set at .45. As noted previously, these values are in a ratio of 1 to 30, so that if any one of the 14 features extracted is inconsistent with a particular letter, that letter receives net inhibition from the features and is rapidly driven into an inactive state.

For simplicity, the features were treated as a constant input, which remained on while letter and word activations (if any) were allowed to take place. At the end of 50 processing cycles, which is virtually asymptotic, output was sampled. Sampling results in the selection of one letter to fill each position; the selected letter is assumed to be all the subject takes away from the target display. As described previously, the forced choice is assumed to be based only on this letter identity information. The subject compares the letter selected for the appropriate position against the forced-choice alternatives. If the letter selected is one of the alternatives, then that alternative is selected. If it is not one of the alternatives, then one of the two alternatives is simply picked at random.

The simulation produced a 10% advantage for letters in words over letters embedded in number signs. Probability-correct forced choice for letters embedded in words was 78% correct, whereas for letters in number signs, performance was 68% correct.

The simulated results for the no-mask condition clearly show a smaller word advantage than for the patterned-mask case. However, the model produces a larger word advantage, which is observed in the experiment (Table 2). As Johnston (1978) has pointed out, there are a number of reasons why an account such as the one we have offered would overestimate the size of the word advantage. First, subjects may occasionally be able to retain an impression of the actual visual information they have been able to extract. On such occasions, feedback from the word level will be of no further benefit. Second, even if subjects only retain a letter identity code, they may tend to choose the forced-choice alternative that is most similar to the letter encoded—instead of simply guessing—when the letter encoded is not one of the two choices. This would tend to result in a greater probability of correct choices and less of a chance for feedback to increase accuracy of performance. It is hard to know exactly how much these factors should be expected to reduce the size of the word advantage under these conditions, but they would certainly bring it more closely in line with the results.

Perception of Letters in Regular Nonwords

One of the most important findings in the literature on word perception is that an item need not be a word in order to produce facilitation with respect to unrelated letter or single letter stimuli. The advantage for pseudowords over unrelated letters has been obtained in a very large number of studies (Aderman & Smith, 1971; Baron & Thurston, 1973; Carr et al., 1978; McClelland, 1976; Spoehr & Smith, 1975). The pseudoword advantage over single letters has been obtained in three studies (Carr et al., 1978; Massaro & Klitzke, 1979; McClelland & Johnston, 1977).

Our model produces the facilitation for pseudowords by allowing them to activate nodes for words that share more than one letter in common with the display. When they occur, these activations produce feedback which strengthens the letters that gave rise to them just as in the case of words. These activations occur in the model if the strength of letter-to-word inhibition is reasonably small compared to the strength of letter-to-word excitation.

To see how this takes place in detail, consider a brief presentation of the pseudoword *MAVE* followed by a patterned mask. (The pseudoword is one used by Glushko, 1979, in developing the idea that partial activa-

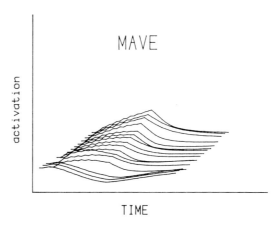

Figure 9. Activation at the word level upon presentation of the nonword *MAVE*.

tions of words are combined to derive pronunciations of pseudowords.) As illustrated in Figure 9, presentation of *MAVE* results in the initial activation of 16 different words. Most of these words, like *have* and *gave*, share three letters with *MAVE*. By and large, these words steadily gain in strength while the target is on and produce feedback to the letter level, sustaining the letters that supported them.

Some of the words are weakly activated for a brief period of time before they fall back below zero. These typically are words like *more* and *many*, which share only two letters with the target but are very high in frequency, so they need little excitation before they exceed threshold. But soon after they exceed threshold, the total activation at the word level becomes strong enough to overcome the weak excitatory input, causing them to drop down just after they begin to rise. Less frequent words sharing two letters with the word displayed have a worse fate still. Since they start out initially at a lower value, they generally fail to receive enough excitation to reach threshold. Thus, when there are several words that have three letters in common with the target, words that share only two letters with the target tend to exert little or no influence. In general then, with pronounceable pseudoword stimuli, the amount of feedback—and hence the amount of facilitation—depends primarily on the activation of nodes for words that share three letters with a displayed pseudoword. It is the nodes for these words that primarily interact with the activations generated by the presentation of the actual target display. In what follows we will call the words that have three letters in common with the target letter string the neighbors of that string.

The amount of feedback a particular letter in a nonword receives depends, in the model, on two primary factors and two secondary factors. The two primary factors are the number of words in the neighborhood that contain the target letter and the number of words that do not. In the case of the *M* in *MAVE*, for example, there are seven words in the neighborhood of *MAVE* that begin with *M*, so the *m* node gets excitatory feedback from all of these. These words are called the "friends" of the *m* node in this case. Because of competition at the word level, the amount of activation that these words receive depends on the total number of words that have three letters in common with the target. Those that share three letters with the target but are inconsistent with the *m* node (e.g., *have*) produce inhibition that tends to limit the activation of the friends of the *m* node, and can thus be considered its "enemies." These words also produce feedback that tends to activate letters that were not actually presented. For example, activation from *have* produces excitatory input to the *h* node, thereby producing some competition with the *m* node. These activations, however, are usually not terribly strong. No one word gets very active, and so letters not in the actual display tend to get fairly weak excitatory feedback. This weak excitation is usually insufficient to overcome the bottom–up inhibition acting on nonpresented letters. Thus, in most cases, the harm done by top–down activation of letters that were not shown is minimal.

A part of the effect we have been describing is illustrated in Figure 10. Here, we compare the activations of the nodes for the letters in *MAVE*. Without feedback, the four curves would be identical to the one single-letter curve included for comparison. So although there is facilitation for all four letters, there are definitely differences in the amount, depending on the number of friends and enemies of each letter. Note that within a given pseudoword, the total number of friends and enemies (i.e., the total number

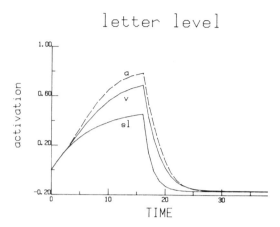

Figure 10. Activation functions for the letters *a* and *v* on presentation of *MAVE*. (Activation function for *e* is indistinguishable from function for *a*, and that for *m* is similar to that for *v*. The activation function for a single letter (s1), or a letter in an unrelated context is included for comparison.)

of words with three letters in common) is the same for all the letters.

There are two other factors that affect the extent to which a particular word will become active at the word level when a particular pseudoword is shown. Although the effects of these factors are only weakly reflected in the activations at the letter level, they are nevertheless interesting to note, since they indicate some synergistic effects that emerge from the interplay of simple excitatory and inhibitory influences in the neighborhood. These are the *rich-get-richer effect* and the *gang effect*. The rich-get-richer effect is illustrated in Figure 11, which compares the activation curves for the nodes for *have, gave,* and *save* under presentation of *MAVE*. The words differ in frequency, which gives the words slight differences in baseline activation. What is interesting is that the difference gets magnified; so that at the point of peak activation, there is a much larger difference. The reason for the amplification can be seen by considering a system containing only two nodes, *a* and *b*, starting at different initial positive activation levels, *a* and *b* at time *t*. Let us suppose that *a* is stronger than *b* at *t*. Then at *t* + 1, *a* will exert more of an inhibitory influence on *b*, since inhibition of a given node is determined by the sum of the activations of all nodes other than itself. This

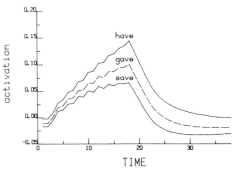

Figure 11. The rich-get-richer effect. (Activation functions for the nodes for *have*, *gave*, and *save* under presentation of *MAVE*.)

advantage for the initially more active nodes is compounded further in the case of the effect of word frequency by the fact that more frequent words creep above threshold first, thereby exerting an inhibitory effect on the lower frequency words when the latter are still too weak to fight back at all.

Even more interesting is the gang effect, which depends on the coordinated action of a related set of word nodes. This effect is depicted in Figure 12. Here, the activation curves for the *move, male, and save* nodes are compared. In the language, *move* and *make* are of approximately equal frequency, so their activations start out at about the same level. But they soon pull apart. Similarly, *save* starts out below *move* but soon reaches a higher activation. The reason for these effects is that *male* and *save* are both members of gangs with several members, whereas *move* is not. Consider first the difference between *male* and *move*. The reason for the difference is that there are several words that share the same three letters with *MAVE* as *male* does. In the list of words used in our simulations, there are six. These words all work together to reinforce the *m*, and *a*, and the *e* nodes, thereby producing much stronger reinforcement for themselves. Thus, these words make up a gang called the *ma_e* gang. In this example, there is also a *_ave* gang consisting of 6 other words, of which *save* is one. All of these work together to reinforce the *a*, *v*, and *e*. Thus, the *a* and *e* are reinforced by two gangs, whereas the

letters *v* and *m* are reinforced by only one each. Now consider the word *move*. This word is a loner; there are no other words in its gang, the *m_ve* gang. Although two of the letters in *move* receive support from one gang each, and one receives support from both other gangs, the letters of *move* are less strongly enhanced by feedback than the letters of the members of the other two gangs. Since continued activation of one word in the face of the competition generated by all of the other partially activated words depends on the activations of the component letter nodes, the words in the other two gangs eventually gain the upper hand and drive *move* back below the activation threshold.

As our study of the *MAVE* example illustrates, the pattern of activation produced by a particular pseudoword is complex and idiosyncratic. In addition to the basic friends and enemies effects, there are also the rich-get-richer and the gang effects. These effects are primarily reflected in the pattern of activation at the word level, but they also exert subtle influences on the activations at the letter level. In general though, the main result is that when the letter-to-word inhibition is low, all four letters in the pseudoword receive some feedback reinforcement. The result, of course, is greater accuracy of reporting letters in pseudowords compared to single letters.

Comparison of performance on words and pseudowords. Let us now consider the fact that the word advantage over pseudowords

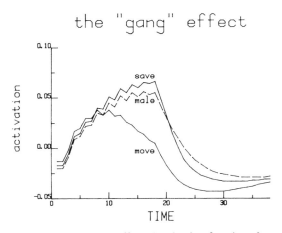

Figure 12. The gang effect. (Activation functions for *move, male,* and *save* under presentation of *MAVE*.)

Table 3
Actual and Simulated Results of the
McClelland & Johnston (1977) Experiments
(Proportion of Correct Forced Choice)

	Target type		
Result class	Word	Pseudoword	Single letter
Actual data			
High BF	.81	.79	.67
Low BF	.78	.77	.64
Average	.80	.78	.66
Simulation			
High BF	.81	.79	.67
Low BF	.79	.77	.67
Average	.80	.78	.67

Note. BF = bigram frequency.

is generally rather small in experiments where the subject knows that the stimuli include pseudowords. Some fairly representative results, from the study of McClelland and Johnston (1977), are illustrated in Table 3. The visual conditions of the study were the same as those used in the patterned-mask condition in Johnston and McClelland (1973). Trials were blocked, so subjects could adopt the optimum strategy for each type of material. The slight word–pseudoword difference, though representative, is not actually statistically reliable in this study.

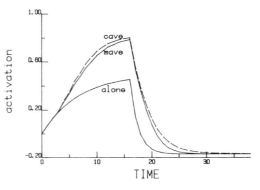

Figure 14. Activation functions for the letter *a*, under presentation of *CAVE* and *MAVE* and alone.

Words differ from pseudowords in that a word strongly activates one node at the word level, whereas a pseudoword does not. While we would tend to think of this as increasing the amount of feedback for words as opposed to pseudowords, there is the word-level inhibition that must be taken into account. This inhibition tends to equalize the total amount of activation at the word level between words and pseudowords. With words, the word shown tends to dominate the pattern of activity, thereby keeping all the words that have three letters in common with it from achieving the activation level they would reach in the absence of a node activated by all four letters. This situation is illustrated for the word *CAVE* in Figure 13. The result is that the sum of the activations of all the active units at the word level is not much different between the two cases. Thus, *CAVE* produces only slightly more facilitation for its constituent letters than *MAVE*, as illustrated in Figure 14.

In addition to the leveling effect of competition at the word level, it turned out that in our model, one of the common design features of studies comparing performance on words and pseudowords would operate to keep performance relatively good on pseudowords. In general, the stimulus materials used in most of these studies are designed by beginning with a list of pairs of words that differ by one letter (e.g., *PEEL–PEEP*). From each pair of words, a pair of nonwords is generated, differing from the orig-

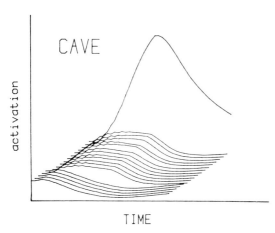

Figure 13. Activity at the word level upon presentation of *CAVE*, with weak letter-to-word inhibition.

inal word pair by just one of the context letters and thereby keeping the actual target letters—and as much of the context as possible—the same between word and pseudoword items (e.g., *TEEL–TEEP*). A previously unnoticed side effect of this matching procedure is that it ensures that the critical letter in each pseudoword has at least one friend, namely the word from the matching pair that differs from it by one context letter. In fact, most of the critical letters in the pseudowords used by McClelland and Johnston (1977) tended to have relatively few enemies, compared to the number of friends. In general, a particular letter should be expected to have three times as many friends as enemies. In the McClelland and Johnston stimuli, the great majority of the stimuli had much larger differentials. Indeed, more than half of the critical letters had no enemies at all.

The puzzling absence of cluster frequency effects. In the account we have just described, facilitation of performance on letters in pseudowords was explained by the fact that pseudowords tend to activate a large number of words, and these words tend to work together to reinforce the activations of letters. This account might seem to suggest that pseudowords that have common letter clusters, and therefore have several letters in common with many words, would tend to produce the greatest facilitation. However, this factor has been manipulated in a number of studies, and little has been found in the way of an effect. The McClelland and Johnston (1977) study is one case in point. As Table 3 illustrates, there is only a slight tendency for superior performance on high cluster frequency words. This slight tendency is also observed in single letter control stimuli, suggesting that the difference may be due to differences in perceptibility of the target letters in the different positions, rather than cluster frequency per se. In any case, the effect is very small. Other studies have likewise failed to find any effect of cluster frequency (Spoehr & Smith, 1975; Manelis, 1974). The lack of an effect is most striking in the McClelland and Johnston study, since the high and low cluster frequency items differed widely in cluster frequency as measured in a number of ways.

In our model, the lack of a cluster frequency effect is due to the effect of mutual inhibition at the word level. As we have seen, this mutual inhibition tends to keep the total activity at the word level roughly constant over a variety of different input patterns, thereby greatly reducing the advantage for high cluster frequency items. Items containing infrequent clusters tend to activate few words, but there is less competition at the word level, so that the words that do become active reach higher activation levels.

The situation is illustrated for the nonwords *TEEL* and *HOET* in Figure 15. Although *TEEL* activates many more words, the total activation is not much different in the two cases.

The total activation is not, of course, the whole story. The ratio of friends to enemies is also important. And it turns out that this ratio is working against the high cluster items more than the low cluster items. In McClelland and Johnston's stimuli, only one of the low cluster frequency nonword pairs had critical letters with any enemies at all! For 23 out of 24 pairs, there was at least one friend (by virtue of the method of stimulus construction) and no enemies. In contrast, for the high cluster frequency pairs, there was a wide range, with some items having several more enemies than friends.

To simulate the McClelland and Johnston (1977) results, we had to select a subset of their stimuli, since some of the words they used were not in our word list. The stimuli had been constructed in sets containing a word pair, a pseudoword pair, and a single letter pair that differed by the same letters in the same position (e.g., *PEEL–PEEP TEEL–TEEP*; ___L ___P). We simply selected all those sets in which both words in the pair appeared in our list. This resulted in a sample of 10 high cluster frequency sets and 10 low cluster frequency sets. The single letter stimuli derived from the high and low cluster frequency pairs were also run through the simulation. Both members of each pair were tested.

Since the stimuli were presented in the actual experiment blocked by material type, we separately selected an optimal time for readout for words, pseudowords, and single letters. Readout time was the same for high and low cluster frequency items of the same type, since these were presented in a mixed list in the actual experiment. As in the simulation of the Johnston and McClelland (1973) results, the display was presented for a duration of 15 cycles.

The simulation results, shown in Table 3, reveal the same general pattern as the actual data. The magnitude of the pseudoword advantage over single letters is just slightly smaller than the word advantage, and the effect of cluster frequency is very slight.

We have yet to consider how the model deals with unrelated letter strings. This depends a little on the exact characteristics of the strings. First let us consider truly ran-

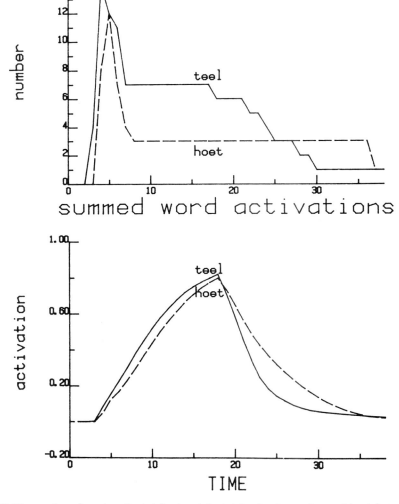

Figure 15. The number of words activated (top) and the total activation at the word level (bottom) upon presentation of the nonwords *TEEL* and *HOET*.

domly generated consonant strings. Such items typically produce some activation at the word level in our model, since they tend to share two letters with several words (one letter out of four is insufficient to activate a word, since three inhibitory inputs outweigh one excitatory input). These strings rarely have three letters in common with any one word. Thus, they only tend to activate a few words very weakly, and because of the weakness of the bottom–up excitation, competition among partially activated words keep any one word from getting very active. So, little benefit results. When we ran our simulation on randomly generated consonant strings, there was only a 1% advantage over single letters.

Some items which have been used as unpronounceable nonwords or unrelated letter strings do produce a weak facilitation. We ran the nonwords used by McClelland and Johnston (1977) in their Experiment 2. These items contain a large number of vowels in positions that vowels typically occupy in words, and they therefore activate more words than, for example, random strings of consonants. The simulation was run under the same conditions as the one reported above for McClelland and Johnston's Experiment 1. The simulation produced a slight advantage for letters in these nonwords, compared to single letters, as did the experiment. In both the simulation and the actual experiment, forced-choice performance was 4% more accurate for letters in these unrelated letter strings than in single letter stimuli.

On the basis of this characteristic of our model, the results of one experiment on the importance of vowels in reading may be reinterpreted. Spoehr and Smith (1975) found that subjects were more accurate when reporting letters in unpronounceable nonwords that contained vowels than in those composed of all consonants. They interpreted the results as supporting the view that subjects parse letter strings into "vocalic center groups." However, an alternative possible account is that the strings containing vowels had more letters in common with actual words than the all consonant strings.

In summary, the model provides a good account of the perceptual advantage for let-

ters in pronounceable nonwords, and for the lack of such an advantage in unrelated letter strings. In addition, it accounts for the small difference between performance on words and pseudowords and for the absence of any really noticeable cluster frequency effect in the McClelland and Johnston (1977) experiment.

The Role of Lexical Constraints

The Johnston (1978) experiment. Several models that have been proposed to account for the word advantage rely on the idea that the context letters in a word facilitate performance by constraining the set of possible letters that might have been presented in the critical letter position. According to models of this class, contexts that strongly constrain what the target letter should be result in greater accuracy of perception than more weakly constraining contexts. For example, the context _HIP should facilitate the perception of an initial S more than the context _INK. The reason is that _HIP is more strongly constraining, since only three letters (S, C, and W) fit in the context to make a word, compared to _INK, where nine letters (D, F, K, L, M, P, R, S, and W) fit in the context to make a word. In a test of such models, Johnston (1978) compared accuracy of perception of letters occurring in high- and low-constraint contexts. The same target letters were tested in the same positions in both cases. For example, the letters S and W were tested in the high-constraint _HIP context and the low-constraint _INK context. Using bright-target/patterned-mask conditions, Johnston found no difference in accuracy of perception between letters in the high- and low-constraint contexts. The results of this experiment are shown in Table 4. Johnston measured letter perception in two ways. He not only asked the subjects to decide which of two letters had been presented (the forced-choice measure), but he also asked subjects to report the whole word and recorded how often they got the critical letter correct. No significant difference was observed in either case. In the forced choice there was a slight difference favoring low-constraint items,

Table 4
Actual and Simulated Results (Probability Correct) From Johnston (1978) Experiments

	Constraint	
Result class	High	Low
Actual data		
Forced choice	.77	.79
Free report	.54	.54
Simulation		
Forced choice	.77	.76
Free report	.56	.54

but in the free report there was no difference at all.

Although our model does use contextual constraints (as they are embodied in specific lexical items), it turns out that it does not predict that highly constraining contexts will facilitate perception of letters much more than weakly constraining contexts under bright-target/patterned-mask conditions. Under such conditions, the role of the word level is not to help the subject select among alternatives left open by an incomplete feature analysis process, as most constraint-based models have assumed, but rather to help strengthen the activation of the nodes for the letters presented. Contextual constraints, at least as manipulated by Johnston, do not have much effect on the magnitude of this strengthening effect.

In detail, what happens in the model when a word is shown is that the presentation results in weak activation of the words that share three letters with the target. Some of these words are friends of the critical letter in that they contain the actual critical letter shown, as well as two of the letters from the context (e.g., *shop* is a friend of the initial *S* in *SHIP*). Some of the words, however, are enemies of the critical letter in that they contain the three context letters of the word but a different letter in the critical letter position (e.g., *chip* and *whip* are enemies of the *S* in *SHIP*). From our point of view, Johnston's (1978) constraint manipulation is essentially a manipulation of the number of enemies the critical letter has in the given context. Johnston's high- and low-constraint stimuli have equal numbers of friends, on the average, but (by design) the high-con-

straint items have fewer enemies, as shown in Table 5.

In the simulation, the friends and enemies of the target word receive some activation. The greater number of enemies in the low-constraint condition is responsible for the small effect of constraint that the model produces. What happens is that the enemies of the critical letter tend to keep nodes for the presented word and for the friends of the critical letter from being quite as strongly activated as they would otherwise be. The effect is quite small for two reasons. First, the node for the word presented receives four excitatory inputs from the letter level, and all other words can only receive at most three excitatory inputs and at least one inhibitory input. As we saw in the case of the word *CAVE*, the node for the correct word dominates the activations at the word level and is predominantly responsible for any feedback to the letter level. Second, while the high-constraint items have fewer enemies, by more than a two-to-one margin, both high- and low-constraint items have, on the average, more friends than enemies. The friends of the target letter work with the actual word shown to keep the activations of the enemies in check, thereby reducing the extent of their inhibitory effect still further. The ratio of the number of friends over the total number of neighbors is not very different in the two conditions, except in the first serial position.

This discussion may give the impression that contextual constraint is not an important variable in our model. In fact, it is quite powerful. But its effects are obscured in the Johnston (1978) experiment because of the strong dominance of the target word when all the features are extracted and the fact that we are concerned with the likelihood of perceiving a particular letter rather than performance in identifying correctly what whole word was shown. We will now consider an experiment in which contextual constraints played a strong role, because the characteristics just mentioned were absent.

The Broadbent and Gregory (1968) experiment. Up to now we have found no evidence that either bigram frequency or lexical constraints have any effect on performance. However, in experiments using

Table 5

Friends and Enemies of the Critical Letters in the Stimuli Used by Johnston (1978)

Critical letter position	High constraint			Low constraint		
	Friends	Enemies	Ratio	Friends	Enemies	Ratio
1	3.33	2.22	.60	3.61	6.44	.36
2	9.17	1.00	.90	6.63	2.88	.70
3	6.30	1.70	.79	7.75	4.30	.64
4	4.96	1.67	.75	6.67	3.50	.66
Average	5.93	1.65		6.17	4.27	

the traditional whole report method, these variables have been shown to have substantial effects. Various studies have shown that recognition thresholds are lower, or recognition accuracy at threshold higher, when relatively unusual words are used (Bouwhuis, 1979; Havens & Foote, 1963; Newbigging, 1961). Such items tend to be low in bigram frequency and at the same time high in lexical constraint.

In one experiment, Broadbent and Gregory (1968) investigated the role of bigram frequency at two different levels of word frequency and found an interesting interaction. We now consider how our model can account for their results. To begin, it is important to note that the visual conditions of their experiment were quite different from those of McClelland and Johnston (1977), in which the data and our model failed to show a bigram frequency effect, and of Johnston (1978), in which the data and the model showed little or no constraint effect. The conditions were like the dim-target/blank-mask conditions discussed above, in that the target was shown briefly against an illuminated background, without being followed by any kind of mask. The dependent measure was the probability of correctly reporting the whole word. The results are indicated in Table 6. A slight advantage for high bigram frequency items over low bigram frequency was obtained for frequent words, although it was not consistent over different subsets of items tested. The main finding was that words of low bigram frequency had an advantage among infrequent words. For these stimuli, higher bigram frequency actually resulted in a lower percent correct.

Unfortunately, Broadbent and Gregory used five-letter words, so we were unable to run a simulation on their actual stimuli. However, we were able to select a subset of the stimuli used in the McClelland and Johnston (1977) experiment that fit the requirements of the Broadbent and Gregory design. We therefore presented these stimuli to our model, under the presentation parameters used in simulating the blank-mask condition of the Johnston and McClelland (1973) experiment above. The only difference was that the output was taken, not from the letter level, as in all of our other simulations, but directly from the word level. The results of the simulation, shown in Table 6, replicate the obtained pattern very nicely. The simulation produced a large advantage for the low bigram items, among the infrequent words, and produced a slight advantage for high bigram items among the frequent words.

In our model, low-frequency words of high bigram frequency are most poorly recognized, because these are the words that have the largest number of neighbors. Under conditions of incomplete feature extraction, which we expect to prevail under these visual

Table 6

Actual and Simulated Results of the Broadbent and Gregory (1968) Experiment (Proportion of Correct Whole Report)

Result class	Word frequency	
	High	Low
Actual data		
High BF	.64	.43
Low BF	.64	.58
Simulation		
High BF	.41	.21
Low BF	.39	.37

Note. BF = bigram frequency.

conditions, the more neighbors a word has the more likely it is to be confused with some other word. This becomes particularly important for lower frequency words. As we have seen, if both a low-frequency word and a high-frequency word are equally compatible with the detected portion of the input, the higher frequency word will tend to dominate. When incomplete feature information is extracted, the relative activation of the target and the neighbors is much lower than when all the features have been seen. Indeed, some neighbors may turn out to be just as compatible with the features extracted as the target itself. Under these circumstances, the word of the highest frequency will tend to gain the upper hand. The probability of correctly reporting a low-frequency word will therefore be much more strongly influenced by the presence of a high-frequency neighbor compatible with the input than the other way around.

But why does the model actually produce a slight reversal with high-frequency words? Even here, it would seem that the presence of numerous neighbors would tend to hurt instead of facilitate performance. However, we have forgotten the fact that the activation of neighbors can be beneficial as well as harmful. The active neighbors produce feedback that strengthens most or all of the letters, and these in turn increase the activation of the node for the word shown. As it happens, there turns out to be a delicate balance for high-frequency words between the negative and positive effects of neighbors, which only slightly favors the words with more neighbors. Indeed, the effect only holds for some of these items. We have not yet had the opportunity to explore all the factors that determine whether the effect of neighbors in individual cases will on balance be positive or negative.

Different effects in different experiments. This discussion of the Broadbent and Gregory (1968) experiment indicates once again that our model is something of a chameleon. The model produces no effect of constraint or bigram frequency under the visual conditions and testing procedures used in the Johnston (1978) and McClelland and Johnston (1977) experiments but does produce such effects under the conditions of the

Broadbent and Gregory (1968) experiment. This flexibility of the model, of course, is fully required by the data. While there are other models of word perception that can account for one or the other type of result, to our knowledge the model presented here is the only scheme that has been worked out to account for both.

Discussion

The interactive activation model does a good job of accounting for the results in the literature on the perception of letters in words and nonwords. The model provides a unified explanation of the results of a variety of experiments and provides a framework in which the effects of manipulations of the visual display characteristics used may be analyzed. In addition, as we shall see in Part 2 (Rumelhart & McClelland, in press), the model readily accounts for a variety of additional phenomena. Moreover, as we shall also show, it can be extended beyond its current domain of applicability with substantial success. In Part 2 we will report a number of experiments demonstrating what we call "context enhancement effects" and show how the model can explain the major findings in the experiments.

One issue that deserves some consideration is the robustness of the model. To what extent do the simulations depend upon particular parameter values? What are the effects of changes of the parameter values? These are extremely complex questions, and we do not have complete answers. However, we have made some observations. First, the basic Reicher (1969) effect can be obtained under a very wide range of different parameters, though of course its exact size will depend on the ensemble of parameter values. However, one thing that seems to be important is the overpowering effect of one incompatible feature in suppressing activations at the letter level. Without this strong bottom-up inhibition, the mask would not effectively drive out the activations previously established by the stimulus. Second, performance on pronounceable nonwords depends on the relative strength of letter-word excitation compared to inhibition and on the strength of the competition among word units. Parameter values can be found which produce no advantage for any multiletter strings except words, whereas other values can be found that produce large advantages for words, pseudowords, and even many nonword strings. The effects (or rather the lack of effects) of letter-cluster frequency and constraints likewise depend on these parameters.

It thus appears that relatively strong feature–letter inhibition is necessary, but at the same time, relatively weak letter–word inhibition is necessary. This discrepancy is a bit puzzling, since we would have thought that the same general principles of operation would have applied to both the letter and the word levels. A possible way to resolve the discrepancy might be to introduce a more sophisticated account of the way masking works. It is quite possible that new inputs act as position-specific "clear signals," disrupting activations created by previous patterns in corresponding locations. Some possible physiological mechanisms that would produce such effects at lower processing levels have been described by Weisstein, Ozog, and Szoc (1975) and by Breitmeyer and Ganz (1976), among others. If we used such a mechanism to account for the basic effect of masking, it might well be possible to lower the feature–letter inhibition considerably. Lowering feature–letter inhibition would then necessitate strong letter–letter inhibition, so that letters that exactly match the input would be able to dominate those with only partial matches. With these changes the letter and word levels would indeed operate by the same principles.

Perhaps it is a bit premature to discuss such issues as robustness, since there are a number of problems that we have not yet resolved. First, we have ignored the fact that there is a high degree of positional uncertainty in reports of letters—particularly letters in unrelated strings, but occasionally also in reports of letters in words and pseudowords (Estes, 1975; McClelland, 1976; McClelland & Johnston, 1977). Another thing that we have not considered very fully is the serial position curve. In general, it appears that performance is more accurate on the end letters in multiletter strings, particularly the first letter. In Part 2 we consider

ways of extending the model to account for both of these aspects of perceptual performance.

Third, there are some effects of set on word perception that we have not considered. Johnston and McClelland (1974) found that perception of letters in words was actually hurt if subjects focused their attention on a single letter position in the word (see also Holender, 1979, and Johnston, 1974). In addition, Aderman and Smith (1971) found that the advantage for pseudowords over unrelated letters only occurs if the subject expects that pseudowords will be shown; and more recently, Carr et al. (1978) have replicated this finding, while at the same time showing that it is apparently not necessary to be prepared for presentations of actual words. Part 2 considers how our model is compatible with this effect also. We will also consider how our model might be extended to account for some recent findings demonstrating effects of letter and word masking on perception of letters in words and other contexts.

In all but one of the experiments we have simulated, the primary (if not the only) data for the experiments were obtained from forced choices between pairs of letters, or strings differing by a single letter. In these cases, it seemed to us most natural to rely on the output of the letter level as the basis for responding. However, it may well be that subjects often base their responses on the output of the word level. Indeed, we have assumed that they do in experiments like the Broadbent and Gregory (1968) study, in which subjects were told to report what word they thought they had seen. This may also have happened in the McClelland and Johnston (1977) and Johnston (1978) studies, in which subjects were instructed to report all four letters before the forced choice on some trials. Indeed, both studies found that the probability of reporting all four letters correctly for letters in words was greater than we would expect given independent processing of each letter position. It seems natural to account for these completely correct reports by assuming that they often occurred on occasions where the subject encoded the item as a word. Even in experiments where only a forced choice is obtained, on many occasions subjects may still come away with a word, rather than a sequence of letters.

In the early phases of the development of our model, we explicitly included the possibility of output from the word level as well as the letter level. We assumed that the subject would either encode a word, with some probability dependent on the activations at the word level or, failing that, would encode some letter for each letter position dependent on the activations at the letter level. However, we found that simply relying on the letter level permitted us to account equally well for the results. In essence, the reason is that the word-level information is incorporated into the activations at the letter level because of the feedback, so that the word level is largely redundant. In addition, of course, readout from the letter level is necessary to the model's account of performance with nonwords. Since it is adequate to account for all of the forced-choice data, and since it is difficult to know exactly how much of the details of free-report data should be attributed to perceptual processes and how much to such things as possible biases in the readout processes and so forth, we have stuck for the present with readout from the letter level.

Another decision that we adopted in order to keep the model within bounds was to exclude the possibility of processing interactions between the visual and phonological systems. However, in the model as sketched at the outset (Figure 1), activations at the letter level interacted with a phonological level as well as the word level. Perhaps the most interesting feature of our model is its ability to account for performance on letters in pronounceable nonwords without assuming any such interactions. We will also see in Part 2 (Rumelhart & McClelland, in press) that certain carefully selected unpronounceable consonant strings produce quite large contextual facilitation effects, compared to other sequences of consonants, which supports our basic position that pronounceability per se is not an important feature of the perceptual facilitation effects we have accounted for.

Another simplification we have adopted in Part 1 has been to consider only cases in which individual letters or strings of letters were presented in the absence of a linguistic context. In Part 2 we will consider the effects of introducing contextual inputs to the word level, and we will explore how the model might work in processing spoken words in context as well.

Reference Notes

1. Hinton, G. E. Relaxation and its role in vision. Unpublished doctoral dissertation, University of Edinburgh, Scotland, 1977.
2. Rumelhart, D. E. *A multicomponent theory of confusion among briefly exposed alphabetic characters* (Tech. Rep. 22). San Diego: University of California, San Diego, Center for Human Information Processing, 1971.

References

Adams, M. J. Models of word recognition. *Cognitive Psychology*, 1979, *11*, 133–176.

Aderman, D., & Smith, E. E. Expectancy as a determinant of functional units in perceptual recognition. *Cognitive Psychology*, 1971, *2*, 117–129.

Anderson, J. A. Neural models with cognitive implications. In D. LaBerge & S. J. Samuels (Eds.), *Basic processes in reading: Perception and comprehension*. Hillsdale, N.J.: Erlbaum, 1977.

Anderson, J. A., Silverstein, J. W., Ritz, S. A., & Jones, R. S. Distinctive features, categorical perception, and probability learning: Some applications of a neural model. *Psychological Review*, 1977, *84*, 413–451.

Baron, J., & Thurston, I. An analysis of the word-superiority effect. *Cognitive Psychology*, 1973, *4*, 207–228.

Bouwhuis, D. G. *Visual recognition of words*. Eindhoven, The Netherlands: Greve Offset B. V., 1979.

Breitmeyer, B. G., & Ganz, L. Implications of sustained and transient channels for theories of visual pattern masking, saccadic suppression, and information processing. *Psychological Review*, 1976, *83*, 1–36.

Broadbent, D. E. Word-frequency effect and response bias. *Psychological Review*, 1967, *74*, 1–15.

Broadbent, D. E., & Gregory, M. Visual perception of words differing in letter digram frequency. *Journal of Verbal Learning and Verbal Behavior*, 1968, *7*, 569–571.

Bruner, J. S. On perceptual readiness. *Psychological Review*, 1957, *64*, 123–152.

Carr, T. H., Davidson, B. J., & Hawkins, H. L. Perceptual flexibility in word recognition: Strategies affect orthographic computation but not lexical access. *Journal of Experimental Psychology: Human Perception and Performance*, 1978, *4*, 674–690.

Cattell, J. M. The time taken up by cerebral operations. *Mind*, 1886, *11*, 220–242.

Estes, W. K. The locus of inferential and perceptual processes in letter identification. *Journal of Experimental Psychology: General*, 1975, *1*, 122–145.

Glushko, R. J. The organization and activation of orthographic knowledge in reading words aloud. *Journal of Experimental Psychology: Human Perception and Performance*, 1979, *5*, 674–691.

Grossberg, S. A theory of visual coding, memory, and development. In E. L. J. Leeuwenberg & H. F. J. M. Buffart (Eds.), *Formal theories of visual perception*. New York: Wiley, 1978.

Havens, L. L., & Foote, W. E. The effect of competition on visual duration threshold and its independence of stimulus frequency. *Journal of Experimental Psychology*, 1963, *65*, 5–11.

Hinton, G. E., & Anderson, J. A. (Eds.), *Parallel models of associative memory*. Hillsdale, N.J.: Erlbaum, 1981.

Holender, D. Identification of letters in words and of single letters with pre- and postknowledge vs. postknowledge of the alternatives. *Perception & Psychophysics*, 1979, *25*, 213–318.

Huey, E. B. *The psychology and pedagogy of reading*. New York: Macmillan, 1908.

Johnston, J. C. *The role of contextual constraint in the perception of letters in words*. Unpublished doctoral dissertation, University of Pennsylvania, 1974.

Johnston, J. C. A test of the sophisticated guessing theory of word perception. *Cognitive Psychology*, 1978, *10*, 123–154.

Johnston, J. C., & McClelland, J. L. Visual factors in word perception. *Perception & Psychophysics*, 1973, *14*, 365–370.

Johnston, J. C., & McClelland, J. L. Perception of letters in words: Seek not and ye shall find. *Science*, 1974, *184*, 1192–1194.

Johnston, J. C., & McClelland, J. L. Experimental tests of a hierarchical model of word identification. *Journal of Verbal Learning and Verbal Behavior*, 1980, *19*, 503–524.

Juola, J. F., Leavitt, D. D., & Choe, C. S. Letter identification in word, nonword, and single letter displays. *Bulletin of the Psychonomic Society*, 1974, *4*, 278–280.

Kohonen, T. *Associative memory: A system-theoretic approach*. West Berlin: Springer-Verlag, 1977.

Kucera, H., & Francis, W. *Computational analysis of present-day American English*. Providence, R.I.: Brown University Press, 1967.

LaBerge, D., & Samuels, S. Toward a theory of automatic information processing in reading. *Cognitive Psychology*, 1974, *6*, 293–323.

Levin, J. A. *Proteus: An activation framework for cognitive process models (ISI/WP-2)*. Marina del Rey, Calif.: Information Sciences Institute, 1976.

Luce, R. D. *Individual choice behavior*. New York: Wiley, 1959.

Manelis, L. The effect of meaningfulness in tachistoscopic word perception. *Perception & Psychophysics*, 1974, *16*, 182–192.

Massaro, D. W., & Klitzke, D. The role of lateral masking and orthographic structure in letter and word recognition. *Acta Psychologica*, 1979, *43*, 413–426.

McClelland, J. L. Preliminary letter identification in the perception of words and nonwords. *Journal of Experimental Psychology: Human Perception and Performance*, 1976, *1*, 80–91.

McClelland, J. L. On the time relations of mental processes: An examination of systems of processes in cascade. *Psychological Review*, 1979, *86*, 287–330.

McClelland, J. L., & Johnston, J. C. The role of familiar units in perception of words and nonwords. *Perception & Psychophysics*, 1977, *22*, 249–261.

Morton, J. Interaction of information in word recognition. *Psychological Review*, 1969, *76*, 165–178.

Neisser, U. *Cognitive psychology*. New York: Appleton-Century-Crofts, 1967.

Newbigging, P. L. The perceptual reintegration of frequent and infrequent words. *Canadian Journal of Psychology*, 1961, *15*, 123–132.

Reicher, G. M. Perceptual recognition as a function of meaningfulness of stimulus material. *Journal of Experimental Psychology*, 1969, *81*, 274–280.

Rumelhart, D. E. A multicomponent theory of the perception of briefly exposed visual displays. *Journal of Mathematical Psychology*, 1970, *7*, 191–218.

Rumelhart, D. E. Toward an interactive model of reading. In S. Dornic (Ed.), *Attention and performance IV*. Hillsdale, N.J.: Erlbaum, 1977.

Rumelhart, D. E., & McClelland, J. L. An interactive activation model of context effects in letter perception: Part 2. The contextual enhancement effect and some tests and extensions of the model. *Psychological Review*, in press.

Rumelhart, D. E., & Siple, P. The process of recognizing tachistoscopically presented words. *Psychological Review*, 1974, *81*, 99–118.

Spoehr, K., & Smith, E. The role of orthographic and phonotactic rules in perceiving letter patterns. *Journal of Experimental Psychology: Human Perception and Performance*, 1975, *1*, 21–34.

Szentagothai, J., & Arbib, M. A. *Conceptual models of neural organization*. Cambridge, Mass.: MIT Press, 1975.

Taylor, G. A., & Chabot, R. J. Differential backward masking of words and letters by masks of varying orthographic structure. *Memory & Cognition*, 1978, *6*, 629–635.

Thompson, M. C., & Massaro, D. W. Visual information and redundancy in reading. *Journal of Experimental Psychology*, 1973, *98*, 49–54.

Turvey, M. On peripheral and central processes in vision: Inferences from an information-processing analysis of masking with patterned stimuli. *Psychological Review*, 1973, *80*, 1–52.

Weisstein, N., Ozog, G., & Szoc, R. A comparison and elaboration of two models of metacontrast. *Psychological Review*, 1975, *82*, 325–343.

Wheeler, D. Processes in word recognition. *Cognitive Psychology*, 1970, *1*, 59–85.

Appendix
Computer Simulation of the Model

The computer program for simulating the interactive activation model was written in the C programming language to run on a Digital PDP 11/45 computer under the UNIX (Trade Mark of Bell Laboratories) operating system. There is now a second version, also in C, which runs under

UNIX on a VAX 11/780. When no other jobs are running on the VAX, a simulation of a single experimental trial takes approximately 15–30 sec.

Data Structures

The simulation relies on several arrays for each of the processing levels in the model. The input is held in an array that contains slots for each of the line segments in the Rumelhart-Siple font in each position. Segments can be present or absent, or their status can be indeterminate (as when the input is made deliberately incomplete). There is another array that holds the information the model has detected about the display. Each element of this array represents a detector for the presence or absence of a feature. When the corresponding feature is detected, the detector's value is set to 1 (remember that both absence and presence must be detected).

At the letter level, one array (the activation array) stores the current activation of each node. A second array (the excitatory buffer) is used to sum all of the excitatory influences reaching each node on a given tick of the clock, and a third array (the inhibitory buffer) is used to sum all of the inhibitory influences reaching each node. In addition there is an output array, containing the current output strength of each letter level node. At the word level, there is an activation array for the current activation of each node, as well as an excitatory buffer and an inhibitory buffer.

Knowledge of Letters and Words

The links among the nodes in the model are stored in a set of tables. There is a table in the program that lists which features are present in each letter and which are absent. Another table contains the spellings of each of the 1,179 words known to the program.

Input

Simulated visual input is entered from a computer terminal or from a text file. Several successive displays within a single "trial" may be specified. Each display is characterized by an onset time (tick number from the start of the trial—see below) and some array of visual information. Each lowercase letter stands for the array of features making up the corresponding letter. Other characters stand for particular mask characters, blanks, and so forth. As examples, "_" stands for a blank, and "0" stands for the ⊠ mask character. Thus the specification:

0 mav–
12 mave
24 0000

instructs the program to present the visual array

consisting of the letters *M*, *A*, and *V* in the first, second, and third letter positions, respectively, at Cycle 0; to present the letter *E* in the fourth position at Cycle 12; and to present an ⊠ mask at Cycle 24. It is also possible to specify any arbitrary feature array to occur in any letter position.

Processing Occurring During Each Cycle

During each cycle, the values of all of the nodes are updated. The activations of letter and word nodes, which were determined on Cycle $t - 1$, are used to determine the activations of these nodes on Cycle t. Activations of feature nodes are updated first, so that they begin to influence letter nodes right away.

The first thing the program does on each cycle is update the input array to reflect any new changes in the display. On cycles when a new display is presented, detectors for features in letter positions in which there has been a change in the input are subject to resetting. A random number generator is used to determine whether each new feature is detected or not. When the new value of a particular feature (present or absent) is detected, the old value is erased. Probability of detection can be set to any probability (in many cases it is simply set to 1.0, so that all of the features are detected).

For each letter in each position, the program then checks the current activation value (i.e., the value computed on the previous cycle) in the activation array. If the node is active (i.e., if its activation is above threshold), its excitatory and inhibitory effects on each node at the word level are computed. To determine whether the letter in question excites or inhibits a particular word node, the program simply examines the spelling of each word to see if the letter is in the word in the appropriate position. If so, excitation is added to the word's excitatory buffer. If not, inhibition is added to the word's inhibitory buffer. The magnitudes of these effects are the product of the driving letter's activation and the appropriate rate parameters. Word-to-letter influences are computed in a similar fashion.

The next step is the computation of the word–word inhibition and the determination of the new word activation values. First, the activations of all the active word nodes are summed. The inhibitory buffer of each word node is incremented by an amount proportional to the summed activation of all other word nodes (i.e., by the product of the total word level activation minus its own activation, if it is active, times the word–word inhibition rate parameter). This completes the influences acting on the word nodes. The value in the inhibitory buffer is subtracted from the value in the excitatory buffer. The result is then subjected to floor and ceiling effects, as described in the article, to determine the net effect of the excitatory and inhibitory input. This net effect is then added to the current activation of the node, and the decay of the current value is subtracted to give a new current value, which is stored in the activation array. Finally, the excitatory and inhibitory buffers are cleared for new input on the next cycle.

Next is the computation of the feature-to-letter influences. For each feature in each letter position, if that feature has been detected, the program checks each letter to see if it contains the feature. If it does, the excitatory buffer for that letter in that position is incremented. If not, the corresponding inhibitory buffer is incremented. After this, the letter–letter inhibition is added into the inhibitory buffers following a similar procedure as was used in computing the word–word inhibitory effects. (Actually, this step is skipped in the reported simulations, since the value of letter–letter inhibition has been set to zero.)

Next is the computation of the new activation values at the letter level. These are computed in just the same way as the new activation values at the word level. Finally, the effect of the current activation is added into the letter's output strength, and the excitatory and inhibitory buffers are cleared for the next cycle.

The order of some of the preceding steps is arbitrary. What is important is that at the end of each cycle, the activations of all the word nodes have been updated to reflect letter activations of the previous cycle and vice versa. The fact that newly detected input influences the letter detectors immediately is not meaningful, since waiting until the next cycle would just add a fixed delay to all of the activations in the system.

Output

To simulate forced-choice performance, the program must be told when to read out the results of processing at the letter level, what position is being tested, and what the two alternatives are. In fact the user actually gives the program the full target display and the full alternative display (e.g., *LEAD-LOAD*), and the program compares them to figure out the critical letter position and the two choice alternatives. Various options are available for monitoring readout performance of the simulation. First, it is possible to have the program print out what the result of readout would be at each time cycle. Second, the user may specify a particular cycle for readout. Third, the user may tell the program to figure out the optimal time for readout and to print both the time and the resulting percent correct performance. This option is used in preliminary runs to determine what readout time to use in the final simulation runs for each experiment.

On each cycle for which output is requested, the program computes the probability that the correct alternative is read out and the probability that the incorrect alternative is read out, based on their response strengths as described in the text. Probability-correct forced choice is then simply the probability that the correct alternative was read out, plus .5 times the probability that neither the correct nor the incorrect alternative was read out.

Observation and Manipulation

It is possible to examine the activation of any node at the end of each cycle. A few useful summaries are also available, such as the number of active word nodes and the sum of their activations, the number of active letter nodes in each position, and so on. It is also possible to alter any of the parameters of the model between cycles or to change a parameter and then start again at Time 0 in order to compare the response of the model under different parameter values.

Running a Simulation

When simulating an experiment with a number of different trials (i.e., a number of different stimulus items in each experimental condition), the information the computer needs about the input and the forced-choice alternatives can be specified in a file, with one line containing all of the necessary information for each trial of the simulation. Typically a few test runs are carried out to choose an optimal exposure duration and readout time. Then the simulation is run with a single specified readout time for each display condition (when different display types are mixed within the same block of trials in the experiment being simulated, a single readout time is used for all display conditions). Note that when the probability of feature detection is set to 1.0, the model is completely deterministic. That is, it computes readout and forced-choice probabilities on the basis of response strengths. These are determined completely by the knowledge stored in the system (e.g., what the system knows about the appearance of the letters and the spellings of the words), by the set of features extracted, and by the values of the various parameters.

26
Introduction

(1982)
Elie L. Bienenstock, Leon N. Cooper, and Paul W. Munro

Theory for the development of neuron selectivity:
orientation specificity and binocular interaction in visual cortex
Journal of Neuroscience 2:32–48

This paper analyzes the development of selective cells in cat visual cortex. Grossberg (paper 24) and von der Malsburg (paper 17) also discuss this problem.

Bienenstock, Cooper, and Munro propose a learning rule that is related to the familiar Hebb rule, but that contains a term related to the average activity of the cell. They show that this new rule is capable of developing selectivity in neurons in the network. Specifically, they show that the synapses of the cells develop a stable set of values corresponding to high selectivity of the cell in a particular environment.

The model neurons used in this paper are the same simple linear integrators seen in the linear associator: they sum up their synaptic inputs—that is, they form the sum of the presynaptic activity times the synaptic weights. To use the notation in the paper, the input pattern is a vector, $\mathbf{d}$, and the synaptic weights on a single cell form a vector, $\mathbf{m}$; then the output of the cell, $\mathbf{c}$, is simply the inner product (dot product) of $\mathbf{m}$ and $\mathbf{d}$, that is, $\mathbf{c} = [\mathbf{m}, \mathbf{d}]$. There is no threshold or other complexities in the neuron response. Activity of a neuron is assumed to be a short term temporal average of the frequency of discharge.

Selectivity is referred to an environment. If we are interested in the development of orientation selectivity, then we must have an environment that provides many possible orientations. The response of their learning system to abnormal environments, where a range of options is not possible, allows the authors to make experimental predictions and to explain the neurophysiological data. However, the constraints involved in defining an environment may be more restrictive than desirable since nature clearly has more options available for inputs than oriented line segments. The ultimate form of development of high selectivity is the 'grandmother cell.' Real neurons in normal environments seem to develop an intermediate level of selectivity, so they show considerable selectivity, but not to the exclusion of responses to similar patterns. And in fact, the simulations presented by Bienenstock, Cooper, and Munro displayed some generalization to similar inputs. This question becomes an important representational issue. Feldman and Ballard (paper 29) discuss this matter in some detail.

The details of the synaptic change rule are quite simple. Bienenstock, Cooper, and Munro propose, first, that there is a simple exponential decay of the synapse so that in the absence of inputs the synaptic strength will decay to zero. Second, they assume something like the traditional Hebb rule. Their rule says the amount of modification of the synapse is a product of the presynaptic activity, and a function, ϕ, of the postsynaptic activity. They assume that there is a modification threshold, $\theta_\mathbf{M}$, which changes the sign of the modification. When postsynaptic activity is above threshold, the synapse increases in strength. When it is below threshold, it decreases in strength. This means that only if the response of the cell to *the entire pattern* drives the cell above threshold, will the synapses get strengthened and response to the pattern increase. The authors refer to this as *temporal competition*, because there is a competition between successive presentations of different patterns. A certain minimum response is necessary for cell response to grow; otherwise suppression of response to the pattern occurs, since when the sign of Hebbian learning is changed, it will become 'antilearning'; that is, response will decrease in the future to that pattern.

Temporal competition is in contrast to *spatial competition*, where there is competition between synapses for resources. There it was assumed that the sum of the synaptic strengths had to add up to a constant, so if one synapse grows, others are decreased. The learning rule proposed by von der Malsburg (in paper 17) is of this type, as was a perceptron variant (Rosenblatt, paper 8).

Setting the modification threshold can be hard. Bienenstock, Cooper and Munro suggest that the best way to do it is to make it a nonlinear function of the average response of the cell. This turns out to work well in both mathematical analysis and computer simulation. The average response of the cell is taken over a long time period. In the simulations, the average of the response of the cell to all the other patterns in the environment was used.

It is quite easy to see that such a modification rule favors selective patterns. Suppose that every possible pattern gave the same response of the cell, that is, the state of minimum selectivity. Let us assume, for illustration, that the modification threshold was set to be the average activity. (The actual function used for setting the modification threshold is nonlinear and tends to drive the modification threshold to a preset value.) The average activity would be the same as the response to any pattern, so there would be no modification of the synaptic strength.

However, this condition is unstable. Suppose, for some reason, random noise perhaps, that the cell responded to one pattern, one time, slightly more strongly than to the other patterns. Then the next response to that pattern would be slightly enhanced because learning would have occurred. Also, the average response of the cell increases slightly, and patterns that were at modification threshold are now below it. This causes 'antilearning' to take place, so future presentations of those patterns give even weaker responses. [This is the synaptic version of the well known Matthew principle: "Unto every one that hath shall be given ... but from him that hath not shall be taken away ..." (Matthew 25:29).] The synapses will tend toward a selective state, with one pattern giving maximum response, and others giving less or no response.

Proving development of selectivity rigorously is difficult, and the bulk of the paper is devoted to the proofs and a set of computer simulations. The simulations apply the model to a number of the effects seen in visual cortex with environmental modification: monocular deprivation, binocular deprivation, and situations where the two eyes see uncorrelated images. A reasonable amount of physiological data is available for these conditions, and it was possible to show good agreement of the simulations with experimental results.

(1982)

Elie L. Bienenstock,[2] Leon N. Cooper,[3] and Paul W. Munro

Theory for the development of neuron selectivity: orientation specificity and binocular interaction in visual cortex[1]

Journal of Neuroscience 2:32–48

Abstract

The development of stimulus selectivity in the primary sensory cortex of higher vertebrates is considered in a general mathematical framework. A synaptic evolution scheme of a new kind is proposed in which incoming patterns rather than converging afferents compete. The change in the efficacy of a given synapse depends not only on instantaneous pre- and postsynaptic activities but also on a slowly varying time-averaged value of the postsynaptic activity. Assuming an appropriate nonlinear form for this dependence, development of selectivity is obtained under quite general conditions on the sensory environment. One does not require nonlinearity of the neuron's integrative power nor does one need to assume any particular form for intracortical circuitry. This is first illustrated in simple cases, e.g., when the environment consists of only two different stimuli presented alternately in a random manner. The following formal statement then holds: the state of the system converges with probability 1 to points of maximum selectivity in the state space. We next consider the problem of early development of orientation selectivity and binocular interaction in primary visual cortex. Giving the environment an appropriate form, we obtain orientation tuning curves and ocular dominance comparable to what is observed in normally reared adult cats or monkeys. Simulations with binocular input and various types of normal or altered environments show good agreement with the relevant experimental data. Experiments are suggested that could test our theory further.

It has been known for some time that sensory neurons at practically all levels display various forms of stimulus selectivity. They may respond preferentially to a tone of a given frequency, a light spot of a given color, a light bar of a certain length, retinal disparity, orientation, etc. We might, therefore, regard stimulus selectivity as a general property of sensory neurons and conjecture that the development of such selectivity obeys some general rule. Most attractive is the idea that some of the mechanisms by which selectivity develops in embryonic or early postnatal life are sufficiently general to allow a unifying theoretical treatment.

In the present paper, we attempt to construct such a mathematical theory of the development of stimulus selectivity in cortex. It is based on (*1*) an elementary definition of a general index of selectivity and (*2*) stochastic differential equations proposed as a description of the evolution of the strengths of all synaptic junctions onto a given cortical neuron.

The ontogenetic development of the visual system, particularly of higher vertebrates, has been studied very extensively. Since the work of Hubel and Wiesel (1959, 1962), it has been known that almost all neurons in the primary visual cortex (area 17) of the normally reared adult cat are selective; they respond in a precise and sometimes highly tuned fashion to a variety of features—in particular, to bars or edges of a given orientation and/or those moving in a given direction through their receptive fields. Further work has shown that the response characteristics of these cortical cells strongly depend on the visual environment experienced by the animal during a *critical period* extending roughly from the 3rd to the 15th week of postnatal life (see, for example, Hubel and Wiesel, 1965; Blakemore and Van Sluyters, 1975; Buis-

[1] This work was supported in part by United States Office of Naval Research Contract N00014-81-K-0136, the Fondation de France, and the Ittleson Foundation, Inc. We would like to express our appreciation to our colleagues at the Brown University Center for Neural Science for their interest and helpful advice. In particular, we thank Professor Stuart Geman for several useful discussions.

[2] Present address: Laboratoire de statistique appliquée, Batiment 425, Université de Paris Sud, 91405 Orsay, France.

[3] To whom correspondence should be addressed at Center for Neural Science, Brown University, Providence, RI 02912.

Copyright © Society for Neuroscience

seret and Imbert, 1976; Frégnac and Imbert, 1978; Frégnac, 1979). Although these experiments show that visual experience plays a determining role in the development of selectivity, the precise nature of this role is still a matter of controversy.

Applying our general ideas to the development of *orientation selectivity* and *binocular interaction* in area 17 of the cat visual cortex, we obtain a theory based on a single mechanism of synaptic modification that accounts for the great variety of experimental results on monocular and binocular experience in normal and various altered visual environments. In addition, we obtain some new predictions.

It is known that various algorithms related to Hebb's principle of synaptic modification (Hebb, 1949) can account for the formation of associative and distributed memories (see, for example, Marr, 1969; Brindley, 1969; Anderson, 1970, 1972; Cooper, 1973; Kohonen, 1977). We therefore suggest that it may be the same fundamental mechanism, accessible to detailed experimental investigation in primary sensory areas of the nervous system, which is also responsible for some of the higher forms of central nervous system organization.

In sections I to III, our ideas are presented in general form, section IV is devoted to the development of orientation selectivity primarily in a normal visual environment, whereas in section V, it is shown that our assumptions also account for normal or partial development of orientation selectivity and binocularity in various normal or altered visual environments.

I. Preliminary Remarks and Definitions

Notation. We simplify the description of the dynamics of a neuron by choosing as variables not the instantaneous incoming time sequence of spikes in each afferent fiber, the instantaneous membrane potential of the neuron, or the time sequence of outgoing spikes but rather the pre- and postsynaptic firing frequencies. These may be thought of as moving time averages of the actual instantaneous variables,[4] where the length of the averaging interval is of the order of magnitude of the membrane time constant, τ. Throughout this paper, these firing frequencies are used as instantaneous variables. This formal neuron is thus a device that performs spatial integration (it integrates the signals impinging all over the soma and dendrites) rather than spatiotemporal integration: the output at time t is a function of the input and synaptic efficacies at t, independent of past history.

A synaptic efficacy m_j characterizes the *net effect* of the presynaptic neuron j on the postsynaptic neuron (in most of the paper, only one postsynaptic neuron is considered). This effect may be mediated through a complex system including perhaps several interneurons, some of which are excitatory and others inhibitory. The resulting "ideal synapse" (Nass and Cooper, 1975) thus may be of either sign, depending on whether the net effect is excitatory or inhibitory; it also may change sign during development.

---

[4] The precise form of the averaging integral (i.e., of the convolution kernel) is not essential. Exponential kernels $K(t) = \exp(-t/\tau)$ often are used in this context (see, e.g., Nass and Cooper, 1975; Uttley, 1976).

A further simplification is to assume that the *integrative power* of the neuron is a linear function, that is:

$$c(t) = \Sigma_j m_j(t) d_j(t) \tag{1}$$

where $c(t)$ is the output at time t, $m_j(t)$ is the efficacy of the jth synapse at time t, $d_j(t)$ is the jth component of the input at time t (the firing frequency of the jth presynaptic neuron), and Σ_j denotes summation over j (i.e., over all presynaptic neurons). We can then write:

$$m(t) = (m_1(t), m_2(t), \ldots, m_N(t))$$

$$d(t) = (d_1(t), d_2(t), \ldots, d_N(t))$$

$$c(t) = m(t) \cdot d(t) \tag{2}$$

$m(t)$ and $d(t)$ are real-valued vectors, of the same dimension, N (i.e., the number of ideal synapses onto the neuron), and $c(t)$ is the inner product (or "dot product") of $m(t)$ and $d(t)$. The vector $m(t)$ (i.e., the array of synaptic efficacies at time t) is called the *state* of the neuron at time t. (Note that $c(t)$ as well as all components of $d(t)$ represent firing frequencies that are measured from the level of average spontaneous activity; thus, they might take negative as well as positive values; $m_j(t)$ is dimensionless.)

The precise form of the integrative power is not essential: our results remain unchanged if, for instance, $c(t) = S(m(t) \cdot d(t))$, with S being a positive-valued sigmoid-shaped function (see Bienenstock, 1980). This is in contrast to other work (e.g., von der Malsburg, 1973) that does require nonlinear integrative power (see "Appendix B").

Selectivity. It is common usage to estimate the orientation selectivity of a single visual cortical neuron by measuring the half-width at half-height—or an equivalent quantity—of its orientation tuning curve. The selectivity then is measured with respect to a parameter of the stimulation, namely the orientation, which takes on values over an interval of 180°. In the present study, various kinds of inputs are considered, e.g., formal inputs with a parameter taking values on a finite set of points rather than a continuous interval. It will be useful then to have a convenient general index of selectivity, defined in all cases. We propose the following:

$$\mathrm{Sel}_d(\mathcal{N}) = 1 - \frac{\text{mean response of } \mathcal{N} \text{ with respect to } \boldsymbol{d}}{\text{maximum response of } \mathcal{N} \text{ with respect to } \boldsymbol{d}} \tag{3}$$

With this definition, selectivity is estimated *with respect to* or *in* an *environment for the neuron*, that is, a random variable $\boldsymbol{d}$ that takes on values in the space of inputs to the neuron $\mathcal{N}$. The variable $\boldsymbol{d}$ represents a random input to the neuron; it is characterized by its probability distribution that may be discrete or continuous. (During normal development, the input to the neuron (or neuronal network) is presumably distributed uniformly over all orientations. In abnormal rearing conditions (e.g., dark reared), the input during development could be different from the input for measuring selectivity. How this should be translated in the formal space R^N will be discussed in section IV.) This distribution defines an environment, mathematically a random variable $\boldsymbol{d}$. Selectivity is estimated (before or after development) with respect to this same environment.[5] Obviously, $\text{Sel}_d(\mathcal{N})$ always falls between 0 and 1 and the higher the selectivity of $\mathcal{N}$ in d, the closer $\text{Sel}_d(\mathcal{N})$ is to 1.

When applied to the formal neuron in state m, definition 3 gives:

$$\text{Sel}_d(m) = 1 - \frac{E[m \cdot d]}{\text{ess sup}(m \cdot d)}$$

where d is any R^N-valued random variable (the formal environment for the neuron). The symbol $E[\ldots]$ stands for "expected value of $\ldots$" (i.e., the mean value with respect to the distribution of d) and "ess sup of $\ldots$" (essential supremum) is equivalent to "maximum of $\ldots$" in most common applications. This is illustrated in Figure 1.

II. Modification of Cortical Synapses

The various factors that influence synaptic modification may be divided broadly into two classes—those dependent on global and those dependent on local information. Global information in the form of chemical or electrical signaling presumably influences in the same way most (or all) modifiable junctions of a given type in a given area. Evidence for the existence of global factors that affect development may be found, for instance, in the work of Kasamatsu and Pettigrew (1976, 1979), Singer (1979, 1980), and Buisseret et al. (1978). On the other hand, local information available at each modifiable synapse can influence each junction in a different manner. In this paper, we are interested primarily in the effect of local information on the development of selectivity.

An early proposal as to how local information could affect synaptic modification was made by Hebb (1949). His, now classical, principle was suggested as a possible neurophysiological basis for operant conditioning: "when an axon of cell A is near enough to excite a cell B and repeatedly or persistently takes part in firing it, some growth process or metabolic change takes place in one or both cells such that A's efficiency, as one of the cells firing B, is increased." Thus, the increase of the synaptic strength connecting A to B is dependent upon the correlated firing of A and B. Such a correlation principle has inspired the work of many theoreticians on various topics related to learning, associative memory, pattern recognition, the organization of neural mappings (retinotopic projections), and the development of selectivity of cortical neurons.

It is fairly clear that, in order to actually use Hebb's principle, one must state conditions for synaptic decrease as specific as those for synaptic increase: if synapses are allowed only to increase, all synapses will eventually saturate; no information will be stored and no selectivity

[5] The mathematical concept that is needed in order to represent the environment, d, during the development period is that of a *stationary stochastic process*, $d(t)$, that is (roughly), a time-dependent random variable whose distribution is invariant in time. For example, d could represent an elongated bar in the receptive field of the neuron, rotating in some random manner around its center. At each instant, the probability of finding the bar in any given orientation is the same as at any other: the distribution of $d(t)$ is time invariant, uniform over the interval $(0, 180°)$.

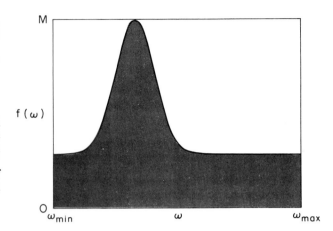

Figure 1. Computing the selectivity with respect to an environment uniformly distributed between ω_{min} and ω_{max}. The abscissa displays a parameter of the stimulus (e.g., orientation $(\omega_{max} - \omega_{min} = 180°)$) and on the ordinate, the neuron's response 0 is the level of the average spontaneous activity; M is the maximum response. The selectivity of the neuron then is given by

$$\text{Sel}_d(\mathcal{N}) = 1 - \frac{1}{M(\omega_{max} - \omega_{min})} \int_{\omega_{min}}^{\omega_{max}} f(\omega)\, d\omega$$

$$= \frac{\text{light area}}{\text{total box area}}$$

This is a simple measure of the breadth of the peak: curves of same selectivity have approximately the same half-width at half-height. (Think, for instance, of triangularly shaped tuning curves.) Typical values for orientation selectivity of adult cortical cells vary between 0.7 and 0.85 ("specific" cells). Selectivity of broadly tuned but still unimodal cells (e.g., those termed "immature" by Buisseret and Imbert (1976) and Frégnac and Imbert (1978)) lies between 0.5 and 0.7. Obviously, 0 is the selectivity of an absolutely flat curve, whereas 1 is the selectivity of a Dirac δ function.

will develop (see, for example, Sejnowski, 1977a, b). What is required is thus a complementary statement to Hebb's principle giving conditions for synaptic decrease.[6]

Such statements usually have resulted in a form of synaptic competition. Consider, for example, one that was proposed by Stent (1973): "when the presynaptic axon of cell A repeatedly and persistently fails to excite the postsynaptic cell B while cell B is firing under the influence of other presynaptic axons, metabolic changes take place in one or both cells such that A's efficiency, as one of the cells firing B, is decreased." According to Stent's principle, the increase of the strength of certain synapses onto neuron B is accompanied by simultaneous decrease of the strength of other synapses onto the same

[6] Nonspecific conditions for synaptic decrease, such as uniform exponential decay, are clearly insufficient too: in Nass and Cooper (1975) for instance, no selectivity is achieved without lateral intracortical inhibition. Other models (von der Malsburg, 1973; Perez et al., 1975) use a normalization rule in conjunction with a hebbian scheme for synaptic increase, which actually results in decrease as well as increase. This normalization rule is discussed in "Appendix B."

neuron B. There thus occurs a *spatial competition between convergent afferents*. A competition mechanism of this kind provides a qualitative explanation of some experimental results on cortical development (e.g., monocularly deprived animals (Stent, 1973)) as well as some aspects of certain more complex deprivation paradigms such as those recently reported by Rauschecker and Singer (1981).

In the present work, we present a mechanism of synaptic modification that results in a temporal competition between input patterns rather than a spatial competition between different synapses. With this mechanism, whether synaptic strength increases or decreases depends upon the magnitude of the postsynaptic response as compared with a variable modification threshold. We show that this can account quantitatively in a more powerful way for increases and decreases in selectivity as well as for a great variety of other experimental results in diverse rearing conditions.

We propose that the change of the jth synapse's strength at the time t obeys the following rule:

$$\dot{\boldsymbol{m}}_j(t) = \phi(\boldsymbol{c}(t))\boldsymbol{d}_j(t) - \epsilon \boldsymbol{m}_j(t) \tag{4}$$

where $\phi(\boldsymbol{c})$ is a scalar function of the postsynaptic activity, $\boldsymbol{c}(t)$, that changes sign at a value, θ_M, of the output called the modification threshold:

$$\phi(\boldsymbol{c}) < 0 \quad \text{for} \quad \boldsymbol{c} < \theta_M; \qquad \phi(\boldsymbol{c}) > 0 \quad \text{for} \quad \boldsymbol{c} > \theta_M$$

The term, $-\epsilon m(t)$, produces a uniform decay of all junctions; this, in most cases, does not affect the behavior of the system if ϵ is small enough. However, as will be seen later, it is important in some situations. Other than this uniform decay, the vector $\boldsymbol{m}$ is driven in the direction of the input $\boldsymbol{d}$ if the output is large (above θ_M) or opposite to the direction of the input if the output is small (below θ_M). As required by Hebb's principle, when $\boldsymbol{d}_j > 0$ and $\boldsymbol{c}$ is large enough, $\boldsymbol{m}_j$ increases. However, when $\boldsymbol{d}_j > 0$ and $\boldsymbol{c}$ is *not large enough*, $\boldsymbol{m}_j$ decreases. We may regard this as a form of *temporal competition between incoming patterns*.

The idea of such a modification scheme was introduced by Cooper et al. (1979). Their use of a constant threshold θ_M, however, resulted in a certain lack of robustness of the system: the response to all patterns could slip below θ_M and then decrease to zero. In the absence of lateral inhibition between neurons, the response might increase to more than one pattern, leading to stable states with a maximal response to more than one pattern.

In this paper, we will see that making an appropriate choice for $\theta_M(t)$ allows correct functioning under quite general conditions and provides remarkable noise tolerance properties.

In our threshold modification scheme, the change of the jth synapse's strength is written as a product of two terms, the presynaptic activity, $\boldsymbol{d}_j(t)$, and a function, $\phi(\boldsymbol{c}(t), \bar{\boldsymbol{c}}(t))$, of the postsynaptic variables, the output, $\boldsymbol{c}(t)$, and the average output, $\bar{\boldsymbol{c}}(t)$. Making use of $\bar{\boldsymbol{c}}(t)$ in the evolutive power of the neuron is a new and essential feature of this work. It is necessary in order to allow both boundedness of the state and efficient threshold modification.

Neglecting the uniform decay term, for the moment

($\epsilon = 0$), in vector notation, we have

$$\dot{\boldsymbol{m}}(t) = \phi(\boldsymbol{c}(t), \bar{\boldsymbol{c}}(t))\,\boldsymbol{d}(t) \tag{5}$$

This, together with equation 2, yields:

$$\dot{\boldsymbol{m}}(t) = \phi(\boldsymbol{m}(t)\cdot\boldsymbol{d}(t), \boldsymbol{m}(t)\cdot\bar{\boldsymbol{d}})\,\boldsymbol{d}(t) \tag{6}$$

The crucial point in the choice of the function $\phi(c, \bar{c})$ is the determination of the threshold $\theta_M(t)$ (i.e., the value of c at which $\phi(c, \bar{c})$ changes sign). A candidate for $\theta_M(t)$ is the average value of the postsynaptic firing rate, $\bar{c}(t)$. The time average is meant to be taken over a period T preceding t much longer than the membrane time constant τ so that $\bar{c}(t)$ evolves on a much slower time scale than $c(t)$. This usually can be approximated[7] by averaging over the distribution of inputs for a given state $\boldsymbol{m}(t)$

$$\bar{\boldsymbol{c}}(t) = \boldsymbol{m}(t)\cdot\bar{\boldsymbol{d}}$$

This results in an essential feature, the *instability of low selectivity points*. (This can be most easily seen at 0 selectivity equilibrium points, where, with any perturbation, the state is driven away from this equilibrium, whatever the input.)

Therefore, if stable equilibrium points exist in the state space, they are of high selectivity. However, do such points exist at all? The answer is generally yes provided that the state is *bounded from the origin and from infinity*. These conditions, instability of low selectivity equilibria as well as boundedness, are fulfilled by a single function $\phi(c, \bar{c})$ if we define $\theta_M(t)$ to be a *nonlinear function* of $c(t)$ (for example, a power with an exponent larger than 1). The final requirement on $\phi(c, \bar{c})$ thus reads:

$$\text{sign } \phi(c, \bar{c}) = \text{sign } \left(c - \left(\frac{\bar{c}}{c_0} \right)^p \bar{c} \right) \quad \text{for} \quad c > 0 \tag{7}$$

$$\phi(0, \bar{c}) = 0 \quad \text{for all} \quad \bar{c}$$

where c_0 and p are two fixed positive constants.[8] The

---

[7] Replacing the time average by an average over the distribution of $\boldsymbol{d}$ is allowed provided that (1) the process $\boldsymbol{d}(t)$ is stationary, (2) the interval, T, of time integration is short with respect to the process of synaptic evolution (i.e., $\boldsymbol{m}(t)$ changes very little during an interval of length T), (3) T is long compared to the mixing rate of the process $\boldsymbol{d}$ (i.e., during a period of length T, the relative time spent by the process $\boldsymbol{d}(t)$ at any point d in the input space is nearly proportional to the weight of the distribution of $\boldsymbol{d}$ at d). Now, synaptic modification of the type involved in changes of selectivity is probably a slow process, requiring minutes or hours (if not days) to be significant, whereas elementary sensory patterns (e.g., oriented stimuli in the receptive field of a given cortical neuron) are normally all experienced in an interval of the order of 1 min or less. Thus, we are able to choose T so that a good estimate of $\bar{c}(t)$ can be available to the neuron. In some experimental situations in which the environment is altered, there are subtle dependences of the sequence by which the final state is reached depending on how rapidly $\bar{c}$ adjusts to the changed environment.

[8] The sign of $\phi(c, \bar{c})$ for $c < 0$ is not crucial since c is essentially a positive quantity: cortical cells in general have low spontaneous activity and, at any rate, are rarely inhibited much below their spontaneous activity level. For the sake of mathematical completeness, one may, however, wish to define $\phi(c, \bar{c})$ for negative c; $\phi(c, \bar{c}) > 0$ is then the most convenient for it allows us to state theorems 1 to 3 below under the most general initial conditions. In addition, the form of ϕ for $c < 0$ can affect calculations such as those of "Appendix C."

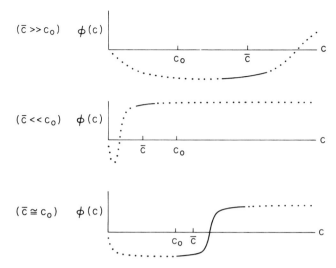

Figure 2. A function satisfying condition 7. The three diagrams show the behavior of $\phi(c, \bar{c})$ as a function of c for three different constant values of $\bar{c}$. In each diagram, the *solid part* of the *curve* represents $\phi(c, \bar{c})$ in the vicinity of $\bar{c}$, which of course is the relevant part of this function. In the *upper diagram* $(\bar{c} \gg c_0)$, although $\phi(c, \bar{c})$ is not negative for all c values as was formally required (see text), the probability that $\phi(c, \bar{c}) > 0$ is small and gets even smaller as $\bar{c}$ increases. The important point in the definition of ϕ is the nonlinearity of $\theta_M(\bar{c})$ which makes it increase or decrease faster than $\bar{c}$, while $\theta_M(\bar{c})$ is of the same order as $\bar{c}$, if $\bar{c}$ itself is of the same order as c_0.

threshold $\theta_M(\bar{c}) = (\bar{c}/c_0)^p \bar{c}$ thus serves two purposes: allowing threshold modification when $\bar{c} \simeq c_0$ as well as driving the state from regions such that $\bar{c} \ll c_0$ or $\bar{c} \gg c_0$. Equation 7 is illustrated in Figure 2.

The process of synaptic growth, starting near zero to eventually end in a stable selective state, may be described as follows. Initially, $\bar{c} \ll c_0$; hence, $\phi(c, \bar{c}) > 0$ for all inputs in the environment: the responses to all inputs grow. With this growth, $\bar{c}$ increases, thus increasing θ_M. Now some inputs result in postsynaptic responses that exceed θ_M, while others—those whose direction is far away from (close to orthogonal to) the favored inputs— give a response less than θ_M. The response to the former continues to grow, while the response to the latter decays. This results in a form of *competition between incoming patterns* rather than competition between synapses. The response to unfavored patterns decays until it reaches zero, where it stabilizes for $\phi(0, \bar{c}) = 0$ for any $\bar{c}$ (equation 7). The response to favored patterns grows until the mean response $\bar{c}$ is high enough, and the state stabilizes. This occurs in spite of the fact that many complicated geometrical relationships may exist between different patterns (i.e., that they are not orthogonal since different patterns may and certainly do share common synapses).

Any function, ϕ, that satisfies equation 7 will give the results that we describe below. The precise form of this function (e.g., the numerical values of p and c_0) will affect the detailed behavior of the system, such as rate of convergence, the height of the maximum response for a selective cell, etc., and would have to be determined by experiment.

III. Mathematical Results

The behavior of system 6 depends critically on the environment, that is, on the distribution of the stationary stochastic process, d. Two classes of distributions may be considered—discrete distributions and continuous distributions. Discrete distributions include K possible inputs $d^1, \ldots, d^K$. These will generally be assumed to occur with the same probability $1/K$. The process d is then a jump process which randomly assumes new values at each time increment. The vector m is (roughly) a Markov process. In the present work, the only continuous distribution that will be considered is a uniform distribution d over a closed one-parameter curve in the input space R^N (section IV).

Although the principles underlying the convergence to selective states are intuitively fairly simple (see the preceding section), mathematical analysis of the system is not entirely straightforward, even for the simplest d. Mathematical results, obtained only for certain discrete distributions, are of two types: (1) equilibrium points are locally stable if and only if they are of the highest available selectivity with respect to the given distribution of d and (2) given any initial value of m in the state space, the probability that $m(t)$ converges to one of the maximum selectivity fixed points as t goes to infinity is 1. Results of the second type are much stronger and require a tedious geometrical analysis. Results are stated here in a somewhat simplified form (obvious requirements of a very mathematical character are omitted). For exact statements and proofs, the reader is referred to Bienenstock (1980).

We first study the simple case where d takes on values on only two possible input vectors, d^1 and d^2, that occur with the same probability:

$$P[d = d^1] = P[d = d^2] = \tfrac{1}{2}$$

Whatever the actual dimension N of the system, it reduces to two dimensions. (Any component of m outside of the linear subspace spanned by d^1 and d^2 will eventually decay to zero due to the uniform decay term.)

It follows immediately from the definition that the maximum value of $\mathrm{Sel}_d(m)$ in the state space is $\tfrac{1}{2}$. It is reached for states m which give a null response when d^1 comes in (i.e., are orthogonal to d^1) but a positive response for d^2—or vice versa. Minimum selectivity, namely zero, is obtained for states m such that $m \cdot d^1 = m \cdot d^2$. Equilibrium states of both kinds indeed exist.

Lemma 1. Let d^1 and d^2 be linearly independent and d satisfy $P[d = d^1] = P[d = d^2] = \tfrac{1}{2}$. Then for any value of ϕ satisfying equation 7, equation 6 admits exactly four fixed points, m^0, m^1, m^2, and $m^{1,2}$ with: $\mathrm{Sel}_d(m^0) = \mathrm{Sel}_d(m^{1,2}) = 0$ and $\mathrm{Sel}_d(m^1) = \mathrm{Sel}_d(m^2) = \tfrac{1}{2}$. (Here the superscripts indicate which of the d^i are *not* orthogonal to m. (m^0 is the origin.) Thus, for instance, $m^1 \cdot d^1 > 0$, $m^1 \cdot d^2 = 0$.)

The behavior of equation 6 depends on the geometry of the inputs, in the present case, on $\cos(d^1, d^2)$. The crucial assumption needed here is that $\cos(d^1, d^2) \geqslant 0$. This is a reasonable assumption which is obviously satisfied if all components of the inputs are positive, as is assumed in some models (von der Malsburg, 1973; Perez et al., 1975). We then may state the following.

Theorem 1. Assume that, in addition to the conditions of

lemma 1, $\cos(d^1, d^2) \geqslant 0$. Then m^0 and $m^{1,2}$ are unstable, m^1 and m^2 are stable, and whatever its initial value, the state of the system converges almost surely (i.e., with probability 1) either to m^1 or to m^2.

Theorem 1 is the basic result in the two-dimensional setting: it characterizes evolution schemes based on *competition between patterns* and states that the state eventually reaches maximal selectivity even when the two input vectors are very close to one another. Obviously this requires that some of the synaptic strengths be negative since the neuron has linear integrative power. Inhibitory connections are thus necessary to obtain selectivity (see also section IV below). Some selectivity is also realizable with no inhibitory connections—not even "intracortical" ones—if the integrative power is appropriately nonlinear. However, whatever the nonlinearity of the integrative power, theorem 1 could not hold for evolution equations based on *competition between converging afferents* (see "Appendix B").

In theorem 1, we have a discrete sensory environment which consists of exactly two different stimuli—a situation, although simple mathematically, not often encountered in nature. It may, however, very well correspond to a visual environment restricted to only horizontally and vertically oriented contours present with equal probability. Theorem 1 then predicts that cortical cells will develop a selective response to one of the two orientations, with no preference for either (other than what may result from initial connectivity). Thus, on a large sample of cortical cells, one should expect as many cells tuned to the horizontal orientation as to the vertical one. (So far, no assumption is made on intracortical circuitry. See "Appendix D.")

The proof of theorem 1 is based on the existence of *trap regions* around each of the selective fixed points.

Theorem 2. Under the same conditions as in theorem 1, there exists around $m^1 (m^2)$ a region $F^1 (F^2)$ such that, once the state enters $F^1 (F^2)$, it converges almost surely to $m^1 (m^2)$.

The meaning of theorem 2 is the following: once $\boldsymbol{m}(t)$ has reached a certain selectivity, it cannot "switch" to another selective region. Applied to cortical cells in a patterned visual environment, this means that, once they become sufficiently committed to certain orientations, they will remain committed to those orientations (provided that the visual environment does not change), becoming more selective as they stabilize to some maximal selectivity. Theorems 1 and 2 are illustrated in Figure 3.

It is worth mentioning that, when $\cos(d^1, d^2) < 0$, the situation is much more complicated: trap regions do not necessarily exist and periodic asymptotic behavior (i.e., limit cycles) may occur, bifurcating from the stable fixed points when $\cos(d^1, d^2)$ becomes too negative (see Bienenstock, 1980).

We now turn to the case where $\boldsymbol{d}$ takes on K values. The following is easily obtained.

Lemma 2. Let $d^1, d^2, \ldots, d^K$ be linearly independent and $\boldsymbol{d}$ satisfy $P[\boldsymbol{d} = d^1] = \ldots = P[\boldsymbol{d} = d^K] = 1/K$. Then, for any function ϕ satisfying equation 7, equation 6 admits exactly 2^K fixed points with selectivities $0, 1/K, 2/K, \ldots, (K-1)/K$. There are K fixed points $m^1, \ldots, m^K$ of selectivity $(K-1)/K$.

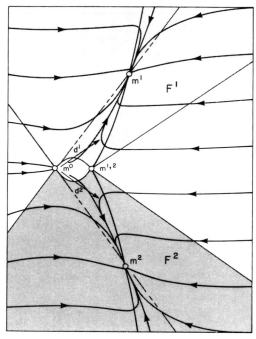

Figure 3. The phase portrait of equation 6 in an environment consisting of two inputs, d^1 and d^2 (theorems 1 and 2). The diagram shows the trajectories of the state of the system, starting from different initial points. This is a computer simulation performed with one given function ϕ satisfying condition 7. Using a different function may slightly change the shape of the trajectories without any essential change in the behavior. The unstable fixed points are $m^{1,2}$ and m^0; the stable ones are m^1 and m^2. The system is a *stochastic* one, which means that the trajectories depend, in fact, on the precise sequence of inputs. As long as the state is in the *unshaded region*, it is not yet known whether it will eventually be attracted to m^1 or m^2. This is determined as the state enters one of the trap (*shaded*) regions, F^1 or F^2. The trajectories shown here are deterministic ones, obtained by alternating $\boldsymbol{d}$ regularly between d^1 and d^2. They are, in fact, the averaged trajectories of the state and are much more regular and smooth than the actual stochastic ones.

Obviously, $(K-1)/K$ is also the maximum possible selectivity with respect to $\boldsymbol{d}$. It means a positive response for one and only one of the inputs. The situation is now much more complicated than what it was with only two inputs: it is not obvious whether, in all cases, the assumption that all of the cosines between inputs are positive is sufficient to yield stability of the maximum selectivity fixed points. However, we may state the following.

Theorem 3. Assume, in addition to the conditions of lemma 2, that $d^1, \ldots, d^K$ are all mutually orthogonal or close to orthogonal. Then the K fixed points of maximum selectivity are stable, and whatever its initial value, the state of the system converges almost surely to one of them.

The proof of theorem 3 also involves trap regions around the K maximally selective fixed points, and the analog of theorem 2 is true here.

Although the general case has not yet been solved analytically, as will be seen in the next section, computer simulations suggest that, for a fairly broad range of environments, if $d^i \cdot d^j \geq 0$, even if $d^1, \ldots, d^K$ are far from being mutually orthogonal, the K fixed points of maximum selectivity are stable.

Simulations suggest further (see, for instance, Fig. 4b) that, even if the $d^1, \ldots, d^K$ are *not* linearly independent and are very far from being mutually orthogonal, the asymptotic selectivity is close to its maximum value with respect to d.

IV. Orientation Selectivity and Binocular Interaction in Visual Cortex

We now apply what has been done to a concrete example, orientation selectivity and binocular interaction in the primary visual cortex. The ordinary development of these properties in mammals depends to a large extent on normal functioning of the visual system (i.e., normal visual experience) during the first few weeks or months of postnatal life. This has been demonstrated many times by various experiments, based mainly on the paradigm of rearing the animal in a restricted sensory environment. In the next two sections, it is shown how equations 4 to 7 account for both normal development as well as development in restricted visual environments.

Consider first a classical test environment used to construct the tuning curve of cortical neurons. This environment consists of an elongated light bar successively presented or moved in all orientations—preferably in a random sequence—in the neuron's receptive field. Thus, all of the parameters of the stimulus are constant except one, the orientation, which is distributed uniformly on a circularly symmetric closed path. We assume that the retinocortical pathway maps this family of stimuli to the cortical neuron's space of inputs in such a way as to preserve the circular symmetry (as defined below). Thus, the typical theoretical environment that will be used for constructing the formal neuron's tuning curve is a random variable d uniformly distributed on a circularly symmetric closed one-parameter family of points in the space R^N. The parameter coding orientation in the receptive field is, in principle, continuous. However, for the purpose of numerical simulations, the distribution is made discrete. Thus, d takes on values on the points $d^1, \ldots, d^K$.

The requirement of circular symmetry is expressed mathematically as follows: the matrix of inner products of the vectors $d^1, \ldots, d^K$ is circular (i.e., each row is obtained from its nearest upper neighbor by shifting it one column to the right) and the rows of the matrix are unimodal. A random variable, d, uniformly distributed on such a set of points will be, hereafter, called a *circular environment*. Such a d may be roughly characterized by three parameters: N, K, and a measure of the mutual geometrical closeness of the d^i vectors, for instance, $\min \cos(d^1, d^i)$.

Now we are faced with the difficult problem of specifying the stationary stochastic process that represents the time sequence of inputs to the neuron during development. In a first analysis, there is no choice but to oversimplify the problem by giving the stochastic process exactly the same distribution as the circular d defined above. In doing so, we assume that development of

orientation selectivity is to a large extent independent of other parameters of the stimulus (e.g., contrast, shape, position in the receptive field, retinal disparity for binocular neurons, etc.). The elementary stimulus for a cortical neuron is a rectilinear contrast edge or bar. Any additional pattern present at the same time in the receptive field is regarded as random noise. (A discussion of this point is given in Cooper et al. (1979).)

IVa. Normal Monocular Input

The behavior of a monocular system in circular environments is investigated by numerically simulating equation 6 with a variety of circular environments, d, and functions ϕ satisfying equation 7. In the simulations presented here, the dimension of the input and state space is generally $N = 37$; the number K of input vectors varies from 12 to 60. (Various kinds of functions ϕ were used: some were stepwise constant; others were smooth, bounded, or unbounded.) One may reasonably expect the system's behavior to be fairly independent of N and K if these are high enough. However, the geometry of d may be determining: if the inputs, d^i, are closely packed together in the state space (i.e., if $\min \cos(d^1, d^i)$ is close to 1), convergence to selective states may presumably be difficult to achieve or even impossible.

Simulations show the following behavior:

1. The state converges rapidly to a fixed point or *attractor*.
2. Various such attractors exist. For a given d and ϕ, they all have the same selectivity, which is close to its maximum value in d.
3. The asymptotic tuning curve is always unimodal. Thus, one may talk of the preferred orientation of an attractor.
4. There exists an attractor for each possible orientation.
5. If there is no initial preference, all orientations have equal probability of attracting the state. (Which one will become favored depends on the exact sequence of inputs.) This does not hold for environments which are not perfectly circular, at least for a single neuron system such as the one studied here.

In Figure 4, a and b show, respectively, the progressive buildup of selectivity and the tuning curve when the state has virtually stabilized.

In summary then, the system behaves in circular environments exactly as we might have expected from the results of the preceding section. However, one should note one important difference: the maximum selectivity for a continuous environment cannot be calculated as simply as it was before. It is only when d is distributed uniformly on K linearly independent vectors that we know that $\max \text{Sel}_d(m) = (K - 1)/K$ (lemma 2). Theorem 3 indicates that, if, in addition, the vectors are nearly orthogonal to one another, this selectivity is indeed asymptotically reached. We could not prove that this is also true when the vectors are arranged circularly but are not mutually orthogonal. However, it could not be disproved by any numerical simulation; therefore, we *conjecture* that this is indeed true. (Reasonable selectivity is attained even in most unfavorable environments. As an

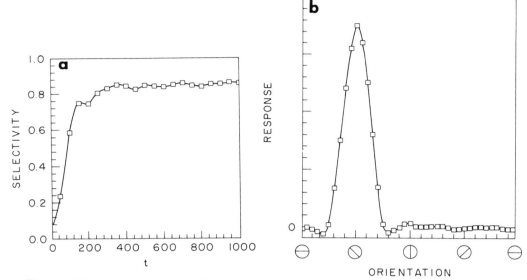

Figure 4. The evolution of a synaptic system in a circular environment. Here, $K = 40$ and $N = 37$ so that the vectors are linearly dependent. The value of the maximum selectivity with respect to d is therefore not precisely calculable. The asymptotic selectivity is approximately 0.9, perhaps the maximum selectivity. *a* demonstrates the progressive buildup of the selectivity in a circular environment d, while *b* shows the resulting tuning curve at $t = 1000$.

example, in a circular d such that all cosines fall between 0.94 and 1, a selectivity of 0.68 was reached after 12,000 iterations.) Notice that, in the present context, this question is only of theoretical interest, since naturally occurring environments are continuous rather than discrete. The behavior of our system in such an environment is very well approximated by a discrete circular d, provided that K is large enough. K is then presumably larger than N, the K inputs are linearly dependent, and we have no explicit formula for max $Sel_d(m)$.

The system thus functions well in a large class of environments. It should be stressed that the numerical value of the parameters that appear explicitly in the evolution equation, namely, c_0 and the exponent p, are not at all critical. Simulations performed with a constant d, with p being varied from 0.01 to 10, yield the same asymptotic limit for the selectivity; the height of the asymptotic tuning curve (i.e., max($m \cdot d$)) is, however, highly dependent on p. This invariance property validates in a sense the definition of $Sel_d(x)$.

Inhibitory synapses are essential here exactly as they are in the two-dimensional case. One way to show this is to substitute 0 for all negative components in the state once it has become selective. This typically results in a drastic drop of selectivity (e.g., from 0.81 to 0.55) although a slight preference generally remains for the original orientation. This may be related to the experimental finding that local pharmacological deactivation of inhibitory connections strongly impairs orientation selectivity by rendering all orientations effective in triggering the cell's response (Sillito, 1975).

Finally, it should be mentioned that the system displays a good noise tolerance, particularly when the state has already reached a selective region. The system then resists presynaptic additive noise with a signal-to-noise

ratio of the order of 1 and postsynaptic noise with a signal-to-noise ratio as small as ¼.

IVb. Restricted Monocular Input

To discuss this situation, we now must include the exponential decay term, $-\epsilon_m(t)$, previously neglected (equation 4). It is clear that the results stated above will be preserved if ϵ is sufficiently smaller than the average of $|\phi(c,\bar{c})|$ (i.e., competition mechanisms are faster than decay). However, exponential decay does become crucial in some situations. One of these is the response of the cell to patterns that were not represented in the environment during development.

Consider, for instance, an environment consisting of a single stimulus d^1. It is then easily shown that system 6 with condition 7 admits one attractor m^1 that satisfies $m^1 \cdot d^1 \simeq c_0$ for small ϵ ($m^1 \cdot d^1 = c_0$ for $\epsilon = 0$). Obviously, for $\epsilon > 0$, m^1 will satisfy $m^1 \cdot d = 0$ for any d orthogonal to d^1. However, the response to a pattern d not orthogonal to d^1 will depend both on ϵ and on $\cos(d, d^1)$. One may for instance find that $m^1 \cdot d \simeq \frac{1}{2}(m^1 \cdot d^1)$ for $\cos(d, d^1) = 0.5$. The selectivity of the neuron in state m^1 with respect to a circular environment $(d^1, \ldots, d^K)$, such that min $\cos(d^1, d^i) = 0.5$, is then lower than 0.5. This should be contrasted with the high selectivity reached by a neuron exposed to all inputs, $d^1 \ldots d^K$.

The one-stimulus environment may be regarded as a case corresponding to rearing the animal in a visual world where only one orientation is present. No controversy remains at present that rearing in such a visual environment results in a cortex in which all visually responsive cells are tuned to the experienced (or nearby) orientations (Blakemore and Cooper, 1970; Hirsch and Spinelli, 1970, 1971; see also Stryker et al., 1978). We see that our theory is in agreement with these findings; moreover, we predict that, in such a cortex, the average selectivity of these cells should be lower than normal. Although there is so far no detailed quantitative study on this point, in a recent study, there is some indication that this may indeed be true: "more neurons with normal orientation tuning were found in the kittens that could see all orientations, or at least horizontal and vertical, than in the kittens that had experienced only one orientation" (Rauschecker and Singer, 1981).

IVc. Binocular Input

We now consider a binocularly driven cell. The firing rate of the neuron at time t becomes

$$c(t) = \boldsymbol{m}_r(t) \cdot \boldsymbol{d}_r(t) + \boldsymbol{m}_l(t) \cdot \boldsymbol{d}_l(t) \qquad (8)$$

with evolution schemes for "right" and "left" states $\boldsymbol{m}_r$ and $\boldsymbol{m}_l$ straightforward generalizations of equation 4. Various possibilities now exist for the input $(\boldsymbol{d}_r, \boldsymbol{d}_l)$: one may wish to simulate normal rearing (both $\boldsymbol{d}_r$ and $\boldsymbol{d}_l$ circular and presumably highly correlated), monocular deprivation, binocular deprivation, etc.

Detailed discussion of the results of simulations under various conditions is given in the next section. The main results are summarized here:

1. In an environment simulating normal binocular rearing, the cell becomes orientation selective and binocular, preferring the same orientation through both eyes.
2. In an environment simulating monocular deprivation, the cell becomes monocular and orientation selective, whatever its initial state.
3. In an environment simulating binocular deprivation, the cell loses whatever orientation selectivity it had but does not lose its responsiveness and, in general, remains driven by both eyes.

V. Development under Different Rearing Conditions: Comparison of Theory with Classical Experimental Data

Related Experimental Data

This brief summary is restricted to area 17 of kitten's cortex. Most kittens first open their eyes at the end of the 1st week after birth. It is not easy to assess whether orientation-selective cells exist at that time in the striate cortex: few cells are visually responsive, and the response's main characteristics are generally "sluggishness" and fatigability. However, it is agreed quite generally that, as soon as cortical cells are reliably visually stimulated (e.g., at 2 weeks), some are orientation selective, whatever the previous visual experience of the animal (cf., Hubel and Wiesel, 1963; Blakemore and Van Sluyters, 1975; Buisseret and Imbert, 1976; Frégnac and Imbert, 1978).

Orientation selectivity develops and extends to all visual cells in area 17 if the animal is reared, and behaves freely, in a normal visual environment (NR): complete "specification" and normal binocularity (about 80% of responsive cells) are reached at about 6 weeks of age (Frégnac and Imbert, 1978). However, if the animal is reared in total darkness from birth to the age of 6 weeks (DR), then none or few orientation-selective cells are recorded (from 0 to 15%, depending on the authors and

the classification criteria); however, the distribution of ocular dominance seems unaffected (Blakemore and Mitchell, 1973; Imbert and Buisseret, 1975; Blakemore and Van Sluyters, 1975; Buisseret and Imbert, 1976; Leventhal and Hirsch, 1980; Frégnac and Imbert, 1978). In animals whose eyelids have been sutured at birth and which are thus binocularly deprived of pattern vision (BD), a somewhat higher proportion (from 12 to 50%) of the visually excitable cells are still orientation selective at 6 weeks (and even beyond 24 months of age) and the proportion of binocular cells is less than normal (Wiesel and Hubel, 1965; Blakemore and Van Sluyters, 1975; Kratz and Spear, 1976; Leventhal and Hirsch, 1977; Watkins et al., 1978).

Of all visual deprivation paradigms, putting one eye in a competitive advantage over the other has probably the most striking consequences: monocular lid suture (MD), if it is performed during a "critical" period (ranging from about 3 to about 12 weeks), results in a rapid loss of binocularity, to the profit of the open eye (Wiesel and Hubel, 1963, 1965); then, opening the closed eye and closing the experienced one may result in a complete reversal of ocular dominance (Blakemore and Van Sluyters, 1974). A disruption of binocularity that does not favor one of the eyes may be obtained, for example, by provoking an artificial strabismus (Hubel and Wiesel, 1965) or by an alternating monocular occlusion, which gives both eyes an equal amount of visual stimulation (Blakemore, 1976). In what follows, we call this uncorrelated rearing (UR).

Theoretical Results

The aim of this section is to show that the experimental results briefly reviewed above follow from our assumptions if one chooses the appropriate distribution for $\boldsymbol{d}$. The model system now consists of a single binocular neuron. The firing rate of the neuron at time t is given by

$$c(t) = \boldsymbol{m}_r(t) \cdot \boldsymbol{d}_r(t) + \boldsymbol{m}_l(t) \cdot \boldsymbol{d}_l(t) \qquad (8)$$

where the indices r and l refer to right and left eyes, respectively. $\boldsymbol{m}_r$ (or $\boldsymbol{m}_l$) obeys the evolution scheme described by equations 4 to 6, where $\boldsymbol{d}_r$ (or $\boldsymbol{d}_l$) is substituted for $\boldsymbol{d}$. The two equations are, of course, coupled, since $c(t)$ depends at each t on both $\boldsymbol{m}_r(t)$ and $\boldsymbol{m}_l(t)$.

The vector $(\boldsymbol{d}_r, \boldsymbol{d}_l)$ is a stationary stochastic process, whose distribution is one of the following, depending on the experimental situation one wishes to simulate.

Normal rearing (NR)

$\boldsymbol{d}_r(t) = \boldsymbol{d}_l(t)$ for all t, and $\boldsymbol{d}_r$ is circular. (Noise terms that may be added to the inputs may or may not be stochastically independent.)

Uncorrelated rearing (UR)

$\boldsymbol{d}_r$ and $\boldsymbol{d}_l$ are i.i.d. (independent identically distributed): they have the same circular distribution, but no statistical relationship exists between them.

Binocular deprivation

Total light deprivation (DR). The $2N$ components of $(\boldsymbol{d}_r, \boldsymbol{d}_l)$ are i.i.d.: $\boldsymbol{d}_r$ and $\boldsymbol{d}_l$ are uncorrelated noise terms, $(\boldsymbol{d}_r, \boldsymbol{d}_l) = (\boldsymbol{n}_r, \boldsymbol{n}_l)$.

Binocular pattern deprivation (BD). $d_r(t) = \lambda_r(t)e$, $d_l(t) = \lambda_l(t)e$, where e is an arbitrary normalized fixed vector with positive components, and λ_r and λ_l are scalar positive valued and i.i.d.

Monocular deprivation (MD)

d_r is circular; d_l is a noise term: $d_l = n$.

In the NR case, the inputs from the two eyes to a binocular cell are probably well correlated. We therefore assume that they are equal, which is mathematically equivalent. The DR distribution represents dark discharge. The BD distribution deserves a more detailed explanation. In this distribution, it is only the *length* λ_r and λ_l of the vectors d_r and d_l that varies in time. This length is thought to correspond to the intensity of light coming through each closed eyelid, whereas the direction of the vector in the input space is determined by the constant "unpatterned" vector e (e.g., $e = (1/\sqrt{N}) \times (1, 1, \ldots, 1)$). One may indeed assume that, when light falls on the retina through the closed lids, there is, at any instant of time, high correlation between the firing rates of all retinal ganglion cells on a relatively large region of the retina. Inputs from the two eyes, however, are probably to some extent asynchronous (cf., Kratz and Spear, 1976); hence the BD distribution.

Simulations of the behavior of the system in these different environments give the following.

NR (Fig. 5a). All asymptotic states are selective and binocular, with matching preferred orientations for stimulation through each eye.

DR (Fig. 5b). The motion of the state (m_r, m_l) resembles a random walk. (The small exponential decay term is necessary here, too, in order to prevent large fluctuations.) The two tuning curves[9] therefore undergo random fluctuations that are essentially determined by the second order statistics of the input d. As can be seen from the figure, these fluctuations may result sometimes in a weak orientation preference or unbalanced ocular dominance. However, the system never stays in such states very long; its average state on the long run is perfectly binocular and non-oriented. Moreover, whatever the second order statistics of d and the circular environment in which tuning curves are assessed, a regular unimodal orientation tuning curve is rarely observed, and selectivity has never exceeded 0.6. Thus, we may conclude that orientation selectivity as observed in the NR case (both experimental and theoretical) cannot be obtained from purely random synaptic weights. It is worth mentioning here that prolonged dark rearing has been reported to increase response variability (Leventhal and Hirsch, 1980); a similar observation was made by Frégnac and Bienenstock (1981).[10]

BD (Fig. 5c). Unlike the DR case, the state *converges*

---

[9] The circular environment which serves to assess the orientation tuning curves is now, in a sense, arbitrary, since it is not at all used in the development period. The same remark applies, of course, to the BD case.

[10] In Figure 1*B* of Frégnac and Bienenstock (1981), which shows averaged orientation tuning curves of a cell recorded in an 86-day-old DR cat, the selectivity is 0.58 at the beginning of the recording session and 0.28 at the end.

(as may easily be proved mathematically). Although there exist both monocular and binocular stable equilibrium points, the asymptotic state is generally monocular if the initial state is taken as 0. The orientation tuning curve then is determined essentially by the relative geometry of the fixed arbitrary vector e and the arbitrary circular environment which serves to assess the tuning curve. Fine unimodal tuning, therefore, is not to be expected.

MD (Fig. 5d). The only stable equilibrium points are monocular and selective. The system converges to such states whatever the initial conditions. In particular, this accounts for reverse suture experiments (Blakemore and Van Sluyters, 1974; Movshon, 1976).

UR (Fig. 5e). This situation is, in a sense, similar to the BD one: the state converges, but monocular as well as binocular equilibria exist. As in the BD case, the asymptotic state generally observed with $m_r(0) = m_l(0) = 0$ is monocular. (This should be attributed to the mismatched inputs from the two eyes, as is done by most authors.) In this case, however, asymptotic states are selective, and when they are binocular, preferred orientations through each eye do not necessarily coincide. It should be mentioned here that Blakemore and Van Sluyters (1974) report that, after a period of alternating monocular occlusion, the remaining binocular cells may differ in their preferred orientations for stimulation through each eye.

These results are in agreement with the classical experimental data in the domain of visual cortex development. Most of them can be obtained fairly easily, with no need of further simulations, as a consequence of the convergence to selective states in the case of a monocularly driven neuron in a circular environment (section IVa).

Some intriguing properties of our theory are more subtle, however, and, in addition to contributing to the results above, provide the opportunity for applications to more complicated experimental paradigms and for new tests. As an example, it is shown in "Appendix C" that, in the MD case, the degree of monocularity of the cortical cell is correlated with its orientation selectivity as well as the diversity of inputs to the open eye. These unexpected predictions agree well with the observation by Cynader and Mitchell (1980) and Trotter et al. (1981) that, after a brief period of monocular exposure, oriented cells are more monocular than non-oriented ones as well as the observation of Rauschecker and Singer (1981) that an open eye with restricted inputs leads to cells oriented to the restricted input that are driven less monocularly than usual. A summary of theoretical results is given below.

VI. Discussion

We have proposed a new mathematical form for synaptic modification and have investigated its consequences on the development of selectivity in cortical neurons. In addition, we have provided a definition of the notion of selectivity with respect to a random variable that might be applied in many different situations (in the domain of development of sensory systems, for example, selectivity of binocular neurons to retinal disparity, etc.). In its application to visual cortex, our theory is in agree-

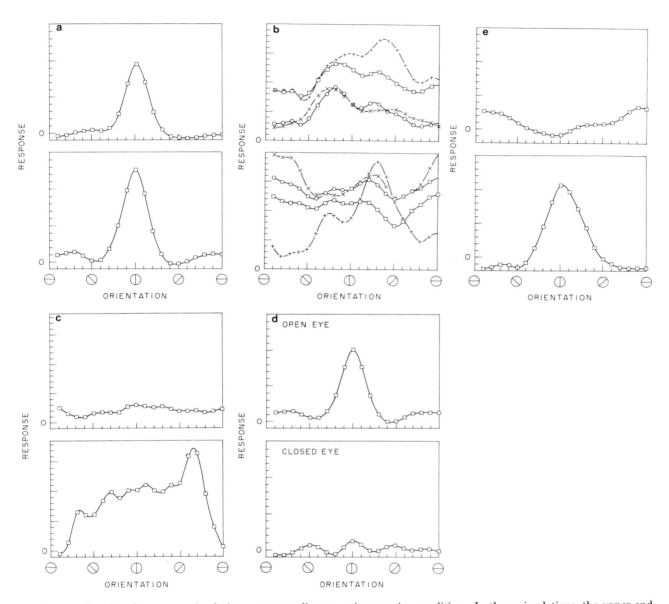

Figure 5. Results of computer simulations corresponding to various rearing conditions. In these simulations, the *upper* and *lower panels* show cell responses to stimuli from the two eyes. *a*, Normal (NR). The cell's response is binocular and selective. *b*, Dark rearing (DR). There is no stable selectivity in the cell's response. The response curve fluctuates randomly. The cell is, on the average, driven binocularly. *c*, Binocular deprivation (BD). The cell reaches a final state corresponding to the arbitrary vector which corresponds to a diffuse input to the retina. The cells sometimes are driven monocularly. This is somewhat analogous to *e* below. *d*, Monocular deprivation (MD). The cell's response is monocular and selective. *e*, Uncorrelated rearing (UR). Both binocular and monocular selective final states are observed.

ment with the classical experimental results obtained over the last generation and offers a number of new predictions, some of which can be tested experimentally. This may lead to the identification of the parameters of the theory and provide indications as to the biochemical mechanisms underlying cortical plasticity.

In a broader context, we may regard our form of synaptic modification as a specific correlation modifica-

tion of a hebbian type. The great majority of models on a synaptic level in domains such as pattern recognition, task learning, or associative memory[11] (which are less accessible to direct neurophysiological experimentation)

---

[11] Notice, for instance, the analogy between states of maximum selectivity as defined here and the optimal associative mappings of Kohonen (1977).

use schemes of a hebbian type with some success. Thus, we are led to conjecture that some form of correlation modification is a very general organizational principle that manifests itself in visual cortex in a manner that is accessible to experiment.

Although synaptic competition is a natural consequence of Hebb's principle, it may be given various mathematical forms. A distinction was made in section II between spatial competition—the form commonly accepted by theoreticians as well as experimentalists—and temporal competition—a new form proposed in this work. Competition is said to be *purely spatial*, or to take place between converging afferents, if the sign of $\dot{m}_j(t)$ is determined by a comparison of the firing rate $d_j(t)$ with firing rates $d_k(t)$ in the other afferents to the neuron at the same time, t. In some schemes (e.g., von der Malsburg, 1973), the sign of $\dot{m}_j(t)$ also depends on the value of the synaptic efficacy $m_j(t)$ relative to $m_k(t)$, $1 \le k \le N$. Formally then, we might characterize competition as purely spatial if

$$\text{sign}\,(\dot{m}_j(t)) = F\left(\frac{d_j(t)}{d_1(t)}, \ldots, \frac{d_j(t)}{d_N(t)}, \frac{m_j(t)}{m_1(t)}, \ldots, \frac{m_j(t)}{m_N(t)}\right) \quad (9)$$

On the other hand, we say that competition is *purely temporal*, or takes place between incoming patterns, if the sign of $\dot{m}_j(t)/d_j(t)$ is independent of j and is determined by a relationship between the postsynaptic neuron's firing rates, $c(t)$ and $c(t')$, $t' < t$:

$$\text{sign}\left(\frac{\dot{m}_j(t)}{d_j(t)}\right) = F(c(t); c(t')), \quad t' < t, \quad j = 1 \ldots N \quad (10)$$

In this work, the modification threshold, θ_M, is given as a function of $\bar{c}$, an average of $c(t')$ over a relatively long time period preceding t, thus satisfying equation 10.

One may, of course, imagine hebbian schemes of a mixed type, involving both spatial and temporal competition.[12] However, the distinction is useful since the performance of the scheme seems to be highly dependent on which of the two classes it is in. This is most clearly seen in the development of selectivity.

In the temporal version used here, asymptotic states are of maximum selectivity with respect to the experienced environment d, independent of the geometry of d. This was rigorously proven analytically in some cases (theorem 1; section III) and conjectured on the basis of numerical results in other cases (circular environments; section IV). In contrast with this, we conjecture that, in any model using pure spatial competition, behavior will usually depend on the geometry of the environment. In any case, maximum selectivity is not reached if the patterns in the environment are not sufficiently separated from one another. This is illustrated in "Appendix B" for one particular model using spatial competition between converging afferents.

We further note that selectivity, as was shown in section V, does not develop in a "pure noise" environment (the distribution termed DR). Some kind of patterned input is required.[13] It follows that, at this level of organization of connectivity, information is being transferred from the environment to the system. This may shed some light on what has been known for a long time as the innate/learned controversy in visual cortex. Our results suggest that this dichotomy is, at best, misleading. The system's potential developmental ability—its evolutive power—may indeed be determined genetically; yet selectivity has no meaning if it does not refer to a given structured environment that determines the final organization of the system.[14]

The present work, however, makes no assumption concerning the initial state of cortex (e.g., the presence or absence of selectivity at eye opening). This question, still a subject of controversy (see Pettigrew, 1978), must be settled experimentally. Further, although we here assume that all synapses are equally modifiable, it could easily be the case that there is variation in modifiability—even one that is time dependent—and that, for example, some of the initial state information including some orientation selectivity is contained in a skeleton of synapses that is less modifiable.[15] Such assumptions can easily be incorporated in fairly obvious extensions of our theory and would, of course, result in the modification of some details of our results. The principal results of our theory, applied to visual cortical neurons and assuming that they are all equally modifiable according to equations 4 to 7, are summarized now. These are either in agreement with existing experimental data or are new and somewhat unexpected consequences of our theory.

Summary of Theoretical Results—New Predictions

Monocularly driven neurons

(1) A monocularly driven neuron in a "normal" (patterned) environment becomes selective. The precise pattern to which it becomes selective is determined at random if the initial selectivity is 0 or may be biased toward a particular pattern if there is a built-in preference for this pattern.

(2) This same neuron in various deprived environments evolves as follows.

Pure noise. The neuron becomes less selective but

---

[12] More complicated temporal or mixed spatiotemporal schemes are possible and some such have been proposed. For example, Sejnowski (1977a) has suggested a form of modification in which the change of the jth synaptic strength involves the co-variance between the jth fiber and postsynaptic activities. In addition, interaction between neurons (such as lateral inhibition) can increase selectivity (see, for example, Nass and Cooper, 1975 and "Appendix D").

[13] Pure noise and circular environments may be regarded as two extreme cases: the first totally lacks structure, whereas the second is highly organized. Intermediate cases (i.e., environments consisting of the sum of a noise process and of a circular process) also have been investigated (see, for example, Bienenstock, 1980). There it is shown that the asymptotic selectivity directly depends on a parameter that measures the degree of structure of the environment.

[14] We note, further, that the mechanism of synaptic modification that we have proposed leads both to what are sometimes called "selective" and "instructive" effects (depending on the structure of the environment and the genetic initial state). Thus, as is already suggested by Rauschecker and Singer (1981), this dichotomy is obscured, or does not appear at all, at the synaptic level.

[15] This skeleton might consist primarily of the contralateral pathway and favor the development of the orientation preference for horizontally and vertically oriented stimuli (see, for example, Frégnac, 1979).

continues to be (somewhat) responsive. It may show an orientation preference, but this is relatively unstable.

Exposure to a single pattern (such as vertical lines). The neuron comes to respond preferentially to the single pattern *but with less selectivity (less sharply tuned) than if all orientations were present in the environment.* This last is a natural consequence of temporal competition between incoming patterns and can provide a good test of our theory.[16]

(3) Inhibitory synapses are required to produce maximum selectivity. If such inhibitory connections are arbitrarily set equal to 0, selectivity diminishes.

Binocularly driven neurons

(1) A binocularly driven neuron in a "normal" (patterned) environment becomes selective and binocular. It is driven selectively by the same pattern from both eyes.

(2) This same binocularly driven neuron in various deprived environments evolves as follows.

Uncorrelated patterned inputs to both eyes. The neuron becomes selective, often monocularly driven; if the neuron is binocular, sometimes it is driven by different patterns from the two eyes.

Patterned input to one eye, noise to the other (monocular deprivation). The neuron becomes selective and generally driven only by the open eye. *There is a correlation between selectivity and binocularity. The more selective the neuron becomes, the more it is driven only by the open eye. A non-selective neuron tends to remain binocularly driven.* This correlation is due, in part, to the fact that it is the same mechanism of synaptic change that serves to increase both the selectivity and ocular dominance of the open eye. However (as shown in "Appendix C"), there is also a subtler connection: it is the non-preferred inputs from the open eye accompanied by noise from the closed eye that drive the neuron's response to the closed eye to 0. Thus, for example, if the visual environment were such that there were mostly preferred inputs to the open eye, *even a selective cell would remain less monocular.* (It should prefer the open eye but remain somewhat driven by the closed eye.) As another example, a kitten dark-reared to the age of about 42 days (when there remain few or no specific cells) and then given monocular exposure to nonpatterned input would retain more binocularly driven cells than a similar animal given patterned input.[17]

Noise input to both eyes (dark rearing or binocular deprivation). The neuron remains non-selective (or loses its selectivity) and diminishes its responsiveness but remains binocularly driven (in contrast to the situation in monocular deprivation).

These theoretical conclusions are consistent with experimental data on increases and decreases in selectivity, data concerning changes in ocular dominance in various rearing conditions, as well as data from more complicated paradigms. Although there are indications in recent work that some of the new predictions are in agreement with experimental results, they provide the opportunity for tests of subtler aspects of the theory.

In conclusion, we note that a precise application of our theory to certain complicated experimental situations would probably require inclusion of some anatomical details, interneuronal interactions, as well as a statement of what information is innate and which synapses are modifiable.[18]

Appendix A: Biochemical Mechanism for Temporal Competition

It is probably premature to propose a detailed physiological mechanism for a mathematical synaptic modification algorithm: too many possibilities exist with no present experimental test to decide among them. However, we propose the following as a possible example.

The dependence of our modification threshold upon the mean postsynaptic activity, which regulates the individual neuron modification in an overall manner, might be the result of a physiological mechanism within the framework proposed by Changeux et al. (1973). Their basic hypothesis is that receptor protein on the postsynaptic membrane exists in two states, one *labile* and the other *stable*; *selective stabilization* of the receptor takes place during development in an activity-dependent fashion. The quantity of labile receptor available for stabilization is determined by the neuron's average activity; that is, labile receptor is not synthesized anymore when the neuron's activity is high for a relatively long period of time ($\bar{c} \gg c_0$) (cf., Changeux and Danchin (1976): "The activity of the postsynaptic cell is expected to regulate the synthesis of receptor.").

Our hypothesis that, during the period when competition really takes place (i.e., when $\bar{c}$ is of the order of c_0 in equation 7), the sign of the modification is determined by the instantaneous activity, c, relative to its mean, $\bar{c}$, requires that a single message, the instantaneous activity, be fed back from the site of integration of the incoming message to the individual synaptic sites, on a rapid time scale (i.e., much faster than the one involved in the overall regulation mechanism). This might be contrasted with the assumption implicit in most spatial competition models, namely, that a chemical substance is redistributed between all subsynaptic sites (cf., the principle of conservation of total synaptic strength (von der Malsburg, 1973)).

---

[16] In addition, the principle of temporal competition suggests an experimental paradigm that could be used to increase the selectivity of a cortical neuron while recording from the same neuron. The paradigm consists of controlling the postsynaptic activity of the neuron while presenting sequentially in its receptive field two stimuli, A and B. Stimulus A (or B) should be associated with a high (or low) instantaneous firing rate in such a way as to keep the cell's mean firing rate at its original value. We predict that the cell will prefer stimulus A eventually (i.e., exhibit selectivity with respect to the discrete environment consisting of A and B). Moreover, we predict that presentation of stimulus A alone will lead to less selectivity. An experiment based on this paradigm is currently being undertaken by one of us (E. L. B.) in collaboration with Yves Frégnac.

[17] In this situation, one might have to distinguish between short and long monocular exposures. In very long monocular exposures, the decay term of equation 4 ($-\epsilon m(t)$) eventually could produce decay of junctions from the closed eye independent of the effect discussed above.

---

[18] This last might be treated as, for example, in the work of Cooper et al. (1979).

Appendix B: von der Malsburg's Model of Development of Orientation Selectivity

A model of development of orientation selectivity using an evolution scheme of the spatial type may be found in the work of von der Malsburg (1973). We present here a brief analysis of this model in view of the definition given in section II. We first show that the type of competition implied by this model is indeed, formally, the spatial one. Next, we investigate the behavior of the system in the simple situation of theorem 1 in section III (i.e., for a two-pattern environment, with the dimension of the system being $N = 2$). We will show why the assumption that is made of nonlinearity of the integrative power is a necessary one. Finally, we prove that the class of two-pattern environments d in which the system behaves nicely (i.e., the state is asymptotically selective with respect to d) is defined by a condition of the type $0 < \cos(d^1, d^2) < a$, where d^1 and d^2 are the two patterns in d, and a is a constant strictly less than 1, which actually depends on the nonlinearity of the integrative power (i.e., on its threshold Θ).

For the purpose of our analysis, we consider a single "cortical" neuron whose integrative and evolutive power are, in our notation, the following:

$$c(t) = (m(t) \cdot d(t))^* \tag{B1}$$

with

$$u^* = \begin{cases} u - \Theta & \text{if } u > \Theta \\ 0 & \text{if } u < \Theta \end{cases} \tag{B2}$$

$$m_j(t + 1) = \gamma(t + 1)(m_j(t) + hc(t)d_j(t)) \qquad j = 1, \ldots, N \tag{B3}$$

with h a small positive constant and $\gamma(t + 1)$ such that:

$$\sum_{j=1}^{N} m_j(t + 1) = \sum_{j=1}^{N} m_j(t) = s \tag{B4}$$

The integrative power is thus nonlinear with threshold Θ. The normalizing factor $\gamma(t + 1)$ in the evolution equation B3 keeps the sum of synaptic weights constant and equal to s. All variables are positive.

Our analysis will be carried out on this *reduced version* of von der Malsburg's model: we simply ignore the fixed intracortical connections assumed there, for these are clearly not sufficient to tune the system to a selective state if individual neurons do not display this property already. As is clearly stated by the author himself, the ability to develop selectivity is an intrinsic property of individual neurons, the intracortical connections being there to organize orientation preference in a coherent way in cortex. (This is also the viewpoint in the present work: see "Appendix D.") Notice that this is by no means a contradiction to the fact that, *in the final state*, intracortical connections, particularly the inhibitory ones, significantly contribute to the selectivity of each neuron.

A straightforward calculation shows that equations B3 and B4 are equivalent to the following.

$$\begin{cases} m_j(t + 1) - m_j(t) = K(t)(d_j(t)/d(t) - m_j(t)/s) & j = 1, \ldots, N \\ K(t) = shc(t)d(t)/(s + hc(t)d(t)) \end{cases} \tag{B5}$$

where

$$d(t) = \sum_{j=1}^{N} d_j(t)$$

(In the simulations, $d(t)$ is actually a constant.)

Thus, according to equation B5, the sign of the change of m_j at time t does not depend on the postsynaptic activity $c(t)$ but on the jth fiber activity $d_j(t)$. This is clearly spatial competition as is suggested by the conservation law (equation B4).

We now investigate the behavior of system B5 in a two-pattern environment: $P[d = d^1] = P[d = d^2] = 0.5$. For this purpose, we slightly modify the original setup: there, the dimension is relatively high ($N = 19$), but the firing frequencies in the afferent fibers are discretely valued (i.e., $d_j = 0$ or 1, $j = 1, \ldots, N$). Here, we take $N = 2$, with $d_{1,2}$ allowed to take any value between 0 and 1. By doing so, we still get a broad range of environments ($\cos(d^1, d^2)$ may assume any value between 0 and 1), but the analysis is made considerably easier. To further simplify, we characterize d by a single parameter $0 < \delta < 1$ by writing $d^1 = (1, \delta)$, $d^2 = (\delta, 1)$. Thus, $\cos(d^1, d^2) = 2\delta/(1 + \delta^2)$. We also set $s = 1$.

Under these circumstances, averaging the evolution equation B5 with respect to d leads to the following:

$$E[m_j(t + 1) - m_j(t)] = \tag{B6}$$

$$\tfrac{1}{2}h(2m_j(t) - 1)(\Theta(1 + \delta) - 2\delta), \qquad j = 1, 2$$

To obtain equation B6, it has been assumed that both inputs yield above threshold responses (i.e., $m \cdot d^1$ and $m \cdot d^2 > \Theta$). Higher order terms in h have been ignored.

We see that the behavior of the system is determined by the sign of the quantity $\Theta(1 + \delta) - 2\delta$. Notice that, since $s = 1$, Θ cannot be arbitrarily high: in order that states m exist such that $m \cdot d^1$ and $m \cdot d^2 > \Theta$, one has to assume that $\Theta < (1 + \delta)/2$.

It follows from equation B6 that, for δ such that $\Theta(1 + \delta) - 2\delta < 0$, there is one attractor of selectivity 0, namely (0.5, 0.5). When δ gets smaller and $\Theta(1 + \delta) - 2\delta$ becomes positive, the solution bifurcates into two attractors of maximum selectivity. We thus conclude that:

1. If the neuron's integrative power is linear (i.e., $\Theta = 0$), the asymptotic state is non-selective. (When $\Theta = 0$ and d^1 and d^2 are orthogonal (i.e., $\delta = 0$), the first order term in h vanishes, yet the second order term also leads to the non-selective fixed point.)

2. Given a fixed $0 < \Theta < 1$, the environments d that are acceptable to the system are those which satisfy $\delta < \Theta/(2 - \Theta)$, which is equivalent to a condition of the type $\cos(d^1, d^2) < a$ with a strictly less than 1. (Notice that, in the actual simulations, d consists of nine stimuli that are indeed well separated from one another, since $\min_i \cos(d^1, d^i) = \frac{1}{7}$.)

Appendix C: Correlation between Ocular Dominance and Selectivity in the Monocular Deprived Environment

Consider the MD environment in section V: it is defined by (d_r, n), where d_r is circular and n is a "pure noise" vector. We will prove that the state $(m_r^*, 0)$ is stable in this environment provided that m_r^* is a stable selective state in the environment d_r.

Let (x_r, x_l) be a small perturbation from equilibrium. The motion at point $(m_r^* + x_r, x_l)$ is given by:

$$\dot{x}_r = \phi(m_r^* \cdot d_r + x_r \cdot d_r + x_l \cdot n, \; m_r^* \cdot \bar{d}_r + x_r \cdot \bar{d}_r) \, d_r \quad \text{(C1r)}$$

$$\dot{x}_l = \phi(m_r^* \cdot d_r + x_r \cdot d_r + x_l \cdot n, \; m_r^* \cdot \bar{d}_r + x_r \cdot \bar{d}_r) n \quad \text{(C1l)}$$

where we assume that the noise has 0 mean.

We analyze separately, somewhat informally, the behavior of the two equations. The stability of equation C1r is immediate from the stability of the selective state m_r^* in the circular environment d_r. To analyze equation C1l, we divide the range of the right eye input d_r into three classes:

1. d_r is such that $m_r \cdot d_r$ is either far above threshold, θ_M, and therefore $\phi(m_r \cdot d_r, \; m_r \cdot \bar{d}_r) > 0$, or far below threshold, θ_M (but still positive), and therefore $\phi(m_r \cdot d_r, \; m_r \cdot \bar{d}_r) < 0$. This case might occur before m_r has reached a stable selective state, m_r^*.
2. d_r is such that $m_r^* \cdot d_r$ is near threshold, θ_M, and therefore, $\phi(m_r^* \cdot d_r, \; m_r^* \cdot \bar{d}_r) \simeq 0$.
3. d_r is such that $m_r^* \cdot d_r \simeq 0$, again resulting in $\phi(m_r^* \cdot d_r, \; m_r^* \cdot \bar{d}_r) \simeq 0$.

For the first class of inputs, the sign of ϕ is determined by d_r alone, hence equation C1l is the equation of a random walk. To investigate the behavior of equation C1l in the two other cases, we neglect the term x_r and linearize ϕ around the relevant one of its two zeros. It is easy to see that case 2 yields

$$\dot{x}_l \simeq \epsilon_1 (x_l \cdot n) n \quad \text{(C2)}$$

whereas, in case 3, one obtains

$$\dot{x}_l \simeq -\epsilon_2 (x_l \cdot n) n \quad \text{(C3)}$$

where ϵ_1 and ϵ_2 are *positive* constants, measuring, respectively, the absolute value of the slope of ϕ at the modification threshold and at zero.[8]

Since n is a noise-like term, its distribution is presumably symmetric with respect to x_l so that averaging equations C2 and C3 yields, respectively

$$\dot{x}_l \simeq \overline{\epsilon_1 n_0^2} x_l \quad \text{(C4)}$$

$$\dot{x}_l \simeq -\overline{\epsilon_2 n_0^2} x_l \quad \text{(C5)}$$

where $\overline{n_0^2}$ is the average squared magnitude of the noise input to a single synaptic junction from the closed eye.

We thus see that input vectors from the first class move x_l randomly, inputs from the second class drive it away from 0, whereas inputs from the third drive it toward 0. In the case where the range of d_r is a set of K linearly independent vectors and m_r^* is of maximum selectivity, $(K-1)/K$, case 1 does not occur at all. (The random contribution occurs only before the synaptic strengths from the open eye have settled to one of their fixed points.) Case 2 occurs only for one input, e.g., d_r^1, with $m_r^* \cdot d_r^1$ exactly equal to threshold, θ_M, and case 3 occurs for the other $K-1$ vectors which are all orthogonal to m_r^*. In the general case (d_r any circular environment), the more selective m_r^* with respect to d_r, the higher the proportion of inputs belonging to class 3, the class that yields equation C5 (i.e., that brings x_l back to 0).

The stability of the global system still depends on the ratio of the quantities ϵ_1 and ϵ_2 as well as on the statistics of the noise term n (e.g., its mean square norm). We may, however, formulate two general conclusions. First, under reasonable assumptions (ϵ_1 of the order of ϵ_2 and the mean square norm of n of the same order as that of d_r), $x_l = 0$ is stable on the average for a selective m_r^*. Second, the residual fluctuation of x_l around zero, essentially due to inputs d_r in classes 1 and 2, is smaller for highly selective m_r^* values than it is for mildly selective ones.

Thus, one should expect that, in a monocularly deprived environment, non-selective neurons tend to remain binocularly driven. In addition, since it is the non-preferred inputs from the open eye accompanied by noise from the closed eye (case 3) that drive the response to the closed eye to 0, if inputs to the open eye were restricted to preferred inputs (case 2), even a selective cell would remain less monocular.

Appendix D: Many-neuron Systems

It is very likely that interactions between cortical neurons play an important role in overall cortical function as well, perhaps, as in selectivity of individual cortical cells (Creutzfeldt et al., 1974; Sillito, 1975). The development of selectivity then might be regarded as a many-neuron problem. Since the underlying principles put forward in this work are stated most clearly at the single unit level (where a more complete analysis is also possible), we have chosen this description. However, the methods employed are also applicable to a system of many cortical neurons interacting with one another. Most important, the result that stable equilibria in a stationary environment are selective with respect to their environment can be taken over to the many-neuron system.

Consider such a system in a stationary external environment. The state of each cortical neuron now has two parts: one relative to the geniculocortical synapses, the other to the cortico-cortical ones. The environment of the neuron is no longer stationary, for the states of all other cortical neurons in the system evolve. Yet, when the system reaches global equilibrium, which will occur under reasonable assumptions, each individual environment becomes stationary. The single unit study then allows us to state that, at least in principle (we do not know a priori that each environment is circular), the state of each neuron is selective with respect to its own individual environment.

In practice, formulation of the many-neuron problem poses two questions. First, the integrative power of the system should be specified. Since the system includes cortico-cortical loops, it is not obvious what the response to a given afferent message should be. The two major alternatives are: (a) stationary cortical activity is reached rapidly (i.e., before the afferent message changes) and (b) relevant cortical activity is transitory. The second question concerns the evolution of cortico-cortical synaptic strengths: should these synapses be regarded as modifiable at all, and if yes, how? von der Malsburg (1973) assumes *alternative a* above and proposes fixed connectivity patterns, short range excitatory and longer range inhibitory.

We have performed a simulation of a many-neuron system using the much simpler (and probably more nat-

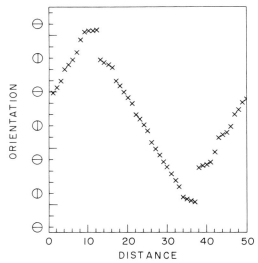

Figure 6. A regular distribution of preferred orientation in a one-dimensional cortex. The system is an array of 50 "cortical" cells arranged in a cyclic way (cell 1 and cell 50 are neighbors) and interconnected according to a fixed short-range-excitation-long-range-inhibition pattern. (Coefficients of interactions are, as a function of increasing intercell distance: 0.4, 0.4, −0.2, −0.4, 0, . . .). The environment d of the system is the usual circular one. Integrative and evolutive powers are described in the text. When the system reaches equilibrium, one has $0.73 \leq \mathrm{Sel}_d(m_i) \leq 0.77$ for all i values between 1 and 50. The diagram shows preferred orientation as a function of cortical coordinate.

ural) *assumption b* above. (*1*) Only monosynaptically and disynaptically mediated components of the afferent message are taken into account for the computation of each cortical neuron's activity before modification is performed and a new stimulus is presented. (*2*) Intracortical connections are fixed and spatially organized as in the work of von der Malsburg (1973). (*3*) The state of each neuron evolves according to equations 6 and 7 of the present work. The results are the following. (*1*) The system's state converges. (*2*) At equilibrium, each neuron stands in a selective state with respect to the environment. (*3*) Preferred orientation—when the environment is a circular one—is a piecewise continuous function of cortical distance (Fig. 6). (*4*) In the final equilibrium state, the intracortical synapses contribute along with the geniculocortical ones to produce the selectivity.

References

Anderson, J. A. (1970) Two models for memory organization using interacting traces. Math. Biosci. *8:* 137–160.

Anderson, J. A. (1972) A simple neural network generating an interactive memory. Math. Biosci. *14:* 197–200.

Bienenstock, E. (1980) A theory of development of neuronal selectivity. Doctoral thesis, Brown University, Providence, RI.

Blakemore, C. (1976) The conditions required for the maintenance of binocularity in the kitten's visual cortex. J. Physiol. (Lond.) *261:* 423–444.

Blakemore, C., and G. F. Cooper (1970) Development of the brain depends on the visual environment. Nature *228:* 477–478.

Blakemore, C., and D. E. Mitchell (1973) Environmental modification of the visual cortex and the neural basis of learning and memory. Nature *241:* 467–468.

Blakemore, C., and R. C. Van Sluyters (1974) Reversal of the physiological effects of monocular deprivation in kittens. Further evidence for a sensitive period. J. Physiol. (Lond.) *237:* 195–216.

Blakemore, C., and R. C. Van Sluyters (1975) Innate and environmental factors in the development of the kitten's visual cortex. J. Physiol. (Lond.) *248:* 663–716.

Brindley, G. S. (1969) Nerve net models of plausible size that perform many simple learning tasks. Proc. R. Soc. Lond. (Biol.) *174:* 173–191.

Buisseret, P., and M. Imbert (1976) Visual cortical cells. Their developmental properties in normal and dark reared kittens. J. Physiol. (Lond.) *255:* 511–525.

Buisseret, P., E. Gary-Bobo, and M. Imbert (1978) Ocular motility and recovery of orientational properties of visual cortical neurones in dark-reared kittens. Nature *272:* 816–817.

Changeux, J. P., and A. Danchin (1976) Selective stabilization of developing synapses as a mechanism for the specification of neuronal networks. Nature *264:* 705–712.

Changeux, J. P., P. Courrège, and A. Danchin (1973) A theory of the epigenesis of neuronal networks by selective stabilization of synapses. Proc. Natl. Acad. Sci. U. S. A. *70:* 2974–2978.

Cooper, L. N. (1973) A possible organization of animal memory and learning. In *Proceedings of the Nobel Symposium on Collective Properties of Physical Systems*, B. Lindquist and S. Lindquist, eds., Vol., 24, pp. 252–264, Academic Press, New York.

Cooper, L. N., F. Lieberman, and E. Oja (1979) A theory for the acquisition and loss of neuron specificity in visual cortex. Biol. Cybern. *33:* 9–28.

Creutzfeldt, O. D., U. Kuhnt, and L. A. Benevento (1974) An intracellular analysis of visual cortical neurones to moving stimuli: Responses in a cooperative neuronal network. Exp. Brain Res. *21:* 251–274.

Cynader, M., and D. E. Mitchell (1980) Prolonged sensitivity to monocular deprivation in dark-reared cats. J. Neurophysiol. *43:* 1026–1040.

Frégnac, Y. (1979) Development of orientation selectivity in the primary visual cortex of normally and dark reared kittens. Biol. Cybern. *34:* 187–204.

Frégnac, Y., and E. Bienenstock (1981) Specific functional modifications of individual cortical neurons, triggered by vision and passive eye movement in immobilized kittens. In *Pathophysiology of the Visual System: Documenta Ophthalmologica*, L. Maffei, ed., Vol. 30, pp. 100–108, Dr. W. Junk, The Hague.

Frégnac, Y., and M. Imbert (1978) Early development of visual cortical cells in normal and dark-reared kittens. Relationship between orientation selectivity and ocular dominance. J. Physiol. (Lond.) *278:* 27–44.

Hebb, D. O. (1949) *Organization of Behavior*, John Wiley and Sons, New York.

Hirsch, H. V. B., and D. N. Spinelli (1970) Visual experience modifies distribution of horizontally and vertically oriented receptive fields in cats. Science *168:* 869–871.

Hirsch, H. V. B., and D. N. Spinelli (1971) Modification of the distribution of receptive field orientation in cats by selective visual exposure during development. Exp. Brain Res. *13:* 509–527.

Hubel, D. H., and T. N. Wiesel (1959) Receptive fields of single neurons in the cat striate cortex. J. Physiol. (Lond.) *148:* 574–591.

Hubel, D. H., and T. N. Wiesel (1962) Receptive fields, binocular interaction and functional architecture in the cat's visual cortex. J. Physiol. (Lond.) *160:* 106–154.

Hubel, D. H., and T. N. Wiesel (1963) Receptive fields of cells in striate cortex of very young, visually inexperienced kittens. J. Neurophysiol. *26:* 994–1002.

Hubel, D. H., and T. N. Wiesel (1965) Binocular interaction in striate cortex of kittens with artificial squint. J. Neurophysiol. *28:* 1041–1059.

Imbert, M., and P. Buisseret (1975) Receptive field characteristics and plastic properties of visual cortical cells in kittens reared with or without visual experience. Exp. Brain Res. *22:* 2–36.

Kasamatsu, T., and J. D. Pettigrew (1976) Depletion of brain catecholamines: Failure of ocular dominance shift after monocular occlusion in kittens. Science *194:* 206–209.

Kasamatsu, T., and J. D. Pettigrew (1979) Preservation of binocularity after monocular deprivation in the striate cortex of kittens treated with 6-hydroxydopamine. J. Comp. Neurol. *185:* 139–181.

Kohonen, T. (1977) *Associative Memory: A System Theoretical Approach,* Springer, Berlin.

Kratz, K. E., and P. D. Spear (1976) Effects of visual deprivation and alterations in binocular competition on responses of striate cortex neurons in the cat. J. Comp. Neurol. *170:* 141–152.

Leventhal, A. G., and H. V. B. Hirsch (1977) Effects of early experience upon orientation selectivity and binocularity of neurons in visual cortex of cats. Proc. Natl. Acad. Sci. U. S. A. *74:* 1272–1276.

Leventhal, A. G., and H. V. B. Hirsch (1980) Receptive field properties of different classes of neurons in visual cortex of normal and dark-reared cats. J. Neurophysiol. *43:* 1111–1132.

Marr, D. (1969) A theory of cerebellar cortex. J. Physiol. (Lond.) *202:* 437–470.

Movshon, J. A. (1976) Reversal of the physiological effects of monocular deprivation in the kitten's visual cortex. J. Physiol. (Lond.) *261:* 125–174.

Nass, M. M., and L. N. Cooper (1975) A theory for the development of feature detecting cells in visual cortex. Biol. Cybern. *19:* 1–18.

Perez, R., L. Glass, and R. J. Shlaer (1975) Development of specificity in the cat visual cortex. J. Math. Biol. *1:* 275–288.

Pettigrew, J. D. (1978) The paradox of the critical period for striate cortex. In *Neuronal Plasticity,* C. W. Cotman, ed., pp. 311–330, Raven Press, New York.

Rauschecker, J. P., and W. Singer (1981) The effects of early visual experience on the cat's visual cortex and their possible explanation by Hebb synapses. J. Physiol. (Lond.) *310:* 215–240.

Sejnowski, T. J. (1977a) Storing covariance with nonlinearly interacting neurons. J. Math. Biol. *4:* 303–321.

Sejnowski, T. J. (1977b) Statistical constraints on synaptic plasticity. J. Theor. Biol. *69:* 385–389.

Sillito, A. M. (1975) The contribution of inhibitory mechanisms to the receptive field properties of neurons in the striate cortex of the cat. J. Physiol. (Lond.) *250:* 305–329.

Singer, W. (1979) Central-core control of visual functions. In *Neuroscience Fourth Study Program,* F. Schmitt and F. Worden, eds., pp. 1093–1109, MIT Press, Cambridge.

Singer, W. (1980) Central gating of developmental plasticity in the cat striate cortex. Verh. Dtsch. Zool. Ges. 268–274.

Stent, G. S. (1973) A physiological mechanism for Hebb's postulate of learning. Proc. Natl. Acad. Sci. U. S. A. *70:* 997–1001.

Stryker, M. P., H. Sherk, A. G. Leventhal, and H. V. Hirsch (1978) Physiological consequences for the cat's visual cortex of effectively restricting early visual experience with oriented contours. J. Neurophysiol. *41:* 896–909.

Trotter, Y., Y. Frégnac, P. Buisseret (1981) Gating control of developmental plasticity by extraocular proprioception in kitten area 17. In *Fourth European Conference on Visual Perception, Gouzieux, France.*

Uttley, A. M. (1976) A two pathway theory of conditioning and adaptive pattern recognition. Brain Res. *102:* 23–35.

von der Malsburg, C. (1973) Self-organization of orientation sensitive cells in the striate cortex. Kybernetik *14:* 85–100.

Watkins, D. W., J. R. Wilson, and S. M. Sherman (1978) Receptive field properties of neurons in binocular and monocular segments of striate cortex in cats raised with binocular lid suture. J. Neurophysiol. *41:* 322–337.

Wiesel, T. N., and D. H. Hubel (1963) Single-cell responses in striate cortex of kittens deprived of vision in one eye. J. Neurophysiol. *26:* 1003–1017.

Wiesel, T. N., and D. H. Hubel (1965) Extent of recovery from the effects of visual deprivation in kittens. J. Neurophysiol. *28:* 1060–1072.

Introduction

(1982)
J. J. Hopfield

Neural networks and physical systems with emergent collective computational abilities
Proceedings of the National Academy of Sciences 79:2554–2558

As far as public visibility goes, the modern era in neural networks dates from the publication of this paper by John Hopfield. This paper is short, clearly written, and brings together a number of strands that up to this point were somewhat separated in the neural network literature. Much of the reason for its impact was simply this: it presented a sophisticated, coherent theoretical picture of how a neural network could work, and what it could do.

We cannot avoid making a comment about the sociology of science, it is hoped without causing offense. John Hopfield is a distinguished physicist. When he talks, people listen. Theory in his hands becomes respectable. Neural networks became instantly legitimate, whereas before, most developments in networks had been the province of somewhat suspect psychologists and neurobiologists, or by those removed from the hot centers of scientific activity.

Other well known physicists represented in this volume had also done distinguished work in neural networks: Leon Cooper, Gordon Shaw, and William Little. Although their work was noted and respected, it was subliminal, as far as the scientific world at large was concerned. The models they proposed were brain models first and useful devices a distant second. Practical implications, though clearly present, were not emphasized.

The one thing that really seems to have made the Hopfield work take fire in terms of public notice was the immediate and strong contact he made with the new chip building technology that was finally capable of constructing the devices he was proposing. The first attempts to make chips followed within a couple of years of this 1982 paper, and by early 1987, AT&T Bell Laboratories had announced successful development of neural net chips, largely based on the Hopfield networks (*Electronics*, March 5, 1987, p. 21) and Carver Mead and coworkers (paper 43) were making artificial sensory systems using VLSI technology, inspired by initial contact with Hopfield. The Caltech environment and the potential usefulness of the neural networks Hopfield discussed made the engineering connection immediate.

The criticism has been made by some old-timers in the neural network field that there was nothing fundamentally new in the model proposed by Hopfield. We have collected in this volume a number of earlier papers, and can let readers draw their own conclusions on this point. Although many of the ideas in this paper have precursors, as Hopfield would be the first to admit (see his list of references!), bringing them all together, with detailed, clear, and powerful mathematical analysis, is creative work of the first order, and the paper richly merits the attention and respect it has received.

There are a number of technical points that are worth mentioning. The order of presentation of ideas that Hopfield uses is the opposite of that used by most network modelers. The standard approach to a neural network is to propose a learning rule, usually based on synaptic modification, and then to show that a number of interesting effects arise from it. Hopfield starts by saying that the function of the nervous system is to develop a number of locally stable points in state space. Other points in state space flow into the stable points (called attractors). This allows a mechanism for correcting errors, since deviations from the stable points disappear. It can also reconstruct missing information since the stable point will appropriately complete missing parts of an incomplete initial state vector.

Hopfield then proceeds to develop a network that shows this desired behavior. He assumes that the basic elements of the network are threshold logic units, which sum synaptic inputs, compare the sum with a threshold, and then respond 1 if the sum is at or above threshold and 0 otherwise. In a later paper (paper 35) Hopfield discusses networks of neurons that can show graded intermediate states. The network is recurrent, in that the neurons connect to each other, with the exception that a neuron does not connect to itself; that is, the connection matrix has zeros down the main diagonal.

Hopfield assumes that the system wants to learn a set of states, $\{V^s\}$, with individual element activities V_i. He suggests a learning rule for constructing elements of the connectivity matrix, which is the Hebb rule, combined with scaling terms, for placing the point for zero connection modification at an activity of one-half. This ensures symmetry of modification magnitude for the two allowable output states of a cell, 0 and 1.

In one paragraph Hopfield suggests one of the most important new techniques to have been proposed in neural networks. He considers the special case of a symmetric matrix, i.e., ones where $\mathbf{T}_{ij} = \mathbf{T}_{ji}$. Then he defines a quantity, called E, which is the sum of all the terms:

$$E = -\tfrac{1}{2} \sum_{i \neq j} \sum \mathbf{T}_{ij} V_i V_j.$$

This term is equivalent to physical *energy*. As the system evolves, due to the feedback dynamics, the energy decreases until it reaches a (perhaps local) minimum. Hopfield next makes the portentous comment, "This case is isomorphic with an Ising model," thereby allowing a deluge of physical theory (and physicists) to enter network modeling. This flood of new participants has transformed the field of neural networks.

The dynamics of evolution of the system state follows a simple rule and is asynchronous. An element, chosen at random, looks at its inputs, and changes state, depending on whether or not the sum of its input is above or below threshold. It can be seen from the form of the energy term that a state change leads either to a decrease in energy or to the energy remaining the same. The updating rule is, therefore, an energy minimizing rule. Modifications of element activities continue until a stable state is reached, that is, a energy minimum is reached.

A number of computer simulations and some analysis led Hopfield to conclude that the number of 'memories' that could be stored accurately by a network was about 15% of the dimensionality. This number agrees well with experience of others. It has also led to a number of attempts to increase storage capacity by various techniques, as well as some more accurate definitions of and computations with storage capacity. However, *most* reasonable models, with *most* reasonable definitions of capacity, end up having a capacity of 10–20% of the number of elements. A number of estimates of capacity can be found in a volume of papers growing out of the 1986 Neural Networks for Computation Conference (Denker, 1986).

Reference

J. S. Denker (Ed.) (1986), *Neural Networks for Computation*, AIP Conference Proceedings 151, New York: American Institute of Physics.

(1982)
J. J. Hopfield

Neural networks and physical systems with emergent collective computational abilities

Proceedings of the National Academy of Sciences 79:2554–2558

ABSTRACT   Computational properties of use to biological organisms or to the construction of computers can emerge as collective properties of systems having a large number of simple equivalent components (or neurons). The physical meaning of content-addressable memory is described by an appropriate phase space flow of the state of a system. A model of such a system is given, based on aspects of neurobiology but readily adapted to integrated circuits. The collective properties of this model produce a content-addressable memory which correctly yields an entire memory from any subpart of sufficient size. The algorithm for the time evolution of the state of the system is based on asynchronous parallel processing. Additional emergent collective properties include some capacity for generalization, familiarity recognition, categorization, error correction, and time sequence retention. The collective properties are only weakly sensitive to details of the modeling or the failure of individual devices.

Given the dynamical electrochemical properties of neurons and their interconnections (synapses), we readily understand schemes that use a few neurons to obtain elementary useful biological behavior (1–3). Our understanding of such simple circuits in electronics allows us to plan larger and more complex circuits which are essential to large computers. Because evolution has no such plan, it becomes relevant to ask whether the ability of large collections of neurons to perform "computational" tasks may in part be a spontaneous collective consequence of having a large number of interacting simple neurons.

In physical systems made from a large number of simple elements, interactions among large numbers of elementary components yield collective phenomena such as the stable magnetic orientations and domains in a magnetic system or the vortex patterns in fluid flow. Do analogous collective phenomena in a system of simple interacting neurons have useful "computational" correlates? For example, are the stability of memories, the construction of categories of generalization, or time-sequential memory also emergent properties and collective in origin? This paper examines a new modeling of this old and fundamental question (4–8) and shows that important computational properties spontaneously arise.

All modeling is based on details, and the details of neuroanatomy and neural function are both myriad and incompletely known (9). In many physical systems, the nature of the emergent collective properties is insensitive to the details inserted in the model (e.g., collisions are essential to generate sound waves, but any reasonable interatomic force law will yield appropriate collisions). In the same spirit, I will seek collective properties that are robust against change in the model details.

The model could be readily implemented by integrated circuit hardware. The conclusions suggest the design of a delo-calized content-addressable memory or categorizer using extensive asynchronous parallel processing.

The general content-addressable memory of a physical system

Suppose that an item stored in memory is "H. A. Kramers & G. H. Wannier *Phys. Rev.* **60**, 252 (1941)." A general content-addressable memory would be capable of retrieving this entire memory item on the basis of sufficient partial information. The input "& Wannier, (1941)" might suffice. An ideal memory could deal with errors and retrieve this reference even from the input "Vannier, (1941)". In computers, only relatively simple forms of content-addressable memory have been made in hardware (10, 11). Sophisticated ideas like error correction in accessing information are usually introduced as software (10).

There are classes of physical systems whose spontaneous behavior can be used as a form of general (and error-correcting) content-addressable memory. Consider the time evolution of a physical system that can be described by a set of general coordinates. A point in state space then represents the instantaneous condition of the system. This state space may be either continuous or discrete (as in the case of N Ising spins).

The equations of motion of the system describe a flow in state space. Various classes of flow patterns are possible, but the systems of use for memory particularly include those that flow toward locally stable points from anywhere within regions around those points. A particle with frictional damping moving in a potential well with two minima exemplifies such a dynamics.

If the flow is not completely deterministic, the description is more complicated. In the two-well problems above, if the frictional force is characterized by a temperature, it must also produce a random driving force. The limit points become small limiting regions, and the stability becomes not absolute. But as long as the stochastic effects are small, the essence of local stable points remains.

Consider a physical system described by many coordinates $X_1 \cdots X_N$, the components of a state vector X. Let the system have locally stable limit points X_a, X_b, $\cdots$. Then, if the system is started sufficiently near any X_a, as at $X = X_a + \Delta$, it will proceed in time until $X \approx X_a$. We can regard the information stored in the system as the vectors X_a, X_b, $\cdots$. The starting point $X = X_a + \Delta$ represents a partial knowledge of the item X_a, and the system then generates the total information X_a.

Any physical system whose dynamics in phase space is dominated by a substantial number of locally stable states to which it is attracted can therefore be regarded as a general content-addressable memory. The physical system will be a potentially useful memory if, in addition, any prescribed set of states can readily be made the stable states of the system.

The model system

The processing devices will be called neurons. Each neuron i has two states like those of McCullough and Pitts (12): $V_i = 0$

The publication costs of this article were defrayed in part by page charge payment. This article must therefore be hereby marked "*advertisement*" in accordance with 18 U. S. C. §1734 solely to indicate this fact.

("not firing") and $V_i = 1$ ("firing at maximum rate"). When neuron i has a connection made to it from neuron j, the strength of connection is defined as T_{ij}. (Nonconnected neurons have $T_{ij} \equiv 0$.) The instantaneous state of the system is specified by listing the N values of V_i, so it is represented by a binary word of N bits.

The state changes in time according to the following algorithm. For each neuron i there is a fixed threshold U_i. Each neuron i readjusts its state randomly in time but with a mean attempt rate W, setting

$$\begin{array}{ll} V_i \to 1 & \\ V_i \to 0 \end{array} \quad \text{if} \quad \sum_{j \neq i} T_{ij} V_j \begin{array}{l} > U_i \\ < U_i \end{array} \qquad [1]$$

Thus, each neuron randomly and asynchronously evaluates whether it is above or below threshold and readjusts accordingly. (Unless otherwise stated, we choose $U_i = 0$.)

Although this model has superficial similarities to the Perceptron (13, 14) the essential differences are responsible for the new results. First, Perceptrons were modeled chiefly with neural connections in a "forward" direction $A \to B \to C \to D$. The analysis of networks with strong backward coupling $A \rightleftarrows B \rightleftarrows C$ proved intractable. All our interesting results arise as consequences of the strong back-coupling. Second, Perceptron studies usually made a random net of neurons deal directly with a real physical world and did not ask the questions essential to finding the more abstract emergent computational properties. Finally, Perceptron modeling required synchronous neurons like a conventional digital computer. There is no evidence for such global synchrony and, given the delays of nerve signal propagation, there would be no way to use global synchrony effectively. Chiefly computational properties which can exist in spite of asynchrony have interesting implications in biology.

The information storage algorithm

Suppose we wish to store the set of states V^s, $s = 1 \cdots n$. We use the storage prescription (15, 16)

$$T_{ij} = \sum_s (2V_i^s - 1)(2V_j^s - 1) \qquad [2]$$

but with $T_{ii} = 0$. From this definition

$$\sum_j T_{ij} V_j^{s'} = \sum_s (2V_i^s - 1) \left[\sum_j V_j^{s'} (2V_j^s - 1) \right] \equiv H_i^{s'} . \qquad [3]$$

The mean value of the bracketed term in Eq. 3 is 0 unless $s = s'$, for which the mean is $N/2$. This pseudoorthogonality yields

$$\sum_j T_{ij} V_j^{s'} \equiv \langle H_i^{s'} \rangle \approx (2V_i^{s'} - 1) N/2 \qquad [4]$$

and is positive if $V_i^{s'} = 1$ and negative if $V_i^{s'} = 0$. Except for the noise coming from the $s \neq s'$ terms, the stored state would always be stable under our processing algorithm.

Such matrices T_{ij} have been used in theories of linear associative nets (15–19) to produce an output pattern from a paired input stimulus, $S_1 \to O_1$. A second association $S_2 \to O_2$ can be simultaneously stored in the same network. But the confusing simulus $0.6\,S_1 + 0.4\,S_2$ will produce a generally meaningless mixed output $0.6\,O_1 + 0.4\,O_2$. Our model, in contrast, will use its strong nonlinearity to make choices, produce categories, and regenerate information and, with high probability, will generate the output O_1 from such a confusing mixed stimulus.

A linear associative net must be connected in a complex way with an external nonlinear logic processor in order to yield true computation (20, 21). Complex circuitry is easy to plan but more difficult to discuss in evolutionary terms. In contrast, our model obtains its emergent computational properties from simple properties of many cells rather than circuitry.

The biological interpretation of the model

Most neurons are capable of generating a train of action potentials—propagating pulses of electrochemical activity—when the average potential across their membrane is held well above its normal resting value. The mean rate at which action potentials are generated is a smooth function of the mean membrane potential, having the general form shown in Fig. 1.

The biological information sent to other neurons often lies in a short-time average of the firing rate (22). When this is so, one can neglect the details of individual action potentials and regard Fig. 1 as a smooth input–output relationship. [Parallel pathways carrying the same information would enhance the ability of the system to extract a short-term average firing rate (23, 24).]

A study of emergent collective effects and spontaneous computation must necessarily focus on the nonlinearity of the input–output relationship. The essence of computation is nonlinear logical operations. The particle interactions that produce true collective effects in particle dynamics come from a nonlinear dependence of forces on positions of the particles. Whereas linear associative networks have emphasized the linear central region (14–19) of Fig. 1, we will replace the input–output relationship by the dot-dash step. Those neurons whose operation is dominantly linear merely provide a pathway of communication between nonlinear neurons. Thus, we consider a network of "on or off" neurons, granting that some of the interconnections may be by way of neurons operating in the linear regime.

Delays in synaptic transmission (of partially stochastic character) and in the transmission of impulses along axons and dendrites produce a delay between the input of a neuron and the generation of an effective output. All such delays have been modeled by a single parameter, the stochastic mean processing time $1/W$.

The input to a particular neuron arises from the current leaks of the synapses to that neuron, which influence the cell mean potential. The synapses are activated by arriving action potentials. The input signal to a cell i can be taken to be

$$\sum_j T_{ij} V_j \qquad [5]$$

where T_{ij} represents the effectiveness of a synapse. Fig. 1 thus

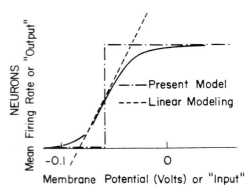

FIG. 1. Firing rate versus membrane voltage for a typical neuron (solid line), dropping to 0 for large negative potentials and saturating for positive potentials. The broken lines show approximations used in modeling.

becomes an input–output relationship for a neuron.

Little, Shaw, and Roney (8, 25, 26) have developed ideas on the collective functioning of neural nets based on "on/off" neurons and synchronous processing. However, in their model the relative timing of action potential spikes was central and resulted in reverberating action potential trains. Our model and theirs have limited formal similarity, although there may be connections at a deeper level.

Most modeling of neural learning networks has been based on synapses of a general type described by Hebb (27) and Eccles (28). The essential ingredient is the modification of T_{ij} by correlations like

$$\Delta T_{ij} = [V_i(t)V_j(t)]_{average} \qquad [6]$$

where the average is some appropriate calculation over past history. Decay in time and effects of $[V_i(t)]_{avg}$ or $[V_j(t)]_{avg}$ are also allowed. Model networks with such synapses (16, 20, 21) can construct the associative T_{ij} of Eq. 2. We will therefore initially assume that such a T_{ij} has been produced by previous experience (or inheritance). The Hebbian property need not reside in single synapses; small groups of cells which produce such a net effect would suffice.

The network of cells we describe performs an abstract calculation and, for applications, the inputs should be appropriately coded. In visual processing, for example, feature extraction should previously have been done. The present modeling might then be related to how an entity or *Gestalt* is remembered or categorized on the basis of inputs representing a collection of its features.

Studies of the collective behaviors of the model

The model has stable limit points. Consider the special case $T_{ij} = T_{ji}$, and define

$$E = -\frac{1}{2}\sum_{i \neq j}\sum T_{ij}V_iV_j \ . \qquad [7]$$

ΔE due to ΔV_i is given by

$$\Delta E = -\Delta V_i \sum_{j \neq i} T_{ij}V_j \ . \qquad [8]$$

Thus, the algorithm for altering V_i causes E to be a monotonically decreasing function. State changes will continue until a least (local) E is reached. This case is isomorphic with an Ising model. T_{ij} provides the role of the exchange coupling, and there is also an external local field at each site. When T_{ij} is symmetric but has a random character (the spin glass) there are known to be many (locally) stable states (29).

Monte Carlo calculations were made on systems of $N = 30$ and $N = 100$, to examine the effect of removing the $T_{ij} = T_{ji}$ restriction. Each element of T_{ij} was chosen as a random number between -1 and 1. The neural architecture of typical cortical regions (30, 31) and also of simple ganglia of invertebrates (32) suggests the importance of 100–10,000 cells with intense mutual interconnections in elementary processing, so our scale of N is slightly small.

The dynamics algorithm was initiated from randomly chosen initial starting configurations. For $N = 30$ the system never displayed an ergodic wandering through state space. Within a time of about $4/W$ it settled into limiting behaviors, the commonest being a stable state. When 50 trials were examined for a particular such random matrix, all would result in one of two or three end states. A few stable states thus collect the flow from most of the initial state space. A simple cycle also occurred occasionally—for example, $\cdots A \rightarrow B \rightarrow A \rightarrow B \cdots$.

The third behavior seen was chaotic wandering in a small region of state space. The Hamming distance between two binary states A and B is defined as the number of places in which the digits are different. The chaotic wandering occurred within a short Hamming distance of one particular state. Statistics were done on the probability p_i of the occurrence of a state in a time of wandering around this minimum, and an entropic measure of the available states M was taken

$$\ln M = -\sum p_i \ln p_i \ . \qquad [9]$$

A value of $M = 25$ was found for $N = 30$. *The flow in phase space produced by this model algorithm has the properties necessary for a physical content-addressable memory* whether or not T_{ij} is symmetric.

Simulations with $N = 100$ were much slower and not quantitatively pursued. They showed qualitative similarity to $N = 30$.

Why should stable limit points or regions persist when $T_{ij} \neq T_{ji}$? If the algorithm at some time changes V_i from 0 to 1 or vice versa, the change of the energy defined in Eq. 7 can be split into two terms, one of which is always negative. The second is identical if T_{ij} is symmetric and is "stochastic" with mean 0 if T_{ij} and T_{ji} are randomly chosen. The algorithm for $T_{ij} \neq T_{ji}$ therefore changes E in a fashion similar to the way E would change in time for a symmetric T_{ij} but with an algorithm corresponding to a finite temperature.

About $0.15 N$ states can be simultaneously remembered before error in recall is severe. Computer modeling of memory storage according to Eq. 2 was carried out for $N = 30$ and $N = 100$. n random memory states were chosen and the corresponding T_{ij} was generated. If a nervous system preprocessed signals for efficient storage, the preprocessed information would appear random (e.g., the coding sequences of DNA have a random character). The random memory vectors thus simulate efficiently encoded real information, as well as representing our ignorance. The system was started at each assigned nominal memory state, and the state was allowed to evolve until stationary.

Typical results are shown in Fig. 2. The statistics are averages over both the states in a given matrix and different matrices. With $n = 5$, the assigned memory states are almost always stable (and exactly recallable). For $n = 15$, about half of the nominally remembered states evolved to stable states with less than 5 errors, but the rest evolved to states quite different from the starting points.

These results can be understood from an analysis of the effect of the noise terms. In Eq. 3, H_i^s is the "effective field" on neuron i when the state of the system is s', one of the nominal memory states. The expectation value of this sum, Eq. 4, is $\pm N/2$ as appropriate. The $s \neq s'$ summation in Eq. 2 contributes no mean, but has a rms noise of $[(n - 1)N/2]^{1/2} \equiv \sigma$. For nN large, this noise is approximately Gaussian and the probability of an error in a single particular bit of a particular memory will be

$$P = \frac{1}{\sqrt{2\pi\sigma^2}} \int_{N/2}^{x} e^{-x^2/2\sigma^2} \, dx \ . \qquad [10]$$

For the case $n = 10$, $N = 100$, $P = 0.0091$, the probability that a state had no errors in its 100 bits should be about $e^{-0.91} \approx 0.40$. In the simulation of Fig. 2, the experimental number was 0.6.

The theoretical scaling of n with N at fixed P was demonstrated in the simulations going between $N = 30$ and $N = 100$. The experimental results of half the memories being well retained at $n = 0.15 N$ and the rest badly retained is expected to

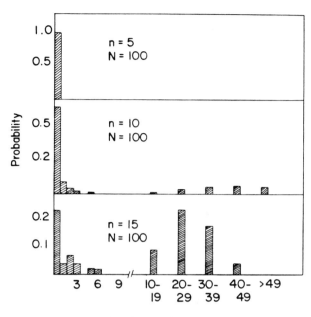

FIG. 2. The probability distribution of the occurrence of errors in the location of the stable states obtained from nominally assigned memories.

be true for all large N. The information storage at a given level of accuracy can be increased by a factor of 2 by a judicious choice of individual neuron thresholds. This choice is equivalent to using variables $\mu_i = \pm 1$, $T_{ij} = \Sigma_s \mu_i^s \mu_j^s$, and a threshold level of 0.

Given some arbitrary starting state, what is the resulting final state (or statistically, states)? To study this, evolutions from randomly chosen initial states were tabulated for $N = 30$ and $n = 5$. From the (inessential) symmetry of the algorithm, if $(101110\cdots)$ is an assigned stable state, $(010001\cdots)$ is also stable. Therefore, the matrices had 10 nominal stable states. Approximately 85% of the trials ended in assigned memories, and 10% ended in stable states of no obvious meaning. An ambiguous 5% landed in stable states very near assigned memories. There was a range of a factor of 20 of the likelihood of finding these 10 states.

The algorithm leads to memories near the starting state. For $N = 30$, $n = 5$, partially random starting states were generated by random modification of known memories. The probability that the final state was that closest to the initial state was studied as a function of the distance between the initial state and the nearest memory state. For distance ≤ 5, the nearest state was reached more than 90% of the time. Beyond that distance, the probability fell off smoothly, dropping to a level of 0.2 (2 times random chance) for a distance of 12.

The phase space flow is apparently dominated by attractors which are the nominally assigned memories, each of which dominates a substantial region around it. The flow is not entirely deterministic, and *the system responds to an ambiguous starting state by a statistical choice* between the memory states it most resembles.

Were it desired to use such a system in an Si-based content-addressable memory, the algorithm should be used and modified to hold the known bits of information while letting the others adjust.

The model was studied by using a "clipped" T_{ij}, replacing T_{ij} in Eq. 3 by ± 1, the algebraic sign of T_{ij}. The purposes were to examine the necessity of a linear synapse supposition (by making a highly nonlinear one) and to examine the efficiency of storage. Only $N(N/2)$ bits of information can possibly be stored in this symmetric matrix. Experimentally, for $N = 100$, $n = 9$, the level of errors was similar to that for the ordinary algorithm at $n = 12$. The signal-to-noise ratio can be evaluated analytically for this clipped algorithm and is reduced by a factor of $(2/\pi)^{1/2}$ compared with the unclipped case. For a fixed error probability, the number of memories must be reduced by $2/\pi$.

With the μ algorithm and the clipped T_{ij}, both analysis and modeling showed that the maximal information stored for $N = 100$ occurred at about $n = 13$. Some errors were present, and the Shannon information stored corresponded to about $N(N/8)$ bits.

New memories can be continually added to T_{ij}. The addition of new memories beyond the capacity overloads the system and makes all memory states irretrievable unless there is a provision for forgetting old memories (16, 27, 28).

The saturation of the possible size of T_{ij} will itself cause forgetting. Let the possible values of T_{ij} be 0, ± 1, ± 2, ± 3, and T_{ij} be freely incremented within this range. If $T_{ij} = 3$, a next increment of $+1$ would be ignored and a next increment of -1 would reduce T_{ij} to 2. When T_{ij} is so constructed, only the recent memory states are retained, with a slightly increased noise level. Memories from the distant past are no longer stable. How far into the past are states remembered depends on the digitizing depth of T_{ij}, and 0, $\cdots$, ± 3 is an appropriate level for $N = 100$. Other schemes can be used to keep too many memories from being simultaneously written, but this particular one is attractive because it requires no delicate balances and is a consequence of natural hardware.

Real neurons need not make synapses both of $i \rightarrow j$ and $j \rightarrow i$. Particular synapses are restricted to one sign of output. We therefore asked whether $T_{ij} = T_{ji}$ is important. Simulations were carried out with only one ij connection: if $T_{ij} \neq 0$, $T_{ji} = 0$. The probability of making errors increased, but the algorithm continued to generate stable minima. A Gaussian noise description of the error rate shows that the signal-to-noise ratio for given n and N should be decreased by the factor $1/\sqrt{2}$, and the simulations were consistent with such a factor. This same analysis shows that the system generally fails in a "soft" fashion, with signal-to-noise ratio and error rate increasing slowly as more synapses fail.

Memories too close to each other are confused and tend to merge. For $N = 100$, a pair of random memories should be separated by 50 $\pm$ 5 Hamming units. The case $N = 100$, $n = 8$, was studied with seven random memories and the eighth made up a Hamming distance of only 30, 20, or 10 from one of the other seven memories. At a distance of 30, both similar memories were usually stable. At a distance of 20, the minima were usually distinct but displaced. At a distance of 10, the minima were often fused.

The algorithm categorizes initial states according to the similarity to memory states. With a threshold of 0, the system behaves as a forced categorizer.

The state $00000\cdots$ is always stable. For a threshold of 0, this stable state is much higher in energy than the stored memory states and very seldom occurs. Adding a uniform threshold in the algorithm is equivalent to raising the effective energy of the stored memories compared to the 0000 state, and 0000 also becomes a likely stable state. The 0000 state is then generated by any initial state that does not resemble adequately closely one of the assigned memories and represents positive recognition that the starting state is not familiar.

Familiarity can be recognized by other means when the memory is drastically overloaded. We examined the case $N = 100$, $n = 500$, in which there is a memory overload of a factor of 25. None of the memory states assigned were stable. The initial rate of processing of a starting state is defined as the number of neuron state readjustments that occur in a time $1/2W$. Familiar and unfamiliar states were distinguishable most of the time at this level of overload on the basis of the initial processing rate, which was faster for unfamiliar states. This kind of familiarity can only be read out of the system by a class of neurons or devices abstracting average properties of the processing group.

For the cases so far considered, the expectation value of T_{ij} was 0 for $i \neq j$. A set of memories can be stored with average correlations, and $\bar{T}_{ij} = C_{ij} \neq 0$ because there is a consistent internal correlation in the memories. If now a partial new state X is stored

$$\Delta T_{ij} = (2X_i - 1)(2X_j - 1) \quad i,j \leq k < N \quad [11]$$

using only k of the neurons rather than N, an attempt to reconstruct it will generate a stable point for all N neurons. The values of $X_{k+1} \cdots X_N$ that result will be determined primarily from the sign of

$$\sum_{j=1}^{k} c_{ij} x_j \quad [12]$$

and X is completed according to the mean correlations of the other memories. The most effective implementation of this capacity stores a large number of correlated matrices weakly followed by a normal storage of X.

A nonsymmetric T_{ij} can lead to the possibility that a minimum will be only metastable and will be replaced in time by another minimum. Additional nonsymmetric terms which could be easily generated by a minor modification of Hebb synapses

$$\Delta T_{ij} = A \sum_s (2V_i^{s+1} - 1)(2V_j^s - 1) \quad [13]$$

were added to T_{ij}. When A was judiciously adjusted, the system would spend a while near V_s and then leave and go to a point near V_{s+1}. But sequences longer than four states proved impossible to generate, and even these were not faithfully followed.

Discussion

In the model network each "neuron" has elementary properties, and the network has little structure. Nonetheless, collective computational properties spontaneously arose. Memories are retained as stable entities or *Gestalts* and can be correctly recalled from any reasonably sized subpart. Ambiguities are resolved on a statistical basis. Some capacity for generalization is present, and time ordering of memories can also be encoded. These properties follow from the nature of the flow in phase space produced by the processing algorithm, which does not appear to be strongly dependent on precise details of the modeling. This robustness suggests that similar effects will obtain even when more neurobiological details are added.

Much of the architecture of regions of the brains of higher animals must be made from a proliferation of simple local circuits with well-defined functions. The bridge between simple circuits and the complex computational properties of higher nervous systems may be the spontaneous emergence of new computational capabilities from the collective behavior of large numbers of simple processing elements.

Implementation of a similar model by using integrated circuits would lead to chips which are much less sensitive to element failure and soft-failure than are normal circuits. Such chips would be wasteful of gates but could be made many times larger than standard designs at a given yield. Their asynchronous parallel processing capability would provide rapid solutions to some special classes of computational problems.

The work at California Institute of Technology was supported in part by National Science Foundation Grant DMR-8107494. This is contribution no. 6580 from the Division of Chemistry and Chemical Engineering.

1. Willows, A. O. D., Dorsett, D. A. & Hoyle, G. (1973) *J. Neurobiol.* **4**, 207–237, 255–285.
2. Kristan, W. B. (1980) in *Information Processing in the Nervous System*, eds. Pinsker, H. M. & Willis, W. D. (Raven, New York), 241–261.
3. Knight, B. W. (1975) *Lect. Math. Life Sci.* **5**, 111–144.
4. Smith, D. R. & Davidson, C. H. (1962) *J. Assoc. Comput. Mach.* **9**, 268–279.
5. Harmon, L. D. (1964) in *Neural Theory and Modeling*, ed. Reiss. R. F. (Stanford Univ. Press, Stanford, CA), pp. 23–24.
6. Amari, S.-I. (1977) *Biol. Cybern.* **26**, 175–185.
7. Amari, S.-I. & Akikazu, T. (1978) *Biol. Cybern.* **29**, 127–136.
8. Little, W. A. (1974) *Math. Biosci.* **19**, 101–120.
9. Marr, J. (1969) *J. Physiol.* **202**, 437–470.
10. Kohonen, T. (1980) *Content Addressable Memories* (Springer, New York).
11. Palm, G. (1980) *Biol. Cybern.* **36**, 19–31.
12. McCulloch, W. S. & Pitts, W. (1943) *Bull. Math Biophys.* **5**, 115–133.
13. Minsky, M. & Papert, S. (1969) *Perceptrons: An Introduction to Computational Geometry* (MIT Press, Cambridge, MA).
14. Rosenblatt, F. (1962) *Principles of Perceptrons* (Spartan, Washington, DC).
15. Cooper, L. N. (1973) in *Proceedings of the Nobel Symposium on Collective Properties of Physical Systems*, eds. Lundqvist, B. & Lundqvist, S. (Academic, New York), 252–264.
16. Cooper, L. N., Liberman, F. & Oja, E. (1979) *Biol. Cybern.* **33**, 9–28.
17. Longuet-Higgins, J. C. (1968) *Proc. Roy. Soc. London Ser. B* **171**, 327–334.
18. Longuet-Higgins, J. C. (1968) *Nature (London)* **217**, 104–105.
19. Kohonen, T. (1977) *Associative Memory—A System-Theoretic Approach* (Springer, New York).
20. Willwacher, G. (1976) *Biol. Cybern.* **24**, 181–198.
21. Anderson, J. A. (1977) *Psych. Rev.* **84**, 413–451.
22. Perkel, D. H. & Bullock, T. H. (1969) *Neurosci. Res. Symp. Summ.* **3**, 405–527.
23. John, E. R. (1972) *Science* **177**, 850–864.
24. Roney, K. J., Scheibel, A. B. & Shaw, G. L. (1979) *Brain Res. Rev.* **1**, 225–271.
25. Little, W. A. & Shaw, G. L. (1978) *Math. Biosci.* **39**, 281–289.
26. Shaw, G. L. & Roney, K. J. (1979) *Phys. Rev. Lett.* **74**, 146–150.
27. Hebb, D. O. (1949) *The Organization of Behavior* (Wiley, New York).
28. Eccles, J. G. (1953) *The Neurophysiological Basis of Mind* (Clarendon, Oxford).
29. Kirkpatrick, S. & Sherrington, D. (1978) *Phys. Rev.* **17**, 4384–4403.
30. Mountcastle, V. B. (1978) in *The Mindful Brain*, eds. Edelman. G. M. & Mountcastle, V. B. (MIT Press, Cambridge, MA). pp. 36–41.
31. Goldman, P. S. & Nauta, W. J. H. (1977) *Brain Res.* **122**, 393–413.
32. Kandel, E. R. (1979) *Sci. Am.* **241**, 61–70.

28
Introduction

(1982)
David Marr

Vision, San Francisco: W. H. Freeman, pp. 19–38, 54–61

David Marr was a brilliant scientist who has had an important influence on several areas of theoretical neurobiology. Adding to his legend was his untimely death at the height of his productivity. He started his career with several impressive papers modeling the cerebellum and cerebral cortex (Marr, 1969, 1970). His best known work is *Vision*, which has been very influential in the field of computational vision. We have included only brief sections here, but the book contains many important ideas. It is hard to excerpt because of the complexity of some of the arguments, and, since it is easily available, the only way to do justice to the ideas in it is to read a copy.

Marr suggested that the way to approach the analysis of an information processing device was to realize that there are three separate levels of function to be understood. First was "computational theory," which was the goal of the computation; second was the "representation and algorithm" used to realize the goal; and third was the "hardware implementation" used to realize the algorithm that realized the goal. An example might be tic-tac-toe, which Marr mentions briefly. The goal is to win the game by getting three of your markers in a line, or, if that is not possible, not to allow your opponent to do so. The representation might include the game board and the markers on it, but might also include some important features of the board: empty squares in critical locations or lines of two markers with an empty square between them. There are a number of algorithms known for tic-tac-toe, and formally describing the internal representation of the board and the computation required would be the second level. Once the algorithm was written in the form of a LISP or Pascal program, it could be run on any convenient machine that would go fast enough for the desired purpose, from a Cray to, as Marr mentions, a special-purpose device made out of Tinkertoys. As long as the machine could implement the algorithm, exactly what kind of hardware was used was a matter of convenience—for example, matters of cost and efficiency would become relevant.

Analysis of perceptual phenomena into these levels has become an article of faith in some quarters, and the levels are construed perhaps more rigidly than Marr might have wanted. Also, although Marr suggested that the levels were "loosely coupled," many connectionist modelers would argue that the connection is tight, especially between the algorithmic level and the hardware level. There are strong interlevel interactions: some desirable algorithms cannot be run efficiently or even at all on some machines. The algorithms chosen and even their goals may be largely hardware driven. Biological systems are concerned with efficiency and speed; it is generally better to be right most of the time and be fast than to be right all of the time and be slow. Settling for second best might be *best* for a real system, a constraint which would affect the goals, the algorithms, and the hardware. (It has been pointed out that in biology, "The best is the enemy of the good.")

Although engineers who can start fresh might have the functional goals in view from the beginning, evolution has to work from the existing hardware. A well known example of this is bird flight, where it is a matter of hot debate as to what was the function of wings before they were large enough to be used for flying. Were they radiators to be used for dumping excess heat, or perhaps bug catching membranes that also became useful in locomotion? Evolution works first from the hardware to the computation; designers of devices or algorithms often work the other way.

Once the hardware performs a useful function, however, evolution can also work the other way, to make an existing computation more efficient. Marr mentions, in a section called "The purpose of vision," a series of biological examples of the impressive way biological systems do just as much as they have to do, and very efficiently: flies are little mobile automata with some specialized representations of relevant aspects of the visual world.

The notion of *representation* will become one of the key issues in neural networks in the next few years. Marr was aware of the importance of this. To him, "A *representation* is a formal system making explicit certain entities or types of information, together with a specification of how the system does this" (p. 20). In human vision, this involves a series of successively more complex analyses of the image at higher and higher levels of abstraction from the image (see table 1.1, p. 37). To recognize an object, for example, we start from a series of intensity values at points on a retinal image and finish with an "object-centered" representation, where our mental description is decoupled from the exact image that gave rise to it. Much of the usefulness of the idea that the world can be represented as separate objects comes from this: that objects can be at least partially understood and described independently of the point of view of a particular observer, or even of the particular example of the object seen. This level of representation is similar in some ways to concepts in cognitive science. Since this amount of abstraction is so difficult to do practically, it must be done in a number of steps, some of which are described in detail in the book.

The last excerpt from the book (section 2.2) describes briefly one of the first steps in the long process of reaching the final object centered representation: to find where the edges in an image are. It is surprising that something we do so easily is so difficult to do artifically. Marr suggests that the best way to look for intensity changes that might correspond to edges is, first, to look at several different scales of spatial frequency, since the same event occurring in several channels might correspond to a true edge, and not random noise. Second, an edge is very likely to correspond to a sudden change in intensity; i.e., the derivative of intensity will be large across the edge. Such a change can be picked out by computing the second derivative of the image intensities and noting the places where the second derivative becomes zero, that is, a *zero crossing*. Marr describes a number of technical criteria that must be met for a good filter capable of detecting such places, and argues that the Laplacian operator operating on the Gaussian distribution is the most practical choice. It is easy to change the scale of such an operator, and it responds to the second derivative appropriately. It also gives good results when used on an image, and several examples of its use are given in the short excerpt.

The most detailed and impressive parts of *Vision* are found in its discussions of low level vision. It is unfortunate that Marr did not have enough time to extend his approach to higher levels of analysis of the image in comparable detail.

References

D. Marr (1969), "A theory of cerebellar cortex," *Journal of Physiology* 202:437–470.

D. Marr (1970), "A theory for cerebral neocortex," *Proceedings of the Royal Society of London, Series B* 176:161–234.

(1982)
David Marr

Vision, San Francisco: W. H. Freeman, pp. 19–38, 54–61

1.2 Understanding Complex Information-Processing Systems

Almost never can a complex system of any kind be understood as a simple extrapolation from the properties of its elementary components. Consider, for example, some gas in a bottle. A description of thermodynamic effects—temperature, pressure, density, and the relationships among these factors—is not formulated by using a large set of equations, one for each of the particles involved. Such effects are described at their own level, that of an enormous collection of particles; the effort is to show that in principle the microscopic and macroscopic descriptions are consistent with one another. If one hopes to achieve a full understanding of a system as complicated as a nervous system, a developing embryo, a set of metabolic pathways, a bottle of gas, or even a large computer program, then one must be prepared to contemplate different kinds of explanation at different levels of description that are linked, at least in principle, into a cohesive whole, even if linking the levels in complete detail is impractical. For the specific case of a system that solves an information-processing problem, there are in addition the twin strands of process and representation, and both these ideas need some discussion.

Representation and Description

A *representation* is a formal system for making explicit certain entities or types of information, together with a specification of how the system does this. And I shall call the result of using a representation to describe a given entity a *description* of the entity in that representation (Marr and Nishihara, 1978).

For example, the Arabic, Roman, and binary numeral systems are all formal systems for representing numbers. The Arabic representation consists of a string of symbols drawn from the set (0, 1, 2, 3, 4, 5, 6, 7, 8, 9), and the rule for constructing the description of a particular integer n is that one decomposes n into a sum of multiples of powers of 10 and unites these multiples into a string with the largest powers on the left and the smallest on the right. Thus, thirty-seven equals $3 \times 10^1 + 7 \times 10^0$, which becomes 37, the Arabic numeral system's description of the number. What this description makes explicit is the number's decomposition into powers of 10. The binary numeral system's description of the number thirty-seven is 100101, and this description makes explicit the number's decomposition into powers of 2. In the Roman numeral system, thirty-seven is represented as XXXVII.

This definition of a representation is quite general. For example, a representation for shape would be a formal scheme for describing some aspects of shape, together with rules that specify how the scheme is applied to any particular shape. A musical score provides a way of representing a symphony; the alphabet allows the construction of a written representation of words; and so forth. The phrase "formal scheme" is critical to the definition, but the reader should not be frightened by it. The reason is simply that we are dealing with information-processing machines, and the way such machines work is by using symbols to stand for things—to represent things, in our terminology. To say that something is a formal scheme means only that it is a set of symbols with rules for putting them together—no more and no less.

A representation, therefore, is not a foreign idea at all—we all use representations all the time. However, the notion that one can capture some aspect of reality by making a description of it using a symbol and that to do so can be useful seems to me a fascinating and powerful idea. But even the simple examples we have discussed introduce some rather general and important issues that arise whenever one chooses to use one particular representation. For example, if one chooses the Arabic numeral representation, it is easy to discover whether a number is a power of 10 but difficult to discover whether it is a power of 2. If one chooses the binary representation, the situation is reversed. Thus, there is a trade-off; any particular representation makes certain information explicit at the expense of information that is pushed into the background any may be quite hard to recover.

This issue is important, because how information is represented can greatly affect how easy it is to do

Copyright © 1982 W. H. Freeman and Company. Reprinted with permission.

different things with it. This is evident even from our numbers example: It is easy to add, to subtract, and even to multiply if the Arabic or binary representations are used, but it is not at all easy to do these things—especially multiplication—with Roman numerals. This is a key reason why the Roman culture failed to develop mathematics in the way the earlier Arabic cultures had.

An analogous problem faces computer engineers today. Electronic technology is much more suited to a binary number system than to the conventional base 10 system, yet humans supply their data and require the results in base 10. The design decision facing the engineer, therefore, is, Should one pay the cost of conversion into base 2, carry out the arithmetic in a binary representation, and then convert back into decimal numbers on output; or should one sacrifice efficiency of circuitry to carry out operations directly in a decimal representation? On the whole, business computers and pocket calculators take the second approach, and general purpose computers take the first. But even though one is not restricted to using just one representation system for a given type of information, the choice of which to use is important and cannot be taken lightly. It determines what information is made explicit and hence what is pushed further into the background, and it has a far-reaching effect on the ease and difficulty with which operations may subsequently be carried out on that information.

Process

The term *process* is very broad. For example, addition is a process, and so is taking a Fourier transform. But so is making a cup of tea, or going shopping. For the purposes of this book, I want to restrict our attention to the meanings associated with machines that are carrying out information-processing tasks. So let us examine in depth the notions behind one simple such device, a cash register at the checkout counter of a supermarket.

There are several levels at which one needs to understand such a device, and it is perhaps most useful to think in terms of three of them. The most abstract is the level of *what* the device does and *why*. What it does is arithmetic, so our first task is to master the theory of addition. Addition is a mapping, usually denoted by $+$, from pairs of numbers into single numbers; for example, $+$ maps the pair $(3, 4)$ to 7, and I shall write this in the form $(3 + 4) \to 7$. Addition has a number of abstract properties, however. It is commutative: both $(3 + 4)$ and $(4 + 3)$ are equal to 7; and associative: the sum of $3 + (4 + 5)$ is the same as the sum of $(3 + 4) + 5$. Then there is the unique distinguished element, zero, the adding of which has no effect: $(4 + 0) \to 4$. Also, for

every number there is a unique "inverse," written (-4) in the case of 4, which when added to the number gives zero: $[4 + (-4)] \to 0$.

Notice that these properties are part of the fundamental *theory* of addition. They are true no matter how the numbers are written—whether in binary. Arabic, or Roman representation—and no matter how the addition is executed. Thus part of this first level is something that might be characterized as *what* is being computed.

The other half of this level of explanation has to do with the question of *why* the cash register performs addition and not, for instance, multiplication when combining the prices of the purchased items to arrive at a final bill. The reason is that the rules we intuitively feel to be appropriate for combining the individual prices in fact define the mathematical operation of addition. These can be formulated as *constraints* in the following way:

1. If you buy nothing, it should cost you nothing; and buying nothing and something should cost the same as buying just the something. (The rules for zero.)

2. The order in which goods are presented to the cashier should not affect the total. (Commutativity.)

3. Arranging the goods into two piles and paying for each pile separately should not effect the total amount you pay. (Associativity; the basic operation for combining prices.)

4. If you buy an item and then return it for a refund, your total expenditure should be zero. (Inverses.)

It is a mathematical theorem that these conditions define the operation of addition, which is therefore the appropriate computation to use.

This whole argument is what I call the *computational theory* of the cash register. Its important features are (1) that it contains separate arguments about what is computed and why and (2) that the resulting operation is defined uniquely by the constraints it has to satisfy. In the theory of visual processes, the underlying task is to reliably derive properties of the world from images of it; the business of isolating constraints that are both powerful enough to allow a process to be defined and generally true of the world is a central theme of our inquiry.

In order that a process shall actually run, however, one has to realize it in some way and therefore choose a representation for the entities that the process manipulates. The second level of the analysis of a process, therefore, involves choosing two things: (1) a *representation* for the input and for the output of the process and (2) an *algorithm* by which the transformation may actually be accomplished. For addition, of course, the input and output representations can both be the same, because they both consist of numbers. However this is not true is general. In the case of a Fourier transform, for example, the input representation may be the time domain, and the output, the frequency domain. If the first of our levels specifies what and why, this second level specifies *how*. For addition, we might choose Arabic numerals for the representations, and for the algorithm we could follow the usual rules about adding the least significant digits first and "carrying" if the sum exceeds 9. Cash registers, whether mechanical or electronic, usually use this type of representation and algorithm.

There are three important points here. First, there is usually a wide choice of representation. Second, the choice of algorithm often depends rather critically on the particular representation that is employed. And third, even for a given fixed representation, there are often several possible algorithms for carrying out the same process. Which one is chosen will usually depend on any particularly desirable or undesirable characteristics that the algorithms may have; for example, one algorithm may be much more efficient than another, or another may be slightly less efficient but more robust (that is, less sensitive to slight inaccuracies in the data on which it must run). Or again, one algorithm may be parallel, and another, serial. The choice, then, may depend on the type of hardware or machinery in which the algorithm is to be embodied physically.

This brings us to the third level, that of the device in which the process is to be realized physically. The important point here is that, once again, the same algorithm may be implemented in quite different technologies. The child who methodically adds two numbers from right to left, carrying a digit when necessary, may be using the same algorithm that is implemented by the wires and transistors of the cash register in the neighborhood supermarket, but the physical realization of the algorithm is quite different in these two cases. Another example: Many people have written computer programs to play tic-tac-toe, and there is a more or less standard algorithm that cannot lose. This algorithm has in fact been implemented by W. D. Hillis and B. Silverman in a quite different technology, in a computer made out of Tinkertoys, a children's wooden building set. The whole monstrously ungainly engine, which actually works, currently resides in a museum at the University of Missouri in St. Louis.

Some styles of algorithm will suit some physical substrates better than others. For example, in conventional digital computers, the number of connections is

Computational theory	Representation and algorithm	Hardware implementation
What is the goal of the computation, why is it appropriate, and what is the logic of the strategy by which it can be carried out?	How can this computational theory be implemented? In particular, what is the representation for the input and output, and what is the algorithm for the transformation?	How can the representation and algorithm be realized physically?

Figure 1-4 The three levels at which any machine carrying out an information-processing task must be understood.

comparable to the number of gates, while in a brain, the number of connections is much larger ($\times 10^4$) than the number of nerve cells. The underlying reason is that wires are rather cheap in biological architecture, because they can grow individually and in three dimensions. In conventional technology, wire laying is more or less restricted to two dimensions, which quite severely restricts the scope for using parallel techniques and algorithms; the same operations are often better carried out serially.

The Three Levels

We can summarize our discussion in something like the manner shown in Figure 1-4, which illustrates the different levels at which an information-processing device must be understood before one can be said to have understood it completely. At one extreme, the top level, is the abstract computational theory of the device, in which the performance of the device is characterized as a mapping from one kind of information to another, the abstract properties of this mapping are defined precisely, and its appropriateness and adequacy for the task at hand are demonstrated. In the center is the choice of representation for the input and output and the algorithm to be used to transfer one into the other. And at the other extreme are the details of how the algorithm and representation are realized physically—the detailed computer architecture, so to speak. These three levels are coupled, but only loosely. The choice of an algorithm is influenced for example, by what is has to do and by the hardware in which it must run. But there is a wide choice available at each level, and the explication of each level involves issues that are rather independent of the other two.

Each of the three levels of description will have its place in the eventual understanding of perceptual information processing, and of course they are logically and causally related. But an important point to note is that since the three levels are only rather loosely

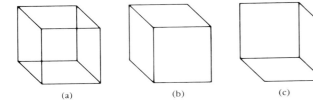

(a)   (b)   (c)

Figure 1-5 The so-called Necker illusion, named after L. A. Necker, the Swiss naturalist who developed it in 1832. The essence of the matter is that the two-dimensional representation (a) has collapsed the depth out of a cube and that a certain aspect of human vision is to recover this missing third dimension. The depth of the cube can indeed be perceived, but two interpretations are possible, (b) and (c). A person's perception characteristically flips from one to the other.

related, some phenomena may be explained at only one or two of them. This means, for example, that a correct explanation of some psychophysical observation must be formulated at the appropriate level. In attempts to relate psychophysical problems to physiology, too often there is confusion about the level at which problems should be addressed. For instance, some are related mainly to the physical mechanisms of vision— such as afterimages (for example, the one you see after staring at a light bulb) or such as the fact that any color can be matched by a suitable mixture of the three primaries (a consequence principally of the fact that we humans have three types of cones). On the other hand, the ambiguity of the Necker cube (Figure 1-5) seems to demand a different kind of explanation. To be sure, part of the explanation of its perceptual reversal must have to do with a bistable neural network (that is, one with two distinct stable states) somewhere inside the brain, but few would feel satisfied by an account that failed to mention the existence of two different but perfectly plausible three-dimensional interpretations of this two-dimensional image.

For some phenomena, the type of explanation required is fairly obvious. Neuroanatomy, for example, is clearly tied principally to the third level, the physical realization of the computation. The same holds for synaptic mechanisms, action potentials, inhibitory interactions, and so forth. Neurophysiology, too, is related mostly to this level, but it can also help us to understand the type of representations being used, particularly if one accepts something along the lines of Barlow's views that I quoted earlier. But one has to exercise extreme caution in making inferences from neurophysiological findings about the algorithms and representations being used, particularly until one has a clear idea about what information needs to be represented and what processes need to be implemented.

Psychophysics, on the other hand, is related more directly to the level of algorithm and representation. Different algorithms tend to fail in radically different ways as they are pushed to the limits of their performance or are deprived of critical information. As we shall see, primarily psychophysical evidence proved to Poggio and myself that our first stereo-matching algorithm (Marr and Poggio, 1976) was not the one that is used by the brain, and the best evidence that our second algorithm (Marr and Poggio, 1979) *is* roughly the one that is used also comes from psychophysics. Of course, the underlying computational theory remained the same in both cases, only the algorithms were different.

Psychophysics can also help to determine the nature of a representation. The work of Roger Shepard (1975), Eleanor Rosch (1978), or Elizabeth Warrington (1975) provides some interesting hints in this direction. More specifically, Stevens (1979) argued from psychophysical experiments that surface orientation is represented by the coordinates of slant and tilt, rather than (for example) the more traditional (p, q) of gradient space (see Chapter 3). He also deduced from the uniformity of the size of errors made by subjects judging surface orientation over a wide range of orientations that the representational quantities used for slant and tilt are pure angles and not, for example, their cosines, sines, or tangents.

More generally, if the idea that different phenomena need to be explained at different levels is kept clearly in mind, it often helps in the assessment of the validity of the different kinds of objections that are raised from time to time. For example, one favorite is that the brain is quite different from a computer because one is parallel and the other serial. The answer to this, of course, is that the distinction between serial and parallel is a distinction at the level of algorithm; it is not fundamental at all—anything programmed in parallel can be rewritten serially (though not necessarily vice versa). The distinction, therefore, provides no grounds for arguing that the brain operates so differently from a computer that a computer could not be programmed to perform the same tasks.

Importance of Computational Theory
Although algorithms and mechanisms are empirically more accessible, it is the top level, the level of computational theory, which is critically important from an information-processing point of view. The reason for this is that the nature of the computations that underlie perception depends more upon the computational problems that have to be solved than upon the particular hardware in which their solutions are implemented.

To phrase the matter another way, an algorithm is likely to be understood more readily by understanding the nature of the problem being solved than by examining the mechanism (and the hardware) in which it is embodied.

In a similar vein, trying to understand perception by studying only neurons is like trying to understand bird flight by studying only feathers: It just cannot be done. In order to understand bird flight, we have to understand aerodynamics; only then do the structure of feathers and the different shapes of birds wings make sense. More to the point, as we shall see, we cannot understand why retinal ganglion cells and lateral geniculate neurons have the receptive fields they do just by studying their anatomy and physiology. We can understand how these cells and neurons behave as they do by studying their wiring and interactions, but in order to understand *why* the receptive fields are as they are—why they are circularly symmetrical and why their excitatory and inhibitory regions have characteristic shapes and distributions—we have to know a little of the theory of differential operators, band-pass channels, and the mathematics of the uncertainty principle (see Chapter 2).

Perhaps it is not surprising that the very specialized empirical disciplines of the neurosciences failed to appreciate fully the absence of computational theory; but it is surprising that this level of approach did not play a more forceful role in the early development of artificial intelligence. For far too long, a heuristic program for carrying out some task was held to be a theory of that task, and the distinction between what a program did and how it did it was not taken seriously. As a result, (1) a style of explanation evolved that invoked the use of special mechanisms to solve particular problems, (2) particular data structures, such as the lists of attribute value pairs called property lists in the LISP programing language, were held to amount to theories of the representation of knowledge, and (3) there was frequently no way to determine whether a program would deal with a particular case other than by running the program.

Failure to recognize this theoretical distinction between *what* and *how* also greatly hampered communication between the fields of artificial intelligence and linguistics. Chomsky's (1965) theory of transformational grammar is a true computational theory in the sense defined earlier. It is concerned solely with specifying what the syntactic decomposition of an English sentence should be, and not at all with how that decomposition should be achieved. Chomsky himself was very clear about this—it is roughly his distinction between competence and performance,

though his idea of performance did include other factors, like stopping in midutterance—but the fact that his theory was defined by transformations, which look like computations, seems to have confused many people. Winograd (1972), for example, felt able to criticize Chomsky's theory on the grounds that it cannot be inverted and so cannot be made to run on a computer; I had heard reflections of the same argument made by Chomsky's colleagues in linguistics as they turn their attention to how grammatical structure might actually be computed from a real English sentence.

The explanation is simply that finding algorithms by which Chomsky's theory may be implemented is a completely different endeavor from formulating the theory itself. In our terms, it is a study at a different level, and both tasks have to be done. This point was appreciated by Marcus (1980), who was concerned precisely with how Chomsky's theory can be realized and with the kinds of constraints on the power of the human grammatical processor that might give rise to the structural constraints in syntax that Chomsky found. It even appears that the emerging "trace" theory of grammar (Chomsky and Lasnik, 1977) may provide a way of synthesizing the two approaches—showing that, for example, some of the rather ad hoc restrictions that form part of the computational theory may be consequences of weaknesses in the computational power that is available for implementing syntactical decoding.

The Approach of J. J. Gibson

In perception, perhaps the nearest anyone came to the level of computational theory was Gibson (1966). However, although some aspects of his thinking were on the right lines, he did not understand properly what information processing was, which led him to seriously underestimate the complexity of the information-processing problems involved in vision and the consequent subtlety that is necessary in approaching them.

Gibson's important contribution was to take the debate away from the philosophical considerations of sense-data and the affective qualities of sensation and to note instead that the important thing about the senses is that they are channels for perception of the real world outside or, in the case of vision, of the visible surfaces. He therefore asked the critically important question, How does one obtain constant perceptions in everyday life on the basis of continually changing sensations? This is exactly the right question, showing that Gibson correctly regarded the problem of perception as that of recovering from sensory information "valid" properties of the external world. His problem was that he had a much oversimplified view of how this should be done. His approach led him to consider higher-order variables—stimulus energy, ratios, proportions, and so on—as "invariants" of the movement of an observer and of changes in stimulation intensity.

"These invariants," he wrote, "correspond to permanent properties of the environment. They constitute, therefore, information about the permanent environment." This led him to a view in which the function of the brain was to "detect invariants" despite changes in "sensations" of light, pressure, or loudness of sound. Thus, he says that the "function of the brain, when looped with its perceptual organs, is not to decode signals, nor to interpret messages, nor to accept images, not to *organize* the sensory input or to *process* the data, in modern terminology. It is to seek and extract information about the environment from the flowing array of ambient energy," and the thought of the nervous system as in some way "resonating" to these invariants. He then embarked on a broad study of animals in their environments, looking for invariants to which they might resonate. This was the basic idea behind the notion of ecological optics (Gibson, 1966, 1979).

Although one can critize certain shortcomings in the quality of Gibson's analysis, its major and, in my view, fatal shortcoming lies at a deeper level and results from a failure to realize two things. First, the detection of physical invariants, like image surfaces, is exactly and precisely an information-processing problem, in modern terminology. And second, he vastly underrated the sheer difficulty of such detection. In discussing the recovery of three-dimensional information from the movement of an observer, he says that "in motion, perspective information alone can be used" (Gibson, 1966, p. 202). And perhaps the key to Gibson is the following:

The detection of non-change when an object moves in the world is not as difficult as it might appear. It is only made to seem difficult when we assume that the perception of constant dimensions of the object must depend on the correcting of sensations of inconstant form and size. The information for the constant dimension of an object is normally carried by invariant relations in an optic array. Rigidity is *specified*. (Emphasis added)

Yes, to be sure, but *how?* Detecting physical invariants is just as difficult as Gibson feared, but nevertheless we can do it. And the only way to understand how is to treat it as an information-processing problem.

The underlying point is that visual information processing is actually very complicated, and Gibson was not the only thinker who was misled by the apparent simplicity of the act of seeing. The whole tradition of philosophical inquiry into the nature of perception seems not to have taken seriously enough the complexity of the information processing involved. For example, Austin's (1962) *Sense and Sensibilia* entertainingly demolishes the argument, apparently favored by earlier philosophers, that since we are sometimes deluded by illusions (for example, a straight stick appears bent if it is partly submerged in water), we see sense-data rather than material things. The answer is simply that usually our perceptual processing does run correctly (it delivers a true description of what is there), but although evolution has seen to it that our processing allows for many changes (like inconstant illumination), the perturbation due to the refraction of light by water is not one of them. And incidentally, although the example of the bent stick has been discussed since Aristotle, I have seen no philosphical inquiry into the nature of the perceptions of, for instance, a heron, which is a bird that feeds by pecking up fish first seen from above the water surface. For such birds the visual correction might be present.

Anyway, my main point here is another one. Austin (1962) spends much time on the idea that perception tells one about real properties of the external world, and one thing he considers is "real shape," (p. 66), a notion which had cropped up earlier in his discussion of a coin that "looked elliptical" from some points of view. Even so,

it had a real shape which remained unchanged. But coins in fact are rather special cases. For one thing their outlines are well defined and very highly stable, and for another they have a known and a nameable shape. But there are plenty of things of which this is not true. What is the real shape of a cloud? ... or of a cat? Does its real shape change whenever it moves? If not, in what posture *is* its real shape on display? Furthermore, is its real shape such as to be fairly smooth outlines, or must it be finely enough serrated to take account of each hair? *It is pretty obvious that there is no answer to these questions—no rules according to which, no procedure by which, answers are to be determined.* (Emphasis added; p. 67)

But there *are* answers to these questions. There are ways of describing the shape of a cat to an arbitrary level of precision (see Chapter 5), and there are rules and procedures for arriving at such descriptions. That is exactly what vision is about, and precisely what makes it complicated.

1.3 A Representational Framework for Vision

Vision is a process that produces from images of the external world a description that is useful to the viewer and not cluttered with irrelevant information (Marr,

1976; Marr and Nishihara, 1978). We have already seen that a process may be thought of as a mapping from one representation to another, and in the case of human vision, the initial representation is in no doubt—it consists of arrays of image intensity values as detected by the photoreceptors in the retina.

It is quite proper to think of an image as a representation; the items that are made explicit are the image intensity values at each point in the array, which we can conveniently denote by $I(x, y)$ at coordinate (x, y). In order to simplify our discussion, we shall neglect for the moment the fact that there are several different types of receptor, and imagine instead that there is just one, so that the image is black-and-white. Each value of $I(x, y)$ thus specifies a particular level of gray; we shall refer to each detector as a picture element or *pixel* and to the whole array I as an image.

But what of the output of the process of vision? We have already agreed that it must consist of a useful description of the world, but that requirement is rather nebulous. Can we not do better? Well, it is perfectly true that, unlike the input, the result of vision is much harder to discern, let alone specify precisely, and an important aspect of this new approach is that it makes quite concrete proposals about what that end is. But before we begin that discussion, let us step back a little and spend a little time formulating the more general issues that are raised by these questions.

The Purpose of Vision
The usefulness of a representation depends upon how well suited it is to the purpose for which it is used. A pigeon uses vision to help it navigate, fly, and seek out food. Many types of jumping spider use vision to tell the difference between a potential meal and a potential mate. One type, for example, has a curious retina formed of two diagonal strips arranged in a V. If it detects a red V on the back of an object lying in front of it, the spider has found a mate. Otherwise, maybe a meal. The frog, as we have seen, detects bugs with its retina; and the rabbit retina is full of special gadgets, including what is apparently a hawk detector, since it responds well to the pattern made by a preying hawk hovering overhead. Human vision, on the other hand, seems to be very much more general, although it clearly contains a variety of special-purpose mechanisms that can, for example, direct the eye toward an unexpected movement in the visual field or cause one to blink or otherwise avoid something that approaches one's head to quickly.

Vision, in short, is used in such a bewildering variety of ways that the visual systems of different animals must differ significantly from one another. Can the

type of formulation that I have been advocating, in terms of representations and processes, possibly prove adequate for them all? I think so. The general point here is that because vision is used by different animals for such a wide variety of purposes, it is inconceivable that all seeing animals use the same representations; each can confidently be expected to use one or more representations that are nicely tailored to the owner's purposes.

As an example, let us consider briefly a primitive but highly efficient visual system that has the added virtue of being well understood. Werner Reichardt's group in Tübingen has spend the last 14 years patiently unraveling the visual flight-control system of the housefly, and in a famous collaboration, Reichardt and Tomaso Poggio have gone far toward solving the problem (Reichardt and Poggio, 1976, 1979; Poggio and Reichardt, 1976). Roughly speaking, the fly's visual apparatus controls its flight through a collection of about five independent, rigidly inflexible, very fast responding systems (the time from visual stimulus to change of torque is only 21 ms). For example, one of these systems is the landing system; if the visual field "explodes" fast enough (because a surface looms nearby), the fly automatically "lands" toward its center. If this center is above the fly, the fly automatically inverts to land upside down. When the feet touch, power to the wings is cut off. Conversely, to take off, the fly jumps; when the feet no longer touch the ground, power is restored to the wings, and the insect flies again.

In-flight control is achieved by independent systems controlling the fly's vertical velocity (through control of the lift generated by the wings) and horizontal direction (determined by the torque produced by the asymmetry of the horizontal thrust from the left and right wings). The visual input to the horizontal control system, for example, is completely described by the two terms

$$r(\psi)\dot{\psi} + D(\psi)$$

where r and D have the form illustrated in Figure 1-6. This input describes how the fly tracks an object that is present at angle ψ in the visual field and has angular velocity $\dot{\psi}$. This system is triggered to track objects of a certain angular dimension in the visual field, and the motor strategy is such that if the visible object was another fly a few inches away, then it would be intercepted successfully. If the target was an elephant 100 yd away, interception would fail because the fly's built-in parameters are for another fly nearby, not an elephant far away.

Thus, fly vision delivers a representation in which at least these three things are specified: (1) whether the

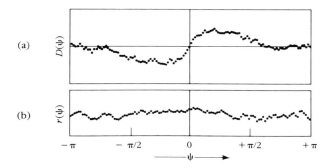

(a) $D(\psi)$

(b) $r(\psi)$

$-\pi$ $-\pi/2$ 0 $+\pi/2$ $+\pi$

$\psi \longrightarrow$

Figure 1-6 The horizontal component of the visual input R to the fly's flight system is described by the formula $R = D(\psi) - r(\psi)\dot{\psi}$, where ψ is the direction of the stimulus and $\dot{\psi}$ is its angular velocity in the fly's visual field. $D(\psi)$ is an odd function, as shown in (a), which has the effect of keeping the target centered in the fly's visual field; $r(\psi)$ is essentially constant as shown in (b).

visual field is looming sufficiently fast that the fly should contemplate landing; (2) whether there is a small patch—it could be a black speck or, it turns out, a textured figure in front of a textured ground—having some kind of motion relative to its background; and if there is such a patch, (3) ψ and $\dot{\psi}$ for this patch are delivered to the motor system. And that is probably about 60% of fly vision. In particular, it is extremely unlikely that the fly has any explicit representation of the visual world around him—no true conception of a surface, for example, but just a few triggers and some specifically fly-centered parameters like ψ and $\dot{\psi}$.

It is clear that human vision is much more complex than this, although it may well incorporate subsystems not unlike the fly's to help with specific and rather low-level tasks like the control of pursuit eye movements. Nevertheless, as Poggio and Reichardt have shown, even these simple systems can be understood in the same sort of way, as information-processing tasks. And one of the fascinating aspects of their work is how they have managed not only to formulate the differential equations that accurately describe the visual control system of the fly but also to express these equations, using the Volterra series expansion, in a way that gives direct information about the minimum possible complexity of connections of the underlying neuronal networks.

Advanced Vision
Visual systems like the fly's serve adequately and with speed and precision the needs of their owners, but they are not very complicated; very little objective information about the world is obtained. The information is all very much subjective—the angular size of the stimulus as the fly sees it rather than the objective size

of the object out there, the angle that the object has in the fly's visual field rather than its position relative to the fly or to some external reference, and the object's angular velocity, again in the fly's visual field, rather than any assessment of its true velocity relative to the fly or to some stationary reference point.

One reason for this simplicity must be that these facts provide the fly with sufficient information for it to survive. Of course, the information is not optimal and from time to time the fly will fritter away its energy chasing a falling leaf a medium distance away or an elephant a long way away as a direct consequence of the inadequacies of its perceptual system. But this apparently does not matter very much—the fly has sufficient excess energy for it to be able to absorb these extra costs. Another reason is certainly that translating these rather subjective measurements into more objective qualities involves much more computation. How, then, should one think about more advanced visual systems—human vision, for example. What are the issues? What kind of information is vision really delivering, and what are the representational issues involved?

My approach to these problems was very much influenced by the fascinating accounts of clinical neurology, such as Critchley (1953) and Warrington and Taylor (1973). Particularly important was a lecture that Elizabeth Warrington gave at MIT in October 1973, in which she described the capacities and limitations of patients who had suffered left or right parietal lesions. For me, the most important thing that she did was to draw a distinction between the two classes of patient (see Warrington and Taylor, 1978). For those with lesions on the right side, recognition of a common object was possible *provided* that the patient's view of it was in some sense straightforward. She used the words *conventional* and *unconventional*—a water pail or a clarinet seen from the side gave "conventional" views but seen end-on gave "unconventional" views. If these patients recognized the object at all, they knew its name and its semantics—that is, its use and purpose, how big it was, how much it weighed, what it was made of, and so forth. If their view was unconventional—a pail seen from above, for example—not only would the patients fail to recognize it, but they would vehemently deny that it *could* be a view of a pail. Patients with left parietal lesions behaved completely differently. Often these patients had no language, so they were unable to name the viewed object or state its purpose and semantics. But they could convey that they correctly perceived its geometry—that is, its shape—even from the unconventional view.

Warrington's talk suggested two things. First, the representation of the shape of an object is stored in a different place and is therefore a quite different kind of thing from the representation of its use and purpose. And second, vision alone can deliver an internal description of the shape of a viewed object, even when the object was not recognized in the conventional sense of understanding its use and purpose.

This was an important moment for me for two reasons. The general trend in the computer vision community was to believe that recognition was so difficult that it required every possible kind of information. The results of this point of view duly appeared a few years later in programs like Freuder's (1974) and Tenenbaum and Barrow's (1976). In the latter program, knowledge about offices—for example, that desks have telephones on them and that telephones are black—was used to help "segment" out a black blob halfway up an image and recognize" it as a telephone. Freuder's program used a similar approach to "segment" and "recognize" a hammer in a scene. Clearly, we do use such knowledge in real life; I once saw a brown blob quivering amongst the lettuce in my garden and correctly identified it as a rabbit, even though the visual information alone was inadequate. And yet here was this young woman calmly telling us not only that her patients could convey to her that they had grasped the shapes of things that she had shown them, even though they could not name the objects or say how they were used, but also that they could happily continue to do so even if she made the task extremely difficult visually by showing them peculiar views or by illuminating the objects in peculiar ways. It seemed clear that the intuitions of the computer vision people were completely wrong and that even in difficult circumstances shapes could be determined by vision alone.

The second important thing, I thought, was that Elizabeth Warrington had put her finger on what was somehow the quintessential fact of human vision—that it tells about shape and space and spatial arrangement. Here lay a way to formulate its purpose—building a description of the shapes and positions of things from images. Of course, that is by no means all that vision can do; it also tells about the illumination and about the reflectances of the surfaces that make the shapes—their brightnesses and colors and visual textures—and about their motion. But these things seemed secondary; they could be hung off a theory in which the main job of vision was to derive a representation of shape.

To the Desirable via the Possible
Finally, one has to come to terms with cold reality. Desirable as it may be to have vision deliver a com-

pletely invariant shape description from an image (whatever that may mean in detail), it is almost certainly impossible in only one step. We can only do what is possible and proceed from there toward what is desirable. Thus we arrived at the idea of a sequence of representations, starting with descriptions that could be obtained straight from an image but that are carefully designed to facilitate the subsequent recovery of gradually more objective, physical properties about an object's shape. The main stepping stone toward this goal is describing the geometry of the visible surfaces, since the information encoded in images, for example by stereopsis, shading, texture, contours, or visual motion, is due to a shape's local surface properties. The objective of many early visual computations is to extract this information.

However, this description of the visible surfaces turns out to be unsuitable for recognition tasks. There are several reasons why, perhaps the most prominent being that like all early visual processes, it depends critically on the vantage point. The final step therefore consists of transforming the viewer-centered surface description into a representation of the three-dimensional shape and spatial arrangement of an object that does not depend upon the direction from which the object is being viewed. This final description is object centered rather than viewer centered.

The overall framework described here therefore divides the derivation of shape information from images into three representational stages: (Table 1-1): (1) the representation of properties of the two-dimensional image, such as intensity changes and local two-dimensional geometry; (2) the representation of properties of the visible surfaces in a viewer-centered coordinate system, such as surface orientation, distance from the viewer, and discontinuities in these quantities; surface reflectance; and some coarse description of the prevailing illumination; and (3) an object-centered representation of the three-dimensional structure and of the organization of the viewed shape, together with some description of its surface properties.

This framework is summarized in Table 1-1. Chapters 2 through 5 give a more detailed account.

2.2 Zero-Crossings and the Raw Primal Sketch

Zero-Crossings
The first of the three stages described above concerns the detection of intensity changes. The two ideas underlying their detection are (1) that intensity changes occur at different scales in an image, and so their optimal detection requires the use of operators of

Table 1-1 Representational framework for deriving shape information from images

Name	Purpose	Primitives
Image(s)	Represents intensity.	Intensity value at each point in the image
Primal sketch	Makes explicit important information about the two-dimensional image, primarily the intensity changes there and their geometrical distribution and organization.	Zero-crossings Blobs Terminations and discontinuities Edge segments Virtual lines Groups Curvilinear organization Boundaries
$2\frac{1}{2}$-D sketch	Makes explicit the orientation and rough depth of the visible surfaces, and contours of discontinuities in these quantities in a viewer-centered coordinate frame.	Local surface orientation (the "needles" primitives) Distance from viewer Discontinuities in depth Discontinuities in surface orientation
3-D model representation	Describes shapes and their spatial organization in an object-centered coordinate frame, using a modular hierarchical representation that includes volumetric primitives (i.e., primitives that represent the volume of space that a shape occupies) as well as surface primitives.	3-D models arranged hierarchically, each one based on a spatial configuration of a few sticks or axes, to which volumetric or surface shape primitives are attached

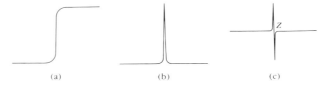

Figure 2-8 The notion of a zero-crossing. The intensity change (a) gives rise to a peak (b) in its first derivative and to a (steep) zero-crossing $Z(c)$ in its second derivative.

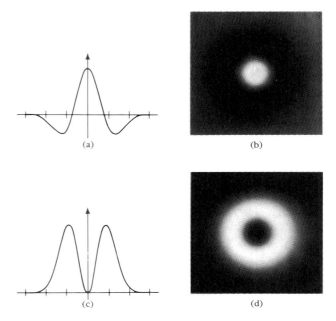

Figure 2-9 $\nabla^2 G$ is shown as a one-dimensional function (a) and in two-dimensions (b) using intensity to indicate the value of the function at each point. (c) and (d) show the Fourier transforms for the one- and two-dimensional cases respectively. (Reprinted by permission from D. Marr and E. Hildreth, "Theory of edge detection." *Proc. R. Soc. Lond. B 207*, pp. 187–217.)

different sizes; and (2) that a sudden intensity change will give rise to a peak or trough in the first derivative or, equivalently, to a *zero-crossing* in the second derivative, as illustrated in Figure 2-8. (A zero-crossing is a place where the value of a function passes from positive to negative).

These ideas suggest that in order to detect intensity changes efficiently, one should search for a filter that has two salient characteristics. First and foremost, it should be a differential operator, taking either a first or second spatial derivative of the image. Second, it should be capable of being tuned to act at any desired scale, so that large filters can be used to detect blurry shadow edges, and small ones to detect sharply focused fine detail in the image.

Marr and Hildreth (1980) argued that the most satisfactory operator fulfilling these conditions is the filter $\nabla^2 G$, where ∇^2 is the Laplacian operator ($\partial^2/\partial x^2 + \partial^2/\partial y^2$) and G stands for the two-dimensional Gaussian distribution

$$G(x,y) = e^{-(x^2+y^2)/2\sigma^2}$$

which has standard deviation σ. $\nabla^2 G$ is a circularly symmetric Mexican-hat-shaped operator whose distribution in two dimensions may be expressed in terms of the radial distance r from the origin by the formula

$$\nabla^2 G(r) = \frac{-1}{\pi\sigma^4}\left(1 - \frac{r^2}{2\sigma^2}\right)e^{-r^2/2\sigma^2} \quad .$$

Figure 2-9 illustrates the one- and two-dimensional forms of this operator, as well as their Fourier transforms.

There are two basic ideas behind the choice of the filter $\nabla^2 G$. The first is that the Gaussian part of it, G, blurs the image, effectively wiping out all structure at scales much smaller than the space constant σ of the Gaussian. To illustrate this, Figure 2-10 shows an

(a)   (b)

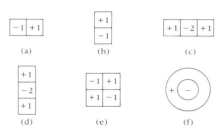

(c)

Figure 2-10 Blurring images is the first step in detecting intensity changes in them. (a) In the original image, intensity changes can take place over a wide range of scales, and no single operator will be very efficient at detecting all of them. The problem is much simplified in an image that has been blurred with a Gaussian filter, because there is, in effect, an upper limit to the rate at which changes can take place. The first part of the edge detection process can be thought of as decomposing the original image into a set of copies, each filtered with a different-sized Gaussian, and then detecting the intensity changes separately in each. (b) The image filtered with a Gaussian having $\sigma = 8$ pixels; in (c), $\sigma = 4$. The image is 320 by 320 elements. (Reprinted by permission from D. Marr and E. Hildreth, "Theory of edge detection," *Proc. R. Soc. Lond. B 207*, pp. 187–217.)

Figure 2-11 The spatial configuration of low-order differential operators. Operators like $\partial/\partial x$ can be roughly realized by filters with the receptive fields illustrated in the figure. (a) $\partial/\partial x$ can be thought of as measuring the difference between the values at two neighboring points along the x-axis. Similarly, (b) shows $\partial/\partial y$. The operator $\partial^2/\partial x^2$ can be thought of as the difference between two neighboring values of $\partial/\partial x$, and so it takes the form shown in (c). The other two second-order operators, $\partial^2/\partial y^2$ and $\partial^2/\partial x\partial y$, appear in (d) and (e), respectively. Finally, the lowest-order isotropic operator, the Laplacian ($\partial^2/\partial x^2 + \partial^2/\partial y2$), which we denote by ∇^2, has the circularly symmetric form shown in (f).

image that has been convolved with two different-sized Gaussians whose space constants σ were 8 pixels (Figure 2-10b) and 4 pixels (Figure 2-10c). The reason why one chooses the Gaussian for this purpose, rather than blurring with a cylindrical pillbox function (for instance), is that the Gaussian distribution has the desirable characteristic of being smooth and localized in both the spatial and frequency domains and, in a strict sense, being the unique distribution that is simultaneously optimally localized in both domains. And the reason, in turn, why this should be a desirable property of our blurring function is that if the blurring is as smooth as possible, both spatially and in the frequency domain, it is least like to introduce any changes that were not present in the original image.

The second idea concerns the derivative part of the filter, ∇^2. The great advantage of using it is economy

of computation. First-order directional derivatives, like $\partial/\partial x$ or $\partial/\partial y$, could be used, in which case one would subsequently have to search for their peaks or troughs at each orientation (as illustrated in Figure 2-8b); or, second-order directional derivatives, like $\partial^2/\partial x^2$ or $\partial^2/\partial y^2$, could be used, in which case intensity changes would correspond to their zero-crossings (see Figure 2-8c). However, the disadvantage of all these operators is that they are directional; they all involve an orientation (see Figure 2-11, which illustrates the spatial organizations, or "receptive fields," in neurophysiological terms of the various first- and second-order differential operators). In order to use the first derivatives, for example, both $\partial I/\partial x$ and $\partial I/\partial y$ have to be measured, and the peaks and troughs in the overall amplitude have to be found. This means that the signed quantity $[(\partial I/\partial x)^2 + (\partial I/\partial y)^2]^{-1/2}$ must also be computed.

Using second-order directional derivative operators involves problems that are even worse than the ones involved in using first-order derivatives. The only way of avoiding these extra computational burdens is to try to choose an orientation-independent operator. The lowest-order isotropic differential operator is the Laplacian ∇^2, and fortunately it so happens that this operator can be used to detect intensity changes provided the blurred image satisfies some quite weak requirements (Marr and Hildreth, 1980).* Images on the whole do satisfy these requirements locally, so in

---

* The mathematical notation for blurring in image intensity function $I(x, y)$ with a Gaussian function G is $G * I$ which is read G convolved with I. The Laplacian of this is denoted by $\nabla^2 (G * I)$ and a mathematical identity allows us to move the ∇^2 operator inside the convolution giving $\nabla^2 (G * I) = (\nabla^2 G) * I$.

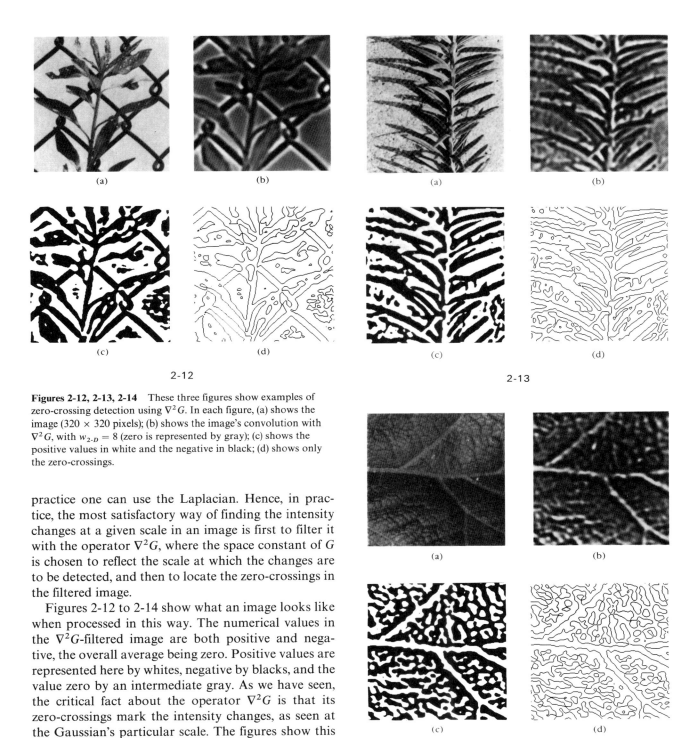

2-12

2-13

Figures 2-12, 2-13, 2-14 These three figures show examples of zero-crossing detection using $\nabla^2 G$. In each figure, (a) shows the image (320 × 320 pixels); (b) shows the image's convolution with $\nabla^2 G$, with $w_{2\text{-}D} = 8$ (zero is represented by gray); (c) shows the positive values in white and the negative in black; (d) shows only the zero-crossings.

2-14

practice one can use the Laplacian. Hence, in practice, the most satisfactory way of finding the intensity changes at a given scale in an image is first to filter it with the operator $\nabla^2 G$, where the space constant of G is chosen to reflect the scale at which the changes are to be detected, and then to locate the zero-crossings in the filtered image.

Figures 2-12 to 2-14 show what an image looks like when processed in this way. The numerical values in the $\nabla^2 G$-filtered image are both positive and negative, the overall average being zero. Positive values are represented here by whites, negative by blacks, and the value zero by an intermediate gray. As we have seen, the critical fact about the operator $\nabla^2 G$ is that its zero-crossings mark the intensity changes, as seen at the Gaussian's particular scale. The figures show this well. In Figure 2-12(c), for instance, the filtered image has been "binarized"—that is, positive values were all set to $+1$ and negative values to -1, and in Figure 2-12(d) the zero-crossings alone are shown. The advantage of the binarized representation is that it also shows the sign of the zero-crossing—which side in the image is the darker.

(a)   (b)

Figure 2-15   Another example of zero-crossings; here, the intensity of the lines has been made to vary with the slope of the zero-crossing, so that it is easier to see which lines correspond to the greater contrast. (Courtesy BBC Horizon.)

In addition, the slope of the zero-crossing depends on the contrast of the intensity change, though not in a very straightforward way. This is illustrated by Figure 2-15, which shows an original image together with zero-crossings that have been marked with curves of varying intensity. The more contrasty the curve, the greater the slope of the zero-crossing at that point, measured perpendicularly to its local orientation.

Zero-crossings like those of Figures 2-12 to 2-15 can be represented symbolically in various ways. I choose to represent them by a set of oriented primitives called *zero-crossing segments*, each describing a piece of the contour whose intensity slope (rate at which the convolution changes across the segment) and local orientation are roughly uniform. Because of their eventual physical significance, it is also important to make explicit those places at which the orientation of a zero-crossing changes "discontinuously." The quotation marks are necessary because one can in fact prove that the zero-crossings of $\nabla^2 G * I$ can never change orientation discontinuously, but one can nevertheless construct a practical definition of discontinuity. In addition, small, closed contours are represented as blobs, each also with an associated orientation, average intensity slope, and size defined by its extent along a major and minor axis. Finally, in keeping with the overall plan, several sizes of operator will be needed to cover the range of scales over which intensity changes occur.

Introduction

(1982)
J. A. Feldman and D. H. Ballard

Connectionist models and their properties
Cognitive Science 6:205–254

This paper is written from a perspective slightly different from those that we have seen. Feldman and Ballard are computer scientists and are concerned with networks as computing devices, even though their primary interest is in neural networks as cognitive models. This difference in background gives an invigorating twist to the problems they address and the devices they propose.

A major problem Feldman and Ballard discuss is the nature of the representation. In the early days of networks it was so amazing to get a network that learned, that exactly what it was learning was irrelevant. The simple statement that a state vector corresponded to something interesting was adequate. Sometimes (e.g., perceptrons) there was an attempt to provide a certain degree of respectability by assuming a topographically organized retina. Marr (paper 28) pointed out how difficult the questions of representation actually are and how many layers and different analyses, all cooperating, are necessary in order to do anything significant with a visual image.

One representational problem that has arisen before in this collection is the distinction between 'grandmother cell' representations and 'distributed representations.' In grandmother cell systems, a single active unit represents a complex concept. The R-units in the early perceptron (Rosenblatt, paper 8) were of this type, as are the word nodes in the Interactive Activation model of McClelland and Rumelhart (paper 25). However, models such as the linear associator and its variants associated patterns of activity with other patterns of activity. If a concept or an event is irreducibly represented as activities in many units, then the representation is called *distributed*.

A familiar example of a distributed representation would be a byte representing a character in a computer. In an 8-bit byte, 256 characters can be represented. The value of 1 bit does not uniquely determine a character; the values of the other 7 are equally necessary. This is a distributed coding. One could equally well represent characters by grandmother cells. If one had a word of 256 bits, then activity in 1 bit could represent a single character.

Both representations have advantages. Distributed codes require fewer units than grandmother cell systems, but they suffer from the problem of cross-talk, as Feldman and Ballard point out. This is a difficulty that was clearly present in pattern associators: if the stored patterns are not orthogonal, then different associations can mix together, causing potential errors of association. In some contexts, this interaction between stored associations can be used to do information processing (Knapp and Anderson, paper 36), but there is no question that cross-talk makes the information processing system fundamentally unpredictable in its behavior, and difficult to analyze. For those interested in practical applications of neural networks, this may be intolerable.

Feldman and Ballard also suggest a novel neuron model, which has similarities to what has been seen before, but which has some unusual features. Their model contains a potential, which can take on values from -10 to $+10$. The neuron sums its inputs, adds it to the potential, and responds to the sum with an integer value, from 1 to 9 if the final potential is above zero. One interesting suggestion is to make the dendrites of the neuron do individual logic computation. It is assumed that each dendrite is capable of doing sums of its own inputs and then responding. Feldman and Ballard suggest that it might be a useful device to compute logic functions of the individual dendritic computations: for example, they suggest that if the cell response takes on the value of the maximum of the dendritic computations, it makes a useful computing unit. Logically, each dendrite computes an alternative representation (for example, the same letter in different type fonts) and the overall response represents recognition of the letter that is independent of font. (There is physiological evidence for complex computation at dendrites; there seems to be agreement that cerebellar Purkinje cells and some cortical cells have dendrites that conduct action potentials; see Shepherd, 1979). They also suggest that it might be convenient on occasion to have units that simply compute logic functions, like McCulloch-Pitts neurons, which they call Q-units.

These complicated individual computing units, coupled with the author's bias toward very selective units, allow Feldman and Ballard to make complex calculations by connecting together small sets of them. For example, one important representational principle is what they call the unit/value principle, where a single unit corresponds to a value on a sensory dimension. Thus a unit might code a particular light intensity value, or a particular location in visual space. As one example, they mention the multiplication tables, where values might correspond to answers. Many different patterns of input, each with its own tuned dendrite, would activate the same output value unit.

Another important idea introduced in this paper is the "Winner-Take-All" (WTA) network. This is a system that lets the unit with the largest activity suppress all others in a group. Suppose that there are multiple possible units activated by an input, corresponding to different values of a sensory parameter. (This might happen because of noise, ambiguity, or from a host of reasonable causes.) Since only one output value is allowed, the WTA network selects the maximum by having the most active unit suppress the others (see Grossberg, paper 24, for some discussions of related networks).

To represent a single seen object, we might need unit values for position, color, depth, size, and shape. Each set of value units would have its own WTA network for selecting the single output value. The unit values representing the object would represent a *stable coalition* of units as the final representation. Note that this final representation is distributed (more than one unit active) but still composed of highly selective units. It could activate a true grandmother cell representing the object.

The issue of selectivity of representation is important, but it is clear that all realistic models partake of aspects of both: real neurons seem to have considerable selectivity; at the same time, even systems with very selective units have multiple active units doing much of the representation.

The rest of this paper gives a number of special cases and illustrative examples of the use of networks and selective units applied to various problems.

One trend seems clear in the development of neural networks, and which is demonstrated by this paper. One focus of network research is shifting from learning algorithms to details of the representations. Some of the learning algorithms are designed specifically to have the ability to do some statistically relevant formation of appropriate representations—for example, self-organization in visual cortex or back propagation. Aesthetically, this development is not for the better, since learning algorithms tend to be general and elegant, and representations tend to be specific and often incorporate unattractive hacks and kludges. But since the nervous system seems to be full of special-purpose hardware, perhaps this is the right way to use neural networks: a few, perhaps not even very powerful, learning algorithms and a number of powerful and highly developed specific representations evolved over millions of years.

Reference

G. M. Shepherd (1979), *The Synaptic Organization of the Brain*, New York: Oxford.

(1982)
J. A. Feldman and D. H. Ballard

Connectionist models and their properties
Cognitive Science 6:205–254

Much of the progress in the fields constituting cognitive science has been based upon the use of explicit information processing models, almost exclusively patterned after conventional serial computers. An extension of these ideas to massively parallel, connectionist models appears to offer a number of advantages. After a preliminary discussion, this paper introduces a general connectionist model and considers how it might be used in cognitive science. Among the issues addressed are: stability and noise-sensitivity, distributed decision-making, time and sequence problems, and the representation of complex concepts.

1. Introduction

Much of the progress in the fields constituting cognitive science has been based upon the use of concrete information processing models (IPM), almost exclusively patterned after conventional sequential computers. There are several reasons for trying to extend IPM to cases where the computations are carried out by a parallel computational engine with perhaps billions of active units. As an introduction, we will attempt to motivate the current interest in massively parallel models from four different perspectives: anatomy, computational complexity, technology, and the role of formal languages in science. It is the last of these which is of primary concern here. We will focus upon a particular formalism, connectionist models (CM), which is based explicitly on an abstraction of our current understanding of the information processing properties of neurons.

Animal brains do not compute like a conventional computer. Comparatively slow (millisecond) neural computing elements with complex, parallel connections form a structure which is dramatically different from a high-speed, predominantly serial machine. Much of current research in the neurosciences is concerned with tracing out these connections and with discovering how they transfer information. One purpose of this paper is to suggest how connectionist theories of the brain can be used to produce stable, detailed models of interesting behaviors. The distributed nature of information processing in the brain is not a new discovery. The traditional view (which we shared) is that

conventional computers and languages were Turing universal and could be made to simulate any parallelism (or analog values) which might be required. Contemporary computer science has sharpened our notions of what is "computable" to include bounds on time, storage, and other resources. It does not seem unreasonable to require that computational models in cognitive science be at least plausible in their postulated resource requirements.

The critical resource that is most obvious is time. Neurons whose basic computational speed is a few milliseconds must be made to account for complex behaviors which are carried out in a few hundred milliseconds (Posner, 1978). This means that *entire complex behaviors are carried out in less than a hundred time steps*. Current AI and simulation programs require millions of time steps. It may appear that the problem posed here is inherently unsolvable and that there is an error in our formulation. But recent results in computational complexity theory (Ja'Ja', 1980) suggest that networks of active computing elements can carry out at least simple computations in the required time range. In subsequent sections we present fast solutions to a variety of relevant computing problems. These solutions involve using massive numbers of units and connections, and we also address the questions of limitations on these resources.

Another recent development is the feasibility of building parallel computers. There is currently the capability to produce chips with 100,000 gates at a reproduction cost of a few cents each, and the technology to go to 1,000,000 gates/chip appears to be in hand. This has two important consequences for the study of CM. The obvious consequence is that it is now feasible to fabricate massively parallel computers, although no one has yet done so (Fahlman, 1980; Hillis, 1981). The second consequence of this development is the renewed interest in the basic properties of highly parallel computation. A major reason why there aren't yet any of these CM machines is that we do not yet know how to design, assemble, test, or program such engines. An important motivation for the careful study of CM is the hope that we will learn more about how to do parallel computing, but we will say no more about that in this paper.

The most important reason for a serious concern in cognitive science for CM is that they might lead to better science. It is obvious that the choice of technical language that is used for expressing hypotheses has a profound influence on the form in which theories are formulated and experiments undertaken. Artificial intelligence and articulating cognitive sciences have made great progress by employing models based on conventional digital computers as theories of intelligent behavior. But a number of crucial phenomena such as associative memory, priming, perceptual rivalry, and the remarkable recovery ability of animals have not yielded to this treatment. A major goal of this paper is to lay a foundation for the systematic use of massively parallel connectionist models in the cognitive sciences, even where these are not yet reducible to physiology or silicon.

Over the past few years, a number of investigators in different fields have begun to employ highly parallel models (idiosyncratically) in their work. The general idea has been advocated for animal models by Arbib (1979) and for cognitive models by Anderson (Anderson et al., 1977) and Ratcliff (1978). Parallel search of semantic memory and various "spreading activation" theories have become common (though not quite consistent) parts of information processing modeling. In machine perception research, massively parallel, cooperative computational theories have become a dominant paradigm (Marr & Poggio, 1976; Rosenfeld et al., 1976) and many of our examples come from our own work in this area (Ballard, 1981; Sabbah, 1981). Scientists looking at performance errors and other nonrepeatable behaviors have not found conventional IPM to be an adequate framework for their efforts. Norman (1981) has recently summarized arguments from cognitive psychology, and Kinsbourne and Hicks (1979) have been led to a similar view from a different perspective. It appears to us that all of these efforts could fit within the CM paradigm outlined here.

One of the most interesting recent studies employing CM techniques is the partial theory of reading developed in (McClelland & Rumelhart, 1981). They were concerned with the word superiority effect and related questions in the perception of printed words, and had a large body of experimental data to explain. One major finding is that the presence of a printed letter in a brief display is easier to determine when the letter is presented in the context of a word than when it is presented alone. The model they developed (cf. Figure 1) explicitly represents three levels of processing: visual features of printed letters, letters, and words. The

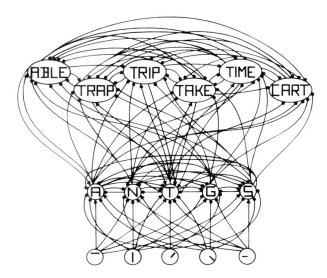

Figure 1 A few of the neighbors of the node for the letter "*t*" in the first position in a word, and their interconnections (McClelland & Rumelhart, 1981).

model assumes that there are positive and negative (circular tipped) connections from visual features to the letters that they can (respectively, cannot) be part of. The connections between letters and words can go in either direction and embody the constraints of English. The model assumes that many units can be simultaneously active, that units form algebraic sums of their inputs and output values proportionally. The activity of a unit is bounded from above and below, has some memory and decays with time. All of these features, and several more, are captured in the abstract unit described in Section 2.

This idea of simultaneously evaluating many hypotheses (here words) has been successfully used in machine perception for some time (Hanson & Riseman, 1978). What has occurred to us relatively recently is that this is a natural model of computation for widely interconnected networks of active elements like those envisioned in connectionist models. The generalization of these ideas to the connectionist view of brain and behavior is that all important encodings in the brain are in terms of the relative strengths of synaptic connections. The fundamental premise of connectionism is that individual neurons *do not transmit large amounts of symbolic information*. Instead they compute by being *appropriately connected* to large numbers of similar units. This is in sharp contrast to the conventional computer model of intelligence prevalent in computer science and cognitive psychology.

The fundamental distinction between the conventional and connectionist computing models can be

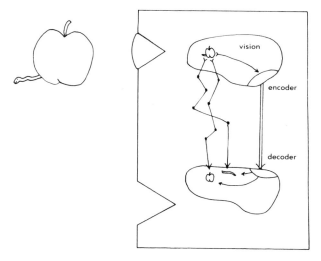

Figure 2 Connectionism vs. symbolic encoding. ⇒ assumes some general encoding. → assumes individual connections.

conveyed by the following example. When one sees an apple and says the phrase "wormy apple," some information must be transferred, however indirectly, from the visual system to the speech system. Either a sequence of special *symbols* that denote a wormy apple is transmitted to the speech system, or there are special *connections* to the speech command area for the words. Figure 2 is a graphic presentation of the two alternatives. The path on the right described by double-lined arrows depicts the situation (as in a computer) where the information that a wormy apple has been seen is encoded by the visual system and sent as an abstract message (perhaps frequency-coded) to a general receiver in the speech system which decodes the message and initiates the appropriate speech act. Notice that a complex message would presumably have to be transmitted sequentially on this channel, and that each end would have to learn the common code for every new concept. No one has yet produced a biologically and computationally plausible realization of this conventional computer model.

The only alternative that we have been able to uncover is described by the path with single-width arrows. This suggests that there are (indirect) links from the units (cells, columns, centers, or what-have-you) that recognize an apple to some units responsible for speaking the word. The connectionist model requires only very simple messages (e.g. stimulus strength) to cross a channel but puts strong demands on the availability of the right connections. Questions concerning the learning and reinforcement of connections are addressed in Feldman, (1981b).

For a number of reasons (including redundancy for reliability), it is highly unlikely that there is exactly one neuron for each concept, but the point of view taken here is that the activity of a small number of neurons (say 10) encodes a concept like apple. An alternative view (Hinton & Anderson, 1981) is that concepts are represented by a "pattern of activity" in a much larger set of neurons (say 1,000) which also represent many other concepts. We have not seen how to carry out a program of specific modeling in terms of these diffuse models. One of the major problems with diffuse models as a parallel computation scheme is cross-talk among concepts. For example, if concepts using units (10, 20, 30, . . .) and (5, 15, 25, . . .) were simultaneously activated, many other concepts, e.g., (20, 25, 30, 35, . . .) would be active as well. In the example of Figure 2, this means that diffuse models would be more like the shared sequential channel. Although a single concept could be transmitted in parallel, complex concepts would have to go one at a time. Simultaneously transmitting multiple concepts that shared units would cause cross-talk. It is still true in our CM that many related units will be triggered by spreading activation, but the representation of each concept is taken to be compact.

Most cognitive scientists believe that the brain appears to be massively parallel and that such structures can compute special functions very well. But massively parallel structures do not seem to be usable for general purpose computing and there is not nearly as much knowledge of how to construct and analyze such models. The common belief (which may well be right) is that there are one or more intermediate levels of computational organization layered on the neuronal structure, and that theories of intelligent behavior should be described in terms of these higher-level languages, such as Production Systems, Predicate Calculus, or LISP. We have not seen a reduction (interpreter, if you will) of any higher formalism which has plausible resource requirements, and this is a problem well worth pursuing.

Our attempts to develop cognitive science models directly in neural terms might fail for one of two reasons. It may be that there really is an interpreted symbol system in animal brains. In this case we would hope that our efforts would break down in a way that could shed light on the nature of this symbol system. The other possibility is that CM techniques are directly applicable but we are unable to figure out how to model some important capacity, e.g., planning. Our program is to continue the CM attack on problems of increasing difficulty (and to induce some of you to join us) until we encounter one that is intractable in our terms. There are a number of problems that are known to be difficult for systems without an interpreted symbolic representation, including complex concepts, learning, and natural language understanding. The current paper is mainly concerned with laying out the formalism and showing how it applies in the easy cases, but we do address the problem of complex concepts in Section 4. We have made some progress on the problem of learning in CM systems (Feldman, 1981b) and are beginning to work seriously on natural language processing and on higher-level vision. Our efforts on planning and long-term memory reorganization have not advanced significantly beyond the discursive presentation in (Feldman, 1980).

We will certainly not get very far in this program without developing some systematic methods of attacking CM tasks and some building-block circuits whose properties we understand. A first step towards a systematic development of CM is to define an abstract computing unit. Our unit is rather more general than previous proposals and is intended to capture the current understanding of the information processing capabilities of neurons. Some useful special cases of our general definition and some properties of very simple networks are developed in Section 2. Among the key ideas are local memory, non-homogeneous and non-linear functions, and the notions of mutual inhibition and stable coalitions.

A major purpose of the rest of the paper is to describe building blocks which we have found useful in constructing CM solutions to various tasks. The constructions are intended to be used to make specific models but the examples in this paper are only suggestive. We present a number of CM solutions to general problems arising in intelligent behavior, but *we are not suggesting that any of these are necessarily employed by nature.* Our notion of an adequate model is one that accounts for *all* of the established relevant findings and this is not a task to be undertaken lightly. We are developing some preliminary sketches (Ballard & Sabbah, 1981; Sabbah, 1981) for a serious model of low and intermediate level vision. As we develop various building blocks and techniques we will also be trying to bury some of the contaminated debris of past neural modeling efforts. Many of our constructions are intended as answers to known hard problems in CM computation. Among the issues addressed are: stability and noise-sensitivity, distributed decision-making, time and sequence problems, and the representation of complex concepts. The crucial questions of learning and change in CM systems are discussed elsewhere (Feldman, 1981b).

2. Neuron-Like Computing Units

As part of our effort to develop a generally useful framework for connectionist theories, we have developed a standard model of the individual unit. It will turn out that a "unit" may be used to model anything from a small part of a neuron to the external functionality of a major subsystem. But the basic notion of unit is meant to loosely correspond to an information processing model of our current understanding of neurons. The particular definitions here were chosen to make it easy to specify detailed examples of relatively complex behaviors. There is no attempt to be minimal or mathematically elegant. The various numerical values appearing in the definitions are arbitrary, but fixed finite bounds play a crucial role in the development. The presentation of the definitions will be in stages, accompanied by examples. A compact technical specification for reference purposes is included as Appendix A. Each unit will be characterized by a small number of discrete states plus:

p—a continuous value in $[-10, 10]$, called *potential* (accuracy of several digits)
v—an *output value*, integers $0 \leqslant v \leqslant 9$
i—a vector of *inputs* $i_1, \ldots, i_n$

P-Units

For some applications, we will be able to use a particularly simple kind of unit whose output v is proportional to its potential p (rounded) when $p > 0$ and which has only one state. In other words

$$p \leftarrow p + \beta \sum w_k i_k \qquad [0 \leq w_k \leq 1]$$
$$v \leftarrow if\ p > \theta\ then\ \text{round}\ (p - \theta)\ else\ 0 \qquad [v = 0 \ldots 9]$$

where β, θ are constants and w_k are weights on the input values. The weights are the sole locus of change with experience in the current model. Most often, the potential and output of a unit will be encoding its *confidence*, and we will sometimes use this term. The "$\leftarrow$" notation is borrowed from the assignment statement of programming languages. This notation covers both continuous and discrete time formulations and allows us to talk about some issues without any explicit mention of time. Of course, certain other questions will inherently involve time and computer simulation of any network of units will raise delicate questions of discretizing time.

The restriction that output take on small integer values is central to our enterprise. The firing frequencies of neurons range from a few to a few hundred impulses per second. In the 1/10 second needed for basic mental events, there can only be a limited amount of information encoded in frequencies. Then ten output values are an attempt to capture this idea. A more accurate rendering of neural events would be to allow 100 discrete values with noise on transmission (cf. Sejnowski, 1977). Transmission time is assumed to be negligible; delay units can be added when transit time needs to be taken into account.

The p-unit is somewhat like classical linear threshold elements (Minsky & Papert, 1972), but there are several differences. The potential, p, is a crude form of memory and is an abstraction of the instantaneous membrane potential that characterizes neurons; it greatly reduces the noise sensitivity of our networks. Without local memory in the unit, one must guarantee that all the inputs required for a computation appear simultaneously at the unit.

One problem with the definition above of a p-unit is that its potential does not decay in the absence of input. This decay is both a physical property of neurons and an important computational feature for our highly parallel models. One computational trick to solve this is to have an inhibitory connection from the unit back to itself. Informally, we identify the negative self feedback with an exponential decay in potential which is mathematically equivalent. With this addition, p-units can be used for many CM tasks of intermediate difficulty. The Interactive Activation models of McClelland and Rumelhart can be described naturally with p-units, and some of our own work (Ballard, 1981) and that of others (Marr & Poggio, 1976) can be done with p-units. But there are a number of additional features which we have found valuable in more complex modeling tasks.

Disjunctive Firing Conditions and Conjunctive Connections

It is both computationally efficient and biologically realistic to allow a unit to respond to one of a number of alternative conditions. One way to view this is to imagine the unit having "dendrites" each of which depicts an alternative enabling condition (Figure 3). For example, one could extend the network of Figure

1 to allow for several different type fonts activating the same letter node, with the higher connections unchanged. Biologically, the firing of a neuron depends, in many cases, on local spatio-temporal summation involving only a small part of the neuron's surface. So-called dendritic spikes transmit the activation to the rest of the cell.

In terms of our formalism, this could be described in a variety of ways. One of the simplest is to define the potential in terms of the maximum of the separate computations, e.g.,

$$p \leftarrow p + \beta \, Max(i_i + i_2 - \varphi, i_3 + i_4 - \varphi, i_5 + i_6 - i_7 - \varphi)$$

where β is a scale constant as in the p-unit and φ is a constant chosen (usually > 10) to suppress noise and require the presence of multiple active inputs (Sabbah, 1981). The minus sign associated with i_7 corresponds to its being an inhibitory input.

It does not seem unreasonable (given current data, Kuffler & Nicholls, 1976) to model the firing rate of some units as the maximum of the rates at its active sites. Units whose potential is changed according to the maximum of a set of algebraic sums will occur frequently in our specific models. One advantage of keeping the processing power of our abstract unit close to that of a neuron is that it helps inform our counting arguments. When we attempt to model a particular function (e.g., stereopsis), we expect to require that the number of units and connections as well as the execution time required by the model are plausible.

The max-of-sum unit is the continuous analog of a logical OR-of-AND (disjunctive normal form) unit and we will sometimes use the latter as an approximate version of the former. The OR-of-AND unit corresponding to Figure 3 is:

$$p \leftarrow p + \alpha \, OR \, (i_1 \& i_2, i_3 \& i_4, i_5 \& i_6 \& (not \, i_7))$$

This formulation stresses the importance that nearby spatial connections *all* be firing before the potential is affected. Hence, in the above example, i_3 and i_4 make a *conjunctive connection* with the unit. The effect of a conjunctive connection can always be simulated with more units but the number of extra units may be very large.

Q-Units and Compound Units

Another useful special case arises when one suppresses the numerical potential, p, and relies upon a finite-state set $\{q\}$ for modeling. If we also identify each input of **i** with a separate named input signal, we can get classical finite automata. A simple example would be a unit that could be started or stopped from firing.

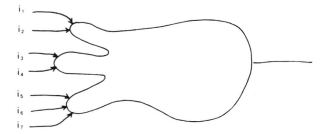

Figure 3 Conjunctive connections and disjunctive input sites.

One could describe the behavior of this unit by a table, with rows corresponding to states in {q} and columns to possible inputs, e.g.,

	i_1 (start)	i_2 (stop)
Firing	Firing	Null
Null	Firing	Null

One would also have to specify an output function, giving output values required by the rest of the network, e.g.,

$v \leftarrow$ if $q =$ Firing then 6 else 0.

This could also be added to the table above. An equivalent notation would be transition networks with states as nodes and inputs and outputs on the arcs.

In order to build models of interesting behaviors we will need to employ many of the same techniques used by designers of complex computers and programs. One of the most powerful techniques will be encapsulation and abstraction of a subnetwork by an individual unit. For example, a system that had separate motor abilities for turning left and turning right (e.g., fins) could use two start-stop units to model a turn-unit, as shown in Figure 4.

Note that the compound unit here has two distinct outputs, when basic units have only one (which can branch, of course). In general, compound units will differ from basic ones only in that they can have several distinct outputs.

The main point of this example is that the turn-unit can be described abstractly, independent of the details of how it is built. For example, using the tabular conventions described above,

	Left	Right	Values Output
a gauche	a gauche	a droit	$v_1 = 7, \quad v_2 = 0$
a droit	a gauche	a droit	$v_1 = 0, \quad v_2 = 8$

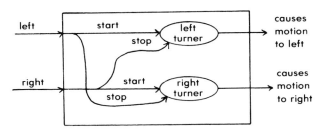

Figure 4 A Turn Unit.

where the right-going output being larger than the left could mean that we have a right-finned robot. There is a great deal more that must be said about the use of states and symbolic input names, about multiple simultaneous inputs, etc., but the idea of describing the external behavior of a system only in enough detail for the task at hand is of great importance. This is one of the few ways known of coping with the complexity of the magnitude needed for serious modeling of biological functions. It is not strictly necessary that the same formalism be used at each level of functional abstraction and, in the long run, we may need to employ a wide range of models. For example, for certain purposes one might like to expand our units in terms of compartmental models of neurons like those of (Perkel, 1979). The advantage of keeping within the same formalism is that we preserve intuition, mathematics, and the ability to use existing simulation programs. With sufficient care, we can use the units defined above to represent large subsystems without giving up the notion that each unit can stand for an abstract neuron. The crucial point is that a subsystem must be elaborated into its neuron-level units for timing and size calculations, but can (hopefully) be described much more simply when only its effects on other subsystems are of direct concern.

Units Employing p and q

It will already have occurred to the reader that a numerical value, like our p, would be useful for modeling the amount of turning to the left or right in the last example. It appears to be generally true that a single numerical value and a small set of discrete states combine to provide a powerful yet tractable modeling unit. This is one reason that the current definitions were chosen. Another reason is that the mixed unit seems to be a particularly convenient way of modeling the information processing behavior of neurons, as generally described. The discrete states enable one to model the effects in neurons of polypeptide modulators, abnormal chemical environments, fatigue, etc. Although these effects are often continuous functions of unit parameters, there are several advantages to using discrete states in our models. Scientists and laymen alike often give distinct names (e.g., cool, warm, hot) to parameter ranges that they want to treat differently. We also can exploit a large literature on understanding loosely-coupled systems as finite-state machines (Sunshine, 1979). It is also traditional to break up a function into separate ranges when it is simpler to describe that way. We have already employed all of these uses of discrete states in our detailed work (Feldman, 1981b; Sabbah, 1981). One example of a unit employing both p and q non-trivially is the following crude neuron model. This model is concerned with saturation and assumes that the output strength, v, is something like average firing frequency. It is not a model of individual action potentials and refractory periods.

We suppose the distinct states of the unit q ∈ {normal, recover}. In *normal* state the unit behaves like a p-unit, but while it is *recovering* it ignores inputs. The following table captures almost all of this behavior.

		$-1 < p < 9$	$p > 9$	Output Value
(incomplete)	normal	$p \leftarrow p + \Sigma i$	$p \leftarrow -p/$ recover	$v \leftarrow \alpha p - \theta$
	recover	normal	⟨impossible⟩	$v \leftarrow 0$

Here we have the change from one state to the other depending on the value of the potential, p, rather than on specific inputs. The recovering state is also characterized by the potential being set negative. The unspecified issue is what determines the duration of the recovering state—there are several possibilities. One is an explicit dishabituation signal like those in Kandel's experiments (Kandel, 1976). Another would be to have the unit sum inputs in the recovering state as well. The reader might want to conside how to add this to the table.

The third possibility, which we will use frequently, is to assume that the potential, p, decays toward zero (from both directions) unless explicitly changed. This implicit decay $p \leftarrow p_0 e^{-kt}$ can be modeled by self inhibition; the decay constant, k, determines the length of the recovery period.

The general definition of our abstract neural computing unit is just a formalization of the ideas presented above. To the previous notions of p, v, and *i* we formally add

{q}—a set of *discrete states*, < 10

and functions from old to new values of these

$p \leftarrow f(i, p, q)$
$q \leftarrow g(i, p, q)$
$v \leftarrow h(i, p, q)$

which we assume, for now, to compute continuously. The form of the f, g, and h functions will vary, but will generally be restricted to conditionals and simple functions. There are both biological and computational reasons for allowing units to respond (for example) logarithmically to their inputs and we have already seen important uses of the maximum function.

The only other notion that we will need is modifiers associated with the inputs of a unit. We elaborate the input vector *i* in terms of received values, weights, and modifiers:

$$\forall j, \, i_j = r_j \cdot w_j \cdot m_j \qquad j = 1, \ldots, n$$

where r_j is the *value* received from a predecessor [r = 0...9]; w_j is a changeable *weight*, unsigned [$0 \leq w_j \leq 1$] (accuracy of several digits); and m is a synapto-synaptic *modifier* which is either 0 or 1.

The weights are the only thing in the system which can change with experience. They are unsigned because we do not want a connection to change from excitatory to inhibitory. The modifier or gate simplifies many of our detailed models. Learning and change will not be treated technically in this paper, but the definitions are included in the Appendix for completeness (Feldman, 1981b).

We conclude this section with some preliminary examples of networks of our units, illustrating the key idea of mutual (lateral) inhibition (Fig. 5). Mutual inhibition is widespread in nature and has been one of the basic computational schemes used in modeling. We will present two examples of how it works to help aid in intuition as well as to illustrate the notation. The basic situation is symmetric configurations of p-units which mutually inhibit one another. Time is broken into discrete intervals for these examples. The examples are too simple to be realistic, but do contain ideas which we will employ repeatedly.

Two P-Units Symmetrically Connected
Suppose

$w_1 = 1, \quad w_2 = -.5$
$p(t + 1) = p(t) + r_1 - (.5)r_2 \qquad r_j = \text{received}$
$v = \text{round}(p) [0 \ldots 9]$

Referring to Figure 5a, suppose the initial input to the unit A.1 is 6, then 2 per time step, and the initial input to B.1 is 5, then 2 per time step. At each time step, each unit changes its potential by adding the external value (r_1) and substracting half the output value of its rival. This system will stabilize to the side of the larger of two instantaneous inputs.

Two Symmetric Coalitions of 2-Units

$w_1 = 1$
$w_2 = .5$
$w_3 = -.5$
$p(t + 1) = p(t) + r_1 + .5(r_2 - r_3)$
$v = \text{round}(p)$
A, C start at 6; B, D at 5;
A, B, C, D have no external input for $t > 1$

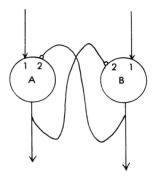

Suppose A₁ received an input of 6 units, then 2 per time step
Suppose B₁ received an input of 5 units, then 2 per time step

t	P(A)	P(B)
1	6	5
2	5.5	4
3	5.5	3.5
4	6	3
5	6.5	2
6	7.5	1
7	9.5	0
8	Sat	0

(a)

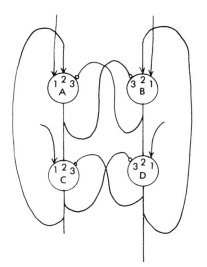

t	P(A)	P(B)	P(C)	P(D)
1	6	5	6	5
2	6.5	4.5	6.5	4.5
3	7.5	3.5	7.5	3.5
4	9.5	1.5	9.5	1.5
	Sat	0	Sat	0

(b)

Figure 5 Small Symmetric Networks.

The connections for this system are shown in Figure 5b. This system converges faster than the previous example. The idea here is that units A and C form a "coalition" with mutually reinforcing connections. The competing units are A vs. B and C vs. D. The last example is the smallest network depicting what we believe to be the basic mode of operation in connectionist systems. The faster convergence is not an artifact; the *positive feedback* among members of a coalition will generally lead to faster convergence than in separate competitions. It is the amount of positive feedback rather than just the size of the coalition that determines the rate of convergence (Feldman & Ballard, 1982). In terms of Figure 1, this could represent the behavior of the rival letters A and T in conjunction with the rival words ABLE and TRAP, in the absence of other active nodes.

Competing coalitions of units will be the organizing principle behind most of our models. Consider the two alternative readings of the Necker cube shown in Figure 6. At each level of visual processing, there are mutually contradictory units representing alternative possibilities. The dashed lines denote the boundaries of coalitions which embody the alternative interpretations of the image. A number of interesting phenomena (e.g., priming, perceptual rivalry, filling, subjective contour) find natural expression in this formalism. We are engaged in an ongoing effort (Ballard, 1981; Sabbah, 1981) to model as much of visual processing as possible within the connectionist framework. The next section describes in some detail a variety of simple networks which we have found to be useful in this effort.

3. Networks of Units

The main restriction imposed by the connectionist paradigm is that no symbolic information is passed from unit to unit. This restriction makes it difficult to employ standard computational devices like parameterized functions. In this section, we present connectionist solutions to a variety of computational problems. The sections address two principal issues. One is: Can the networks be connected up in a way that is sufficient to represent the problem at hand? The other is: Given these connections, how can the networks exhibit appropriate dynamic behavior, such as making a decision at an appropriate time?

Using a Unit to Represent a Value

One key to many of our constructions is the dedication of a separate unit to each value of each parameter of interest, which we term the unit/value principle. We

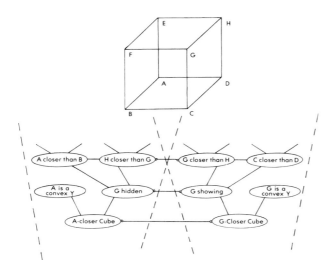

Figure 6   The Necker Cube.

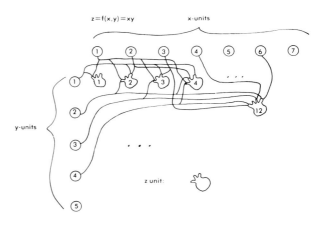

Figure 7   Multiplication Units.

will show how to compute using unit/value networks and present arguments that the number of units required is not unreasonable. In this representation the output of a unit may be thought of as a confidence measure. Suppose a network of depth units encodes the distance of some object from the retina. Then if the unit representing depth = 2 saturates, the network is expressing confidence that the distance is two units. Similarly, the "G-hidden" node in Figure 6 expresses confidence in its assertion. There is much neurophysiological evidence to suggest unit/value organizations in less abstract cortical maps. Examples are edge sensitive units (Hubel & Wiesel, 1979) and perceptual color units (Zeki, 1980), which are relatively insensitive to illumination spectra. Experiments with cortical motor control in the monkey and cat (Wurtz & Albano, 1980) suggest a unit/value organization. Our hypothesis is that the unit/value organization is widespread, and is a fundamental design principle.

Although many physical neurons do seem to follow the unit/value rule and respond according to the reliability of a particular configuration, there are also other neurons whose output represents the range of some parameter, and apparently some units whose firing frequency reflects both range and strength information (Scientific American, 1979). Both of the latter types can be accommodated within our definition of a unit, but we will employ only unit/value networks in the remainder of this paper.

In the unit/value representation, much computation is done by table look-up. As a simple example, let us consider the multiplication of two variables, i.e., $z = xy$. In the unit/value formalism there will be units for *every*

value of x and y that is important. Appropriate pairs of these will make a conjunctive connection with another unit cell representing a specific value for the product. Figure 7 shows this for a small set of units representing values for x and y. Notice that the confidence (expressed as output value) that a particular product is an answer can be a linear function of the maximum of the sums of the confidences of its two inputs. A major problem with function tables (and with CM in general) is the potential combinatorial explosion in the number of units required for a computation. A naive approach would demand N^2 units to represent all products of numbers from 1 to N. The network of Figure 7 requires many fewer units because each product is represented only once, another advantage of conjunctive connections. We could use even fewer units by exploiting positional notation and replacing each output connection with a conjunction of outputs from units representing multiples of 1, 10, 100, etc. The question of efficient ways of building connection networks is treated in detail in Section 4 (cf. also Hinton, 1981a; 1981b).

Modifiers and Mappings

The idea of function tables (Fig. 7) can be extended through the use of *variable mappings*. In our definition of the computational unit, we included a binary modifier, m, as an option on every connection. As the definition specifies, if the modifier associated with a connection is zero, the value v sent along that connection is ignored. Thus the modifier denotes inhibition, or blocking. There is considerable evidence in nature for synapses on synapses (Kandel, 1976) and the modifiers add greatly to the computational simplicity of our networks. Let us start with an initial informal example of the use of modifiers and mappings. Suppose that one has a model of grass as green except in California where

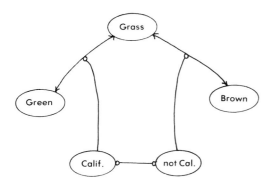

Figure 8 Gross is Green connection modified by California.

it is brown (golden), as shown in Figure 8. Here we can see that grass and green are potential members of a coalition (can reinforce one another) except when the link is blocked. This use is similar to the cancellation link of (Fahlman, 1979) and gives a crude idea of how context can effect perception in our models. Note that in Figure 8 we are using a shorthand notation. A modifier touching a double-ended arrow actually blocks two connections. (Sometimes we also omit the arrowheads when connection is double-ended.)

Mappings can also be used to select among a number of possible values. Consider the example of the relation between depth, physical size, and retinal size of a circle. (For now, assume that the circle is centered on and orthogonal to the line of sight, that the focus is fixed, etc.) Then there is a fixed relation between the size of retinal image and the size of the physical circle for any given depth. That is, each depth specifies a *mapping* from retinal to physical size (see Fig. 9). Here we suppose the scales for depth and the two sizes are chosen so that unit depth means the same numerical size. If we knew the depth of the object (by touch, context, or magic) we would know its physical size. The network above allows retinal size 2 to reinforce physical size 2 when depth = 1 but inhibits this connection for all other depths. Similarly, at depth 3, we should interpret retinal size 2 as physical size 8, and inhibit other interpretations. Several remarks are in order. First, notice that this network implements a function phys = f(ret, dep) that maps from retinal size and depth to physical size, providing an example of how to replace functions with parameters by mappings. For the simple case of looking at one object perpendicular to the line of sight, there will be one consistent coalition of units which will be stable. The work does something more, and this is crucial to our enterprise; the network can represent the consistency relation R among the three quantities: depth, retinal size, and physical size. It embodies not only the function f, but its two in-

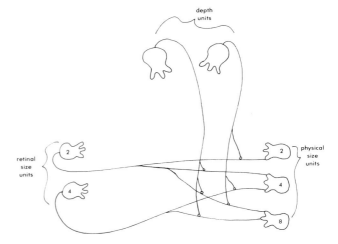

Figure 9 Depth Network using Modifiers.

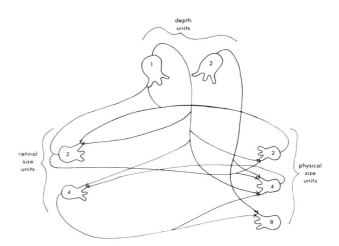

Figure 10 Depth Network using Conjunctive Connections.

verse functions as well (dep = f_1(ret, phys), and ret = f_2(phys, dep)). (The network as shown does not include the links for f_1 and f_2, but these are similar to those for f.) Most of Section 5 is devoted to laying out networks that embody theories of particular visual consistency relations.

The idea of modifiers is, in a sense, complementary to that of conjunctive connections. For example, the network of Figure 9 could be transformed into the following network (Fig. 10). In this network the variables for physical size, depth, and retinal size are all given equal weight. For example, physical size = 4 and depth = 1 make a *conjunctive connection* with retinal size = 4. Each of the value units in a competing row could be connected to all of its competitors by inhibitory links and this would tend to make the network activate only one value in each category. The general issue of rivalry and coalitions will be discussed in the next two sub-sections.

When should a relation be implemented with modifiers and when should it be implemented with conjunctive connections? A simple, nonrigorous answer to this question can be obtained by examining the size of two sets of units: (1) the number of units that would have to be inhibited by modifiers; and (2) the number of units that would have to be reinforced with conjunctive connections. If (1) is larger than (2), then one should choose modifiers; otherwise choose conjunctive connections. Sometimes the choice is obvious: to implement the brown Californian grass example of Figure 8 with conjunctive connections, one would have to reinforce all units representing places that had green grass! Clearly in this case it is easier to handle the exception with modifiers. On the other hand, the depth relation $R(phy, dep, ret)$ is more cheaply implemented with conjunctive connections. Since our modifiers are strictly binary, conjunctive connections have the additional advantage of continuous modulation.

To see how the conjunctive connection strategy works in general, suppose a constraint relation to be satisfied involves a variable x, e.g., $f(x, y, z, w) = 0$. For a particular value of x, there will be triples of values of y, z, and w that satisfy the relation f. Each of these triples should make a conjunctive connection with the unit representing the x-value. There could also be 3-input conjunctions at each value of y, z, w. Each of these four different kinds of conjunctive connections corresponds to an interpretation of the *relation* $f(x, y, z, w) = 0$ as a *function*, i.e., $x = f_1(y, z, w)$, $y = f_2(x, z, w)$, $z = f_3(x, y, w)$, or $w = f_4(x, y, z)$. Of course, these functions need not be single-valued. This network connection pattern could be extended to more than four variables, but high numbers of variables would tend to increase its sensitivity to noisy inputs. Hinton has suggested a special notation for the situation where a network exactly captures a consistency relation. The

mutually consistent values are all shown to be centrally linked (Fig. 11). This notation provides an elegant way of presenting the interactions among networks, but must be used with care. Writing down a triangle diagram does not insure that the underlying mappings can be made consistent or computationally well-behaved.

Winner-Take-All Networks and Regulated Networks

A very general problem that arises in any distributed computing situation is how to get the entire system to make a decision (or perform a coherent action, etc.). Biologically necessary examples of this behavior abound; ranging from turning left or right, through fight-or-flight responses, to interpretations of ambiguous words and images. Decision-making is a particularly important issue for the current model because of its restrictions on information flow and because of the almost linear nature of the p-units used in many of our specific examples. Decision-making introduces the notions of *stable states* and *convergence* of networks.

One way to deal with the issue of coherent decisions in a connectionist framework is to introduce *winner-take-all* (WTA) networks, which have the property that only the unit with the highest potential (among a set of contenders) will have output above zero after some setting time (Fig. 12). There are a number of ways to construct WTA networks from the units described above. For our purposes it is enough to consider one example of a WTA network which will operate in one time step for a set of contenders each of whom can read the potential of all of the others. Each unit in the network computes its new potential according to the rule:

$$p \leftarrow if\ p > \max(i_j, .1)\ then\ p\ else\ 0.$$

That is, each unit sets itself to zero if it knows of a higher input. This is fast and simple, but probably a little too complex to be plausible as the behavior of a single neuron. There is a standard trick (apparently widely used by nature) to convert this into a more plausible scheme. Replace each unit above with two

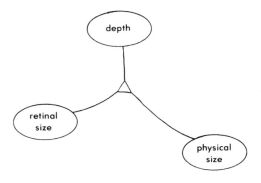

Figure 11 Notation for consistency relations.

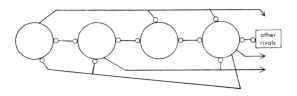

Figure 12 Winner-Take-All. Each unit stops if it sees a higher value.

units; one computes the maximum of the competitor's inputs and inhibits the other. The circuit above can be strengthened by adding a reverse inhibitory link, or one could use a modifier on the output, etc. Obviously one could have a WTA layer that got intputs from some set of competitors and settled to a winner when triggered to do so by some downstream network. This is an exact analogy of strobing an output buffer in a conventional computer.

One problem with previous neural modeling attempts is that the circuits proposed were often unnaturally delicate (unstable). Small changes in parameter values would cause the networks to oscillate or converge to incorrect answers. We will have to be careful not to fall into this trap, but would like to avoid detailed analysis of each particular model for delicacy in this paper. What appears to be required are some building blocks and combination rules that preserve the desired properties. For example, the WTA subnetworks of the last example will not oscillate in the absence of oscillating inputs. This is also true of any symmetric mutually inhibitory subnetwork. This is intuitively clear and could be proven rigorously under a variety of assumptions (cf. Grossberg, 1980). If every unit receives inhibition proportional to the activity (potential) of each of its rivals, the instantaneous leader will receive less inhibition and thus not lose its lead unless the inputs change significantly.

Another useful principle is the employment of lower-bound and upper-bound cells to keep the total activity of a network within bounds (Fig. 13). Suppose that we add two extra units, LB and UB, to a network which has coordinated output. The LB cell compares the total (sum) activity of the units of the network with a lower bound and sends positive activation uniformly to all members if the sum is too low. The UB cell inhibits all units equally if the sum of activity is too high. Notice that LB and UB can be parameters set from outside the nework. Under a wide range of conditions (but not all), the LB-UB augmented network can be designed to preserve order relationships among the outputs v_j

of the original network while keeping the sum between LB and UB.

We will often assume that LB-UB pairs are used to keep the sum of outputs from a nework within a given range. This same mechanism also goes far towards eliminating the twin perils of uniform saturation and uniform silence which can easily arise in mutual inhibition networks. Thus we will often be able to reason about the computation of a network assuming that it stays active and bounded.

Stable Coalitions

For a massively parallel system to actually make a decision (or do something), there will have to be states in which some activity strongly dominates. Such stable, connected, high confidence units are termed *stable coalitions*. A stable coalition is our architecturally-biased term for the psychological notions of percept, action, etc. We have shown some simple instances of stable coalitions, in Figure 5b and the WTA network. In the depth networks of Figures 9 and 10, a stable coalition would be three units representing consistent values of retinal size, depth, and physical size. But the general idea is that a very large complex subsystem must stabilize, e.g., to a fixed interpretation of visual input, as in Figure 1. The way we believe this to happen is through mutually reinforcing coalitions which dominate all rival activity when the decision is required. The simplest case of this is Figure 5b, where the two units A and B form a coalition which suppresses C and D. Formally, *a coalition will be called stable when the output of all its members is non-decreasing*. Notice that a coalition is not a particular anatomical structure, but an instantaneously mutually reinforcing set of units, in the spirit of Hebb's cell assemblies (Jusczyk & Klein, 1980).

What can we say about the conditions under which coalitions will become and remain stable? We will begin informally with an almost trivial condition. Consider a set of units $\{a, b, \ldots\}$ which we wish to examine as a possible coalition, π. For now, we assume that the units in π are all p-units and are in the non-saturated range and have no decay. Thus for each u in π,

$$p(u) \leftarrow p(u) + \text{Exc} - \text{Inh},$$

where Exc is the weighted sum of excitatory inputs and Inh is the weighted sum of inhibitory inputs. Now suppose that $\text{Exc}|\pi$, the excitation from the coalition π only, were greater than INH, the largest possible inhibition receivable by u, for each unit u in π, i.e.,

(SC) $\forall u \in \pi; \text{Exc}|\pi > \text{INH}$

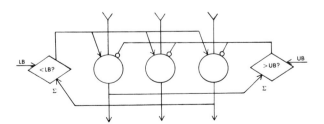

Figure 13 Regulated Network. If sum exceeds UB all units get uniform inhibition.

Then it follows that

$$\forall\, u \in \pi; p(u) \leftarrow p(u) + \delta \text{ where } \delta > 0.$$

That is, the potential of every unit in the coalition will increase. This is not only true instantaneously, but remains true as long as nothing external changes (we are ignoring state change, saturation, and decay). This is because $Exc|\pi$ continues to increase as the potential of the members of π increases. Taking saturation into account adds no new problems; if all of the units in π are saturated, the change, δ, will be zero, but the coalition will remain stable.

The condition that the excitation from other coalition members alone, $Exc|\pi$, be greater than any possible inhibition INH for each unit may appear to be too strong to be useful. It is certainly true that coalitions can be stable without condition (SC) being met. The condition (SC) is useful for model building because it may be relatively easy to establish. Notice that INH is directly computable from the description of the unit; it is the largest negative weighted sum possible. If inhibition in our networks is mutual, the upper-bound possible after a fixed time τ, INHτ, will depend on the current value of potential in each unit u. The simplest case of this is when two units are "deadly rivals"— each gets all its inhibition from the other. In such cases, it may well be feasible to show that after some time τ, the stable coalition condition will hold (in the absence of decay, fatigue, and changes external to the network). Often, it will be enough to show that the coalition has a stable "frontier," the set of units with outputs to some system under investigation.

There are a number of interesting properties of the stable coalition principle. First notice that it does not prohibit multiple stable coalitions nor single coalitions which contain units which mutually inhibit one another (although excessive mutual inhibition is precluded). If the units in the coalition had non-zero decay, the coalition excitation $Exc|\pi$ would have to exceed both INH and decay for the coalition to be stable. We suppose that a stable coalition yields control when its input elements change (fatigue and explicit resets are also feasible). To model coalitions with changeable inputs, we add boundary elements, which also had external "Input" and thus whose condition for being part of a stable coalition, π, would be:

$$Exc|\pi + \text{Input} > \text{INH}.$$

This kind of unit could disrupt the coalition if its Input went too low. The mathematical analysis of CM networks and stable coalitions continues to be a problem of interest. We have achieved some understanding of special cases (Feldman & Ballard, 1982) and these results have been useful in designing CM too complex to analyze in closed form.

4. Conserving Connections

It is currently estimated that there are about 10^{11} neurons and 10^{15} connections in the human brain and that each neuron receives intput from about 10^3–10^4 other neurons. These numbers are quite large, but not so large as to present no problems for connectionist theories. It is also important to remember that neurons are not switching devices; the same signal is propagated along all of the outgoing branches. For example, suppose some model called for a separate, dedicated path between all possible pairs of units in two layers in size N. It is easy to show that this requires N^2 intermediate sites. This means, for example, that there are not enough neurons in the brain to provide such a cross-bar switch for substructures of a million elements each. Similarly, there are not enough neurons to provide one to represent each complex object at every position, orientation, and scale of visual space. Although the development of connectionist models is in its perinatal period, we have been able to accumulate a number of ideas on how some of the required computations can be carried out without excessive resource requirements. Five of the most important of these are described below: (1) functional decomposition; (2) limited precision computation; (3) coarse and coarse-fine coding; (4) tuning; and (5) spatial coherence.

Functional Decomposition

When the number of variables in the function becomes large, the fan-in or number of input connections could become unrealistically large. For example, with the function $t = f(u, v, w, x, y, z)$ implemented with 100 values of t, when each of its arguments can have 100 distinct values, would require an average number of inputs per unit of $10^{12}/10^2$, or 10^{10}. However, there are simple ways of trading units for connections. One is to replicate the number of units with each value. This is a good solution when the inputs can be partitioned in some natural way as in the vision examples in the next section. A more powerful technique is to use intermediate units when the computation can be decomposed in some way. For example, if $f(u, v, w, x, y, z) = g(u, v) o\, h(w, x, y, z)$, where o is some composition, then separate networks of value units for $f(g, h)$, $g(u, v)$, and $h(w, x, y, z)$ can be used. The outputs from the g and h units can be combined in conjunctive connections according to the composition operator o in a third network representing f. An example is the case of word recognition. Letter-feature units would have to

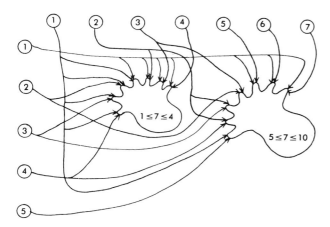

Figure 14 Modified Multiplication Table using Less Units.

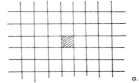

Figure 15a Coarse coding example. In a two-dimensional measurement space, the presence of a measurement can be encoded by making a single unit in the fine resolution space have a high confidence value. The same measurement can be encoded by making overlapping coarse units in three distinct coarse arrays have high confidence values.

connect to vastly more word units without the imposition of the intermediate level of letter units. The letter units limit the ways letter-feature units can appear in a word.

Limited Precision Computation

In the multiplication example $z = xy$, the number of z units required is proportional to $N_x N_y$ even when redundant value units are eliminated, and in general the number of units could grow exponentially with the number of arguments. However, there are several refinements which can drastically reduce the number of required units. One way to do this is to fix the number of units at the *precision* required for the computation. Figure 14 shows the network of Figure 7 modified when less computational accuracy is required.

This is the same principle that is incorporated in integer calculations in a sequential computer: computations are rounded to within the machine's accuracy. Accuracy is related to the number of bits and the number representation. The main difference is that since the sequential computer is general purpose, the number representations are conservative, involving large number of bits. The neural units need only represent sufficient accuracy for the problem at hand. This will generally vary from network to network, and may involve very inhomogeneous, special purpose number representations.

Coarse and Coarse-Fine Coding

Coarse coding is a general technical device for reducing the number of units needed to represent a range of values with some fixed precision, due to Hinton (1980). As Figure 15a suggests, one can represent a more precise value as the simultaneous activation of several (here 3) overlapping coarse-valued units. In general, D

simultaneous activations of coarse cells of diameter D precise units suffice. For a parameter space of dimension k, a range of F values can be captured by only F^k/D^{k-1} units rather than F^k in the naive method. The coarse coding trick and the related coarse-fine trick to be described next both depend on the input at any given time being sparse relative to the set of all values expressible by the network.

The coarse-fine coding technique is useful when the space of values to be represented has a natural structure which can be exploited. Suppose a set of units represents a vector parameter v which can be thought of as partitioned into two components (r, s). Suppose further that the number of units required to represent the subspace r is N_r, and that required to represent s is N_s. Then the number of units required to represent v is $N_r N_s$. It is easy to construct examples in vision where the product $N_r N_s$ is too close to the upper bound of 10^{11} units to be realistic. Consider the case of trihedral (v) vertices, an important visual cue. Three angles and two position coordinates are necessary to uniquely define every possible trihedral vertex. (Two angles define the types of vertex (arrow, y-joint); the third specifies the rotation of the joint in space.) If we use 5 degree angle sensitivity and 10^5 spatial sample points, the number of units is given by $N_r \approx 3.6 \times 10^5$ and $N_s = 10^5$ so that $N_r N_s \approx 3.6 \times 10^{10}$. How can we achieve the required representation accuracy with less units?

In many instances, one can take advantage of the fact that the *actual occurrence* of parameters is sparse. In terms of trihedral vertices, one assumes that in an image, such vertices will rarely occur in tight spatial clusters. (If they do, they cannot be resolved as individuals simultaneously.) Given that simultaneous proximal values of parameters are unlikely, they can be represented accurately for other computations, without excessive cost.

The solution is to decompose the space v into two

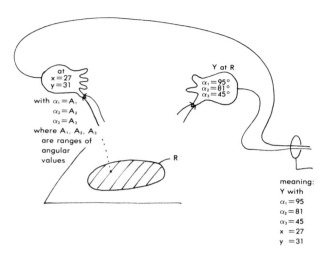

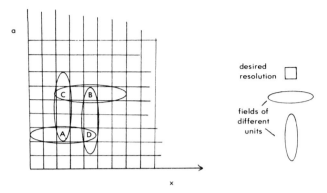

Figure 16 Inputs at A & B cause ghosts at C & D.

Figure 15b Coarse angle—fine position and coarse position—fine angle units combine to yield precise values of all five parameters.

subspaces, r and s, each with unilaterally reduced resolution.

Instead of $N_r N_s$ units, we represent v with two spaces, one with $N_{r'} N_s$ units where $N_{r'} \ll N_r$ and another with $N_r N_{s'}$ units where $N_{s'} \ll N_s$.

To illustrate this technique with the example of trihedral vertices we choose

$N_{s'} = 0.01 N_s$ and $N_{r'} = 0.01 N_r$.

Thus the dimensions of the two sets of units are:

$N_{s'} N_r = 3.6 \times 10^8$

and

$N_s N_{r'} = 3.6 \times 10^8$.

The choices result in one set of units which accurately represent the angle measurements and fire for a specific trihedral vertex anywhere in a fairly broad visual region, and another set of units which fire only if a general trihedral vertex is present at the precise position. The coarse-fine technique can be viewed as replacing the square coarse-valued covering in Figure 15a with rectangular (multi-dimensional) coverings, like those shown in Figure 16. In terms of our value units, the coarse-fine representation of trihedral vertices is shown in Figure 15b.

If the trihedral angle enters into another relation, say $R(v, \alpha)$, where both its angle and position are required accurately, one conjunctively connects pairs of appropriate units from each of the reduced resolution spaces to appropriate R-units. The conjunctive connection represents the *intersection* of each of its components' *fields*. Essentially the same mechanism will suffice for

conjoining (e.g.) accurate color with coarse velocity information.

An important limitation of these techniques, however, is that the input must be sparse. If inputs are too closely spaced, "ghost" firings will occur. In Figure 16, two sets of overlapping fields are shown, each with unilaterally reduced resolution. Actual input at points A and B will produce an erroneous indication of an input at C, in addition to the correct signals. The sparseness requirement has been shown to be satisfied in a number of experiments with visual data (Ballard & Kimball, 1981a, 1981b; Ballard & Sabbah, 1981).

The resolution device involves a units/connections tradeoff, but in general, the tradeoff is attractive. To see this, consider a unit that receives input from a network representing a vector parameter v. If n is the number of places where the output is used, and conjunctive connections are used to conjoin the D firing units, then Dn synapses are required. Thus if A is the number of non-coarse coded units to achieve a given acuity, then coarse coding is attractive when $A/D^{k-1} > Dn$, assuming connections and units are equally scarce. This result is optimistic in that, when other uses of conjunctive connections are taken into account, the number of conjunctive units could be unrealistically large.

Tuning

The idea of tuning further exploits networks composed of coarsely- and finely-grained units. Suppose there are n fine resolution units of a feature A and n fine resolutions for a feature B. To have explicit units for feature values AB, n^2 units would be required. This is an untenable solution for large feature spaces (the number of units grows exponentially with the number of features), so alternatives must be sought. One solution to this problem is to vary the grain of the AB units so that they are only coarsely represented. This solution has its attendant disadvantages in that separate stimuli within the limits of the coarse resolution grain cannot be distinguished. Also, a set of weak stimuli can be misinterpreted. A better solution is to have a coarse unit that would respond only to a single saturated unit within its input range. In that way a collection of weak inputs is not misinterpreted.

This situation can be achieved by having the units in each finely-tuned network that are in the field of a coarse unit laterally inhibit each other, e.g., in the WTA network of Figure 5a. The outputs of these individual feature units then form disjunctive connections with appropriate coarse resolution multiple feature units. If m is the grain of the coarse resolution units along with each feature dimension, the number of disjunctions per coarse unit is $(n/m)^2$. The result of this connection strategy is that a coarse unit responds with a strength that varies as the strengths of the largest maximum in the subnetwork of each of the finely-tuned units that correspond to its field. The response of a coarse-tuned unit is the maximum of the sums of the conjunctive inputs from the finely tuned units which connect to it. In terms of Figure 15, a tuned coarse-angle cell would respond only to one high-confidence pair of angles in its range, and not to several weak ones (which couldn't correctly appear all at one position). This is a better property than just having unstructured coarse units and it will be exploited in the next section, when we deal with perceiving complex objects.

Spatial Coherence

The most serious problem which requires conserving connections is the representation of complex concepts. The obvious way of representing concepts (sets of properties) is to dedicate a separate unit to each conjunction of features. In fact, it first appears that one would need a separate unit for each combination at each location in the visual field. We will present here a simple way around the problem of separate units for each location and deal with the more general problem in the next section.

The basic problem can be readily seen in the example of Figure 17. Suppose there were one unit each for finally recognizing concepts like colored circles and squares. Now consider the case when a red circle (at x = 7) and a blue square (at x = 11) simultaneously appear in the visual field. If the various "colored figure" units simply summed their inputs, the incorrect "blue circle" unit would see two active inputs, just like the correct "red circle" and "blue square" units. This problem is known as cross-talk, and is always a potential hazard in CM networks. The solution presented in Figure 17 is quite general. Each unit is assumed to have a separate conjunctive connection site for each

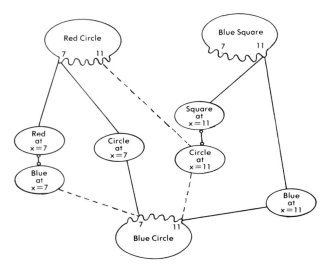

Figure 17 Spatial coherence on inputs can represent complex concepts without cross-talk. Solid lines show active inputs and dashed lines (some of the) inactive inputs.

position of the visual field. In our example, the correct units get dual inputs to a single site (and are activated) while the partially matched units receive separated inputs and are not activated. Only sets of properties which are spatially coherent can serve to activate concept units. This example was meant to show how spatial coherence could be used with conjunctive connections to eliminate cross-talk. There are a number of additional ways of using spatial coherence, each of which involves different tradeoffs. These are discussed in the next section, which considers some sample applications in more detail.

5. Applications

This section illustrates the power of the CM paradigm via two groups of examples. The first shows how the various techniques for conserving connections can be used in an idealized form of perception of a complex object. Here the point is that an object has multiple features which are computed in parallel via the transform methodology. The second group of examples starts with a relatively simple problem, that of vergence eye movements, to illustrate motor control using value units. In this example, control is immediate; a visual signal produces an instantaneous output (within the settling time constants of the units). Extensions of this idea use space as a buffer for time. For motor output, space allows the incoporation of more complex motor commands. For speech input, spatial buffering allows for phoneme recognition based on *subsequent* information.

These examples were chosen to show that CM can provide a unified representation for both perception and motor control. This is important since an animal is hardly ever passively responding to its environment. Instead, it seems involved in what Arbib has called a perception-action cycle (Arbib, 1979). Perceptions result in actions which in turn cause new perceptions, and so on. Massive parallelism changes the way the perception-action cycle is viewed. In the traditional view, one would convert the input to a language which uses variables, and then use these variables to direct motor commands. CM suggests that we think of accomplishing the same actions via a transformation: sensory input is transformed (connected to) to abstract representational units, which in turn are transformed (connected to) to motor units. This will obviously work for reflex actions. The examples are intended to suggest how more flexible command and control structures can also be represented by systems of value units.

Object Recognition

The examples of Figures 1 and 6 are representative of the problem of gestalt perception: that of seeing parts of an image as a single percept (object). An "object" is indicated by the "simultaneous" appearance of a number of 'visual features" in the correct relative spatial positions. In any realistic case, this will involve a variety of features at several different levels of abstraction and complex interaction among them. A comprehensive model of this process would be a prototype theory of visual perception and is well beyond the scope of this paper. What we will do here is consider the prerequisite task of constructing CM solutions to the problems of detecting non-punctate visual features and of forming sets of the features which could help characterize a percept. We will refer throughout to the prototype problem of detecting Fred's frisbee, which is known to be round, baby-blue, and moving fairly fast. The development suppresses many important issues such as hierarchical descriptions, perspective, occlusion, and the integration of separate fixations, not to mention learning. A brief discussion of how these might be tackled follows the technical material.

The first problem is to develop a general CM technique for detecting features and properties of images, given that these features are not usually detectable at a single point in some retinotopic map. The basic idea is to find parameters which characterize the feature in question and connect each retinotopic detector to the parameter values consistent with its detectand.

Consider the problem of detecting lines in an image from short edge segments. Different lines can be represented by units having different discrete parameter

values, e.g. in the line equation $p = x \cos \theta + y \sin \theta$, the parameters are p and θ. Thus edge units at (x, y, α) could be connected to appropriate line units. Note that this example is analogous to the word recognition example (Fig. 1). Edges are analogous to letters and lines to words. As in the words-letter example, "top-down" connections allow the existence of a line to raise the confidence of a local edge. In our line detection example, lines in the image are high potential (confidence) units in a slope-intercept (θ, p) parameter space. High confidence edge units produce high confidence line units by virtue of the newtork connectivity. This general way of describing this relationship between parts of an image (e.g., edges) and the associated parameters (e.g., p, θ for a line) is a connectionist interpretation of the *Hough transform* (Duda & Hart, 1972). Since each parameter value is determined by a large number of inputs, the method is inherently noise-resistant and was invented for this purpose. A Hough transform network for circles (like Fred's frisbee) would involve one parameter for size plus two for spatial location, and exactly this method has been used for tumor detection in chest radiographs (Kimme et al., 1975). Notice that the circle parameter space is itself retinotopic in that the centers of circles have specified locations; this will be important in registering multiple features.

The Hough transform is a formalism for specifying excitatory links between units. The general requirements are that part of an image representation can be represented by a parameter vector **a** in an image space A and a feature can be represented by a vector **b** which is an element of a feature space B. *Physical constraints* f(**a**, **b**) = 0 relate **a** and **b**. The space A represents spatially indexed units, and each individual element $\mathbf{a}_k$ is only consistent with certain elements in the space B, owing to the constraint imposed by the relation f. Thus for each $\mathbf{a}_k$ it is impossible to compute the set

$$B_k = \{\mathbf{b} | \mathbf{a}_k \text{ and } f(\mathbf{a}_k, \mathbf{b}) \leq \delta_b\}$$

where B_k is the set of units in the feature space network B that the $\mathbf{a}_k$ unit must connect to, and the constant δ_b is related to the quantization in the space B. Let H(**b**) be the number of active connections the value unit **b** receives from units in A. H(**b**) is the number of image measurements which are consistent with the parameter value **b**. The potential of units in B is given by p(**b**) ← H(**b**)/Σ_bH(**b**). The value p(**b**) can stand for the confidence that segment with feature value **b** is present in the image. If the measurement represented by **a** is realized as groups of units, e.g., $\mathbf{a} = (\mathbf{a}_1, \mathbf{a}_2)$, then conjunctive connections are required to implement the constraint relation.

Implementing these networks often results in a set of *very sparsely distributed* high-confidence feature space units. In implementations of the line detection example, only approximately 1% of the units have maximum confidence values. This figure is also typical of other modalities. In general, each $\mathbf{a}_k$ and the relationship f will not determine a single unit in $\mathbf{B}_k$ as in the line detection example, but there still will be isolated high-confidence units. Figure 1 shows why this is the case: different $\mathbf{a}_k$ letter-feature units connect to common units in the letter space B.

We have found that parameter spaces combine with the growing body of knowledge on specific physical constraints to provide a powerful and robust model for the simultaneous computation of invariant object properties such as reflectance, curvature, and relative motion (Ballard, 1981).

Of course segmentation must involve ways of associating peaks in several different feature spaces and methods for doing this are discussed presently, but the cornerstone of the techniques are high-confidence units in the individual-modality feature spaces. In extending the single feature case to multiple features, the most serious problem is the immense size of the cross product of the spatial dimensions with those of interesting features such as color, velocity, and texture. Thus to explain how image-like input such as color and optical flow are related to abstract objects such as "a blue, fast-moving thing," it becomes necessary to use all the techniques of the previous sections.

Even if we assume that there is a special unit for recognizing images of Fred's frisbee, it cannot be the case that there is a separate one of these units for each point in the visual field. One weak solution to this kind of problem was given in Figure 17 of the last section. There could conceivably be a separate 3-way conjunctive connection on the Fred's frisbee unit for each position in space. Activation of one conjunct would require the simultaneous activation of circle, baby-blue, and fairly-fast in the same part of the visual field. The solution style with separate conjunctions for every point in space becomes increasingly implausible as we consider more complex objects with hierarchical and multiple descriptions. The spatially registered conjunctions would have to be preserved throughout the structure.

The problem of going from a set of descriptors (features) to the object which is the best match to the set is known in artificial intelligence as the *indexing problem*. The feature set is viewed as an index (as in a data base). There have been several proposed parallel hierarchical network solutions to the indexing problem (Fahlman, 1979; Hillis, 1981) and these can be

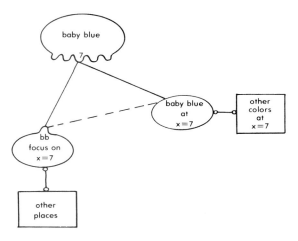

Figure 18 Spatial focus unit can gate only input from attended positions.

mapped into CM terms. But these designs assume that the network is presented with sets of descriptors which are already partitioned; precisely the vision problem we are trying to solve. There are three additional mechanisms that seem to be necessary, two of which have already been discussed. Coarse coding and tuning (as discussed in Section 4) make it much less costly to represent conjunctions. In addition, some general concepts (e.g., blue frisbee) might be indexed more efficiently through less precise units. The new idea is an extension of spatial coherence that exploits the fact that the networks respond to activity that occurs together in time. If there were a way to focus the activity of the network on one area at a time, only properties detected in that area would compete to index objects.

The obvious way to focus attention on one area of the visual field is with eye movements, but there is evidence that focus can also be done within a fixation. The general idea of internal spatial focus is shown in Figure 18. In this network, the general "baby-blue" unit is configured to have separate conjunctive inputs for each point in space, like the blue-square units of Figure 17. The difference is that the second input to the conjunction comes from a "focus" unit, and this makes a much more general network. The idea of making a unit (e.g., baby blue) more responsive to inputs from a given spatial position can be implemented in different ways. The conjunctive connection at the $x = 7$ lobe of the baby-blue unit is the most direct way. But treating this conjunct as a strict AND would mean that all spatial units would have to be active when there was no focus. An alternative would be to have the "focus on 7" unit boost the output of the "baby blue at 7" unit (and all of its rivals) as shown by the dashed line; this would eliminate the need for separate spatial conjunctions on the baby-blue unit, but would alter the potential of all the units at the position being attended. The trade-offs become even trickier when goal-directed input is taken into account, but both methods have the same effect on indexing. If the system has its attention directed only to x = 7, then the only feature units activated at all will be those whose local representatives are dominant (in their WTA) at x = 7. In such a case, there would be a time when the only concept units active in the entire network would be those for x = 7. This does not "solve" the problem of identifying objects in a visual scene, but it does suggest that sequentially focusing attention on separate places can help significantly. There is considerable reason to suppose (Posner, 1978; Triesman, 1980) that people do this even in tasks without eye movement.

There are other ways of looking at the network of Figure 18. Suppose the system had reason to focus on some particular property (e.g., baby-blue). If we make bi-directional the links from "focus on x = 7" to "baby-blue" and "baby-blue at 7," a nice possibility arises. The "focus on 7" unit could have a conjunctive connection for each separate property at its position. If, for example, baby-blue was chosen for focus and was the dominant color at x = 7, then the "focus on x = 7" unit would dominate its rivals. This suggests another way in which the recognition of complex objects could be helped by spatial focus. Figure 19 depicts the fairly general situation.

In Figure 19, the units representing baby-blue, circular, and fairly-fast are assumed to be for the entire visual field and moderately precise. The dotted arrows to the "Fred's frisbee" node suggest that there might be more levels of description in a realistic system. The spatial focus links involving baby-blue are the same as in Figure 18, and are replicated for the other two properties. Notice that the position-specific sensing units do not have their potentials affected by spatial focus units, so that the sensed data can remain intact. The network of Figure 19 can be used in several ways.

If attention has been focused on x = 7 for any reason, the various space-independent units whose representatives are most active at x = 7 will become most active, presumably leading to the activation (recognition) of Fred's frisbee. If a top-down goal of looking for Fred's frisbee (or even just something baby-blue) is active, then the "focus on x = 7" will tend to defeat its WTA rivals, leading to the same result. A third possibility is a little more complicated, but quite powerful. Suppose that a given image, even in context, activates too many property units so that no objects are effectively indexed. One strategy would be to systematically scan each area of the visual field, eliminating confounding activity from other areas. But it is also possible to be more efficient. If some property unit (say baby-blue) were strongly activated, the network could focus attention on all the positions with that property. In this case it is like putting a baby-blue filter in front of the scene, and should often lead to better convergence in the networks for shape, speed, etc.

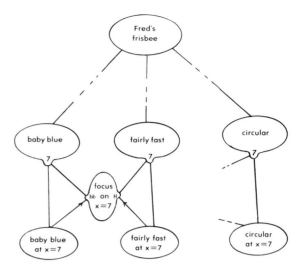

Figure 19 Spatial focus and indexing.

One should compare the network of Figure 17 with Figures 18 and 19. In the former, parallel co-existing concepts are possible if we assume delicate arrangements of conjunctive connections. The latter networks are more robust but use sequentiality to eliminate cross-talk.

Time and Sequence

Connectionist models do not initially appear to be well-suited to representing changes with time. The network for computing some function can be made quite fast, but it will be fixed in functionality. There are two quite different aspects of time variability of connectionist structures. One is time-varying responses, i.e., long-term modification of the networks (through changing weights) and short-term changes in the behavior of a fixed network with time. The second aspect is sequence: the problem of analyzing inherently sequential output (such as speech) or producing inherently sequential output (such as motor commands) with parallel models. The problem of change will be deferred to (Feldman, 1981b). The problem of sequence is discussed here.

There are a number of biologically suggested mechanisms for changing the weight (w_j) of synaptic

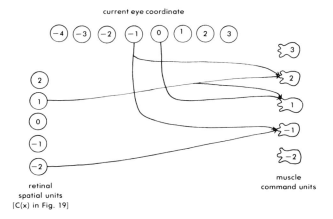

current eye coordinate

retinal
spatial units
[C(x) in Fig. 19]

muscle
command units

Figure 20   Distributed Control of Eye Fixations.

connections, but none of them are nearly rapid enough to account for our ability to hear, read, or speak. The ability to perceive a time-varying signal like speech or to integrate the images from successive fixations must be achieved (according to our dogma) by some dynamic (electrical) activity in the networks. As usual, we will present computational solutions to the problems of sequence that appear to be consistent with known structural and performance constraints. These are, again, too crude to be taken literally but do suggest that connectionist models can describe the phenomena.

Motor Control of the Eye   To see how the transform notion of distributed units might work for motor control, we present a simplistic model of vergence eye movements. (The same idea may be valid for fixations, but control probably takes place at higher levels of abstraction.) In this model retinotopic (spatial) units are connected directly to muscle control units. Each retinotopic unit can if saturated cause the appropriate contraction so that the new eye position is centered on that unit. When several retinotopic units saturate, each enables a muscle control unit independently and the muscle itself contracts an average amount.

Figure 20 shows the idea for a one-dimensional retina. For example, with units at positions 2, 4, 5, and 6 saturated, the net result is that the muscle is centered at 17/4 or 4.25. (This idea can be extended to the case where the retinotopic units have overlapping fields.) This kind of organization could be extended to more complex movement models such as that of the organization of the superior colliculus in the monkey (Wurtz & Albano, 1980).

Notice that each retinotopic unit is capable of enabling different muscle control units. The appropriate

one is determined by the enabled x-origin unit which inhibits commands to the inappropriate control units via modifiers.

One problem with this simple network arises when disparate groups of retinotopic units are saturated. The present configuration can send the eye to an average position if the features are truly identical. The newtork can be modified with additional connections so that only a single connected component of saturated units is enabled by using additional object primitives. A version of this WTA motor control idea has already been used in a computer model of the frog tectum (Didday, 1976).

There are still many details to be worked out before this could be considered a realistic model of vergence control, but it does illustrate the basic idea: local spatially separate sensors have *distinct, active* connections which could be averaged at the muscle for fine motor control or be fed to some intermediate network for the control of more complex behaviors.

Converting Space to Time   Consider the problem of controlling a simple physical motion, such as throwing a ball. It is not hard to imagine that in a skilled motor performance unit-groups fire each other in a fixed succession, leading to the motor sequence. The computational problem is that there is a unique set of effector units (say at the spinal level) that must receive input from each group at the right time. Figure 21a depicts a simple case in which there are two effector units (e_1, e_2) that must be activated alternatively. The circles marked 1–4 represent units (or groups of units) which activate their successor and inhibit their predecessor (cf. Delcomyn, 1980). The main point is that a succession of outputs to a single effector set can be modeled as a sequence of time-exclusive groups representing instantaneous coordinate signals. Moving from one time step to the next could be controlled by pure timing for ballistic movements, or by a proprioceptive feedback signal. There is, of course, an enormous amount more than this to motor control, and realistic models would have to model force control, ballistic movements, gravity compensation, etc.

The second part of Figure 21 depicts a somewhat fanciful notion of how a variety of output sequences could share a collection of lower level response units. The network shown has a single "Dixie" unit which can start a sequence and which joins in conjunctive connections with each note to specify its successor. At each time step, a WTA network decides what note gets sounded. One can imagine adding the rhythm network and transposition networks to other keys and to other modalities of output.

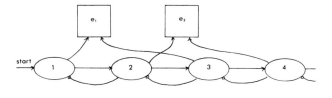

a. Sequence and Suppression

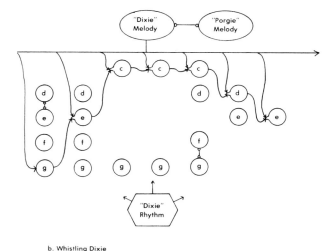

b. Whistling Dixie

Figure 21 Mapping Space to Time.

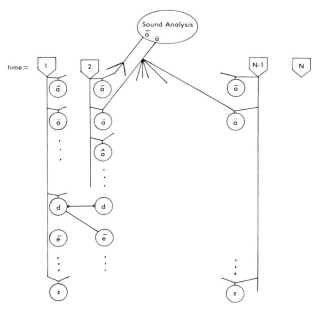

Figure 22 Mapping Time to Space.

Converting Time to Space The sequencer model for skilled movements was greatly simplified by the assumption that the sequence of activities was pre-wired. How could one (still crudely, of course) model a situation like speech perception where there is a largely unpredictable time-varying computation to be carried out? One solution is to combine the sequencer model of Figure 21 with a simple vision-like scheme. We assume that speech is recognized by being sequenced into a buffer of about the length of a phrase and then is relaxed against context in the way described above for vision. For simplicity, assume that there are two identical buffers, each having a pervasive modifier (m_j) innervation so that either one can be switched into or out of its connections. We are particularly concerned with the process of going from a sequence of potential phonetic features into an interpreted phrase. Figure 22 gives an idea of how this might happen.

Assume that there is a separate unit for each potential feature for each time step up to the length of the buffer. The network which analyzes sound is connected identically to each column, but conjunction allows only the connections to the active column to transmit values. Under ideal circumstances, at each time step

exactly one feature unit would be active. A phrase would then be laid out on the buffer like an image on the "mind's eye," and the analogous kind of relaxation cones (cf. Figure 1, 6) involving morphemes, words, etc., could be brought to bear. The more realistic case where sounds are locally ambiguous presents no additional problems. We assume that, at each time step, the various competing features get varying activation. Diphone constraints could be captured by (+ or −) links to the next column as suggested by Figure 22. The result is a multiple possibility relaxation problem— again exactly like that in visual perception. The fact that each potential feature could be assigned a row of units is essential to this solution; we do not know how to make an analogous model for a sequence of sounds which cannot be clearly categorized and combined. Recall that the purpose of this example is to indicate how time-varying input could be treated in connectionist models. The problem of actually laying out detailed models for language skills is enormous and our example may or may not be useful in its current form. Some of the considerations that arise in distributed modeling of language skills are presented in Arbib and Caplan, (1979).

Conclusions

The CM paradigm advanced in this paper has been applied successfully only to relatively low-level tasks. There is no reason, as yet, to be confident that an intermediate symbolic representation will not be required for modeling higher cognitive processes. There is, however, the beginning of a collection of efforts which can be interpreted as attempting CM approaches to higher level tasks. These include work which explicitly uses parallelism in planning (Stefik, 1981) and deduction, and work which incorporates more connectionist architectural notions of value units (Forbus, 1981) and coarse coding (Garvey, 1981).

We have now completed six years of intensive effort on the development of connectionist models and their application to the description of complex tasks. While we have only touched the surface, the results to data are very encouraging. Somewhat to our surprise, we have yet to encounter a challenge to the basic formulation. Our attempts to model in detail particular computations (Ballard & Sabbah, 1981; Sabbah, 1981) have led to a number of new insights (for us, at least) into these specific tasks. Attempts like this one to formulate and solve general computational problems in realistic connectionist terms have proven to be difficult, but less so than we would have guessed. There appear to be a number of interesting technical problems within the theory and a wide range of questions about brains and behavior which might benefit from an approach along the lines suggested in this paper.

Appendix: Summary of Definitions and Notation

A **unit** is a computational entity comprising:

$\{q\}$—a set of *discrete states*, < 10

p—a continuous value in $[-10, 10]$, called *potential* (accuracy of several digits)

v—an *output value*, integers $0 \leq v \leq 9$

i—a vector of *inputs* $i_1, \ldots, i_n$

and functions from old to new values of these

$p \leftarrow f(i, p, q)$
$q \leftarrow g(i, p, q)$
$v \leftarrow h(i, p, q)$

which we assume to compute continuously. The form of the f, g, and h functions will vary, but will generally be restricted to conditionals and simple functions.

P-Units

For some applications, we will use a particularly simple kind of unit whose output v is proportional to its potential p (rounded) (when $p > 0$) and which has only one state. In other words

$p \leftarrow p + \beta \Sigma w_k i_k$ $[0 \leq w_k \leq 1]$
$v \leftarrow$ *if* $p > \theta$ *then* round $(p - \theta)$ *else* 0 $[v = 0 \ldots 9]$

where β, θ are constants and w_k are weights on the input values.

Conjunctive Connections

In terms of our formalism, this could be described in a variety of ways. One of the simplest is to define the potential in terms of the maximum, e.g.,

$p \leftarrow p + \beta \text{Max}(i_1 + i_2 - \varphi, i_3 + i_4 - \varphi, i_5 + i_6 - i_7 - \varphi)$

where β is a scale constant as in the p-unit and φ is a constant chosen (usually > 10) to suppress noise and require the presence of multiple active inputs. The minus sign associated with i_7 corresponds to its being an inhibitory input. The max-of-sum unit is the continuous analog of a logical OR-of-AND (disjunctive normal form) unit and we will sometimes use the latter as an approximate version of the former. The OR-of-AND unit corresponding to the above is:

$p \leftarrow p + \alpha \text{ OR} (i_1 \& i_2, i_3 \& i_4, i_5 \& i_6 \& (\text{not } i_7))$

Winner-take-all (WTA) networks have the property that only the one with the highest potential (among a set of contenders) will have output above zero after some settling time.

A coalition will be called stable when the output of all of its members is non-decreasing.

Change

For our purposes, it is useful to have all the adaptability of networks be confined to changes in weights. While there is known to be some growth of new connections in adults, it does not appear to be fast or extensive enough to play a major role in learning. For technical reasons, we consider very local growth or decay of connections to be changes in existing connection patterns. Obviously, models concerned with developing systems would need a richer notion of change in connectionist networks (cf. von der Malsburg & Willshaw, 1977). We provide each unit with a memory vector μ which can be updated:

$\mu \leftarrow c(i, p, q, x, w, \mu)$

where μ is the intermediate-term memory vector, w is the weight vector, i, p, and q are as always, and x is an additional single integer intput ($0 \leq x \leq 1$) which captures the notion of the importance and value of the current behavior. Instantaneous establishment of long-term memory imprinting would be equivalent to having $\mu = w$. The assumption is that the consolidation of long-term changes is a separate process.

We postulate that important, favorable or unfavorable, behaviors can give rise to faster learning. The rationale for this is given in (Feldman, 1980, 1981a), which also lays out informally our views on how short- and long-term learning could occur in connectionist networks. A detailed technical discussion of this material, along the lines of this paper, is presented in (Feldman, 1981b). Obviously enough, a plausible model of learning and memory is a prerequisite for any serious scientific use of connectionism.

References

Anderson, J. A., Silverstein, J. W., Ritz, S. A., & Jones, R. S. Distinctive features, categorical perception, and probability learning: Some applications of a neural model. *Psychological Review*, September 1977, *84*(5), 413–451.

Arbib, M. A. *Perceptual structures and distributed motor control.* COINS (Tech. Rep. 79-11). University of Massachusetts, Computer and Information Science, and Center for Systems Neuroscience, June 1979.

Arbib, M. A., & Caplan, D. Neurolinguistics must be computational. *The Brain and Behavioral Sciences*, 1979, *2*, 449–483.

Ballard, D. H. Parameter networks: Towards a theory of low-level vision. *Proceedings of the 7th IJCAI*, Vancouver, BC, August 1981.

Ballard, D. H., & Kimball, O. A. *Rigid body motion from depth and optical flow* (Tech. Rep. 70). New York: University of Rochester, Computer Science Department, in press, 1981. (a)

Ballard, D. H., & Kimball, O. A. *Shape and light source direction from shading* (Tech. Rep.). Rochester, NY: University of Rochester, Computer Science Department, in press, 1981. (b)

Ballard, D. H., & Sabbah, D. On shapes. *Proceedings of the 7th IJCAI*, Vancouver, BC, August 1981.

Collins, A. M., & Loftus, E. F. A spreading-activation theory of semantic processing. *Psychological Review*, November 1975, *82*, 407–429.

Delcomyn, F. Neural basis of rhythmic behavior in animals. *Science*, October 1980, *210*, 492–498.

Dell, G. S., & Reich, P. A. Toward a unified model of slips of the tongue. In V. A. Fromkin (Ed.), *Errors in Linguistic Performance: Slips of the Tongue, Ear, Pen, and Hand.* New York: Academic Press, 1980.

Didday, R. L. A model of visuomotor mechanisms in the frog optic tectum. *Mathematical Bioscience*, 1976, *30*, 169–180.

Duda, R. O., & Hart, P. E. Use of the Hough transform to detect lines and curves in pictures. *Communications of the ACM 15*(1), January 1972, 11–15.

Edelman, G., & Mountcastle, B. *The Mindful Brain.* Boston, MA: MIT Press, 1978.

Fahlman, S. E. *NETL, A System for Representing and Using Real Knowledge.* Boston, MA: MIT Press, 1979.

Fahlman, S. E. The Hashnet interconnection scheme. Computer Science Department, Carnegie-Mellon University, June 1980.

Feldman, J. A. *A distributed information processing model of visual memory* (Tech. Rep. 52). Rochester, NY: University of Rochester, Computer Science Department, 1980.

Feldman, J. A. A connectionist model of visual memory. In G. E. Hinton & J. A. Anderson (Eds.), *Parallel Models of Associative Memory.* Hillsdale, NJ: Lawrence Erlbaum Associates, 1981. (a)

Feldman, J. A. *Memory and change in connection networks* (Tech. Rep. 96). Rochester, NY: University of Rochester, Computer Science Department, October 1981. (b)

Feldman, J. A. *Four frames suffice* (Tech. Rep. 99). Rochester, NY: University of Rochester, Computer Science Department, in press, 1982.

Feldman, J. A., & Ballard, D. H. *Computing with connections* (Tech. Rep. 72). Rochester, NY: University of Rochester, Computer Science Department, 1981; to appear in book by A. Rosenfeld & J. Beck (Eds.), 1982.

Forbus, K. D. Qualitative reasoning about physical processes. *Proceedings of the 7th IJCAI*, Vancouver, BC, August 1981, 326–330.

Freuder, E. C. Synthesizing constraint expressions. *Communications of the ACM*, November 1978, *21*(11), 958–965.

Garvey, T. D., Lowrance, J. D., & Fischler, M. A. An inference technique for integrating knowledge from disparate sources. *Proceedings of the 7th IJCAI*, Vancouver, BC, August 1981, 319–325.

Grossberg, S. Biological competition: Decision rules, pattern formation, and oscillations. *Proc. National Academy of Science USA*, April 1980, *77*(4), 2238–2342.

Hanson, A. R., & Riseman, E. M., (Eds.). *Computer Vision Systems.* New York: Academic Press, 1978.

Hillis, W. D. The connection machine (Computer architecture for the new wave). AI Memo 646, M.I.T., September 1981.

Hinton, G. E. Relaxation and its role in vision. (Ph.D. thesis, University of Edinburgh, December 1977.)

Hinton, G. E. Draft of Technical Report. La Jolla, CA: University of California at San Diego, 1980.

Hinton, G. E. The role of spatial working memory in shape perception. *Proceeding of the Cognitive Science Conference*, Berkeley, CA, August 1981. (a) 56–60.

Hinton, G. E., & Anderson, J. A. (Eds.). *Parallel Models of Associative Memory.* Hillsdale, NJ: Lawrence Erlbaum Associates, 1981.

Horn, B. K. P., & Schunck, B. G. Determining Optical Flow. AI Memo 572, AI Lab, MIT, April 1980.

Hubel, D. H., & Wiesel, T. N. Brain mechanisms of vision. *Scientific American*, September 1979, 150–162.

Ja'Ja', J., & Simon, J. Parallel algorithms in graph theory: Planarity testing. CS 80-14, Computer Science Department, Pennsylvania State University, June 1980.

Jusczyk, P. W., & Klein, R. M. (Eds.). *The Nature of Thought: Essays in Honor of D. O. Hebb.* Hillsdale, NJ: Lawrence Erlbaum Associates, 1980.

Kandel, E. R. *The Cellular Basis of Behavior.* San Francisco, CA: Freeman, 1976.

Kimme, C., Sklansky, J., & Ballard, D. Finding circles by an array of accumulators. *Communications of the ACM*, February 1975.

Kinsbourne, M., & Hicks, R. E. Functional cerebral space: A model for overflow, transfer and interference effects in human performance: A tutorial review. In J. Requin (Ed.), *Attention and Performance 7.* Hillsdale, NJ: Lawrence Erlbaum Associates, 1979.

Kosslyn, S. M. *Images and Mind.* Cambridge, MA: Harvard University Press, 1980.

Kuffler, S. W., & Nicholls, J. G. *From Neuron to Brain: A Cellular Approach to the Function of the Nervous System.* Sunderland, MA: Sinauer Associates, Inc., Publishers, 1976.

Marr, D. C., & Poggio, T. Cooperative computation of stereo disparity *Science*, 1976, *194*, 283–287.

McClelland, J. L., & Rumelhart, D. E. An interactive activation model of the effect of context in perception: Part 1. *Psychological Review*, 1981.

Minsky, M., & Papert, S. *Perceptrons.* Cambridge, MA: The MIT Press, 1972.

Norman, D. A. A psychologist views human processing: Human errors and other phenomena suggest processing mechanisms. *Proceedings of the 7th IJCAI*, Vancouver, BC, August 1981, 1097–1101.

Perkel, D. H., & Mulloney, B. Calibrating compartmental models of neurons. *American Journal of Physiology* 1979, *235*(1), R93–R98.

Posner, M. I. *Chronometric Explorations of Mind.* Hillsdale, NJ: Lawrence Erlbaum Associates, 1978.

Prager, J. M. Extracting and labeling boundary segments in natural scenes. *IEEE Trans. PAMI*, January 1980, *2*(1), 16–27.

Ratcliff, R. A theory of memory retrieval. *Psychological Review*, March 1978, *85*(2), 59–108.

Rosenfeld, A., Hummel, R. A., & Zucker, S. W. Scene labelling by relaxation operations *IEEE Trans. SMC 6*, 1976.

Sabbah, D. Design of a highly parallel visual recognition system. *Proceedings of the 7th IJCAI* Vancouver, BC, August, 1981.

Scientific American. *The Brain.* San Francisco, CA.: W. H. Freeman and Company, 1979.

Sejnowski, T. J. Strong covariance with nonlinearly interacting neurons. *Journal of Mathematical Biology*, 1977, *4*(4), 303–321.

Smith, E. E., Shoben, E. J., & Rips, L. J. Structure and process in semantic memory: A featural model for semantic decisions. *Psychological Review*, 1974, *81*(3), 214–241.

Stefik, M. Planning with Constraints (MOLGEN: Part 1). *Artificial Intelligence*, *16*(2), 1981.

Stent, G. S. A physiological mechanism for Hebb's postulate of learning. *Proc. National Academy of Science USA*, April 1973, *70*(4), 997–1001.

Sunshine, C. A. Formal techniques for protocol specification and verification. *IEEE Computer*, August 1979.

Torioka, T. Pattern separability in a random neural net with inhibitory connections. *Biological Cybernetics*, 1979, *34*, 53–62.

Triesman, A. M., & Gelade, G. A feature-integration theory of attention. *Cognitive Psychology*, 1980, *12*, 97–136.

Ullman, S. Relaxation and constrained optimization by local processes. *Computer Graphics and Image Processing*, 1979, *10*, 115–125.

von der Malsburg, Ch., & Willshaw, D. J. How to label nerve cells so that they can interconnect in an ordered fashion. *Proc. National Academy of Science USA*, November 1977, *74*(11), 5176–5178.

Wickelgren, W. A. Chunking and consolidation: A theoretical synthesis of semantic networks, configuring in conditioning, S-R versus cognitive learning, normal forgetting, the amnesic syndrome, and the hippocampal arousal system. *Psychological Review*, 1979, *86*(1), 44–60.

Wurtz, R. H., & Albano, J. E. Visual-motor function of the primate superior colliculus. *Annual Review of Neuroscience*, 1980, *3*, 189–226.

Zeki, S. The representation of colours in the cerebral cortex. *Nature*, April 1980, *284*, 412–418.

30
Introduction

(1982)
Teuvo Kohonen

Self-organized formation of topologically correct feature maps
Biological Cybernetics 43:59–69

The brain is organized in many places so that aspects of the sensory environment are represented in the form of two-dimensional maps. For example, in the visual system, there are several topographic mappings of visual space onto the surface of the visual cortex. There are organized mappings of the body surface onto the cortex in both motor (the motor homunculus) and somatosensory areas, and tonotopic mappings of frequency in the auditory cortex. The use of topographic representations, where some important aspect of a sensory modality is related to the physical locations of the cells on a surface, is so common that it obviously serves an important information processing function. (For a review, see Knudsen, du Lac, and Esterly, 1987.)

Kohonen is attempting to construct an artificial system that can show the same behavior. This is not quite the same as developing selective cells in response to particular input patterns. It is a related question: developing topographically organized systems. Clearly, there is development of cell selectivity in such a system as well, since cells develop maximum responsiveness to particular values of a parameter. But the emphasis here is on the global organization.

The fundamental fact that must hold true for a topographically organized system is that nearby units respond similarly. The essential mechanism of the Kohonen scheme is to cause the system to modify itself so this occurs. The way it is done is straightforward. Units start off by responding randomly to the parameter of interest—frequency, spatial location, whatever. An input signal, with some value of the parameter, is provided. One unit responds *best* to that input. This unit is located. The neighbors of this unit, defined to be some region around it, and the unit itself have their synaptic weights changed, so they also now respond more like the best unit than they did before. The synapses are subject to normalization of some form, so that the sum of the weights is roughly constant; increasing one strength diminishes the others on that unit. These simple assumptions seem adequate to do the topographic ordering: neighborhoods become similar in their response properties, and global order follows from the smooth gradation of one neighborhood into the next. Computer simulations show that the organization is fast and reliable. It is possible to prove directly in a few simple cases that the systems will organize properly. Kohonen describes this model in more detail in a recent book (Kohonen, 1984).

Kohonen suggests several biological mechanisms that could give rise to such a system. He suggests that it is possible to form the clusters by assuming a system where nearby cells excite each other, with inhibition for longer distances—very similar to the kind of center-surround organization seen in a number of brain regions. Excitatory interactions among nearby cells tend to form compact regions that are highly excited.

Such regions, coupled with a simple modification of the Hebb learning rule, could implement the system.

Kohonen describes a number of simulations of various forms of the model. The conclusion is that the self-organizing effects seem to be relatively insensitive to parameters. Often the size of the excitatory region is changed during learning, starting out quite large and shrinking as organization proceeds.

This paper is interesting for a number of reasons. It suggests that topographic self-organization can be done by several rather simple mechanisms. There is no doubt that this kind of organization is useful for representing information in a real nervous system, because it is so frequently found in the brain. Using Kohonen's technique, it is possible to have artificial systems topographically self-organize in useful and effective ways. Kohonen has used it to make a speech recognition device that can recognize a large vocabulary of isolated words. It works by forming an organized map of the acoustic cues of phonemes on a surface, and then recognizes words by using their trajectories across the phoneme space (Kohonen, 1987).

References

E. I. Knudsen, S. du Lac, and S. D. Esterly (1987), "Computational maps in the brain," *Annual Review of Neuroscience, 1987* 10:41–65.

T. Kohonen (1984), *Self-Organization and Associative Memory*, Berlin: Springer.

T. Kohonen (1987), "Representation of sensory information in self-organizing feature maps, and relation of these maps to distributed memory networks," *Optical and Hybrid Computing*, H. H. Szu (Ed.), SPIE Vol. 634, Bellingham, WA: SPIE.

(1982)
Teuvo Kohonen

Self-organized formation of topologically correct feature maps
Biological Cybernetics 43:59–69

Abstract. This work contains a theoretical study and computer simulations of a new self-organizing process. The principal discovery is that in a simple network of adaptive physical elements which receives signals from a primary event space, the signal representations are automatically mapped onto a set of output responses in such a way that the responses acquire the same topological order as that of the primary events. In other words, a principle has been discovered which facilitates the automatic formation of topologically correct maps of features of observable events. The basic self-organizing system is a one- or two-dimensional array of processing units resembling a network of threshold-logic units, and characterized by short-range lateral feedback between neighbouring units. Several types of computer simulations are used to demonstrate the ordering process as well as the conditions under which it fails.

1. Introduction

The present work has evolved from a recent discovery by the author (Kohonen, 1981), i.e. that topologically correct maps of structured distributions of signals can be formed in, say, a one- or two-dimensional array of processing units which did not have this structure initially. This principle is a generalization of the formation of direct topographic projections between two laminar structures known as *retinotectal mapping* (Willshaw and Malsburg, 1976, 1979; Malsburg and Willshaw, 1977; Amari, 1980). It will be introduced here in a general form in which signals of any modality may be used. There are no restrictions on the automatic formation of maps of completely abstract or conceptual items provided their signal representations or feature values are expressible in a metric or topological space which allows their ordering. In other words, we shall not restrict ourselves to topographical maps but consider *maps of patterns relating to an arbitrary feature or attribute space, and at any level of abstraction.*

The processing units by which these mappings are implemented can be identified with concrete physical adaptive components of a type similar to the Perceptrons (Rosenblatt, 1961). There is a characteristic feature in these new models, namely, a *local feedback* which makes map formation possible. The main objective of this work has been to demonstrate that *external signal activity alone, assuming a proper structural and functional description of system behavior, is sufficient for enforcing mappings of the above kind into the system.*

The present work is related to an idealized *neural* structure. However, the intention is by no means to assert that self-organization is mainly of neural origin; on the contrary, there are good reasons to assume that the basic state of readiness is often determined genetically. This does not exclude the possibility, however, that self-organization may significantly be affected and sometimes even completely determined by sensory experiences. On the other hand, the logic underlying this model is readily generalizable to mechanisms other than neural.

There are indeed many kinds of maps or images of sensory experiences in the brain; the most familiar ones are the retinotopic, somatotopic, and tonotopic projections in the primary sensory areas, as well as the somatotopic order of cells in the motor cortex. There is some evidence (Lynch et al., 1978) that topographic maps of the exterior environment are formed in the hippocampus. These observations suggest that the brains of different species would also more generally be able to produce maps of occurrences that are only indirectly related to the sensory inputs; notice that the signals received by the sensory areas have also been transformed by sensory organs, ganglia, and relay

© Springer-Verlag 1982

nuclei. If the ability to form maps were ubiquitous in the brain, then one could easily explain its power to operate on *semantic* items: some areas of the brain could simply create and order specialized cells or cell groups in conformity with high-level features and their combinations.

The possibility of constructing spatial maps for attributes and features in fact revives the old question of *how symbolic representations for concepts could be formed automatically*; most of the models of automatic problem solving and representation of knowledge have simply skipped this question.

2. Preliminary Simulations

In order to elucidate the self-organizing processes discussed in this paper, their operation is first demonstrated by means of ultimately simplified system models. The essential constituents of these systems are: 1. An array of processing units which receive coherent inputs from an event space and form simple discriminant functions of their input signals. 2. A mechanism which compares the discriminant functions and selects the unit with the greatest function value. 3. Some kind of local interaction which simultaneously activates the selected unit and its nearest neighbours. 4. An adaptive process which makes the parameters of the activated units increase their discriminant function values relating to the present input.

2.1. Definition of Ordered Mappings

Consider Fig. 1 which delineates a simple one-level self-organizing system. Information about the events $A_1, A_2, A_3, \ldots$ taking place in the exterior world is mediated in the form of *sensory signals* to a set of

processing units (shown here as a one-dimensional array for simplicity) via a *relaying network*. The sets of sensory signals S_i distributed to each processing unit i may be nonidentical and the number of signals in each S_i may be different; however, these signals are assumed to be *coherent* in the sense that they are uniquely determined by the same events A_k. Assume that the events A_k can be *ordered* in some metric or topological way such that $A_1 \mathbf{R} A_2 \mathbf{R} A_3 \ldots$ where $\mathbf{R}$ stands for a general ordering relation which is transitive (the above implies, e.g., that $A_1 \mathbf{R} A_3$). Assume further that the processing units produce output responses to the events with scalar values $\eta_i(A_1), \eta_i(A_2), \ldots$.

Definition. The system of Fig. 1 is said to implement a *one-dimensional ordered mapping* if for $i_1 > i_2 > i_3 > \ldots$,

$$\eta_{i_1}(A_1) = \max_j \{\eta_j(A_1) | j = 1, 2, \ldots, n\}$$

$$\eta_{i_2}(A_2) = \max_j \{\eta_j(A_2) | j = 1, 2, \ldots, n\}$$

$$\eta_{i_3}(A_3) = \max_j \{\eta_j(A_3) | j = 1, 2, \ldots, n\}$$

etc.

The above definition is readily generalizable to two- and higher-dimensional arrays of processing units; in this case some *topological order* must be definable for the events A_k, induced by more than one ordering relation with respect to different attributes. On the other hand, the topology of the array is simply defined by the definition of neighbours to each unit. If the unit with the maximum response to a particular event is regarded as the image of the latter, then *the mapping is said to be ordered if the topological relations of the images and the events are similar.*

2.2. Formation of Topological Maps in a Two-Dimensional Array with Identical Inputs to all Units

Consider Fig. 2 which delineates a rectangular array of processing units. In the first experiment, the relaying network was neglected, and the same set of input

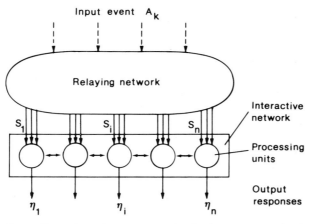

Fig. 1. Illustration of a system which implements an ordered mapping

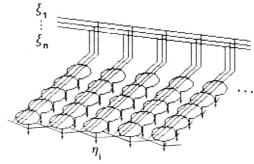

Fig. 2. Two-dimensional array of processing units

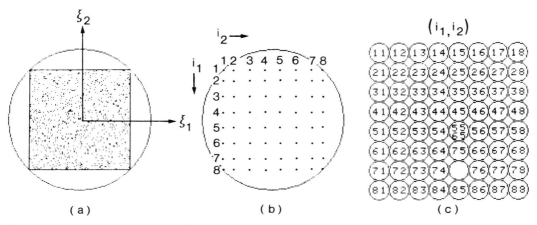

Fig. 3. a Distribution of training vectors (front view of the surface of a unit sphere in R^3). The distribution had edges, each of which contained as many vectors as the inside. **b** Test vectors which are mapped into the outputs of the processing unit array. **c** Images of the test vectors at the outputs

signals $\{\xi_1, \xi_2, ..., \xi_n\}$ was connected to all units. In accordance with notations used in mathematical system theory, this set of signals is expressed as a column vector $x = [\xi_1, \xi_2, ..., \xi_n]^T \in R^n$ where T denotes the transpose. Unit i shall have *input weights* or parameters $\mu_{i1}, \mu_{i2}, ..., \mu_{in}$ which are expressible as another column vector $m_i = [\mu_{i1}, \mu_{i2}, ..., \mu_{in}]^T \in R^n$. The unit shall form the discriminant function

$$\eta_i = \sum_{j=1}^{n} \mu_{ij}\xi_j = m_i^T x. \tag{1}$$

A discrimination mechanism (Sect. 3) shall further operate by which the maximum of the η_i is singled out:

$$\eta_k = \max_i \{\eta_i\}. \tag{2}$$

For unit k and all the eight of its nearest neighbours (except at the edges of the array where the number of neighbours was different) the following adaptive process is then assumed to be active:

$$m_i(t+1) = \frac{m_i(t) + \alpha x(t)}{\|m_i(t) + \alpha x(t)\|_E}, \tag{3}$$

where the variables have been labelled by a discrete-time index t (an integer), α is a "gain parameter" in adaptation, and the denominator is the Euclidean norm of the numerator. Equation (3) otherwise resembles the well-known teaching rule of the Perceptron, except that the direction of the corrections is always the same as that of x (no decision process or supervision is involved), and the weight vectors are normalized. Normalization improves selectivity in discrimination, and it is also beneficial in maintaining the "memory resources" at a certain level. Notice that the process of Eq. (3) does not change the length of m_i but only *rotates* m_i *towards* x. Nonetheless it is not always

necessary that the norm be Euclidean as in Eq. (3) (Sect. 3.3).

Simulation 1. A sequence of *training vectors* $\{x(t)\}$ was derived from the *structured* distribution shown in Fig. 3a. Without much loss of generality, the lengths of the $x(t)$ were normalized to unity whereby their distribution lies on the surface of the unit sphere in R^3. The "training vectors" were picked up noncyclically, in a completely random fashion from this distribution. The initial values for the parameters μ_{ij} were also defined as random numbers. The gain parameter α was made a function of the iteration step, e.g., proportional to $1/t$. (A decreasing sequence was necessary for stabilization, and this choice complies with that frequently used in mathematical models of learning systems.)

To test the final state of the system after many iterations, a set of test vectors from the distribution of Fig. 3a, as shown in Fig. 3b, was defined. The images of these vectors (i.e. those units which gave the largest responses to particular input vectors) are shown in Fig. 3c. It may be clearly discernible that an *ordered mapping* has been formed in the process. The map has also *formatted* itself along the sides of the array.

What actually caused the self-ordering? Some fundamental properties of this process can be determined by means of the following argumentation. The corrective process of Eq. (3) increases the parallelism of the activated (neighbouring) vectors. Thus the *differential order* all over the array will be increased on the average. However, differential ordering steps of the above kind cannot take place independently of each other. As all units in the array have neighbours which they affect during adaptation, changes in individual units cannot be *compatible* unless they result in a *global order*. The boundary effects in the array delimit the

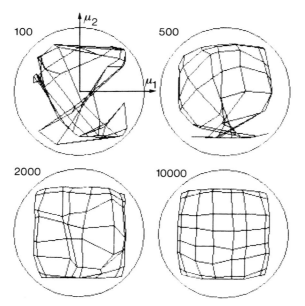

Fig. 4. Distribution of the weight vectors $m_i(t)$ at different times. The number of training steps is shown above the distribution. Interaction between nearest neighbours only

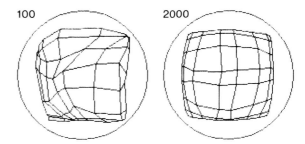

Fig. 5. Same as Fig. 4 except that a longer interaction range was used

format of the map in a manner somewhat similar to the way boundary conditions determine the solution of a differential equation.

Simulation 2. A clear conception of the ordering process is obtainable if the sequence of the weight vectors is illustrated using computer graphics. For this purpose, the vectors were assumed to be *three-dimensional*. Obviously the distribution of the weight vectors tends to imitate that of the training vectors $x(t)$. Since the vectors are normalized, they lie on the surface of a unit sphere in R^3. The order of the weight vectors in this distribution can be indicated simply by a lattice of lines which conforms with the topology of the processing unit array. A line connecting two weight vectors m_i and m_j is used only to indicate that the two corresponding units i and j are adjacent in the array. Figure 4 now shows a typical development of the vectors $m_i(t)$ in time; the illustration may be self-explanatory.

Simulation 3. This experiment was made in order to find out whether the ordering of the weight vectors would proceed faster if Eq. (3) were applied to more than the eight nearest neighbours of the selected unit. A number of experiments were made, and one of the best methods found was to apply Eq. (3) as such to the selected unit and its nearest eight neighbours while using an adaptation gain value of $\alpha/4$ for those 16 units which surrounded the previous ones. A result relating to the previous training vectors is given in Fig. 5. In this case ordering seems to proceed more quickly and

more reliably; on the other hand, the final result is perhaps not as good as before.

Comments. Something should be said here about simulations 1 through 3. There are eight equally probable symmetrical alternatives in which the map may be realized in the ordering process. One way to break the symmetry and to define a particular orientation for the map is to define "seeds", i.e. units with fixed, predetermined input weights. Another possibility is to use nonsymmetrical distributions and arrays which might have the same effect. We shall not take up this question in more detail.

2.3. Formation of Feature Maps in a One-Dimensional Array with Non-Identical but Coherent Inputs to the Units (Frequency Map)

The primary purpose of this experiment was to show that for self-organization *non-identical but coherent* inputs are sufficient.

Simulation 4. Consider Fig. 6, which depicts a one-dimensional array of processing units governed by system equations (1) through (3). In this case each unit except the outermost ones has two nearest neighbours.

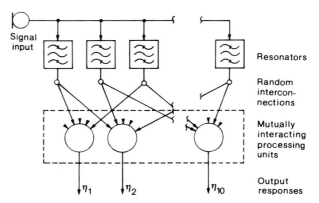

Fig. 6. Illustration of the one-dimensional system used in the self-organized formation of a frequency map

Table 1. Formation of frequency maps in Simulation 4. The resonators (20 in number) corresponded to second-order filters with quality factor $Q = 2.5$ and resonant frequencies selected at random from the range $[1, 2]$. The training frequencies were selected at random from the range $[0.5, 1]$. This table shows two different ordering results. The numbers in the table indicate those test frequencies to which each processing unit became most sensitive

Unit i	1	2	3	4	5	6	7	8	9	10
Frequency map in Experiment 1, 2000 steps	0.55	0.60	0.67	0.70	0.77	0.82	0.83	0.94	0.98	0.83
Frequency map in Experiment 2, 3500 steps	0.99	0.98	0.98	0.97	0.90	0.81	0.73	0.69	0.62	0.59

This system will receive sinusoidal signals and become ordered according to their frequency. Assume a set of resonators or bandpass filters tuned at random to different frequencies. *Five* inputs to each array unit are now picked up at random from the resonator outputs, so that there is no initial correlation or order in any structure or parameters. Next we shall carry out a series of adaptation operations, each time generating a new sinusoidal signal with a randomly chosen frequency. After a number of iteration steps the array units start to become sensitized to different frequencies in an ascending or descending order. The results of a few experiments are shown in Table 1.

Although this model system was a completely fictive one, a striking resemblance to the *tonotopic maps* formed in the auditory cortices of mammals (e.g., Reale and Imig, 1980) can be discerned; the extent of the disorders in natural maps is also similar.

3. A Possible Embodiment of Self-Organization in a Neural Structure

The models in Sect. 2 were set up without any reference to physical realizability. This section will discuss assumptions which lead to essentially similar self-organization in a *physical*, possibly *neural* system. It will be useful to realize that the complete process, in the earlier as well as present models, always consists of two phases which can be implemented, studied, and adjusted independently: 1. Formation of an activity cluster in the array around the unit at which activation was maximum. 2. Adaptive change in the input weights of those units where activity was confined.

It is salient that many structures of the central nervous system (CNS) are essentially two-dimensional, let alone the stratification of cells in several laminae. On the other hand, it is also rather generally agreed that in the neocortex, for instance, which has a pronounced vertical texture, the cell responses are very similar in the vertical direction. Many investigators

even hold the view that the cortical cell mass is functionally organized in *vertical columns*. It seems that such columns are organized around specific afferent axons so that they perform the basic input-output transformation of signals (Mountcastle, 1957; Towe, 1975).

There is both anatomical and physiological evidence for the following type of *lateral interaction* between cells: 1. Short-range lateral excitation reaching laterally up to a radius of 50 to 100 μm (in

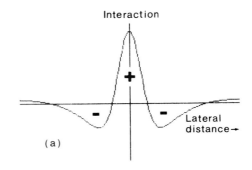

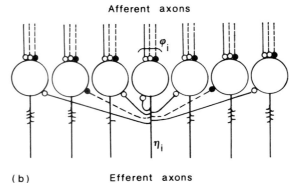

Fig. 7. a Lateral interaction around an arbitrary point of excitation, as a function of distance. Positive value: excitation. Negative value: inhibition. **b** Schematic representation of lateral connectivity which may implement the function shown at **a**. Open (small) circle: excitatory synapse. Solid circle: inhibitory synapse. Dotted line: polysynaptic connection. The variables φ_i and η_i: see simulations

primates); 2. The excitatory area is surrounded by a penumbra of inhibitory action reaching up to a radius of 200 to 500 μm; 3. A weaker excitatory action surrounds the inhibitory penumbra and reaches up to a radius of several centimeters.

The form of the lateral interaction function is depicted in Fig. 7a, and a schematic representation of a laminar network model that has the type of lateral interconnectivity possibly underlying the observed interactions, is delineated in Fig. 7b.

This particular network model is first used to demonstrate, in an ultimately simplified configuration, that the activity of neighbouring cells, due to the lateral interactions, can become clustered in small groups of roughly the same lateral dimension as the diameter of the excitatory or inhibitory region. The second step in modelling is then to show that if changes in the synaptic efficacies of the input connections are changed adaptively in proportion to presynaptic as well as postsynaptic activity, the process will be very similar to that already discussed in Sect. 2.

3.1. Dynamic Behaviour of the Network Activity

The CNS neurons usually fire rather regularly at a rate which depends on the integrated presynaptic transmission. It is a good approximation to assume that differential changes in the postsynaptic potential add up linearly. The overall average triggering frequency η of the neuron is then expressible as

$$\eta = \sigma\left[\sum_j \beta_j \xi_j\right], \tag{4}$$

where $\sigma[\cdot]$ defines a characteristic functional form which we shall study in a few cases, the ξ_j are the presynaptic impulse frequencies of all synapses, and the β_j now correspond to the synaptic efficacies. The β_j are positive for excitatory synapses and negative for the inhibitory ones.

Consider an array of principal neurons as depicted in Fig. 7; the interneurons have not been shown explicitly but manifest themselves through the lateral couplings. In accordance with Eq. (4) we will write for every output

$$\eta_i(t) = \sigma\left[\varphi_i(t) + \sum_{k \in S_i} \gamma_k \eta_k(t - \Delta t)\right], \tag{5}$$

where $\varphi_i(t)$ is the integrated depolarization caused by all afferent (external) inputs, and the second term represents inputs due to lateral couplings. Here S_i is the set of cells connected to cell i. The coefficients γ_k depend not only on synaptic efficacies but also on the type of lateral interconnections, and the γ_k around cell i shall roughly depend on the distance according to

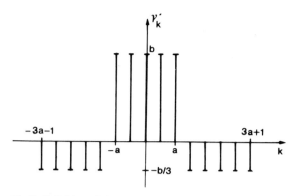

Fig. 8. Definition of the γ'_k coefficients used in Simulation 5

Fig. 7a. The synaptic transmission and latency delays Δt of the lateral couplings are assumed to be identical since their variations are of no interest here. It is essential to first study the recurrent process in which the $\eta_i(t)$ settle down to their asymptotic values in time.

Simulation 5. Most of the characteristic features of the above process will already be found in a one-dimensional model in which a row of cells is interconnected to yield a lateral excitability qualitatively similar to that given in Fig. 7a. (The long-range excitatory interaction is neglected.) Let us write

$$\eta_i(t) = \sigma\left[\varphi_i + \sum_{k=-L}^{+L} \gamma'_k \eta_{i+k}(t-1)\right], \tag{6}$$

where we have made $\Delta t = 1$; the coefficients γ'_k have been defined in Fig. 8. Moreover, $\sigma[a] = 0$ was chosen for $a < 0$, $\sigma[a] = a$ for $0 \leq a \leq A$, and $\sigma[a] = A$ for $a > A$.

In Fig. 9a, the outputs stabilized to values proportional to those of the φ_i. The form of the distribution changed due to the lateral connections. In Fig. 9b the $\eta_i(t)$ tend to *stable clusters* which have a lateral extension of the same order of magnitude as the excitatory center in the connectivity function. Such clusters may have a relation to the physiological "columns" of the cortical organization, although here they simply follow from lateral interactions.

It ought to be realized that these clusters are usually *self-resetting*; they tend to decay due to habituation, fluctuation of activation, etc.

It should be pointed out that for good clustering the *width* of the interaction function of Figs. 7 or 8 cannot be very small in relation to the curvature of the input activation. Otherwise, the lateral interaction tends only to enhance the borders of the input activation. Symptoms of such an effect are discernible in Fig. 9b (also Wilson and Cowan, 1973) (see also Sect. 4.3).

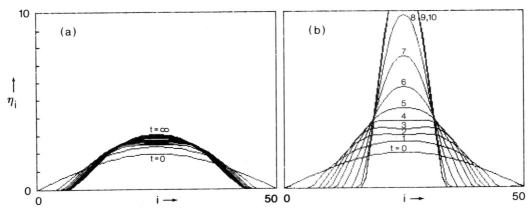

Fig. 9a and b. Development of activity in time over a one-dimensional interconnected array, vs. unit position. Input excitation: $\varphi_i = 2\sin(\pi i/50)$. **a** The lateral feedback was below a certain critical value (the parameters relating to Fig. 8 were: $a = 5$, $b = 0.024$, and the saturation limit was $= 10$). **b** Same as in **a** except that the lateral feedback exceeded the critical value ($b = 0.039$)

3.2. On the Analytical Model of Synaptic Plasticity

The synaptic efficacy can adaptively depend on signal values in many ways; potentiation and habituation effects can have widely different durations and be proportional to presynaptic or postsynaptic activity. In complex learning situations in which associative learning is present, the more permanent synaptic changes seem to need both presynaptic and postsynaptic activation. This is roughly the law usually referred to as the *Hebbian hypothesis* (Hebb, 1949). Some experimental evidence for the presence of both presynaptic and postsynaptic factors in the plasticity of cells in the visual cortex has recently been provided (Singer, 1977; Rauschecker and Singer, 1979) (Levy, 1980).

However, the original hypothesis of Hebb, which in effect stated that the efficacy of a synapse is increased with simultaneous presynaptic and postsynaptic triggering, is unsatisfactory for at least the following reasons: 1. Changes occur in one direction only. 2. There would be changes with background activity. Therefore, a more natural possibility which also satisfies the essential requirements is that *the sum of synaptic resources of a cell during a relatively short time span is approximately constant, and changes are induced only in the relative efficacies of the synapses. This is possible if the synaptic efficacy is mainly determined by one or more postsynaptic factors which are redistributed between the synaptic sites in proportion to their use* (Kohonen, 1977). Notice that for synaptic transmission various chemical agents and energy must be supplied. Their reserves are also limited.

It seems necessary to express the principle of limited synaptic resources in one form or another, not only for physical reasons but also since it seems to be very favourable for many learning processes, including the one discussed in this paper.

One of the simplest analytical expressions for changing synaptic efficacy of the above type follows from rather simple dynamics (Kohonen, 1977):

$$d\mu/dt = \alpha(\xi - \xi_b)\eta\,, \qquad (7)$$

where μ is the efficacy which corresponds to the input weight expressed in Eqs. (1) through (3), ξ is the presynaptic input (triggering frequency of the presynaptic neuron), ξ_b is an effective background value, η is the postsynaptic triggering frequency, and α is a proportionality constant (which depends on the type and location of the synapse). *Notice that all plastic synapses in the present model can be excitatory.*

3.3. Demonstration of Self-Organization in Networks of Neuron-Like Elements

We shall restrict ourselves below to physical models which have been constructed to implement the two partial processes mentioned at the beginning of this subsection: clustering of activity (Phase 1), and adaptation of the input weights (Phase 2).

Phase 1. Many simulations performed on the complete system models reported below have shown convincingly that *it is immaterial how the activity cluster is formed in Phase 1, as long as it attains the proper form* (Sect. 4); consequently, one may experiment with many differential equations for the system description. Since the numerical integration of these equations is usually rather tedious, it was considered permissible to speed up this phase as much as possible. Several simplifications for the simulation of the dynamic process were suggested. The most straighforward method, without losing much fidelity with respect to the original processes, was to make the increments of $\eta_i(t)$ fairly large at every interval of t, in fact on the order of one decade. Such a speed-up is normal in the discrete-time computing models applied in system theory.

It was then concluded that since the discrimination process is in any case a well-established phenomenon, its accurate modelling might not contribute anything essential when contrasted with the more interesting Phase 2. Relying on the results achieved in modelling Phase 1, the solution in each discrimination process (training step) was simply *postulated* to be some static function of the input excitations $\varphi_i = \sum_{j=1}^{n} \mu_{ij}\xi_j$. A simple, although not quite equivalent way, is to introduce a threshold by defining a *floating bias function* δ common for all units, and then by putting the system equations into a form in which the solution is determined implicitly:

$$\eta_i = \sigma\left[\varphi_i + \sum_{k \in S_i} \gamma_k \eta_k - \delta\right],$$
$$\delta = \max_i \{\eta_i\} - \varepsilon. \tag{8}$$

The nonlinearity $\sigma[\cdot]$ might be similar to that applied in Eq. (6), and ε is a small positive constant.

Perhaps the simplest model which can still be regarded as physical *first* performs thresholding of the input excitation and then lets the short-range local feedback amplify the activity and spread it to neighbouring units.

$$\varphi_i' = \sigma[\varphi_i - \delta], \qquad \delta = \max_i \{\varphi_i\} - \varepsilon,$$
$$\eta_i = \sum_{k \in S_i} \gamma_k \varphi_k'. \tag{9}$$

All the above equations have been tested to yield roughly similar activity clusters.

Phase 2. The wanted operation in the self-organizing process would be to rotate the weight vectors at each training step in the proper direction. Straightforward application of modifiability law of the type expressed in Eq. (7) would yield an expression

$$\mu_{ij}(t+1) = \mu_{ij}(t) + \alpha\eta_i(t)(\xi_j - \xi_b). \tag{10}$$

This, however, does not yet involve any *normalization*. Therefore it must be pointed out that Eq. (7) is already an approximation; *it was in fact derived by postulating that $\sum \mu_{ij}$ is constant.* Therefore, a more accurate version of Eq. (10) is the following

$$\mu_{ij}(t+1) = \frac{\mu_{ij}(t) + \alpha\eta_i(t)(\xi_j - \xi_b)}{\sum_j [\mu_{ij}(t) + \alpha\eta_i(t)(\xi_j - \xi_b)]}, \tag{11}$$

where normalization based on the conservation of "memory" resources (their linear sum) has been made.

Notice that the factor $\eta_i(t)$ in Eqs. (10) and (11) in fact corresponds to the selection rule relating to Eq. (3); the input weights of only the activated units change. Proportionality to $\eta_i(t)$ further means that the correction becomes a function of distance from the maximum response. However, in addition to rotating the m_i vectors, this process still affects their lengths.

It has been pointed out (Oja, 1981) that the factor which is assumed to be redistributed in the "memory" process actually need not be directly proportional to the input weight; for instance, if the input efficacy were proportional to the square root of this factor (a weaker function than the linear one!), then the denominator of Eq. (11) would already become similar to that applied in the teaching rule Eq. (3). Another interesting fact is that the Euclidean norm follows from a simple forgetting law. One may note further that a particular norm, and a particular form of the discrimination function should be related to each other; the most important requirement is to achieve good discrimination between neighbouring responses in one way or another.

Many simulations with physical process models of the above type were carried out; to make them comparable to those performed on the more fictive models of Sect. 2, three-dimensional vectors alone were used. The following adaptive law was then applied:

$$\mu_{ij}(t+1) = \frac{\mu_{ij}(t) + \alpha\eta_i(t)(\xi_j - \xi_b)}{\left\{\sum_{j=1}^{n} [\mu_{ij}(t) + \alpha\eta_i(t)(\xi_j - \xi_b)]^2\right\}^{1/2}}. \tag{12}$$

The results were not particularly sensitive to the value of ξ_b, which could be made zero.

Simulation 6. These experiments were performed with the simplest physical system model expressed in Eqs. (9), and in general they yielded very good results for many parameter values. The parameter ε can be used to control selectivity of the responses, and it also affects the width of the activity cluster. In this simulation ε was 0.05; on the other hand, if it was made equal to or greater than, say, 0.1, an interesting "collapse" phenomenon (Sect. 4.4) occurred. A "contraction" phenomenon (Sect. 4.2) due to boundary effects has also taken place (Fig. 10).

With the more complicated system models simulations with varying degrees of success have been performed. The reasons for different kinds of outcomes are discussed in Sect. 4 in more detail.

Conclusion. The conditions described in this section are in general *favourable* for the implementation of self-organization in a physical system. Most of these functions are also *realizable* with relatively simple components. This raises an intriguing question about

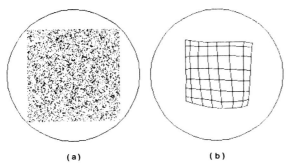

Fig. 10. a Distribution of training vectors used in a simple physical system model. b Distribution of weight vectors m_i after 4000 training steps

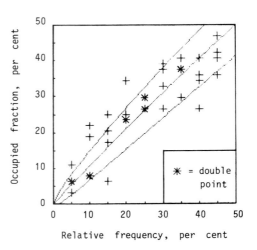

Fig. 11. Diagram showing the fraction of the processing unit array occupied by pattern vectors of one half of the distribution, vs. their relative frequency. Middle curve: optimal third-degree weighted least-square fit. The other curves: standard deviation

the realizability of this phenomenon in the *neural networks*. At least if one does not stipulate that the *exact* mathematical expression should be valid, but allows some variance, which in any case retains certain necessary conditions (good discrimination between neighbouring responses and formation of activity clusters in one way or another), the conditions met in neural circuits could also be conducive to this phenomenon.

4. Some Special Effects

4.1. The Magnification Factor

In this subsection an important property characteristic of biological organisms will be demonstrated: the *scale* or *magnification factor* of the map is usually not constant but a function of location in the array which depends on the frequency of input events that have mapped into that location during adaptation. It seems that the magnification factor is approximately proportional to the above frequency; this means that *the network resources are utilized optimally in accordance with need*.

The optimal allocation of resources in the mapping was demonstrated by means of the following series of experiments. The distribution of the training vectors, which was uniform over an area, was made variable; in a given experiment one half, say the right-hand one, had a different contingent of the total distribution. In other words the relative frequency of vectors drawn from this half was variable, and in every experiment the mapping was permitted to settle down to the asymptotic value. The relative fraction of the map into which these vectors were then mapped, or the "occupation of memory", was evaluated vs. the relative frequency of the vectors, and plotted into Fig. 11 (Kohonen, 1981).

4.2. Boundary Effects

Another effect which may deform the maps is caused by the fact that the arrays of processing units had borders. The outermost units had no neighbours on one side with which they could interact. Since the distribution of training vectors is usually bordered, too, some kind of "boundary effects" can then be observed. The most typical is "*contraction*" of the distribution of weight vectors: on average the outermost vectors are rotated more often inwards than outwards. This effect will iteratively spread to other vectors with the result that the whole distribution will be contracted from that of the training vectors. The "contraction" effect was clearly discernible in Simulation 6.

It must be realized that the distribution of weight vectors on the one hand, and the output map on the other have scales which are *reciprocal* to each other: if the distribution of weight vectors is contracted, the corresponding output map is *expanded*, and may even *overflow* the array. In fact, this is the phenomenon mentioned above, which, when occurring in modest amounts, is useful and even essential for effective map formation. It will *format* the map automatically along the borders. In still other words, there seems to be a kind of "pressure" in the map which tends to "mould" it into the given form.

Further it has to be remarked that brain networks do not have abrupt borders; accordingly, such boundary effects need not be considered in this particular form. Brain networks are often *parcelled* into subareas each of which receives signals with different origin.

Ordering within each subarea may thus occur independently and be only slightly affected at the demarcation zones.

4.3. The "Pinch" Phenomenon

Although the type of self-organization reported in this paper is not particularly "brittle" with respect to parameters, conditions in which this process fails do exist. One of the typical effects encountered is termed the "pinch" phenomenon. This means simply that the distribution of the memory vectors does not spread out into a planar configuration, but is instead concentrated onto a *ring*. Some kind of one-dimensional order of vectors may be discernible along the perimeter of the ring which, however, does not produce a meaningful map in the processing unit array.

A typical condition for the "pinch" phenomenon is *poor selectivity* in the discrimination process, especially *when the range of lateral interaction is too short*. This results in several "peaks" of activity, usually at the opposite edges of the array. In consequence, contradictory corrections are imposed on the weight vectors. Figure 12 exemplifies a typical distribution of weight vectors (shown without lattice lines) when this effect was fully developed.

4.4. The "Collapse" Phenomenon

Another pitfall related to the "pinch" phenomenon, and not much different in principle from the contraction effect discussed in Sect. 4.2, should be mentioned. This is the outcome in which all weight vectors tend to attain the same value. This is termed "collapse" (of the distribution, not of the map). This phenomenon was observed *when the range of lateral interaction was too long*. For instance, a too low threshold in the models of Sect. 3.4 resulted in the "collapse".

This effect is manifested in the output responses so that large groups of units give the same response.

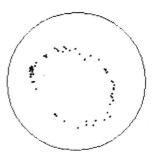

Fig. 12. An example of a weight vector distribution when the "pinch" phenomenon was due

4.5. The "Focusing" Phenomenon

This case is opposite to the "collapse". It may happen that one or more array units take over, i.e., they become sensitized to large portions of the distribution of the training vectors. In this case there is usually no order in the mapping. This failure is mainly due to poor general design or malfunction of the system, *especially if the lateral interaction between neighbouring elements is too weak*. Another common reason is that *the normalization of the weight vectors is not done in a proper way* whereby discrimination between the responses of neighbouring units becomes impossible.

5. On the Possible Roles of Various Neural Circuits Made by Interneurons

In view of the effects reported in Sects. 4.1 through 4.4 now seems possible to draw conclusions about the meaning of certain structures met in neural networks. Throughout the history of neuroanatomy and neurophysiology there has been much speculation about the purpose of the polysynaptic circuits made by various interneurons in CNS structures. Some investigators seem to be searching for an explanation in terms of complex computational operations while others see the basic implementation of feature detectors for sensory signals in these circuits. There are, e.g., quite specific circuits made by some cells such as the bipolar, chandelier, and basket cells in the cortex, for which a more detailed explanation is needed.

However, the roles of the above-mentioned cells, as well as the characteristic ramification of the axon collaterals of most cortical cell types would become quite obvious if the purpose was to implement a neural network with a capacity for predominantly two-dimensional self-organization. The bipolar cells, the recurrent axon collaterals of various cells, and the general vertical texture of the cortex warrant a high degree of conductance and spreading of signal activity in the *vertical* direction. On the other hand, if it were necessary to implement a *lateral interaction*, such as that described by the excitation function of the type delineated in Fig. 8a, then the stellate, basket, chandelier, etc. interneurons, and the horizontal intracortical axon collaterals would account for the desired lateral coupling.

The effects reported in Sects. 4.1 through 4.4 indicate that although self-organization of this type is not a particularly "brittle" phenomenon, the form of the lateral interaction function needs some adjustment. In a neural tissue this must be done by active circuits for which the interneurons are needed.

It might even be said that *the fraction of a certain cell type in the composition of all neurons is a tuning parameter by which an optimal form of local interaction can be defined.* The characteristic branchings of a particular cell type may roughly serve a similar purpose to that of the different basis functions in mathematical functional expansions. This conception is also in agreement with the fact that all cell types are not present in different species, or even in different parts of the brain: for some purposes it may be sufficient to "tune" a certain interaction with fewer types of cell ("basis functions"), while for more exacting tasks a richer variety of cells would be needed. Such recruitment of new forms according to need would be in complete agreement with the general principles of evolution.

References

Amari, S.-I.: Topographic organization of nerve fields. Bull. Math. Biol. **42**, 339–364 (1980)

Hebb, D.: Organization of behavior. New York: Wiley 1949

Kohonen, T.: Associative memory – a system-theoretical approach. Berlin, Heidelberg, New York: Springer 1977, 1978

Kohonen, T.: Automatic formation of topological maps of patterns in a self-organizing system. In: Proc. 2nd Scand. Conf. on Image Analysis, pp. 214–220, Oja, E., Simula, O. (eds.). Espoo: Suomen Hahmontunnistustutkimuksen Seura 1981

Levy, W.: Limiting characteristics of a candidate elementary memory unit: LTP studies of entorhinal-dentate synapses. (To appear in a book based on the workshop "Synaptic modification, neuron selectivity, and nervous system organization", Brown University, Rhode Island, Nov. 16–19, 1980)

Lynch, G.S., Rose, G., Gall, C.M.: In: Functions of the septo-hippocampal system, pp. 5–19. Amsterdam: Ciba Foundation, Elsevier 1978

Malsburg, Ch. von der: Self-organization of orientation sensitive cells in the striate cortex. Kybernetik **14**, 85–100 (1973)

Malsburg, Ch. von der, Willshaw, D.J.: How to label nerve cells so that they can interconnect in an ordered fashion. Proc. Natl. Acad. Sci. USA **74**, 5176–5178 (1977)

Mountcastle, V.B.: Modality and topographic properties of single neurons of cat's somatic sensory cortex. J. Neurophys. **20**, 408–434 (1957)

Oja, E.: A simplified neuron model as a principal component analyzer (1981) (to be published)

Rauschecker, J.P., Singer, W.: Changes in the circuitry of the kitten's visual cortex are gated by postsynaptic activity. Nature **280**, 58–60 (1979)

Reale, R.A., Imig, T.J.: Tonotopic organization in auditory cortex of the cat. J. Comp. Neurol. **192**, 265–291 (1980)

Rosenblatt, F.: Principles of neurodynamics: Perceptrons and the theory of brain mechanisms. Washington, D.C.: Spartan Books 1961

Singer, W., Rauschecker, J., Werth, R.: The effect of monocular exposure to temporal contrasts on ocular dominance in kittens. Brain Res. **134**, 568–572 (1977)

Swindale, N.V.: A model for the formation of ocular dominance stripes. Proc. R. Soc. B**208**, 243–264 (1980)

Towe, A.: Notes on the hypothesis of columnar organization in somatosensory cerebral cortex. Brain Behav. Evol. **11**, 16–47 (1975)

Willshaw, D.J., Malsburg, Ch. von der: How patterned neural connections can be set up by self-organization. Proc. R. Soc. B**194**, 431–445 (1976)

Willshaw, D.J., Malsburg, Ch. von der: A marker induction mechanism for the establishment of ordered neural mappings; its application to the retino-tectal problem. Phil. Trans. R. Soc. Lond. B**287**, 203–243 (1979)

Wilson, H.R., Cowan, J.D.: A mathematical theory of the functional dynamics of cortical and thalamic nervous tissue. Kybernetik **13**, 55–80 (1973)

31
Introduction

(1983)
Kunihiko Fukushima, Sei Miyake, and Takayuki Ito

Neocognitron: a neural network model for a mechanism of visual pattern recognition
IEEE Transactions on Systems, Man, and Cybernetics SMC-13:826–834

This paper and the one following (Barto, Sutton, and Anderson, paper 32) are from a special issue of the *IEEE Transactions on Systems, Man, and Cybernetics* on neural models in 1983. This special issue provides a snapshot of the accomplishments of the field at the time of the beginning of the current explosion of interest in it. There are many excellent papers in the issue and it is highly recommended.

Fukushima had proposed earlier (Fukushima, 1975) a multilayered neural network with strong self-organizing properties, which he called the "Cognitron." The more recent paper reprinted here has an interesting engineering orientation, which was unusual in the field of neural networks in 1983. It is a modification of the Cognitron architecture—hence Neocognitron.

Multilayer networks are hard to analyze. One of the reasons for this is what is called the 'credit assignment problem.' Barto, Sutton, and Anderson (paper 32) discuss this from the point of view of reinforcement delayed in time, which gives rise to some of the same technical problems as in multilayer networks.

In a learning system, we often have privileged access to the inputs and the outputs of the system. We might change connections between input and output elements by using a Hebb rule, say, or provide an error signal by subtracting the desired output from the actual output, as in the Widrow-Hoff rule. But given model neurons in an inner layer that are neither input nor output units ('hidden units'), it is hard to know how to adjust their synaptic weights so as to improve accuracy. Boltzmann machines (paper 38) and the 'back propagation' algorithm (paper 41) allow an explicit solution to the problem of learning in multilayer networks in a number of cases.

The multilayer learning system proposed by Fukushima, Miyake, and Ito takes a somewhat different point of view. It is assumed that the builder of the network knows roughly what kind of result is wanted and the information processing strategy to be employed. For the problem described in the paper, they were interested in the specific problem of recognizing handwritten characters. This meant they had to constrain the system considerably because of the nature of the inputs; that is, the inputs could only be configurations of lines of varying degrees of complexity. They wanted their system to be able to recognize characters presented anywhere on the field of view of the system. They modeled their system after the anatomy and physiology of the visual system, as described by Hubel and Weisel. They assumed a number of modules in series, which were driven by an initial retina of photoreceptors. Each module had a number of "S-cells," based on the properties of simple cells in the primary visual cortex, and "C-cells," based on the properties of complex cells.

As used in the paper, the S-cells responded to particular features in the driving layer, and the C-cells responded to the features picked up by the S-cells. The features learned were largely determined by the experimenter. The C-cell layers were driven by the S-cell layers. Therefore, by design, the C-cells abstracted over position the features that the S-cells picked up.

It should be pointed out, if only as an aside, that the physiology of simple and complex cells is somewhat more complicated than this. It is true that in primary visual cortex, simple cells pick up oriented line segments in particular locations, and complex cells tend to pick up orientation somewhat independently of location. The initial model proposed by Hubel and Wiesel in 1962 was strictly hierarchical, with spatially dependent features followed by a stage of spatially independent abstracted representations of the same features, with simple cells feeding complex cells. It seems, however, that simple and complex cells are partially parallel systems, and not strictly hierarchical. Their properties are reflections of two different cell populations, one largely linear, with receptive fields spatially localized, sensitive to low angular velocities of motion, and high spatial frequencies, and the other nonlinear, with receptive fields spatially less localized, and sensitive to high velocities of motion and low spatial frequencies. (In the cat these two classes of cells have become known as X and Y cells.) Complex cells not only receive inputs from simple cells, as a simple hierarchy would suggest, but also direct inputs. It is possible, therefore, for complex cells to respond before simple cells, due to the higher conduction velocity of the large Y cell axons. The idea of a simple generalizing information processing hierarchy, though still partially correct, is only a fragment of the physiological story. Henry (1985) gives a summary of this fascinating, evolving bit of neurophysiology.

In any case, the idea of layered feature extraction has a lot to recommend it from an engineering, as well as a physiological, point of view. Fukushima, Miyake, and Ito make artful use of the idea. They solve the credit assignment problem by assuming that the "teacher" knows roughly what features will be required and that learning progresses sequentially from the retina to the last stage, moving up the hierarchy. Because their system is organized in successive stages of response (S-cell) and spatial abstraction (C-cell), followed by other stages of response and spatial abstraction, they need only teach one S-cell to pick up a useful feature. Then the synaptic strengths can be copied to other cells. Their paper contains some discussions of the engineering trade-offs required to allow the cells to develop sufficient selectivity to be useful, but at the same time allow tolerance to distortion and noise. Features provided as inputs and taught to the system range from lines of various orientations at low levels to curves and corners at higher levels, to fragments of characters and characters at the highest levels.

After learning is completed, the final Neocognitron system is capable of recognizing handwritten numerals presented in any visual field location, even with considerable distortion. In this paper, Fukushima, Miyake, and Ito are quite modest in describing their results, which are actually very impressive. They have made a videotape of the Neocognitron in operation that demonstrates its capabilities and succeeds in giving an intuitive feeling for the operation of the system.

This paper gives rise to a number of interesting procedural points, which will become more and more important as the field of neural networks matures. By taking ideas from physiology, engineering, and network theory, and by combining them with some handcrafted fine tuning, Fukushima, Miyake, and Ito created a system capable of a task that is both difficult and highly practical. There is a large commercial market for recognizers of handwritten characters in Japan, where because of nature of the written language there is a great deal of handwritten communication. Some of the techniques used are not general. Some are. However, a glance at any book in neurophysiology will quickly convince a reader that there is some very special-purpose hardware inside the head as well. Carrying through practical tasks to completion may be a promising scientific research strategy, as well as being useful in itself, because it requires the researcher to make the trade-offs between generality and specificity in light of the task and the available resources that the nervous system must also make.

References

K. Fukushima (1975), "Cognitron: a self-organizing multilayered neural network," *Biological Cybernetics* 20:121–136.

G. H. Henry (1985), "Physiology of the cat striate cortex," *Cerebral Cortex*, Vol. 3: *Visual Cortex*, A. Peters and E. G. Jones (Eds.), New York: Plenum, pp. 119–155.

(1983)
Kunihiko Fukushima, Sei Miyake, and Takayuki Ito

Neocognitron: a neural network model for a mechanism of visual pattern recognition
IEEE Transactions on Systems, Man, and Cybernetics SMC-13:826–834

Abstract—A neural network model, called a "neocognitron," for a mechanism of visual pattern recognition was proposed earlier, and the result of computer simulation for a small-scale network was shown. A neocognitron with a larger-scale network is now simulated on a digital computer and is shown to have a great capability for visual pattern recognition: The neocognitron's ability to recognize handwritten Arabic numerals, even with considerable deformations in shape, is demonstrated. The neocognitron is a multilayered network consisting of a cascaded connection of many layers of cells. The information of the stimulus pattern given to the input layer is processed step by step in each stage of the multilayered network. A cell in a deeper layer generally has a tendency to respond selectively to a more complicated feature of the stimulus patterns and, at the same time, has a larger receptive field and is less sensitive to shifts in position of the stimulus patterns. Thus each cell of the deepest layer of the network responds selectively to a specific stimulus pattern and is not affected by the distortion in shape or the shift in position of the pattern. The synapses between the cells in the network are modifiable, and the neocognitron has a function of learning. A learning-with-a-teacher process is used to reinforce these modifiable synapses in the new model, instead of the learning-without-a-teacher process which was applied to the previous small-scale model.

I. INTRODUCTION

THE NEURAL mechanism of visual pattern recognition in the brain is little known, and revealing it by conventional physiological experiments alone seems to be almost impossible. So, we take a slightly different approach to this problem. If we could make a neural network model which has the same capability for pattern recognition as a human being, it would give us a powerful clue to the understanding of the neural mechanism in the brain. In this paper, we discuss how to synthesize a neural network model in order to endow it with pattern recognition capability like that of a human being.

Several models were proposed with this intention [1]–[6]. In synthesizing such models, one of the most difficult problems is to design the networks so as to show position- and deformation-invariant responses. Some of these conventional models fail to recognize patterns which are shifted in position or deformed in shape. Although the four-layer perceptron [2] shows a kind of position-invariant responses, it works correctly only when the distance of shift is equal to one of the several specific values which are determined during the training of the network.

A few years ago, the authors [7], [8] proposed a multilayered neural network model, called a "neocognitron," which is capable of recognizing stimulus patterns correctly without being affected by any shift in position or even by considerable distortion in shape of the patterns. The result of computer simulation of a neocognitron with a small-scale network was reported there.

In this present paper, a neocognitron with a larger-scale network is simulated on a minicomputer PDP-11/34 and is shown to have a great capability for visual pattern recognition. The new model consists of nine layers of cells, while

Copyright © 1983 IEEE. Reprinted with permission.

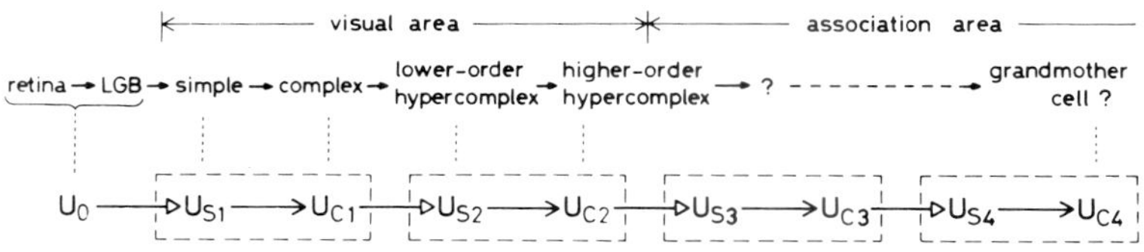

Fig. 1. Comparison between hierarchical model by Hubel and Wiesel and structure of neural network of neocognitron.

the previous model consisted of seven layers. We demonstrate that the new model can be trained to recognize handwritten Arabic numerals even with considerable deformations in shape.

We use a learning-with-a-teacher process for the reinforcement of the modifiable synapses in the new large-scale model, instead of the learning-without-a-teacher process applied to the previous model. In this paper, we focus on the mechanism for pattern recognition rather than that for self-organization.

II. STRUCTURE OF THE NETWORK

The neocognition is a multilayered network with a hierarchical structure similar to the hierarchical model for the visual system proposed by Hubel and Wiesel [9], [10]. As shown in Fig. 1, the neocognitron is composed of a cascaded connection of a number of modular structures preceded by an input layer U_0 consisting of photoreceptor array. Each of the modular structures is composed of two layers of cells, namely, a layer U_S consisting of S cells, and a layer U_C consisting of C cells. The layers U_S and U_C in the lth module are denoted by U_{Sl} and U_{Cl}, respectively. An S cell has a response characteristic similar to a simple cell or a lower order hypercomplex cell according to the classification by Hubel and Wiesel, while a C cell resembles a complex cell or a higher order hypercomplex cells. In this network, a cell in a higher stage generally has a tendency to respond selectively to a more complicated feature of the stimulus pattern and, at the same time, has a larger receptive field and is more insensitive to the shift in position of the stimulus pattern.

Each S cell has modifiable input synapses which are reinforced with learning and acquires an ability to extract a specific stimulus feature. That is, an S cell comes to respond only to a specific stimulus feature and not to respond to other features.

Each C cell has afferent synapses leading from a group of S cells which have receptive fields of similar characteristics at approximately the same position on the input layer. This means that all of the presynaptic S cells are to extract almost the same stimulus feature but from slightly different positions on the input layer. The efficiencies of the synapses are determined in such a way that the C cell will be activated whenever at least one of its presynaptic S cells is active. Hence, even if a stimulus pattern which has elicited a large response from the C cell is shifted a little in position, the C cell will keep responding as before, because another presynaptic S cell will become active instead of the first one. In other words, a C cell responds to the same

stimulus feature as its presynaptic S cells do but is less sensitive to the shift in position of the stimulus feature.

S cells or C cells in any single layer are sorted into subgroups according to the optimum stimulus features of their receptive fields. Since the cells in each subgroup are set in a two-dimensional array, we call the subgroup as a "cell plane." We will also use the terminology S plane and C plane to represent the cell planes consisting of S cells and C cells, respectively. All the cells in a single cell plane have input synapses of the same spatial distribution, and only the positions of the presynaptic cells are shifted in parallel depending on the position of the postsynaptic cells. Even in the process of learning, in which the efficiencies of the synapses are modified, the modification is performed always under this restriction.

Fig. 2 is a schematic diagram illustrating the synaptic connections between layers. Each tetragon drawn with heavy lines represents an S plane or a C plane, and each vertical tetragon drawn with thin lines, in which S planes or C planes are enclosed, represents an S layer or a C layer.

In Fig. 2, for the sake of simplicity, only one cell is shown in each cell plane. Each of these cells receives input synapses from the cells within the area enclosed by the ellipse in its preceding layer. All the other cells in the same cell plane have input synapses of the same spatial distribution, and only the positions of the presynaptic cells are shifted in parallel from cell to cell. Hence all the cells in a single cell plane have receptive fields of the same function but at different positions.

Since the cells in the network are interconnected in a cascade as shown in Fig. 2, the deeper the layer is, the larger becomes the receptive field of each cell of that layer. The density of the cells in each cell plane is so determined as to decrease in accordance with the increase of the size of the receptive fields. The number of cells in each layer is shown at the bottom of Fig. 2. In the deepest module, the receptive field of each C cell becomes so large as to cover the whole input layer, and each C plane is so determined as to have only one C cell. Fig. 3 illustrates concretely how the cells of each cell plane are interconnected to the cells of other cell planes.

S cells and C cells are excitatory cells. Although it is not shown in Figs. 2 and 3, we have inhibitory V_C cells in C layers.

Here, we will describe the outputs of these cells with numerical expressions. All the cells employed in the neocognitron are of analog type; that is, the input and output signals of the cells take nonnegative analog values proportional to the instantaneous firing frequencies of

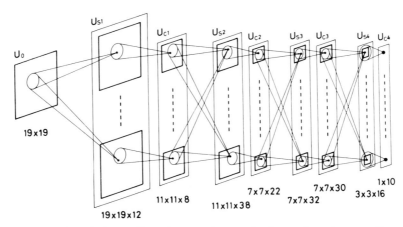

Fig. 2. Schematic diagram illustrating synaptic connections between layers in neocognitron.

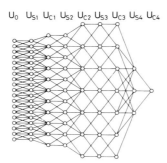

Fig. 3. One-dimensional view of interconnections between cells of different cell planes. Only one cell plane is drawn in each layer.

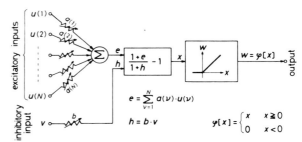

Fig. 4. Input-to-output characteristics of S cell: typical example of cells employed in neocognitron.

actual biological neurons. The output of a photoreceptor is denoted by $u_0(\boldsymbol{n})$ where $\boldsymbol{n}$ represents the two-dimensional coordinates indicating the location of the cell. We will use notations $U_{Sl}(k, \boldsymbol{n})$ to represent the output of an S cell in the kth S plane in the lth module, and $u_{Cl}(k, \boldsymbol{n})$ to represent the output of a C cell in the kth C plane in that module, where $\boldsymbol{n}$ is the two-dimensional coordinates representing the position of these cells' receptive fields on the input layer.

As shown in Fig. 4, S cells have inhibitory inputs with shunting mechanism. Incidentally, S cells have the same characteristics as the excitatory cells employed in the conventional cognitron [5], [6]. The output of an S cell of the kth S plane in the lth module is given by

argument representing the position $\boldsymbol{n}$ of the receptive field of cell $u_{Sl}(k, \boldsymbol{n})$.

Parameter r_l in (1) controls the intensity of the inhibition. The larger the value of r_l is, the more selective becomes the cell's response to its specific feature. Their values are $r_1 = 1.7$, $r_2 = 4.0$, $r_3 = 1.5$, and $r_4 = 1.0$. (A detailed discussion on the response of S cells will be given in Section III-B.)

The inhibitory cell $v_{Cl-1}(\boldsymbol{n})$, which is sending an inhibitory signal to cell $U_{Sl}(k, \boldsymbol{n})$, receives afferent synapses from the same group of cells as $u_{Sl}(k, \boldsymbol{n})$ does and yields an output proportional to the weighted root mean square of its inputs:

$$v_{Cl-1}(\boldsymbol{n}) = \sqrt{\sum_{\kappa=1}^{K_{Cl-1}}\sum_{\boldsymbol{v}\in A_l}c_{l-1}(\boldsymbol{v}) \cdot u_{Cl-1}^2(\kappa, \boldsymbol{n} + \boldsymbol{v})} . \quad (2)$$

$$U_{Sl}(k, \boldsymbol{n}) = r_l \cdot \phi \left(\frac{1 + \sum_{\kappa=1}^{K_{Cl-1}}\sum_{\boldsymbol{v}\in A_l}a_l(\kappa, \boldsymbol{v}, k) \cdot u_{Cl-1}(\kappa, \boldsymbol{n} + \boldsymbol{v})}{1 + \dfrac{r_l}{1 + r_l} \cdot b_l(k) \cdot v_{Cl-1}(\boldsymbol{n})} - 1 \right), \qquad k = 1, 2, \cdots, K_{Sl} \quad (1)$$

where $\phi[x] = \max(x, 0)$. In the case of $l = 1$ in (1), $u_{Cl-1}(\kappa, \boldsymbol{n})$ stands for $u_0(\boldsymbol{n})$, and we have $K_{Cl-1} = 1$.

Here, $a_l(\kappa, \boldsymbol{v}, k)$ and $b_l(k)$ represent the efficiencies of the excitatory and inhibitory modifiable synapses, respectively. As described before, all the S cells in the same S plane are assumed to have an identical set of afferent synapses. Hence $a_l(\kappa, \boldsymbol{v}, k)$ and $b_l(k)$ do not contain any

The efficiencies of the unmodifiable synapses $c_{l-1}(\boldsymbol{v})$ are determined so as to decrease monotonically with respect to $|\boldsymbol{v}|$ and to satisfy

$$K_{Cl-1} \cdot \sum_{\boldsymbol{v}\in A_l} c_{l-1}(\boldsymbol{v}) = 1. \quad (3)$$

The size of the connection area A_l of these cells is set to be small in the first module and to increase with the depth l as illustrated in Fig. 3.

The output of a C cell of the kth C plane in the lth module is given by

$$u_{Cl}(k, \boldsymbol{n}) = \psi\left(\sum_{\kappa=1}^{K_{Sl}} j_l(\kappa, k) \sum_{\boldsymbol{v} \in D_l} d_l(\boldsymbol{v}, k) \cdot u_{Sl}(\kappa, \boldsymbol{n} + \boldsymbol{v})\right),$$
$$\cdot k = 1, 2, \cdots, K_{Cl}, \quad (4)$$

where

$$\psi[x] = \begin{cases} x/(\alpha_l + x), & (x \geqq 0); \\ 0, & (x < 0). \end{cases}$$

The parameter α_l is a positive constant which determines the degree of saturation of the output. Their values are $\alpha_1 = \alpha_2 = \alpha_3 = 0.25$, and $\alpha_4 = 1.0$.

In (4), $d_l(\boldsymbol{v}, k)$ represents the efficiencies of the excitatory synapses leading from S cells, and $j_l(\kappa, k)$ takes value one or zero depending on whether synaptic connections really exist from the κth S plane to the kth C plane. The value of $d_l(\boldsymbol{v}, k)$ is determined so as to decrease monotonically with respect to $|\boldsymbol{v}|$ and is independent of k except for $l = 1$. The size of the connection area D_l is set to be small in the first module and to increase with depth l as illustrated in Fig. 3.

The process of pattern recognition in this multilayered network can be briefly summarized as follows. The stimulus pattern is first observed within a narrow range by each of the S cells in the first module, and several features of the stimulus pattern are extracted. In the next module, these features are combined by observation over a little larger range, and higher order features are extracted. Operations of this kind are repeatedly applied through a cascaded connection of a number of modules. In each stage of these operations, a small amount of positional error is tolerated. The operation by which positional errors are tolerated little by little, not at a single stage, plays an important role in endowing the network with an ability to recognize even distorted patterns.

III. SYNAPTIC CONNECTIONS BETWEEN CELLS

The synaptic connections in the new model of the neocognitron are reinforced by means of a supervised learning, that is, a learning-with-a-teacher process. During the training process, the network is presented with a set of training patterns to the input layer, together with the instructions which cells in the network should come to respond to each of the training patterns. This algorithm is different from that used in the previous model [7], [8]. In the new model, the algorithm for the reinforcement of synapses is determined from a standpoint of an engineering application to a design of a pattern recognizer rather than from that of pure biological modeling. That is, the algorithm is determined with the criterion of obtaining a better performance in handwritten character recognition.

A. Reinforcement of the Input Synapses of S Cells

The reinforcement of the synaptic connections are performed in sequence from the distal to the deeper layers. That is, the reinforcement of the input synapses of the lth layer is performed after completion of the reinforcement of up to the $(l - 1)$th layer.

A number of cell planes are in an S layer. These cell planes are reinforced one at a time. In order to reinforce a cell plane, the "teacher" presents a training pattern to the input layer, and at the same time chooses one S cell which should work as the "representative" from that cell plane. The representative cell works like a seed in the crystal growth. The input synapses to the representative cell are reinforced depending on the stimuli given to these synapses. That is, only the synapses through which nonzero signals are coming are reinforced. As the result, the representative cell acquires a selective responsiveness to the training pattern which is now presented to the input layer. All the other cells in that cell plane have their input synapses reinforced in the identical manner as their representative.

This algorithm can be expressed as follows. Let cell $u_{Sl}(\hat{k}, \hat{\boldsymbol{n}})$ be the representative. The modifiable synapses $a_l(\kappa, \boldsymbol{v}, \hat{k})$ and $b_l(\hat{k})$, which are afferent to the S cells of this S plane, are reinforced by the amount shown below:

$$\Delta a_l(\kappa, \boldsymbol{v}, \hat{k}) = q_l \cdot c_{l-1}(\boldsymbol{v}) \ u_{Cl-1}(\kappa, \hat{\boldsymbol{n}} + \boldsymbol{v}), \quad (5)$$
$$\Delta b_l(\hat{k}) = q_l \cdot v_{Cl-1}(\hat{\boldsymbol{n}}), \quad (6)$$

where q_l is a positive constant which determines the amount of reinforcement. The initial values of these modifiable synapses are all zero.

We can choose any cell of a cell plane as the representative, and the choice of the representative does not have so much effect on the result of training, provided that the training pattern is presented at a proper position in the respective field of the representative. Hence, in the computer simulation discussed later, we always choose the cell situated at the center of each cell plane as the representative.

In the computer simulation, the number of training patterns given to each cell plane is from one to four, depending on the required allowance to the deformation of the stimulus features. (See the following section for more discussions.)

B. Analysis of the Response of an S Cell

In this section, we discuss how each S cell is trained to respond selectively to differences in stimulus patterns. Since the structure between two adjoining modules is similar in all parts of the network, we observe the response of an arbitrary S cell $U_{S1}(k, \boldsymbol{n})$ of layer U_{S1} as a typical example. Fig. 5 shows the synaptic connections converging to such a cell. For the sake of simplicity, we will omit the suffixes S, $l = 1$ and the arguments $k, \boldsymbol{n}$ and represent the response of this cell simply by u. Similarly, we will use the notation v for the output of the inhibitory cell $v_{C0}(\boldsymbol{n})$, which sends an

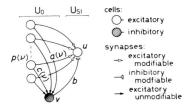

Fig. 5. Synaptic connections converging to S cell.

inhibitory signal to cell u. For the other variables, the arguments k and n and suffixes S, C, l, and $l - 1$ will also be omitted.

Let $p(v)$ be the response of the cells of layer U_0 situated in the connection area of cell u, so that

$$p(v) = u_0(n + v). \tag{7}$$

In other words, $p(v)$ is the stimulus pattern (or feature) presented to the receptive field of cell u.

With this notation, (1) and (2) can be written

$$u = r \cdot \phi \left(\frac{1 + \Sigma_v a(v) \cdot p(v)}{1 + \dfrac{r}{1 + r} \cdot b \cdot v} - 1 \right) \tag{8}$$

$$v = \sqrt{\Sigma_v c(v) \cdot p^2(v)}. \tag{9}$$

When cell u is chosen as the representative, the amounts of reinforcement of the modifiable synapses are derived from (5) and (6), that is,

$$\Delta a(v) = q \cdot c(v) \cdot p(v), \tag{10}$$

$$\Delta b = q \cdot v. \tag{11}$$

Let s be defined by

$$s = \frac{\Sigma_v a(v) \cdot p(v)}{b \cdot v}. \tag{12}$$

Then (8) reduces approximately to

$$u \simeq r \cdot \phi \left(\frac{r + 1}{r} \cdot s - 1 \right), \tag{13}$$

provided that $a(v)$ and b are sufficiently large.

Let a stimulus pattern $p(v) = P(v)$ be presented, and let cell u be chosen as the representative. Then, from (5) and (6), we obtain

$$a(v) = q \cdot c(v) \cdot P(v), \tag{14}$$

$$b = q\sqrt{\Sigma_v c(v) \cdot P^2(v)}. \tag{15}$$

Substituting (9), (14), and (15) into (12), we obtain

$$s = \frac{\Sigma_v c(v) \cdot P(v) \cdot p(v)}{\sqrt{\Sigma_v c(v) \cdot P^2(v)} \cdot \sqrt{\Sigma_v c(v) \cdot p^2(v)}}. \tag{16}$$

If we regard $p(v)$ and $P(v)$ as vectors, (16) can be interpreted as the (weighted) inner product of the two vectors normalized by the norms of both vectors. In other words, s gives the cosine of the angle between the two vectors $p(v)$ and $P(v)$ in the multidimensional vector space. Therefore, we have $s = 1$ only when $p(v) = P(v)$, and we have $s < 1$ for all patterns such as $p(v) \neq P(v)$. This

means that s becomes maximum for the training pattern and becomes smaller for any other patterns.

If parameter q is large enough, (13) holds. When an arbitrary pattern $p(v)$ is presented, and if it satisfies $s > r/(r + 1)$, we have $u > 0$ by (13). Conversely, for a pattern which makes $s \leq r/(r + 1)$, cell u does not respond. We can interpret by saying that cell u judges the similarity between patterns $p(v)$ and $P(v)$ using the criterion defined by (16) and that it responds only to patterns judged to be similar to $P(v)$. Incidentally, if $p(v) = P(v)$, we have $s = 1$ and consequently $u \simeq 1$.

Since the value $r/(r + 1)$ tends to one with increase of r, a larger value of r makes the cell's response more selective to one specific pattern or feature. In other words, a large value of r endows the cell with a high ability to discriminate patterns of different classes. However, a higher selectivity of the cell's response is not always desirable, because it decreases the ability to tolerate the deformation of patterns. Hence the value of r should be determined at a point of compromise between these two contradictory conditions.

In the above analysis, we supposed that cell u is trained only for one particular pattern $P(v)$. When cell u has been trained to two patterns, say, to $P_1(v)$ and $P_2(v)$, $P(v)$ in the above discussions should be replaced with $\{P_1(v) + P_2(v)\}$. Hence cell u acquires a tendency to respond equally to both $P_1(v)$ and $P_2(v)$. This, however, depends on the value of r, and also on the similarity between $P_1(v)$ and $P_2(v)$. If the difference between $P_1(v)$ and $P_2(v)$ is too large, or if the value of r is too large, cell u comes to respond neither to $P_1(v)$ nor to $P_2(v)$.

The above discussion is not restricted to S cells of layer U_{S1}. Each S cell in succeeding modules shows a similar type of response, if we regard the response of the C cells in its connection area in the preceding layer as its input pattern.

C. Layers U_{S1} and U_{C1}

Layer U_{S1} has 12 cell planes, and each cell plane contains the same number of cells as layer U_0, that is, 19×19 (see Figs. 2 and 3). These S cells have their input synapses reinforced so as to acquire the ability to detect line components of various orientations.

The training patterns which are used for training the 12 cell planes are displayed in column a_1 in Fig. 6. This figure shows, for example, that the cells of the first cell plane are trained to detect a horizontal line component. We can also interpret this by saying that the patterns in column a_1 show the structure of the receptive fields of the cells of layer U_{S1}.

Since the spread of the excitatory input synapses $a_1(\kappa, v, k)$ of each S cell (i.e., the connection area A_1 in (1) and (2)) is as small as 3×3, cases exist where two different cell planes should be prepared for detecting a line of a particular orientation. For example, in Fig. 6, the second and third cell planes of layer U_{S1} have the receptive fields of the same preferred orientation but of different structures. Hence the outputs from such pairs of cell planes are

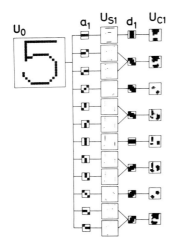

Fig. 6. Example of response of cells of layers U_0, U_{S1}, and U_{C1}, and synaptic connections between them.

joined together at the input stage of layer U_{C1} as shown in Fig. 6. The parameter $j_1(\kappa, k)$ in (4) takes value one or zero, depending on this joining condition. For instance, $j_1(\kappa, 2) = 1$ for $\kappa = 2$ and 3, and $j_1(\kappa, 2) = 0$ for other κ. Because of this joining process, layer U_{C1} contains only eight cell planes.

Parameter r_1 in (1), which determines the selectivity of an S cell, is set at value of 1.7. Since the stimulus feature which is used for training an S cell contains three active elements, the S cell with $r_1 = 1.7$ yields nonzero output for a stimulus feature contaminated with up to two additive elements of noise, or one additive and one subtractive elements. However, it does not respond to a stimulus feature with two subtractive elements of noise or more. (These results can be obtained from the analysis in Section III-B as well as from the computer simulation.)

The spatial distribution of the input synapses $d_1(v, k)$ of a C cell (i.e., the connection area D_1 in (4)) is 5×5 in size, but all of these 5×5 synapses are not effective. As shown in Fig. 6, the effective part of the distribution is elongated to the direction perpendicular to its preferred orientation and is compressed in the direction of its preferred orientation.

Since each C cell receives excitatory signals from a number of S cells, it usually responds similarly as its neighboring C cells. Hence it is possible to reduce the number of C cells in each cell plane compared to that of S cells. The density of cells in each cell plane of layer U_{C1} are thinned out by two to one compared to that of layer U_{S1} both in horizontal and vertical directions. Thus as shown in Fig. 3, the number of cells in each cell plane is reduced to 11×11 in layer U_{C1}.

D. Layers U_{S2} and U_{C2}

Layer U_{S2} has 38 cell planes, and each cell plane contains 11×11 S cells. Layer U_{C2} has only 22 cell planes, because the outputs from some of the cell planes of layer U_{S2} are joined together at the input stage of layer U_{C2}.

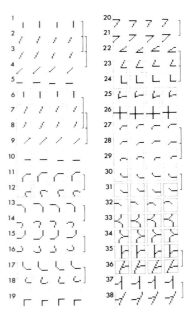

Fig. 7. Training patterns used to train 38 cell planes of layer U_{S2}. Way of joining at input stage of layer U_{C2} is also shown to right of each group of training patterns.

Each cell plane of layer U_{C2} contains 7×7 C cells because of the two to one thinning-out of the cell density as was shown in Fig. 3.

Each S cell of layer U_{S2} has modifiable excitatory input synapses of 3×3 spatial distribution. Since the preceding layer U_{C1} has eight cell planes, the total number of the excitatory input synapses to each S cell is $3 \times 3 \times 8$. All of these synapses are not reinforced by learning, but most of them usually stay at the initial value of zero. The input synapses to each C cell of layer U_{C2} have spatial distribution of 5×5.

Figure 7 shows the training patterns used for training the 38 cell planes of layer U_{S2}. Four training patterns, in which the same stimulus feature is shifted in parallel to each other by one element in both horizontal and vertical directions, are used to train each cell plane. The reason why the use of four patterns are necessary is discussed below.

Because the cells of this layer are thinned out by two to one compared to those of layer U_0 in both horizontal and vertical directions, each S cell should take charge of extracting a specific stimulus feature from four different positions on layer U_0. In this network, the two to one thinning-out of the cells is already made at the stage of layer U_{C1}, from which the relevant S cells receive synaptic connections. The effect of this thinning-out is not so small, and a somewhat different spatial response might appear in the preceding layer U_{C1} when the stimulus pattern is shifted in position by one element. Hence each cell-plane of layer U_{S2} should be trained with four different patterns beforehand so as to come to respond equally to them.

In this experiment, we intend to train the neocognitron so as to recognize handwritten Arabic numerals. When patterns are written by hand, the stimulus features in the patterns usually suffer from considerable deformations depending on the writers. However, the way of deformation is not at random but usually has some tendency. Some of such deformed features are detected separately in a number of cell planes of layer U_{S2} and are combined together at the input stage of layer U_{C2}. In Fig. 7, the lines drawn to the right of the 36 groups of training patterns indicate how this joining is made.

E. Layers U_{S3} and U_{C3}

Layer U_{S3} has 32 cell planes, and U_{C3} has 30 cell planes. The number of cells in each cell plane is 7×7 for both layers U_{S3} and U_{C3}. Thinning-out is not performed between these layers. Each S cell has $3 \times 3 \times 22$ modifiable excitatory input synapses, and the input synapses of each C cell have spatial distribution of 3×3.

Fig. 8 shows the training patterns used for training the 32 cell planes of layer U_{S3} and also shows how the outputs from these cell planes are joined together at the input stage of layer U_{C3}. Most of these training patterns consist of some parts of the standard numeral patterns which are to be taught to this network.

As is seen in Fig. 8, only two or three different patterns are used to train each cell plane of layer U_{S3}. They are deformed in shape or varied in size to each other. In the case of this layer, it is not necessary to present all of the deformed patterns which should be detected by the cell plane. Presentation of only a few number of typical patterns is enough for the training of each cell plane, because a considerable amount of deformation has already been absorbed before this stage.

F. Layers U_{S4} and U_{C4}

Layer U_{S4} has 16 cell planes, and each cell plane has 3×3 S cells. Each of these S cells has $5 \times 5 \times 30$ modifiable input synapses. Although the number of cells in each cell plane is reduced in layer U_{S4} from that in the preceding layer U_{C3}, no thinning out is made between these layers. Only the cells near the periphery of the cell planes of layer U_{S4} are omitted, because they are of little use for the recognition of the whole input pattern (see Fig. 3).

The 16 cell planes of layer U_{S4} are trained with the 16 sets of patterns as shown in Fig. 9 and are joined together into ten cell planes at layer U_{C4}. Each cell plane of layer U_{C4} has only one cell which has input synapses of 3×3 spatial distribution.

The ten cells of layer U_{C4} have one-to-one correspondence with ten Arabic numerals. In Figs. 10–12, which will be discussed later, they are arranged vertically from zero to nine in order at the rightmost column. Among these cells, a mechanism of lateral inhibition exists, although it is omitted in (4).

For some of the numerals, more than one quite different styles of writing are accustomed to be used. For each of these numerals, two S planes are prepared and are trained independently with two typical patterns of different styles

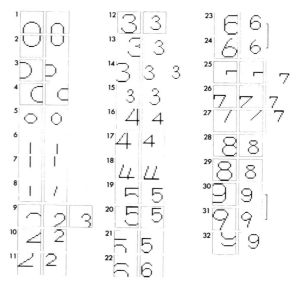

Fig. 8. Training patterns used to train 32 cell planes of layer U_{S3} and way of joining at input stage of layer U_{C3}.

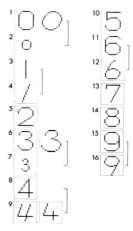

Fig. 9. Training patterns used to train 16 cell planes of layer U_{S4} and way of joining at input stage of layer U_{C4}.

as shown in Fig. 9, and their outputs are joined together at the input stage of layer U_{C4}.

IV. RESPONSE OF THE NETWORK

The neocognitron, which has been trained with the procedure discussed in the previous chapter, is tested with various input patterns. Fig. 10 shows the response of the cells in the network to one of the patterns used for training the network. It is seen that only cell 2 of layer U_{C4} yields an output. This means that the neocognitron recognizes the input pattern correctly. Even if the input pattern is deformed from the training pattern as much as shown in Fig. 11, the neocognitron recognizes it correctly.

In the case of Fig. 12, two of the cells of layer U_{C4} respond; that is, a large output is obtained from cell 5 and

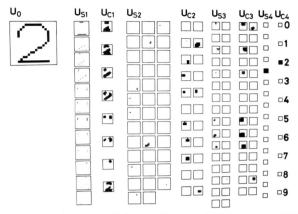

Fig. 10. Response of cells in network to one of training patterns "2."

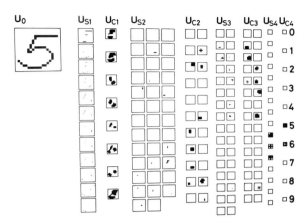

Fig. 12. Response of cells in network to deformed pattern, which elicits response from two cells of layer U_{C4}.

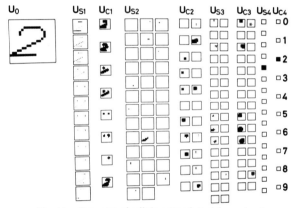

Fig. 11. Response of cells in network to deformed pattern.

a small output from cell 6. This means that the neocognitron correctly judges that the input pattern is 5, but also admits that the input pattern slightly resembles 6.

Fig. 13 shows some examples of the stimulus patterns which the neocognitron correctly recognizes. On the other hand, Fig. 14 shows some examples of the patterns which cannot be correctly recognized. Some of these patterns elicit no response from any of the cells of layer U_{C4}, and the others elicit responses from wrong cells of layer U_{C4}.

V. Discussion

We have demonstrated that the neocognitron recognizes handwritten numerals of various styles of penmanship correctly, even if they are considerably distorted in shape. Although the result is shown for the recognition of Arabic numerals, the neocognitron can be trained to recognize other set of patterns such as alphabet, geometrical shapes, or others.

The number of cell planes of each layer should be changed adaptively, depending on the set of patterns which

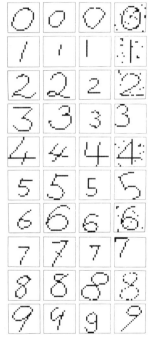

Fig. 13. Some examples of deformed numerals which neocognitron recognizes correctly.

the neocognitron should learn to recognize. The program for the computer simulation is made in such a way that the number of cell planes can be chosen freely and can readily be increased when necessary.

Although each S cell has a large number of modifiable input synapses, all of them are not generally reinforced by learning. On the contrary, most of them remain at the initial state in which their efficiencies are zero. Furthermore, the modifiable synapses tend to be reinforced in clusters. In the computer program, we made full use of these characteristics of the synapses and reduced the re-

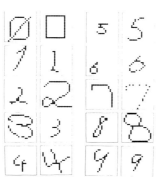

Fig. 14. Some examples of distorted patterns which are not correctly recognized.

quired memory capacity and increased the computation speed by eliminating unnecessary calculations.

In the simulated model, we made two to one thinning-out in several parts of the network in order to increase the computation speed. The thinning-out between layers U_{S1} and U_{C1}, however, was too coarse compared to the 5×5 spread of the input synapses of the cells of layer U_{C1}. As a result, we felt a little difficulty in training the network, and we had to use four different training patterns for each cell plane of layer U_{S2}. If we do not make the thinning-out at this stage, we can possibly improve the capability of the network further.

ACKNOWLEDGMENT

The authors are very grateful to Mr. Toshinori Hirano for his assistance in making the computer program.

REFERENCES

[1] F. Rosenblatt, *Principles of Neurodynamics*. Washington, DC: Spartan, 1962.
[2] H. D. Block, B. W. Knight, and F. Rosenblatt, "Analysis of a four-layer series-coupled perceptron. II," *Rev. Mod. Phys.*, vol. 34, pp. 135–152, Jan. 1962.
[3] H. Marko and H. Giebel, "Recognition of handwritten characters with a system of homogeneous layers," *Nachrichtentechnische Zeitschrift*, vol. 23, pp. 455–459, Sept. 1970.
[4] H. Marko, "A biological approach to pattern recognition," *IEEE Trans. Syst., Man, Cybern.*, vol. SMC-4, pp. 34–39, Jan. 1974.
[5] K. Fukushima, "Cognitron: A self-organizing mulitlayered neural network," *Biol. Cybern.*, vol. 20, pp. 121–136, Nov. 1975.
[6] _____, "Cognitron: A self-organizing multilayered neural network model," NHK Tech. Monograph, no. 30, Jan. 1981.
[7] _____, "Neocognitron: A self-organizing neural network model for a mechanism of pattern recognition unaffected by shift in position," *Biol. Cybern.*, vol. 36, pp. 193–202, Apr. 1980.
[8] K. Fukushima and S. Miyake, "Neocognitron: A new algorithm for pattern recognition tolerant of deformations and shifts in position," *Pattern Recognition*, vol. 15, pp. 455–469, 1982.
[9] D. H. Hubel and T. N. Wiesel, "Receptive fields, binocular interaction and functional architecture in cat's visual cortex," *J. Physiol.* (London), vol. 160, pp. 106–154, Jan. 1962.
[10] D. H. Hubel and T. N. Wiesel, "Receptive fields and functional architecture in two nonstriate visual area (18 and 19) of the cat," *J. Neurophysiol.*, vol. 28, pp. 229–289, 1965.

32
Introduction

(1983)
Andrew G. Barto, Richard S. Sutton, and Charles W. Anderson

Neuronlike adaptive elements that can solve difficult learning control problems
IEEE Transactions on Systems, Man, and Cybernetics SMC-13:834–846

One of the hardest and most important problems for both brains and learning systems to solve is called the "credit assignment" problem, that is, knowing how to change connection strengths in complicated networks so as to produce a network that can do what is wanted. There are a number of forms of this problem. In multilayer networks it is hard to give rules for specifying appropriate connections for the inner layers, that is, layers that contain "hidden units" that are neither input nor output units. The Neocognitron (paper 31) learned in a multilayer network, but by a sequential directed learning procedure, where only a single layer at a time is plastic, and where the types of features that the layer should respond to were known. It was not until the development of the generalized error correcting rule now called back propagation (papers 41 and 42) that it was possible to give efficient general learning rules for multilayer networks.

Barto, Sutton, and Anderson discuss a different aspect of the credit assignment problem. Error correction techniques such as the Widrow-Hoff rule and back propagation require detailed knowledge about the nature of the error. In general, the system must know exactly what the appropriate response was and what the system response was, so that a detailed error signal can be formed and used to make corrections in the network. There are many places where detailed information about the error is not available—only knowledge that an error was made. An example of a complex system of this type would be a chess game, where a loss might have been caused by an error at any one of many earlier moves. Trying to find where the mistake was and when it was made gives substance to many chess arguments.

The example used by Barto, Sutton, and Anderson in their paper analyzes and suggests a solution to another class of credit assignment problems. They use a cart, which holds a pole that is free to pivot. The cart can move on a track between two stops. The object of the system is to keep the pole from falling over by pushing the cart back and forth between the stops. An error is made when the pole falls over or the cart hits the stop. The physics of this task is familiar to anyone who has balanced a broom or a baseball bat on the hand. It is quite easy with a little practice.

The problem here is that the error feedback is not very informative. When the pole falls over or the cart hits the stop, all it means is that a mistake was made some time in the past. Even with perfect control after the initial mistake, the error could not be prevented. There is no idea of when the mistake was made or how large the mistake was.

The strategy used by Barto, Sutton, and Anderson is to assume that there are two adaptive devices involved: first is the *associative search element*, which takes the current state of the physical system (i.e., the position and velocity of the cart) and, by an associative learning rule, gives an output specifying a control action—the force to be applied to the cart; second is what they call the *adaptive critic element*. The critic is looking ahead and predicting the expected reinforcement from the environment, given a particular action from the associative search element. The learning rule of the associative search element incorporates a prediction of reinforcement from the critic, so the two elements are tightly coupled.

Since the critic element is constantly trying to predict the expected reinforcement associated with particular input states, and modifying itself appropriately with experience, the associative search element is constantly modifying its weights, even when an explicit error has not been made. In human terms, the critic element is acting something like a parent predicting direction of future reinforcement—either negative ("If you eat that, you will get a stomach ache.") or positive ("If you clean up your room, Santa Claus will remember you."). The associative search element is taking the predicted reinforcement into account when it learns what to do.

For convenience in simulation and analysis, Barto, Sutton, and Anderson only use a single adaptive element critic and a single associative search element. However, they point out that their system could easily be turned into a more traditional multineuron network with only minor modifications.

An interesting aspect of the approach taken in this paper is that it is consistent with a large body of psychological literature on animal learning. In fact, use of the adaptive critic element is close to the Rescorla-Wagner model of classical conditioning, probably the most successful current model of classical conditioning. A highly recommended paper by Sutton and Barto in *Psychological Review* (1981) discusses the connections between adaptive system theory and psychological theory in great detail. This paper by Sutton and Barto is also notable for bringing the Widrow-Hoff error correction technique and related work in adaptive control theory to the attention of psychologically oriented neural modelers, where it had a major impact.

Reference

R. S. Sutton and A. G. Barto (1981), "Toward a modern theory of adaptive networks: expectation and prediction," *Psychological Review* 88:135–171.

(1983)
Andrew G. Barto, Richard S. Sutton, and Charles W. Anderson

Neuronlike adaptive elements that can solve difficult learning control problems

IEEE Transactions on Systems, Man, and Cybernetics SMC-13:834–846

Abstract—It is shown how a system consisting of two neuronlike adaptive elements can solve a difficult leaning control problem. The task is to balance a pole that is hinged to a movable cart by applying forces to the cart's base. It is assumed that the equations of motion of the cart–pole system are not known and that the only feedback evaluating performance is a failure signal that occurs when the pole falls past a certain angle from the vertical, or the cart reaches an end of a track. This evaluative feedback is of much lower quality than is required by standard adaptive control techniques. It is argued that the learning problems faced by adaptive elements that are components of adaptive networks are at least as difficult as this version of the pole-balancing problem. The learning system consists of a single *associative search element* (ASE) and a single *adaptive critic element* (ACE). In the course of learning to balance the pole, the ASE constructs associations between input and output by searching under the influence of reinforcement feedback, and the ACE constructs a more informative evaluation function than reinforcement feedback alone can provide. The differences between this approach and other attempts to solve problems using neuronlike elements are discussed, as is the relation of this work to classical and instrumental conditioning in animal learning studies and its possible implications for research in the neurosciences.

Manuscript received August 1, 1982; revised April 20, 1983. This work was supported by AFOSR and the Air Force Wright Aeronautical Laboratory under Contract F33615-80-C-1088.

Copyright © 1983 IEEE. Reprinted with permission.

I. Introduction

MATHEMATICALLY formulated networks of neuronlike elements have been studied both as models of specific neural circuits and as abstract, though biologically inspired, computational architectures. As models of specific neural circuits, network models can provide theories to explain anatomical and physiological data. As computational architectures, they represent attempts to explore possible substrates for intelligent behavior, both natural and artificial. Networks of this second category are relevant to brain and behavioral science to the extent that their behavior can be related to phenomena of animal behavior for which no plausible mechanisms are known, thereby suggesting novel lines of empirical research. They are relevant to artificial intelligence to the extent that they exhibit forms of problem solving, knowledge acquisition, or data storage that are difficult to achieve by more conventional means.

In this article we illustrate an abstract neural network approach that we believe can have relevance for both neuroscience and computer science. Advances in our appreciation of the complexity of biological cells make it clear that the 35-year old metaphor that places the neuron at the level of the computer logic gate is inadequate. Neurons and synapses have information processing capabilities that make use of both short- and long-term information storage, locally implemented by complex biochemical mechanisms. Biochemical networks within cells are known to perform functions that had previously been attributed to networks of interacting cells. These facts call for new neural metaphors. Moreover, advances in computer science suggest the possibility of achieving sophisticated problem-solving capacity through networks of interacting components that are themselves powerful problem-solving systems (e.g., [1] and [2]). In our approach, network components are neuronlike in their basic structure and behavior and communicate by means of excitatory and inhibitory signals rather than by symbolic messages, but they are much more complex than neuronlike adaptive elements studied in the past. Rather than asking how very primitive components can be interconnected in order to solve problems, we are pursuing questions about how components that are themselves capable of solving relatively difficult problems can interact in order to solve problems that are even more difficult.

This article is devoted to the justification of the design of two types of neuronlike adaptive elements and an illustration of the problem-solving capacities of a system consisting of a single element of each type. We call one element an *associative search element* (ASE) and the other an *adaptive critic element* (ACE). As a vehicle for introducing our adaptive elements, we describe an earlier adaptive problem-solving system, called "boxes," developed by Michie and Chambers [3], [4]. We show that a learning strategy similar to theirs can be implemented by a *single* ASE, and we show how its learning performance can be improved by the addition of a *single* ACE. To illustrate the problem-solving capabilities of these elements, we use the pole-balancing control problem posed by Michie and Chambers to illustrate their boxes algorithm, and we compare the performance of their system with that of our own. We conclude with a brief discussion of behavioral interpretations of our adaptive elements and their possible implications for neuroscience. A strong analogy exists between the behavior of the ACE and animal behavior in classical conditioning experiments, and parallels can be seen between the behavior of the ASE/ACE system and animal behavior in instrumental learning experiments. The adaptive elements we describe are refinements of those we have discussed previously [5]–[10] and were suggested by the work of Klopf [11], [12]. Our approach also has similarities with the work of Widrow and colleagues [13], [14] on what they called "bootstrap adaptation."

The significance of endowing single adaptive elements with this level of problem-solving capability is twofold. First, we wish to suggest neural metaphors, constrained by the computational demands of problem-solving, that postulate functions for the complex cellular mechanisms that are rapidly being elucidated as the study of the cellular basis of learning progresses. Second, we wish to suggest that if adaptive elements are to learn effectively *as network components*, then they must possess adaptive capabilities at least as robust as those of the elements discussed here. As we argue in the following, the learning problem faced by an adaptive element that is deeply embedded in the interior of a network is characterized by some of the same types of complexities that are present in the pole-balancing task considered here.

Thus, although the algorithms that we implement by means of single adaptive elements can obviously be implemented by networks of many simpler elements, we are attempting to delineate those properties required of components if they are to learn how to function as interconnected, cooperating components of networks. The extensive history of attempts to construct powerful adaptive networks and the generally acknowledged failure of these attempts suggest that network components as simple as those usually considered are not adequate. This lesson from previous theoretical studies, together with our contention that the view of neural function that constrained these studies was too limited, leads us to study elements as complex as the ASE and ACE. Despite our ultimate interest in networks, we do not present results in this paper that show that the elements discussed here are able to learn as components of powerful adaptive networks. However, previous simulation experiments with networks of similar elements have provided preliminary support for our approach to adaptive networks [5], [6], [8], and the research discussed here represents an initial attempt to move toward more difficult learning problems.

Although we intend to raise questions about the level in the functional hierarchy of the nervous system at which neurons can be said to act, we are not claiming that there is necessarily a strict correspondence between single neurons and ACE's and ASE's. Some of the features of these elements clearly are not neuronlike but can be implemented in standard ways by elements more faithful to neural limitations. For example, the ASE can "fire" with both negative and positive output values, but it can be implemented by a pair of reciprocally inhibiting elements, each capable only of positive "spikes." Consequently, by the term "neuronlike element" we do not mean a literal neuron model, and we purposefully exclude well-known neuron properties which would have no clear functional role in the present problem.

Our interest in the pole-balancing problem arises from its convenience as a test bed for exploring a variety of algorithms that may enable elements to learn effectively when embedded in networks. We are not interested in pole balancing *per se*, and our formulation of the problem, following that of Michie and Chambers [3], [4], makes it much more difficult than it would need to be if one were simply interested in controlling this type of dynamical system. We assume that the controller's design must be based upon very little knowledge of the controlled system's dynamics and that the evaluative feedback provided to the controller is of much lower quality than is required by standard adaptive control methods. These constraints produce a difficult learning control problem and reflect some of the conditions that we believe characterize the tasks faced by network components. While a variety of well-developed adaptive control methods can be (and have been) successfully applied to pole balancing, we know of none that are directly applicable to the problem subject to the constraints we impose. Additionally, the algorithm we describe can be applied to nonnumerical problems as well as to problems requiring the control of dynamical systems.

II. Learning within Networks

Many of the previous studies of adaptive networks of neuronlike elements focused on adaptive elements that are capable of solving certain types of pattern classification problems. Elements such as the ADALINE (adaptive linear element [16]) and those employed in the Perceptron [15] perform supervised learning pattern classification (see, for example, [17]). These elements form linear discrimination rules by adjusting a set of "synaptic" weights in an attempt to match their response to each training input pattern with a desired response, or correct classification, that is provided by a "teacher." The resulting discrimination rule can be used to classify new pattern instances (perhaps incorrectly), thereby providing a form of generalization. The algorithms implemented by these adaptive elements are closely related to iterative regression methods used in adaptive control for the identification of unknown system parameters [17].

Unfortunately, a network composed of these types of adaptive elements can only learn if its environment contains a teacher that can supply *each* component adaptive element with its individual desired response for each pattern in a training sequence. This is the Achilles' heel of supervised learning pattern classifiers as network compo-

nents. In many problem-solving tasks, the network's environment may be able to provide assessments of certain *consequences* of the *collective activity* of all of the network components but the environment cannot know the desired responses of individual elements or even evaluate the behavior of individual elements. To use terms encountered in the artificial intelligence literature (e.g., [18]), the network's internal mechanism is not very "transparent" to the "critic."

Other approaches to the problem of learning within adaptive networks rely on adaptive elements that require neither teachers nor critics. These elements employ some form of unsupervised learning, or clustering, algorithm, often based on Hebb's [19] hypothesis that repeated pairing of pre- and postsynaptic activity strengthens synaptic efficacy. While clustering is likely to play an important role in sophisticated problem-solving systems, it does not by itself provide the necessary means for a system to improve performance in tasks determined by factors external to the system, such as, for example, the task of controlling an environment having initially unknown dynamics. For these types of tasks, a learning system must not just cluster information but must form those clusters that are useful in terms of the system's interaction with its environment. Thus it seems necessary to consider networks that learn under the influence of some sort of evaluative feedback, but this feedback cannot be so informative as to provide individualized instruction to each adaptive element.

These considerations have led us to study adaptive elements that are capable of learning to improve performance with respect to an evaluation function that assesses the consequences, which may be quite indirect, of element actions but does not directly specify these actions. Further, these elements are capable of improving performance under conditions of considerable uncertainty. Since evaluative feedback, or reinforcement feedback, will generally assess the performance of the entire network rather than the performance of individual elements, a high degree of uncertainty is necessarily present in the optimization problem faced by any individual component. Additional uncertainty arises from any delay that might exist between the time of an element's action and the time it receives the resulting reinforcement. The reinforcement feedback received by a network component at any time will generally depend upon factors other than its own action taken some fixed time earlier; it will additionally depend upon the actions of a large number of components taken at a variety of earlier times.

The ASE implements one part of our approach to these problems. Since we assume its environment is unable to provide desired responses, the ASE must *discover* what responses lead to improvements in performance. It employs a trial-and-error, or generate-and-test, search process. In the presence of input signals, it generates actions by a random process. Based on feedback that evaluates the problem-solving consequences of the actions, the ASE "tunes in" input signals to bias the action generation process, conditionally on the input, so that it will more likely generate the actions leading to improved performance. Different actions can be optimal when taken in the presence of different input signals. Actions that lead to improved performance when taken in the presence of certain input signals become associated with those signals in a developing input–output mapping. This type of stochastic search allows the ASE to improve performance under conditions of uncertainty. We have called this general process *associative search* [8] to emphasize both its association formation and generate-and-test search aspects.

In providing elements with these capabilities, we have been guided by the hypothesis of Klopf [11], [12] that neurons implement a strategy for attempting to maximize the frequency of occurrence of one type of input signal and minimize the frequency of occurrence of another type. According to this hypothesis, in other words, neurons can be conditioned in an operant or instrumental manner, where certain types of inputs act as rewarding stimuli and others act as punishing stimuli. A neuron learns how to attain certain types of inputs and avoid others by adjusting the transmission efficacy of its synapses according to the consequences of its discharges as fed back through pathways both internal to the nervous system and external to the animal. The ASE departs in several ways from Klopf's hypothesis, but his underlying idea remains the same.

III. Error Correction Versus Reinforcement Learning

Considerable misunderstanding is evident in the literature about how this type of "reinforcement learning" differs from supervised learning pattern classification as performed, for example, by Perceptrons and ADALINE's. It is important to emphasize these differences before we describe our adaptive elements. Supervised learning pattern classification elements are sometimes formulated in such a manner that the training process occurs as follows. A training pattern is presented to the element which responds as directed by its current set of weights; based on knowledge of the correct response, the element's environment feeds back an error signal giving the difference between the actual and correct resonses; the element uses this error signal to update its weight values. This sequence is repeated for all of the training patterns until the error signals become zero. These error signals are response-contingent feedback to the adaptive element, but it is misleading to view this process as a general form of reinforcement learning.

One important difference between the error-correction process just described and reinforcement learning as implemented by the ASE is that the latter does not rely exclusively on its weight values to determine its actions. Instead, it generates actions by a random process that is merely *biased* by the combination of its weight values and the input patterns. Actions are thus not appropriately viewed strictly as *responses* to input patterns. The random component of the generation process introduces the variety that is necessary to serve as the basis for subsequent selection by

evaluative feedback. The ASE therefore searches in its action space in a manner that supervised learning pattern classification machines do not.

Additionally, significant differences exist between general performance evaluation signals and the signed error signals required by supervised learning pattern classification elements. To supply a signed error signal, the environment must know both what the actual action was and what it should have been.[1] Evaluation of performance, on the other hand, may be based on a relative assessment of certain consequences of the element's actions rather than on knowledge of both the correct and actual actions. Widrow *et al.* [13] used the phrase "learning with a critic" to distinguish this type of process from learning with a teacher, as supervised learning pattern classification is sometimes called.

Very few studies have been made of neuronlike elements capable of learning under reinforcement feedback that is less informative than are signed error signals (Farley and Clark [20]; Minsky [21]; and Widrow *et al.* [13]). Indeed, considerable confusion arises from an unfortunate inconsistency in the usage of the term "error." What psychologists mean by trial-and-error learning is not the same as the error-correction process used by supervised learning pattern classification machines. Like the process employed by our ASE, trial-and-error learning is a "selectional" rather than an "instructional" process (cf. the usage of these terms by Edelman [22], although the selectional mechanism of the ASE is quite different from the one he proposes). Much more could be said about these issues, but we shall let the following example further clarify them. It will be apparent that elements such as Perceptrons and ADALINE's cannot by themselves solve the control problem we will consider.

IV. The Credit Assignment Problem

One can view the uncertainty discussed in the foregoing as a result of a fundamental problem that faces any learning system, whether it is natural or artificial, that has been called the "credit-assignment" problem by artificial intelligence researchers [18], [23]. This is the problem of determining what parts of a complex interacting or interlocking set of mechanisms, decisions, or actions deserve credit (blame) for improvements (decrements) in the overall performance of the system. The credit-assignment problem is especially acute when evaluative feedback to the learning system occurs infrequently, for example, upon the completion of a long series of decisions or actions.

Given the widely acknowledged importance of the credit-assignment problem for adaptive problem-solving systems, it is surprising that techniques for its solution have not been more intensely studied. The most successful,

---

[1] It is thus possible to formulate this training paradigm as one in which the learning machine's environment provides training patterns together with their desired responses (as we have done in Section II), and the system itself determines its error. This formulation does not involve feedback that passes through the machine's environment and more clearly reveals the limited nature of this type of process.

and perhaps the most extensible, solution to date was used in the checkers-playing program written by Samuel [24] more than two decade ago. A few isolated studies using similar techniques have been undertaken (Doran [25]; Holland [26]; Minsky [21], [23]; and Witten [27]), but the current approaches to the credit-assignment problem in artificial intelligence largely rely on providing the critic with domain-specific knowledge [18], [27]. Samuel's method, on the other hand, is one by which the system improves its own internal critic by a learning process.

The ACE implements a strategy most closely related to the methods of Samuel [24] and Witten [27] for reducing the severity of the credit-assignment problem. It adaptively develops an evaluation function that is more informative than the one directly available from the learning system's environment. This reduces the uncertainty under which the ASE must learn. The ACE was developed primarily by Sutton as a refinement of the adaptive element model of classical conditioning introduced by Sutton and Barto [9].

V. A Learning Control Problem: Pole Balancing

Fig. 1 shows a schematic representation of a cart to which a rigid pole is hinged. The cart is free to move within the bounds of a one-dimensional track. The pole is free to move only in the vertical plane of the cart and track. The controller can apply an impulsive "left" or "right" force F of fixed magnitude to the cart at discrete time intervals. The cart–pole system was simulated by digital computer using a very detailed model that includes all of the nonlinearities and reactive forces of the physical system (the Appendix provides details of the cart–pole model and simulations). The cart–pole model has four state variables:

x position of the cart on the track,
θ angle of the pole with the vertical,
$\dot{x}$ cart velocity, and
$\dot{\theta}$ rate of change of the angle.

Parameters specify the pole length and mass, cart mass, coefficients of friction between the cart and the track and at the hinge between the pole and the cart, the impulsive control force magnitude, the force due to gravity, and the simulation time step size.

The control problem we pose is identical to the one studied by Michie and Chambers. We assume that the equations of motion of the cart–pole system are not known and that there is no preexisting controller that can be imitated. At each time step, the controller receives a vector giving the cart–pole system's state at that instant. If the pole falls or the cart hits the track boundary, the controller receives a failure signal, the cart–pole system (but not the controller's memory) is reset to its initial state, and another learning trial begins. The controller must attempt to generate controlling forces in order to avoid the failure signal for as long as possible. No evaluative feedback other than the failure signal is available.

Learning to avoid the failure signal under these constraints is a very different problem than learning to balance

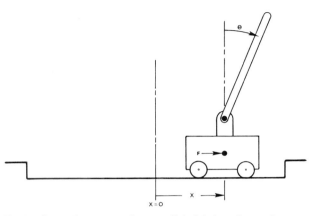

Fig. 1. Cart–pole system to be controlled. Solution of system's equations of motion approximated numerically (see Appendix).

the pole under the conditions usually assumed by control theorists. Since the failure signal will occur only after a long sequence of individual control decisions, a difficult credit-assignment problem arises in the attempt to determine which decisions were responsible for the failure. Neither a continuously available error signal nor a continuously available performance evaluation signal exists, as is the case in more conventional formulations of pole balancing. For example, Widrow and Smith [14] used a linear regression method, implemented by an ADALINE, to approximate the bang-bang control law required for balancing the pole. In order to use this method, however, they had to supply the controller with a signed error signal at each time step whose determination required external knowledge of the correct control decision for that time step. The present formulation of the problem, on the other hand, requires the learning system to discover for itself which control decisions are correct, and in so doing, solve a difficult credit-assignment problem that is completely absent in the usual versions of this problem.

VI. The Boxes System

By first describing Michie and Chambers' [3], [4] boxes system, we can provide much of the justification for the design of our adaptive elements. The strategy of these authors was to decompose the pole-balancing problem into a number of independent subproblems and to use an identical generate-and-test rule for learning to solve each subproblem. They divided the four-dimensional cart–pole state space into disjoint regions (or boxes) by quantizing the four state variables. They distinguished three grades of cart position, six of the pole angle, three of cart velocity, and three of pole angular velocity [4]. We use a similar partition of the state space based on the following quantization thresholds:

1) x: ± 0.8, ± 2.4 m,
2) θ: 0, ± 1, ± 6, $\pm 12°$,
3) $\dot{x}$: ± 0.5, $\pm \infty$ m/s,
4) $\dot{\theta}$: ± 50, $\pm \infty°$/s.

This yields $3 \times 3 \times 6 \times 3 = 162$ regions corresponding to all of the combinations of the intervals. The physical units of these thresholds differ from those used in [3] and [4]. We chose these values and units to produce what seemed like a physically realistic control problem, given our parameterization of the cart–pole simulation (Michie and Chambers did not publish the parameters of their cart–pole simulation. See the Appendix for our parameter values). At present we assume, as Michie and Chambers did, that this quantization is provided from the start (see Section X).

Each box is imagined to contain a *local demon* whose job is to choose a control action (left or right) whenever the system state enters its box. The local demon must learn to choose the action that will tend to be correlated with long system lifetime, that is, a long time until the occurrence of the failure signal. A *global demon* inspects the incoming state vector at each time step and alerts the local demon whose box contains that system state. When a failure signal is received, the global demon distributes it to all local demons. Each local demon maintains estimates of the expected lifetimes of the system following a left decision and following a right decision. A local demon's estimate of the expected lifetime for left is a weighted average of actual system lifetimes over all past occasions that the system state entered the demon's box and the decision left was made. The expected lifetime for the decision right is determined in the same way for occasions in which a right decision was made.

More specifically, upon being signaled by the global demon that the system state has entered its box, a local demon does the following.

1) It chooses the control action left or right according to which has the longest lifetime estimate. The control system emits the control action as soon as the decision is made.

2) It remembers which action was just taken and begins to count time steps.

3) When a failure signal is received, it uses its current count to update the left or right lifetime estimate, depending on which action it chose when its box was entered.

Michie and Chambers' actual algorithm is somewhat more complicated than this, but this description is sufficient for our present purposes. Details are provided in [3] where it is shown that the system is capable of learning to balance the pole for extended periods of time (in one reported run, the pole was balanced for a time approximately corresponding to one hour of real time). Notice that since the effect of a demon's decision will depend on the decisions made by other demons whose boxes are visited during a trial (where a trial is the time period from reset to failure), the environment of a local demon, consisting of the other demons as well as the cart–pole system, does not consistently evaluate the demon's actions.

VII. THE ASSOCIATIVE SEARCH ELEMENT (ASE)

Obviously, many possibilities exist for implementing a system like boxes using neuronlike elements. We know, for example, that any algorithm can be implemented by a network of McCulloch–Pitts abstract neurons acting as logic gates and delay units. Such an implementation would illustrate the neural metaphor resulting from the very earliest contact between neuroscience and digital technology [29]. More recent neural metaphors suggest that each local demon might be implemented by a network of adaptive neurons that would be set into reverberatory activity under conditions corresponding to the demon's box being entered by the state vector. Upon receipt of the failure signal, the magnitude of this reverberatory activity would somehow alter synapses used for triggering control actions. The global demon might be implemented by a neural network responsible for quantizing the system state vectors, conjunctively combining the results, and activating appropriate local demon networks (a neural decoder—see Section X). Finally, an element or network of elements would be required for channeling the action of each local demon network to a common efferent pathway.

In the neuronlike implementation we are pursuing, however, a local demon corresponds to the mechanism of a single synapse (to use the language of neural metaphor), and the output pathway of the postsynaptic element (the ASE) provides the common efferent pathway for control signals. At each synapse of the ASE are both a long-term memory trace that determines control actions and a short-term memory trace that is required to update the long-term trace, a role similar to that of a local demon's counter in the boxes algorithm. To accomplish the global demon's job of activating the appropriate local demon, we assume the existence of a decoder that has four real-valued input pathways (for the system state vector) and 162 binary valued output pathways corresponding to the boxes of Michie and Chambers' system (Fig. 2). The decoder transforms each state vector into a 162-component binary vector whose components are all zeros except for a single one in the position corresponding to the box containing the state vector. This vector is provided as input to the ASE and effectively selects the synapse corresponding to the appropriate box. For the other job of the global demon, that of distributing a failure signal to all of the local demons, we just let the adaptive element receive the failure signal via its reinforcement pathway and distribute the information to all of of its afferent synapses. In this way the entire boxes algorithm can be implemented by a single neuronlike ASE and an appropriate decoder.

In more detail, an ASE is defined as follows. The element has a reinforcement input pathway, n pathways for nonreinforcement input, and a single output pathway (Fig. 2). Let $x_i(t), 1 \le i \le n$, denote the real-valued signal on the ith nonreinforcement input pathway at time t, and let $y(t)$ denote the output at time t. Associated with each nonreinforcement input pathway i is a real-valued weight with value at time t denoted by $w_i(t)$.

The element's output $y(t)$ is determined from the input vector $X(t) = (x_1(t), \cdots, x_n(t))$ as follows:

$$y(t) = f\left[\sum_{i=1}^{n} w_i(t) x_i(t) + \text{noise}(t) \right] \qquad (1)$$

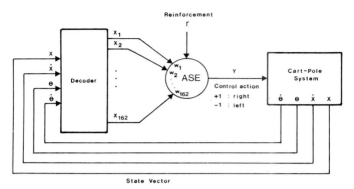

Fig. 2. ASE controller for cart–pole system. ASE's input is determined from current cart–pole state vector by decoder that produces output vector consisting of zeros with single one indicating which of 162 boxes contains state vector. ASE's output determines force applied to cart. Reinforcement is constant throughout trial and becomes -1 to signal failure.

where noise (t) is a real random variable with probability density function d and f is either a threshold, sigmoid, or identity function. For the pole-balancing illustration, d is the mean zero Gaussian distribution with variance σ^2, and f is the following threshold function:

$$f(x) = \begin{cases} +1, & \text{if } x \geq 0 \quad \text{(control action right)} \\ -1, & \text{if } x < 0 \quad \text{(control action left)}. \end{cases}$$

This follows the usual linear threshold convention common in adaptive network studies, but our approach does not depend strongly on the specifics of the input/output function of the element.

According to (1), actions are emitted even in the absence of nonzero input signals. The element's output is determined by chance, with a probability biased by the weighted sum of the input signals. If that sum is zero, the left and right control actions are equally probable. Assuming the decoder input shown in Fig. 2, a positive weight w_i, for example, would make the decision right more probable than left when box i is entered by the system state vector. The value of a weight, therefore, plays a role corresponding to the difference between the expected lifetimes for the left and right actions stored by a local demon in the boxes system. However, unlike the boxes system, the weight only determines the probability of an action rather than the action itself. The learning process updates the action probabilities. Also note that an input vector need not be of the restricted form produced by the decoder in order for (1) and the equations that follow to the meaningful.

The weights w_i, $1 \leqslant i \leqslant n$, change over (discrete) time as follows:

$$w_i(t+1) = w_i(t) + \alpha r(t) e_i(t) \qquad (2)$$

where

α positive constant determining the rate of change of w_i,

$r(t)$ real-valued *reinforcement* at time t, and

$e_i(t)$ *eligibility* at time t of input pathway i.

The basic idea expressed by (2) is that whenever certain conditions (to be discussed later) hold for input pathway i, then that pathway becomes eligible to have its weight

modified, and it remains eligible for some period of time after the conditions cease to hold. How w_i changes depends on the reinforcement received during periods of eligibility. If the reinforcement indicates improved performance, then the weights of the eligible pathways are changed so as to make the element more likely to do whatever it did that made those pathways eligible. If reinforcement indicates decreased performance, then the weights of the eligible pathways are changed to make the element more likely to do something else. The term "eligibility" and this weight update scheme are derived from the theory of Klopf [11], [12] and have precursors in the work of Farley and Clark [20], Minsky [21], and others. This general approach to reinforcement learning is related to the theory of stochastic learning automata [30], [31], which has its roots in the work of Bush and Mosteller [32] and Tsetlin [33].

Reinforcement: Positive r indicates the occurrence of a rewarding event and negative r indicates the occurrence of a punishing event.[2] It can be regarded as a measure of the *change* in the value of a performance criterion as commonly used in control theory. For the pole-balancing problem, r remains zero throughout a trial and becomes -1 when failure occurs.

Eligibility: Klopf [11] proposed that a pathway should reach maximum eligibility a short time after the occurrence of a pairing of a nonzero input signal on that pathway with the "firing" of the element. Eligibility should decay thereafter toward zero. Thus, when the consequences of the element's firing are fed back to the element, credit or blame can be assigned to the weights that will alter the firing probability when a similar input pattern occurs in the future. More generally, the eligibility of a pathway reflects the extent to which input activity on that pathway was paired in the past with element output activity. The eligibility of pathway i at time t is therefore a *trace* of the product $y(\tau)x_i(\tau)$ for times τ preceding t. If either or both of the quantities $y(\tau)$ and $x_i(\tau)$ are negative (as they can be for the ASE defined earlier), then credit is assigned

---

[2] A negative value of r is not the same as a psychologists's "negative reinforcement." In psychology, negative reinforcement is reinforcement due to the cessation of an aversive stimulus.

appropriately via (2) if eligibility is a trace of the signed product $y(\tau)x_i(\tau)$.

For computational simplicity, we generate exponentially decaying eligibility traces e_i using the following linear difference equation:

$$e_i(t + 1) = \delta e_i(t) + (1 - \delta)y(t)x_i(t), \qquad (3)$$

where $\delta, 0 \leq \delta < 1$, determines the trace decay rate. Note that each synapse has its own local eligibility trace.

Eligibility plays a role analogous to the part of the boxes local-demon algorithm that, when the demon's box is entered and an action has been chosen, remembers what action was chosen and begins to count. The factor $x_i(t)$ in (3) triggers the eligibility trace, a kind of count, or contributes to an ongoing trace, whenever box i is entered ($x_i(t) = 1$). Instead of explicitly remembering what action was chosen, our system contributes a different amount to the eligibility trace depending on what action was chosen (via the term $y(t)$ in (3)). Thus the trace contains information not only about how long ago a box was entered but also about what decision was made when it was entered.

Unlike the count initiated by a local demon in the boxes system, however, the eligibility trace effectively counts down rather than up (more precisely, its magnitude decays toward zero). Recall that reinforcement r remains zero until a failure occurs, at which time it becomes -1. Thus whatever control decision was made when a box was visited will always be made *less* likely when the failure occurs, but the longer the time interval between the decision and the occurrence of the failure signal, the less this decrease in probability will be. From one perspective, this process seems appropriate. Since the failure signal always eventually occurs, the action that was taken may deserve some of the blame for the failure. However, this view misses the point that even though both actions inevitably lead to failure, one action is probably *better* than the other. The learning process defined by (1)–(3) needs to be more subtle to ensure convergence to the actions that yield the least punishment in cases in which only punishment is available. In the present article, we build this subtlety into the ACE rather than into the ASE. Among its other functions, the ACE constructs predictions of reinforcement so that if punishment is less than its expected level, it acts as reward. For the pole-balancing task, the ASE as defined here must operate in conjunction with the ACE.

Although the boxes system and the version of the pole-balancing problem described earlier serve well to make an ASE's design understandable, the ASE does not represent an attempt to duplicate the boxes algorithm in neuronlike form. We are interested in tasks more general than the pole-balancing problem and in learning systems that are more general than the boxes system. An ASE is less restricted than the boxes system in several ways. First, the boxes system is based on the subdivision of the problem space into a finite number of nonoverlapping regions, and no generalization is attempted between regions. It develops a control rule that is effectively specified by means of a lookup table. Although a form of generalization can be easily added to the boxes algorithm by using an averaging process over neighboring boxes (see Section X) it is not immediately obvious how to extend the algorithm to take advantage of the other forms of generalization that would be possible if the controlled system's states could be represented by arbitrary vectors rather than only by the standard unit basis vectors which are produced by a suitable decoder. The ASE can accept arbitrary input vectors and, although we do not illustrate it in this article, can be regarded as a step toward extending the type of generalization produced by error-correction supervised learning pattern classification methods to the less restricted reinforcement learning paradigm (see Section III).

The boxes system is also restricted in that its design was based on the *a priori* knowledge that the time until failure was to serve as the evaluation criterion and that the learning process would be divided into distinct trials that would always end with a failure signal. This knowledge permitted Michie and Chambers to reduce the uncertainty in the problem by restricting each local demon to choosing the same action each time its box was entered during any given trial. The ASE, on the other hand, is capable of working to achieve rewarding events and to avoid punishing events which might occur at any time. It is not exclusively failure-driven, and its operation is specified without reference to the notion of a trial.

VIII. THE ADAPTIVE CRITIC ELEMENT (ACE)

Fig. 3 shows an ASE together with an ACE configured for the pole-balancing task. The ACE receives the externally supplied reinforcement signal which it uses to determine how to compute, on the basis of the current cart–pole state vector, an improved reinforcement signal that it sends to the ASE. Expressed in terms of the boxes system, the job of the ACE is to store in each box a prediction or expectation of the reinforcement that can eventually be obtained from the environment by choosing an action for that box. The ACE uses this prediction to determine a reinforcement signal that it delivers to the ASE whenever the box is entered by the cart–pole state, thus permitting learning to occur throughout the pole-balancing trials rather than solely upon failure. This greatly decreases the uncertainty faced by the ASE. The central idea behind the ACE algorithm is that predictions are formed that predict not just reinforcement but also future predictions of reinforcement.

Like the ASE, the ACE has a reinforcement input pathway, n pathways for nonreinforcement input, and a single output pathway (Fig. 3). Let $r(t)$ denote the real-valued reinforcement at time t; let $x_i(t), 1 \leq i \leq n$, denote the real-valued signal on the ith nonreinforcement input pathway at time t; and let $\hat{r}(t)$ denote the real-valued output signal at time t. Each nonreinforcement input pathway i has a weight with real value $v_i(t)$ at time t. The output $\hat{r}$ is the improved reinforcement signal that is used by the ASE in place of r in (2).

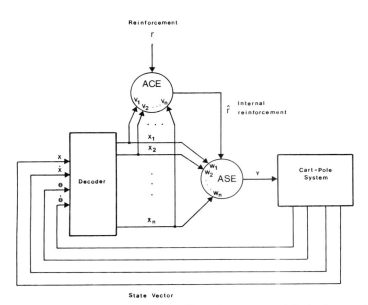

Fig. 3. ASE and ACE configured for pole-balancing task. ACE receives same nonreinforcing input as ASE and uses it to compute an improved or internal reinforcement signal to be used by ASE.

In order to produce $\hat{r}(t)$, the ACE must determine a prediction $p(t)$ of eventual reinforcement that is a function of the input vector $X(t)$ (which in the boxes paradigm, simply selects a box). We let

$$p(t) = \sum_{i=1}^{n} v_i(t) x_i(t) \qquad (4)$$

and seek a means of updating the weights v_i so that $p(t)$ converges to an accurate prediction. The updating rule we use is

$$v_i(t + 1) = v_i(t) + \beta [r(t) + \gamma p(t) - p(t - 1)] \bar{x}_i(t), \qquad (5)$$

where β is a positive constant determining the rate of change of v_i; $\gamma, 0 < \gamma \leqslant 1$, is a constant to be explained below; $r(t)$ is the reinforcement signal supplied by the environment at time t; and $\bar{x}_i(t)$ is the value at time t of a *trace* of the input variable x_i.

It is beyond the scope of the present paper to explain the derivation of this learning rule fully (see [7] and [9]). Very briefly, the trace $\bar{x}_i$ acts much like the eligibility trace e_i defined by (3). Here, however, an input pathway gains positive eligibility whenever a nonzero signal is present on that pathway, irrespective of what the element's action is. We compute $\bar{x}_i$ using the following linear difference equation (cf. (3)):

$$\bar{x}_i(t + 1) = \lambda \bar{x}_i(t) + (1 - \lambda) x_i(t), \qquad (6)$$

where $\lambda, 0 \leqslant \lambda < 1$ determines the trace decay rate.

According to (6), an eligible pathways's weight changes whenever the actual reinforcement $r(t)$ plus the current prediction $p(t)$ differs from the value $p(t - 1)$ that was predicted for this sum. Closely related to the ADALINE learning rule and related regression techniques, this rule provides a means of finding weight values such that $p(t - 1)$ approximates $r(t) + \gamma p(t)$, or, equivalently, such that $p(t)$ approximates $r(t + 1) + \gamma p(t + 1)$. By attempting to

predict its own prediction, the learning rule produces predictions that tend to be the earliest possible indictions of eventual reinforcement. The constant γ, related to Witten's [27] "discount factor," provides for eventual extinction of predictions in the absence of external reinforcement. If $\gamma = 1$, predictions will be self-sustaining in the absence of external reinforcement; whereas if $0 < \gamma < 1$, predictions will decay in the absence of external reinforcement. In our simulations, $\gamma = 0.95$.

The ACE's output, the improved or internal reinforcement signal, is computed from these predictions as follows:

$$\hat{r}(t) = r(t) + \gamma p(t) - p(t - 1). \qquad (7)$$

This is the same expression appearing in (5). The reader should note that with $\hat{r}$ substituted for r in (2), the weight updating rules for the ASE and ACE ((2) and (5), respectively) differ only in their forms of eligibility traces. The ASE's traces are conditional on its output, whereas the ACE's are not.

Although this process works for arbitrary input vectors, it is easiest to justify (7) by again specializing to the boxes input representation. According to (7), as the cart–pole state moves between boxes without failure occurring (i.e., $r(t) = 0$), the reinforcement $\hat{r}(t)$ sent to the ASE is the difference between the prediction of reinforcement of the current box (discounted by γ) and the prediction of reinforcement of the previous box. Increases in reinforcement prediction therefore become rewarding events (assuming $\gamma = 1$), and decreases become penalizing events.

When failure occurs, the situation is slightly different. Given the way the control problem is represented, when failure occurs the cart–pole state is not in any box. Thus all $x_i(t)$ are equal to zero at failure, and according to (4), so is $p(t)$. Upon failure, then, the reinforcement sent to the ASE is the externally supplied reinforcement $r(t) = -1$, minus the previous prediction $p(t - 1)$. Consequently, an unpredicted failure results in $\hat{r}(t)$ being negative. This both punishes the actions made preceding the failure and, via (5), increments the predictions of failure (i.e., decrements the p's) of the boxes entered before the failure. A fully predicted failure generates no punishment. However, when a box with such a high prediction of failure is entered from a box with a lower prediction of failure, the recently made actions are punished and the recent predictions of failure are incremented, just as they were initially upon failure. Similarly, if the cart–pole state moves from a box with a higher prediction of failure to a box with a lower prediction, the recent actions are rewarded and recent predictions of failure are decremented (i.e., the p's are incremented). The system thus learns which boxes are "safe" and which are "dangerous." It punishes itself for moving from any box to a more dangerous box and rewards itself for moving from any box to a safer box. In the following we discuss the relation between the behavior of the ACE and that of animals in classical conditioning experiments.

IX. Simulation Results

We implemented the boxes system as described in [3] and [4] as well as our systems shown in Fig. 2 (ASE alone) and Fig. 3 (ASE with ACE). We wanted to determine what kinds of neuronlike elements could attain or exceed the performance of the boxes system. Our results suggest that a system using an ASE with internal reinforcement supplied by an ACE is easily able to outperform the boxes system. We must emphasize at the outset, however, that it is not our intention to criticize Michie and Chambers' program: the boxes system they described was in an initial state of development and clearly could be extended to include a mechanism analogous to our ACE. We make comparisons with the performance of the boxes system because it provides a convenient reference point.

We simulated a series of runs of each learning system attempting to control the same cart–pole simulation (see the Appendix). Each run consisted of a sequence of trials, where each trial began with the cart–pole state $x = 0$, $\dot{x} = 0$, $\theta = 0$, $\dot{\theta} = 0$, and ended with a failure signal indicating that θ left the interval $[-12°, 12°]$ or x left the interval $[-2.4 \text{ m}, 2.4 \text{ m}]$. We also set all the trace variables e_i to zero at the start of each trial. The learning systems were "naive" at the start of each run (i.e., all the weights w_i and v_i were set to zero). At the start of each boxes run, we supplied a different seed value to the pseudorandom number generator that we used to initialize the state of the learning system and to break ties in comparing expected lifetimes in order to choose control actions. We did not reset the cart–pole state to a randomly chosen state at the start of each trial as was done in the experiments reported in [3] and [4]. At the start of each run of an ASE system, we supplied a different seed to the pseudorandom number generator that we used to generate the noise used in (1). We approximated this Gaussian random variable by the usual procedure of summing uniformly distributed random variables (we used an eightfold sum). Since the ASE runs began with weight vectors equal to zero, initial actions for each box were equiprobable, and initial ACE predictions were zero. Except for the random number generator seeds,

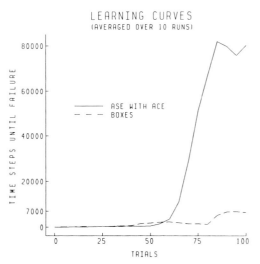

Fig. 4. Simulation results. Performance of boxes system and ASE/ACE system averaged over ten runs. See text for complete explanation.

identical parameter values were used for all runs. Runs consisted of 100 trials unless the run's duration exceeded 500 000 time steps (approximately 2.8 h of simulated real time), in which case the run was terminated. For our implementation of the boxes system, we used the parameter values published in [3]. We experimented with other parameter values without obtaining consistently better performance. We did not attempt to optimize the performance of the systems using the ASE. We picked values that seemed reasonable based on our previous experience with similar adaptive elements.

Figs. 4 and 5 show the results of our simulations of boxes and the ASE/ACE system. The graphs of Fig. 4 are averages of performance over the ten runs that produced the individual graphs shown in Fig. 5. In both figures, a single point is plotted for each bin of five trials giving the number of time steps until failure averaged over those five trials. Almost all runs of the ASE/ACE system, and one run of the boxes system, were terminated after 500 000 time steps before all 100 trials took place (those whose graphs terminate short of 100 trials in Fig. 5). We stopped the simulation before failure on the last trials of these runs. To produce the averages for all 100 trials shown in Fig. 4, we needed to make special provision for the interrupted runs. If the duration of the trial that was underway when the run was interrupted was less than the duration of the immediately preceding (and therefore complete) trial, then we assigned to fictitious remaining trials the duration of that preceding trial. Otherwise, we assigned to fictitious remaining trials the duration of the last trial when it was interrupted. We did this to prevent any short interrupted trials from producing deceptively low averages.

The ASE/ACE system achieved much longer runs than did the boxes system. Fig. 5 shows that the ACE/ASE system tended to solve the problem before it had experienced 100 failures, whereas the boxes system tended not to. Obviously, we cannot make definitive statements about the relative performance of these systems, or about the general utility of the ASE/ACE system solely on the basis of these

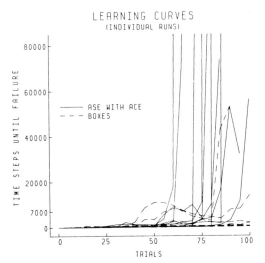

Fig. 5. Simulation results. Performance of boxes system and ASE/ACE system in individual runs that were averaged to produce Fig. 4. See text for complete explanation.

experiments. However, these results encourage us to continue developing the principles upon which the ASE/ACE system is based.

The parameter values used in producing these results were $\alpha = 1000$, $\beta = 0.5$, $\delta = 0.9$, $\gamma = 0.95$, $\lambda = 0.8$, and $\sigma = 0.01$. Except for one set of extreme values, these were the only values we tried. The large value of α was chosen so that large changes in the weights w_i occurred upon reinforcement. This caused the probability of a rewarded action to become nearly one and the probability of a penalized action to become nearly zero. We did this in an attempt to implement in our system the feature of the boxes system that causes each local demon to choose the same action each time its box is entered in any given trial. This greatly reduces the uncertainty in the problem but would be inappropriate, we think, for problems in which other reinforcing events could occur during trials. The parameters δ and λ determine the durations of the eligibility traces. Their values, 0.9 and 0.8, respectively, cause long, slowly decaying traces to form, as seemed appropriate given the nature of the problem.

The good performance of the ASE/ACE system was almost entirely due to the ACE's supplying reinforcement throughout trials. For the boxes system and for an ASE without an ACE, learning occurs only upon failure, an event that becomes less frequent as learning proceeds. With the ACE in place, an ASE can receive feedback on every time step. The learning produced by this feedback causes the system to attempt to enter particular parts of the state space and to avoid others. We simulated the control problem using an ASE without an ACE, using the same parameter settings that worked well for the ASE/ACE experiments. The ASE was not able to attain the level of performance shown by the boxes system. These shortcomings of the ASE are due to difficulties in the convergence process in tasks involving only penalizing feedback, as

discussed in Section VII. The use of reinforcement computed by the ACE markedly changes this property of the pole-balancing problem. At present, we have little experience with ASE-like elements operating without ACE supplied reinforcement in the pole-balancing problem.

X. THE DECODER

We have assumed the existence of a decoder that effectively divides the cart–pole state space into a number of disjoint regions by transforming each state vector into a vector having 162 components, all but one of which is zero. We call this a decoder after a similar device used in computer memory circuits to transform each memory address into a signal on the wire connected to the physical location having that address. With this decoder providing its input, the ASE essentially fills in a lookup table of control actions. Similarly, the ACE fills in a table of reinforcement predictions. Each item of information is stored by the setting of the value of a single synaptic weight at a given location.

As a consequence of this localized storage scheme, no generalization occurs beyond the confines of a given box. Given the relative smoothness of the cart–pole dynamics, learning would be faster if information stored in a box could be extrapolated to neighboring boxes (using the Euclidean metric). This can be accomplished by using a kind of decoder that produces activity on overlapping sets of output pathways. It is interesting to note that in several theories of sensorimotor learning, it is postulated that the granular layer of the cerebellum implements just this kind of decoder and that Purkinje cells are adaptive elements [34], [35].[3]

Localized extrapolation is not the only type of generalization that can be useful. There has lately been increasing interest in "associative memory networks" that use distributed representations in which dispersed rather than localized patterns of activity encode information [36], [37]. Rather than implementing table-lookup storage, associative memories use weighted summations to compute output vectors from input vectors. This style of information storage provides generalization among patterns according to where they lie with respect to a set of linear discriminant functions. Since the ASE and ACE use weighted summations that are defined for arbitrary input vectors, they implement linear discriminant functions and are capable of forming information storage networks having all of the properties that have generated interest in associative memory networks. Unlike the associative memory networks discussed in the literature, however, networks of ASE-like components are capable of discovering via reinforcement learning what information is useful to store. These aspects

---

[3]In these theories, the adaptive elements perform supervised learning pattern classification, with climbing fiber input providing the desired responses, and not the type of reinforcement learning with which the present article is concerned. If the adaptive capacity of a Purkinje cell were limited to that postulated in these theories, then a Purkinje cell would not be able to solve the type of problem illustrated by the pole-balancing task described in this article.

of ASE's are emphasized by Barto *et al.* [8] where *associative search* networks are discussed.

Whether environmental states are represented using localized or distributed patterns, the problem remains of how to choose the specifics of the representations in order to facilitate learning. In this article we followed Michie and Chambers in choosing a state–space partition based on special knowledge of the control task. As they point out, it is easy to choose a partition that makes the task impossible [4]. For the next stage of development of the boxes system, Michie and Chambers planned to give the system the ability to change the boundaries of the boxes by the processes of "splitting" and "lumping" [4]. We are not aware of any results they published on these processes, but we were motivated in part by their comments to experiment with layered networks of ASE-like adaptive elements in order to examine the feasibility of implementing a kind of adaptive decoder. Some preliminary results, reported in [5], were encouraging, and we are continuing our investigation in this direction.

XI. Animal Learning

Minsky has pointed out [23] that methods for reducing the severity of the credit-assignment problem like the one used in Samuel's checkers player are suggestive of secondary or conditioned reinforcement phenomena in animal learning studies. A stimulus acquires reinforcing qualities (i.e., becomes a secondary reinforcer) if it predicts either primary reinforcement (e.g., food or shock) or some other secondary reinforcer. It is generally held that higher order classical conditioning, whereby previously conditioned conditioned stimuli (CS's) can act as unconditioned stimuli (US's) for earlier potential CS's, is the basis for the development of secondary reinforcement [38].

The ACE is a refinement of the model of classical conditioning that was presented in [9]. That model's behavior is consistent with the Rescorla–Wagner model of classical conditioning [39]. While not without certain problems, the Rescorla–Wagner model has been the most influential model of classical conditioning for the last ten years [40]. One interpretation of the basic premise of the Rescorla-Wagner model is that the degree to which an event is "unexpected" or "surprising" determines the degree to which it enters into associations with earlier events. Stimuli lose their reinforcing qualities to the extent that they are expected on the basis of the occurrence of earlier stimuli. The model upon which the ACE is based extends the basic mechanism of the Rescorla–Wagner model to

provide for some of the features of higher order conditioning, the influence of relative event timing within trials, and the occurrence of conditioned responses (CR's) that anticipate the US. In these terms, the failure signal r corresponds to the US, the signals x_i from the decoder correspond to potential CS's, and the prediction p corresponds to a component of the CR. We have not yet thoroughly investigated the extent to which the ACE/ASE system is a valid model of animal behavior in instrumental conditioning experiments.

XII. Conclusion

It should be clear that our approach differs from that of the pioneering adaptive neural-network theorists of the 1950's and 1960's. We have built into single neuronlike adaptive elements a problem-solving capacity that in many respects exceeds that achieved in the past by entire simulated neural networks. The metaphor for neural function suggested by this approach provides, at least to us, the first convincing inkling of how nervous tissue could possibly be capable of its exquisite feats of problem solving and control.

We argued that components of powerful adaptive networks must be at least as sophisticated as the components described in this article. If this were true for biological networks as well as for artificial networks, then it would suggest that parallels might exist between neurons and the adaptive elements described here. It would suggest, for example, that 1) there are single neuron analogs of instrumental conditioning and chemically specialized reinforcing neurons that may themselves be adaptive (see [41]); 2) the random component of an instrumental neuron's behavior is necessary for generating variety to serve as the basis for subsequent selection; and 3) mechanisms exist for maintaining relatively long-lasting synaptically local traces of activity that modulate changes in synaptic efficacy. Although some of these implications are supported in varying degrees by existing data, there are no data that provide direct support for the existence of the specific mechanisms used in our adaptive elements. By showing how neuronlike elements can solve genuinely difficult problems that are solved routinely by many animals, we hope to stimulate interest in the relevant experimental research.

Appendix
Details of the Cart–Pole Simulation

The cart–pole system is modeled by the following nonlinear differential equations (see [42]):

$$\ddot{\theta}_t = \frac{g \sin \theta_t + \cos \theta_t \left[\dfrac{-F_t - ml\dot{\theta}_t^2 \sin\theta_t + \mu_c \operatorname{sgn}(\dot{x}_t)}{m_c + m} \right] - \dfrac{\mu_p \dot{\theta}_t}{ml}}{l \left[\dfrac{4}{3} - \dfrac{m \cos^2 \theta_t}{m_c + m} \right]}$$

$$\ddot{x}_t = \frac{F_t + ml \left[\dot{\theta}_t^2 \sin \theta_t - \ddot{\theta}_t \cos \theta_t \right] - \mu_c \operatorname{sgn}(\dot{x}_t)}{m_c + m}$$

where

$g = -9.8$ m/s^2, acceleration due to gravity,

$m_c = 1.0$ kg, mass of cart

$m = 0.1$ kg, mass of pole,

$l = 0.5$ m, half-pole length,

$\mu_c = 0.0005$, coefficient of friction of cart on track,

$\mu_p = 0.000002$, coefficient of friction of pole on cart,

$F_t = \pm 10.0$ newtons, force applied to cart's center of mass at time t.

We initially used the Adams–Moulton predictor–corrector method to approximate numerically the solution of these equations, but the results reported in this article were produced using Euler's method with a time step of 0.02 s for the sake of computational speed. Comparisons of solutions generated by the Adams–Moulton methods and the less accurate Euler method did not reveal discrepencies that we deemed significant for the purposes of this article.

ACKNOWLEDGMENT

We thank A. H. Klopf for bringing to us a set of ideas filled with possibilities. We are grateful also to D. N. Spinelli, M. A. Arbib, and S. Epstein for their valuable comments and criticisms; to D. Lawton for first making us aware of Michie and Chambers' boxes system; to D. Politis and W. Licata for pointing out an important error in our original cart–pole simulations; and to S. Parker for essential help in preparing the manuscript.

REFERENCES

[1] B. Chandrasekaran, "Natural and social system metaphors for distributed problem solving: Introduction to the issue," *IEEE Trans. Syst., Man., Cybern.*, vol. SMC-11, pp. 1–5, 1981.

[2] V. R. Lesser and D. D. Corkill, "Functionally-accurate, cooperative distributed systems," *IEEE Trans. Syst., Man, Cybern.*, vol. SMC-11, pp. 81–96, 1981.

[3] D. Michie and R. A. Chambers, "BOXES: An experiment in adaptive control," in *Machine Intelligence 2*, E. Dale and D. Michie, Eds. Edinburgh: Oliver and Boyd, 1968, pp. 137–152.

[4] D. Michie and R. A. Chambers, "'Boxes' as a model of pattern-formation," in *Towards a Theoretical Biology*, vol. 1, *Prolegomena*, C. H. Waddington, Ed. Edinburgh: Edinburgh Univ. Press, 1968, pp. 206–215.

[5] A. G. Barto, C. W. Anderson, and R. S. Sutton, "Synthesis of nonlinear control surfaces by a layered associative search network," *Biol. Cybern.*, vol. 43, pp. 175–185, 1982.

[6] A. G. Barto and R. S. Sutton, "Landmark learning: An illustration of associative search," *Biol. Cybern.*, vol. 42, pp. 1–8, 1981.

[7] ———, "Simulation of anticipatory responses in classical conditioning by a neuron-like adaptive element," *Behavioral Brain Res.*, vol. 4, pp. 221–235, 1982.

[8] A. G. Barto, R. S. Sutton, and P. S. Brouwer, "Associative search network: A reinforcement learning associative memory," *Biol. Cybern.*, vol. 40, pp. 201–211, 1981.

[9] R. S. Sutton and A. G. Barto, "Toward a modern theory of adaptive networks: Expectation and prediction," *Psychol. Rev.*, vol. 88, pp. 135–171, 1981.

[10] ———, "An adaptive network that constructs and uses an internal model of its world," *Cognition and Brain Theory*, vol. 4, pp. 213–246, 1981.

[11] A. H. Klopf, "Brain function and adaptive systems—A heterostatic theory," Air Force Cambridge Res. Lab. Res. Rep., AFCRL-72-0164, Bedford, MA, 1972. (A summary appears in *Proc. Int. Conf. Syst., Man, Cybern.*, 1974).

[12] A. H. Klopf, *The Hedonistic Neuron: A Theory of Memory, Learning, and Intelligence.* Washington, DC: Hemisphere, 1982.

[13] B. Widrow, N. K. Gupta, and S. Maitra, "Punish/reward: learning with a critic in adaptive threshold systems," *IEEE Trans. Syst., Man, Cybern.*, vol. SMC-3, pp. 455–465, 1973.

[14] B. Widrow and F. W. Smith, "Pattern-recognizing control systems," in *Computer and Information Sciences*, J. T. Tow and R. H. Wilcox, Eds. Clever Hume Press, 1964, pp. 288–317.

[15] F. Rosenblatt, *Principles of Neurodynamics.* New York: Spartan, 1962.

[16] B. Widrow and M. E. Hoff, "Adaptive switching circuits," in *1960 WESCON Conv. Record*, part IV, 1960, pp. 96–104.

[17] R. O. Duda and P. E. Hart, *Pattern Classification and Scene Analysis.* New York: Wiley, 1973.

[18] P. R. Cohen and E. A. Feigenbaum, *The Handbook of Artificial Intelligence*, vol. 3. Los Altos, CA: Kauffman, 1982.

[19] D. O. Hebb, *Organization of Behavior.* New York: Wiley, 1949.

[20] B. G. Farley and W. A. Clark, "Simulation of self-organizing systems by digital computer," *IRE Trans. Inform. Theory*, vol. PGIT-4, pp. 76–84, 1954

[21] M. L. Minsky, "Theory of neural-analog reinforcement systems and its application to the brain-model problem," Ph.D. dissertation, Princeton Univ., Princeton, NJ, 1954.

[22] G. M. Edelman, "Group selection and phasic reentrant signaling: A theory of higher brain function," in *The Mindful Brain: Cortical Organization and the Group-Selective Theory of Higher Brain Function*, G. M. Edelman and V. B. Mountcastle, Eds. Cambridge, MA: MIT Press, 1978.

[23] M. L. Minsky, "Steps toward artificial intelligence," *Proc. IRE*, vol. 49, pp. 8–30, 1961.

[24] A. L. Samuel, "Some studies in machine learning using the game of checkers," *IBM J. Res. Develop.*, vol. 3, pp. 210–229, 1959.

[25] J. Doran, "An approach to automatic problem solving," in *Machine Intelligence*, vol. 1, E. L. Collins and D. Michie, Eds. Edinburgh: Oliver and Boyd, 1967, pp. 105–123.

[26] J. H. Holland, "Adaptive algorithms for discovering and using general patterns in growing knowledge-bases," *Int. J. Policy Anal. Inform. Syst.*, vol. 4, pp. 217–240, 1980.

[27] I. H. Witten, "An adaptive optimal controller for discrete-time Markov environments," *Inform. Contr.*, vol. 34, pp. 286–295, 1977.

[28] T. D. Dietterich and B. G. Buchanan, "The role of the critic in learning systems," Stanford Univ. Tech. Rep., STAN-CS-81-891, 1981.

[29] W. S. McCulloch and W. H. Pitts, "A logical calculus of the ideas immanent in nervous activity," *Bull. Math. Biophys.*, vol. 5, pp. 115–133, 1943.

[30] K. S. Narendra and M. A. L. Thatachar, "Learning automata—A survey," *IEEE Trans. Syst., Man, Cybern.*, vol. SMC-4, pp. 323–334, 1974.

[31] S. Lakshmivarahan, *Learning Algorithms Theory and Applications.* New York: Springer-Verlag, 1981.

[32] R. R. Bush and F. Mosteller, *Stochastic Models for Learning.* New York: Wiley, 1958.

[33] M. L. Tsetlin, *Automaton Theory and Modelling of Biological Systems.* New York: Academic, 1973.

[34] J. A. Albus, *Brains, Behavior, and Robotics.* Peterborough, NH: BYTE Books, 1981.

[35] D. Marr, "A theory of cerebellar cortex," *J. Physiol.*, vol. 202, pp. 437–470, 1969.

[36] G. E. Hinton and J. A. Anderson, Eds., *Parallel Models of Associative Memory.* Hillsdale, NJ: Erlbaum, 1981.

[37] T. Kohonen, *Associative Memory: A System Theoretic Approach.* Berlin, Germany: Springer, 1977.

[38] R. A. Rescorla, *Pavlovian Second-Order Conditioning: Studies in Associative Learning.* Hillsdale, NJ: Erlbaum, 1980.

[39] R. A. Rescorla and A. R. Wagner, "A theory of Pavlovian conditioning: Variations in the effectiveness of reinforcement and nonreinforcement," in *Classical Conditioning II: Current Research and Theory*, A. H. Black and W. F. Prokasy, Eds. New York: Appleton-Century-Crofts, 1972.

[40] A. Dickinson, *Contemporary Animal Learning Theory.* Cambridge: Cambridge Univ. Press, 1980.

[41] L. Stein and J. D. Belluzzi, "Beyond the reflex arc: A neuronal model of operant conditioning," in *Changing Concepts of the Nervous System.* New York: Academic, 1982.

[42] R. H. Cannon, *Dynamics of Physical Systems.* New York: McGraw-Hill, 1967.

Introduction

(1983)
S. Kirkpatrick, C. D. Gelatt, Jr., and M. P. Vecchi

Optimization by simulated annealing
Science 220:671–680

Selfridge, in this collection (paper 9), pointed out that if we climb a hill a step at a time, always going upward, it might be a way to maximize a function. Selfridge also pointed out some of the problems involved in hill climbing. The major problem addressed by the paper by Kirkpatrick, Gelatt, and Vecchi and the major problem in this literature in general is that we might not be climbing the right hill; that is, we might be finding the maximum of one of the foothills (a local maximum) rather than finding the overall (global) maximum.

Many of the problems that we are concerned with in this collection of papers require finding a minimum value of a function, rather than the maximum values; that is, we are descending into a valley rather than climbing a hill. This is because most of the functions that grow out of neural networks are related to energy in physical systems or to cost, where we want to find the smallest rather than the largest value. However, the mathematical techniques are identical, whether one is trying to go up or down.

If we are only allowed to make decisions about the best direction to move in based on the local topography, we are very restricted. All we can do is see if the ground slopes up or down, and then move in the best direction.

We need some way to get occasional samples of the rest of the terrain. There are a number of ways to do this. As one example, we could do a systematic search over the parameters of the function, but this can take an appalling amount of computer time, even if the problem is small; and unless the samples are taken close enough together, we might completely miss a deep narrow valley.

A question for debate in medieval theology was whether God could create two hills without an intervening valley (or if we are minimizing a function, two valleys without an intervening hill). Unfortunately, when optimizing functions, the answer seems to be no. To get out of a local minimum and into the global minimum, we must climb over the ridge between them. Rules based only on the local topography will not let us do this. This paper by Kirkpatrick, Gelatt, and Vecchi suggests another way of sampling more than a local region of parameters.

The basic idea is very simple, and seems to have been first used as a technique in computer simulation over thirty years ago by Metropolis. Instead of *always* going downhill, we go downhill *most* of the time.

The metaphor used in this technique is physical. There is a system parameter called *temperature* that controls how much random searching is done by the system. The thermodynamic probability of finding the system in a particular configuration with a given energy is proportional to the Boltzmann probability factor,

$$e^{-E(\text{configuration})/kT}.$$

Suppose we have two configurations, say C_1 and C_2, with energy E_1 and E_2. Then the ratio of the probabilities of the two configurations is given by

$$\frac{\text{Prob}(C_1)}{\text{Prob}(C_2)} = e^{-[E(C_1)-E(C_2)]/kT}.$$

For example, a molecule of gas in the earth's atmosphere has its lowest energy when it is sitting on the ground. But because the temperature of the earth is about 300° Kelvin, the probability of finding the molecule above the surface is quite large, because the difference in energy between a configuration with a particular molecule at 0 feet and at 10 feet above ground is very small compared to kT. Then the probability of finding the molecule at one height and the probability of finding it at the other are roughly the same. However, the difference in energy between a molecule at 0 feet and at 100,000 feet is substantial, so molecules are more *likely* to be low in altitude than high. But there will still be some, though fewer, at high altitudes. Low energy configurations are more probable than high energy ones. If we wait a long time, the system will take on many configurations, with their relative probabilities given by the ratio of the two Boltzmann factors.

If the temperature is very high, the system, when it changes, explores a large number of possible configurations. Both high and low energy configurations are almost equally probable because the denominator, kT, is so large that it makes the difference in probabilities between different configurations small. That is, the system is almost as likely to increase its energy when it changes state as to decrease it.

If we drop the temperature, the probabilities of high energy configurations decrease relative to the low energy ones. The state of the system is liable to remain longer in the lower energy states. If the system is at a very low temperature, any configuration beyond the lowest energy one that the system can reach is highly improbable and the system state is frozen.

We cannot just start off at zero temperature for the same reasons that give us trouble with local minima in other problems. If we start at a particular point, and explore new configurations, the system will try to lower the energy if it can, because at zero temperature any lower energy state is infinitely more probable than a higher energy state. So the system can be trapped in a local minimum, even though there are lower minima elsewhere. But at very high energies, the system will not stay in the low energy states for very long.

The idea discussed in the paper by Kirkpatrick, Gelatt, and Vecchi is to start the system off at a high temperature and then gradually to drop the temperature to zero. The system is free to explore different states. When this is done, for example, when a crystal is grown or a metal is cooled, it is referred to as annealing. For example, a crystal is the lowest energy state of many solids. Crystals are often grown by melting a solid and then cooling the system very gradually so that the system can reach this lowest energy state. The slower the cooling, the larger and more regular the final crystal is liable to be.

It is something of an art to know how slowly to cool the system in practice. This paper discusses some successful applications of these *simulated annealing* techniques to a number of practical problems: the layout of computer chips, routing of wires on printed circuit board, and a famous optimization problem (the "Traveling Salesman" problem). The technique in all these cases is to find a function—cost or energy—to be minimized, to choose a starting system temperature, and then to drop the temperature according to a useful annealing schedule. The results are quite good, and this class of algorithms has proved to be of considerable use in practical applications.

Simulated annealing is used in Geman and Geman (paper 37) and in Ackley, Hinton, and Sejnowski (paper 38), where it is combined with a neural network model.

(1983)
S. Kirkpatrick, C. D. Gelatt, Jr., and M. P. Vecchi

Optimization by simulated annealing
Science 220:671–680

There is a deep and useful connection between statistical mechanics (the behavior of systems with many degrees of freedom in thermal equilibrium at a finite temperature) and multivariate or combinatorial optimization (finding the minimum of a given function depending on many parameters). A detailed analogy with annealing in solids provides a framework for optimization of the properties of very large and complex systems. This connection to statistical mechanics exposes new information and provides an unfamiliar perspective on traditional optimization problems and methods.

In this article we briefly review the central constructs in combinatorial optimization and in statistical mechanics and then develop the similarities between the two fields. We show how the Metropolis algorithm for approximate numerical simulation of the behavior of a many-body system at a finite temperature provides a natural tool for bringing the techniques of statistical mechanics to bear on optimization.

We have applied this point of view to a number of problems arising in optimal design of computers. Applications to partitioning, component placement, and wiring of electronic systems are described in this article. In each context, we introduce the problem and discuss the improvements available from optimization.

Of classic optimization problems, the traveling salesman problem has received the most intensive study. To test the power of simulated annealing, we used the algorithm on traveling salesman problems with as many as several thousand cities. This work is described in a final section, followed by our conclusions.

Combinatorial Optimization

The subject of combinatorial optimization (*1*) consists of a set of problems that are central to the disciplines of computer science and engineering. Research in this area aims at developing efficient techniques for finding minimum or maximum values of a function of very many independent variables (*2*). This function, usually

Copyright © 1983 AAAS.

called the cost function or objective function, represents a quantitative measure of the "goodness' of some complex system. The cost function depends on the detailed configuration of the many parts of that system. We are most familiar with optimization problems occurring in the physical design of computers, so examples used below are drawn from that context. The number of variables involved may range up into the tens of thousands.

The classic example, because it is so simply stated, of a combinatorial optimization problem is the traveling salesman problem. Given a list of N cities and a means of calculating the cost of traveling between any two cities, one must plan the salesman's route, which will pass through each city once and return finally to the starting point, minimizing the total cost. Problems with this flavor arise in all areas of scheduling and design. Two subsidiary problems are of general interest: predicting the expected cost of the salesman's optimal route, averaged over some class of typical arrangements of cities, and estimating or obtaining bounds for the computing effort necessary to determine that route.

All exact methods known for determining an optimal route require a computing effort that increases exponentially with N, so that in practice exact solutions can be attempted only on problems involving a few hundred cities or less. The traveling salesman belongs to the large class of NP-complete (nondeterministic polynomial time complete) problems, which has received extensive study in the past 10 years (*3*). No method for exact solution with a computing effort bounded by a power of N has been found for any of these problems, but if such a solution were found, it could be mapped into a procedure for solving all members of the class. It is not known what features of the individual problems in the NP-complete class are the cause of their difficulty.

Since the NP-complete class of problems contains many situations of practical interest, heuristic methods have been developed with computational requirements proportional to small powers of N. Heuristics are rather problem-specific: there is no guarantee that a heuristic procedure for finding near-optimal solutions for one NP-complete problem will be effective for another.

There are two basic strategies for heuristics: "divide-and-conquer" and iterative improvement. In the first, one divides the problem into subproblems of manageable size, then solves the subproblems. The solutions to the subproblems must then be patched back together. For this method to produce very good solutions, the subproblems must be naturally disjoint, and the division made must be an appropriate one, so that errors made in patching do not offset the gains obtained in applying more powerful methods to the subproblems (4).

In iterative improvement (5,6), one starts with the system in a known configuration. A standard rearrangement operation is applied to all parts of the system in turn, until a rearranged configuration that improves the cost function is discovered. The rearranged configuration then becomes the new configuration of the system, and the process is continued until no further improvements can be found. Iterative improvement consists of a search in this coordinate space for rearrangement steps which lead downhill. Since this search usually gets stuck in a local but not a global optimum, it is customary to carry out the process several times, starting from different randomly generated configurations, and save the best result.

There is a body of literature analyzing the results to be expected and the computing requirements of common heuristic methods when applied to the most popular problems (1–3). This analysis usually focuses on the worst-case situation—for instance, attempts to bound from above the ratio between the cost obtained by a heuristic method and the exact minimum cost for any member of a family of similarly structured problems. There are relatively few discussions of the average performance of heuristic algorithms, because the analysis is usually more difficult and the nature of the appropriate average to study is not always clear. We will argue that as the size of optimization problems increases, the worst-case analysis of a problem will become increasingly irrelevant, and the average performance of algorithms will dominate the analysis of practical applications. This large number limit is the domain of statistical mechanics.

Statistical Mechanics

Statistical mechanics is the central discipline of condensed matter physics, a body of methods for analyzing aggregate properties of the large numbers of atoms to be found in samples of liquid or solid matter (7). Because the number of atoms is of order 10^{23} per cubic centimeter, only the most probable behavior of the system in thermal equilibrium at a given temperature is observed in experiments. This can be characterized by the average and small fluctuations about the average behavior of the system, when the average is taken over the ensemble of identical systems introduced by Gibbs. In this ensemble, each configuration, defined by the set of atomic positions, $\{r_i\}$, of the system is weighted by its Boltzmann probability factor, $\exp(-E(\{r_i\})/k_B T)$, where $E(\{r_i\})$ is the energy of the configuration, k_B is Boltzmann's constant, and T is temperature.

A fundamental question in statistical mechanics concerns what happens to the system in the limit of low temperature—for example, whether the atoms remain fluid or solidify, and if they solidify, whether they form a crystalline solid or a glass. Ground states and configurations close to them in energy are extremely rare among all the configurations of a macroscopic body, yet they dominate its properties at low temperatures because as T is lowered the Boltzmann distribution collapses into the lowest energy state or states.

As a simplified example, consider the magnetic properties of a chain of atoms whose magnetic moments, μ_i, are allowed to point only "up" or "down," states denoted by $\mu_i = \pm 1$. The interaction energy between two such adjacent spins can be written $J\mu_i\mu_{i+1}$. Interaction between each adjacent pair of spins contributes $\pm J$ to the total energy of the chain. For an N-spin chain, if all configurations are equally likely the interaction energy has a binomial distribution, with the maximum and minimum energies given by $\pm NJ$ and the most probable state having zero energy. In this view, the ground state configurations have statistical weight $\exp(-N/2)$ smaller than the zero-energy configurations. A Boltzmann factor, $\exp(-E/k_B T)$, can offset this if $k_B T$ is smaller than J. If we focus on the problem of finding empirically the system's ground state, this factor is seen to drastically increase the efficiency of such a search.

In practical contexts, low temperature is not a sufficient condition for finding ground states of matter. Experiments that determine the low-temperature state of a material—for example, by growing a single crystal from a melt—are done by careful annealing, first melting the substance, then lowering the temperature slowly, and spending a long time at temperatures in the vicinity of the freezing point. If this is not done, and the substance is allowed to get out of equilibrium, the resulting crystal will have many defects, or the substance may form a glass, with no crystalline order and only metastable, locally optimal structures.

Finding the low-temperature state of a system when a prescription for calculating its energy is given is an optimization problem not unlike those encountered in combinatorial optimization. However, the concept of the temperature of a physical system has no obvious equivalent in the systems being optimized. We will introduce an effective temperature for optimization, and show how one can carry out a simulated annealing process in order to obtain better heuristic solutions to combinatorial optimization problems.

Iterative improvement, commonly applied to such problems, is much like the microscopic rearrangement processes modeled by statistical mechanics, with the cost function playing the role of energy. However, accepting only rearrangements that lower the cost function of the system is like extremely rapid quenching from high temperatures to $T = 0$, so it should not be surprising that resulting solutions are usually metastable. The Metropolis procedure from statistical mechanics provides a generalization of iterative improvement in which controlled uphill steps can also be incorporated in the search for a better solution.

Metropolis et al. (8), in the earliest days of scientific computing, introduced a simple algorithm that can be used to provide an efficient simulation of a collection of atoms in equilibrium at a given temperature. In each step of this algorithm, an atom is given a small random displacement and the resulting change, ΔE, in the energy of the system is computed. If $\Delta E \leq 0$, the displacement is accepted, and the configuration with the displaced atom is used as the starting point of the next step. The case $\Delta E > 0$ is treated probabilistically: the probability that the configuration is accepted is $P(\Delta E) = \exp(-\Delta E/k_B T)$. Random numbers uniformly distributed in the interval $(0, 1)$ are a convenient means of implementing the random part of the algorithm. One such number is selected and compared with $P(\Delta E)$. If it is less than $P(\Delta E)$, the new configuration is retained; if not, the original configuration is used to start the next step. By repeating the basic step many times, one simulates the thermal motion of atoms in thermal contact with a heat bath at temperature T. This choice of $P(\Delta E)$ has the consequence that the system evolves into a Boltzmann distribution.

Using the cost function in place of the energy and defining configurations by a set of parameters $\{x_i\}$, it is straightforward with the Metropolis procedure to generate a population of configurations of a given optimization problem at some effective temperature. This temperature is simply a control parameter in the same units as the cost function. The simulated annealing process consists of first "melting" the system being optimized at a high effective temperature, then lower-

ing the temperature by slow stages until the system "freezes" and no further changes occur. At each temperature, the simulation must proceed long enough for the system to reach a steady state. The sequence of temperatures and the number of rearrangements of the $\{x_i\}$ attempted to reach equilibrium at each temperature can be considered an annealing schedule.

Annealing, as implemented by the Metropolis procedure, differs from iterative improvement in that the procedure need not get stuck since transitions out of a local optimum are always possible at nonzero temperature. A second and more important feature is that a sort of adaptive divide-and-conquer occurs. Gross features of the eventual state of the system appear at higher temperatures; fine details develop at lower temperatures. This will be discussed with specific examples.

Statistical mechanics contains many useful tricks for extracting properties of a macroscopic system from microscopic averages. Ensemble averages can be obtained from a single generating function, the partition function, Z,

$$Z = \mathrm{Tr} \exp\left(\frac{-E}{k_B T}\right) \tag{1}$$

in which the trace symbol, Tr, denotes a sum over all possible configurations of the atoms in the sample system. The logarithm of Z, called the free energy, $F(T)$, contains information about the average energy, $\langle E(T) \rangle$, and also the entropy, $S(T)$, which is the logarithm of the number of configurations contributing to the ensemble at T:

$$-k_B T \ln Z = F(T) = \langle E(T) \rangle - TS \tag{2}$$

Boltzmann-weighted ensemble averages are easily expressed in terms of derivatives of F. Thus the average energy is given by

$$\langle E(T) \rangle = \frac{-d \ln Z}{d(1/k_B T)} \tag{3}$$

and the rate of change of the energy with respect to the control parameter, T, is related to the size of typical variations in the energy by

$$C(T) = \frac{d\langle E(T) \rangle}{dT}$$

$$= \frac{[\langle E(T)^2 \rangle - \langle E(T) \rangle^2]}{k_B T^2} \tag{4}$$

In statistical mechanics $C(T)$ is called the specific heat. A large value of C signals a change in the state of order of a system, and can be used in the optimization context to indicate that freezing has begun and hence that very slow cooling is required. It can also be used to determine the entropy by the thermodynamic relation

$$\frac{dS(T)}{dT} = \frac{C(T)}{T} \tag{5}$$

Integrating Eq. 5 gives

$$S(T) = S(T_1) - \int_T^{T_1} \frac{C(T')\,dT}{T} \tag{6}$$

where T_1 is a temperature at which S is known, usually by an approximation valid at high temperatures.

The analogy between cooling a fluid and optimization may fail in one important respect. In ideal fluids all the atoms are alike and the ground state is a regular crystal. A typical optimization problem will contain many distinct, noninterchangeable elements, so a regular solution is unlikely. However, much research in condensed matter physics is directed at sysems with quenched-in randomness, in which the atoms are not all alike. An important feature of such systems, termed "frustration," is that interactions favoring different and incompatible kinds of ordering may be simultaneously present (9). Tne magnetic alloys known as "spin glasses," which exhibit competition between ferromagnetic and antiferromagnetic spin ordering, are the best understood example of frustration (10). It is now believed that highly frustrated systems like spin glasses have many nearly degenerate random ground states rather than a single ground state with a high degree of symmetry. These systems stand in the same relation to conventional magnets as glasses do to crystals, hence the name.

The physical properties of spin glasses at low temperatures provide a possible guide for understanding the possibilities of optimizing complex systems subject to conflicting (frustrating) constraints.

Physical Design of Computers

The physical design of electronic systems and the methods and simplifications employed to automate this process have been reviewed (11, 12). We first provide some background and definitions related to applications of the simulated annealing framework to specific problems that arise in optimal design of computer systems and subsystems. Physical design follows logical design. After the detailed specification of the logic of a system is complete, it is necessary to specify the precise physical realization of the system in a particular technology.

This process is usually divided into several stages. First, the design must be partitioned into groups small enough to fit the available packages, for example, into groups of circuits small enough to fit into a single chip, or into groups of chips and associated discrete components that can fit onto a card or other higher level package. Second, the circuits are assigned specific locations on the chip. This stage is usually called placement. Finally, the circuits are connected by wires formed photolithographically out of a thin metal film, often in several layers. Assigning paths, or routes, to the wires is usually done in two stages. In rough or global wiring, the wires are assigned to regions that represent schematically the capacity of the intended package. In detailed wiring (also called exact embedding), each wire is given a unique complete path. From the detailed wiring results, masks can be generated and chips made.

At each stage of design one wants to optimize the eventual performance of the system without compromising the feasibility of the subsequent design stages. Thus partitioning must be done in such a way that the number of circuits in each partition is small enough to fit easily into the available package, yet the number of signals that must cross partition boundaries (each requiring slow, power-consuming driver circuitry) is minimized. The major focus in placement is on minimizing the length of connections, since this translates into the time required for propagation of signals, and thus into the speed of the finished system. However, the placements with the shortest implied wire lengths may not be wirable, because of the presence of regions in which the wiring is too congested for the packaging technology. Congestion, therefore, should also be anticipated and minimized during the placement process. In wiring, it is desirable to maintain the minimum possible wire lengths while minimizing sources of noise, such as cross talk between adjacent wires. We show in this and the next two sections how these conflicting goals can be combined and made the basis of an automatic optimization procedure.

The tight schedules involved present major obstacles to automation and optimization of large system design, even when computers are employed to speed up the mechanical tasks and reduce the chance of error. Possibilities of feedback, in which early stages of a design are redone to solve problems that became apparent only at later stages, are greatly reduced as the scale of the overall system being designed increases. Optimization procedures that can incorporate, even approximately, information about the chance of success of later stages of such complex designs will be increasingly valuable in the limit of very large scale.

System performance is almost always achieved at the expense of design convenience. The partitioning problem provides a clean example of this. Consider N circuits that are to be partitioned between two chips.

Propagating a signal across a chip boundary is always slow, so the number of signals required to cross between the two must be minimized. Putting all the circuits on one chip eliminates signal crossings, but usually there is no room. Instead, for later convenience, it is desirable to divide the circuits about equally.

If we have connectivity information in a matrix whose elements $\{a_{ij}\}$ are the number of signals passing between circuits i and j, and we indicate which chip circuit i is placed on by a two-valued variable $\mu_i = \pm 1$, then N_c, the number of signals that must cross a chip boundary is given by $\sum_{i>j}(a_{ij}/4)(\mu_i - \mu_j)^2$. Calculating $\sum_i \mu_i$ gives the difference between the numbers of circuits on the two chips. Squaring this imbalance and introducing a coefficient, λ, to express the relative costs of imbalance and boundary crossings, we obtain an objective function, f, for the partition problem:

$$f = \sum_{i>j}\left(\lambda - \frac{a_{ij}}{2}\right)\mu_i\mu_j \tag{7}$$

Reasonable values of λ should satisfy $\lambda \lesssim z/2$, where z is the average number of circuits connected to a typical circuit (fan-in plus fan-out). Choosing $\lambda \simeq z/2$ implies giving equal weight to changes in the balance and crossing scores.

The objective function f has precisely the form of a Hamiltonian, or energy function, studied in the theory of random magnets, when the common simplifying assumption is made that the spins, μ_i, have only two allowed orientations (up or down), as in the linear chain example of the previous section. It combines local, random, attractive ("ferromagnetic") interactions, resulting from the a_{ij}'s, with a long-range repulsive ("antiferromagnetic") interaction due to λ. No configuration of the $\{\mu_i\}$ can simultaneously satisfy all the interactions, so the system is "frustrated," in the sense formalized by Toulouse (9).

If the a_{ij} are completely uncorrelated, it can be shown (13) that this Hamiltonian has a spin glass phase at low temperatures. This implies for the associated magnetic problem that there are many degenerate "ground states" of nearly equal energy and no obvious symmetry. The magnetic state of a spin glass is very stable at low temperatures (14), so the ground states have energies well below the energies of the random high-temperature states, and transforming one ground state into another will usually require considerable rearrangement. Thus this analogy has several implications for optimization of partition:

1) Even in the presence of frustration, significant improvements over a random starting partition are possible.

2) There will be many good near-optimal solutions, so a stochastic search procedure such as simulated annealing should find some.

3) No one of the ground states is significantly better than the others, so it is not very fruitful to search for the absolute optimum.

In developing Eq. 7 we made several severe simplifications, considering only two-way partitioning and ignoring the fact that most signals connect more than two circuits. Objective functions analogous to f that include both complications are easily constructed. They no longer have the simple quadratic form of Eq. 7, but the qualitative feature, frustration, remains dominant. The form of the Hamiltonian makes no difference in the Metropolis Monte Carlo algorithm. Evaluation of the change in function when a circuit is shifted to a new chip remains rapid as the definition of f becomes more complicated.

It is likely that the a_{ij} are somewhat correlated, since any design has considerable logical structure. Efforts to understand the nature of this structure by analyzing the surface-to-volume ratio of components of electronic sysems [as in "Rent's rule" (15)] conclude that the circuits in a typical system could be connected with short-range interactions if they were embedded in a space with dimension between two and three. Uncorrelated connections, by contrast, can be thought of as infinite-dimensional, since they are never short-range.

The identification of Eq. 7 as a spin glass Hamiltonian is not affected by the reduction to a two- or three-dimensional problem, as long as $\lambda N \simeq z/2$. The degree of ground state degeneracy increases with decreasing dimensionality. For the uncorrelated model, there are typically of order $N^{1/2}$ nearly degenerate ground states (14), while in two and three dimensions, $2^{\alpha N}$, for some small value, α, are expected (16). This implies that finding a near-optimum solution should become easier, the lower the effective dimensionality of the problem. The entropy, measurable as shown in Eq. 6, provides a measure of the degeneracy of solutions. $S(T)$ is the logarithm of the number of solutions equal to or better than the average result encountered at temperature T.

As an example of the partitioning problem, we have taken the logic design for a single-chip IBM "370 microprocessor" (17) and considered partitioning it into two chips. The original design has approximately 5000 primitive logic gates and 200 external signals (the chip has 200 logic pins). The results of this study are plotted in Fig. 1. If one randomly assigns gates to the two chips, one finds the distribution marked $T = \infty$ for the number of pins required. Each of the two chips (with about 2500 circuits) would need 3000 pins. The

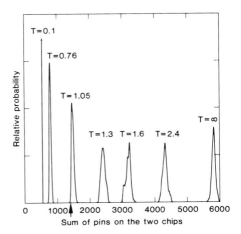

Figure 1 Distribution of total number of pins required in two-way partition of a microprocessor at various temperatures. Arrow indicates best solution obtained by rapid quenching as opposed to annealing.

other distributions in Fig. 1 show the results of simulated annealing.

Monte Carlo annealing is simple to implement in this case. Each proposed configuration change simply flips a randomly chosen circuit from one chip to the other. The new number of external connections, C, to the two chips is calculated (an external connection is a net with circuits on both chips, or a circuit connected to one of the pins of the original single-chip design), as is the new balance score, B, calculated as in deriving Eq. 7. The objective function analogous to Eq. 7 is

$$f = C + \lambda B \qquad (8)$$

where C is the sum of the number of external connections on the two chips and B is the balance score. For this example, $\lambda = 0.01$.

For the annealing schedule we chose to start at a high "temperature," $T_0 = 10$, where essentially all proposed circuit flips are accepted, then cool exponentially, $T_n = (T_1/T_0)^n T_0$, with the ratio $T_1/T_0 = 0.9$. At each temperature enough flips are attempted that either there are ten accepted flips per circuit on the average (for this case, 50,000 accepted flips at each temperature), or the number of attempts exceeds 100 times the number of circuits before ten flips per circuit have been accepted. If the desired number of acceptances is not achieved at three successive temperatures, the system is considered "frozen" and annealing stops.

The finite temperature curves in Fig. 1 show the distribution of pins per chip for the configurations sampled at $T = 2.5$, 1.0, and 0.1. As one would expect from the statistical mechanical analog, the distribution shifts to fewer pins and sharpens as the temperature is

decreased. The sharpening is one consequence of the decrease in the number of configurations that contribute to the equilibrium ensemble at the lower temperature. In the language of statistical mechanics, the entropy of the system decreases. For this sample run in the low-temperature limit, the two chips required 353 and 321 pins, respectively. There are 237 nets connecting the two chips (requiring a pin on each chip) in addition to the 200 inputs and outputs of the original chip. The final partition in this example has the circuits exactly evenly distributed between the two partitions. Using a more complicated balance score, which did not penalize imbalance of less than 100 circuits, we found partitions resulting in chips with 271 and 183 pins.

If, instead of slowly cooling, one were to start from a random partition and accept only flips that reduce the objective function (equivalent to setting $T = 0$ in the Metropolis rule), the result is chips with approximately 700 pins (several such runs led to results with 677 to 730 pins). Rapid cooling results in a system frozen into a metastable state far from the optimal configuration. The best result obtained after several rapid quenches is indicated by the arrow in Fig. 1.

Placement

Placement is a further refinement of the logic partitioning process, in which the circuits are given physical positions (*11,12,18,19*). In principle, the two stages could be combined, although this is not often possible in practice. The objectives in placement are to minimize signal propagation times or distances while satisfying prescribed electrical constraints, without creating regions so congested that there will not be room later to connect the circuits with actual wire.

Physical design of computers includes several distinct categories of placement problems, depending on the packages involved (*20*). The larger objects to be placed include chips that must reside in a higher level package, such as a printed circuit card or fired ceramic "module" (*21*). These chip carriers must in turn be placed on a backplane or "board," which is simply a very large printed circuit card. The chips seen today contain from tens to tens of thousands of logic circuits, and each chip carrier or board will provide from one to ten thousand interconnections. The partition and placement problems decouple poorly in this situation, since the choice of which chip should carry a given piece of logic will be influenced by the position of that chip.

The simplest placement problems arise in designing chips with structured layout rules. These are called "gate array" or "master slice" chips. In these chips, standard logic circuits, such as three- or four-input

NOR's, are preplaced in a regular grid arrangement, and the designer specifies only the signal wiring, which occupies the final, highest, layers of the chip. The circuits may all be identical, or they may be described in terms of a few standard groupings of two or more adjacent cells.

As an example of a placement problem with realistic complexity without too many complications arising from package idiosyncrasies, we consider 98 chips packaged on one multilayer ceramic module of the IBM 3081 processor (21). Each chip can be placed on any of 100 sites, in a 10 × 10 grid on the top surface of the module. Information about the connections to be made through the signal-carrying planes of the module is contained in a "netlist," which groups sets of pins that see the same signal.

The state of the system can be briefly represented by a list of the 98 chips with their x and y coordinates, or a list of the contents of each of the 100 legal locations. A sufficient set of moves to use for annealing is interchanges of the contents of two locations. This results in either the interchange of two chips or the interchange of a chip and a vacancy. For more efficient search at low temperatures, it is helpful to allow restrictions on the distance across which an interchange may occur.

To measure congestion at the same time as wire length, we use a convenient intermediate analysis of the layout, a net-crossing histogram. Its construction is summarized in Fig. 2. We divide the package surface by a set of natural boundaries. In this example, we use the boundaries between adjacent rows or columns of chip sites. The histogram then contains the number of nets crossing each boundary. Since at least one wire must be routed across each boundary crossed, the sum of the entries in the histogram of Fig. 2 is the sum of the horizontal extents of the rectangles bounding each net, and is a lower bound to the horizontal wire length required. Constructing a vertical net-crossing histo-

gram and summing its entries gives a similar estimate of the vertical wire length.

The peak of the histogram provides a lower bound to the amount of wire that must be provided in the worst case, since each net requires at least one wiring channel somewhere on the boundary. To combine this information into a single objective function, we introduce a threshold level for each histogram—an amount of wire that will nearly exhaust the available wire capacity—and then sum for all histogram elements that exceed the threshold the square of the excess over threshold. Adding this quantity to the estimated length gives the objective function that was used.

Figure 3 shows the stages of a simulated annealing run on the 98-chip module. Figure 3a shows the chip locations from the original design, with vertical and horizontal net-crossing histograms indicated. The different shading patterns distinguish the groups of chips that carry out different functions. Each such group was designed and placed together, usually by a single designer. The net-crossing histograms show that the center of the layout is much more congested than the edges, most likely because the chips known to have the most critical timing constraints were placed in the center of the module to allow the greatest number of other chips to be close to them.

Heating the original design until the chips diffuse about freely quickly produces a random-looking arrangement, Fig. 3b. Cooling very slowly until the chips move sluggishly and the objective function ceases to decrease rapidly with change of temperature produced the result in Fig. 3c. The net-crossing histograms have peaks comparable to the peak heights in the original placement, but are much flatter. At this "freezing point," we find that the functionally related groups of chips have reorganized from the melt, but now are spatially separated in an overall arrangement quite different from the original placement. In the final result, Fig. 3d, the histogram peaks are about 30 percent less than in the original placement. Integrating them, we find that total wire length, estimated in this way, is decreased by about 10 percent. The computing requirements for this example were modest: 250,000 interchanges were attempted, requiring 12 minutes of computation on an IBM 3033.

Between the temperature at which clusters form and freezing starts (Fig. 3c) and the final result (Fig. 3d) there are many further local rearrangements. The functional groups have remained in the same regions, but their shapes and relative alignments continue to change throughout the low-temperature part of the annealing process. This illustrates that the introduction of temperature to the optimization process per-

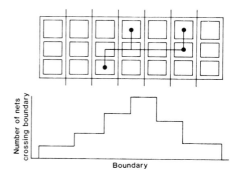

Figure 2 Construction of a horizontal net-crossing histogram.

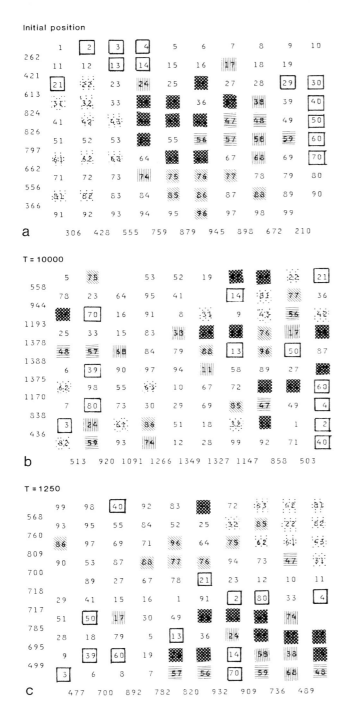

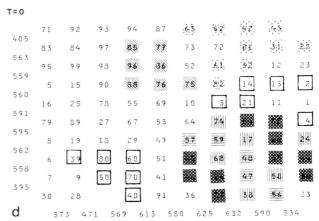

Figure 3 Ninety-eight chips on a ceramic module from the IBM 3081. Chips are identified by number (1 to 100, with 20 and 100 absent) and function. The dark squares comprise an adder, the three types of squares with ruled lines are chips that control and supply data to the adder, the lightly dotted chips perform logical arithmetic (bitwise AND, OR, and so on), and the open squares denote general-purpose registers, which serve both arithmetic units. The numbers at the left and lower edges of the module image are the vertical and horizontal net-crossing histograms, respectively. (a) Original chip placement; (b) a configuration at $T = 10,000$; (c) $T = 1250$; (d) a zero-temperature result.

mits a controlled, adaptive division of the problem through the evolution of natural clusters at the freezing temperature. Early prescription of natural clusters is also a central feature of several sophisticated placement programs used in master slice chip placement (22, 23).

A quantity corresponding to the thermodynamic specific heat is defined for this problem by taking the derivative with respect to temperature of the average value of the objective function observed at a given temperature. This is plotted in Fig. 4. Just as a maximum in the specific heat of a fluid indicates the onset of freezing or the formation of clusters, we find specific heat maxima at two temperatures, each indicating a different type of ordering in the problem. The higher temperature peak corresponds to the aggregation of clusters of functionally related objects, driven apart by the congestion term in the scoring. The lower temperature peak indicates the further decrease in wire length obtained by local rearrangements. This sort of measurement can be useful in practice as a means of determining the temperature ranges in which the important rearrangements in the design are occurring, where slower cooling with be helpful.

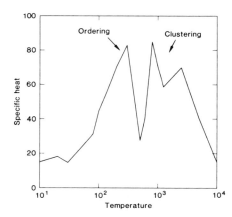

Figure 4 Specific heat as a function of temperature for the design of Fig. 3, a to d.

Wiring

After placement, specific legal routings must be found for the wires needed to connect the circuits. The techniques typically applied to generate such routings are sequential in nature, treating one wire at a time with incomplete information about the positions and effects of the other wires (*11, 24*). Annealing is inherently free of this sequence dependence. In this section we describe a simulated annealing approach to wiring, using the ceramic module of the last section as an example.

Nets with many pins must first be broken into connections—pairs of pins joined by a single continuous wire. This "ordering" of each net is highly dependent on the nature of the circuits being connected and the package technology. Orderings permitting more than two pins to be connected are sometimes allowed, but will not be discussed here.

The usual procedure, given an ordering, is first to construct a coarse-scale routing for each connection from which the ultimate detailed wiring can be completed. Package technologies and structured image chips have prearranged areas of fixed capacity for the wires. For the rough routing to be successful, it must not call for wire densities that exceed this capacity.

We can model the rough routing problem (and even simple cases of detailed embedding) by lumping all actual pin positions into a regular grid of points, which are treated as the sources and sinks of all connections. The wires are then to be routed along the links that connect adjacent grid points.

The objectives in global routing are to minimize wire length and, often, the number of bends in wires, while spreading the wire as evenly as possible to simplify exact embedding and later revision. Wires are to be routed around regions in which wire demand exceeds

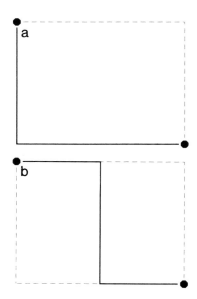

Figure 5 Examples of (a) L-shaped and (b) Z-shaped wire rearrangements.

capacity if possible, so that they will not "overflow," requiring drastic rearrangements of the other wires during exact embedding. Wire bends are costly in packages that confine the north-south and east-west wires to different layers, since each bend requires a connection between two layers. Two classes of moves that maintain the minimum wire length are shown in Fig. 5. In the L-shaped move of Fig. 5a, only the essential bends are permitted, while the Z-shaped move of Fig. 5b introduces one extra bend. We will explore the optimization possible wih these two moves.

For a simple objective function that will reward the most balanced arrangement of wire, we calculate the square of the number of wires on each link of the network, sum the squares for all links, and term the result F. If there are N_L links and N_w wires, a global routing program that deals with a high density of wires will attempt to route precisely the average number of wires, N_w/N_L, along each link. In this limit F is bounded below by N_w^2/N_L. One can use the same objective function for a low-density (or high-resolution) limit appropriate for detailed wiring. In that case, all the links have either one or no wires, and links with two or more wires are illegal. For this limit the best possible value of F will be N_w/N_L.

For the L-shaped moves, F has a relatively simple form. Let $\varepsilon_{iv} = +1$ along the links that connection i has for one orientation, -1 for the other orientation, and 0 otherwise. Let a_{iv} be 1 if the ith connection can run through the vth link in either of its two positions, and 0 othewise. Note that a_{iv} is just ε_{iv}^2. Then if $\mu_i =$

± 1 indicates which route the ith connection has taken, we obtain for the number of wires along the vth link,

$$n_v = \sum_i \frac{a_{iv}(\varepsilon_{iv}\mu_i + 1)}{2} + n_v(0) \qquad (9)$$

where $n_v(0)$ is the contribution from straight wires, which cannot move without increasing their length, or blockages.

Summing the n_v^2 gives

$$F = \sum_{ij} J_{ij}\mu_i\mu_j + \sum_i h_i\mu_i + \text{constants} \qquad (10)$$

which has the form of the Hamiltonian for a random magnetic alloy or spin glass, like that discussed earlier. The "random field," h_i, felt by each movable connection reflects the difference, on the average, between the congestion associated with the two possible paths:

$$h_i = \sum_v \varepsilon_{iv}\left[2n_v(0) + \sum_j a_{jv}\right] \qquad (11)$$

The interaction between two wires is proportional to the number of links on which the two nets can overlap, its sign depending on their orientation conventions:

$$J_{ij} = \sum_v \frac{\varepsilon_{iv}\varepsilon_{jv}}{4} \qquad (12)$$

Both J_{ij} and h_i vanish, on average, so it is the fluctuations in the terms that make up F which will control the nature of the low-energy states. This is also true in spin glasses. We have not tried to exhibit a functional form for the objective function with Z-moves allowed, but simply calculate it by first constructing the actual amounts of wire found along each link.

To assess the value of annealing in wiring this model, we studied an ensemble of randomly situated connections, under various statistical assumptions. Here we consider routing wires for the 98 chips on a module considered earlier. First, we show in Fig. 6 the arrangement of wire that results from assigning each wire to an L-shaped path, choosing orientations at random. The thickness of the links is proportional to the number of wires on each link. The congested area that gave rise to the peaks in the histograms discussed above is seen in the wiring just below and to the right of the center of the module. The maximum numbers of wires along a single link in Fig. 6 are 173 (x direction) and 143 (y direction), so the design is also anisotropic. Various ways of rearranging the wiring paths were studied. Monte Carlo annealing with Z-moves gave the best solution, shown in Fig. 7. In this example, the largest numbers of wires on a single link are 105 (x) and 96 (y).

We compare the various methods of improving the

Random
Grid size 10

Figure 6 Wire density in the 98-chip module with the connections randomly assigned to perimeter routes. Chips are in the original placement.

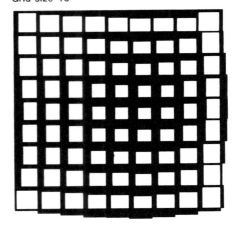

M.C. Z-paths
Grid size 10

Figure 7 Wire density after simulated annealing of the wire routing, using Z-shaped moves.

wire arrangement by plotting (Fig. 8) the highest wire density found in each column of x-links for each of the methods. The unevenness of the density profiles was already seen when we considered net-crossing histograms as input information to direct placement. The lines shown represent random assignment of wires with L-moves; aligning wires in the direction of least average congestion—that is, along h_i—followed by cooling for one pass at zero T; simulated annealing with L-moves only; and annealing with Z-moves. Finally, the light dashed line shows the optimum result, in which the wires are distributed with all links carrying as close

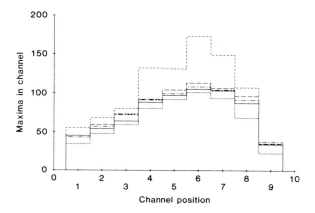

Figure 8 Histogram of the maximum wire densities within a given column of x-links, for the various methods of routing.

to the average weight as possible. The optimum cannot be attained in this example without stretching wires beyond their minimum length, because the connections are too unevenly arranged. Any method of optimization gives a significant improvement over the estimate obtained by assigning wire routings at random. All reduce the peak wire density on a link by more than 45 percent. Simulated annealing with Z-moves improved the random routing by 57 percent, averaging results for both x and y links.

Traveling Salesmen

Quantitative analysis of the simulated annealing algorithm or comparison between it and other heuristics requires problems simpler than physical design of computers. There is an extensive literature on algorithms for the traveling salesman problem (3, 4), so it provides a natural context for this discussion.

If the cost of travel between two cities is proportional to the distance between them, then each instance of a traveling salesman problem is simply a list of the positions of N cities. For example, an arrangement of N points positioned at random in a square generates one instance. The distance can be calculated in either the Euclidean metric or a "Manhattan" metric, in which the distance between two points is the sum of their separations along the two coordinate axes. The latter is appropriate for physical design applications, and easier to compute, so we will adopt it.

We let the side of the square have length $N^{1/2}$, so that the average distance between each city and its nearest neighbor is independent of N. It can be shown that this choice of length units leaves the optimal tour length per step independent of N, when one averages over many instances, keeping N fixed (25). Call this aver-

age optimal step length α. To bound α from above, a numerical experiment was performed with the following "greedy" heuristic algorithm. From each city, go to the nearest city not already on the tour. From the Nth city, return directly to the first. In the worst case, the ratio of the length of such a greedy tour to the optimal tour is proportional to $\ln(N)$ (26), but on average, we find that its step length is about 1.12. The variance of the greedy step length decreases as $N^{-1/2}$, so the situation envisioned in the worst case analysis is unobservably rare for large N.

To construct a simulated annealing algorithm, we need a means of representing the tour and a means of generating random rearrangements of the tour. Each tour can be described by a permuted list of the numbers 1 to N, which represents the cities. A powerful and general set of moves was introduced by Lin and Kernighan (27, 28). Each move consists of reversing the direction in which a section of the tour is traversed. More complicated moves have been used to enhance the searching effectiveness of iterative improvement. We find with the adaptive divide-and-conquer effect of annealing at intermediate temperatures that the subsequence reversal moves are sufficient (29).

An annealing schedule was determined empirically. The temperature at which segments flow about freely will be of order $N^{1/2}$, since that is the average bond length when the tour is highly random. Temperatures less than 1 should be cold. We were able to anneal into locally optimal solutions with $\alpha \leq 0.95$ for N up to 6000 sites. The largest traveling salesman problem in the plane for which a proved exact solution has been obtained and published (to our knowledge) has 318 points (30).

Real cities are not uniformly distributed, but are clumped, with dense and sparse regions. To introduce this feature into an ensemble of traveling salesman problems, albeit in an exaggerated form, we confine the randomly distributed cities to nine distinct regions with empty gaps between them. The temperature gives the simulated annealing method a means of separating out the problem of the coarse structure of the tour from the local details. At temperatures, such as $T = 1.2$ (Fig. 9a), where the small-scale structure of the paths is completely disordered, the longer steps across the gaps are already becoming infrequent and steps joining regions more than one gap are eliminated. The configurations studied below $T = 0.8$ (for instance, Fig. 9b) had the minimal number of long steps, but the detailed arrangement of the long steps continued to change down to $T = 0.4$ (Fig. 9c). Below $T = 0.4$, no further changes in the arrangement of the long steps were seen,

Figure 9 Results at four temperatures for a clustered 400-city traveling salesman problem. The points are uniformly distributed in nine regions. (a) $T = 1.2$, $\alpha = 2.0567$; (b) $T = 0.8$, $\alpha = 1.515$; (c) $T = 0.4$, $\alpha = 1.055$; (d) $T = 0.0$, $\alpha = 0.7839$.

but the small-scale structure within each region continued to evolve, with the result shown in Fig. 9d.

Summary and Conclusions

Implementing the appropriate Metropolis algorithm to simulate annealing of a combinatorial optimization problem is straightforward, and easily extended to new problems. Four ingredients are needed: a concise description of a configuration of the system; a random generator of "moves" or rearrangements of the elements in a configuration; a quantitative objective function containing the trade-offs that have to be made; and an annealing schedule of the temperatures and length of times for which the system is to be evolved. The annealing schedule may be developed by trial and error for a given problem, or may consist of just warming the system until it is obviously melted, then cooling in slow stages until diffusion of the components ceases. Inventing the most effective sets of moves and deciding

which factors to incorporate into the objective function require insight into the problem being solved and may not be obvious. However, existing methods of iterative improvement can provide natural elements on which to base a simulated annealing algorithm.

The connection with statistical mechanics offers some novel perspectives on familiar optimization problems. Mean field theory for the ordered state at low temperatures may be of use in estimating the average results to be obtained by optimization. The comparison with models of disordered interacting systems gives insight into the ease or difficulty of finding heuristic solutions of the associated optimization problems, and provides a classification more discriminating than the blanket "worst-case" assignment of many optimization problems to the NP-complete category. It appears that for the large optimization problems that arise in current engineering practice a "most probable" or average behavior analysis will be more useful in assessing the value of a heuristic than the traditional worst-case arguments. For such analysis to be useful and accurate, better knowledge of the appropriate ensembles is required.

Freezing, at the temperatures where large clusters form, sets a limit on the energies reachable by a rapidly cooled spin glass. Further energy lowering is possible only by slow annealing. We expect similar freezing effects to limit the effectiveness of the common device of employing iterative improvement repeatedly from different random starting configurations.

Simulated annealing extends two of the most widely used heuristic techniques. The temperature distinguishes classes of rearrangements, so that rearrangements causing large changes in the objective function occur at high temperatures, while the small changes are deferred until low temperatures. This is an adaptive form of the divide-and-conquer approach. Like most iterative improvement schemes, the Metropolis algorithm proceeds in small steps from one configuration to the next, but the temperature keeps the algorithm from getting stuck by permitting uphill moves. Our numerical studies suggest that results of good quality are obtained with annealing schedules in which the amount of computational effort scales as N or as a small power of N. The slow increase of effort with increasing N and the generality of the method give promise that simulated annealing will be a very widely applicable heuristic optimization technique.

Dunham (5) has described iterative improvement as the natural framework for heuristic design, calling it "design by natural selection." [See Lin (6) for a fuller discussion.] In simulated annealing, we appear to have found a richer framework for the construction of heuristic algorithms, since the extra control provided by introducing a temperature allows us to separate out problems on different scales.

Simulation of the process of arriving at an optimal design by annealing under control of a schedule is an example of an evolutionary process modeled accurately by purely stochastic means. In fact, it may be a better model of selection processes in nature than is iterative improvement. Also, it provides an intriguing instance of "artifical intelligence," in which the computer has arrived almost uninstructed at a solution that might have been thought to require the intervention of human intelligence.

References and Notes

1. E. L. Lawlor, *Combinatorial Optimization* (Holt, Rinehart & Winston, New York, 1976).

2. A. V. Aho, J. E. Hopcroft, J. D. Ullman, *The Design and Analysis of Computer Algorithms* (Addison-Wesley, Reading, Mass., 1974).

3. M. R. Garey and D. S. Johnson, *Computers and Intractability: A Guide to the Theory of NP-Completeness* (Freeman, San Francisco, 1979).

4. R. Karp, *Math. Oper. Res.* **2**, 209 (1977).

5. B. Dunham, *Synthese* **15**, 254 (1963).

6. S. Lin, *Networks* **5**, 33 (1975).

7. For a concise and elegant presentation of the basic ideas of statistical mechanics, see E. Shrödinger, *Statistical Thermodynamics* (Cambridge Univ. Press, London, 1946).

8. N. Metropolis, A. Rosenbluth, M. Rosenbluth, A. Teller, E. Teller, *J. Chem. Phys.* **21**, 1087 (1953).

9. G. Toulouse, *Commun. Phys.* **2**, 115 (1977).

10. For review articles, see C. Castellani, C. DiCastro, L. Peliti, Eds., *Disordered Systems and Localization* (Springer, New York, 1981).

11. J. Soukup, *Proc. IEEE* **69**, 1281 (1981).

12. M. A. Breuer, Ed., *Design Automation of Digital Systems* (Prentice-Hall, Engelwood Cliffs, N.J., 1972).

13. D. Sherrington and S. Kirkpatrick, *Phys. Rev. Lett.* **35**, 1792 (1975); S. Kirkpatrick and D. Sherrington, *Phys. Rev. B* **17**, 4384 (1978).

14. A. P. Young and S. Kirkpatrick, *Phys. Rev. B* **25**, 440 (1982).

15. B. Mandelbrot, *Fractals: Form, Chance, and Dimension* (Freeman, San Francisco, 1979), pp. 237–239.

16. S. Kirkpatrick, *Phys. Rev. B* **16**, 4630 (1977).

17. C. Davis, G. Maley, R. Simmons, H. Stoller, R. Warren, T. Wohr, in *Proceedings of the IEEE International Conference on Circuits and Computers*, N. B. Guy Rabbat, Ed. (IEEE, New York, 1980), pp. 669–673.

18. M. A. Hanan, P. K. Wolff, B. J. Agule, *J. Des. Autom. Fault-Tolerant Comput.* **2**, 145 (1978).

19. M. Breuer, *ibid.* **1**, 343 (1977).

20. P. W. Case, M. Correia, W. Gianopulos, W. R. Heller, H. Ofek, T. C. Raymond, R. L. Simek, C. B. Steiglitz, *IBM J. Res. Dev.* **25**, 631 (1981).

21. A. J. Blodgett and D. R. Barbout, *ibid.* **26**, 30 (1982); A. J. Blodgett, in *Proceedings of the Electronics and Computers Conference* (IEEE, New York, 1980), pp. 283–285.

22. K. A. Chen, M. Feuer, K. H. Khokhani, N. Nan, S. Schmidt, in *Proceedings of the 14th IEEE Design Automation Conference* (New Orleans, La., 1977), pp. 298–302.

23. K. W. Lallier, J. B. Hickson, Jr., R. K. Jackson, paper presented at the European Conference on Design Automation, September 1981.

24. D. Hightower, in *Proceedings of the 6th IEEE Design Automation Workshop* (Miami Beach, Fla., June 1969), pp. 1–24.

25. J. Beardwood, J. H. Halton, J. M. Hammersley, *Proc. Cambridge Philos. Soc.* **55**, 299 (1959).

26. D. J. Resenkrantz, R. E. Stearns, P. M. Lewis, *SIAM (Soc. Ind. Appl. Math.) J. Comput.* **6**, 563 (1977).

27. S. Lin, *Bell Syst. Tech. J.* **44**, 2245 (1965).

28. ——— and B. W. Kernighan, *Oper. Res.* **21**, 498 (1973).

29. V. Cerny has described an approach to the traveling salesman problem similar to ours in a manuscript received after this article was submitted for publication.

30. H. Crowder and M. W. Padberg, *Manage. Sci.* **26**, 495 (1980).

31. The experience and collaborative efforts of many of our colleagues have been essential to this work. In particular, we thank J. Cooper, W. Donath, B. Dunham, T. Enger, W. Heller, J. Hickson, G. Hsi, D. Jepsen, H. Koch, R. Linsker, C. Mehanian, S. Rothman, and U. Schultz.

Introduction

(1984)
Francis Crick

Function of the thalamic reticular complex: the searchlight hypothesis
Proceedings of the National Academy of Sciences 81:4586–4590

This paper by Francis Crick brings together ideas from many of the areas that a modern neural modeler should know about. One of the entertaining aspects of research in neural networks, from the point of view of participants, is the necessity to be at least aware of results in a large number of fields. It is not the research area for narrow specialists.

In this paper, we see a good deal of detailed neuroanatomy and physiology used to propose a functional brain model using a novel short term synaptic modification suggested by von der Malsburg to explain some suggestive results in cognitive psychology found by Anne Triesman. This eclectic combination is typical of what has become known as Cognitive Science, and neural network modeling is one important facet of contemporary Cognitive Science.

This paper concerns itself with a number of issues that are becoming more important in network research. Up to this point, we have had models for learning that have been implicitly or explicitly concerned with long term memory effects, that is, permanent synaptic changes resulting from events more than a few hours or days old. However, in the actual operation of cognition, many short term effects, some associated with short term memory and some linked to what is usually called attention, occur. Crick discusses in this paper some short term learning effects that have time scales of seconds or less. These can be interpreted as a model for attention, but they also have some short term memory-like effects as well.

These effects are made to occur by a particular neuroanatomical arrangement of cells, discussed at considerable length, and by a particular form of very short term Hebbian synaptic modification proposed by Christoph von der Malsburg. A more detailed and mathematical discussion of the implications of short term modifiability is contained in a recent paper by von der Malsburg and Bienenstock (1986). They describe the modification rules as follows (p. 250):

Rule A: successful synaptic events enhance the transmission efficacy of the synapse.

Rule B: transmission failures such as presynaptic firing without postsynaptic firing, and possibly also 'failures' of the inverse type, i.e., postsynaptic without presynaptic firing, depress synaptic activity.

Both types of plastic change become effective within a few milliseconds....

Short-term plasticity is restricted to a small range of strengths: the absolute value of the strength saturates after a small number of similar events. In case of successful transmission, it reaches a maximal level. In case of failures, it settles at a minimum....

Short-term plasticity lets connectivity and activity evolve on the same time scale, leading to the notion of a joint activity-and-connectivity state. It is proposed that the events which underlie brain function are best described by such compound states....

Traditional Hebbian synaptic modifications are assumed to give rise to slow, long term temporal integration of correlations over very long times. The short term effects operate using very short term correlations. Von der Malsburg and Bienenstock go on to show that activity-and-connectivity states give rise to attractor dynamics, similar in spirit to those we have seen in other contexts in this collection.

Crick draws on the prediction that these short term synaptic effects allow for the formation of flexible, somewhat transient representations of information determined by both the sensory inputs and the structure of long term memory. This notion is embedded in a number of specific suggestions about how the brain might realize such a system and how it might be used psychologically. The general implication for the field is that effects occurring at short time scales may be very important and perhaps can be analyzed by some of the same techniques used in learning models that assume much longer time scales. There is the hope that all these ideas can be brought together (as the brain does it!) in one neatly fitting package.

Reference

C. von der Malsburg and E. Bienenstock (1986), "Statistical coding and short term synaptic plasticity: a scheme for knowledge representation in the brain," *Disordered Systems and Biological Organization*, E. Bienenstock, F. Fogelman Soulié, and G. Weisbuch (Eds.), Berlin: Springer.

(1984)
Francis Crick

Function of the thalamic reticular complex: the searchlight hypothesis
Proceedings of the National Academy of Sciences 81:4586–4590

ABSTRACT   It is suggested that in the brain the internal attentional searchlight, proposed by Treisman and others, is controlled by the reticular complex of the thalamus (including the closely related perigeniculate nucleus) and that the expression of the searchlight is the production of rapid bursts of firing in a subset of thalamic neurons. It is also suggested that the conjunctions produced by the attentional searchlight are mediated by rapidly modifiable synapses—here called Malsburg synapses—and especially by rapid bursts acting on them. The activation of Malsburg synapses is envisaged as producing transient cell assemblies, including "vertical" ones that temporarily unite neurons at different levels in the neural hierarchy.

This paper presents a set of speculative hypotheses concerning the functions of the thalamus and, in particular, the nucleus reticularis of the thalamus and the related perigeniculate nucleus. For ease of exposition I have drawn my examples mainly from the visual system of primates, but I expect the ideas to apply to all mammals and also to other systems, such as the language system in man.

Visual System

It is now well established that in the early visual system of primates there are at least 10 distinct visual areas in the neocortex. [For a recent summary, see Van Essen and Maunsell (1).] If we include all areas whose main concern is with vision, there may be perhaps twice that number. To a good approximation, the early visual areas can be arranged in a branching hierarchy. Each of these areas has a crude "map" of (part of) the visual world. The first visual area (area 17, also called the striate cortex) on one side of the head maps one-half of the visual world in rather fine detail. Its cells can respond to relatively simple visual "features," such as orientation, spatial frequency, disparity (between the two eyes), etc. This particular area is a large one so that the connections between different parts of it are relatively local. Each part therefore responds mainly to the properties of a small local part of the visual field (2).

As one proceeds to areas higher in the hierarchy, the "mapping" becomes more diffuse. At the same time the neurons appear to respond to more complex features in the visual field. Different cortical areas specialize, to some extent, in different features, one responding mainly to motion, another more to color, etc. In the higher areas a neuron hardly knows where in the visual field the stimulus (such as a face) is arising, while the feature it responds to may be so complex that individual neurons are often difficult to characterize effectively (3, 4).

Thus, the different areas analyze the visual field in different ways. This is not, however, how we appear to see the world. Our inner visual picture of the external world has a

The publication costs of this article were defrayed in part by page charge payment. This article must therefore be hereby marked "*advertisement*" in accordance with 18 U.S.C. §1734 solely to indicate this fact.

unity. How then does the brain put together all of these different activities to produce a unified picture so that, for example, for any object the right color is associated with the right shape?

The Searchlight

The pioneer work of Treisman and her colleagues (5–8), supported more recently by the elegant experiments of Julesz (9–11), have revealed a remarkable fact. If only a very short space of time is available, especially in the presence of "distractors," the brain is unable to make these conjunctions reliably. For example, a human subject can rapidly spot an "*S*" mixed in with a randomly arranged set of green *X*s and brown *T*s—it "pops out" at him. His performance is also rapid for a blue letter mixed in with the same set. However, if he is asked to detect a green *T* (which requires that he recognize the *conjunction* of a chosen color with a chosen shape), he usually takes much more time. Moreover, the time needed increases linearly with the number of distractors (the green *X*s and brown *T*s) as if the mind were searching the letters *in series*, as if the brain had an internal attentional searchlight that moved around from one visual object to the next, with steps as fast as 70 msec in favorable cases. In this metaphor the searchlight is not supposed to light up part of a completely dark landscape but, like a searchlight at dusk, it intensifies part of a scene that is already visible to some extent.

If there is indeed a searchlight mechanism in the brain, how does it work and where is it located? To approach this problem we must study the general layout of the brain and, in particular, that of the neocortex and the thalamus. The essential facts we need at this stage are as follows.

Thalamus

The thalamus is often divided into two parts: the dorsal thalamus, which is the main bulk of it, and the ventral thalamus. [For a general account of the thalamus, see the review by Jones (12).] For the moment when I speak of the thalamus I shall mean the dorsal thalamus.

Almost all input to the cortex, with the exception of the olfactory input, passes through the thalamus. For this reason it is sometimes called the gateway to the cortex. There are some exceptions—the diffuse projections from the brainstem, the projections from the claustrum, and also some projections from the amygdala and basal forebrain—that need not concern us here.

This generalization is not true for projections *from* the cortex, which do not need to pass through the thalamus. Nevertheless, for each projection *from* a region of the thalamus there is a corresponding reverse projection from that part of the cortex *to* the corresponding region of the thalamus. In some cases at least this reverse projection has more axons than the forward projection.

Most of the neurons in the thalamus are relay cells—that is, they receive an input from outside the thalamus (for example, the lateral geniculate nucleus of the thalamus gets a major input from the retina) and project directly to the cortex. Their axons form type I synapses and therefore are probably excitatory. There is a minority of small neurons in the thalamus—their exact number is somewhat controversial—that appear to form type II synapses and are therefore probably inhibitory.

While on the face of it the thalamus appears to be a mere relay, this seems highly unlikely. Its size and its strategic position make it very probable that it has some more important function.

Reticular Complex

Much of the rest of the thalamus is often referred to as the ventral thalamus. This includes the reticular complex (part of which is often called the perigeniculate nucleus), the ventral lateral geniculate nucleus, and the zona incerta. In what follows I shall, for ease of exposition, use the term reticular complex to include the perigeniculate nucleus. Again, "thalamus" means the dorsal thalamus. Although much of the following information comes from the cat or the rat, there is no reason to think that it does not also apply to the primate thalamus.

The reticular complex is a thin sheet of neurons, in most places only a few cells thick, which partly surrounds the (dorsal) thalamus (13–32) (see Fig. 1). All axons from the thalamus to the cerebral cortex pass through it, as do all of the reverse projections from the cortex to the thalamus. The intralaminar nuclei of the thalamus, which project very strongly to the striatum, also send their axons through it, as may some of the axons from the globus pallidus that project back to the thalamus.

It is believed that many of the axons that pass in both directions through the reticular complex give off collaterals that make excitatory synaptic contacts in it (15, 18, 21, 29, 30). If the thalamus is the gateway to the cortex, the reticular complex might be described as the guardian of the gateway. Its exact function is unknown.

Not only is its position remarkable, but its structure is also unusual. It consists largely (if not entirely) of neurons whose dendrites often spread rather extensively in the plane of the nucleus (29). The size of these neurons is somewhat different in different parts of the complex (31). Their axons, which project to the thalamus, give off rather extensive collaterals that ramify, sometimes for long distances, within the sheet of the reticular complex (19, 29). This is in marked contrast

with most of the nuclei of the thalamus, the principal cells of which have few, if any, collaterals either within each nucleus or between nuclei. The nuclei of the thalamus (with the exception of the intralaminar nuclei) keep themselves to themselves. The neurons of the reticular complex, on the other hand, appear to communicate extensively with each other. Moreover, it is characteristic of them that they fire in long bursts at a very rapid rate (25).

An even more remarkable property of reticular neurons concerns their output. Whereas all of the output neurons of the thalamus make type I synapses and appear to be excitatory, many (if not all) of the neurons in the reticular complex appear to be GABAergic (GABA = γ-aminobutyric acid) and thus almost certainly inhibitory (26–28). The excitation in the complex must come almost exclusively from the activity of the various axons passing through it.

Both the input and the output of the complex are arranged topographically (16, 29, 30, 32). It seems likely that if a particular group of axons going from the thalamus to the cortex passes through a small region of the reticular complex, the reverse projection probably passes through or near that same region. There may well be a rough map of the whole cortex on the reticular complex, though how precise this map may be is not known. It should be remembered, however, that the spread of the receiving dendrites of the reticular nucleus is quite large.

The projection of the reticular complex to the thalamus is also not random. Though any individual axon may spread fairly widely, there is a very crude topography in the arrangement. The projection from any one part of the reticular complex probably projects to that part of the thalamus from which it receives input as well as other neighboring parts. The exact nature of these various mappings would repay further study.

The neurons of the reticular complex project to the (dorsal) thalamus. The evidence suggests that they mainly contact the principal (relay) cells of the thalamus (22). What effect does the reticular input have on the behavior of the cells in the dorsal thalamus?

Obviously this is a crucial question. Let us consider two oversimplified but contrasting hypotheses. The first is that the main effect of the reticular complex is inhibitory. This would lead to the following general picture. The traffic passing through the reticular complex will produce excitation. Let us assume that one patch of the complex is more excited than the rest because of special activity in the thalamo-cortical pathways. The effect of this will be 2-fold. That region will tend to suppress somewhat the other parts of the reticular complex, because of the many inhibitory collaterals. It will also suppress the corresponding thalamic region. These two effects will damp down the thalamus in its most active region and have the opposite effect (since the inhibition from the reticular complex will be reduced there) on the remaining parts. The total effect will be to even out the activity of the thalamus. This is not a very exciting conclusion. The function of the reticular complex would be to act as an overall thermostat of thalamic activity, making the warm parts cooler and the cool parts warmer.

The second hypothesis is just the opposite. Let us assume that the effect of the reticular complex on the dorsal thalamus is mainly excitation in some form or other. Then we see that, once again, an active patch in the complex will tend to suppress many other parts of the complex. This time, however, the effect will be to heat up the warmer parts of the thalamus and cool down the cooler parts. We shall have positive feedback rather than negative feedback, so that "attention" will be focused on the most active thalamo-cortical regions.

How can GABAergic neurons produce some sort of excitatory effect on the relay cells of the thalamus? One possibili-

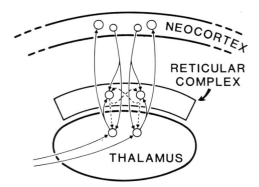

FIG. 1. The main connections of the reticular complex, highly diagramatic and not at all to scale. Solid lines represent excitatory axons. Dashed lines show GABAergic (inhibitory) axons. Arrows represent synapses.

ty is that they might synapse only onto the local inhibitory neurons in the thalamus. By inhibiting these inhibitory cells they would thereby increase the effect of incoming excitation on the relay cells.

This is certainly possible but the anatomic evidence (22) suggests that in the main the neurons of the reticular complex project directly to the thalamic relay neurons. One would expect that this would inhibit these neurons. We must therefore ask if thalamic neurons show any unusual types of behavior.

Properties of Thalamic Neurons

The recent work of Llinás and Jahnsen (33–35) on thalamic slices from the guinea pig confirms that this is indeed the case. Their papers should be consulted for the detailed results, which are complicated, but, broadly, they show that all thalamic neurons display two relatively distinct modes of behavior. When the cell is near its normal resting potential (say, −60 mV) it responds to an injected current by firing (producing axonal spikes) at a fairly modest rate, usually between 25 and 100 spikes per second. The rate increases with the value of the current injected.

If, on the other hand, the negative potential of the membrane is increased somewhat (that is, if the cell is hyperpolarized) to, say, −70 mV, then a neuron responds to an injected current, after a short delay, with a spike or a *short fast burst of spikes*, firing briefly at rates nearer *300 spikes per second*. Moreover, the after-effect of this burst is that, even though the injected current is maintained constant, the cell will not produce a further burst for a time of the order of 80–150 msec. Jahnsen and Llinás (34, 35), by means of many elegant controls, have shown that this behavior depends on a number of special ion channels, including a Ca^{2+}-dependent K^+ conductance.

Thus, it is at least possible that the effect of the GABAergic neurons of the reticular complex on the thalamic relay cells is to produce a brief burst of firing in response to incoming excitations, followed by a more prolonged inhibition. Whether this is actually the effect they produce in natural circumstances remains to be seen, since it is not easy to deduce this with certainty from the results of Jahnsen and Llinás on slices.

The Searchlight Hypothesis

What do we require of a searchlight? It should be able to sample the activity in the cortex and/or the thalamus and decide "where the action is." It should then be able to intensify the thalamic input to that region of the cortex, probably by making the active thalamic neurons in that region fire more rapidly than usual. It must then be able to turn off its beam, move to the next place demanding attention, and repeat the process.

It seems remarkable, to say the least, that the nature of the reticular complex and the behavior of the thalamic neurons fit this requirement so neatly. The extensive inhibitory collaterals in the reticular complex may allow it to select a small region that corresponds to the most active part of the thalamo-cortical maps. Its inhibitory output, by making more negative the membrane potential of the relevant thalamic neurons, could allow them to produce a very rapid, short burst and also effectively turns them off for 100 msec or so. This means that the reticular complex will no longer respond at that patch and its activity can thus move to the next most active patch. We are thus led to two plausible hypotheses:

(i) *The searchlight is controlled by the reticular complex of the thalamus.*

(ii) *The expression of the searchlight is the production of rapid firing in a subset of active thalamic neurons.*

So far I have lumped the perigeniculate nucleus (17–24) in with the reticular nucleus proper which adjoins it. It seems probable that the lateral geniculate nucleus (which in primates projects mainly to the first visual area of the cortex) sends collaterals of its output to the perigeniculate nucleus, while the rest of the dorsal thalamus sends collaterals to the reticular nucleus proper (20). This suggests that there may be at least two searchlights: one for the first visual area and another for all of the rest. Indeed, there may be several separate searchlights. Their number will depend in part on the range and strength of the inhibitory collaterals within the reticular complex. Clearly, much more needs to be known about both the neuroanatomy and the neurophysiology of the various parts of the reticular complex.

Malsburg Synapses

We must now ask: what could the searchlight usefully do? Treisman's results (5–8) suggest that what we want it to do is to form *temporary* "conjunctions" of neurons. One possibility is that the conjunction is expressed merely by the relevant neurons firing simultaneously, or at least in a highly correlated manner. In artificial intelligence the problem would be solved by "creating a line" between the units. There is no way that the searchlight can rapidly produce new dendrites, new axons, or even new axon terminals in the brain. The only plausible way to create a line in a short time is to strengthen an existing synapse in some way. This is the essence of the idea put forward in 1981 by von der Malsburg in a little known but very suggestive paper.* After describing the conjunction problem in general terms he proposed that a synapse could alter its synaptic weight (roughly speaking, the weight is the effect a presynaptic spike has on the potential at the axon hillock of the postsynaptic cell) on a fast time scale ("fractions of a second"). He proposed that when there was a strong correlation between presynaptic and postsynaptic activity, the strength of the synapse was temporarily increased—a dynamic version of Hebb's well-known rule (36)—and that with *un*correlated pre- and postsynaptic signals the strength would be temporarily decreased below its normal resting value.

Notice that we are not concerned here with *long-term* alterations in weight, as we would if we were considering learning, but very short-term *transient* alterations that would occur during the act of visual perception. The idea is not, however, limited to the visual system but is supposed to apply to all parts of the neocortex and possibly to other parts of the brain as well.

Most previous theoretical work on neural nets does not use this idea, though there are exceptions (37, 38). The usual convention is that while a net, or set of nets, is *performing*, the synaptic weights are kept constant. They are only allowed to alter when *learning* is being studied. Thus, von der Malsburg's idea represents a rather radical alteration to the usual assumptions. I propose that such (hypothetical) synapses be called Malsburg synapses. Notice that in the cortex the number of synapses exceeds the number of neurons by at least three orders of magnitude.

Let us then accept for the moment that Malsburg synapses are at least plausible. We are still a long way from knowing the exact rules for their behavior.—How much can their strength be increased? What exactly determines this increase (or decrease) of strength? How rapidly can this happen? How does this temporary alteration decay?—to say nothing of the molecular mechanisms underlying such changes.

---

*von der Malsburg, C. (1981) Internal Report 81-2 (Department of Neurobiology, Max-Planck-Institute for Biophysical Chemistry, Goettingen, F.R.G.).

In spite of all of these uncertainties it seems not unreasonable to assume that the effect of the searchlight is to activate Malsburg synapses. We are thus led to a third hypothesis.

(*iii*) *The conjunctions produced by the searchlight are mediated by Malsburg synapses, especially by rapid bursts acting on them.*

We still have to explain exactly how activated Malsburg synapses form associations of neurons. This is discussed by von der Malsburg in his paper in some detail but most readers may find his discussion hard to follow. His argument depends on the assumption that the system needs to have more than one such association active at about the same time. He describes at some length how correlations, acting on Malsburg synapses, can link cells into groups and thus form what he calls topological networks. What characterizes one such cell assembly is that the neurons in it fire "simultaneously," an idea that goes back to Hebb (36). von der Malsburg suggests that two kinds of signal patterns can exist in a topological network: waves running through the network or groups of cells switching synchronously between an active and a silent state. He next discusses how a set of cells rather than a single cell might form what he calls a "network element." Finally, after an elaborate development of this theme he broaches the "bandwidth problem." In simple terms, how can we avoid these various groups of cells interfering with each other?

Cell Assemblies

The cell assembly idea is a powerful one. Since a neuron can usually be made to fire by several different combinations of its inputs, the *significance* of its firing is necessarily ambiguous. It is thus a reasonable deduction that this ambiguity can be removed, at least in part, by the firing pattern of an *assembly* of cells. This arrangement is more economical than having many distinct neurons, each with very high specificity. This type of argument goes back to Young (39) in 1802.

There has been much theoretical work on what we may loosely describe as associative nets. The nets are usually considered to consist of neurons of a similar type, receiving input, in most cases, from similar sources and sending their output mainly to similar places. If we regard neurons (in, say, the visual system) as being arranged in some sort of hierarchy, then we can usefully refer to such an assembly as a *horizontal* assembly.

von der Malsburg's ideas, however, permit another type of assembly. In his theory a cell at a higher level is associated with one at a lower level (we are here ignoring the direction of the connection), and these, in turn, may be associated with those at a still lower level. (By "associated with" I mean that the cells fire approximately simultaneously.) For example, a cell at a higher level that signified the general idea "face" would be temporarily associated, by Malsburg synapses, with cells that signified the parts of the face, and, in turn, perhaps with their parts. Such an assembly might usefully be called a *vertical* assembly. It is these vertical assemblies that have to be constructed anew for each different visual scene, or for each sentence, etc. Without them it would be a difficult job to unite the higher level concepts with their low level details in a rather short time. This idea is reminiscent of the K lines of Minsky.[†]

The idea of *transient* vertical assemblies is a very powerful one. It solves in one blow the combinational problem—that is, how the brain can respond to an almost unlimited number of distinct sentences, passages of music, visual scenes, etc. The solution is to use *temporary* combinations of a subset of

a much more limited number of units (the 10^{12} or so neurons in the central nervous system), each new combination being brought into action as the circumstances demand and then largely discarded. Without this device the brain would either require vastly more neurons to do the job or its ability to perceive, think, and act would be very severely restricted. This is the thrust of von der Malsburg's arguments.

A somewhat similar set of ideas about simultaneous firing has been put forward by Abeles. His monograph (40) should be consulted for details. He proposes the concept of "synfire chains"—sets of cells, each set firing synchronously, connected in chains, which fire sequentially. He gives a plausible numerical argument, based mainly on anatomical connectivity, which suggests that to establish a functioning synfire chain only a few (perhaps five or so) synapses would be available at any one neuron. Since this is such a small number he deduces that the individual synapses must be strengthened (if these five synapses by themselves are to fire the cell) perhaps by a factor of 5 or so. However, he gives no indication as to how this strengthening might be done.

Abeles' argument stresses the importance to the system of the *exact time of firing* of each spike, rather than the *average* rate of firing, which is often taken to be the more relevant variable. This exact timing is also an important aspect of von der Malsburg's ideas. These arguments can also be supported by considering the probable values of the passive cable constants of cortical dendrites. Very rough estimates (for example, $\tau = 8$ msec, $X = \lambda/5$) suggest that inputs will not add satisfactorily unless they arrive within a few milliseconds of each other (see figure 3.18 in ref. 41).

Notice the idea that a cell assembly consists of neurons firing simultaneously (or at least in a highly correlated manner) is a very natural one, since this means that the impact of their joint firing on *other* neurons, elsewhere in the system, will be large. The content of the cell assembly—the "meaning" of all of the neurons so linked together—can in this way be impressed on the rest of the system in a manner that would not be possible if all of the neurons in it fired at random times, unless they were firing very rapidly indeed. Therefore, our fourth hypothesis follows.

(*iv*) *Conjunctions are expressed by cell assemblies, especially assemblies of cells in different cortical regions.*

It should not be assumed that cell assemblies can only be formed by the searchlight mechanism. Some important ones may well be laid down, or partly laid down, genetically (e.g., faces?) or be formed by prolonged learning (e.g., reading letters or words?).

It is clear that much further theoretical work is needed to develop these ideas and make them more precise. If the members of a vertical assembly fire approximately synchronously, exactly how regular and how close together in time do these firings have to be? Are there special pathways or devices to promote more simultaneous firing? Are dendritic spikes involved? How does one avoid confusion between different cell assemblies? Do neurons in *different* cell assemblies briefly inhibit each other, so that accidental synchrony is made more difficult? Etc.

The idea that the dorsal thalamus and the reticular complex are concerned with attention is not novel (19, 42, 43). What is novel (as far as I know) is the suggestion that they control and express the internal attentional searchlight proposed by Treisman (5–8), Julesz (9, 10), Posner (44, 45), and others. For this searchlight at least two features are required. The first is the rapid movement of the searchlight from place to place while the eyes remain in one position, as discussed above. There is, however, another aspect. The brain must know what it is searching *for* (the green *T* in the example given earlier) so that it may know when its hunt is successful. In other words, the brain must know *what* to attend to. That aspect, which may involve other cortical areas

---

[†]Minsky, M. (1979) Artificial Intelligence Memo No. 516 (Artificial Intelligence Laboratory, Massachusetts Institute of Technology, Cambridge, MA).

such as the frontal cortex, has not been discussed here. The basic searchlight mechanism may depend on several parts of the reticular complex, but these may be influenced by top-down pathways, or by other searchlights in other parts of the reticular complex, which may be partly controlled by which ideas are receiving attention. An important function of the reticular complex may be to limit the number of subjects the thalamus can pay attention to at any one time.

Experimental Tests

These will not be discussed here in detail. It suffices to say that many of the suggestions, such as the behavior of the dorsal thalamus and the reticular complex, are susceptible to fairly direct tests. The exact behavior of reticular neurons and thalamic neurons is difficult to predict with confidence, since they contain a number of very different ion channels. Experiments on slices should therefore be complemented by experiments on animals. Obviously, most of such experiments should be done on alert, behaving animals, if possible with natural stimuli. An animal under an anesthetic can hardly be expected to display all aspects of attention. Various psychophysical tests are also possible.

Other aspects of these ideas, such as the behavior of Malsburg synapses, may be more difficult to test in the immediate future. It seems more than likely that dendritic spines are involved, both the spines themselves and the synapses on them (46, 47).

The existence and the importance of rapid bursts of firing can also be tested. Such bursts, followed by a quiet interval, have been seen in neurons in the visual cortex of a curarized, unanesthetized and artificially respired cat when they respond to an optimal visual signal [see figure 1 in Morrell (48)]. It is unlikely that the two systems—the rapid-burst system and the slow-firing system—will be quite as distinct as implied here. In fact, as von der Malsburg has pointed out, one would expect them to interact.

Thus, all of these ideas, plausible though they may be, must be regarded at the moment as speculative until supported by much stronger experimental evidence. In spite of this, they appear as if they might begin to form a useful bridge between certain parts of cognitive psychology, on the one hand, and the world of neuroanatomy and neurophysiology on the other.

Note Added in Proof. Recent unpublished experimental work suggests that the reticular complex may produce bursts of firing in some thalamic neurons but merely an increase of firing rate in others.

This work originated as a result of extensive discussions with Dr. Christopher Longuet-Higgins. I thank him and many other colleagues who have commented on the idea, in particular, Drs. Richard Anderson, Max Cowan, Simon LeVay, Don MacLeod, Graeme Mitchison, Tomaso Poggio, V. S. Ramachandran, Terrence Sejnowski, and Christoph von der Malsburg. This work has been supported by the J. W. Kieckhefer Foundation and the System Development Foundation.

1. Van Essen, D. C. & Maunsell, J. H. R. (1983) *Trends Neurosci.* **6,** 370–375.
2. Hubel, D. H. & Wiesel, T. H. (1977) *Proc. R. Soc. London Ser. B* **198,** 1–59.
3. Bruce, C., Desimone, R. & Gross, C. G. (1981) *J. Neurophys.* **46,** 369–384.
4. Perrett, D. I., Rolls, E. T. & Caan, W. (1982) *Exp. Brain Res.* **47,** 329–342.
5. Treisman, A. (1977) *Percept. Psychophys.* **22,** 1–11.
6. Treisman, A. M. & Gelade, G. (1980) *Cognit. Psychol.* **12,** 97–136.
7. Treisman, A. & Schmidt, H. (1982) *Cognit. Psychol.* **14,** 107–141.
8. Treisman, A. (1983) in *Physical and Biological Processing of Images,* eds. Braddick, O. J. & Sleigh, A. C. (Springer, New York), pp. 316–325.
9. Julesz, B. (1980) *Philos. Trans. R. Soc. London Ser. B.* **290,** 83–94.
10. Julesz, B. (1981) *Nature (London)* **290,** 91–97.
11. Bergen, J. R. & Julesz, B. (1983) *Nature (London)* **303,** 696–698.
12. Jones, E. G. (1983) in *Chemical Neuroanatomy,* ed. Emson, P. C. (Raven, New York), pp. 257–293.
13. Sumitomo, I., Nakamura, M. & Iwama, K. (1976) *Exp. Neurol.* **51,** 110–123.
14. Dubin, M. W. & Cleland, B. G. (1977) *J. Neurophys.* **40,** 410–427.
15. Montero, V. M., Guillery, R. W. & Woolsey, C. N. (1977) *Brain Res.* **138,** 407–421.
16. Montero, V. M. and Scott, G. L. (1981) *Neuroscience* **6,** 2561–2577.
17. Ahlsén, G., Lindström, S. & Sybirska, E. (1978) *Brain Res.* **156,** 106–109.
18. Ahlsén, G. & Lindström, S. (1982) *Brain Res.* **236,** 477–481.
19. Ahlsén, G. & Lindström, S. (1982) *Brain Res.* **236,** 482–486.
20. Ahlsén, G., Lindström, S. & Lo, F.-S. (1982) *Exp. Brain Res.* **46,** 118–126.
21. Ahlsén, G. & Lindström, S. (1983) *Acta Physiol. Scand.* **118,** 181–184.
22. Ohara, P. T., Sefton, A. J. & Lieberman, A. R. (1980) *Brain Res.* **197,** 503–506.
23. Hale, P. T., Sefton, A. J., Baur, L. A. & Cottee, L. J. (1982) *Exp. Brain Res.* **45,** 217–229.
24. Ide, L. S. (1982) *J. Comp. Neuro.* **210,** 317–334.
25. Schlag, J. & Waszak, M. (1971) *Exp. Neurol.* **32,** 79–97.
26. Houser, C. R., Vaughn, J. E., Barber, R. P. & Roberts, E. (1980) *Brain Res.* **200,** 341–354.
27. Oertel, W. H., Graybiel, A. M., Mugnaini, E., Elde, R. P., Schmechel, D. E. & Kopin, I. J. (1983) *J. Neurosci.* **3,** 1322–1332.
28. Ohara, P. T., Lieberman, A. R., Hunt, S. P. & Wu, J.-Y. (1983) *Neuroscience* **8,** 189–211.
29. Scheibel, M. E. & Scheibel, A. B. (1966) *Brain Res.* **1,** 43–62.
30. Jones, E. G. (1975) *J. Comp. Neurol.* **162,** 285–308.
31. Scheibel, M. E. & Scheibel, A. B. (1972) *Exp. Neurol.* **34,** 316–322.
32. Minderhoud, J. M. (1971) *Exp. Brain Res.* **12,** 435–446.
33. Llinás, R. & Jahnsen, H. (1982) *Nature (London)* **297,** 406–408.
34. Jahnsen, H. & Llinás, R. (1984) *J. Physiol.* **349,** 205–226.
35. Jahnsen, H. & Llinás, R. (1984) *J. Physiol.* **349,** 227–247.
36. Hebb, D. O. (1949) *Organization of Behavior* (Wiley, New York).
37. Little, W. A. & Shaw, G. L. (1975) *Behav. Biol.* **14,** 115–133.
38. Edelman, G. M. & Reeke, G. N. (1982) *Proc. Natl. Acad. Sci. USA* **79,** 2091–2095.
39. Young, T. (1802) *Philos. Trans.* 12–48.
40. Abeles, M. (1982) *Local Cortical Circuits: Studies of Brain Function* (Springer, New York), Vol. 6.
41. Jack, J. J. B., Noble, D. & Tsien, R. W. (1975) *Electric Current Flow in Excitable Cells* (Clarendon, Oxford).
42. Yingling, C. D. & Skinner, J. E. (1977) in *Attention, Voluntary Contraction and Event-Cerebral Potentials,* ed. Desmedt, J. E. (Karger, Basel, Switzerland), pp. 70–96.
43. Skinner, J. E. & Yingling, C. D. (1977) in *Attention, Voluntary Contraction and Event-Cerebral Potentials,* ed. Desmedt, J. E. (Karger, Basel, Switzerland), pp. 30–69.
44. Posner, M. I. (1982) *Am. Psychol.* **37,** 168–179.
45. Posner, M. I., Cohen, Y. & Rafal, R. D. (1982) *Philos. Trans. R. Soc. London Ser. B* **298,** 187–198.
46. Perkel, D. H. (1983) *J. Physiol. (Paris)* **78,** 695–699.
47. Koch, C. & Poggio, T. (1983) *Proc. R. Soc. London Ser. B.* **218,** 455–477.
48. Morrell, F. (1972) in *Brain and Human Behavior,* eds. Karczmar, A. G. & Eccles, J. C. (Springer, New York), pp. 259–289.

Introduction

(1984)
J. J. Hopfield

Neurons with graded response have collective computational properties like those of two-state neurons

Proceedings of the National Academy of Sciences 81:3088–3092

This paper is an extension of the model presented in Hopfield's earlier work (paper 27). The earlier paper, which appeared in 1982, used McCulloch-Pitts neurons, capable of only one of two states—on or off—at their outputs. Although these neurons are easy to analyze, they are not much like neurons in the real nervous system. In real neurons, output seems to be related to the short term average frequency at which the cell fires action potentials, which means that neurons are capable of graded outputs. There are a number of complexities involved in interpreting the behavior of neurons, but the conclusion that neurons are not binary is generally accepted.

The new paper uses the same computational strategy as the earlier paper: the network is developing stable states, the stable states are related to stored information, the network finds stable states because its dynamics minimize energy, and the stable states are energy minima.

Hopfield now proposes a more realistic neuron model and shows that the results found in the previous paper still hold. The neuron contains an internal variable, the linear sum of its excitation and inhibition. This sum takes the form of the sum of the product of presynaptic activity weighted by the appropriate connection strength. (This sum of weighted inputs is used in nearly every model using continuous valued units in this collection.) There is no threshold, but the internal variable is converted into an output activity by a sigmoidal nonlinearity (see Grossberg, paper 19). This sigmoidal nonlinear function is monotone increasing as the sum of inputs increases. However, the slope is very low for large values of the sum of inputs, so large increases in the sum have a small effect on output activity. The slope of the sigmoid is low for small values of the sum as well.

Hopfield then uses the same, very powerful, technique that he used in his first paper: he sets up an energy function and shows that the evolution of the system in time, given the properties of the neurons, will be to decrease energy. (Technically, the formula for energy is a Liapunov function for this system.)

An important point for the computations is that the response of a cell to changes in its inputs is not instantaneous, because neurons have considerable amounts of membrane capacity and resistance. The membrane potential does not change immediately, but the membrane capacitors must be charged, and the time constant of the membrane can be quite long. Energy required to charge the membrane capacitors makes an important contribution to the energy function. Of interest for practical applications is the demonstration by Hopfield that this model is identical to a set of operational amplifiers connected together.

Hopfield is also able to show a connection between the stable points of the discrete model proposed in 1982 and the continuous one. The slope of the sigmoid corresponds to the gain of the system. If the gain is high, then the stable points of the continuous system correspond to the stable points of the discrete system, and are corners of the state space. (These are the stable points found by Anderson et al. in paper 22. This model has a Liaponov function that does not contain a charging term; see Golden, 1986.)

Lower gains cause the energy minima to move inward from the corners but still to stay near them; as gain gets small enough, the minima start to disappear.

There is a brief, and quite fascinating, discussion of the role of action potentials in a network. If we assume that an action potential delivers a quantal charge to the postsynaptic cell, and there are significant integrating time constants involved, then the existence of discrete action potentials merely creates a bit of noise. They do not alter the energy minimum seeking properties of the network. If we believe in the kind of network computations done here, this suggests that the existence of action potentials degrades the system by introducing noise. This is quite a change from the operational bias of a McCulloch-Pitts neuron, where the all or none property of the action potential was paramount. There is rather good evidence from neurophysiology that the function of action potentials is not logic but the conveyance of information about membrane potentials over long distances when it is necessary to use very poor, leaky, high resistance cables, that is, cell axons (Stevens, 1967). If communication distances were very short, one might not see action potentials at all. This is true in the retina, where only the output cells, the retinal ganglion cells, seem to use traditional action potentials.

References

R. M. Golden (1986), "The 'brain-state-in-a-box' neural model is a gradient descent algorithm," *Journal of Mathematical Psychology* 30:73–80.

C. Stevens (1967), *Neurophysiology: A Primer*, New York: Wiley.

(1984)
J. J. Hopfield

Neurons with graded response have collective computational properties like those of two-state neurons

Proceedings of the National Academy of Sciences 81:3088–3092

ABSTRACT   A model for a large network of "neurons" with a graded response (or sigmoid input–output relation) is studied. This deterministic system has collective properties in very close correspondence with the earlier stochastic model based on McCulloch–Pitts neurons. The content-addressable memory and other emergent collective properties of the original model also are present in the graded response model. The idea that such collective properties are used in biological systems is given added credence by the continued presence of such properties for more nearly biological "neurons." Collective analog electrical circuits of the kind described will certainly function. The collective states of the two models have a simple correspondence. The original model will continue to be useful for simulations, because its connection to graded response systems is established. Equations that include the effect of action potentials in the graded response system are also developed.

Recent papers (1–3) have explored the ability of a system of highly interconnected "neurons" to have useful collective computational properties. These properties emerge spontaneously in a system having a large number of elementary "neurons." Content-addressable memory (CAM) is one of the simplest collective properties of such a system. The mathematical modeling has been based on "neurons" that are different both from real biological neurons and from the realistic functioning of simple electronic circuits. Some of these differences are major enough that neurobiologists and circuit engineers alike have questioned whether real neural or electrical circuits would actually exhibit the kind of behaviors found in the model system even if the "neurons" were connected in the fashion envisioned.

Two major divergences between the model and biological or physical systems stand out. Real neurons (and real physical devices such as operational amplifiers that might mimic them) have continuous input–output relations. (Action potentials are omitted until *Discussion*.) The original modeling used two-state McCulloch–Pitts (4) threshold devices having outputs of 0 or 1 only. Real neurons and real physical circuits have integrative time delays due to capacitance, and the time evolution of the state of such systems should be represented by a differential equation (perhaps with added noise). The original modeling used a stochastic algorithm involving sudden 0–1 or 1–0 changes of states of neurons at random times. This paper shows that the important properties of the original model remain intact when these two simplifications of the modeling are eliminated. Although it is uncertain whether the properties of these new continuous "neurons" are yet close enough to the essential properties of real neurons (and/or their dendritic arborization) to be directly applicable to neurobiology, a major conceptual obstacle has been eliminated. It is certain that a CAM constructed on the basic ideas

The publication costs of this article were defrayed in part by page charge payment. This article must therefore be hereby marked "*advertisement*" in accordance with 18 U.S.C. §1734 solely to indicate this fact.

of the original model (1) but built of operational amplifiers and resistors will function.

Form of the Original Model

The original model used two-state threshold "neurons" that followed a stochastic algorithm. Each model neuron i had two states, characterized by the output V_i of the neuron having the values V_i^0 or V_i^1 (which may often be taken as 0 and 1, respectively). The input of each neuron came from two sources, external inputs I_i and inputs from other neurons. The total input to neuron i is then

$$\text{Input to } i = H_i = \sum_{j\neq i} T_{ij}V_j + I_i. \qquad [1]$$

The element T_{ij} can be biologically viewed as a description of the synaptic interconnection strength from neuron j to neuron i.

CAM and other useful computations in this system involve the change of state of the system with time. The motion of the state of a system of N neurons in state space describes the computation that the set of neurons is performing. A model therefore must describe how the state evolves in time, and the original model describes this in terms of a stochastic evolution. Each neuron samples its input at random times. It changes the value of its output or leaves it fixed according to a threshold rule with thresholds U_i.

$$V_i \rightarrow V_i^0 \text{ if } \sum_{j\neq i} T_{ij}V_j + I_i < U_i$$

$$\rightarrow V_i^1 \text{ if } \sum_{j\neq i} T_{ij}V_j + I_i > U_i. \qquad [2]$$

The interrogation of each neuron is a stochastic process, taking place at a mean rate W for each neuron. The times of interrogation of each neuron are independent of the times at which other neurons are interrogated. The algorithm is thus *asynchronous*, in contrast to the usual kind of processing done with threshold devices. This asynchrony was deliberately introduced to represent a combination of propagation delays, jitter, and noise in real neural systems. Synchronous systems might have additional collective properties (5, 6).

The original model behaves as an associative memory (or CAM) when the state space flow generated by the algorithm is characterized by a set of stable fixed points. If these stable points describe a simple flow in which nearby points in state space tend to remain close during the flow (i.e., a nonmixing flow), then initial states that are close (in Hamming distance) to a particular stable state and far from all others will tend to terminate in that nearby stable state.

Abbreviations: CAM, content-addressable memory; RC, resistance–capacitance.

If the location of a particular stable point in state space is thought of as the information of a particular memory of the system, states near to that particular stable point contain partial information about that memory. From an initial state of partial information about a memory, a final stable state with all the information of the memory is found. The memory is reached not by knowing an address, but rather by supplying in the initial state some subpart of the memory. Any subpart of adequate size will do—the memory is truly addressable by *content* rather than location. A given T matrix contains many memories simultaneously, which are reconstructed individually from partial information in an initial state.

Convergent flow to stable states is the essential feature of this CAM operation. There is a simple mathematical condition which guarantees that the state space flow algorithm converges on stable states. Any symmetric T with zero diagonal elements (i.e., $T_{ij} = T_{ji}$, $T_{ii} = 0$) will produce such a flow. The proof of this property followed from the construction of an appropriate energy function that is always decreased by any state change produced by the algorithm. Consider the function

$$E = -\frac{1}{2}\sum_{i \neq j}\sum T_{ij}V_iV_j - \sum_i I_iV_i + \sum_i U_iV_i. \qquad [3]$$

The change ΔE in E due to changing the state of neuron i by ΔV_i is

$$\Delta E = -\left[\sum_{j \neq i} T_{ij}V_j + I_i - U_i\right]\Delta V_i. \qquad [4]$$

But according to the algorithm, ΔV_i is positive only when the bracket is positive, and similarly for the negative case. Thus any change in E under the algorithm is negative. E is bounded, so the iteration of the algorithm must lead to stable states that do not further change with time.

A Continuous, Deterministic Model

We now construct a model that is based on continuous variables and responses but retains all the significant behaviors of the original model. Let the output variable V_i for neuron i have the range $V_i^0 \leq V_i \leq V_i^1$ and be a continuous and monotone-increasing function of the instantaneous input u_i to neuron i. The typical input–output relation $g_i(u_i)$ shown in Fig. 1a is sigmoid with asymptotes V_i^0 and V_i^1. For neurons exhibiting action potentials, u_i could be thought of as the mean soma potential of a neuron from the total effect of its excitatory and inhibitory inputs. V_i can be viewed as the short-term average of the firing rate of the cell i. Other biological interpretations are possible—for example, nonlinear processing may be done at junctions in a dendritic arbor (7), and the model "neurons" could represent such junctions. In terms of electrical circuits, $g_i(u_i)$ represents the input–output characteristic of a nonlinear amplifier with negligible response time. It is convenient also to define the inverse output–input relation, $g_i^{-1}(V)$.

In a biological system, u_i will lag behind the instantaneous outputs V_j of the other cells because of the input capacitance C of the cell membranes, the transmembrane resistance R, and the finite impedance T_{ij}^{-1} between the output V_j and the cell body of cell i. Thus there is a resistance–capacitance (RC) charging equation that determines the rate of change of u_i.

$$C_i(du_i/dt) = \sum_j T_{ij}V_j - u_i/R_i + I_i$$
$$u_i = g_i^{-1}(V_i). \qquad [5]$$

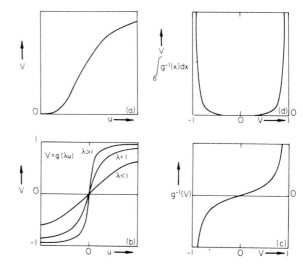

FIG. 1. (a) The sigmoid input–output relation for a typical neuron. All the $g(u)$ of this paper have such a form, with possible horizontal and vertical translations. (b) The input–output relation $g(\lambda u)$ for the "neurons" of the continuous model for three values of the gain scaling parameter λ. (c) The output–input relation $u = g^{-1}(V)$ for the g shown in b. (d) The contribution of g to the energy of Eq. 5 as a function of V.

$T_{ij}V_j$ represents the electrical current input to cell i due to the present potential of cell j, and T_{ij} is thus the synapse efficacy. Linear summing of inputs is assumed. T_{ij} of both signs should occur. I_i is any other (fixed) input current to neuron i.

The same set of equations represents the resistively connected network of electrical amplifiers sketched in Fig. 2. It appears more complicated than the description of the neural system because the electrical problem of providing inhibition and excitation requires an additional inverting amplifier and a negative signal wire. The magnitude of T_{ij} is $1/R_{ij}$, where R_{ij} is the resistor connecting the output of j to the input line i, while the sign of T_{ij} is determined by the choice of the posi-

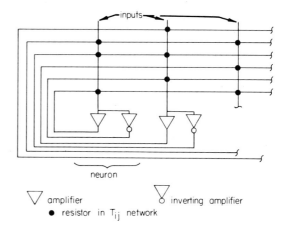

FIG. 2. An electrical circuit that corresponds to Eq. 5 when the amplifiers are fast. The input capacitance and resistances are not drawn. A particularly simple special case can have all positive T_{ij} of the same strength and no negative T_{ij} and replaces the array of negative wires with a single negative feedback amplifier sending a common output to each "neuron."

tive or negative output of amplifier j at the connection site. R_i is now

$$1/R_i = 1/\rho_i + \sum_j 1/R_{ij}, \qquad [6]$$

where ρ_i is the input resistance of amplifier i. C_i is the total input capacitance of the amplifier i and its associated input lead. We presume the output impedance of the amplifiers is negligible. These simplifications result in Eq. 5 being appropriate also for the network of Fig. 2.

Consider the quantity

$$E = -\frac{1}{2} \sum_{i,j} T_{ij} V_i V_j$$

$$+ \sum_i (1/R_i) \int_0^{V_i} g_i^{-1}(V) dV + \sum_i I_i V_i. \qquad [7]$$

Its time derivative for a symmetric T is

$$dE/dt = -\sum_i dV_i/dt \left(\sum_j T_{ij} V_j - u_i/R_i + I_i \right). \qquad [8]$$

The parenthesis is the right-hand side of Eq. 5, so

$$dE/dT = -\sum_i C_i (dV_i/dt)(du_i/dt)$$

$$= -\sum C_i g_i^{-1\prime}(V_i)(dV_i/dt)^2. \qquad [9]$$

Since $g_i^{-1}(V_i)$ is a monotone increasing function and C_i is positive, each term in this sum is nonnegative. Therefore

$$dE/dt \leq 0, \quad dE/dt = 0 \rightarrow dV_i/dt = 0 \text{ for all } i. \qquad [10]$$

Together with the boundedness of E, Eq. 10 shows that the time evolution of the system is a motion in state space that seeks out minima in E and comes to a stop at such points. E is a Liapunov function for the system.

This deterministic model has the same flow properties in its continuous space that the stochastic model does in its discrete space. It can therefore be used in CAM or any other computational task for which an energy function is essential (3). We expect that the qualitative effects of disorganized or organized anti-symmetric parts of T_{ij} should have similar effects on the CAM operation of the new and old system. The new computational behaviors (such as learning sequences) that can be produced by antisymmetric contributions to T_{ij} within the stochastic model will also hold for the deterministic continuous model. Anecdotal support for these assertions comes from unpublished work of John Platt (California Institute of Technology) solving Eq. 5 on a computer with some random T_{ij} removed from an otherwise symmetric T, and from experimental work of John Lambe (Jet Propulsion Laboratory), David Feinstein (California Institute of Technology), and Platt generating sequences of states by using an antisymmetric part of T in a real circuit of a six "neurons" (personal communications).

Relation Between the Stable States of the Two Models

For a given T, the stable states of the continuous system have a simple correspondence with the stable states of the stochastic system. We will work with a slightly simplified instance of the general equations to put a minimum of mathematics in the way of seeing the correspondence. The same basic idea carries over, with more arithmetic, to the general case.

Consider the case in which $V_i^0 < 0 < V_i^1$ for all i. Then the zero of voltage for each V_i can be chosen such that $g_i(0) = 0$ for all i. Because the values of asymptotes are totally unimportant in all that follows, we will simplify notation by taking them as ± 1 for all i. The second simplification is to treat the case in which $I_i = 0$ for all i. Finally, while the continuous case has an energy function with self-connections T_{ii}, the discrete case need not, so $T_{ii} = 0$ will be assumed for the following analysis.

This continuous system has for symmetric T the underlying energy function

$$E = -\frac{1}{2} \sum_{j \neq i} \sum T_{ij} V_i V_j + \sum 1/R_i \int_0^{V_i} g_i^{-1}(V) dV. \qquad [11]$$

Where are the maxima and minima of the *first term* of Eq. 11 in the domain of the hypercube $-1 \leq V_i \leq 1$ for all i? In the usual case, all extrema lie at *corners* of the N-dimensional hypercube space. [In the pathological case that T is a positive or negative definite matrix, an extremum is also possible in the interior of the space. This is not the case for information storage matrices of the usual type (1).]

The discrete, stochastic algorithm searches for minimal states at the corners of the hypercube—corners that are lower than adjacent corners. Since E is a linear function of a single V_i along any cube edge, the energy minima (or maxima) of

$$E = -\frac{1}{2} \sum_{i \neq j} \sum T_{ij} V_i V_j \qquad [12]$$

for the discrete space $V_i = \pm 1$ are exactly the same corners as the energy maxima and minima for the continuous case $-1 \leq V_i \leq 1$.

The second term in Eq. 11 alters the overall picture somewhat. To understand that alteration most easily, the gain g can be scaled, replacing

$$V_i = g_i(u_i) \text{ by } V_i = g_i(\lambda u_i)$$

and

$$u_i = g_i^{-1}(V_i) \text{ by } u_i = (1/\lambda) g_i^{-1}(V_i). \qquad [13]$$

This scaling changes the steepness of the sigmoid gain curve without altering the output asymptotes, as indicated in Fig. 1b. $g_i(x)$ now represents a standard form in which the scale factor $\lambda = 1$ corresponds to a standard gain, $\lambda \gg 1$ to a system with very high gain and step-like gain curve, and λ small corresponds to a low gain and flat sigmoid curve (Fig. 1b). The second term in E is now

$$+ \frac{1}{\lambda} \sum_i 1/R_i \int_0^{V_i} g_i^{-1}(V) dV. \qquad [14]$$

The integral is zero for $V_i = 0$ and positive otherwise, getting very large as V_i approaches ± 1 because of the slowness with which $g(V)$ approaches its asymptotes (Fig. 1d). However, in the high-gain limit $\lambda \rightarrow \infty$ this second term becomes negligible, and the locations of the maxima and minima of the full energy expression become the same as that of Eq. 12 or Eq. 3 in the absence of inputs and zero thresholds. *The only stable points of the very high gain, continuous, deterministic system therefore correspond to the stable points of the stochastic system.*

For large but finite λ, the second term in Eq. 11 begins to contribute. The form of $g_i(V_i)$ leads to a large positive contribution near all surfaces, edges, and corners of the hypercube while it still contributes negligibly far from the surfaces. This leads to an energy surface that still has its maxima at corners but the minima become displaced slightly toward the interior

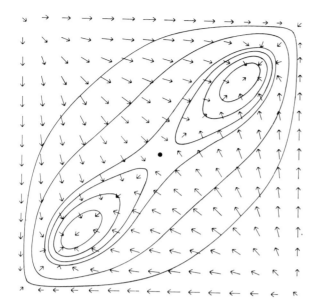

FIG. 3. An energy contour map for a two-neuron, two-stable-state system. The ordinate and abscissa are the outputs of the two neurons. Stable states are located near the lower left and upper right corners, and unstable extrema at the other two corners. The arrows show the motion of the state from Eq. 5. This motion is not in general perpendicular to the energy contours. The system parameters are $T_{12} = T_{21} = 1$, $\lambda = 1.4$, and $g(u) = (2/\pi)\tan^{-1}(\pi\lambda u/2)$. Energy contours are 0.449, 0.156, 0.017, -0.003, -0.023, and -0.041.

of the space. As λ decreases, each minimum moves further inward. As λ is further decreased, minima disappear one at a time, when the topology of the energy surface makes a minimum and a saddle point coalesce. Ultimately, for very small λ, the second term in Eq. 11 dominates, and the only minimum is at $V_i = 0$. When the gain is large enough that there are many minima, each is associated with a well-defined minimum of the infinite gain case—as the gain is increased, each minimum will move until it reaches a particular cube corner when λ → ∞. The same kind of mapping relation holds in general between the continuous deterministic system with sigmoid response curves and the stochastic model.

An energy contour map for a two-neuron (or two operational amplifier) system with two stable states is illustrated in Fig. 3. The two axes are the outputs of the two amplifiers. The upper left and lower right corners are stable minima for infinite gain, and the minima are displaced inward by the finite gain.

There are many general theorems about stability in networks of differential equations representing chemistry, circuits, and biology (8–12). The importance of this simple symmetric system is not merely its stability, but the fact that the correspondence with a discrete system lends it a special relation to elementary computational devices and concepts.

DISCUSSION

Real neurons and real amplifiers have graded, continuous outputs as a function of their inputs (or sigmoid input–output curves of finite steepness) rather than steplike, two-state response curves. Our original stochastic model of CAM and other collective properties of assemblies of neurons was based on two-state neurons. A continuous, deterministic neuron network of interconnected neurons with graded responses has been analyzed in the previous two sections. It functions as a CAM in precisely the same collective way as did the original stochastic model of CAM. A set of memories

can be nonlocally stored in a matrix of synaptic (or resistive) interconnections in such a way that particular memories can be reconstructed from a starting state that gives partial information about one of them.

The convergence of the neuronal state of the continuous, deterministic model to its stable states (memories) is based on the existence of an energy function that directs the flow in state space. Such a function can be constructed in the continuous, deterministic model when T is symmetric, just as was the case for the original stochastic model with two-state neurons. Other interesting uses and interpretations of the behaviors of the original model based on the existence of an underlying energy function will also hold for the continuous ("graded response") model (3).

A direct correspondence between the stable states of the two models was shown. For steep response curves (high gain) there is a 1:1 correspondence between the memories of the two models. When the response is less steep (lower gain) the continuous-response model can have fewer stable states than the stochastic model with the same T matrix, but the existing stable states will still correspond to particular stable states of the stochastic model. This simple correspondence is possible because of the quadratic form of the interaction between different neurons in the energy function. More complicated energy functions, which have occasionally been used in constraint satisfaction problems (13, 14), may have in addition stable states within the interior of the domain of state space in the continuous model which have no correspondence within the discrete two-state model.

This analysis indicates that a real circuit of operational amplifiers, capacitors, and resistors should be able to operate as a CAM, reconstructing the stable states that have been designed into T. As long as T is symmetric and the amplifiers are fast compared with the characteristic RC time of the input network, the system will converge to stable states and cannot oscillate or display chaotic behavior. While the symmetry of the network is essential to the mathematics, a pragmatic view indicates that approximate symmetry will suffice, as was experimentally shown in the stochastic model. Equivalence of the gain curves and input capacitance of the amplifiers is not needed. For high-gain systems, the stable states of the real circuit will be exactly those predicted by the stochastic model.

Neuronal and electromagnetic signals have finite propagation velocities. A neural circuit that is to operate in the mode described must have propagation delays that are considerably shorter than the RC or chemical integration time of the network. The same must be true for the slowness of amplifier response in the case of the electrical circuit.

The continuous model supplements, rather than replaces, the original stochastic description. The important properties of the original model are not due to its simplifications, but come from the general structure lying behind the model. Because the original model is very efficient to simulate on a digital computer, it will often be more practical to develop ideas and simulations on that model even when use on biological neurons or analog circuits is intended. The interesting collective properties transcend the 0–1 stochastic simplifications.

Neurons often communicate through action potentials. The output of such neurons consists of a series of sharp spikes having a mean frequency (when averaged over a short time) that is described by the input–output relation of Fig. 1a. In addition, the delivery of transmitter at a synapse is quantized in vesicles. Thus Eq. 5 can be only an equation for the behavior of a neural network neglecting the quantal noise due to action potentials and the releases of discrete vesicles. Because the system operates by moving downhill on an energy surface, the injection of a small amount of quantal noise will not greatly change the minimum-seeking behavior.

Eq. **5** has a generalization to include action potentials. Let all neurons have the same gain curves $g(u)$, input capacitance C, input impedance R, and maximum firing rate F. Let $g(u)$ have asymptotes 0 and 1. When a neuron has an input u, it is presumed to produce action potentials $V_0\delta(t - t_{\text{firing}})$ in a stochastic fashion with a probability $Fg(u)$ of producing an action potential per unit time. This stochastic view preserves the basic idea of the input signal being transformed into a firing rate but does not allow precise timing of individual action potentials. A synapse with strength T_{ij} will deliver a quantal charge V_0T_{ij} to the input capacitance of neuron i when neuron j produces an action potential. Let $P(u_1, u_2, \ldots u_i, \ldots, u_N, t)du_1, du_2, \ldots, du_N$ be the probability that input potential 1 has the value $u_1, \ldots$. The evolution of the state of the network is described by

$$\partial P/\partial t = \sum_i (1/RC)(\partial(u_iP)/\partial u_i)$$

$$+ \sum_j Fg(u_j)[-P + P(u_1 - T_{1j}V_0/C,\ldots, u_i - T_{ij}V_0/C,\ldots)]. \quad [15]$$

If V_0 is small, the term in brackets can be expanded in a Taylor series, yielding

$$\partial P/\partial t = \sum_i (1/RC)(\partial(u_iP)/\partial u_i)$$

$$- \sum_j (\partial P/\partial u_i)(V_0F/C) \sum_i T_{ij}\, g(u_j)$$

$$+ V_0^2 F/2C^2 \sum_{i,j,k} g(u_k)T_{ik}T_{jk}\, (\partial^2 P/\partial u_i\partial u_j). \quad [16]$$

In the limit as $V_0 \to 0$, $F \to \infty$ such that $FV_0 = $ constant, the second derivative term can be omitted. This simplification has the solutions that are identical to those of the continuous, deterministic model, namely

$$P = \prod \delta(u_i - u_i(t)),$$

where $u_i(t)$ obeys Eq. **5**.

In the model, stochastic noise from the action potentials disappears in this limit and the continuous model of Eq. **5** is recovered. The second derivative term in Eq. **16** produces noise in the system in the same fashion that diffusion produces broadening in mobility–diffusion equations. These equations permit the study of the effects of action potential noise on the continuous, deterministic system. Questions such as the duration of stability of nominal stable states of the continuous, deterministic model Eq. **5** in the presence of action potential noise should be directly answerable from analysis or simulations of Eq. **15** or **16**. Unfortunately the steady-state solution of this problem is *not* equivalent to a thermal distribution—while Eq. **15** is a master equation, it does not have detailed balance even in the high-gain limit, and the quantal noise is not characterized by a temperature.

The author thanks David Feinstein, John Lambe, Carver Mead, and John Platt for discussions and permission to mention unpublished work. The work at California Institute of Technology was supported in part by National Science Foundation Grant DMR-8107494. This is contribution no. 6975 from the Division of Chemistry and Chemical Engineering, California Institute of Technology.

1. Hopfield, J. J. (1982) *Proc. Natl. Acad. Sci. USA* **79**, 2554–2558.
2. Hopfield, J. J. (1984) in *Modeling and Analysis in Biomedicine*, ed. Nicolini, C. (World Scientific Publishing, New York), in press.
3. Hinton, G. E. & Sejnowski, T. J. (1983) in *Proceedings of the IEEE Computer Science Conference on Computer Vision and Pattern Recognition* (Washington, DC), pp. 448–453.
4. McCulloch, W. A. & Pitts, W. (1943) *Bull. Math. Biophys.* **5**, 115–133.
5. Little, W. A. (1974) *Math. Biosci.* **19**, 101–120.
6. Little, W. A. & Shaw, G. L. (1978) *Math. Biosci.* **39**, 281–289.
7. Poggio, T. & Torre, V. (1981) in *Theoretical Approaches to Neurobiology*, eds. Reichardt, W. E. & Poggio, T. (MIT Press, Cambridge, MA), pp. 28–38.
8. Glansdorf, P. & Prigogine, R. (1971) in *Thermodynamic Theory of Structure, Stability, and Fluctuations* (Wiley, New York), pp. 61–67.
9. Landauer, R. (1975) *J. Stat. Phys.* **13**, 1–16.
10. Glass, L. & Kauffman, S. A. (1973) *J. Theor. Biol.* **39**, 103–129.
11. Grossberg, S. (1973) *Stud. Appl. Math.* **52**, 213–257.
12. Glass, L. (1975) *J. Chem. Phys.* **63**, 1325–1335.
13. Kirkpatrick, S., Gelatt, C. D. & Vecchi, M. P. (1983) *Science* **220**, 671–680.
14. Geman, S. & Geman, D. (1984) *IEEE Transactions Pat. Anal. Mech. Intell.*, in press.

36

Introduction

(1984)
Andrew G. Knapp and James A. Anderson

Theory of categorization based on distributed memory storage
Journal of Experimental Psychology: Learning, Memory, and Cognition 10:616–637

There are real cultural differences in the way an engineer would measure the storage capacity of a neural network and the way a psychologist or cognitive scientist would. An engineer might measure capacity by the number of different items a memory could store and retrieve before it started to make errors, that is, when it started retrieving patterns that were not stored or could not find patterns that were stored. If what you got is not what you put in, then an error has occurred. A psychologist might comment that perhaps this is not the right measure of the way a useful, as opposed to an accurate, biological memory works. One way human memory works is by lumping different things together: we see many different kinds of tables each day—large ones, small ones, plastic ones, wooden ones—but we give them all the same name, "table." One function of language is to form and manipulate categories. We often do not care about exact details, but whether an event or thing is like other events and things in some useful way. It can be shown experimentally that humans systematically distort stored items in memory in many situations. These are not errors, but are adaptive ways of learning complex environments that display regularities, but where events seldom or never repeat exactly.

This paper by Knapp and Anderson uses a neural network to form a simple category structure. In this case, the neural model was simple enough so it could be checked against data from a particular set of experiments. We have seen before (McClelland and Rumelhart, paper 25) that neural network models can be used to make contact with experimental data from humans. Even if neural networks have no practical value at all, they have given rise to models that can be used to explain some aspects of human psychology.

Suppose we define capacity in the engineering sense: the number of random state vectors that the system can store before the retrieved vector is too different from the stored vector. Calculations of capacity in simple neural networks give a value that is a fraction of the dimensionality. Hopfield (paper 27) arrives at a value of 15% of the dimensionality, and this is consistent with most simulations. It is possible to increase capacity somewhat by using special storage techniques, but what starts to be lost are the useful information processing abilities of the memory, for example, generalizing and correcting errors in the data.

The experiment analyzed in this paper involves learning patterns of random dots. Patterns, called *prototypes*, are generated. Next, *exemplars* are formed from the prototypes by moving the dots random distances and directions: the average distance moved governs similarity between the prototype and an exemplar formed from it. During the first part of the experiment, subjects are taught to classify exemplars of a particular prototype together by making the appropriate classification, i.e., by pressing a key on a keyboard. Subjects are given feedback as to whether their classification was correct. The experiment is not hard, and after a few presentations, subjects are quite accurate. During the test phase, subjects are given new patterns to classify and given the instructions to make the "best" classification. New patterns can be either what they saw during learning (old exemplars), new patterns generated from the prototype (new exemplars), or, most important, the prototype, which they never saw.

Subjects classify the prototype most accurately and respond most quickly to it; and, if asked, they are more sure that they saw the prototype than what they actually did see. Clearly, memory is showing the systematic distortions mentioned earlier. Experiments showing a prototype effect are easy to do, and the effect has been shown for a very wide range of stimulus materials. It has been shown, as well, in infants a few months old (Bomba and Siqueland, 1983). One interpretation of these experiments is that we are seeing, in microcosm, the formation of a simple "concept." Mental representation of a category or concept by a prototype or best example has considerable psychological support (see Rosch, 1978). It may be an important strategy used to reduce the burden on the rather limited capacity human information processing system.

A computational memory such as a neural network has two distinct modes of operation when it has to store many items. If each item is distinct, and it is necessary to keep them separate, then storage is severely limited—to a fraction of the dimensionality of the system. If, on the other hand, we have a system that forms equivalence classes in storage, by forming "concepts" or "categories" as this one does, then its capacity for discrete items is irrelevant. What matters is correct classification of new items, and new learning may simply confirm the categorization structure of previous learning. The memory has become a memory for concepts and not a memory for particular events.

The neural network model used to explain the data is a straightforward application of the linear associator (papers 14 and 15). If a number of patterns are associated with the same output, then a kind of averaging of the input patterns occurs. It is easy to show that the system will respond best to the average patterns. It is possible to vary many of the experimental details of the system and predict the effects of the variation on the experimental results, and much of the paper is devoted to this.

The much maligned linearity of these early models is in fact what gives them most of their concept forming power. Building a degree of small signal linearity into a network turns out to be an effective way of forming and using concept-like entities.

The other point of modeling interest in this paper is the representation of the dot patterns. It is assumed that a dot gives rise to a little bump of activity on a topographically arranged two-dimensional surface, modeled after the topographic representations found in cortex. This means that if a dot is displaced slightly, its cortical activity pattern remains quite similar, since the bumps of activity still can overlap. This builds in a natural similarity metric, based on the width of the bump representing a dot. Since examples of a prototype involve relatively small dot motions from the prototype, overlaps between representations of different examples are greatest at the prototype location. Thus the entire model rests on getting the representations 'right,' that is, arranging it so that similar things have similar representations, allowing strong interactions between similar things in memory.

We think that this particular paper illustrates a very important general point about neural networks. *A good representation does most of the work.* Neural networks, no matter of what kind, are weak computers, but a proper representation, coupled with the abstracting and cooperative effects of the network, can become extremely fast, powerful, and accurate. Unfortunately, representations are usually problem specific. But as applications of networks get closer, we suspect that the emphasis in modeling will shift from learning algorithms, which are already very powerful, toward getting the representations right, so that the learning and the dynamics can work effectively.

The model presented in this paper could be considered an approximation of a situation where a relatively small number of not very interesting events are stored. Later work on the concept forming power of this and related networks has shown that the incorporation of large signal nonlinearities and error correction into the dynamics and learning maintains most of the interesting category forming abilities of the network (Anderson and Murphy, 1986; McClelland and Rumelhart, 1985). Concept forming abilities in the nonlinear networks correspond, perhaps, to situations where large numbers of events are stored, and where richer and more interesting structures can be formed.

References

J. A. Anderson and G. L. Murphy (1986), "Psychological concepts in a parallel system," *Physica* D22:218–336.

P. C. Bomba and E. R. Siqueland (1983), "The nature and structure of infant form categories," *Journal of Experimental Child Psychology* 35:294–328.

J. L. McClelland and D. E. Rumelhart (1985), "Distributed memory and the representation of general and specific information," *Journal of Experimental Psychology: General* 114:159–188.

E. Rosch (1978), "Principles of categorization," *Cognition and Categorization*, E. Rosch and B. B. Lloyd (Eds.), Hillsdale, NJ: Erlbaum.

(1984)
Andrew G. Knapp and James A. Anderson

Theory of categorization based on distributed memory storage
Journal of Experimental Psychology: Learning, Memory, and Cognition 10:616–637

As an alternative to *probabilistic* and *exemplar* models of categorization, we develop a model based on the assumption of distributed memory storage. We discuss the model in the context of experiments using stimuli composed of random dots. When the number of exemplars of the stimulus patterns is small, new dot patterns are classified according to their similarity to learned exemplars; when the number is large, accuracy depends on a dot pattern's similarity to a prototype pattern. The *distributed memory* model is used to explain a number of aspects of previously reported experimental findings. We also report two new experiments. In the first, the perceived similarity was measured between two dot patterns, one a distortion of the other. In the second, groups of exemplar patterns derived from a category prototype were classified together in a category-learning task. Detailed computer simulations are described for the two experiments.

Because events in the world rarely repeat themselves exactly, organisms must possess ways to relate new information to what has been learned in the past. In particular, the problem faced by a behaving creature is often to reduce the varied inputs it receives into a smaller number of equivalence classes (e.g., friend, foe, or food) and to classify correctly new inputs that are similar to, but not identical with, what has been encountered previously.

Categorization—the ability to organize information into equivalence classes—has fundamental importance for any organism that relies on learning for survival. The purpose of this article is to present a theory of memory in which, under certain conditions, categories are formed automatically by the act of storage. We will show that the proposed storage and retrieval scheme captures some of the phenomena observed when human subjects categorize in the laboratory.

Our approach is derived from a view of memory that regards information as being stored in a distributed fashion, with separate traces interacting in storage.

Theory

Models of Categorization

Recently, a great deal of effort has been directed toward the study of human categorization (see Mervis & Rosch, 1981, and Smith & Medin, 1981, for recent reviews). According to the formulation of Smith and Medin (1981) models of categorization fall broadly into two families, depending on how categories are assumed to be represented in memory.

Probabilistic models assume that the representation of a category consists of a unitary description of valid category members. However, not all properties of the description are necessarily true of all category members, so that category membership is actually a continuous rather than a two-valued function. This class of models includes the spreading activation model of Collins and Loftus (1975), the feature comparison model (Smith,

This article represents a collaborative effort and the order of authors is arbitrary.

The work described was supported in part by Grants BNS-79-23900 and BNS-82-14728 from the National Science Foundation, Memory and Cognitive Processes Section to J.A. Some support was provided by the United States Office of Naval Research under Contract N00014-81-K-0136. Computer facilities were provided by the Center for Cognitive Science, Brown University. Computer facilities were supported with assistance from the National Science Foundation, the Alfred P. Sloan Foundation, and the Digital Equipment Corporation.

The authors thank Richard Millward, director of the center. Special thanks to the referees of a previous version of this article and to Richard Shiffrin for valuable criticisms and comments.

Copyright 1984 by the American Psychological Association, Inc. Reprinted by permission of the publisher and author.

Shoben, & Rips, 1974), and several models that represent categories as points in a multidimensional space.

One set of probabilistic models we are especially concerned with are those that represent a concept by an average (sometimes called a *prototype*) of category instances. A body of empirical evidence has been used to support the idea that, in some tasks, subjects abstract a prototype from items classified together during learning and that novel items are classified according to the prototype they most resemble. This evidence includes the finding that category prototypes are sometimes classified more accurately than other category members, including the exemplars actually seen during learning (Franks & Bransford, 1971; Posner & Keele, 1968, 1970; Strange, Keeney, Kessel, & Jenkins, 1970; also see Robbins et al., 1978), suggesting that the prototype is a main constituent in the category's mental representation.

Exemplar models make up the other main family of categorization models. According to this view, no single description of a category exists. Rather, an aggregate of separate descriptions of some or all category members serves to represent the category (Brooks, 1978; Nelson, 1974). According to this view, stimuli are categorized according to the number and/or type of stored exemplars they retrieve. These proposals include the proximity and best-examples models (Reed, 1972), as well as the context model of Medin and Schaffer (1978). Proponents of exemplar models have pointed out that many of the findings taken to support prototype abstraction can be equally well explained if only the learned instances are assumed to be stored in memory (see, e.g., Hintzman & Ludlam, 1980).

Although they lead to quite different approaches to the problem of categorization, the memory representations postulated by both probabilistic and exemplar models are similar in an important respect. Both views assume the descriptions representing learned categories to be stored separately from one another. In the case of exemplar models, the descriptions of individual category members that combine to make up a category's representation are likewise assumed to be separate, identifiable entities. The idea that items stored in memory reside in unique locations or form

distinct traces is an implicit assumption common to both families of models. Rejecting the assumption of separate storage leads to a third class of models, which we call *distributed memory* models of categorization.

The fundamental assumption behind the distributed memory approach is that remembered items share many or all of the same storage elements, so that one cannot properly point to a single memory trace. Previous theoretical work demonstrated that information can be both stored and retrieved without assuming separate storage of individual items (for fuller descriptions see Anderson & Hinton, 1981; Hinton & Anderson, 1981; Kohonen, 1977). In addition, distributed memory models have been applied to a variety of cognitive processes, including associative learning (Anderson, 1983; Eich, 1982; Murdock, 1982), list learning tasks (Anderson, 1973, 1977), as well as categorical perception, distinctive feature analysis, and probability learning (Anderson, Silverstein, Ritz, & Jones, 1977).

It is worth emphasizing that the assumption of a distributed memory does not preclude more familiar models of categorization. It is possible to implement probabilistic or exemplar model "software" on distributed memory "hardware." However, the converse is not always true: The behavior of some distributed memory models can be mimicked only with ad hoc assumptions or at great computational expense if separate storage is assumed. We illustrate this point by developing a distributed memory model that acts as a simple categorizer and we compare the results of two experiments pertaining to categorization with the predictions of the model.

A Distributed Memory Categorizer: Introduction

The model to be presented was inspired by speculation about how associative learning might occur in the nervous system. A review of some neuroscientific evidence bearing on the model is available (Levy, Anderson, & Lehmkuhle, 1984). More detailed presentation of the material in the following section can be found in Anderson et al. (1977) or Anderson and Hinton (1981).

We begin by assuming that a large number

of richly interconnected but simple elements participate in the storage of information. This is the basic distributed memory assumption. We refer to these elements as *neurons,* and we refer to their connections as *synapses,* while remaining agnostic about their possible realization in the nervous system.

Each neuron has an activity that depends on the synaptic inputs it receives from other neurons, where a *synaptic input* is defined as the activity of the input neuron multiplied by a weighting factor that we call the *strength* of the synapse. In particular, we assume that neurons behave as linear integrators: A neuron's activity is simply the weighted sum of its synaptic inputs. Thus, if a neuron receives inputs from three other neurons whose activities are 20, −10, and 2, with synaptic strengths 0.5, 0.2 and −1.0, respectively, its activity will be as follows:

$$(20)(0.5) + (-10)(0.2) + (2)(-1.0) = 6.$$

Note that both neuronal activities and synaptic strengths can take on negative as well as positive values.

Information is represented in such a system by the pattern of individual neuron activities across a large number of neurons. Formally, we denote these activity patterns by N component vectors, where each element in the vector is the activity of a single neuron, and N is the number of neurons. Now suppose there are two such sets of N neurons, alpha and beta, connected so that every neuron in beta receives an input from every neuron in alpha as illustrated in Figure 1.

We can conveniently represent the N^2 synaptic strengths by an $N \times N$ connectivity matrix **A**, where each entry A(i, j) is the strength of the synapse between neuron i in

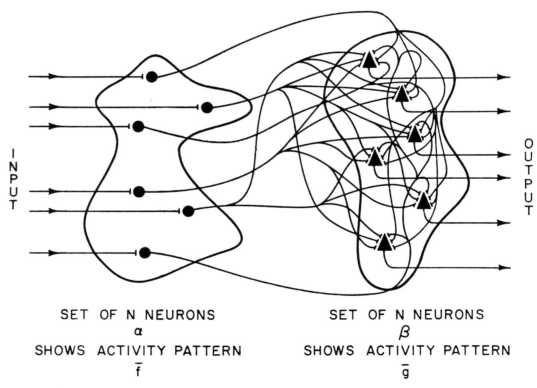

SET OF N NEURONS
α
SHOWS ACTIVITY PATTERN
$\overline{f}$

SET OF N NEURONS
β
SHOWS ACTIVITY PATTERN
$\overline{g}$

Figure 1. Models assume two sets of N, neurons alpha projecting to beta. (Every neuron in alpha projects to every neuron in beta. This drawing has $N = 6$. From "Distinctive Features, Categorical Perception, and Probability Learning: Some Applications of a Neural Model" by J. A. Anderson, J. W. Silverstein, S. A. Ritz, and R. S. Jones, 1977, *Psychological Review, 84,* p. 416. Copyright 1977 by the American Psychological Association, Inc. Reprinted by permission.)

alpha and neuron j in beta. In the absence of other inputs, the activity of each neuron in beta is thus completely determined by the activities of the neurons in alpha and by the synaptic strengths. In vector notation, this relation is as follows: (pattern in beta) = $\mathbf{Af}$.

Information can be stored in such a system by modifying the synaptic strengths. Suppose that all the strengths are initially zero, that activity pattern $\mathbf{f}$ occurs in alpha, and that pattern $\mathbf{g}$ simultaneously occurs in beta. Here, $\mathbf{f}$ may denote a stimulus and $\mathbf{g}$, a response, and it is desirable for some reason to learn the association between them. We assume that in such a situation, the synaptic strengths are able to change according to the following rule:

$$A(i, j) \propto f(i)g(j),$$

that is, the strength of each synapse is changed proportionally to the product of the pre- and postsynaptic neurons (cf. Hebb, 1949).

For illustration, suppose that the proportionality constant in the learning equation just presented is 1 and the vector $\mathbf{f}$ is normalized so that the length of $\mathbf{f}$ is set equal to 1 and $\mathbf{A}$ is initialized to 0 (i.e., the inner product, $[\mathbf{f}, \mathbf{f}]$ is 1, and $\mathbf{A}$ is all 0s). The resulting connectivity matrix becomes the following:

$$\mathbf{A} = \mathbf{gf}^T.$$

The superscript T refers to the matrix operation *transpose*. The transpose of a matrix interchanges rows and columns. We assume as a convention that all of our vectors are column vectors; the transpose of a column vector is a row vector. This convention allows us to write the vectors and matrices arising from this model in a compact and useful form. Of particular importance is the observation that the inner product (dot product) of two vectors $\mathbf{f}$ and $\mathbf{g}$ is given by the following:

$$[\mathbf{f}, \mathbf{g}] = \mathbf{f}^T\mathbf{g},$$

and an outer product matrix (the matrix $\mathbf{A}$ described earlier) is given when the terms $\mathbf{f}$ and $\mathbf{g}$ are reversed in the expression just presented. These observations follow from the familiar rules of matrix multiplication.

Now suppose that after the synapses comprising $\mathbf{A}$ have been modified, pattern $\mathbf{f}$ again occurs in alpha. By the learning rule,

(pattern in beta) = $\mathbf{Af}$;

(pattern in beta) = $\mathbf{gf}^T\mathbf{f}$;

(pattern in beta) = $\mathbf{g}$.

Therefore, the result of this form of synaptic modification is that subsequent occurrences of $\mathbf{f}$ in alpha give rise to $\mathbf{g}$ in beta—the two patterns have become associated.

In general, such a simple system can associate a number, say n, of patterns,

$$\mathbf{f}_1 \rightarrow \mathbf{g}_1, \mathbf{f}_2 \rightarrow \mathbf{g}_2, \ldots, \mathbf{f}_n \rightarrow \mathbf{g}_n,$$

up to a potential capacity of N, though the practical capacity is less. Each association increments the matrix according to the learning rule, so the final matrix becomes the following:

$$\mathbf{A} = \sum_{i=1}^{n} \mathbf{g}_i\mathbf{f}_i^T.$$

Although the n associations are mixed together at the N^2 synapses, information might not have been lost.

When one of the fs, say $\mathbf{f}_i$, occurs in alpha, then we have the following:

(pattern in beta) = $\mathbf{Af}_i$

$$\text{(pattern in beta)} = \mathbf{g}_i\mathbf{f}_i^T\mathbf{f}_i + \sum_{i \neq j}^{n} \mathbf{g}_j(\mathbf{f}_j^T\mathbf{f}_i).$$

Consider the special case where the simulus set (the fs) are orthogonal, that is, the fs are at right angles to each other, and their inner product $[\mathbf{f}_i, \mathbf{f}_j] = 0$ when $i \neq j$, and 1 when $i = j$. Then

$$\text{(pattern in beta)} = \mathbf{g}_i,$$

because all the other inner products are 0. This is a useful approximation, because if N, the dimensionality, is large and the fs have statistically independent components, they will be close to orthogonal.

Categorization

The model can function as a simple categorizer by making one additional assumption. Let us make the fundamental coding assumption that the activity patterns representing similar stimuli are themselves similar, that is, their state vectors are correlated. This means the inner product between two similar patterns is not small.

Now consider the case described earlier, where the model has made the association $\mathbf{f} \rightarrow \mathbf{g}$. Let us restrict our attention to the magnitude of the vector in beta that results when various patterns occur in alpha. We have just shown that when $\mathbf{f}$ occurs in alpha, $\mathbf{g}$ occurs in beta. When a new pattern $\mathbf{f}'$ occurs in alpha, then

$$(\text{pattern in beta}) = \mathbf{g}\mathbf{f}^T\mathbf{f}'.$$

If $\mathbf{f}$ and $\mathbf{f}'$ are not similar, their inner product $[\mathbf{f}, \mathbf{f}']$ is small. If $\mathbf{f}$ is similar to $\mathbf{f}'$, then the inner product will be large. The model responds to input patterns based on similarity to $\mathbf{f}$. The nature of the learning assumption gives us a generalization mechanism. Furthermore, this formulation suggests that the perceived similarity of two stimuli should be systematically related to the inner product $[\mathbf{f}, \mathbf{f}']$ of the two neural codings. We test this prediction in Experiment 1.

Multiple Member Categories

We next apply the distributed-memory model to a more realistic situation where a category contains many similar items. Here, a set of similar activity patterns (representing the category members) becomes associated with the same response, for example, the category name. It is convenient to discuss such a set of vectors with respect to their mean. Let us assume the mean is taken over all potential members of the category.

Specifically, consider a set of correlated vectors $\mathbf{f}_1 \cdots \mathbf{f}_n$, with mean $\mathbf{p}$. Each individual vector in the set can be written as the sum of the mean vector and an additional noise vector, $\mathbf{d}_i$, representing the deviation from the mean, that is,

$$\mathbf{f}_i = \mathbf{p} + \mathbf{d}_i.$$

When these n patterns occur in alpha and are all associated with the same response, $\mathbf{g}$, in beta, the final connectivity matrix will be as follows:

$$\mathbf{A} = \sum_{i=1}^{n} \mathbf{g}\mathbf{f}_i^T;$$

$$\mathbf{A} = \mathbf{g} \sum_{i=1}^{n} (\mathbf{p}^T + \mathbf{d}_i^T);$$

$$\mathbf{A} = n\mathbf{g}\mathbf{p}^T + \mathbf{g} \sum_{i=1}^{n} \mathbf{d}_i^T.$$

We discuss this result quantitatively in the next few sections. Let us point out some qualitative results now. The term containing the sum of the noise vectors (the $\mathbf{d}_i$) is particularly important. Suppose that this term is relatively small, as could happen if the system learned many randomly chosen members of the category (so the $\mathbf{d}_i$s cancel on the average, and their sum is small) and/or if $\mathbf{d}_i$ is not very large. In that case, the connectivity matrix is approximated by the following:

$$\mathbf{A} = n\mathbf{g}\mathbf{p}^T.$$

The system behaves as if it had repeatedly learned only one pattern, $\mathbf{p}$, the mean of the correlated set of vectors it was exposed to. Under these conditions, the simple association model extracts a "noisy signal"—the prototype—just like an average response computer. In this respect, the distributed memory model behaves like a prototype model because the most powerful response will be to the pattern $\mathbf{p}$, which might never have been seen.

However, if the sum of the $\mathbf{d}_i$ terms is not relatively small, as might happen if the system only sees a few patterns from the set and/or if the average $\mathbf{d}_i$ is large, as we shall see, the response of the model will depend on the similarities between the novel input and each of the learned patterns, that is, the system will behave more like an exemplar model.

Finally, we need to consider what happens when members of more than one category can occur. Suppose (for illustration) the system learns items drawn from three categories with means of $\mathbf{p}_1$, $\mathbf{p}_2$, and $\mathbf{p}_3$, respectively, with n exemplars presented from each category. If $\mathbf{g}_1$, $\mathbf{g}_2$, and $\mathbf{g}_3$ are the responses associated with the three categories, then

$$\mathbf{A} = \mathbf{g}_1\mathbf{s}_1^T + \mathbf{g}_2\mathbf{s}_2^T + \mathbf{g}_3\mathbf{s}_3^T,$$

where each sum $\mathbf{s}_i$ is composed of the sum of the n exemplars as before.

When presented with an input $\mathbf{f}$, an exemplar of one of the categories, the output pattern, $\mathbf{g}$, will be given by the following:

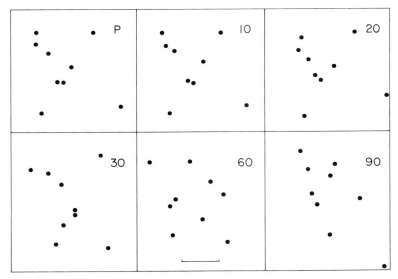

Figure 2. A prototype dot pattern (P) followed by five exemplars at various levels of distortion. (Dots were generated on a 512 × 512 array and were presented to subjects on a CRT screen. The number in the upper right-hand corner of each exemplar refers to the average number of locations moved on the array. A distance of 100 screen units is indicated.)

$$\mathbf{Af} = g_1 \mathbf{s}_1{}^T \mathbf{f} + g_2 \mathbf{s}_2{}^T \mathbf{f} + g_3 \mathbf{s}_3{}^T \mathbf{f}.$$

If the distortions of the prototypes are small, then this sum can be approximated by the following:

$$\mathbf{Af} = n([\mathbf{p}_1, \mathbf{f}]\mathbf{g}_1 + [\mathbf{p}_2, \mathbf{f}]\mathbf{g}_2 + [\mathbf{p}_3, \mathbf{f}]\mathbf{g}_3).$$

Due to superposition (this is a linear system), the actual response pattern, $\mathbf{g}$, is a sum of the responses, $\mathbf{g}_i$, weighted by the inner products, $[\mathbf{p}_i, \mathbf{f}]$. If the $\mathbf{p}_i$ are dissimilar, the inner product between an exemplar of one prototype and the other prototypes is small on the average, and the admixture of outputs associated with the other categories will also be small. Therefore only the appropriate response occurs. This seems to be the case in the experiments to be described. (For computational convenience, if the categories are well separated, we need only inspect the inner products $[\mathbf{f}, \mathbf{p}_i]$ and choose the response associated with the largest inner product, which is how our simulations were done.)

Random Dot-Pattern Stimuli

In order to make contact with experiments that could be related to the distributed memory model, it was necessary to use stimuli whose relations to one another could be readily quantified. These considerations led us to adopt the artificial categories first used by Posner and Keele (1968, 1970).

These stimuli consist of nonmeaningful arrangements of dots. In a typical experiment, several readily distinguishable patterns are produced by randomly distributing nine dots within a display area. These original patterns are called *prototypes*. A family of distortions of each prototype can then be generated by moving the dots random distances in random directions according to various rules. A single distortion of a prototype is called an *exemplar*. By manipulating the motion of the dots, category exemplars can be either grossly distorted versions or only slight variations of the prototype. Each prototype and its progeny constitute an artificial category. Like many natural categories, dot-pattern categories are ill defined (Neisser, 1967) because any pattern can be transformed into another by an appropriate distortion.

Shown in Figure 2 are a prototype and five exemplars at increasing levels of distortion as used in our experiments. Full details of their generation are given later. Each exemplar is the result of moving the prototype dots various distances drawn from a normal distribution. The mean of the distribution (in display screen units) is given above each exemplar.

Similarity

It is apparent from Figure 2 that the resemblance between an exemplar and its prototype decreases as the dot displacements used to create the exemplar increase. As just described, our distributed memory model predicts a general form for this relation. The perceived similarity between two stimuli is related to the inner product [**f**, **f**′], where **f** and **f**′ are the distributed activity patterns representing the two stimuli.

To obtain quantitative similarity predictions from the distributed memory model for these stimuli, we must specify a neural coding of the input patterns, that is, we must describe how the state vectors are generated from the physical input pattern. We know from many sources that there is a powerful topographic mapping of visual space onto the surface of the cerebral cortex. Let us consider the coding due to a single dot in visual space. We assume that this will map onto a "bump" of activity on a hypothetical surface composed of the many elementary neurons that represent the neural coding. Because real neurons have receptive fields of varying width, even in a single cortical region, we assume there is a falloff of activity from the central location corresponding to the exact topographic location of the dots: Some cells have large receptive fields and respond to a dot even if the center of their fields is far away from the dot location; other receptive fields are much smaller and must be precisely centered on the dot location to be activated. Shown in Figure 3 is an exponential falloff of activation. The activity at a point, a(r), is given by the following:

$$a(r) = \exp(-r/\lambda),$$

where λ is the length constant of the exponential, and r is the distance from the center of the distribution to the point whose activity is to be computed.

Interactions Between Activity Patterns

We can compute the inner product between the activity patterns of any pair of dots. We

investigated a number of different activity patterns due to single dots. The best results were obtained with the exponential distribution shown in Figure 3. In this case, we are able to derive a closed-form approximation to the inner product.

To evaluate the dot product, we approximate it by a continuous distribution, though in reality it is made up of discrete neuronlike elements. Dimensionality (i.e., the number of neurons in the system) is extremely large in this approximation. Suppose we have two activity patterns due to single dots separated by a distance d. For convenience, let the length constant equal one, and consider two exponential distributions, a(x, y) and b(x, y) separated by a distance d along the x axis. We then need to evaluate the product of the two functions over the plane to approximate the dot product (which would be a summation of products at many discrete points). Therefore we want to evaluate the integral, I(d), which will be a function of displacement, d. If

$$a(x, y) = \exp\left[-\sqrt{(x + d/2)^2 + y^2}\right]$$

and

$$b(x, y) = \exp\left[-\sqrt{(x - d/2)^2 + y^2}\right],$$

we want to evaluate

$$I(d) = \int\int_{-\infty}^{\infty} a(x, y)b(x, y)dxdy.$$

Let us normalize the function so that I(d) = 1, when d = 0. This integral can be computed exactly and is given by the following:

$$I(d) = (\tfrac{1}{2})d^2 K_2(d),$$

where $K_2(d)$ is a modified Bessel function of order 2 (see Abramowitz & Stegun, 1964). For connoisseurs of integration, this integral can be done by observing that the loci of constant product are ellipses. The equations are converted to elliptical coordinates (Korn & Korn, 1968) and are then integrated with the tables in Gradshteyn and Ryzhik (1965).

Pathologies of Linear Systems

Although many interesting aspects of the model we have discussed arise from its basic

RESPONSE TO A DOT

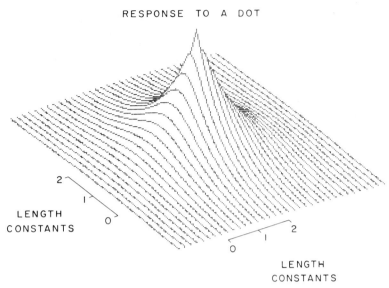

Figure 3. Activity pattern on a hypothetical cortex due to a single dot in the real world. (There is an exponential falloff with distance from a central point. Height corresponds to activity. Length constant of the exponential is shown.)

linearity, linearity has some undesirable properties as well. Hinton (1981) gave examples of stimuli that linear models cannot properly associate. Minsky and Papert (1969) discussed at length limitations in processing power of classes of models closely related to the one we have presented here, in particular, their inability to tell a connected geometrical figure from an unconnected one. A discussion of linear and nonlinear nervous system models with reference to experimental data can be found in Anderson and Silverstein (1978).

Significant nonlinearities exist in the nervous system and are particularly important in some kinds of classification tasks. A nonlinear model specifically for categorization behavior was presented earlier (Anderson et al., 1977; see also Anderson & Silverstein, 1978). Formally, the linear model presented here could be considered the strictly linear portion of the more powerful (but less tractable) nonlinear model. We must point out that the prototype effects are a result of the linearities in the system. We feel it is useful to hold with the assumption of linearity as long as possible, though acknowledging its limitations, because a linear approximation often is a good indicator of the behavior of a more realistic but more complex nonlinear model. Linearizing a complex nonlinear problem is a frequently used analytical technique.

Difficulties Arising From Inadequacies of Representation

Along with the problem of linearity, there is an equally serious problem with the details of representation. The representation of the stimulus suggested here makes no specific use of pair-wise or higher order relations between different dots. We assume single dots in the stimulus are represented as spatially organized bumps of activity and are then added to the memory. Suppose we constructed a group of stimuli comprising single dots in the various positions corresponding to dots in one of our nine dot patterns. If these single dots were presented one at a time, were associated with the proper response, and were learned, one would predict the final matrix memory would be the same as if all nine dots were presented at once, because only the sum of activities is involved. Yet clearly these are very different experimental situations and should not give rise to the same memory representation.

The representation of the stimuli proposed does not differentiate these cases. A response to this problem is to observe that one pattern of nine dots looks grossly like another pattern of nine dots. Though we are only concerned here with the part of the vector that can differ between patterns, the full neural codings contain a significant additional context that is identical between different patterns: dot number, the experimental situation, and so forth. This constant part is assumed to play no part in our computations because it is identical for all the dot patterns, but if we make a significant change in the experimental situation (such as dot number), it can become important. This constant part of the vector could contain elements reflecting gross nonlinearities as long as these elements did not vary across the stimulus sets.

However, the coding proposed is also not adequate for some more reasonable situations. Consider a set of exemplars that are formed from a prototype by translating all the dots in the pattern in the same direction and the same distance, that is, a simple translation of the entire pattern. We then form a memory from a number of such exemplars. Now suppose we consider dot patterns formed by taking dots from different patterns, that is, one dot from an exemplar formed by a translation to the right, one from a translation down, and so forth. The memory formed according to the model from this new set of patterns would be identical to the first because it is composed of the same total set of dots, and the gross context (i.e., dot number, etc.) is the same. Yet the psychological perception of these two cases surely would be very different.

It seems clear that the first set of exemplars would be recognized in an experiment simply as a series of translations of an identical pattern. The second set would be seen and would act as distortions because the patterns could not be perceived as identical to each other.

The model we present here is not capable of differentiating these two sets of exemplars. The best way to extend the model to account for this problem would be to increase the kind of information about the stimulus represented in the state vector, that is, to make the coding richer. If, for example, we not only represent single dots in the state vector but also the pair-wise relations between dots, we could differentiate the exemplars above.

(A pair-wise dot relation could be the length and direction of the line formed between a pair of dots.) In the case of simple translations of the whole pattern, the state vectors would all have identical pair-wise relations, whereas exemplars constructed from dots taken from different translations would have different pair-wise relations. Therefore, the state vectors would no longer be identical, and the memories would not be identical.

In the experiments described here, the simple representation of dot patterns seems adequate, and inclusion of pair-wise and higher order relations seems not to add much to the ability of the representation to discriminate among patterns. More realistic simulations would require more complete representations: Dot patterns seem to be sufficiently unfamiliar and impoverished to be approximated with a simple representation.

This example makes a general point about the representation of information by the nervous system: It is consistent with both neurophysiology and common sense that information is represented by elements that can respond to a great many aspects of the stimulus, some of which are useful for a particular task but most of which are not. Making the representation more complex and more complete requires making the state vector larger. A very rich representation may be conceptually a little untidy and certainly difficult to handle in a computer simulation—problems that do not concern the nervous system but do concern the nervous system modeller. The problem could be rephrased in the language of pattern recognition as that of building a system that makes effective use of feature vectors that are composed of extremely large numbers of poor features. The models presented here are good at working with such representations. More frequent practical use is not made of such techniques because they are slow and inefficient on current computers. Because the algorithms used in the models discussed here are highly parallel, they might become fast, efficient, and practical on a parallel computer.

Properties of the Linear Model

The qualitative behavior of the model is more understandable if we begin by considering the summed representation of a single dot in the nine dot patterns (see Figures 4 and 5). These diagrams may be considered close-up views of the category's representation, with similar summation occuring in the regions of the other eight dots.

Shown in Figure 4 are activity patterns due to four exemplar dots equally spaced from each other and from the location of the prototype dot. As the physical separation between the exemplar dots increases, the peaks of activity in the sum separate. At first, the distribution of activity in the sum as the dots separate simply seems to broaden the peak located at the prototype location. Then bumps due to the individual exemplars appear. Even with dots well separated, there is still substantial representation at the prototype location. It is clear that some representation of variability (the width of the activity pattern) is present, as well as representation of the central tendency. In Figure 4, the left-hand side shows the actual sum; the right-hand side has the maximum values in the pattern drawn as the same height so relative curve shapes can be compared.

Shown in Figure 5 are the formations of sums from two different exemplars (top) and from eight different exemplars (bottom). In the diagrams, the exemplar dots have been chosen to be equally spaced about a central location corresponding to the dot in the prototype pattern. In both sums, the activity at the prototype location is substantial. In the two-exemplar sum, the activity is greatest at the location of the individual exemplars, but in the eight-exemplar sum, the activity is largest at the prototype location, and the overall distribution of activity is more uniform. The smaller number of exemplars produces a "lumpy" sum dominated by learned exemplars, whereas the larger number yields a "smoother" sum with the prototype enhanced.

When is a Prototype Not Extracted (and Why)

This mixed behavior deserves comment. Intuition tells us that different categories demand different mental representations. Let us take an important example from outside the world of dot patterns. Intuitively, we feel some things are too different from each other to form natural equivalence classes or to be represented by a single prototype, though it is always possible to bind explicitly very different things together to form complex concepts. Consider the letters *A* and *a*. In some contexts, these two forms represent the same noises and have identical interpretations. In other contexts, they are clearly separated, having different names, such as *Capital A* and *small a*. It seems unlikely and undesirable that presentation of *A* and *a* would lead to formation of their average as the best example of the written letter *ay*. In this case, it seems best to assume that there are two simple prototypes formed that in some contexts are associated with the same name.

Formation of a complicated concept from two simple ones is quite straightforward. Suppose **f** and **f'** are associated with the same response, **g**, and **f** and **f'** are sufficiently dissimilar to be orthogonal to one another. Then, we form the overall **A** as follows:

$$\mathbf{A} = \mathbf{g}\mathbf{f}^T + \mathbf{g}\mathbf{f'}^T.$$

Presentation of either **f** or **f'** would evoke the correct output, **g**. If it were possible to present the sum (**f** + **f'**), we would also evoke the associated output, with larger amplitude. Often it is not possible to present the two items simultaneously: Suppose *A* gives rise to coding **f**, and *a* gives rise to coding **f'**. Presentation of the image of *A* superimposed on the image of *a* does not generally give the sum **f** + **f'** as the resulting input neural representation because of a number of effects in the initial stages of the visual system. If, however, the letters are presented in sequence or are spatially separated, generating a representation with less mutual interference, it seems intuitive that the name *ay* is indeed likely to be evoked as the strongest common association of two different state vectors, that is, the common response will be enhanced. We can use this technique computationally to disambiguate ambiguous or complicated concepts (Elio & Anderson, 1981; Hayes-Roth & Hayes-Roth, 1977), an application described elsewhere.

Therefore, almost as interesting a phenomenon as the enhancement of prototypes are the conditions when prototypes are not enhanced. We can predict when this occurs

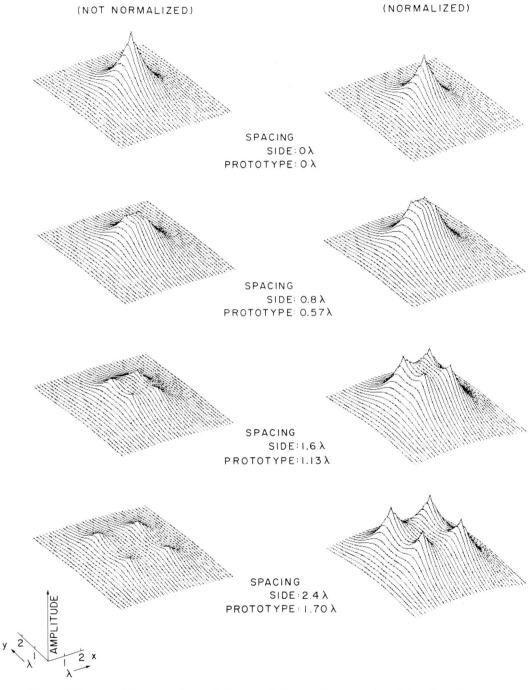

Figure 4. The sum of four exemplars each due to a single dot. (Demonstration of the effect of increasing spacing between dots. There is relatively greater representation of individual exemplars as spacing increases but still a significant buildup at the prototype location. The left-hand column gives the simple sum of the individual exemplar activity patterns; the right-hand column equates the maximum activity of each displacement, so shapes of curves can be compared. λ = the length constant).

with our stimuli with our model. Let us consider the surfaces presented in Figure 4. Note that in large distortions, though there is enhancement at the prototype location, the peaks representing the individual exemplars are larger than the value of the sum at the prototype location. In the smaller distortions, there may be an enhancement of the prototype over the individual exemplars.

Let us analyze the simple case presented in Figure 4. Four exemplars are presented equally spaced from the prototype location (0, 0) and from each other at locations (+x, 0), (0, +x), (0, −x), and (−x, 0). We assume the falloff function is exp(−r), where r is the distance from the peak, as before, with length constant at 1. Let us consider only a single dot location. Two dots are separated from it by a distance $2^{1/2}x$. The dot on the other side of the square is $2x$ distant. At the prototype location (a distance x from each exemplar),

the amplitude of the sum, s, is given by the following:

$$s(0, 0) = 4 \exp (-x).$$

At the location of, say, the exemplar at (0, x), the value of the sum is given by the following:

$$s(0, x) = 1 + 2 \exp (-2^{1/2}x) + \exp (-2x).$$

The sums at the other three dot locations are identical.

If x is greater than 0 and less than 0.692, there is actual enhancement of the sum at the prototype location, that is, the amplitude of the sum at (0, 0) is greater than at any dot location.

We are interested, though, in a more complex quantity: the relative sizes of the inner product between the prototype or an exemplar and the sum of old exemplars. This can be a complicated function because the shape of

MEMORY FORMATION

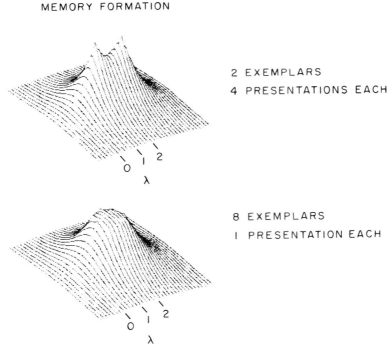

2 EXEMPLARS

4 PRESENTATIONS EACH

8 EXEMPLARS

1 PRESENTATION EACH

Figure 5. Demonstration of the qualitative difference between a "memory" sum constructed from two different exemplars and one constructed from eight different exemplars. (Exemplars in both cases had the same average separation from the prototype location. Note the very "lumpy" memory produced when only a few exemplars are stored. In this case, there is representation of item information; when many exemplars are stored, the prototype gives the largest response. λ = the length constant.)

the activity patterns due to a single dot are involved.

We can simulate a simple version of this situation. Let us assume that a number of exemplars are located equally spaced on the unit circle, with the prototype at the center of the circle. The number of exemplars in the simulations were varied from two to a large number. We varied the length constant and computed inner products between an input pattern located at the prototype location and at the location of one of the dots on the circle. Presented in Figure 6 is the ratio of the memory inner product at the prototype location and at the location of an exemplar.

It can be seen that when the distance to the exemplars is very small (in length constants), the values at prototype and exemplar locations are almost identical, because everything adds without significant falloff. This case corresponds to very small, average dot movement.

When the distance to the exemplars is very large (in length constants), activity from one exemplar falls to almost zero before it encounters activity from another exemplar. There is essentially no representation at the prototype location. This case corresponds to storage of single exemplars with no prototype formation. This corresponds in the experiments to very large, average movement of dots.

There is also a region of optimal prototype enhancement, which reaches a peak value around 20% greater than the value at the location of any individual exemplar. (On Figure 6, we plotted the best length constant, 11.3 screen units, derived from the similarity measures of Experiment 1.)

Category Formation Without Distinct Responses

It has been reported in the literature (Fried, 1979) that it is possible to detect the presence of categories without specifically associating each set of exemplars with a different response. This can easily be demonstrated with a program to generate random dot prototypes and exemplars of the kind used in our experiments. Informally, after looking at patterns for a while, it becomes obvious that sets of patterns belong together.

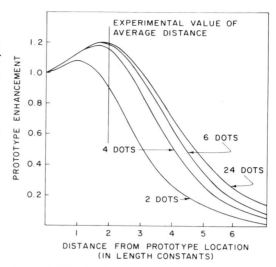

Figure 6. This simulation assumed varying numbers of dots (from 2 to 24) were arranged on the unit circle. (The prototype was located at the center of the circle. The graph gives the ratio of the system response to a dot at the prototype location and the system response to a dot at the location of an old exemplar. The y axis gives the radius of the circle in length constants. The average displacement of patterns used in Experiment 2 is indicated.)

Let us assume that the response for all of the patterns consists of some nonzero output pattern $\mathbf{g}$. Suppose we have three different clusters, each constructed from n exemplars of a prototype, where $\mathbf{p}_1$, $\mathbf{p}_2$, and $\mathbf{p}_3$ are the three prototypes. We form the matrix $\mathbf{A}$ exactly as described previously, and as long as the distortions are not very large, we have shown the matrix $\mathbf{A}$ is approximated by the following:

$$\mathbf{A} = n\mathbf{g}(\mathbf{p}_1 + \mathbf{p}_2 + \mathbf{p}_3)^T.$$

The amplitude of response (length of the output vector) to an input vector $\mathbf{f}$ will be proportional to the sum of the inner products:

$$\mathbf{Af} = n([\mathbf{p}_1, \mathbf{f}] + [\mathbf{p}_2, \mathbf{f}] + [\mathbf{p}_3, \mathbf{f}])\mathbf{g}.$$

Response classes may therefore become detectable as local maxima in the amplitude of response.

Local maxima can be seen to develop for dot patterns using the calculation we performed in the previous section. Consider the storage of very highly distorted exemplars, that is, when the average displacement of a dot from its prototype is very large compared with the length constant. If we identify highly distorted exemplars with different prototypes, all of these prototypes are now associated with the same output activity pattern, by assumption, a consistent nonzero activity pattern. As we have just shown (in Figure 6), the system now only responds strongly to the well-separated prototypes and responds poorly to other (new) patterns.

The two-exemplar example in Figure 5 and the largest displacements displayed in Figure 4 show such local maxima developing.

Let us make two additional comments. Although we do not discuss it further here, unsupervised development of response selectivity can be demonstrated in refinements of the models described earlier and forms the basis of some of the neural models of the development of visual cortical selectivity that have been discussed at length elsewhere (see Cooper, 1981; Cooper, Liberman, & Oja, 1979).

Also, the nonlinear categorizer mentioned earlier as an extension of the simple linear model presented (Anderson et al., 1977) will develop response selectivity based on similarity relations between members of a stimulus set without supervision and without requiring different responses for each different member of the set. This has been demonstrated in numerous computer simulations. (See Anderson & Mozer, 1981, for a detailed description of such a simulation.)

General Comments on Experiments 1 and 2

We used stimuli composed of random patterns of dots. The studies took place in Brown University's Human Learning Laboratory (Millward, Aikin, & Wickens, 1972). The experiments were controlled by a Digital Equipment Corporation PDP-8/e minicomputer. The dot-pattern stimuli appeared on Tektronix 502 oscilloscopes. The display area of each oscilloscope screen measured 10 × 10 cm and was divided into a 512 × 512 unit grid. This grain was determined by the digital-to-analog converters used to transmit the stimuli from the computer to the oscilloscopes. The intensity of the oscilloscope beam was adjusted so that the stimulus dots

were as small as possible while remaining clearly visible.

All of the experiments involved two types of stimuli: prototypes and exemplars. Each prototype pattern was created by randomly placing nine dots within a 300 × 300 unit grid centered in the 512 × 512 unit display area. Exemplars of a prototype were constructed by displacing the prototype dots short distances. For each dot, the direction of motion was chosen at random from a uniform distribution on the range 0°–360°. The distance moved by a dot was determined by a probability distribution specified by the experimenters.

Shown in Figure 2 are a prototype and five exemplars at increasing levels of distortion. Each exemplar is the result of moving the prototype dots various distances drawn from a normal distribution. The mean of the distribution (in display screen units) is given above each exemplar. The standard deviation of the distribution was one third of the mean in all cases. Because the prototypes were generated toward the middle of the display screen, the dots had ample room to move. The computer halted at the border any dots that attempted to stray beyond the confines of the display area, but such events were rare, even for the largest displacements used.

The subjects were paid volunteers from the Brown University Psychology Department's pool of students and staff. The subjects sat in dimly lit booths. The display screens were viewed through openings in the front wall of each booth, approximately 75 cm from the subjects; the dot patterns thus subtended about 6° of visual angle. Subjects indicated their responses by pressing keys on teletype keyboards located below and in front of the display windows. A row of computer-operated lights mounted on the keyboards provided feedback when required.

Experiment 1: Similarity Between Two Dot Patterns

It is apparent from Figure 2 that the similarity between an exemplar and its prototype decreases as the dot displacements used to create the exemplar increase. In Experiment 1, subjects judged the similarity of prototype–exemplar pairs at several levels of distortion.

Method

Subjects. Ten members of the Brown University community received $3 for participating in a single, hour-long session. Subjects were unfamiliar with the stimuli and were unaware of the purpose of the experiment. All subjects were tested individually.

Stimuli and design. Subjects viewed pairs of sequentially presented dot patterns. The first member of each pair was a prototype generated as described previously, and the second was an exemplar derived from that prototype. The terms *prototype* and *exemplar* are arbitrary in this context because a new prototype was generated for each trial, and no prototype was repeated. A new set of stimuli was created for each subject.

We manipulated the distortion used to create the exemplar patterns as follows: Several levels of distortion were obtained by moving all the dots in the prototype pattern the same average distance—15, 30, 45, 60, or 90 units—in random directions. The average movement of a dot varied slightly from pattern to pattern.

Each subject rated 40 different prototype–exemplar pairs at each level of average distortion and rated 20 patterns in which the two patterns were identical. Trials were divided into two blocks of 100, separated by a short rest period. Within a trial block, there were 20 pairs at each distortion level, ordered randomly for each subject.

Procedure. Subjects were told they would be shown a series of dot-pattern pairs, presented sequentially. They were instructed to rate each pair on an 8-point similarity scale, ranging from *highly similar* (1) to *very different* (8), by pressing the appropriate teletype key.

The first member of each pair was presented for 5 s, followed by a 1-s interstimulus interval (ISI), followed by the second pattern, which remained visible until the subject responded. A 2-s interval followed each response, during which time a keyboard light indicated that the subject should prepare for the next trial.

At the start of the experimental session, each subject rated 33 practice trials, 3 at each distortion level in random order, to become familiar with the procedure and with the range of similarities.

Results

Plotted in Figure 7 are the mean similarity ratings for each level of average dot movement in the distortion. The identical patterns received a significantly larger rating than the minimum possible value of 1.0, $z(200) = 2.02$, although the largest mean rating (6.76 for the most distorted variable-distance pairs) was substantially less than the theoretical maximum of 8.0.

Discussion and Theory

Computer simulations of the similarity experiments. In the computer simulation, we generated patterns like those used in the

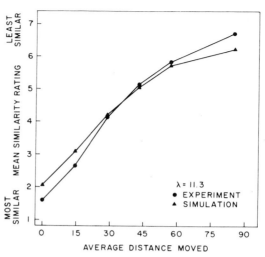

Figure 7. Mean similarity ratings as a function of average displacement. (λ = the length constant.)

experiments, distorted the patterns, computed the inner product (measure of similarity), and then computed a measure of goodness of fit of the inner product to the experimental data.

We had 81 pairs of dots (nine dots taken two at a time, one from a pattern and the other from its distortion) between which we computed inner products. In the computer program, we formed a table of distances between dots. Note that 9 members of the table of distances represent distances between a dot and the displacement of that dot in the distortion, and the other 72, the distance between a dot and a different dot (cross terms). If average displacement were small, we would expect most of the contribution to the inner product to be concentrated among particular dots and their displacements. As the displacement increased, we should expect more contribution from the cross terms representing interactions between different dots. (That is, the pattern is starting to mix up as the dot displacement gets large.) The largest displacements used in our experiments had a significant contribution from terms due to different dots. Using the best-fitting length constants computed from the similarity experiments, the component of the inner product due to a dot and its shifted counterpart was roughly equal to the sum of the inner products of the dot with the other eight dots in the distortion for the largest distortions.

By assumption, the magnitude of the inner product between activity patterns is directly related to similarity. The only free parameter in the simulation was the length constant of the exponential falloff of activity.

Mathematical representations of the patterns were constructed that were statistically identical to those actually presented. For the similarity simulation, a dot pattern and a distortion were generated, and the inner product of the neural codings of the two patterns were computed. The experimental and theoretical values were compared.

For reasons now obscure, experimental subjects were required to use low ratings for high similarity, and high ratings for low similarity. We assumed a linear transformation existed between experimental rating and computed inner product. Several measures of goodness of fit were used in the computations. First, the best-fitting straight line (in the sense of linear regression) was computed between predicted similarity and experimental similarity. The length constant that minimized the mean square distance between this line and the experimental data was found. Second, the correlation between predicted and experimental values was computed.

The results of the two measures of fit were close to each other: The length constant was 11.3 for the minimum mean square distance and was 14.5 for best correlation. The maxima of the relations between length constant and the measures of goodness of fit were quite broad. The average, best-fitting length constant for minimizing distance between computed and experimental similarity was 11.3 screen units (mean square distance of 0.210), producing an average correlation of 0.97.

Several functions representing falloff of activity due to a single dot were investigated at various times: Gaussians, exponentials, laterally inhibited functions, and others. The best fits were obtained with simple exponentials, which is why this function is used in the simulations and in the figures. Exact shape of the falloff was not critical, that is, the pattern of fits obtained was similar for different falloffs, but best fits were not quite as good numerically.

Experimental and theoretical fits are plotted on the graph in Figure 7. Values of parameters were those used for the smallest mean square distance between simulated and experimental values. The best-fitting length constant was computed for five different sequences of patterns. Each sequence was generated from a different, random-number generator seed.

Inspection of Figure 7 shows the fit is quite close, but there are systematic deviations between predicted and experimental data. We feel that by modifying the shape of the falloff function, we could improve fit somewhat, but to do this and test it properly would require more ambitious experiments.

Experiment 2: Number of Exemplars

Experiment 1 demonstrated reasonable agreement between the distributed-memory model's predictions and subjects' behavior when asked to rate the similarity between two dot patterns. We next sought to test this same model's predictions for a true categorization task.

In Experiment 2, subjects learned three categories containing 1, 6, and 24 members, respectively. The learning stimuli were exemplars of one of three prototypes, and the degree of distortion between each exemplar and its prototype was held constant. Only the number of learned stimuli varied across the categories. Subjects learned the categories by classifying the exemplars and by receiving feedback about their decisions. During this process, the categories were presented with equal frequency to avoid biasing subjects' classifications. Following the learning phase of the experiment, subjects classified (without feedback) a series of dot patterns that included the training stimuli again, new exemplars created from the same category prototypes, and the prototypes themselves.

Method

Subjects. Twenty-one paid subjects from the same pool drawn on in Experiment 1 participated in a single, 20-min session. To ease data analysis, the subjects were tested in groups of 3. Within each group, all 3 subjects saw the same stimuli in the same random order.

Stimuli and design. The experiment had a learning phase, in which subjects classified category exemplars with feedback, and a testing phase, in which they classified old, new, and prototype patterns without receiving feedback. For each group of subjects, three prototypes (designated A, B, and C) were constructed. Exemplars of the prototype were created by the method described earlier. The dot displacements were drawn from a normal distribution, with $M = 24$ and $SD = 8$. This distance was chosen in pilot studies because it seemed to maximize the experimental prototype enhancement. The conditions giving rise to the greatest prototype enhancement were discussed in the section "When is a Prototype Not Extracted (and Why)," and this distance was close to the maximum.

The learning stimuli consisted of 1 exemplar of Prototype A displayed 24 times, 6 exemplars of Prototype B displayed 4 times each, and 24 exemplars of Prototype C shown only once each, for a total of 72 learning trials, with each category represented 24 times during learning. In the testing phase, the single, learned exemplar from Category A was presented 8 times, 4 of the old exemplars from Category B were presented twice each, and 8 of the old C exemplars were each presented once. In addition, 8 newly constructed exemplars of each prototype were presented once each, and each prototype was shown 8 times. The subjects had encountered neither the prototypes nor the new exemplars during learning. Finally, 9 unrelated control patterns (newly generated prototypes) were presented once each for a total of 81 test trials: 8 old exemplar trials, 8 new exemplar trials, and 8 prototype trials for each category, plus the 9 control trials.

Procedure. The subjects were told that they would be shown a series of dot patterns and that their task would be to determine which patterns were to be grouped together under the same response. Subjects were instructed to classify each pattern by pressing one of three teletype keys. The keys were randomly paired with the categories at the start of the session.

Stimuli were presented one at a time until all 3 subjects had responded or for a maximum of 5 s. In the learning phase, subjects were urged to respond during the display interval, if possible, but not to rush their responses. Each stimulus presentation was followed by a 5-s feedback interval, during which the correct response for that trial was indicated by illuminating a light above the appropriate response key.

In the testing phase, a 5-s blank ISI followed each stimulus presentation. Subjects were instructed to respond as quickly as possible, based on what they had learned in the first part of the experiment, and were told that no feedback would be given.

Results

The data from two subject groups had to be discarded due to equipment failure.

The mean number of errors made in the learning phase was 15.1 (range = 4–43). For Categories A, B, and C, these means were 2.8, 6.2, and 6.1, respectively (ranges = 0–16, 0–17, and 1–10). Subjects made significantly fewer errors learning the one-exemplar category than learning the other two categories ($S = 7.3$, on a Friedman test); several subjects made only 1 error on this category during the entire learning phase. This advantage did not transfer to the testing phase, where the overall classification accuracy was 88%

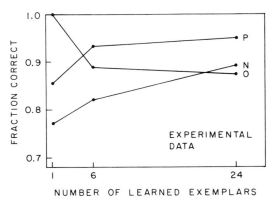

Figure 8. Percentage of correct test trial classifications for Experiment 2. (N = new exemplars, O = old exemplars, and P = prototypes.)

for Categories A and B, and 91% for Category C.

Shown in Figure 8 is the percentage correct classification for each test trial combination of stimulus type (old, new, prototype) and number of learned exemplars (1, 6, 24) collapsed across subject groups. A ceiling effect is apparent for the single, old exemplar in the one-instance category: All the subjects classified this pattern correctly all the time.

An analysis of variance (ANOVA) was performed on the classification data, treating stimulus type and number of learned exemplars as within-subjects variables and treating group as a between-subjects variable. The main effect of type was significant, $F(2, 20) = 7.21$, $MS = 0.21$, $p < .05$, but the main effects of number and group were not (both Fs < 1). The Type × Number interaction reached significance, $F(4, 40) = 3.99$, $MS = 0.22$, $p < .05$; performance on the old exemplars decreased with larger numbers of learned items, whereas performance on the new and prototype instances increased. The prototypes were always classified more accurately than were other novel exemplars by about the same margin. No other interactions approached significance.

Reaction time (RT) data were collected to provide convergent support for conclusions drawn from the classification data. Previous studies using dot-pattern stimuli (Homa, Cross, Cornell, Goldman, & Shwartz, 1973; Posner & Keele, 1968; see also Omohundro & Homa, 1981) have reported that response speed tends to be correlated with classification accuracy. The RT data, summarized in Figure 9, generally confirm the pattern of the error data.

Subjects assigned the unrelated control patterns to the 1-, 6-, and 24-instance categories 5.9%, 31.1%, and 63.0% of the time, respectively. This corroborates the finding of Homa et al. (1973) that subjects tend to assign random patterns to the category containing the largest number of learned instances. The mean RT for the control patterns was 1,556 ms, about 350 ms longer than for the other stimuli. Clearly, these patterns are recognized as not belonging to any of the three learned categories. This finding provides reassurance that prototypes chosen at random differ from one another a good deal, so there is little category overlap. This was the subjective impression of the experimenters and seems to be consistent with the data.

Discussion and Theory

The results of Experiment 2 support the qualitative predictions of the distributed memory model. Although varying the number of learned exemplars did not substantially affect subjects' overall accuracy in classifying test stimuli, it did affect their relative accuracies for the different types of category instances. Old exemplars were classified more accurately for the smallest category, and a prototype advantage was apparent for the largest category. We proposed a specific form for the activity patterns representing dot patterns and experimentally derived a length constant in Experiment 1, so we can model Experiment 2 without making any additional assumptions.

Computer simulation of prototype extraction. Our computer program is a straightforward realization of the distributed memory model. Prototype dot patterns were generated randomly, with parameters identical to those used in the experiments. Exemplars were formed according to the rules followed in the experiments. Sums of exemplars were constructed, and a "memory" was formed. If there were three prototypes in a particular experiment, then three sums were constructed and were kept separate. This rule was used instead of generating a response with the full matrix model because (a) the selectivity of

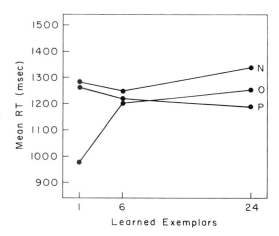

Figure 9. Correct reaction times (RT) for classification for Experiment 2. (N = new exemplars, O = old exemplars, and P = prototypes.)

the system was given by the similarities of the inputs, (b) we did not know how best to represent the outputs, (c) both experimental and theoretical results suggested that different prototypes were sufficiently different from each other so that they did not interact significantly, and (d) we could greatly ease our computational demands by only looking at the inner products. An input was classified as to which sum it was most similar to by computing the inner product with each sum in turn and by choosing the largest. The sum with greatest similarity was the classification of the input, and the associated response was assumed to occur.

In the experiments described previously, we used groups containing 1, 6, and 24 exemplars of particular prototypes. It may be pointed out, with justice, that a prototype cannot be formed when only a single exemplar is presented. This is, of course, correct. However the model will generate similarities for new exemplars and the prototype for this case just as it will when many exemplars are presented, so we should be able to predict the responses in this special case with the same model we used for multiple exemplars. A model for generalization is part of the distributed memory model and should be predictable as a special condition of a prototype experiment.

The statistics of the patterns used in the simulations were identical in all respects to those in the real experiments. The only parameter to vary was the length constant of the falloff of the individual, dot-activity patterns. A total of 50 different sets of 24 patterns was used for each value of length constant.

The length constant found for best fit in the similarity experiment was 11.3. Shown in Figure 10 are the results when this value was used in the prototype simulation program. Although there is no way to make a direct mapping of inner product into percentage of correct responses without numerous additional assumptions, there should be a monotonic relation between them. The simulations shown in Figure 10 seem to show qualitative similarity to the experimental results in Figure 8. Note, in both figures, the crossing over of the responses to old exemplars shown for the 6-exemplar case and the coming together of the responses to old and new exemplars in the 24-exemplar case. There is a strong response to the single, old exemplar in the 1-exemplar case, but the responses to the prototype and new exemplars for this (seemingly different) case fall in line with the other data.

As discussed earlier, prototypes are not necessarily enhanced relative to old and new exemplars. By changing the length constant, we can demonstrate this effect. If the length constant is very small relative to the dot displacement, there is little prototype enhancement, and the system primarily responds to old exemplars, even when many exemplars are learned. If the length constant is large relative to dot displacement, the system responds to old, new, and prototype exemplars almost identically, and again, there is little differential enhancement of the prototype. Maximum prototype enhancement is found for intermediate length constants. Shown in Figure 11 are simulations of Experiment 2 for various length constants from small (5) to large (20).

General Discussion

We have attempted to account for some aspects of human categorization with a model whose primary assumption holds that memory traces can add together in storage. The agreement obtained between theory and data suggests that a neurally inspired explanation

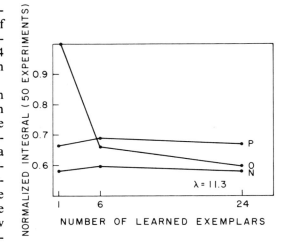

Figure 10. Simulation of Experiment 2 using the best-fitting length constant (λ) determined in the similarity experiment, Experiment 1. (cf. Figure 8.)

of categorization deserves consideration along with rival models. However, two difficulties lie in the way of attempts to compare the distributed memory model with other models. First, in order for the distributed memory model to generate specific predictions, one must make assumptions about how stimuli might be represented. We have done this for dot patterns but have not as yet done so for more commonly used stimuli. Second, it is notoriously difficult to get the various categorization models to make differential predictions (Hintzman & Ludlam, 1980; Smith & Medin, 1981, p. 182). Indeed, several different models can be made to predict the experimental results reported here.

Consider a pure prototype model that stores only the average of learned instances and that classifies novel stimuli according to their similarity to the average. Such a model predicts the results of Experiment 2 as follows. For the one-item category, the stored average is simply the lone, learned item. Naturally, this pattern will be classified most accurately on a subsequent test. The objective prototype will have an advantage over other new patterns, being more similar to the learned pattern than a randomly selected exemplar (in our experimental patterns, 25.2 screen units average displacement vs. 38.5 screen units). As category size increases, the pure prototype model predicts that classification of the prototype should improve, because the computed average matches the actual prototype more and more closely. Meanwhile, accuracy for the old and new exemplars should converge because they are equally similar to the prototype.

Although it is possible to make differential predictions between the prototype and exemplar models simply based on consideration of some special cases, the most telling differences arise when one considers the effective, dual representation of exemplars and prototype found in Figure 5, say. Here, the response to a dot at the prototype location will be identical for two exemplars presented four times each and for eight exemplars presented once. The inner product in this case will depend only on the eight distances between the center and the exemplar and will not depend on the exact arrangement of the exemplars. However, the response to an old exemplar will be greatly different in the two

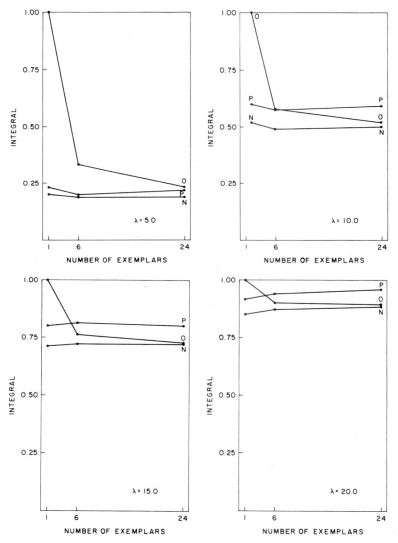

Figure 11. Simulations of Experiment 2 with various length constants (λ), showing the shift from learning of specifics about old exemplars (small length constants) to prototype enhancement (with more presented exemplars and intermediate length constants) to lack of differentiation between old examples, new examples, and the prototype (very large length constant).

cases: relatively larger when the number of exemplars is smaller. A simple prototype model would not predict this because the average location is identical in both cases, and the displacement from the average location of an old exemplar is the same.

Exemplar models, too, can be used to explain the present findings. As a trivial example, one can construct a model that stores learned exemplars separately yet classifies test items using a similarity computation equivalent to the inner product function proposed here. It is difficult to conceive of a theoretical motivation for such a model. One well-motivated exemplar model that also predicts some of the present results is the context model (Medin & Schaffer, 1978). The context model computes the similarity between two items as the product of their similarity values along one or more featural dimensions. It can judge two nearly identical stimuli as being dissimilar if they differ in one highly salient attribute. For the distributed-memory model, the analogous situation occurs when corresponding elements in two activity patterns are large compared with other elements and have opposite signs, thus contributing negatively to the inner product. It is difficult to compare the two models further, because they assume such different forms of mental representation.

The context model was prompted by Medin and Schaffer's finding that a test item's similarity to learned exemplars was a more important determinant of subjects' classification accuracy than the item's similarity to the category prototype, a result that argues against a pure prototype model. Because average old–new similarity in Experiment 2 was correlated with category size, the context model can explain the observed effects of manipulating the number of learned exemplars, too. However, it is not presently clear that old–new similarity is always a determining factor of accuracy in categorization tasks. Homa, Sterling, and Trepel (1981) trained subjects to categorize highly distorted exemplars and explicitly manipulated the similarity between old and new exemplars. Although classification accuracy for new patterns depended strongly on their similarity to old exemplars, the effect of old–new similarity was greater for small categories (5 old exemplars) than for large ones (20 old exemplars), a prediction

of the distributed memory model. This effect was especially pronounced for the category prototypes, which were progressively enhanced in the larger categories, as was also found in our Experiment 2. This finding was not well predicted by the context model.

Homa et al. (1981) interpreted their results as support for a mixed model of categorization, in which a category's representation includes both abstracted prototypes and individual exemplars, with their relative weights determined by a number of aspects of the experimental situation. Two of these aspects, category size and within-category variability, are the primary determinants of exemplar versus prototype dominance in the distributed memory model. Perhaps the distributed memory model may best be considered as a particular embodiment of the mixed model proposed by Homa et al. (1981), because it both stores instances, and given suitable input, abstracts a prototype as a direct consequence of the memory storage operation.

Even without definitive experimental validation, the approach we have outlined can draw support from the following observations: (a) The same model has been successfully applied to a variety of other cognitive behaviors, (b) the model is robust with respect to the details of the representation of stimuli, (c) the model correctly predicts the quantitative (Experiment 1) and qualitative (Experiment 2) results of the experiments presented here and does so using a single value of its one adjustable parameter. We hope this discussion raises the possibility that theoretical investigations of neural processes may shed light on psychological theory.

References

Abramowitz, M., & Stegun, I. A. (Eds.). (1964). Handbook of mathematical functions. *National Bureau of Standards applied mathematics series: No. 55.* Washington, DC: U.S. Government Printing Office.

Anderson, J. A. (1973). A theory for the recognition of items from short memorized lists. *Psychological Review, 80,* 417–438.

Anderson, J. A. (1977). Neural models with cognitive implications. In D. LaBerge & S. J. Samuels (Eds.), *Basic processes in reading* (pp. 27–90). Hillsdale, NJ: Erlbaum.

Anderson, J. A. (1983). Neural models for cognition. *IEEE transactions on system, man, and cybernetics, SMC-13,* 799–815.

Anderson, J. A., & Hinton, G. E. (1981). Models of information processing in the brain. In G. E. Hinton

& J. A. Anderson (Eds.), *Parallel models of associative memory* (pp. 9–48). Hillsdale, NJ: Erlbaum.

Anderson, J. A., & Mozer, M. C. (1981). Categorization and selective neurons. In G. E. Hinton & J. A. Anderson (Eds.), *Parallel models of associative memory* (pp. 213–236). Hillsdale, NJ: Erlbaum.

Anderson, J. A., & Silverstein, J. W. (1978). Reply to Grossberg. *Psychological Review, 85,* 597–603.

Anderson, J. A., Silverstein, J. W., Ritz, S. A., & Jones, R. S. (1977). Distinctive features, categorical perception, and probability learning: Some applications of a neural model. *Psychological Review, 84,* 413–451.

Brooks, L. (1978). Non-analytical concept formation and memory for instances. In E. Rosch & B. Lloyd (Eds.), *Cognition and categorization* (pp. 169–211). Hillsdale, NJ: Erlbaum.

Collins, A. M., & Loftus, E. F. (1975). A spreading-activation theory of semantic processing. *Psychological Review, 82,* 407–428.

Cooper, L. (1981). Distributed memory in the central nervous system: Possible test of assumptions in visual cortex. In F. O. Schmitt, F. G. Worden, G. Adelman, & S. G. Dennis (Eds.), *The organization of cerebral cortex* (pp. 479–503). Cambridge, MA: MIT Press.

Cooper, L. N., Liberman, F., & Oja, E. (1979). A theory for the acquisition and loss of neuron specificity in visual cortex. *Biological Cybernetics, 33,* 9–28.

Eich, J. M. (1982). A composite holographic associative recall model. *Psychological Review, 89,* 627–661.

Elio, R., & Anderson, J. R. (1981). The effects of category generalizations and instance similarity on schema abstraction. *Journal of Experimental Psychology: Human Learning and Memory, 7,* 397–417.

Franks, J. J., & Bransford, J. D. (1971). Abstraction of visual patterns. *Journal of Experimental Psychology, 90,* 65–74.

Fried, L. S. (1979). *Perceptual learning and classification with ill-defined categories* (Tech. Rep. No. MMPP 79-6). Ann Arbor: University of Michigan, Psychology Department.

Gradshteyn, I. S., & Ryzhik, I. M. (1965). *Table of integrals, series, and products* (4th ed.). New York: Academic Press.

Hayes-Roth, B., & Hayes-Roth, F. (1977). Concept learning and the recognition and classification of exemplars. *Journal of Verbal Learning and Verbal Behavior, 16,* 321–338.

Hebb, D. O. (1949). *The organization of behavior.* New York: Wiley.

Hinton, G. E. (1981). Implementing semantic networks in parallel hardware. In G. E. Hinton & J. A. Anderson (Eds.), *Parallel models of associative memory* (pp. 161–187). Hillsdale, NJ: Erlbaum.

Hinton, G. E., & Anderson, J. A. (Eds.). (1981). *Parallel models of associative memory.* Hillsdale, NJ: Erlbaum.

Hintzman, D. L., & Ludlam, G. (1980). Differential forgetting of prototypes and old instances: Simulation by an exemplar-based classification model. *Memory & Cognition, 8,* 378–382.

Homa, D., Cross, J., Cornell, D., Goldman, D., & Shwartz, S. (1973). Prototype abstraction and classification of new instances as a function of the number

of instances defining the prototype. *Journal of Experimental Psychology, 101,* 116–122.

Homa, D., Sterling, S., & Trepel, L. (1981). Limitations of exemplar-based generalization and the abstraction of categorical information. *Journal of Experimental Psychology: Human Learning and Memory, 7,* 418–439.

Kohonen, T. (1977). *Associative memory: A system theoretic approach.* Berlin: Springer-Verlag.

Korn, G. A., & Korn, T. M. (1968). *Mathematical handbook for scientists and engineers* (2nd ed.). New York: McGraw-Hill.

Levy, W., Anderson, J. A., & Lehmkuhle, W. (Eds.). (1984). *Synaptic change in the nervous system.* Hillsdale, NJ: Erlbaum.

Medin, D. L., & Schaffer, M. M. (1978). Context theory of classification learning. *Psychological Review, 85,* 207–238.

Mervis, C. B., & Rosch, E. (1981). Categorization of natural objects. *Annual Review of Psychology, 32,* 89–115.

Millward, R. B., Aikin, J., & Wickens, T. D. (1972). The Human Learning Laboratory at Brown University. In, *Computers in the psychological laboratory* (Vol. 2, pp. 35–48). Maynard, MA: Digital Equipment Corporation.

Minsky, M., & Papert, S. (1969). *Perceptrons.* Cambridge, MA: MIT Press.

Murdock, B. B., Jr. (1982). A theory for the storage and retrieval of item and associative information. *Psychological Review, 89,* 609–626.

Neisser, U. (1967). *Cognitive psychology.* New York: Appleton-Century-Crofts.

Nelson, K. (1974). Concept, word and sentences: Interrelations in acquisition and development. *Psychological Review, 81,* 267–285.

Omohundro, J., & Homa, D. (1981). Search for abstracted information. *American Journal of Psychology, 9,* 324–331.

Posner, M. I., & Keele, S. W. (1968). On the genesis of abstract ideas. *Journal of Experimental Psychology, 77,* 353–363.

Posner, M. I., & Keele, S. W. (1970). Retention of abstract ideas. *Journal of Experimental Psychology, 83,* 304–308.

Reed, S. K. (1972). Pattern recognition and categorization. *Cognitive Psychology, 3,* 382–407.

Robbins, D., Barresi, J., Compton, P., Furst, A., Russo, M., & Smith, M. A. (1978). The genesis and use of exemplar vs. prototype knowledge in abstract category learning. *Memory & Cognition, 6,* 473–480.

Smith, E. E., & Medin, D. L. (1981). *Categories and concepts.* Cambridge, MA: Harvard University Press.

Smith, E. E., Shoben, E. J., & Rips, L. J. (1974). Structure and processing in semantic memory: A feature model for semantic decision. *Psychological Review, 81,* 214–241.

Strange, W., Keeney, T., Kessel, F. S., & Jenkins, J. J. (1970). Abstraction over time from distortions of random dot patterns—a replication. *Journal of Experimental Psychology, 83,* 508–510.

37
Introduction

(1984)
Stuart Geman and Donald Geman

Stochastic relaxation, Gibbs distributions, and the Bayesian restoration of images
IEEE Transactions on Pattern Analysis and Machine Intelligence PAMI-6:721–741

There are a number of significant ideas in this difficult but important paper by two brothers, Stuart and Donald Geman. It is frequently cited in the literature on simulated annealing, and also in the Boltzmann machine literature on neural networks. It is not, strictly speaking, a neural network paper, though the theoretical connections are very close.

Geman and Geman use the problem of image processing as a basis for their examples, but start the paper with an abstract discussion of how to work with complex, probabilistic systems.

One of the most important ideas in the study of the nervous system, and in neural networks, and one that we have seen before, is that the brain is designed to function in a particular environment. As William James put it in *Psychology* (*Briefer Course*), "... our inner faculties are *adapted* in advance to the world in which we dwell" (p. 3). This makes the nervous system sometimes quite special-purpose in its architecture, and makes the specific representation of the information to be processed important, and problem dependent. In Marr and Poggio (paper 20) we saw how the incorporation of assumptions about the real world allowed a connection matrix to be constructed and used for combining images from the two eyes to form a stereoscopic depth map. Some configurations of objects in the world were much more likely than others, and the structure of the nervous system should reflect that structure, and use it.

Geman and Geman base their analysis on this idea: the world has structure, and some configurations of the image are more likely than other configurations. Since images in computers are represented as pixels (picture elements), many of these constraints can be expressed as constraints on local areas of the image. For example, suppose we know that objects tend to have smooth surfaces. This implies that nearby points will tend to have the same value of brightness, although at edges there may be discontinuities. But since there are many more pixels involved in representing the surface than the edges, similarity between adjacent pixels is more probable than big differences.

Geman and Geman also suggest a *line process*, which falls "in between" the processes that govern relations between individual pixels. Nearby pixels tend to have the same properties, since they lie on the same surface, *unless* an edge comes between them. If an edge is between them, then they need have (in this model) no necessary relation at all. So the line process probabilistically decouples pixels that lie on opposite sides of the line. Lines or edges have their own constraints as well; lines rarely end blindly and frequently run straight for long distances.

Geman and Geman assume that, through experience, or design, the probabilistic relations between a pixel and its neighbors are known. Suppose we are given a noisy image. The algorithms proposed in the paper suggest how the underlying noise-free image can be estimated. The noise is assumed to be random. So the problem is to find the image that is most probable, given the input. This is called the "maximum a posteriori" (MAP) estimate. The probabilistic constraints on the neighborhoods are referred to as a "Markov Random Field" (MRF). Markov Random Fields capture prior knowledge about the world by describing probabilities of particular patterns of relations between the pixels.

The difficult computational problem is to use the local probabilities to generate the globally most probable image. Some pixels have unlikely values, given the rules assumed for images, and are probably noise; others agree with their neighbors and are probably correct. This problem becomes very much like the situation found in simulated annealing (paper 33), where the totality of interactions among individual units determines a global energy and the problem is to find the minimum global energy. Here, the problem is to find the image with maximum probabilities, given the initial image. A search is impossible because of the huge number of possible states of the system.

Suppose we consider a single pixel. If we changed the value of the pixel so that it increased its local probabilities, then we would have a strict gradient ascent procedure in probability. This would be subject to the problems with local maxima that we have seen before. If, however, there existed a global parameter, T (for Temperature) which had the same effect as temperature in a physical system, we would be able to use exactly the same technique that was used in simulated annealing. We would then have the ability to choose *less* probable configurations occasionally and avoid local maxima.

There is a powerful theorem that makes an explicit connection between the probabilities in the Markov Random Field formulation and the Gibbs distribution, which characterizes physical systems with energies and temperatures. The theorem shows that it is possible to choose an *energy* function to realize the desired local probabilistic relations if the system is a Markov Random Field. The Gibbs distribution is followed by thermodynamic physical systems and is the same as what was called the "Boltzmann probability factor" in Kirkpatrick, Gelatt, and Vecchi (paper 33). Once the energy function is used, the close connection with the 'energies' in neural networks of the type used by Hopfield (paper 27) or by Ackley, Hinton, and Sejnowski (paper 38) becomes clear. In Geman and Geman's words, the system using energy "... represents an *imaginary* physical system whose lowest energy states are exactly the MAP estimates of the original image, given the degraded 'data'" (p. 734). It is generally much easier and more intuitive to construct energy relations than to construct equivalent probabilistic relations.

Once an energy is defined, annealing can be used; that is, the process starts at a high temperature, where choosing less probable configurations occurs often, and gradually decreases in temperature until changes in configuration always increase the local probabilities.

Besides pointing out the important connection between the MAP estimates and the Gibbs distribution, and how to exploit it, an important result about annealing is proved in this paper. Until Geman and Geman were able to do the detailed analysis, annealing schedules were ad hoc. They usually worked, but it was not clear why.

Geman and Geman showed that if the temperature at the kth step, $T(k)$, was kept above the value

$$T(k) \geqslant \frac{c}{\log(1 + k)},$$

then the system will converge to the minimum energy distribution.

There is a rich variety of other topics in this paper, treated with a degree of mathematical rigor unlike other papers in this collection. The computer simulations show how effective this approach can be in working with image data.

(1984)
Stuart Geman and Donald Geman

Stochastic relaxation, Gibbs distributions, and the Bayesian restoration of images
IEEE Transactions on Pattern Analysis and Machine Intelligence PAMI-6:721–741

Abstract—We make an analogy between images and statistical mechanics systems. Pixel gray levels and the presence and orientation of edges are viewed as states of atoms or molecules in a lattice-like physical system. The assignment of an energy function in the physical system determines its Gibbs distribution. Because of the Gibbs distribution, Markov random field (MRF) equivalence, this assignment also determines an MRF image model. The energy function is a more convenient and natural mechanism for embodying picture attributes than are the local characteristics of the MRF. For a range of degradation mechanisms, including blurring, nonlinear deformations, and multiplicative or additive noise, the posterior distribution is an MRF with a structure akin to the image model. By the analogy, the posterior distribution defines another (imaginary) physical system. Gradual temperature reduction in the physical system isolates low energy states ("annealing"), or what is the same thing, the most probable states under the Gibbs distribution. The analogous operation under the posterior distribution yields the maximum *a posteriori* (MAP) estimate of the image given the degraded observations. The result is a highly parallel "relaxation" algorithm for MAP estimation. We establish convergence properties of the algorithm and we experiment with some simple pictures, for which good restorations are obtained at low signal-to-noise ratios.

Index Terms—Annealing, Gibbs distribution, image restoration, line process, MAP estimate, Markov random field, relaxation, scene modeling, spatial degradation.

I. Introduction

THE restoration of degraded images is a branch of digital picture processing, closely related to image segmentation and boundary finding, and extensively studied for its evident practical importance as well as theoretical interest. An analysis of the major applications and procedures (model-based and otherwise) through approximately 1980 may be found in [47]. There are numerous existing models (see [34]) and algorithms and the field is currently very active. Here we adopt a Bayesian approach, and introduce a "hierarchical," stochastic model for the original image, based on the *Gibbs distribution*, and a new restoration algorithm, based on stochastic relaxation and *annealing*, for computing the maximum *a posteriori* (MAP) estimate of the original image given the degraded image. This algorithm is highly parallel and exploits the equivalence between Gibbs distributions and *Markov random fields* (MRF).

The essence of our approach to restoration is a stochastic relaxation algorithm which generates a sequence of images that converges in an appropriate sense to the MAP estimate. This sequence evolves by *local* (and potentially *parallel*) changes in pixel gray levels and in locations and orientations of boundary elements. Deterministic, iterative-improvement methods generate a sequence of images that monotonically increase the posterior distribution (our "objective function"). In contrast, stochastic relaxation permits changes that *decrease* the posterior distribution as well. These are made on a *random* basis, the effect of which is to avoid convergence to *local maxima*. This should not be confused with "probabilistic relaxation" ("relaxation labeling"), which is deterministic; see Section X.

The stochastic relaxation algorithm can be informally described as follows.

1) A local change is made in the image based upon the current values of pixels and boundary elements in the immediate "neighborhood." This change is *random*, and is generated by sampling from a local conditional probability distribution.

2) The local conditional distributions are dependent on a global control parameter T called "temperature." At *low* temperatures the local conditional distributions concentrate on states that *increase* the objective function, whereas at high temperatures the distribution is essentially uniform. The limiting cases, $T = 0$ and $T = \infty$, correspond respectively to greedy algorithms (such as gradient ascent) and undirected (i.e., "purely random") changes. (High temperatures induce a loose coupling between neighboring pixels and a chaotic appearance to the image. At low temperatures the coupling is tighter and the images appear more regular.)

3) Our image restorations avoid local maxima by beginning at high temperatures where many of the stochastic changes will actually decrease the objective function. As the relaxation proceeds, temperature is gradually lowered and the process behaves increasingly like iterative improvement. (This gradual reduction of temperature simulates "annealing," a procedure by which certain chemical systems can be driven to their low energy, highly regular, states.)

Our "annealing theorem" prescribes a schedule for lowering temperature which guarantees convergence to the global maxima of the posterior distribution. In practice, this schedule may be too slow for application, and we use it only as a guide in choosing the functional form of the temperature-time dependence. Readers familiar with Monte Carlo methods in statistical physics will recognize our stochastic relaxation algorithm as a "heat bath" version of the *Metropolis algorithm* [42]. The idea of introducing temperature and simulating an-

Copyright © 1984 IEEE. Reprinted with permission.

nealing is due to Černý [8] and Kirkpatrick *et al.* [40], both of whom used it for combinatorial optimization, including the traveling salesman problem. Kirkpatrick also applied it to computer design.

Since our approach is Bayesian it is model-based, with the "model" captured by the prior distribution. Our models are "hierarchical," by which we mean layered processes reflecting the type and degree of *a priori* knowledge about the class of images under study. In this paper, we regard the original image as a pair $X = (F, L)$ where F is the matrix of observable pixel intensities and L denotes a (dual) matrix of unobservable edge elements. Thus the usual gray levels are considered a marginal process. We refer to F as the *intensity process* and L as the *line process*. In future work we shall expand this model by adjoining other, mainly geometric, attribute processes.

The degradation model allows for noise, blurring, and some nonlinearities, and hence is characteristic of most photochemical and photoelectric systems. More specifically, the degraded image G is of the form $\phi(H(F)) \odot N$, where H is the blurring matrix, ϕ is a possibly nonlinear (memoryless) transformation, N is an independent noise field, and $\odot$ denotes any suitably invertible operation, such as addition or multiplication. Surprisingly, these nonlinearities do not affect the computational burden.

To pin things down, let us briefly discuss the Markovian nature of the intensity process; similar remarks apply to the line process, the pair (F, L), and the distribution of (F, L) conditional on the "data" G. Of course, all of this will be discussed in detail in the main body of the paper.

Let $Z_m = \{(i,j) : 1 \leqslant i, j \leqslant m\}$ denote the $m \times m$ integer lattice; then $F = \{F_{i,j}\}$, $(i,j) \in Z_m$, denotes the gray levels of the original, digitized image. Lowercase letters will denote the values assumed by these (random) variables; thus, for example, $\{F = f\}$ stands for $\{F_{i,j} = f_{i,j}, (i,j) \in Z_m\}$. We regard F as a sample realization of a random field, usually isotropic and homogeneous, and with significant correlations well beyond nearest neighbors. Specifically, we model F as an MRF, or, what is the same (see Section IV), we assume that the probability law of F is a Gibbs *distribution*. Given a *neighborhood system* $\mathcal{F} = \{\mathcal{F}_{i,j}, (i,j) \in Z_m\}$, where $\mathcal{F}_{i,j} \subseteq Z_m$ denotes the neighbors of (i,j), an MRF over $(Z_m, \mathcal{F})$ is a stochastic process indexed by Z_m for which, for every (i,j) and every f,

$$P(F_{i,j} = f_{i,j} \mid F_{k,l} = f_{k,l}, (k,l) \neq (i,j))$$
$$= P(F_{i,j} = f_{i,j} \mid F_{k,l} = f_{k,l}, (k,l) \in \mathcal{F}_{i,j}). \quad (1.1)$$

The MRF–Gibbs equivalence provides an explicit formula for the *joint* probability distribution $P(F = f)$ in terms of an *energy function*, the choice of which, together with $\mathcal{F}$, supplies a powerful mechanism for modeling spatial continuity and other scene features.

The relaxation algorithm is designed to maximize the conditional probability distribution of (F, L) given the data $G = g$, i.e., find the mode of the *posterior distribution* $P(X = x \mid G = g)$. This form of Bayesian estimation is known as *maximum a posteriori* or MAP estimation, or sometimes as *penalized maximum likelihood* because one seeks to maximize $\log P(G = g \mid X = x) + \log P(X = x)$ as a function of x; the second term is

the "penalty term." MAP estimation has been successfully employed in special settings (see, e.g., Hunt [31] and Hansen and Elliott [25]) and we share the opinion of many that the MAP formulation (and a Bayesian approach in general; see also [24], [43], [45]) is well-suited to restoration, particularly for handling general forms of spatial degradation. Moreover, the distribution of G itself need not be known, which is fortunate due to its usual complexity. On the other hand, MAP estimation clearly presents a formidable computational problem. The number of possible intensity images is L^{m^2}, where L denotes the number of allowable gray levels, which rules out any direct search, even for small ($m = 64$), binary ($L = 2$) scenes. Consequently, one is usually obliged to make simplifying assumptions about the image and degradation models as well as compromises at the computational stage. Here, the computational problem is overcome by exploiting the pivotal observation that the posterior distribution is again Gibbsian with approximately the same neighborhood system as the original image, together with a sampling method which we call the *Gibbs Sampler*. Indeed, our principal theoretical contribution is a general, practical, and mathematically coherent approach for investigating MRF's by sampling (Theorem A), and by computing modes (Theorem B) and expectations (Theorem C).

The Gibbs Sampler generates realizations from a given MRF by a "relaxation" technique akin to site-replacement algorithms in statistical physics, such as "spin-flip" and "exchange" systems. The prototype is due to Metropolis *et al.* [42]; see also [7], [18], and Section X. Cross and Jain [12] use one of these algorithms invented for studying binary alloys. ("Relaxation labeling" in the sense of [13], [30], [46], [47] is different; see Section X.) The Markov property (1.1) permits parallel updating of the line and pixel sites, each of which is "refreshed" according to a simple recipe determined by the governing distribution. Thus, both parts of the MRF–Gibbs equivalence are exploited, for computing and modeling, respectively. Moreover, minimum mean-square error (MMSE) estimation is also feasible by using the (temporal) ergodicity of the relaxation chain to compute *means* w.r.t. the posterior distribution. However, we shall not pursue this approach.

We have used a comparatively slow, raster scan-serial version of the Gibbs Sampler to generate images and restorations (see Section XIII). But the algorithm is parallel; it could be executed in essentially one-half the time with two processors running simultaneously, or in one-third the time with three, and so on. The full parallel potential is realized by assigning one (simple) processor to each site of the intensity process and to each site of the line process. Whatever the number of processors, parallel implementation is made feasible by a small communications requirement among processors. The communications burden is related to the neighborhood size of the graph associated with the image model, and herein lies much of the power of the hierarchical structure: although the field model $X = (F, L)$ has a local graph structure, the *marginal* distribution on the observable intensity process F has a *completely connected graph*. The introduction of a hierarchy dramatically expands the richness of the model of the observed process while only moderately adding to the computational burden. We shall return to these points in Sections IV and XI.

The MAP algorithm depends on an *annealing schedule*, which refers to the (sufficiently) slow decrease of a ("control") parameter T that corresponds to *temperature* in a physical system. As T decreases, samples from the posterior distribution are forced towards the minimal energy configurations; these correspond to the mode(s) of the distribution. Theorem B makes this precise, and is, to our knowledge, the first theoretical result of this nature. Roughly speaking, it says that if the temperature $T(k)$ employed in executing the kth site replacement (i.e., the kth image in the iteration scheme) satisfies the bound

$$T(k) \geqslant \frac{c}{\log (1 + k)}$$

for every k, where c is a constant independent of k, then with probability converging to one (as $k \to \infty$), the configurations generated by the algorithm will be those of minimal energy. Put another way, the algorithm generates a Markov chain which converges *in distribution* to the uniform measure over the minimal energy configurations. (It should be emphasized that *pointwise* convergence, i.e., convergence *with probability one*, is in general not possible.) These issues are discussed in Section XII, and the algorithm is demonstrated in Section XIII on a variety of degraded images. We also discuss the nature of the constant c in regard to practical convergence rates. Basically, we believe that the logarithmic rate is best possible. However, the best (i.e., smallest) value of c that we have obtained to date (see the Appendix) is far too large for computational value and our restorations are actually performed with small values of c. As yet, we do not know how to bring the theory in line with experimental results in this regard.

The role of the Gibbs (or Boltzmann) distribution, and other notions from statistical physics, in the construction of "expert systems" is expanding. To begin with, we refer the reader to [21] for the original formulation of our computational method and of a general approach to expert systems based on maximum entropy extensions. As previously mentioned, Černý [8] and Kirkpatrick *et al.* [40] introduced annealing into combinatorial optimization. Other examples include the work of Cheeseman [9] on maximum entropy and diagnosis and of Hinton and Sejnowski [29] on neural modeling of inference and learning.

This paper is organized as follows. The degradation model is described in the next section, and the undegraded image models are presented in Section IV after preliminary material on graphs and neighborhood systems in Section III. In particular, Section IV contains the definitions of MRF's, Gibbs distributions, and the equivalence theorem. Due to the plethora of Markovian models in the literature, we pause in Section V to compare ours to others, and in Section VI to explain some connections with maximum entropy methods. In Section VII we raise the issues of parameter estimation and model selection, and indicate why we are avoiding the former for the time being. The posterior distribution is computed in Section VIII and the corresponding optimization problem is addressed in Section IX. The concept of stochastic relaxation is reviewed

in Section X, including its origins in physics. Sections XI and XII are devoted to the Gibbs Sampler, dealing, respectively, with its mechanical and mathematical workings. Our experimental results appear in Section XIII, followed by concluding remarks.

II. Degraded Image Model

We follow the standard modeling of the (intensity) image formation and recording processes, and refer the reader to [31] or [47] for better accounts of the physical mechanisms.

Let H denote the "blurring matrix" corresponding to a shift-invariant point-spread function. The formation of F gives rise to a blurred image $H(\mathrm{F})$ which is recorded by a sensor. The latter often involves a nonlinear transformation of $H(\mathrm{F})$, denoted here by ϕ, in addition to random sensor noise $\mathrm{N} = \{\eta_{i,j}\}$, which we assume to consist of independent, and for definiteness, Gaussian variables with mean μ and standard deviation σ.

Our methods apply to essentially arbitrary noise processes $\mathrm{N} = \{\eta_{i,j}\}$, discrete or continuous. However, computational feasibility requires that the description of N as an MRF (this can always be done; see Section IV) has an associated graph structure that is approximately "local"; the same requirement is applied to the image process $\mathrm{X} = (\mathrm{F}, \mathrm{L})$. For clarity, we forgo full generality and focus on the traditional Gaussian white noise case. Extension to a general noise process is mostly a matter of notation.

The degraded image is then a function of $\phi(H(\mathrm{F}))$ and N, say $\psi(\phi(H(\mathrm{F})), \mathrm{N})$, for example, addition or multiplication. (To compute the posterior distribution, we only need to assume that $b \to \psi(a, b)$ is invertible for each a.) For notational ease, we will write

$$\mathrm{G} = \phi(H(\mathrm{F})) \odot \mathrm{N}. \tag{2.1}$$

At the pixel level, for each $(i, j) \in Z_m$,

$$G_{i,j} = \phi\left(\sum_{(k, l)} H(i - k, j - l) F_{k, l} \right) \odot \eta_{i,j}. \tag{2.2}$$

The mathematical results require an additional assumption, namely, that F and N be independent as stochastic processes (and likewise for L and N) and we assume this henceforth. This is customary, although we recognize the limitation in certain contexts, e.g., for nuclear scan pictures.

For computational purposes, the degree of locality of F should be approximately preserved by (2.1), so that the neighborhood systems for the prior and posterior distributions on (F, L) are comparable. This is achieved when H is a simple convolution over a small window. For instance, take

$$H(k, l) = \begin{cases} \frac{1}{2}, & k = 0, l = 0 \\ \frac{1}{16}, & |k|, |l| \leq 1, (k, l) \neq (0, 0) \end{cases} \tag{2.3}$$

so that the intensity at (i, j) is weighted equally with the average of the eight nearest neighbors. The function ϕ is unrestricted, bearing in mind that the true noise level depends on ϕ, $\odot$, and σ. Typically, ϕ is logarithmic (film) or algebraic (TV).

An important special case, which occurs in two-dimensional (2-D) signal theory, is the segmentation of noisy images into coherent regions. The usual model is

$$G = F + N \tag{2.4}$$

where N is white noise and the number of intensity levels is small. This is the model entertained by Hansen and Elliott [25] for simple, binary MRF's F, and by many other workers with varying assumptions about F; see [14], [16], [17]. In this case, namely (2.4), we can extract simple images under extremely low signal-to-noise ratios.

The full degraded image is (G, L); that is, the "line process" is not transformed.

III. GRAPHS AND NEIGHBORHOODS

Here and in Section IV we present the general theory of MRF's on graphs, focusing on the aspects and examples which figure in the experimental restorations. The level of abstraction is warranted by the variety of MRF's, graphs, and probability distributions simultaneously under discussion.

Let $S = \{s_1, s_2, \cdots, s_N\}$ be a set of *sites* and let $\mathcal{G} = \{\mathcal{G}_s, s \in S\}$ be a *neighborhood system* for S, meaning any collection of subsets of S for which 1) $s \notin \mathcal{G}_s$ and 2) $s \in \mathcal{G}_r \Leftrightarrow r \in \mathcal{G}_s$. Obviously, $\mathcal{G}_s$ is the set of *neighbors* of s and the pair $\{S, \mathcal{G}\}$ is a graph in the usual way. A subset $C \subseteq S$ is a *clique* if every pair of distinct sites in C are neighbors; $\mathcal{C}$ denotes the set of cliques.

The special cases below are especially relevant.

Case 1: $S = Z_m$. This is the set of pixel sites for the intensity process F; $\{s_1, s_2, \cdots, s_N\}$, $N = m^2$, is any ordering of the lattice points. We are interested in homogeneous neighborhood systems of the form

$$\mathcal{G} = \mathcal{F}_c = \{\mathcal{F}_{i,j}, (i,j) \in Z_m\}; \mathcal{F}_{i,j}$$
$$= \{(k,l) \in Z_m : 0 < (k-i)^2 + (l-j)^2 \leqslant c\}.$$

Notice that sites at or near the boundary have fewer neighbors than interior ones; this is the so-called "free boundary" and is more natural for picture processing than torodial lattices and other periodic boundaries. Fig. 1(a), (b), (c) shows the (interior) neighborhood configurations for $c = 1, 2, 8$; $c = 1$ is the first-order or nearest-neighbor system common in physics, in which $\mathcal{F}_{i,j} = \{(i, j-1), (i, j+1), (i-1, j), (i+1, j)\}$, with adjustments at the boundaries. In each case, (i, j) is at the center, and the symbol $\circ$ stands for a neighboring pixel. The cliques for $c = 1$ are all subsets of Z_m of the form $\{(i, j)\}$, $\{(i, j), (i, j+1)\}$ or $\{(i, j), (i+1, j)\}$, shown in Fig. 1(d). For $c = 2$, we have the cliques in Fig. 1(d) as well as those in Fig. 1(e). Obviously, the number of clique types grows rapidly with c. However, only small cliques appear in the model for F actually employed in this paper; indeed, the degree of progress with only *pair* interactions is somewhat surprising. Nonetheless, more complex images will likely necessitate more complex energies. Our experiments (see Section XIII) suggest that much of this additional complexity can be accommodated while maintaining modest neighborhood sizes by further developing the hierarchy.

Case 2: $S = D_m$, the "dual" $m \times m$ lattice. Think of these sites as placed midway between each vertical or horizontal pair of pixels, and as representing the possible locations of "edge

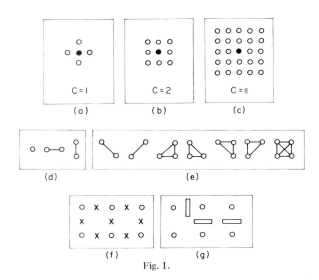

Fig. 1.

elements." Shown in Fig. 1(f) are six pixel sites together with seven line sites denoted by an X. The six surrounding X's are the neighbors of the middle X for the neighborhood system we denote by $\mathcal{L} = \{\mathcal{L}_d, d \in D_m\}$. Fig. 1(g) is a segment of a realization of a binary line process for which, at each line site, there may or may not be an edge element. We also consider line processes with more than two levels, corresponding to edge elements with varying orientations.

Case 3: $S = Z_m \cup D_m$. This is the setup for the field (F, L). Z_m has neighborhood system $\mathcal{F}_1$ (nearest-neighbor lattice) and D_m has the above-described system. The pixel neighbors of sites in D_m are the two pixels on each side, and hence each (interior) pixel has four line site neighbors.

IV. MARKOV RANDOM FIELDS AND GIBBS DISTRIBUTIONS

We now describe a class of stochastic processes that includes both the prior and posterior distribution on the original image. In general, this class of processes (namely, MRF's) is neither homogeneous nor isotropic, assuming the index set S has enough geometric structure to even *define* a suitable family of translations and rotations. However, the *particular* models we choose for prior distributions on the original image are in fact both homogeneous and isotropic in an appropriate sense. (This is not the case for the *posterior* distribution.) We refer the reader to Section XIII for a precise description of the prior models employed in our experiments, and in particular for specific examples of the role of the line elements.

As in Section III, $\{S, G\}$ denotes an arbitrary graph. Let $X = \{X_s, s \in S\}$ denote *any* family of random variables indexed by S. For simplicity, we can assume a common state space, say $\Lambda \doteq \{0, 1, 2, \cdots, L - 1\}$, so that $X_s \in \Lambda$ for all s; the extension to site-dependent state spaces, appropriate when S consists of both line and pixel sites, is entirely straightforward (although not merely a notational matter due to the "positivity condition" below). Let Ω be the set of all possible *configurations*:

$$\Omega = \{\omega = (x_{s_1}, \cdots, x_{s_N}) : x_{s_i} \in \Lambda, 1 \leqslant i \leqslant N\}.$$

As usual, the event $\{X_{s_1} = x_{s_1}, \cdots, X_{s_N} = x_{s_N}\}$ is abbreviated $\{X = \omega\}$.

X is an MRF with respect to $\mathcal{G}$ if

$$P(X = \omega) > 0 \qquad \text{for all} \quad \omega \in \Omega; \qquad (4.1)$$

$$P(X_s = x_s | X_r = x_r, r \neq s) = P(X_s = x_s | X_r = x_r, r \in \mathcal{G}_s) \qquad (4.2)$$

for every $s \in S$ and $(x_{s_1}, \cdots, x_{s_N}) \in \Omega$. Technically, what is meant here is that the pair $\{X, P\}$ satisfies (4.1) and (4.2) relative to some probability measure on Ω. The collection of functions on the left-hand side of (4.2) is called the *local characteristics* of the MRF and it turns out that the (joint) probability distribution $P(X = \omega)$ of *any* process satisfying (4.1) is *uniquely* determined by these conditional probabilities; see, e.g., [6, p. 195].

The concept of an MRF is essentially due to Dobrushin [15] and is one way of extending Markovian dependence from 1-D to a general setting; there are, of course, many others, some of which will be reviewed in Section V.

Notice that *any* X satisfying (4.1) is an MRF if the neighborhoods are large enough to encompass the dependencies. The utility of the concept, at least in regard to image modeling, is that priors are available with neighborhoods that are small enough to ensure feasible computational loads and yet still rich enough to model and restore interesting classes of images (and textures: [12]).

Ordinary 1-D Markov chains are MRF's relative to the nearest-neighbor system on $S = \{1, 2, \cdots, N\}$ (i.e., $\mathcal{G}_1 = \{2\}$, $\mathcal{G}_i = \{i - 1, i + 1\}$ $2 \leq i \leq N - 1$, $\mathcal{G}_N = \{N - 1\}$) if we assume all positive transitions and the chain is started in equilibrium. In other words, the "one-sided" Markov property

$$P(X_k = x_k | X_j = x_j, j \leq k - 1) = P(X_k = x_k | X_{k-1} = x_{k-1})$$

and the "two-sided" Markov property

$$P(X_k = x_k | X_j = x_j, j \neq k) = P(X_k = x_k | X_j = x_j, j \in \mathcal{G}_k)$$

are equivalent. Similarly for an rth order Markov process on the line with respect to the r nearest neighbors on one side and on both sides. (This appears to be doubted in [1] but follows, eventually, from straightforward calculations or immediately from the Gibbs connection.)

Gibbs models were introduced into image modeling by Hassner and Sklansky [28], although the treatment there is mostly expository and limited to the binary case.

A *Gibbs distribution* relative to $\{S, \mathcal{G}\}$ is a probability measure π on Ω with the following representation:

$$\pi(\omega) = \frac{1}{Z} e^{-U(\omega)/T} \qquad (4.3)$$

where Z and T are constants and U, called the *energy function*, is of the form

$$U(\omega) = \sum_{C \in \mathcal{C}} V_C(\omega). \qquad (4.4)$$

Recall that $\mathcal{C}$ denotes the set of cliques for $\mathcal{G}$. Each V_C is a function on Ω with the property that $V_C(\omega)$ depends only on those coordinates x_s of ω for which $s \in C$. Such a family $\{V_C, C \in \mathcal{C}\}$ is called a *potential*. Z is the normalizing constant:

$$Z \doteq \sum_\omega e^{-U(\omega)/T} \qquad (4.5)$$

and is called the *partition function*. Finally, T stands for "temperature"; for our purposes, T controls the degree of "peaking" in the "density" π. Choosing T "small" exaggerates the mode(s), making them easier to find by sampling; this is the principle of annealing, and will be applied to the posterior distribution $\pi(f, l) = P(F = f, L = l | G = g)$ in order to find the MAP estimate. Of course, we will show that $\pi(f, l)$ is Gibbsian and identify the energy and neighborhood system in terms of those for the priors. The *choice* of the prior distributions, i.e., of the particular functions V_C for the image model $\pi(\omega) = P(X = \omega)$, will be discussed later on; see Section VII for some general remarks and Section XIII for the particular models employed in our experiments.

The terminology obviously comes from statistical physics, wherein such measures are "equilibrium states" for physical systems, such as ferromagnets, ideal gases, and binary alloys. The V_C functions represent contributions to the total energy from external fields (singleton cliques), pair interactions (doubletons), and so forth. Most of the interest there, and in the mathematical literature, centers on the case in which S is an *infinite*, 2-D or 3-D lattice; singularities in Z may then occur at certain ("critical") temperatures and are associated with "phase transitions."

Typically, several free parameters are involved in the specification of U, and Z is then a function of those parameters— notoriously intractable. For more information see [3], [5], [6], [23], [32], and [39].

The best-known of these lattice systems is the Ising model, invented in 1925 by E. Ising [33] to help explain ferromagnetism. Here, $S = Z_m$ and $\mathcal{G} = \mathcal{F}_1$, the nearest-neighbor system. The most general form of U is then

$$U(\omega) = \sum V_{\{i, j\}}(x_{i, j}) + \sum V_{\{(i, j), (i+1, j)\}}(x_{i, j}, x_{i+1, j})$$
$$+ \sum V_{\{(i, j), (i, j+1)\}}(x_{i, j}, x_{i, j+1}) \qquad (4.6)$$

where the sums extend over all $(i, j) \in Z_m$ for which the indicated cliques make sense. The Ising model is the special case of (4.6) in which X is binary ($L = 2$), homogeneous (= strictly stationary), and isotropic (= rotationally invarient):

$$U(\omega) = \alpha \sum x_{i, j} + \beta \left(\sum x_{i, j} x_{i+1, j} + \sum x_{i, j} x_{i, j+1} \right) \quad (4.7)$$

for some parameters α and β, which measure, respectively, the external field and bonding strengths.

Returning to the general formulation, recall that the local characteristics

$$\pi(x_s | x_r, r \neq s) = \frac{\pi(\omega)}{\sum_{x_s \in \Lambda} \pi(\omega)} \qquad s \in S, \omega \in \Omega$$

uniquely determine π for any probability measure π on Ω, $\pi(\omega) > 0$ for all ω. The difficulty with the MRF formulation *by itself* is that

i) the *joint* distribution of the X_s is not apparent;

ii) it is extremely difficult to spot local characteristics, i.e., to determine when a given set of functions $\psi(x_s \mid x_r, r \neq s)$, $s \in S$, $(x_{s_1}, \cdots, x_{s_N}) \in \Omega$, are conditional probabilities for some (necessarily unique) distribution on Ω.

For example, Chellappa and Kashyap [10] allude to i) as a disadvantage of the "conditional Markov" models. See also the discussion in [6]. In fact, these apparent limitations to the MRF formulation have been noted by a number of authors, many of whom were obviously not aware of the following theorem.

Theorem: Let $\mathcal{G}$ be a neighborhood system. Then X is an MRF with respect to $\mathcal{G}$ if and only if $\pi(\omega) = P(\mathrm{X} = \omega)$ is a Gibbs distribution with respect to $\mathcal{G}$.

Among other benefits, this equivalence provides us with a simple, practical way of specifying MRF's, namely by specifying potentials, which is easy, instead of local characteristics, which is nearly impossible. In fact, with some experience, one can choose U's in accordance with the desired *local* behavior, at least at the intensity level. In short, the modeling and consistency problems of i) and ii) are eliminated.

Proofs may be found in many places now; see, e.g., [39] and the references therein, or the approach via the Hammersley-Clifford expansion in [6]. An influential discussion of this correspondence appears in Spitzer's work, e.g., [48]. Explicit formulas exist for obtaining U from the local characteristics. Conversely, the local characteristics of π are obtained in a straightforward way from the potentials: use the defining ratios and make the allowable cancellations. Fix $s \in S$, $\omega = (x_{s_1}, \cdots, x_{s_N}) \in \Omega$, and let ω^x denote the configuration which is x at site s and agrees with ω everywhere else. Then if $\pi(\omega) = P(\mathrm{X} = \omega)$ is Gibbsian,

$$P(X_s = x_s \mid X_r = x_r, r \neq s) = Z_s^{-1} \exp - \frac{1}{T} \sum_{C \,:\, s \in C} V_C(\omega) \tag{4.8}$$

$$Z_s \doteq \sum_{x \in \Lambda} \exp - \frac{1}{T} \sum_{C \,:\, s \in C} V_C(\omega^x). \tag{4.9}$$

Notice that the right-hand side of (4.8) only depends on x_s and on x_r, $r \in \mathcal{G}_s$, since any site in a clique containing s must be a neighbor of s. These formulas will be used repeatedly to program the Gibbs Sampler for local site replacements.

For the Ising model, the conditional probability that $X_{i,j} = x_{i,j}$, given the states at $S \setminus \{i, j\}$, or equivalently, just the four nearest neighbors, reduces to

$$\frac{e^{-x_{ij}(\alpha + \beta v_{i,j})}}{1 + e^{-(\alpha + \beta v_{i,j})}}$$

where $v_{i,j} = x_{i,j-1} + x_{i-1,j} + x_{i,j+1} + x_{i+1,j}$. This is also known as the autologistic model and has been used for texture modeling in [12]. More generally, if the local characteristics are given by an exponential family and if $V_C(\omega) \equiv 0$ for $|C| > 2$, then the pair potentials always "factor" into a product of two like terms; see [6].

We conclude with some further discussion of a remark made in Section I: that the hierarchical structure introduced with the line process L expands the graph structure of the *marginal* distribution of the intensity process F. Consider first an arbitrary MRF X with respect to a graph $\{S, \mathcal{G}\}$. Fix $r \in S$ and let $\hat{\mathrm{X}} = \{X_s, s \in S, s \neq r\}$. The marginal distribution $\hat{P}$ of $\hat{\mathrm{X}}$ is derived from the distribution P of X by summing over the range of X_r. Use the Gibbs representation for P and perform this summation: the resulting expression for $\hat{P}$ can be put in the Gibbs form, and from this the neighborhood system on $\hat{S} \doteq S \setminus \{r\}$ can be inferred. The conclusion of this exercise is that $s_1, s_2 \in \hat{S}$ are, in general, neighbors if either i) they were neighbors in S under $\mathcal{G}$ or ii) each is a neighbor of $r \in S$ under $\mathcal{G}$. Now let $\mathrm{X} = (\mathrm{F}, \mathrm{L})$, with neighborhood system defined at the end of Section III. Successive summations of the distribution of X over the ranges of the elements of L yields the marginal distribution of the observable intensity process F. Each summation leaves a graph structure associated with the marginal distribution of the remaining variables, and this can be related to the original neighborhood system by following the preceding discussion of the general case. It is easily seen that when all of the summations are performed, the remaining graph is completely connected; under the marginal distribution of F, all sites are neighbors. This calculation suggests that significant long-range interactions can be introduced through the development of hierarchical structures without sacrificing the computational advantages of local neighborhood systems.

V. RELATED MARKOV IMAGE MODELS

The use of neighborhoods is, of course, pervasive in the literature: they offer a geometric framework for the clustering of pixel intensities and for many types of statistical models. In particular, the Markov property is a natural way to formalize these notions. The result is a somewhat bewildering array of Markov-type image models and it seems worthwhile to pause to relate these to MRF's. The process under consideration is $\mathrm{F} = \{F_{i,j}, (i, j) \in Z_m\}$, the gray levels, or really any pixel attribute.

An early work in this direction is Abend, Harley and Kanal [1] about pattern classification. Among many novel ideas, there is the notion of a *Markov mesh* (MM) process, in which the Markovian dependence is *causal*: generally, one assumes that, for all (i, j) and f,

$$\begin{aligned} P(F_{i,j} &= f_{i,j} \mid F_{k,l} = f_{k,l}, (k, l) \in A_{i,j}) \\ &= P(F_{i,j} = f_{i,j} \mid F_{k,l} = f_{k,l}, (k, l) \in B_{i,j}) \end{aligned} \tag{5.1}$$

where $B_{i,j} \subseteq A_{i,j} \subseteq \{(k, l) : k < i \text{ or } l < j\}$. A common example is $B_{i,j} = \{(i - 1, j), (i - 1, j - 1), (i, j - 1)\}$. Besag [6], Kanal [37], and Pickard [44] also discuss such "unilateral" processes, which are usually a subclass of MRF's, although the resulting (bilateral) neighborhoods can be irregular. Anyway, for MM models the emphasis is on the causal, iterative aspects, including a recursive representation for the joint probabilities. Incidentally, a Gibbs type description of rth order Markov chains is given in [1]; of course, the full Gibbs-MRF equivalence is not perceived and was not for about five years. Derin *et al.* [14] model F as an MM process and use recursive Bayes smoothing to recover F from a noisy version $F + N$; the algorithms exploit the causality to maximize the univariate posterior distribution at each pixel based on the data over a strip containing it, and are very effective at low S/N ratios for some simple images.

Motivated by a paper of Lévy [41], Woods [51] defined "P-Markov" processes for the resolution of wavenumber spectra. The definition involves two spatial regions separated by a "boundary" of width P, and correspond to the past, future, and present in 1-D. Woods also considers a family of "wide-sense" Markov fields of the form

$$F_{i,j} = \sum_{(k,\,l)\in W_p} \theta_{k,\,l}F_{i-k,\,j-l} + U_{i,j} \qquad (5.2)$$

where $W_p = \{(k,\,l):0 < k^2 + l^2 \leqslant P\}$, $\theta_{k,\,l}$ are the MMSE coefficients for projecting $F_{i,j}$ on $\{F_{k,\,l},\,(k,\,l)\in(i,\,j)+W_p\}$, and $\{U_{i,j}\}$ is the error, generally nonwhite. The main theoretical result is that if $\{U_{i,j}\}$ is homogeneous, Gaussian, and satisfies a few other assumptions, then F is Gaussian, P-Markov and vice-versa. In general, there are consistency problems and the P-Markov property is hard to verify. In the nearest-neighbor case, one gets a Gaussian MRF.

Other "wide-sense" Markov processes appear in Jain and Angel [35] and Stuller and Kurz [49]. The assumptions in [35] are a nearest-neighbor system, white noise, and no blur; restoration is achieved by recursively filtering the rows $\{F_{i,j}\}_{j=1}^{m}$, which form a vector-valued, second-order Markov chain, to find the optimal interpolator of each row. In [49], causality is introduced and earlier work is generalized by considering an arbitrary "scanning pattern."

The "spatial interaction models" in Chellappa and Kashyap [10], [38] satisfy (5.2) for general coefficients and W's. The model is causal if W lies in the third quadrant. The authors consider "simultaneous autoregressive" (SAR) models, wherein the noise is white, and "conditional Markov" (CM) models, wherein the "bilateral" Markov property holds (i.e., (1.1) with $\mathcal{F}_{i,j} = (i,\,j) + W$) in addition to (5.2), and the noise is nonwhite. Thus, the CM models are MRF's, although in [10], [38] the boundary of Z_m is periodic, and hence boundary conditions must be adjoined to (5.2). Given any (homogeneous) SAR process there exists a unique CM process with the same spectral density, although different neighborhood structure. The converse holds in the Gaussian case but is generally false (see the discussion in Besag [6]). MMSE restoration of blurred images with additive Gaussian noise is discussed in [10]; the original image is SAR or CM, usually Gaussian.

Finally, Hansen and Elliott [25] and Elliott et al. [17] design MAP algorithms for the segmentation of remotely sensed data with high levels of additive noise. The image model is a nearest-neighbor, binary MRF. However, the autologistic form of the joint distribution is not recognized due to the lack of the Gibbs formulation. The conditional probabilities are approximated by the product of four 1-D transitions, and segmentation is performed by dynamic programming, first for each row and then for the entire images. More recent work in Elliott et al. [16] is along the same lines, namely MAP estimation, via dynamic programming, of very noisy but simple images; the major differences are the use of the Gibbs formulation and improvements in the algorithms. Similar work, applied to boundary finding, can be found in Cooper and Sung

[11], who use a Markov boundary model and a deterministic relaxation scheme.

VI. Maximum Entropy Restoration

There are several contact points. The Gibbs distribution can be derived (directly from physical principles in statistical mechanics) by maximizing entropy: basically, it has maximal entropy among all probability measures (equilibrium states) on Ω with the same *average* energy. Thus it is no accident that, like maximum entropy (ME) methods, ours are well-suited to nonlinear problems; see [50]. Moreover, based on the success of ME restoration (along the lines suggested by Jaynes [36]) for recovering randomly pulsed objects (cf. Frieden [19]), we intend in the future to analyze such data (e.g., starfield photographs) by our methods.

We should also like to mention the interesting observation of Trussell [50] that conventional ME restoration is a special case of MAP estimation in which the prior distribution on F is

$$P(F = f) = \exp\left(-\beta \sum f_{i,j}\log f_{i,j}\right)\Big/(\text{normalizing constant}).$$

By "conventional ME," we refer to maximizing the entropy $\sum f_{i,j} \log f_{i,j}$ subject to $\sum \eta_{i,j}^2 = \text{constant}$ ($\eta_{i,j}$ is here again the noise process); see [2]. Other ME methods (e.g., [19]) do not appear to be MAP-related.

VII. Model Selection and Parameter Estimation

The quality of the restoration will clearly depend on choices made at the modeling stage, in our case about specific energy types, attribute processes, and parameters. Cross and Jain [12] use maximum likelihood estimation in the context of Besag's [6] "coding scheme," as well as standard goodness-of-fit tests, for matching realizations of autobinomial MRF's to real textures. Kashyap and Chellappa [38] introduce some new methods for parameter estimation and the choice of neighborhoods for the SAR and CM models, mostly in the Gaussian case. These are but two examples.

For uncorrupted, simple MRF's, the coding methods do finesse the problem of the partition function. However, for more complex models and for corrupted data, we feel that the coding methods are ultimately inadequate due to the complexity of the distribution of G. This view seems to be shared by other authors, although in different contexts. Of course, for MRF's, the obstacles facing conventional statistical inference due to Z have often been noted. Even for the Ising model, analytical results are rare; a famous exception is Onsager's work on the correlational structure.

At any rate, we have developed a new method [20] for estimating clique parameters from the "noisy" data, and this will be implemented in a forthcoming paper. For now, we are obliged to choose the parameters on an ad hoc basis (which is common), but hasten to add that the quality of restoration does not seem to have been adversely affected, probably due to the relative simplicity of the MRF's we actually use for the line and intensity processes; see Section XIII.

One should also address the *general* choice of π and G. This is really quite different than parameter estimation and somewhat related to "image understanding": how does one incorporate "real-world knowledge" into the modeling process? In image interpretation systems, various semantical and hierarchical models have been proposed (see, e.g., [26]). We have begun our study of hierarchical Gibbs models in this paper. A *general theory* of interactive, self-adjusting models that is practical and mathematically coherent may lie far ahead.

VIII. Posterior Distribution

We now turn to the posterior distribution $P(F = f, L = l \mid G = g)$ of the original image given the "data" g. In this section we take $S = Z_m \cup D_m$, the collection of pixel and line sites, with some neighborhood system $\mathcal{G} = \{\mathcal{G}_s, s \in S\}$; an example of such a "mixed" graph was given in Section III. The configuration space is the set of all pairs $\omega = (f, l)$ where the components of f assume values among the allowable gray levels and those of l among the (coded) line states.

We assume that X is an MRF relative to $\{S, \mathcal{G}\}$ with corresponding energy function U and potentials $\{V_C\}$:

$$P(F = f, L = l) = e^{-U(f, l)/T}/Z$$

$$U(f, l) = \sum_C V_C(f, l).$$

For convenience, take $T = 1$

Recall that $G = \phi(H(F)) \odot N$, where N is white Gaussian noise with mean μ and variance σ^2 and is independent of X.

We emphasize that what follows is easily extended to processes N that are more general MRF's, although we still require that N be independent of X. The operation $\odot$ is assumed invertible and we will write $N = \Phi(G, \phi(H(F))) = \{\Phi_s, s \in Z_m\}$ to indicate this inverse.

Let $\mathcal{H}_s$, $s \in Z_m$, denote the pixels which affect the blurred image $H(F)$ at s. For instance, for the H in (2.3), $\mathcal{H}_s$ is the 3×3 square centered at s. Observe that Φ_s, $s \in Z_m$, depends only on g_s and $\{f_t, t \in \mathcal{H}_s\}$. By the shift-invariance of H, $\mathcal{H}_{r+s} = s + \mathcal{H}_r$ where $\mathcal{H}_r \subseteq Z_m$, $s + r \in Z_m$, and $s + \mathcal{H}_r$ is understood to be intersected with Z_m, if necessary. In addition, we will assume that $\{\mathcal{H}_s\}$ is "symmetric" in that $r \in \mathcal{H}_0 \Rightarrow -r \in \mathcal{H}_0$. Then the collection $\{\mathcal{H}_s \backslash \{s\}, s \in Z_m\}$ is a neighborhood system over Z_m. Let $\mathcal{H}^2$ denote the second-order system, i.e.,

$$\mathcal{H}_s^2 = \bigcup_{r \in \mathcal{H}_s} \mathcal{H}_r, \quad s \in Z_m.$$

Then it is not hard to see that $\{\mathcal{H}_s^2 \backslash \{s\}, s \in Z_m\}$ is also a neighborhood system. Finally, set $\mathcal{G}^P = \{\mathcal{G}_s^P, s \in S\}$ where

$$\mathcal{G}_s^P = \begin{cases} \mathcal{G}_s, & s \in D_m \\ \mathcal{G}_s \cup \mathcal{H}_s^2 \backslash \{s\}, & s \in Z_m. \end{cases} \quad (8.1)$$

The "P" stands for "posterior"; some thought shows that $\mathcal{G}^P$ is a neighborhood system on S.

Let $\mu \in \mathbb{R}^M (M = N^2)$ have all components $= \mu$ and let $\| \cdot \|$ denote the usual norm in $\mathbb{R}^M$: $\|V\|^2 = \Sigma_1^M V_i^2$.

Theorem: For each g fixed, $P(X = \omega \mid G = g)$ is a Gibbs distribution over $\{S, \mathcal{G}^P\}$ with energy function

$$U^P(f, l) = U(f, l) + \|\mu - \Phi(g, \phi(H(f)))\|^2/2\sigma^2. \quad (8.2)$$

Proof: Using standard results about "regular conditional expectations," we can and do assume that

$$P(X = \omega \mid G = g) = \frac{P(G = g \mid X = \omega) P(X = \omega)}{P(G = g)} \quad (8.3)$$

for all $\omega = (f, l)$, for each g.

Since $P(G = g)$ is a constant and $P(X = \omega) = e^{-U(\omega)}/Z$, the key term is

$$P(G = g \mid X = \omega) = P(\phi(H(F)) \odot N = g \mid F = f, L = l)$$
$$= P(N = \Phi(g, \phi(H(f))) \mid F = f, L = l)$$
$$= P(N = \Phi(g, \phi(H(f))))$$

(since N is independent of F and L)

$$= (2\pi\sigma^2)^{-M/2} \exp - \left(\frac{1}{2\sigma^2}\right) \|\mu - \Phi\|^2.$$

We will write Φ for $\Phi(g, \phi(H(f)))$. Collecting constants we have, from (8.3),

$$P(X = \omega \mid G = g) = e^{-U^P(\omega)}/Z^P$$

for U^P as in (8.2); Z^P is the usual normalizing constant (which will depend on g). It remains to determine the neighborhood structure.

Intuitively, the line sites should have the *same* neighbors whereas the neighbors $\mathcal{G}_s$ of a pixel site $s \in Z_m$ should be augmented in accordance with the blurring mechanism.

Take $s \in D_m$. The local characteristics at s for the posterior distribution are, by (8.2),

$$P(L_s = l_s \mid L_r = l_r, r \neq s, r \in D_m, F = f, G = g)$$
$$= \frac{e^{-U^P(f, l)}}{\sum_{l_s} e^{-U^P(f, l)}} = \frac{e^{-U(f, l)}}{\sum_{l_s} e^{-U(f, l)}}$$

where the sum extends over all possible values of L_s. Hence $\mathcal{G}_s^P = \mathcal{G}_s$.

For $s \in Z_m$, the term in (8.2) involving Φ does not cancel out. Now $\Phi(g, \phi(H(f))) = \{\Phi_s, s \in Z_m\}$ and let us denote the dependencies in Φ_s by writing $\Phi_s = \Phi_s(g_s; f_t, t \in \mathcal{H}_s)$. Then

$$P(F_s = f_s \mid F_r = f_r, r \neq s, r \in Z_m, L = l, G = g)$$
$$= \frac{e^{-U^P(f, l)}}{\sum_{f_s} e^{-U^P(f, l)}}; U^P(f, l)$$
$$= U(f, l) + \sum_{r \in Z_m} (\Phi_r - \mu)^2/2\sigma^2. \quad (8.4)$$

Decompose U^P as follows:

$$U^P(f, l) = \sum_{C: s \in C} V_C(f, l)$$
$$\quad + (2\sigma^2)^{-1} \sum_{r: s \in \mathcal{H}_r} (\Phi_r(g_r; f_t, t \in \mathcal{H}_r) - \mu)^2$$
$$\quad + \sum_{C: s \notin C} V_C(f, l)$$
$$\quad + (2\sigma^2)^{-1} \sum_{r: s \notin \mathcal{H}_r} (\Phi_r(g_r; f_t, t \in \mathcal{H}_r) - \mu)^2.$$

Since the last two terms do not involve f_s (remember that V_C only depends on the sites in C), the ratio in (8.4) depends only on the first two terms above. The first term depends only on coordinates of (f, l) for sites in $\mathcal{G}_s (s \in C \Rightarrow C \subseteq \mathcal{G}_s)$ and the second term only on sites in

$$\bigcup_{r : s \in \mathcal{H}_r} \mathcal{H}_r = \bigcup_{r \in \mathcal{H}_s} \mathcal{H}_r \doteq \mathcal{H}_s^2 .$$

Hence, $\mathcal{G}_s^P = \mathcal{G}_s \cup \mathcal{H}_s^2 \setminus \{s\}$, as asserted in the theorem. $\square$

IX. The Computational Problem

The posterior distribution $P(X = \omega | g)$ is a powerful tool for image analysis; in principle, we can construct the optimal (Bayesian) estimator for the original image, examine images sampled from $P(X = \omega | g)$, estimate parameters, design near-optimal statistical tests for the presence or absence of special objects, and so forth. But a conventional approach to any of these involves prohibitive computations. Specifically, our job here is to find the value(s) of ω which maximize the posterior distribution for a fixed g, i.e., *minimize*

$$U(f, l) + \| \mu - \Phi(g, \phi(H(f)))\|^2 / 2\sigma^2, (f, l) \in \Omega \quad (9.1)$$

where (see Section VIII) Φ is defined by $\phi(H(f)) \odot \Phi = g$. Even without L, the size of Ω is at least 2^{4000}, corresponding to a binary image on a small (64×64) lattice. Hence, the identification of even near-optimal solutions is extremely difficult for such a relatively complex function.

In Sections XI and XII we will describe our stochastic relaxation method for this kind of optimization. The same method works for sampling and for computing expectations (and hence forming likelihood ratios), as will be explained in Section XI. The algorithm is highly parallel, but our current implementation is serial: it uses a single processor. The restoration of more complex images than those in Section XIII, probably involving more levels in the hierarchy, may necessitate *some* parallel processing.

X. Stochastic Relaxation

There are many types of "relaxation," two of them being the type used in statistical physics and the type developed in image processing called "relaxation labeling" (RL), or sometimes "probabilistic relaxation." Basically, ours is of the former class, referred to here as SR, although there are some common features with RL.

The "Metropolis algorithm" (Metropolis *et al.* [42]) and others like it [7], [18] were invented to study the equilibrium properties, especially ensemble averages, time-evolution, and low-temperature behavior, of very large systems of essentially identical, interacting components, such as molecules in a gas or atoms in binary alloys.

Let Ω denote the possible configurations of the system; for example, $\omega \in \Omega$ might be the molecular positions or site configuration. If the system is in thermal equilibrium with its surroundings, then the probability (or "Boltzmann factor") of ω is given by

$$\pi(\omega) = e^{-\beta \mathcal{E}(\omega)} \Big/ \sum_{\omega} e^{-\beta \mathcal{E}(\omega)}, \quad \omega \in \Omega$$

where $\mathcal{E}(\omega)$ is the potential energy of ω and $\beta = 1/KT$ where K is Boltzmann's constant and T is absolute temperature. We have already seen an example in the Ising model (4.7). Usually, one needs to compute ensemble averages of the form

$$\langle Y \rangle = \int_{\Omega} Y(\omega) \, d\pi(\omega) = \frac{\sum_{\omega} Y(\omega) \, e^{-\beta \mathcal{E}(\omega)}}{\sum_{\omega} e^{-\beta \mathcal{E}(\omega)}}$$

where Y is some variable of interest. This cannot be done analytically. In the usual Monte Carlo method, one restricts the sums above to a *sample* of ω's drawn uniformly from Ω. This, however, breaks down in the situation above: the exponential factor puts most of the mass of π over a very small part of Ω, and hence one tends to choose samples of very low probability. The idea in [42] is to choose the samples from π instead of uniformly and then weight the samples evenly instead of by $d\pi$. In other words, one obtains $\omega_1, \omega_2, \cdots, \omega_R$ from π and $\langle Y \rangle$ is approximated by the usual ergodic averages:

$$\langle Y \rangle \approx \frac{1}{R} \sum_{r=1}^{R} Y(\omega_r). \quad (10.1)$$

Briefly, the sampling algorithm in [42] is as follows. Given the state of the system at "time" t, say $X(t)$, one randomly chooses another configuration η and computes the energy change $\Delta \mathcal{E} = \mathcal{E}(\eta) - \mathcal{E}(X(t))$ and the quantity

$$q = \frac{\pi(\eta)}{\pi(X(t))} = e^{-\beta \Delta \mathcal{E}}. \quad (10.2)$$

If $q > 1$, the move to η is allowed and $X(t+1) = \eta$, whereas if $q \leqslant 1$, the transition is made *with probability* q. Thus we choose $0 \leqslant \xi \leqslant 1$ uniformly and set $X(t+1) = \eta$ if $\xi \leqslant q$ and $X(t+1) = X(t)$ if $\xi > q$. (A "parallel processing variant" of this for simulating certain binary MRF's is given by Berger and Bonomi [4].)

In binary, "single-flip" studies, $\eta = X(t)$ except at one site, whereas in "spin-exchange" [18] systems, a pair of neighboring sites is selected. In either case, the "flip" or "exchange" is made with probability $q/(1 + q)$, where q is given in (10.2). In special cases, the single-flip system is equivalent to our Gibbs Sampler. The exchange algorithm in Cross and Jain [12] is motivated by work on the evolution of binary alloys. The samples generated are used for visual inspection and statistical testing, comparing the real and simulated textures. The model is an autobinomial MRF; see [6] or [12]. The algorithm is not suitable (nor intended) for restoration: for one thing, the intensity histogram is constant throughout the iteration process. This is necessarily the case with exchange systems which depend heavily on the initial configuration.

The algorithm in Hassner and Sklansky [28] is apparently a modification of one in Bortz *et al.* [7]. Another application of these ideas outside statistical mechanics appears in Hinton and Sejnowski [29], a paper about neural modeling but a spiritual cousin of ours. In particular, the parallel nature of these algorithms is emphasized.

The essence of every SR scheme is that changes ($\omega \rightarrow \eta$) which *increase* energy, i.e., *lower* probability, are permitted.

By contrast, deterministic algorithms only allow jumps to states of lower energy and invariably get "stuck" in *local* minima. To get to samples from π, we must occasionally "backtrack."

All of these algorithms can be cast in a general theory involving Markov chains with state space Ω. See Hammersley and Handscomb [27] for a readable treatment. The goal is an irreducible, aperiodic chain with equilibrium measure π. If $\omega_1, \omega_2, \cdots, \omega_R$ is a realization of such a chain, then standard results yield (10.1), in fact at a rate $O(R^{-1/2})$ as $R \to \infty$. In this setup an auxiliary transition matrix is used to go from ω to η, and the general replacement recipe involves the same ratio $\pi(\eta)/\pi(\omega)$. The Markovian properties of the Gibbs Sampler will be described in the following sections.

Chemical annealing is a method for determining the low energy states of a material by a gradual lowering of temperature. The process is delicate: if T is lowered too rapidly and insufficient time is spent at temperatures near the freezing point, then the process may bog down in nonequilibrium states, corresponding to flaws in the material, etc. In *simulated* annealing, Kirkpatrick *et al.* [40] identify the solution of an optimal (computer) design problem with the ground state of an imaginary physical system, and then employ the Metropolis algorithm to reach "steady-state" at each of a decreasing sequence of temperatures $\{T_n\}$. This sequence, and the time spent at each temperature, is called an "annealing schedule." In [40], this is done on an ad hoc basis using guidelines developed for chemical annealing. Here, we prove the existence of annealing schedules which guarantee convergence to minimum energy states (see Section XII for formal definitions), and we identify the *rate* of decrease relative to the number of full sweeps.

Turning to RL, there are many similarities with SR, both in purpose and, at least abstractly, in method. RL was designed for the assignment of numeric or symbolic labels to objects in a visual system, such as intensity levels to pixels or geometric labels to cube edges, in order to achieve a "global interpretation" that is consistent with the context and certain "local constraints." Ideally, the process evolves by a series of *local* changes, which are intended to be simple, homogeneous, and performed in parallel. The local constraints are usually so-called "compatibility functions," which are much like statistical correlations, and often defined in reference to a graph. We refer the reader to Davis and Rosenfeld [13] for an expository treatment, to Rosenfeld *et al.* [46] for the origins, to Hummel and Zucker [30] for recent work on the logical and mathematical foundations, and to Rosenfeld and Kak [47] for applications to iterative segmentation.

But there are also fundamental differences. First, most variants of RL are rather ad hoc and heuristic. Second, and more importantly, RL is essentially a *nonstochastic* process, both in the interaction model and in the updating algorithms. (Indeed, various probabilistic analogies are often avoided as misleading; see [30], for example.) There is nothing in RL corresponding to an equilibrium measure or even a joint probability law over configurations, whereas there is no analogue in SR of the all-important, iterative updating *formulas* and corresponding sequence of "probability estimates" for various hypotheses involving pixel or object classification.

In summary, there are shared goals and shared features (lo-cality, parallelism, etc.) but SR and RL are quite distinct, at least as practiced in the references made here.

XI. GIBBS SAMPLER: GENERAL DESCRIPTION

We return to the general notation of Section IV: $\backslash = \{X_s, s \in S\}$ is an MRF over a graph $\{\mathcal{G}_s, s \in S\}$ with state spaces Λ_s, configuration space $\Omega = \Pi_s \Lambda_s$, and Gibbs distribution $\pi(\omega) = e^{-U(\omega)/T}/Z$, $\omega \in \Omega$.

The general computational problems are

A) sample from the distribution π;
B) minimize U over Ω;
C) compute expected values.

Of course, we are most concerned with B), which corresponds to MAP estimation when π is the posterior distribution. The most basic problem is A), however, because A) together with annealing yields B) and A) together with the ergodic theorem yields C). We will state three theorems corresponding to A), B), and C) above. Theorem C is not used here and will be proven elsewhere; we state it because of its potential importance to other methods of restoration and to hypothesis testing.

Let us imagine a simple processor placed at each site s of the graph. The connectivity relation among the processors is determined by the bonds: the processor at s is connected to each processor for the sites in $\mathcal{G}_s$. In the cases of interest here (and elsewhere) the number of sites N is very large. However, the size of the neighborhoods, and thus the number of connections to a given processor, is modest, only eight in our experiments, including line, pixel and mixed bonds.

The state of the machine evolves by discrete changes and it is therefore convenient to discretize time, say $t = 1, 2, 3, \cdots$. At time t, the state of the processor at site s is a random variable $X_s(t)$ with values in Λ_s. The total configuration is $X(t) \doteq (X_{s_1}(t), X_{s_2}(t), \cdots, X_{s_N}(t))$, which evolves due to state changes of the individual processors. The starting configuration, $X(0)$, is arbitrary. At each epoch, only *one* site undergoes a (possible) change, so that $X(t-1)$ and $X(t)$ can differ in at most one coordinate. Let $n_1, n_2, \cdots$ be the sequence in which the sites are "visited" for replacement; thus, $n_t \in S$ and $X_{s_i}(t) = X_{s_i}(t-1)$, $i \neq n_t$. Each processor is programmed to follow the same algorithm: at time t, a sample is drawn *from the local characteristics* of π for $s = n_t$ and $\omega = X(t-1)$. In other words, we choose a state $x \in \Lambda_{n_t}$ from the conditional distribution of X_{n_t} given the observed states of the neighboring sites $X_r(t-1)$, $r \in \mathcal{G}_{n_t}$. The new configuration $X(t)$ has $X_{n_t}(t) = x$ and $X_s(t) = X_s(t-1)$, $s \neq n_t$.

These are *local* computations, and *identical* in nature when π is homogeneous. Moreover, the actual calculation is *trivial* since the local characteristics are generally very simple. These conditional probabilities were discussed in Section IV and we refer the reader again to formulas (4.8) and (4.9). Notice that Z does not appear.

Given an initial configuration $X(0)$, we thus obtain a sequence $X(1)$, $X(2)$, $X(3)$, $\cdots$ of configurations whose convergence properties will be described in Section XII. The limits obtained do not depend on $X(0)$. The sequence (n_t) we actually use is simply the one corresponding to a raster scan, i.e., repeatedly visiting all the sites in some "natural" fixed order. Of course, in this case one does not actually need a processor at each site. But the theorems are valid for very general (not necessarily periodic) sequences (n_t) allowing for *asychronous* schemes in which each processor could be driven *by its own clock*. Let us briefly discuss such a parallel implementation of the Gibbs Sampler and its advantage over the serial version.

Computation is parallel in the sense that it is realized by simple and alike units operating largely independently. Units are dependent only to the extent that each must transmit its current state to its neighbors. Most importantly, the amount of time required for one complete update of the entire system is *independent of the number of sites*. In the raster version, we simply "move" a processor from site to site. Upon arriving at a site, this processor must first load the local neighborhood relations and state values, perform the replacement, and move on. The time required to refresh S grows linearly with $N = |S|$. Thus, for example, for the purposes at hand, the parallel procedure is potentially at least 10^4 times faster than the raster version we used, and which required considerable CPU time on a VAX 780. Of course, we recognize that the fully parallel version will require extremely sophisticated new hardware, although we understand that small prototypes of similar machines are underway at several places.

A more modest degree of parallelism can be simply implemented. Since the convergence theorems are independent of the details of the site replacement scheme $n_1, n_2, \cdots$ the graph associated with the MRF X can be divided into collections of sites with each collection assigned to an independently running (asynchronous) processor. Each such processor would execute a raster scan updating of its assigned sites. Communication requirements will be small if the division of the graph respects the natural topology of the scene, provided, of course, that the neighborhood systems are reasonably local. Such an implementation, with five or ten micro- or minicomputers, represents a straightforward application of available technology.

XII. GIBBS SAMPLER: MATHEMATICAL FOUNDATIONS

As in Section XI, (n_t), $t = 1, 2, \cdots$, is the sequence in which the sites are visited for updating, and $X_s(t)$ denotes the state of site s after t replacement opportunities, of which only those for which $n_\tau = s$, $1 \leqslant \tau \leqslant t$, involve site s. For simplicity, we will assume a common state space $\Lambda_s \equiv \Lambda = \{0, 1, \cdots, L - 1\}$, and as usual that $0 < \pi(\omega) < 1$ for all $\omega \in \Omega$ or, what is the same, that $\sup_\omega |U(\omega)| < \infty$. The initial configuration is $X(0)$.

We now investigate the statistical properties of the random process $\{X(t), t = 0, 1, 2, \cdots\}$. The evolution $X(t - 1) \to X(t)$ of the system was explained in Section XI. In mathematical terms,

$$P(X_s(t) = x_s, s \in S)$$
$$= \pi(X_{n_t} = x_{n_t} | X_s = x_s, s \neq n_t) P(X_s(t - 1)$$
$$= x_s, s \neq n_t) \tag{12.1}$$

where, of course, $\pi = e^{-U/T}/Z$ is the Gibbs measure which drives the process. Our first result states that the distribution of $X(t)$ converges to π as $t \to \infty$ regardless of $X(0)$. The only

assumption is that we continue to visit every site, obviously a necessary condition for convergence.

Theorem A (Relaxation): Assume that for each $s \in S$, the sequence $\{n_t, t \geqslant 1\}$ contains s infinitely often. Then for every starting configuration $\eta \in \Omega$ and every $\omega \in \Omega$,

$$\lim_{t \to \infty} P(X(t) = \omega | X(0) = \eta) = \pi(\omega). \tag{12.2}$$

The proof appears in the Appendix, along with that of Theorem B. Like the Metropolis algorithm, the Gibbs Sampler produces a Markov chain $\{X(t), t = 0, 1, 2, \cdots\}$ with π as equilibrium distribution. The only complication is that the transition probabilities associated with the Gibbs Sampler are nonstationary, and their matrix representations do not commute. This precludes the usual algebraic treatment. These issues are discussed in more detail at the beginning of the Appendix.

We now turn to annealing. Hitherto the temperature has been fixed. Theorem B is an "annealing schedule" or rate of temperature decrease which forces the system into the lowest energy states. The necessary programming modification in the relaxation process is trivial, and the *local* nature of the calculations is preserved.

Let us indicate the dependence of π on T by writing π_T, and let $T(t)$ denote the temperature at stage t. The annealing procedure generates a different process $\{X(t), t = 1, 2, \cdots\}$ such that

$$P(X_s(t) = x_s, s \in S)$$
$$= \pi_{T(t)}(X_{n_t} = x_{n_t} | X_s = x_s, s \neq n_t)$$
$$\cdot P(X_s(t - 1) = x_s, s \neq n_t). \tag{12.3}$$

Let

$$\Omega_0 = \{\omega \in \Omega : U(\omega) = \min_\eta U(\eta)\}, \tag{12.4}$$

and let π_0 be the uniform distribution on Ω_0. Finally, define

$$U^* = \max_\omega U(\omega),$$
$$U_* = \min_\omega U(\omega),$$
$$\Delta = U^* - U_*. \tag{12.5}$$

Theorem B (Annealing): Assume that there exists an integer $\tau \geqslant N$ such that for every $t = 0, 1, 2, \cdots$ we have

$$S \subseteq \{n_{t+1}, n_{t+2}, \cdots, n_{t+\tau}\}.$$

Let $T(t)$ be any decreasing sequence of temperatures for which

a) $T(t) \to 0$ as $t \to \infty$;

b) $T(t) \geqslant N\Delta/\log t$
     for all $t \geqslant t_0$ for some integer $t_0 \geqslant 2$.

Then for any starting configuration $\eta \in \Omega$ and for every $\omega \in \Omega$,

$$\lim_{t \to \infty} P(X(t) = \omega | X(0) = \eta) = \pi_0(\omega). \tag{12.6}$$

The first condition is that the individual "clocks" do not slow to an arbitrarily low frequency as the system evolves, and imposes no limitations in practice. For raster replacement,

$\tau = N$. The major practical weakness is b); we cannot truly follow the "schedule" $N\Delta/\log t$. For example, with $N = 20,000$ and $\Delta = 1$, it would take $e^{40,000}$ site visits to reach $T = 0.5$. We single out this temperature because we have obtained good results by making T decrease from approximately $T = 4$ to $T = 0.5$ over 300-1000 sweeps (= $300N - 1000N$ replacements), using a schedule of the form $C/\log(1 + k)$, where k is the number of full sweeps. (Notice that the condition in b) is then satisfied provided C is sufficiently large.) Apparently, the bound in b) is far from optimal, at least as concerns the constant $N\Delta$. (In fact, the proof of Theorem B does establish something stronger, namely that Δ can be taken as the largest absolute difference in energies associated with pairs ω and ω^* which differ at only one coordinate. But this improvement still leaves $N\Delta$ too large for actual practice.) On the other hand, the logarithmic rate is not too surprising in view of the widespread experience of chemists that T must be lowered very slowly, particularly near the freezing point. Otherwise one encounters undesirable physical embodiments of *local* energy minima.

Concerning ergodicity, in statistical physics one attempts to predict the observable quantities of a system in equilibrium; these are the "time averages" of functions on Ω. Under the "ergodic hypothesis," one assuumes that (10.1) is in force, so that time averages approach the corresponding "phase averages" or expected values. The analog for our system is the assertion that, in some suitable sense,

$$\lim_{n \to \infty} \frac{1}{n} \sum_{t=1}^{n} Y(X(t)) = \int_{\Omega} Y(\omega) \, d\pi(\omega). \qquad (12.7)$$

(Here again T is fixed.) As we have already stated, a direct calculation of the righthand side of (12.7), namely,

$$\sum_{\omega} Y(\omega) e^{-U(\omega)/T} \Big/ \sum_{\omega} e^{-U(\omega)/T}$$

is impossible in general. The left-hand side of (12.7) suggests that we use the Gibbs Sampler and compute a time average of the function Y. For most physical systems, the ergodic hypothesis is just that—a *hypothesis*—which can rarely be verified in practice. Fortunately, for our system it is not too difficult to directly establish ergodicity.

Theorem C (Ergodicity): Assume that there exists a τ such that $S \subseteq \{n_{t+1}, \cdots, n_{t+\tau}\}$ for all t. Then for every function Y on Ω and for every starting configuration $\eta \in \Omega$, (12.7) holds with probability one.

XIII. Experimental Results

There are three groups of pictures. Each contains an original image, several degraded versions, and the corresponding restorations, usually at two stages of the annealing process to illustrate its evolution. The degradations are formed from combinations of

i) ϕ absent or $\phi(x) = \sqrt{x}$;
ii) multiplicative or additive noise;
iii) signal-to-noise levels.

The signal-to-noise ratios are all very low. For blurring, we always took the convolution H in (2.3). The restorations are

all MAP estimates generated by the serial Gibbs Sampler with annealing schedule

$$T(k) = \frac{C}{\log(1 + k)}, \qquad 1 \leqslant k \leqslant K$$

where $T(k)$ is the temperature during the kth *iteration* (= full sweep of S), so that K is the total number of iterations. In each case, $C = 3.0$ or $C = 4.0$. No pre- or postfiltering, nor anything else was done. The models for the intensity and line processes were kept as simple as possible; indeed, only cliques of size two appear in the intensity model.

Group 1: The original image [Fig. 2(a)] is a sample of an MRF on Z_{128} with $L = 5$ intensities and the eight-neighbor system (Fig. 1, $c = 2$). The potentials $V_C = 0$ unless $C = \{r, s\}$, in which case

$$V_C(f) = \begin{cases} \frac{1}{3}, & f_s = f_r \\ -\frac{1}{3}, & f_s \neq f_r. \end{cases}$$

Two hundred iterations (at $T \equiv 1$) were made to generate Fig. 2(a).

The first degraded version is Fig. 2(b), which is simply Fig. 2(a) plus Gaussian noise with $\sigma = 1.5$ *relative to* gray levels f, $1 \leqslant f \leqslant 5$. Fig. 2(c) is the restoration of Fig. 2(b) with $K = 25$ iterations only, i.e., early in the annealing process. In Fig. 2(d), $K = 300$.

The second degraded image [Fig. 3(b)] uses the model

$$G = H(F)^{1/2} \cdot N \qquad (13.1)$$

where $\mu = 1$ and $\sigma = 0.1$, again relative to intensities $1 \leqslant f \leqslant 5$. Fig. 3(c) and 3(d) shows the restorations of Fig. 3(b) with $K = 25$ and $K = 300$, respectively.

Group 2: Fig. 4(a) is "hand-drawn." The lattice size is 64×64 and there are three gray levels. Gaussian noise ($\mu = 0$, $\sigma = 0.7$) was added to produce Fig. 4(b). We tried two types of restoration on Fig. 4(b). First, we used the "blob process" which generated Fig. 2(a) for the F-model. There was no line process and $K = 1000$. Obviously these are flaws; see Fig. 4(c).

A line process L was then adjoined to F for the original image model, and the corresponding restoration after 1000 iterations is shown in Fig. 4(d). L itself was described in Case 2 of Section III and the neighborhood system for (F, L) on $Z_{64} \cup D_{64}$ was discussed in Case 3 of Section III. The (prior) distribution on $X = (F, L)$ was as follows. The range of F is $\{0, 1, 2\}$ ($L = 3$ intensities). The energy $U(f, l)$ consists of two terms, say $U(f|l) + U(l)$. To understand the interaction term $U(f|l)$, let d denote a line site, say between pixels r and s. If $L_d = 1$, i.e., an edge element is "present" at d, then the bond between s and r is "broken" and we set $V_{\{r, s\}}(f_r, f_s) = 0$ regardless of f_r, f_s; otherwise ($L_d = 0$) $V_{\{r, s\}}$ is as before except that $\pm\frac{1}{3}$ are replaced by ± 1. As for $U(l)$, only cliques of size four are nonzero, of which there are six distinct types up to rotations. These are shown in Fig. 5(a) with their associated energy values.

Then we corrupted the hand-drawn figure using (13.1) with the same noise parameters as Fig. 3(b), obtaining Fig. 6(b), which is restored in Fig. 6(c) using the same prior on (F, L) as above and with $K = 1000$ iterations.

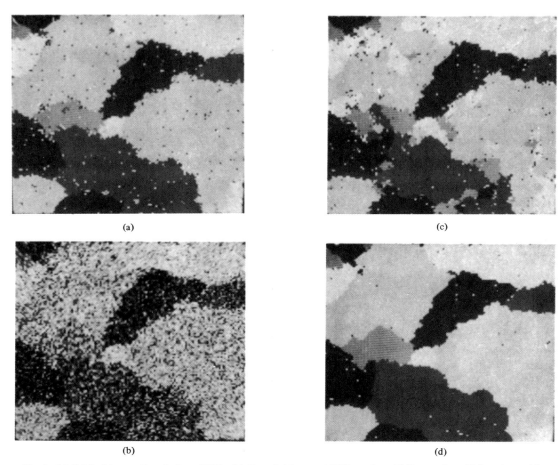

Fig. 2. (a) Original image: Sample from MRF. (b) Degraded image: Additive noise. (c) Restoration: 25 iterations. (d) Restoration: 300 iterations.

Group 3: The results in Group 2 suggest a boundary-finding algorithm for general shapes: allow the line process more directional freedom. Group 3 is an exercise in boundary finding at essentially 0 dB. Fig. 7(a) is a 64 × 64 segment of a roadside photograph that we obtained from the Visions Research Group at the University of Massachusetts. The levels are scaled so that the (existing) two peaks in the histogram occur at $f = 0$ and $f = 1$. We regard Fig. 7(a) as the *blurred image* $H(\mathsf{F})$. Noise is added in Fig. 7(b); the standard error is $\sigma = 0.5$ *relative to the two main gray levels $f = 0, 1$.*

Figs. 7(c) and 7(d) are "restorations" of Fig. 7(b) for $K = 100$ and $K = 1000$ iterations, respectively. The outcome of the line process is indicated by painting black any pixels to the left of or above a "broken bond." The two main regions, comprising the sign and the arrow, are perfectly circumscribed by a continuous sequence of line elements.

The model for X is more complex than the one in Group 2. There are now four possible states for each line site corresponding to "off" ($l = 0$) and three directions, shown in Fig. 5(b). The $U(f|\boldsymbol{l})$ term is the same as before in that the pixel bond between r and s is broken whenever $l_d \neq 0$. The range of F is $\{0, 1\}$ ($L = 2$).

Only cliques of size four are nonzero in $U(\boldsymbol{l})$, as before. However, there are now many combinations for $(l_{d_1}, l_{d_2}, l_{d_3}, l_{d_4})$ given such a clique $C = \{d_1, d_2, d_3, d_4\}$ of line sites, although the number is substantially reduced by assuming rotational invariance, which we do. Fig. 5(c) shows the convention we will use for the ordering and an example of the notation. The energies for the possible configurations $(l_{d_i}, 1 \leqslant i \leqslant 4)$ range from 0 to 2.70. (Remember that high energies correspond to low probability, and that the exponential exaggerates differences.) We took $V(0, 0, 0, 0) = 0$ and $V(l_{d_i}, 1 \leqslant i \leqslant 4) = 2.70$ otherwise, except when exactly two of the l_{d_i} are nonzero. Parallel segments [e.g., $(1, 0, 1, 0)$] receive energy 2.70; sharp turns [e.g., $(0, 2, 1, 0)$] and other "corner" types get 1.80; mild turns [e.g., $(0, 2, 3, 0)$] are 1.35; and continuations [e.g., $(2, 0, 2, 0)$ or $(0, 1, 3, 0)$] are 0.90.

XIV. Concluding Remarks

We have introduced some new theoretical and processing methods for image restoration. The models and estimates are noncausal and nonlinear, and do not represent extensions into two dimensions of one-dimensional filtering and smoothing

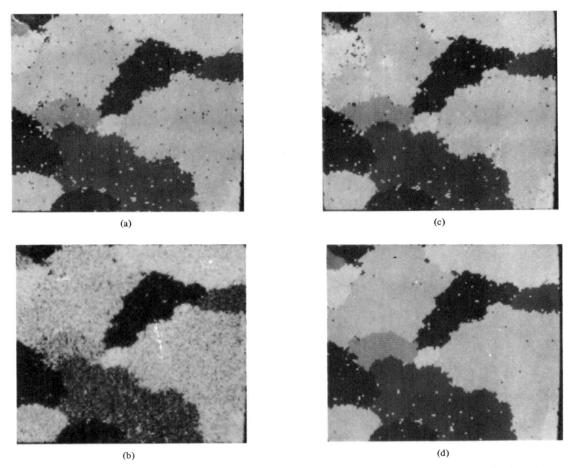

(a)

(c)

(b)

(d)

Fig. 3. (a) Original image: Sample from MRF. (b) Degraded image: Blur, nonlinear transformation, multiplicative noise. (c) Restoration: 25 iterations. (d) Restoration: 300 iterations.

algorithms. Rather, our work is largely inspired by the methods of statistical physics for investigating the time-evolution and equilibrium behavior of large, lattice-based systems.

There are, of course, *many* well-known and remarkable features of these massive, homogeneous physical systems. Among these is the evolution to minimal energy states, regardless of initial conditions. In our work posterior (Gibbs) distribution represents an *imaginary* physical system whose lowest energy states are exactly the MAP estimates of the original image given the degraded "data."

The approach is very flexible. The MRF–Gibbs class of models is tailor-made for representing the dependencies among the intensity levels of nearby pixels as well as for augmenting the usual, pixel-based process by other, unobservable attribute processes, such as our "line process," in order to bring exogenous information into the model. Moreover, the degradation model is almost unrestricted; in particular, we allow for deformations due to the image formation and recording processes. All that is required is that the posterior distribution have a "reasonable" neighborhood structure as a MRF, for in that case the computational load can be accommodated by appro-

priate variants (such as the Gibbs Sampler) of relaxation algorithms for dynamical systems.

APPENDIX
PROOFS OF THEOREMS

Background and Notation

Recall that $\Lambda = \{0, 1, 2, \cdots, L - 1\}$ is the common state space, that η, η', ω, etc. denote elements of the configuration space $\Omega = \Lambda^N$, and that the sites $S = \{s_1, s_2, \cdots, s_N\}$ are visited for updating in the order $\{n_1, n_2, \cdots\} \subset S$. The resulting stochastic process is $\{X(t), t = 0, 1, 2, \cdots\}$, where $X(0)$ is the initial configuration.

For Theorem A, the transitions are governed by the Gibbs distribution $\pi(\omega) = e^{-U(\omega)/T}/Z$ in accordance with (12.1), whereas, for Theorem B (annealing), we use $\pi_{T(t)}$ (see Section XII) for the transition $X(t - 1) \rightarrow X(t)$ [see (12.3)].

Let us briefly discuss the process $\{X(t), t \geqslant 0\}$, restricting attention to constant temperature; the annealing case is essentially the same. To begin with, $\{X(t), t \geqslant 0\}$ is indeed a Markov chain; this is apparent from its construction. Fix t and

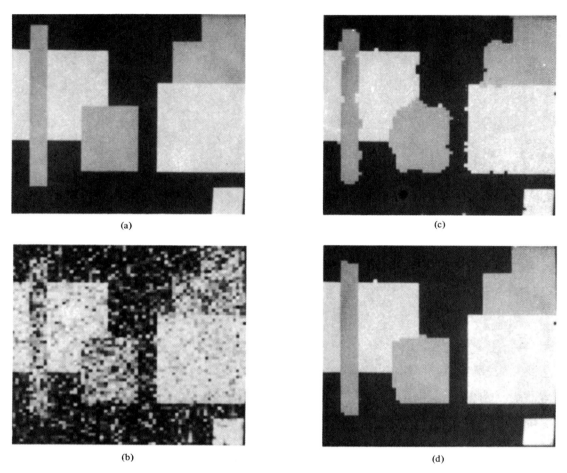

Fig. 4. (a) Original image: "Hand-drawn." (b) Degraded image: Additive noise. (c) Restoration: Without line process; 1000 iterations. (d) Restoration: Including line process; 1000 iterations.

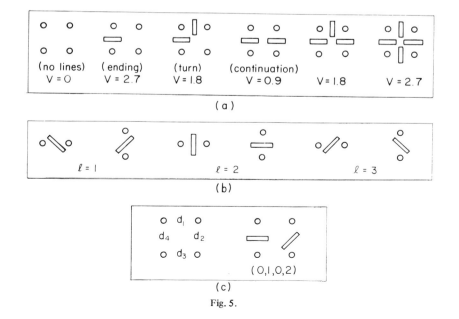

Fig. 5.

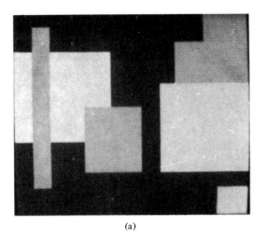

(a)

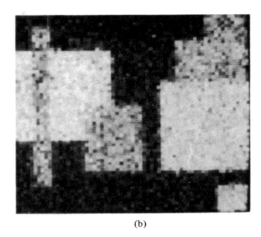

(b)

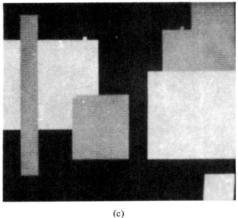

(c)

Fig. 6. (a) Original image: "Hand-drawn." (b) Degraded image: Blur, nonlinear transformation, multiplicative noise. (c) Restoration: including line process; 1000 iterations.

$\omega \in \Omega$. For any $x \in \Lambda$, let ω^x denote the configuration which is x at site n_t and agrees with ω elsewhere. The transition matrix *at time t* is

$$(M_t)_{\eta, \omega} = \begin{cases} \pi(X_{n_t} = x_{n_t} | X_s = x_s, s \neq n_t), \\ \qquad \text{if} \quad \eta = \omega^x \quad \text{for some} \quad x \in \Lambda \\ 0, \quad \text{otherwise} \end{cases}$$

where $(M_t)_{\eta, \omega}$ denotes the row η, column ω entry of M_t, and $\omega = (x_{s_1}, x_{s_2}, \cdots, x_{s_N})$. In particular, the chain is *nonstationary*, although clearly *aperiodic* and *irreducible* (since $\pi(\omega) > 0 \ \forall \ \omega$). Moreover, given any starting vector (distribution) μ_0, the distribution of $X(t)$ is given by the vector $\mu_0 \Pi^t_{j=1} M_j$, i.e.,

$$P_{\mu_0}(X(t) = \omega) = \left(\mu_0 \times \prod_{j=1}^t M_j \right)_\omega$$

$$= \sum_\eta P(X(t) = \omega | X(0) = \eta) \mu_0(\eta).$$

Notice that π is the (necessarily) unique invariant vector, i.e., for every $t = 1, 2, \cdots,$

$$\pi(\omega) = (\pi M_t)_\omega = \sum_\eta P(X(t) = \omega | X(0) = \eta) \pi(\eta). \qquad \text{(A.1)}$$

To see this, fix t and $\omega = \{x_s\}$, and write

$$(\pi M_t)_\omega = \sum_\eta \pi(\eta)(M_t)_{\eta, \omega}$$

$$= \sum_{x \in \Lambda} \pi(\omega^x)(M_t)_{\omega^x, \omega}$$

$$= (M_t)_{\omega^{x'}, \omega} \sum_{x \in \Lambda} \pi(\omega^x) \quad (\text{for } any \quad x' \in \Lambda)$$

$$= \pi(X_{n_t} = x_{n_t} | X_s = x_s, s \neq n_t) \ \pi(X_s = x_s, s \neq n_t)$$

$$= \pi(\omega).$$

It will be convenient to use the following, semistandard notation for transitions. For nonnegative integers $r < t$ and $\omega, \eta \in \Omega$, set

$$P(t, \omega | r, \eta) = P(X(t) = \omega | X(r) = \eta)$$

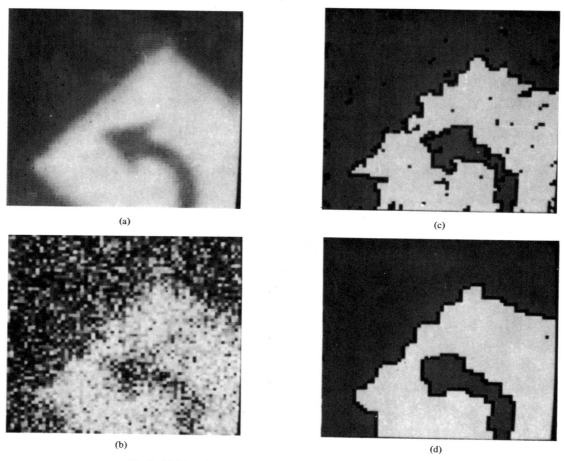

Fig. 7. (a) Blurred image (roadside scene). (b) Degraded image: Additive noise. (c) Restoration including line process; 100 iterations. (d) Restoration including line process; 1000 iterations.

and, for any *distribution* μ on Ω, set

$$P(t, \omega \mid r, \mu) = \sum_{\eta} P(t, \omega \mid r, \eta) \, \mu(\eta).$$

Finally, $\|\mu - \nu\|$ denotes the L^1 distance between two distributions on Ω:

$$\|\mu - \nu\| = \sum_{\omega} |\mu(\omega) - \nu(\omega)|.$$

Obviously, $\mu_n \rightarrow \mu(n \rightarrow \infty)$ in distribution (i.e., $\mu_n(\omega) \rightarrow \mu(\omega)$ $\forall \omega$) if and only if $\|\mu_n - \mu\| \rightarrow 0$, $n \rightarrow \infty$. (Remember that Ω is finite.)

Proof of Theorem A: Set $T_0 = 0$ and define $T_1 < T_2 < \cdots$ such that

$$S \subseteq \{n_{T_{k-1}+1}, n_{T_{k-1}+2}, \cdots, n_{T_k}\}, \quad k = 1, 2, \cdots.$$

This is possible since every site is visited infinitely often. Clearly (at least) k iterations or full sweeps have been completed by "time" T_k. In particular, $kN \leqslant T_k < \infty \; \forall k$. Let

$$K(t) = \sup \{k : T_k < t\}.$$

Obviously $K(t) \rightarrow \infty$ at $t \rightarrow \infty$. The proof of Theorem A is based on the following lemma, which also figures in the proof of the annealing theorem.

Lemma 1: There exists a constant r, $0 \leqslant r < 1$, such that for every $t = 1, 2, \cdots$,

$$\sup_{\omega, \, \eta', \, \eta''} |P(X(t) = \omega \mid X(0) = \eta') - P(X(t)$$

$$= \omega \mid X(0) = \eta'')| \leqslant r^{K(t)}.$$

Assume for now that the lemma is true. Since π is an invariant vector for the chain:

$$\overline{\lim_{t \rightarrow \infty}} \sup_{\omega, \, \eta} |P(X(t)$$

$$= \omega \mid X(0) = \eta) - \pi(\omega)|$$

$$= \overline{\lim_{t \rightarrow \infty}} \sup_{\omega, \, \eta} \left| \sum_{\eta'} \pi(\eta') \{P(X(t)$$

$$= \omega \mid X(0) = \eta) - P(X(t) = \omega \mid X(0) = \eta')\} \right|$$

[by (A.1)]

$$\leqslant \varlimsup_{t \to \infty} \sup_{\omega, \eta', \eta''} \left| P(X(t) = \omega \,|\, X(0) = \eta') \right.$$

$$\left. - P(X(t) = \omega \,|\, X(0) = \eta'') \right|$$

$$= 0, \text{ by Lemma 1.}$$

So it suffices to prove Lemma 1.

Proof of Lemma 1: For each $k = 1, 2, \cdots$ and $1 \leqslant i \leqslant N$, let m_i be the time of the last replacement of site s_i before $T_k + 1$, i.e.,

$$m_i = \sup \{t : t \leqslant T_k, n_t = s_i\}.$$

We can assume, without loss of generality, that $m_1 > m_2 > \cdots > m_N$; otherwise, relabel the sites. For any $\omega = (x_{s_1}, \cdots, x_{s_N})$ and ω',

$$P(X(T_k) = \omega \,|\, X(T_{k-1}) = \omega')$$

$$= P(X_{s_1}(m_1) = x_{s_1}, \cdots, X_{s_N}(m_N)$$

$$= x_{s_N} \,|\, X(T_{k-1}) = \omega')$$

$$= \prod_{j=1}^{N} P(X_{s_j}(m_j) = x_{s_j} \,|\, X_{s_{j+1}}(m_{j+1})$$

$$= x_{s_{j+1}}, \cdots, X_{s_N}(m_N) = x_{s_N}, X(T_{k-1}) = \omega').$$

Let δ be the smallest probability among the local characteristics:

$$\delta = \inf_{\substack{(x_{s_1}, \cdots, x_{s_N}) \in \Omega \\ 1 \leqslant i \leqslant N}} \pi(X_{s_i} = x_{s_i} \,|\, X_{s_j} = x_{s_j}, j \neq i).$$

Then $0 < \delta < 1$ and a little reflection shows that every term in the product above is at least δ. Hence,

$$\inf_{\substack{k = 1, 2, \cdots \\ \omega, \omega'}} P(X(T_k) = \omega \,|\, X(T_{k-1}) = \omega') \geqslant \delta^N. \qquad \text{(A.2)}$$

Consider now the inequality asserted in Lemma 1. It is trivial for $t \leqslant T_1$ since in this case $K(t) = 0$. For $t > T_1$,

$$\sup_{\omega, \eta', \eta''} \left| P(X(t) = \omega \,|\, X(0) = \eta') - P(X(t) = \omega \,|\, X(0) = \eta'') \right|$$

$$= \sup_{\omega} \{ \sup_{\eta} P(X(t) = \omega \,|\, X(0) = \eta)$$

$$- \inf_{\eta} P(X(t) = \omega \,|\, X(0) = \eta) \}$$

$$= \sup_{\omega} \{ \sup_{\eta} \sum_{\omega'} P(X(t) = \omega \,|\, X(T_1)$$

$$= \omega') P(X(T_1) = \omega' \,|\, X(0) = \eta)$$

$$- \inf_{\eta} \sum_{\omega'} P(X(t) = \omega \,|\, X(T_1)$$

$$= \omega') P(X(T_1) = \omega' \,|\, X(0) = \eta) \}$$

$$\doteq \sup_{\omega} Q(t, \omega).$$

Certainly, for each $\omega \in \Omega$,

$$\sup_{\eta} \sum_{\omega'} P(X(t) = \omega \,|\, X(T_1) = \omega') \, P(X(T_1) = \omega' \,|\, X(0) = \eta)$$

$$\leqslant \sup_{\mu} \sum_{\omega'} P(X(t) = \omega \,|\, X(T_1) = \omega') \, \mu(\omega')$$

where the supremum is over all probability measures μ on Ω which, by (A.2), are subject to $\mu(\omega') \geqslant \delta^N \; \forall \, \omega'$. Suppose $\omega' \to P(X(t) = \omega \,|\, X(T_1) = \omega')$ is maximized at $\omega' = \omega^*$ (which depends on ω). Then the last supremum is attained by placing mass δ^N on each ω' and the remaining mass, namely, $1 - |\Omega| \, \delta^N = 1 - L^N \delta^N$, on ω^*. The value so obtained is

$$(1 - (L^N - 1) \, \delta^N) \, P(X(t)$$

$$= \omega \,|\, X(T_1) = \omega^*)$$

$$+ \delta^N \sum_{\omega' \neq \omega^*} P(X(t) = \omega \,|\, X(T_1) = \omega').$$

Similarly,

$$\inf_{\eta} \sum_{\omega'} P(X(t) = \omega \,|\, X(T_1) = \omega') \, P(X(T_1) = \omega' \,|\, X(0) = \eta)$$

$$\geqslant (1 - (L^N - 1) \, \delta^N) \, P(X(t)$$

$$= \omega \,|\, X(T_1) = \omega_*)$$

$$+ \delta^N \sum_{\omega' \neq \omega_*} P(X(t) = \omega \,|\, X(T_1) = \omega_*)$$

where $\omega' \to P(X(t) = \omega \,|\, X(T_1) = \omega')$ is *minimized* at ω_*. It follows immediately that

$$Q(t, \omega) \leqslant (1 - L^N \delta^N) \, \{ P(X(t)$$

$$= \omega \,|\, X(T_1) = \omega^*) - P(X(t) = \omega \,|\, X(T_1) = \omega_*) \},$$

and hence,

$$\sup_{\omega, \eta', \eta''} \left| P(X(t) = \omega \,|\, X(0) = \eta') - P(X(t) = \omega \,|\, X(0) = \eta'') \right|$$

$$\leqslant (1 - L^N \delta^N) \sup_{\omega, \eta', \eta''} \left| P(X(t) \right.$$

$$= \omega \,|\, X(T_1) = \eta') - P(X(t)$$

$$\left. = \omega \,|\, X(T_1) = \eta'') \right|.$$

Proceeding in this way, we obtain the bound

$$(1 - L^N \delta^N)^{K(t)} \sup_{\omega, \eta', \eta''} \left| P(X(t) \right.$$

$$= \omega \,|\, X(T_{K(t)}) = \eta') - P(X(t) = \omega \,|\, X(T_{K(t)}) = \eta'') \Big|$$

and the lemma now follows with $r = 1 - L^N \delta^N$. Notice that $r = 0$ corresponds to the (degenerate) case in which $\delta = L^{-1}$, i.e., all the local characteristics are uniform on Λ. Q.E.D.

Proof of Theorem B: We first state two lemmas.

Lemma 2: For every $t_0 = 0, 1, 2, \cdots$,

$$\lim_{t \to \infty} \sup_{\omega, \eta', \eta''} \left| P(X(t) \right.$$

$$= \omega \,|\, X(t_0) = \eta') - P(X(t) = \omega \,|\, X(t_0) = \eta'') \Big| = 0.$$

Lemma 3:

$$\lim_{t_0 \to \infty} \sup_{t \geqslant t_0} \|P(t, \cdot \,|\, t_0, \pi_0) - \pi_0\| = 0.$$

Recall that π_0 is the uniform probability measure over the minimal energy states $\Omega_0 = \{\omega : U(\omega) = \min_\eta U(\eta)\}$.

First we show how these lemmas imply Theorem B, which states that $P(X(t) = \cdot \,|\, X(0) = \eta)$ converges to π_0 as $t \to \infty$. For any $\eta \in \Omega$,

$$\varlimsup_{t \to \infty} \|P(X(t) = \cdot \,|\, X(0) = \eta) - \pi_0\|$$

$$= \varlimsup_{t_0 \to \infty} \varlimsup_{\substack{t \to \infty \\ t \geqslant t_0}} \left\| \sum_{\eta'} P(t, \cdot \,|\, t_0, \eta') \right.$$

$$\left. \cdot P(t_0, \eta' \,|\, 0, \eta) - \pi_0 \right\|$$

$$\leqslant \varlimsup_{t_0 \to \infty} \varlimsup_{\substack{t \to \infty \\ t \geqslant t_0}} \left\| \sum_{\eta'} P(t, \cdot \,|\, t_0, \eta') \right.$$

$$\left. \cdot P(t_0, \eta' \,|\, 0, \eta) - P(t, \cdot \,|\, t_0, \pi_0) \right\|$$

$$+ \varlimsup_{t_0 \to \infty} \varlimsup_{\substack{t \to \infty \\ t \geqslant t_0}} \|P(t, \cdot \,|\, t_0, \pi_0) - \pi_0\|.$$

The last term is zero by Lemma 3. Furthermore, since $P(t_0, \cdot \,|\, 0, \eta)$ and π_0 have total mass 1, we have

$$\left\| \sum_{\eta'} P(t, \cdot \,|\, t_0, \eta') P(t_0, \eta' \,|\, 0, \eta) - P(t, \cdot \,|\, t_0, \pi_0) \right\|$$

$$= \sum_\omega \sup_{\eta''} \left| \sum_{\eta'} (P(t, \omega \,|\, t_0, \eta') - P(t, \omega \,|\, t_0, \eta'')) \right.$$

$$\times (P(t_0, \eta' \,|\, 0, \eta) - \pi_0(\eta')) \Big|$$

$$\leqslant 2 \sum_\omega \sup_{\eta', \eta''} |P(t, \omega \,|\, t_0, \eta') - P(t, \omega \,|\, t_0, \eta'')|.$$

Finally, then,

$$\varlimsup_{t \to \infty} \|P(X(t) = \cdot \,|\, X(0) = \eta) - \pi_0\|$$

$$\leqslant 2 \sum_\omega \varlimsup_{t_0 \to \infty} \varlimsup_{\substack{t \to \infty \\ t \geqslant t_0}} \sup_{\eta', \eta''} |P(t, \omega \,|\, t_0, \eta')$$

$$- P(t, \omega \,|\, t_0, \eta'')|$$

$$= 0 \quad \text{by Lemma 2.} \qquad \text{Q.E.D.}$$

Proof of Lemma 2: We follow the proof of Lemma 1. Fix $t_0 = 0, 1, \cdots$ and define $T_k = t_0 + k\tau$, $k = 0, 1, 2, \cdots$. Recall that $S \subseteq \{n_{t+1}, \cdots, n_{t+\tau}\}$ for all t by hypothesis, that $\pi_{T(t)}(\omega) = e^{-U(\omega)/T(t)}/Z$ and that U^*, U_* are the maximum and minimum of $U(\omega)$, respectively, the range being $\Delta = U^* - U_*$. Let

$$\delta(t) = \inf_{\substack{1 \leqslant i \leqslant N \\ (x_{s_1}, \cdots, x_{s_N}) \in \Omega}} \pi_{T(t)}(X_{s_i} = x_{s_i} \,|\, X_{s_j} = x_{s_j}, j \neq i).$$

Observe that

$$\delta(t) \geqslant \frac{e^{-U^*/T(t)}}{L e^{-U_*/T(t)}} = \frac{1}{L} e^{-\Delta/T(t)}.$$

Now fix k for the moment and define the m_i as before:

$$m_i = \sup\{t : t \leqslant T_k, n_t = s_i\}, \quad 1 \leqslant i \leqslant N.$$

We again assume that $m_1 > m_2 > \cdots > m_N$. Then

$$P(X(T_k) = \omega \,|\, X(T_{k-1}) = \omega')$$

$$= P(X_{s_1}(m_1) = x_{s_1}, \cdots, X_{s_N}(m_N)$$

$$= x_{s_N} \,|\, X(T_{k-1}) = \omega')$$

$$= \prod_{j=1}^N P(X_{s_j}(m_j) = x_{s_j} \,|\, X_{s_{j+1}}(m_{j+1})$$

$$= x_{s_{j+1}}, \cdots, X_{s_N}(m_N) = x_{s_N}, X(T_{k-1}) = \omega')$$

$$\geqslant \prod_{j=1}^N \delta(m_j) \quad \text{(using (12.3) and the definition of δ)}$$

$$\geqslant L^{-N} \prod_{j=1}^N e^{-\Delta/T(m_j)}$$

$$\geqslant L^{-N} \exp\left\{ -\frac{\Delta N}{T(t_0 + k\tau)} \right\} \quad \text{(since $m_j \leqslant T_k$}$$

$$= t_0 + k\tau, j = 1, 2, \cdots, N, \text{ and } T(\cdot) \text{ is decreasing)}$$

$$\geqslant L^{-N}(t_0 + k\tau)^{-1}$$

wherever $t_0 + k\tau$ is sufficiently large. In fact, for a sufficiently small constant C, we can and do assume that

$$\inf_{\omega, \omega'} P(X(T_k) = \omega \,|\, X(T_{k-1}) = \omega') \geqslant \frac{CL^{-N}}{t_0 + k\tau} \tag{A.3}$$

for every $t_0 = 0, 1, 2, \cdots$ and $k = 1, 2, \cdots$, bearing in mind that T_k depends on t_0.

For each $t > t_0$, define $K(t) = \sup\{k : T_k < t\}$ so that $K(t) \to \infty$ as $t \to \infty$. Fix $t > T_1$ and continue to follow the argument in Lemma 1, but using (A.3) in place of (A.2), obtaining

$$\sup_{\omega, \eta', \eta''} |P(X(t) = \omega \,|\, X(t_0) = \eta') - P(X(t) = \omega \,|\, X(t_0) = \eta'')|$$

$$\leqslant \prod_{k=1}^{K(t)} \left(1 - \frac{C}{t_0 + k\tau}\right).$$

Hence it will be sufficient to show that

$$\lim_{m \to \infty} \prod_{k=1}^m \left(1 - \frac{C}{t_0 + k\tau}\right) = 0 \tag{A.4}$$

for every t_0. However, (A.4) is a well-known consequence of the divergence of the series $\sum_k (t_0 + k\tau)^{-1}$ for all t_0, τ. This completes the proof of Lemma 2.

Proof of Lemma 3: The probability measures $P(t, \cdot \,|\, t_0, \pi_0)$ figure prominently in the proof, and for notational ease we prefer to write $P_{t_0, t}(\cdot)$, so that for any $t \geqslant t_0 > 0$ we have

$$P_{t_0, t}(\omega) = \sum_\eta P(X(t) = \omega \,|\, X(t_0) = \eta) \pi_0(\eta).$$

To begin with, we claim that for any $t > t_0 \geqslant 0$,

$$\|P_{t_0, t} - \pi_{T(t)}\| \leqslant \|P_{t_0, t-1} - \pi_{T(t)}\|. \tag{A.5}$$

Assume for convenience that $n_t = s_1$. Then

$$\|P_{t_0,t} - \pi_{T(t)}\|$$

$$= \sum_{(x_{s_1}, \cdots, x_{s_N})} |\pi_{T(t)}(X_{s_1} = x_{s_1} | X_s = x_s, s \neq s_1)$$

$$\cdot P_{t_0,t-1}(X_s = x_s, s \neq s_1)$$

$$- \pi_{T(t)}(X_s = x_s, s \in S)|$$

$$= \sum_{x_{s_2}, \cdots, x_{s_N}} \left\{ \sum_{x_{s_1} \in \Lambda} \pi_{T(t)}(X_{s_1} = x_{s_1} | X_s = x_s, s \neq s_1) \right.$$

$$\times |P_{t_0,t-1}(X_s = x_s, s \neq s_1)$$

$$\left. - \pi_{T(t)}(X_s = x_s, s \neq s_1)| \right\}$$

$$= \sum_{x_{s_2}, \cdots, x_{s_N}} |P_{t_0,t-1}(X_s = x_s, s \neq s_1)$$

$$- \pi_{T(t)}(X_s = x_s, s \neq s_1)|$$

$$= \sum_{x_{s_2}, \cdots, x_{s_N}} \left| \sum_{x_{s_1}} \{P_{t_0,t-1}(X_s = x_s, s \in S) \right.$$

$$\left. - \pi_{T(t)}(X_s = x_s, s \in S)\} \right|$$

$$\leq \sum_{(x_{s_1}, \cdots, x_{s_N}) \in \Omega} |P_{t_0,t-1}(X_s = x_s, s \in S)$$

$$- \pi_{T(t)}(X_s = x_s, s \in S)|$$

$$= \|P_{t_0,t-1} - \pi_{T(t)}\|.$$

Observe that $\|\pi_0 - \pi_{T(t)}\| \to 0$ as $t \to \infty$. To see this, let $|\Omega_0|$ be the size of Ω_0. Then

$$\pi_{T(t)}(\omega) = \frac{e^{-U(\omega)/T(t)}}{\displaystyle\sum_{\omega' \in \Omega_0} e^{-U(\omega')/T(t)} + \sum_{\omega' \in \Omega \setminus \Omega_0} e^{-U(\omega')/T(t)}}$$

$$= \frac{e^{-(U(\omega) - U_*)/T(t)}}{|\Omega_0| + \displaystyle\sum_{\omega' \in \Omega \setminus \Omega_0} e^{-(U(\omega') - U_*)/T(t)}}$$

$$\xrightarrow{t \to \infty} \begin{cases} 0, & \omega \notin \Omega_0 \\ \dfrac{1}{|\Omega_0|}, & \omega \in \Omega_0. \end{cases} \tag{A.6}$$

Next, we claim that

$$\sum_{t=1}^{\infty} \|\pi_{T(t)} - \pi_{T(t+1)}\| < \infty. \tag{A.7}$$

Since

$$\sum_{t=1}^{\infty} \|\pi_{T(t)} - \pi_{T(t+1)}\| = \sum_{\omega} \sum_{t=1}^{\infty} |\pi_{T(t)}(\omega) - \pi_{T(t+1)}(\omega)|$$

and since $\pi_{T(t)}(\omega) \to \pi_0(\omega)$ for every ω, it will be enough to show that, for every ω, $\pi_T(\omega)$ is monotone (increasing or decreasing) in T for all T sufficiently small. But this is clear from (A.6): if $\omega \notin \Omega_0$, then a little calculus shows that $\pi_T(\omega)$ is strictly increasing for $T \in (0, \epsilon)$ for some ϵ, whereas if $\omega \in \Omega_0$, then $\pi_T(\omega)$ is strictly decreasing for all $T > 0$.

Lemma 3 can now be obtained from (A.5) and (A.7) in the following way. Fix $t > t_0 \geq 0$:

$$\|P_{t_0,t} - \pi_0\|$$

$$\leq \|P_{t_0,t} - \pi_{T(t)}\| + \|\pi_{T(t)} - \pi_0\|$$

$$\leq \|P_{t_0,t-1} - \pi_{T(t)}\| + \|\pi_{T(t)} - \pi_0\|, \quad \text{by (A.5)}$$

$$\leq \|P_{t_0,t-1} - \pi_{T(t-1)}\| + \|\pi_{T(t-1)} - \pi_{T(t)}\| + \|\pi_{T(t)} - \pi_0\|$$

$$\leq \|P_{t_0,t-2} - \pi_{T(t-1)}\| + \|\pi_{T(t-1)} - \pi_{T(t)}\| + \|\pi_{T(t)} - \pi_0\|$$

$$\leq \|P_{t_0,t-2} - \pi_{T(t-2)}\| + \|\pi_{T(t-2)} - \pi_{T(t-1)}\| + \|\pi_{T(t-1)} - \pi_{T(t)}\| + \|\pi_{T(t)} - \pi_0\|.$$

Proceeding in this way,

$$\|P_{t_0,t} - \pi_0\| \leq \|P_{t_0,t_0} - \pi_{T(t_0)}\| + \sum_{k=t_0}^{t-1} \|\pi_{T(k)} - \pi_{T(k+1)}\| + \|\pi_{T(t)} - \pi_0\|.$$

Since $P_{t_0,t_0} = \pi_0$ and $\|\pi_{T(t)} - \pi_0\| \to 0$ as $t \to \infty$, we have,

$$\overline{\lim_{t_0 \to \infty}} \sup_{t \geq t_0} \|P_{t_0,t} - \pi_0\|$$

$$\leq \overline{\lim_{t_0 \to \infty}} \sup_{t > t_0} \sum_{k=t_0}^{t-1} \|\pi_{T(k)} - \pi_{T(k+1)}\|$$

$$= \overline{\lim_{t_0 \to \infty}} \sum_{k=t_0}^{\infty} \|\pi_{T(k)} - \pi_{T(k+1)}\|$$

$$= 0 \quad \text{due to (A.7).} \qquad \text{Q.E.D.}$$

ACKNOWLEDGMENT

The authors would like to acknowledge their debt to U. Grenander for a flow of ideas; his work on pattern theory [23] prefigures much of what is here. They also thank D. E. McClure and S. Epstein for their sound advice and technical assistance, and V. Mirelli for introducing them to the practical side of image processing as well as arguing for MRF scene models.

REFERENCES

[1] K. Abend, T. J. Harley, and L. N. Kanal, "Classification of binary random patterns," *IEEE Trans. Inform. Theory*, vol. IT-11, pp. 538–544, 1965.

[2] H. C. Andrews and B. R. Hunt, *Digital Image Restoration*. Englewood Cliffs, NJ, Prentice-Hall, 1977.

[3] M. S. Bartlett, *The Statistical Analysis of Spatial Pattern*. London: Chapman and Hall, 1976.

[4] T. Berger and F. Bonomi, "Parallel updating of certain Markov random fields," preprint.

[5] J. Besag, "Nearest-neighbor systems and the auto-logistic model for binary data," *J. Royal Statist. Soc.*, series B, vol. 34, pp. 75–83, 1972.

[6] ——, "Spatial interaction and the statistical analysis of lattice systems (with discussion)," *J. Royal Statist. Soc.*, series B, vol. 36, pp. 192–326, 1974.

[7] A. B. Bortz, M. H. Kalos, and J. L. Lebowitz, "A new algorithm

for Monte Carlo simulation of Ising spin systems," *J. Comp. Phys.*, vol. 17, pp. 10–18, 1975.

[8] V. Cerný, "A thermodynamical approach to the travelling salesman problem: an efficient simulation algorithm," preprint, Inst. Phys. & Biophys., Comenius Univ., Bratislava, 1982.

[9] P. Cheeseman, "A method of computing maximum entropy probability values for expert systems," preprint.

[10] R. Chellappa and R. L. Kashyap, "Digital image restoration using spatial interaction models," *IEEE Trans. Acoust., Speech, Signal Processing*, vol. ASSP-30, pp. 461–472, 1982.

[11] D. B. Cooper and F. P. Sung, "Multiple-window parallel adaptive boundary finding in computer vision," *IEEE Trans. Pattern Anal. Machine Intell.*, vol. PAMI-5, pp. 299–316, 1983.

[12] G. C. Cross and A. K. Jain, "Markov random field texture models," *IEEE Trans. Pattern Anal. Machine Intell.*, vol. PAMI-5, pp. 25–39, 1983.

[13] L. S. Davis and A. Rosenfeld, "Cooperating processes for low-level vision: A survey," 1980.

[14] H. Derin, H. Elliott, R. Christi, and D. Geman, "Bayes smoothing algorithms for segmentation of images modelled by Markov random fields," Univ. Massachusetts Tech. Rep., Aug. 1983.

[15] R. L. Dobrushin, "The description of a random field by means of conditional probabilities and conditions of its regularity," *Theory Prob. Appl.*, vol. 13, pp. 197–224, 1968.

[16] H. Elliott, H. Derin, R. Christi, and D. Geman, "Application of the Gibbs distribution to image segmentation," Univ. Massachusetts Tech. Rep., Aug. 1983.

[17] H. Elliott, F. R. Hansen, L. Srinivasan, and M. F. Tenorio, "Application of MAP estimation techniques to image segmentation," Univ. Massachusetts Tech. Rep., 1982.

[18] P. A. Flinn, "Monte Carlo calculation of phase separation in a 2-dimensional Ising system," *J. Statist. Phys.*, vol. 10, pp. 89–97, 1974.

[19] B. R. Frieden, "Restoring with maximum likelihood and maximum entropy," *J. Opt. Soc. Amer.*, vol. 62, pp. 511–518, 1972.

[20] D. Geman and S. Geman, "Parameter estimation for some Markov random fields," Brown Univ. Tech. Rep., Aug. 1983.

[21] S. Geman, "Stochastic relaxation methods for image restoration and expert systems," in *Proc. ARO Workshop: Unsupervised Image Analysis*, Brown Univ., 1983; to appear in *Automated Image Analysis: Theory and Experiments*, D. B. Cooper, R. L. Launer, and D. E. McClure, Eds. New York: Academic, 1984.

[22] U. Grenander, *Lectures in Pattern Theory*, Vols. I–III. New York: Springer-Verlag, 1981.

[23] D. Griffeath, "Introduction to random fields," in *Denumerable Markov Chains*, Kemeny, Knapp and Snell, Eds. New York: Springer-Verlag, 1976.

[24] A. Habibi, "Two-dimensional Bayesian estimate of images," *Proc. IEEE*, vol. 60, pp. 878–883, 1972.

[25] F. R. Hansen and H. Elliott, "Image segmentation using simple Markov field models," *Comput. Graphics Image Processing*, vol. 20, pp. 101–132, 1982.

[26] A. R. Hanson and E. M. Riseman, "Segmentation of natural scenes," in *Computer Visions Systems*. New York: Academic, 1978.

[27] J. M. Hammersley and D. C. Handscomb, *Monte Carlo Methods*. London: Methuen, 1964.

[28] M. Hassner and J. Sklansky, "The use of Markov random fields as models of texture," *Comput. Graphics Image Processing*, vol. 12, pp. 357–370, 1980.

[29] G. E. Hinton and T. J. Sejnowski, "Optimal perceptual inference," in *Proc. IEEE Conf. Comput. Vision Pattern Recognition*, 1983.

[30] R. A. Hummel and S. W. Zucker, "On the foundations of relaxation labeling processes," *IEEE Trans. Pattern Anal. Machine Intell.*, vol. PAMI-5, pp. 267–287, 1983.

[31] B. R. Hunt, "Bayesian methods in nonlinear digital image restoration," *IEEE Trans. Comput.*, vol. C-23, pp. 219–229, 1977.

[32] V. Isham, "An introduction to spatial point processes and Markov random fields," *Int. Statist. Rev.*, vol. 49, pp. 21–43, 1981.

[33] E. Ising, *Zeitschrift Physik*, vol. 31, p. 253, 1925.

[34] A. K. Jain, "Advances in mathematical models for image processing," *Proc. IEEE*, vol. 69, pp. 502–528, 1981.

[35] A. K. Jain and E. Angel, "Image restoration, modeling and reduction of dimensionality," *IEEE Trans. Comput.*, vol. C-23, pp. 470–476, 1974.

[36] E. T. Jaynes, "Prior probabilities," *IEEE Trans. Syst. Sci. Cybern.*, vol. SSC-4, pp. 227–241, 1968.

[37] L. N. Kanal, "Markov mesh models," in *Image Modeling*. New York: Academic, 1980.

[38] R. L. Kashyap and R. Chellappa, "Estimation and choice of neighbors in spatial interaction models of images," *IEEE Trans. Inform. Theory*, vol. IT-29, pp. 60–72, 1983.

[39] R. Kinderman and J. L. Snell, *Markov Random Fields and Their Applications*. Providence, RI: Amer. Math. Soc., 1980.

[40] S. Kirkpatrick, C. D. Gellatt, Jr., and M. P. Vecchi, "Optimization by simulated annealing," IBM Thomas J. Watson Research Center, Yorktown Heights, NY, 1982.

[41] P. A. Levy, "A special problem of Brownian motion and a general theory of Gaussian random functions," in *Proc. 3rd Berkeley Symp. Math. Statist. and Prob.*, vol. 2, 1956.

[42] N. Metropolis, A. W. Rosenbluth, M. N. Rosenbluth, A. H. Teller, and E. Teller, "Equations of state calculations by fast computing machines," *J. Chem. Phys.*, vol. 21, pp. 1087–1091, 1953.

[43] N. E. Nahi and T. Assefi, "Bayesian recursive image estimation," *IEEE Trans. Comput.*, vol. C-21, pp. 734–738, 1972.

[44] D. K. Pickard, "A curious binary lattice process," *J. Appl. Prob.*, vol. 14, pp. 717–731, 1977.

[45] W. H. Richardson, "Bayesian-based iterative method of image restoration," *J. Opt. Soc. Amer.*, vol. 62, pp. 55–59, 1972.

[46] A. Rosenfeld, R. A. Hummel, and S. W. Zucker, "Scene labeling by relaxation operations," *IEEE Trans. Syst., Man, Cybern.*, vol. SMC-6, pp. 420–433, 197.

[47] A. Rosenfeld and A. C. Kak, *Digital Picture Processing*, vols. 1, 2, 2nd ed. New York: Academic, 1982.

[48] F. Spitzer, "Markov random fields and Gibbs ensembles," *Amer. Math. Mon.*, vol. 78, pp. 142–154, 1971.

[49] J. A. Stuller and B. Kruz, "Two-dimensional Markov representations of sampled images," *IEEE Trans. Commun.*, vol. COM-24, pp. 1148–1152, 1976.

[50] H. J. Trussell, "The relationship between image restoration by the maximum a posteriori method and a maximum entropy method," *IEEE Trans. Acoust., Speech, Signal Processing*, vol. ASSP-28, pp. 114–117, 1980.

[51] J. W. Woods, "Two-dimensional discrete Markovian fields," *IEEE Trans. Inform. Theory*, vol. IT-18, pp. 232–240, 1972.

38
Introduction

(1985)
David H. Ackley, Geoffrey E. Hinton, and Terrence J. Sejnowski

A learning algorithm for Boltzmann machines
Cognitive Science 9:147–169

There are two separate aspects to most neural networks: the dynamics of the network and the learning algorithm for determining the strengths of connections in the network. Forming the connections is quite a different process from using them after they are formed.

In this paper we see a dynamics that is similar to others we have seen before—in particular, to Hopfield (paper 27), Kirkpatrick, Gelatt, and Vecchi (paper 33), and Geman and Geman (paper 37). In addition, the authors propose a learning rule that is capable of giving rise to network weights that can solve some difficult problems, as shown by computer simulations. Solving the problems correctly requires the system to form interesting internal representations of the input data.

The "Boltzmann machine" is a neural network whose basic elements are the same as those used by Hopfield in paper 27. The computational units are binary, and can be on or off. Units add up their inputs and make a decision based on the sum of the synaptic weights times the input activity for all the units connected to a particular unit. If the sum is greater than a threshold, θ, the unit takes on the value one; otherwise, the unit takes on the value zero. The authors define an "energy" of the system (equation 1) that is effectively the same as the energy term used by Hopfield: it is a quadratic energy function, the sum of products of the synaptic weights and the values of the two elements of the state vector coupled together by that weight plus a term involving the thresholds of the units. In the original Hopfield model, the thresholds were all zero. If we assume asynchronous updating, given the rule for neurons changing state, it is easy to show that the system always reduces energy or keeps it constant.

This was the result that Hopfield obtained. The energy minimum that the system finds may be a local minimum, rather than the global energy minimum. In some applications, the local minima are associated with particular stored items, and there may be no need to reach a global minimum. However, in many cases the global minimum is required. The global minimum may be just over the ridge, but the state change rules in the simple model only allow the energy of the system to drop or stay the same.

Ackley, Hinton, and Sejnowski introduce a probabilistic component into the system by making the individual units stochastic rather than deterministic. The difference in energy, ΔE, is computed for the two different states of the unit. Then the unit is switched to the on state with probability p given by

$$p = \frac{1}{1 + e^{-\Delta E/T}}.$$

This means the relative probabilities of two states that differ in energy by ΔE will be given by the Boltzmann distribution—hence the name Boltzmann machine. This means that the unit can assume a new state value even if the result is an overall increase in energy, allowing, in conjunction with simulated annealing, a good chance of finding the global energy minimum.

The clever joining of a neural network with simulated annealing has created a system with an effective dynamics: it is capable of searching a state space for the pattern giving the global energy minimum. The next problem for a general neural network is to construct a learning algorithm that can put the minimum where you want it, i.e., so the minimum is the solution to the problem the network is designed to solve.

The rest of the paper is devoted to structuring and teaching the network so that it is capable of coming up with answers to interesting questions. Many questions (for example, Exclusive-OR of two inputs; see Minksy and Papert, paper 13) cannot be solved by a simple associative system. There are several ways of solving a problem like this. One is to expand the initial representation, so it includes units that correspond to higher order relations between the inputs. The other is to let the system contain what are called "hidden" units, that is, units that are neither input units nor output units. In a multilayer system the hidden units could correspond to the middle layers. It is this multilayer architecture that is at the heart of what Ackley, Hinton, and Sejnowski discuss and simulate.

In any connectionist learning scheme, it is necessary to specify a rule to change the weights coupling units together. But, with hidden units, we are faced with the "credit assignment" problem that we have seen before—for example, in Barto, Sutton, and Anderson (paper 32)—since it is not obvious what the properties of the hidden units and their connections are, nor how to change these properties.

One useful definition of learning for Boltzmann machines involves matching probabilities between the environment and the network. All states of the network are possible at thermal equilibrium, with their relative probabilities given by the Boltzmann distribution. If the probabilities of the states in the network are the same as the probabilities of states of the environment, then the network has an accurate model of the environment. This is a strange definition of learning, but useful in many situations. It can be shown that there is a simple local learning rule for adjusting the weights in the Boltzmann machine so as to increase the fit between the network and the environment, although the derivation of the rule is not trivial. The rule involves first letting the system "free run." The probabilities of the states taken by each unit can be estimated. Then the visible (input and output) units are "clamped," that is, forced to take appropriate values. Again, values of the probabilities of the states of the units are estimated. Then local weight changes are possible, the change being proportional to the difference in the probabilities of the units coupled by that weight.

This process is slow, since it is necessary to estimate probabilities, but it works, as shown by a number of intriguing computer simulations. The best known problem is called the "encoder problem" and has become a benchmark for many learning networks.

Suppose we have 4 input and 4 output units, and 2 hidden units (the 4-2-4 encoder). The network is strictly feedforward: the input units connect to the hidden units and the hidden units connect to the output units. We want to connect a unit in the input layer to a unit in the output layer, so that if, say, the third unit turns on in the input layer, then the third unit turns on in the output layer. If there were 4 units in the hidden layer, there would be a trivial solution: to make a direct connection between an input unit and an output unit. To code which of 4 units is active with only 2 hidden units is not so easy. Binary counting is how humans can solve this problem. In simulations, the learning system is reliably able to achieve this solution: it sets the coupling coefficients so that the hidden layer is effectively a binary counter.

One could argue that this is not a very interesting problem. But the fact that it was solved successfully in such an elegant way is impressive. One can view the hidden layers as developing an effective representation of the inputs, based on the task to be done. The representations that networks can develop are effective—and optimal in some cases where optimality can be checked. We shall see more examples of this in papers 41 and 42, where a newer learning algorithm, back propagation, is presented.

The Boltzmann machine presented in this paper is impressive. The entire package is an effective and important *learning system*. It can find global energy minima using a technique that can be analyzed theoretically, and it has an effective, though slow, learning algorithm, which, in some architectures, can develop unit responses that look very much like effective abstract representations. Sejnowski, Kienker, and Hinton (1986) present some other examples of the powers of the Boltzmann machine—for example, detection of symmetry—as do Hinton and Sejnowski (1986). Paul Smolensky (1986) has developed a closely related theory, called Harmony Theory, which has been thoroughly analyzed and applied to a number of interesting simulations.

References

G. E. Hinton and T. J. Sejnowski (1986), "Learning and relearning in Boltzmann machines," *Parallel Distributed Processing*, Vol. 1, D. E. Rumelhart and J. L. McClelland (Eds.), Cambridge, MA: MIT Press, pp. 282–317.

T. J. Sejnowksi, P. K. Kienker, and G. E. Hinton (1986), "Learning symmetry groups with hidden units: beyond the perceptron," *Physica* D22:260–275.

P. Smolensky (1986), "Information processing in dynamical systems: foundations of Harmony Theory," *Parallel Distributed Processing*, Vol. 1, D. E. Rumelhart and J. L. McClelland (Eds.), Cambridge, MA: MIT Press, pp. 194–281.

(1985)

David H. Ackley, Geoffrey E. Hinton, and Terrence J. Sejnowski

A learning algorithm for Boltzmann machines
Cognitive Science 9:147–169

The computational power of massively parallel networks of simple processing elements resides in the communication bandwidth provided by the hardware connections between elements. These connections can allow a significant fraction of the knowledge of the system to be applied to an instance of a problem in a very short time. One kind of computation for which massively parallel networks appear to be well suited is large constraint satisfaction searches, but to use the connections efficiently two conditions must be met: First, a search technique that is suitable for parallel networks must be found. Second, there must be some way of choosing internal representations which allow the preexisting hardware connections to be used efficiently for encoding the constraints in the domain being searched. We describe a general parallel search method, based on statistical mechanics, and we show how it leads to a general learning rule for modifying the connection strengths so as to incorporate knowledge about a task domain in an efficient way. We describe some simple examples in which the learning algorithm creates internal representations that are demonstrably the most efficient way of using the preexisting connectivity structure.

1. Introduction

Evidence about the architecture of the brain and the potential of the new VLSI technology have led to a resurgence of interest in "connectionist" systems (Feldman & Ballard, 1982; Hinton & Anderson, 1981) that score their long-term knowledge as the strengths of the connections between simple neuron-like processing elements. These networks are clearly suited to tasks like vision that can be performed efficiently in parallel networks which have physical connections in just the places where processes need to communicate. For problems like surface interpolation from sparse depth data (Grimson, 1981; Terzopoulos, 1984) where the necessary decision units and communication paths can be determined in advance, it is relatively easy to see how to make good use of massive parallelism. The more difficult problem is to discover parallel organizations that do not require so much problem-dependent information to be built into the architecture of the network. Ideally, such a system would adapt a given

structure of processors and communication paths to whatever problem it was faced with.

This paper presents a type of parallel constraint satisfaction network which we call a "Boltzmann Machine" that is capable of learning the underlying constraints that characterize a domain simply by being shown examples from the domain. The network modifies the strengths of its connections so as to construct an internal *generative* model that produces examples with the same probability distribution as the examples it is shown. Then, when shown any particular example, the network can "interpret" it by finding values of the variables in the internal model that would generate the example. When shown a partial example, the network can complete it by finding internal variable values that generate the partial example and using them to generate the remainder. At present, we have an interesting mathematical result that guarantees that a certain learning procedure will build internal representations which allow the connection strengths to capture the underlying constraints that are implicit in a large ensemble of examples taken from a domain. We also have simulations which show that the theory works for some simple cases, but the current version of the learning algorithm is very slow.

The search for general principles that allow parallel networks to learn the structure of their environment has often begun with the assumption that networks are randomly wired. This seems to us to be just as wrong as the view that *all* knowledge is innate. If there are connectivity structures that are good for particular tasks that the network will have to perform, it is much more efficient to build these in at the start. However, not all tasks can be foreseen, and even for ones that can, fine-tuning may still be helpful.

Another common belief is that a general connectionist learning rule would make sequential "rule-based" models unnecessary. We believe that this view stems from a misunderstanding of the need for multiple levels of description of large systems, which can be usefully viewed as either parallel or serial depending on the grain of the analysis. Most of the key issues and questions that have been studed in the context of sequential models do not magically disappear in connectionist models. It is still necessary to perform searches for good solutions to problems or good interpretations of perceptual input, and to create complex internal representations. Ultimately it will be necessary to bridge the gap between hardware-oriented connectionist descriptions and the more abstract symbol manipulation models that have proved to be an extremely powerful and pervasive way of describing human information processing (Newell & Simon, 1972).

2. The Boltzmann Machine

The Boltzmann Machine is a parallel computational organization that is well suited to constraint satisfaction tasks involving large numbers of "weak" constraints. Constraint-satisfaction searches (e.g., Waltz, 1975; Winston, 1984) normally use "strong" constraints that *must* be satisfied by any solution. In problem domains such as games and puzzles, for example, the goal criteria often have this character, so strong constraints are the rule.[1] In some problem domains, such as finding the most plausible interpretation of an image, many of the criteria are not all-or-none, and frequently even the best possible solution violates some constraints (Hinton, 1977). A variation that is more appropriate for such domains uses weak constraints that incur a cost when violated. The quality of a solution is then determined by the total cost of all the constraints that it violates. In a perceptual interpretation task, for example, this total cost should reflect the implausibility of the interpretation.

The machine is composed of primitive computing elements called *units* that are connected to each other by bidirectional *links*. A unit is always in one of two states, *on* or *off*, and it adopts these states as a probabilistic function of the states of its neighboring units and the *weights* on its links to them. The weights can take on real values of either sign. A unit being on or off is taken to mean that the system currently accepts or rejects some elemental hypothesis about the domain. The weight on a link represents a weak pairwise constraint between two hypotheses. A positive weight indicates that the two hypotheses tend to support one another; if one is currently accepted, accepting the other should be more likely. Conversely, a negative weight suggests, other things being equal, that the two hypotheses should not both be accepted. Link weights

are *symmetric*, having the same strength in both directions (Hinton & Sejnowski, 1983).[2]

The resulting structure is related to a system described by Hopfield (1982), and as in his system, each global state of the network can be assigned a single number called the "energy" of that state. With the right assumptions, the individual units can be made to act so as to *minimize the global energy*. If *some* of the units are externally forced or "clamped" into particular states to represent a particular input, the system will then find the minimum energy configuration that is compatible with that input. The energy of a configuration can be interpreted as the extent to which that combination of hypotheses violates the constraints implicit in the problem domain, so in minimizing energy the system evolves towards "interpretations" of that input that increasingly satisfy the constraints of the problem domain.

The energy of a global configuration is defined as

$$E = -\sum_{i<j} w_{ij} s_i s_j + \sum_i \theta_i s_i \qquad (1)$$

where w_{ij} is the strength of connection between units i and j, s_i is 1 if unit i is on and 0 otherwise, and θ_i is a threshold.

2.1. Minimizing Energy

A simple algorithm for finding a combination of truth values that is a *local* minimum is to switch each hypothesis into whichever of its two states yields the lower total energy given the current states of the other hypotheses. If hardware units make their decisions asynchronously, and if transmission times are negligible, then the system always settles into a local energy minimum (Hopfield, 1982). Because the connections are symmetric, the difference between the energy of the whole system with the kth hypothesis rejected and its energy with the kth hypothesis accepted can be determined locally by the kth unit, and this "energy gap" is just

$$\Delta E_k = \sum_i w_{ki} s_i - \theta_k \qquad (2)$$

Therefore, the rule for minimizing the energy contributed by a unit is to adopt the *on* state if its total input from the other units and from outside the system

[1] But, see (Berliner & Ackley, 1982) for argument that, even in such domains, strong constraints must be used only where absolutely necessary for legal play, and in particular must not propagate into the determination of *good* play.

[2] Requiring the weights to be symmetric may seem to restrict the constraints that can be represented. Although a constraint on boolean variables A and B such as "$A \equiv B$ with a penalty of 2 points for violation" is obviously symmetric in A and B, "$A \Rightarrow B$ with a penalty of 2 points for violation" appears to be fundamentally asymmetric. Nevertheless, this constraint can be represented by the combination of a constraint on A alone and a symmetric pairwise constraint as follows: "Lose 2 points if A is true" and "Win 2 points if both A and B are true."

exceeds its threshold. This is the familiar rule for binary threshold units.

The threshold terms can be eliminated from Eqs. (1) and (2) by making the following observation: the effect of θ_i on the global energy or on the energy gap of an individual unit is identical to the effect of a link with strength $-\theta_i$ between unit i and a special unit that is by definition always held in the *on* state. This "true unit" need have no physical reality, but it simplifies the computations by allowing the threshold of a unit to be treated in the same manner as the links. The value $-\theta_i$ is called the *bias* of unit i. If a permanently active "true unit" is assumed to be part of every network, then Eqs. (1) and (2) can be written as:

$$E = -\sum_{i<j} w_{ij}s_i s_j \tag{3}$$

$$\Delta E_k = \sum_i w_{ki}s_i \tag{4}$$

2.2. Using Noise to Escape from Local Minima

The simple, deterministic algorithm suffers from the standard weakness of gradient descent methods: It gets stuck in *local* minima that are not globally optimal. This is not a problem in Hopfield's system because the local energy minima of his network are used to store "items": If the system is started near some local minimum, the desired behavior is to fall into that minimum, not to find the global minimum. For constraint satisfaction tasks, however, the system must try to escape from local minima in order to find the configuration that is the global minimum given the current input.

A simple way to get out of local minima is to occasionally allow jumps to configurations to higher energy. An algorithm with this property was introduced by Metropolis, Rosenbluth, Rosenbluth, Teller, & Teller (1953) to study average properties of thermodynamic systems (Binder, 1978) and has recently been applied to problems of constraint satisfaction (Kirkpatrick, Gelatt, & Vecchi, 1983). We adopt a form of the Metropolis algorithm that is suitable for parallel computation: If the energy gap between the *on* and *off* states of the kth unit is ΔE_k then regardless of the previous state set $s_k = 1$ with probability

$$p_k = \frac{1}{(1 + e^{-\Delta E_k/T})} \tag{5}$$

where T is a parameter that acts like temperature (see Figure 1).

The decision rule in Eq. (5) is the same as that for a particle which has two energy states. A system of such particles in contact with a heat bath at a given temperature will eventually reach thermal equilibrium and the

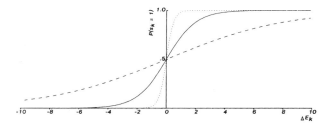

Figure 1 Eq. (5) at $T = 1.0$ (*solid*), $T = 4.0$ (*dashed*), and $T = 0.25$ (*dotted*).

probability of finding the system in any global state will then obey a Boltzmann distribution. Similarly, a network of units obeying this decision rule will eventually reach "thermal equilibrium" and the relative probability of two global states will follow the Boltzman distribution:

$$\frac{P_\alpha}{P_\beta} = e^{-(E_\alpha - E_\beta)/T} \tag{6}$$

where P_α is the probability of being in the αth global state, and E_α is the energy of that state.

The Boltzmann distribution has some beautiful mathematical properties and it is initimately related to information theory. In particular, the difference in the log probabilities of two global states is just their energy difference (at a temperature of 1). The simplicity of this relationship and the fact that the equilibrium distribution is independent of the path followed in reaching equilibrium are what make Boltzmann machines interesting.

At low temperatures there is a strong bias in favor of states with low energy, but the time required to reach equilibrium may be long. At higher temperatures the bias is not so favorable but equilibrium is reached faster. A good way to beat this trade-off is to start at a high temperature and gradually reduce it. This corresponds to annealing a physical system (Kirkpatrick, Gelatt, & Vecchi, 1983). At high temperatures, the network will ignore small energy differences and will rapidly approach equilibrium. In doing so, it will perform a search of the coarse overall structure of the space of global states, and will find a good minimum at that coarse level. As the temperature is lowered, it will begin to respond to smaller energy differences and will find one of the better minima within the coarse-scale minimum it discovered at high temperature. Kirkpatrick et al. have shown that this way of searching the coarse structure before the fine is very effective for combinatorial problems like graph partitioning, and we believe it will also prove useful when trying to satisfy multiple weak constraints, even though it will clearly fail in cases where the best solution corresponds to a minimum that is deep, narrow, and isolated.

3. A Learning Algorithm

Perhaps the most interesting aspect of the Boltzmann Machine formulation is that it leads to a domain-independent learning algorithm that modifies the connection strengths between units in such a way that the whole network develops an internal model which captures the underlying structure of its environment. There has been a long history of failure in the search for such algorithms (Newell, 1982), and many people (particularly in Artificial Intelligence) now believe that no such algorithms exist. The major technical stumbling block which prevented the generalization of simple learning algorithms to more complex networks was this: To be capable of interesting computations, a network must contain nonlinear elements that are not directly constrained by the input, and when such a network does the wrong thing it appears to be impossible to decide which of the many connection strengths is at fault. This "credit-assignment" problem was what led to the demise of perceptrons (Minsky & Papert, 1968; Rosenblatt, 1961). The perceptron convergence theorem guarantees that the weights of a single layer of decision units can be trained, but it could not be generalized to networks of such units when the task did not directly specify how to use all the units in the network.

This version of the credit-assignment problem can be solved within the Boltzmann Machine formulation. By using the right stochastic decision rule, and by running the network until it reaches "thermal equilibrium" at some finite temperature, we achieve a mathematically simple relationship between the probability of a global state and its energy. For a network that is running freely without any input from the environment, this relationship is given by Eq. (6). Because the energy is a *linear* function of the weights (Eq. 1) this leads to a remarkably simple relationship between the log probabilities of global states and the individual connection strengths:

$$\frac{\partial \ln P_\alpha}{\partial w_{ij}} = \frac{1}{T}[s_i^\alpha s_j^\alpha - p'_{ij}] \qquad (7)$$

where s_i^α is the state of the ith unit in the αth global state (so $s_i^\alpha s_j^\alpha$ is 1 only if units i and j are both on in state α), and p'_{ij} is just the probability of finding the two units i and j on at the same time when the system is at equilibrium.

Given Eq. (7), it is possible to manipulate the log probabilities of global states. If the environment directly specifies the required probabilities P_α for each global state α, there is a straightforward way of converging on a set of weights that achieve those probabilities, provided any such set exists (for details, see Hinton & Sejnowski, 1983a). However, this is not a particularly interesting kind of learning because the system has to be given the required probabilities of *complete* global states. This means that the central question of what internal representation should be used has already been decided by the environment. The interesting problem arises when the environment implicitly contains high-order constraints and the network must choose internal representations that allow these constraints to be expressed efficiently.

3.1. Modeling the Underlying Structure of an Environment

The units of a Boltzmann Machine partition into two functional groups, a nonempty set of *visible* units and a possibly empty set of *hidden* units. The visible units are the interface between the network and the environment; during training all the visible units are clamped into specific states by the environment; when testing for completion ability, any subset of the visible units may be clamped. The hidden units, if any, are never clamped by the environment and can be used to "explain" underlying constraints in the ensemble of input vectors that cannot be represented by pairwise constraints among the visible units. A hidden unit would be needed, for example, if the environment demanded that the states of three visible units should have even parity—a regularity that cannot be enforced by pairwise interactions alone. Using hidden units to represent more complex hypotheses about the states of the visible units, such higher-order constraints among the visible units can be reduced to first and second-order constraints among the whole set of units.

We assume that each of the environmental input vectors persists for long enough to allow the network to approach thermal equilibrium, and we ignore any structure that may exist in the *sequence* of environmental vectors. The structure of an environment can then be specified by giving the probability distribution over all 2^v states of the v visible units. The network will be said to have a perfect model of the environment if it achieves exactly the same probability distribution over these 2^v states when it is running freely at thermal equilibrium with all units unclamped so there is no environmental input.

Unless the number of hidden units is exponentially large compared to the number of visible units, it will be impossible to achieve a *perfect* model because even if the network is totally connected the $(v + h - 1) \times$ $(v + h)/2$ weights and $(v + h)$ biases among the v visible and h hidden units will be insufficient to model the 2^v probabilities of the states of the visible units specified by the environment. However, if there are regularities in the environment, and if the network uses its hidden units to capture these regularities, it may achieve a good match to the environmental probabilities.

An information-theoretic measure of the discrepancy between the network's internal model and the environment is

$$G = \sum_\alpha P(V_\alpha) \ln \frac{P(V_\alpha)}{P'(V_\alpha)} \tag{8}$$

where $P(V_\alpha)$ is the probability of the αth state of the visible units when their states are determined by the environment, and $P'(V_\alpha)$ is the corresponding probability when the network is running freely with no environmental input. The G metric, sometimes called the asymmetric divergence or information gain (Kullback, 1959; Renyi, 1962), is a measure of the distance from the distribution given by the $P'(V_\alpha)$ to the distribution given by the $P(V_\alpha)$. G is zero if and only if the distributions are identical; otherwise it is positive.

The term $P'(V_\alpha)$ depends on the weights, and so G can be altered by changing them. To perform gradient descent in G, it is necessary to know the partial derivative of G with respect to each individual weight. In most cross-coupled nonlinear networks it is very hard to derive this quantity, but because of the simple relationships that hold at thermal equilibrium, the partial derivative of G is straightfoward to derive for our networks. The probabilities of global states are determined by their energies (Eq. 6) and the energies are determined by the weights (Eq. 1). Using these equations the partial derivative of G (see the appendix) is:

$$\frac{\partial G}{\partial w_{ij}} = -\frac{1}{T}(p_{ij} - p'_{ij}) \tag{9}$$

where p_{ij} is the average probability of two units both being in the *on* state when the environment is clamping the states of the visible units, and p'_{ij}, as in Eq. (7), is the corresponding probability when the environmental input is not present and the network is running freely. (Both these probabilities must be measured at equilibrium.) Note the similarity between this equation and Eq. (7), which shows how changing a weight affects the log probability of a single state.

To minimize G, it is therefore sufficient to observe p_{ij} and p'_{ij} when the network is at thermal equilibrium, and to change each weight by an amount proportional to the difference between these two probabilities:

$$\Delta w_{ij} = \varepsilon(p_{ij} - p'_{ij}) \tag{10}$$

where ε scales the size of each weight change.

A surprising feature of this rule is that it uses only *locally available* information. The change in a weight depends only on the behavior of the two units it connects, even though the change optimizes a global measure, and the best value for each weight depends on the values of all the other weights. If there are no hidden units, it can be shown that G-space is concave (when viewed from above) so that simple gradient descent will not get trapped at poor local minima. With hidden units, however, there can be local minima that correspond to different ways of using the hidden units to represent the higher-order constraints that are implicit in the probability distribution of environmental vectors. Some techniques for handling these more complex G-spaces are discussed in the next section.

Once G has been minimized the network will have captured as well as possible the regularities in the environment, and these regularities will be enforced when performing completion. An alternative view is that the network, in minimizing G, is finding the set of weights that is most likely to have generated the set of environmental vectors. It can be shown that maximizing this likelihood is mathematically equivalent to minimizing G (Peter Brown, personal communication, 1983).

3.2. Controlling the Learning

There are a number of free parameters and possible variations in the learning algorithm presented above. As well as the size of ε, which determines the size of each step taken for gradient descent, the lengths of time over which p_{ij} and p'_{ij} are estimated have a significant impact on the learning process. The values employed for the simulations presented here were selected primarily on the basis of empirical observations.

A practical system which estimates p_{ij} and p'_{ij} will necessarily have some noise in the estimates, leading to occasional "uphill steps" in the value of G. Since hidden units in a network can create local minima in G, this is not necessarily a liability. The effect of the noise in the estimates can be reduced, if desired, by using a small value for ε or by collecting statistics for a longer time, and so it is relatively easy to implement an annealing search for the minimum of G.

The objective function G is a metric that specifies how well two probability distributions match. Problems arise if an environment specifies that only a small subset of the possible patterns over the visible units ever occur. By default, the unmentioned patterns must occur with probability zero, and the only way a Boltzmann Machine running at a non-zero temperature can guarantee that certain configurations *never* occur is to give those configurations infinitely high energy, which requires infinitely large weights.

One way to avoid this implicit demand for infinite weights is to occasionally provide "noisy" input vectors. This can be done by filtering the "correct" input vectors through a process that has a small probability of reversing each of the bits. These noisy vectors are then clamped on the visible units. If the noise is small, the correct vectors will dominate the statistics, but every vector will have some chance of occuring and so infinite energies will not be needed. This "noisy clamping" technique was used for all the examples presented here. It works quite well, but we are not entirely satisfied with it and have been investigating other methods of preventing the weights from growing too large when only a few of the possible input vectors ever occur.

The simulations presented in the next section employed a modification of the obvious steepest descent method implied by Eq. (10). Instead of changing w_{ij} by an amount proportional to $p_{ij} - p'_{ij}$, it is simply incremented by a fixed "weight-step" if $p_{ij} > p'_{ij}$ and decremented by the same amount if $p_{ij} < p'_{ij}$. The advantage of this method over steepest descent is that it can cope with wide variations in the first and second derivatives of G. It can make significant progress on dimensions where G changes gently without taking very large divergent steps on dimensions where G falls rapidly and then rises rapidly again. There is no suitable value for the ε in Eq. (10) in such cases. Any value large enough to allow progress along the gently sloping floor of a ravine will cause divergent oscillations up and down the steep sides of the ravine.[3]

4. The Encoder Problem

The "encoder problem" (suggested to us by Sanjaya Addanki) is a simple abstraction of the recurring task of communicating information among various components of a parallel network. We have used this problem to test out the learning algorithm because it is clear what the optimal solution is like and it is nontrivial to discover it. Two groups of visible units, designated V_1 and V_2, represent two systems that wish to communicate their states. Each group has v units. In the simple

[3] The problem of finding a suitable value for ε disappears if one performs a line search for the lowest value of G along the current direction of steepest descent, but line searches are inapplicable in this case. *Only* the local gradient is available. There are bounds on the second derivative that can be used to pick conservative values of ε (Mark Derthick, personal communication, 1984), and methods of this kind are currently under investigation.

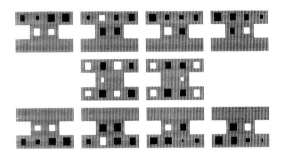

Figure 2 A solution to an encoder problem. The link weights are displayed using a recursive notation. Each unit is represented by a shaded 1-shaped box; from top to bottom the rows of boxes represent groups V_1, H, and V_2. Each shaded box is a map of the entire network, showing the strengths of that unit's connections to other units. At each position in a box, the size of the white (positive) or black (negative) rectangle indicates the magnitude of the weight. In the position that would correspond to a unit connecting to itself (the second position in the top row of the second unit in the top row, for example), the bias is displayed. All connections between units appear twice in the diagram, once in the box for each of the two units being connected. For example, the black square in the top right corner of the left-most unit of V_1 represents the same connection as the black square in the top left corner of the rightmost unit of V_1. This connection has a weight of -30.

formulation we consider here, each group has only one unit on at a time, so there are only v different states of each group. V_1 and V_2 are not connected directly but both are connected to a group of h hidden units H, with $h < v$ so H may acts as a limited capacity bottleneck through which information about the states of V_1 and V_2 must be squeezed. Since all simulations began with all weights set of zero, finding a solution to such a problem requires that the two visible groups come to agree upon the meanings of a set of codes without any *a priori* conventions for communication through H.

To permit perfect communication between the visible groups, it must be the case that $h \geqslant log_2 v$. We investigated minimal cases in which $h = log_2 v$, and cases when h was somewhat larger than $log_2 v$. In all cases, the environment for the network consisted of v equiprobable vectors of length $2v$ which specified that one unit in V_1 and the corresponding unit in V_2 should be on together with all other units off. Each visible group is completely connected internally and each is completely connected to H, but the units in H are not connected to each other.

Because of the severe speed limitation of simulation on a sequential machine, and because the learning requires many annealings, we have primarily experimented with small versions of the encoder problem. For example, Figure 2 shows a good solution to a "4-2-4" encoder problem in which $v = 4$ and $h = 2$. The interconnections between the visible groups and

H have developed a binary coding—each visible unit causes a different pattern of *on* and *off* states in the units of *H*, and corresponding units in V_1 and V_2 support identical patterns in *H*. Note how the bias of the second unit of V_1 and V_2 is possible to compensate for the fact that the code which represents that unit has all the *H* units turned off.

4.1. The 4-2-4 Encoder

The experiments on neworks with $v = 4$ and $h = 2$ were performed using the following learning cycle:

1. *Estimation of p_{ij}*: Each environmental vector in turn was clamped over the visible units. For each environmental vector, the network was allowed to reach equilibrium twice. Statistics about how often pairs of units were both on together were gathered at equilibrium. To prevent the weights from growing too large we used the "noisy" clamping technique described in Section 3.2. Each *on* bit of a clamped vector was set to *off* with a probability of 0.15 and each *off* bit was set to *on* with a probability of 0.05.

2. *Estimation of p'_{ij}*: The network was completely unclamped and allowed to reach equilibrium at a temperature of 10. Statistics about co-occurrences were then gathered for as many annealings as were used to estimate p_{ij}.

3. *Updating the weights*: All weights in the network were incremented or decremented by a fixed weight-step of 2, with the sign of the increment being determined by the sign of $p_{ij} - p'_{ij}$.

When a settling to equilibrium was required, all the unclamped units were randomized with equal probability on or off (corresponding to raising the temperature to infinity), and then the network was allowed to run for the following times at the following temperatures: $[2@20, 2@15, 2@12, 4@10]$.[4] After this annealing schedule it was assumed that the network had reached equilibrium, and statistics were collected at a temperature of 10 for 10 units of time.

We observed three main phases in the search for the global minimum of *G*, and found that the occurrence of these phases was relatively insensitive to the precise parameters used. The first phase begins with all the weights set to zero, and is characterized by the development of negative weights throughout most of the network, implementing two winner-take-all networks that model the simplest aspect of the environmental

[4] One unit of time is defined as the time required for each unit to be given, on average, one chance to change its state. This means that if there are *n* unclamped units, a time period of 1 involves *n* random probes in which some unit is given a chance to change its state.

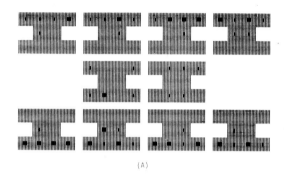

(A)

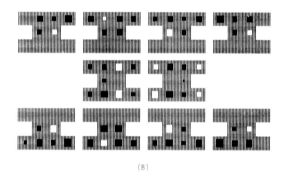

(B)

Figure 3 Two phases in the development of the perfect binary encoding shown in Figure 2. The weights are shown (A) after 4 learning trials and (B) after 60 learning trials.

structure—only one unit in each visible group is normally active at a time. In a *4-2-4* encoder, for example, the number of possible patterns over the visible units is 2^8. By implementing a winner-take-all network among each group of four this can be reduced to 4×4 low energy patterns. Only the final reduction from 2^4 to 2^2 low energy patterns requires the hidden units to be used for communicating between the two visible groups. Figure 3a shows a *4-2-4* encoder network after four learning cycles.

Although the hidden units are exploited for inhibition in the first phase, the lateral inhibition task can be handled by the connections within the visible group alone. In the second phase, the hidden units begin to develop positive weights to some of the units in the visible groups, and they tend to maintain symmetry between the sign and approximate magnitude of a connection to a unit in V_1 and the corresponding unit in V_2. The second phase finishes when every hidden unit has significant connection weights to each unit in V_1 and analogous weights to each unit in V_2, and most of the different codes are being used, but there are some codes that are used more than once and some not at all. Figure 3b shows the same network after 60 learning cycles.

Occasionally, all the codes are being used at the end of the second phase in which case the problem is solved. Usually, however, there is a third and longest phase during which the learning algorithm sorts out the remaining conflicts and finds a global minimum. There are two basic mechanisms involved in the sorting out process. Consider the conflict between the first and fourth units in Figure 3b, which are both employing the code $\langle -, + \rangle$. When the system is running without environmental input, the two units will be on together quite frequently. Consequently, $p'_{1,4}$ will be higher than $p_{1,4}$ because the environmental input tends to prevent the two units from being on together. Hence, the learning algorithm keeps decreasing the weight of the connection between the first and fourth units in each group, and they come to inhibit each other strongly. (This effect explains the variations in inhibitory weights in Figure 2. Visible units with similar codes are the ones that inhibit each other strongly.) Visible units thus compete for "territory" in the space of possible codes, and this repulsion effect causes codes to migrate away from similar neighbors. In addition to the repulsion effect, we observed another process that tends to eventually bring the unused codes adjacent (in terms of hamming distance) to codes that are involved in a conflict. The mechanics of this process are somewhat subtle and we do not take the time to expand on them here.

The third phase finishes when all the codes are being used, and the weights then tend to increase so that the solution locks in and remains stable against the fluctuations caused by random variations in the co-occurrence statistics. (Figure 2 is the same network shown in Figure 3, after 120 learning cycles.)

In 250 different tests of the *4-2-4* encoder, it always found one of the global minima, and once there it remained there. The median time required to discover four different codes was 110 learning cycles. The longest time was 1810 learning cycles.

4.2. The 4-3-4 Encoder

A variation on the binary encoder problem is to give H more units than are absolutely necessary for encoding the patterns in V_1 and V_2. A simple example is the *4-3-4* encoder which was run with the same parameters as the *4-2-4* encoder. In this case the learning algorithm quickly finds four different codes. Then it always goes on to modify the codes so that they are optimally spaced out and no pair differ by only a single bit, as shown in Figure 4. The median time to find four well-spaced codes was 270 learning cycles and the maximum time in 200 trials was 1090.

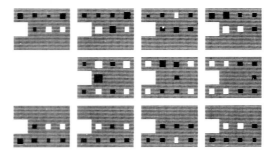

Figure 4 A 4-3-4 encoder that has developed optimally spaced codes.

4.3. The 8-3-8 Encoder

With $v = 8$ and $h = 3$ it took many more learning cycles to find all 8 three-bit codes. We did 20 simulations, running each for 4000 learning cycles using the same parameters as for the *4-2-4* case (but with a probability of 0.02 of reversing each *off* unit during noisy clamping). The algorithm found all 8 codes in 16 out of 20 simulations and found 7 codes in the rest. The median time to find 7 codes was 210 learning cycles and the median time to find all 8 was 1570 cycles.

The difficulty of finding all 8 codes is not surprising since the fraction of the weight space that counts as a solution is much smaller than in the *4-2-4* case. Sets of weights that use 7 of the 8 different codes are found fairly rapidly and they constitute local minima which are far more numerous than the global minima and have almost as good a value of G. In this type of G-space, the learning algorithm must be carefully tuned to achieve a global minimum, and even then it is very slow. We believe that the G-spaces for which the algorithm is well-suited are ones where there are a great many possible solutions and it is not essential to get the very best one. For large networks to learn in a reasonable time, it may be necessary to have enough units and weights and a liberal enough specification of the task so that no single unit or weight is essential. The next example illustrates the advantages of having some spare capacity.

4.4. The 40-10-40 Encoder

A somewhat larger example is the *40-10-40* encoder. The 10 units in H are almost twice the theoretical minimum, but H still acts as a limited bandwidth bottleneck. The learning algorithm works well on this problem. Figure 5 shows its performance when given a pattern in V_1 and required to settle to the corresponding pattern in V_2. Each learning cycle involved annealing once with each of the 40 environmental vectors clamped, and the same number of times without

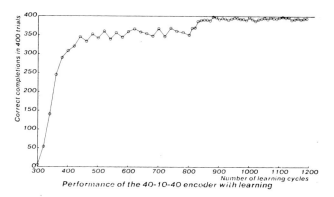

Performance of the 40-10-40 encoder with learning

Figure 5 Completion accuracy of a 40-10-40 encoder during learning. The network was tested by clamping the states of the units in V_1 and letting the remainder of the network reach equilibrium. If just the correct unit was on in V_2, the test was successful. This was repeated 10 times for each of the 40 units in V_1. For the first 300 learning cycles the network was run without connecting up the hidden units. This ensured that each group of 40 visible units developed enough lateral inhibition to implement an effective winner-take-all network. The hidden units were then connected up and for the next 500 learning cycles we used "noisy" clamping switching *on* bits to *off* with a probability of 0.1 and *off* bits to *on* with a probability of 0.0025. After this we removed the noise and this explains the sharp rise in performance after 800 cycles. The final performance asymptotes at 98.6% correct.

clamping. The final performance asymptotes at 98.6% correct.

The codes that the network selected to represent the patterns in V_1 and V_2 were all separated by a hamming distance of at least 2, which is very unlikely to happen by chance. As a test, we compared the weights of the connections between visible and hidden units. Each visible unit has 10 weights connecting it to the hidden units, and to avoid errors, the 10 dimensional weight vectors for two different visible units should not be too similar. The cosine of the angle between two vectors was used as a measure of similarity, and no two codes had a similarity greater than 0.73, whereas many pairs had similarities of 0.8 or higher when the same weights were randomly rearranged to provide a control group for comparison.

To achieve good performance on the completion tests, it was necessary to use a very gentle annealing schedule during testing. The schedule spent twice as long at each temperature and went down to half the final temperature of the schedule used during learning. As the annealing was made faster, the error rate increased, thus giving a very natural speed/accuracy trade-off. We have not pursued this issue any further, but it may prove fruitful because some of the better current models of the speed/accuracy trade-off in human reaction time experiments involve the idea of a biased random walk (Ratcliff, 1978), and the annealing search gives rise to similar underlying mathematics.

5. Representation in Parallel Networks

So far, we have avoided the issue of how complex concepts would be represented in a Boltzmann machine. The individual units stand for "hypotheses," but what is the relationship between these hypotheses and the kinds of concepts for which we have words? Some workers suggest that a concept should be represented in an essentially "local" fashion: The activation of one or a few computing units is the representation for a concept (Feldman & Ballard, 1982); while others view concepts as "distributed" entities: A particular pattern of activity over a large group of units represents a concept, and different concepts corresponds to *alternative* patterns of activity over the same group of units (Hinton, 1981).

One of the better arguments in favor of local representations is their inherent modularity. Knowledge about relationships between concepts is localized in specific connections and is therefore easy to add, remove, and modify, if some reasonable scheme for forming hardware connections can be found (Fahlman, 1980; Feldman, 1982). With distributed representations, however, the knowledge is diffuse. This is good for tolerance to local hardware damage, but it appears to make the design of modules to perform specific functions much harder. It is particularly difficult to see how new distributed representations of concepts could originate spontaneously.

In a Boltzmann machine, a distributed representation corresponds to an energy minimum, and so the problem of creating a good collection of distributed representations is equivalent to the problem of creating a good "energy landscape." The learning algorithm we have presented is capable of solving this problem, and it therefore makes distributed representations considerably more plausible. The diffuseness of any one piece of knowledge is no longer a series objection, because the mathematical simplicitly of the Boltzmann distribution makes it possible to manipulate all the diffuse local weights in a coherent way on the basis of purely local information. The formation of a simple set of distributed representations is illustrated by the encoder problems.

5.1. Communicating Information between Modules

The encoder problem examples also suggest a method for communicating symbols between various components of a parallel computational network. Feldman and Ballard (1982) present sketches of two implementations for this task; using the example of the transmission of the concept "wormy apple" from where it is recognized in the perceptual system to where the phrase "wormy apple" can be generated by the speech system. They argue that there appears to be only two ways that this could be accomplished. In the first method, the perceptual information is encoded into a set of symbols that are then transmitted as messages to the speech system, where they are decoded into a form suitable for utterance. In this case, there would be a set of general-purpose communication lines, analogous to a bus in a conventional computer, that would be used as the medium for all such messages from the visual system to the speech system. Feldman and Ballard describe the problems with such a system as:

• Complex messages would presumably have to be transmitted sequentially over the communication lines.

• Both sender and receiver would have to learn the common code for each new concept.

• The method seems biologically implausible as a mechanism for the brain.

The alternative implementation they suggest requires an individual, dedicated hardware pathway for each concept that is communicated from the perceptual system. The idea is that the simultaneous activation of "apple" and "worm" in the perceptual system can be transmitted over private links to their counterparts in the speech system. The critical issues for such an implementation are having the necessary connections available between concepts, and being able to establish new connection pathways as new concepts are learned in the two systems. The main point of this approach is that the links between the computing units carry simple, nonsymbolic information such as a single activation level.

The behavior of the Boltzmann machine when presented with an encoder problem demonstrates a way of communicating concepts that largely combines the best of the two implementations mentioned. Like the second approach, the computing units are small, the links carry a simple numeric value, and the computational and connection requirements are within the range of biological plausibility. Like the first approach, the architecture is such that many different concepts can be transmitted over the same communication lines, allowing for effective use of limited connections. The learning of new codes to present new concepts emerges automatically as a cooperative process from the G-minimization learning algorithm.

6. Conclusion

The application of statistical mechanics to constraint satisfaction searches in parallel networks is a promising new area that has been discovered independently by several other groups (Geman & Geman, 1983; Smolensky, 1983). There are many interesting issues that we have only mentioned in passing. Some of these issues are discussed in greater detail elsewhere: Hinton and Sejnowski (1983b) and Geman and Geman (1983) describe the relation to Bayesian inference and to more conventional relaxation techniques; Fahlman, Hinton, and Sejnowski (1983) compare Boltzmann machines with some alternative parallel schemes, and discuss some knowledge representation issues. An expanded version of this paper (Hinton, Sejnowski, & Ackley, 1984) presents this material in greater depth and discusses a number of related issues such as the relationship to the brain and the problem of sequential behavior. It also shows how the probabilistic decision function could be realized using gaussian noise, how the assumptions of symmetry in the physical connections and of no time delay in transmission can be relaxed, and describes results of simulations on some other tasks.

Systems with symmetric weights form an interesting class of computational device because their dynamics is governed by an energy function.[5] This is what makes it possible to analyze their behavior and to use them for iterative constraint satisfaction. In their influential exploration of perceptrons, Minsky and Papert (1968, p. 231) concluded that: "Multilayer machines with loops clearly open up all the questions of the general theory of automata." Although this statement is very plausible, recent developments suggest that it may be misleading because it ignores the symmetric case, and it seems to have led to the general belief that it would be impossible to find powerful learning algorithms for networks of perceptron-like elements.

We believe that the Boltzmann Machine is a simple example of a class of interesting stochastic models that exploit the close relationship between Boltzmann distributions and information theory.

All of this will lead to theories [of computation] which are much less rigidly of an all-or-none nature than past and present formal logic. They will be of a much less combinatorial, and much more analytical, character. In fact, there are numerous indications to make us believe that this new system of formal logic will move closer to another discipline which

---

[5] One can easily write down a similar energy function for asymmetric newtorks, but this energy function does not govern the behavior of the network when the links are given their normal causal interpretation.

has been little linked in the past with logic. This is thermodynamics, primarily in the form it was received from Boltzmann, and is that part of theoretical physics which comes nearest in some of its aspects to manipulating and measuring information. (John Von Neumann, *Collected Works* Vol. 5, p. 304)

Appendix: Derivation of the Learning Algorithm

When a network is free-running at equilibrium the probability distribution over the visible units is given by

$$P'(V_\alpha) = \sum_\beta P'(V_\alpha \wedge H_\beta) = \frac{\sum_\beta e^{-E_{\alpha\beta}/T}}{\sum_{\lambda\mu} e^{-E_{\lambda\mu}/T}} \qquad (11)$$

where V_α is a vector of states of the visible units, H_β is a vector of states of the hidden units, and $E_{\alpha\beta}$ is the energy of the system in state $V_\alpha \wedge H_\beta$

$$E_{\alpha\beta} = -\sum_{i<j} w_{ij} s_i^{\alpha\beta} s_j^{\alpha\beta}.$$

Hence,

$$\frac{\partial e^{-E_{\alpha\beta}/T}}{\partial w_{ij}} = \frac{1}{T} s_i^{\alpha\beta} s_j^{\alpha\beta} e^{-E_\alpha/T}.$$

Differentiating (11) then yields

$$\frac{\partial P'(V_\alpha)}{\partial w_{ij}} = \frac{\frac{1}{T} \sum_\beta e^{-E_{\alpha\beta}/T} s_i^{\alpha\beta} s_j^{\alpha\beta}}{\sum_{\alpha\beta} e^{-E_{\alpha\beta}/T}}$$

$$- \frac{\sum_\beta e^{-E_{\alpha\beta}/T} \frac{1}{T} \sum_{\lambda\mu} e^{-E_{\lambda\mu}/T} s_i^{\lambda\mu} s_j^{\lambda\mu}}{\left(\sum_{\lambda\mu} e^{-E_{\lambda\mu}/T}\right)^2}$$

$$= \frac{1}{T}\left[\sum_\beta P'(V_\alpha \wedge H_\beta) s_i^{\alpha\beta} s_j^{\alpha\beta}\right.$$

$$\left. - P'(V_\alpha) \sum_{\lambda\mu} P'(V_\lambda \wedge H_\mu) s_i^{\lambda\mu} s_j^{\lambda\mu}\right].$$

This derivative is used to compute the gradient of the *G*-measure

$$G = \sum_\alpha P(V_\alpha) \ln \frac{P(V_\alpha)}{P'(V_\alpha)}$$

where $P(V_\alpha)$ is the clamped probability distribution over the visible units and is independent of w_{ij}. So

$$\frac{\partial G}{\partial w_{ij}} = -\sum_\alpha \frac{P(V_\alpha)}{P'(V_\alpha)} \frac{\partial P'(V_\alpha)}{\partial w_{ij}}$$

$$= -\frac{1}{T} \sum_\alpha \frac{P(V_\alpha)}{P'(V_\alpha)} \left[\sum_\beta P'(V_\alpha \wedge H_\beta) s_i^{\alpha\beta} s_j^{\alpha\beta}\right.$$

$$\left. - P'(V_\alpha) \sum_{\lambda\mu} P'(V_\lambda \wedge H_\mu) s_i^{\lambda\mu} s_j^{\lambda\mu}\right].$$

Now,

$$P(V_\alpha \wedge H_\beta) = P(H_\beta|V_\alpha) P(V_\alpha),$$

$$P'(V_\alpha \wedge H_\beta) = P'(H_\beta|V_\alpha) P'(V_\alpha),$$

and

$$P'(H_\beta|V_\alpha) = P(H_\beta|V_\alpha). \qquad (12)$$

Equation (12) holds because the probability of a hidden state given some visible state must be the same in equilibrium whether the visible units were clamped in that state or arrived there by free-running. Hence,

$$P'(V_\alpha \wedge H_\beta) \frac{P(V_\alpha)}{P'(V_\alpha)} = P(V_\alpha \wedge H_\beta).$$

Also,

$$\sum_\alpha P(V_\alpha) = 1.$$

Therefore,

$$\frac{\partial G}{\partial w_{ij}} = -\frac{1}{T}[p_{ij} - p'_{ij}]$$

where

$$p_{ij} \overset{def}{=} \sum_{\alpha\beta} P(V_\alpha \wedge H_\beta) s_i^{\alpha\beta} s_j^{\alpha\beta}$$

and

$$p'_{ij} \overset{def}{=} \sum_{\lambda\mu} P'(V_\lambda \wedge H_\mu) s_i^{\lambda\mu} s_j^{\lambda\mu}$$

as given in (9).

The Boltzmann Machine learning algorithm can also be formulated as an input-output model. The visible units are divided into an input set I and an output set O, and an environment specifies a set of conditional probabilities of the form $P(O_\beta|I_\alpha)$. During the "training" phase the environment clamps both the input and output units, and p_{ij}s are estimated. During the "testing" phase the input units are clamped and the output units and hidden units free-run, and p'_{ij}s are estimated. The appropriate *G* measure in this case is

$$G = \sum_{\alpha\beta} P(I_\alpha \wedge O_\beta) \ln \frac{P(O_\beta|I_\alpha)}{P'(O_\beta|I_\alpha)}$$

Similar mathematics apply in this formulation and $\partial G/\partial w_{ij}$ is the same as before.

Acknowledgment

*The research reported here was supported by grants from the System Development Foundation. We thank Peter Brown, Francis Crick, Mark Derthick, Scott Fahlman, Jerry Feldman, Stuart Geman, Gail Gong, John Hopfield, Jay McClelland, Barak Pearlmutter, Harry Printz, Dave Rumelhart, Tim Shallice, Paul Smolensky, Rich Szeliski, and Venkataraman Venkatasubramanian for helpful discussions.

Reprint requests should be addressed to David Ackley, Computer Science Department, Carnegie-Mellon University, Pittsburgh, PA 15213.

References

Berliner, H. J., & Ackley, D. H. (1982, August). The QBKG system: Generating explanations from a non-discrete knowledge representation. *Proceedings of the National Conference on Artificial Intelligence AAAI-82*, Pittsburgh, PA, 213–216.

Binder, K. (Ed.) (1978). *The Monte-Carlo method in statistical physics.* New York: Springer-Verlag.

Fahlman, S. E. (1980, June). The Hashnet Interconnection Scheme. (Tech. Rep. No. CMU-CS-80-125), Carnegie-Mellon University, Pittsburgh, PA.

Fahlman, S. E., Hinton, G. E., & Sejnowski, T. J. (1983, August). Massively parallel architectures for AI: NETL, Thistle, and Boltzmann Machines. *Proceedings of the National Conference on Artificial Intelligence AAAI-83*, Washington, DC, 109–113.

Feldman, J. A. (1982). Dynamic connections in neural networks. *Biological Cybernetics, 46,* 27–39.

Feldman, J. A., & Ballard, D. H. (1982). Connectionist models and their properties. *Cognitive Science, 6,* 205–254.

Geman, S., & Geman, D. (1983). Stochastic relaxation, Gibbs distributions, and the Bayesian restoration of images. Unpublished manuscript.

Grimson, W. E. L. (1981). *From images to surfaces.* Cambridge, MA: MIT Press.

Hinton, G. E. (1977). *Relaxation and its role in vision.* Unpublished doctoral dissertation, University of Edinburgh. Described in D. H. Ballard & C. M. Brown (Eds.), *Computer Vision.* Englewood Cliffs, NJ: Prentice-Hall, 408–430.

Hinton, G. E. (1981). Implementing semantic networks in parallel hardware. In G. E. Hinton & J. A. Anderson (Eds.), *Parallel Models of Associative Memory.* Hillsdale, NJ: Erlbaum.

Hinton, G. E., & Anderson, J. A. (1981). *Parallel models of associative memory.* Hillsdale, NJ: Erlbaum.

Hinton, G. E., Sejnowski, T. J. (1983a, May). Analyzing cooperative computation. *Proceedings of the Fifth Annual Conference of the Cognitive Science Society.* Rochester, NY.

Hinton, G. E., & Sejnowski, T. J. (1983b, June). Optimal perceptual inference. *Proceedings of the IEEE Computer Society Conference on Computer Vision and Pattern Recognition.* Washington, DC, pp. 448–453.

Hinton, G. E., Sejnowski, T. J., & Ackley, D. H. (1984, May). *Boltzmann Machines: Constraint satisfaction networks that learn.* (Tech. Rep. No. CMU-CS-84-119). Pittsburgh, PA: Carnegie-Mellon University.

Hopfield, J. J. (1982). Neural networks and physical systems with emergent collective computational abilities. *Proceedings of the National Academy of Sciences USA, 79,* 2554–2558.

Kirkpatrick, S., Gelatt, C. D., & Vecchi, M. P. (1983). Optimization by simulated annealing. *Science, 220,* 671–680.

Kullback, S. (1959). *Information theory and statistics.* New York: Wiley.

Metropolis, N., Rosenbluth, A., Rosenbluth, M., Teller, A., & Teller, E. (1953). Equation of state calculations for fast computing machines. *Journal of Chemical Physics, 6,* 1087.

Minsky, M., & Papert, S. (1968). *Perceptrons.* Cambridge, MA: MIT Press.

Newell, A. (1982). *Intellectual issues in the history of artificial intelligence.* (Tech. Rep. No. CMU-CS-82-142). Pittsburgh, PA: Carnegie-Mellon University.

Newell, A., & Simon, H. A. (1972). *Human problem solving.* Englewood Cliffs, NJ: Prentice-Hall, 1972.

Ratcliff, R. (1978). A theory of memory retrieval. *Psychological Review, 85,* 59–108.

Renyi, A. (1962). *Probability theory.* Amsterdam: North-Holland.

Rosenblatt, F. (1961). *Principles of neurodynamics: Perceptrons and the theory of brain mechanisms.* Washington, DC: Spartan.

Smolensky, P. (1983, August). Schema selection and stochastic inference in modular environments. *Proceedings of the National Conference on Artificial Intelligence AAAI-83*, Washington, DC. 109–113.

Terzopoulos, D. (1984). *Multiresolution computation of visible-surface representations.* Unpublished doctoral dissertation, MIT, Cambridge, MA.

Waltz, D. L. (1975). Understanding line drawings of scenes with shadows. In P. Winston (Ed.), *The Psychology of Computer Vision.* New York: McGraw-Hill.

Winston, P. H. (1984). *Artificial Intelligence.* (2nd ed.) Reading, MA: Addison-Wesley.

Introduction

(1985)
Nabil H. Farhat, Demetri Psaltis, Aluizio Prata, and Eung Paek

Optical implementation of the Hopfield model
Applied Optics 24:1469–1475

One of the most exciting things about the current renaissance of neural network modeling is the real possibility of building extremely fast, powerful special-purpose hardware. Neural networks are intrinsically parallel, highly interconnected, low precision, and, in many versions, partially linear for small signal interactions. Programs to simulate networks tend to be short and quite simple. However, they use an immense amount of computer time when they run because a highly interconnected parallel system must be modeled with serial hardware. Even with reasonable access to super-computers, it is hard to study large networks. Networks will have to be large and fast to be practical.

It is easy to design special-purpose hardware to realize networks because the basic structure of most network models is so simple and repetitive. There are currently a number of projects underway to build custom VLSI implementations of networks, and as of early 1987, the first devices have been announced. There are also several "neural network accelerators" commercially available, or nearly so, to do fast network computations.

All the neural network systems that are close to commercial availability use traditional silicon VLSI technology of various types. However, there are alternative ways of realizing networks that have great promise for the future. There has been interest in optical computing for a number of years. At first glance, optical computation has tremendous advantages: it is parallel, very fast, and rugged. However, in spite of a good deal of effort, there has not been much progress in making practical computing devices using optics. (An SPIE Institute volume edited by Szu, 1986, gives more details.)

One reason for this is that computation as it has evolved over the past forty years is highly nonlinear. It operates with logical operations, and the elements that realize the logic are two-state devices. It is difficult to build such nonlinear optical components. This approach to optical computing assumes that one would end with a machine that was like a traditional computer, except that much of the hardware would be optical, not electronic.

It is hard to escape the uneasy feeling that this is not the right way to use the strengths of optical hardware.

It turns out that it is quite easy to make neural networks optically. This paper by Farhat, Psaltis, Prata, and Paek describes how the 1982 Hopfield model (paper 27) can be realized using straightforward optical hardware and describes the operation of such a device. Other neural networks, for example, the linear associator, have also been implemented optically.

To realize the Hopfield model what one needs is a parallel matrix multiplier, a limiter, and feedback. The essential nonlinearity is a thresholding operation, which can be built into the feedback hardware. The connection matrix is a set of dark and light areas on a photographic transparency. The required apparatus is simple. The assumed asynchronous updating in the Hopfield model, where the state of one neuron at a time is changed, is not specifically used. Finite response times and random differences between elements in the hardware seem to accomplish the same function.

It is easy to see why optical hardware is so well suited to neural networks. Optical systems are fully parallel, and allow a very high degree of interconnection, because crossing light beams do not interfere. When optical interactions do occur they are graded, and sometimes linear, or close to it for small signals. This is one of the reasons why it is difficult to make digital hardware with optical devices. However, small signal linearity is used for many neural network interactions, so this problem becomes a virtue in the right environment. It is easy to build in large signal nonlinearities, such as clipping. Since optical interactions usually involve the measurement of a continuous quantity (intensity or phase), they are of limited precision, which is also not a problem for neural networks.

Practical hardware questions are important for the future of neural network research. Simulations with the connectivity of the brain are now within reach. The average cortical cell has on the order of thousands, perhaps tens of thousands, of synaptic connections. Almost every laboratory doing active neural network research has access to significant computational power in the form of scientific workstations or shared use of a mainframe. With this kind of computing power, it is routine to simulate systems with hundreds of connections per unit. Use of a supercomputer, which is nearly routine in many environments, raises the simulation connectivity to the low thousands, which is comparable to the biological connectivity. However, simulations of the size of the number of neurons in the nervous system are still many orders of magnitudes beyond our reach. There are millions or tens of millions of cells in a cortical structure. The largest foreseeable network simulation, even projecting into the future with current or proposed supercomputer architecture, will be far below this level. Yet many organizing principles for real brains may become important at these sizes. If we look at the ratio of the number of synapses on a cell to the number of cells in a system, ratios for current simulations usually are of order of magnitude one; that is, there are about as many synapses per cell as the total number of cells. In real brains, the ratio is a very small fraction—say, one in a thousand or one in ten thousand. Organizing a system with such limited connectivity relative to the number of elements may be hard. Scaling our current simulations may be more difficult than usually presumed.

The only way to perform simulations using realistic numbers of elements and connections will be to use special-purpose hardware. Inexpensive special-purpose hardware will allow many practical applications of neural networks to become feasible. It is possible that inexpensive, powerful optical computers, organized using brain-like principles, may be common in the twenty-first century, as an adjunct or supplement to traditional digital computers.

Reference

H. Szu (Ed.) (1986), *Optical and Hybrid Computing*, SPIE Vol. 634, Bellingham, WA: SPIE.

(1985)

Nabil H. Farhat, Demetri Psaltis, Aluizio Prata, and Eung Paek

Optical implementation of the Hopfield model

Applied Optics 24:1469–1475

Optical implementation of content addressable associative memory based on the Hopfield model for neural networks and on the addition of nonlinear iterative feedback to a vector–matrix multiplier is described. Numerical and experimental results presented show that the approach is capable of introducing accuracy and robustness to optical processing while maintaining the traditional advantages of optics, namely, parallelism and massive interconnection capability. Moreover a potentially useful link between neural processing and optics that can be of interest in pattern recognition and machine vision is established.

I. Introduction

It is well known that neural networks in the eye–brain system process information in parallel with the aid of large numbers of simple interconnected processing elements, the neurons. It is also known that the system is very adept at recognition and recall from partial information and has remarkable error correction capabilities.

Recently Hopfield described a simple model[1] for the operation of neural networks. The action of individual neurons is modeled as a thresholding operation and information is stored in the interconnections among the neurons. Computation is performed by setting the state (on or off) of some of the neurons according to an external stimulus and, with the interconnections set according to the recipe that Hopfield prescribed, the state of all neurons that are interconnected to those that are externally stimulated spontaneously converges to the stored pattern that is most similar to the external input. The basic operation performed is a nearest-neighbor search, a fundamental operation for pattern recognition, associative memory, and error correction. A remarkable property of the model is that powerful global computation is performed with very simple, identical logic elements (the neurons). The interconnections provide the computation power to these simple logic elements and also enhance dramatically the stor-age capacity; approximately $N/4 \ln N$ bits/neuron can be stored in a network in which each neuron is connected to N others.[2] Another important feature is that synchronization among the parallel computing elements is not required, making concurrent, distributed processing feasible in a massively parallel structure. Finally, the model is insensitive to local imperfections such as variations in the threshold level of individual neurons or the weights of the interconnections.

Given these characteristics we were motivated to investigate the feasibility of implementing optical information processing and storage systems that are based on this and other similar models of associative memory.[3,4] Optical techniques offer an effective means for the implementation of programmable global interconnections of very large numbers of identical parallel logic elements. In addition, emerging optical technologies such as 2-D spatial light modulators, optical bistability, and thin-film optical amplifiers appear to be very well suited for performing the thresholding operation that is necessary for the implementation of the model.

The principle of the Hopfield model and its implications in optical information processing have been discussed earlier.[5,6] Here we review briefly the main features of the model, give as an example the results of a numerical simulation, describe schemes for its optical implementation, then present experimental results obtained with one of the schemes and discuss their implications as a content addressable associative memory (CAM).

II. Hopfield Model

Given a set of **M** bipolar, binary $(1,-1)$ vectors $\mathbf{v}_i^{(m)}$, $i = 1,2,3\ldots N$, $m = 1,2,3\ldots\mathbf{M}$, these are stored in a synaptic matrix in accordance with the recipe

$$T_{ij} = \sum_{m=1}^{\mathbf{M}} \mathbf{v}_i^{(m)}\mathbf{v}_j^{(m)}, \qquad i,j = 1,2,3\ldots N, \qquad T_{ii} = 0, \qquad (1)$$

$\mathbf{v}_i^{(m)}$ are referred to as the nominal state vectors of the

© 1985 Optical Society of America.

memory. If the memory is addressed by multiplying the matrix T_{ij} with one of the state vectors, say $v_i^{(mo)}$, it yields the estimate

$$\hat{v}_i^{(mo)} = \sum_j^N T_{ij} v_j^{(mo)} \tag{2}$$

$$= \sum_{j \neq i}^N \sum_m^{\mathbf{M}} v_i^{(m)} v_j^{(m)} v_j^{(mo)}$$

$$= (N-1)v_i^{(mo)} + \sum_{m \neq m_0} \alpha_{m,mo} v_i^{(m)}, \tag{3}$$

where

$$\alpha_{m,mo} = \sum_j^N v_j^{(mo)} v_j^{(m)}.$$

$\hat{v}_i^{(mo)}$ consists of the sum of two terms: the first is the input vector amplified by $(N-1)$; the second is a linear combination of the remaining stored vectors and it represents an unwanted cross-talk term. The value of the coefficients $\alpha_{m,mo}$ is equal to $\sqrt{N-1}$ on the average (the standard deviation of the sum of $N-1$ random bits), and since $(\mathbf{M}-1)$ such coefficients are randomly added, the value of the second term will on the average be equal to $\sqrt{(\mathbf{M}-1)(N-1)}$. If N is sufficiently larger than $\mathbf{M}$, with high probability the elements of the vector $\hat{v}_i^{(mo)}$ will be positive if the corresponding elements of $v_i^{(mo)}$ are equal to $+1$ and negative otherwise. Thresholding of $\hat{v}_i^{(mo)}$ will therefore yield $v_i^{(mo)}$:

$$v_i^{(mo)} = \text{sgn}[\hat{v}_i^{(mo)}] = \begin{cases} +1 \text{ if } \hat{v}_i^{(mo)} > 0 \\ -1 \text{ otherwise.} \end{cases} \tag{4}$$

When the memory is addressed with a binary valued vector that is not one of the stored words, the vector–matrix multiplication and thresholding operation yield an output binary valued vector which, in general, is an approximation of the stored word that is at the shortest Hamming distance from the input vector. If this output vector is fed back and used as the input to the memory, the new output is generally a more accurate version of the stored word and continued iteration converges to the correct vector.

The insertion and readout of memories described above are depicted schematically in Fig. 1. Note that in Fig. 1(b) the estimate $\hat{v}_i^{(mo)}$ can be viewed as the weighted projection of T_{ij}. Recognition of an input vector that corresponds to one of the state vectors of the memory or is close to it (in the Hamming sense) is manifested by a stable state of the system. In practice unipolar binary $(0,1)$ vectors or words $b_i^{(m)}$ of bit length N may be of interest. The above equations are then applicable with $[2b_i^{(m)} - 1]$ replacing $v_i^{(m)}$ in Eq. (1) and $b_i^{(mo)}$ replacing $v_i^{(mo)}$ in Eq. (2). For such vectors the SNR of the estimate $\hat{v}_i^{(mo)}$ can be shown to be lower by a factor of $\sqrt{2}$.[1]

An example of the T_{ij} matrix formed from four binary unipolar vectors, each being $N = 20$ bits long, is given in Fig. 2 along with the result of a numerical simulation of the process of initializing the memory matrix with a partial version of $b_i^{(4)}$ in which the first eight digits of $b_i^{(4)}$ are retained and the remainder set to zero. The Hamming distance between the initializing vector and $b_i^{(4)}$ is 6 bits and it is 9 or more bits for the other three

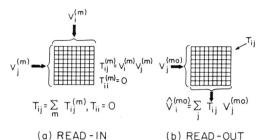

(a) READ-IN   (b) READ-OUT

Fig. 1.   (a) Insertion and (b) readout of memories.

stored vectors. It is seen that the partial input is recognized as $b_i^{(4)}$ in the third iteration and the output remains stable as $b_i^{(4)}$ thereafter. This convergence to a stable state generally persists even when the T_{ij} matrix is binarized or clipped by replacing negative elements by minus ones and positive elements by plus ones evidencing the robustness of the CAM. A binary synaptic matrix has the practical advantage of being more readily implementable with fast programmable spatial light modulators (SLM) with storage capability such as the Litton Lightmod.[7] Such a binary matrix, implemented photographically, is utilized in the optical implementation described in Sec. III and evaluated in Sec. IV of this paper.

Several schemes for optical implementation of a CAM based on the Hopfield model have been described earlier.[5] In one of the implementations an array of light emitting diodes (LEDs) is used to represent the logic elements or neurons of the network. Their state (on or off) can represent unipolar binary vectors such as the state vectors $b_i^{(m)}$ that are stored in the memory matrix T_{ij}. Global interconnection of the elements is realized as shown in Fig. 3(a) through the addition of nonlinear feedback (thresholding, gain, and feedback) to a conventional optical vector–matrix multiplier[8] in which the array of LEDs represents the input vector and an array of photodiodes (PDs) is used to detect the output vector. The output is thresholded and fed back in parallel to drive the corresponding elements of the LED array. Multiplication of the input vector by the T_{ij} matrix is achieved by horizontal imaging and vertical smearing of the input vector that is displayed by the LEDs on the plane of the T_{ij} mask [by means of an anamorphic lens system omitted from Fig. 3(a) for simplicity]. A second anamorphic lens system (also not shown) is used to collect the light emerging from each row of the T_{ij} mask on individual photosites of the PD array. A bipolar T_{ij} matrix is realized in incoherent light by dividing each row of the T_{ij} matrix into two subrows, one for positive and one for negative values and bringing the light emerging from each subrow to focus on two adjacent photosites of the PD array that are electrically connected in opposition as depicted in Fig. 3(b). In the system shown in Fig. 3(a), feedback is achieved by electronic wiring. It is possible and preferable to dispose of electronic wiring altogether and replace it by optical feedback. This can be achieved by combining the PD and LED arrays in a single compact hybrid or

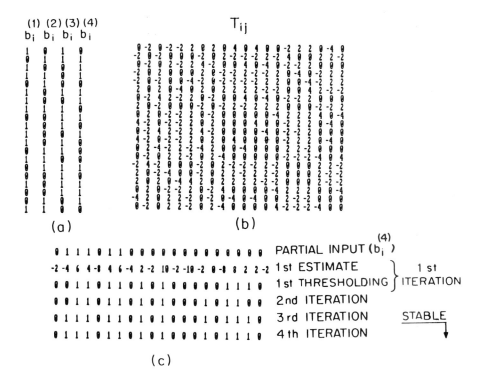

(a)   (b)

PARTIAL INPUT ($b_i^{(4)}$)

1st ESTIMATE   } 1st

1st THRESHOLDING   } ITERATION

2nd ITERATION

3rd ITERATION   STABLE

4th ITERATION

(c)

Fig. 2. Numerical example of recovery from partial input; $N = 20$, $M = 4$. (a) Stored vectors, (b) memory or (synaptic) matrix, (c) results of initializing with a partial version of $b_i^{(4)}$.

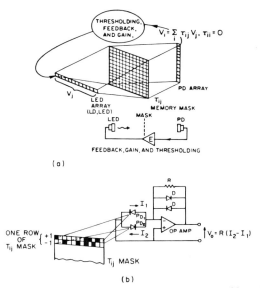

Fig. 3. Concept for optical implementation of a content addressable memory based on the Hopfield model. (a) Matrix–vector multiplier incorporating nonlinear electronic feedback. (b) Scheme for realizing a binary bipolar memory mask transmittance in incoherent light.

monolithic structure that can also be made to contain all ICs for thresholding, amplification, and driving of LEDs. Optical feedback becomes even more attractive when we consider that arrays of nonlinear optical light amplifiers with internal feedback[9] or optical bistability

devices (OBDs)[10] can be used to replace the PD/LED arrays. This can lead to simple compact CAM structures that may be interconnected to perform higher-order computations than the nearest-neighbor search performed by a single CAM.

We have assembled a simple optical system that is a variation of the scheme presented in Fig. 3(a) to simulate a network of $N = 32$ neurons. The system, details of which are given in Figs. 5–8, was constructed with an array of thirty-two LEDs and two multichannel silicon PD arrays, each consisting of thirty-two elements. Twice as many PD elements as LEDs are needed in order to implement a bipolar memory mask transmittance in incoherent light in accordance with the scheme of Fig. 3(b). A bipolar binary T_{ij} mask was prepared for M = 3 binary state vectors. The three vectors or words chosen, their Hamming distances from each other, and the resulting T_{ij} memory matrix are shown in Fig. 4. The mean Hamming distance between the three vectors is 16. A binary photographic transparency of 32 × 64 square pixels was computer generated from the T_{ij} matrix by assigning the positive values in any given row of T_{ij} to transparent pixels in one subrow of the mask and the negative values to transparent pixels in the adjacent subrow. To insure that the image of the input LED array is uniformly smeared over the memory mask it was found convenient to split the mask in two halves, as shown in Fig. 5, and to use the resulting submasks in two identical optical arms as shown in Fig. 6. The size of the subrows of the memory submasks was made exactly equal to the element size of the PD arrays in the vertical direction which were placed in register

Stored words:

```
Word 1 : 1 1 1 0 0 0 0 1 0 1 0 1 1 1 0 1 1 0 1 1 1 1 0 1 1 0 0 0 0 0 1 0
Word 2 : 0 1 1 0 0 0 0 0 0 0 1 0 0 1 0 1 0 1 0 0 1 1 1 1 0 1 0 1 1 0 1 0
Word 3 : 1 0 1 1 0 0 1 1 1 1 1 1 1 1 1 1 0 0 0 1 0 1 1 0 0 0 0 1 1 0 0 0 0
```

Hamming distance from word to word:

WORD	1	2	3
1	0	15	14
2	15	0	19
3	14	19	0

Clipped memory matrix:

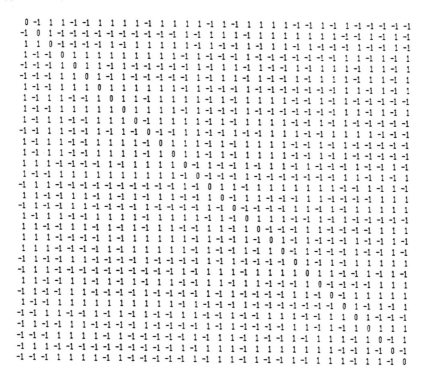

```
 0 -1  1  1 -1 -1  1  1  1  1 -1  1  1  1  1 -1  1 -1  1  1  1  1 -1 -1  1 -1  1 -1 -1 -1 -1 -1
-1  0  1 -1 -1 -1 -1 -1 -1 -1 -1  1 -1  1  1  1  1  1  1  1  1  1 -1 -1  1 -1  1 -1  1 -1  1 -1
 1  1  0 -1 -1 -1 -1  1 -1  1  1  1  1  1  1 -1  1 -1 -1  1 -1  1  1  1 -1 -1 -1 -1  1 -1  1 -1
 1 -1 -1  0  1  1  1  1  1  1  1  1 -1  1 -1 -1 -1  1 -1 -1 -1 -1 -1 -1 -1  1  1 -1  1 -1  1  1
-1 -1 -1  1  0  1  1 -1  1 -1 -1 -1 -1 -1  1  1  1 -1  1 -1 -1  1  1  1  1 -1  1  1 -1  1  1  1
-1 -1 -1  1  1  0  1 -1  1 -1 -1 -1 -1 -1  1  1  1 -1  1 -1 -1  1  1  1  1 -1  1  1 -1  1  1  1
 1 -1 -1  1  1  1  0  1  1  1 -1  1 -1  1  1 -1  1 -1 -1 -1 -1 -1 -1 -1 -1  1  1 -1  1 -1  1  1
 1 -1  1  1 -1 -1  1  0  1  1 -1  1  1  1  1 -1  1 -1  1  1  1  1 -1 -1  1 -1  1 -1 -1 -1 -1 -1
 1 -1 -1  1  1  1  1  1  0  1  1  1 -1  1 -1 -1  1 -1 -1 -1 -1 -1 -1 -1 -1  1  1 -1  1 -1  1  1
 1 -1  1  1 -1 -1  1  1  1  0 -1  1  1  1  1 -1  1 -1  1  1  1  1 -1 -1  1 -1  1 -1 -1 -1 -1 -1
-1 -1 -1  1  1  1 -1 -1  1 -1  0 -1 -1  1  1  1 -1  1  1  1  1  1  1 -1  1  1  1  1 -1  1 -1 -1
 1 -1  1  1 -1 -1  1  1  1  1 -1  0  1  1 -1  1 -1  1  1  1  1  1 -1 -1  1 -1  1 -1 -1 -1 -1 -1
 1 -1  1  1 -1 -1  1  1  1  1 -1  1  0  1  1 -1  1 -1  1  1  1  1 -1 -1  1 -1  1 -1 -1 -1 -1 -1
 1  1  1 -1 -1 -1 -1  1 -1  1  1  1  1  0 -1  1 -1 -1  1 -1  1  1 -1 -1  1 -1 -1  1 -1 -1  1 -1
-1 -1  1  1  1  1  1  1  1  1  1 -1  1 -1  0 -1 -1  1  1 -1 -1  1 -1 -1 -1 -1 -1  1 -1  1 -1  1
-1  1  1 -1 -1 -1 -1 -1 -1 -1 -1  1 -1  1 -1  0  1  1 -1  1  1 -1  1 -1  1  1  1  1 -1  1 -1  1
 1  1 -1 -1  1  1  1  1  1  1 -1 -1  1 -1 -1  1  0 -1  1  1  1 -1 -1 -1  1 -1 -1 -1 -1  1  1  1
-1  1 -1  1  1  1 -1 -1 -1  1 -1 -1 -1 -1  1 -1  0 -1 -1 -1 -1  1 -1  1 -1  1 -1  1  1  1  1  1
 1 -1  1  1 -1 -1  1  1  1  1  1  1  1  1 -1  1 -1  0  1  1  1 -1  1 -1  1 -1 -1 -1 -1 -1 -1 -1
 1  1 -1 -1  1  1 -1  1 -1  1 -1  1  1 -1 -1  1  1 -1  0 -1 -1  1  1 -1  1  1 -1 -1 -1 -1  1  1
 1  1  1 -1 -1 -1 -1  1 -1  1  1  1  1  1 -1  1  1 -1  1  0 -1  1 -1 -1  1 -1 -1 -1 -1 -1  1 -1
 1  1 -1 -1  1  1 -1  1 -1  1  1 -1 -1  1 -1  1  1  1 -1  1  0  1 -1 -1 -1  1 -1 -1  1 -1  1  1
 1  1  1 -1 -1 -1 -1  1 -1  1  1  1  1 -1 -1  1 -1  1  1  1 -1  0  1 -1  1 -1 -1 -1  1 -1  1 -1
-1 -1 -1  1 -1 -1 -1 -1 -1 -1  1 -1 -1 -1 -1 -1 -1 -1 -1 -1  1 -1  0  1 -1  1  1  1  1  1  1  1
-1  1  1 -1 -1 -1 -1 -1 -1 -1  1 -1 -1 -1 -1  1 -1  1  1 -1  1  1 -1  0  1 -1  1 -1  1  1 -1  1
 1 -1 -1  1  1  1  1  1  1  1 -1  1  1 -1 -1  1 -1  1  1 -1 -1 -1  1  1  0 -1 -1 -1 -1  1  1  1
-1  1 -1  1  1  1 -1 -1 -1  1 -1 -1 -1 -1  1  1 -1  1 -1  1 -1  1 -1  1 -1  0 -1  1  1  1  1  1
 1 -1 -1  1  1  1  1  1  1  1 -1  1  1  1 -1  1 -1 -1 -1 -1 -1 -1  1  1 -1 -1  0 -1 -1  1  1  1
 1 -1 -1  1 -1 -1  1 -1  1 -1 -1 -1 -1 -1 -1 -1 -1  1 -1 -1  1 -1  1  1 -1  1 -1  0  1 -1  1  1
-1 -1  1  1  1  1 -1 -1  1 -1  1 -1 -1  1  1  1  1  1 -1 -1 -1 -1  1  1  1  1  1  1 -1  0 -1  1
-1  1  1 -1 -1 -1 -1 -1 -1 -1  1 -1 -1  1 -1  1  1  1 -1  1  1  1  1 -1  1  1 -1  1  1 -1  0 -1
-1 -1 -1  1  1  1  1  1  1  1 -1  1  1 -1  1 -1  1 -1 -1  1 -1  1 -1  1  1  1  1  1 -1  1 -1  0
```

Fig. 4. Stored words, their Hamming distances, and their clipped T_{ij} memory matrix.

Fig. 5. Two halves of T_{ij} memory mask.

against the masks. Light emerging from each subrow of a memory submask was collected (spatially integrated) by one of the vertically oriented elements of the multichannel PD array. In this fashion the anamorphic optics required in the output part of Fig. 3(a) are disposed of, resulting in a more simple and compact system. Pictorial views of the input LED array and the two submask/PD array assemblies are shown in Figs. 7(a) and (b), respectively. In Fig. 7(b) the left memory submask/PD array assembly is shown with the submask removed to reveal the silicon PD array situated behind it. All electronic circuits (amplifiers, thresholding comparators, LED drivers, etc.) in the thirty-two parallel feedback channels are contained in the electronic amplification and thresholding box shown in Fig. 6(a) and in the boxes on which the LED array and the two submask/PD array assemblies are mounted (see Fig. 7). A pictorial view of a composing and display box is shown in Fig. 8. This contains an arrangement of thirty-two switches and a thirty-two element LED display panel whose elements are connected in parallel to the input LED array. The function of this box is to compose and

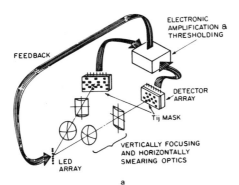

Fig. 6. Arrangement for optical implementation of the Hopfield model: (a) optoelectronic circuit diagram, (b) pictorial view.

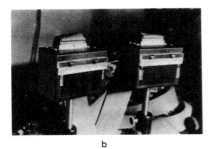

Fig. 7. Views of (a) input LED array and (b) memory submask/PD array assemblies.

display the binary input word or vector that appears on the input LED array of the system shown in Fig. 7(a). Once an input vector is selected it appears displayed on the composing box and on the input LED box simultaneously. A single switch is then thrown to release the system into operation with the composed vector as the

Fig. 8. Word composer and display box.

initializing vector. The final state of the system, the output, appears after a few iterations displayed on the input LED array and the display box simultaneously. The above procedure provides for convenient exercising of the system in order to study its response vs stimulus behavior. An input vector is composed and its Hamming distance from each of the nominal state vectors stored in the memory is noted. The vector is then used to initialize the CAM as described above and the output vector representing the final state of the CAM appearing, almost immediately, on the display box is noted. The response time of the electronic feedback channels as determined by the 3-dB roll-off of the amplifiers was ~60 msec. Speed of operation was not an issue in this study, and thus low response time was chosen to facilitate the experiment.

IV. Results

The results of exercising and evaluating the performance of the system we described in the preceding section are tabulated in Table I. The first run of initializing vectors used in exercising the system were error laden versions of the first word $b_i^{(1)}$. These were obtained from $b_i^{(1)}$ by successively altering (switching) the states of 1,2,3 . . . up to N of its digits starting from the Nth digit. In doing so the Hamming distance between the initializing vector and $b_i^{(1)}$ is increased linearly in unit steps as shown in the first column of Table I whereas, on the average, the Hamming distance between all these initializing vectors and the other two state vectors remained approximately the same, about $N/2 = 16$. The final states of the memory, i.e., the steady-state vectors displayed at the output of the system (the composing and display box) when the memory is prompted by the initializing vectors, are listed in column 2 of Table I. When the Hamming distance of the initializing vector from $b_i^{(1)}$ is <11, the input is always recognized correctly as $b_i^{(1)}$. The CAM is able therefore to recognize the input vector as $b_i^{(1)}$ even when up to 11 of its digits (37.5%) are wrong. This performance is identical to the results obtained with a digital simulation shown in parenthesis in column 2 for comparison. When the Hamming distance is increased further to values lying between 12 and 22, the CAM is confused and identifies erroneously other state vectors, mostly $b_i^{(3)}$, as the input. In this range, the Hamming distance of the initializing vectors from any of the stored vectors is approximately equal making it more difficult for the CAM to decide. Note that the performance of

Table I. Optical CAM Performance

Hamming distance of initializing vector from $b_i^{(m)}$	Recognized vector ($m = 1$)	Recognized vector ($m = 2$)	Recognized vector ($m = 3$)
0	1 (1)	2 (2)	3 (3)
1	1 (1)	2 (2)	3 (3)
2	1 (1)	2 (2)	3 (3)
3	1 (1)	2 (2)	3 (3)
4	1 (1)	2 (2)	3 (3)
5	1 (1)	2 (2)	3 (3)
6	1 (1)	2 (2)	3 (3)
7	1 (1)	2 (2)	3 (3)
8	1 (1)	2 (2)	3 (3)
9	1 (1)	2 (2)	3 (3)
10	1 (1)	1 (1)	3 (3)
11	1 (1)	2 (2)	3 (3)
12	3 (3)	3,2 ($\overline{3}$)	3 (3)
13	3 (3)	$\overline{3}$ (3)	3 ($\overline{2}$)
14	3 (3)	1,$\overline{3}$ (1)	3 ($\overline{2}$)
15	1 (OSC)	1 (1)	2,3 ($\overline{2}$)
16	3 (OSC)	1 (1)	$\overline{2}$ ($\overline{2}$)
17	3 (OSC)	1 (OSC)	$\overline{2}$ ($\overline{2}$)
18	3 (3)	1 ($\overline{2}$)	3 (OSC)
19	3 ($\overline{2}$)	$\overline{2}$ ($\overline{2}$)	$\overline{2}$ ($\overline{2}$)
20	3 ($\overline{1}$)	$\overline{2}$ ($\overline{2}$)	$\overline{2}$ (OSC)
21	1,2 ($\overline{1}$)	$\overline{2}$ ($\overline{2}$)	$\overline{3}$ (OSC)
22	3 ($\overline{1}$)	$\overline{2}$ ($\overline{2}$)	$\overline{3}$ (OSC)
23	$\overline{1}$ ($\overline{1}$)	$\overline{2}$ ($\overline{2}$)	$\overline{3}$ (OSC)
24	$\overline{1}$ ($\overline{1}$)	$\overline{2}$ ($\overline{2}$)	$\overline{3}$ ($\overline{3}$)
25	$\overline{1}$ ($\overline{1}$)	$\overline{2}$ ($\overline{2}$)	$\overline{3}$ ($\overline{3}$)
26	$\overline{1}$ ($\overline{1}$)	$\overline{2}$ ($\overline{2}$)	$\overline{3}$ ($\overline{3}$)
27	$\overline{1}$ ($\overline{1}$)	$\overline{2}$ ($\overline{2}$)	$\overline{3}$ ($\overline{3}$)
28	$\overline{1}$ ($\overline{1}$)	$\overline{2}$ ($\overline{2}$)	$\overline{3}$ ($\overline{3}$)
29	$\overline{1}$ ($\overline{1}$)	$\overline{2}$ ($\overline{2}$)	$\overline{3}$ ($\overline{3}$)
30	$\overline{1}$ ($\overline{1}$)	$\overline{2}$ ($\overline{2}$)	$\overline{3}$ ($\overline{3}$)
31	$\overline{1}$ ($\overline{1}$)	$\overline{2}$ ($\overline{2}$)	$\overline{3}$ ($\overline{3}$)
32	$\overline{1}$ ($\overline{1}$)	$\overline{2}$ ($\overline{2}$)	$\overline{3}$ ($\overline{3}$)

the CAM and results of digital simulation in this range of Hamming distance are comparable except for the appearance of oscillations (designated by OSC) in the digital simulation when the outcome oscillated between several vectors that were not the nominal state vectors of the CAM. Beyond a Hamming distance of 22 both the optical system and the digital simulation identified the initializing vectors as the complement $\overline{b}_i^{(1)}$ of $b_i^{(1)}$. This is expected because it can be shown using Eq. (1) that the T_{ij} matrix formed from a set of vectors $b_i^{(m)}$ is identical to that formed by the complementary set $\overline{b}_i^{(m)}$. The complementary vector can be viewed as a contrast reversed version of the original vector in which zeros and ones are interchanged. Recognition of a complementary state vector by the CAM is analogous to our recognizing a photographic image from the negative.

Similar results of initializing the CAM with error laden versions of $b_i^{(2)}$ and $b_i^{(3)}$ were also obtained. These are presented in columns 2 and 3 of Table I. Here again we see when the Hamming distance of the initializing vector from $b_i^{(3)}$, for example, ranged between 1 and 14, the CAM recognized the input correctly as $b_i^{(3)}$ as shown in column 3 of the table and as such it did slightly better than the results of digital simulation. Oscillatory behavior is also observed here in the digital simulation when the range of Hamming distance between the ini-

tializing vector from all stored vectors approached the mean Hamming distance between the stored vectors. Beyond this range the memory recognizes the input as the complementary of $b_i^{(3)}$.

In studying the results presented in Table I several observations can be made: The optically implemented CAM is working as accurately as the digital simulations and perhaps better if we consider the absence of oscillations. These are believed to be suppressed in the system because of the nonsharp thresholding performed by the smoothly varying nonlinear transfer function of electronic circuits compared with the sharp thresholding in digital comptations. The smooth nonlinear transfer function and the finite time constant of the optical system provide a relaxation mechanism that substitutes for the role of asynchronous switching required by the Hopfield model. Generally the system was able to conduct successful nearest-neighbor search when the inputs to the system are versions of the nominal state vectors containing up to ~30% error in their digits. It is worth noting that this performance is achieved in a system built from off-the-shelf electronic and optical components and with relatively little effort in optimizing and fine tuning the system for improved accuracy, thereby confirming the fact that accurate global computation can be performed with relatively inaccurate individual components.

V. Discussion

The number **M** of state vectors of length N that can be stored at any time in the interconnection matrix T_{ij} is limited to a fraction of N. An estimate of $\mathbf{M} \simeq 0.1N$ is indicated in simulations involving a hundred neurons or less[1] and a theoretical estimate of $\mathbf{M} \simeq N/4 \ln N$ has recently been obtained.[2] It is worthwhile to consider the number of bits that can be stored per interconnection or per neuron. The number of pixels required to form the interconnection matrix is N^2. Since such a T_{ij} memory matrix can store up to $\mathbf{M} \simeq N/4 \ln N$ (N-tuples), the number of bits stored is $\mathbf{M}N = N^2/4 \ln N$. The number of bits stored per memory matrix element or interconnection is $\mathbf{M}N/N^2 = (4 \ln N)^{-1}$, while the number of bits stored per neuron is $\mathbf{M}N/N = \mathbf{M}$.

The number of stored memories that can be searched for a given initializing input can be increased by using a dynamic memory mask that is rapidly addressed with different T_{ij} matrices each corresponding to different sets of **M** vectors. The advantage of programmable SLMs for realizing this goal are evident. For example, the Litton Lightmod (magnetooptic light modulator), which has nonvolatile storage capability and can provide high frame rates, could be used. A frame rate of 60 Hz is presently specified for commercially available units of 128 × 128 pixels which are serially addressed.[7] Units with 256 × 256 pixels are also likely to be available in the near future with the same frame rate capability. Assuming a memory mask is realized with a Litton Lightmod of 256 × 256 pixels we have $N = 256$, $\mathbf{M} \simeq 0.1N \simeq 26$ and a total of 26 × 60 = 1560 vectors can be searched or compared per second against an initializing input vector. Speeding up the frame rate of the Litton

Lightmod to increase memory throughput beyond the above value by implementing parallel addressing schemes is also possible. Calculations show that the maximum frame rate possible for the device operating in reflection mode with its drive lines heat sunk is 10 kHz.[7] This means the memory throughput estimated above can be increased to search 2.6×10^5 vectors/sec, each being 256 bits long, or a total of 6.7×10^8 bits/sec. This is certainly a respectable figure, specially when we consider the error correcting capability and the associative addressing mode of the Hopfield model; i.e., useful computation is performed in addition to memory addressing.

The findings presented here show that the Hopfield model for neural networks and other similar models for content addressable and associative memory fit well the attributes of optics, namely, parallel processing and massive interconnection capabilities. These capabilities allow optical implementation of large neural networks based on the model. The availability of nonlinear or bistable optical light amplifiers with internal feedback, optical bistability devices, and nonvolatile high speed spatial light modulators could greatly simplify the construction of optical CAMs and result in compact modules that can be readily interconnected to perform more general computation than nearest-neighbor search. Such systems can find use in future generation computers, artificial intelligence, and machine vision.

The work described in this paper was performed while one of the authors, N.F., was on scholarly leave at the California Institute of Technology. This author wishes to express his appreciation to CIT and the University of Pennsylvania for facilitating his sabbatical leave. The work was supported in part by the Army Research Office and in part by the Air Force Office of Scientific Research.

The subject matter of this paper is based on a paper presented at the OSA Annual Meeting, San Diego, Oct. 1984.

References

1. J. J. Hopfield, "Neural Networks and Physical Systems with Emergent Collective Computational Abilities," Proc. Natl. Acad. Sci. USA **79**, 2554 (1982).
2. R. J. McEliece, E. C. Posner, and S. Venkatesh, California Institute of Technology, Electrical Engineering Department; private communication.
3. G. E. Hinton and J. A. Anderson, *Parallel Models of Associative Memory* (LEA Publishers, Hillsdale, N.J., 1981).
4. T. Kohonen, *Content Addressable Memories* (Springer, New York, 1980).
5. D. Psaltis and N. Farhat, "A New Approach to Optical Information Processing Based On the Hopfield Model," in *Technical Digest, ICO-13 Conference, Sapporo* (1984), p. 24.
6. D. Psaltis and N. Farhat, "Optical Information Processing Based on an Associative-Memory Model of Neural Nets with Thresholding and Feedback," Opt. Lett. **10**, 98 (1985).
7. W. Ross, D. Psaltis, and R. Anderson, "Two-Dimensional Magneto-Optic Spatial Light Modulator For Signal Processing," Opt. Eng. **22**, 485 (1983).
8. J. W. Goodman, A. R. Dias, and L. M. Woody, "Fully Parallel, High-Speed Incoherent Optical Method for Performing Discrete Fourier Transforms," Opt. Lett. **2**, 1 (1978).
9. Z. Porada, "Thin Film Light Amplifier with Optical Feedback," Thin Solid Films **109**, 213 (1983).
10. H. M. Gibbs *et al.*, "Optical Bistable Devices: The Basic Components of All-Optical Circuits," Proc. Soc. Photo-Opt. Instrum. Eng. **269**, 75 (1981).

40

Introduction

(1986)
Terrence J. Sejnowski and Charles R. Rosenberg

NETtalk: a parallel network that learns to read aloud
The Johns Hopkins University Electrical Engineering and Computer Science Technical Report
JHU/EECS-86/01, 32 pp.

English spelling is notoriously inconsistent. However, it does follow rules, although weak rules with exceptions and qualifications. This structure makes it well suited to a neural network, since networks are good at picking up statistical regularities, and are not bothered by occasional inconsistencies, as would be a more rigid rule based system.

As an example, consider the sentence "This is a test." The two letters "is" occur in both the first and second words, each time followed by a space, so the immediate context of 's' is identical. However, in the word "this" the 's' is unvoiced, that is, pronounced /s/. In the word "is" the 's' is voiced, that is, pronounced /z/.

In spite of its inconveniences, some have defended inconsistent spelling as making English related to both phonetic and an ideographic representations. Since there are so many dialects in English, it would be impossible to come up with a universal phonetic spelling scheme that would fit them all. Each dialect having its own phonetic spelling might signal the start of the fragmentation of English into separate languages. One of the editors of this collection of papers still recalls with dismay and amusement an article on rationalized English spelling read a number of years ago in a British publication. The author of the article proposed eliminating the terminal 'r' in words like "butter" or "corner" since it was not pronounced in modern English, ignoring, of course, the fact that most English speaking inhabitants of North America pronounce it. One man's phonetic spelling is another man's arbitrary representation.

NETtalk is a neural network that has learned how to read: specifically, it takes strings of characters forming English text and converts them into strings of phonemes that can serve as input to a speech synthesizer. The Digital Equipment Corporation sells a commercial product named "DECtalk" that also turns text into speech—hence the similarity of names. DECtalk is a complex rule based expert system developed over a number of years. Since NETtalk was a learning system, it potentially could learn to 'read' quickly and mechanically.

Sejnowski and Rosenberg used a three-layer feedforward network, with an input and output layer of units, as well as a layer of hidden units between the input and output layers. They have used both the Boltzmann machine learning algorithm (paper 38) and back propagation (papers 41 and 42) on this problem, with comparable results, though back propagation was faster to learn.

The input layer looked at a seven-character window of text. The network generated an output corresponding to the center character in the window. A window of this size seems to be enough to pronounce all except a few unusually difficult characters. Some pronounciation decisions require detailed knowledge of English, for example, to differentiate between the verb 'to lead' and the metal 'lead.'

The representation of characters at the input layer is localized, with one active unit representing a character, space, or punctuation, for a total of 29 units, times 7 character positions, for a total of 203 input units. The number of hidden units varied from simulation to simulation, but for continuous speech 80 hidden units were used. There were 26 features (23 articulatory features, 3 for stress) represented in the output layer. Phonemes were represented by multiple simultaneously active output units. Simulations used about 300 units total with around 20,000 connections. The authors also did some experiments with distributed input representations, with results ultimately comparable to the local representation. In one simulation they used the system to develop its own preferred input representation.

Sejnowski and Rosenberg taught the system both isolated words and continuous speech. Error correction was used and the network was trained to give the correct output phoneme when presented with input character strings. The authors were able to find samples of transcribed speech in the literature on children's language and used this as the training set.

After roughly 12 CPU hours of training on a DEC VAX, NETtalk was producing phonemes from the training set correctly 95% of the time. New samples of similar text showed a drop in accuracy, but still at around the 80% level. Sejnowski has played a "developmental" tape of NETtalk's output at a number of meetings. It passes through a babbling phase, then starts approximating speech more and more closely. By the time learning stops, it is quite understandable, though it still makes occasional errors. The errors are usually close enough to the correct output so that the speech is still comprehensible.

Sejnowski and Rosenberg were interested in what was going on in the hidden layers, because of the theoretical reasons for believing that the hidden layer units are developing efficient representations. Units in the hidden layers each seem to respond to several input characters, and have no obvious interpretations. Later work has pursued this point. Some units seem to respond more strongly to some classes of characters (for example, vowels) than others, but nowhere is there strict localization. Representations in the hidden layers are distributed to some degree.

One of the reasons for the resurgence of interest in neural networks is their potential for practical applications. NETtalk is significant, because it seems to be close to a real application. It is a good example of a network being applied to a problem it can handle well: a system with regularities, but with exceptions to the regularities. In short, it is a cognitive problem of the messy kind that humans work with all the time and are good at coping with.

An expanded version of this technical report was recently published in the journal *Complex Systems* (Sejnowski and Rosenberg, 1987).

Reference

T. J. Sejnowski and C. R. Rosenberg (1987), "Parallel networks that learn to pronounce English text," *Complex Systems* 1:145–168.

(1986)
Terrence J. Sejnowski and Charles R. Rosenberg

NETtalk: a parallel network that learns to read aloud
The Johns Hopkins University Electrical Engineering and Computer Science Technical Report
JHU/EECS-86/01, 32 pp.

Unrestricted English text can be converted to speech by applying phonological rules and handling exceptions with a look-up table. However, this approach is highly labor intensive since each entry and rule must be hand-crafted. NETtalk is an alternative approach that is based on an automated learning procedure for a parallel network of deterministic processing units. After training on a corpus of informal continuous speech, it achieves good performance and generalizes to novel words. The distributed internal representations of the phonological regularities discovered by the network are damage resistant.

Introduction

English is amongst the most difficult languages to read aloud. Most phonological rules for transforming letters to speech sounds have exceptions that are often context-sensitive (1). For example, the "a" in almost all words ending in "ave", such as "brave" and "gave", is a long vowel, but-not in "have", and there are some words such as "survey" that can vary in pronunciation with their syntactic role. The problem of reconciling rules and exceptions in converting text to speech shares some characteristics with difficult problems in artificial intelligence that have traditionally been approached with rule-based knowledge representations (2), such as natural language translation and abduction in expert systems (3).

DECtalk (4) is a commercial product that can produce intelligible speech synthesis in a restricted domain. DECtalk uses two methods for first converting text to phonemes (5): A word is first looked up in a pronunciation dictionary of common words; if it is not found there, then a set of phonological rules is applied. Phonemes and stress assignments are then converted into speech sounds using transition rules and digital speech synthesis. For novel words that are not correctly pronounced, this approach requires explicit intervention to update the dictionary and rules (6).

An alternative approach to knowledge representation is based on massively-parallel network models (7, 8). Knowledge in these models is distributed over many processing units and the behavior of the network in response to a particular input pattern is a collective decision based on the exchange of information between the processing units. These are called connectionist models because they are "programmed" by specifying the connectivity of the network and the strength or weight of each connection. In some cases the connections can be determined using insights from the problem domain, particularly when the networks have a regular pattern (9). It would be desireable to generate networks automatically, and one method is to "compile" a network from a description of a problem, such as a parser for a grammar (10). Another automatic method is incremental learning, which allows networks to be created by repetitive training on examples.

Rumelhart and McClelland (11) have successfully taught a one-layer network model to produce the past tenses of English verbs using the perceptron learning algorithm (12). The verb-learning network is rule-following but not rule-based in the sense that no rules are explicitly programmed into the network, but after training, the network behaves as if it were following rules. This is a consequence of the learning algorithm, which takes advantage of regularities of past tense endings to minimize the number of weights needed in the network, as shown by the ability of the network to generalize to novel verbs and pseudoverbs. However, a network with one layer of modifiable weights is severely restricted in its ability to discover higher-order features (13, 14).

Until recently, learning in multilayered networks was an unsolved problem and considered by some impossible (13, p. 231). In a multilayered machine the internal, or hidden, units can be used as feature detectors which perform a mapping between input units and output units, and the difficult problem is to discover the proper features. The Boltzmann machine learning algorithm is capable of finding features that allow a network to generalize from examples (8, 14, 15), and several other learning algorithms are now known for multilayered networks which can also discover good features (16, 17).

In this paper we explore the applicability of network learning algorithms with one to three layers of modifiable connections to the problem of converting text to

a

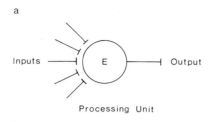

Inputs ——⊣〔 E 〕⊢—— Output

Processing Unit

b

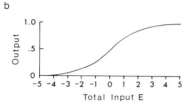

Figure 1 (a) Schematic model of a processing unit receiving inputs from other processing units. (b) Transformation between summed inputs and output of a processing unit, as given by Eq. 2.

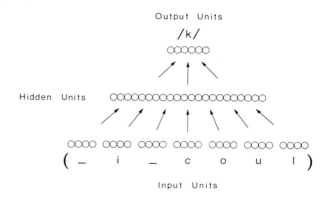

Figure 2 Schematic drawing of the network architecture. Input units are shown on the bottom of the pyramid, with 7 groups of 29 units in each group. Each hidden unit in the intermediate layer receives inputs from all of the input units on the bottom layer, and in turn sends its output to all 26 units in the output layer. An example of an input string of letters is shown below the inputs groups, and the correct output phoneme for the middle letter is shown above the output layer. For 80 hidden units, which were used for the corpus of continuous informal speech, there was a total of 309 units and 18,629 weights in the network, including a variable threshold for each unit.

speech. The model, which we call NETtalk, demonstrates that a relatively small network can capture most of the significant regularities in English pronunciation as well as absorb many of the irregularities. NETtalk can be trained on any dialect of any language and the resulting network can be implemented directly in hardware.

We will first describe the network architecture and the learning algorithm that we used, and then present the results obtained from simulations. Finally, we discuss the computational complexity of NETtalk and some of its biological implications.

Network Architecture

Processing Units

The network is composed of processing units that non-linearly transform their summed, continuous-valued inputs, as illustrated in Fig. 1. The connection strength, or weight, linking one unit to another unit can be a positive or negative real value, representing either an excitatory or an inhibitory influence of the first unit on the output of the second unit. Each unit also has a threshold which is subtracted from the summed input. The threshold is implemented as weight from a unit that has a fixed value of 1 so that the same notation and learning algorithm can also be applied to the thresholds as well as the weights. The output of the ith unit is determined by first summing all of its inputs

$$E_i = \sum_j w_{ij} p_j \qquad (1)$$

where w_{ij} is the weight from the jth to the ith unit, and then applying a sigmoidal transformation

$$p_i = P(E_i) = \frac{1}{1 + e^{-E_i}} \qquad (2)$$

as shown in Fig. 1.

The network used in NETtalk is hierarchically arranged into three layers of units: an input layer, an output layer and an intermediate or "hidden" layer, as illustrated in Fig. 2. Information flows through the network from bottom to top. First the letters units at the base are clamped, then the states of the hidden units are determined by Eqs. 1 & 2, and finally, the states of the phoneme units at the top are determined (30).

Representations of Letters and Phonemes

There are seven groups of units in the input layer, and one group of units in each of the other two layers. Each input group encodes one letter of the input text, so that strings of seven letters are presented to the input units at any one time. The desired output of the network is the correct phoneme, or contrastive speech sound, associated with the center, or fourth, letter of this seven letter "window". The other six letters (three on either side of the center letter) provide a partial context for this decision. The test is stepped through the window letter-by-letter. At each step, the network computes a phoneme, and after each word the weights are adjusted according to how closely the computed pronunciation matches the correct one.

The letters and phonemes are represented in different ways. The letters are represented locally within each group by 26 dedicated units, one for each letter of the alphabet, plus an additional 3 units to encode punctuation and word boundaries. The phonemes, in contrast, are represented in terms of 23 articulatory features, such as point of articulation, voicing, vowel height, and so on, as summarized in Table 1. Three additional units encode stress and syllable boundaries. This is a distributed representation since each output unit participates in the encoding of several phonemes (18).

The hidden units neither receive direct input nor have direct output, but are used by the network to form internal representations that are appropriate for solving the mapping problem of letters to phonemes. The goal of the learning algorithm is to adjust the weights between the units in the network in order to make the hidden units good feature detectors.

Learning Algorithm

Two texts were used to train the network: Phonetic transcriptions from informal, continuous speech of a child (19) and a 20,012 word corpus from a dictionary (20). A subset of 1000 words was chosen from this dictionary taken from the Brown corpus of the most common words in English (21). The corresponding letters and phonemes were aligned and a special symbol for continuation, "-", was inserted whenever a letter was silent or part of a graphemic letter combination, as in the conversion from "phone" to the phonemes /f-on-/ (see Table 1). Two procedures were used to move the text through the window of 7 input groups. For the corpus of informal, continuous speech the text was moved through continuously with word boundary symbols between the words. Several words or word fragments could be within the window at the same time. For the dictionary, the word were placed in random order and were moved through the window individually.

The weights were incrementally adjusted during the training according to the discrepancy between the desired and actual values of the output units. For each phoneme, this error was "back-propagated" from the output to the input layer using the learning algorithm introduced by Rumelhart, et al. (17). Each weight in the network is adjusted to minimize its contribution to the total mean square error between the desired and actual outputs. Briefly, the weights were updated according to:

$$w_{ij}{}^{(n)}(t+1) = \alpha w_{ij}{}^{(n)}(t) + (1-\alpha)\varepsilon\delta_i{}^{(n+1)}P_j{}^{(n)} \tag{3}$$

where $w_{ij}{}^{(n)}$ is the weight from the jth unit in layer n to the ith unit in layer n + 1, the parameter α smooths the gradient by over-relaxation (typically 0.9), ε controls

the rate of learning (typically 2.0). The error signal $\delta_i{}^{(n)}$ for layer n was calculated starting from the output layer N:

$$\delta_i{}^{(N)} = (p_i{}^* - p_i{}^{(N)})P'(E_i{}^{(N)}) \tag{4}$$

and recursively back-propagating the differences to lower layers

$$\delta_i{}^{(n)} = \sum_j \delta_j{}^{(n+1)} w_{ji}{}^{(n)} P'(E_i{}^{(n)}),$$

where $P'(E)$ is the first derivative of $P(E)$, $p_i{}^*$ was the desired value of the ith unit in the output layer, and $p_i{}^{(N)}$ was the actual value obtained from the network. For most of the simulations the error signal was back-propagated only when the difference between the actual and desired values were greater than a margin of 0.1. The gradients in Eq. 4 were accumulated over several letters and Eq. 3 was applied only once for each word. The weights in the network were always initialized to small random values uniformly distributed between -0.3 and 0.3; this was necessary to differentiate the hidden units.

Performance

A simulator was written in the C programming language for configuring a network with arbitrary connectivity, training it on a corpus and collecting statistics on its performance. A network of 10,000 weights had a throughput during learning of about 2 letters/sec on a VAX 780 FPA. Two measures of performance were computed. The output was considered a "perfect match" if the value of each articulatory feature was within a margin of 0.1 of its correct value. This was a much stricter criterion than the "best guess", which was the phoneme making the smallest angle with the output vector. The performance was also assayed by "playing" the output string of phonemes and stresses through DECtalk, bypassing the front end that converts letters to phonemes.

Continuous Informal Speech

This corpus was a difficult one because the same word was often pronounced several different ways; phonemes were commonly modified or elided at word boundaries. The learning curve for 1024 words from the informal speech corpus is shown in Fig. 3. The percentage of correct best guesses for the phonemes rose rapidly at first and continued to rise at slower rate throughout the learning, reaching 95% after 50,000 words. Perfect matches were rarer, but were at 55% and still rising at the termination of the simulation. Primary and secondary stresses and syllable boundaries

Table 1 Articulatory representation of phonemes and punctuations[a]

Symbol	Phoneme	Articulatory features
/a/	father	Low, Tensed, Central2
/b/	bet	Voiced, Labial, Stop
/c/	bought	Unvoiced, Velar, Medium
/d/	debt	Voiced, Alveolar, Stop
/e/	bake	Medium, Tensed, Front2
/f/	fin	Unvoiced, Labial, Fricative
/g/	guess	Voiced, Velar, Stop
/h/	head	Unvoiced, Glottal, Glide
/i/	Pete	High, Tensed, Front1
/k/	Ken	Unvoiced, Velar, Stop
/l/	let	Voiced, Dental, Liquid
/m/	met	Voiced, Labial, Nasal
/n/	net	Voiced, Alveolar, Nasal
/o/	boat	Medium, Tensed, Back2
/p/	pet	Unvoiced, Labial, Stop
/r/	red	Voiced, Palatal, Liquid
/s/	sit	Unvoiced, Alveolar, Fricative
/t/	test	Unvoiced, Alveolar, Stop
/u/	lute	High, Tensed, Back2
/v/	vest	Voiced, Labial, Fricative
/w/	wet	Voiced, Labial, Glide
/x/	about	Medium, Central2
/y/	yet	Voiced, Palatal, Glide
/z/	zoo	Voiced, Alveolar, Fricative
/A/	bite	Medium, Tensed, Front2 + Central1
/C/	chin	Unvoiced, Palatal, Affricative
/D/	this	Voiced, Dental, Fricative
/E/	bet	Medium, Front1 + Front2
/G/	sing	Voiced, Velar, Nasal
/I/	bit	High, Front1
/J/	gin	Voiced, Velar, Nasal
/K/	sexual	Unvoiced, Palatal, Fricative + Velar, Affricative (Compound: [k] + [S])
/L/	bottle	Voiced, Alveolar, Liquid
/M/	absym	Voiced, Dental, Nasal
/N/	button	Voiced, Palatal, Nasal
/O/	boy	Medium, Tensed, Central1 + Central2
/Q/	quest	Voiced, Labial + Velar, Affricative, Stop
/R/	bird	Voiced, Velar, Liquid
/S/	shin	Unvoiced, Palatal, Fricative
/T/	thin	Unvoiced, Dental, Fricative
/U/	book	High, Back1
/W/	bout	High + Medium, Tensed, Central2 + Back1
/X/	excess	Unvoiced, Affricative, Front2 + Central1
/Y/	cute	High, Tensed, Front1 + Front2 + Central1
/Z/	leisure	Voiced, Palatal, Fricative
/@/	bat	Low, Front2
/!/	Nazi	Unvoiced, Labial + Dental, Affricative (Compound: [t] + [s])
/#/	examine	Voiced, Palatal + Velar, Affricative (Compound: [g] + [z])
/*/	one	Voiced, Glide, Front1 + Low, Central1 (Compound: [w] + [^])
/:/	logic	High, Front1 + Front2
/^/	but	Low, Central1
/-/	Continuation	Silent, Elide
/–/	Word Boundary	Pause, Elide
/./	Period	Pause, Full Stop

a. Output representations for phonemes and punctuations. The symbols for phonemes in square brackets are a superset of ARPAbet (4) and are associated with the sound of the italicized part of the adjacent word. Compound phonemes were introduced when a single letter was associated with more than one primary phoneme. The continuation symbol was used when a letter was silent. Two or more of the following 17 articulatory feature units were used to represent each phoneme: *Position in mouth*: Labial = Front1, Dental = Front2, Alveolar = Central1, Palatal = Central2, Velar = Back1, Glottal = Back2; *Phoneme Type*: Stop, Nasal, Fricative, Affricative, Glide, Liquid, Voiced, Tensed; *Vowel Frequency*: High, Medium, Low. Four additional output units were used to represent each punctuation: Silent, Elide, Pause, Full Stop.

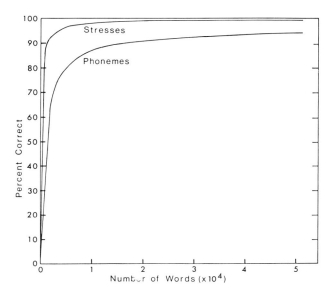

Figure 3 Learning curves for phonemes and stresses during training on the 1024 word corpus of continuous informal speech. The percent of correct best guesses are shown as functions of the number of training words.

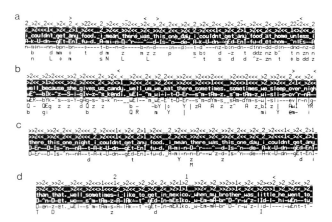

Figure 4 Examples of raw output from the simulator during learning on a corpus of 1024 words of continuous informal speech: (a) after the first 200 words of training starting from random weights, (b) after one pass through the corpus, and (c) after 25 passes through the corpus. (d) Output of the network during testing on a continuation of the corpus. The letters within the black stripe are the text (middle row), phonemes (bottom row) and stresses (top row) from the training corpus. The symbols for the phonemes are given in Table 1. The stresses are represented by a number (primary = 1, secondary = 0, tertiary = 2), and the syllable boundaries are indicated by a reversal in the direction of the arrows: "$\langle$ $\rangle$". The output of the network is shown above and below the black stripe. The phonemes making the smallest angle with the output vector are shown in rank order below the black stripe with the best guess at the top. The stresses are similarly listed in rank order above the black stripe with the best guess at the bottom.

were learned very quickly for all words and achieved nearly perfect performance by 5,000 words, as shown in Fig. 3.

Representative examples of output at the beginning and near the end of the training are shown in Fig. 4. The distinction between vowels and consonants was made early; however, the network substituted the same vowel for all vowels and the same consonant for all consonants, which resulted in a babbling sound. A second stage occured when word boundaries are recognized, and the output then resembled pseudowords. After just a few passes through the network many of the words were intelligible, and by 10 passes the text was understandable.

Errors in the best guesses were far from random. For example, few errors in a well-trained network were confusions between vowels and consonants: most confusions were between phonemes that were very similar, such as the difference in voicing between the "th" sounds in "thesis" and "these". Some errors were due to inconsistencies in the original training corpus. Nevertheless, the intelligibility of the speech was quite good.

A network trained on a 1024 word corpus of informal speech was tested without training on a 439 word continuation from the same speaker. The performance was 78% best guesses and 35% perfect matches, which indicates that much of the learning was transferred to novel words. An excerpt from the new corpus is shown in Fig. 4d.

A graphical summary of the weights between the letter units and some of the hidden units is shown in Fig. 5. The pattern of excitatory and inhibitory weights to a hidden unit can be considered its "receptive field", in analogy with the receptive fields of sensory neurons. Most hidden units responded to more than one pattern of letters. We examined performance of a highly-trained network to random perturbations of the weights. As shown in Fig. 6, random perturbations of the weights uniformly distributed on the interval $[-0.5, 0.5]$ had little effect on the performance of the network, and degradation was gradual with increasing damage. Since the distribution of the weights had a standard deviation of 1.2, the amount of information conveyed by each weight is only a few bits. Relearning was about ten times faster than the original learning starting from the same level of performance. Similar fault tolerance and fast recovery from damage has also been observed using the Boltzmann learning algorithm (15).

Dictionary

We used the 1000 most common English words to study how the performance of the network and learn-

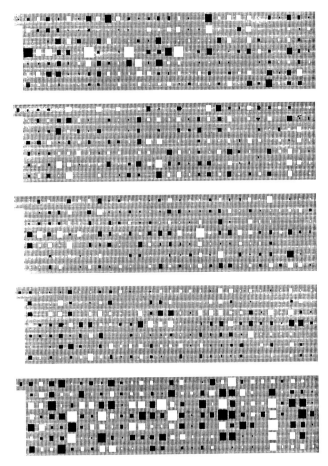

Figure 5 Hinton diagram showing weights from the layer of input units to 5 representative hidden units taken from a network with 80 hidden units that was trained on a corpus of continuous informal speech. Each gray rectangle represents one hidden unit and each square within a rectangle represents a weight. The area of the square is proportional to the magnitude of the weight and the sign of the weight is indicated by the color: white for positive, or excitatory weights, and black for negative, or inhibitory weights (28). The largest weights have magnitudes of about 4. Each row of squares within a gray rectangle represents the weights from one group of input units, with the leftmost input group on the top row and the rightmost input group on the bottom row. The isolated weight in the upper left corner of each array is the bias (negative threshold) of the unit, which was treated like a weight to a true unit. Out of 29 squares in each row, the first 26 represent the weights from the letters of the alphabet, from *a* to *z*, and the last three represent the punctuations, including word boundaries. Thus, the square in the lower left corner of a hidden unit is the weight it receives when the letter "a" is present in the rightmost input group. For most hidden units more than one combination of letters will cause it to produce a large output. These are called distributed representations. However, a few hidden units, such as the third from the top, had a more restricted pattern of weights that could be called a local representation.

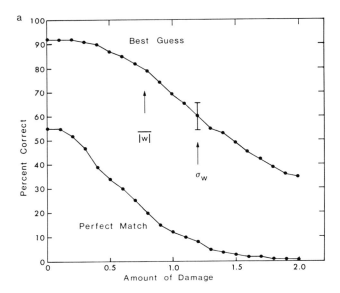

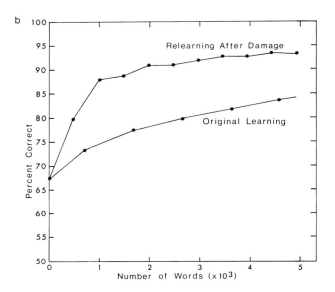

Figure 6 Damage to the network and recovery from damage. a) Performance of a network as a function of the amount of damage to the weights. The network had been previously trained on 50 passes though the corpus of continuous informal speech. The weights were then damaged by adding a random component to each weight uniformly distributed on the interval $[-d, d]$, where d is the amount of damage plotted on the abscissa. The performance shown is the average of at least two disrupted networks for each value of d. For $d = 1.2$, 22 disrupted networks were tested to obtain a standard deviation of 6%. The average absolute value of the weights in the network was $|w| = 0.77$ and the standard deviation was $\sigma = 1.2$, as indicated by the arrows. The best guesses were more resistant to damage than the perfect matches. There was little degradation of the best guesses until $d = 0.5$, and the falloff with increasing damage was gentle. (b) Retraining of a damaged network compared with the original learning curve starting from the same level of performance. The network was damaged with $d = 1.2$ and was retrained using the same corpus and learning parameters that were used to train it. There is a rapid recovery phase during the first pass through the network followed by a slower healing process similar in time course to the later stages of the original training. These two phases can be accounted for by the shape of the error metric in weight space, which typically has deep ravines (15).

ing rate scale with the number of hidden units. The most common English words are also amongst the most irregular, so this was also a test of the capacity of the network to absorb exceptions. With no hidden units, the performance rose quickly and saturated at 82% best guesses as shown in Fig. 7a. This represents the part of the mapping that can be accomplished by linearly separable partitioning of the input space (13). The pattern of errors was different from that observed in networks with a layer of hidden units in that many were stereotyped and inappropriate. Hidden units allow more contextual influence by recognizing higher-order features amongst combinations of input units (14).

The rate of learning and final performance increased with the number of hidden units, as shown in Fig. 7a. The best performance achieved with 120 hidden units was 98% best guesses, better than the performance achieved with continuous informal speech, which was more difficult because of the variability in real-world speech. Examples of two letter-to-sound correspondences are shown in Fig. 7b. The network with 120 hidden units was tested on the randomized dictionary of 20,012 words. Without learning, the average performance was 77% best guesses and 28% perfect matches. Following 5 training passes through the dictionary, the performance increased to 90% best guesses and 48% perfect matches.

Letters and punctuations were represented by single units in the input groups; this is a local representation and had the advantage that the receptive fields of the hidden units were more easily interpreted in terms of letters. Simulations were also performed with distributed representations, similar in spirit to the articulatory representation used for the output units. For a particular distributed representation with 16 units per input group we found that the general level of performance was comparable to that with the local representation, but there was a consistent difficulty with several correspondences between letters and phonemes.

We used the learning algorithm to discover a good distributed input representation by introducing an additional group of 10 units between each input letter group and the group of hidden units. The resulting network had three layers of modifiable weights. The performance of this network with 160 hidden units in the layer after training on 5 passes through the 20,012 word dictionary was 89% best guesses and 49% perfect matches. The number of weights in this network was comparable to a network with a local input representation and 76 hidden units. None of the difficulties experienced with previous hand-crafted distributed input representations occurred. As expected,

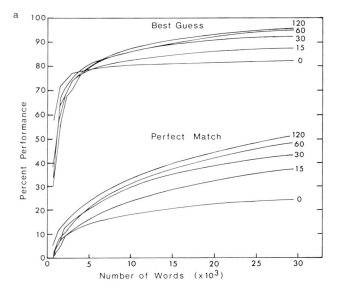

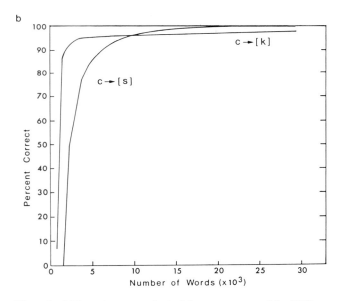

Figure 7 (a) Learning curves for training on a corpus of the 1000 most common words in English using different numbers of hidden units, as indicated beside each curve. For the case with no hidden units, the input units were directly connected to the output units. Both the percent correct best guesses and perfect matches are shown. (b) Performance during learning of two representative phonological rules, the hard and soft pronunciation of the letter "c". Note that the soft "c" takes longer to learn, but eventually achieves perfect accuracy. The hard "c" occurs about twice as often as the soft "c" in the training corpus. Children show a similar difficulty with learning to read words with the soft "c" (29).

each group of 10 units developed a highly distributed representation.

Computational Complexity

The translation of letters to phonemes can be analyzed as a mapping problem. Consider a domain of 29 symbols for letters and punctuations taken in strings of length 7. We would like to construct a deterministic mapping from these strings to a range of 51 symbols representing phonemes (23). Only a subset of all possible mappings actually occur in English speech and the problem is to find a compact description of this mapping which takes advantage of the regularities and also captures the exceptions (24).

For a restricted text this problem can be solved by specifying entries in a look-up table determined by letter strings in English words. For a text of 1000 words this would consist of about 5,000 entries since there are 5 letters on average per word and there would be at most one entry in the table per letter. However, this look-up table generalizes poorly when applied to new words in an unrestricted text. One way to generalize is to look for partial matches; this could be implemented by compiling frequency tables of letter pairs, triples, etc. for all combinations of positions within the window. There are two practical problems with this method: First the size of these tables grows exponentially with the size of the window, with about 500,000 entries needed for a text of 1000 words and a window of 7 letters. Second, some weighting scheme is required to combine evidence from different partial matches.

In NETtalk, the weighting of input letters is performed by the weights between the letter units and the hidden units, and the weighting of the features is performed by the weights between the hidden units and the output units. The learning algorithm discovers those combinations of letters that are particularly efficient at implementing the correspondences between letters and phonemes (14). The mapping is distributed in that each significant combination of letters is encoded by several hidden units, and each hidden unit recognizes more than one sequence of letters; as a consequence, the performance of the network is highly resistant to both localized and diffuse damage. Exceptions to regularities are also recognized by their features so that a separate look-up table such as that used in DECtalk is unnecessary. We are currently examining assemblies of units that appear to be related to particular letter-to-sound correspondence rules. Learning algorithms make it possible to design these efficient mappings without direct human intervention.

Biological Implications

The processing units used in the network share some properties with neurons, such as a high degree of connectivity, summation of excitatory and inhibitory influences through synaptic weights, and a nonlinear input-output function that resembles the firing rate of a neuron as a function of integrated synaptic inputs, but there are also many differences, such as the absence of explicit action potentials and an integration time constant. However, insights may be gained concerning the representation of information in large populations of neurons by examining the way that these simple network models solve problems like text-to-speech. Although the detailed implementations may be different, similar principles may apply to both neural networks and massively-parallel network models (25).

During the early stages of learning in NETtalk, the sounds produced by the network are uncannily similar to early speech sounds in children (26). However, our model of text-to-speech combines two different processes that occur at different stages of human development: learning to talk and learning to read. By the time that a human child learns to read, phonetic representations for words are already well developed. Nonetheless, the phonological mappings produced by NET-talk are efficient encodings for a parallel network and may be comparable to those used by humans.

NETtalk can be used to study the importance of particular phonological rules in the context of a particular corpus by presenting the network with nonsense words that are constructed to critically test a proposed rule. The performance of the network can also be studied following damage of the network. The patterns of errors following simulated "lesions" in the network by either removing units or by disrupting the weights can be compared with reading errors observed in humans suffering from acquired dyslexia (27).

NETtalk is clearly limited in its ability to handle ambiguities that require syntactic and semantic levels of analysis. It is perhaps surprising that the network was capable of reaching a significant level of performance using a window of only seven letters. A human level of performance would require information from larger parts of sentences to control intonation, stress contours and prosody. It should be possible to incorporate these variables into a structured network and apply the learning algorithm to them as well.

References

1. N. Chomsky and M. Halle, M., *The Sound Pattern of English*, (Harper & Row, New York, 1968); R. L. Venezky, *The Structure of English Orthography*, (Mouton, The Hague, 1970); L. Henderson,

Orthography and World Recognition in Reading, (Academic Press, New York, 1982).

2. For a discussion of exceptions in knowledge representations based on inheritance hierarchies see D. W. Etherington and R. Reiter, [*Int. Joint Conf. on Artificial Intelligence*, (William Kauffman, Inc., Los Altos, California, 1983), pp. 104–108] and D. S. Touretzky, D. S. [*Proc. National Conf. Artif. Intelligence*, (William Kauffman, Inc., Los Altos, California, 1984), pp. 322–325]. An alternative network formulation of this problem based on evidence theory is given by L. Shastri, and J. A. Feldman, [*Proc. 9th International Joint Conf. on Artificial Intelligence*, (William Kauffman, Inc., Los Altos, California, 1985)].

3. W. Haas, *Phonographic Translation*, (Manchester University Press: Manchester, 1970). For an account of abduction in plausible inference, see Ch. 8 in E. Charniak and D. McDermott, *Artificial Intelligence*, (Addison Wesley, Reading, MA, 1985).

4. Digital Equipment Corporation, *DTC-01-AA* For a study of speech synthesis intelligibility, see D. B. Pisoni, H. C. Nusbaum, B. G. Greene, *Proc. IEEE* **73**, 1665 (1985).

5. J. Allen, *Proc. IEEE* **64**, 433 (1976); D. Klatt, D. *J. Acoust. Soc. Am.* **67**, 971 (1980).

6. S. R. Hertz, J. Kadin, K. J. Karplus, *Proc. IEEE* **73**, 1589, (1985).

7. D. J. Amit, H. Gutfreund, H. Sompolinsky, *Phys. Rev. A* **32**, 1007 (1985); J. A. Anderson, *IEEE Trans. Systems, Man, Cybernetics* **13**, 799 (1983); M. A. Arbib, *Annals Biomed. Eng.* **3**, 238 (1975); D. H. Ballard, G. E. Hinton, T. J. Sejnowski, *Nature* **306**, 21 (1983); A. G. Barto, R. S. Sutton and C. W. Anderson, *IEEE Trans. Systems, Man, Cybernetics* **13**, 835 (1983); M. A. Cohen and S. Grossberg, *IEEE Trans. Systems, Man, Cybernet.* **13**, 875 (1983); S. E. Fahlman, G. E. Hinton, T. J. Sejnowski, in Proceedings of the National Conference on Artificial Intelligence, (William Kauffman, Inc., Los Altos, Proceedings of the National Conference on Artificial Intelligence, 1983), pp. 109–113; G. E. Hinton and J. A. Anderson (Eds), *Parallel Models of Associative Memory*, (Erlbaum Associates, Hillsdale, NJ, 1981); J. A. Feldman, D. H. Ballard, *Cognitive Science* **6**, 205 (1982); T. Hogg and B. A. Huberman, *Proc. National Acad. Sci. USA* **81**, 6871 (1984); J. J. Hopfield, *Proc. National Acad. Sci. USA* **79**, 2554 (1982); C. Koch, J. Marroquin and A. Yuille, *Proc. National Acad. Sci. USA* (in press); T. Kohonen, *Self-Organization and Associative Memory*, (Springer-Verlag, New York, 1984); J. A. McClelland and D. E. Rumelhart, *Psych. Rev.* **88**, 375 (1981); P. Peretto, *Biological Cybernetics* **50**, 51 (1984); P. Smolensky in *Proc. of the National Conference on Artificial Intelligence*, (William Kauffman, Inc., Los Altos, California, 1983), pp. 378–382; G. Tolouse, S. Dehaene and J.-P. Changeux, *Proc. National Acad. Sci.* (in press); C. von der Malsburg and E. Bienenstock, In: *Disordered Systems and Biological Organization*, F. Fogelman, F. Weisbuch and E. Bienenstock, Eds. (Springer-Verlag, Berlin, 1986); D. K. Waltz and J. B. Pollack, *Cog. Sci.* **9**, 51 (1985); S. Wolfram, *Physica D* in press (1986).

8. G. E. Hinton, T. J. Sejnowski, *Proceedings of the IEEE Computer Society Conference on Computer Vision & Pattern Recognition*, Washington, D.C., pp. 448–453 (1983); D. H. Ackley, G. E. Hinton, T. J. Sejnowski, *Cognitive Science* **9**, 147 (1985).

9. For example, D. Marr and T. Poggio [*Science* **194**, 283 (1976)] designed a network that could compute depth from random-dot stereograms; T. J. Sejnowski and G. E. Hinton, [in *Vision, Brain and Cooperative Computation*, M. A. Arbib and A. R. Hanson (Eds.) (MIT Press, Cambridge, MA, 1986)] designed a network to separate figure from ground in images; J. J. Hopfield & D. Tank, [*Biolog. Cybernetics* **52**, 1 (1985)] designed a network that finds moderately good tours for the traveling salesman problem quickly: Their network model is a symmetric version of the continuous nonlinear model analyzed by T. J. Sejnowski in *Parallel Models of Associative Memory*, G. E. Hinton and J. A. Anderson (Eds.) (Erlbaum Associates, Hillsdale, NJ, 1981), pp. 189–212.

10. B. Selman, and G. Hirst, *Proc. 7th Annual Conf. of the Cognitive Science Soc.* (1985); M. Fanty, University of Rochester Computer Science Technical Report TR-174 (1985); see also A. S. Lapedes and R. M. Farber, *Physica D*, (in press).

11. D. E. Rumelhart and J. L. McClelland, On Learning the Past Tenses of English Verbs, in: D. E. Rumelhart and J. L. McClelland, (Eds.) *Parallel Distributed Processing: Explorations in the Microstructure of Cognition*. (MIT Press, Cambridge, 1986)

12. The perceptron, which was introduced by F. Rosenblatt [*Principles of Neurodynamics*, (Spartan Books, New York, 1959)], uses binary threshold units that are deterministic, but Rumelhart and McClelland use a probabilistic update rule that turns their network into a one-layer Boltzmann machine (8). It can be shown that the perceptron learning algorithm in this case is identical to the Boltzmann learning algorithm where co-occurrence statistics are only collected for one iteration.

13. Perceptrons can only learn first-order predicates. [M. Minsky & S. Papert, *Perceptrons*, (MIT Press, Cambridge, 1969)]. McClelland and Rumelhart used a third-order coding scheme to pre-process the inputs and post-process the outputs, which made it possible to learn the mapping with only one layer of modifiable weights.

14. T. J. Sejnowski, P. K. Kienker, and G. E. Hinton, *Physica D*, (in press).

15. G. E. Hinton and T. J. Sejnowski, Learning and Relearning in Boltzmann Machines in: D. E. Rumelhart and J. L. McClelland, (Eds.) *Parallel Distributed Processing: Explorations in the Microstructure of Cognition*. (MIT Press, Cambridge, 1986)

16. A. G. Barto and C. W. Andersen, *Proc. 7th Annual Conf. Cognitive Science Sc.* (1985); D. B. Parker, MIT Center for Computational Research in Economics and Management Science, TR-47 (1985); Y. LeCun, In: *Disordered Systems and Biological Organization*, F. Fogelman, F. Weisbuch and E. Bienenstock, Eds. (Springer-Verlag, Berlin, 1986);

17. D. E. Rumelhart, G. E. Hinton, and R. J. Williams, Learning Internal Representations by Error Propagation, in: D. E. Rumelhart and J. L. McClelland, (Eds.) *Parallel Distributed Processing: Explorations in the Microstructure of Cognition*. (MIT Press, Cambridge, 1986).

18. G. E. Hinton, J. L. McClelland, and D. E. Rumelhart, Distributed Representations, in: D. E. Rumelhart and J. L. McClelland, (Eds.) *Parallel Distributed Processing: Explorations in the Microstructure of Cognition*. (MIT Press, Cambridge, 1986).

19. E. C. Carterette, and M. G. Jones, *Informal Speech*. (University of California Press, Los Angeles, 1974).

20. *Merriam Webster's Pocket Dictionary*, 1974.

21. H. Kuchera and W. N. Francis, *Computational Analysis of Modern-Day American English*, (Brown University Press, Providence, RI, 1967).

22. This supervised learning algorithm only uses positive examples and requires that the teacher correct every error for every output unit. Although children are corrected while they are learning to read, some learning also occurs through non-supervised observation of correct reading. Non-supervised learning algorithms would be worth exploring in the context of this problem [S. Grossberg, *Biolog. Cybernetics* **23**, 121 (1976); E. L. Bienenstock, L. N. Cooper, and P. W. Munro, *J. Neuroscience* **2**, 32 (1982); D. E. Rumelhart and D. Zipser, *Cog. Sci.* **9**, 75 (1985)].

23. A. N. Kolmogorov [*Dokl. Akad. Nauk SSSR* **114**, 953; *AMS Translation* **2**, 55 (1957)] has studied the class of functions that can be computed with a layered network of nonlinear processing units, and these results have been extended by G. Palm [*Biol. Cybernetics* **31**, 119 (1978)]. S. Wolfram [*Nature* **311**, 419 (1984)] has studied the

computational complexity of cellular automata, a related architecture with discrete states. The complexity analysis of what can be learned as opposed to what can be computed has only recently been addressed. L. Valiant [*Communications of the ACM*, **27**, 1134 (1984); *Proc. 9th International Joint Conf. on Artifical Intelligence*, (William Kauffman, Inc., Los Altos, California, 1985) pp. 560–566] has analyzed the learning of disjunctions of conjunctons and found a subclass that can be learned in polynomial time. See also E. M. Gold, *Inform & Control* **16**, 447 (1967).

24. A Wijk [in: *Alphabets for English*, W. Haas (Ed.), (Manchester University Press, Manchester, England, 1969)] has estimated that there are 102 graphemes or basic letter groupings in English, and P. R. Hanna, J. S. Hanna, R. E. Hodges, R. E. and E. H. Rudorf, [in: *Phoneme-Grapheme Correspondences as Cues to Spelling Improvement*, (US Department of Health, Education and Welfare, Washington, D.C., 1966)] used 170 graphemic patterns in designing phoneme-grapheme correspondence rules for spelling. However, additional rules are also needed to specify the segmentation of unrestricted English text, which is an unsolved problem. Henderson (1, p. 82) states that "... in the absence of morphological constraints it seems clear that no segmenting procedure can be formulated so as to result in a correct translation of all English words."

25. T. J. Sejnowski, Open Questions About Computation in Cerebral Cortex, In: D. E. Rumelhart, and J. L. McClelland, (Eds.) *Parallel Distributed Processing: Explorations in the Microstructure of Cognition*. (MIT Press, Cambridge, 1986); D. H. Ballard, *Behavioral and Brain Sciences* (in press); D. Tank and J. J. Hopfield, *Science* (in press).

26. Rumelhart and McClelland (11) have observed stages during the network learning of past tenses of English verbs that resemble the learning of past tenses by children. These developmental patterns may be a general property of incremental learning in networks with distributed representations.

27. T. Shallice, E. K. Warrington, and R. McCarthy, *Quarterly Journal of Experimental Psychology*, **35A**, 111 (1983); A. Caramazza, G. Miceli, M. C. Silveri, and A. Laudanna, *Cognitive Neuropsychol.* **2**, 81 (1985).

28. This visual representation of the weights is much more efficient than a traditional wiring diagram. For a more extensive description of this representation of a network, see (8, 14).

29. R. L. Venezky and D. Johnson, *J. Educ. Psychol.* **64**, 109 (1973).

30. Note that the output of the network only depends on letters and not on neighboring phonemes. As a consequence, the network cannot take advantage of phonotactic regularities.

31. We are grateful to Drs. Alfonso Caramazza, Stephen Hanson, James McClelland, Geoffrey Hinton, George Miller, David Rumelhart, Timothy Shallice, and Stephen Wolfram for useful insights and helpful discussions on language and learning. Bell Communications Research at Morristown, N.J. provided computational support. We also wish to thank Drs. Peter Brown, Edward Carterette, Howard Nusbaum and Alexander Weibel for their help in obtaining corpora. C.R.R. was supported by a grant to Dr. Michael Gazzaniga at the Division of Cognitive Neuroscience at the Cornell Medical Center, and T.J.S. was supported by grants from the National Science Foundation, System Development Foundation, Sloan Foundation, General Electric Corporation, Exxon Education Foundation, Allied Corporation Foundation, Westinghouse, and Smith, Kline & French Laboratories.

Introduction

(1986)
D. E. Rumelhart, G. E. Hinton, and R. J. Williams
Learning internal representations by error propagation
Parallel Distributed Processing: Explorations in the Microstructures of Cognition, Vol. 1,
D. E. Rumelhart and J. L. McClelland (Eds.), Cambridge, MA: MIT Press, pp. 318–362

(1986)
David E. Rumelhart, Geoffrey E. Hinton, and Ronald J. Williams
Learning representations by back-propagating errors
Nature 323:533–536

These two papers describe the same model: what has become known in the neural network world as "back propagation," or, informally, "backprop." It is, as of mid-1987, the most popular learning algorithm for working with multilayer networks. It is considerably faster than the other learning algorithm that is successful with multilayer networks, the Boltzmann machine (paper 38).

In addition to the work described here, back propagation was discovered independently in two other places about the same time, an interesting case of simultaneous independent discovery of a solution, once the nature of the learning problem was made explicit (Le Cun, 1986; Parker, 1985). It also turned out that the algorithm had been described earlier by Paul J. Werbos in his Harvard Ph.D. thesis in August 1974 (Werbos, 1974).

We have included two papers on back propagation. The mathematical proof of back propagation involves repeated applications of the chain rule for partial derivatives, and can be confusing. The calculations are done in slightly different ways in the two papers. The short *Nature* paper also contains an elegant computer simulation.

Back propagation is a generalization of the Widrow-Hoff error correction rule (see paper 10). The original Widrow-Hoff technique formed an error signal, which is the difference between what the output is and what it was supposed to be. Synaptic strengths were changed in proportion to the error times the input signal, which diminishes the error in the direction of the gradient (the direction of most rapid change in the error). This algorithm can be realized locally in a neural network. It is not hard to suggest a system where the input, output, and desired output (together forming the error signal) are present at the same synapse. This does not mean that it actually exists—just that it is easy to propose a plausible circuit!

In a multilayer network containing hidden units, that is, units that are neither input nor output units, the problem is much more difficult. The error signal can be formed as before, but many synapses can give rise to the error, not just the ones at the output layers. Since we usually do not know what the hidden units are supposed to do, we cannot directly compute the error signal for hidden units.

The "generalized delta rule" gives a recipe for adjusting the strengths of synapses on internal units based on the error at the output. However, the internal units must be told how large the error is, and how strongly the internal units are connected to the output units in error. (If an internal unit can make no contribution to the error, it obviously need not modify its weights.) This involves running the synapses 'backward' so that the internal unit knows both how strongly it is connected to an output unit and the error at that unit. The internal unit then sums up all its weighted error contributions. It knows the strengths of the inputs it receives, and can modify its synaptic weights according to a rule very similar to that used by the output units:

according to the product of input and summed, weighted errors from higher layers going backward. The implementation of back propagation involves a forward pass through the layers to estimate the error, and then a backward pass modifying the synapses to decrease the error.

Practical implementations of the back propagation algorithm are not difficult, but without modification it is still rather slow, especially for systems with many layers. There is currently a great deal of work being devoted to various ways of speeding up learning, some of which are ingenious, theoretically sound ways of improving the algorithm.

Real synapses do not run backward. Making a plausible physiological model of back propagation is not easy. There are many examples of reciprocal connections between higher and lower cortical levels, but it is unclear that such projections have the correct properties, either anatomical or physiological, to serve in a back propagation network.

One of the most important aspects of both Boltzmann machines and back propagation involves the nature of the representation of information that is formed in the hidden units. The title of paper 41 in the book *Parallel Distributed Processing* places "learning internal representations" first. In the solution to the encoder problem found by the Boltzmann machine (see paper 38) the representations of the input data found by the learning algorithm, when it was successful, looked like binary counting. If a good representation exists, and use of it is necessary in order to solve a problem, then back propagation seems to be able to find it in some situations.

The representations formed in the hidden units may be effective ways of representing the important information in the input signal. It has been proposed by several groups that back propagation can be used to develop algorithms for data compression. Suppose we arrange it so that the input and output layers contain the same number of units, but that there are fewer units in the hidden layers, and that input and output layers have no direct connections. Suppose we wish to train the system so that given an input, it reproduces as accurately as possible the input pattern at the output. If it can successfully do this, then the internal representation at the hidden layers contains adequate information to reconstruct the output to some degree of accuracy. So to communicate the input pattern, we need send only the values of the hidden units.

Understanding the nature of optimal representations is a matter of great interest in both cognitive science and engineering. We have seen other papers in this book where effective internal representations were formed by neural networks, for example learning in visual cortex and topographic organization of cortex. Back propagation suggests a new and powerful way to explore good representations; it is also the most effective current learning algorithm for complex, multilayer systems.

References

Y. Le Cun (1986), "Learning processes in an asymmetric threshold network," *Disordered Systems and Biological Organization*, E. Bienenstock, F. Fogelman Souli, and G. Weisbuch (Eds.), Berlin: Springer.

D. Parker (1985), "Learning Logic," Technical report TR-87, Center for Computational Research in Economics and Management Science, MIT, Cambridge, MA.

P. J. Werbos (1974), "Beyond regression: new tools for prediction and analysis in the behavioral sciences," Ph.D. thesis, Harvard University, Cambridge, MA.

(1986)
D. E. Rumelhart, G. E. Hinton, and R. J. Williams

Learning internal representations by error propagation
Parallel Distributed Processing: Explorations in the Microstructures of Cognition, Vol. 1,
D. E. Rumelhart and J. L. McClelland (Eds.), Cambridge, MA: MIT Press, pp. 318–362

The Problem

We now have a rather good understanding of simple two-layer associative networks in which a set of input patterns arriving at an input layer are mapped directly to a set of output patterns at an output layer. Such networks have no *hidden* units. They involve only *input* and *output* units. In these cases there is no *internal representation*. The coding provided by the external world must suffice. These networks have proved useful in a wide variety of applications (cf. Chapters 2, 17, and 18). Perhaps the essential character of such networks is that they map similar input patterns to similar output patterns. This is what allows these networks to make reasonable generalizations and perform reasonably on patterns that have never before been presented. The similarity of patterns in a PDP system is determined by their overlap. The overlap in such networks is determined outside the learning system itself—by whatever produces the patterns.

The constraint that similar input patterns lead to similar outputs can lead to an inability of the system to learn certain mappings from input to output. Whenever the representation provided by the outside world is such that the similarity structure of the input and output patterns are very different, a network without internal representations (i.e., a network without hidden units) will be unable to perform the necessary mappings. A classic example of this case is the *exclusive-or* (XOR) problem illustrated in Table 1. Here we see that those patterns which overlap least are supposed to generate identical output values. This problem and many others like it cannot be performed by networks without hidden units with which to create their own internal representations of the input patterns. It is interesting to note that had the input patterns contained a third input taking the value 1 whenever the first two have value 1 as shown in Table 2, a two-layer system would be able to solve the problem.

Minsky and Papert (1969) have provided a very careful analysis of conditions under which such systems are capable of carrying out the required mappings. They show that in a large number of interesting cases, networks of this kind are incapable of solving the problems. On the other hand, as Minsky and Papert also pointed out, if there is a layer of simple perceptron-like hidden units, as shown in Figure 1, with which the original input pattern can be augmented, there is always a recoding (i.e., an internal representation) of the input patterns in the hidden units in which the similarity of the patterns among the hidden units can support any required mapping from the input to the output units. Thus, if we have the right connections from the input units to a large enough set of hidden units, we can always find a representation that will perform any mapping from input to output through these hidden units. In the case of the XOR problem, the addition of a feature that detects the conjunction of the input units changes the similarity structure of the patterns sufficiently to allow the solution to be learned. As illustrated in Figure 2, this can be done with a single hidden unit. The numbers on the arrows represent the strengths of the connections among the units. The numbers written in the circles represent the thresholds of the units. The value of $+1.5$ for the threshold of the hidden unit insures that it will be turned on only when both input units are on. The value 0.5 for the output unit insures that it will turn on only when it receives a net positive input greater than 0.5. The weight of -2 from the hidden unit to the output unit insures that the output unit will not come on when both input units are on. Note that from the point of view of the output unit, the hidden unit is treated as simply another input unit. It is as if the input patterns consisted of three rather than two units.

The existence of networks such as this illustrates the potential power of hidden units and internal representations. The problem, as noted by Minsky and Papert, is that whereas there is a very simple guaranteed learning rule for all problems that can be solved without hidden units, namely, the perceptron convergence procedure (or the variation due originally to Widrow and Hoff, 1960, which we call the delta rule; see Chapter 11), there is no equally powerful rule for learning in networks with hidden units. There have been three

Table 1

Input patterns		Output patterns
00	→	0
01	→	1
10	→	1
11	→	0

Table 2

Input patterns		Output patterns
000	→	0
010	→	1
100	→	1
111	→	0

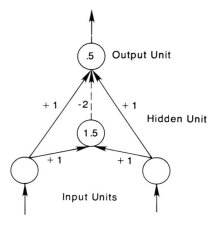

Figure 2 A simple XOR network with one hidden unit. See text for explanation.

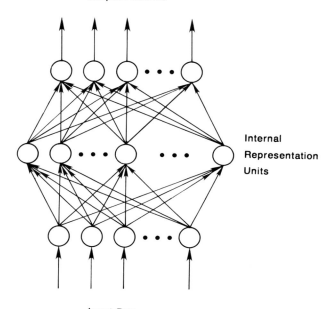

Figure 1 A multilayer network. In this case the information coming to the input units is *recoded* into an internal representation and the outputs are generated by the internal representation rather than by the original pattern. Input patterns can always be encoded, if there are enough hidden units, in a form so that the appropriate output pattern can be generated from any input pattern.

basic responses to this lack. One response is represented by competitive learning (Chapter 5) in which simple *unsupervised* learning rules are employed so that useful hidden units develop. Although these approaches are promising, there is no external force to *insure* that hidden units appropriate for the required mapping are developed. The second response is to simply *assume* an internal representation that, on some a priori grounds, seems reasonable. This is the tack taken in the chapter on verb learning (Chapter 18) and in the interactive activation model of word perception (McClelland & Rumelhart, 1981; Rumelhart & McClelland, 1982). The third approach is to attempt to *develop* a learning procedure capable of learning an internal representation adequate for performing the task at hand. One such development is presented in the discussion of Boltzmann machines in Chapter 7. As we have seen, this procedure involves the use of stochastic units, requires the network to reach equilibrium in two different phases, and is limited to symmetric networks. Another recent approach, also employing stochastic units, has been developed by Barto (1985) and various of his colleagues (cf. Barto & Anandan, 1985). In this chapter we present another alternative that works with deterministic units, that involves only local computations, and that is a clear generalization of the delta rule. We call this the *generalized delta rule*. From other considerations, Parker (1985) has independently derived a similar generalization, which he calls *learning-logic*. Le Cun (1985) has also studied a roughly similar learning scheme. In the remainder of this chapter we first derive the generalized delta rule, then we illustrate its use by providing some results of our simulations, and finally we indicate some further generalizations of the basic idea.

The Generalized Delta Rule

The learning procedure we propose involves the presentation of a set of pairs of input and output patterns. The system first uses the input vector to produce its own output vector and then compares this with the *desired output*, or *target* vector. If there is no difference, no learning takes place. Otherwise the weights are changed to reduce the difference. In this case, with no hidden units, this generates the standard delta rule as described in Chapters 2 and 11. The rule for changing weights following presentation of input/output pair p is given by

$$\Delta_p w_{ji} = \eta(t_{pj} - o_{pj})i_{pi} = \eta\delta_{pj}i_{pi} \tag{1}$$

where t_{pj} is the target input for jth component of the output pattern for pattern p, o_{pj} is the jth element of the actual output pattern produced by the presentation of input pattern p, i_{pi} is the value of the ith element of the input pattern, $\delta_{pj} = t_{pj} - o_{pj}$, and $\Delta_p w_{ij}$ is the change to be made to the weight from the ith to the jth unit following presentation of pattern p.

The delta rule and gradient descent   There are many ways of derivating this rule. For present purposes, it is useful to see that for linear units it minmizes the squares of the differences between the actual and the desired output values summed over the output units and all pairs of input/output vectors. One way to show this is to show that the derivative of the error measure with respect to each weight is proportional to the weight change dictated by the delta rule, with negative constant of proportionality. This corresponds to performing steepest descent on a surface in weight space whose height at any point in weight space is equal to the error measure. (Note that some of the following sections are written in italics. These sections constitute informal derivations of the claims made in the surrounding text and can be omitted by the reader who finds such derivations tedious.)

To be more specific, then, let

$$E_p = \frac{1}{2}\sum_j (t_{pj} - o_{pj})^2 \tag{2}$$

be our measure of the error on input/output pattern p and let $E = \sum E_p$ be our overall measure of the error. We wish to show that the delta rule implements a gradient descent in E when the units are linear. We will proceed by simply showing that

$$-\frac{\partial E_p}{\partial w_{ji}} = \delta_{pj}i_{pi},$$

which is proportional to $\Delta_p w_{ji}$ as prescribed by the delta rule. When there are no hidden units it is straightforward to compute the relevant derivative. For this purpose we use the chain rule to write the derivative as the product of two parts: the derivative of the error with respect to the output of the unit times the derivative of the output with respect to the weight.

$$\frac{\partial E_p}{\partial w_{ji}} = \frac{\partial E_p}{\partial o_{pj}}\frac{\partial o_{pj}}{\partial w_{ji}}. \tag{3}$$

The first part tells how the error changes with the output of the jth unit and the second parts tells how much changing w_{ji} changes that output. Now, the derivatives are easy to compute. First, from Equation 2

$$\frac{\partial E_p}{\partial o_{pj}} = -(t_{pj} - o_{pj}) = -\delta_{pj}. \tag{4}$$

Not surprisingly, the contribution of unit u_j to the error is simply proportional to δ_{pj}. Moreover, since we have linear units,

$$0_{pj} = \sum_i w_{ji}i_{pi}, \tag{5}$$

from which we conclude that

$$\frac{\partial o_{pj}}{\partial w_{ji}} = i_{pi}.$$

Thus, substituting back into Equation 3, we see that

$$-\frac{\partial E_p}{\partial w_{ji}} = \delta_{pj}i_{pi} \tag{6}$$

as desired. Now, combining this with the observation that

$$\frac{\partial E}{\partial w_{ji}} = \sum_p \frac{\partial E_p}{\partial w_{ji}}$$

should lead us to conclude that the net change in w_{ji} after one complete cycle of pattern presentations is proportional to this derivative and hence that the delta rule implements a gradient descent in E. In fact, this is strictly true only if the values of the weights are not changed during this cycle. By changing the weights after each pattern is presented we depart to some extent from a true gradient descent in E. Nevertheless, provided the learning rate (i.e., the constant of proportionality) is sufficiently small, this departure will be negligible and the delta rule will implement a very close approximation to gradient descent in sum-squared error. In particular, with small enough learning rate, the delta rule will find a set of weights minimizing this error function.

The delta rule for semilinear activation functions in feedforward neworks   We have shown how the standard delta rule essentially implements gradient descent in sum-squared error for linear activation functions. In this case, without hidden units, the error surface is shaped like a bowl with only one minimum, so gradient descent is guaranteed to find the best set of weights. With hidden units, however, it is not so obvious how to compute the derivatives, and the error surface is not concave upwards, so there is the danger of getting stuck in local minima. The main theoretical contribution of this chapter is to show that there is an efficient way of computing the derivatives. The main empirical contribution is to show that the apparently fatal problem of local minima is irrelevant in a wide variety of learning tasks.

At the end of the chapter we show how the generalized delta rule can be applied to arbitrary networks, but, to begin with, we confine ourselves to *layered feedforward* networks. In these networks, the input units are the bottom layer and the output units are the top layer. There can be many layers of hidden units in between, but every unit must send its output to higher layers than its own and must receive its input from lower layers than its own. Given an input vector, the output vector is computed by a forward pass which computes the activity levels of each layer in turn using the already computed activity levels in the earlier layers.

Since we are primarily interested in extending this result to the case with hidden units and since, for reasons outlined in Chapter 2, hidden units with linear activation functions provide no advantage, we begin by generalizing our analysis to the set of nonlinear activation functions which we call *semilinear* (see Chapter 2). A semilinear activation function is one in which the output of a unit is a nondecreasing and differentiable function of the net total output,

$$net_{pj} = \sum_i w_{ji} o_{pi}, \tag{7}$$

where $o_i = i_i$ if unit i is an input unit. Thus, a semilinear activation function is one in which

$$o_{pj} = f_j(net_{pj}) \tag{8}$$

and f is differentiable and nondecreasing. The generalized delta rule works if the network consists of units having semilinear activation functions. Notice that linear threshold units do not satisfy the requirement because their derivative is infinite at the threshold and zero elsewhere.

To get the correct generalization of the delta rule, we must set

$$\Delta_p w_{ji} \propto -\frac{\partial E_p}{\partial w_{ji}},$$

where E is the same sum-squared error function defined earlier. As in the standard delta rule it is again useful to see this derivative as resulting from the product of two parts: one part reflecting the change in error as a function of the change in the net input to the unit and one part representing the effect of changing a particular weight on the net input. Thus we can write

$$\frac{\partial E_p}{\partial w_{ji}} = \frac{\partial E_p}{\partial net_{pj}} \frac{\partial net_{pj}}{\partial w_{ji}}. \tag{9}$$

By Equation 7 we see that the second factor is

$$\frac{\partial net_{pj}}{\partial w_{ji}} = \frac{\partial}{\partial w_{ji}} \sum_k w_{jk} o_{pk} = o_{pi}. \tag{10}$$

Now let us define

$$\delta_{pj} = -\frac{\partial E_p}{\partial net_{pj}}.$$

(By comparing this to Equation 4, note that this is consistent with the definition of δ_{pj} used in the original delta rule for linear units since $o_{pj} = net_{pj}$ when unit u_j is linear.) Equation 9 thus has the equivalent form

$$-\frac{\partial E_p}{\partial w_{ji}} = \delta_{pj} o_{pi}.$$

This says that to implement gradient descent in E we should make our weight changes according to

$$\Delta_p w_{ji} = \eta \delta_{pj} o_{pi}, \tag{11}$$

just as in the standard delta rule. The trick is to figure out what δ_{pj} should be for each unit u_j in the network. The interesting result, which we now derive, is that there is a simple recursive computation of these δ's which can be implemented by propagating error signals backward through the network.

To compute $\delta_{pj} = -\partial E_p/\partial net_{pj}$, we apply the chain rule to write this partial derivative as the product of two factors, one factor reflecting the change in error as a function of the output of the unit and one reflecting the change in the output as a function of changes in the input. Thus, we have

$$\delta_{pj} = -\frac{\partial E_p}{\partial net_{pj}} = -\frac{\partial E_p}{\partial o_{pj}} \frac{\partial o_{pj}}{\partial net_{pj}}. \tag{12}$$

Let us compute the second factor. By Equation 8 we see that

$$\frac{\partial o_{pj}}{\partial net_{pj}} = f'_j(net_{pj}),$$

which is simply the derivative of the squashing function f_j for the jth unit, evaluated at the net input net_{pj} to that unit. To compute the first factor, we consider two cases. First, assume that unit u_j is an output unit of the network. In this case, it follows from the definition of E_p that

$$\frac{\partial E_p}{\partial o_{pj}} = -(t_{pj} - o_{pj}),$$

which is the same result as we obtained with the standard delta rule. Substituting for the two factors in Equation 12, we get

$$\delta_{pj} = (t_{pj} - o_{pj})f'_j(net_{pj}) \tag{13}$$

for any output unit u_j. If u_j is not an output unit we use the chain rule to write

$$\sum_k \frac{\partial E_p}{\partial net_{pk}} \frac{\partial net_{pk}}{\partial o_{pj}} = \sum_k \frac{\partial E_p}{\partial net_{pk}} \frac{\partial}{\partial o_{pj}} \sum_i w_{ki} o_{pi}$$

$$= \sum_k \frac{\partial E_p}{\partial net_{pk}} w_{kj} = -\sum_k \delta_{pk} w_{kj}.$$

In this case, substituting for the two factors in Equation 12 yields

$$\delta_{pj} = f'_j(net_{pj}) \sum_k \delta_{pk} w_{kj} \qquad (14)$$

whenever u_j is not an output unit. Equations 13 and 14 give a recursive procedure for computing the δ's for all units in the network, which are then used to compute the weight changes in the network according to Equation 11. This procedure constitutes the generalized delta rule for a feedforward network of semilinear units.

These results can be summarized in three equations. First, the generalized delta rule has exactly the same form as the standard delta rule of Equation 1. The weight on each line should be changed by an amount proportional to the product of an error signal, δ, available to the unit receiving input along that line and the output of the unit sending activation along that line. In symbols,

$$\Delta_p w_{ji} = \eta \delta_{pj} o_{pi}.$$

The other two equations specify the error signal. Essentially, the determination of the error signal is a recursive process which starts with the output units. If a unit is an output unit, its error signal is very similar to the standard delta rule. It is given by

$$\delta_{pj} = (t_{pj} - o_{pj}) f'_j(net_{pj})$$

where $f'_j(net_{pj})$ is the derivative of the semilinear activation function which maps the total input to the unit to an output value. Finally, the error signal for hidden units for which there is no specified target is determined recursively in terms of the error signals of the units to which it directly connects and the weights of those connections. That is,

$$\delta_{pj} = f'_j(net_{pj}) \sum_k \delta_{pk} w_{kj}$$

whenever the unit is not an output unit.

The application of the generalized delta rule, thus, involves two phases: During the first phase the input is presented and propagated forward through the network to compute the output value o_{pj} for each unit. This output is then compared with the targets, resulting in an error signal δ_{pj} for each output unit. The second phase involves a backward pass through the network (analogous to the initial forward pass) during which the error signal is passed to each unit in the network and the appropriate weight changes are made.

This second, backward pass allows the recursive computation of δ as indicated above. The first step is to compute δ for each of the output units. This is simply the difference between the actual and desired output values times the derivative of the squashing function. We can then compute weight changes for all connections that feed into the final layer. After this is done, then compute δ's for all units in the penultimate layer. This propagates the errors back one layer, and the same process can be repeated for every layer. The backward pass has the same computational complexity as the forward pass, and so it is not unduly expensive.

We have now generated a gradient descent method for finding weights in any feedforward network with semilinear units. Before reporting our results with these networks, it is useful to note some further observations. It is interesting that not all weights need be variable. Any number of weights in the network can be fixed. In this case, error is still propagated as before; the fixed weights are simply not modified. It should also be noted that there is no reason why some output units might not receive inputs from other output units in earlier layers. In this case, those units receive two different kinds of error: that from the direct comparison with the target and that passed through the other output units whose activation it affects. In this case, the correct procedure is to simply add the weight changes dictated by the direct comparison to that propagated back from the other output units.

Simulation Results

We now have a learning procedure which could, in principle, evolve a set of weights to produce an arbitrary mapping from input to output. However, the procedure we have produced is a gradient descent procedure and, as such, is bound by all of the problems of any hill climbing procedure—namely, the problem of local maxima or (in our case) minima. Moreover, there is a question of how long it might take a system to learn. Even if we could guarantee that it would eventually find a solution, there is the question of whether our procedure could learn in a reasonable period of time. It is interesting to ask what hidden units the system actually develops in the solution of particular problems. This is the question of what kinds of internal representations the system actually creates. We do not yet have definitive answers to these questions. However, we have carried out many simulations which lead us to be optimistic about the local minima and time questions and to be surprised by the kinds of representations our learning mechanism discovers. Before proceeding with our results, we must describe our simulation system in more detail. In particular, we must specify an activation function and show how the system can compute the derivative of this function.

A useful activation function   In our above derivations the derivative of the activation function of unit u_j, $f'_j(net_j)$, always played a role. This implies that we need an activation function for which a derivative exists. It is interesting to note that the linear threshold function, on which the perceptron is based, is discontinuous and hence will not suffice for the generalized delta rule. Similarly, since a linear system achieves no advantage from hidden units, a linear activation function will not suffice either. Thus, we need a continuous, nonlinear activation function. In most of our experiments we have used the *logistic* activation function in which

$$o_{pj} = \frac{1}{1 + e^{-(\sum_i w_{ji}o_{pi} + \theta_j)}} \tag{15}$$

where θ_j is a bias similar in function to a threshold.[1] In order to apply our learning rule, we need to know the derivative of this function with respect to its total input, net_{pj}, where $net_{pj} = \sum w_{ji}o_{pi} + \theta_j$. It is easy to show that this derivative is given by

$$\frac{\partial o_{pj}}{\partial net_{pj}} = o_{pj}(1 - o_{pj}).$$

Thus, for the logistic activation function, the error signal, δ_{pj}, for an output unit is given by

$$\delta_{pj} = (t_{pj} - o_{pj})o_{pj}(1 - o_{pj}),$$

and the error for an arbitrary hidden u_j is given by

$$\delta_{pj} = o_{pj}(1 - o_{pj}) \sum_k \delta_{pk} w_{kj}.$$

It should be noted that the derivative, $o_{pj}(1 - o_{pj})$, reaches its maximum for $o_{pj} = 0.5$ and, since $0 \leqslant o_{pj} \leqslant 1$, approaches its minimum as o_{pj} approaches zero or one. Since the amount of change in a given weight is proportional to this derivative, weights will be changed most for those units that are near their midrange and, in some sense, not yet committed to being either on or off. This feature, we believe, contributes to the stability of the learning of the system.

One other feature of this activation function should be noted. The system can not actually reach its extreme values of 1 or 0 without infinitely large weights. Therefore, in a practical learning situation in which the desired

outputs are binary $\{0, 1\}$, the system can never actually achieve these values. Therefore, we typically use the values of 0.1 and 0.9 as the targets, even though we will talk as if values of $\{0, 1\}$ are sought.

The learning rate   Our learning procedure requires only that the change in weight be proportional to $\partial E_p/\partial w$. True gradient descent requires that infinitesimal steps be taken. The constant of proportionality is the learning rate in our procedure. The larger this constant, the larger the changes in the weights. For practical purposes we choose a learning rate that is as large as possible without leading to oscillation. This offers the most rapid learning. One way to increase the learning rate without leading to oscillation is to modify the generalized delta rule to include a *momentum* term. This can be accomplished by the following rule:

$$\Delta w_{ji}(n + 1) = \eta(\delta_{pj}o_{pi}) + \alpha\Delta w_{ji}(n) \tag{16}$$

where the subscript n indexes the presentation number, η is the learning rate, and α is a constant which determines the effect of past weight changes on the current direction of movement in weight space. This provides a kind of momentum in weight space that effectively filters out high-frequency variations of the error-surface in the weight space. This is useful in spaces containing long ravines that are characterized by sharp curvature across the ravine and a gently sloping floor. The sharp curvature tends to cause divergent oscillations across the ravine. To prevent these it is necessary to take very small steps, but this causes very slow progress along the ravine. The momentum filters out the high curvature and thus allows the effective weight steps to be bigger. In most of our simulations α was about 0.9. Our experience has been that we get the same solutions by setting $\alpha = 0$ and reducing the size of η, but the system learns much faster overall with larger values of α and η.

Symmetry breaking   Our learning procedure has one more problem that can be readily overcome and this is the problem of symmetry breaking. If all weights start out with equal values and if the solution requires that unequal weights be developed, the system can never learn. This is because error is propagated back through the weights in proportion to the values of the weights. This means that all hidden units connected directly to the output inputs will get identical error signals, and, since the weights changes depend on the error signals, the weights from those units to the output units must always be the same. The system is starting out at a kind of *local maximum*, which keeps the weights equal, but it is a maximum of the error function, so once it escapes it will never return. We counteract this

---

[1] Note that the values of the bias, θ_j, can be learned just like any other weights. We simply imagine that θ_j is the weight from a unit that is always on.

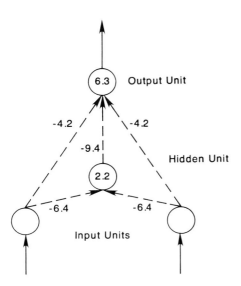

Figure 3 Observed XOR network. The connection weights are written on the arrows and the biases are written in the circles. Note a positive bias means that the unit is on unless turned off.

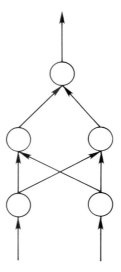

Figure 4 A simple architecture for solving XOR with two hidden units and no direct connections from input to output.

problem by starting the system with small random weights. Under these conditions symmetry problems of this kind do not arise.

The XOR Problem

It is useful to begin with the exclusive-or problem since it is the classic problem requiring hidden units and since many other difficult problems involve an XOR as a subproblem. We have run the XOR problem many times and with a couple of exceptions discussed below, the system has always solved the problem. Figure 3 shows one of the solutions to the problem. This solution was reached after 558 sweeps through the four stimulus patterns with a learning rate of $\eta = 0.5$. In this case, both the hidden unit and the output unit have *positive biases* so they are on unless turned off. The hidden unit turns on if neither input unit is on. When it is on, it turns off the output unit. The connections from input to output units arranged themselves so that they turn off the output unit whenever both inputs are on. In this case, the network has settled to a solution which is a sort of mirror image of the one illustrated in Figure 2.

We have taught the system to solve the XOR problem hundreds of times. Sometimes we have used a single hidden unit and direct connections to the output unit as illustrated here, and other times we have allowed two hidden units and set the connections from the input units to the outputs to be zero, as shown in Figure 4. In only two cases has the system encountered a *local minimum* and thus been unable to solve the

problem. Both cases involved the two hidden units version of the problem and both ended up in the same local minimum. Figure 5 shows the weights for the local minimum. In this case, the system correctly responds to two of the patterns—namely, the patterns 00 and 10. In the cases of the other two patterns 11 and 01, the output unit gets a net input of zero. This leads to an output value of 0.5 for both of these patterns. This state was reached after 6,587 presentations of each pattern with $\eta = 0.25$.[2] Although many problems require more presentations for learning to occur, further trials on this problem merely increase the magnitude of the weights but do not lead to any improvement in performance. We do not know the frequency of such local minima, but our experience with this and other problems is that they are quite rare. We have found only one other situation in which a local minimum has occurred in many hundreds of problems of various sorts. We will discuss this case below.

The XOR problem has proved a useful test case for a number of other studies. Using the architecture illustrated in Figure 4, a student in our laboratory, Yves Chauvin, has studied the effect of varying the number of hidden units and varying the learning rate on time to solve the problem. Using as a learning criterion an error of 0.01 per pattern, Yves found that the average number of presentations to solve the problem with $\eta = 0.25$ varied from about 245 for the case with two hidden units to about 120 presentations for 32 hidden

[2] If we set $\eta = 0.5$ or above, the system escapes this minimum. In general, however, the best way to avoid local minima is probably to use very small values of η.

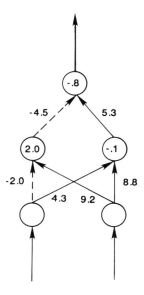

Figure 5 A network at a local minimum for the exclusive-or problem. The dotted lines indicate negative weights. Note that whenever the right most input unit is on it turns on *both* hidden units. The weights connecting the hidden units to the output are arranged so that when both hidden units are on, the output unit gets a net input of zero. This leads to an output value of 0.5. In the other cases the network provides the correct answer.

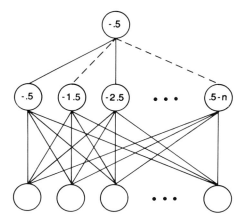

Figure 6 A paradigm for the solutions to the parity problem discovered by the the learning system. See text for explanation.

Table 3

Number of *on* input units		Hidden unit patterns		Output value
0	→	1111	→	0
1	→	1011	→	1
2	→	1010	→	0
3	→	0010	→	1
4	→	0000	→	0

units. The results can be summarized by $P = 280 - 33 \log_2 H$, where P is the required number of presentations and H is the number of hidden units employed. Thus, the time to solve XOR is reduced linearly with the logarithm of the number of hidden units. This result holds for values of H up to about 40 in the case of XOR. The general result that the time to solution is reduced by increasing the number of hidden units has been observed in virtually all of our simulations. Yves also studied the time to solution as a function of learning rate for the case of eight hidden units. He found an average of about 450 presentations with $\eta = 0.1$ to about 68 presentations with $\eta = 0.75$. He also found that learning rates larger than this led to unstable behavior. However, within this range larger learning rates speeded the learning substantially. In most of our problems we have employed learning rates of $\eta = 0.25$ or smaller and have had no difficulty.

Parity

One of the problems given a good deal of discussion by Minsky and Papert (1969) is the parity problem, in which the output required is 1 if the input pattern contains an odd number of 1s and 0 otherwise. This is a very difficult problem because the most similar patterns (those which differ by a single bit) require different answers. The XOR problem is a parity problem with

input patterns of size two. We have tried a number of parity problems with patterns ranging from size two to eight. Generally we have employed layered networks in which direct connections from the input to the output units are not allowed, but must be mediated through a set of hidden units. In this architecture, it requires at least N hidden units to solve parity with patterns of length N. Figure 6 illustrates the basic paradigm for the solutions discovered by the system. The solid lines in the figure indicate weights of $+1$ and the dotted lines indicate weights of -1. The numbers in the circles represent the biases of the units. Basically, the hidden units arranged themselves so that they count the number of inputs. In the diagram, the one at the far left comes on if one or more input units are on, the next comes on if two or more are on, etc. All of the hidden units come on if all of the input lines are on. The first m hidden units come on whenever m bits are on in the input pattern. The hidden units then connect with alternately positive and negative weights. In this way the net input from the hidden units is zero for even numbers and $+1$ for odd numbers. Table 3 shows the actual solution attained for one of our simulations with four input lines and four hidden units. This solution was reached after 2,825 presentations of each of the

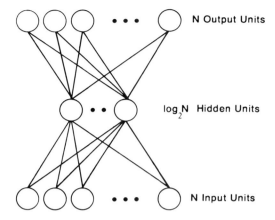

Figure 7 A network for solving the encoder problem. In this problem there are N orthogonal input patterns each paired with one of N orthogonal output patterns. There are only $\log_2 N$ hidden units. Thus, if the hidden units take on binary values, the hidden units must form a binary number to encode each of the input patterns. This is exactly what the system learns to do.

sixteen patterns with $\eta = 0.5$. Note that the solution is roughly a mirror image of that shown in Figure 6 in that the number of hidden units turned on is equal to the number of zero input values rather than the number of ones. Beyond that the principle is that shown above. It should be noted that the internal representation created by the learning rule is to arrange that the number of hidden units that come on is equal to the number of zeros in the input and that the particular hidden units that come on depend *only* on the number, not on which input units are on. This is exactly the sort of recoding *required* by parity. It is not the kind of representation readily discovered by unsupervised learning schemes such as competitive learning.

The Encoding Problem

Ackley, Hinton, and Sejnowski (1985) have posed a problem in which a set of orthogonal input patterns are mapped to a set of orthogonal output patterns through a small set of hidden units. In such cases the internal representations of the patterns on the hidden units must be rather efficient. Suppose that we attempt to map N input patterns onto N output patterns. Suppose further that $\log_2 N$ hidden units are provided. In this case, we expect that the system will learn to use the hidden units to form a binary code with a distinct binary pattern for each of the N input patterns. Figure 7 illustrates the basic architecture for the encoder problem is to learn an encoding of an N bit pattern into a $\log_2 N$ bit pattern and then learn to decode this representation into the output pattern. We have presented the system with a number of these problems. Here we

Table 4

Input patterns		Output patterns
10000000	→	10000000
01000000	→	01000000
00100000	→	00100000
00010000	→	00010000
00001000	→	00001000
00000100	→	00000100
00000010	→	00000010
00000001	→	00000001

Table 5

Input patterns		Hidden unit patterns				Output patterns
10000000	→	.5	0	0	→	10000000
01000000	→	0	1	0	→	01000000
00100000	→	1	1	0	→	00100000
00010000	→	1	1	1	→	00010000
00001000	→	0	1	1	→	00001000
00000100	→	.5	0	1	→	00000100
00000010	→	1	0	.5	→	00000010
00000001	→	0	0	.5	→	00000001

Table 6

Input patterns		Output patterns
00	→	1000
01	→	0100
10	→	0010
11	→	0001

present a problem with eight input patterns, eight output patterns, and three hidden units. In this case the required mapping is the identity mapping illustrated in Table 4. The problem is simply to turn on the same bit in the output as in the input. Table 5 shows the mapping generated by our learning system on this example. It is of some interest that the system employed its ability to use intermediate values in solving this problem. It could, of course, have found a solution in which the hidden units took on only the values of zero and one. Often it does just that, but in this instance, and many others, there are solutions that use the intermediate values, and the learning system finds them even though it has a bias toward extreme values. It is possible to set up problems that *require* the system to make use of intermediate values in order to solve a problem. We now turn to such a case.

Table 6 shows a very simple problem in which we have to convert from a *distributed representation* over two units into a *local representation* over four units. The similarity structure of the distributed input

Table 7

Input patterns		Singleton hidden unit		Remaining hidden units					Output patterns
10	→	0	→	1	1	1	0	→	0010
11	→	2	→	1	1	0	0	→	0001
00	→	6	→	.5	1	1	.3	→	1000
01	→	1	→	0	0	0	1	→	0100

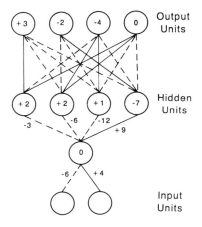

Figure 8 The network illustrating the use of intermediate values in solving a problem. See text for explanation.

patterns is simply not preserved in the local output representation.

We presented this problem to our learning system with a number of constraints which made it especially difficult. The two input units were only allowed to connect to a single hidden unit which, in turn, was allowed to connect to four more hidden units. Only these four hidden units were allowed to connect to the four output units. To solve this problem, then, the system must first convert the distributed representation of the input patterns into various intermediate values of the singleton hidden unit in which different activation values correspond to the different input patterns. These continuous values must then be converted back through the next layer of hidden units—first to another distributed representation and then, finally, to a local representation. This problem was presented to the system and it reached a solution after 5,226 presentations with $\eta = 0.05$.[3] Table 7 shows the sequence of representations the system actually developed in order to transform the patterns and solve the problem. Note each of the four input patterns was mapped into a particular activation value of the singleton hidden unit. These values were then mapped onto distributed patterns at the next layer of hidden units which were finally mapped into the required local representation at the output level. In principle, this trick of mapping patterns into activation values and then converting those activation values back into patterns could be done for any number of patterns, but it becomes increasingly difficult for the system to make the necessary distinctions as ever smaller differences among activation values must be distinguished. Figure 8 shows the network the system developed to do this job. The connection weights from the hidden units to the output units have been suppressed for clarity. (The sign of the connection, however, is indicated by the form of the connection—e.g., dashed lines mean inhibitory connections). The four different activation values were generated by having relatively large weights of opposite sign. One input line turns the hidden unit full on, one turns it full off. The two differ by a relatively small

[3] Relatively small learning rates make units employing intermediate values easier to obtain.

amount so that when both turn on, the unit attains a value intermediate between 0 and 0.5. When neither turns on, the near zero bias causes the unit to attain a value slightly over 0.5. The connections to the second layer of hidden units is likewise interesting. When the hidden unit is full on, the right-most of these hidden units is turned on and all others turned off. When the hidden unit is turned off, the other three of these hidden units are on and the left-most unit off. The other connections from the singleton hidden unit to the other hidden units are graded so that a distinct pattern is turned on for its other two values. Here we have an example of the flexibility of the learning system.

Our experience is that there is a propensity for the hidden units to take on extreme values, but, whenever the learning problem calls for it, they can learn to take on graded values. It is likely that the propensity to take on extreme values follows from the fact that the logistic is a sigmoid so that increasing magnitudes of its inputs push it toward zero or one. This means that in a problem in which intermediate values are required, the incoming weights must remain of moderate size. It is interesting that the derivation of the generalized delta rule does not depend on all of the units having identical activation functions. Thus, it would be possible for some units, those required to encode information in a graded fashion, to be linear while others might be logistic. The linear unit would have a much wider dynamic range and could encode more different values. This would be a useful role for a linear unit in a network with hidden units.

Symmetry

Another interesting problem we studied is that of classifying input strings as to whether or not they are

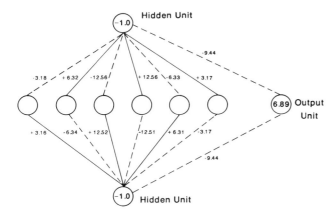

Figure 9 Network for solving the symmetry problem. The six open circles represent the input units. There are two hidden units, one shown above and one below the input units. The output unit is shown to the far right. See text for explanation.

symmetric about their center. We used patterns of various lengths with various numbers of hidden units. To our surprise, we discovered that the problem can always be solved with only two hidden units. To understand the derived representation, consider one of the solutions generated by our system for strings of length six. This solution was arrived at after 1,208 presentations of each six-bit pattern with $\eta = 0.1$. The final network is shown in Figure 9. For simplicity we have shown the six input units in the center of the diagram with one hidden unit above and one below. The output unit, which signals whether or not the string is symmetric about its center, is shown at the far right. The key point to see about this solution is that for a given hidden unit, weights that are symmetric about the middle are equal in magnitude and opposite in sign. That means that if a symmetric pattern is on, both hidden units will receive a net input of zero from the input units, and, since the hidden units have a negative bias, both will be off. In this case, the output unit, having a positive bias, will be on. The next most important thing to note about the solution is that the weights on each side of the midpoint of the string are in the ratio of $1:2:4$. This insures that each of the eight patterns that can occur on each side of the midpoint sends a unique activation sum to the hidden unit. This assures that there is no pattern on the left that will exactly balance a non-mirror-image pattern on the right. Finally, the two hidden units have identical patterns of weights from the input units except for sign. This insures that for every nonsymmetric pattern, at least one of the two hidden units will come on and turn on the output unit. To summarize, the network is arranged so that both hidden units will receive exactly

zero activation from the input units when the pattern is symmetric, and at least one of them will receive positive input for every nonsymmetric pattern.

This problem was interesting to us because the learning system developed a much more elegant solution to the problem than we had previously considered. This problem was not the only one in which this happened. The parity solution discovered by the learning procedure was also one that we had not discovered prior to testing the problem with our learning procedure. Indeed, we frequently discover these more elegant solutions by giving the system more hidden units than it needs and observing that it does not make use of some of those provided. Some analysis of the actual solutions discovered often leads us to the discovery of a better solution involving fewer hidden units.

Addition
Another interesting problem on which we have tested our learning algorithm is the simple binary addition problem. This problem is interesting because there is a very elegant solution to it, because it is the one problem we have found where we can reliably find local minima and because the way of avoiding these local minima gives us some insight into the conditions under which local minima may be found and avoided. Figure 10 illustrates the basic problem and a minimal solution to it. There are four input units, three output units, and two hidden units. The output patterns can be viewed as the binary representation of the sum of two two-bit binary numbers represented by the input patterns. The second and fourth input units in the diagram correspond to the low-order bits of the two binary numbers and the first and third units correspond to the two higher order bits. The hidden units correspond to the *carry bits* in the summation. Thus the hidden unit on the far right comes on when both of the lower order bits in the input pattern are turned on, and the one on the left comes on when both higher order bits are turned on or when one of the higher order bits and the other hidden unit is turned on. In the diagram, the weights on all lines are assumed to be $+1$ except where noted. Inhibitory connections are indicated by dashed lines. As usual, the biases are indicated by the numbers in the circles. To understand how this network works, it is useful to note that the lowest order output bit is determined by an exclusive-or among the two low-order input bits. One way to solve this XOR problem is to have a hidden unit come on when both low-order input bits are on and then have it inhibit the output unit. Otherwise either of the low-order input units can turn on the low-order output bit. The middle bit is somewhat more difficult. Note that the middle bit

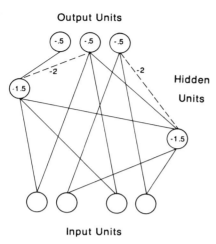

Output Units

Hidden

Units

Input Units

Figure 10 Minimal network for adding two two-bit binary numbers. There are four input units, three output units, and two hidden units. The output patterns can be viewed as the binary representation of the sum of two two-bit binary numbers represented by the input patterns. The second and fourth input units in the diagram correspond to the low-order bits of the two binary numbers, and the first and third units correspond to the two higher order bits. The hidden units correspond to the carry bits in the summation. The hidden unit on the far right comes on when both of the lower order bits in the input pattern are turned on, and the one on the left comes on when both higher order bits are turned on or when one of the higher order bits and the other hidden unit is turned on. The weights on all lines are assumed to be +1 except where noted. Negative connections are indicated by dashed lines. As usual, the biases are indicated by the numbers in the circles.

should come on whenever an odd number of the set containing the two higher order input bits and the lower order carry bit is turned on. Observation will confirm that the network shown performs that task. The left-most hidden unit receives inputs from the two higher order bits and from the carry bit. Its bias is such that it will come on whenever two or more of its inputs are turned on. The middle output unit receives positive inputs from the same three units and a negative input of −2 from the second hidden unit. This insures that whenever just one of the three are turned on, the second hidden unit will remain off and the output bit will come on. Whenever exactly two of the three are on, the hidden unit will turn on and counteract the two units exciting the output bit, so it will stay off. Finally, when all three are turned on, the output bit will receive −2 from its carry bit and +3 from its other three inputs. The net is positive, so the middle unit will be on. Finally, the third output bit should turn on whenever the second hidden unit is on—that is, whenever there is a carry from the second bit. Here then we have a minimal network to carry out the job at hand. More-

over, it should be noted that the concept behind this network is generalizable to an arbitrary number of input and output bits. In general, for adding two m bit binary numbers we will require $2m$ input units, m hidden units, and $m + 1$ output units.

Unfortunately, this is the one problem we have found that reliably leads the system into local minima. At the start in our learning trials on this problem we allow any input unit to connect to any output unit and to any hidden unit. We allow any hidden unit to connect to any output unit, and we allow one of the hidden units to connect to the other hidden unit, but, since we can have no loops, the connection in the opposite direction is disallowed. Sometimes the system will discover essentially the same network shown in the figure.[4] Often, however, the system ends up in a local minimum. The problem arises when the XOR problem on the low-order bits is not solved in the way shown in the diagram. One way it can fail is when the "higher" of the two hidden units is "selected" to solve the XOR problem. This is a problem because then the other hidden unit cannot "see" the carry bit and therefore cannot finally solve the problem. This problem seems to stem from the fact that the learning of the second output bit is always dependent on learning the first (because information about the carry is necessary to learn the second bit) and therefore lags behind the learning of the first bit and has no influence on the selection of a hidden unit to solve the first XOR problem. Thus, about half of the time (in this problem) the wrong unit is chosen and the problem cannot be solved. In this case, the system finds a solution for all of the sums except the $11 + 11 \rightarrow 110 \, (3 + 3 = 6)$ case in which it misses the carry into the middle bit and gets $11 + 11 \rightarrow 100$ instead. This problem differs from others we have solved in as much as the hidden units are not "equipotential" here. In most of our other problems the hidden units have been equipotential, and this problem has not arisen.

It should be noted, however, that there is a relatively simple way out of the problem—namely, add some extra hidden units. In this case we can afford to make a mistake on one or more selections and the system can still solve the problems. For the problem of adding two-bit numbers we have found that the system always solves the problem with one extra hidden unit. With larger numbers it may require two or three more. For purposes of illustration, we show the results of one of our runs with three rather than the minimum two

[4] The network is the same except for the highest order bit. The highest order bit is always on whenever three or more of the input units are on. This is always learned first and always learned with direct connections to the input units.

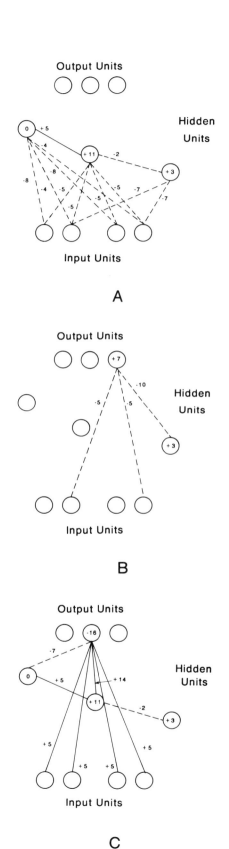

A

B

C

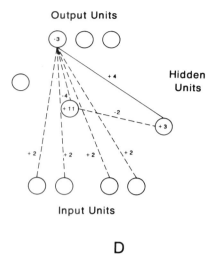

D

Figure 11 Network found for the summation problem. *A*: The connections from the input units to the three hidden units and the connections among the hidden units. *B*: The connections from the input and hidden units to the lowest order output unit. *C*: The connections from the input and hidden units to the middle output unit. *D*: The connections from the input and hidden units to the highest order output unit.

hidden units. Figure 11 shows the state reached by the network after 3,020 presentations of each input pattern and with a learning rate of $\eta = 0.5$. For convenience, we show the network in four parts. In Figure 11A we show the connections to and among the hidden units. This figure shows the internal representation generated for this problem. The "lowest" hidden unit turns off whenever either of the low-order bits are on. In other words it detects the case in which no low-order bit is turn on. The "highest" hidden unit is arranged so that it comes on whenever the sum is less than two. The conditions under which the middle hidden unit comes on are more complex. Table 8 shows the patterns of hidden units which occur to each of the sixteen input patterns. Figure 11B shows the connections to the lowest order output unit. Noting that the relevant hidden unit comes on when neither low-order input unit is on, it is clear how the system computes XOR. When both low-order inputs are off, the output unit is turned off by the hidden unit. When both low-order input units are on, the output is turned off directly by the two input units. If just one is on, the positive bias on the output unit keeps it on. Figure 11C gives the connections to the middle output unit, and in Figure 11D we show those connections to the leftmost, highest order output unit. It is somewhat difficult to see how these connections always lead to the correct output answer, but, as can be verified from the figures, the network is balanced so that this works.

Table 8

Input patterns		Hidden unit patterns		Output patterns
00 + 00	→	111	→	000
00 + 01	→	110	→	001
00 + 10	→	011	→	010
00 + 11	→	010	→	011
01 + 00	→	110	→	001
01 + 01	→	010	→	010
01 + 10	→	010	→	011
01 + 11	→	000	→	100
10 + 00	→	011	→	010
10 + 01	→	010	→	011
10 + 10	→	001	→	100
10 + 11	→	000	→	101
11 + 00	→	010	→	011
11 + 01	→	000	→	100
11 + 10	→	000	→	101
11 + 11	→	000	→	110

Table 9

Input patterns		Output patterns
0000	→	000
0001	→	001
0010	→	010
0011	→	011
0100	→	100
0101	→	101
0110	→	110
0111	→	111
1000	→	111
1001	→	110
1010	→	101
1011	→	100
1100	→	011
1101	→	010
1110	→	001
1111	→	000

It should be pointed out that most of the problems described thus far have involved hidden units with quite simple interpretations. It is much more often the case, especially when the number of hidden units exceeds the minimum number required for the task, that the hidden units are not readily interpreted. This follows from the fact that there is very little tendency for *localist* representations to develop. Typically the internal representations are distributed and it is the *pattern* of activity over the hidden units, not the meaning of any particular hidden unit that is important.

The Negation Problem

Consider a situation in which the input to a system consists of patterns of $n + 1$ binary values and an output of n values. Suppose further that the general rule is that n of the input units should be mapped directly to the output patterns. One of the input bits, however, is special. It is a negation bit. When that bit is off, the rest of the pattern is supposed to map straight through, but when it is on, the complement of the pattern is to be mapped to the output. Table 9 shows the appropriate mapping. In this case the left element of the input pattern is the negation bit, but the system has no way of knowing this and must learn which bit is the negation bit. In this case, weights were allowed from any input unit to any hidden or output unit and from any hidden unit to any output unit. The system learned to set all of the weights to zero except those shown in Figure 12. The basic structure of the problem and of the solution is evident in the figure. Clearly the problem was reduced to a set of three XORs between the negation bit and each input. In the case of the two right-most input units, the XOR problems were solved

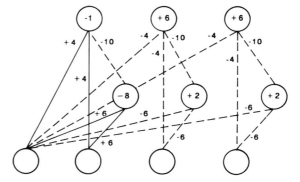

Figure 12 The solution discovered for the negation problem. The left-most unit is the negation unit. The problem has been reduced and solved as three exclusive-ors between the negation unit and each of the other three units.

by recruiting a hidden unit to detect the case in which *neither* the negation unit *nor* the corresponding input unit was on. In the third case, the hidden unit detects the case in which *both* the negation unit *and* relevant input were on. In this case the problem was solved in less than 5,000 passes through the stimulus set with $\eta = 0.25$.

The T-C Problem

Most of the problems discussed so far (except the symmetry problem) are rather abstract mathematical problems. We now turn to a more geometric problem —that of discriminating between a *T* and a *C*— independent of translation and rotation. Figure 13 shows the stimulus patterns used in these experiments. Note, these patterns are each made of five squares and differ from one another by a single square. Moreover, as Minsky and Papert (1969) point out, when con-

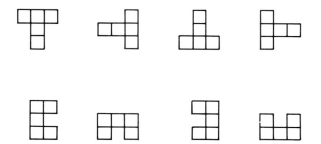

Figure 13 The stimulus set for the T-C problem. The set consists of a block *T* and a block *C* in each of four orientations. One of the eight patterns is presented on each trial.

sidering the set of patterns over all possible translations and rotations (of 90°, 180°, and 270°), the patterns do not differ in the set of distances among their pairs of squares. To see a difference between the sets of patterns one must look, at least, at configurations of triplets of squares. Thus Minsky and Papert call this a problem of *order three*.[5] In order to facilitate the learning, a rather different architecture was employed for this problem. Figure 14 shows the basic structure of the network we employed. Input patterns were now conceptualized as two-dimensional patterns superimposed on a rectangular grid. Rather than allowing each input unit to connect to each hidden unit, the hidden units themselves were organized into a two-dimensional grid with each unit receiving input from a square 3 × 3 region of the input space. In this sense, the overlapping square regions constitute the predefined *receptive field* of the hidden units. Each of the hidden units, over the entire field, feeds into a single output unit which is to take on the value 1 if the input is a *T* (at any location or orientation) and 0 if the input is a *C*. Further, in order that the learning that occurred be independent of where on the field the pattern appeared, we constrained all of the units to learn exactly the same pattern of weights. In this way each unit was constrained to compute exactly the same function over its receptive field—the receptive fields were constrained to all have the same shape. This guarantees translation independence and avoids any possible "edge effects" in the learning. The learning can readily be extended to arbitrarily large fields of input units. This constraint was accomplished by simply adding together the weight changes dictated by the delta rule for each unit and then changing all weights exactly the same amount. In this way, the whole field of hidden units consists simply of replications of a single feature detector centered on different regions of

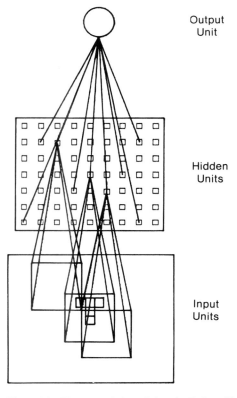

Figure 14 The network for solving the T-C problem. See text for explanation.

the input space, and the learning that occurs in one part of the field is automatically generalized to the rest of the field.[6]

We have run this problem in this way a number of times. As a result, we have found a number of solutions. Perhaps the simplest way to understand the system is by looking at the form of the receptive field for the hidden units. Figure 15 shows several of the receptive fields we have seen.[7] Figure 15A shows the most local representation developed. This *on-center-off-surround* detector turns out to be an excellent *T* detector. Since, as illustrated, a *T* can extend into the on-center and achieve a net input of +1, this detector will be turned on for a *T* at any orientation. On the other hand, any *C* extending into the center must cover at least *two* inhibitory cells. With this detector the bias can be set so that only one of the whole field of inhibitory units will come on whenever a *T* is presented and none of the hidden units will be turned on by any *C*. This

[6] A similar procedure has been employed by Fukushima (1980) in his *neocognitron* and by Kienker, Sejnowski, Hinton, and Schumacher (1985).
[7] The ratios of the weights are about right. The actual values can be larger or smaller than the values given in the figure.

[5] Terry Sejnowski pointed out to us that the T-C problem was difficult for models of this sort to learn and therefore worthy of study.

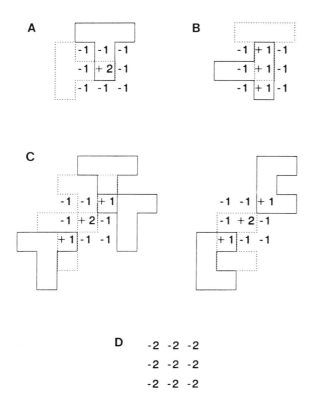

Figure 15 Receptive fields found in different runs of the T-C problem. *A*: An on-center–off-surround receptive field for detecting *T*'s. *B*: A vertical bar detector which responds to *T*'s more strongly than *C*'s. *C*: A diagonal bar detector. A *T* activates five such detectors whereas a *C* activates only three such detectors. *D*: A compactness detector. This inhibitory receptive field turns off whenever an input covers any region of its receptive field. Since the *C* is more compact than the *T* it turns off 20 such detectors whereas the *T* turns off 21 of them.

is a kind of *protrusion* detector which differentiates between a *T* and *C* by detecting the protrusion of the *T*.

The receptive field shown in Figure 15B is again a kind of *T* detector. Every *T* activates one of the hidden units by an amount +2 and none of the hidden units receives more than +1 from any of the *C*'s. As shown in the figure, *T*'s at 90° and 270° send a total of +2 to the hidden units on which the crossbar lines up. The *T*'s at the other two orientations receive +2 from the way it detects the vertical protrusions of those two characters. Figure 15C shows a more distributed representation. As illustrated in the figure, each *T* activities five different hidden units whereas each *C* excites only three hidden units. In this case the system again is differentiating between the characters on the basis of the protruding end of the *T* which is not shared by the *C*.

Finally, the receptive field shown in Figure 15D is

even more interesting. In this case every hidden unit has a positive bias so that it is on unless turned off. The strength of the inhibitory weights are such that if a character overlaps the receptive field of a hidden unit, that unit turns off. The system works because a *C* is more compact than a *T* and therefore the *T* turns off more units that the *C*. The *T* turns off 21 hidden units, and the *C* turns off only 20. This is a truly distributed representation. In each case, the solution was reached in from about 5,000 to 10,000 presentations of the set of eight patterns.[8]

It is interesting that the inhibitory type of receptive field shown in Figure 15D was the most common and that there is a predominance of inhibitory connections in this and indeed all of our simulations. This can be understood by considering the trajectory through which the learning typically moves. At first, when the system is presented with a difficult problem, the initial random connections are as likely to mislead as to give the correct answer. In this case, it is best for the output units to take on a value of 0.5 than to take on a more extreme value. This follows from the form of the error function given in Equation 2. The output unit can achieve a constant output of 0.5 by turning off those units feeding into it. Thus, the first thing that happens in virtually every difficult problem is that the hidden units are turned off. One way to achieve this is to have the input units inhibit the hidden units. As the system begins to sort things out and to learn the appropriate function some of the connections will typically go positive, but the majority of the connections will remain negative. This *bias* for solutions involving inhibitory inputs can often lead to nonintuitive results in which hidden units are often on unless turned off by the input.

More Simulation Results

We have offered a sample of our results in this section. In addition to having studied our learning system on the problems discussed here, we have employed back propagation for learning to multiply binary digits, to play tic-tac-toe, to distinguish between vertical and horizontal lines, to perform sequences of actions, to recognize characters, to associate random vectors, and a host of other applications. In all of these applications we have found that the generalized delta rule was capable of generating the kinds of internal representations required for the problems in question. We have found local minima to be very rare and that the system learns in a reasonable period of time. Still more studies

---

[8] Since translation independence was built into the learning procedure, it makes no difference *where* the input occurs; the same thing will be learned wherever the pattern is presented. Thus, there are only eight distinct patterns to be presented to the system.

of this type will be required to understand precisely the conditions under which the system will be plagued by local minima. Suffice it to say that the problem has not been serious to date. We now turn to a pointer to some future developments.

Some Further Generalizations

We have intensively studied the learning characteristics of the generalized delta rule on feedforward networks and semilinear activations functions. Interestingly these are not the most general cases to which the learning procedure is applicable. As yet we have only studied a few examples of the more fully generalized system, but it is relatively easy to apply the same learning rule to sigma-pi units and to recurrent networks. We will simply sketch the basic ideas here.

The Generalized Delta Rule and Sigma-Pi Units

It will be recalled from Chapter 2 that in the case of sigma-pi units we have

$$o_j = f_j\left(\sum_i w_{ji} \prod_k o_{i_k}\right) \qquad (17)$$

where i varies over the set of conjuncts feeding into unit j and k varies over the elements of the conjuncts. For simplicity of exposition, we restrict ourselves to the case in which no conjuncts involve more than two elements. In this case we can notate the weight from the conjunction of units i and j to unit k by w_{kij}. The weight on the direct connection from unit i to unit j would, thus, be w_{jii}, and since the relation is multiplicative, $w_{kij} = w_{kji}$. We can now rewrite Equation 17 as

$$o_j = f_j\left(\sum_{i,h} w_{jhi} o_h o_i\right).$$

We now set

$$\Delta_p w_{kij} \propto -\frac{\partial E_p}{\partial w_{kij}}.$$

Taking the derivative and simplifying, we get a rule for sigma-pi units strictly analogous to the rule for semilinear activation functions:

$$\Delta_p w_{kij} = \delta_k o_i o_j.$$

We can see the correct form of the error signal, δ, for this case by inspecting Figure 16. Consider the appropriate value of δ_i for unit u_i in the figure. As before, the correct value of δ_i is given by the sum of the δ's for all of the units into which u_i feeds, weighted by the amount of effect due to the activation of u_i times the derivative of the activation function. In the case of semilinear

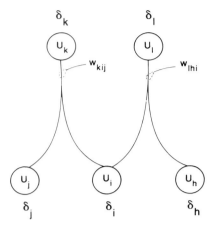

Figure 16   The generalized delta rule for sigma-pi units. The products of activation values of individual units activate output units. See text for explanation of how the δ values are computed in this case.

functions, the measure of a unit's effect on another unit is given simply by the weight w connecting the first unit to the second. In this case, the u_i's effect on u_k depends not only on w_{kij}, but also on the value of u_j. Thus, we have

$$\delta_i = f'_i(net_i) \sum_{j,k} \delta_k w_{kij} o_j$$

if u_i is not an output unit and, as before,

$$\delta_i = f'_i(net_i)(t_i - o_i)$$

if it is an output unit.

Recurrent Nets

We have thus far restricted ourselves to *feedforward* nets. This may seem like a substantial restriction, but as Minsky and Papert point out, there is, for every recurrent network, a feedforward network with identical behavior (over a finite period of time). We will now indicate how this construction can proceed and thereby show the correct form of the learning rule for the recurrent network. Consider the simple recurrent network shown in Figure 17A. The same network in a feedforward architecture is shown in Figure 17B. The behavior of a recurrent network can be achieved in a feedforward network at the cost of duplicating the hardware many times over for the feedforward version of the network.[9] We have distinct units and distinct weights for each point in time. For naming convenience, we subscript each unit with its unit number in the corresponding recurrent network and the time it repre-

[9] Note that in this discussion, and indeed in our entire development here, we have assumed a discrete time system with synchronous update and with each connection involving a unit delay.

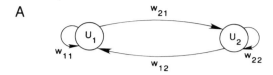

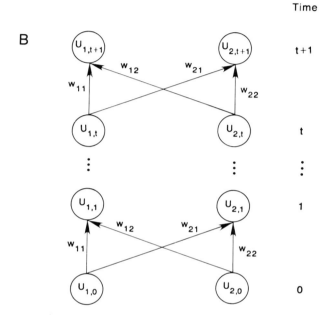

Figure 17 A comparison of a recurrent network and a feedforward network with identical behavior. *A*: A completely connected recurrent network with two units. *B*: A feedforward network which behaves the same as the recurrent network. In this case, we have a separate unit for each time step and we require that the weights connecting each layer of units to the next be the same for all layers. Moreover, they must be the same as the analogous weights in the recurrent case.

sents. As long as we constrain the weights at each level of the feedforward network to be the same, we have a feedforward network which performs identically with the recurrent network of Figure 17A. The appropriate method for maintaining the constraint that all weights be equal is simply to keep track of the changes dictated for each weight at each level and then change each of the weights according to the *sum* of these individually prescribed changes. Now, the general rule for determining the change prescribed for a weight in the system for a particular time is simply to take the product of an appropriate error measure δ and the input along the relevant line both for the appropriate times. Thus, the problem of specifying the correct learning rule for recurrent networks is simply one of determining the appropriate value of δ for each time. In a feedforward

network we determine δ by multiplying the derivative of the activation function by the sum of the δ's for those units it feeds into weighted by the connection strengths. The same process works for the recurrent network— except in this case, the value of δ associated with a particular unit changes in time as a unit passes error back, sometimes to itself. After each iteration, as error is being passed back through the network, the change in weight for that iteration must be added to the weight changes specified by the preceding iterations and the sum stored. This process of passing error through the network should continue for a number of iterations equal to the number of iterations through which the activation was originally passed. At this point, the appropriate changes to all of the weights can be made.

In general, the procedure for a recurrent network is that an input (generally a sequence) is presented to the system while it runs for some number of iterations. At certain specified times during the operation of the system, the output of certain units are compared to the target for that unit at that time and error signals are generated. Each such error signal is then passed back through the network for a number of iterations equal to the number of iterations used in the forward pass. Weight changes are computed at each iteration and a sum of all the weight changes dictated for a particular weight is saved. Finally, after all such error signals have been propagated through the system, the weights are changed. The major problem with this procedure is the memory required. Not only does the system have to hold its summed weight changes while the error is being propagated, but each unit must somehow record the sequence of activation values through which it was driven during the original processing. This follows from the fact that during each iteration while the error is passed back through the system, the current δ is relevant to a point earlier in time and the required weight changes depend on the activation levels of the units at that time. It is not entirely clear how such a mechanism could be implemented in the brain. Nevertheless, it is tantalizing to realize that such a procedure is potentially very powerful, since the problem it is attempting to solve amounts to that of finding a sequential program (like that for a digital computer) that produces specified input-sequence/output-sequence pairs. Furthermore, the interaction of the teacher with the system can be quite flexible, so that, for example, should the system get stuck in a local minimum, the teacher could introduce "hints" in the form of desired output values for intermediate stages of processing. Our experience with recurrent networks is limited, but we have carried out some experiments.

We turn first to a very simple problem in which the system is induced to invent a shift register to solve the problem.

Learning to be a shift register Perhaps the simplest class of recurrent problems we have studied is one in which the input and output units are one and the same and there are no hidden units. We simply present a pattern and let the system process it for a period of time. The state of the system is then compared to some target state. If it hasn't reached the target state at the designated time, error is injected into the system and it modifies its weights. Then it is shown a new input pattern and restarted. In these cases, there is no constraint on the connections in the system. Any unit can connect to any other unit. The simplest such problem we have studied is what we call the *shift register* problem. In this problem, the units are conceptualized as a circular shift register. An arbitrary bit pattern is first established on the units. They are then allowed to process for two time-steps. The target state, after those two time-steps, is the original pattern shifted two spaces to the left. The interesting question here concerns the state of the units between the presentation of the start state and the time at which the target state is presented. One solution to the problem is for the system to become a shift register and shift the pattern exactly one unit to the left during each time period. If the system did this then it would surely be shifted two places to the left after two time units. We have tried this problem with groups of three or five units and, if we constrain the biases on all of the units to be negative (so the units are off unless turned on), the system always learns to be a shift register of this sort.[10] Thus, even though in principle any unit can connect to any other unit, the system actually learns to set all weights to zero except the ones connecting a unit to its left neighbor. Since the target states were determined on the assumption of a circular register, the left-most unit developed a strong connection to the right-most unit. The system learns this relatively quickly. With $\eta = 0.25$ it learns perfectly in fewer than 200 sweeps through the set of possible patterns with either three- or five-unit systems.

The tasks we have described so far are exceptionally simple, but they do illustrate how the algorithm works with unrestricted networks. We have attempted a few more difficult problems with recurrent networks. One of the more interesting involves learning to complete

Table 10 25 sequences to be learned

AA1212	AB1223	AC1231	AD1221	AE1213
BA2312	BB2323	BC2331	BD2321	BE2313
CA3112	CB3123	CC3131	CD3121	CE3113
DA2112	DB2123	DC2131	DD2121	DE2113
EA1312	EB1323	EC1331	ED1321	EE1313

sequences of patterns. Our final examples comes from this domain.

Learning to complete sequences Table 10 shows a set of 25 sequences which were chosen so that the first two items of a sequence uniquely determine the remaining four. We used this set of sequences to test out the learning abilities of a recurrent network. The network consisted of five input units (A, B, C, D, E), 30 hidden units, and three output units (1, 2, 3). At time 1, the input corresponding to the first item of the sequence is turned on and the other input units are turned off. At Time 2, the input unit for the second item in the sequence is turned on and the others are all turned off. Then all the input units are turned off and kept off for the remaining four steps of the forward iteration. The network must learn to make the output units adopt states that represent the rest of the sequence. Unlike simple feedforward networks (or their iterative equivalents), the errors are not only assessed at the final layer or time. The output units must adopt the appropriate states *during* the forward iteration, and so during the back-propagation phase, errors are injected at each time-step by comparing the remembered actual states of the output units with their desired states.

The learning procedure for recurrent nets places no constraints on the allowable connectivity structure.[11] For the sequence completion problem, we used one-way connections from the input units to the hidden units and from the hidden units to the output units. Every hidden unit had a one-way connection to every other hidden unit and to itself, and every output unit was also connected to every other output unit and to itself. All the connections started with small random weights uniformly distributed between -0.3 and $+0.3$. All the hidden and output units started with an activity level of 0.2 at the beginning of each sequence.

We used a version of the learning procedure in which the gradient of the error with respect to each weight is computed for a whole set of examples before the weights are changed. This means that each connection

---

[10] If the constraint that biases be negative is not imposed, other solutions are possible. These solutions can involve the units passing through the complements of the shifted pattern or even through more complicated intermediate states. These trajectories are interesting in that they match a simple shift register on all even numbers of shifts, but do not match following an odd number of shifts.

[11] The constraint in feedforward networks is that it must be possible to arrange the units into layers such that units do not influence units in the same or lower layers. In recurrent networks this amounts to the constraint that during the forward iteration, future states must not affect past ones.

Table 11 Performance of the network on five novel test sequences

Input sequence	A	D	—	—	—	—
Desired outputs	—	—	1	2	2	1
Actual states of						
Output unit 1	0.2	0.12	0.90	0.22	0.11	0.83
Output unit 2	0.2	0.16	0.13	0.82	0.88	0.03
Output unit 3	0.2	0.07	0.08	0.03	0.01	0.22
Input sequence	B	E	—	—	—	—
Desired outputs	—	—	2	3	1	3
Actual states of						
Output unit 1	0.2	0.12	0.20	0.25	0.48	0.26
Output unit 2	0.2	0.16	0.80	0.05	0.04	0.09
Output unit 3	0.2	0.07	0.02	0.79	0.48	0.53
Input sequence	C	A	—	—	—	—
Desired outputs	—	—	3	1	1	2
Actuals state of						
Output unit 1	0.2	0.12	0.19	0.80	0.87	0.11
Output unit 2	0.2	0.16	0.19	0.00	0.13	0.70
Output unit 3	0.2	0.07	0.80	0.13	0.01	0.25
Input sequence	D	B	—	—	—	—
Desired outputs	—	—	2	1	2	3
Actual states of						
Output unit 1	0.2	0.12	0.16	0.79	0.07	0.11
Output unit 2	0.2	0.16	0.80	0.15	0.87	0.05
Output unit 3	0.2	0.07	0.20	0.01	0.13	0.96
Input sequence	E	C	—	—	—	—
Desired outputs	—	—	1	3	3	1
Actual states of						
Output unit 1	0.2	0.12	0.80	0.09	0.27	0.78
Output unit 2	0.2	0.16	0.20	0.13	0.01	0.02
Output unit 3	0.2	0.07	0.07	0.94	0.76	0.13

Table 12 Six variations of the sequence EA1312 produced by presenting the first two items at variable times[a]

EA--1312	E-A-1312	E--A1312
-EA-1312	-E-A1312	--EA1312

a. With these temporal variations, the 25 sequences shown in Table 10 can be used to generate 150 different sequences.

must accumulate the sum of the gradients for all the examples and for all the time steps involved in each example. During training, we used a particular set of 20 examples, and after these were learned almost perfectly we tested the network on the remaining examples to see if it had picked up on the obvious regularity that relates the first two items of a sequence to the subsequent four. The results are shown in Table 11. For four out of the five test sequences, the output units all have the correct values at all times (assuming we treat values above 0.5 as 1 and values below 0.5 as 0). The network has clearly captured the rule that the first item of a sequence determines the third and fourth, and the second determines the fifth and sixth. We repeated the simulation with a different set of random initial weights, and it got all five test sequences correct.

The learning required 260 sweeps through all 20 training sequences. The errors in the output units were computed as follows: For a unit that should be on,

there was no error if its activity level was above 0.8, otherwise the derivative of the error was the amount below 0.8. Similarly, for output units that should be off, the derivative of the error was the amount above 0.2. After each sweep, each weight was decremented by .02 times the total gradient accumulated on that sweep plus 0.9 times the previous weight change.

We have shown that the learning procedure can be used to create a network with interesting sequential behavior, but the particular problem we used can be solved by simply using the hidden units to create "delay lines" which hold information for a fixed length of time before allowing it to influence the output. A harder problem that cannot be solved with delay lines of fixed duration is shown in Table 12. The output is the same as before, but the two input items can arrive at variable times so that the item arriving at time 2, for example, could be either the first or the second item and could therefore determine the states of the output units at either the fifth and sixth or the seventh and eighth times. The new task is equivalent to requiring a buffer that receives two input "words" at variable times and outputs their "phonemic realizations" one after the other. This problem was solved successfully by a network similar to the one above except that it had 60 hidden units and half of their possible interconnections were omitted at random. The learning was much slower, requiring thousands of sweeps through all 136 training examples. There were also a few more errors on the 14 test examples, but the generalization was still good with most of the test sequences being completed perfectly.

Conclusion

In their pessimistic dicussion of perceptrons, Minsky and Papert (1969) finally discuss multilayer machines near the end of their book. They state:

The perceptron has shown itself worthy of study despite (and even because of!) its severe limitations. It has many features that attract attention: its linearity; its intriguing learning theorem; its clear paradigmatic simplicity as a kind of parallel computation. There is no reason to suppose that any of these virtues carry over to the many-layered version. Nevertheless, we consider it to be an important research problem to elucidate (or reject) our intuitive judgement that the extension is sterile. Perhaps some powerful convergence theorem will be discovered, or some profound reason for the failure to produce an interesting "learning theorem" for the multilayered machine will be found. (P. 231–232)

Although our learning results do not *guarantee* that we can find a solution for all solvable problems, our analyses and results have shown that as a practical matter, the error propagation scheme leads to solutions in virtually every case. In short, we believe that we have answered Minsky and Papert's challenge and *have* found a learning result sufficiently powerful to demonstrate that their pessimism about learning in multilayer machines was misplaced.

One way to view the procedure we have been describing is as a parallel computer that, having been shown the appropriate input/output exemplars specifying some function, programs itself to compute that function in general. Parallel computers are notoriously difficult to program. Here we have a mechanism whereby we do not actually have to know how to write the program in order to get the system to do it. Parker (1985) has emphasized this point.

On many occasions we have been surprised to learn of new methods of computing interesting functions by observing the behavior of our learning algorithm. This also raised the question of generalization. In most of the cases presented above, we have presented the system with the entire set of exemplars. It is interesting to ask what would happen if we presented only a subset of the exemplars at training time and then watched the system generalize to remaining exemplars. In small problems such as those presented here, the system sometimes finds solutions to the problems which do not properly generalize. However, preliminary results on larger problems are very encouraging in this regard. This search is still in progress and cannot be reported here. This is currently a very active interest of ours.

Finally, we should say that this work is not yet in a finished form. We have only begun our study of recurrent networks and sigma-pi units. We have not yet applied our learning procedure to many very complex problems. However, the results to date are encouraging and we are continuing our work.

(1986)
David E. Rumelhart, Geoffrey E. Hinton, and Ronald J. Williams

Learning representations by back-propagating errors
Nature 323:533–536

We describe a new learning procedure, back-propagation, for networks of neurone-like units. The procedure repeatedly adjusts the weights of the connections in the network so as to minimize a measure of the difference between the actual output vector of the net and the desired output vector. As a result of the weight adjustments, internal 'hidden" units which are not part of the input or output come to represent important features of the task domain, and the regularities in the task are captured by the interactions of these units. The ability to create useful new features distinguishes back-propagation from earlier, simpler methods such as the perceptron-convergence procedure[1].

There have been many attempts to design self-organizing neural networks. The aim is to find a powerful synaptic modification rule that will allow an arbitrarily connected neural network to develop an internal structure that is appropriate for a particular task domain. The task is specified by giving the desired state vector of the output units for each state vector of the input units. If the input units are directly connected to the output units it is relatively easy to find learning rules that iteratively adjust the relative strengths of the connections so as to progressively reduce the difference between the actual and desired output vectors[2]. Learning becomes more interesting but more difficult when we introduce hidden units whose actual or desired states are not specified by the task. (In perceptrons, there are "feature analysers' between the input and output that are not true hidden units because their input connections are fixed by hand, so their states are completely determined by the input vector: they do not learn representations.) The learning procedure must decide under what circumstances the hidden units should be active in order to help achieve the desired input–output behaviour. This amounts to deciding what these units should represent. We demonstrate that a general purpose and relatively simple procedure is powerful enough to construct appropriate internal representations.

The simplest form of the learning procedure is for layered networks which have a layer of input units at the bottom; any number of intermediate layers; and a layer of output units at the top. Connections within a layer or from higher to lower layers are forbidden, but connections can skip intermediate layers. An input vector is presented to the network by setting the states of the input units. Then the states of the units in each layer are determined by applying equations (1) and (2) to the connections coming from lower layers. All units within a layer have their states set in parallel, but different layers have their states set sequentially, starting at the bottom and working upwards until the states of the output units are determined.

The total input, x_j, to units j is a linear function of the outputs, y_i, of the units that are connected to j and of the weights, w_{ji}, on these connections

$$x_j = \sum_i y_i w_{ji} \tag{1}$$

Units can be given biases by introducing an extra input to each unit which always has a value of 1. The weight on this extra input is called the bias and is equivalent to a threshold of the opposite sign. It can be treated just like the other weights.

A unit has a real-valued output, y_j, which is a non-linear function of its total input

$$y_j = \frac{1}{1 + e^{-x_j}} \tag{2}$$

It is not necessary to use exactly the functions given in equations (1) and (2). Any input–output function which has a bounded derivative will do. However, the use of a linear function for combining the inputs to a unit before applying the nonlinearity greatly simplifies the learning procedure.

The aim is to find a set of weights that ensure that for each input vector the output vector produced by the network is the same as (or sufficiently close to) the desired output vector. If there is a fixed, finite set of input–output cases, the total error in the performance of the network with a particular set of weights can be computed by comparing the actual and desired output vectors for every case. The total error, E, is defined as

$$E = \tfrac{1}{2} \sum_c \sum_j (y_{j,c} - d_{j,c})^2 \tag{3}$$

Copyright © 1986 Macmillan Magazines Limited. Reprinted by permission.

where c is an index over cases (input–output pairs), j is an index over output units, y is the actual state of an output unit and d is its desired state. To minimize E by gradient descent it is necessary to compute the partial derivative of E with respect to each weight in the network. This is simply the sum of the partial derivatives for each of the input–output cases. For a given case, the partial derivatives of the error with respect to each weight are computed in two passes. We have already described the forward pass in which the units in each layer have their states determined by the input they receive from units in lower layers using equations (1) and (2). The backward pass which propagates derivatives from the top layer back to the bottom one is more complicated.

The backward pass starts by computing $\partial E/\partial y$ for each of the output units. Differentiating equation (3) for a particular case, c, and suppressing the index c gives

$$\partial E/\partial y_j = y_j - d_j \tag{4}$$

We can then apply the chain rule to compute $\partial E/\partial x_j$

$$\partial E/\partial x_j = \partial E/\partial y_j \cdot \mathrm{d}y_j/\mathrm{d}x_j$$

Differentiating equation (2) to get the value of $\mathrm{d}y_j/\mathrm{d}x_j$ and substituting gives

$$\partial E/\partial x_j = \partial E/\partial y_j \cdot y_j(1 - y_j) \tag{5}$$

This means that we know how a change in the total input x to an output unit will affect the error. But this total input is just a linear function of the states of the lower level units and it is also a linear function of the weights on the connections, so it is easy to compute how the error will be affeced by changing these states and weights. For a weight w_{ji}, from i to j the derivative is

$$\partial E/\partial w_{ji} = \partial E/\partial x_j \cdot \partial x_j/\partial w_{ji}$$
$$= \partial E/\partial x_j \cdot y_i \tag{6}$$

and for the output of the ith unit the comtribution to $\partial E/\partial y_i$ resulting from the effect of i on j is simply

$$\partial E/\partial x_j \cdot \partial x_j/\partial y_i = \partial E/\partial x_j \cdot w_{ji}$$

so taking into account all the connections emanating from unit i we have

$$\partial E/\partial y_i = \sum_j \partial E/\partial x_j \cdot w_{ji} \tag{7}$$

We have now seen how to compute $\partial E/\partial y$ for any unit in the penultimate layer when given $\partial E/\partial y$ for all units in the last layer. We can therefore repeat this procedure to compute this term for successively earlier layers, computing $\partial E/\partial w$ for the weights as we go.

One way of using $\partial E/\partial w$ is to change the weights after every input–output case. This has the advantage that no separate memory is required for the derivatives. An alternative scheme, which we used in the research reported here, is to accumulate $\partial E/\partial w$ over all the input–output cases before changing the weights. The simplest version of gradient descent is to change each weight by an amount proportional to the accumulated $\partial E/\partial w$

$$\Delta w = -\varepsilon \partial E/\partial w \tag{8}$$

This method does not converge as rapidly as methods which make use of the second derivatives, but it is much simpler and can easily be implemented by local computations in parallel hardware. It can be significantly improved, without sacrificing the simplicity and locality, by using an acceleration method in which the current gradient is used to modify the velocity of the point in weight space instead of its position

$$\Delta w(t) = \varepsilon \partial E/\partial w(t) + \alpha \Delta w(t - 1) \tag{9}$$

where t is incremented by 1 for each sweep through the whole set of input–output cases, and α is an exponential decay factor between 0 and 1 that determines the relative contribution of the current gradient and earlier gradients to the weight change.

To break symmetry we start with small random weights. Variants on the learning procedure have been discovered independently by David Parker (personal communication) and by Yann Le Cun[3].

One simple task that cannot be done by just connecting the input units to the output units is the detection of symmetry. To detect whether the binary activity levels of a one-dimensional array of input units are symmetrical about the centre point, it is essential to use an intermediate layer because the activity in an individual input unit, considered alone, provides no evidence about the symmetry or non-symmetry of the whole input vector, so simply adding up the evidence from the individual input units is insufficient. (A more formal proof that intermediate units are required is given in ref. 2.) The learning procedure discovered an elegant solution using just two intermediate units, as shown in Fig. 1.

Another interesting task is to store the information in the two family trees (Fig. 2). Figure 3 shows the network we used, and Fig. 4 shows the "receptive fields' of some of the hidden units after the network was trained on 100 of the 104 possible triples.

So far, we have only dealt with layered, feed-forward networks. The equivalence between layered networks and recurrent networks that are run iteratively is shown in Fig. 5.

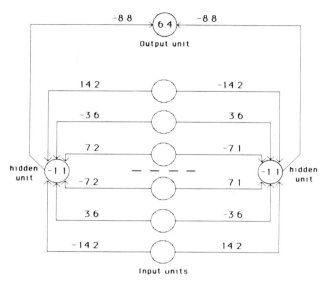

Figure 1 A network that has learned to detect mirror symmetry in the input vector. The numbers on the arcs are weights and the numbers inside the nodes are biases. The learning required 1,425 sweeps through the set of 64 possible input vectors, with the weights being adjusted on the basis of the accumulated gradient after each sweep. The values of the parameters in equation (9) were $\varepsilon = 0.1$ and $\alpha = 0.9$. The initial weights were random and were uniformly distributed between -0.3 and 0.3. The key property of this solution is that for a given hidden unit, weights that are symmetric about the middle of the input vector are equal in magnitude and opposite in sign. So if a symmetrical pattern is presented, both hidden units will receive a net input of 0 from the input units, and, because the hidden units have a negative bias, both will be off. In this case the output unit, having a positive bias, will be on. Note that the weights on each side of the midpoint are in the ratio $1:2:4$. This ensures that each of the eight patterns that can occur above the midpoint sends a unique activation sum to each hidden unit, so the only pattern below the midpoint that can exactly balance this sum is the symmetrical one. For all non-symmetrical patterns, both hidden units will receive non-zero activations from the input units. The two hidden units have identical patterns of weights but with opposite signs, so for every non-symmetric pattern one hidden unit will come on and suppress the output unit.

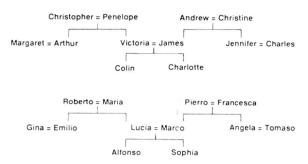

Figure 2 Two isomorphic family trees. The information can be expressed as a set of triples of the form ⟨person 1⟩⟨relationship⟩ ⟨person 2⟩, where the possible relationships are {father, mother, husband, wife, son, daughter, uncle, aunt, brother, sister, nephew, niece}. A layered net can be said to 'know' these triples if it can produce the third term of each triple when given the first two. The first two terms are encoded by activating two of the input units, and the network must then complete the proposition by activating the output unit that represents the third term.

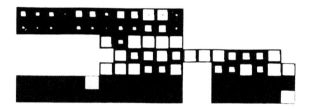

Figure 3 Activity levels in a five-layer network after it has learned. The bottom layer has 24 input units on the left for representing ⟨person 1⟩ and 12 input units on the right for representing the relationship. The white squares inside these two groups show the activity levels of the units. There is one active unit in the first group representing Colin and one in the second group representing the relationship 'has-aunt'. Each of the two input groups is totally connected to its own group of 6 units in the second layer. These groups learn to encode people and relation-ships as distributed patterns of activity. The second layer is totally connected to the central layer of 12 units, and thes   ‸e connected to the penultimate layer of 6 units. The activity in the penultimate layer must activate the correct output units, each of which stands for a particular ⟨person 2⟩. In this case, there are two correct answers (marked by black dots) because Colin has two aunts. Both the input units and the output units are laid out spatially with the English people in one row and the isomorphic Italians immedi-ately below.

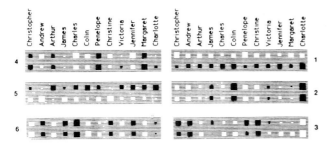

Figure 4 The weights from the 24 input units that represent people to the 6 units in the second layer that learn distributed representations of people. White rectangles, excitatory weights; black rectangles, inhibitory weights; area of the rectangle encodes the magnitude of the weight. The weights from the 12 English people are in the top row of each unit. Unit 1 is primarily concerned with the distinction between English and Italian and most of the other units ignore this distinction. This means that the representation of an English person is very similar to the representation of their Italian equivalent. The network is making use of the isomorphism between the two family trees to allow it to share structure and it will therefore tend to generalize sensibly from one tree to the other. Unit 2 encodes which generation a person belongs to, and unit 6 encodes which branch of the family they come from. The features captured by the hidden units are not at all explicit in the input and output encodings, since these use a separate unit for each person. Because the hidden features capture the underlying structure of the task domain, the network generalizes correctly to the four triples on which it was not trained. We trained the network for 1500 sweeps, using $\varepsilon = 0.005$ and $\alpha = 0.5$ for the first 20 sweeps and $\varepsilon = 0.01$ and $\alpha = 0.9$ for the remaining sweeps. To make it easier to interpret the weights we introduced 'weight-decay' by decrementing every weight by 0.2% after each weight change. After prolonged learning, the decay was balanced by $\partial E/\partial w$, so the final magnitude of each weight indicates its usefulness in reducing the error. To prevent the network needing large weights to drive the outputs to 1 or 0, the error was considered to be zero if output units that should be on had activities above 0.8 and output units that should be off had activities below 0.2.

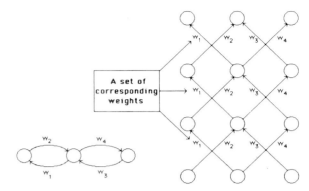

Figure 5 A synchronous iterative net that is run for three iterations and the equivalent layered net. Each time-step in the recurrent net corresponds to a layer in the layered net. The learning procedure for layered nets can be mapped into a learning procedure for iterative nets. Two complications arise in performing this mapping: first, in a layered net the output levels of the units in the intermediate layers during the forward pass are required for performing the backward pass (see equations (5) and (6)). So in an iterative net it is necessary to store the history of output states of each unit. Second, for a layered net to be equivalent to an iterative net, corresponding weights between different layers must have the same value. To preserve this property, we average $\partial E/\partial w$ for all the weights in each set of corresponding weights and then change each weight in the set by an amount proportional to this average gradient. With these two provisos, the learning procedure can be applied directly to iterative nets. These nets can then either learn to perform iterative searches or learn sequential structures[4].

The most obvious drawback of the learning procedure is that the error-surface may contain local minima so that gradient descent is not guaranteed to find a global minimum. However, experience with many tasks shows that the network very rarely gets struck in poor local minima that are significantly worse than the global minimum. We have only encountered this undesirable behaviour in networks that have just enough connections to perform the task. Adding a few more connections creates extra dimensions in weight-space and these dimensions provide paths around the barriers that create poor local minima in the lower dimensional subspaces.

The learning procedure, in its current form, is not a plausible model of learning in brains. However, applying the procedure to various tasks shows that interesting internal representations can be constructed by gradient descent in weight-space, and this suggests that it is worth looking for more biologically plausible ways of doing gradient descent in neural networks.

We thank the System Development Foundation and the Office of Naval Research for financial support.

Notes and References

1. Rosenblatt, F. *Principles of Neurodynamics* (Spartan, Washington, DC, 1961).

2. Minsky, M. L. & Papert, S. *Perceptrons* (MIT, Cambridge, 1969).

3. Le Cun, Y. *Proc. Cognitiva* **85**, 599–604 (1985).

4. Rumelhart, D. E., Hinton, G. E. & Williams, R. J. in *Parallel Distributed Processing: Explorations in the Microstructure of Cognition*. Vol. 1: *Foundations* (eds Rumelhart, D. E. & McClelland, J. L.) 318–362 (MIT, Cambridge, 1986).

43
Introduction

(1987)
Massimo A. Sivilotti, Michelle A. Mahowald, and Carver A. Mead

Real-time visual computations using analog CMOS processing arrays
Advanced Research in VLSI: Proceedings of the 1987 Stanford Conference, P. Losleben (Ed.), Cambridge, MA: MIT Press, pp. 295–312

Currently a number of groups in private industry and the universities are trying to make devices to implement neural networks. Some groups are using optical techniques (see paper 39); others are using VLSI. At present, there are design projects underway to implement many of the models we have presented here: dynamical attractor networks, associative networks, Boltzmann machines. For the last paper in our collection, we have chosen a description of one of the most interesting and highly developed approaches to using VLSI technology for realizing a neural network. We have done so even though technology advances so quickly that any paper on hardware will be out of date by the time it is published.

A group at the California Institute of Technology, directed by Carver Mead, is systematically trying to design VLSI devices to act as sensory systems. The circuitry is modeled as closely as possible on the structure of mammalian sensory systems. This paper describes their progress in modeling the retina, but other work in this laboratory is trying to model the cochlea, to be used as the first stage of an artificial auditory system.

It is sometimes not appreciated what a complex information processing system is contained in the retina of the higher animals. In the frog many cells are specialized to detect objects of interest to frogs—most prominently, bugs. The visual world of the frog must be nearly blank until it detects one of the few objects it was designed to see.

The retina of primates is less specialized. However, it is still a formidable information processing device. Most cells in the retina respond most strongly to moving stimuli, and respond weakly or not at all to stationary objects. There is also strong lateral inhibition, which enhances contrast and provides some gain control. There are mechanisms that shift the system from one class of receptors (rods, for night vision) to another class (cones, for day vision) as the light level changes. The size of receptive fields of the retinal output cells changes with light intensity, so that at very low intensities cells integrate light over a large spatial region and over a long period of time, trading resolution for sensitivity. As intensities increase, the retina rewires itself so as to give higher spatial resolution, but lower sensitivity. There are several different classes of output cells in cats: somewhat linear cells with small receptive cells (X-cells), less linear cells with larger receptive fields (Y-cells), and group of complex, highly nonlinear cells that look almost like a piece of specialized frog retina mixed into the retina (W-cells).

What the cortical visual areas are receiving does not look anything like a point-for-point copy of light intensities; rather, it has been highly preprocessed. This early preprocessing is surely very effective at reducing the computational burden of higher levels.

Sivilotti, Mahowald, and Mead are trying to build devices that incorporate as much of this peripheral computing power as they can. As one example, used by this group as an important starting point in their work, most retinal cells respond very strongly to motion. For retinal cells, changes in light intensity are the single most potent aspect of the image. If one is building a traditional computer vision system, motion could be detected by looking for changes in intensity computed from a succession of stationary images. However, the computational problems become extremely difficult. It is a hard correspondence problem to discover which points in the two different views of the moving object are the same.

The authors view the problem of building a neural network as a branch of analog circuit design. This is certainly correct. The bulk of the paper is a description of the construction techniques that can be used to obtain a large dynamic range in the receptor units, to obtain sensitivity to changes in light intensity, and the practical problems involved in putting a number of such devices on a single chip.

Figures 18 and 19 in their paper give an idea of their success in capturing some aspects of retinal processing. Carver Mead shows a dramatic videotape of this demonstration at lectures. When the artificial retina is looking at a stationary set of vanes, there is barely an image on the screen. When the vanes start to rotate, the retina explodes into life, and its responses to the edges of the vanes become large, outlining them clearly. Having higher layers analyze a dynamic image of this type is considerably different from having higher layers analyze a static image, and changes processing requirements considerably. The approach described in this paper has a great deal to recommend it if we want to make systems that are, like us (or so we flatter ourselves to be), intelligent perceivers.

The high level associative networks discussed in many other places in this collection, if present in the brain, are operating with carefully and effectively preprocessed inputs. The highest levels can do only a few kinds of simple computation. In the brain most of the work has been done for them by special-purpose hardware, tried and tested by evolution, at the lower levels. It is a good strategy for the brain to follow and will be a good strategy for artifical devices to use as well.

(1987)
Massimo A. Sivilotti, Michelle A. Mahowald, and Carver A. Mead

Real-time visual computations using analog CMOS processing arrays
Advanced Research in VLSI: Proceedings of the 1987 Stanford Conference, P. Losleben (Ed.),
Cambridge, MA: MIT Press, pp. 295–312

Integration of photosensors and processing elements provides a mechanism to concurrently perform computations previously intractable in real-time. We have used this approach to model biological early vision processes. A set of VLSI "retina" chips have been fabricated, using large scale analog circuits (over 100K transistors in total). Analog processing provides sophisticated, compact functional elements, and avoids some of the aliasing problems encountered in conventional sampled-data artificial vision systems.

Integration of Photoreception and Processing

By their very two-dimensional nature, images constitute a high bandwidth interface with the real world. The most powerful supercomputers are incapable of even rudimentary analysis of static images. Real-time analysis of motion information, requiring computation over several images, is completely infeasible. Yet biological early vision processes are clearly able to perform these computations, by exploiting the inherent parallelism of visual inputs in a truly concurrent fashion. Computations are spatially localized, and computing elements are replicated as required. Only significant information is transmitted along the optic nerve.

Artificial vision systems are further limited by the early sampling performed by television camera frontends. Because each point is sampled only every 1/30 second, an object can easily move several pixels between samples. Motion interpretation has thus been converted from a local problem to the much more difficult *correspondence problem*. In signal processing terms, high frequency information is irretrievably lost due to *aliasing*.

Modeled on biological architectures, our approach is to spatially interleave integrated photoreceptors and processing on a VLSI die [16]. To obtain a sufficiently rich, yet compact, set of computing elements, analog micropower CMOS circuits are used. The resulting high density permits complex systems to be built, that demonstrate powerful collective behaviors [20]. Finally, by performing temporal operations on continuous data, prior to sampling for transmission off-chip, susceptibility to aliasing is reduced.

Models for Retinal Computations

The retina performs the first step in visual processing and provides the data for all subsequent stages of the visual system. Although various species perform slightly different sets of retinal computations, there are several aspects of visual processing that are common to many different organisms [6]. These ubiquitous features include logarithmic compression of the incoming intensity at the detector level, and the extensive use of lateral and temporal inhibition in the retinal computations. These functions are computed using smoothly varying (continuous) analog potentials, rather than neuronal action potentials.

Several explanations have been proposed for why the visual system performs these computations in the retina. As is often the case when investigating biological systems, it is not possible to determine *the* reason that the system adopted a particular strategy; rather, these systems are optimzied with respect to multiple constraints [18].

One particular set of advantages to retinal processing can be observed by assuming that the function of the retina is the "neat packaging of information" [2] to be sent to higher visual areas. Retinal ganglion cells transmit information to the brain by propagating action potentials along their axons in the optic nerve. There are a finite number of discriminable signal levels coming over the optic nerve due to intrinsic noise. The visual information must be encoded in such a way that the full dynamic range of the neuron is used. Automatic gain control mechanisms, such as logarithmic compression of intensity and center-surround antagonistic receptive fields allow the system to encode detail over a large range of ambient light levels. In addition to the limited resolution of signal levels, the discrete-time nature of the action potential limits the temporal resolution of events. A large fraction of retinal processing is dedicated to extracting motion events. If retinal ganglion cells encoded simple intensity, then any change in intensity would be encoded as a change in pulse rate. Even assuming that such an encoding allowed no statistical fluctuation in pulse interval, the time at which such a change had occurred could be determined only to the time between pulses. In signal processing terms, the derivative information would have been aliased away by temporal sampling of the image. For this reason, motion detecting ganglion cells produce action potentials that correspond to changes in intensity, rahter than intensity itself. In this way, a pulse burst corresponds to an important feature moving over that particular place on the retina. Higher-level correlations among events can then be reconstructed without loss of information due to temporal aliasing.

Light-Level Independence

A vision system intended for operation in an unconstrained environment must include automatic gain control (AGC) with respect to absolute ambient light level. Taking the logarithm of the incident light intensity is a simple local AGC mechanism. Receptors with logarithmic response have the additional advantage of providing output voltages whose difference is proportional to the *contrast ratios* within the image, which are the perceptually important parameters.

An integrated photoreceptor with an output that is logarithmic over 5 orders of magnitude in light intensity is shown in Figure 1a. Its operation is similar to that of one previously described [9]; a large-area bipolar transistor is formed using the n-well for the base and p-type diffusion as the emitter. The substrate forms the collector, and hence the device is operated in a common-collector configuration. The output voltage biases the gate of a p-channel MOS feedback transistor operating in subthreshold. In this regime, the channel current is exponential in the gate voltage with a slope of about 1 decade per 100 mV. If a second subthreshold transistor is used for source degeneration, the slope can be decreased to about 1 decade per 300 mV. This

arrangement provides a larger output voltage swing, and sets the output voltage in the 1.0 to 2.5 V range below V_{dd}, making direct coupling to subsequent stages possible. The feedback current is generated by mirroring the load current for the emitter of the phototransistor. If the feedback to the phototransistor base is omitted (Figure 1b), the receptor is sensitive to lower light levels (by a factor of h_{fe}), but will saturate at bright levels, as the MOS loads leave subthreshold. These receptors operate at light levels comparable to the useful range of cones in human retinas, and form the basis for the RET10 chip discussed later.

A Simple Local Computation: Discrete-Time Derivative

We perceive motion when a point in an image displays non-zero spatial and temporal derivatives. In other words, an edge (space derivative) that is moving causes a change in brightness at that point in the image. Thus, a local time derivative is the simplest computation to highlight areas of an image that are moving. Because this computation is purely local, no interpixel communication is required. The derivative can be approximated by comparing the present photoreceptor output with some suitably delayed version of the output.

A discrete-time derivative based on the circuit in Figure 2 forms the core of the RET20 chip. When switch S_1 is closed, the amplifier forms a unity-gain follower stage that stores the current state of the system on capacitor C_1. The switch is then opened, and any evolution of the input away from the sampled value is amplified by the open-loop voltage gain of the (wide-range) transconductance amplifier [21].

If switch S_1 is implemented with a MOS pass transistor, transient switching charge is injected onto C_1 [15]. To minimize this effect, a transconductance amplifier was used for the switch (Figure 3). When the bias current in A_2 is reduced to zero, the capacitor is effectively isolated from the input. Less noise charge is

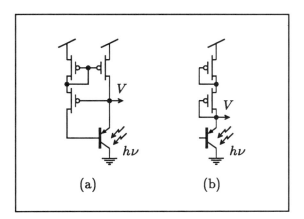

(a)   (b)

Figure 1 Logarithmic photoreceptors.

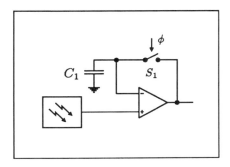

Figure 2 Photoreceptor with discrete-time differentiator.

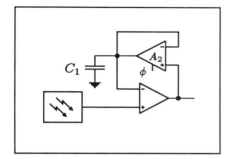

Figure 3 Transconductance amplifier used as low-noise switch.

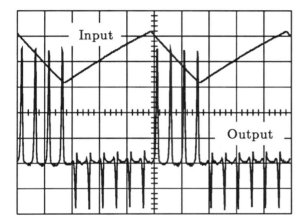

Figure 4 Discrete-time derivative: sample operation.

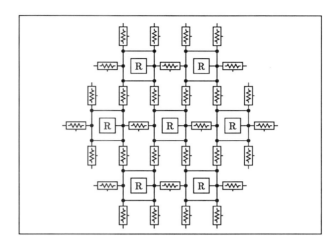

Figure 5 RET30 tesselation.

generated because the channel charge is symmetrically divided between both branches of the differential pair (which have identical operating points in the follower configuration), and because any capacitive clock feedthrough is decreased by the cascode connection of the differential pair. In addition, the clock need not be rail-to-rail (hence decreasing dV/dt even further), and it can be single phase and unipolar. Figure 4 illustrates the output of the circuit when presented with an asymmetrical 100 Hz triangle wave input (ϕ clock frequency 1000 Hz).

RET30: Continuous-Time Derivatives and Local Space Derivatives

Biological retinas contain horizontal cells that provide lateral conductance and can be loosely thought of as providing an average of the signal values in the neighborhood with which the local signal can be compared. Inspired by this model, the RET30 (Figure 5) consists of an array of receptors, R, interconnected by a hexagonal resistive network [11]. To provide temporal smoothing, a capacitor to ground is located at each junction of 6 neighboring horizontal resistors.

Each local processing element takes the difference between the potential of the horizontal network and that of the receptor output, and drives the local potential of the horizontal network toward the local receptor output potential. The "derivative" computed is the difference between the input signal and a spatially and temporally smoothed version of that signal. The spatial part of the processing emphasizes areas in the image containing the most information. The emphasis corresponds to a discrete approximation to a Laplacian operator applied to the image. The temporal part of the processing is a finite-gain, single time-constant differentiation.

Horizontal Resistors To construct a practical space–time derivative system, we must be able to create time constants of the same order as the time scale of motion events, without using enormous area for capacitors. The horizontal network spreads the potential at one point outward through a resistive sheet. To keep the time constant (τ_h) of the spreading on the same scale as others in the system, enormous resistor values are required, (10^{11}–10^{13} Ω)—larger than the resistance of any integrated device we can build. A circuit that implements a very high-value resistance in a controlled way is shown in Figure 6. V_1 and V_2 represent potentials V_i of two neighboring locations in the network. The current I_0 into the upper node is constrained to be the same as that out of the lower node by a current-mirror arrangement. Hence, any current out of node 1 must flow into node 2. By symmetry, the magnitude of this current must be zero when the voltages are equal, and will be monotonically related to their difference. However, it can never exceed I_0. The I–V relationship is shown in Figure 7. The limiting current I_0 is set by the current mirror input, and controls the value of τ_h.

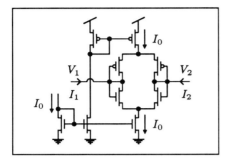

Figure 6 Horizontal resistor.

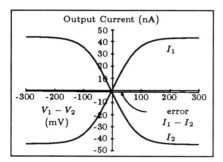

Figure 7 I-V characteristic.

As can be seen from the figure, the region of linear operation of the horizontal circuits is about ± 100 mV, corresponding to an illumination contrast ratio of about 2:1 at the photoreceptors.

Time Differentiators The differentiator of the RET30 chip is implemented using the same circuit as that shown in Figure 3, except the A_2 amplifier is not clocked. Because the ϕ input on the second amplifier controls the maximum current that may flow into or out of the storage capacitor, it determines the rate at which the capacitor is charged, and hence the time-constant τ_r. The net current into the capacitor is of the same form as that shown in Figure 7, for the same reasons. This circuit has unity gain at DC, and a gain at short times set by the open-circuit gain of the first amplifier, as in the discrete-time case. Experimental data illustrating the operation of the continuous-time differentiator are shown in Figure 8.

The advantage of this arrangement is that any input offset voltage of the first amplifier is not multiplied by the gain of the amplifier in its effect on the output voltage.

It can be argued that the saturating characteristic of all these circuits is desirable, as it prevents one extreme input (or faulty circuit element) from paralyzing the entire network. Thus, even for these simple operations,

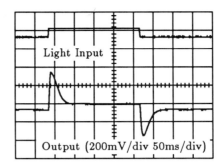

Figure 8 Continuous-time differentiator.

many of the properties of collective circuits can be preserved.

As shown, maximum outputs will occur when high contrast features move over the retina. If only time derivative information is desired, the horizontal network is unnecessary, and can be disabled by setting the I_0 input current to zero. When enabled, the horizontal network computes the extent to which the light received by an individual receptor differs from the average level in its neighborhood. It is thus most sensitive to a point, less to a corner, even less to an edge, and not at all to a uniform gradient. The system can be made to display a sustained response to one of its preferred stimuli even if that stimulus is stationary.

CMOS Design Frame for Scanning Arrays

Fundamental limits on the number of pads imposed by available VLSI packaging technologies, coupled with the high area cost of dedicated wiring within the imaging array, necessitate a scalable communication architecture. Time multiplexing of sampled data signals requires a minimal number of pads, and simplifies the external system design by reducing the number of external components needed to compensate for the different electrical nature of the off-chip environment.

To facilitate experimentation with a wide variety of processing core cells, a "design frame" approach was adopted, with a standard peripheral frame providing all the communication support functions, as well as the data sampling mechanism. By separating the computation and communication tasks, the responsibility of the core cell designer is simplified, and consists of providing, as the result of some computation, an analog voltage to be sampled and transmitted. An additional benefit is the independence of the tiling topology from the computation topology. For example, our design frames support true hexagonal tiling (Figure 9), hexagonal tiling using offset rectangles (Figure 10), and pure

Figure 9 Hexagonal tiling with Boston geometry (HEXRET).

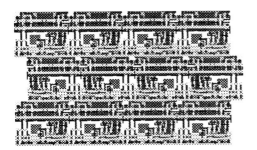

Figure 10 Hexagonal tiling with Manhattan geometry (RET30).

rectangular tiling; in all cases, the pixel stream is (offset) rectangular.

Considerable work has been done on charge-transfer systems, particularly for CCD image arrays [5]. Recently, bipolar phototransducer arrays have received some attention due to their more favorable saturation and antiblooming properties [7,1,3]. In general, the non-CCD scanning techniques consist of switching some charge basin onto a column line, then switching that column's charge packet onto a global output line. Often, these switches are implemented with specially fabricated low-threshold MOS devices [14]. The primary source of fixed-pattern noise on the output is due to inversion charge variation in the switching transistors [13,23]. Mechanisms to reduce this noise include integration over the entire pixel time, and sampling at some constant point during the pixel event.

Our RETxx chips have taken a somewhat different approach to obtaining an acceptable signal-to-noise ratio (SNR). First, photocharge is integrated at each pixel site. It is not then destructively dumped on the output line, but rather is stored on a local capacitor, which is nondestructively sampled using a single MOS charge–sense transitor. The *current* in this "bit line" is sensed, eventually by an external amplifier. To minimize signal propagation delays due to $C(dV/dt)$ losses (*i.e.* charging/discharging the highly capacitive bit lines, output line, and output pad/off-chip wiring), current-steering sensing is employed throughout.

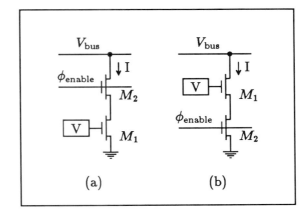

Figure 11 Current transducers.

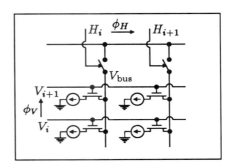

Figure 12 Horizontal switches.

Pixels are enabled on a row-by-row basis by a switched pass transistor in series with the transduction transistor (Figure 11). Two configurations are possible: Figure 11a shows a conventional cascode arrangement that minimizes the dependence of I on V_{bus}; Figure 11b shows a configuration designed to maximize the linear range of I with V, by operating M_2 in the ohmic region to provide source degeneration for M_1. In either case, linear operation can be guaranteed if V_{bus} is maintained sufficiently close to ground, so M_1 is operating in its ohmic regime.

One entire row of the scanning array is enabled simultaneously, at the line clock rate ϕ_V. Within each scan line, the pixel clock ϕ_H sequences the connection of the different "bit" lines onto the single output line (Figure 12). To maintain control of V_{bus}, and to keep this value independent of the pixel current I, the $N - 1$ bit lines not connected to the output bus are instead connected to a dummy bus, biased at the desired value of V_{bus} (Figure 13). Thus, the pixel current I flows at all times.

Off-chip, the current in the output line is converted to a voltage by a current–sense amplifier (Figure 14). This configuration implicitly biases the output line at

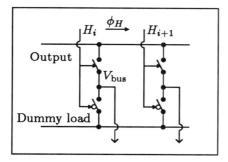

Figure 13 Dummy load line.

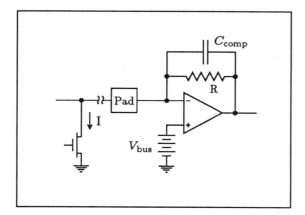

Figure 14 External current-sense amplifier.

V_{bus}. External compensation in the form of C_{comp} is required to counteract the highly capacitive input seen by the opamp.

The last circuit elements to consider in the analog signal path are the multiplexing switches in Figure 13. These can inject noise charge onto the output lines in two ways: (1) as a MOS transistor shuts off, the mobile charge in the channel region is divided (nearly equally) between the source and the drain, and (2) the switching clock itself can couple through the overlap capacitance of the gate with the source-drain regions. To minimize these problems, CMOS transmission gates driven by complementary phases derived from the horizontal clock were used as switches. Any injected parasitic charge is offset by the opposite physical process occurring in the complementary transistor, as well as by the same process occuring in reverse in an adjacent pass gate (as that column becomes enabled). More complicated charge compensation schemes are possible [8,19], but our simulation results indicated that these schemes did not significantly improve the clock noise suppression in this case.

Compensation of charge injection using comple-

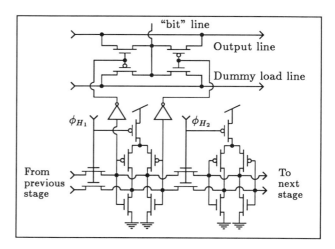

Figure 15 CSRL stage with multiplexer.

mentary devices depends on the availability of a clock and its logical complement with *minimal skew* between these signals. In the presence of appreciable skew, the transmission gate switch actually can be noisier than a single pass transistor switch. For this reason, the clock signal and its complement are generated simultaneously by a CMOS set−reset logic (CSRL) [12] shift register, in which both signals are propagated together. This shift register is clocked with a two-phase nonoverlapping clock running at the pixel rate. Because the CSRL outputs are not fully restored during one clock phase, they are buffered by a pair of inverters. The full circuit, including multiplexing switches, is shown in Figure 15. The input to the shift register chain is brought off-chip. During one line period, a single "1" is shifted into the register, and is shifted sequentially through all the pixel columns.

The same CSRL design is used in the vertical shift register to sequentially enable successive rows. The buffer inverters serve the additional function of driving the highly capacitive "word line" (ϕ_{enable} in Figure 11).

A functional diagram of the entire system is shown in Figure 16. An objective of the design frame was to simplify the system-level interface with the pixel array. There are only eight signal lines (four clocks—ϕ_{V_1}, ϕ_{V_2}, ϕ_{H_1}, ϕ_{H_2}; two shift register inputs—H_{in}, V_{in}; and two analog outputs—I_{out}, $I_{dummyload}$) plus power and any signals specific to the core cells in the pixel array.

Implementation and Experimental Results

We fabricated three different chips, incoporating the RET10, RET20, and RET30 core cells. Pixel array sizes and chip dimensions are shown in Table 1. The first

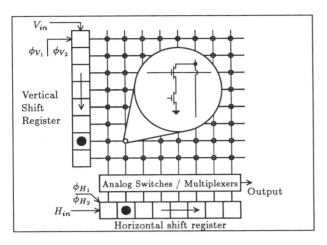

Figure 16 Design frame: system level interface.

Table 1 Performance of core cells (3 μm feature size)

Chip	Pixel size ($\mu m \times \mu m$)	Array size	Measured SNR (dB)
Phototransistor	33 × 33	N/A	
RET10	92 × 80	88 × 88	62
RET20	113 × 98	64 × 64	70
RET30	164 × 143	48 × 48	60

Figure 17 RET10 sample output.

versions were fabricated on MOSIS [4] run M57Q in August 1985. Except for the HEXRET, they all share the same design frame, implemented in Manhattan geometry using the WOL design tool set [10,17]; the HEXRET implemented the RET10 circuit using arbitrary angle geometry, and was layed out with the TIGGER/POOH design system [22].

The chips were tested with two different setups: (1) a workstation based tester capable of running all the chip clocks, and of digitizing and displaying the output signal at a rate of 60,000 pixels/sec (approximately 8 frames/sec for RET10), and (2) a TTL clock generator board that also produced a raster on which the chip output could be superposed and displayed on an oscilloscope at rates of more than 400,000 pixels/sec (over 50 frames/sec).

Figure 17 shows sample output from the RET10 chip, and clearly illustrates the logarithmic compression performed by the photoreceptors. Figures 18 and 19 demonstrate the operation of the RET30 chip. In both cases, the stimulus is a dark cross mounted on a rotating axis. In Figure 18 the cross is stationary and, with the horizontal resistor network disabled, no image is seen. In Figure 19 the cross is rotating at approximately 10 rpm, and is clearly visible.

Figure 18 RET30—stationary image.

Figure 19 RET30—moving image.

Conclusions

Although many models have been proposed for the visual system, it is not possible to simulate enough cases to gain real confidence in the model, even on our most powerful computers. For this reason, we will not really understand visual processing, especially with respect to demanding problems such as motion analysis, until we succeed in building a system that is capable of doing visual processing in real-time. Until recently, we have not had a technology in which such fundamental synthetic investigations could be carried out. With the evolution of high-density VLSI technology, we have a means for conducting these extremely important investigations. The work will not be trivial. Previously, the most massive application of large-scale circuits has been in digital systems. Although analog integrated circuit techniques have developed along with digital ones, no comparable methods exist for managing the complexity of extremely large analog systems. This paper not only has described a prototype vision system, but also has illustrated an approach to problems of this class.

Acknowledgments

This research was supported by a grant from the System Development Foundation, and an equipment grant from Hewlett-Packard Corp. The authors are indebted to Dick Lyon, Lars Neilsen, Michael Emerling, and John Tanner for many useful discussions.

References

1 M. Aoki, H. Ando, S. Ohba, I. Takemoto, S. Nagahara, T. Nakano, M. Kubo, and T. Fujita. 2/3-inch format MOS single-chip color imager. *IEEE Journal of Solid State Circuits*, SC-17(2):375–380, April 1982.

2 H. B. Barlow. Three points about lateral inhibition. In W. A. Rosenblith, editor, *Sensory Communication*, pages 782–786, M.I.T. Press, Cambridge Mass., 1961.

3 Savvas G. Chamberlain and Jim P. Y. Lee. A novel wide dynamic range silicon photodetector and linear imaging array. *IEEE Journal of Solid State Circuits*, SC-19(1):41–48, February 1984.

4 D. Cohen and G. Lewicki. MOSIS—the ARPA silicon broker. In *Proceedings from the Second Caltech Conference on VLSI*, pages 29–44, California Institute of Technology, Pasadena, CA, 1981.

5 P. L. P. Dillon, D. M. Lewis, and F. G. Kasper. Color imaging system using a single CCD area array. *IEEE Trans. Electron Devices*, ED-25:102–107, February 1978.

6 Akimichi Kaneko. Physiology of the retina. *Ann. Rev. Neurosci*, 2:169–191, 1979.

7 N. Koike, I. Takemoto, K. Satoh, S. Hanamura, S. Nagahara, and M. Kubo. MOS area sensor: part I—design consideration and performance of an *n-p-n* structure 484 × 384 element color MOS imager. *IEEE Journal of Solid State Circuits*, SC-15(4):741–746, August 1980.

8 J. L. McCreary and P. R. Gray. All-MOS charge redistribution analog-to-digital conversion techniques—part I. *IEEE Journal of Solid State Circuits*, SC-10:371–379, December 1975.

9 Carver Mead. A sensitive electronic photoreceptor. In *1985 Chapel Hill Conference on Very Large Scale Integration*, pages 463–471, 1985.

10 Carver Mead. *The WOLery*. Technical Report 5113:TR:84, California Institute of Technology, June 1984.

11 Carver Mead and Michelle Mahowald. An electronic model of the Y-system of mammalian retina. California Institute of Technology, Internal Technical Memo 5144:DF:84, June 1984.

12 Carver Mead and John Wawrsynek. A new discipline for CMOS design: an architecture for sound synthesis. In *1985 Chapel Hill Conference on Very Large Scale Integration*, pages 87–104, 1985.

13 S. Ohba, M. Nakai, H. Ando, S. Hanamura, S. Shimada, K. Satoh, K. Takahashi, M. Kubo, and T. Fujita. MOS area sensor: part II—low-noise MOS area sensor with antiblooming photodiodes. *IEEE Journal of Solid State Circuits*, SC-15(4):747–752, August 1980.

14 Yoshio Ohkubo. An analysis of fixed pattern noise for MOS-CCD type image sensors under quasi-stationary conditions. *IEEE Journal of Solid State Circuits*, SC-21(4):555–560, August 1986.

15 Bing J. Sheu and Chenming Hu. Switch-induced error voltage on a switched capacitor. *IEEE Journal of Solid State Circuits*, SC-19(4):519–525, August 1984.

16 Massimo A. Sivilotti. *Toward a Motion-Based VLSI Vision System*. Master's thesis, California Institute of Technology, 1986. 5225:TR:86.

17 Massimo A. Sivilotti. *A User's Guide to the WOL Design Tools*. Technical Report 5237:TR:86, California Institute of Technology, 1986.

18 M. V. Srivivasan, S. B. Laughlin, and A. Dubs. Predictive coding: a fresh view of inhibition in the retina. *Proc. R. Soc. London B*, 216:427–459, 1982.

19 R. E. Suarez, P. R. Gray, and D. A. Hodges. All-MOS charge redistribution analog-to-digital conversion techniques—part II. *IEEE Journal of Solid State Circuits*, SC-10:379–385, December 1975.

20 John Edward Tanner. *Integrated Optical Motion Detection.* PhD thesis, California Institute of Technology, 1986. 5223:TR:86.

21 Eric A. Vittoz. Micropower techniques. In Yannis Tsividis and Paolo Antognetti, editors, *Design of MOS VLSI Circuits for Telecommunications*, pages 104–144, Prentice-Hall, Englewood Cliffs, NJ, 1985.

22 Telle E. Whitney. *Hierarchical Composition of VLSI Circuits.* PhD thesis, California Institute of Technology, 1985. 5189:TR:85.

23 W. B. Wilson, H. Z. Massoud, E. J. Swanson, R. T. George Jr., and R. B. Fair. Measurement and modeling of charge feedthrough in *n*-channel MOS analog switches. *IEEE Journal of Solid State Circuits*, SC-20(6):1206–1213, December 1985.

Afterword

What is the current state of connectionist and neural network modeling in 1987? It is an area of intense activity, growing and realistic interest in practical applications, and, unfortunately, escalating "hype," as those with commercial interests become involved. Those of us with some experience in the field are keenly aware of the perceptron debacle, and fervently hope that claims will be kept modest, justifiable, and under control.

Let us conclude by giving a series of overall impressions of the current status of the field, the power of the systems, and some possible future research directions.

1. Connectionist models have already had an important influence on cognitive science. A whole new class of theories of the mind is in the process of emerging. There is serious talk about a "paradigm shift" away from rule based or symbol manipulation theories to theories based on, or at least consistent with, neural networks. Connectionism is a prominent feature of recent issues of *Cognitive Science* and the annual meetings of the Cognitive Science Society.

2. Neuroscientists are much more conservative. Historically, the roots of neural networks are in neuroscience. In the past twenty years, neuroscience has become more and more empirical, and more and more concerned with molecular mechanisms. Interest in large scale theories, common in earlier days, has waned, and the field now focuses on biological minutiae. However, there is still a small, but now perhaps increasing, interest in "systems neuroscience." There are currently several examples of groups that have taken the ideas found in network models and made use of them to explain or interpret experimental data. Good examples of work combining experimental neuroscience and network theory might include the work coming from the laboratory of Walter Freeman (see Baird, 1986), the work on tensor models of the cerebellum (see Pellionisz and Llinas, 1985), the work described by Gluck and Thompson (1987), and experiments on synaptic learning specifically designed to check the assumptions of network theory (see Levy, 1985; Kelso, Ganong, and Brown, 1986). These are only a few of the better known examples; there are encouraging signs of influences going in both directions between neuroscience and network modeling.

3. For practical systems, cognitive applications are initially the most promising—that is, applications with a strongly psychological flavor, such as generalizing, guessing, or working with noisy and partially rule governed systems. Attempts to make neural networks more like traditional computers—i.e., to make them do logic, or to store and retrieve very large amounts of information with high accuracy—are not making good use of the virtues of networks.

4. The standards of accuracy and precision that traditional computers have accustomed us to will have to be discarded. If computers using neural networks are implemented, they will intrinsically make mistakes. This is the dark side of "creativity" and "generalization."

5. Simple networks have weak formal computing power—basically only associative or similarity operations. They can do more, though sometimes with difficulty, and it will be of interest to explore their limits, partially because study of networks in this context may give us insights into how humans perform such 'unnatural' functions.

6. Most of the power of networks is in the representational details of the vector coding. The systems are strongly memory driven and somewhat inflexible. They are computing memories, not computers. Constructing the details of the representation so that it learns and computes what you want it to will be more important in the future than deciding what learning algorithm to use.

7. Knowledge networks and structures can be implemented so that they evolve serially in the time domain, but this has not been studied often. One of the most difficult problems in networks at present is formulation of systems that are oriented toward the time domain. To solve problems such as speech recognition, natural language understanding, and many aspects of visual perception, it will be necessary to make networks that evolve continuously in time. We are only starting to study such systems. Most work on networks up to this point has been strongly oriented toward spatial representations, though sometimes with interesting temporal properties.

8. Hybrid systems may be practical quite soon and are potentially very powerful. Traditional Artificial Intelligence is "high level"; networks as currently implemented are "low level." There is an obvious potential synergy that can be exploited in practical systems.

9. Neural networks will not be practically useful until cheap special-purpose parallel hardware is available. A small digital computer, for example, a calculator, is useful. A neural network of equivalent small size is not of much value.

10. Many neural networks require low precision matrix multiplications, parallel computation, a degree of linearity, and high interconnectivity. They are, therefore, excellent matches to optical computers or to special-purpose analog or digital hardware. Many such devices are being built. However, hardware will prove relatively easy to build. How to use the hardware will be the tough part.

11. The software for neural networks will be by far the most difficult, important, and painful aspect of practical applications. Most research in industry has ignored this problem, concentrating instead on hardware problems that are well defined, tractable, and solvable. When software is considered at all, it is usually discussed in a vague way, incorporating the current buzz words: fault tolerance, noise immunity, graceful degradation, content addressable memory, massive parallelism, and so on. This may impress the company executives or venture capitalists, but software and applications will be many times more difficult than hardware.

12. In terms of real software development for connectionist systems, cognitive scientists are designing some of the software and suggesting applications for the parallel hard-

ware. Cognitive scientists are studying the functions of the mind—what those in the field fondly believe is the "existence proof" of the powers of neural networks. Those interested in practical applications of neural networks would be well advised to look carefully at the construction and functions of the one that already works.

References

B. Baird (1986), "Non-linear dynamics of pattern formation and pattern recognition in the rabbit olfactory bulb," *Evolution, Games and Learning: Models for Adaptation in Machines and Nature*, D. Farmer, A. Lapedes, N. Packard, and B. Wendroff (Eds.), Amsterdam: North Holland.

M. A. Gluck and R. F. Thompson (1987), "Modelling the neural substrates of associative learning and memory: a computational approach," *Psychological Review* 94:176–191.

S. R. Kelso, A. H. Ganong, and T. H. Brown (1986), "Hebbian synapses in hippocampus," *Proceedings of the National Academy of Sciences* 83:5326–5330.

W. B. Levy (1985), "Associative changes at the synapse: LTP in the hippocampus," *Synaptic Modification, Neuron Selectivity and Nervous System Organization*, W. B. Levy, J. A. Anderson, and S. Lehmkuhle (Eds.), Hillsdale, NJ: Erlbaum.

A. Pellionisz and R. Llinas (1985), "Tensor network theory of the metaorganization of functional geometries in the central nervous system" *Neuroscience* 16:245–273.

Name Index

Italicized page numbers indicate papers appearing in this volume.

Subject Index